Quantum Theory of the Solid State

Student Edition

Quantum Theory of the Solid State

Student Edition

JOSEPH CALLAWAY
Department of Physics and Astronomy
Louisiana State University
Baton Rouge, Louisiana

ACADEMIC PRESS New York San Francisco London
A Subsidiary of Harcourt Brace Jovanovich, Publishers

To Mary

ACADEMIC PRESS, INC.
111 Fifth Avenue, New York, New York 10003

United Kingdom Edition published by
ACADEMIC PRESS, INC. (LONDON) LTD.
24/28 Oval Road, London NW1

LIBRARY OF CONGRESS CATALOG CARD NUMBER: 72-12200

ISBN 0-12-155256-X

PRINTED IN THE UNITED STATES OF AMERICA

Contents

Preface

The purpose of this book is to describe the concepts and methods, and to introduce some of the central problems of the quantum theory of solids. It should be suitable as a textbook for students who have completed a one-year course in quantum mechanics and have some familiarity with the experimental facts of solid state physics. It should also be useful as a reference work. I have attempted a moderately comprehensive coverage: The physics of solids is, in fact, a rather diverse subject.

A book with these aims must develop both principles and mathematical techniques; in addition, it should assist the reader in making his way through the more specialized periodical literature. To this end, fairly lengthy bibliographies have been included at the end of each chapter, although these are not intended to be complete in any area. If these references are used in conjunction with the Science Citation Index, it should be possible to follow many specific subjects to the frontier of present research.

This work is intended to be a single intellectual unit, although for reasons of convenience it has been divided into two parts. Part A contains much of the formalism required for the theoretical study of solids; Part B is oriented toward more specific problems. Thus, Part A includes phenomenological treatments of lattice vibrations and magnetic order, a discussion of symmetry groups, and a description of the properties of one-electron wave functions and the principal techniques for calculating energy levels. In Part B the machinery developed previously is applied to impurities, disordered systems, the effects of external fields, and transport phenomena (including superconductivity). The book concludes with an introduction to many-body theory, including some applications.

The specific selection of topics is obviously a personal one, and some areas of considerable importance, such as mechanical properties, surfaces, electron diffraction, and amorphous materials, have not been included. Experimental results are used occasionally and illustratively. There is no detailed confrontation of specific approximations with experimental data. MKS, cgs, and atomic units have been used rather interchangeably. Some problems have been included.

I am indebted to my colleague, Dr. John Kimball, and to several students (W. Y. Ching, M. Eswaran, G. S. Grest, W. Y. Hsia, M. Singh, and C. S. Wang) for critical readings of portions of the manuscript.

Note to the Reader

In Part B the superscript "plus" ($^+$) is used to indicate a Hermitian adjoint instead of the superscript "dagger" (†) used in Part A.

Part A

CHAPTER 1

Lattice Dynamics

In this chapter we will present some portions of the theory of the vibrations of crystal lattices. We will also describe some of the geometrical relations and constructions which are useful in almost all branches of the theory of solids. Our point of view with respect to the lattice vibrations will be phenomenological, that is, we will assume that the forces between atoms are known and can be described by a set of force constants, which are the second derivatives of the interatomic potentials with respect to atomic displacements. The displacements themselves are assumed to be small, so that for the most part, the forces may be regarded as linear functions of the atomic displacements. This is the harmonic approximation; the lattice is treated as a collection of coupled simple harmonic oscillators. It is the task of a more fundamental theory to determine the interatomic potential and thus the force constants which we have regarded as disposable parameters, apart from general restrictions imposed by symmetry and invariance considerations.

1.1 Equations of Motion and Their Solution

In this section we will obtain the general equations of motion for a vibrating lattice and indicate the method of solution with respect to a simple example. The treatment is largely based on the presentation of Maradudin *et al.* (1971).

1.1.1 *The Dynamical Matrix*

The periodicity of a crystal is described by a set of vectors $\mathbf{R}_i$ which locate each unit cell in the crystal. We assume the crystal contains a large or even an infinite number of unit cells and neglect any effects due to boundary surfaces. Each $\mathbf{R}_i$ may be expressed in terms of three independent (noncoplanar) primitive translation vectors ($\mathbf{a}_1$, $\mathbf{a}_2$, $\mathbf{a}_3$) in the following way:

$$\mathbf{R}_i = n_{i1}\mathbf{a}_1 + n_{i2}\mathbf{a}_2 + n_{i3}\mathbf{a}_3 \tag{1.1.1}$$

where the n_{ij} are integers. Each unit cell contains r atoms. The locations of the r atoms are given by the vectors $\mathbf{d}_\kappa$ where κ indicates the different atoms in the unit cell, and takes the values 1, 2, . . . , r. The general position of the κth atom in the ith unit cell is then

$$\mathbf{X}_{i\kappa} = \mathbf{R}_i + \mathbf{d}_\kappa. \tag{1.1.2}$$

We now suppose that each atom is displaced from its equilibrium position by an amount $\mathbf{u}_{i\kappa}$ (the αth rectangular component of $\mathbf{u}_{i\kappa}$ is denoted by $u_{\alpha,i\kappa}$). The mass of the κth atom is M_κ. The atoms are heavy enough so that their behavior can, in most instances, be described classically.

The total kinetic energy of the lattice is therefore

$$T = \tfrac{1}{2} \sum_{\alpha,\, i,\, \kappa} M_\kappa \dot{u}^2{}_{\alpha,i\kappa} \tag{1.1.3}$$

where the dot indicates derivative with respect to time.

Let the total potential energy of the crystal be denoted by Φ. This quantity will be a function of the atomic positions. When the atoms are displaced from their equilibrium positions, Φ will differ from its equilibrium value Φ_0. This difference may be expressed as a Taylor series in the atomic displacements

$$\Phi = \Phi_0 + \sum_{i\kappa\alpha} \Phi_{\alpha,i\kappa} u_{\alpha,i\kappa} + \tfrac{1}{2} \sum_{i\kappa\alpha,\, j\nu\beta} \Phi_{\alpha\beta,i\kappa,j\nu} u_{\alpha,i\kappa} u_{\beta,j\nu} + \cdots. \tag{1.1.4}$$

In this chapter we will make the harmonic approximation, that is, we will neglect terms in (1.1.4) higher than second order in the atomic displacements. The quantities Φ_α are derivatives of the potential energy.

$$\Phi_{\alpha,i\kappa} = [\partial\Phi/\partial u_{\alpha,i\kappa}]_0, \qquad \Phi_{\alpha\beta,i\kappa,j\nu} = [\partial^2\Phi/\partial u_{\alpha,i\kappa}\, \partial u_{\beta,j\nu}]_0. \tag{1.1.5}$$

The subscript 0 indicates that the derivatives are evaluated in the equilibrium configuration of the crystal.

It is apparent that $\Phi_{\alpha,i\kappa}$ in the negative of the αth component of the net force on atom $i\kappa$ in its equilibrium position. However, this notion is contredictory in that if there were a net force on an atom, it would move, and so the original position would not have been one of equilibrium. Thus we must have

$$\Phi_{\alpha,i\kappa} = 0. \tag{1.1.6}$$

The Hamiltonian of the system is therefore

$$H = \Phi_0 + \tfrac{1}{2}\sum M_\kappa \dot{u}^2_{\alpha,i\kappa} + \tfrac{1}{2}\sum_{\alpha i\kappa,\beta j\nu} \Phi_{\alpha\beta,i\kappa,j\nu} u_{\alpha,i\kappa} u_{\beta,j\nu}. \tag{1.1.7}$$

The equations of motion of the lattice are then easily found to be

$$M_\kappa \ddot{u}_{\alpha,i\kappa} = -\partial\Phi/\partial u_{\alpha,i\kappa} = -\sum_{\beta j\nu} \Phi_{\alpha\beta,i\kappa,j\nu} u_{\beta,j\nu}. \tag{1.1.8}$$

There are a number of restrictions that may be imposed by general physical considerations on the coefficients $\Phi_{\alpha\beta,i\kappa,j\nu}$. Some of these include:

(1) $\Phi_{\alpha\beta,i\kappa,j\nu}$ depends on $\mathbf{R}_i$ and $\mathbf{R}_j$ only through their vector difference $\mathbf{R}_i - \mathbf{R}_j$. This follows because we may displace the origin of coordinates arbitrarily without altering the $\Phi_{\alpha\beta}$.

(2) Suppose the lattice is displaced rigidly (all $u_{\beta,j\nu}$ are made independent of j and ν). No acceleration can result. Thus

$$\sum_{j\nu} \Phi_{\alpha\beta,i\kappa,j\nu} = 0. \tag{1.1.9}$$

Other restrictions are derived by Maradudin *et al.* (1971).

Let us obtain periodic solutions to (1.1.8). We write

$$u_{\alpha,i\kappa} = M_\kappa^{-1/2} u_{\alpha,\kappa}(\mathbf{k}) \exp[-i\omega t + i\mathbf{k}\cdot\mathbf{R}_i]. \tag{1.1.10}$$

Here $u_{\alpha,\kappa}(\mathbf{k})$ is assumed to be independent of $\mathbf{R}_i$. This is to be substituted into (1.1.8). We obtain

$$-M_\kappa^{1/2}\omega^2 \exp[-i\omega t + i\mathbf{k}\cdot\mathbf{R}_i] u_{\alpha,\kappa}(\mathbf{k}) = -\sum_{\beta j\nu} M_\nu^{-1/2} \Phi_{\alpha\beta,i\kappa,j\nu} u_{\beta,\nu}(\mathbf{k}) \exp[-i\omega t + i\mathbf{k}\cdot\mathbf{R}_i + i\mathbf{k}\cdot(\mathbf{R}_j - \mathbf{R}_i)]. \tag{1.1.11}$$

Since Φ depends only on $\mathbf{R}_i - \mathbf{R}_j$, we may replace the sum over $\mathbf{R}_j$ by one on $\mathbf{R}_i - \mathbf{R}_j$. Thus we have the set of simultaneous equations

$$\omega^2 u_{\alpha,\kappa}(\mathbf{k}) = \sum_{\nu\beta} D_{\alpha\beta,\kappa\nu}(\mathbf{k}) u_{\beta,\nu}(\mathbf{k}) \tag{1.1.12}$$

where

$$D_{\alpha\beta,\kappa\nu}(\mathbf{k}) = (M_\kappa M_\nu)^{-1/2} \sum_{\mathbf{R}_i - \mathbf{R}_j} \Phi_{\alpha\beta,i\kappa,j\nu} \exp[-ik\cdot(\mathbf{R}_i - \mathbf{R}_j)]. \tag{1.1.13}$$

D is frequently referred to as the "dynamical matrix," and $\mathbf{k}$ is the wave vector of the vibrational wave. The condition for the set of linear homogeneous equations to possess a nontrivial solution is

$$\det[\omega^2 \delta_{\alpha\beta}\, \delta_{\kappa\nu} - D_{\alpha\beta,\kappa\nu}(\mathbf{k})] = 0. \tag{1.1.14}$$

The matrix D is of dimension $3r \times 3r$ (recall that r is the number of atoms in the unit cell). Furthermore, it is Hermitian:

$$D^*_{\beta\alpha,\nu\kappa}(\mathbf{k}) = \sum_{\mathbf{R}_i-\mathbf{R}_j} \Phi_{\beta\alpha,i\nu,j\kappa} \exp[-i\mathbf{k}\cdot(\mathbf{R}_j - \mathbf{R}_i)]$$

$$= \sum \Phi_{\alpha\beta,j\kappa,i\nu} \exp[-i\mathbf{k}\cdot(\mathbf{R}_j - \mathbf{R}_i)] = D_{\alpha\beta,\kappa\nu}(\mathbf{k}). \quad (1.1.15)$$

We have used the symmetry property of the derivatives of Φ that $\Phi_{\beta\alpha,i\nu,j\kappa} = \Phi_{\alpha\beta,j\kappa,i\nu}$.

Thus we see that there are $3r$ real eigenvalues to be determined. We denote these eigenvalues by $\omega_j^2(\mathbf{k})$ ($j = 1, \ldots, r$). They are the squares of the normal mode frequencies for the crystal. The index j designates a branch; and within a branch ω^2 will be a continuous function of $\mathbf{k}$ (out to a certain limit, as we will find later).

The equation

$$\omega = \omega_j(\mathbf{k}) \qquad (j = 1, 2, \ldots, 3r) \quad (1.1.16)$$

is known as the *dispersion relation* for the crystal.

For each of the $3r$ values of ω for given $\mathbf{k}$, there is an eigenvector of D which we denote by $\mathbf{e}^{(j)}$, or $e_{\alpha,\nu}{}^{(j)}(\mathbf{k})$. This satisfies

$$\omega_j^2(\mathbf{k}) e_{\alpha,\nu}{}^{(j)}(\mathbf{k}) = \sum_{\beta\kappa} D_{\alpha\beta,\nu\kappa}(\mathbf{k}) e_{\beta,\kappa}{}^{(j)}(\mathbf{k}). \quad (1.1.17)$$

These vectors are determined by (1.1.17) only up to a constant factor; however, they may be normalized conveniently. The e's are in fact elements of a unitary matrix which diagonalizes D. As a result, we have both orthonormality and completeness relations in the form

$$\text{orthonormality:} \quad \sum_{\alpha,\kappa} e_{\alpha,\kappa}{}^{(j)*}(\mathbf{k}) e_{\alpha,\kappa}{}^{(i)}(\mathbf{k}) = \delta_{ji}, \quad (1.1.18a)$$

$$\text{completeness:} \quad \sum_{j} e_{\alpha,\kappa}{}^{(j)*}(\mathbf{k}) e_{\beta,\nu}{}^{(j)}(\mathbf{k}) = \delta_{\alpha\beta}\, \delta_{\kappa\nu}. \quad (1.1.18b)$$

Since the $\Phi_{\alpha\beta}$ are real, it follows from (1.1.13) that

$$D_{\alpha\beta,\kappa\nu}(-\mathbf{k}) = D^*_{\alpha\beta,\kappa\nu}(\mathbf{k}). \quad (1.1.19)$$

Since D is Hermitian, the eigenvalues are always real; it therefore follows that

$$\omega_j^2(\mathbf{k}) = \omega_j^2(-\mathbf{k}). \quad (1.1.20)$$

This relation can also be shown to be a consequence of time reversal symmetry.

If we now take the complex conjugate of Eq. (1.1.17), we see that the eigenvectors must be proportional:

$$e_{\alpha,\kappa}{}^{(j)*}(\mathbf{k}) = c e_{\alpha,\kappa}{}^{(j)}(-\mathbf{k}).$$

We require that completeness and orthonormality relations hold for $(-\mathbf{k})$ as well as for $(\mathbf{k})$ so that c must be a complex number of modulus unity. We will choose c to be 1, so that

$$e_{\alpha,\kappa}^{(j)*}(\mathbf{k}) = e_{\alpha,\kappa}^{(j)}(-\mathbf{k}). \tag{1.1.21}$$

1.1.2 *Some Properties of the Vibrational Spectrum*

Three of the branches of the spectrum are such that ω goes to 0 as $\mathbf{k}$ goes to 0. To see this, set $\mathbf{k} = 0$ in (1.1.17), which then becomes

$$\omega_j^2(0)e_{\alpha\kappa}^{(j)}(0) = \sum_{\beta j\nu} [\Phi_{\alpha\beta,i\kappa,j\nu}/(M_\kappa M_\nu)^{1/2}]e_{\beta\nu}^{(j)}(0). \tag{1.1.22}$$

Now we can solve (1.1.22) trivially by supposing that, for each β, $(e_{\beta\nu}(0)/M_\nu^{1/2})$ is independent of ν. Then the right side of (1.1.22) vanishes because of (1.1.9) and we have a solution with $\omega = 0$.

The modes which have this property are called *acoustic modes.* The remaining $3r - 3$ modes are called *optical modes.* The atomic displacements corresponding to the $\omega = 0$ acoustic modes are, from (1.1.10)

$$\mathbf{u}_{i\kappa} = M_\kappa^{-1/2}\mathbf{e}_\kappa(0) = \text{const.}$$

All r particles in each unit cell move in parallel with equal amplitudes. This is characteristic of an elastic wave of infinite wavelength.

Let us now consider the case of $r = 2$, corresponding to an ionic crystal with two atoms in each unit cell. We apply Eq. (1.1.18a) and let j refer to one of the optical branches while i refers to any of the acoustic branches. Further, we allow κ to take the values $+$, $-$ which may be considered to refer to the ions of positive and negative charge, respectively. We may now write Eq. (1.1.18a)

$$\mathbf{e}_+^{(j)}(0)\cdot\mathbf{e}_+^{(i)}(0) + \mathbf{e}_-^{(j)}(0)\cdot\mathbf{e}_-^{(i)}(0) = 0. \tag{1.1.23}$$

We have already seen that for the acoustic branch

$$\mathbf{e}_+^{(i)}(0)/M_+^{1/2} = \mathbf{e}_-^{(i)}(0)/M_-^{1/2}.$$

Thus we have

$$\mathbf{e}_+^{(i)}(0)\cdot[\mathbf{e}_+^{(j)}(0) + (M_-/M_+)^{1/2}\mathbf{e}_-^{(j)}(0)] = 0.$$

Since the three polarization vectors $\mathbf{e}^{(i)}$ $(i = 1, 2, 3)$ for the acoustic modes are independent, it follows that

$$M_+^{1/2}\mathbf{e}_+^{(j)}(0) + M_-^{1/2}\mathbf{e}_-^{(j)}(0) = 0 \tag{1.1.24}$$

which implies, through (1.1.10), that

$$M_+\mathbf{u}_{i+} + M_-\mathbf{u}_{i-} = 0. \tag{1.1.25}$$

This result may be interpreted as meaning that the two ions in any unit cell vibrate 180° out of phase with each other, but in such a way that the center of mass of the cell remains fixed. Because the two ions have opposite charges, there will be a net oscillating dipole moment in the crystal. Recall that from (1.1.10), since we are concerned with $k = 0$, each cell will vibrate in phase with every other cell. Such a dipole moment can interact with an external electric field, and this gives the modes their name. In the case of more than two atoms per unit cell, the interpretation will not be so simple.

We continue with the case $r = 2$, and now consider a cubic crystal. We will determine the frequency of the vibrations at $k = 0$. We multiply (1.1.17) by $e_{\alpha\nu}^{(j)*}(\mathbf{k})$ and sum over $\alpha\nu$. With the use of (1.1.18a) we have

$$\omega_j^2(\mathbf{k}) = \sum_{\alpha\nu,\beta\kappa} e_{\alpha,\nu}^{(j)*}(\mathbf{k}) D_{\alpha\beta,\nu\kappa}(\mathbf{k}) e_{\beta,\kappa}^{(j)}(\mathbf{k}). \tag{1.1.26}$$

This is very much like a quantum mechanical expectation value, and is valid for any $\mathbf{k}$. Now we substitute for D, using (1.1.12), and put $k = 0$ explicitly:

$$\omega_j^2(0) = \sum_{\alpha\nu,\beta\kappa,\mathbf{R}_i-\mathbf{R}_l} (M_\kappa M_\nu)^{-1/2} e_{\alpha,\nu}^{(j)*}(0)\, \Phi_{\alpha\beta,i\nu,l\kappa} e_{\beta,\kappa}^{(j)}(0). \tag{1.1.27}$$

We now manipulate this expression. Write out the sum over ν and κ explicitly in terms of $+$ and $-$:

$$\begin{aligned}\omega_j^2(0) = \sum_{\alpha\beta,\mathbf{R}_i-\mathbf{R}_l} \{ & e_{\alpha+}^{(j)*}(0)[(M_+)^{-1}\Phi_{\alpha\beta,i+,l+} e_{\beta+}^{(j)}(0) \\ & + (M_+M_-)^{-1/2}\Phi_{\alpha\beta,i+,l-} e_{\beta-}^{(j)}(0)] \\ & + e_{\alpha-}^{(j)*}(0)[(M_+M_-)^{-1/2}\Phi_{\alpha\beta,i-,l+} e_{\beta+}^{(j)}(0) \\ & + (M_-)^{-1}\Phi_{\alpha\beta,i-,l-} e_{\beta-}^{(j)}(0)]\}.\end{aligned}$$

Then we use (1.1.24) and (1.1.9) to obtain

$$\begin{aligned}\omega_j^2(0) = \sum_{\alpha\beta;\mathbf{R}_i-\mathbf{R}_l} \{ & [e_{\alpha+}^{(j)*}(0) e_{\beta+}^{(j)}(0) + e_{\alpha-}^{(j)*} e_{\beta-}^{(j)}(0)] \\ & \times [(\Phi_{\alpha\beta,i+,l+}/M_+) + (\Phi_{\alpha\beta,i-,l-}/M_-)]\}.\end{aligned} \tag{1.1.28}$$

We now assume the crystal is cubic. In this case, if we sum over all lattice sites, terms with $\alpha \neq \beta$ must disappear, since transformations such as $x \to x$, $y \to -y$, $z \to z$ change the sign of the second derivatives but cannot change the potential energy of the system. Similarly, the application of transformations such as $x \to y$, $y \to x$, $z \to z$ shows that all diagonal $(\alpha = \beta)$ terms are equal. Thus

$$\sum_{\mathbf{R}_i-\mathbf{R}_l} \Phi_{\alpha\beta,i\pm,l\pm} = \delta_{\alpha\beta} \sum_{\mathbf{R}_i-\mathbf{R}_l} \Phi_{\alpha\alpha,i\pm,l\pm}, \tag{1.1.29}$$

and the right side is independent of α. We can use this result in combination with (1.1.18a) to obtain

$$\omega_j^2(0) = \sum_{\mathbf{R}_i-\mathbf{R}_l} [(\Phi_{\alpha\alpha,i+,l+}/M_+) + (\Phi_{\alpha\alpha,i-,l-}/M_-)]. \quad (1.1.30)$$

The right side of (1.1.30) is independent of j, and we may therefore conclude that the frequencies of all three optical modes at $k = 0$ in a cubic crystal are equal.

We now want to consider the small k behavior of the acoustic branches in more detail. In this case, ω becomes linearly proportional to k, the proportionality constant being the appropriate velocity of sound. As the algebra can, in the general case, become quite messy we consider here only a monatomic lattice (one atom per unit cell). In this case, we may drop the indices κ and ν, and consider Eq. (1.1.11) in the form

$$\omega^2(\mathbf{k})u_\alpha(\mathbf{k}) = \sum_\beta D_{\alpha\beta}(\mathbf{k})u_\beta(\mathbf{k}). \quad (1.1.31)$$

We expand the dynamical matrix in powers of $\mathbf{k}$, since $\mathbf{k}$ is considered to be small. Thus

$$D_{\alpha\beta}(\mathbf{k}) = D_{\alpha\beta}(0) + \sum_\gamma c_{\alpha\beta,\gamma}k_\gamma + \sum_{\gamma\delta} c_{\alpha\beta,\gamma\delta}k_\gamma k_\delta + \cdots \quad (1.1.32)$$

where

$$D_{\alpha\beta}(0) = (1/M) \sum_{\mathbf{R}_i-\mathbf{R}_j} \Phi_{\alpha\beta}(\mathbf{R}_i - \mathbf{R}_j) = (1/M) \sum_i \Phi_{\alpha\beta}(\mathbf{R}_i); \quad (1.1.33)$$

also

$$c_{\alpha\beta,\gamma} = (1/M) \sum_i \Phi_{\alpha\beta}(\mathbf{R}_i)x_\gamma^{(i)}, \quad (1.1.34)$$

$$c_{\alpha\beta,\gamma\delta} = (-1/2M) \sum_i \Phi_{\alpha\beta}(\mathbf{R}_i)x_\gamma^{(i)}x_\delta^{(i)}. \quad (1.1.35)$$

We have defined $x_\gamma^{(i)} = (\mathbf{R}_i)_\gamma$. It is possible, however, to show that $c_{\alpha\beta,\gamma}$ vanishes. This may be seen if we realize that the second derivatives of the potential energy must be even functions of $\mathbf{R}_i - \mathbf{R}_j$. This follows since the atomic interactions must be assumed to be unchanged by an inversion, that is, if $\mathbf{R}_i \to -\mathbf{R}_i$ and $\mathbf{R}_j \to -\mathbf{R}_j$. Then, there will be compensatory contributions from $\mathbf{R}_i$ and $-\mathbf{R}_i$. Also, we have from (1.1.9) that $D_{\alpha\beta}(0) = 0$. [This leads to $\omega^2(0) = 0$ in this case.] Hence our small k formula is

$$\omega^2(\mathbf{k})u_\alpha(\mathbf{k}) = \sum_{\nu\delta\beta} [c_{\alpha\beta,\nu\delta}k_\nu k_\delta]u_\beta(\mathbf{k}). \quad (1.1.36)$$

The frequency ω can be expanded as a power series in the components of $\mathbf{k}$.

Since $\omega_j(0) = 0$, the leading term in this is of order k, and, therefore, the leading dependence of ω^2 on the magnitude of $\mathbf{k}$ is of second order. This being so, we may neglect the dependence of u_α on $|\mathbf{k}|$. It will, however, continue to depend on direction so we will denote it by $u_\alpha(\hat{\mathbf{k}})$. Thus we have

$$\omega^2 u_\alpha(\hat{\mathbf{k}}) = \sum_{\nu\delta\beta} c_{\alpha\beta,\nu\delta} k_\nu k_\delta u_\beta(\hat{\mathbf{k}}). \tag{1.1.37}$$

This equation has the same form as the equation determining the vibration frequencies for an elastic continuum. We must consider (1.1.37) as an eigenvalue equation. This equation determines the proportionality constant between ω and $|\mathbf{k}|$ for the three possible acoustic waves. This proportionality constant is the relevant sound velocity. The direction cosines of the polarization vectors are also determined.

1.1.3 *Example: A Simple Cubic Lattice*

Let us illustrate these considerations by discussing a relatively simple case: a monatomic, simple cubic lattice of lattice constant a in which forces are assumed to act between one atom and its first and second nearest neighbors. The forces are assumed to be central, that is, the potential energy is a function of the distance between atomic pairs only (and not angle). Only those displacements which change the distance between atoms (in first order) will contribute to Φ. Such displacements must be along the vector connecting the atoms in equilibrium. Let $\mathbf{u}_i$ be the displacement vector for atom i. We have the potential function

$$\Phi = \Phi_0 + (\alpha/2a^2) \sum_{ij,\,\mathrm{nn}} [(\mathbf{R}_i - \mathbf{R}_j)\cdot(\mathbf{u}_i - \mathbf{u}_j)]^2$$

$$+ (\gamma/2a^2) \sum_{ij,\,\mathrm{sn}} [(\mathbf{R}_i - \mathbf{R}_j)\cdot(\mathbf{u}_i - \mathbf{u}_j)]^2 \tag{1.1.38}$$

where nn indicates nearest neighbors, sn second neighbors, and the constants α and γ are second derivatives of the potential energy. Let a be the lattice constant, and x_i be the x component of u_i with respect to the crystal axes, etc. Then

$$\alpha = -(\partial^2\Phi/\partial x_i\,\partial x_j) \qquad \text{for} \quad |\mathbf{R}_i - \mathbf{R}_j| = a, \tag{1.1.39}$$

$$\gamma = -(\partial^2\Phi/\partial x_i\,\partial x_j) \qquad \text{for} \quad |\mathbf{R}_i - \mathbf{R}_j| = 2^{1/2}a. \tag{1.1.40}$$

The sum in (1.1.38) includes different nearest neighbor pairs. In order to make the notation more transparent in the following, we replace the single indices $i, j, \ldots$ by triples, e.g., lmn. We have, dropping the Φ_0 term,

$$\Phi = (\alpha/2) \sum_{lmn} [(x_{lmn} - x_{l+1,m,n})^2 + (y_{lmn} - y_{l,m+1,n})^2 + (z_{lmn} - z_{l,m,n+1})^2]$$
$$+ (\gamma/2) \sum_{lmn} [(x_{lmn} - x_{l+1,m+1,n} + y_{lmn} - y_{l+1,m+1,n})^2$$
$$+ (x_{lmn} - x_{l+1,m-1,n} - y_{lmn} + y_{l+1,m-1,n})^2$$
$$+ (x_{lmn} - x_{l+1,m,n+1} + z_{lmn} - z_{l+1,m,n+1})^2$$
$$+ (x_{lmn} - x_{l+1,m,n-1} - z_{lmn} + z_{l+1,m,n-1})^2$$
$$+ (y_{lmn} - y_{l,m+1,n+1} + z_{lmn} - z_{l,m+1,n+1})^2$$
$$+ (y_{lmn} - y_{l,m+1,n-1} - z_{lmn} + z_{l,m+1,n-1})^2]. \qquad (1.1.41)$$

We now calculate the elements of the dynamical matrix according to (1.1.13). The sums over equivalent atoms can be performed easily. We obtain

$$MD_{11}(k) = 2\alpha(1 - \cos k_x a) + 4\gamma(2 - \cos k_x a \cos k_y a - \cos k_x a \cos k_z a),$$
$$MD_{22}(k) = 2\alpha(1 - \cos k_y a) + 4\gamma(2 - \cos k_y a \cos k_x a - \cos k_y a \cos k_z a),$$
$$MD_{33}(k) = 2\alpha(1 - \cos k_z a) + 4\gamma(2 - \cos k_z a \cos k_x a - \cos k_z a \cos k_y a),$$
$$MD_{12}(k) = 4\gamma \sin k_x a \sin k_y a,$$
$$MD_{13}(k) = 4\gamma \sin k_x a \sin k_z a,$$
$$MD_{23}(k) = 4\gamma \sin k_y a \sin k_z a. \qquad (1.1.42)$$

When these values are substituted into (1.1.14) we obtain a secular equation of third degree in ω^2. The roots of this equation are the characteristic frequencies for which we are looking.

In general, this equation must be solved numerically for some assumed values of α and γ. Here, we will investigate two simple directions:

(1) $k_y = k_z = 0$. In this case, we have the solution

$$M\omega_1^2 = (2\alpha + 8\gamma)(1 - \cos k_x a), \qquad M\omega_2^2 = M\omega_3^2 = 4\gamma(1 - \cos k_x a).$$

Solution 1 corresponds to a longitudinal wave (displacement and propagation in the x direction); solutions 2 and 3 are transverse with displacements in the yz plane. For small values of $k_x a$, we expand the cosine and have

$$\omega_1^2 = \omega_l^2 = (\alpha + 4\gamma) k_x^2 a^2/M, \qquad \omega_2^2 = \omega_3^2 = \omega_t^2 = 2\gamma k_x^2 a^2/M.$$

If we write $\omega = kc$, where c is the speed of sound, we see that

$$\text{longitudinal branch:} \quad c_l = a[(\alpha + 4\gamma)/M]^{1/2},$$

$$\text{transverse branch:} \quad c_t = a[(2\gamma/M)]^{1/2}.$$

Now let us recall from elasticity theory (Kittel, 1966, Chapter 4; Huntington, 1957)

$$c_l = (c_{11}/\rho_0)^{1/2}, \qquad c_t = (c_{44}/\rho_0)^{1/2}$$

where ρ_0 is the mass density and c_{11}, c_{44} are elastic constants. In this case $\rho_0 = M/a^3$, so we have

$$c_{11} = (\alpha + 4\gamma)/a, \qquad c_{44} = 2\gamma/a.$$

(2) Suppose $k_x = k_y = k_z = k_0$ (111 axis). In this case the longitudinal frequency branch has

$$M\omega_l^2 = 2\alpha(1 - \cos k_0 a) + 16\gamma \sin^2 k_0 a \approx (\alpha + 16\gamma) k_0^2 a^2.$$

The transverse branches have

$$M\omega_t^2 = 2\alpha(1 - \cos k_0 a) + 4\gamma \sin^2 k_0 a \approx (\alpha + 4\gamma) k_0^2 a^2.$$

Again, using elasticity theory, the longitudinal sound velocity is, in this direction $[(c_{11} + 2c_{12} + 4c_{44})/3\rho]^{1/2}$. Combining this with our other results, we see that $c_{12} = 2\gamma/a = c_{44}$. A simple solution can also be obtained for the 110 direction.

1.2 The Reciprocal Lattice and the Brillouin Zone

1.2.1 *Reciprocal Lattice Vectors*

We must now consider the specification of the **k** vector which was introduced in the previous section. The allowed values of **k** are determined in principle by the boundary conditions imposed on the displacement vector **u**. If the results depend critically on the precise boundary conditions which would specify the behavior of **u** on the surface of the specimen, lattice vibration theory would be in serious trouble. Fortunately, this is not the case, since the number of atoms on the surface of a crystal is very small compared to the total number, the ratio being typically of the order of 10^{-7}.

It turns out that, in a large crystal, we may regard the values of **k** as forming a dense and uniform distribution; in fact, we may treat **k** as a continuous variable. The justification for this procedure is the following: We impose periodic boundary conditions on a crystal, and subsequently pass to the limit in which the periodic volume becomes infinite. Periodic boundary conditions imply that the crystal repeats itself in all respects after translation through vectors $\mathbf{R}_1 = (2N_1 + 1)\mathbf{a}_1$, $\mathbf{R}_2 = (2N_2 + 1)\mathbf{a}_2$, $\mathbf{R}_3 = (2N_3 + 1)\mathbf{a}_3$ where $\mathbf{a}_1$, $\mathbf{a}_2$, and $\mathbf{a}_3$ are the three primitive translation vectors of the lattice. The quantities N_1, N_2, N_3 are (large) integers and the particular form is chosen for later convenience. In this way, all surface effects are discarded. We require that the allowed values of $\mathbf{k}$ be such that

$$\exp[i\mathbf{k}\cdot(2N_1 + 1)\mathbf{a}_1] = \exp[i\mathbf{k}\cdot(2N_2 + 1)\mathbf{a}_2] = \exp[i\mathbf{k}\cdot(2N_3 + 1)\mathbf{a}_3] = 1. \tag{1.2.1}$$

To see what this implies, we define a set of vectors by the relation

$$\mathbf{b}_j\cdot\mathbf{a}_i = 2\pi\,\delta_{ij}. \tag{1.2.2}$$

The factor of 2π is introduced for later convenience. Vectors $\mathbf{b}_j$ which satisfy this relation can always be found:

$$\mathbf{b}_1 = \frac{2\pi(\mathbf{a}_2 \times \mathbf{a}_3)}{\mathbf{a}_1\cdot(\mathbf{a}_2 \times \mathbf{a}_3)}; \qquad \mathbf{b}_2 = \frac{2\pi(\mathbf{a}_3 \times \mathbf{a}_1)}{\mathbf{a}_1\cdot(\mathbf{a}_2 \times \mathbf{a}_3)}; \qquad \mathbf{b}_3 = \frac{2\pi(\mathbf{a}_1 \times \mathbf{a}_2)}{\mathbf{a}_1\cdot(\mathbf{a}_2 \times \mathbf{a}_3)}. \tag{1.2.3}$$

The vectors $\mathbf{b}_j$ are used as primitive vectors to construct a lattice which we will call the *reciprocal lattice.*

Now we will define a set of vectors $\mathbf{K}_j$ by

$$\mathbf{K}_j = g_{j1}\mathbf{b}_1 + g_{j2}\mathbf{b}_2 + g_{j3}\mathbf{b}_3 \tag{1.2.4}$$

where the g_{ji} are any integers. These vectors satisfy (for all i, j)

$$\mathbf{K}_j\cdot\mathbf{R}_i = 2\pi n_{ij} \tag{1.2.5}$$

where the quantity n_{ij} is an integer. The vectors $\mathbf{K}_j$ can be considered as defining a lattice of points, which in view of (1.2.5) is the reciprocal lattice. The $\mathbf{K}_j$ are referred to as *reciprocal lattice vectors.*

Now let us consider the equation defining the dynamical matrix, Eq. (1.1.13):

$$D_{\alpha\beta,\kappa\lambda}(\mathbf{k}) = (M_\kappa M_\lambda)^{-1/2} \sum_{\mathbf{R}_i-\mathbf{R}_j} \Phi_{\alpha\beta,i\kappa,j\lambda} \exp[i\mathbf{k}\cdot(\mathbf{R}_j - \mathbf{R}_i)]. \tag{1.1.13$'$}$$

If we increase the wave vector $\mathbf{k}$ above by any reciprocal lattice vector $\mathbf{K}_l$, we see from (1.2.5) that the same dynamical matrix, and therefore the same frequency spectrum, will be obtained. Consequently, it is not necessary to consider both $\mathbf{k}$ and $\mathbf{k} + \mathbf{K}_l$.

Two vectors $\mathbf{k}$ and $\mathbf{k}'$ such that $\mathbf{k} = \mathbf{k}' + \mathbf{K}_l$ (for any $\mathbf{K}_l$) are said to be *equivalent.* In view of the above paragraph, it is necessary to consider only **k** vectors which are not equivalent to each other. Such a region of nonequivalent **k** may be constructed geometrically: the region of space so enclosed is referred to as a *Brillouin zone.*

1.2.2 *Brillouin Zones*

The construction is made as follows. Choose any reciprocal lattice point as origin. Draw the vectors connecting this point with (all) other lattice points. Next, construct a set of planes which are the perpendicular bieectors of these vectors. The smallest solid figure containing the origin is the "Brillouin zone" (Brillouin, 1931). The Brillouin zone for the simple cubic lattice is the cube shown in Fig. 1.2.1.

Points **k** on the surface of the zone must satisfy the condition

$$\mathbf{k}^2 = (\mathbf{k} - \mathbf{K}_n)^2 \qquad \text{or} \qquad \mathbf{K}_n^2 - 2\mathbf{k}\cdot\mathbf{K}_n = 0 \tag{1.2.6}$$

for some reciprocal lattice vector $\mathbf{K}_n$. In calculating the lattice vibration spectrum, we need consider only **k** values lying within the zone.

In the literature one occasionally encounters so-called "higher" Brillouin zones. These are figures formed in the construction procedure by planes bisecting vectors to more distant neighbors. In the present work, only a single Brillouin zone will be considered for any lattice. It must also be noted that our restriction of the definition of $\omega(\mathbf{k})$ to values inside the Brillouin zone is only a convention. Since **k** space may be filled by a set of Brillouin zones, one centered on each point of the reciprocal lattice, an alternative convention for defining the frequency as a function of **k** throughout **k** space is possible in which it is required to be a periodic function of **k** which repeats its values in each zone. If this is done, however, one must be careful not to count repetitions of a given $\omega(\mathbf{k})$ as different vibrational modes.

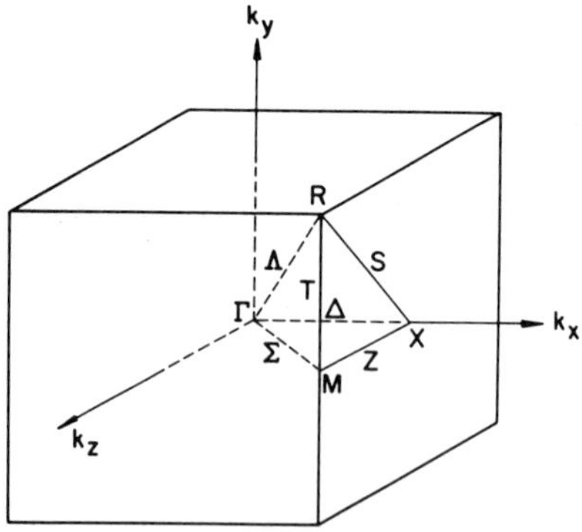

Fig. 1.2.1. Brillouin zone for the simple cubic lattice. Points and lines of symmetry are indicated.

From what has been said, it follows that any **k** vector may be written

$$\mathbf{k} = [h_1/(2N_1 + 1)]\mathbf{b}_1 + [h_2/(2N_2 + 1)]\mathbf{b}_2 + [h_3/(2N_3 + 1)]\mathbf{b}_3 \tag{1.2.7}$$

where the h_i are integers in the range $-N_i \leq h_i \leq N_i$. The values of **k** which are given by the above formula may not all be within the Brillouin zone. However, it is always possible to bring any such **k** outside the zone back into it by translation by a reciprocal lattice vector, and we will subsequently always assume this has been done.

The values of **k** form a uniform and dense distribution. In the limit in which the N_i are allowed to become infinite, we may convert sums over possible **k** values into integrals over the Brillouin zone through the relation

$$\sum_{\mathbf{k}} = (\mathfrak{N}/V) \int_{\mathrm{BZ}} d^3k \tag{1.2.8}$$

where $\mathfrak{N} = (2N_1 + 1)(2N_2 + 1)(2N_3 + 1)$ is the number of unit cells and V is the volume of the Brillouin zone. This volume is

$$V = (2\pi)^3/\Omega \tag{1.2.9}$$

where Ω is the volume of the unit cell in real space. Hence

$$\sum_{\mathbf{k}} = [\mathfrak{N}\Omega/(2\pi)^3] \int d^3k \cdots = [V/(2\pi)^3] \int d^3k \tag{1.2.10}$$

where V is the volume of the portion of crystal used in defining the periodic boundary conditions.

As examples of the foregoing considerations, let us examine the body-centered cubic and face-centered cubic lattices. These are of particular importance since they are the most symmetric structures assumed by single elements. In the former case, possible choices for the three primitive translation vectors are $(a/2)(\mathbf{i} + \mathbf{j} + \mathbf{k})$, $(a/2)(\mathbf{i} + \mathbf{j} - \mathbf{k})$, and $(a/2)(\mathbf{i} - \mathbf{j} + \mathbf{k})$, where a is the lattice parameter. For the face-centered cubic lattice we have vectors $(a/2)(\mathbf{j} + \mathbf{k})$, $(a/2)(\mathbf{i} - \mathbf{k})$, and $(a/2)(\mathbf{i} - \mathbf{j})$. If the basis vectors of the reciprocal lattice are now constructed according to Eq. (1.2.3), it is immediately seen that the lattice reciprocal to the body-centered cubic structure is face-centered cubic. The basic reciprocal lattice vectors are those given for the face-centered cubic structure with $a/2$ replaced by $2\pi/a$, and the lattice reciprocal to the face-centered cubic is body-centered cubic. The Brillouin zones for these structures may be constructed according to the procedures given. The zones are shown in

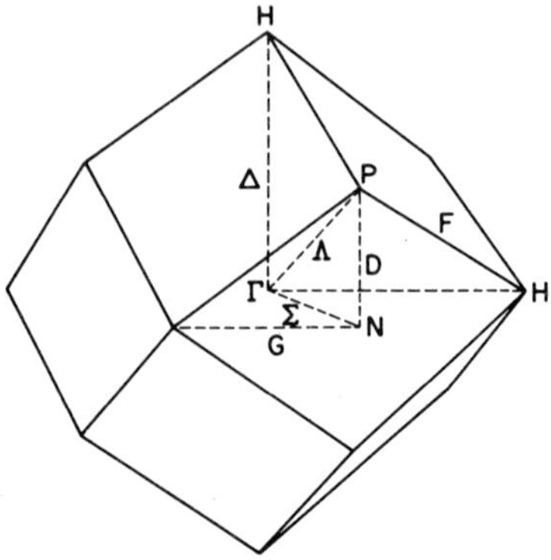

Fig. 1.2.2.

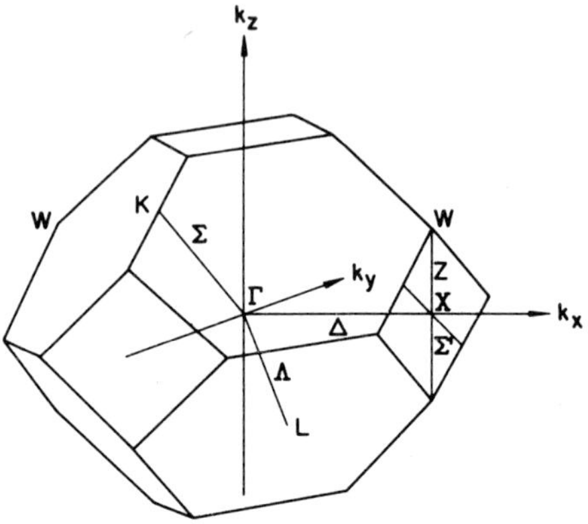

Fig. 1.2.3.

Fig. 1.2.2. Brillouin zone for the body-centered cubic lattice. Points and lines of symmetry are indicated.

Fig. 1.2.3. Brillouin zone for the face-centered cubic lattice. Points and lines of symmetry are indicated.

Figs. 1.2.2 and 1.2.3. The symmetry points of the zone have been labeled according to the notation introduced by Bouckaert *et al.* (1936).

In both cases the center of the zone is designated by Γ. The 100, 111, and 110 axes are labeled Δ, Λ, and Σ, respectively. In the zone for the body-centered cubic lattice, the principal "symmetry points" of the zone are Γ, H, P, and N. The last three points are the intersections of the Δ, Λ, and Σ axes, respectively, with the faces of the zone. A particular point H has coordinates $(2\pi/a)(1, 0, 0)$. All six points H can be obtained from the original one by adding reciprocal lattice vectors of the type $(2\pi/a)(-1, 1, 0)$ or $(2\pi/a)(-2, 0, 0)$, etc. All these points are equivalent. A particular point P has a $\mathbf{k}$ vector $(2\pi/a)(\frac{1}{2}, \frac{1}{2}, \frac{1}{2})$. It is easily seen that three other points P are equivalent to it: $(2\pi/a)(-\frac{1}{2}, -\frac{1}{2}, \frac{1}{2})$, $(2\pi/a)(-\frac{1}{2}, \frac{1}{2}, -\frac{1}{2})$, and $(2\pi/a)(\frac{1}{2}, -\frac{1}{2}, -\frac{1}{2})$. However, the points with coordinates $(2\pi/a)(-\frac{1}{2}, -\frac{1}{2}, -\frac{1}{2})$, $(2\pi/a)(-\frac{1}{2}, \frac{1}{2}, \frac{1}{2})$, $(2\pi/a)(\frac{1}{2}, -\frac{1}{2}, \frac{1}{2})$, and $(2\pi/a)(\frac{1}{2}, \frac{1}{2}, -\frac{1}{2})$, while equivalent to each other, are not equivalent to any of the points in the first group. There are two inequivalent points P. The 12 points N have $\mathbf{k}$ vectors of the type $(2\pi/a)(\frac{1}{2}, \frac{1}{2}, 0)$. These points are equivalent in pairs, the point defined being equivalent to the point $(2\pi/a)$ $(-\frac{1}{2}, -\frac{1}{2}, 0)$. There are six inequivalent points N.

In the Brillouin zone for the face-centered cubic lattice, the points X have the same coordinates as the points H discussed previously, but since the reciprocal lattice vectors are now of the type $(2\pi/a)(1, 1, 1)$, $(2\pi/a)$

$(2, 0, 0)$, etc., each point X is equivalent to only one other such point. There are three inequivalent points X. Similarly, the point L with coordinates $(2\pi/a)(\frac{1}{2}, \frac{1}{2}, \frac{1}{2})$ is equivalent to point $(2\pi/a)(-\frac{1}{2}, -\frac{1}{2}, -\frac{1}{2})$, but not to any others; thus, there are four inequivalent points L. One also finds there are 24 corner points W with coordinates $2\pi(1, \frac{1}{2}, 0)$, but these are equivalent in groups of four. For instance, the points $(2\pi/a)(1, \frac{1}{2}, 0)$, $(2\pi/a)(-1, \frac{1}{2}, 0)$, $(2\pi/a)(0, -\frac{1}{2}, 1)$, and $(2\pi/a)(0, -\frac{1}{2}, -1)$ are equivalent. There are six inequivalent points. Finally, each of the 12 points K of the type $(2\pi/a)$ $(\frac{3}{4}, \frac{3}{4}, 0)$ is equivalent to two of the 24 points U, whose coordinates are (for the particular K given) $(2\pi/a)(-\frac{1}{4}, -\frac{1}{4}, 1)$ and $(2\pi/a)(-\frac{1}{4}, -\frac{1}{4}, -1)$. Portions of the lattice vibration spectrum for diamond, which has the same Brillouin zone as the face-centered cubic lattice, are shown in Fig. 1.2.4.

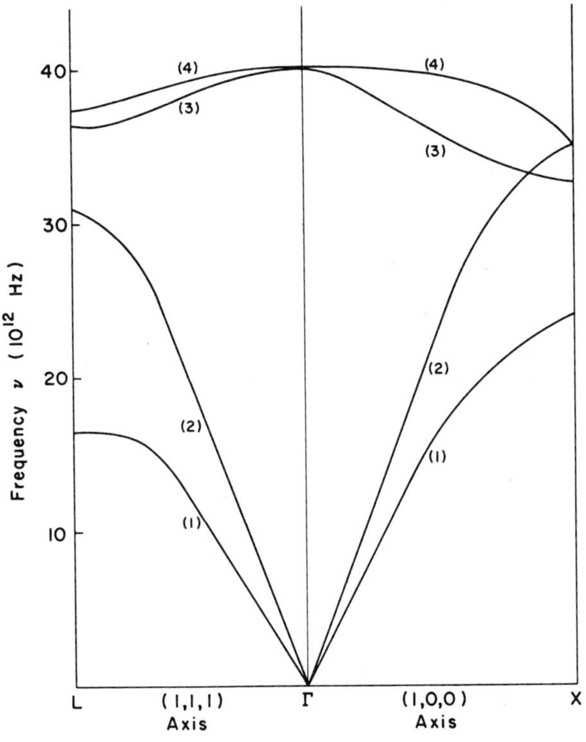

Fig. 1.2.4. Lattice vibration spectrum of diamond (roughly according to Dolling and Cowley, 1966). The Brillouin zone for this crystal is shown in Fig. 1.2.3. The order of branches is: (1) transverse acoustic (doubly degenerate), (2) longitudinal acoustic, (3) transverse optical (doubly degenerate), and (4) longitudinal optical.

1.3 Optical Properties: Classical Theory

In the infrared region of the electromagnetic spectrum, the optical properties of many insulating crystals are controlled by the lattice vibrations. We will consider here the classical theory of the interaction of an electric field with a vibrating ionic lattice. The ions are assumed to have some effective charge q, and not to deform (or polarize) internally in the vibration. We are interested in frequencies corresponding to optical vibrational modes. For light, $k = \omega/c$ is very small at such frequencies (corresponding to infrared radiation) compared to the dimensions of a typical Brillouin zone, and we need, therefore, to consider only the interaction with the optical modes of $k = 0$. Other modes cannot stay in phase with the electric field over an extended region of space. The corrections due to distortion of the ions have been determined by Szigeti (1949).

1.3.1 *The Dielectric Function*

We will first write the equation of motion for an optical mode in the presence of an alternating electric field (Born and Huang, 1954). For simplicity, we consider only a crystal with two atoms per unit cell. The extension to more complex lattices has been given by Barker (1964).

We use (1.1.8) to write out the equations of motion explicitly, (λ, $\kappa = +, -$), adding a term $\pm qE$ to the equations

$$M_+\ddot{u}_{\alpha,i+} = -\sum_{\beta j}[\Phi_{\alpha\beta,i+,j+}u_{\beta,j+} + \Phi_{\alpha\beta,i+,j-}u_{\beta,j-}] + qE,$$

$$M_-\ddot{u}_{\alpha,i-} = -\sum_{\beta j}[\Phi_{\alpha\beta,i-,j+}u_{\beta,j+} + \Phi_{\alpha\beta,i-,j-}u_{\beta,j-}] - qE. \quad (1.3.1)$$

The theory can be improved by the addition of a phenomenological damping term, proportional to $\dot{u}$. We will not do this here, as the present treatment neglecting damping is sufficiently accurate for our purposes. We are concerned with the $k = 0$ mode for which [from (1.1.10)] the u's are independent of i and j. We then apply (1.1.9) and obtain, after rearrangement,

$$\ddot{u}_{\alpha,+} - \ddot{u}_{\alpha,-} = -\sum_{\beta j}[\Phi_{\alpha\beta,i+,j+}/M_+ + \Phi_{\alpha\beta,i-,j-}/M_-](u_{\beta,+} - u_{\beta,-})$$
$$+ qE(1/M^+ + 1/M^-). \quad (1.3.2)$$

Now define $\mathbf{w} = \mathbf{u}_+ - \mathbf{u}_-$. We have, after using (1.1.29) and (1.1.30),

$$\ddot{\mathbf{w}} = -\omega_0^2\mathbf{w} + q\mathbf{E}/\mu \quad (1.3.3)$$

where ω_0 is the optical mode frequency and μ is the reduced ionic mass [$\mu = M_+M_-/(M_+ + M_-)$].

In the presence of the vibration, there is a dynamic dipole moment in each cell, which is just $q\mathbf{w}$. The dipole moment per unit volume is then $q\mathbf{w}/\Omega$ where Ω is the volume of the unit cell. This gives the contribution of the lattice vibrations to the polarizability of the crystal, in the approximation that the ions do not distort while vibrating,

$$\mathbf{P} = q\mathbf{w}/\Omega. \tag{1.3.4}$$

The displacement $\mathbf{D}$ in the crystal is $\mathbf{D} = \epsilon_\infty \mathbf{E} + \mathbf{P}$. MKS units are used in electromagnetic equations.

The quantity ϵ_∞ is the permittivity of the crystal at frequencies high compared to all lattice frequencies but small compared to those of electronic transitions which occur typically in the visible or ultraviolet range. It differs from ϵ_0, the permittivity of free space, by a contribution from virtual electronic transitions. Thus

$$D = \epsilon_\infty \mathbf{E} + q\mathbf{w}/\Omega. \tag{1.3.5}$$

Let us consider an applied electric field $\mathbf{E} = \mathbf{E}_0 e^{-i\omega t}$. Put $\mathbf{w} = \mathbf{w}_0 e^{-i\omega t}$. We obtain from (1.3.3)

$$\mathbf{w}_0 = q\mathbf{E}_0/\mu(\omega_0^2 - \omega^2). \tag{1.3.6}$$

Substitute (1.3.6) in (1.3.5) and obtain $\mathbf{D}$ given by

$$\mathbf{D} = [\epsilon_\infty + (q^2/\mu\Omega)(\omega_0^2 - \omega^2)^{-1}]\mathbf{E}. \tag{1.3.7}$$

Let us write $\mathbf{D} = \epsilon\mathbf{E} = \epsilon_0 \kappa \mathbf{E}$ where ϵ is the permittivity and κ is the dielectric function,

$$\kappa = (\epsilon_\infty/\epsilon_0) + (q^2/\mu\Omega\epsilon_0)(\omega_0^2 - \omega^2)^{-1}. \tag{1.3.8}$$

At high frequencies $(\omega \gg \omega_0)$ $\kappa = \kappa_\infty = \epsilon_\infty/\epsilon_0$.

It will be observed that κ is negative in a region of frequencies

$$\omega_0^2 \leq \omega^2 \leq \omega_L^2. \tag{1.3.9}$$

Let $S = q^2/\mu\Omega\epsilon_0$. We find ω_L from Eq. (1.3.8):

$$\kappa_\infty + [S/(\omega_0^2 - \omega_L^2)] = 0, \qquad \omega_L^2 = \omega_0^2 + (S/\kappa_\infty). \tag{1.3.10}$$

In the region between ω_0 and ω_L, the crystal is perfectly reflecting: no electromagnetic wave will propagate. Inclusion of damping would remove the divergence at ω_0 and introduce absorption. We can eliminate S from the relation between ω_L and ω_0 by introducing the zero-frequency dielectric constant

$$\kappa_0 = \kappa_\infty + (S/\omega_0^2). \tag{1.3.11}$$

Note that the low frequency permittivity of the crystal is not ϵ_0 but $\kappa_0\epsilon_0$.

We have

$$\omega_L^2/\omega_0^2 = \kappa_0/\kappa_\infty. \tag{1.3.12}$$

This is Lyddane–Sachs–Teller relation (1941). With the use of (1.3.11) we may rewrite (1.3.8) in the form

$$\kappa(\omega) = \kappa_\infty + [\omega_0^2(\kappa_0 - \kappa_\infty)/(\omega_0^2 - \omega^2)]. \tag{1.3.13}$$

This is independent of the poorly known effective charge q. All the quantities in (1.3.13) can be determined experimentally.

1.3.2 *Polaritons*

Let us now consider the general dispersion curve for the propagation of a photon of wave vector $\mathbf{k}$ coupled to an optical mode in a crystal. From electromagnetic theory, we find this is given by

$$k^2 = \omega^2 \epsilon \mu_0 \tag{1.3.14}$$

where μ_0 is the permeability of free space. Since $c^2 = 1/\mu_0\epsilon_0$, we have

$$k^2c^2 = \omega^2\kappa(\omega). \tag{1.3.15}$$

Let us insert (1.3.8) for $\kappa(\omega)$ and solve for $\omega(k)$. The equation obtained is

$$\omega^2\{\kappa_\infty + [(\kappa_0 - \kappa_\infty)/(\omega_0^2 - \omega^2)]\omega_0^2\} = k^2c^2. \tag{1.3.16}$$

The solutions are

$$\omega^2 = (1/2\kappa_\infty)(k^2c^2 + \kappa_0\omega_0^2) \pm (1/2\kappa_\infty)[(k^2c^2 + \kappa_0\omega_0^2)^2 - 4\kappa_\infty\omega_0^2k^2c^2]^{1/2}. \tag{1.3.17}$$

As $k \to 0$, the solutions are

$$\omega^2 = (\kappa_0/\kappa_\infty)\omega_0^2 = \omega_L^2 \tag{1.3.18}$$

and

$$\omega^2 = (1/\kappa_0)k^2c^2. \tag{1.3.19}$$

These correspond to the phonon and photon modes. For large k, the two solutions are

$$\omega^2 = k^2c^2/\kappa_\infty \qquad \text{and} \qquad \omega^2 = \omega_0^2. \tag{1.3.20}$$

If the solutions of (1.3.17) are plotted as functions of k, it will be seen that the curves do not cross. An example computed with numbers appropriate to gallium phosphide is shown in Fig. 1.3.1 (see Henry and Hopfield, 1965). The two solutions of (1.3.17) correspond to transverse electromagnetic waves and lattice vibrations separately at $k = 0$; for $k \neq 0$, these disturbances are coupled. In the region of k in which the curves would try to cross, there is strong coupling between the optical and lattice waves and the actual normal modes of the coupled system contain both light-like

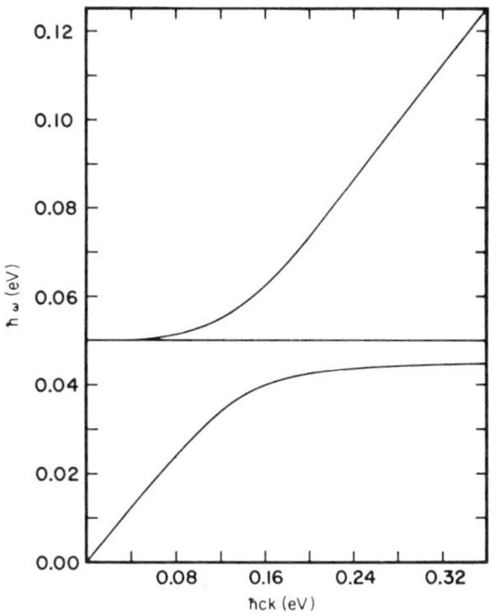

Fig. 1.3.1. Polariton dispersion relation for GaP. These curves are calculated assuming $\kappa_0 = 10.182$, $\kappa_\infty = 8.457$, $\omega_0 = 0.0455$.

and lattice-like components. The coupled system of optical modes plus transverse electromagnetic waves is referred to as a *polariton* (Hopfield, 1958, 1969). For k larger than in the region of avoided crossing, the dispersion curves again resemble optical modes and light, separately, but the character in this respect of the two curves has been exchanged.

The system has an additional mode, corresponding to a longitudinal vibration, which does not couple to the electromagnetic field. This vibration appears as the horizontal line in the figure. Its frequency is determined by the condition

$$\epsilon(\omega) = 0. \tag{1.3.21}$$

This condition, as it refers to a normal mode, is a consequence of the requirement that the particular one of Maxwell's equations

$$\nabla \cdot \mathbf{D} = \nabla \cdot (\epsilon \mathbf{E}) = 0 \tag{1.3.22}$$

be satisfied. Evidently there are two possibilities of satisfying (1.3.21); either through a transverse mode in which $\mathbf{k} \cdot \mathbf{E} = 0$, or through the case of vanishing dielectric constant (1.3.21), in which a longitudinal disturbance is possible. The solution of (1.3.21) is $\omega = \omega_L$ where ω_L is given by (1.3.10) or (1.3.12).

1.4 Quantization of Lattice Vibrations

In this section we will develop the formal machinery for the quantization of the lattice vibrations. The initial step is to uncouple the coupled lattice oscillators of the classical theory and then to express the Hamiltonian for the vibrating lattice as a sum of terms for the separate oscillators. To this end, we introduce the normal coordinate transformation.

1.4.1 *Normal Coordinates and the Hamiltonian*

Any lattice displacement can be expressed as a combination of the eigenvectors of the dynamical matrix. We write

$$u_{\alpha,l\kappa} = (\Omega/8\pi^3)(\mathfrak{N}/M_\kappa)^{1/2} \sum_j \int d^3k\, e_{\alpha,\kappa}^{(j)}(\mathbf{k}) Q_j(\mathbf{k}, \mathbf{t}) \exp(i\mathbf{k}\cdot\mathbf{R}_l). \quad (1.4.1a)$$

The constants standing in front of the summation above are chosen so as to insure the correct normalization of the expression for the total energy which will be obtained subsequently [Eq. (1.4.18)]. Since (1.4.1a) employs an integration over **k**, we will describe this and similar equations in terms of continuous **k**. It is, however, frequently convenient to consider **k** to be a discrete quantity. In this case, we have instead

$$u_{\alpha,l\kappa} = (\mathfrak{N}M_\kappa)^{-1/2} \sum_{j\mathbf{k}} e_{\alpha,\kappa}^{(j)}(\mathbf{k}) Q_j(\mathbf{k}, t) \exp(i\mathbf{k}\cdot\mathbf{R}_l). \quad (1.4.1b)$$

In this section, **k** will be treated as continuous, but a discrete representation will be employed in some of the following sections. The $u_{\alpha,l\kappa}$ must be real: $u^*_{\alpha,l\kappa} = u_{\alpha,l\kappa}$. Consequently

$$Q_j^*(\mathbf{k}, t) = Q_j(-\mathbf{k}, t). \quad (1.4.2)$$

The Q_j are referred to as *normal coordinates*. The lattice kinetic energy was given by (1.1.3). After substituting (1.4.1a) we have

$$T = \tfrac{1}{2}\mathfrak{N} \sum_{\alpha,l,\kappa,j,j'} (\Omega/8\pi^3)^2 \iint d^3k\, d^3q\, e_{\alpha,\kappa}^{(j)}(\mathbf{k}) e_{\alpha,\kappa}^{(j')}(\mathbf{q})$$
$$\times \exp[i(\mathbf{k}+\mathbf{q})\cdot\mathbf{R}_l]\dot{Q}_j(\mathbf{k}, t)\dot{Q}_{j'}(\mathbf{q}, t). \quad (1.4.3)$$

As before, the dot indicates derivative with respect to time.

Now we will use the summation relation (Appendix A)

$$\sum_l \exp[i(\mathbf{k}+\mathbf{q})\cdot\mathbf{R}_l] = [(2\pi)^3/\Omega] \sum_s \delta(\mathbf{k}+\mathbf{q}-\mathbf{K}_s). \quad (1.4.4)$$

However, **k** and **q** above are, by our previously discussed convention, required to be inside the Brillouin zone; so that $\mathbf{k}+\mathbf{q}$ cannot equal a nonzero reciprocal lattice vector, and we have, with the use of (1.1.21),

$$T = \tfrac{1}{2}[V/(2\pi)^3] \sum_{\alpha\kappa jj'} \int e_{\alpha,\kappa}^{(j)}(\mathbf{k}) e_{\alpha\kappa}^{(j')}(-\mathbf{k}) \dot{Q}_j(\mathbf{k}, t) Q_{j'}(-\mathbf{k}, t)\, d^3k$$

$$= \tfrac{1}{2}[V/(2\pi)^3] \sum_{\alpha\kappa jj'} \int e_{\alpha\kappa}^{(j')*}(\mathbf{k}) e_{\alpha\kappa}^{(j)}(\mathbf{k}) \dot{Q}^*_{j'}(\mathbf{k}, t) Q_j(\mathbf{k}, t)\, d^3k.$$

We now use the orthonormality relation (1.1.18a) to reduce this further to

$$T = \tfrac{1}{2}[V/(2\pi)^3] \sum_j \int d^3k \mid \dot{Q}_j(\mathbf{k}, t) \mid^2. \tag{1.4.5}$$

We also treat the potential energy in the same manner. Substitute (1.4.1a) into (1.1.13). The constant term Φ_0 is discussed

$$\Phi = \tfrac{1}{2} \sum_{i\kappa\alpha,\, l\lambda\beta} \Phi_{\alpha\beta,i\kappa,l\lambda}[\mathfrak{N}\Omega^2/(2\pi)^6(M_\kappa M_\lambda)^{1/2}] \sum_{jj'} \int d^3q\, d^3k$$
$$\times\, e_{\alpha\kappa}{}^{(j)}(\mathbf{k}) e_{\beta\lambda}{}^{(j')}(\mathbf{q}) Q_j{}^*(\mathbf{k}, t) Q_{j'}(\mathbf{q}, t) \exp[i(\mathbf{k}\cdot\mathbf{R}_i + \mathbf{q}\cdot\mathbf{R}_l)].$$

To reduce this, we recall that Φ is a function of $\mathbf{R}_i - \mathbf{R}_l$ only. Then let $\mathbf{k}\cdot\mathbf{R}_i = \mathbf{k}\cdot(\mathbf{R}_i - \mathbf{R}_l) + \mathbf{k}\cdot\mathbf{R}_l$. Insert this in the exponential. The sum over i and l becomes a sum over $\mathbf{R}_i - \mathbf{R}_l$ and the sum on l alone can be done, yielding $[(2\pi)^3/\Omega]\,\delta(\mathbf{k} + \mathbf{q})$. Then we use the definition of the dynamical matrix to yield

$$\Phi = \tfrac{1}{2}[V/(2\pi)^3] \sum_{\kappa\lambda,\, \alpha\beta,\, jj'} \int d^3k\, D_{\alpha\beta,\kappa\lambda}(\mathbf{k}) e_{\alpha\kappa}{}^{(j)}(\mathbf{k})$$
$$\times e_{\beta\lambda}{}^{(j')}(-\mathbf{k}) Q_j(\mathbf{k}, t) Q_j(-\mathbf{k}, t). \tag{1.4.6}$$

However, the e's are eigenvectors of the dynamical matrix. We use (1.1.17) and (1.1.18a) to reduce this still further to

$$\Phi = \tfrac{1}{2}[V/(2\pi)^3] \sum_j \int d^3k\, \omega_j{}^2(\mathbf{k}) \mid Q_j(\mathbf{k}, t) \mid^2. \tag{1.4.7}$$

The Lagrangian for the lattice is

$$L = T - V = [V/(2\pi)^3] \int \mathfrak{L}\, d^3k$$

$$= \tfrac{1}{2}[V/(2\pi)^3] \sum_j \int d^3k\, [\dot{Q}_j{}^*(\mathbf{k}) \dot{Q}_j(\mathbf{k}) - \omega_j{}^2(\mathbf{k}) Q_j{}^*(\mathbf{k}) Q_j(\mathbf{k})] \tag{1.4.8}$$

where $\mathfrak{L}$ is the Lagrangian density. The $Q_j(\mathbf{k})$ are regarded as cannonical coordinates. The momentum conjugate to $Q_j(\mathbf{k})$ is

$$P_j(\mathbf{k}) = \partial\mathfrak{L}/\partial\dot{Q}_j{}^*(k) = \dot{Q}_j(k). \tag{1.4.9a}$$

[Note that the factor of $\tfrac{1}{2}$ disappears since $Q_j(\mathbf{k})$ effectively occurs twice on account of (1.4.2).] Now we can write the Hamiltonian

$$H = \tfrac{1}{2}[V/(2\pi)^3]$$
$$\times \sum_j \int d^3k\, [P_j{}^*(\mathbf{k}) P_j(\mathbf{k}) + \omega_j{}^2(\mathbf{k}) Q_j{}^*(\mathbf{k}) Q_j(\mathbf{k})]. \tag{1.4.9b}$$

This Hamiltonian is simply the sum of Hamiltonians for uncoupled simple harmonic oscillators. Hamilton's cannonical equations now yield the classical equations of motion in the form

$$\ddot{Q}_j(\mathbf{k}) + \omega_j^2(\mathbf{k})Q_j(\mathbf{k}) = 0. \tag{1.4.10}$$

1.4.2 *Phonons*

The transition to quantum mechanics is made by regarding the $Q_j(\mathbf{k})$ and $P_j(\mathbf{k})$ as quantum mechanical operators which have the commutation relations

$$[Q_j^*(\mathbf{k}), P_{j'}(\mathbf{q})] = i\hbar\, \delta(\mathbf{k} - \mathbf{q})\, \delta_{jj'}. \tag{1.4.11a}$$

Here it is assumed that $\mathbf{k}$ and $\mathbf{q}$ are inside the Brillouin zone. To include surface points, or to extend this to $\mathbf{k}$, $\mathbf{q}$, outside the BZ, we write

$$[Q_j^*(\mathbf{k}), P_{j'}(\mathbf{q})] = i\hbar\, \delta_{jj'} \sum_s \delta(\mathbf{k} - \mathbf{q} - \mathbf{K}_s). \tag{1.4.11b}$$

If we take the complex conjugate of (1.4.11a), we have

$$[Q_j(\mathbf{k}), P^*_{j'}(\mathbf{q})] = -i\hbar\, \delta(\mathbf{k} - \mathbf{q})\, \delta_{jj'}.$$

Other commutators involving the Q's and P's vanish:

$$[Q_j(\mathbf{k}), Q^*_{j'}(\mathbf{q})] = [P_j(\mathbf{k}), P^*_{j'}(\mathbf{q})] = [Q_j(\mathbf{k}), P_{j'}(\mathbf{q})] = 0. \tag{1.4.12}$$

We will now introduce creation and annihilation operators a_{kj}, $a^\dagger_{kj}$ by

$$a_j(\mathbf{k}) = \{[\omega_j(\mathbf{k})/2\hbar]^{1/2}Q_j(\mathbf{k}) + i[2\hbar\omega_j(\mathbf{k})]^{-1/2}P_j(\mathbf{k})\}, \tag{1.4.13a}$$

$$a_j^\dagger(\mathbf{k}) = \{[\omega_j(\mathbf{k})/2\hbar]^{1/2}Q_j^*(\mathbf{k}) - i[2\hbar\omega_j(\mathbf{k})]^{-1/2}P_j^*(\mathbf{k})\}. \tag{1.4.13b}$$

The inverse of these relations is

$$Q_j(\mathbf{k}) = [\hbar/2\omega_j(\mathbf{k})]^{1/2}[a_j(\mathbf{k}) + a_j^\dagger(-\mathbf{k})], \tag{1.4.14a}$$

$$P_j(\mathbf{k}) = i^{-1}[\hbar\omega_j(\mathbf{k})/2]^{1/2}[a_j(\mathbf{k}) - a_j^\dagger(-\mathbf{k})]. \tag{1.4.14b}$$

We then obtain the commutation relations

$$[a_j(\mathbf{k}), a^\dagger_{j'}(\mathbf{q})] = \delta_{jj'}\, \delta(\mathbf{k} - \mathbf{q}) \tag{1.4.15a}$$

and

$$[a_j(\mathbf{k}), a_{j'}(\mathbf{q})] = [a_j^\dagger(\mathbf{k}), a^\dagger_{j'}(\mathbf{q})] = 0. \tag{1.4.15b}$$

When Eqs. (1.4.14) are substituted into H [Eq. (1.4.9)], we obtain

$$H = [V/(2\pi)^3] \sum_j \int d^3k\, \hbar\omega_j(\mathbf{k})[a_j^\dagger(\mathbf{k})a_j(\mathbf{k}) + \tfrac{1}{2}]. \tag{1.4.16}$$

Equations (1.4.15) and (1.4.16) are fundamental results. The Hamiltonian has been expressed as a sum of independent quantum harmonic oscillators, whose frequencies are those of the classical normal modes. Henceforth, these quantum oscillators will be called *phonons*. Standard procedures may now be employed to deduce the eigenvalues of H and the matrix elements of a and $a^\dagger$ (Messiah, 1961). This argument will not be repeated here. For our future use, the essential results are as follows: The number of phonons in a state of wave vector $\mathbf{k}$ belonging to branch j is the eigenvalue of the number operator

$$n_j(\mathbf{k}) = a_j^\dagger(\mathbf{k})a_j(\mathbf{k}). \tag{1.4.17}$$

We will use the symbol $n_j(\mathbf{k})$ interchangeably for the operator and its eigenvalues (which are positive integers). The energy is

$$E = [V/(2\pi)^3] \sum_j \int d^3k\ \hbar\omega_j(\mathbf{k})[n_j(\mathbf{k}) + \tfrac{1}{2}]. \tag{1.4.18}$$

This formula will form the basis for the discussion of thermodynamics in Section 1.5. We will also require the matrix elements of the operators a and $a^\dagger$ between states which are eigenstates of $n_j(\mathbf{k})$. These may be determined by inspection from the basic equations below. Suppose $|\, n_j(\mathbf{k})\rangle$ is an eigenstate of the number operator. Then

$$a_j(\mathbf{k})\,|\, n_j(\mathbf{k})\rangle = [n_j(\mathbf{k})]^{1/2}\,|\, n_j(\mathbf{k}) - 1\rangle, \tag{1.4.19}$$

$$a_j^\dagger(\mathbf{k})\,|\, n_j(\mathbf{k})\rangle = [n_j(\mathbf{k}) + 1]^{1/2}\,|\, n_j(\mathbf{k}) + 1\rangle. \tag{1.4.20}$$

The operators a and $a^\dagger$ will be referred to as *annihilation* and *creation operators* since they decrease or increase, respectively, the number of phonons in a given mode by 1. In the harmonic approximation, there is no interaction between phonon states. An arbitrary number of phonons may be present in any mode.

A state of the vibrating crystal may be specified by listing the number of phonons present in each mode. If $|\, 0\rangle$ denotes the state in which no phonons are present, the n phonon state, properly normalized is

$$|\,\{n_j(\mathbf{k})\}\rangle = \{\prod_{k_i j_i} [n_{j_i}(\mathbf{k}_i)!]\}^{-1/2} a^\dagger_{j_1}(\mathbf{k}_1)a^\dagger_{j_2}(\mathbf{k}_2)\cdots a^\dagger_{j_n}(\mathbf{k}_n)\,|\, 0\rangle. \tag{1.4.21}$$

In this state, there are n_{j_i} phonons present in the ith mode (wave vector $\mathbf{k}_i$, branch j_i). The total number of phonons present, n, is the sum over all modes,

$$\sum_{j,i} n_{j_i}(\mathbf{k}_i) = n.$$

For convenience, we have considered the modes to be discrete. A detailed discussion of the second quantization procedure employed here is given by Schweber (1961).

For future use, we require an expression for an atomic displacement in terms of phonon creation and annihilation operators. To obtain this we combine (1.4.1a) and (1.4.14a):

$$u_{\alpha,l\kappa} = (\Omega/8\pi^3)(\hbar\mathfrak{N}/M_\kappa)^{1/2} \sum_j \int \{d^3k/[2\omega_j(\mathbf{k})]^{1/2}\} e_{\alpha,\kappa}{}^{(j)}(\mathbf{k})[a_j(\mathbf{k}) + a_j{}^\dagger(-\mathbf{k})] \exp(i\mathbf{k}\cdot\mathbf{R}_l). \tag{1.4.22a}$$

This is the operator in the Schrödinger picture: it is time-independent. An alternative form is

$$u_{\alpha,l\kappa} = (\Omega/8\pi^3)(\hbar\mathfrak{N}/M_\kappa)^{1/2} \sum_j \int \{d^3k/[2\omega_j(\mathbf{k})]^{1/2}\}[e_{\alpha,\kappa}{}^{(j)}(\mathbf{k}) \times\exp(i\mathbf{k}\cdot\mathbf{R}_l)a_j(\mathbf{k}) + e_{\alpha,\kappa}{}^{(j)*}(\mathbf{k}) \exp(-i\mathbf{k}\cdot\mathbf{R}_l)a_j{}^\dagger(\mathbf{k})]. \tag{1.4.22b}$$

It is also desirable to have an expression for $u_{\alpha,l\kappa}$ when $\mathbf{k}$ is treated as a discrete quantity. Instead of (1.4.22b), one finds

$$u_{\alpha,l\kappa} = (\hbar/\mathfrak{N}M_\kappa)^{1/2} \sum_{j\mathbf{k}} [2\omega_j(\mathbf{k})]^{-1/2}\{e_{\alpha,\kappa}{}^{(j)}(\mathbf{k}) \exp(i\mathbf{k}\cdot\mathbf{R}_l)a_j(\mathbf{k}) + e_{\alpha,\kappa}{}^{(j)}(\mathbf{k}) \exp(-i\mathbf{k}\cdot\mathbf{R}_l)a_j{}^\dagger \mathrm{I}(\mathbf{k})\} \tag{1.4.23}$$

1.5 Thermodynamics and the Density of States

For many applications in thermodynamics and transport theory, it is desirable to introduce the density of states (or frequency distribution function).

1.5.1 *Definitions and General Properties*

We define $g(\omega)\, d\omega$ to be the fractional number of frequencies in the range between ω and $\omega + d\omega$. Similarly, $G(\omega^2)$ is the fractional number of squared frequencies in the range $(\omega^2 \to \omega^2 + d\omega^2)$. These functions are related by

$$g(\omega) = 2\omega G(\omega^2). \tag{1.5.1}$$

We will normalize the density of states by requiring that for each branch

$$\int_0^{\omega_L^{(j)}} g_j(\omega)\, d\omega = 1 \tag{1.5.2}$$

where $\omega_L^{(j)}$ is the largest frequency in this branch. Thus the entire distribution function satisfies

$$\int_0^{\omega_L} g(\omega)\, d\omega = \sum_{j=1}^{3r} \int_0^{\omega_L} g_j(\omega)\, d\omega = 3r \tag{1.5.3}$$

where r is, as before, the number of atoms in the unit cell and ω_L is the maximum vibration frequency of the crystal. The computation and the use of this object will be investigated subsequently.

It was shown in Section 1.4 that the energy in each mode of vibration $(j, \mathbf{k})$ of the crystal is $\hbar\omega_j(\mathbf{k})[n_j(\mathbf{k}) + \frac{1}{2}]$ where $n_j(\mathbf{k})$ is the number of quanta in the mode (or occupation number). The total energy of the crystal is, for some assumed set of occupation numbers,

$$E_{\{n\}} = \sum_{j\mathbf{k}} [n_j(\mathbf{k}) + \tfrac{1}{2}]\hbar\omega_j(\mathbf{k}). \tag{1.5.4}$$

The calculation of the equilibrium thermodynamic properties of a crystal begins with the partition function Z. This can be written in terms of a sum over all possible choices of occupation numbers

$$Z = \sum_{\{n_j(\mathbf{k})\}} \exp(-\beta E_{\{n\}}) \tag{1.5.5}$$

where $\beta = 1/KT$, K being Boltzmann's constant. Substitute Eq. (1.5.4):

$$Z = \prod_{j\mathbf{k}} \exp[-\tfrac{1}{2}\beta\hbar\omega_j(\mathbf{k})] \sum_{n_j(\mathbf{k})} \exp(-\beta n_j\hbar\omega_j).$$

Now $n_j(\mathbf{k})$ takes all positive integral values including zero; the sum is therefore a geometrical series, and we have

$$\sum_{n_j(\mathbf{k})} \exp[-\beta n_j\hbar\omega_j(\mathbf{k})] = 1/\{1 - \exp[-\beta\hbar\omega_j(\mathbf{k})]\}.$$

Thus

$$Z = \prod_{j\mathbf{k}} \exp[-\tfrac{1}{2}\beta\hbar\omega_j(\mathbf{k})]/\{1 - \exp[-\beta\hbar\omega_j(\mathbf{k})]\}. \tag{1.5.6}$$

This enables us to start thermodynamics. We now obtain the Helmholtz free energy F:

$$\begin{aligned} F = -KT\ln Z &= KT \sum_{j\mathbf{k}} \ln\{[1 - \exp(-\beta\hbar\omega_j(\mathbf{k}))]/\exp[-\tfrac{1}{2}\beta\hbar\omega_j(\mathbf{k})]\} \\ &= KT \sum_{j\mathbf{k}} \ln[2\sinh(\hbar\omega_j(\mathbf{k})/2KT)]. \end{aligned} \tag{1.5.7}$$

The entropy S is

$$\begin{aligned} S = -(\partial F/\partial T)_V = K \sum_{j\mathbf{k}} (&[\hbar\omega_j(\mathbf{k})/2KT]\,\mathrm{ctnh}[\hbar\omega_j(\mathbf{k})/2KT] \\ &- \ln\{2\sinh[\hbar\omega_j(\mathbf{k})/2KT]\}). \end{aligned} \tag{1.5.8}$$

All quantities refer to constant volume. Then the internal energy U is

$$U = F + TS = \tfrac{1}{2} \sum_{j\mathbf{k}} \hbar\omega_j(\mathbf{k}) \operatorname{ctnh}[\hbar\omega_j(\mathbf{k})/2KT]$$
$$= \sum_{j\mathbf{k}} ([\hbar\omega_j(\mathbf{k})/2] + \{\hbar\omega_j(\mathbf{k})/[\exp[\hbar\omega_j(\mathbf{k})/KT] - 1]\}). \tag{1.5.9}$$

The heat capacity at constant volume is

$$C = (\partial U/\partial T)_V$$
$$= K \sum_{j\mathbf{k}} [\hbar\omega_j(\mathbf{k})/KT]^2 (\exp[\hbar\omega_j(\mathbf{k})/KT]/\{\exp[\hbar\omega_j(\mathbf{k})/KT] - 1\}^2)$$
$$= K \sum_{j\mathbf{k}} [\hbar\omega_j(\mathbf{k})/2KT]^2/\sinh^2[\hbar\omega_j(\mathbf{k})/2KT]. \tag{1.5.10}$$

These expressions may be converted to integrals over the frequency by introducing the density of states. This may be done by direct use of the definition, recognizing that the only dependence on $\mathbf{k}$ is through $\omega_j(\mathbf{k})$, or more formally, as will be discussed below. For example,

$$U = \mathfrak{N} \int g(\omega) \{(\hbar\omega/2) + [\hbar\omega/(e^{\hbar\omega/KT} - 1)]\}\, d\omega, \tag{1.5.11}$$

$$C = \mathfrak{N} K \int g(\omega) (\hbar\omega/2KT)^2 \operatorname{csch}^2(\hbar\omega/2KT)\, d\omega. \tag{1.5.12}$$

According to the definition of the density of states, we are required, essentially, to count frequencies which are equal to ω. The counting is done by means of the Kronecker delta $\delta_{\omega,\omega_j(\mathbf{k})}$ or by the delta function $\delta[\omega - \omega_j(\mathbf{k})]$, which gives a unit contribution if $\omega_j(\mathbf{k}) = \omega$, and zero otherwise. Thus

$$g_j(\omega) \approx \sum_{j\mathbf{k}} \delta_{\omega,\omega_j(\mathbf{k})} \approx \int d^3k\, \delta[\omega - \omega_j(\mathbf{k})].$$

The required constant of proportionality is determined by normalization. Recall that $g(\omega)$ is the fractional number of frequencies; that is, the ratio of the number of frequencies in the particular range to the number in the entire zone. The total number of states in a given branch is just $\mathfrak{N}$ [from Eq. (1.2.10)]. Hence, again using (1.2.10),

$$g_j(\omega) = [\Omega/(2\pi)^3] \int d^3k\, \delta[\omega - \omega_j(\mathbf{k})]. \tag{1.5.13}$$

This relation was first introduced by Bowers and Rosenstock (1950). Equations (1.2.10) and (1.3.13) may be used to introduce the density of states into thermodynamic functions. Thus, considering the specific heat, for example, we derive (1.5.12) by setting

$$C = [\mathfrak{N}\Omega/(2\pi)^3]K \sum_j \int d^3k\ (\hbar\omega_j/2KT)^2 \operatorname{csch}^2(\hbar\omega_j(\mathbf{k})/2KT)$$

$$= [\mathfrak{N}\Omega/(2\pi)^3]K \sum_j \int d^3k \int d\omega\ (\hbar\omega/2KT)^2 \operatorname{csch}^2(\hbar\omega/2KT)\ \delta[\omega - \omega_j(\mathbf{k})]$$

$$= \mathfrak{N}K \int d\omega \sum_j g_j(\omega)\,(\hbar\omega/2KT)^2 \operatorname{csch}^2(\hbar\omega/2KT)$$

$$= \mathfrak{N}K \int d\omega\, g(\omega)\,(\hbar\omega/2KT)^2 \operatorname{csch}^2(\hbar\omega/2KT).$$

1.5.2 *Calculation of the Density of States*

We will base our discussion of the density of states on Eq. (1.5.13). Let us note first that in the one-dimensional case (in which the volume or more properly the length of the Brillouin zone is $2\pi/a$, a being the distance between like atoms), Eq. (1.5.13) gives

$$g(\omega) = (a/\pi)\,|dk/d\omega|. \tag{1.5.14}$$

For a linear chain of like atoms with nearest neighbor interactions, this formula yields

$$g(\omega) = 2[\pi(\omega_m{}^2 - \omega^2)^{1/2}]^{-1} \qquad (\omega < \omega_m)$$

where ω_m is the highest frequency of the chain. The inverse square root singularity is characteristic of one-dimensional problems.

The calculation of the density of states in a three-dimensional system is frequently a difficult task, requiring sophisticated numerical methods (for a discussion of such methods, see Gilat, 1972). Our considerations will be limited to simple cases to which analytical procedures may be applied, at least in part.

The density of states can be formally reduced to a surface integral by the replacement

$$d^3k \to (dS_\omega/|\nabla_\mathbf{k}\omega|)d\omega$$

where dS_ω is an element of area on a surface of constant frequency. Thus

$$g_j(\omega) = (\Omega/8\pi^3) \int dS_{\omega j}/|\nabla_\mathbf{k}\omega_j|. \tag{1.5.15}$$

The surface specified in (1.5.15) is that for which $\omega_j(\mathbf{k}) = \omega$.

The complexity of the surface of integration makes (1.5.15) difficult to apply. Another approach of some interest is to introduce the integral representation of the delta function

$$\delta(x) = (1/2\pi) \int_{-\infty}^{\infty} e^{ixt}\, dt. \tag{1.5.16}$$

This enables us to write the density of states as a Fourier transform

$$g_j(\omega) = (1/2\pi) \int e^{i\omega t} f_j(t)\, dt \tag{1.5.17}$$

where

$$f_j(t) = [\Omega/(2\pi)^3] \int d^3k \exp[-i\omega_j(\mathbf{k})t]. \tag{1.5.18}$$

As an example of this we will consider the calculation of $G(\omega^2)$ for a monatomic simple cubic lattice with nearest neighbor interactions only. However, in order not to have an entirely trivial problem we will include noncentral forces: In this case, we consider a potential energy which has the form [instead of (1.1.41)]

$$\begin{aligned}\Phi = (\alpha/2) \sum_{lmn} [&(x_{lmn} - x_{l+1,m,n})^2 + (y_{lmn} - y_{l,m+1,n})^2 + (z_{lmn} - z_{l,m,n+1})^2] \\ + (\beta/2) \sum_{lmn} [&(x_{lmn} - x_{l,m+1,n})^2 + (x_{lmn} - x_{l,m,n+1})^2 \\ &+ (y_{lmn} - y_{l+1,m,n})^2 + (y_{lmn} - y_{l,m,n+1})^2 \\ &+ (z_{lmn} - z_{l+1,m,n})^2 + (z_{lmn} - z_{l,m+1,n})^2].\end{aligned} \tag{1.5.19}$$

The parameter β describes noncentral forces, that is, forces on an atom produced by a displacement of a neighboring atom which is perpendicular to the line joining them. The dynamical matrix is diagonal in this model (not a very realistic situation). We find

$$\begin{aligned}\omega_1^2 &= 2\alpha(1 - \cos k_x a) + 2\beta(2 - \cos k_y a - \cos k_z a), \\ \omega_2^2 &= 2\alpha(1 - \cos k_y a) + 2\beta(2 - \cos k_x a - \cos k_z a), \\ \omega_3^2 &= 2\alpha(1 - \cos k_z a) + 2\beta(2 - \cos k_x a - \cos k_y a).\end{aligned} \tag{1.5.20}$$

The frequency spectrum is evidently of the same form for each branch. In this case $f(t)$ can be written as the product of three single integrals; a typical one of which is (we consider ω^2 rather than ω)

$$(a/2\pi) \int_{-\pi/a}^{\pi/a} dk \exp[-2i\alpha t(1 - \cos k_x a)].$$

This integral may be evaluated using the formula

$$(1/2\pi) \int_{-\pi}^{\pi} \exp(2i\alpha t \cos q)\, dq = J_0(2\alpha t)$$

where J_0 is a Bessel function of order zero. Then

$$G_j(\omega^2) = (1/2\pi) \int_{-\infty}^{\infty} dt \exp[i(\omega^2 - 2\alpha - 4\beta)t] J_0(2\alpha t) J_0^2(2\beta t)$$

$$= (1/\pi) \int_0^{\infty} dt \cos[(\omega^2 - 2\alpha - 4\beta)t]\, J_0(2\alpha t) J_0^2(2\beta t). \qquad (1.5.21)$$

A graph of this function is shown in Fig. 1.5.1 for the particular case $\alpha = \beta$. It will be noticed that G has some sharp corners. Such sharp corners are not just special features of the particular approximate model we are using.

In two dimensions, the corresponding integral may be evaluated in closed form. We obtain for the case $\alpha = \beta$,

$$G(\omega^2) = (4/\pi^2\omega_m^2) K[(4\omega/\omega_m)(1 - \omega/\omega_m)] \qquad (0 < \omega < \omega_m)$$

$$= 0 \qquad (\omega > \omega_m) \qquad (1.5.22)$$

where ω_m is the maximum frequency $\omega_m^2 = 8\alpha$ and K is the complete elliptic

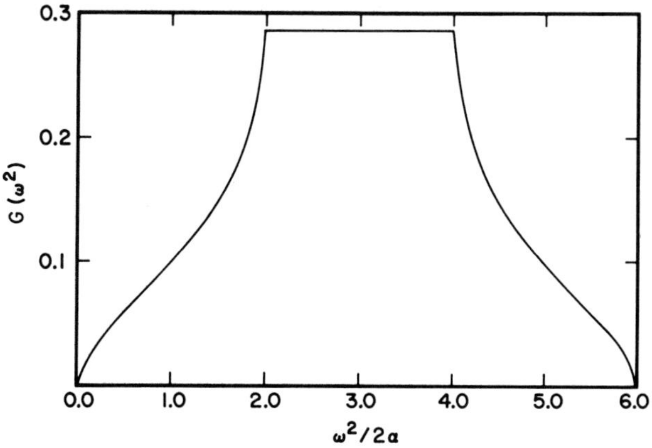

Fig. 1.5.1. Density of states $G(\omega^2)$ for a simple cubic lattice according to Eq. (1.5.21).

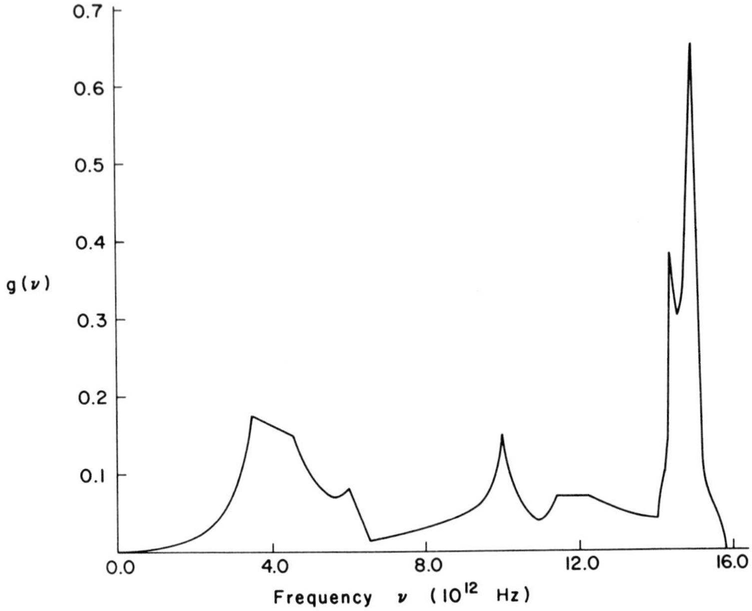

Fig. 1.5.2. Density of states for silicon. [After Dolling and Cowley (1966).]

integral of the first kind:

$$K(k) = \int_0^{\pi/2} d\phi/(1 - k^2 \sin^2 \phi)^{1/2}. \qquad (1.5.23)$$

This density of states has a logarithmic singularity for $\omega = \frac{1}{2}\omega_m$. Such logarithmic singularities are characteristic of two-dimensional systems.

In general, all densities of states exhibit sharp structure. In Fig. 1.5.2 we show a computed density of states for silicon. The rich structure is apparent.

1.5.3 *Van Hove Singularities*

This structure has its origins in the following considerations. There are many points in the Brillouin zone where $|\nabla_k \omega(\mathbf{k})|$ is required to vanish by reasons of symmetry. Such arguments will be examined in Chapter 3. If this happens, we see from Eq. (1.5.15) that the density of states may be singular. Suppose $\mathbf{k}_0$ is a point at which $|\nabla_k \omega| = 0$ (for simplicity we assume there is no degeneracy). We will investigate what happens to $g(\omega)$.

Let us consider the expansion of the frequency in powers of $\mathbf{s} = \mathbf{k} - \mathbf{k}_0$. The frequency is evidently a quadratic form in the components of $\mathbf{s}$; we may without loss of generality imagine we have chosen our coordinate system to coincide with the principal axes of this form, so that we have

$$\omega(\mathbf{k}) = \omega(\mathbf{k}_0) + \sum_{i=1}^{3} \alpha_i s_i^2. \qquad (1.5.24)$$

A point $\mathbf{k}_0$, where such an expansion is possible, is called an "analytic critical point." There are four possible types of analytic critical points (designated P_i according to the values of the coefficients α). These are enumerated as follows:

P_0: $\alpha_1, \alpha_2, \alpha_3$ all negative (maximum),
P_1: α_1, α_2 negative, α_3 positive (saddle point),
P_2: α_1 negative, α_2, α_3 positive (saddle point),
P_3: $\alpha_1, \alpha_2, \alpha_3$ all positive (minimum).

There are fundamental considerations of a topological nature which relate the number of critical points of the several kinds. These relations, first obtained by Morse (1938), were applied to the frequency distribution function for lattice vibrations by Van Hove (1953) and by Phillips (1956). The frequency may be considered to be a multiply periodic function of wave vector in the reciprocal lattice. For the purposes of illustrating the argument in a simple manner, let us consider (Montroll, 1954) a two-dimensional square reciprocal lattice which is shown in Fig. 1.5.3. Suppose

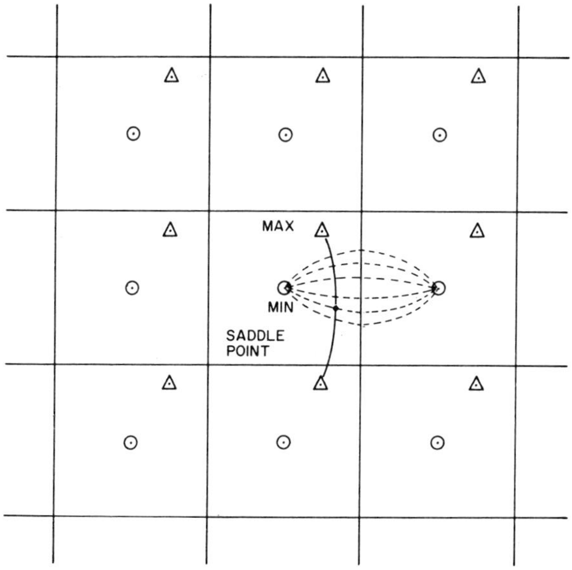

Fig. 1.5.3. Critical points in a square lattice.

that a simple dispersion relation exists in this system, which has a maximum and a minimum of frequency for the points in the cell shown in the diagram. Let us imagine a set of curves connecting the minima in adjacent cells as shown. On each curve there is a point at which the frequency is a relative maximum. The locus of such relative maxima may be obtained: we suppose it is the solid curve passing through the maxima. On this curve, there is a lowest relative maximum: this point is a saddle point. Similarly, on drawing curves connecting the absolute maxima in two cells, one obtains a locus of relative minima, and thereby finds another saddle point which is the highest relative minimum. Evidently two saddle points must exist. (Actually, if the absolute maximum occurred at a point of low symmetry as shown in the diagram, it would have to be repeated at seven other points inside the cell.)

The fundamental result of the previous discussion is that the numbers of the critical points of the various type are not independent if the $\omega(\mathbf{k})$ function is multiply periodic. Let N_j be the number of critical points of type P_j (we are still considering only analytic critical points). The relations which must hold between the N_j are (for a three-dimensional situation), according to Morse,

$$N_0 \geq 1, \qquad N_1 - N_0 \geq 2, \qquad N_2 - N_1 + N_0 \geq 1,$$
$$N_3 - N_2 + N_1 - N_0 = 0. \tag{1.5.25}$$

The minimum set of critical points in a given band is thus one maximum, one minimum, and three saddle points of each kind (P_1 and P_2). The simple expression

$$\omega^2 = 2\alpha[3 - \cos k_x a - \cos k_y a - \cos k_z a]$$

exhibits the minimum number of critical points. The minimum occurs at the zone center Γ, three saddle points of type P_2 occur at the face center X, three saddle points of type P_1 occur at the middle of an edge M, and the maximum occurs at the corner R.

We will now study the behavior of the density of states near a critical point following the treatment of Wannier (1959). It is convenient to define new variables $r_i = |\alpha_i|^{1/2} s_i$ in (1.5.24) so that

$$\omega(\mathbf{k}) - \omega(\mathbf{k}_0) = \sum_{i=1}^{3} \epsilon_i r_i^2 \qquad \text{with} \quad \epsilon_i = \pm 1. \tag{1.5.26}$$

The behavior in the vicinity of a minimum or maximum can be determined easily. Suppose we have a maximum. In this case, the ϵ_i will all be negative. We substitute in (1.5.13), change the variable of integration from k to r, and find immediately that

$$g(\omega) \approx [\omega(\mathbf{k}_0) - \omega]^{1/2}. \tag{1.5.27}$$

Near a minimum, we find similarly that

$$g(\omega) \approx [\omega - \omega(\mathbf{k}_0)]^{1/2}. \tag{1.5.28}$$

However, this does not describe the behavior of the acoustic branches near $\mathbf{k} = 0$ since $|\nabla_k\omega|$ is independent of $|\mathbf{k}|$. In that case, we see that for small k, $g(\omega) \approx \omega^2$. More explicitly, let

$$\omega_j(\mathbf{k}) = v_j(\theta, \phi)k \tag{1.5.29}$$

for some branch j where v_j is the speed of sound in the direction specified by θ, ϕ, and $k = |\mathbf{k}|$. Then we find directly from (1.5.13), that

$$g_j(\omega) = \omega^2[\Omega/(2\pi)^3] \int d\sigma/v_j^3(\theta, \phi) \tag{1.5.30}$$

where we integrate over solid angles ($d\sigma = \sin\theta\, d\theta\, d\phi$). The total density of states for the acoustic branches for frequencies low enough for (1.5.29) to apply is then

$$g_{\mathrm{A}}(\omega) = \sum_j g_j(\omega) = 3\Omega\omega^2/2\pi^2 v_{\mathrm{A}}^3 \tag{1.5.31}$$

where, for future convenience, we have defined

$$3/v_{\mathrm{A}}^3 = (1/4\pi) \sum_j \int d\sigma/v_j^3(\theta, \phi). \tag{1.5.32}$$

Next, we consider in detail the behavior of the density of states near a saddle point. These introduce discontinuities in the derivative of the density of states with respect to frequency. In order to have a concrete example, let us assume

$$\omega(\mathbf{k}) - \omega(\mathbf{k}_0) = r_1^2 + r_2^2 - r_3^2. \tag{1.5.33}$$

When $\omega(\mathbf{k}) > \omega(\mathbf{k}_0)$, the surfaces of constant frequency may be represented as hyperboloids of one sheet; when $\omega(\mathbf{k}) = \omega(\mathbf{k}_0)$, the surface becomes a cone passing through the saddle point; for $\omega(\mathbf{k}) < \omega(\mathbf{k}_0)$, the energy surfaces are hyperboloids of two sheets. The area of the surfaces of constant energy changes very rapidly when $\omega(\mathbf{k})$ passes through $\omega(\mathbf{k}_0)$; this increase in area is responsible for the discontinuity of the derivative of the density of states. To determine $g(\omega)$, put

$$|\omega(\mathbf{k}) - \omega(\mathbf{k}_0)| = a^2,$$

$$r_1 = a\cosh\xi\cos\phi, \qquad r_2 = a\cosh\xi\sin\phi, \qquad r_3 = a\sinh\xi$$
$$\text{for}\quad \omega(\mathbf{k}) > \omega(\mathbf{k}_0), \tag{1.5.34a}$$

$$r_1 = a\sinh\xi\cos\phi, \qquad r_2 = a\sinh\xi\sin\phi, \qquad r_3 = a\cosh\xi$$
$$\text{for}\quad \omega(\mathbf{k}) < \omega(\mathbf{k}_0). \tag{1.5.34b}$$

The surfaces of constant frequency are then surfaces of constant a. The density of states is determined by expressing the volume element d^3k in the coordinates specified in (1.5.34) with the use of the Jacobian determinant. The integrals must be limited to a finite region in the vicinity of the saddle point. This may be done by requiring that

$$r_1^2 + r_2^2 + r_3^2 = a^2(\sinh^2\xi + \cosh^2\xi) = R^2 \tag{1.5.35}$$

where R is some fixed number. Thus, for $\omega(\mathbf{k}) > \omega(\mathbf{k}_0)$, $0 \le \sinh\xi \le [\frac{1}{2}(R^2a^{-2} - 1)]^{1/2}$ and for $\omega(\mathbf{k}) < \omega(\mathbf{k}_0)$, $1 \le \cosh\xi \le [\frac{1}{2}(R^2a^{-2} + 1)]^{1/2}$. Thus, for $\omega(\mathbf{k}) > \omega(\mathbf{k}_0)$,

$$g(\omega) \propto (R^2 - a^2)^{1/2} = [R^2 + \omega(\mathbf{k}_0) - \omega(\mathbf{k})]^{1/2}, \tag{1.5.36a}$$

while, for $E(\mathbf{k}) < E(\mathbf{k}_0)$,

$$g(\omega) \propto (R^2 + a^2)^{1/2} - a = [(R^2 + \omega(\mathbf{k}_0) - \omega(\mathbf{k})]^{1/2} - [\omega(\mathbf{k}_0) - \omega(\mathbf{k})]^{1/2}. \tag{1.5.36b}$$

The function $[R^2 + \omega(\mathbf{k}_0) - \omega(\mathbf{k})]^{1/2}$ is continuous and has continuous derivatives at the saddle point, so it is evident that $g(\omega)$ has a discontinuity in its first derivative. Similar considerations may easily be applied near a P_1 point.

The frequency surfaces may be much more complicated than allowed by (1.5.24) in the vicinity of a point of degeneracy. We will not analyze these cases in detail. Phillips has shown that an index i and a topological weight q may be assigned to each such nonanalytic critical point so that the relations (1.5.25) remain valid [q is the number of times the critical point is to be counted in applying (1.5.25)]. These relations are used in the following manner: one determines the critical points of each type required by the symmetry considerations of the previous section, and tests to see if the relations (1.5.25) are satisfied. If they are, no further critical points are required; if not, additional critical points must exist, and have to be located. Once the critical points are located, the behavior of the density of states as a function of frequency can be determined in the neighborhood of the point. The construction of the density of states is greatly simplified with this information (see, for instance, Callaway and Hughes, 1962).

1.5.4 *The Debye Theory of Specific Heat*

We will conclude this section with a brief discussion of the Debye theory of specific heat. This is obtained if the expression we have derived for the density of states of the low frequency acoustic modes (1.5.31) is used in the expression (1.5.12) for the specific heat. Even if optical modes are present, only the acoustic modes will contribute to the specific heat at low temperatures. Before making this substitution, it is convenient to normalize the density of states by assuming that (1.5.31) is valid for all frequencies up to some maximum ω_m, and that $g(\omega) = 0$ if $\omega > \omega_m$. Since there are three acoustic branches with one mode per atom per branch, ω_m is defined by

$$\int_0^{\omega_m} g_A(\omega)\, d\omega = (3\Omega/2\pi^2 v_A{}^3) \int_0^{\omega_m} \omega^2\, d\omega = 3 \qquad (1.5.37)$$

or

$$\omega_m = (6\pi^2/\Omega)^{1/3} v_A \qquad (1.5.38)$$

where v_A is given by (1.5.32). We define a quantity with the dimensions of temperature (the Debye temperature) by

$$K\Theta = \hbar\omega_m.$$

Thus

$$\Theta = (2\pi\hbar/K)(3/4\pi\Omega)^{1/3} v_A = \left[(\Omega/9) \sum_j \int d\sigma/v_j{}^3(\theta, \phi)\right]^{-1/3}. \qquad (1.5.39)$$

The normalized density of states is

$$\begin{aligned} g_A(\omega) &= 9\omega^2/\omega_m{}^3 && (\omega \leq \omega_m) \\ &= 0 && (\omega > \omega_m). \end{aligned} \qquad (1.5.40)$$

Now substitute this into (1.5.12), and introduce a dimensionless variable $x = \hbar\omega/KT_B$. Then

$$C = (\mathfrak{N}K/4)(9T^3/\Theta^3) \int_0^{\Theta/T} x^4 \operatorname{csch}^2(x/2)\, dx. \qquad (1.5.41)$$

For high temperatures, we expand $\operatorname{csch} x/2 \approx 2/x$. This gives

$$C = 3\mathfrak{N}K, \qquad (1.5.42)$$

which is simply a consequence of the correct normalization of the density of states. However, we are mainly interested in low temperatures ($T \ll \Theta$)

for which the upper limit may be allowed to become infinite. Then

$$\tfrac{1}{4}\int_0^\infty x^4 \operatorname{csch}^2(x/2)\, dx = \int_0^\infty [x^4e^x/(e^x - 1)^2]\, dx = 4\pi^4/15.$$

Thus the low temperature specific heat is

$$C = \mathfrak{N}K(12\pi^4/5)(T/\Theta)^3. \tag{1.5.43}$$

This is the Debye T^3 law.

For temperatures lying in between the regions in which the limiting forms (1.5.42) and (1.5.43) will be valid, the integral in (1.5.41) must be evaluated numerically.

The effect of considering a more realistic representation of the density of states is to produce departures from the simple forms obtained here. However, (1.5.43) will always be valid for sufficiently low temperatures; however, as the temperature increases, the observed specific heat will fail to fit the Debye function with constant Θ. Experimentally, Θ will appear to be a function of temperature. Usually, numerical evaluation of (1.5.12) will be required with an accurate $g(\omega)$ in order to fit the data.

1.6 Scattering of Thermal Neutrons by a Vibrating Crystal Lattice

Measurements of the scattering of long wavelength neutrons (energies of 0.1 eV or less) by solids have proved to be of great significance. In this section, we will study the interaction between neutrons and phonons; in Section 2.5 the topic of magnetic scattering of neutrons by a system of spins will be considered. The subject has been reviewed by Marshall and Lovesey (1971); brief accounts which are close to the present point of view may be found in Maradudin *et al.* (1971), and Kittel (1963). An earlier review, mainly concerned with structure determinations, was given by Shull and Wollan (1956). The mathematical treatment used here was developed by Van Hove (1954).

1.6.1 *Consequences of Conservation Relations*

Since the mathematical treatment of neutron scattering is fairly complex, we begin with an elementary description of the process. Neutrons in the energy region of interest have wavelengths of the order of a few angstroms. Thus they are able to interact effectively with several atoms. The scattering may be either elastic or inelastic, and either coherent or incoherent. Our interest here concerns coherent processes, in which the neutron may be considered to be scattered by the crystal as a whole. Incoherent scattering furnishes a diffuse background, and results from effects of disorder, such as a random distribution of nuclear isotopes with different scattering properties, or a disordered distribution of atomic spins. Such effects cause local fluctuations in the scattering. In a coherent scattering process energy is conserved, and the change in wave vector is restricted to a reciprocal lattice vector. In an incoherent process, there is no conservation of wave vector.

Let the initial and final wave vectors of the neutron be $\mathbf{k}_i$ and $\mathbf{k}_f$, respectively. In the case of coherent elastic scattering, the energy of the neutron remains the same

$$\mathbf{k}_i^2 = \mathbf{k}_f^2. \tag{1.6.1}$$

However, as in the case of X-ray diffraction, the neutron wave vector may change by a reciprocal lattice vector $\mathbf{K}_s$, say, if this is consistent with (1.6.1):

$$\mathbf{k}_i - \mathbf{k}_f = \mathbf{K}_s. \tag{1.6.2}$$

These equations determine possible values for θ, the angle between $\mathbf{k}_f$ and $\mathbf{k}_i$:

$$k_i \sin(\theta/2) = \tfrac{1}{2}K_s. \tag{1.6.3}$$

This is the Bragg condition.

Coherent elastic scattering furnishes information about the crystal structure, but not about lattice dynamics. For this we must study inelastic events in which a phonon is emitted or absorbed. Such processes were first observed by Brockhouse and Stewart (1955). We restrict our attention to coherent inelastic scattering involving a single phonon of wave vector $\mathbf{q}$ and energy $\hbar\omega_j(\mathbf{q})$. We now have

$$(\hbar^2/2m)(\mathbf{k}_i^2 - \mathbf{k}_f^2) = \pm\hbar\omega_j(\mathbf{q}) \tag{1.6.4}$$

where m is the mass of the neutron, and the plus $(+)$ sign corresponds to emission of the phonon, the minus $(-)$ to absorption. Conservation of wave vector gives the relation

$$\mathbf{k}_i - \mathbf{k}_f = \mathbf{K}_s \pm \mathbf{q}. \tag{1.6.5}$$

Since $\omega_j(\mathbf{K}_s \pm \mathbf{q}) = \omega_j(\mathbf{q})$, we can combine (1.6.4) and (1.6.5) to give

$$(\hbar^2/2m)(\mathbf{k}_i^2 - \mathbf{k}_f^2) = \pm\hbar\omega_j(\mathbf{k}_i - \mathbf{k}_f). \tag{1.6.6}$$

To understand the implication of these relations (Placzek and Van Hove, 1954), we begin by supposing that the neutrons are initially very cold, so that $\mathbf{k}_i \approx 0$. Only absorption is possible. The final neutron wave vector is restricted to the range of magnitudes

$$0 \leq k_f \leq (2m\omega_m/\hbar)^{1/2} \tag{1.6.7}$$

where ω_m is the maximum frequency in the crystal. Equation (1.6.6) becomes

$$\mathbf{k}_f^2 = 2m\omega_j(\mathbf{k}_f)/\hbar. \tag{1.6.8}$$

This equation defines a scattering surface S_j such that the end points of allowed values of $\mathbf{k}_f$ lie on the surface. There are as many surfaces S_j as there are branches in the vibrational spectrum. Suppose we select arbitrarily a particular direction for $\mathbf{k}_f$. Observation of the scattering in that direction will reveal peaks of neutrons of different energy corresponding to the allowed values of $\mathbf{k}_f$ determined from (1.6.8).

For finite $\mathbf{k}_i$, in the case of absorption, the range of final wave vectors is evidently

$$k_i \leq k_f \leq [\mathbf{k}_i^2 + (2m\omega_m/\hbar)]^{1/2}. \tag{1.6.9}$$

Phonon emission may also be possible. The conditions for emission are, however, more stringent than for absorption. It is easily seen that Eq. (1.6.3), which describes elastic scattering, has no nontrivial solution if $|\mathbf{k}_i|$ is smaller than half the magnitude of the smallest nonzero reciprocal lattice vector, which we call $\mathbf{K}_1$. By the same argument, phonon emission is not possible unless $|\mathbf{k}_i|$ is larger than $\frac{1}{2}|\mathbf{K}_1|$: then it may occur in at least some directions. Emission is possible for all directions of $\mathbf{k}_i$ if $\mathbf{k}_i$ is larger than a value approximately $(2m\omega_m/\hbar)^{1/2}$. Further discussion of these points is given by Placzek and Van Hove (1954).

1.6.2 *Formulation in Terms of Correlation Functions*

We will now discuss the quantum mechanical theory of neutron scattering. To this end, it is desirable to have a simple description of the interaction potential. In order to avoid the complications of nuclear physics, we adopt the Fermi pseudopotential. This is an interaction between a neutron and a nucleus of the form

$$v(\mathbf{r}) = (2\pi a\hbar^2/m)\,\delta(\mathbf{r}) \tag{1.6.10}$$

where a is the neutron–nucleus scattering length which can be determined experimentally. The scattering length characterizes low energy neutron–single nucleus scattering if there are no complicating resonances in the energy region considered; that is, the total cross section is just

$$\sigma = 4\pi a^2. \tag{1.6.11}$$

The interaction (1.6.10) is intended for use in a Born approximation calculation, and ensures that the Born approximation will give the correct scattering. The treatment of the nucleus as a point, implied by the delta function, is legitimate when the neutron wavelength is much larger than the nuclear size, which certainly is the case in the circumstances considered here.

The interaction of a neutron with the crystal can now be represented as

$$V(\mathbf{r}) = \sum_{\mu} v(\mathbf{r} - \mathbf{x}_{\mu}) = (2\pi a\hbar^2/m) \sum_{\mu} \delta(\mathbf{r} - \mathbf{x}_{\mu}). \qquad (1.6.12)$$

For simplicity, we consider a monatomic crystal. The atoms are assumed to be located at positions $\mathbf{x}_{\mu}$ (μ designates a unit cell). The $\mathbf{x}_{\mu}$ are related to the atomic displacements $\mathbf{u}_{\mu}$ by

$$\mathbf{x}_{\mu} = \mathbf{R}_{\mu} + \mathbf{u}_{\mu}$$

where $\mathbf{R}_{\mu}$ is a direct lattice vector. The interaction (1.6.12) is to be treated in the first Born approximation as is the case for (1.6.10). From a semiclassical point of view, the $\mathbf{x}_{\mu}$ depend on time, and this time-dependent interaction can lead to inelastic processes. We consider the $\mathbf{x}_{\mu}$ to be operators on the phonon system, and will ultimately express them in terms of creation and annihilation operators.

The initial state of the system has n_{i} phonons in assorted states and a neutron with wave vector $\mathbf{k}_{\text{i}}$; the final state has n_{f} phonons and a neutron of wave vector $\mathbf{k}_{\text{f}}$. The transition matrix element is proportional to

$$\langle n_{\text{f}} \mid \int \exp(-i\mathbf{k}_{\text{f}}\cdot\mathbf{r}) V(\mathbf{r}) \exp(i\mathbf{k}_{\text{i}}\cdot\mathbf{r})\, d^3r \mid n_{\text{i}}\rangle$$

$$= (2\pi a\hbar^2/m) \langle n_{\text{f}} \mid \int \sum_{\mu} \exp[i(\mathbf{k}_{\text{i}} - \mathbf{k}_{\text{f}})\cdot\mathbf{x}_{\mu}] \mid n_{\text{i}}\rangle.$$

The differential cross section for scattering into energy interval $d\epsilon$ is given, in the Born approximation, by the square modulus of this quantity multiplied by

$$(m^2/4\pi^2\hbar^4)(k_{\text{f}}/k_{\text{i}})\, \delta(E_{\text{f}} - E_{\text{i}}).$$

Therefore, we obtain

$$\partial^2\sigma/\partial\Omega\, \partial\epsilon = a^2(k_{\text{f}}/k_{\text{i}}) \mid \langle n_{\text{f}} \mid \sum_{\mu} \exp[-i(\mathbf{k}_{\text{f}} - \mathbf{k}_{\text{i}})\cdot\mathbf{x}_{\mu}] \mid n_{\text{i}}\rangle \mid^2 \delta(E_{\text{f}} - E_{\text{i}}). \qquad (1.6.13)$$

We write $E_{\text{f}} - E_{\text{i}} = E_{n\text{f}} - E_{n\text{i}} - \hbar\omega$ where

$$\hbar\omega = (\hbar^2/2m)(k_{\text{i}}^2 - k_{\text{f}}^2) \qquad (1.6.14)$$

represents the energy transfer, and $E_{n\text{f}}$ ($E_{n\text{i}}$) is the energy of the phonons present in the final (initial) state of the system. However, we do not know exactly what the initial state of the system is and, therefore, must average over a thermal equilibrium distribution of phonon populations. We must sum over all possible final states n_f as well. We write the cross section in the form

$$d^2\sigma/d\Omega\, d\epsilon = (a^2/\hbar)(k_f/k_i)S(\boldsymbol{\kappa}, \omega) \tag{1.6.15}$$

where $\boldsymbol{\kappa} = \mathbf{k}_i - \mathbf{k}_f$ and

$$S(\boldsymbol{\kappa}, \omega) = \sum_{n_i} P_{ni} \sum_{n_f} |\langle n_f | \sum_{\mu} \exp(i\boldsymbol{\kappa}\cdot\mathbf{x}_\mu) | n_i\rangle|^2 \delta[\omega + (E_{ni} - E_{nf})/\hbar]. \tag{1.6.16}$$

The quantity P_{ni} is the probability distribution function (or statistical weight) for the initial state. Our attention will be concentrated on the function $S(\boldsymbol{\kappa}, \omega)$, which contains all the solid state physics of the problem.

It is convenient to introduce its Fourier transform $F(\boldsymbol{\kappa}, t)$

$$S(\boldsymbol{\kappa}, \omega) = 1/(2\pi) \int_{\infty}^{\infty} e^{-i\omega t} F(\boldsymbol{\kappa}, t)\, dt, \tag{1.6.17}$$

$$F(\boldsymbol{\kappa}, t) = \int_{\infty}^{\infty} e^{i\omega t} S(\boldsymbol{\kappa}, \omega)\, d\omega. \tag{1.6.18}$$

We insert (1.6.16) into (1.6.18) and obtain

$$\begin{aligned} F(\boldsymbol{\kappa}, t) &= \sum_{n_i} P_{ni} \sum_{n_f} |\langle n_f | \sum_{\mu} \exp(i\boldsymbol{\kappa}\cdot\mathbf{x}_\mu) | n_i\rangle|^2 \exp[i(E_{nf} - E_{ni})t/\hbar] \\ &= \sum_{n_i} P_{ni} \sum_{n_f} \sum_{\mu\nu} \langle n_i | \exp(-i\boldsymbol{\kappa}\cdot\mathbf{x}_\nu) | n_f\rangle \exp(iE_{nf}t/\hbar) \\ &\quad \times \langle n_f | \exp(i\boldsymbol{\kappa}\cdot\mathbf{x}_\mu) | n_i\rangle \exp(-iE_{ni}t/\hbar). \end{aligned} \tag{1.6.19}$$

The $\mathbf{x}_\mu$ in these formulas are quantum operators in the Schrödinger picture. It turns out to be convenient to introduce Heisenberg operators $\mathbf{x}_\mu(t)$:

$$\mathbf{x}_\mu(t) = \exp(iHt/\hbar)\mathbf{x}_\mu(0) \exp(-iHt/\hbar) \tag{1.6.20}$$

where $\mathbf{x}_\mu(0)$ is the Schrödinger operator. We also have

$$\exp(iHt/\hbar) \exp(i\boldsymbol{\kappa}\cdot\mathbf{x}_\mu) \exp(-iHt/\hbar) = \exp[i\boldsymbol{\kappa}\cdot\mathbf{x}_\mu(t)]. \tag{1.6.21}$$

This substitution is easily accomplished since

$$\begin{aligned} &\exp(iE_{nf}t/\hbar) \langle n_f | \exp(i\boldsymbol{\kappa}\cdot\mathbf{x}_\mu) | n_i\rangle \exp(-iE_{ni}t/h) \\ &\quad = \langle n_f | \exp(iHt/\hbar) \exp(i\boldsymbol{\kappa}\cdot\mathbf{x}_\mu) \exp(-iHt/\hbar) | n_i\rangle \\ &\quad = \langle n_f | \exp[i\boldsymbol{\kappa}\cdot\mathbf{x}_\mu(t)] | n_i\rangle. \end{aligned}$$

It is now possible to perform the sum over final phonon states by closure. This gives

$$F(\boldsymbol{\kappa}, t) = \sum_{n_i} P_{ni} \sum_{\mu\nu} \langle n_i \mid \exp[-i\boldsymbol{\kappa}\cdot\mathbf{x}_\nu(0)] \exp[i\boldsymbol{\kappa}\cdot\mathbf{x}_\mu(t)] \mid n_i\rangle. \tag{1.6.22}$$

We denote the statistical average over phonon states by $\langle\ \rangle_T$:

$$\sum_{n_i} P_{ni}\langle n_i \mid \exp[-i\boldsymbol{\kappa}\cdot\mathbf{x}_\nu(0)] \exp[i\boldsymbol{\kappa}\cdot\mathbf{x}_\mu(t)] \mid n_i\rangle = \langle \exp[-i\boldsymbol{\kappa}\cdot\mathbf{x}_\nu(0)] \exp[i\boldsymbol{\kappa}\cdot\mathbf{x}_\mu(t)]\rangle_T.$$

We now have the simple form

$$S(\boldsymbol{\kappa}, \omega) = (1/2\pi) \int_{\infty}^{\infty} dt \exp(-i\omega t) \sum_{\mu\nu} \langle \exp[-i\boldsymbol{\kappa}\cdot\mathbf{x}_\nu(0)] \exp[i\boldsymbol{\kappa}\cdot\mathbf{x}_\mu(t)]\rangle_T. \tag{1.6.23}$$

Equation (1.6.23) enables us to express the scattering cross section as a thermal average of a correlation function of two Heisenberg operators at different times. Unfortunately, the operators $\mathbf{x}_\mu(0)$ and $\mathbf{x}_\mu(t)$ do not commute. Even when Schrödinger picture operators commute, the same operators in the Heisenberg picture do not since they refer to different times. This is a serious complication.

1.6.3 *Evaluation of the Correlation Function*

We now consider the evaluation of $S(\boldsymbol{\kappa}, \omega)$. This can be done with a substantial degree of rigor within the basic framework of the harmonic approximation. To proceed, we first separate the constant part of the $\mathbf{x}_\mu$ by writing $\mathbf{x}_\mu = \mathbf{R}_\mu + \mathbf{u}_\mu(t)$, where $\mathbf{u}_\mu$ is the displacement from equilibrium

$$\langle \exp[-i\boldsymbol{\kappa}\cdot\mathbf{x}_\nu(0)] \exp[i\boldsymbol{\kappa}\cdot\mathbf{x}_\mu(t)]\rangle_T = \exp[i\boldsymbol{\kappa}\cdot(\mathbf{R}_\mu - \mathbf{R}_\nu)]\langle \exp[-i\boldsymbol{\kappa}\cdot\mathbf{u}_\nu(0)] \exp[i\boldsymbol{\kappa}\cdot\mathbf{u}_\mu(t)]\rangle.$$

The product of exponentials of noncommuting displacement operators can be simplified using the following (Baker–Hausdorff) theorem

$$e^A e^B = e^{(A+B)} e^{1/2[A,B]} = e^{(B+A)} e^{1/2[A,B]} \tag{1.6.24}$$

in which $[\cdots]$ denotes the commutator. This expression is correct provided that $[A, B]$ commutes with A and with B. Then

$$\exp[-i\boldsymbol{\kappa}\cdot\mathbf{u}_\nu(0)] \exp[i\boldsymbol{\kappa}\cdot\mathbf{u}_\mu(t)] = \exp[i\boldsymbol{\kappa}\cdot(\mathbf{u}_\mu(t) - \mathbf{u}_\nu(0))] \times \exp\tfrac{1}{2}[\boldsymbol{\kappa}\cdot\mathbf{u}_\nu(0), \boldsymbol{\kappa}\cdot\mathbf{u}_\mu(t)]. \tag{1.6.25}$$

We will now compute the commutator. It will be found to be a constant so that use of Eq. (1.6.24) is justified. The operator $\mathbf{u}_\nu$ is given in the Schrödinger picture in Eq. (1.4.23). This is the same as the Heisenberg operator at time $t = 0$. To obtain the Heisenberg operator at arbitrary t, we must recall that a_j destroys and $a_j^\dagger$ creates a phonon. Then

$$a_j(\mathbf{k}, t) = \exp(iHt/\hbar)a_j(\mathbf{k}) \exp(-iHt/\hbar) = \exp[-i\omega_j(\mathbf{k})t]a_j(\mathbf{k}),$$

$$a_j^\dagger(\mathbf{k}, t) = \exp[i\omega_j(\mathbf{k})t]a_j^\dagger(\mathbf{k}).$$

Thus [Eq. (1.5.22b)],

$$\mathbf{u}_\mu(t) = (\hbar/\mathfrak{N}M)^{1/2} \sum_{j\mathbf{k}} [2\omega_j(\mathbf{k})]^{-1/2}[\mathbf{e}^{(j)}(\mathbf{k}) \times \exp\{i[\mathbf{k}\cdot\mathbf{R}_\mu - \omega_j(\mathbf{k})t]\}a_j(\mathbf{k}) + \mathbf{e}^{(j)}(\mathbf{k}) \exp\{-i[\mathbf{k}\cdot\mathbf{R}_\mu - \omega_j(\mathbf{k})t]\}a_j^\dagger(\mathbf{k})]. \quad (1.6.26)$$

The $\mathbf{e}$'s may be assumed to be real in a monatomic crystal. We can now work out the commutator

$$[u_{\alpha,\nu}(0), u_{\beta,\mu}(t)] = (2i\hbar/\mathfrak{N}M) \sum_{j\mathbf{k}} [2\omega_j(\mathbf{k})]^{-1/2} e_\alpha^{(j)}(\mathbf{k}) e_\beta^{(j)}(\mathbf{k}) \sin\theta_{\mu\nu}(\mathbf{k}, t) \quad (1.6.27)$$

where

$$\theta_{\mu\nu}(\mathbf{k}, t) = \mathbf{k}\cdot(\mathbf{R}_\mu - \mathbf{R}_\nu) + \omega_j(\mathbf{k})t. \quad (1.6.28)$$

At the moment, all we require of this complicated expression is that it be a number. Thus we may take the commutator outside the thermal average. We have

$$\langle \exp[-i\boldsymbol{\kappa}\cdot\mathbf{u}_\nu(0)] \exp[i\boldsymbol{\kappa}\cdot\mathbf{u}_\mu(t)]\rangle_T = \langle \exp\{i\boldsymbol{\kappa}\cdot[\mathbf{u}_\mu(t) - \mathbf{u}_\nu(0)]\}\rangle_T \times \exp\{-\tfrac{1}{2}[\boldsymbol{\kappa}\cdot\mathbf{u}_\mu(t), \boldsymbol{\kappa}\cdot\mathbf{u}_\nu(0)]\}. \quad (1.6.29)$$

To evaluate this, we require the result

$$\langle \exp\{i\boldsymbol{\kappa}\cdot(\mathbf{u}_\mu(t) - \mathbf{u}_\nu(0))\}\rangle_T = \exp\{-\tfrac{1}{2}\langle[\boldsymbol{\kappa}\cdot(\mathbf{u}_\mu(t) - \mathbf{u}_\nu(0))]^2\rangle_T\}. \quad (1.6.30)$$

The proof is quite lengthy and succeeds only because the system is equivalent to a set of uncoupled harmonic oscillators. We start by defining the thermal average in terms of a trace:

$$\langle f \rangle_T = \mathrm{tr}[fe^{-\beta H}]/\mathrm{tr}[e^{-\beta H}], \qquad \beta = 1/KT. \tag{1.6.31}$$

The trace may be evaluated on the eigenstates of $H \mid n_j(\mathbf{k}) \rangle$, which satisfy

$$H \mid n_j(\mathbf{k}) \rangle = \hbar\omega_j(\mathbf{k})[n_j(\mathbf{k}) + \tfrac{1}{2}] \mid n_j(\mathbf{k}) \rangle. \tag{1.6.32}$$

The operators of interest to us have the form

$$f = \exp\{i \sum_{j\mathbf{k}} [c_j(\mathbf{k})a_j(\mathbf{k}) + c_j^*(\mathbf{k})a_j^\dagger(\mathbf{k})]\} \tag{1.6.33}$$

where the $c_j(\mathbf{k})$ are coefficients. Since the $a_j(\mathbf{k})$ and $a_j^\dagger(\mathbf{k})$ commute with all creation and annihilation operators for other modes of different j and different $\mathbf{k}$, the expression for $\langle f \rangle$ factors into a produce of averages for a single mode:

$$\langle f \rangle_T = \prod_{j\mathbf{k}} \langle f_{j\mathbf{k}} \rangle_T \tag{1.6.34}$$

where

$$\langle f_{j\mathbf{k}} \rangle_T = Z^{-1} \sum_n \langle n_j(\mathbf{k}) \mid \exp\{-i[c_j(\mathbf{k})a_j(\mathbf{k}) + c_j^*(\mathbf{k})a_j^\dagger(\mathbf{k})]\} \mid n_j(\mathbf{k}) \rangle$$
$$\times \exp\{-\beta[n_j(\mathbf{k}) + \tfrac{1}{2}]\hbar\omega_j(\mathbf{k})\} \tag{1.6.35}$$

and

$$Z = \sum_n \exp\{-\beta[n_j(\mathbf{k}) + \tfrac{1}{2}]\hbar\omega_j(\mathbf{k})\}$$
$$= \exp[-\beta\hbar\omega_j(\mathbf{k})/2]/\{1 - \exp[-\beta\hbar\omega_j(\mathbf{k})]\}. \tag{1.6.36}$$

At this point, we drop subscripts and arguments since we are considering only a single mode. We use the Baker–Hausdorff theorem (1.6.24) to write

$$\exp[i(ca + c^*a^\dagger)] = \exp(ica)\exp(ic^*a^\dagger)\exp(-\tfrac{1}{2}[ica, ic^*a^\dagger]).$$

However,

$$[ca, c^*a^\dagger] = |c|^2[a, a^\dagger] = |c|^2.$$

The matrix element of interest becomes

$$\langle n \mid \exp(ica)\exp(ic^*a^\dagger) \mid n \rangle \exp(\tfrac{1}{2} |c|^2). \tag{1.6.37}$$

To evaluate this, we introduce a complete set of states $\mid N \rangle$ so that (1.6.37) becomes

$$\exp(\tfrac{1}{2} |c|^2) \sum_N \langle n \mid \exp(ica) \mid N \rangle \langle N \mid \exp(ic^*a^\dagger) \mid n \rangle. \tag{1.6.38}$$

The matrix elements of the exponentials can be determined by expanding each exponential in a power series. The individual terms are evaluated with the aid of the formula

$$(a^\dagger)^l \mid n\rangle = [(n+l)!/n!]^{1/2} \mid n+l\rangle.$$

Thus

$$\langle n \mid (a^\dagger)^l \mid N\rangle = \delta_{N,n+l}[(n+l)!/n!]^{1/2} = (N!/n!)^{1/2}\,\delta_{N,n+l},$$

and similarly,

$$\langle n \mid (a)^l \mid N\rangle = \delta_{N-l,n}(N!/n!)^{1/2}.$$

Thus

$$\langle N \mid \exp(ic^*a^\dagger) \mid n\rangle = \sum_l (i^l c^{*l}/l!)\,\langle N \mid (a^\dagger)^l \mid n\rangle$$
$$= (N!/n!)^{1/2}(i)^{N-n}[c^{*N-n}/(N-n)!].$$

We use this expression to rewrite (1.6.38) (with $s = N - n$):

$$\exp(\tfrac{1}{2} \mid c\mid^2) \sum_{s=0} (-1)^s\{[(s+n)!]/n!\}\{\mid c\mid^{2s}/[s!]^2\}. \quad (1.6.39)$$

We substitute (1.6.39) into (1.6.35) and obtain the double sum, which we denote as $\mathcal{S}$:

$$\mathcal{S} = \exp(\tfrac{1}{2} \mid c\mid^2)/Z \sum_{ns} (-1)^s$$
$$\times \exp[-\beta(n+\tfrac{1}{2})\hbar\omega]\{[(s+n)!]/n!\}\{\mid c\mid^{2s}/[s!]^2\}. \quad (1.6.40)$$

This sum can be evaluated exactly. Let $x = \mid c\mid^2$, $y = \exp(-\beta\hbar\omega)$. We are concerned with an expression of the form

$$\sum_{n=0}^{\infty}\sum_{s=0}^{\infty} (-1)^s[(n+s)!/n!(s!)^2]x^s y^n.$$

We use the identity

$$1/(1-y)^{s+1} = \sum_{n=0}^{\infty} [(s+n)!/s!n!]y^n,$$

and the formula (1.6.36) for Z to rewrite $\mathcal{S}$ as

$$\mathcal{S} = e^{x/2} \sum_s (1/s!)[(-1)^s x^s/(1-y)^s] = \exp\{x[\tfrac{1}{2} - (1-y)^{-1}]\}.$$

We return to the original variables and obtain

$$\mathcal{S} = \exp\{-(\mid c\mid^2/2)\ \mathrm{ctnh}\ \beta\hbar\omega/2\}. \quad (1.6.41)$$

We compare this with the result of the simple average (for a single mode)

$$R = \langle (ca + c^*a^\dagger)^2 \rangle$$
$$= (1/Z) \sum_n \exp[-\beta\hbar\omega(n + \tfrac{1}{2})] \langle n \mid (ca + c^*a^\dagger)^2 \mid n \rangle. \quad (1.6.42)$$

Terms containing aa and $a^\dagger a^\dagger$ vanish. Also $aa^\dagger + a^\dagger a = 2n + 1$, so we have

$$R = \mid c \mid^2 [1 - \exp(-\beta\hbar\omega/2)] \sum_n (2n + 1) \exp(-\beta\hbar\omega n).$$

The sum is easy to evaluate and we obtain

$$R = \mid c \mid^2 \operatorname{ctnh}(\beta\hbar\omega/2). \quad (1.6.43)$$

Thus

$$\mathcal{S} = \exp[-R/2]. \quad (1.6.44)$$

We have now shown that for a single mode

$$\langle \exp[i(ca + c^*a^\dagger)] \rangle_T = \exp[-\tfrac{1}{2} \langle (ca + c^*a^\dagger)^2 \rangle_T]. \quad (1.6.45)$$

However, each mode enters the averaging process independently. Consequently,

$$\langle \exp\{i \sum_{j\mathbf{k}} [c_j(\mathbf{k}) a_j(\mathbf{k}) + c_j^*(\mathbf{k}) a_j^\dagger(\mathbf{k})]\} \rangle_T$$
$$= \exp\{-\tfrac{1}{2} \sum_{j\mathbf{k}} \langle [c_j(\mathbf{k}) a_j(\mathbf{k}) + c_j^*(\mathbf{k}) a_j^\dagger(\mathbf{k})]^2 \rangle_T\}. \quad (1.6.46)$$

This implies the relation we want

$$\langle \exp[i\boldsymbol{\kappa} \cdot (\mathbf{u}_\mu(t) - \mathbf{u}_\nu(0))] \rangle_T = \exp\{-\tfrac{1}{2} \langle [\boldsymbol{\kappa} \cdot (\mathbf{u}_\mu(t) - \mathbf{u}_\nu(0))]^2 \rangle_T\}. \quad (1.6.30')$$

Equation (1.6.30′) is substituted into (1.6.29). The result is written in the form

$$\langle \exp[-i\boldsymbol{\kappa} \cdot \mathbf{u}_\nu(0)] \exp[i\boldsymbol{\kappa} \cdot \mathbf{u}_\mu(t)] \rangle_T = \exp[-Q_{\mu\nu}(\boldsymbol{\kappa}, t)] \quad (1.6.47)$$

where

$$Q_{\mu\nu}(\boldsymbol{\kappa}, t) = \tfrac{1}{2}\{ \langle [\boldsymbol{\kappa} \cdot (\mathbf{u}_\mu(t) - \mathbf{u}_\nu(0))]^2 \rangle_T - [\boldsymbol{\kappa} \cdot \mathbf{u}_\mu(t), \boldsymbol{\kappa} \cdot \mathbf{u}_\nu(0)]\}. \quad (1.6.48)$$

The remaining average is of the type considered in (1.6.42). First, it is necessary to express $\mathbf{u}_\mu(t) - \mathbf{u}_\mu(0)$ in terms of phonon operators

$$\mathbf{u}_\mu(t) - \mathbf{u}_\nu(0) = (\hbar/\mathfrak{N}M)^{1/2} \sum_{j\mathbf{k}} [2\omega_j(\mathbf{k})]^{-1/2}$$
$$\times (\mathbf{e}^{(j)}(\mathbf{k}) \{\exp[i(\mathbf{k} \cdot \mathbf{R}_\mu - \omega_j(\mathbf{k})t)]$$
$$- \exp(i\mathbf{k} \cdot \mathbf{R}_\nu)\} a_j(\mathbf{k}) + \mathbf{e}^{(j)}(\mathbf{k}) \{\exp[-i(\mathbf{k} \cdot \mathbf{R}_\mu - \omega_j(\mathbf{k})t)]$$
$$- \exp(-i\mathbf{k} \cdot \mathbf{R}_\nu)\} a_j^\dagger(\mathbf{k})).$$

We have to square this. This process introduces terms like $a_j(\mathbf{k})a_j(\mathbf{q})$, etc. which have zero average value in a stationary state. We discard all terms of this type in the square, retaining only those proportional to the phonon number operator. The sum is converted into an integral according to Eq. (1.2.10). This gives

$$\{\boldsymbol{\kappa}\cdot[\mathbf{u}_\mu(t) - \mathbf{u}_\nu(0)]\}^2 = (\Omega\hbar/8\pi^3 M)\,\kappa_\alpha\kappa_\beta \sum_j \int [d^3k/2\omega_j(\mathbf{k})]e_\alpha^{(j)}(\mathbf{k})e_\beta^{(j)}(\mathbf{k})$$
$$\times \{2 - \exp[i\theta_{\mu\nu}(\mathbf{k},t)] - \exp[-i\theta_{\mu\nu}(\mathbf{k},t)]\}$$
$$\times (2n_j(\mathbf{k}) + 1). \qquad (1.6.49)$$

A summation over the repeated Cartesian indices (α, β) is understood. The result is combined with Eq. (1.6.27) to yield

$$Q_{\mu\nu}(\boldsymbol{\kappa}, t) = (\Omega/16\pi^3 M)\,\kappa_\alpha\kappa_\beta \sum_j \int [d^3k/\omega_j(\mathbf{k})]e_\alpha^{(j)}(\mathbf{k})e_\beta^{(j)}(\mathbf{k})$$
$$\times [(1 - \cos\theta_{\mu\nu}(\mathbf{k},t))(2\langle n_j(\mathbf{k})\rangle + 1) - i\sin\theta_{\mu\nu}(\mathbf{k},t)]. \qquad (1.6.50)$$

The thermal average of (1.6.49) is performed simply by replacing $n_j(\mathbf{k})$ by its average.

It proves to be convenient to extract from $Q_{\mu\nu}$ the part which is independent of time:

$$Q_{\mu\nu}(\boldsymbol{\kappa}, t) = 2W(\boldsymbol{\kappa}) - G_{\mu\nu}(\boldsymbol{\kappa}, t) \qquad (1.6.51)$$

where (the factor of 2 is a convention)

$$2W = (\Omega\hbar/16\pi^3 M)\,\kappa_\alpha\kappa_\beta \sum_j \int [d^3k/\omega_j(\mathbf{k})]e_\alpha^{(j)}(\mathbf{k})e_\beta^{(j)}(\mathbf{k})[2\langle n_j(\mathbf{k})\rangle + 1]. \qquad (1.6.52)$$

This is the so-called "Debye–Waller" factor which governs the temperature dependence of the scattering cross section. It will be studied in detail below. We now have

$$\exp[-Q_{\mu\nu}(\boldsymbol{\kappa}, t)] = \exp[-2W(\boldsymbol{\kappa})]\exp[G_{\mu\nu}(\boldsymbol{\kappa}, t)]$$

where

$$G_{\mu\nu}(\boldsymbol{\kappa}, t) = (\Omega\hbar/16\pi^3 M)\,\kappa_\alpha\kappa_\beta \sum_j \int [d^3k/\omega_j(\mathbf{k})]e_\alpha^{j}(\mathbf{k})e_\beta^{j}(\mathbf{k})$$
$$\times [(2\langle n_j(\mathbf{k})\rangle + 1)\cos\theta_{\mu\nu} + i\sin\theta_{\mu\nu}]. \qquad (1.6.53)$$

The function $\exp(G_{\mu\nu})$ describes neutron scattering events in which an arbitrary number of phonons may be absorbed or emitted. Usually, only one-phonon processes are of interest, and these may be extracted by an expansion

$$\exp[G_{\mu\nu}(\boldsymbol{\kappa}, t)] = 1 + G_{\mu\nu}(\boldsymbol{\kappa}, t) \tag{1.6.54}$$

in which only first-order terms are retained. We now have for the quantity F defined by (1.6.22)

$$\begin{aligned} F(\boldsymbol{\kappa}, t) = \exp[-2W(\boldsymbol{\kappa})] \sum_{\mu\nu} \Big(&\exp[i\boldsymbol{\kappa}\cdot(\mathbf{R}_\mu - \mathbf{R}_\nu)] \\ &+ \kappa_\alpha\kappa_\beta(\Omega\hbar/16\pi^3 M) \sum_j \int [d^3k/\omega_j(\mathbf{k})] e_\alpha^{(j)}(\mathbf{k}) e_\beta^{(j)}(\mathbf{k}) \\ &\times \{(\langle n_j(\mathbf{k})\rangle + 1) \exp[i\omega_j(\mathbf{k})t] \exp[i(\boldsymbol{\kappa} + \mathbf{k})\cdot(\mathbf{R}_\mu - \mathbf{R}_\nu)] \\ &+ \langle n_j(\mathbf{k})\rangle \exp[-i\omega_j(\mathbf{k})t] \exp[i(\boldsymbol{\kappa} - \mathbf{k})\cdot(\mathbf{R}_\mu - \mathbf{R}_\nu)]\}\Big). \end{aligned} \tag{1.6.55}$$

We may now do the sum on $\mathbf{R}_\mu$ and $\mathbf{R}_\nu$. We have

$$\sum_{\mu\nu} \exp[i\boldsymbol{\kappa}\cdot(\mathbf{R}_\mu - \mathbf{R}_\nu)] = [(2\pi)^3/\Omega]\mathfrak{N} \sum_l \delta(\boldsymbol{\kappa} - \mathbf{K}_l)$$

where $\mathfrak{N}$ is the number of cells in the crystal and the $\mathbf{K}_l$ are reciprocal lattice vectors. Thus

$$\begin{aligned} F(\boldsymbol{\kappa}, t) = \mathfrak{N} \exp[-2W(\boldsymbol{\kappa})] \Big\{&[(2\pi)^3/\Omega] \sum_l \delta(\boldsymbol{\kappa} - \mathbf{K}_l) \\ &+ \kappa_\alpha\kappa_\beta(\hbar/2M) \sum_{jl} \int [d^3k/\omega_j(\mathbf{k})] e_\alpha^{(j)}(\mathbf{k}) e_\beta^{(j)}(\mathbf{k}) \\ &\times [(\langle n_j(\mathbf{k})\rangle + 1) \exp(i\omega_j(\mathbf{k})t)\, \delta(\boldsymbol{\kappa} + \mathbf{k} - \mathbf{K}_l) \\ &+ \langle n_j(\mathbf{k})\rangle \exp[-i\omega_j(\mathbf{k})t]\, \delta(\boldsymbol{\kappa} - \mathbf{k} - \mathbf{K}_l)]\Big\}. \end{aligned}$$

The integration over $\mathbf{k}$ is performed easily:

$$\begin{aligned} F(\boldsymbol{\kappa}, \mathbf{t}) = \mathfrak{N} \exp[-2W(\boldsymbol{\kappa})] \{&[(2\pi)^3/\Omega] \sum_l \delta(\boldsymbol{\kappa} - \mathbf{K}_l) \\ &+ \kappa_\alpha\kappa_\beta \sum_j [[\hbar/2M\omega_j(\boldsymbol{\kappa})] e_\alpha^{(j)}(\boldsymbol{\kappa}) e_\beta^{(j)}(\boldsymbol{\kappa}) \\ &\times \{(\langle n_j(\boldsymbol{\kappa})\rangle + 1) \exp[i\omega_j(\boldsymbol{\kappa})t] + \langle n_j(\boldsymbol{\kappa})\rangle \exp[-i\omega_j(\boldsymbol{\kappa})t]\}]\}. \end{aligned} \tag{1.6.56}$$

It is assumed in the second term above that the actual neutron momentum transfer vector $\boldsymbol{\kappa}$ is reduced by an appropriate $\mathbf{K}_l$ to lie in the Brillouin zone. The quantities $\omega_j(\boldsymbol{\kappa})$, $n_j(\boldsymbol{\kappa})$, and $\mathbf{e}^{(j)}(\boldsymbol{\kappa})$ are unaffected by this. We can now complete the calculation by substituting (1.6.56) into (1.6.23), and executing the integration:

$$
\begin{aligned}
S(\boldsymbol{\kappa}, \omega) = \mathfrak{N} e^{-2W(\kappa)} \{ & [(2\pi)^3/\Omega]\, \delta(\omega) \sum_l \delta(\boldsymbol{\kappa} - \mathbf{K}_l) \\
& + (\kappa_\alpha \kappa_\beta / 2M) \hbar \sum_j [1/\omega_j(\boldsymbol{\kappa})] e_\alpha^{(j)}(\boldsymbol{\kappa}) e_\beta^{(j)}(\boldsymbol{\kappa}) \\
& \times \{ [\langle n_j(\boldsymbol{\kappa}) \rangle + 1]\, \delta[\omega - \omega_j(\boldsymbol{\kappa})] + \langle n_j(\boldsymbol{\kappa}) \rangle \delta[\omega + \omega_j(\boldsymbol{\kappa})] \} \}.
\end{aligned}
\tag{1.6.57}
$$

This result contains three parts. The first, the term involving $\delta(\omega)$, evidently describes scattering without transfer of energy. This is the coherent elastic scattering. The momentum conserving delta function in this term gives rise to the Bragg peaks mentioned at the beginning of this section. The second and third terms describe scattering in which a single phonon is involved. From (1.6.14), we see that a positive value of ω corresponds to loss of energy by the neutron to the crystal, that is, phonon emission; while a negative value of ω corresponds to phonon absorption. The delta functions express conservation of energy. In an actual experiment, the neutron beam will not be strictly monoenergetic, but will have a spread of energies. In addition, anharmonic forces provide a means for phonon–phonon interactions so that single-phonon states are not exact eigenstates of the actual Hamiltonian. A phonon then has a finite lifetime. In consequence of these considerations, the delta function spikes will be replaced by finite peaks.

Inspection of (1.6.57) shows that phonon emission involves a factor

$$\langle n_j(\boldsymbol{\kappa}) \rangle + 1 = [1 - \exp(-\hbar\omega_j(\boldsymbol{\kappa})/KT)]^{-1}, \tag{1.6.58}$$

while phonon absorption is governed by

$$\langle n_j(\boldsymbol{\kappa}) \rangle = \{\exp[h\omega_j(\boldsymbol{\kappa})/KT] - 1\}^{-1}. \tag{1.6.59}$$

The presence of these factors is predicted in perturbation theory, since the amplitude for single-phonon emission is proportional to the square root of (1.6.58) and the amplitude for absorption to the square root of (1.6.59). These factors determine the relative temperature dependences of emission and absorption; thus for a phonon of energy large compared to KT, emission will be much more probable than absorption provided, of course, that the neutron has adequate energy.

1.6.4 The Debye–Waller Factor

Another, and vitally important source of temperature dependence, is the factor e^{-2W} which multiplies the entire function. This is the Debye–Waller factor, which has the effect of reducing the probability of all coherent processes as temperature increases. It is also important in the theory of the Mössbauer effect, which we will consider in the next section. We consider the definition of $2W$, as given by (1.6.52). It is interesting to compare this with $\langle |\mathbf{u}_0|^2\rangle_T$, the average amplitude of vibration. As a consequence of translation invariance this must be the same for all atoms: it cannot depend on the choice of origin. We therefore calculate $\langle |\mathbf{u}_0|\rangle_T^2$. The operator u_0 is given by

$$\mathbf{u}_0 = (\hbar/\mathfrak{N}M)^{1/2} \sum_{j\mathbf{k}} [2\omega_j(\mathbf{k})]^{-1/2}[\mathbf{e}^{(j)}(\mathbf{k})a_j(\mathbf{k}) + \mathbf{e}^{(j)}(\mathbf{k})a_j^{\dagger}(\mathbf{k})]. \tag{1.6.60}$$

We square and average over a thermal distribution. Using the same procedures as before, we need retain only the diagonal terms involving $aa^{\dagger} + a^{\dagger}a$. The result is

$$\langle \mathbf{u}_0^2\rangle = (\Omega\hbar/16\pi^3M) \sum_j \int [d^3k/\omega_j(\mathbf{k})][2\langle n_j\rangle + 1] \tag{1.6.61}$$

where we have used the fact that the $\mathbf{e}^{(j)}$ are orthogonal unit vectors. This is quite similar to $2W$. In that case, we have in each mode the factor $(\boldsymbol{\kappa}\cdot\mathbf{e}^{(j)})^2$. The result of integration over angles will be approximated by replacing this factor by $\frac{1}{3}$ which will be done at the beginning. Thus

$$2W = [\Omega\hbar/3(16\pi^3M)]\kappa^2 \sum_j \int [d^3k/\omega_j(\mathbf{k})][2\langle n_j\rangle + 1]. \tag{1.6.62}$$

So

$$2W = \tfrac{1}{3}\kappa^2\langle |u_0|^2\rangle. \tag{1.6.63}$$

We will now evaluate this explicitly. From (1.6.58) and (1.6.59) we have

$$\langle 2n_j(\mathbf{k}) + 1\rangle = \operatorname{ctnh} \beta h\omega_j(\mathbf{k})/2.$$

Thus

$$2W = (\Omega\hbar/48\pi^3M)\kappa^2 \sum_j \int [d^3k/\omega_j(\mathbf{k})]\operatorname{ctnh}[\beta h\omega_j(\mathbf{k})/2]. \tag{1.6.64}$$

Let us consider low temperatures where $\beta = 1/KT$ is quite large. For large x, ctnh $x \to 1$, and we separate this term

$$2W = (\Omega\hbar/48\pi^3 M)\kappa^2 \sum_j \int [d^3k/\omega_j(\mathbf{k})]\{1 + [\text{ctnh}(\beta\hbar\omega_j(\mathbf{k})/2) - 1]\}. \tag{1.6.65}$$

This is still exact. To obtain a more explicit result, we must specify $\omega_j(\mathbf{k})$. For simplicity, we consider a Debye spectrum in which $\omega_j(\mathbf{k}) = v_j\mathbf{k}$; $d^3k = 4\pi\mathbf{k}^2\,dk = (4\pi/v_j^3)\omega_j^2\,d\omega$. Thus

$$\begin{aligned} 2W &= (\Omega\hbar/12\pi^2 M)\kappa^2 \sum_j \int_0^{\omega_{mj}} \omega_j\{1 + [\text{ctnh}(\beta\hbar/2)\omega_j - 1]\}\,d\omega \\ &= (\Omega\hbar/12\pi^2 M)\kappa^2 \sum_j (1/v_j^3)[(\omega^2_{mj}/2) + (\pi^2/3\beta^2\hbar^2)] \end{aligned} \tag{1.6.66}$$

in which ω_{mj} is the maximum frequency in branch j. To achieve additional simplicity, suppose all the v_j are equal ($v_j = v_s$ the average speed of sound). Put $h\omega_m = K\Theta$, where Θ is the Debye temperature. Also $\hbar\omega_m = \hbar v_s k_m$ where k_m is the radius of the sphere whose volume equals that of the Brillouin zone. So

$$K\Theta/\hbar k_m = v_s \qquad \text{and} \qquad (4\pi/3)k_m^3 = 8\pi^3/\Omega.$$

The final result is

$$2W = \tfrac{3}{2}(\hbar^2\kappa^2/2M)(1/K\Theta)[1 + (2\pi^2/3)(T/\Theta)^2]. \tag{1.6.67}$$

The magnitude of the Debye–Waller factor depends on the ratio of two energies. If we replaced the atomic mass M by the neutron mass m, the factor $\hbar^2\kappa^2/2m$ would represent the energy of a neutron with momentum $\hbar\kappa$. We see that this is reduced by the factor m/M, and compared to the Debye energy $K\Theta$. This κ dependence has the effect of reducing the intensity of coherent processes involving large wave vectors compared to those with small wave vectors. We also notice a term proportional to $(T/\Theta)^2$, so that all coherent processes will decrease in intensity with increasing temperature. At high temperatures,

$$\text{ctnh}[\beta\hbar\omega_j(\mathbf{k})/2] = 2/\beta\hbar\omega_j(\mathbf{k}),$$

and we see that $2W$ is directly proportional to temperature in this case. In the Debye approximation, the high temperature limit is

$$2W = 3(\hbar^2\kappa^2/2M)(1/K\Theta)(T/\Theta). \tag{1.6.68}$$

1.7 The Mössbauer Effect

When radioactive nuclei which emit low energy γ rays (less than 200 keV) are bound in a solid, a significant portion of the decays may occur without appreciable transfer of energy to the lattice. The resulting γ-ray spectrum is extremely sharp, and it is possible to observe resonant absorption of these γ rays by unexcited atoms. The effect was first observed by Mössbauer (1958a, b; 1959) although the essential physics was given much earlier by Lamb (1939); it has since become a tool of significant importance in solid state physics for the determination of hyperfine interactions (interactions between a nucleus and surrounding electrons).

The hyperfine interaction always involves the product of a nuclear quantity; such as the nuclear magnetic dipole moment or electric quadrupole moment, and an atomic quantity such as the electron density at the nucleus; or the electric field gradient at the nuclear site. Consequently, the Mössbauer effect furnishes useful information in nuclear physics as well, and the extremely narrow width of the line has made possible a laboratory observation of the gravitational red shift (Pound and Rebka, 1959; Pound and Snider, 1965).

There is extensive literature on the Mössbauer effect, including several reviews. No attempt will be made to give a complete list even of the reviews. Useful references include Frauenfelder (1962), Wertheim (1964), and Danon (1968).

1.7.1 *Elementary Theory*

The physical problem can be described quite simply: Consider first a free nucleus at rest of mass M, which has an excited state of energy E_x above the ground state. The nucleus is initially in the excited state, and emits a γ ray of energy E_γ, and momentum E_γ/c, c being the speed of light. However, E_γ does not quite equal E_x, since some recoil energy is transferred to the nucleus. The momentum of the nucleus is $p = E_\gamma/c$; and the recoil energy associated with this is R:

$$R = p^2/2M = E_\gamma^2/2Mc^2. \tag{1.7.1}$$

Thus the energy of the γ ray is

$$E_\gamma = E_x - R = E_x - (E_\gamma^2/2Mc^2). \tag{1.7.2a}$$

Since R is much smaller than E_x, a good approximation to (1.7.2a) is

$$E_\gamma \approx E_x[1 - (E_x/2Mc^2)]. \tag{1.7.2b}$$

Typically, E_x is in the range of 10 to 100 keV; while R ranges from 2×10^{-4} to 5×10^{-2} eV. One of the most commonly employed nuclei is ^{57}Fe, in which $E_x = 14$ keV and $R = 0.002$ eV. As small as R is, it is still much larger than the natural linewidth Γ, which in the case of ^{57}Fe, is 4.6×10^{-9} eV.

In order for a nucleus in its ground state to absorb a γ-ray and make a transition to an excited state, an energy $E_\gamma + R$ must be supplied. Since R is much larger than the linewidth, emitted γ rays will not be reabsorbed with appreciable probability. The situation here is quite different from that in atomic physics where Γ/R is normally much larger than unity and resonant absorption is easily observed.

If the radioactive nucleus is not free, but is rigidly bound in a solid, the situation is different. Crudely, it is as if the (macroscopic) mass of the solid appeared in (1.7.2) instead of the nuclear mass, in which case the recoil energy is effectively zero. This is an oversimplification because the binding of an atom to a solid is not rigid; the crystal lattice possesses vibrational degrees of freedom, the phonons, which a displaced nucleus can excite. However, there is a some probability that in a particular decay no phonon will be emitted, although when a large number of atoms are considered, the average energy transferred to the lattice in the emission process is R (Lipkin, 1960). The fraction of transitions without recoil (called the *recoilless fraction* and denoted by f) will be calculated below; this is a problem in lattice dynamics. A detailed discussion of this problem has been given by Maradudin (1964).

1.7.2 *Calculation of the Recoilless Fraction*

We consider the interaction of a radioactive nucleus bound in a crystal with a radiation field. For our purposes, it is not necessary to consider an explicitiy quantum description of the radiation field: we may use the semiclassical theory. The interaction Hamiltonian is, for a single nucleus,

$$-(e/m_{\mathrm{p}}) \sum_i \mathbf{A}(\mathbf{x}_i) \cdot \mathbf{p}_i \tag{1.7.3}$$

where $\mathbf{x}_i$ and $\mathbf{p}_i$ are the proton coordinate and momentum, m_{p} is the proton mass, and the sum includes all the protons in the nucleus. We may neglect interaction with the magnetic moment for our purposes. Suppose the center of mass of the nucleus is located at position $\mathbf{X}_\mu$ ($\mathbf{X}_\mu = \mathbf{R}_\mu + \mathbf{u}_\mu$) in a crystal. To keep the notation reasonably simple, we consider a monatomic crystal. We introduce relative coordinates for the protons in a given nucleus with respect to the center of mass: $\mathbf{x}_i = \mathbf{X}_\mu + \mathbf{r}_i$. Consider a plane electromagnetic wave of wave vector $\mathbf{k}_\gamma$. The vector potential $A(\mathbf{x}_i)$ is proportional to $\exp(i\mathbf{k}_\gamma \cdot \mathbf{x}_i)$. We require matrix elements of the interaction between initial and final nuclear states $|\,\mathrm{i}\rangle$ and $|\,\mathrm{f}\rangle$. When the nucleus makes the transition i → f, the state of the lattice may change, going from n_i to n_f phonons in the various modes. The matrix element for the transition is

$$M = \langle \{n_\mathrm{f}(\mathbf{q})\} \mid \exp(i\mathbf{k}_\gamma \cdot \mathbf{X}_\mu) \mid \{n_\mathrm{i}(\mathbf{q})\}\rangle\langle \mathrm{f} \mid a(\mathbf{k}_\gamma) \mid \mathrm{i}\rangle \qquad (1.7.4)$$

where

$$a(\mathbf{k}_\gamma) \sim \sum_j \exp(i\mathbf{k}_\gamma \cdot \mathbf{r}_j)\,\mathbf{A}_0 \cdot \mathbf{p}_j, \qquad (1.7.5)$$

and $\mathbf{A}_0$ is the amplitude of the radiation field. Thus $a(\mathbf{k}_\gamma)$ contains all the nuclear physics of the problem and will not concern us further here. Note that $\langle \mathrm{f} \mid a(\mathbf{k}_\gamma) \mid \mathrm{i}\rangle$ is independent of lattice position. We are interested in the matrix element for a transition between phonon states. The transition rate is proportional to $P(n_\mathrm{f}, n_\mathrm{i})$ where

$$P(n_\mathrm{f}, n_\mathrm{i}) = |\langle n_\mathrm{f} \mid \exp(i\mathbf{k}_\gamma \cdot \mathbf{X}_\mu) \mid n_\mathrm{i}\rangle|^2. \qquad (1.7.6)$$

Since the probability of a nuclear transition is independent of the phonon states, we may therefore consider only the lattice matrix element explicitly.

The recoilless fraction f is the relative probability that a transition will occur without change in the energy of the phonon distribution. This is written as

$$f = \sum_\mathrm{f} |\langle n_\mathrm{i} \mid \exp(i\mathbf{k}_\gamma \cdot \mathbf{X}_\mu) \mid n_\mathrm{f}\rangle|^2\,\delta(E_\mathrm{f} - E_\mathrm{i}). \qquad (1.7.7)$$

The normalization is correct, since if we sum over all possible final states including those with an arbitrary transfer of energy to the phonons (this deletes the delta function), we find

$$\sum_{n_\mathrm{f}} |\langle n_\mathrm{i} \mid \exp(i\mathbf{k}_\gamma \cdot \mathbf{X}_\mu) \mid n_\mathrm{f}\rangle|^2 = \sum_{n_\mathrm{f}} \langle n_\mathrm{i} \mid \exp(i\mathbf{k}_\gamma \cdot \mathbf{X}_\mu) \mid n_\mathrm{f}\rangle$$
$$\times \langle n_\mathrm{f} \mid \exp(-i\mathbf{k}_\gamma \cdot \mathbf{X}_\mu) \mid n_\mathrm{i}\rangle = \langle n_\mathrm{i} \mid n_\mathrm{i}\rangle = 1. \qquad (1.7.8)$$

We must, however, allow for our ignorance of the actual phonon distribution by averaging over the probability of a particular distribution of occupation numbers. Hence, we should multiply (1.7.7) by $P_{n\mathrm{i}}(T)$, the probability of obtaining a particular phonon distribution at temperature T and sum over all states n_i. We may also assume the equilibrium position of the radioactive atom to be at the origin: thus $\mathbf{X}_0 = \mathbf{u}_0$, the displacement of this atom, and

$$f = \sum_\mathrm{if} P_{n\mathrm{i}}(T)\,|\langle n_\mathrm{i} \mid \exp(i\mathbf{k}_\gamma \cdot \mathbf{u}_0) \mid n_\mathrm{f}\rangle|^2\,\delta(E_\mathrm{f} - E_\mathrm{i}). \qquad (1.7.9)$$

The reader will recognize this object as a special case of the function

$S(\boldsymbol{\kappa}, \omega)$ defined in (1.6.16), in which the energy transfer is zero, and with the restriction that only $\mu = 0$ is considered in the sum.

From this, we see that we may obtain f simply from the first term of the expression (1.6.57) for $S(\boldsymbol{\kappa}, \omega)$ by the following process: Delete the factor of $\mathfrak{N}$ and the momentum-conserving delta function, and integrate over an infinitesimal range of frequencies. The result is just

$$f = \exp(-\tfrac{1}{3}k_\gamma^2\langle \mathbf{u}_0^2\rangle) = \exp[-2W(\mathbf{k}_\gamma)] \tag{1.7.10}$$

where $2W$ is just the Debye–Waller factor given formally by Eq. (1.6.52) and approximately evaluated in Eq. (1.6.67). This means that at low temperatures

$$f = \exp\{-\tfrac{3}{2}(\hbar^2\mathbf{k}_\gamma^2/2M)(1/K\Theta)[1 + (2\pi^2/3)(T/\Theta)^2]\}. \tag{1.7.11}$$

We notice the presence of the ratio of the recoil energy to the Debye energy $K\Theta$. For this reason, the Mössbauer effect is observed only in low energy γ transitions.

These arguments are easily extended to determine the shape of the absorption or emission cross section, including processes in which any number of phonons are involved. Since the natural width of the γ-ray line is negligible compared to typical phonon energies, no harm results if we consider the transition to be sharp. Thus if the final state in emission differs from the initial state by an energy $\hbar\omega$, we have instead of (1.7.8)

$$f(\omega) = \sum_{\mathrm{if}} P_{n\mathrm{i}}(T) \mid \langle n_\mathrm{i} \mid \exp(i\mathbf{k}_\gamma\cdot\mathbf{u}_0) \mid n_\mathrm{f}\rangle \mid^2 \delta(E_\mathrm{f} - E_\mathrm{i} - \hbar\omega). \tag{1.7.12}$$

The shape of the line will be characterized by the function $f(\omega)$. The calculation clearly is nearly identical to that of Section 1.6 and leads to the presence of single, double, etc. phonon wings on the line.

1.7.3 *Average Energy Transferred*

We will now derive Lipkin's (1960) result mentioned previously: The average energy transferred to the phonon system is equal to the expected recoil energy $\hbar^2\mathbf{k}_\gamma^2/2M$. To this end, we consider the commutator of the Hamiltonian for the crystal with $\exp(i\mathbf{k}_\gamma\cdot\mathbf{u}_0)$. The eigenstates of this Hamiltonian are states containing a definite number of phonons. The only portion of the Hamiltonian which does not commute with $\exp(i\mathbf{k}_\gamma\cdot\mathbf{u}_0)$ is the kinetic energy operator for the radioactive nucleus $\mathbf{p}^2/2M$.

$$\begin{aligned}[H, \exp(i\mathbf{k}_\gamma\cdot\mathbf{u}_0)] &= [\mathbf{p}^2/2M, \exp(i\mathbf{k}_\gamma\cdot\mathbf{u}_0)] \\ &= \exp(i\mathbf{k}_\gamma\cdot\mathbf{u}_0)[(\hbar^2\mathbf{k}_\gamma^2/2M) + (\hbar/M)\mathbf{k}_\gamma\cdot\mathbf{p}].\end{aligned}$$

We now form the double commutator

$$[[H, \exp(i\mathbf{k}_\gamma\cdot\mathbf{u}_0)], \exp(-i\mathbf{k}_\gamma\cdot\mathbf{u}_0)] = 2H - \exp(i\mathbf{k}_\gamma\cdot\mathbf{u}_0)\, H \exp(-i\mathbf{k}_\gamma\cdot\mathbf{u}_0)$$
$$- \exp(-i\mathbf{k}_\gamma\cdot\mathbf{u}_0)\, H \exp(i\mathbf{k}_\gamma\cdot\mathbf{u}_0)$$
$$= -\hbar^2\mathbf{k}_\gamma{}^2/M. \qquad (1.7.13)$$

Consider the expectation value of the double commutator in the initial state $|n_i\rangle$, characterized by energy E_i:

$$\langle n_i | [[H, \exp(i\mathbf{k}_\gamma\cdot\mathbf{u}_0)], \exp(-i\mathbf{k}_\gamma\cdot\mathbf{u}_0)] | n_i\rangle$$
$$= 2E_i - \langle n_i | \exp(i\mathbf{k}_\gamma\cdot\mathbf{u}_0) H \exp(-i\mathbf{k}_\gamma\cdot\mathbf{u}_0) | n_i\rangle$$
$$- \langle n_i | \exp(-i\mathbf{k}_\gamma\cdot\mathbf{u}_0) H \exp(i\mathbf{k}_\gamma\cdot\mathbf{u}_0) | n_i$$
$$= 2E_i - \sum_f \{\langle n_i | \exp(i\mathbf{k}_\gamma\cdot\mathbf{u}_0) | n_f\rangle\langle n_f | H \exp(i\mathbf{k}_\gamma\cdot\mathbf{u}_0) | n_i\rangle$$
$$+ \langle n_i | \exp(-i\mathbf{k}_\gamma\cdot\mathbf{u}_0) | n_f\rangle\langle n_f | H \exp(i\mathbf{k}_\gamma\cdot\mathbf{u}_0) | n_i\rangle\}$$
$$= 2E_i - 2\sum_f E_f | \langle n_i | \exp(i\mathbf{k}_\gamma\cdot\mathbf{u}_0) | n_f\rangle |^2. \qquad (1.7.14)$$

We have inserted a complete set of final states $|n_f\rangle$. With the use of (1.7.6) and (1.7.11), we may write this in the form

$$\sum_f (E_f - E_i)P(n_f, n_i) = \hbar^2\mathbf{k}_\gamma{}^2/2M. \qquad (1.7.15)$$

This is the desired result.

One of the reasons for the great utility of the Mössbauer effect in solid state physics is that, through the Zeeman effect, magnetic fields at the nuclear site can be measured. If the nucleus has spin I different from zero, there will be a magnetic moment $\boldsymbol{\mu}_n = \gamma_n\hbar\boldsymbol{I}$ (where γ_n is the nuclear gyromagnetic ratio), and this magnetic moment will interact with an effective magnetic field H_{eff} if it exists. That interaction is

$$H_Z = -\boldsymbol{\mu}_n\cdot\boldsymbol{H}_{eff}. \qquad (1.7.16)$$

This interaction completely removes the $(2I + 1)$-fold degeneracy of spin orientation. The splitting, although small, is easily observable.

1.8 Lattice Thermal Conductivity

A heat current in an electrically insulating crystal is carried by phonons. This situation contrasts with that in metals, where electrons are the primary carriers. Suppose one end of a sample is connected to a hot reservoir; the other to a cold reservoir. An excess of phonons is generated at the hot end and diffuses to the cold end. However, if the harmonic approximation were exactly valid, and the crystal were perfect, there would be no resistance to the heat flow and the thermal conductivity would be infinite. It is necessary to introduce anharmonic forces, or imperfections (or both) to obtain a finite thermal conductivity. Thus the study of lattice thermal conductivity is of considerable importance in furnishing information about anharmonic effects. [For reviews, see Klemens (1969, 1956), Berman (1967), Carruthers (1961), and Liebfried (1955).]

1.8.1 *The Boltzmann Equation*

The heat current carried by a phonon of wave vector $\mathbf{q}$ in branch j is $\hbar\omega_j(\mathbf{q})\mathbf{v}_j(\mathbf{q})$ where $\mathbf{v}_j(\mathbf{q})$ is the (group) velocity appropriate to the mode. We multiply by the average number of phonons in $j\mathbf{q}$, and sum over all modes. Thus we have for the heat flow per unit volume of crystal

$$\mathbf{Q} = (1/\mathcal{V}) \sum_{j\mathbf{q}} n_j(\mathbf{q})\hbar\omega_j(\mathbf{q})\mathbf{v}_j(\mathbf{q})$$

$$= [1/(2\pi)^3] \sum_{j} \int d^3q\, n_j(\mathbf{q})\hbar\omega_j(\mathbf{q})\mathbf{v}_j(\mathbf{q}). \tag{1.8.1}$$

The group velocity $\mathbf{v}_j(\mathbf{q})$ is given by

$$\mathbf{v}_j(\mathbf{q}) = \nabla_{\mathbf{q}}\omega_j(\mathbf{q}). \tag{1.8.2}$$

At very low temperatures, $\mathbf{Q}$ may be approximately related to the crystal momentum of the phonon system since only low energy modes will be excited. For such modes, $\omega_j(\mathbf{q}) = |\mathbf{v}_j(\theta, \phi)|\, q$. Suppose for simplicity that the direction dependence of v_j can be ignored. Then we may write

$$Q = (1/\mathcal{V}) \sum_{j\mathbf{q}} n_j(\mathbf{q})\hbar\mathbf{q}v_j^2 = \sum_j \mathbf{P}_j v_j^2 \tag{1.8.3}$$

where

$$\mathbf{P}_j = (1/\mathcal{V}) \sum_{\mathbf{q}} n_j(\mathbf{q})\hbar\mathbf{q}. \tag{1.8.4}$$

This quantity is the total wave vector for phonons in branch j.

The thermal conductivity is obtained from $\mathbf{Q}$. The heat current is proportional to the temperature gradient

$$\mathbf{Q} = -\kappa\,\nabla T \tag{1.8.5a}$$

where κ is the thermal conductivity. In a cubic crystal, only a single quantity κ describes the flow of heat. In an anisotropic crystal, we have

$$Q_\alpha = -\sum_\beta \kappa_{\alpha\beta}\, \partial T/\partial x_\beta. \tag{1.8.5b}$$

In this case, the $\kappa_{\alpha\beta}$ are the elements of a second-rank (symmetric) tensor.

We now consider the flow of phonons of mean wave vector $\mathbf{q}$ belonging to branch j through a small volume element. The number of these phonons changes as a result of two effects: (1) The number of phonons depends on the temperature gradient, and phonons tend to flow down the gradient. (2) Phonons collide with each other and with imperfections. The number in $j\mathbf{q}$ is considered to be a function of position $\mathbf{r}$, in accord with (1.8.1).

The rate of change of the mean occupation number due to the motion of the phonons is $-\nabla_{\mathbf{r}} n_j(\mathbf{q}) \cdot \mathbf{v}_j(\mathbf{q})$. This is proportional to the net flow of phonons out of a small volume element due to the variation of temperature with position. The change due to collisions is written as $[\partial n_j(\mathbf{q})/\partial t]_c$. The sum of these terms must vanish in a steady state situation

$$[\partial n_j(\mathbf{q})/\partial t]_c - \nabla_{\mathbf{r}} n_j(\mathbf{q}) \cdot \mathbf{v}_j(\mathbf{q}) = 0. \tag{1.8.6}$$

This is the *Boltzmann equation.* The quantity $n_j(\mathbf{q})$ will depend on $\mathbf{r}$ only through its dependence on temperature. Thus

$$\nabla_{\mathbf{r}} n_j(\mathbf{q}) = dn_j(\mathbf{q})/dT\ \nabla_{\mathbf{r}} T. \tag{1.8.7}$$

We therefore have

$$[\partial n_j(\mathbf{q})/\partial t]_c = \mathbf{v}_j(\mathbf{q}) \cdot \nabla_{\mathbf{r}} T\ dn_j(\mathbf{q})/dT. \tag{1.8.8}$$

This equation forms the basis for our study of thermal conductivity. A similar equation is obtained for electron transport processes (see Part B, Chapter 7) except that effects of external fields must be included.

1.8.2 *Phonon Scattering*

It is necessary to evaluate the collision term $[\partial n_j(\mathbf{q})/\partial t]_c$. To do this we have to study the processes which scatter phonons. Two principal contributors to this scattering are (1) anharmonic forces and (2) imperfections. We consider first anharmonic forces. The theory of anharmonic effects in crystals is reviewed by Liebfried and Ludwig (1961). The leading contribution is of third order in the atomic displacements. We call this V_3, and write

$$V_3 = (1/3!) \sum B \begin{pmatrix} i & l & n \\ \kappa & \lambda & \mu \\ \alpha & \beta & \gamma \end{pmatrix} u_{\alpha,i\kappa} u_{\beta,l\lambda} u_{\gamma,n\mu}. \tag{1.8.9}$$

The B coefficients are third derivatives of the interatomic potential and are to be regarded at the present as known parameters. Fourth- and higher-order terms exist, but will be ignored. We need to consider the effect of such a term on the phonon population. To do this, we must substitute (1.4.23). To simplify the results, we will consider here a monatomic crystal, so that we may drop the indices κ, λ, μ. Then

$$V_3 = \frac{1}{3!}\left(\frac{\hbar}{2\mathfrak{N}M}\right)^{3/2} \sum_{jj'j''} \sum_{\mathbf{k}\mathbf{k}'\mathbf{k}''} [\omega_j(\mathbf{k})\omega_{j'}(\mathbf{k}')\omega_{j''}(\mathbf{k}'')]^{-1/2} \sum B\begin{pmatrix} i & l & n \\ \alpha & \beta & \gamma \end{pmatrix}$$
$$\times \{[e_\alpha^{(j)}(\mathbf{k}) \exp(i\mathbf{k}\cdot\mathbf{R}_i)a_j(\mathbf{k}) + e_\alpha^{(j)}(\mathbf{k}) \exp(-i\mathbf{k}\cdot\mathbf{R}_i)a_j^\dagger(\mathbf{k})]$$
$$\times [e_\beta^{(j')}(\mathbf{k}') \exp(i\mathbf{k}'\cdot\mathbf{R}_l)a_{j'}(\mathbf{k}') + e_\beta^{(j')}(\mathbf{k}') \exp(-i\mathbf{k}'\cdot\mathbf{R}_l)a_j^\dagger(\mathbf{k}')]$$
$$\times [e_\gamma^{(j'')}(\mathbf{k}'') \exp(i\mathbf{k}''\cdot\mathbf{R}_n)a_{j''}(\mathbf{k}'')$$
$$+ e_\gamma^{(j'')}(\mathbf{k}'') \exp(-i\mathbf{k}''\cdot\mathbf{R}_n)a^\dagger_{j''}(\mathbf{k}'')]\}. \qquad (1.8.10)$$

The product contains eight terms. However, the two involving aaa and $a^+a^+a^+$ do not contribute to scattering in first order since they cannot conserve wave vector. Of the remaining six terms, only two are essentially different. Note that the commutator of two a's can be discarded since it leads to a term with only one a or a^+ and hence does not conserve energy. We make use of the invariance of B with respect to permutations to reduce V_3 to

$$V_3 = \frac{1}{2}\left(\frac{\hbar}{2\mathfrak{N}M}\right)^{3/2} \sum_{jj'j''} \sum_{\mathbf{k}\mathbf{k}'\mathbf{k}''} [\omega_j(\mathbf{k})\omega_{j'}(\mathbf{k}')\omega_{j''}(\mathbf{k}'')]^{-1/2} \sum B\begin{pmatrix} i & l & n \\ \alpha & \beta & \gamma \end{pmatrix}$$
$$\times \{e_\alpha^{(j)*}(\mathbf{k})e_\beta^{(j')}(\mathbf{k}')e_\gamma^{(j'')}(\mathbf{k}'') \exp[i(\mathbf{k}'\cdot\mathbf{R}_l + \mathbf{k}''\cdot\mathbf{R}_n - \mathbf{k}\cdot\mathbf{R}_i)]$$
$$\times a_j^\dagger(\mathbf{k})a_j(\mathbf{k}')a_{j''}(\mathbf{k}'') + \text{h.c.}\} \qquad (1.8.11)$$

where h.c. stands for the Hermitian conjugate and therefore involves terms $a_j^\dagger(\mathbf{k})a_j^\dagger(\mathbf{k}')a_{j''}(\mathbf{k}'')$.

Equation (1.8.11) shows that V_3 contains terms describing the collision of two phonons to produce a third phonon; or conversely, the splitting of a single phonon into two others. Conservation of energy in such processes is enforced by the usual delta function multiplying the square of a matrix element in time-dependent perturbation theory. We also expect wave vector to be conserved (up to a reciprocal lattice vector). This comes about as follows:

In an infinite crystal the coefficients B cannot depend on the absolute position of the atomic cells, but only on their relative position. Thus we may introduce vectors

$$\mathbf{R}_\mu = \mathbf{R}_l - \mathbf{R}_i, \qquad \mathbf{R}_\nu = \mathbf{R}_n - \mathbf{R}_i$$

and consider B to depend on $\mathbf{R}_\mu$ and $\mathbf{R}_\nu$ only. We make this substitution in Equation (1.8.11). Then the sum cn i stands separately and may be performed:

$$\sum_i \exp[i(\mathbf{k}' + \mathbf{k}'' - \mathbf{k})\cdot\mathbf{R}_i] = \mathfrak{N}\delta_{\mathbf{k}'+\mathbf{k}'',\mathbf{k}+\mathbf{K}_s} \tag{1.8.12}$$

We have

$$V_3 = \frac{\hbar}{4M}\left(\frac{\hbar}{2\mathfrak{N}M}\right)^{1/2} \sum_{jj'j''s} \sum_{\mathbf{k}\mathbf{k}'\mathbf{k}''} [\omega_j(\mathbf{k})\omega_j(\mathbf{k}')\omega_j(\mathbf{k}'')]^{-1/2}$$

$$\times \delta_{\mathbf{k}'+\mathbf{k}'',\mathbf{k}+\mathbf{K}_s}$$

$$\times \sum_{\mu\nu,\alpha\beta\gamma} B\begin{pmatrix} \mu & \nu & \\ \alpha & \beta & \gamma \end{pmatrix} e_\alpha^{(j)}(\mathbf{k}) e_\beta^{(j')}(\mathbf{k}') e_\gamma^{(j'')}(\mathbf{k}'')$$

$$\times \exp[i(\mathbf{k}'\cdot\mathbf{R}_\mu + \mathbf{k}''\cdot\mathbf{R}_\nu)] a_j^\dagger(\mathbf{k}) a_{j'}(\mathbf{k}') a_{j''}(\mathbf{k}'') + \cdots. \tag{1.8.13}$$

The notation may be simplified somewhat by defining

$$b_{jj'j''}(\mathbf{k}, \mathbf{k}', \mathbf{k}'') = \sum_{\mu\nu,\alpha\beta\gamma} B\begin{pmatrix} \mu & \nu & \\ \alpha & \beta & \gamma \end{pmatrix} e_\alpha^{(j)}(\mathbf{k}) e_\beta^{(j')}(\mathbf{k}') e_\gamma^{(j'')}(\mathbf{k}'')$$

$$\times \exp[i(\mathbf{k}'\cdot\mathbf{R}_\mu + \mathbf{k}''\cdot\mathbf{R}_\nu)].$$

Note that if $\mathbf{k}'$ or $\mathbf{k}''$ is increased by $\mathbf{K}_s$, the result is unchanged. Also $b(\mathbf{k}, \mathbf{k}', \mathbf{k}'') = b(\mathbf{k}, -\mathbf{k}', \mathbf{k}'')$ for real $\mathbf{e}$ which we assume. Thus

$$V_3 = \frac{\hbar}{4M}\left(\frac{\hbar}{2\mathfrak{N}M}\right)^{1/2} \sum_{jj'j''s} \sum_{\mathbf{k}\mathbf{k}''} \frac{b_{jj'j''}(\mathbf{k}, \mathbf{k}', \mathbf{k} - \mathbf{k}')}{[\omega_j(\mathbf{k})\omega_{j'}(\mathbf{k}')\omega_{j''}(\mathbf{k}' - \mathbf{k} + \mathbf{K}_s)]}$$

$$\times a_j^\dagger(\mathbf{k}) a_{j'}(\mathbf{k}') a_{j''}(\mathbf{k} - \mathbf{k}' + \mathbf{K}_s) + \cdots]. \tag{1.8.14}$$

The delta function of Eq. (1.8.12) indicates that the scattering process does not occur unless

$$\mathbf{k} = \mathbf{k}' + \mathbf{k}'' - \mathbf{K}_s. \tag{1.8.15}$$

Suppose $\mathbf{K}_s = 0$. Then the wave vector of the outgoing phonon ($\mathbf{k}$) is equal to the sum of the wave vectors of the incoming phonons. Such events are called *normal* processes. If $\mathbf{K}_s \neq 0$, the event is called an *umklapp* process. In an umklapp process, the wave vector of the outgoing phonon

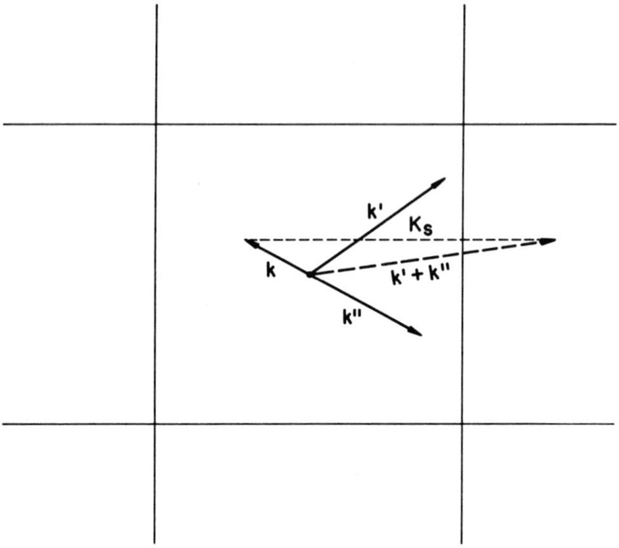

Fig. 1.8.1. An umklapp process.

is quite different from the sum of the incoming ones. This is illustrated in Fig. 1.8.1. Umklapp processes are of great importance in the theory of lattice thermal conductivity.

The significance of umklapp processes is made apparent by the following argument. Suppose only normal processes occurred. Then the sum of the wave vectors of all the phonon states $\mathbf{P} = \sum_j \mathbf{P}_j$ would remain constant. Hence, collisions could not lead from a state of nonzero $\mathbf{P}$ to a state of zero $\mathbf{P}$. However, in equilibrium we must have $\mathbf{P} = 0$. In a state of nonzero $\mathbf{P}$, a heat current is obtained [see Eq. (1.8.3)]. Therefore once a heat current was established in a crystal, it would continue to flow even in the absence of a thermal gradient, if only normal processes were available to scatter phonons. Such a condition indicates zero thermal resistance. Umklapp processes lead to a change in the total wave vector of the phonon system and thus are essential to the existence of a thermal resistance in pure crystals.

Additional restrictions on possible processes are imposed by conservation of energy. We must have

$$\omega_j(\mathbf{k}) = \omega_{j'}(\mathbf{k}') + \omega_{j''}(\mathbf{k} - \mathbf{k}' + \mathbf{K}_s). \qquad (1.8.16)$$

The nature of the restrictions can be visualized for simple $\omega(\mathbf{k})$ relations through the following construction (Peierls, 1955). For simplicity consider a direction in which two phonon branches are degenerate.

We suppose both $\mathbf{k}$ and $\mathbf{k}'$ to lie in this direction and draw curves representing $\omega(\mathbf{k})$. Suppose we choose j' to be the lowest (transverse) branch (see Fig. 1.8.2). At $(\omega_{j'}, \mathbf{k}')$, repeat the construction of the $\omega(\mathbf{k})$ curves. The intersection of the reconstructed curves with the original ones indicates possible three-phonon processes. The curves which would extend outside the zone are brought back into the zone in accord with $\omega_{j''}(\mathbf{k}) = \omega_{j''}(\mathbf{k} + \mathbf{K}_s)$. The point of intersection shows ω and $\mathbf{k}$; the horizontal dashed line gives $\mathbf{k} - \mathbf{k}' + \mathbf{K}_s$. In the figure, intersections 1 and 2 correspond to normal three-phonon processes while intersections 3 and 4 represent umklapps.

In general, considerations of the possible scattering processes are quite complex because of the necessity of considering the detailed form of the $\omega(\mathbf{k})$ relation. The existence of lines of degeneracy plays a very important role (Herring, 1954).

We will now follow Peierls in writing an expression for the rate of change of $n_j(\mathbf{k})$ due to three-phonon interactions. The rate of transitions between specified initial and final states is

$$W_{\mathrm{fi}} = (2\pi/\hbar) \mid \langle \mathrm{f} \mid V_3 \mid \mathrm{i} \rangle \mid^2 \delta(E_{\mathrm{f}} - E_{\mathrm{i}}). \tag{1.8.17}$$

Consider a transition in which the number of phonons in $j\mathbf{k}$ increases by 1.

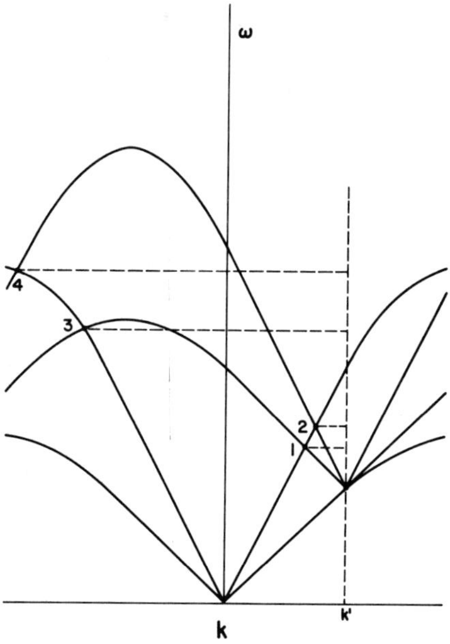

Fig. 1.8.2. Energetics of umklapp processes (see text for a description).

We will obtain the contribution of the first term of (1.8.14) to the transition rate. The sum over $\mathbf{k}'$ is replaced by an integral in accord with (1.2.10).

$$\frac{2\pi}{\hbar}\frac{\hbar^2}{16M^2}\frac{\Omega\hbar}{16\pi^3 M}\sum_{j'j''s}\int d^3k' \frac{|b_{jj'j''}(\mathbf{k},\mathbf{k}',\mathbf{k}-\mathbf{k}')|^2}{\omega_j(\mathbf{k})\omega_{j'}(\mathbf{k}')\omega_{j''}(\mathbf{k}'-\mathbf{k}+\mathbf{K}_s)}$$
$$\times (n_j(\mathbf{k})+1)n_{j'}(\mathbf{k}')n_{j''}(\mathbf{k}-\mathbf{k}'+\mathbf{K}_s)$$
$$\times \delta[(\hbar)(\omega_j(\mathbf{k})-\omega_{j'}(\mathbf{k}')-\omega_{j''}(\mathbf{k}-\mathbf{k}'+\mathbf{K}_s))]. \qquad (1.8.18)$$

To simplify the notation, we will abbreviate $n_j(\mathbf{k})$ by n, $n_{j'}(\mathbf{k}')$ by n', and $n_{j''}(\mathbf{k}-\mathbf{k}'+\mathbf{K}_s)$ by n''. Similarily $\omega_j(\mathbf{k}) \equiv \omega$, $\omega_{j'}(\mathbf{k}') \equiv \omega'$, and $\omega_{j''}(\mathbf{k}'-\mathbf{k}+\mathbf{K}_s) = \omega''$.

To obtain the total transition rate of the phonon population in $j\mathbf{k}$ we have to include three other processes. These are: (2) Phonon $j\mathbf{k}$ splits into two others. This gives factors $n(n'+1)(n''+1)$, and enters with a minus sign since it represents a loss of phonons from the mode. (3) Phonon $j''\mathbf{k}''$ splits into $j'\mathbf{k}'$ and $j\mathbf{k}$. In this case, we have number factors $(n+1)(n'+1)n''$ and an energy conservation factor $\delta(\omega+\omega'-\omega'')$. (4) Phonon $j'\mathbf{k}'$ absorbs $j\mathbf{k}$ and produces $j''\mathbf{k}''$. This is the opposite of (3) and yields $nn'(n''+1)$.

Processes (2) and (3) require an additional factor of 2 since they may go with $j''\mathbf{k}''$ replaced by $j'\mathbf{k}'$. Altogether there are six processes which agrees with the number of possible energy conserving terms in V_3. For a detailed analysis of the terms contributing to V_3, see Liebfried (1955). Then we obtain

$$\left(\frac{\partial n_j(\mathbf{k})}{\partial t}\right)_c = \frac{\hbar}{128\pi^2 M^3}\sum_{j'j''s}\int d^3k' \frac{|b_{jj'j''}(\mathbf{k},\mathbf{k}',\mathbf{k}-\mathbf{k}')|^2}{\omega\omega'\omega''}$$
$$\times \{\delta(\omega-\omega'-\omega'')[(n+1)n'n''-n(n'+1)(n''+1)]$$
$$+ 2\,\delta(\omega+\omega'-\omega'')[(n+1)(n'+1)n''-nn'(n''+1)]\}. \qquad (1.8.19)$$

We now write the distribution function as the sum of two parts, $n = n_0 + n_1$, where n_0 is the mean phonon number in thermal equilibrium and n_1 is the difference due to the thermal conduction process:

$$n_0 = 1/[\exp(\hbar\omega/KT) - 1]. \qquad (1.8.20)$$

We have as an identity, that if $\omega = \omega' + \omega''$,

$$(n_0+1)n_0'n_0'' = n_0(n_0'+1)(n_0''+1). \qquad (1.8.21)$$

(The proof is immediate on substitution.) This shows that with n replaced by n_0, $(\partial n/\partial t)_c$ is zero: the equilibrium distribution is unchanged by collisions, as is apparent from physical reasoning.

Let us write n_1 in the form

$$n_1 = n_0(n_0 + 1)g \tag{1.8.22}$$

where g is to be determined. We retain only first-order terms in g, since g will be of first order in the thermal gradient. Higher-order terms would involve $(\nabla T)^2$, which is neglected. Frequent use of (1.8.21) then leads to the expression

$$\begin{aligned}\left(\frac{\partial n_j(\mathbf{k})}{\partial t}\right)_c = \frac{\Omega\hbar}{128\pi^2 M^3} \sum_{j'j''s} \int d^3k' \frac{|b_{jj'j''}(\mathbf{k}, \mathbf{k}', \mathbf{k} - \mathbf{k}')|^2}{\omega\omega'\omega''} \\ \times \{\delta(\omega - \omega' - \omega'')[(n_0 + 1)n_0'n_0''(g' + g'' - g)] \\ + 2\,\delta(\omega + \omega' - \omega'')[(n_0 + 1)(n_0' + 1)n_0''(g'' - g' - g)]\}.\end{aligned} \tag{1.8.23}$$

We return to the Boltzmann equation (1.8.8). In calculating (dn_j/dT), we retain only the contribution from n_0 as this already includes ∇T. Thus we obtain

$$dn_0/dT = n_0(n_0 + 1)(\hbar\omega/KT^2).$$

The Boltzmann equation can now be written correct through terms of first order in ∇T. The subscript 0 is dropped, but arguments are restored.

$$\begin{aligned}&\frac{\hbar\omega_j(\mathbf{q})}{KT^2} n_j(\mathbf{q})[n_j(\mathbf{q}) + 1]\mathbf{v}_j(\mathbf{q})\cdot\nabla T \\ &= \frac{\hbar}{128\pi^2 M^3} \sum_{j'j''s} \int d^3k \frac{|b_{jj'j''}(\mathbf{q}, \mathbf{k}, \mathbf{q} - \mathbf{k})|^2}{\omega_j(\mathbf{q})\omega_{j'}(\mathbf{k})\omega_{j''}(\mathbf{q} - \mathbf{k} + \mathbf{K}_s)} \\ &\quad \times \{\delta[\omega_j(\mathbf{q}) - \omega_{j'}(\mathbf{k}) - \omega_{j''}(\mathbf{q} - \mathbf{k} + \mathbf{K}_s)] \\ &\quad \times (n_j(\mathbf{q}) + 1)n_{j'}(\mathbf{k})n_{j''}(\mathbf{q} - \mathbf{k} + \mathbf{K}_s) \\ &\quad \times [g_{j'}(\mathbf{k}) + g_{j''}(\mathbf{q} - \mathbf{k} + \mathbf{K}_s) - g_j(\mathbf{q})] \\ &\quad + 2[\delta(\omega_j(\mathbf{q}) + \omega_{j'}(\mathbf{k}) - \omega_{j''}(\mathbf{q} - \mathbf{k} + \mathbf{K}_s))(n_j(\mathbf{q}) + 1) \\ &\quad \times (n_{j'}(\mathbf{k}) + 1)n_{j''}(\mathbf{q} - \mathbf{k} + \mathbf{K}_s)(g_{j''}(\mathbf{q} - \mathbf{k} + \mathbf{K}_s) \\ &\quad - g_j(\mathbf{q}) - g_{j'}(\mathbf{k}))]\}.\end{aligned} \tag{1.8.24}$$

Equation (1.8.24) is to be regarded as an integral equation to be solved for the g_j. Its solution would describe the conduction of heat in a perfect crystal (no impurities or defects) at temperatures low enough so that four-phonon processes could be neglected.

This equation is quite difficult to solve. We can, however, draw rigorous conclusions applicable to the high temperature region ($T > \Theta$). Then 1 may be neglected compared to n, and n is itself proportional to temperature. The left side of (1.8.24) becomes independent of T so that g must be proportional to T^{-3}. We now consider the calculation of the heat current by (1.8.1). The term n_0 gives no contribution and since $g \approx T^{-3}$, $n_1 \approx T^{-1}$ [by (1.8.22)]. Therefore $Q \approx T^{-1}$ and we arrive at the conclusion that in the high-temperature region the lattice thermal conductivity of all crystals should be proportional to $1/T$. This is observed.

1.8.3 *The Relaxation Time Approximation*

Equation (1.8.24) is far too difficult to solve in practical cases, and in any event, its solution would not include scattering processes other than three-phonon events. As a practical matter, one must make approximations, and the most common of these is the relaxation time approximation. In this procedure, we write

$$(dn_j(\mathbf{q})/dt)_c = [n_{0j}(\mathbf{q}) - n_j(\mathbf{q})]/\tau_j(\mathbf{q}) = -n_{1j}(\mathbf{q})/\tau_j(\mathbf{q}) \quad (1.8.25)$$

where $\tau_j(\mathbf{q})$ is the relaxation time for mode $j\mathbf{q}$.

There are some important difficulties with the relaxation time approximation, as will be described subsequently; however, let us accept it for the moment. We insert this into the Boltzmann equation (1.8.24) and solve for $n_{1j}(\mathbf{q})$:

$$n_{1j}(\mathbf{q}) = -[\hbar\omega_j(\mathbf{q})/KT^2]n_{0j}(\mathbf{q})[n_{0j}(\mathbf{q}) + 1]\mathbf{v}_j(\mathbf{q}) \cdot \nabla T\, \tau_j(\mathbf{q}). \quad (1.8.26)$$

This leads to a heat current, from (1.8.1),

$$\begin{aligned} Q_\alpha &= [1/(2\pi)^3] \sum_j \int d^3q\, n_{1j}(\mathbf{q}) v_\alpha(\mathbf{q}) \hbar\omega_j(\mathbf{q}) \\ &= [-1/(2\pi)^3] \sum_j \int d^3q\, [\hbar\omega_j(\mathbf{q})/KT^2] n_{0j}(\mathbf{q})[n_{0j}(\mathbf{q}) + 1] \\ &\quad \times v_{\alpha j}(\mathbf{q}) v_{\beta j}(\mathbf{q}) \tau_j(\mathbf{q})\, \partial T/\partial x_\beta, \end{aligned} \quad (1.8.27)$$

and thus to a thermal conductivity tensor

$$\begin{aligned} \kappa_{\alpha\beta} &= [1/(2\pi)^3] \sum_j \int d^3q\, [\hbar\omega_j(\mathbf{q})/KT^2] n_{0j}(\mathbf{q})[n_{0j}(\mathbf{q}) + 1] \\ &\quad \times \tau_j(\mathbf{q}) v_{\alpha j}(\mathbf{q}) v_{\beta j}(\mathbf{q}). \end{aligned} \quad (1.8.28)$$

In order to interpret this equation, let us recall from Eq. (1.3.10) that the contribution of mode $j\mathbf{q}$ to the crystal specific heat per unit volume is $C_j(\mathbf{q})$,

$$C = [1/(2\pi)^3] \sum_j \int C_j(\mathbf{q})\, d^3q$$

where

$$C_j(\mathbf{q}) = [(\hbar\omega_j(\mathbf{q}))^2/KT^2] n_{0j}(\mathbf{q})[n_{0j}(\mathbf{q}) + 1].$$

Thus

$$\kappa_{\alpha\beta} = [1/(2\pi)^3] \sum_j \int d^3q\, C_j(\mathbf{q})\tau_j(\mathbf{q})v_{\alpha j}(\mathbf{q})v_{\beta j}(\mathbf{q}). \qquad (1.8.29)$$

The functions C_j and τ_j should have the full symmetry of the crystal. The phonon velocities have the symmetry of vectors. In a cubic crystal, this implies that the average over directions of $v_{\alpha j}v_{\beta j}$ is zero unless $\alpha = \beta$ and that the average of $v^2{}_{\alpha j}$ is independent of α; equal to $\frac{1}{3}$ the directional average of $v_j{}^2$. Then

$$\kappa_{\alpha\beta} = \kappa\,\delta_{\alpha\beta}, \qquad \kappa = [1/3(2\pi)^3] \sum_j \int d^3q\, C_j(\mathbf{q})\tau_j(\mathbf{q})v_j{}^2(\mathbf{q}). \qquad (1.8.30)$$

Suppose that the phonon speeds and relaxation times were in fact independent of q. Then (1.8.30) would reduce to

$$\kappa = \tfrac{1}{3}Cv^2\tau = \tfrac{1}{3}CvL \qquad (1.8.31)$$

where C is the crystal specific heat and $L = v\tau$ is the phonon mean free path. This result is characteristic of elementary kinetic theory.

It is, however, a very poor approximation to ignore the frequency dependence of the relaxation time except at the very lowest temperatures for which scattering by crystal boundaries is dominant. It would take us too long to attempt to calculate these at the present. We will therefore consider a simple phenomenological model in which we neglect dispersion altogether, and treat the speed of sound as a constant v_s, as in the Debye theory of specific heat. Furthermore, we will neglect the difference between branches of the vibration spectrum. We replace the sum over j by a factor of 3, counting only acoustic branches since, because of the large excitation energy, optic modes will not contribute at low temperatures. Simple functional forms are assumed for the relaxation times. We have

$$\kappa = [v_s{}^2/(2\pi)^3] \int d^3q\, C(q)\tau(q). \qquad (1.8.32)$$

We will suppose that phonons are scattered by point defects with a relaxation time

$$\tau_D^{-1} = A\omega^4 \tag{1.8.33}$$

where A is proportional to $c(1 - c)$, c being the concentration of defects (and is independent of temperature) and by phonon–phonon processes with a relaxation time

$$\tau_{ph}^{-1} = BT^3\omega^2. \tag{1.8.34}$$

Equation (1.8.33) is due to Klemens (1955). It can be obtained by the methods of scattering theory which will be discussed in Section 5.2 (Callaway, 1963). The validity of (1.8.34) is more questionable; it is an approximation based on the work of Herring (1954), which seems to be reasonably adequate empirically. We will also ignore the difference between normal and umklapp processes which will be valid if umklapp processes are not too rare. Finally, to handle very low temperatures we will consider a boundary scattering process with a frequency- and temperature-independent relaxation time

$$\tau_B^{-1} = v_s/L \tag{1.8.35}$$

where L is of the order of the macroscopic dimensions of the specimen (Casimir, 1938). The scattering processes are assumed to be independent so that the probabilities are additive. This means that the reciprocal relaxation times add. Therefore, put (Callaway, 1959)

$$\tau^{-1} = \tau_D^{-1} + \tau_{ph}^{-1} + \tau_B^{-1} = A\omega^4 + BT^3\omega^2 + v_s/L.$$

This is substituted into (1.8.32). We obtain, with $\omega = v_s q$,

$$\kappa = (K/2\pi^2 v_s) \int_0^{K\Theta/\hbar} \tau(\omega)(\hbar\omega/KT)^2[e^{\hbar\omega/KT}/(e^{\hbar\omega/KT} - 1)^2]\,\omega^2\,d\omega. \tag{1.8.36}$$

The upper limit is the maximum frequency of the vibrational spectrum, which we assume to be given by the Debye temperature. It is convenient to introduce the dimensionless variable $x = \hbar\omega/KT$ so that we have

$$\kappa = (K/2\pi^2 v_s)(KT/\hbar)^3 \int_0^{\Theta/T} [x^4/(Dx^4 + Ex^2 + v_s/L)][e^x/(e^x - 1)^2]\,dx \tag{1.8.37}$$

where

$$D = A(KT/\hbar)^4, \qquad E = BT^3(KT/\hbar)^2.$$

Even with all the approximations we have made, the result is too complicated to evaluate analytically. If (1.8.37) is integrated numerically, it is found to be possible to fit observed thermal conductivity curves for many materials over a substantial temperature range with reasonable choices of A, B, and C.

A limiting case may be examined analytically. Boundary scattering is ignored. We approximate $x^2e^x/(e^x - 1)^2$ by the leading term in its expansion for small x, 1. Then

$$\kappa = [K/(2\pi^2 v_s)](KT/\hbar)^3 \int_0^{\Theta/T} dx/(Dx^2 + E)$$

$$= [K/(2\pi^2 v_s)][1/(DE)^{1/2}](KT/\hbar)^3 \tan^{-1}[(\Theta/T)(D/E)^{1/2}]. \quad (1.8.38)$$

At low temperatures ($T \ll \Theta$), and for strong defect scattering,

$$\kappa = K[4\pi v_s (AB)^{1/2} T^{3/2}]^{-1}. \quad (1.8.39)$$

We note that the thermal conductivity turns out to be inversely proportional to the square root of the defect concentration. This is observed in some cases.

On the other hand, if the defect scattering is small compared to umklapp processes, and if the temperature is still low enough so that (1.8.34) may be used and the upper limit of the integral (1.8.37), Θ/T, may be made infinite; we may expand the denominator of (1.8.37), and evaluate the integral explicitly. Boundary scattering is neglected. The result is

$$\kappa = (K^2/6v_s\hbar BT^2)[1 - (4\pi^2/5)(K^2/\hbar^2)(A/BT)]. \quad (1.8.40)$$

The principal interest here is in the effect of the point defect scattering. The thermal resistivity $W = 1/\kappa$ can be expressed as

$$W = W_p + W_D \quad (1.8.41)$$

where W_p is the resistivity of "pure" material (no defects) and W_D is the specific contribution from the defects. We find

$$W_D = (24\pi^2/5)v_s(AT/\hbar). \quad (1.8.42)$$

The defects thus contribute an additional term to the resistivity which is proportional to temperature and is independent of the phonon–phonon scattering (except through the numerical constant, which depends on the particular power law used to express the frequency dependence).

1.8.4 *Role of Normal Three-Phonon Processes*

Let us now investigate the effect of considering the conservation of total wavevector by normal processes (Callaway, 1959). This is a complex problem and a full treatment requires the solution of (1.8.24). However, an approximate method can be developed which seems to be reasonably successful. The procedure is intuitive in nature, and consists in modifying

(1.8.25) as follows:

$$(\partial n(\mathbf{q})/\partial t)_c = [n_0(\mathbf{q}) - n(\mathbf{q})]/\tau_u(\mathbf{q}) + [n_0(\mathbf{q}, \boldsymbol{\lambda}) - n(\mathbf{q})]/\tau_N(\mathbf{q}). \tag{1.8.43}$$

We are not considering the different branches of the vibrational spectrum, so the index j has been dropped. The first term on the right side of (1.8.43) is just that included in (1.8.25), and represents the effect of wave vector-destroying processes which tend to return the distorted distribution caused by the thermal gradient to equilibrium. The quantity $\tau_u(q)$ includes all such processes: point defects, umklapps, crystal boundaries, etc.

The second term on the right of (1.8.43) is intended to account for the normal processes, and τ_N is the relaxation time for them. However, we note that normal processes do not tend to relax the phonon distribution to the equilibrium distribution n_0, but to a displaced or flowing distribution $n_0(\mathbf{q}, \boldsymbol{\lambda})$ (Klemens, 1955).

$$n_0(\mathbf{q},\boldsymbol{\lambda}) = \{\exp[(\hbar\omega(\mathbf{q}) - \boldsymbol{\lambda}\cdot\mathbf{q})/KT] - 1\}^{-1}.$$

Let us write $n_1 = n(\mathbf{q}) - n_0(\mathbf{q})$. For simplicity of notation we do not write the arguments of the functions from here on. In using (1.8.43), we may expand, treating $\boldsymbol{\lambda}$ as small since as will be seen, $\boldsymbol{\lambda}$ is proportional to the temperature gradient. The Boltzmann equation, (1.8.8) may now be written as

$$[(-\hbar\omega/KT^2)\mathbf{v}_s\cdot\nabla T + (\boldsymbol{\lambda}\cdot\mathbf{q}/\tau_N KT)][n_0(n_0+1)] - (1/\tau_N + 1/\tau_u)n_1 = 0. \tag{1.8.44}$$

We can now define a combined relation time τ_c by

$$\tau_c^{-1} = \tau_N^{-1} + \tau_u^{-1}. \tag{1.8.45}$$

The quantity τ_c is the relaxation time which was previously used in (1.8.36). To simplify (1.8.44) we write n_1 as in (1.8.26),

$$n_1 = -\tau_T \mathbf{v}_s\cdot\nabla T(\hbar\omega/KT^2)n_0(n_0+1) \tag{1.8.46}$$

where τ_T is some total relaxation time to be determined. This leads to a thermal conductivity

$$\kappa = [v_s^2/(2\pi)^3]\int \tau_T(\mathbf{q})C(\mathbf{q})\, d^3q. \tag{1.8.47}$$

We have here employed the same arguments as used previously to obtain (1.8.30). To determine τ_T we first substitute (1.8.46) and (1.8.45) into (1.8.44) which becomes

$$(\hbar\omega/\tau_c T)\tau_T v_s\cdot\nabla T + (\boldsymbol{\lambda}\cdot\mathbf{q}/\tau_N) = (\hbar\omega/T)\mathbf{v}_s\cdot\nabla T. \tag{1.8.48}$$

Since $\boldsymbol{\lambda}$ determines the total crystal wave vector of the phonon distribution, we expect $\boldsymbol{\lambda}$ to be a constant vector in the direction of the temperature gradient. We can define a parameter β with the dimensions of a relaxation time by

$$\boldsymbol{\lambda} = -(\hbar/T)\beta v_s^2 \nabla T. \qquad (1.8.49)$$

In the Debye approximation to the phonon spectrum which we are using $\mathbf{q} = \mathbf{v}_s\omega/v_s^2$. Thus

$$\boldsymbol{\lambda}\cdot\mathbf{q} = -(\hbar\omega/T)\beta\mathbf{v}_s\cdot\nabla T, \qquad (1.8.50)$$

and (1.8.48) gives

$$\tau_T = \tau_c(1 + \beta/\tau_N). \qquad (1.8.51)$$

The factor $(1 + \beta/\tau_N)$ which multiplies τ_c in (1.8.51) evidently expresses the correction to the simple relaxation time approximation due to the particular nature of the normal processes. A further condition is neccesary to determine β. We obtain this by requiring that the rate of change of the total phonon wave vector due to N processes should vanish. We express this as

$$\int (\partial n/\partial T)_N\mathbf{q}\, d^3q = \int \{[n_0(\mathbf{q}, \boldsymbol{\lambda}) - n(\mathbf{q})]/\tau_N(\mathbf{q})\}\mathbf{q}\, d^3q = 0. \qquad (1.8.52)$$

We substitute (1.8.43), (1.8.46), and (1.8.50) into (1.8.52) and have, for small $\boldsymbol{\lambda}$,

$$\int n_0(n_0 + 1)(\hbar\omega/KT^2)\mathbf{v}_s(\mathbf{v}_s\cdot\nabla T)(\tau_T - \beta)(\omega/\tau_N v_s^2)\, d^3q = 0.$$

We insert (1.8.51) for τ_T and solve for β. The result is (in terms of the dimensionless variable $x = \hbar\omega/KT$)

$$\beta = \int_0^{\Theta/T} \frac{\tau_c}{\tau_N}\frac{x^4e^x}{(e^x - 1)^2}\,dx \Bigg/ \int_0^{\Theta/T} \frac{1}{\tau_N}\left(1 - \frac{\tau_c}{\tau_N}\right)\frac{x^4e^x}{(e^x - 1)^2}\,dx. \qquad (1.8.53)$$

The resulting thermal conductivity is

$$\kappa = (K/2\pi^2 v_s)(I_1 + \beta I_2) \qquad (1.8.54)$$

where

$$I_1 = (KT/\hbar)^3 \int_0^{\Theta/T} \tau_c(x)[x^4e^x/(e^x - 1)^2]\,dx, \qquad (1.8.55)$$

$$I_2 = (KT/\hbar)^3 \int_0^{\Theta/T} (\tau_c/\tau_N)[x^4e^x/(e^x - 1)^2]\,dx. \qquad (1.8.56)$$

There are two questions that arise naturally at this point: (1) How important is the correction βI_2? and (2) Is the simple approach we have taken to the complications of the full Boltzmann equation adequate?

The first question can be answered as long as we stay within the framework of the approximations already introduced (Callaway, 1961). We expect a significant correction only when the wave-vector-conserving processes are much stronger than those contained in τ_u which reduce the total wave vector. This is unusual, but it can happen, and if it does, the results are quite interesting. Suppose, to be specific, that only point defect scattering and normal processes occur and that $\tau_D \gg \tau_N$. Then the leading term in the expression for the thermal resistivity W is

$$W = \frac{2\pi^2 v_s}{\hbar}\left(\frac{\hbar}{KT}\right)^3 \int_0^{\Theta/T} \frac{1}{\tau_D} \frac{x^4 e^x}{(e^x - 1)^2}\, dx \Bigg/ \left[\int_0^{\Theta/T} \frac{x^4 e^x}{(e^x - 1)^2}\, dx\right]^2. \quad (1.8.57)$$

To appreciate (1.8.57) let us assume in addition that the temperature is low enough so that Θ/T may be made infinite, and let us use (1.8.33) for τ_D. Then we find

$$W = 120\pi^2 v_s A T/\hbar. \quad (1.8.58)$$

This should be compared to (1.8.42). It will be seen that the defect resistance is 25 times greater than that predicted for the situation in which umklapp processes dominate the phonon scattering. The result (1.8.58) was first obtained by Ziman (1956), using quite different methods. Equation (1.8.58) must, of course, be regarded as an idealized limiting case, as some umklapp processes will always be present. It is interesting that this limit is approached to some extent in the case of solid helium, at very low temperatures, as in the experiments of Walker and Fairbank (1960) and Berman *et al.* (1965). The I_2 term also appears to be quite significant in lithium fluoride (Berman and Brock, 1965).

The second question concerning the general validity of the approach is more difficult to answer. The derivation presented here follows the intuitive approach of the original papers. A more formal and more adequate treatment has been given by Krumhansl (1965) leading to essentially the same result. The choice of relaxation times assumed to facilitate evaluation of the integrals in particular, the formula (1.8.34) for normal processes, is more questionable (Nettleton, 1963; Holland, 1963). It appears that, to a certain extent, more adequate approximations to the relaxation times can be developed within the basic theoretical framework.

1.9 Quantum Theory of the Interaction of Lattice Vibrations with Electromagnetic Radiation

1.9.1 *The Interaction Hamiltonian*

In Section 1.3, a classical description was given of the interaction of light with a vibrating crystal. We will now discuss the quantum theory of such processes. The interaction of electromagnetic radiation, characterized by a vector potential A with a crystal lattice is described by the interaction Hamiltonian

$$H_{\text{int}} = \sum_{\mu\kappa} [-(q_\kappa/M_\kappa)\mathbf{A}(\mathbf{x}_{\mu\kappa})\cdot\mathbf{p}_{\mu\kappa} + (q^2_\kappa/2M_\kappa)\mathbf{A}^2] \tag{1.9.1}$$

in which q_κ is the charge of the κth atom and M_κ its mass. We confine our attention to a hypothetical system of atoms which do not deform (rigid ions). The quantity $\mathbf{x}_{\mu\kappa}$ is the position of the κth atom in the μth unit cell; $\mathbf{p}_{\mu\kappa}$ is the corresponding momentum and $\mathbf{A}(\mathbf{x}_{\mu\kappa})$ is the vector potential of the radiation field evaluated at the actual position of the atom. A more complete theory must consider the detailed electronic structure of the crystal, which the present theory grossly oversimplifies and will be developed subsequently (Section 6.5).

We use Eqs. (1.4.1), (1.4.9), and (1.4.14b) to write

$$p_{\alpha,\mu\kappa} = M_\kappa \dot{u}_{\alpha,\mu\kappa} = (1/i)[\Omega/(2\pi)^3][\hbar\mathfrak{N}/2M_\kappa]^{1/2} \sum_j \int d^3k\, [\omega_j(\mathbf{k})^{1/2}]$$
$$\times [e_{\alpha,\kappa}^{(j)}(\mathbf{k}) \exp(i\mathbf{k}\cdot\mathbf{R}_\mu) a_j(\mathbf{k}) - e_{\alpha,\kappa}^{(j)*}(\mathbf{k}) \exp(-i\mathbf{k}\cdot\mathbf{R}_\mu) a_j^\dagger(\mathbf{k})]. \tag{1.9.2}$$

For most problems in solid state physics, a semiclassical treatment of the radiation field is sufficient, and will generally be employed in this book. It is, however, of some intrinsic intellectual interest to consider an example of a fully quantum theory, and that will be done in this section. Here, the electromagnetic field is represented in second quantized form according to the procedures described in Appendix B. MKS units are employed.

$$A(\mathbf{x}_{\mu\kappa}) \doteq c(\hbar\mu_0/\kappa_\infty)^{1/2}(2\pi)^{-3} \sum_\rho \int d^3s\, \{\boldsymbol{\varepsilon}_\rho(\mathbf{s})/[2\nu(\mathbf{s})]^{1/2}\}$$
$$\times [\exp(i\mathbf{s}\cdot\mathbf{x}_{\mu\kappa}) b_\rho(\mathbf{s}) + \exp(-i\mathbf{s}\cdot\mathbf{x}_{\mu\kappa}) b_\rho^\dagger(\mathbf{s})]. \tag{1.9.3}$$

where $b_\rho^\dagger(\mathbf{s})$ and $b_\rho(\mathbf{s})$ are creation and annihilation operators for a phonon of wave vector $\mathbf{s}$ and polarization ρ. The quantity μ_0 is the permeability of free space, $\boldsymbol{\varepsilon}_\rho$ is the polarization vector, κ_∞ is the high frequency dielectric constant, and $\nu = cs/\kappa_\infty^{1/2}$.

The $b_\rho(\mathbf{s})$ satisfy commutation relations appropriate to a free Bose field

$$[b_\rho(\mathbf{s}), b^\dagger_{\rho'}(\mathbf{s}')] = \delta_{\rho'\rho}\,\delta(\mathbf{s} - \mathbf{s}'), \tag{1.9.4a}$$

$$[b_\rho(\mathbf{s}), b_{\rho'}(\mathbf{s}')] = 0. \tag{1.9.4b}$$

Since the b's and a's represent independent physical quantities, they commute with each other:

$$[b_\rho(\mathbf{s}), a_j^\dagger(\mathbf{k})] = 0, \quad \text{etc.} \tag{1.9.5}$$

Note that $\mathbf{s}$ is not restricted to the Brillouin zone.

We will discuss the absorption (or emission) of infrared radiation by a diatomic lattice of oppositely charged ions. Only single-photon, single-phonon processes will be considered. This means we may consider $\mathbf{A}$ to be evaluated at the equilibrium position of the ions. We put $x_{\mu\kappa} = \mathbf{R}_\mu + \mathbf{d}_\kappa$ in Eq. (1.9.3) and neglect the $\mathbf{A}^2$ term in (1.9.1). The term retained is

$$\begin{aligned} H_{\text{int}} = \frac{1}{2} i \frac{\Omega c}{(2\pi)^6} \left(\frac{\hbar^2 \mu_0 \mathfrak{N}}{\kappa_\infty}\right)^{1/2} \sum_{\mu\kappa, j\rho} (q_\kappa / M_\kappa^{1/2}) \iint d^3k\, d^3s\, [\omega_j(\mathbf{k})/\nu(\mathbf{s})]^{1/2} \\ \times (\mathbf{e}_\kappa^{(j)}(\mathbf{k}) \cdot \boldsymbol{\varepsilon}_\rho(\mathbf{s}) \exp(i\mathbf{k} \cdot \mathbf{R}_\mu) \{\exp[i\mathbf{s} \cdot (\mathbf{R}_\mu + \mathbf{d}_\kappa)] a_j(\mathbf{k}) b_\rho(\mathbf{s}) \\ + \exp[-i\mathbf{s} \cdot (\mathbf{R}_\mu + \mathbf{d}_\kappa)] a_j(\mathbf{k}) b_\rho^\dagger(\mathbf{s})\} \\ - \mathbf{e}_\kappa^{(j)*}(\mathbf{k}) \cdot \boldsymbol{\varepsilon}_\rho(\mathbf{s}) \exp(-i\mathbf{k} \cdot \mathbf{R}_\mu) \{\exp[i\mathbf{s} \cdot (\mathbf{R}_\mu + \mathbf{d}_\kappa)] a_j^\dagger(\mathbf{k}) b_\rho(\mathbf{s}) \\ + \exp[-i\mathbf{s} \cdot (\mathbf{R}_\mu + \mathbf{d}_\kappa)] a_j^\dagger(\mathbf{k}) b_\rho^\dagger(\mathbf{s})\}). \end{aligned} \tag{1.9.6}$$

1.9.2 *Single-Photon Absorption*

We will consider a process in which a single photon is absorbed and a single phonon is emitted. Initially the system is in a state $|i\rangle$ with $n_{\rho'}(\mathbf{s}')$ photons of polarization ρ' with wave vector $\mathbf{s}'$ and $n_{j'}(\mathbf{k}')$ phonons of wave vector $\mathbf{k}'$ and polarization j'. In the final state $|\mathrm{f}\rangle$ there are $n_{\rho'}(\mathbf{s}') - 1$ photons in $\rho'\mathbf{s}'$ and $n_{j'}(\mathbf{k}') + 1$ phonons present. The essential matrix element is

$$\begin{aligned} \langle n_{\rho'}(\mathbf{s}') - 1, n_{j'}(\mathbf{k}') + 1 \mid a_j^\dagger(\mathbf{k}) b_\rho(\mathbf{s}) \mid n_{j'}(\mathbf{k}'), n_{\rho'}(\mathbf{s}') \rangle \\ = [n_{j'}(\mathbf{k}') + 1]^{1/2} [n_{\rho'}(\mathbf{s}')]^{1/2}\, \delta(\mathbf{k} - \mathbf{k}')\, \delta(\mathbf{s} - \mathbf{s}')\, \delta_{jj'}\, \delta_{\rho\rho'}. \end{aligned}$$

Other light quanta and phonons which do not take part in the transition may be present. The matrix element of H_{int} is

$$\begin{aligned} \langle \mathrm{f} \mid H_{\text{int}} \mid \mathrm{i} \rangle &= \langle n_{\rho'}(\mathbf{s}') - 1, n_j(\mathbf{k}') + 1 \mid H_{\text{int}} \mid n_{j'}(\mathbf{k}') n_{\rho'}(\mathbf{s}') \rangle \\ &= -\frac{1}{2} i \frac{\Omega c}{(2\pi)^6} \left(\frac{\hbar^2 \mu_0 \mathfrak{N}}{\kappa_\infty}\right)^{1/2} [(n_{j'}(\mathbf{k}') + 1) n_\rho(\mathbf{s}')]^{1/2} \\ &\quad \times \sum_{\mu\kappa} [\omega_j(\mathbf{k}')/\nu(\mathbf{s}')]^{1/2} q_\kappa / M_\kappa^{1/2} \mathbf{e}_\kappa^{(j')*}(\mathbf{k}') \cdot \boldsymbol{\varepsilon}_{\rho'}(\mathbf{s}') \\ &\quad \times \exp[i(\mathbf{s}' - \mathbf{k}') \cdot \mathbf{R}_\mu] \exp(i\mathbf{s}' \cdot \mathbf{d}_\kappa). \end{aligned} \tag{1.9.7}$$

The sum over μ can now be done, yielding $[(2\pi)^3/\Omega] \sum_l \delta(\mathbf{k}' - \mathbf{s}' + \mathbf{K}_l)$. To simplify the notation, we will drop the prime designation.

The energy of a photon with a wave vector of the order of a nonzero reciprocal lattice vector is quite large, far outside the phonon spectrum, so conservation of energy forbids transitions with $\mathbf{K}_l \neq 0$. (Recall that a delta function of energy conservation multiplies the expression for the transition probability). Thus $\mathbf{s} = \mathbf{k}$,

$$\langle \mathrm{f} \mid H_{\mathrm{int}} \mid \mathrm{i} \rangle = -[ic/2(2\pi)^3](\hbar^2\mu_0\mathfrak{M}/\kappa_\infty)^{1/2} \times [(n_j(\mathbf{k}) + 1)n_p(\mathbf{s})]^{1/2}[\omega_j(\mathbf{k})/\nu(\mathbf{k})]^{1/2} \times \sum_\kappa q_\kappa/M_\kappa^{1/2}\mathbf{e}_\kappa^{(j)}(\mathbf{k}) \cdot \boldsymbol{\varepsilon}_p(\mathbf{k}) \exp(i\mathbf{k}\cdot\mathbf{d}_\kappa)\,\delta(\mathbf{k} - \mathbf{s}). \quad (1.9.8)$$

The allowed values of $\mathbf{k}$ are so small that the exponential factor $\exp(i\mathbf{k}\cdot\mathbf{d}_\kappa)$ may be replaced by unity.

Our considerations will be limited to a diatomic crystal with oppositely charged atoms ($\kappa = +, -$; $q_+ = -q_- = q$, as in Section 1.1) such as NaCl, ZnS. For small k, if j represents an acoustic mode, $\mathbf{e}_+/M_+^{1/2} = \mathbf{e}_-/M_-^{1/2}$. The matrix element vanishes. However, for an optic mode, $\mathbf{e}_- = -(M_+/M_-)^{1/2}\mathbf{e}_+$. These quantities refer to $k = 0$ and are thus independent of $\mathbf{k}$ or $\mathbf{s}$. If the atoms in the unit cell are alike, absorption by optic modes does not occur in first order.

The reduced mass of the atoms in the unit cell is denoted by $\mathfrak{M} = M_+M_-/M_+ + M_-$. The sum over κ in (1.9.8) is performed:

$$\langle \mathrm{f} \mid H_{\mathrm{int}} \mid \mathrm{i} \rangle = -[ic/2(2\pi)^3](\hbar^2\mu_0\mathfrak{M}/\kappa_\infty)^{1/2}[(n_j(\mathbf{k}) + 1)n_p(\mathbf{s})]^{1/2} \times [\omega_j(\mathbf{k})/\nu(\mathbf{k})]^{1/2}(qM_+^{1/2}/\mathfrak{M})\mathbf{e}_+^{(j)} \cdot \boldsymbol{\varepsilon}_p\,\delta(\mathbf{k} - \mathbf{s}), \quad (1.9.9)$$

which is valid for optic modes. Note that since $\boldsymbol{\varepsilon}_p$ is perpendicular to $\mathbf{s}$, only the transverse optic modes can absorb.

We now must square this matrix element and multiply by

$$(2\pi/\hbar)\,\delta(E_\mathrm{f} - E_\mathrm{i}) = (2\pi/\hbar^2)\,\delta[\omega_j(\mathbf{k}) - \nu(\mathbf{s})].$$

However, in doing this, we encounter the apparently undefined factor $\delta(\mathbf{k} - \mathbf{s})^2$. This problem arises in all transitions involving two continuum states. However, the transition rate is well defined and contains only a single delta function describing momentum conservation [consult Part B, Section 6.5.2 for further discussion of this point]. The transition rate is

$$W = (\pi/2)[c^2\mu_0 V/\Omega\kappa_\infty](2\pi)^{-6}(n_j(\mathbf{k}) + 1)n_p(\mathbf{s}) \times [\mathbf{e}_+^{(j)} \cdot \boldsymbol{\varepsilon}_p]^2(q^2M_+/\mathfrak{M}^2)\,\delta(\mathbf{k} - \mathbf{s})\,\delta[\omega_j(\mathbf{k}) - \nu(\mathbf{k})]. \quad (1.9.10)$$

The factor of ω/ν has disappeared since energy conservation requires $\omega = \nu$.

The rate of energy absorption per unit time per unit volume for a crystal at temperature T will be calculated. This quantity will be denoted by U. To do this multiply (1.9.10) by $\hbar\nu(\mathbf{s})$, divide by the crystal volume, and integrate over both **s** and **k**. We also sum over polarization vectors $\boldsymbol{\varepsilon}_\rho$ and vibrational branches j:

$$U = (\pi/2)(c^2\mu_0/\kappa_\infty\Omega)(q^2M_+/\mathfrak{M}^2)[1/(2\pi)^6] \sum_{\rho j} \int d^3s\, d^3k$$
$$\times [(n_j(\mathbf{k}) + 1)n_\rho(\mathbf{s})]\hbar\nu[\mathbf{e}_+^{(j)} \cdot \boldsymbol{\varepsilon}_\rho]^2 \,\delta[\omega_j(\mathbf{k}) - \nu(\mathbf{k})]\,\delta(\mathbf{k} - \mathbf{s}). \tag{1.9.11}$$

The factor $\hbar\nu n_\rho(\mathbf{s})$ may be related to the intensity of the light as follows: The integral over photon wave vectors is converted to an integral over ν, in which each quantum contributes an amount $\hbar\nu$ to the energy density and $\hbar c\nu/\kappa_\infty$ to the intensity. The total intensity of the light is expressed as the integral of an intensity per unit frequency $I(\nu)$ over all frequencies ν:

$$[1/(2\pi)^3\kappa_\infty] \int \hbar c\nu n_\rho\, d^3s = \int I_\rho(\nu)\, d\nu. \tag{1.9.12}$$

We consider unpolarized light for which $I_\rho = \frac{1}{2}I$, independent of ρ. Note also that the phonon occupation number $n_j(\mathbf{k})$ is a function of energy only:

$$U = [\pi\mu_0 c/4\Omega(2\pi)^3](q^2M_+/\mathfrak{M}^2) \int d\nu\, I(\nu)[n(\nu) + 1]$$
$$\times \sum_{j\rho} [\mathbf{e}_+^{(j)} \cdot \boldsymbol{\varepsilon}_\rho]^2 \,\delta[\omega_j(\mathbf{k}) - \nu]. \tag{1.9.13}$$

A dimensionless effective charge Q_{eff} is now defined by the relation

$$q^2M_+ \sum_{\rho j} [\mathbf{e}_+^{(j)} \cdot \boldsymbol{\varepsilon}_\rho]^2 \equiv \tfrac{4}{3}\mathfrak{M}Q^2_{\text{eff}}\, e^2. \tag{1.9.14}$$

The definition is motivated by the fact that the average of the square of the cosine of an angle between two vectors is $\frac{1}{3}$, and that there are two polarizations and two transverse branches which can contribute to (1.9.14). Hence

$$U = [\pi\mu_0 ce^2/3(2\pi)^3\Omega](Q^2_{\text{eff}}/\mathfrak{M})I(\omega)[n(\omega) + 1]. \tag{1.9.15}$$

Finally, we average over a thermal distribution of phonons at temperature T:

$$\langle n(\omega) + 1\rangle_T = \tfrac{1}{2}[1 + \operatorname{ctnh}(\hbar\omega/2KT)]. \tag{1.9.16}$$

Our result for U is

$$U = (\pi/6)[\mu_0 e^2c/(2\pi)^3\Omega](Q^2_{\text{eff}}/\mathfrak{M})I(\omega)[1 + \operatorname{ctnh}(\hbar\omega/2KT)].$$

This may be simplified since $\mu_0 = (c^2\epsilon_0)^{-1}$ and in MKS units the dimensionless fine structure constant α is given by

$$\alpha = e^2/4\pi\epsilon_0\hbar c. \tag{1.9.17}$$

Thus

$$U = (\alpha/12\pi)(\hbar Q^2_{\text{eff}}/\mathfrak{M}\Omega)I(\omega)[1 + \text{ctnh}(\hbar\omega/2KT)]. \tag{1.9.18}$$

This equation describes single-photon optical absorption.

1.9.3 *Scattering of Photons*

In addition to the absorption (or emission) of light by a phonon system, scattering can occur with a change in frequency. This is the *Raman effect* (Raman, 1928). In this process, a photon of frequency ω_i incident from some light source is absorbed and a scattered photon of frequency ω_s is created, accompanied by the creation or absorption of a lattice phonon of frequency ω. We may have either

$$\omega_s = \omega_i - \omega \tag{1.9.19a}$$

(so-called *Stokes component*), or

$$\omega_s = \omega_i + \omega \tag{1.9.19b}$$

(*anti-Stokes component*). Since the photons involved are usually in the infrared or visible region of the spectrum, their wave vectors are quite small on the scale relevant to phonons; thus the $k = 0$ phonons alone are involved. The Raman effect in crystals has been reviewed by Loudon (1964); the same author has also developed the quantum theory of the process (Loudon, 1963). For a review of earlier work, see Menzies (1953).

The Hamiltonian we have developed, Eq. (1.9.6), does not give an adequate description of the process, but requires extension. According to Loudon, the dominant contribution to the Raman effect may be described as a three-step process. (1) an electron in the crystal absorbs the incident photon making a (virtual) transition to an excited state. (2) In the excited state, the electron interacts with the vibrating lattice with the absorption or emission of a phonon. (3) The scattered electron next interacts with the electromagnetic field, this time emitting a photon. Third-order perturbation theory is required to describe this sequence of events. Other contributions to the Raman effect come from processes in which a phonon is created directly by the photon; the phonon splits into two phonons (or absorbs another phonon) via the anharmonic interaction previously described in connection with thermal conduction, and finally one phonon again interacts with the field, being converted into a photon. Alternately, one of the photon–electromagnetic field interactions may involve second-order (two-phonon) processes which we have ignored. None of these processes will be discussed in mathematical detail here. Since they are of second and third order, it is evident and is observed that the Raman effect is generally weak. However, one should note that optical modes in crystals with two like atoms in the unit cell (such as diamond, silicon, geramanium) which do not absorb light directly, can participate in (and thus their frequencies can be measured by) the Raman effect.

While most interest concerns Raman scattering by optical modes, acoustic modes may also be involved. In this case, the phenomenon is known as *Brillouin scattering* (Brillouin, 1922). The energy relations are the same as in Eq. (1.9.19) for the wave vectors

$$\mathbf{k}_i = \mathbf{k}_s \pm \mathbf{k} \tag{1.9.20}$$

where $\mathbf{k}_i$, $\mathbf{k}_s$, and $\mathbf{k}$ are the wave vectors of the incident and scattered photons and the phonon, respectively. Since the fractional change in photon frequencies is usually small, we can assume that $k_i = k_s$, and then the angle of scattering θ (the angle between $\mathbf{k}_i$ and $\mathbf{k}_s$) is determined from

$$k = 2k_i \sin(\theta/2). \tag{1.9.21}$$

For acoustic phonons, we put $k = \omega/v$, where v is the velocity of sound for the appropriate branch, while for the photon, $k_i = \kappa_\infty^{1/2}\omega_i/c$. Thus in Brillouin scattering, we have

$$|\omega_i \pm \omega_s|/\omega_i = (2\kappa_\infty^{1/2}v/c)\sin(\theta/2). \tag{1.9.22}$$

Because v/c is quite small (of order 10^{-5}), the Brillouin shifts are typically small; of the order of 2 or 3 cm^{-1}. Since three acoustic branches may participate, three Stokes and three anti-Stokes components may be observed.

Problems

1 Calculate the vibrational spectrum for a one-dimensional crystal with two kinds of atoms, of masses M and m, arranged alternately. Assume that only nearest neighbor atoms interact.

2 Obtain the density of states for the model crystal described in Problem 1.

3 Determine an expression for the linear momentum of a simple cubic crystal containing a single phonon of wave vector $\mathbf{K}$. Show that the use of periodic boundary conditions implies that the momentum is zero if $\mathbf{K} \neq 0$. What happens if $\mathbf{K} = 0$?

4 The static dielectric constant of LiF is 9.27 and the high frequency (optical) dielectric constant is 1.93. The restrahl frequency corresponds to

a wavelength of 32.6 μm. Find the frequency of the longitudinal optical mode. Compare your result with the observations of Berreman (1963). For what wavelength will the index of refraction be unity?

5 Derive Eqs. (1.4.15a), (1.4.15b), and (1.4.16).

6 Suppose that a phonon dispersion relation in a material with the face centered cubic lattice has the form

$$\omega^2 = 3(\alpha + \beta) - \alpha(\cos k_x a/2 \cos k_y a/2 + \cos k_x a/2 \cos k_z a/2 + \cos k_y a/2 \cos k_z a/2) - \beta(\cos k_x a + \cos k_y a + \cos k_z a).$$

Locate the critical points for $\beta/\alpha = 0.05$ and show that the Morse relations are satisfied.

7 Determine an expression for the entropy of a crystal in the Debye approximation. Find the low temperature limit of this formula. Assume that the Debye Temperature of Na is 156°K. Find the lattice entropy of Na at 3.9°K, 7.8°K, and 39°K. You may use the tables of the Debye Function (*NBS Handbook of Mathematical Functions*). Express your results in cal/mol-deg.

8 Find the energy eigenvalues of the Hamiltonian

$$H = \hbar\omega(a^\dagger a + b^\dagger b) + \gamma(ab + a^\dagger b^\dagger)$$

in which a and b are annihilation operators for distinguishable bosons, and $|\gamma/\hbar\omega| < 1$.

9 Show that the reciprocal relaxation time resulting from the scattering of phonons by impurity atoms which differ only in their mass from those of the perfect crystal can be written as $\tau_D^{-1} = A\omega^4$. Determine an expression for A in terms of the mass difference. Use first-order perturbation theory.

10 Suppose that the relaxation time for normal three-phonon processes at low temperature is proportional to $(\omega T^4)^{-1}$ instead of $(\omega^2 T^3)^{-1}$. How will the thermal conductivity depend on temperature and defect concentration in the strong defect scattering limit?

11 Consider the total momentum P and energy U in a phonon system according to the simple one-branch Debye-like model used in studying thermal conductivity. Use the Boltzmann equation to obtain expressions for the time and space rates of change of these quantities in the absence of a thermal gradient. Assume that phonons scatter only through normal three-phonon processes so that energy and momentum are conserved. Combine the equations to obtain a wave equation describing periodic variations in energy in the phonon system, and thereby show that the speed of second sound in solids is $v_s/3^{1/2}$, where v_s is the speed of ordinary sound.

References

Barker, A. S. (1964). *Phys. Rev.* **136,** A1290.

Berman, R. (1967). *Sci. Progr. Oxford* **55,** 357.

Berman, R., and Brock, J. C. F. (1965). *Proc. Roy. Soc.* (*London*) **A289,** 46.

Berman, R., Bounds, C. L., and Rogers, S. J. (1965). *Proc. Roy. Soc.* (*London*) **A289,** 66.

Berreman, D. W. (1963). *Phys. Rev.* **130,** 2193.

Born, M., and Huang, K. (1954). "The Dynamical Theory of Crystal Lattices." Oxford Univ. Press (Clarendon), London and New York.

Bouckaert, L. P., Smoluchowski, R., and Wigner, E. (1936). *Phys. Rev.* **50,** 58.

Bowers, W. A., and Rosenstock, H. B. (1950). *J. Chem. Phys.* **18,** 1056.

Brillouin, L. (1922). *Ann. Phys. Paris* **17,** 88.

Brillouin, L. (1931). "Quantenstatistik." Springer, Berlin.

Brockhouse, B. N., and Stewart, A. T. (1955). *Phys. Rev.* **100,** 756.

Callaway, J. (1959). *Phys. Rev.* **113,** 1046.

Callaway, J. (1961). *Phys. Rev.* **122,** 787.

Callaway, J. (1963). *Nuovo Cimento* **29,** 883.

Callaway, J., and Hughes, A. J. (1962). *Phys. Rev.* **128,** 134.

Carruthers, P. (1961). *Rev. Mod. Phys.* **33,** 92.

Casimir, H. B. G. (1938). *Physica* **5,** 495.

Danon, J. (1968). "Lectures on the Mössbauer Effect." Gordon and Breach, New York.

Dolling, G., and Cowley, R. A. (1966). *Proc. Phys. Soc.* **88,** 463.

Frauenfelder, H. (1962). "The Mössbauer Effect." Benjamin, New York.

Gilat, G. (1972). *J. Comp. Phys.* **10,** 432.

Henry, C. H., and Hopfield, J. J. (1965). *Phys. Rev. Lett.* **15,** 964.

Herring, C. (1954). *Phys. Rev.* **95,** 954.

Holland, M. G. (1963). *Phys. Rev.* **132,** 246.

Hopfield, J. J. (1958). *Phys. Rev.* **112,** 1555.

Hopfield, J. J. (1969). Free and Bound Excitons. *In* "Elementary Excitations and Their Interactions in Solids" (A. A. Maradudin and G. F. Nardelli, eds.), p. 413. Plenum Press, New York.

Huntington, H. B. (1957). *Solid State Phys.* **7,** 213.

Kittel, C. (1963). "Quantum Theory of Solids." Wiley, New York.

Kittel, C. (1966). "Introduction to Solid State Physics," Chapter 4. Wiley, New York.

Klemens, P. G. (1955). *Proc. Phys. Soc.* **A68,** 1113.

Klemens, P. G. (1956). Thermal Conductivity of Solids at Low Temperatures. *In* "Encyclopedia of Physics" (S. Flügge, ed.), Vol. 14, p. 198. Springer-Verlag, Berlin.

Klemens, P. G. (1969). "Thermal Conductivity" (R. P. Tye, ed.), Vol. 1, p. 2. Academic Press, New York.

Krumhansl, J. A. (1965). *Proc. Phys. Soc.* **85,** 921.

Lamb, W. E. (1939). *Phys. Rev.* **55,** 190.

Liebfried, G. (1955). Gittertheorie der mechanischen und thermischen Eigenschaffen der Kristalle. *In* "Encyclopedia of Physics" (S. Flügge, ed.), p. 290. Springer-Verlag, Berlin.

Liebfried, G., and Ludwig, W. (1961). *Solid State Phys.* **12,** 276.

Lipkin, H. J. (1960). *Ann. Phys.* (*N. Y.*) **9,** 332.

Loudon, R. (1963). *Proc. Roy. Soc.* (*London*) **A275,** 218.

Loudon, R. (1964). *Advan. Phys.* **13,** 423.

Lyddane, R. H., Sachs, R. G., and Teller, E. (1941). *Phys. Rev.* **59,** 673.

Maradudin, A. A. (1964). *Rev. Mod. Phys.* **36,** 417.
Maradudin, A. A., Montroll, E. W., Weiss, G. H., and Ipatova, I. P. (1971). "Theory of Lattice Dynamics in the Harmonic Approximation" (*Solid State Phys.* Suppl. 3), 2nd ed. Academic Press, New York.
Marshall, W., and Lovesey, S. W. (1971). "Theory of Thermal Neutron Scattering." Oxford Univ. Press, London and New York.
Menzies, A. C. (1953). *Rep. Progr. Phys.* **16,** 83.
Messiah, A. (1961). "Quantum Mechanics," Vol. 1, Chapter 12. North-Holland Publ., Amsterdam.
Montroll, E. W. (1954). *Amer. Math. Monthly* **61,** 46.
Morse, M. (1938). "Functional Topology and Abstract Variational Theory." *Memor. Sci. Math.* **92,** Gauthier-Villars, Paris.
Mossbauer, R. L. (1958a). *Z. Phys.* **151,** 124.
Mossbauer, R. L. (1958b). *Naturwissenschaften* **45,** 538.
Mossbauer, R. L. (1959). *Z. Naturforsch.* **14a,** 211.
Nettleton, R. E. (1963). *Phys. Rev.* **132,** 2032.
Peierls, R. E. (1955). "Quantum Theory of Solids." Oxford Univ. Press, London and New York.
Phillips, J. C. (1956). *Phys. Rev.* **104,** 1263.
Placzek, G., and VanHove, L. (1954). *Phys. Rev.* **93,** 1207.
Pound, R. V., and Rebka, G. A. (1959). *Phys. Rev. Lett.* **3,** 439.
Pound, R. V., and Snider, J. L. (1965). *Phys. Rev.* **140,** B788.
Raman, C. V. (1928). *Ind. J. Phys.* **2,** 387.
Schweber, S. (1961). "An Introduction to Relativistic Quantum Field Theory," Chapter 6. Harper, New York.
Shull, C. G., and Wollan, E. O. (1956). *Solid State Phys.* **2,** 137.
Szigeti, B. (1949). *Trans. Faraday Soc.* **45,** 155.
Van Hove, L. (1953). *Phys. Rev.* **89,** 1189.
Van Hove, L. (1954). *Phys. Rev.* **95,** 249.
Walker, E. J., and Fairbank, H. A. (1960). *Phys. Rev.* **118,** 913.
Wannier, G. (1959). "Elements of Solid State Theory." Cambridge Univ. Press, London and New York.
Wertheim, G. K. (1964). "Mössbauer Effect: Principles and Applications." Academic Press, New York.
Ziman, J. M. (1956). *Can. J. Phys.* **34,** 1256.

CHAPTER 2

Phenomenological Theories of Magnetic Order

2.1 General Description

The problem of magnetic order in solids is of interest for both technological and fundamental reasons. The former require no emphasis here; in regard to the latter, it may be well to point out that one of a very few soluble models of a phase transition is furnished by the (two-dimensional) Ising model of a ferromagnet. Our concern here will be with phenomenological approaches to the problem of magnetic order, that is, we assume the existence of a certain Hamiltonian, which couples the spins of atoms in a crystal, and investigate the consequences.

It is assumed that we can consider a crystal lattice in which the only relevant degrees of freedom are individual atomic spins. Let $\mathbf{S}_j$ represent the spin on lattice site $\mathbf{R}_j$ (for simplicity we consider only one magnetic atom to be present in each unit cell). The $\mathbf{S}_j$ are quantum mechanical operators, whose rectangular components $S_{j\alpha'}$ etc., obey the commutation rules

$$[S_{j\alpha'}S_{l\beta}] = \delta_{jl}\, i\hbar\epsilon_{\alpha\beta\gamma}S_{j\gamma} \tag{2.1.1}$$

in which $\epsilon_{\alpha\beta\gamma}$ is the completely antisymmetric Levi–Civitta symbol ($\epsilon_{123} = \epsilon_{231} = \epsilon_{312} = 1$; $\epsilon_{213} = \epsilon_{321} = \epsilon_{132} = -1$), and summation over γ is understood. In what follows we will invariably set $\hbar = 1$. The operator $\mathbf{S}_j^2$ commutes with the Hamiltonian given below, and its eigenvalue $S(S+1)$ determines the magnitude of the spin on a site, which is independent of j.

The assumed Hamiltonian is

$$H_{\mathrm{H}} = -2\sum_{i>j} J_{ij}\mathbf{S}_i\cdot\mathbf{S}_j \tag{2.1.2}$$

in which the quantities J_{ij} (which are referred to as *exchange parameters*) are functions of the distance between sites i and j,

$$J_{ij} = J_{ij}(\,|\,\mathbf{R}_i - \mathbf{R}_j\,|\,). \tag{2.1.3}$$

This Hamiltonian is called the *Heisenberg Hamiltonian*, although it was actually introduced by Dirac (1929) and first used extensively for the study of magnetically ordered solids by van Vleck (1932).

The algebraic sign is so chosen in (2.1.2) that if the J_{ij} are all positive, the ground state will be ferromagnetic. If some of the J_{ij} are negative, a much more complex state of affairs results. Antiferromagnetic ordering is a possible consequence.

There is an enormous literature concerned with this Hamiltonian, both in regard to derivation and application. Since the Hamiltonian has been rather controversial we will begin with a few remarks justifying its use. A general and rigorous analysis of this problem has been given by Herring (1966). As a result of this work, the theory is on a firm foundation with respect to basic principles. We will summarize Herring's arguments in the next section; it will suffice here to say that with a suitable definition of J_{ij}, Eq. (2.1.2) may be rigorously applied to an assembly of well separated atoms; the terms neglected being smaller in a definite sense than those retained. It is a different, more difficult, and largely unanswered question whether H_{H} is applicable to any particular real solid at the actual equilibrium atomic spacing.

2.2 Interaction of Atomic Spins at Large Distances

2.2.1 *Derivation of the Heisenberg Hamiltonian*

We wish to investigate here the justification for the use of H_{H} [Eq. (2.1.2)]. The arguments are those of Herring (1966). We will limit explicit discussion to atoms with orbitally nondegenerate ground states ($L = 0, J = S$). Orbital degeneracy has also been considered by Herring, but we shall not discuss this here.

It is perhaps desirable to begin by stating the result. For any assembly of weakly interacting atoms or molecules, each of which in isolation has a ground state which is orbitally nondegenerate but has exact degeneracy with respect to orientation of its spin, the eigenvalues of the $\prod_i(2S_i + 1)$ lowest levels are the same as those of the operator

$$H_{\mathrm{eff}} = \mathcal{E}_0 - 2\sum_{i>j} J_{ij}\mathbf{S}_i\cdot\mathbf{S}_j. \tag{2.2.1}$$

Here $\mathcal{E}_0$ is a constant; the J_{ij} are determined by the coordinate wave function of the atoms involved (an explicit formula will be given later) and the corrections are, roughly, of the order of J^2.

We are going to indicate only the essential steps in the proof. Herring's (1966) argument is much more extensive, and the reader is referred to his article for more detail. We suppose that we have an assembly of atoms all of which are at sufficiently large distances so that they interact only weakly. Atom i has n_i electrons. The true wave function for the system is antisymmetric under the interchange of all the electrons; however, we begin by ignoring this and consider electrons on different atoms to be distinguishable. If the atoms are infinitely far apart, an eigenstate of the system under the hypothesis of distinguishable electrons will be described by having the first n_1 electrons on atom 1, electrons $n_1 + 1$ to $(n_1 + n_2)$ on atom 2, and so forth. The wave function of the electrons on each atom is antisymmetric with respect to interchange of the electrons on that atom (only). At this point, we do not have antisymmetry with respect to interchange of electrons on different atoms. Finally, we assume that each atom i is in a definite spin state described by the operator $\mathbf{S}_i$. The collection of all the eigenvalues of S_i^2 and S_{iz} will be denoted by Γ. We can use this to characterize the system. The wave function for the system in this situation will be denoted by $\bar{\Phi}_\Gamma$.

Now allow the atoms to move into finite, but large, separations where there is weak interaction. The state of the system, using the same hypothesis of distinguishability, will not be $\bar{\Phi}_\Gamma$ but will be one which has evolved from $\bar{\Phi}_\Gamma$ under the influence of the interaction. Each possible state of the noninteracting system will evolve into a specific state of the interacting atoms if they are far enough apart, and so we may characterize the states of the interacting atoms by the quantum numbers of the ones from which they evolved. These states are denoted by Φ_Γ.

A state Φ_Γ has the following properties:

(1) It is antisymmetric under all operators of the subgroup G_0 of coordinate–spin permutations which permute the first n_1 electrons among themselves, the next n_2 among themselves, and so on.

(2) The first n_1 electrons are in a spin state described by the eigenvalues S_1^2, S_{1z} of the spin operator $\mathbf{S}_1$, and so on.

Real electrons must obey the exclusion principle. We may antisymmetrize any N electron function by applying the antisymmetrization operator A,

$$A = (1/N!) \sum_P \delta_P \, P^{(r)} P^{(s)} \tag{2.2.2}$$

in which $P^{(r)}$ and $P^{(s)}$ apply a permutation $P = P^{(r)}P^{(s)}$ to coordinates and spins, respectively, and $\delta_P = +1$ or -1 according as the permutation is even or odd. The state $A\bar{\Phi}_\Gamma$ is a physically possible state for noninteracting atoms, and $A\Phi_\Gamma$ will be a possible state for interacting atoms. The states $A\Phi_\Gamma$ are not eigenstates of the Hamiltonian including interaction, but they at least form a complete set of states in a subspace of Hilbert space of degeneracy $\nu = \prod_i(2S_i + 1)$. When the atoms are far apart it is enough to consider diagonalizing the Hamiltonian in this subspace, since states outside this subspace have a finite excitation energy with respect to those in it. Therefore, perturbation theory shows that correct results will be obtained to first order in the interaction by considering only the degenerate states. Thus we have the equation

$$\det[(\Phi_\Gamma \mid HA \mid \Phi_{\Gamma'}) - \mathcal{E}(\Phi_\Gamma \mid A \mid \Phi_{\Gamma'})] = 0 \tag{2.2.3}$$

where H is the full Hamiltonian for the system of interacting atoms, including the interaction between all electron pairs. We now insert A as given by (2.2.2) and consider the terms which arise.

The largest terms arise from those permutations P in A which merely interchange electrons on a single atom. There are $\prod_i(n_i!)$ such terms. Since each atomic function is already antisymmetric $\delta_P P$ leaves Φ_Γ unchanged for such P's. Therefore, in each such case we obtain the same results as for the identity $P = E$; further, these terms are diagonal (proportional to $\delta_{\Gamma\Gamma'}$) since states of different Γ are orthogonal.

The next largest terms are those which interchange a single pair of electrons. We may expect that any Φ_Γ decreases exponentially when any electron is far removed from the site of its own atom. Let the smallest screening constant in an expression representing this behavior be α. Then Φ_Γ decays at least as fast as $\exp[-\alpha \mid \mathbf{r}_m - \mathbf{R}_i(m) \mid]$ when electron m is removed from its "own" atom i. (Powers of r which may multiply this expression are ignored.)

A matrix element describing electron exchange is typically of the form given below in Eqs. (2.2.15) and (2.2.16). Thus a term in (2.2.3) in which a single pair of electrons has been exchanged between atoms at distance R will be of order $e^{-2\alpha R}$. Terms involving the exchange of two pairs will be of order $e^{-4\alpha R}$, and so on. For this reason we neglect interchanges of more than a single pair.

Any permutation which leaves every electron on its own atom except for a single interchange between atoms i and j can be written as

$$P = QP_{ij}Q' \tag{2.2.4}$$

where P_{ij} interchanges the first electron on i with the first electron on j, and Q and Q' are permutations which interchange electrons only on their own atoms. As we saw before, the operators Q, Q' have no effect on the matrix elements. Thus in the summation over all P's the contribution from this type will be

$$\sum \delta_P P = -\nu_{ij} P_{ij} \tag{2.2.5}$$

where ν_{ij} is the number of distinct P's of this type. This number is just the product of the number of electrons on i and j times the number of permutations which interchange electrons on their own atom

$$\nu_{ij} = n_i n_j \prod_l (n_l!). \tag{2.2.6}$$

We now have for the $\Gamma\Gamma'$ element of the matrix contained in (2.2.3):

$$[\prod_l (n_l!)/N!][(\bar{\mathcal{E}}_\phi - \mathcal{E})\delta_{\Gamma\Gamma'} - \sum_{i>j} n_i n_j (\phi_\Gamma \mid (H - \mathcal{E}) P_{ij} \mid \Phi_{\Gamma'})$$
$$+ O(e^{-4\alpha R}) + \cdots]. \tag{2.2.7}$$

We have denoted by $\bar{\mathcal{E}}_\Phi$ the average energy of Φ_Γ:

$$\bar{\mathcal{E}}_\Phi = (\Phi_\Gamma \mid H \mid \Phi_\Gamma), \tag{2.2.8}$$

and this is independent of Γ as long as the Hamiltonian is, as we assume, independent of spin.

We now recall that any permutation P can be expressed as a product of a coordinate permutation $P^{(r)}$ and a spin permutation $P^{(s)}$. The operator $P^{(s)}(m_i, m_j)$ exchanges the spins of electrons m_i and m_j. It may be expressed in terms of electron spin operators ($\mathbf{S}_{mi} = \frac{1}{2}\boldsymbol{\sigma}_{mi}$, the $\boldsymbol{\sigma}$ being Pauli matrices)

$$P^{(s)}(m_i, m_j) = (\tfrac{1}{2} + 2\mathbf{S}_{mi}\cdot\mathbf{S}_{mj}). \tag{2.2.9}$$

This result is central to the development. It may be verified by writing the expression in terms of raising and lowering operators. Thus

$$\tfrac{1}{2} + 2\mathbf{S}_1\cdot\mathbf{S}_2 = \tfrac{1}{2} + 2S_{1z}S_{2z} + S_{1+}S_{2-} + S_{1-}S_{2+}.$$

Then allow the result to operate on products of spin functions (α, β) for the two electrons (with $S_{1z}\alpha_1 = \frac{1}{2}\alpha_1$, $S_{1z}\beta_1 = -\frac{1}{2}\beta_1$) $(\alpha_1\alpha_2)$, $(\alpha_1\beta_2)$, $(\alpha_2\beta_1)$, $(\beta_1\beta_2)$. Thus

$$(\tfrac{1}{2} + 2S_1\cdot S_2)\alpha_1\beta_2 = \beta_1\alpha_2,$$

etc. The products of spin functions given above are a complete set for two-particle spin space, so the result is independent of the use of eigenfunctions of S_z.

We now remove the common factor $\prod(n_l!)/N!$ from (2.2.7), and insert (2.2.9). We obtain

$$(\bar{\mathcal{E}}_\Phi - \mathcal{E})\delta_{\Gamma\Gamma'} - \sum_{i>j} n_i n_j (\Phi_\Gamma \mid (H - \mathcal{E}) P^{(r)}(m_i, m_j)$$
$$\times (\tfrac{1}{2} + 2\mathbf{S}_{mi}\cdot\mathbf{S}_{mj}) \mid \Phi_{\Gamma'}) + O(e^{-4\alpha R}). \tag{2.2.10}$$

The remaining permutation operator $P^{(r)}(m_i, m_j)$ interchanges the coordinates of electrons m_i and m_j.

To complete the derivation of (2.2.1), it remains to express the term involving individual electron spin operators $\mathbf{S}_{mi} \cdot \mathbf{S}_{mj}$ as the scalar product of complete atomic spin operators $\mathbf{S}_i \cdot \mathbf{S}_j$. In the case of one-electron atoms, $\mathbf{S}_{mi} = \mathbf{S}_i$, and the argument is already complete. For the general case we need note only that $\mathbf{S}_{mi}$ and $\mathbf{S}_i$ are vector operators, and according to the Wigner–Eckart theorem (Wigner, 1959, Chapter 21), the matrices representing any two vector operators in a subspace of states of definite J can differ at most by a scalar factor of proportionality. Thus we arrive at (2.2.1), where J_{ij} includes the various coordinate integrals, etc.

2.2.2 *Estimation of the Exchange Parameters*

An explicit expression for J_{ij} can be obtained from (2.2.10) without making additional assumptions about the atomic wave functions. We will, however, consider here only the simplest possible case of two hydrogen atoms as an illustration of the general considerations.

For two electrons, we diagonalize the matrix of (2.2.3) by introducing singlet and triplet spin functions. These functions are eigenfunctions of $\mathbf{S}_1 \cdot \mathbf{S}_2$, which has eigenvalues of $\frac{1}{4}$ in the triplet state and $-\frac{3}{4}$ in the singlet state. The functions Φ_Γ are written as

$$\Phi_\Gamma(1, 2) = \phi_\Gamma(\mathbf{r}_1, \mathbf{r}_2)\theta_\Gamma(s_1, s_2) \tag{2.2.11}$$

where θ_Γ is a singlet or triplet spin function as described above and ϕ_Γ is a coordinate function. On comparing (2.2.10) with (2.2.1) and considering the above argument, we see that

$$J = \tfrac{1}{2}(\mathcal{E}_s - \mathcal{E}_t) = \iint d^3r_1\, d^3r_2\, \phi_\Gamma^*(\mathbf{r}_1, \mathbf{r}_2) \times [(H - \mathcal{E})P^{(r)}(1, 2)\phi_\Gamma(\mathbf{r}_1, \mathbf{r}_2)]. \tag{2.2.12}$$

The wave functions ϕ_Γ are supposed to be linear combinations of actual eigenfunctions of the Hamiltonian, including electron–electron interactions, modified so that electrons on different atoms are regarded as distinguishable. The standard method of evaluating J, which gives the Heitler–London result, is to approximate ϕ by a product of atomic wave functions. Consider the diagram of Fig. 2.2.1. The protons are designated A, B; the electrons 1 and 2. The Hamiltonian is, for fixed separations R of the protons,

$$H = (-\hbar^2/2m)[\nabla_1^2 + \nabla_2^2] - e^2(r_{1A}^{-1} + r_{1B}^{-1} + r_{2A}^{-1} + r_{2B}^{-1} - r_{12}^{-1} - R^{-1}). \tag{2.2.13}$$

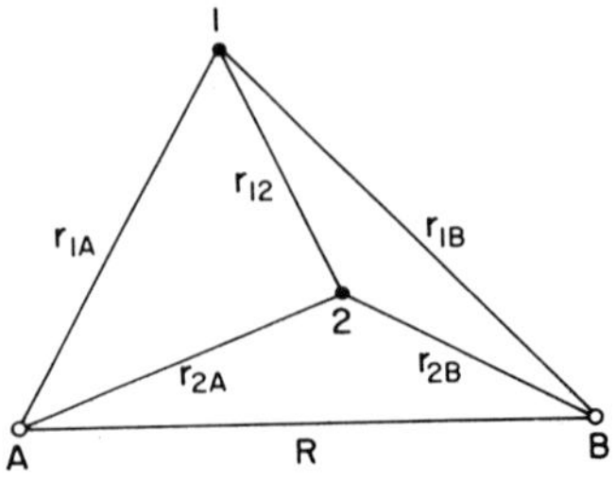

Fig. 2.2.1. Coordinate system for H_2.

The assumed wave function in this approximation is

$$\phi_\Gamma(\mathbf{r}_1, \mathbf{r}_2) \approx u_A(\mathbf{r}_1)u_B(\mathbf{r}_2) \tag{2.2.14}$$

where $u_A(\mathbf{r}_1)$ is a hydrogen atom wave function in which electron 1 is bound to atom A:

$$u_A(\mathbf{r}_1) = (\pi a_0^3)^{-1/2} \exp(-r_{1A}/a_0)$$

in which a_0 is the Bohr radius. Then

$$P(1, 2)\phi_\Gamma(\mathbf{r}_1, \mathbf{r}_2) = u_A(\mathbf{r}_2)u_B(\mathbf{r}_1).$$

The energy $\mathcal{E}$ which appears in (2.2.12) can be approximated to the accuracy we are concerned with at present by 2ϵ, where ϵ is the energy of an isolated ground state hydrogen atom. The corrections here are of order e^{-4R}, which we neglect. We then find

$$J = e^2 \iint d^3r_1\, d^3r_2\, u_A(\mathbf{r}_1)u_B(\mathbf{r}_2)$$
$$\times\, (r_{12}^{-1} + R^{-1} - r_{2A}^{-1} - r_{1B}^{-1})\, u_B(\mathbf{r}_1)u_A(\mathbf{r}_2). \tag{2.2.15}$$

This is the Heitler–London result in the large R limit. The difference in normalization of the wave function in the singlet and triplet states which appears in the usual formula does not appear here. It would contribute a term of order e^{-4R} to the result, which we neglect.

The integrals have been evaluated by Sugiura (1927). In the large R limit, the result is (in rydbergs, 13.6 eV)

$$J = [-(56/45) + (4\gamma/15) + (4/15)\ln(R/a_0)]$$
$$\times\, (R/a_0)^3 \exp(-2R/a_0) \tag{2.2.16}$$

where a_0 is the Bohr radius and γ is Euler's constant ($\gamma = 0.577$). For reasonable values of R, J is negative, that is, the singlet state has lower energy than the triplet. However, at very large R, the logarithmic term forces J to become positive. This results in error as it conflicts with the expectation, which can be justified rigorously, that the ground state wave function must be nodeless. The singlet state must be lower than the triplet for hydrogen at all value of R. The error arises from the neglect of any electron interaction effects in the approximate wave function (2.2.14).

Herring and Flicker (1964) obtained the leading term in J for large R. In order to do this, they multiply (2.2.14) by a function of $\mathbf{r}_1$ and $\mathbf{r}_2$ in the form $\exp[-S_1(\mathbf{r}_1, \mathbf{r}_2)]$ which is obtained by approximate solution of the Schrödinger equation for large separations. They then express J as a surface integral on a hyperplane in six-dimensional space such that $r^2{}_{1A} + r^2{}_{2B} = r^2{}_{1B} + r^2{}_{1A}$. After evaluation, the result is (in rydbergs)

$$J = -1.636R^{5/2} \exp[-2R/a_0] + O(R^2 \exp[-2R/a_0]). \qquad (2.2.17)$$

The result differs from (2.2.16) in that the logarithm is missing and the dominant power of R multiplying the exponential is different. However, the numerical difference is not significant for (R/a_0) between 5 and 12. In this region, there is substantial agreement between the expressions, and the Heitler–London procedure is adequate. This is probably the most important region for applications to solids: We are usually not interested in extraordinarily weak atomic interactions at enormous distances, but in the coupling of atoms at moderate distances, i.e., distances somewhat larger than the radius of atomic orbital of interest but of the order of the distance between nearest neighbor like atoms in an ionic crystal; typically 3–4 Å.

Although the Heitler–London approach should be sufficient to determine the exchange parameters at reasonably large separations, even this procedure is quite difficult to carry out in practice. The reader is referred to Herring (1966) for details. One important conclusion of Herring should be noted: Exchange parameters computed in this way are expected to be negative (antiferromagnetic). In real solids there are, however, many other effects contributing to spin coupling beyond those considered here (see, for instance, Sections 3.4 and 8.4). We will proceed from this point by treating the Heisenberg Hamiltonian phenomenologically: the exchange parameters will be regarded as adjustable, to be determined, if possible, from fits to experimental measurements of magnetic properties.

2.3 Molecular Field Theory

Molecular field theory furnishes the simplest approximate method for studying the properties of a system of spins coupled by a Heisenberg interaction. This approach was first introduced by Pierre Weiss (1907). The essential idea is that the spins behave as if each were acted upon by an effective magnetic field which is proportional to the magnetization of the crystal. The relatively simple mathematics of the method makes extensive development possible. A general exposition has been given by Smart (1966), which we shall follow.

2.3.1 *The Effective Field*

In the case of ferromagnetic crystals, the theory can be adequately presented under the restriction that we consider coupling only between atoms and their nearest neighbors. The number of nearest neighbors is conventionally denoted by Z. From (2.1.2) atom i interacts with its neighbors by a single-atom Hamiltonian $\mathcal{H}_1$, which is given by

$$\mathcal{H}_1 = -2J\mathbf{S}_i \cdot \sum_{j=1}^{Z} \mathbf{S}_j. \tag{2.3.1}$$

We define an effective magnetic field $\mathbf{H}_e$:

$$g\beta\mathbf{H}_e = 2J \sum_{j=1}^{Z} \mathbf{S}_j \tag{2.3.2}$$

in which g is the gyromagnetic ratio and β represents the Bohr magneton. Thus

$$\mathcal{H}_1 = -g\beta\mathbf{S}_i \cdot \mathbf{H}_e. \tag{2.3.3}$$

Equation (2.3.2) cannot rigorously define a simple magnetic field since the $\mathbf{S}_j$ are quantum mechanical operators; however, we assume (and this is essential) that in (2.3.2), each $\mathbf{S}_j$ can be replaced by its average value $\langle \mathbf{S}_j \rangle$. The magnetization of the crystal is

$$\mathbf{M} = Ng\beta\langle \mathbf{S}_j \rangle \tag{2.3.4}$$

where N is the number of atoms per unit volume and all magnetic atoms are assumed to be identical. Thus, on combining (2.3.2) and (2.3.4) we have

$$\mathbf{H}_e = (2ZJ/g\beta)\langle \mathbf{S}_j \rangle = (2ZJ/Ng^2\beta^2)\mathbf{M} = \gamma\mathbf{M} \tag{2.3.5a}$$

where

$$\gamma = 2ZJ/Ng^2\beta^2. \tag{2.3.5b}$$

We may add to the effective field $\mathbf{H}_e$ an external field $\mathbf{H}_0$. The total field on any atom is then

$$\mathbf{H}_T = \mathbf{H}_0 + \mathbf{H}_e. \tag{2.3.6}$$

We will use $\mathbf{H}_0$ to define the z axis. The basic Hamiltonian of (2.1.2) contains nothing which ties the spins to the crystal axes (that is, there is no anisotropy) so the spins will tend to line up along $\mathbf{H}_0$. Then $\mathbf{H}_e$ will be in the same direction as $\mathbf{H}_0$. Since all vectors are in the same direction, their vector nature is not important and the fields may be treated as scalars. We replace H_e by H_T and have, instead of (2.3.3),

$$\mathcal{H}_1 = -g\beta S_z H_T. \tag{2.3.7}$$

The eigenvalues of $\mathcal{H}_1$ are E_m:

$$E_m = -g\beta m H_T \tag{2.3.8}$$

where m is an integer or half integer in the range between $-S$ and S.

The thermodynamic properties of the system are obtained from the partition function:

$$\mathcal{Z} = \sum_{m=-S}^{S} \exp(-E_m/KT) = \sum_{m=-S}^{S} \exp(g\beta H_T m/KT). \tag{2.3.9}$$

The sum is a geometrical series. The result is

$$\mathcal{Z} = \sinh[g\beta H_T(2S+1)/2KT]/\sinh(g\beta H_T/2KT). \tag{2.3.10}$$

The magnetization is given by (2.3.4). The thermal average of any operator $\mathcal{O}$ is given by

$$\langle \mathcal{O} \rangle = \mathrm{tr}[\mathcal{O}e^{-H/KT}]/\mathrm{tr}(e^{-H/KT}) = (1/\mathcal{Z})\ \mathrm{tr}(\mathcal{O}e^{-H/KT}). \tag{2.3.11}$$

Thus

$$\begin{aligned} M &= (Ng\beta/\mathcal{Z})\ \mathrm{tr}(S_{jz}e^{-H/KT}) \\ &= (Ng\beta/\mathcal{Z}) \sum_{m=-S}^{S} m \exp(mg\beta H_T/KT). \end{aligned} \tag{2.3.12}$$

The sum may be evaluated easily by differentiation of (2.3.10) with respect to H_T. The result is

$$M = Ng\beta S B_S(x) \tag{2.3.12a}$$

where

$$\begin{aligned} B_S(x) &= [(2S+1)/2S]\ \mathrm{ctnh}\{[(2S+1)/2S]x\} \\ &\quad - (1/2S)\ \mathrm{ctnh}(x/2S) \end{aligned} \tag{2.3.12b}$$

and

$$x = g\beta S H_T/KT \tag{2.3.12c}$$

with B_S referred to as the *Brillouin function*. Its argument x is the ratio of the energy of a spin S in the field H_T to the thermal energy KT. A graph of $B_S(x)$ for several values of S is shown in Fig. 2.3.1. For $S = \frac{1}{2}$, $B_{1/2}(x) = \tanh x$.

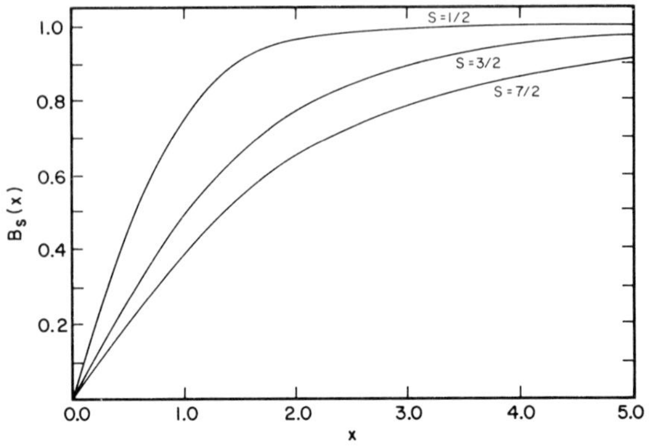

Fig. 2.3.1. Brillouin function $B_S(x)$ for $S = \frac{1}{2}, \frac{3}{2}, \frac{7}{2}$.

2.3.2 *The Ferromagnetic Transition*

We may note in passing that these results can be used directly to describe the behavior of an isolated magnetic ion in a crystal (we simply set $H_T = H_0$). For the prospective ferromagnet we proceed as follows: Eq. (2.3.12a) gives the magnetization as a function of the total effective field, while from (2.3.6) we see that the total field depends on the magnetization. Hence, we must obtain a solution of the pair of equations simultaneously.

We begin by writing the equations in dimensionless form so as to eliminate some of the constants. Since the maximum value of the Brillouin function is 1, the maximum possible value of the magnetization is M_0, where

$$M_0 = Ng\beta S. \tag{2.3.13}$$

We define the reduced magnetization

$$\sigma = M/M_0 = B_S(x). \tag{2.3.14}$$

We define x_0 in analogy with (2.3.12c), $x_0 = g\beta H_0 S/KT$, and we have

$$x = x_0 + (2ZJS^2/KT)\sigma. \tag{2.3.15}$$

The solution for σ may be obtained graphically. We plot $\sigma(x)$ from Eqs. (2.3.14) and (2.3.15). The simultaneous solution of these equations is the point of intersection as shown in Fig. 2.3.2.

Let us consider first the case $H_0 = 0$. For small values of the argument x

$$\operatorname{ctnh} x = (1/x)[1 + (x^2/3) - \tfrac{1}{45}x^4].$$

From this, we obtain the small argument form of the Brillouin function

$$B_S(x) = [(S+1)/S](x/3) - \{[(2S+1)^4 - 1]/(2S)^4\}(x^3/45). \qquad (2.3.16)$$

A spontaneous magnetic field develops at the temperature T_c at which it is first possible to satisfy (2.3.14) and (2.3.15). To determine this, we substitute the leading term of (2.3.16) into (2.3.15). The result is

$$T_c = \tfrac{2}{3}[JZS(S+1)/K]. \qquad (2.3.17)$$

For temperatures $T > T_c$, we obtain a solution only if an external field is present. In this case, the magnetization is proportional to the external field H_0. We find from (2.3.15) still using the small argument expansion

$$x = x_0 + (T_c/T)x.$$

This gives for the susceptibility

$$\chi = M/H_0 = C/(T - T_c) \qquad (2.3.18a)$$

where

$$C = Ng^2\beta^2S(S+1)/3K. \qquad (2.3.19)$$

If Eq. (2.3.18) is written in the form

$$\chi = C/(T - \Theta), \qquad (2.3.18b)$$

we have the Curie–Weiss law. The parameter Θ which appears in (2.3.18b) is called the *paramagnetic Curie temperature*. It is found that many materials obey (2.3.18b) in the high temperature region to a reasonable degree

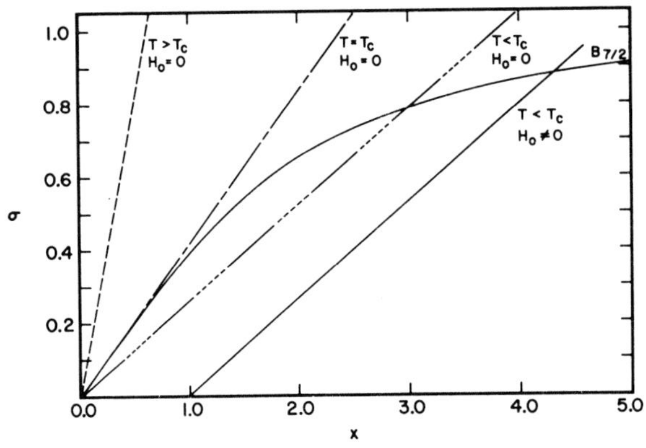

Fig. 2.3.2. Determination of magnetization in molecular field theory. The simultaneous solution of Eqs. (2.3.14) and (2.3.15) is shown for four cases.

of accuracy; but the specific result of the present model ($\Theta = T_c$) may not be satisfied. The constant C is known as the Curie constant.

As the temperature T approaches the Curie temperature from above, the reciprocal susceptibility χ^{-1} approaches zero, and a spontaneous magnetization is established. The spontaneous magnetization in zero field satisfies the transcendental equation

$$\sigma = B_S\{[3S/(S+1)](T_c/T)\sigma\}. \tag{2.3.20}$$

To obtain an expression valid near T_c, we expand the right side of (2.3.20), and this time, we use both terms of (2.3.16). The resulting expression is solved for σ^2 and we find

$$\sigma^2 = (10/3)[(S+1)^2/((S+1)^2+S^2)][T^2(T_c-T)/T_c^3] \tag{2.3.21a}$$

$$\approx (10/3)[(S+1)^2/((S+1)^2+S^2)][(T_c-T)/T_c]. \tag{2.3.21b}$$

The essential conclusion is that, as $T \to T_c$,

$$\sigma \approx (T_c - T)^{1/2}, \qquad d\sigma/dT \approx -(T_c - T)^{-1/2}. \tag{2.3.21c}$$

These results will be of interest later in comparison with better theories of the ferromagnetic phase transition. In particular, the exponents in (2.3.21c) are not in precise agreement either with more complete theories or with experiment.

In the low temperature limit ($T \to 0$), the magnetization M approaches its maximum value M_0. Let us examine σ in this limit, in which the argument of the Brillouin function is large. Since, as $|x| \to \infty$,

$$\operatorname{ctnh} x \to 1 + 2e^{-2x} + \cdots,$$

we find for $|x| \to \infty$

$$B_S(x) \to 1 - (1/S)e^{-x/S}, \tag{2.3.22}$$

which yields immediately

$$\sigma = 1 - \exp\{-[3/(S+1)](T_c/T)\}. \tag{2.3.23}$$

For small T, σ differs from unity by terms which are exponentially small. From the point of view of comparison with experiment, this is a major weakness, as one finds instead that $\sigma \approx 1 - aT^{3/2}$ (a being a numerical constant). The reason for this discrepancy is that molecular field theory does not properly describe spin waves which are the elementary excitations of a system of coupled spins. Spin wave theory does give a proper account of the low temperature behavior of the Heisenberg ferromagnet.

We now consider the specific heat of a ferromagnet in the molecular field approximation. The simplest way to obtain this is to calculate the internal energy directly. The energy of a single atom j is, for zero applied field,

$$E_j = g\beta S_{jz}H_e.$$

To calculate the energy of the system, we take one half the average energy of the atoms in the internal field so that we avoid counting atomic interactions twice, and then add the (full) energy of interaction with the external field.

$$U = -Ng\beta\langle S_{jz}\rangle(\tfrac{1}{2}H_e + H_0) = -M(\tfrac{1}{2}\gamma M + H_0). \qquad (2.3.24)$$

The vector character of M and H can be neglected since M, and therefore H_e, are parallel to H_0. Equation (2.3.24) can be written in several different ways:

$$\begin{aligned} U &= -\tfrac{1}{2}\gamma M_0^2\sigma[\sigma + (2H_0/\gamma M_0)] \\ &= -\tfrac{1}{2}\gamma M_0^2\sigma[\sigma + (g\beta H_0/JSZ)] \\ &= -(NKTS/2)\sigma(x + x_0). \end{aligned} \qquad (2.3.25)$$

The specific heat is

$$C_m = (\partial U/\partial T)_{H_0} = -\gamma M_0^2\,(\partial\sigma/\partial T)_{H_0}\,[\sigma + (H_0/\gamma M_0)]. \qquad (2.3.26)$$

This result may also be obtained from the partition function (2.3.10). We recall that $F = -NKT \ln Z$, and $S = -(\partial F/\partial T)_{H_T}$, $M = -(\partial F/\partial H_T)$. Given S, we may find the specific heat by differentiation with respect to temperature, holding the external field H_0 fixed. We find

$$C_m = T\,(\partial S/\partial T)_{H_0} = -H_T\,(\partial M/\partial T)_{H_0},$$

which will be seen to agree exactly with (2.3.26). The specific heat is a continuous function of temperature unless there is no external field ($H_0 = 0$). Then above T_c, $\sigma = 0$, and we see that $C = 0$ also. For $T < T_c$, we may use (2.3.21a):

$$\begin{aligned} C_m &= -\tfrac{1}{2}\gamma M_0^2\, d(\sigma^2)/dT \\ &= (5/3)(\gamma M_0^2/T_c)[(S+1)^2/((S+1)^2 + S^2)] \\ &= 5NK[S(S+1)/((S+1)^2 + S^2)]. \end{aligned} \qquad (2.3.27)$$

Thus we see that the specific heat has a finite discontinuity at the Curie temperature. This indicates a second-order phase transition. The magnetude of this discontinuity is independent of the strength of the exchange interaction. Experiment shows, instead, approximately a logarithmic singularity at T_c, and a high-temperature tail (for more information, see Section 2.7).

2.3.3 *Antiferromagnetism*

Antiferromagnetic ordering of spins may occur if the coupling of spins is dominated by negative values of J. In an antiferromagnetically ordered system, the spins are aligned so that the net magnetic moment is substantially smaller than if the spins were parallel or ferromagnetically ordered. The simplest example of this occurs if the crystal can be divided into two sublattices, such that each sublattice is aligned ferromagnetically, but the magnetizations of the two sublattices are opposed and so cancel. More complicated situations are not only possible, but frequently observed; however, we will confine our attention here to the simplest case. Molecular field theory can be used to describe this situation, as well as ferromagnetism, and we will sketch this treatment briefly.

We consider a system of two sublattices, consisting of identical atoms. There is an exchange parameter J_{12} connecting neighboring spins on the different lattices and a parameter J_{11} connecting nearest neighbor spins on the same sublattice. The effective magnetic field acting on sublattice i ($i = 1, 2$) is

$$\mathbf{H}_i = \mathbf{H}_0 + \sum_j \gamma_{ij}\mathbf{M}_j. \qquad (2.3.28)$$

The argument which lead to (2.3.5a) and (2.3.5b) in the ferromagnetic case now yields

$$\gamma_{ij} = 2Z_{ij}J_{ij}/(N/n)g^2\beta^2 \qquad (2.3.29a)$$

where n is the number of sublattices and Z_{ij} is the number of neighbors of an atom on sublattice i which are on sublattice j and are connected with i by the exchange parameter J_{ij}. In the present case, $n = 2$, and if we suppose that the nearest neighbors of an atom on 1 are all on 2, and vice versa, then $Z_{12} = Z_1$, the total number of nearest neighbors. We suppose that the second neighbors of a given atom are on the same sublattice, so we put $Z_{11} = Z_{22} = Z_2$, the number of second neighbors. We also relabel $J_{12} = J_1$, the first neighbor exchange parameter, and $J_{11} = J_{22} = J_2$, the second neighbor exchange parameter. Then

$$\gamma_{12} = \gamma_{21} = 4Z_1J_1/Ng^2\beta^2; \qquad \gamma_{11} = \gamma_{22} = 4Z_2J_2/Ng^2\beta^2. \qquad (2.3.29b)$$

We repeat the arguments of the ferromagnetic case. The reduced magnetization on sublattice i is given by

$$\sigma_i = B_s(x_i) \qquad (2.3.30)$$

where $x_i = g\beta H_iS/KT$. Note that $M_{0i} = (N/2)g\beta S$. First we consider the susceptibility in the high temperature region in which the material is paramagnetic. In this case, we expect the material to magnetize in the direction of H_0, so we may again neglect the vector nature of H_0 and M. We also use the high temperature approximation to the Brillouin function (2.3.16), retaining only the first term. This is substituted into (2.3.28), giving

$$M_i - (C/nT) \sum_j \gamma_{ij} M_j = (C/nT) H_0 \tag{2.3.31}$$

where C is the Curie constant defined by (2.3.19). Equation (2.3.31) is valid for an arbitrary number of sublattices if the γ_{ij} are defined appropriately. There are, in fact, n simultaneous equations. We are here concerned with the total magnetization, so we sum (2.3.31) with respect to i and make use of the symmetry properties of the coefficients γ: $\gamma_{ij} = \gamma_{ji}$ for $i \neq j$, and $\gamma_{ii} = \gamma_{jj}$. The result is

$$M[1 - (C/nT) \sum_j \gamma_{ij}] = CH_0/T$$

and

$$\chi = M/H_0 = C/(T + \Theta_A) \tag{2.3.32}$$

where

$$\Theta_A = -(C/n) \sum \gamma_{ij} = -[2S(S+1)/3K] \sum_k Z_k J_k \tag{2.3.33}$$

where Z_k is the number of kth neighbors of a given atom and J_k is the appropriate exchange integral.

Antiferromagnetism is characterized by the dominance of terms with negative J, that is, in an antiferromagnetic crystal, Θ_A as defined by (2.3.33) will be a positive temperature. Thus the susceptibility of an antiferromagnet in the paramagnetic region obeys a Curie–Weiss law with a negative Curie temperature $T_c = -\Theta_A$.

To investigate the possibility of a transition to an ordered state in zero external field, we set $H_0 = 0$ in (2.3.31) and look for a nontrivial solution by requiring the determinant of the coefficients to vanish:

$$\det[(nT/C)\,\delta_{ij} - \gamma_{ij}] = 0. \tag{2.3.34}$$

The roots of this equation give a set of n possible transition temperatures. The eigenvectors give the ratios of the M_i, that is, the relative magnetizations of the various sublattices. Although general methods exist to solve this equation in arbitrary systems (see Smart, 1966), we will confine our attention here to the two-sublattice system described previously. In this case (2.3.34) becomes

$$\begin{vmatrix} 2T/C - \gamma_{11} & \gamma_{12} \\ \gamma_{12} & 2T/C - \gamma_{11} \end{vmatrix} = 0$$

where we have used the symmetry properties of the γ's mentioned previously. The solutions give the transition temperatures

$$T = (C/2)(\gamma_{11} \pm \gamma_{12}) = [2S(S+1)/3K][Z_2J_2 \pm Z_1J_1]. \qquad (2.3.35)$$

The solution with the plus (+) sign has $T = -\Theta_A$. From the previous discussion, we see that this is a transition at a negative temperature, and is, therefore, not observable. This solution corresponds to ferromagnetism ($M_1 = M_2$) as is seen from the eigenvector; our result is that if Θ_A is positive, ferromagnetism is an unstable state of the system.

The other root of (2.3.35) has an eigenvector which gives $M_1 = -M_2$, so that the two sublattices are magnetized oppositely. This is the antiferromagnetic state. The transition temperature is called the *Néel temperature* T_N:

$$T_N = [2S(S+1)/3K](Z_2J_2 - Z_1J_1) \qquad (2.3.36)$$

where T_N will be positive, corresponding to an antiferromagnetic transition at a positive temperature if J_1 is negative and J_2 is not so negative that it compensates for J_1, i.e., [$J_2 > 0$ or if $J_2 < 0$, $|J_2| < (Z_1/Z_2)|J_1|$]. If these conditions are violated, the two-sublattice picture from which we started was incorrect, and a more complicated kind of magnetic ordering will occur.

Many improvements and extensions of molecular field theory are possible. The interested reader is referred to the work by Smart which we have already cited. We will not consider these here, but investigate instead another approach to the Heisenberg Hamiltonian, that of spin waves.

2.4 Spin Waves

We consider the Heisenberg Hamiltonian from a more rigorously quantum mechanical point of view than is afforded by molecular field theory. Specifically we investigate the ground state and low lying excited states of a system of coupled spins. Our attention is devoted initially to the case of ferromagnetic coupling (positive J's) for which the most rigorous results are available. For antiferromagnetic coupling (negative J), not even the ground state is known exactly, in three dimensions. More complete surveys of spin wave theory have been given by Akhiezer *et al.* (1968) and by Keffer (1966).

2.4.1 *The Ground State of the Heisenberg Ferromagnet*

We begin by considering the commutation relations between spin operators $\mathbf{S}_i$. It is convenient to introduce the raising and lowering operators S_i^+, S_i^-:

$$S_i^{\pm} = S_i^x \pm iS_i^y.$$

We have

$$[S_i^z, S_j^\pm] = \pm\delta_{ij}\, S_i^\pm \tag{2.4.1a}$$

and

$$[S_i^+, S_j^-] = 2\,\delta_{ij}\, S_i^z. \tag{2.4.1b}$$

It should be noticed that spin operators for different sites commute.

We now consider the Heisenberg Hamiltonian

$$\begin{aligned} H_{\mathrm{H}} &= -2\sum_{i>j} J_{ij}\mathbf{S}_i\cdot\mathbf{S}_j \\ &= -2\sum_{i>j} J_{ij}(S_i^zS_j^z + \tfrac{1}{2}S_i^+S_j^- + \tfrac{1}{2}S_i^-S_j^+) \\ &= -\sum_{i,j} J_{ij}(S_i^zS_j^z + S_i^-S_j^+). \end{aligned} \tag{2.4.2}$$

The last step is possible since the difference

$$\tfrac{1}{2}S_i^-S_j^+ - \tfrac{1}{2}S_i^+S_j^- = i(S_i^xS_j^y - S_i^yS_j^x)$$

is antisymmetric i and j and so vanishes on summation. We also find it convenient to eliminate the factor of 2 and define $J_{ii} = 0$, and thus extend the sum over all i, j, without restriction.

Let us consider the state of the coupled spins in which each S_i^z attains its maximum value S. We call this state $|\,0\,\rangle$, the completely ferromagnetic state. This state has the property that

$$S_j^+\,|\,0\rangle = 0, \qquad S_j^z\,|\,0\rangle = S\,|\,0\rangle \tag{2.4.3}$$

for any j; that is, no spin can be raised. The energy of this state is

$$H\,|\,0\rangle = E_0\,|\,0\rangle, \qquad E_0 = -S^2\sum_{i,j} J_{ij}. \tag{2.4.4a}$$

In the case in which J_{ij} differs from zero only when i and j are nearest neighbors, the sum on j gives N, the total number of atoms in the crystal while the sum on i for fixed j gives the number of nearest neighbors Z, so

$$E_0 = -NS^2JZ. \tag{2.4.4b}$$

The state $|\,0\rangle$ can easily be seen to be the ground state of the system for positive J. If any spin has less than its maximum z component, then eigenvalue of $\sum \mathbf{S}_i\cdot\mathbf{S}_j$ will be less in such a case. Of course, the ground state is highly degenerate as there is nothing to fix the direction of the z axis (This can be specified only by the addition of an external field.) In the absence of an external field, there are $(2NS + 1)$ possible orientations of the total spin; and all these states have the same energy. The implications of this will be examined subsequently.

2.4.2 *Excited States*

We now look at the low lying excited states of the system. These should consist of states in which a single spin has deviated from the alignment of the completely ferromagnetic state. We begin by defining a set of orthonormal states in which the spin at a particular site, l say, has been lowered,

$$| l\rangle = [1/(2S)^{1/2}]S_l^- | 0\rangle. \tag{2.4.5}$$

The dual states are $\langle i | = [1/(2S)^{1/2}]\langle 0 | S_i^+$. These spin deviation states are orthonormal:

$$\begin{aligned}\langle i | l\rangle &= (1/2S)\langle 0 | S_i^+S_l^- | 0\rangle \\ &= (1/2S)\langle 0 | 2\,\delta_{il}\,S_i^z + S_l^-S_i^+ | 0\rangle \\ &= (1/2S)\langle 0 | 2\,\delta_{il}\,S_i^z | 0\rangle = \delta_{il}.\end{aligned} \tag{2.4.6}$$

The states $| l\rangle$ are localized. They are not eigenfunctions of the Hamiltonian, but do furnish a convenient basis for the solution of the problem of determining eigenstates. For the moment, we will denote an eigenfunction by $| E\rangle$ and will add other quantum numbers as necessary. We can expand $| E\rangle$ as

$$| E\rangle = \sum_l \phi(\mathbf{R}_l) \, | l\rangle. \tag{2.4.7}$$

We require $H | E\rangle = E | E\rangle$. In order to find an equation satisfied by the coefficients $\phi(\mathbf{R}_l)$, we form the scalar product with a typical state $| n\rangle$:

$$\sum_l \langle n | H | l\rangle\phi(\mathbf{R}_l) = E\phi(\mathbf{R}_n). \tag{2.4.8}$$

We must now calculate the matrix element

$$\begin{aligned}\langle n | H | l\rangle &= -(1/2S) \sum_{ij} J_{ij}\langle 0 | S_n^+\mathbf{S}_i\cdot\mathbf{S}_jS_l^- | 0\rangle \\ &= -(1/2S) \sum_{ij} J_{ij}\langle 0 | S_n^+(S_i^zS_j^z + S_i^-S_j^+)S_l^- | 0\rangle.\end{aligned}$$

The matrix elements are evaluated by commuting the S^+ operators through those on their right so that they ultimately act on, and annihilate, the ground state $| 0\rangle$. We obtain for the matrix element

$$\langle n | H | l\rangle = -\delta_{nl}\,[S^2 \sum_{ij} J_{ij} - 2S \sum_j J_{nj}] - 2SJ_{nl}. \tag{2.4.9}$$

Let $\epsilon = E - E_0$ where E_0 is given by (2.4.4a). Then (2.4.9) becomes

$$2S \sum_l [(\sum_j J_{lj})\,\delta_{nl} - J_{nl}]\phi(\mathbf{R}_l) = \epsilon\phi(\mathbf{R}_n). \tag{2.4.10}$$

Equation (2.4.10) can be solved easily. The essential fact which makes solution possible is that the exchange integral coupling the spins on two sites is a function of the distance between sites only:

$$J_{nl} = J(\,|\,\mathbf{R}_n - \mathbf{R}_l\,|\,). \tag{2.4.11}$$

We introduce the *ansatz*

$$\phi(\mathbf{R}_l) \propto \exp(i\mathbf{k}\cdot\mathbf{R}_l). \tag{2.4.12}$$

This yields a solution and gives

$$\epsilon = \epsilon(\mathbf{k}) = 2S\sum_l J(\mathbf{R}_l)[1 - \exp(i\mathbf{k}\cdot\mathbf{R}_l)]. \tag{2.4.13}$$

The solutions, as given by (2.4.12) are wavelike, and are conventionally called either *spin waves* or *magnons*. The wave vector $\mathbf{k}$ is restricted to the Brillouin zone by the argument given in a different connection in Chapter 1: If $\mathbf{k}$ is increased by $\mathbf{K}_s$ where $\mathbf{K}_s$ is any reciprocal lattice vector, both ϕ and $\epsilon(\mathbf{k})$ are unchanged. The function $\epsilon(\mathbf{k})$ gives what is called the *magnon dispersion relation.* Our result expresses the fact that the eigenstates of the system are not spin deviations localized on a single site, but rather are spin deviations spread through the entire system.

We will now introduce the wave vector $\mathbf{k}$ into the designation of ϕ by writing $\phi(\mathbf{k}, \mathbf{R}_l)$. We may adopt the normalization

$$\sum_l \phi^*(\mathbf{k}, \mathbf{R}_l)\phi(\mathbf{q}, \mathbf{R}_l) = \delta(\mathbf{k} - \mathbf{q}) \tag{2.4.14a}$$

which is satisfied by

$$\phi(\mathbf{k}, \mathbf{R}_l) = [\Omega^{1/2}/(2\pi)^{3/2}]\exp(i\mathbf{k}\cdot\mathbf{R}_l). \tag{2.4.15a}$$

Alternately, we may use discrete normalization

$$\sum_l \phi^*(\mathbf{k}, \mathbf{R}_l)\phi(\mathbf{q}, \mathbf{R}_l) = \delta_{\mathbf{kq}}. \tag{2.4.14b}$$

This is attained by

$$\phi(\mathbf{k}, \mathbf{R}_l) = N^{-1/2}\exp(i\mathbf{k}\cdot\mathbf{R}_l) \tag{2.4.15b}$$

where N is the number of unit cells in the crystal.

It follows from (2.4.11) that for a crystal with a center of inversion

$$\epsilon(\mathbf{k}) = 2S\sum_l J(\mathbf{R}_l)(1 - \cos\mathbf{k}\cdot\mathbf{R}_l), \tag{2.4.16}$$

and therefore that

$$\epsilon(\mathbf{k}) = \epsilon(-\mathbf{k}). \tag{2.4.17}$$

It should be emphasized that our result (2.4.13) for the energy of a single spin wave is rigorous. No approximations have been made beyond the initial introduction of the Heisenberg Hamiltonian.

It is enlightening to examine (2.4.16) in a special case. Let us consider a simple cubic lattice and suppose that the exchange integrals J can be neglected beyond third neighbors. The first neighbors are located at sites of the type $(a, 0, 0)$, the second at $(a, a, 0)$, and the third at (a, a, a). Then in this case

$$\begin{aligned}\epsilon(k) = 2S\{&[6J(a) + 12J(2^{1/2}a) + 8J(3^{1/2}a)]\\ &-2J(a)(\cos k_x a + \cos k_y a + \cos k_z a)\\ &- 4J(2^{1/2}a)(\cos k_x a \cos k_y a\\ &+ \cos k_z a \cos k_x a + \cos k_y a \cos k_z a)\\ &- 8J(3^{1/2}a) \cos k_x a \cos k_y a \cos k_z a\}.\end{aligned} \tag{2.4.18}$$

It is indicated here that the energy can be expressed in terms of a symmetrized combination of cosines. Equation (2.4.18) may be easily adapted to other cubic lattices (body-centered and face-centered) by suitably rearranging terms and adjusting the distances.

If the utility of (2.4.16) were really confined to the situation in which there is just one and only one reversed spin in a ferromagnet very little would have been accomplished. It must be supposed instead that our results can be used to describe a crystal containing a small concentration of reversed spins, that is, (2.4.16) should apply the thermodynamics of a ferromagnet at low temperatures where the number of reversed spins is small compared to the total number of spins in the crystal.

Let us determine the maximum energy in the spin wave spectrum from (2.4.18) for the simple, body-centered, and face-centered cubic structures. This is denoted by ϵ_m. For simplicity we use the nearest neighbor approximation and call the exchange integral J in each case. We find the values $24JS$, $32JS$, and $32JS$, respectively. It is interesting to compare these values with the Curie temperature in the molecular field model of the previous section [Eq. (2.3.17)]. The results are

$$\begin{aligned}\epsilon_m/KT_c &= 6/(S+1) \qquad \text{(simple cubic)},\\ &= 6/(S+1) \qquad \text{(body-centered)},\\ &= 4/(S+1) \qquad \text{(face-centered)}.\end{aligned}$$

For small S ($S = \frac{1}{2}$, say), ϵ_m is significantly larger than T_c; but for large S

($S = \frac{7}{2}$), as found in the Europium chalcogenides (EuO, EuS, etc.), for example, and in the face-centered cubic structure, the ratio approaches unity. The inference here is that we do not expect to find spin waves of energies near the top of the band excited with appreciable amplitude under circumstances in which the theory might be expected to be valid except possibly for large S.

A spin wave state of wave vector $|\mathbf{k}\rangle$ may be constructed as follows [we use discrete normalization (2.4.14b), (2.4.15b) from here on]:

$$|\mathbf{k}\rangle = N^{-1/2} \sum_l \exp(i\mathbf{k}\cdot\mathbf{R}_l)\;|l\rangle. \tag{2.4.19}$$

These states are eigenstates of the $\mathbf{S}^2$, the square of the total spin. We establish this as follows. Consider

$$\mathbf{S}^2 = (\sum_i \mathbf{S}_i)^2 = \sum_{i,j} \mathbf{S}_i\cdot\mathbf{S}_j.$$

Let $\mathbf{S}^2$ operate on a localized spin deviation state $|l\rangle$. We find

$$\begin{aligned}\mathbf{S}^2\,|l\rangle &= [1/(2S)^{1/2}] \sum_{ij} [S_i^zS_j^z + \tfrac{1}{2}S_i^-S_j^+ + \tfrac{1}{2}S_i^+S_j^-]S_l^-\,|0\rangle \\ &= NS(NS-1)\,|l\rangle + 2S \sum_j |j\rangle.\end{aligned}$$

Note that in this case we must include both $S_i^-S_j^+$ and $S_i^+S_j^-$ since terms with $i = j$ do occur here. Then

$$\mathbf{S}^2\,|\mathbf{k}\rangle = NS(NS-1)\,|\mathbf{k}\rangle + (2S/N^{1/2})[\sum_l \exp(i\mathbf{k}\cdot\mathbf{R}_l)] \sum_j |j\rangle,$$

but

$$\sum_l \exp(i\mathbf{k}\cdot\mathbf{R}_l) = \mathbf{N}\,\delta_{\mathbf{k},0} \qquad \text{and} \qquad \sum_j |j\rangle = N^{1/2}\,|\mathbf{k}=0\rangle,$$

so

$$\mathbf{S}^2\,|\mathbf{k}\rangle = [NS(NS-1) + 2NS\,\delta_{\mathbf{k},0}]\,|\mathbf{k}\rangle. \tag{2.4.20}$$

Thus a spin wave of wave vector $\mathbf{k}$ is an eigenstate of $\mathbf{S}^2$ for the entire crystal

$$\mathbf{S}^2\,|\mathbf{k}\rangle = \mathcal{S}(\mathcal{S}+1)\,|\mathbf{k}\rangle.$$

If $k = 0$, $\mathcal{S} = NS$, but if $k \neq 0$, $\mathcal{S} = NS - 1$. From this, we see that a spin wave of infinite wavelength ($k = 0$), has the same value of $\mathcal{S}$ as does the ground state. In the absence of an external magnetic field, the energy of the system cannot depend on the orientation of the spin with respect to an arbitrary z axis fixed in space. Excitation of a spin wave of zero wave vector corresponds to such a rotation, and therefore cannot change the energy of the system.

There is another way of looking at this result, which is a consequence of the Goldstone theorem on broken symmetry (Goldstone *et al.*, 1962). Their conclusion, which is also of considerable importance in elementary particle physics, may be stated as follows: If there is a continuous symmetry transformation under which the Lagrangian of a field is invariant, then either the vacuum state is also invariant under the transformation, or there must exist spinless particles of zero mass. In the present case, the Hamiltonian is invariant under rotations about any axis; however, the ferromagnetic ground state assumes a definite value for S_z and is thus not invariant. Consequently states of zero excitation energy must exist; in our case, there are the spin waves of $k = 0$.

2.4.3 *External Magnetic Fields*

Let us consider the dependence of the spin wave on an external magnetic field H_0. We add to the Hamiltonian (2), a term $\mathcal{H}_m$:

$$\mathcal{H}_m = -g\beta H_0 \sum_i S_{iz}. \tag{2.4.21}$$

The field has been assumed to define the z direction. The preceding calculation is easily repeated. A term $-g\beta H_0 NS$ is added to E_0, while $\epsilon(\mathbf{k})$ becomes

$$\epsilon(\mathbf{k}) = 2S \sum_l J(\mathbf{R}_l)(1 - \cos \mathbf{k}\cdot\mathbf{R}_l) + g\beta H_0. \tag{2.4.22}$$

In the presence of H_0, a finite energy $g\beta H_0$ is required to excite a spin wave. This can be described by saying that the excitation spectrum is separated from the ground state by an energy gap which is proportional to the applied field. If no field is present, a spin wave of wave vector $k = 0$ may be excited without any change in energy, that is, the excitation spectrum touches the ground state energy.

The dependence of the spin wave energy on an external field that we have calculated raises the following problem: What should be done with the magnetization $\mathbf{M}$ of the ferromagnet? Should it be accounted for as part of H_0 in determining $\epsilon(\mathbf{k})$? The answer to this question turns out to be rather complicated, and was first given by Holstein and Primakoff (1940). To account for $\mathbf{M}$, it is necessary to add to the Heisenberg Hamiltonian (plus H_m) a term which allows for the dipole–dipole interaction of the spins. This term will be called $\mathcal{H}_d$, and is given by

$$\mathcal{H}_d = \tfrac{1}{2}(g\beta)^2 \sum_{i,j} (1/\mathbf{R}^3_{ij})\{\mathbf{S}_i\cdot\mathbf{S}_j - [(\mathbf{S}_i\cdot\mathbf{R}_{ij})(\mathbf{S}_j\cdot\mathbf{R}_{ij})/\mathbf{R}^2_{ij}]\}. \tag{2.4.23}$$

The calculation of spin wave energies, here denoted by $E(\mathbf{k})$, when the Hamiltonian consists of $\mathcal{H}_H + \mathcal{H}_d + \mathcal{H}_m$, is rather long and the details will not be given. The result may be put in the form (following Charap and Boyd, 1964)

$$E(\mathbf{k}) = [\epsilon(\mathbf{k}) + g\beta H_0][1 + \phi(\mathbf{k}) \sin^2\theta_k]^{1/2}. \tag{2.4.24}$$

In this expression, $\epsilon(\mathbf{k})$ is the spin wave energy from the Heisenberg Hamiltonian in the absence of an external field, that is, $\epsilon(\mathbf{k})$ is given by (2.4.16). The quantity $\phi(\mathbf{k})$ is given by

$$\phi(\mathbf{k}) = 4\pi g\beta M_0/[\epsilon(\mathbf{k}) + g\beta H_0] \tag{2.4.25}$$

where M_0 is the saturation magnetization ($M_0 = Ng\beta S$) and θ_k is the angle between the wave vector of the spin wave $\mathbf{k}$ and the magnetization. If $\phi(\mathbf{k})$ is small, that is if $\epsilon(\mathbf{k})$ or $g\beta H_0$ or both are larger than $4\pi g\beta M_0$, the square root in (2.4.24) can be expanded to yield

$$E(\mathbf{k}) = \epsilon(\mathbf{k}) + g\beta H_0 + 2\pi g\beta M_0 \sin^2\theta_k.$$

Suppose we then replace $\sin\theta_k$ by its average value of $\frac{2}{3}$. The result is

$$E(\mathbf{k}) = \epsilon(\mathbf{k}) + g\beta[H_0 + (4\pi M_0/3)]. \tag{2.4.26}$$

Under these approximations, the effect of dipole–dipole interactions on the spin wave energy is to justify the treatment of the sample magnetization as an effective magnetic field in the fashion that one would expect from a glance at (2.4.22). In most cases, this approximation is adequate for the calculation of thermodynamic properties, which depend on integrals over the spin wave spectrum. However, suppose that the external magnetic field H_0 vanishes, and that we consider the limit of very small $\mathbf{k}$ so that $\epsilon(\mathbf{k}) \ll 4\pi g\beta M_0$. Then

$$E(\mathbf{k}) = [4\pi g\beta M_0\epsilon(\mathbf{k})]^{1/2} \sin\theta_k.$$

This expression goes to zero as $\mathbf{k}$ approaches zero indicating that, in the absence of an external field, there is no gap separating spin waves from the ground state. Consistency with the argument involving broken symmetries is maintained.

2.4.4 *Statistics of Spin Waves*

At this point, we wish to consider the calculation of some thermodynamic properties of the Heisenberg ferromagnet using spin wave theory. Such calculations are rather simple at low temperatures. One problem which has to be faced initially is the question of what statistics to use: Spins are not exactly bosons or fermions, that is, the commutation rules for spin operators are not those of a Bose or Fermi field. However, it turns out that at low temperatures where only a few spin waves are excited, spin waves may be approximately treated as independent bosons.

To see what is involved in this, we attempt to make a correspondence between the localized spin operators $\mathbf{S}_l$ and boson operators a_l as follows:

$$S_l^+ \to (2S)^{1/2} a_l, \qquad S_l^- \to (2S)^{1/2} a_l^\dagger, \qquad S_l^z \to S - a_l^\dagger a_l. \tag{2.4.27a}$$

The operators a_l, $a_l^\dagger$ are required to obey the commutation rules

$$[a_j, a_l^\dagger] = \delta_{jl}, \qquad [a_j^\dagger, a_l^\dagger] = [a_j, a_l] = 0. \tag{2.4.28}$$

The Hamiltonian can be expressed in terms of the new operators as

$$H_{\mathrm{H}} = -\sum_{i,j} J_{ij}[(S - a_i^\dagger a_i)(S - a_j^\dagger a_j) + 2S a_i^\dagger a_j]. \tag{2.4.29a}$$

The difficulty here is that the assumed commutation rules for the a's are not exactly consistent with the actual rules obeyed by the S's [Eq. (2.4.1)]. For this reason, (2.4.27a) is not an equality. On comparison, we see that (2.4.1a) is satisfied but (2.4.1b) is not. However, (2.4.1b) is satisfied "on the average," that is, if we form the expectation of (2.4.1b) using the completely ferromagnetic state, the equation is consistent with (2.4.27). Another point of difference between spins and free bosons arises from the fact that there is a minimum possible value of S_l^z, namely $-S$; whereas any number of free bosons can be in a given state. This discussion does suggest that the correspondence (2.4.27) should be useful at low temperatures where the number of spin deviations is small, and this expectation is born out by more complete analysis.

In a fundamental paper, Holstein and Primakoff (1940) proposed a different correspondence between spin deviations and bosons. Their relations are

$$\begin{aligned} S_l^z &\to S - a_l^\dagger a_l, \\ S_l^+ &\to (2S)^{1/2}[1 - (a_l^\dagger a_l/2S)]^{1/2} a_l, \\ S_l^- &\to (2S)^{1/2} a_l^\dagger [1 - (a_l^\dagger a_l/2S)]^{1/2}. \end{aligned} \tag{2.4.27b}$$

The a_l are assumed to satisfy the commutation rules (2.4.28). The Holstein–Primakoff correspondence is reduced to (2.4.27a) if the square roots $[1 - (a_l^\dagger a_l/2S)]^{1/2}$ are replaced by unity. A short calculation shows that this correspondence is exactly consistent with the fundamental commutation rules. However, this success has been achieved at the high price of complicating the expression for the Hamiltonian, which becomes

$$\begin{aligned} H = -\sum_{i,j} J_{ij}\{&(S - a_i^\dagger a_i)(S - a_j^\dagger a_j) \\ &+ 2S a_i^\dagger [1 - (a_i^\dagger a_i/2S)]^{1/2}[1 - (a_j^\dagger a_j/2S)]^{1/2} a_j\}. \end{aligned} \tag{2.4.29b}$$

This cannot be expressed in terms of integral powers of the a's, and is consequently extremely difficult to handle. In fact, one is usually forced to approximate the expressions $[1 - (a_l^\dagger a_l/2S)]^{1/2}$ by unity, as should by valid at low temperatures. However, if this is done, it appears that there is no practical advantage to the Holstein–Primakoff substitution, as we have returned to (2.4.29a).

Other substitutions have been proposed. In particular, Dyson (1956a,b) and Maleev (1958) have proposed a procedure which is satisfactory except that the relation $(S_i^+)^\dagger = S_i^-$ is not satisfied. They choose

$$
\begin{aligned}
S_l^z &\to a_l^\dagger a_l - S, \\
S_l^+ &\to (2S)^{1/2} a_l^\dagger, \\
S_l^- &\to (2S)^{1/2}[1 - (a_l^\dagger a_l/2S)]a_l.
\end{aligned}
\tag{2.4.27c}
$$

Here the ferromagnetic ground state is chosen to have $S_i^z = -S$, so that creation of a spin deviation is accomplished by S^+ rather than S_l^-. This convention is different from the one used here previously, but equally acceptable. This transformation may be somewhat easier to use than the Holstein–Primakoff substitution, but it also involves a price; which is that the resulting Hamiltonian is not Hermitian. This complication can be handled; however, we shall not pursue this topic further here. The interested reader is referred to the articles by Dyson for further information.

We now form the "Fourier transform" of the a_l: Define

$$
\begin{aligned}
a(\mathbf{k}) &= N^{-1/2} \sum_l \exp(i\mathbf{k}\cdot\mathbf{R}_l) a_l, \\
a^\dagger(\mathbf{k}) &= N^{-1/2} \sum_l \exp(-i\mathbf{k}\cdot\mathbf{R}_l) a_l^\dagger.
\end{aligned}
\tag{2.4.30}
$$

The inverse relations are

$$
\begin{aligned}
a_i &= N^{-1/2} \sum_{\mathbf{k}} \exp(-i\mathbf{k}\cdot\mathbf{R}_i) a(\mathbf{k}), \\
a_i^\dagger &= N^{-1/2} \sum_{\mathbf{k}} \exp(i\mathbf{k}\cdot\mathbf{R}_i) a^\dagger(\mathbf{k}).
\end{aligned}
\tag{2.4.31}
$$

The wave vector $\mathbf{k}$ is confined to the Brillouin zone. It now follows from (2.4.24) that

$$
[a(\mathbf{k}), a^\dagger(\mathbf{q})] = \delta_{\mathbf{kq}}, \quad [a(\mathbf{k}), a(\mathbf{q})] = [a^\dagger(\mathbf{k}), a^\dagger(\mathbf{q})] = 0. \tag{2.4.32}
$$

The operators $a^\dagger(\mathbf{k})$, $a(\mathbf{k})$ are creation and destruction operators for spin waves. This is legitimate since the single-spin wave state $|\,\mathbf{k}\rangle$ given by (2.4.19) may be expressed as

$$
|\,\mathbf{k}\rangle = a^\dagger(\mathbf{k})\,|\,0\rangle. \tag{2.4.33}
$$

This expression is valid in all the correspondences (2.4.27) and the state $| \mathbf{k} \rangle$ is as we have seen, an eigenstate of the Hamiltonian. However, the states in which more than one spin is excited, such as

$$| \mathbf{k}, \mathbf{q} \rangle = a^{\dagger}(\mathbf{k}) a^{\dagger}(\mathbf{q}) \mid 0 \rangle \tag{2.4.34}$$

which contains two spin waves, are not eigenstates. There is, in fact, an effective interaction between spin waves which has been discussed by several authors: Dyson (1956a,b), Wortis (1963, 1965), Boyd and Callaway (1965), Silberglitt and Harris (1968).

The effective interaction between spin waves may be handled in different ways. One straightforward approach is to take the Hamiltonian (2.4.29b) involving the Holstein–Primakoff transformation, expand the square roots, and collect all terms of fourth order in the a's. Then one introduces the transformation (2.4.30) and obtains an effective interaction, of order k^2 for small k, between spin waves. A different procedure which avoids operator substitutions of the sort of (2.4.27a)–(2.4.27c) altogether has been devised (Boyd and Callaway, 1965). This involves the construction of a set of orthonormal two-spin deviation states which are most conveniently characterized by a total wave vector and relative position. The matrix of the Heisenberg Hamiltonian on this basis can be constructed (it is diagonal with respect to the total wave vector). In this picture there is an interaction between two spin deviations on neighboring sites, as is evident physically. The scattering cross section for this problem can be obtained by the scattering techniques which will be described in Part B, Chapter 5. It is interesting to note that the scattering cross section is of fourth order in $\mathbf{k}$ for small $\mathbf{k}$, as was first obtained by Dyson (1956a). If we consider two spin waves with wave vectors $\mathbf{k}$, $\mathbf{k}'$, both small, the cross section is

$$\sigma = (\mathbf{k} \cdot \mathbf{k}')^2 / 8\pi S^2$$

which is quite small. However, the interaction between large $\mathbf{k}$ spin waves is by no means small: it is found that both bound states (Wortis, 1963; Hanus, 1963) and scattering resonances exist for suitably large values of the total wave vector. These techniques can also be applied to determine the effect of spin wave interactions on properties such as the specific heat and spontaneous magnetization.

The conclusion of the previous discussion is that at low temperatures, spin waves are to be treated as free bosons. This means that the average number of spin waves of wave vector $\mathbf{k}$ at temperature T is given by the Bose distribution function

$$\langle a^{\dagger}(\mathbf{k}) a(\mathbf{k}) \rangle_{\mathrm{T}} = \bar{n}(\mathbf{k}, T) = \{\exp[\epsilon(\mathbf{k})/KT] - 1\}^{-1} \tag{2.4.35}$$

where $\epsilon(\mathbf{k})$ is given by (2.4.16). If an external field or dipole–dipole interactions are included, we must use (2.4.24) instead, but we will neglect such effects here.

2.4.5 *Spontaneous Magnetization, Specific Heat, and Thermal Conductivity*

The first application of spin wave theory will be to determine the temperature dependence of the spontaneous magnetization. As in Section 2.3 we let

$$\sigma(T) = M(T)/M_0$$

where $M_0 = Ng\beta S$ is the maximum value for the magnetization, and is obtained at $T = 0°\text{K}$. The magnetization at temperature T is

$$M(T) = g\beta\langle \sum_i S_{iz}\rangle_{\text{T}} = g\beta\langle (NS - \sum_l a_l^\dagger a_l)\rangle_{\text{T}} \tag{2.4.36a}$$

where $\langle\cdots\rangle_{\text{T}}$ indicates a thermal average. Thus

$$\sigma(T) = 1 - \langle(1/NS) \sum_l a_l^\dagger a_l\rangle_{\text{T}}. \tag{2.4.36b}$$

Spin wave operators are introduced according to (2.4.31):

$$\begin{aligned}\sigma(T) &= 1 - (1/N^2S)\langle \sum_{\mathbf{kq}} \sum_l \exp[i(\mathbf{k} - \mathbf{q})\cdot\mathbf{R}_l]a^\dagger(\mathbf{k})a(\mathbf{q})\rangle_{\text{T}} \\ &= 1 - (1/NS) \sum_{\mathbf{k}} \langle a^\dagger(\mathbf{k})a(\mathbf{k})\rangle_{\text{T}}.\end{aligned} \tag{2.4.37}$$

In order to perform the computation in (2.4.33), we replace the sum by an integral in accord with the usual rule

$$(1/N)\sum_{\mathbf{k}} \to [\Omega/(2\pi)^3]\int d^3k.$$

Thus we require

$$\begin{aligned}&(1/NS)\sum_{\mathbf{k}} \langle a^\dagger(\mathbf{k})a(\mathbf{k})\rangle_{\text{T}} \\ &\qquad = [\Omega/(2\pi)^3S]\int d^3k\{\exp[\epsilon(\mathbf{k})/KT] - 1\}^{-1}.\end{aligned} \tag{2.4.38}$$

If dipolar interactions are involved, a more complicated equation for σ results (Holstein and Primakoff, 1940). The integral in (2.4.34) is actually rather difficult because it is restricted to the Brillouin zone, whose detailed shape must be taken into account and because the actual energy expression (2.4.16) must be used. This means that numerical computation is required.

However, at temperatures which are very low compared to the Curie temperature, these complications may be neglected. For such temperatures, the principal contribution to the integral will come from small values of $\epsilon(\mathbf{k})$, and this implies small $\mathbf{k}$. The energy function $\epsilon(\mathbf{k})$ given by (2.4.16) is expanded in powers of $\mathbf{k}$, and only the leading term is retained.

$$\epsilon(\mathbf{k}) = S \sum_l J(\mathbf{R}_l)(\mathbf{k}\cdot\mathbf{R}_l)^2. \tag{2.4.39}$$

The value of the sum depends on the exchange integrals and the particular lattice structure involved. For simplicity, we confine our attention to cubic lattices. Then sum over all the atoms which are in the nth shell from the origin. These have the same value of J:

$$\sum_{n_{\text{nbs}}} J(\mathbf{R}_l)(\mathbf{k}\cdot\mathbf{R}_l)^2 = \tfrac{1}{3}Z_n k^2 \mid \mathbf{R}_n \mid^2 J_n$$

where n_{nbs} denotes nth neighbors, Z_n is the number of nth neighbors, and J_n is the exchange parameter for these neighbors. Thus (2.4.39) becomes

$$\epsilon(\mathbf{k}) = \tfrac{1}{3}k^2 S \sum_n Z_n J_n \mid \mathbf{R}_n \mid^2 \tag{2.4.40}$$

where the sum now runs over the different sets of neighbors.

The integral in (2.4.38) can now be evaluated using (2.4.40). An answer whose accuracy is consistent with the use of the expansion (2.4.40) is obtained by replacing the Brillouin zone by a sphere of equal volume, whose radius is k_m:

$$4\pi k_m{}^3/3 = (2\pi)^3/\Omega.$$

Then we have

$$\int d^3k\{\exp[\epsilon(\mathbf{k})/KT] - 1\}^{-1}$$
$$= 4\pi \int_0^{k_m} k^2\, dk\, [\exp(Ak^2/KT) - 1]^{-1}$$

where

$$A = \tfrac{1}{3}S \sum_n Z_n J_n \mid \mathbf{R}_n \mid^2. \tag{2.4.41}$$

Now introduce dimensionless variables $x = Ak^2/KT$. The integral becomes

$$2\pi(KT/A)^{3/2} \int_0^{x_m} x^{1/2}\, dx/(e^x - 1)$$

where $x_m = Ak_m{}^2/KT$. For the very low temperatures we are considering, the upper limit may be made infinite. The errors are of order $\exp(-x_m)$

which is negligible. The integral is then

$$\int_0^\infty x^{1/2}/(e^x - 1)\,dx = \Gamma(\tfrac{3}{2})\zeta(\tfrac{3}{2})$$

where ζ is the Riemann zeta function

$$\zeta(s) = \sum_{n=1}^{\infty} n^{-s}.$$

Thus, finally,

$$\sigma(T) = 1 - (\Omega/4\pi^2 S)(KT/A)^{3/2}\Gamma(\tfrac{3}{2})\zeta(\tfrac{3}{2}). \qquad (2.4.42)$$

The essential result is that the spontaneous magnetization decreases with temperature at $T^{3/2}$ when T is small. This result is rigorous and contrasts with molecular field theory which predicts an exponentially small decrease of the spontaneous magnetization for small T. Equation (2.4.42) can be written in a somewhat simpler form if we consider a cubic crystal (simple, body-centered, or face-centered) with interactions between nearest neighbor spins only. In each case

$$A = 2JSa^2 \qquad (2.4.43)$$

where the subscripts have been dropped and a is the cubic lattice constant. Also, $\Omega = a^3/n$, where $n = 1, 2, 4$ in simple-cubic, body-centered, and face-centered cubic crystals, respectively. We then find

$$\sigma(T) = 1 - (1/nS)(KT/8\pi JS)^{3/2}\zeta(\tfrac{3}{2}). \qquad (2.4.44)$$

Corrections to this simple result are of higher order in temperature. One source is additional terms (k^4, k^6, . . .) in the expansion of $\epsilon(\mathbf{k})$ in powers of k. It is easy to see that each additional power of k^2 in ϵ contributes an additional power of T in σ. There result additional contributions to σ of order $T^{5/2}$, $T^{7/2}$, etc. The coefficients of these terms can be computed in a rather straightforward way, but we shall not do this. When such terms are important, it is probably necessary to consider the actual shape of the Brillouin zone, and this requires a numerical calculation.

Additional corrections arise from the interaction of spin waves with each other. Dyson (1956b) showed that the leading contribution from this effect to the magnetization is of order T^4.

Next, we calculate the specific heat of the Heisenberg ferromagnet at low temperatures. This proceeds quite similarly. The internal energy of the crystal U at temperature T is

$$U = E_0 + \sum_{\mathbf{k}} \epsilon(\mathbf{k})\bar{n}(\mathbf{k}, T) \qquad (2.4.45)$$

where E_0 is the energy of the ground state. The sum is transformed into an integral in the usual manner

$$U = E_0 + [N\Omega/(2\pi)^3] \int d^3k\, \epsilon(\mathbf{k})/\{\exp[\epsilon(\mathbf{k})/KT] - 1\}. \tag{2.4.46}$$

As is the case for the magnetization, a simple result is obtained in the low temperature limit. We use the expansion (2.4.36), and find with the same substitutions,

$$\begin{aligned} U &= E_0 + (N\Omega/2\pi^2)A \int k^4\, dk/[\exp(Ak^2/KT) - 1] \\ &= E_0 + (N\Omega/4\pi^2)A(KT/A)^{5/2} \int_0^\infty [x^{3/2}/(e^x - 1)]\, dx \\ &= E_0 + (N\Omega A/4\pi^2)(KT/A)^{5/2}\Gamma(\tfrac{5}{2})\zeta(\tfrac{5}{2}). \end{aligned} \tag{2.4.47}$$

The internal energy departs from the ground state energy by terms of order $T^{5/2}$. The specific heat is

$$C = dU/dT = NK(5\Omega/8\pi^2)(KT/A)^{3/2}\Gamma(\tfrac{5}{2})\zeta(\tfrac{5}{2}). \tag{2.4.48}$$

For the case of cubic crystals with nearest neighbor interactions, (2.4.48) becomes

$$C = NK(15/4n)(KT/8\pi JS)^{3/2}\zeta(\tfrac{5}{2}) \tag{2.4.49}$$

[and $n = 1, 2$, or 4 as specified just below (2.4.43)].

The specific heat of a Heisenberg ferromagnet at low temperatures is seen to be proportional to $T^{3/2}$. This compares with the specific heat due to lattice vibrations proportional to T^3. The difference is due to the fact that $\epsilon(\mathbf{k}) \approx k^2$ for spin waves while $\epsilon(\mathbf{k}) \approx k$ for phonons. Consequently, for small k, less energy is required to excite spin waves than phonons, leading to a weaker T dependence of the specific heat at low temperatures. Thus, in an insulating ferromagnet, spin wave excitation dominates the specific heat at low temperatures. This effect is particularly marked if $T_c \ll \Theta$. (Θ is the Debye temperature), and makes possible an estimation of J. The insulating ferromagnet EuS is a good example (McCollum and Callaway, 1962; Charap and Boyd, 1964).

Corrections to the $T^{3/2}$ dependence of the specific heat may be obtained as described in connection with the magnetization. Specifically, the next term is of order $T^{5/2}$ and results from inclusion of the k^4 term in $\epsilon(\mathbf{k})$. The contribution from spin wave interactions is of order T^4.

Since spin wave excitation may dominate the specific heat, it is natural also to inquire about the thermal conductivity. In order to estimate this we will follow the simple approach of Section 1.8 and assume the existence of a time of relaxation $\tau(\mathbf{k})$ for spin waves of wave vector $\mathbf{k}$. The thermal conductivity can be obtained by generalizing the simple formula

$$\kappa = \tfrac{1}{3}Cvl \tag{2.4.50}$$

where C is the specific heat, v the velocity, and l the mean free path, so that one considers a sum over all the spin wave modes

$$\kappa = \tfrac{1}{3}\sum_{\mathbf{k}} C(\mathbf{k})v(\mathbf{k})l(\mathbf{k}) = [N\Omega/3(2\pi)^3]\int C(\mathbf{k})v(\mathbf{k})l(\mathbf{k})\, d^3k. \tag{2.4.51}$$

In this formula $C(\mathbf{k})$ is the contribution of spin waves of wave vector $\mathbf{k}$ to the specific heat:

$$\begin{aligned} C(\mathbf{k}) &= dU(\mathbf{k})/dT = d/dT[\epsilon(\mathbf{k})/(e^{\epsilon(\mathbf{k})/KT} - 1)] \\ &= K(\epsilon/KT)^2[1/(e^{\epsilon/KT} - 1)^2]. \end{aligned} \tag{2.4.52}$$

The other terms in (2.4.46) are the group velocity $\hbar v(\mathbf{k}) = \nabla_{\mathbf{k}}\epsilon(\mathbf{k})$ and the mean free path $l(\mathbf{k})$. Spin waves may be scattered by other spin waves, by phonons, by impurities, by free electrons, by grain boundaries, etc. Since experimental information on spin wave thermal conductivity is sparse, we will consider here only the simplest possibility, that of a constant mean free path, $l(\mathbf{k}) = L_s$.

The thermal conductivity per unit volume κ_V is then

$$\kappa_V = [L_sK/3(2\pi)^3\hbar]\int d^3k\ (\epsilon/KT)^2[e^{\epsilon/KT}/(e^{\epsilon/KT} - 1)^2]\, d\epsilon/dk.$$

We will evaluate this integral in the same approximations as used for the magnetization and the specific heat. The usual substitutions lead to

$$\begin{aligned} \kappa_V &= (L_sK/6\pi^2)[(KT)^2/A\hbar]\int_0^\infty [x^3e^x/(e^x - 1)^2]\, dx \\ &= (L_sK/\pi^2)[(KT)^2/A]\zeta(3). \end{aligned} \tag{2.4.53}$$

The thermal conductivity is proportional to T^2 at low temperatures in the approximation of a constant mean free path. This contrasts with the $T^{3/2}$ behavior one might expect from the elementary formula (2.4.50). The difference results from the fact that v is not constant as may be assumed for phonons at low temperatures. Let us compare the spin wave thermal conductivity from (2.4.53) with the corresponding result for phonons, assuming a constant mean free path L_{ph} in that case,

$$\kappa_{\text{ph}} = (2K\pi^2/15)(L_{\text{ph}}/v_s^2)(KT/\hbar)^3,$$

in which v_s is the appropriately averaged speed of sound. The ratio of the thermal conductivities is

$$K_{\text{ph}}/K_{\text{sw}} = [2\pi^4/15\zeta(3)][KTA/(\hbar v_s)^2](L_{\text{ph}}/L_s). \tag{2.4.54}$$

If the mean free path of phonons and spin waves were nearly equal in a material at low temperature (as might occur if boundary scattering was dominant for both), we see that spin wave conduction would be more important. Unfortunately, experimental information is meager.

The calculated magnetization, specific heat, and thermal conductivity are sensitive to the presence of an external magnetic field. Since an external field produces a gap in the spin wave spectrum by introducing a minimum energy for the excitation of a spin wave, one expects a field-dependent reduction in the specific heat and thermal conductivity, and an increase in the magnetization. Estimates of these effects can be made by using the spin wave energy as given by (2.4.26) in the integrals; for more accurate work, one must use (2.4.24). At low temperatures [but not so low that (2.4.27) should be used for the energies], one finds an exponential dependence of those quantities on $(g\beta H)/KT$ where here $H = H_0 + 4\pi M_0/3$. The dependence of specific heat and thermal conductivity on magnetic field furnishes a useful means to determine experimentally whether spin waves are the dominant excitations. For example, measurements of the field dependence of the specific heat of EuS are in substantial agreement with theory (Passenheim *et al.*, 1966).

2.4.6 *Spin Waves in Antiferromagnets*

We now turn to a consideration of the spin wave spectrum of an antiferromagnet. This problem is somewhat more difficult than for ferromagnets because, as was pointed out in Section 2.3, the ground state of the Heisenberg Hamiltonian for negative (antiferromagnetic) exchange parameters is not known exactly. The following calculations are based on the two-sublattice model discussed in Section 2.3. When the magnetic order is more complicated than this, as frequently is the case, the calculation of the spin wave spectrum is correspondingly more complex. We will consider only the two-sublattice case and neglect dipole–dipole interactions here. The reader who wishes a more complete treatment should consult the book by Turov (1965). It is also useful to consult a fundamental paper by Anderson (1952), on which much of the present treatment is based. A modern treatment has been presented by Harris *et al.* (1971).

We begin with a two-sublattice model and try to work from there to the physical ground and excited states. Let $\mathbf{S}_a$ represent any atomic spin on one sublattice and $\mathbf{S}_b$ any spin on the other. All of the a spins are up in the two-sublattice state, and all of the b spins are down, that is,

$$\mathbf{S}_a^+ \mid 0\rangle = \mathbf{S}_b^- \mid 0\rangle = 0. \tag{2.4.55}$$

The Hamiltonian we consider is the usual Heisenberg Hamiltonian H_{H} supplemented by an external field. In the case of an antiferromagnet, it is customary to introduce a fictitious magnetic field which has different directions on each sublattice. If the spins are in the positive z direction on sublattice a and in the negative z direction on sublattice b, we add a term to H_{H} which is

$$-g\beta H_{\mathrm{A}} \sum_a S_a{}^z + g\beta H_{\mathrm{A}} \sum_b S_b{}^z. \tag{2.4.56a}$$

The field H_{A} is called the "anisotropy field." Its function is to stabilize the assumed sublattice arrangement of the spins. The total Hamiltonian that we consider is then

$$H = -\sum_{m,n} J_{mn}\mathbf{S}_m\cdot\mathbf{S}_n - g\beta H_{\mathrm{A}} \sum_a S_a{}^z + g\beta H_{\mathrm{A}} \sum_b S_b{}^z. \tag{2.4.56b}$$

The anisotropy field is a useful device, but it is not very satisfying from a more fundamental point of view. However, (2.4.56b) is to be regarded as only an approximation to the full spin Hamiltonian referred to in Section 2.1, and which will be discussed in more detail in Chapter 3. There are in this anisotropy terms resulting from crystal field effects which have consequences similar to (2.4.56a) in that the spins have preferred orientations with respect to the crystal axes. This problem will not be discussed here. Instead we will use a phenomenological anisotropy field. The exchange interaction is assumed to be limited to nearest neighbors; none of the nearest neighbors of a given atom being nearest neighbors of each other.

Two sets of creation and annihilation operators for spin deviations at an atomic site are introduced. Consider first atom i on sublattice a:

$$S^+{}_{ai} \to (2S)^{1/2}a_i, \qquad S^-{}_{ai} \to (2S)^{1/2}a_i{}^\dagger, \qquad S^z{}_{ai} \to S - a_i{}^\dagger a_i. \tag{2.4.57}$$

The Holstein–Primakoff substitution would be more satisfactory in terms of commutation relations, but we should have to approximate it by (2.4.57). For atom j on sublattice b we put

$$S^+{}_{bj} \to (2S)^{1/2}b_j{}^\dagger, \qquad S^-{}_{bj} \to (2S)^{1/2}b_j, \qquad S^z{}_{bj} \to b_j{}^\dagger b_j - S. \tag{2.4.58}$$

The operators a, b satisfy the usual commutation relations (2.4.28) and the a's and b's commute with each other:

$$[a_i, b_j] = [a_i^\dagger, b_j] = [a_i, b_j^\dagger] = 0.$$

These operators are substituted into the Hamiltonian (2.4.56b). Next, we introduce a transformation to spin wave variables through

$$a(\mathbf{k}) = (2/N)^{1/2} \sum_i \exp(i\mathbf{k}\cdot\mathbf{R}_i) a_i,$$

$$b(\mathbf{k}) = (2/N)^{1/2} \sum_j \exp(-i\mathbf{k}\cdot\mathbf{R}_j)\, b_j, \tag{2.4.59}$$

and the corresponding conjugates. The summations are restricted to the sublattices a and b, respectively, each of which contain $N/2$ atoms.

The expressions (2.4.58) and (2.4.59) are substituted into the Hamiltonian (2.4.56b). Fourth-order terms are neglected. The result is

$$H = E_0 + \sum_k \{(g\beta H_A - 2JSZ)[a^\dagger(\mathbf{k})a(\mathbf{k}) + b^\dagger(\mathbf{k})b(\mathbf{k})] - 2JS\gamma(\mathbf{k})[a(\mathbf{k})b(\mathbf{k}) + a^\dagger(\mathbf{k})b^\dagger(\mathbf{k})]\}. \tag{2.4.60}$$

Here Z is the number of nearest neighbors,

$$E_0 = N(JS^2Z - g\beta H_A S), \tag{2.4.61}$$

and

$$\gamma(\mathbf{k}) = \sum_{nn} \exp(i\mathbf{k}\cdot\mathbf{\Delta}) \tag{2.4.62}$$

where $\mathbf{\Delta}$ is a vector connecting an atom with one of its nearest neighbors, and the sum includes all such vectors. It will be observed that unlike the ferromagnet, the Hamiltonian is not diagonal after the substitution of the spin wave operators. It can be diagonalized, however, by a simple transformation.

The transformation amounts to the determination of new creation and annihilation operators $A(\mathbf{k})$, $B(\mathbf{k})$ which are linear combinations of the $a(\mathbf{k})$, $b(\mathbf{k})$:

$$a(\mathbf{k}) = \alpha(\mathbf{k})A(\mathbf{k}) + \beta(\mathbf{k})B^\dagger(\mathbf{k}), \qquad b(\mathbf{k}) = \beta(\mathbf{k})A^\dagger(\mathbf{k}) + \alpha(\mathbf{k})B(\mathbf{k}). \tag{2.4.63a}$$

The coefficients in the combination α, β are real. The operators A, B must satisfy the same commutation relations as the a, b:

$$[A(\mathbf{k}), A^\dagger(\mathbf{q})] = [B(\mathbf{k}), B^\dagger(\mathbf{q})] = \delta_{\mathbf{kq}}, \qquad [A(\mathbf{k}), B(\mathbf{q})] = 0. \tag{2.4.63b}$$

This imposes one condition on the α, β:

$$\alpha^2(\mathbf{k}) - \beta^2(\mathbf{k}) = 1. \tag{2.4.64}$$

We substitute (2.4.62) into (2.4.60). The result is

$$\begin{aligned} H = E_0' + \sum_{\mathbf{k}} \{ & [A^\dagger(\mathbf{k})A(\mathbf{k}) \\ & + B^\dagger(\mathbf{k})B(\mathbf{k})][(\alpha^2 + \beta^2)(g\beta H_A - 2JSZ) - 4\alpha\beta JS\gamma(\mathbf{k})] \\ & + [A^\dagger(\mathbf{k})B^\dagger(\mathbf{k}) + A(\mathbf{k})B(\mathbf{k})][2\alpha\beta(g\beta H_A - 2JSZ) \\ & - 2(\alpha^2 + \beta^2)JS\gamma(\mathbf{k})]\} \end{aligned} \tag{2.4.65}$$

with

$$E_0' = E_0 + 2\sum_{\mathbf{k}} [\beta^2(\mathbf{k})(g\beta H_A - 2JSZ) - 2JS\gamma(\mathbf{k})\alpha(\mathbf{k})\beta(\mathbf{k})]. \tag{2.4.66}$$

The coefficients α, β are now chosen so that the off-diagonal term in (2.4.65) vanishes:

$$2\alpha\beta(g\beta H_A - 2JSZ) - (\alpha^2 + \beta^2)2JS\gamma(\mathbf{k}) = 0. \tag{2.4.67}$$

Equations (2.4.64) and (2.4.66) must be satisfied simultaneously. The solutions are

$$\begin{aligned} \alpha(\mathbf{k}) &= \rho(\mathbf{k})[\rho^2(\mathbf{k}) - 4J^2S^2\gamma^2(\mathbf{k})]^{-1/2}, \\ \beta(\mathbf{k}) &= 2JS\gamma(\mathbf{k})[\rho(\mathbf{k})^2 - 4J^2S^2\gamma^2(\mathbf{k})]^{-1/2} \end{aligned} \tag{2.4.68}$$

where

$$\rho(\mathbf{k}) = (g\beta H_A - 2JSZ) + [(g\beta H_A - 2JSZ)^2 - 4J^2S^2\gamma^2(\mathbf{k})]^{1/2}. \tag{2.4.69}$$

These results are substituted back into the expression for H. The result may be expressed as

$$H = E_0' + \sum_{\mathbf{k}} \epsilon(\mathbf{k})[A^\dagger(\mathbf{k})A(\mathbf{k}) + B^\dagger(\mathbf{k})B(\mathbf{k})]. \tag{2.4.70}$$

The Hamiltonian has now been diagonalized. The spin wave excitation energy $\epsilon(\mathbf{k})$ is given by

$$\begin{aligned} \epsilon(\mathbf{k}) &= [\alpha^2(\mathbf{k}) + \beta^2(\mathbf{k})](g\beta H_A - 2JSZ) - 4\alpha(\mathbf{k})\beta(\mathbf{k})JS\gamma(\mathbf{k}) \\ &= [(g\beta H_A - 2JSZ)^2 - 4J^2S^2\gamma^2(\mathbf{k})]^{1/2}. \end{aligned} \tag{2.4.71a}$$

The negative sign n in Eq. (2.4.71a) suggests that we might obtain an imaginary $\epsilon(\mathbf{k})$; however, J is negative for an antiferromagnet, $J = -|J|$, so that

$$\epsilon(\mathbf{k}) = [(g\beta H_A + 2\,|\,J\,|\,SZ)^2 - 4J^2S^2\gamma^2(\mathbf{k})^2]^{1/2}. \qquad (2.4.71b)$$

Since $\gamma(\mathbf{k}) \leq Z$, $\epsilon(\mathbf{k})$ is always real. As $\mathbf{k} \to 0$, $\gamma(\mathbf{k}) \to Z$. Thus

$$\epsilon(0) = [g\beta H_A(g\beta H_A + 4JSZ)]^{1/2}. \qquad (2.4.72a)$$

A finite energy is required to excite a spin wave of $\mathbf{k} = 0$. This is a consequence of the introduction of the anisotropy field. The quantity $g\beta H_A$ is called the *anisotropy energy* ϵ_A and $2JSZ$ the exchange energy ϵ_X, so that

$$\epsilon(0) = [\epsilon_A(\epsilon_A + 2\epsilon_X)]^{1/2}. \qquad (2.4.72b)$$

Next, consider $k > 0$. We suppose for simplicity that the nearest neighbors have cubic symmetry, so that, to second order

$$\gamma(\mathbf{k}) = Z - \mathbf{k}^2a^2 \qquad (2.4.73)$$

where a is the cubic lattice constant. Thus, to this order,

$$\epsilon(\mathbf{k}) = [(g\beta H_A)^2 + 4(g\beta H_A)\,|\,J\,|\,SZ + 8J^2S^2Zk^2a^2]^{1/2}. \qquad (2.4.74a)$$

This is interesting in the limit that $JSZka \gg gH_A$. Provided that this limit is compatible with $ka \ll 1$, this simplifies to

$$\epsilon(\mathbf{k}) \approx (8Z)^{1/2}JSka. \qquad (2.4.74b)$$

Unlike the ferromagnet, an antiferromagnet has an approximately linear $\epsilon(\mathbf{k})$ curve for small k, similar to that encountered in the lattice vibration problem. The analogy indicates that the specific heat and the thermal conductivity will both be proportional to T^3 at low temperatures. However, if the temperature is low enough so that $KT < g\beta H_A$, there will be an exponential dependence of these quantities due to the existence of a gap in the energy spectrum.

In the present picture, each unit cell of the crystal contains two atoms; there are $N/2$ such unit cells, and as many allowed values of $\mathbf{k}$: $(N/2)$. There are two spin wave modes of the same energy for each value of $\mathbf{k}$. One mode corresponds to the A operators [the term $A^\dagger(\mathbf{k})A(\mathbf{k})$ in (2.4.70)], the other to the B operators. These modes are independent but degenerate excitations; so that there are N spin wave modes in all.

The zero point energy E_0' can be computed from (2.4.61) and (2.4.66). This becomes, with the aid of (2.4.71a),

$$E_0' = NJZS(S+1) - Ng\beta H_A(S+\tfrac{1}{2}) + \sum_{\mathbf{k}} \epsilon(\mathbf{k}). \qquad (2.4.75)$$

Let us look at this in an approximation in which H_A is neglected. This gives for $\epsilon(\mathbf{k})$ [remember $\epsilon(\mathbf{k})$ must be positive although J is negative]

$$\epsilon(\mathbf{k}) = -2JSZ[1 - (1/Z^2)\gamma^2(\mathbf{k})]^{1/2}.$$

In this limit, we have

$$E_0' = NJZS\{(S+1) - (2/N)\sum_{\mathbf{k}}[1-\gamma^2(\mathbf{k})/Z^2]^{1/2}\}. \qquad (2.4.76a)$$

In order to compare with the result of Anderson (1952), we rewrite (2.4.76a) in the form

$$E_0' = NJZS^2(1+\lambda/S) \qquad (2.4.76b)$$

where

$$\lambda = 1 - (2/N)\sum_{\mathbf{k}}[1-\gamma^2(\mathbf{k})/Z^2]^{1/2}. \qquad (2.4.76c)$$

The sum can be converted to an integral over a Brillouin zone whose volume corresponds to $N/2$ values of $\mathbf{k}$. To obtain accurate results, the integral must be done numerically. For a simple cubic lattice, Anderson finds $\lambda = 0.097$.

The diagonalization of the Hamiltonian (2.4.60) led to a ground state energy E_0', which we have just calculated. The character of the ground state is also different from the two-sublattice model insofar as the transformation (2.4.63) leads to a ground state in which the spins on each sublattice are not perfectly aligned. Let us consider the z component of the total spin of all atoms on sublattice a. Call this $\mathcal{S}_a^z$:

$$\begin{aligned}\mathcal{S}_a^z &= \sum_{i \text{ on } a} S_i^z = \sum_i (S - a_i^\dagger a_i)\\ &= (N/2)[S - (2/N)\sum a^\dagger(\mathbf{k})a(\mathbf{k})]\\ &= (N/2)\{S - (2/N)\sum_{\mathbf{k}}\{\alpha^2(\mathbf{k})A^\dagger(\mathbf{k})A(\mathbf{k}) + \beta^2(\mathbf{k})B(\mathbf{k})B^\dagger(\mathbf{k})\\ &\quad + \alpha(\mathbf{k})\beta(\mathbf{k})[A(\mathbf{k})B(\mathbf{k}) + A^\dagger(\mathbf{k})B^\dagger(\mathbf{k})]\}. \end{aligned} \qquad (2.4.77)$$

This operator is not diagonal. However, we may find the expectation value in the ground state readily enough and this will describe the situation at $T = 0$. Only the term $BB^\dagger$ contributes from the summation, and this gives

$$\langle \mathcal{S}_a^z \rangle = (N/2)\{S - (2/N)\sum_{\mathbf{k}}\beta^2(\mathbf{k})\}. \qquad (2.4.78)$$

A similar result is obtained for sublattice B. Evaluation is facilitated by writing $\beta^2 = \frac{1}{2}(\alpha^2+\beta^2) - \frac{1}{2}$. To compute this, we use Eq. (2.4.68) and consider only the case $H_A = 0$. Let ΔS_a^z represent the fractional departure of the spins on sublattice a from perfect alignment, that is,

$$\begin{aligned}\Delta S_a^z &= (2/N)[(N/2)S - \langle \mathcal{S}_a^z\rangle]\\ &= -\tfrac{1}{2} + (2/N)\sum_{\mathbf{k}}\tfrac{1}{2}[\alpha^2(\mathbf{k}) + \beta^2(\mathbf{k})]\\ &= \tfrac{1}{2}\{-1 + (2/N)\sum_{\mathbf{k}}[1-\gamma^2(\mathbf{k})/Z^2]^{-1/2}\}. \end{aligned} \qquad (2.4.79)$$

The sum may be converted to an integral as over the Brillouin zone. Again, it must be evaluated numerically. For a simple cubic lattice, Anderson obtains $\Delta S_a{}^z = 0.078$.

The preceding calculations have yielded spin wave dispersion relations containing only a single branch. The reader who recalls the discussion of lattice vibrations in Chapter 1 may wonder why we have not obtained spin wave spectra with more than one branch. In fact, such spectra are found when the magnetic ordering is more complicated than we have assumed. In general there should be as many branches in the spin wave spectrum as there are magnetic sublattices. Under some circumstances, as in the simple antiferromagnet we have discussed, the two branches are degenerate, but this is usually not the case. For example, one may consider magnetite (Fe_3O_4). When ordered, the magnetic structure involves six interpenetrating face-centered cubic lattices and gives a spin wave spectrum with six branches (Glasser and Milford, 1963).

2.4.7 *Absorption of Light by a System of Coupled Spins*

This discussion of spin waves will be concluded by considering the interactions between spin waves and electromagnetic radiation. This has been a topic of considerable recent interest. Both absorption and scattering of light by spin waves have been observed; in addition so-called *magnon side bands* may be found in the vicinity of other transitions involving a more obvious change of electronic state on the part of magnetic atoms in a crystal (See, for instance, Sell, 1968; Fleury and Porto, 1968; Moriya, 1968; Loudon, 1968), and references contained therein]. In general, photons, magnons, and phonons will be coupled to form mixed propogating modes (Huberman *et al.*, 1972). In this section, only a single process will be described: the absorption of light accompanied by the emission of one or more magnons. The treatment follows that of Moriya. Such absorption is observed in antiferromagnets of the rutile structure, in particular, MnF_2 (Allen *et al.*, 1966).

The explanation of the absorption cannot be entirely based on the Heisenberg Hamiltonian, as this does not include any coupling between spins and an external electric field. The observed transitions are electric dipole in character and it is necessary to suppose the existence of an effective interaction Hamiltonian between spins and the radiation field of the form

$$H_{\mathrm{R}} = -\boldsymbol{\mathcal{E}}\cdot\mathbf{P}_{\mathrm{eff}} \tag{2.4.80}$$

where $\boldsymbol{\mathcal{E}}$ is the incident electric field and $\mathbf{P}_{\mathrm{eff}}$ is a dipole moment operator which is a function of the arrangement of spins in the system and is expressed as a combination of spin operators involving (1) single atomic sites, (2) pairs of sites, and so on:

$$\mathbf{P}_{\text{eff}} = \sum_j \mathbf{P}_j + \sum_{jl} \mathbf{P}_{jl} + \cdots \tag{2.4.81}$$

(j, l refer to atomic sites). The operators P_j, P_{jl}, ... are assumed to be quadratic in the spins; thus if α, β, γ denote Cartesian components, general expressions are

$$P_j^\alpha = \sum_{\beta\gamma} K_j^{\alpha\beta\gamma} S_j^\beta S_j^\gamma$$

$$P^\alpha{}_{jl} = \Pi^\alpha{}_{jl} \mathbf{S}_j \cdot \mathbf{S}_l + \sum_{\beta\gamma} \Gamma_{jl}{}^{\alpha,\beta\gamma}(S_j^\beta S_l^\gamma + S_j^\gamma S_l^\beta)$$

$$+ \sum_\beta d_{jl}{}^{\alpha,\beta}[\mathbf{S}_j \times \mathbf{S}_l]^\beta. \tag{2.4.82}$$

The quantities K, Π, Γ, d are tensors which can, in principle, be calculated, but are most conveniently regarded as phenomenological parameters, which are not zero unless symmetry considerations intervene (as they often do). It turns out that the term involving $\mathbf{S}_j \cdot \mathbf{S}_l$ appears to be most important in compounds like MnF_2 (where $\mathbf{P}_j$ is required to vanish by symmetry), and we will concentrate our attention on this.

Let us consider briefly how a term coupling spins can arise, as described by Tanabe *et al.* (1965). We are concerned with a process in which a pair of atoms flip their spins. For simplicity, let us suppose that we have a single electron in state ϕ on each of two atoms, a and b, with either $\uparrow$ or $\downarrow$ spin possible. We are concerned with a process connecting an initial state $|\phi_{b\uparrow}\phi_{a\downarrow}\rangle$ to $|\phi_{a\uparrow}\phi_{b\downarrow}\rangle$. Since two electrons are involved and since the interaction between at atomic system and an external electromagnetic field is represented by a sum of one-particle operators, it is evidently necessary to consider a second-order process. The electromagnetic field can combine with an exchange matrix element of the coulomb interaction between electrons, for example. This gives a matrix element of the form

$$\langle \phi_{a\uparrow}\phi_{b\downarrow} | \mathbf{P}_{\text{eff}} | \phi_{b\uparrow}\phi_{a\downarrow} \rangle$$

$$= \sum_\mu \frac{\langle \phi_{a\uparrow}\phi_{b\downarrow} | \mathbf{P} | \phi_{\mu\uparrow}\phi_{b\downarrow} \rangle}{E(\phi_a) - E(\phi_\mu)} \langle \phi_{\mu\uparrow}\phi_{b\downarrow} | V | \phi_{b\uparrow}\phi_{a\downarrow} \rangle + \text{interchanges.} \tag{2.4.83}$$

Here $\mathbf{P}$ is the actual electric dipole operator, ϕ_μ is any state on atoms a and b, and the interchanges include the exchange of the order of P and V plus the use of $|\phi_{a\uparrow}\phi_{\mu\downarrow}\rangle$ as the intermediate state. It turns out that if the arrangement of the functions ϕ_a and ϕ_b has a center of symmetry, $\mathbf{P}_{\text{eff}}$ as given by (2.4.83) vanishes.

It is convenient to introduce electron spin operators for the electrons in ϕ_a and ϕ_b so that an operator equivalent of (2.4.83) can be formed:

$$\mathbf{P}_{\text{eff}} = \mathbf{\Pi}_{ab}(\mathbf{s}_a \cdot \mathbf{s}_b). \tag{2.4.84}$$

Summation over occupied orbitals then gives the term involving $\mathbf{\Pi}_{jl}$ in (2.4.82).

We now return to (2.4.82) and retain only the term containing $\mathbf{\Pi}$. The creation and annihilation operators for spin deviations are substituted from (2.4.57) and (2.4.58). A constant term is discarded, yielding

$$\mathbf{P}_{\text{eff}} = S \sum_{jl} \mathbf{\Pi}_{jl}(-a_j^\dagger a_j + b_l^\dagger b_l + a_j b_l + a_j^\dagger b_l^\dagger). \tag{2.4.85}$$

Equation (2.4.85) contains terms involving the simultaneous creation (or absorption) of two magnons. This part ($-a_j^\dagger a_j + b_l^\dagger b_l$ terms can be neglected) gives, after substitution of (2.4.59)

$$\mathbf{P}_{\text{eff}} = (2S/N) \sum_{jl,\mathbf{kq}} \mathbf{\Pi}_{jl} \exp[i(\mathbf{q}\cdot\mathbf{R}_l - \mathbf{k}\cdot\mathbf{R}_j)]$$
$$\times [a(\mathbf{k})b(\mathbf{q}) + a^\dagger(\mathbf{q})b^\dagger(\mathbf{k})].$$

We should expect $\mathbf{\Pi}_{jl}$ to depend on $(\mathbf{R}_j - \mathbf{R}_l)$ only. Then it is useful to define

$$\mathbf{\Pi}(\mathbf{k}) = \sum_j \mathbf{\Pi}_{jl} \exp[-i\mathbf{k}\cdot(\mathbf{R}_j - \mathbf{R}_l)].$$

Equation (2.4.85) becomes

$$\mathbf{P}_{\text{eff}} = S \sum_{\mathbf{k}} \mathbf{\Pi}(\mathbf{k})[a(\mathbf{k})b(\mathbf{k}) + a^\dagger(\mathbf{k})b^\dagger(\mathbf{k})]. \tag{2.4.86}$$

We can now proceed to develop an expression for the absorption constant. The physical magnon operators $A(\mathbf{k})$, $B(\mathbf{k})$ are introduced by Eq. (2.4.63). In first order of time-dependent perturbation theory, Eqs. (2.4.80) and (2.4.86) contain terms which describe the absorption of a photon and the creation of two magnons. Let s be the photon wave vector and $\hbar\omega$ its energy. Since $s = \hbar\omega/c$ (where c is the speed of light) and since the photons of interest are in the infrared, s is very small compared to the dimensions of the Brillouin zone. Just as in the problem of the absorption of light by a vibrating lattice, discussed in Section 1.9, it is legitimate to neglect $\mathbf{s}$ entirely. The usual procedures of the semiclassical theory of radiation yield a formula for the transition rate W which specifies the number of transitions per unit time,

$$W = (2\pi/\hbar) \,|\langle \mathrm{f} \,|\, H_{\mathrm{R}} \,|\, \mathrm{i}\rangle|^2 \,\delta(E_{\mathrm{f}} - E_{\mathrm{i}}). \tag{2.4.87}$$

This formula applies to precisely specified initial and final states $|\,\mathrm{i}\rangle$, $|\,\mathrm{f}\rangle$. In the present case, these states are members of a continuous spectrum, so that it is necessary to sum (integrate) over a range of final states. The temperature is assumed to be zero. Then

$$W = (2\pi S^2/\hbar)\cdot\tfrac{1}{2}\sum_{\mathbf{k}} |\,\boldsymbol{\mathcal{E}}\cdot\boldsymbol{\Pi}(\mathbf{k})\,|^2\,\delta[\hbar\omega - 2\epsilon(\mathbf{k})]. \qquad (2.4.88)$$

The factor of $\frac{1}{2}$ multiplies the sum since the two-magnon state with magnon 1 of type "A" and magnon 2 of type "B" is not distinct from that with magnon 1 of type "B" and magnon 2 of type "A." The presence of the delta function indicates that the absorption will exhibit the characteristic features of the density of spin wave states.

The absorption constant α is defined as the ratio of the energy removed from the incident beam per unit volume and time to the incident flux

$$\alpha = (\hbar\omega) \times \frac{\text{number of transitions per unit volume and time}}{\text{incident flux}}. \qquad (2.4.89)$$

The energy flux is interpreted as the product of the energy density times the speed of flow. In cgs units, the energy density in the medium for an incident plane wave is $\epsilon\mathcal{E}^2/4\pi$ instantaneously, where ϵ is the dielectric constant or $\epsilon\mathcal{E}^2/8\pi$ when averaged over a cycle, ϵ now being the amplitude. The speed of flow is c/n, where n is the index of refraction. Note that $\epsilon = n^2$. Thus we find

$$\alpha = (4\pi^2 S^2\omega/cn\mathcal{V}) \sum_{\mathbf{k}} [\,|\,\boldsymbol{\mathcal{E}}\cdot\boldsymbol{\Pi}(\mathbf{k})\,|^2/\mathcal{E}^2]\,\delta[(\hbar\omega/2) - \epsilon(\mathbf{k})] \qquad (2.4.90)$$

in which $\mathcal{V}$ is the volume of the crystal. The density of spin wave states of energy E per unit volume, $G(E)$, is

$$G(E) = (1/\mathcal{V}) \sum_{\mathbf{k}} \delta[E - \epsilon(\mathbf{k})]. \qquad (2.4.91)$$

This function can be introduced into (2.4.90) if we average the matrix element over a surface of constant energy. The average of a quantity over such a surface is denoted by $\langle\cdots\rangle_E$. Equation (2.4.90) becomes

$$\alpha = (4\pi^2 S^2\omega/cn)G(\hbar\omega/2)(1/\mathcal{E}^2)\langle\,|\,\boldsymbol{\mathcal{E}}\cdot\boldsymbol{\Pi}(\mathbf{k})\,|^2\rangle_{(\hbar\omega/2)}. \qquad (2.4.92)$$

To the extent to which it is legitimate to consider the average matrix element to be independent of energy, we see that the two-magnon absorption measures the density of spin wave states at an energy equal to one-half the photon energy.

2.5 Scattering of Slow Neutrons by Magnetically Ordered Systems

The scattering of slow neutrons by a system of atoms with magnetic moments furnishes an important means of obtaining information about the magnetic order in a crystal. Direct measurement of the energies and lifetimes of spin waves is possible. The disordering of spins near the Curie temperature can also be investigated. Detailed reviews of the theory have been given by DeGennes (1963) and Izyumov (1963). Our treatment will be based on the Heisenberg model for the coupling of atomic spins. Many of the results are generally valid. For a treatment based on an itinerant electron model, see Izuyama *et al.* (1963).

2.5.1 *The Interaction Hamiltonian*

The scattering process results from the interaction of the neutron and electron magnetic moments. For a single atom, the magnetic scattering is of the same general order of magnitude as nuclear scattering. The Hamiltonian $\mathcal{H}$ includes the energy of an electron of magnetic moment $\boldsymbol{\mu}_e$ in the magnetic field $\mathbf{H}_n$ produced by the neutron

$$\mathcal{H} = -\boldsymbol{\mu}_e \cdot \mathbf{H}_n. \tag{2.5.1}$$

The reader should consult the classic paper of Halpern and Johnson (1939) for a more comprehensive discussion of the basic approximations. Only the contribution of the electron spin to the electron magnetic moment is included. Let $\mathbf{r}$ be the separation of electron and neutron. The field produced by the neutron at the position of the electron is

$$\mathbf{H}_n = -\nabla \times [\boldsymbol{\mu}_n \times \nabla(1/r)] \tag{2.5.2}$$

where $\boldsymbol{\mu}_n$ is the neutron magnetic moment. This can be rewritten, using appropriate vector identities, as

$$\mathbf{H}_n = -\boldsymbol{\mu}_n \nabla^2(1/r) + (\boldsymbol{\mu}_n \cdot \nabla)[\nabla(1/r)].$$

However, $\nabla^2(1/r) = -4\pi\,\delta(\mathbf{r})$, so

$$\mathbf{H}_n = 4\pi\boldsymbol{\mu}_n\,\delta(\mathbf{r}) - (\boldsymbol{\mu}_n/r^3) + [3(\boldsymbol{\mu}_n \cdot \mathbf{r})\mathbf{r}/r^5].$$

This gives for the Hamiltonian

$$\mathcal{H} = (\boldsymbol{\mu}_e \cdot \boldsymbol{\mu}_n/r^3) - [3(\boldsymbol{\mu}_n \cdot \mathbf{r})(\boldsymbol{\mu}_e \cdot \mathbf{r})/r^5] - 4\pi\boldsymbol{\mu}_e \cdot \boldsymbol{\mu}_n\,\delta(\mathbf{r}). \tag{2.5.3}$$

It is important to note that the first two terms make a contribution at $r = 0$. Consider the integral of $\mathbf{H}_n$ through a sphere of very small radius.

This is

$$\int \mathbf{H}_n\, d^3r = -\int \nabla \times [\boldsymbol{\mu}_n \times \nabla(1/r)]\, d^3r = \int [\boldsymbol{\mu}_n \times \nabla(1/r)] \times d\mathbf{S}$$

where the last integral is over the surface of the sphere. Since $d\mathbf{S} = \mathbf{r}/\mid r \mid da$ where da is a scalar element of area, we obtain

$$\int \mathbf{H}_n\, d^3r = \boldsymbol{\mu}_n \int da/r^2 - \int \mathbf{r}(\boldsymbol{\mu}\cdot\mathbf{r})\, da/r^4 = 8\pi\boldsymbol{\mu}_n/3.$$

This remains finite as the sphere radius goes to zero. Thus, for $r = 0$, we have the contact interaction

$$\mathcal{H}_c = (-8\pi/3)\, \boldsymbol{\mu}_e\cdot\boldsymbol{\mu}_n\, \delta(\mathbf{r}), \tag{2.5.4}$$

whereas for $r \neq 0$, we use the dipole–dipole form

$$\mathcal{H}_d = \boldsymbol{\mu}_e\cdot\boldsymbol{\mu}_n/r^3 - 3(\boldsymbol{\mu}_n\cdot\mathbf{r})(\boldsymbol{\mu}_e\cdot\mathbf{r})/r^5. \tag{2.5.5}$$

The dipole–dipole term $\mathcal{H}_d$ was already mentioned in Section 2.4 as contributing to the interaction of atomic spins on different atoms; the contact term is of vital importance in the theory of hyperfine structure. It is also permissible to use (2.5.3) as long as the $r \to 0$ limit is treated carefully.

2.5.2 *Magnetic Scattering by a Single Atom*

Let us now consider the scattering of the neutron by a single atom using this Hamiltonian. The transition matrix element is computed in the Born approximation. The initial state of the system has the neutron in a plane wave state of wave vector $\mathbf{k}_i$ (coordinates $\mathbf{x}_n$) and spin state $\mid m_i\rangle$ (with respect to some axis of quantization). The state of the atom is described by a wave function $\psi_i(\mathbf{x}_e, \mathbf{s}_e)$ where $\mathbf{x}_e$, $\mathbf{s}_e$ represent the totality of electron position and spin coordinates. Similarly, in the final state we have the neutron with wave vector $\mathbf{k}_f$, spin state $\mid m_f\rangle$, and atomic state $\psi_f(\mathbf{x}_e, \mathbf{s}_e)$. Thus we require the matrix element

$$\langle \mathrm{f} \mid \mathcal{H} \mid \mathrm{i}\rangle = \int d^3x_n\, d\tau_e \exp[-i(\mathbf{k}_f - \mathbf{k}_i)\cdot\mathbf{x}_n] \times \psi_f^*(\mathbf{x}_e, \mathbf{s}_e)\langle m_f \mid H \mid m_i\rangle\psi_i(\mathbf{x}_e, \mathbf{s}_e). \tag{2.5.6}$$

Summation over electron spin coordinates is understood.

The Hamiltonian in (2.5.6) is the sum of terms like (2.5.3) referring to the interaction between each electron and the neutron. We use the form

$$\mathcal{H} = \sum_j \mathcal{H}(\mathbf{r}_{jn}) = -\sum_j [(\boldsymbol{\mu}_n \cdot \nabla_j)(\boldsymbol{\mu}_{ej} \cdot \nabla_j)(1/r_{jn}) + 4\pi \boldsymbol{\mu}_{ej} \cdot \boldsymbol{\mu}_n \, \delta(\mathbf{r}_{jn})] \tag{2.5.7}$$

where $\boldsymbol{\mu}_{ej}$ is the magnetic moment operator for the jth electron and $\mathbf{r}_{jn} = \mathbf{x}_j - \mathbf{x}_n$ is the position of electron j with respect to the neutron. Equation (2.5.6) is rewritten as (with $\mathbf{K} = \mathbf{k}_f - \mathbf{k}_i$).

$$\langle f \mid \mathcal{H} \mid i \rangle = \langle m_f \mid \sum_j \int d\tau_e \, \psi_f^*(\mathbf{x}_e, \mathbf{s}_e) \int d^3x_n \exp(-i\mathbf{K} \cdot \mathbf{x}_n) \times \mathcal{H}(\mathbf{r}_{jn}) \psi_i(\mathbf{x}_e, \mathbf{s}_e) \mid m_i \rangle.$$

The variables of integration in the integral over neutron coordinates are changed from $\mathbf{x}_n$ to $\mathbf{r}_{nj}$. Then consider a single term, which is

$$\int d^3x_n \exp(-i\mathbf{K} \cdot \mathbf{x}_n) \mathcal{H}(\mathbf{r}_{jn})$$

$$= \exp(-i\mathbf{K} \cdot \mathbf{x}_j) \int d^3r_{jn} \exp(i\mathbf{K} \cdot \mathbf{r}_{jn}) \mathcal{H}(\mathbf{r}_{jn})$$

$$= -\exp(-i\mathbf{K} \cdot \mathbf{x}_j) \int d^3r_{jn} \exp(i\mathbf{K} \cdot \mathbf{r}_{jn}) [(\boldsymbol{\mu}_n \cdot \nabla_j)(\boldsymbol{\mu}_e \cdot \nabla_j)(1/r_{jn}) + 4\pi \boldsymbol{\mu}_{ej} \cdot \boldsymbol{\mu}_n \, \delta(\mathbf{r}_{jn})]$$

$$= \exp(-i\mathbf{K} \cdot \mathbf{x}_j) \{ -4\pi \boldsymbol{\mu}_{ej} \cdot \boldsymbol{\mu}_n + (\boldsymbol{\mu}_n \cdot \mathbf{K})(\boldsymbol{\mu}_{ej} \cdot \mathbf{K}) \int d^3r_{jn} \exp(i\mathbf{K} \cdot \mathbf{r}_{jn})(1/r_{jn}) \}$$

$$= -4\pi \exp(-i\mathbf{K} \cdot \mathbf{x}_j) [\boldsymbol{\mu}_{ej} \cdot \boldsymbol{\mu}_n - [(\boldsymbol{\mu}_n \cdot \mathbf{K})(\boldsymbol{\mu}_{ej} \cdot \mathbf{K})/K^2]]. \tag{2.5.8}$$

We have integrated by parts twice. It is convenient to define an operator **M** as follows:

$$\mathbf{M}(\mathbf{K}) = \sum_j \exp(-i\mathbf{K} \cdot \mathbf{x}_j) \boldsymbol{\mu}_{ej}. \tag{2.5.9}$$

The basic matrix element $\langle f \mid \mathcal{H} \mid i \rangle$ now becomes

$$\langle f \mid \mathcal{H} \mid i \rangle = -4\pi \langle m_f \mid \int d\tau_e \, \psi_f^*(\mathbf{x}_e, \mathbf{s}_e) \times [\mathbf{M} \cdot \boldsymbol{\mu}_n - [(\mathbf{M} \cdot \mathbf{K})(\boldsymbol{\mu}_n \cdot \mathbf{K})/K^2]] \psi_i(\mathbf{x}_e, \mathbf{s}_e) \mid m_i \rangle. \tag{2.5.10}$$

Further simplification is possible if the atomic transition does not involve a change of electronic state (apart from rearrangement of the spins). This is the usual case, as we are concerned with neutrons whose energy is low compared to the separation between different atomic states. Let **J** be the total angular momentum of the state considered. Its eigenvalues are $j(j+1)$. The states ψ_f, ψ_i are assumed to have the same j, but different values of m_j: $m_j^{(f)}$, $m_j^{(i)}$. Then it is a standard result for any vector operator **T** which obeys the commutation rule

$$[J_\alpha, T_\beta] = i\epsilon_{\alpha\beta\gamma}T_\gamma \tag{2.5.11}$$

that (Condon and Shortley, 1964)

$$j(j+1)(m_j^{(f)} \mid \mathbf{T} \mid m_j^{(i)}) = (m_j^{(f)} \mid \mathbf{J} \mid m_j^{(i)})(m_j^{(i)} \mid \mathbf{J}\cdot\mathbf{T} \mid m_j^{(i)}). \tag{2.5.12}$$

The operator **M** defined above satisfies (2.5.11) so that

$$\begin{aligned}\int \psi_f(\mathbf{x}_e, \mathbf{s}_e)[\mathbf{M}\cdot\boldsymbol{\mu}_n - [(\mathbf{M}\cdot\mathbf{K})(\boldsymbol{\mu}_n\cdot\mathbf{K})/K^2]]\psi_i(\mathbf{x}_e, \mathbf{s}_e)\, d\tau_e \\ = [1/j(j+1)]\langle m_j^{(f)} \mid \mathbf{J}\cdot\boldsymbol{\mu}_n \\ - [(\mathbf{J}\cdot\mathbf{K})(\boldsymbol{\mu}_n\cdot\mathbf{K})/K^2] \mid m_j^{(i)}\rangle\langle m_j^{(i)} \mid \mathbf{J}\cdot\mathbf{M} \mid m_j^{(i)}\rangle.\end{aligned} \tag{2.5.13}$$

It is frequently permissible to replace **J** by **S** and j by s in the matrix element since the orbital part of the angular momentum is usually quenched in the solid. We have separated the matrix element into one part involving the coupled neutron and atomic spins (and with no integration over atomic space coordinates) and another part, independent of the neutron spin, containing the electron coordinates. This is the atomic form factor $F(\mathbf{K})$. Define

$$\begin{aligned}(-e\hbar/mc)F(\mathbf{K}) &= [1/j(j+1)]\langle m_j^{(i)} \mid \mathbf{J}\cdot\mathbf{M} \mid m_j^{|i)}\rangle \\ &= [1/j(j+1)]\int d\tau_e\, \psi_i^*(\mathbf{x}_e, \mathbf{s}_e) \sum_j \exp(-i\mathbf{K}\cdot\mathbf{x}_j) \\ &\quad \times \boldsymbol{\mu}_{ej}\cdot\mathbf{J}\psi_i(\mathbf{x}_e, \mathbf{s}_e).\end{aligned} \tag{2.5.14}$$

The factor $e\hbar/mc$ is included explicitly so that $F(\mathbf{K})$ will be dimensionless (here **J** and **S** have dimensionless eigenvalues and e is positive). If we replace **J** by **S** as mentioned above, and consider only the contribution of spin to $\boldsymbol{\mu}$, we have

$$F(\mathbf{K}) = [1/s(s+1)]\int d\tau_e\psi_i^*(\mathbf{x}_e, \mathbf{s}_e)[\sum_j \exp(-i\mathbf{K}\cdot\mathbf{x}_j)]\mathbf{s}_j\cdot\mathbf{S}\psi_i(\mathbf{x}_e, \mathbf{s}_e). \tag{2.5.15}$$

The form factor is normalized so that $F(0) = 1$. We have put $\boldsymbol{\mu}_{ej} = -(e\hbar/mc)\mathbf{s}_j$; and now write $\boldsymbol{\mu}_n = -(ge\hbar/Mc)\mathbf{s}_n$. Then we have

$$\langle \mathrm{f} \mid \mathcal{H} \mid \mathrm{i} \rangle = -(4\pi g e^2 \hbar^2/mMc^2) F(\mathbf{K}) \langle m_j^{(\mathrm{f})} m_j \mid \mathbf{s}_n \cdot \mathbf{s} - [(\mathbf{S}\cdot\mathbf{K})(\mathbf{s}_n\cdot\mathbf{K})/K^2] \mid m_j^{(\mathrm{i})} m_\mathrm{i} \rangle. \tag{2.5.16}$$

Here M is the mass of the neutron and $g\,(=1.91)$ the neutron magnetic moment (with the nuclear magneton factored out). Equation (2.5.16) presents the transition matrix element for spin flip scattering by a single atom.

2.5.3 *Scattering by Magnetic Ions in a Lattice*

We now turn to the case of a crystal composed of atoms with magnetic moments. The present treatment is based on that of Van Hove (1954). The atoms are located at equilibrium positions $\mathbf{R}_\mu$ (for simplicity, consider a monatomic crystal with one atom per unit cell), and neglect the lattice vibrations. In this case, for each atomic site $\mathbf{R}_\mu$, (2.5.6) is modified by referring the neutron coordinate to $\mathbf{R}_\mu$. The variable of integration is $\mathbf{x}_n - \mathbf{R}_\mu$. A factor $\exp(-i\mathbf{K}\cdot\mathbf{R}_\mu)$ appears outside the integral which now depends on μ through the atomic spin operators $\mathbf{S}_\mu$. We find instead of (2.5.16)

$$\langle \mathrm{f} \mid \mathcal{H} \mid \mathrm{i} \rangle = -(4\pi g e^2 \hbar^2/mMc^2) \sum_\mu \exp(-i\mathbf{K}\cdot\mathbf{R}_\mu) F(\mathbf{K}) \times \langle m_{j\mu}^{(\mathrm{f})} m_\mathrm{f} \mid \mathbf{s}_n\cdot\mathbf{S}_\mu - [(\mathbf{S}_\mu\cdot\mathbf{K})(\mathbf{s}_n\cdot\mathbf{K})/K^2] \mid m_{j\mu}^{(\mathrm{i})} m_\mathrm{i} \rangle. \tag{2.5.17}$$

The differential cross section is calculated by squaring (2.5.17); multiplying by $k_\mathrm{f}/k_\mathrm{i}(M/2\pi\hbar^2)^2$. A delta function is inserted to represent energy conservation.

Suppose that the neutron beam is unpolarized. Then it is necessary to average over the orientation of the neutron spin in the initial state. We also average over the initial atomic spin distribution at the temperature of measurement, and sum over the final spin orientations of the neutron and the atoms. The states of the atomic spin system are denoted by labels s_f, s_i:

$$\begin{aligned} d^2\sigma/d\Omega\, d\epsilon = {} & (2ge^2/mc^2)^2 (k_\mathrm{f}/k_\mathrm{i}) \mid F(\mathbf{K}) \mid^2 \sum_{\mathrm{i,f}} P_\mathrm{i} \langle s_\mathrm{i}, m_\mathrm{i} \mid \\ & \times \sum_\mu \exp(i\mathbf{K}\cdot\mathbf{R}_\mu) [\mathbf{s}_n\cdot\mathbf{S}_\mu - [(\mathbf{S}_\mu\cdot\mathbf{K})(\mathbf{s}_n\cdot\mathbf{K})/K^2]] \\ & \times \mid s_\mathrm{f}, m_\mathrm{f} \rangle \langle m_\mathrm{f}, s_\mathrm{f} \mid \sum_\nu \exp(-i\mathbf{K}\cdot\mathbf{R}_\nu) \\ & \times [\mathbf{s}_n\cdot\mathbf{S}_\nu - [(\mathbf{S}_\nu\cdot\mathbf{K})(\mathbf{s}_n\cdot\mathbf{K})/K^2]] \mid s_\mathrm{i}, m_\mathrm{i} \rangle\, \delta(E_\mathrm{f} - E_\mathrm{i}). \end{aligned} \tag{2.5.18}$$

Here, P_i is the probability of finding a particular initial state. We write the energy difference in (2.5.18) as

$$E_f - E_i = E_{sf} - E_{si} - \hbar\omega \tag{2.5.19a}$$

where

$$\hbar\omega = (\hbar^2/2M)(\mathbf{k}_i^2 - \mathbf{k}_f^2). \tag{2.5.19b}$$

Here E_{sf}, E_{si} are the energies of the final and initial states of the atomic spin system, and $\hbar\omega$ is the energy lost or gained by the neutron.

The neutron spins can be eliminated. Since a complete set of neutron spin states is involved, the neutron spin operators in the two matrix elements may be combined. Averages of the combination $s_n^\alpha s_n^\beta$ where α and β are rectangular coordinates, must be considered. It is easy to see, using the spin density matrix formalism (Dicke and Wittke, 1960) that such an average gives $\frac{1}{4}\,\delta_{\alpha\beta}$. We now rewrite (2.5.18) in a conventional form

$$d^2\sigma/d\Omega\, d\epsilon = (ge^2/mc^2)^2(k_f/k_i)(1/\hbar)\,|\,F(\mathbf{K})\,|^2 \times \sum_{\alpha\beta}[\delta_{\alpha\beta} - (K_\alpha K_\beta/K^2)]S_{\alpha\beta}(\mathbf{K},\omega) \tag{2.5.20}$$

where

$$S_{\alpha\beta}(\mathbf{K},\omega) = \sum_{i,f} P_i \langle s_i \,|\, \sum_\mu \exp(i\mathbf{K}\cdot\mathbf{R}_\mu) S_\mu^\alpha \,|\, s_f\rangle \times \langle s_f \,|\, \sum_\nu \exp(-i\mathbf{K}\cdot\mathbf{R}_\nu) S_\nu^\beta \,|\, s_i\rangle\, \delta[\omega - (E_{sf} - E_{si})/\hbar]. \tag{2.5.21}$$

From this point, our attention focuses on the function $S_{\alpha\beta}(\mathbf{K},\omega)$, which can be expressed in terms of a correlation function in a fashion similar to that of Section 1.6. The Fourier transform of $S_{\alpha\beta}$ denoted by $\Gamma_{\alpha\beta}$ is introduced:

$$\Gamma_{\alpha\beta}(\mathbf{K},t) = \int_{-\infty}^{\infty} e^{i\omega t} S_{\alpha\beta}(\mathbf{K},\omega)\, d\omega. \tag{2.5.22a}$$

The inverse relation is

$$S_{\alpha\beta}(\mathbf{K},\omega) = (1/2\pi)\int_{-\infty}^{\infty} \Gamma_{\alpha\beta}(\mathbf{K},t) e^{-i\omega t}\, dt. \tag{2.5.22b}$$

Equation (2.5.21) is substituted into (2.5.22a), which yields

$$\Gamma_{\alpha\beta}(\mathbf{K},t) = \sum_{if} P_i \exp(-iE_{si}t/\hbar)\langle s_i \,|\, \sum_\mu \exp(i\mathbf{K}\cdot\mathbf{R}_\mu) S_\mu^\alpha \,|\, s_f\rangle \times \exp(iE_{sf}t/\hbar)\langle s_f \,|\, \sum_\nu \exp(-i\mathbf{K}\cdot\mathbf{R}_\nu) S_\nu^\beta \,|\, s_i\rangle. \tag{2.5.23}$$

It is convenient to introduce time-dependent atomic spin operators in the Heisenberg picture by

$$S_\mu{}^\alpha(t) = \exp(-iHt/\hbar)\, S_\mu{}^\alpha \exp(iHt/\hbar), \qquad S_\mu{}^\alpha(0) = S_\mu{}^\alpha. \tag{2.5.24}$$

Here H is the Hamiltonian of the spin system. Note that

$$\langle s_\mathrm{i} \mid \sum_\mu \exp(i\mathbf{K}\cdot\mathbf{R}_\mu) S_\mu{}^\alpha(t) \mid s_\mathrm{f}\rangle$$
$$= \exp(-iE_{s\mathrm{i}}t/\hbar) \langle s_\mathrm{i} \mid \sum_\mu \exp(i\mathbf{K}\cdot\mathbf{R}_\mu) S_\mu{}^\alpha(0) \mid s_\mathrm{f}\rangle \exp(iE_{s\mathrm{f}}t/\hbar).$$

Introduction of the time-dependent operators enables us to absorb the exponential factors. The sum over the complete set of final states of the spin system $\mid s_\mathrm{f}\rangle$ gives

$$\Gamma_{\alpha\beta}(\mathbf{K}, t) = \sum_\mathrm{i} P_\mathrm{i} \langle s_\mathrm{i} \mid \sum_{\mu\nu} \exp[i\mathbf{K}\cdot(\mathbf{R}_\mu - \mathbf{R}_\nu)] S_\mu{}^\alpha(t) S_\nu{}^\beta(0) \mid s_\mathrm{i}\rangle. \tag{2.5.25}$$

The factor P_i expresses the probability that a particular initial state of energy E_i will occur at temperature T:

$$P_\mathrm{i} = \exp(-E_{s\mathrm{i}}/KT)/Z \tag{2.5.26}$$

where Z is the partition function of the spin system

$$Z = \sum_i \exp(-E_{s\mathrm{i}}/KT). \tag{2.5.27}$$

It is apparent that $\Gamma_{\alpha\beta}(\mathbf{K}, t)$ is the thermal average of a spin correlation function. We denote this by $\langle\cdots\rangle_\mathrm{T}$ as in Section 1.6. The formulas may be simplified in either of two ways. First, we note that the summation in (2.5.25) involves $\mathbf{R}_\mu - \mathbf{R}_\nu$ only. Thus we may set $\mathbf{R}_\nu = 0$ and multiply the result by a factor of N:

$$\Gamma_{\alpha\beta}(\mathbf{K}, t) = N \sum_\mu \exp(i\mathbf{K}\cdot\mathbf{R}_\mu) \langle S_\mu{}^\alpha(t) S_0{}^\beta(0)\rangle_\mathrm{T}. \tag{2.5.28}$$

In this form, we see that the correlation between the spin at site 0 and time zero and that at site μ and time t is involved. Alternately, we may introduce Fourier transformed spin operators

$$S^\alpha(\mathbf{K}, t) = N^{-1/2} \sum \exp(i\mathbf{K}\cdot\mathbf{R}_\mu) S_\mu{}^\alpha(t), \tag{2.5.29}$$

etc. We then obtain

$$\Gamma_{\alpha\beta}(\mathbf{K}, t) = N \langle S^\alpha(\mathbf{K}, t) S^\beta(-\mathbf{K}, 0)\rangle_\mathrm{T}. \tag{2.5.30}$$

To extract some results from this formalism, consider first the simplest case: that of uncorrelated paramagnetic ions. This situation obtains both if a small concentration of magnetic ions is embedded in an inert host, and at temperatures very high compared to the ordering temperature in a material which is magnetically ordered at low temperatures. In this case, all states of the spin system are equally probable, and there is no correlation between spins on different sites. There is then effectively no coupling between spins and, in the absence of external fields, no way in which the neutron can transfer energy to the spin system. The scattering will be elastic and we must have

$$S_{\alpha\beta}(\mathbf{K}, \omega) = \tfrac{1}{3}S(S+1)\,\delta_{\alpha\beta}\,\delta(\omega). \tag{2.5.31}$$

We integrate over energies and sum over α, β using

$$\sum_{\alpha\beta} \delta_{\alpha\beta}[\delta_{\alpha\beta} - (K_\alpha K_\beta/K^2)] = 2.$$

Integration over ω then gives

$$d\sigma/d\mu = \tfrac{2}{3}(gr_0)^2 S(S+1)\,|\,F(\mathbf{K})\,|^2 \tag{2.5.32}$$

where $r_0 = e^2/mc^2$ is the classical electron radius. The scattering depends on the angle through the atomic form factor $F(\mathbf{K})$, and the process may therefore be used to measure the form factor. In this way it is possible to determine experimentally the distribution of magnetically active electrons in a paramagnetic atom or ion.

We will now consider elastic scattering from a ferromagnetic crystal. An inelastic component will also be present due to the excitation or absorption of spin waves by the neutrons. This will be discussed subsequently. A physical argument determines the elastic scattering: Suppose the spins in the ferromagnet are aligned along an axis $\hat{\boldsymbol{\sigma}}$. For large times, the correlation function will be zero except for α, $\beta = \sigma$. In this case, each S will contribute its thermal average at the relevant temperature. We then have a time-independent component of Γ:

$$\begin{aligned}\Gamma_{\alpha\beta}(\mathbf{K}) &= N\langle S\rangle_{\mathrm{T}}^2\,\delta_{\alpha\sigma}\,\delta_{\beta\sigma}\sum_{\mu}\exp(i\mathbf{K}\cdot\mathbf{R}_\mu)\\ &= N\langle S\rangle_{\mathrm{T}}^2[(2\pi)^3/\Omega]\,\delta_{\alpha\sigma}\,\delta_{\beta\sigma}\sum_{l}\delta(\mathbf{K}-\mathbf{K}_l)\end{aligned} \tag{2.5.33}$$

where the $\mathbf{K}_l$ are reciprocal lattice vectors and $\langle S\rangle_{\mathrm{T}}$ is the thermal average of an atomic spin at temperature T. This result is inserted into (2.5.22b), and a factor of $\delta(\omega)$ is obtained from the integration. Finally, the result is inserted into (2.5.20), and an integration over energy is performed. We find

$$d\sigma/d\Omega = (gr_0)^2 N[(2\pi)^3/\Omega]\langle S\rangle_{\mathrm{T}}^2 \sum_{l} |\,F(\mathbf{K})\,|^2[1 - \hat{\mathbf{K}}\cdot\hat{\boldsymbol{\sigma}})^2]\,\delta(\mathbf{K}-\mathbf{K}_l) \tag{2.5.34}$$

where $\hat{\boldsymbol{\sigma}}$ is a unit vector in the direction of the magnetization, and $\hat{\mathbf{K}}$ is a unit vector in the direction of $\mathbf{K}$. The result describes magnetic Bragg scattering: it occurs for the same reciprocal lattice vectors as in the case of nuclear scattering.

The results for ferromagnetic ordering can be readily generalized to more complicated spin arrangements. The essential result is that peaks in the scattering can occur for reciprocal lattice vectors characteristic of the actual magnetic order. For example, suppose that antiferromagnetic order is present in a system with only one kind of magnetic ion, and that this order can be described on a sublattice model. We apply (2.5.33) with two modifications:

(1) $$\sum_{\mu} \exp(i\mathbf{K}\cdot\mathbf{R}_{\mu}) = [(2\pi)^3/\Omega_{\mathrm{m}}] \sum_{l} \delta(\mathbf{K}-\mathbf{K}_l)$$

where Ω_{m} is the volume of the magnetic unit cell and the $\mathbf{K}_l$ refer to the magnetic order rather than to the ordinary crystal lattice;

(2) It is necessary to take account of the different spin directions associated with the different sublattices. If this is described by vectors $\hat{\boldsymbol{\sigma}}_i$ ($i = 1, \ldots, n$), where n is the number of sublattices, we have, instead of (2.5.34),

$$d\sigma/d\Omega = (gr_0)^2 N[(2\pi)^3/\Omega_{\mathrm{m}}]\langle S\rangle_{\mathrm{T}}^2 \times \sum_{i,l} |F(\mathbf{K})|^2 [1-(\hat{\mathbf{K}}\cdot\hat{\boldsymbol{\sigma}}_{\mathrm{i}})^2]\, \delta(\mathbf{K}-\mathbf{K}_l). \tag{2.5.35}$$

Neutron scattering is commonly used to determine the particular magnetic order existing in an antiferromagnet.

Lattice vibrations affect magnetic scattering. This will not be discussed in detail. The essential result is that lattice vibrations reduce the scattering cross section (Izyumov, 1963). This may be accounted for by inserting into the summations in (2.5.33) and (2.5.35) [and multiplying (2.5.32) by] a factor $\exp[-2W(\mathbf{K})]$, which is the temperature dependent Debye–Waller factor computed in Section 1.6.

In addition to the elastic scattering described above, there will be an inelastic component which results from those events in which a neutron looses or gains energy from the spin system. At temperatures low compared to the Curie temperature, such processes can be described in terms of the emission and absorption of spin waves. We will conclude this section with a description of this process in ferromagnets following the treatment of Izyumov (1963).

Let us suppose that, as usual, the equilibrium direction of the spins is the z axis. The sum over α and β in Eq. (2.5.20) is a fairly complicated object, but if we retain only the largest components of $S_{\alpha\beta}$, it becomes rather simple. The components to retain include S_{zz}, which gives the elastic scattering, and the components S_{+-} and S_{-+} which correspond to one-quantum emission and absorption. The cross section is expressed in terms of correlation functions by

$$d^2\sigma/d\Omega\, d\epsilon = (ge^2/mc^2)^2(k_f/k_i) \mid F(\mathbf{K}) \mid^2 (1/2\pi) \int e^{-i\omega t}\{[1 - (\hat{\mathbf{K}}\cdot\hat{\boldsymbol{\eth}})^2]\Gamma_{zz} + \tfrac{1}{4}[1 + (\hat{\mathbf{K}}\cdot\hat{\boldsymbol{\eth}})^2](\Gamma_{+-} + \Gamma_{-+})\}\, dt. \tag{2.5.36}$$

We will now evaluate Γ_{+-}: This quantity is defined by (2.5.25)

$$\Gamma_{+-} = \sum_{\mu\nu} \exp[i\mathbf{K}\cdot(\mathbf{R}_\mu - \mathbf{R}_\nu)]\, \langle S_\mu^+(t) S_\nu^-(0)\rangle. \tag{2.5.37}$$

The operators appearing in (2.5.37) are related to spin wave operators through (2.4.27a) and (2.4.30):

$$S_\nu^-(0) = (2S/N)^{1/2} \sum_{\mathbf{p}} \exp(i\mathbf{p}\cdot\mathbf{R}_\nu) a^\dagger(\mathbf{p}).$$

Thus

$$\Gamma_{+-} = (2S/N) \sum_{\mu\nu,\mathbf{qp}} \exp[i(\mathbf{K} - \mathbf{q})\cdot\mathbf{R}_\mu] \exp[i(\mathbf{p} - \mathbf{K})\cdot\mathbf{R}_\nu] \times \langle \exp(-iHt/\hbar) a(\mathbf{q}) \exp(iHt/\hbar) a^\dagger(\mathbf{p})\rangle_T.$$

However,

$$\langle \exp(-iHt/\hbar) a(\mathbf{q}) \exp(iHt/\hbar) a^\dagger(\mathbf{p})\rangle = \exp[i\epsilon(\mathbf{p})t/\hbar]\, \delta_{\mathbf{q},\mathbf{p}}[1 + \bar{n}(\mathbf{p}, T)] \tag{2.5.38}$$

where $\epsilon(\mathbf{p})$ is the spin wave energy given by (2.4.13) and $\bar{n}(\mathbf{p}, T)$ is the average occupation number for spin waves of wave vector $\mathbf{p}$ at temperature T:

$$\bar{n}(\mathbf{p}, T) = \{\exp[\epsilon(\mathbf{p})/KT] - 1\}^{-1}. \tag{2.5.39}$$

Three of the sums are performed to give

$$\Gamma_{+-} = (2S)[(2\pi)^3/\Omega] \sum_{\mathbf{q},l} \delta(\mathbf{K} - \mathbf{q} - \mathbf{K}_l) \exp[i\epsilon(\mathbf{q})t/\hbar][1 + \bar{n}(\mathbf{q}, T)]. \tag{2.5.40}$$

Similarly we have

$$\Gamma_{-+} = (2S)[(2\pi)^3/\Omega] \sum_{\mathbf{q},l} \delta(\mathbf{K} + \mathbf{q} - \mathbf{K}_l) \exp[-i\epsilon(\mathbf{q})t/\hbar]\bar{n}(\mathbf{q}, T). \tag{2.5.41}$$

The time integrals yield a delta function for energy conservation;

$\delta[\omega \pm \epsilon(\mathbf{q})/\hbar]$. The sum of $\mathbf{q}$ is converted to an integral:

$$(1/N) \sum_{q} \rightarrow [\Omega/(2\pi)^3] \int d^3q.$$

The following result is obtained for the inelastic cross section (resulting from Γ_{-+} and Γ_{+-} only):

$$\begin{aligned} d^2\sigma/d\Omega\, d\epsilon = {} & (NS/2)(ge^2/mc^2)^2(1/\hbar)\, |F(\mathbf{K})|^2 \\ & \times [1 + (\hat{\mathbf{K}}\cdot\hat{\delta})^2] \sum_{l} \int d^3q(k_f/k_i) \\ & \times \{\delta[\omega - \epsilon(\mathbf{q})/\hbar]\, \delta(\mathbf{K} - \mathbf{q} - \mathbf{K}_l)\, [1 + \bar{n}(\mathbf{q}, T)] \\ & + \delta[\omega + \epsilon(\mathbf{q})/\hbar]\, \delta(\mathbf{K} + \mathbf{q} - \mathbf{K}_l)\, \bar{n}(\mathbf{q}, T)\}. \end{aligned} \tag{2.5.42}$$

The other correlation function Γ_{zz} is independent of time and gives the elastic scattering which we have already described in (2.5.33). We obtain the additional result that the quantity $\langle S\rangle_T$ used there is

$$\langle S\rangle_T = S - (1/N) \sum_{\mathbf{k}} \bar{n}(\mathbf{k}, T). \tag{2.5.43}$$

The physical content of (2.5.42) which describes one-quantum inelastic scattering may be summarized as follows: One should see peaks whose location is determined by the conservation laws. Since $\epsilon(\mathbf{K} + \mathbf{K}_l) = \epsilon(\mathbf{K})$, this implies

$$E_n(\mathbf{k}_i) - E_n(\mathbf{k}_f) = \pm\epsilon(\mathbf{K}) \tag{2.5.44}$$

where E_n is the neutron energy. These peaks are repeated at all values of $\mathbf{K} \pm \mathbf{K}_l$, where $\mathbf{K}_l$ is any reciprocal lattice vector. The intensity of each peak is governed both by the atomic form factor $F(\mathbf{K})$ and the Bose function $n(\mathbf{K}, T)$. Thus, at low temperature, emission of a spin wave will be far more probable than absorption.

Inelastic neutron scattering was first used by Brockhouse (1957) to study spin wave dispersion relations. Since that time, the technique has been refined, and results are now available for many important systems, including metallic ferromagnets (see, for instance, Minkiewicz *et al.*, 1969). We will not attempt to review experimental data here.

2.6 The Ising Model

2.6.1 *General Discussion*

In this section we will consider a model of a ferromagnetic system that is somewhat more simple and correspondingly less realistic than that furnished by the Heisenberg Hamiltonian. This model is obtained by considering only z components of the spin operators in the Heisenberg Hamiltonian; the x, y components are discarded. This means that our spins are classical quantities: there can be no problem of noncommutativity of operators when only one component is retained. Further, the spins are restricted to have two values ± 1: This is essentially a spin $\frac{1}{2}$ system, and the exchange parameters are redefined appropriately. Let σ_i, σ_j be classical variables taking the value ± 1. The Hamiltonian is

$$\mathcal{H} = -\tfrac{1}{2} \sum_{i<j} J(\,|\,\mathbf{R}_i - \mathbf{R}_j\,|\,)\sigma_i\sigma_j - H \sum_i \sigma_i \tag{2.6.1}$$

where

$$H = \tfrac{1}{2} g\beta H_{\mathrm{M}},$$

H_{M} being the magnetic field strength. It is conventional in the theory of the Ising model to rescale the exchange parameters to absorb the factor of $\frac{1}{2}$ in (2.6.1); thus put $J' = \frac{1}{2}J$; then drop the prime. We consider in detail only positive values of J, corresponding to ferromagnetic ordering. If J is negative, antiferromagnetism results. Unfortunately, the Hamiltonian is too complicated for complete solution unless certain additional and rather drastic restrictions are imposed on the model: Only nearest neighbor interactions are included and the lattice is assumed to be two-dimensional and rectangular. The Hamiltonian reduces to

$$\mathcal{H} = -J_1 \sum \sigma_{n,m}\sigma_{n+1,m} - J_2 \sum \sigma_{n,m}\sigma_{n,m+1} - H \sum \sigma_{n,m}. \tag{2.6.2}$$

The integers n, m, etc. refer to the nth row and mth column, respectively. The lattice has N rows and M columns; these quantities ultimately are allowed to become infinite in such a way that the ratio M/N remains constant. Periodic boundary conditions are required $\sigma_{n,M+1} = \sigma_{n,1}$ and $\sigma_{N+1,m} = \sigma_{1,m}$. The impositions of these boundary conditions is equivalent to wrapping the lattice on a torus.

The reader may question the value of a model so severely restricted. The system, though apparently simple, is actually soluble only with difficulty (and then only if $H = 0$). Interest focuses on the phase transition. The Ising model gives an exactly soluble example of a phase transition; and this is of great significance. In partial defense of the utility of the model despite the restrictions, one may observe that phase transitions are a common feature of bulk matter, appearing in all sorts of systems in which the interactions are quite different. Perhaps the details of the interaction are not too important. The Ising model has been adapted to describe order–disorder transitions (Domb, 1960), and has been used as a model for liquid–gas condensation (Lee and Yang, 1952). Certain magnetic salts, such as $Tb(OH)_3$ have such strongly anisotropic exchange as to resemble Ising ferromagnets (Scott and Wolf, 1969).

The calculation proceeds by evaluation of the partition function Z defined by

$$Z = \sum_{\sigma_{11}=\pm 1} \cdots \sum_{\sigma_{NM}=\pm 1} \exp[-\mathcal{H}(\sigma_{11}\cdots\sigma_{NM})/K_B T] \qquad (2.6.3)$$

(K_B is Boltzmann's constant).

The thermodynamic properties are found from derivatives of the partition function in a standard way. Determination of the partition function requires, in principle, knowledge of all the eigenvalues of the Hamiltonian, even though the present calculation of Z is not explicitly performed in that manner. Recall that in the case of the Heisenberg Hamiltonian, we were able to obtain exact energies for the ferromagnetic ground state and some of the low lying excited states. It would have been necessary to have solutions for states with an arbitrary number of spin deviations in order to describe the region around the transition temperature. Exact solutions for such states are not available at present for the Heisenberg Hamiltonian. We are therefore restricted to the much simpler Ising model.

The Ising model was introduced in 1925 (Ising, 1925). The exact solution for the two-dimensional case in the absence of a magnetic field was first obtained by Onsager (1944). Interest in the problem has remained strong to the present. Reviews have been published by Newell and Montroll (1953) and Domb (1960). We will follow the method of Schultz *et al.* (1964) in the evaluation of the partition function. The following notation is introduced:

$$K_1 = J_1/K_B T, \qquad K_2 = J_2/K_B T; \qquad \mathcal{H}' = H/K_B T. \qquad (2.6.4)$$

2.6.2 *The One-Dimensional Problem*

It is useful to consider first an Ising system in one dimension with periodic boundary conditions ($\sigma_1 = \sigma_{N+1}$). The techniques used to treat the two-dimensional case are easier to understand if they are first applied in one dimension. The partition function is

$$Z = \sum_{\sigma_1\ldots\sigma_N} \exp(K_1 \sum_n \sigma_n\sigma_{n+1}) \exp(\mathcal{H}' \sum \sigma_n). \qquad (2.6.5)$$

This multiple sum can be written as the trace of a matrix product: It is useful to double the number of variables by introducing a set $\sigma_1', \ldots, \sigma_N'$:

$$Z = \sum_{\sigma_1\ldots\sigma_n\sigma_1'\cdots\sigma_n'} \{\exp(\mathcal{H}'\sigma_1)\,\delta_{\sigma_1\sigma_1'}\exp(K_1\sigma_1'\sigma_2)$$
$$\times \exp(\mathcal{H}'\sigma_2)\,\delta_{\sigma_2\sigma_2'}\exp(K_1\sigma_2'\sigma_3)\cdots\}. \qquad (2.6.6)$$

Next, introduce two 2×2 matrices

$$(V_1)_{\sigma_i\sigma_j} = \exp(K_1\sigma_i\sigma_j), \qquad (V_2)_{\sigma_i\sigma_j} = \exp(\mathcal{H}'\sigma_i)\,\delta_{\sigma_i\sigma_j}. \tag{2.6.7}$$

The rows and columns of these matrices correspond to the different possible values (± 1) for σ_i and σ_j. Thus

$$V_1 = \begin{pmatrix} \exp K_1 & \exp(-K_1) \\ \exp(-K_1) & \exp K_1 \end{pmatrix}, \qquad V_2 = \begin{pmatrix} \exp \mathcal{H}' & 0 \\ 0 & \exp(-\mathcal{H}') \end{pmatrix}. \tag{2.6.8}$$

Then

$$Z = \operatorname{tr}(V_2V_1 \cdots V_2V_1) = \operatorname{tr}(V_2V_1)^N. \tag{2.6.9}$$

The value of the trace is unchanged if we replace V_2V_1 by the symmetric matrix V called the *transfer matrix*:

$$V = V_2^{1/2}V_1V_2^{1/2}. \tag{2.6.10a}$$

Thus

$$Z = \operatorname{tr}(V_2^{1/2}V_1V_2^{1/2})^N = \operatorname{tr}(V^N). \tag{2.6.10b}$$

An explicit expression for V is

$$V = \begin{pmatrix} \exp(K + \mathcal{H}') & \exp(-K) \\ \exp(-K) & \exp(K - \mathcal{H}') \end{pmatrix}. \tag{2.6.11}$$

The evaluation of the partition function is conveniently considered in terms of the eigenvalues of V. Let us denote these as Λ_1 and Λ_2, where $\Lambda_1 > \Lambda_2$. Then

$$Z = \Lambda_1^N + \Lambda_2^N = \Lambda_1^N[1 + (\Lambda_2/\Lambda_1)^N]. \tag{2.6.12}$$

As $N \to \infty$ the contribution of the second term goes to zero. The partition function is seen to be determined by the largest eigenvalue of the transfer matrix. In the present case, this is

$$\Lambda = e^K \cosh \mathcal{H}' + [e^{2K} \sinh^2 \mathcal{H}' + e^{-2K}]^{1/2}. \tag{2.6.13}$$

The free energy per spin is

$$F = -(KT/N) \ln Z = -KT \ln \Lambda_1 \tag{2.6.14}$$

(the last step applies in the limit $N \to \infty$). The magnetization M per spin is

$$M = -\partial F/\partial H_M = (g\beta/2)\, \partial \ln \Lambda_1/\partial\mathcal{H}'. \tag{2.6.15}$$

We find

$$M = (g\beta/2)e^K \sinh \mathcal{H}' \,\{e^{2K} \sinh^2\mathcal{H}' + e^{-2K}\}^{-1/2}. \tag{2.6.16}$$

For any temperature $T > 0$, zero magnetic field corresponds to $\mathcal{H}' = 0$. Equation (2.6.16) asserts that in this limit $M \to 0$, so that there is no spontaneous magnetization at finite temperatures. Thus the one-dimensional Ising system does not exhibit a phase transition. However, it is easy to see that the magnetic susceptibility

$$\chi = \partial M/\partial H_M = (g\beta/2KT)\, \partial M/\partial \mathcal{H}' \qquad (2.6.17)$$

approaches infinity for $H_M = 0$ as $T \to 0$ if K is positive, corresponding to a ferromagnetic interaction.

Note that if $H_M = 0$,

$$Z = 2^N \cosh^N K, \qquad (2.6.18a)$$

whereas if $K = 0$,

$$Z = 2^N \cosh N\mathcal{H}'. \qquad (2.6.18b)$$

These results are the same as would be obtained for a system of N non-interacting particles with two energy levels. However, if both H_M and $K \neq 0$, the problem is less trivial, as we have seen. Note that the absence of a phase transition in the one-dimensional Ising system is consistent with the general result that in a system in which the forces are short range, no phase exhibiting long-range order can exist (Landau and Lifshitz, 1958).

2.6.3 *Two Dimensions*

The two-dimensional case is much more difficult. Since this problem has not been solved otherwise, we will now put $H_M = 0$. It will first be shown that the partition function can be written in the form of (2.6.10) after suitable generalization of the elements involved.

The matrices V_1 and V_2 must be generalized. We continue to use V_1 to describe interactions within a given row (of M atoms), whereas V_2 will now be employed to describe the interaction of different rows with each other.

To this end, we introduce two sets of matrices τ_α and $\boldsymbol{\tau}$. The three matrices τ_α ($\alpha = x, y, z$) are just the usual 2×2 Pauli spin matrices. This notation, rather than the more usual σ, has been introduced in order to avoid confusion with the site spin matrices. The $\boldsymbol{\tau}$ are certain direct products involving the τ_α which will be defined in detail below. We first note that V_1 given by (2.6.7) can be expressed as

$$V_1 = \exp(K_1)I + \exp(-K_1)\tau_x = \exp(K_1)[I + \exp(-2K_1)\tau_x]$$

where I is the 2×2 unit matrix. Since $\tau_i^2 = I$ for each i, it follows that

$$\exp(a\tau_\alpha) = I \cosh a + \tau_\alpha \sinh a = \cosh a\, (I + \tau_\alpha \tanh a).$$

Now define k_1 by

$$\tanh k_1 = \exp(-2K_1). \tag{2.6.19}$$

We obtain

$$V_1 = (2/\sinh 2k_1)^{1/2} \exp(k_1\tau_x) = (2 \sinh 2K_1)^{1/2} \exp(k_1\tau_x) \tag{2.6.20}$$

where the identity

$$\sinh 2k_1 \sinh 2K_1 = 1 \tag{2.6.21}$$

has been used.

The Hamiltonian in the exponential of the partition function can now be rearranged so that one first sums over all the interactions between one row and the next. However, to describe the state of affairs in a given row we need a $(2M \times 2M)$-dimensional matrix instead of a 2×2. The matrices $\boldsymbol{\tau}$ are constructed for this purpose. Consider the jth site in a given row. We define, using the direct product notation,

$$\boldsymbol{\tau}_{j,\alpha} = I \times I \times \cdots \times I \times \tau_\alpha \times I \times \cdots \times I \tag{2.6.22}$$

in which $\alpha = x, y, z$; with τ_α in the jth spot and I as a 2×2 unit matrix. For a definition of the direct product of matrices, see Arfken (1970, p. 164). The generalized matrix $\mathbf{V}_1$ is the direct product of M matrices of the form (2.6.20) (Newell and Montroll, 1953):

$$\mathbf{V}_1 = (2 \sinh 2K_1)^{M/2} \exp(k_1\tau_x) \times \exp(k_1\tau_x) \times \cdots \times \exp(k_1\tau_x). \tag{2.6.23}$$

It is a theorem concerning the multiplication of direct product matrices that

$$(A_1 \times B_1)(A_2 \times B_2) = (A_1A_2) \times (B_1B_2) \tag{2.6.24}$$

(where $\times$ represents the direct product, and otherwise we have ordinary matrix multiplication). From this, it is easy to show that

$$\begin{aligned} \mathbf{V}_1 &= (2 \sinh 2K_1)^{M/2} \prod_j \exp(k_1\boldsymbol{\tau}_{j,x}) \\ &= (2 \sinh 2K_1)^{M/2} \exp(k_1 \sum_j \boldsymbol{\tau}_{j,x}). \end{aligned} \tag{2.6.25}$$

Now consider the interaction between adjacent rows. Given one particular row described by matrices $\boldsymbol{\tau}_{j,\alpha}$; let us denote the appropriate matrices for an adjacent row by $\boldsymbol{\tau}'_{j,\alpha}$. The interaction is such that the nonvanishing terms connecting the adjacent rows involve the same value of j. To describe the contribution of the interaction of two adjacent rows to the partition function, we introduce a matrix V_2:

$$V_2 = \exp(K_2 \sum \boldsymbol{\tau}_{j,z}\boldsymbol{\tau}'_{j,z}). \tag{2.6.26}$$

The partition function can now be written as in (2.6.9); we have

$$Z = \operatorname{tr}(V_2^{1/2} V_1 V_2^{1/2})^N = \operatorname{tr} V^N \tag{2.6.27}$$

where V is the transfer matrix for the two-dimensional Ising model.

2.6.4 *Transformation to Fermion Operators*

We now address the problem of determining the eigenvalues of V.

It is a consequence of the multiplication rule for direct product matrices that the matrices $\boldsymbol{\tau}_{j,\alpha}$ obey commutation rules of the same form as the τ_α (or the Pauli matrices σ). Thus, we may construct "raising" and "lowering" operators $\boldsymbol{\tau}_{j,\pm}$,

$$\boldsymbol{\tau}_{j,\pm} = \tfrac{1}{2}[\boldsymbol{\tau}_{j,x} \pm i\boldsymbol{\tau}_{j,y}], \tag{2.6.28}$$

which obey

$$[\boldsymbol{\tau}_{j,\pm}, \boldsymbol{\tau}_{l,\pm}] = 0 \qquad (j \neq l), \tag{2.6.29a}$$

$$\{\boldsymbol{\tau}_{j,+}, \boldsymbol{\tau}_{j,-}\} = 1, \tag{2.6.29b}$$

$$(\boldsymbol{\tau}_{j,+})^2 = (\boldsymbol{\tau}_{j,-})^2 = 0 \tag{2.6.29c}$$

where the brackets $[\cdots]$ designate the commutator and the braces $\{\ \}$ indicate the anticommutator: $\{A, B\} = AB + BA$.

Unfortunately, these rules are not characteristic of either a Boson system or a Fermion system. It is possible, however, to introduce a transformation to Fermion operators. Define

$$C_j = \left[\exp\left(\pi i \sum_{l=1}^{j-1} \boldsymbol{\tau}_{l,+}\boldsymbol{\tau}_{l,-}\right)\right]\boldsymbol{\tau}_{j,-}, \qquad C_j^+ = \left[\exp\left(\pi i \sum_{l=1}^{j-1} \boldsymbol{\tau}_{l,+}\boldsymbol{\tau}_{l,-}\right)\right]\boldsymbol{\tau}_{j,+}. \tag{2.6.30}$$

These operators obey the Fermion (anticommutation) rules

$$\{C_j^+, C_l\} = \delta_{jl}, \qquad \{C_j, C_l\} = \{C_j^+, C_l^+\} = 0. \tag{2.6.31}$$

To see that (2.6.31) is true, we first observe that the transformation has no effect on (2.6.29b) and (2.6.29c), which are already consistent with (2.6.31). The C's characterized by different site indices anticommute. We may suppose, without loss of generality, that $l < j$. Then we have, for instance,

$$C_j C_l^+ = \exp\left[\pi i\left(\sum_{k=1}^{j-1} + \sum_{k=1}^{l-1}\right)\boldsymbol{\tau}_{k,+}\boldsymbol{\tau}_{k,-}\right]\boldsymbol{\tau}_{j,-}\boldsymbol{\tau}_{l,+}.$$

However,

$$C_l^+ C_j = \exp\left\{\left[2\pi i \sum_{k=1}^{l-1} + \pi i \sum_{k=l+1}^{j-1}\right]\boldsymbol{\tau}_{k,+}\boldsymbol{\tau}_{k,-}\right\}\boldsymbol{\tau}_{l,+}\exp(\pi i \boldsymbol{\tau}_{l+}\boldsymbol{\tau}_{l-})\,\boldsymbol{\tau}_{j,-}$$

$$= \exp\left\{\left[2\pi i \sum_{k=1}^{l-1} + \pi i \sum_{k=l+1}^{j-1}\right]\boldsymbol{\tau}_{k,+}\boldsymbol{\tau}_{k,-}\right\}\boldsymbol{\tau}_{l,+}\boldsymbol{\tau}_{j,-} = -C_j C_l^+$$

where we have used $\boldsymbol{\tau}_{l,+}\exp(\pi i \boldsymbol{\tau}_{l,+}\boldsymbol{\tau}_{l,-}) = \boldsymbol{\tau}_{l,+}$, and the fact that since the sums in the exponentials in (2.6.30) do not contain $\boldsymbol{\tau}_j$, the exponentials commute with $\boldsymbol{\tau}_j$. Also,

$$C_j^+ C_j = \boldsymbol{\tau}_{j,+}\boldsymbol{\tau}_{j,-}. \tag{2.6.32}$$

The combination $C_j^+ C_j$ has eigenvalues 1 or 0 and may be considered as a number operator.

It then follows that the transformation can be inverted:

$$\boldsymbol{\tau}_{j,-} = \left[\exp \pi i \sum_{l=1}^{j-1} C_l^+ C_l\right] C_j, \qquad \boldsymbol{\tau}_{j,+} = \left[\exp \pi i \sum_{l=1}^{j-1} C_l^+ C_l\right] C_j^+. \tag{2.6.33}$$

The preceding derivation of V associated the matrices $\boldsymbol{\tau}_x$ with the usual Pauli matrix σ_x. However, the results for the partition function cannot be altered if we make the change $\boldsymbol{\tau}_{j,x} \to -\boldsymbol{\tau}_{j,z}$; $\boldsymbol{\tau}_{j,z} \to \boldsymbol{\tau}_{j,x}$. This amounts to a unitary transformation of the $\boldsymbol{\tau}$, and in effect diagonalizes $\boldsymbol{\tau}_x$, while $\boldsymbol{\tau}_z$ loses its diagonal form. We have instead of (2.6.25)

$$V_1 = (2 \sinh 2K_1)^{M/2} \exp[-2k_1 \sum_j (\boldsymbol{\tau}_{j,+}\boldsymbol{\tau}_{j,-} - \tfrac{1}{2})]$$

$$= (2 \sinh 2K_1)^{M/2} \exp[-2k_1 \sum_l (C_l^+ C_l - \tfrac{1}{2})]$$

where, in the last step, (2.6.32) has been used.

After considerable algebra involving the commutation rules, the following relations are obtained for $j < M$:

$$\boldsymbol{\tau}_{j,+}\boldsymbol{\tau}_{j+1,-} = C_j^+ C_{j+1}, \qquad \boldsymbol{\tau}_{j,-}\boldsymbol{\tau}_{j+1,-} = -C_j C_{j+1},$$

$$\boldsymbol{\tau}_{j,+}\boldsymbol{\tau}_{j+1,+} = C_j^+ C^+_{j+1}, \qquad \boldsymbol{\tau}_{j,-}\boldsymbol{\tau}_{j+1,+} = -C_j C^+_{j+1}. \tag{2.6.34}$$

Thus, if each row has free ends,

$$V_2 = \exp\left[K_2 \sum_1^{M-1} (C_j^+ - C_j)(C^+_{j+1} + C_{j+1})\right]. \tag{2.6.35}$$

It is, however, desirable to impose cyclic boundary conditions so that the system will acquire translational invariance. This complicates matters somewhat at present, because a bond between the last and the first spins is introduced. Then, instead of (2.6.34),

$$\begin{aligned} \tau_{M,+}\tau_{1,-} &= -(-1)^n C_M^+ C_1 \neq C_M^+ C_1, \\ \tau_{M,+}\tau_{1,+} &= -(-1)^n C_M^+ C_1^+ \neq C_M^+ C_1^+, \\ \tau_{M,-}\tau_{1,-} &= (-1)^n C_M C_1 \neq -C_M C_1 \end{aligned} \tag{2.6.36}$$

where n is the total number operator,

$$n = \sum_1^M \tau_{l,+}\tau_{l,-} = \sum_1^M C_l^+ C_l. \tag{2.6.37}$$

The expression for V_2 is now

$$V_2 = \exp\left\{K_2\left[\sum_{j=1}^{M-1}(C_j^+ - C_j)(C^+_{j+1} + C_{j+1}) - (-1)^n(C_M^+ - C_M)(C_1^+ - C_1)\right]\right\}. \tag{2.6.38}$$

In order to calculate the partition function, states of the system are classified according to the eigenvalues of the total number operator n. Because all terms in V_2 and V_1 are bilinear in the fermion operators, we may consider separately states of even or odd n.

When V_2 acts on even states, it is equivalent to the operator

$$V_2^+ = \exp\left[K_2 \sum_{j=1}^{M}(C_j^+ - C_j)(C^+_{j+1} + C_{j+1})\right] \tag{2.6.39a}$$

where

$$C_{M+1} = -C_1 \quad \text{and} \quad C^+_{M+1} = -C_1^+. \tag{2.6.40a}$$

When V_2 acts on states of odd n, we use V_2^-:

$$V_2^- = \left[K_2 \sum_{j=1}^{M}(C_j^+ - C_j)(C^+_{j+1} + C_{j+1})\right] \tag{2.6.39b}$$

with

$$C_{M+1} = C_1, \qquad C^+_{M+1} = C_1^+. \tag{2.6.40b}$$

2.6.5 *Eigenvalues of the Transfer Matrix*

The result of these considerations is that we must obtain the eigenvalues of the operators

$$V^{\pm} = (2\sinh 2K_1)^{M/2} \exp\left[\tfrac{1}{2}K_2 \sum_{j=1}^{M} (C_j^+ - C_j)(C^+_{j+1} + C_{j+1})\right]$$

$$\times \exp\left[-2k_1 \sum_{j=1}^{M} (C_j^+C_j - \tfrac{1}{2})\right]$$

$$\times \exp\left[\tfrac{1}{2}K_2 \sum_{j=1}^{M} (C_j^+ + C_j)(C^+_{j+1} + C_{j+1})\right]. \tag{2.6.41}$$

The $\pm$ sign refers to the definition (2.6.40a, b) of the operators C_{M+1}, $C^+{}_{M+1}$.

We accept only those eigenvalues of V^+ corresponding to states with an even number of fermions; and only those eigenvalues of V^- for an odd number of fermions. Others are discarded.

The problem of obtaining the eigenvalues of V is solved by constructing a linear transformation from the operators C_j, C_j^+ to new operators which leaves V in a form resembling

$$V \propto \exp\left[-\sum_q \epsilon_q(\chi_q^+\chi_q + \text{const})\right]. \tag{2.6.42}$$

with a suitable definition of the operators. To determine this transformation, it is convenient to introduce running waves. This is now possible because of the efforts that have been made to ensure translational symmetry. We put

$$C_j = M^{-1/2}e^{-i\pi/4} \sum_q e^{ijq}\eta_q. \tag{2.6.43}$$

The phase factor $e^{-i\pi/4}$ was introduced to simplify some formulas to be obtained below. The allowed values of q are chosen so that the conditions (2.6.40a, b) are satisfied. Suppose for convenience that M is even. The condition (2.6.40a) requires that q must be an odd multiple of π/M; thus the allowed values are

$$\pm\pi/M, \pm 3\pi/M, \ldots, \pm(M-1)\pi/M. \tag{2.6.44a}$$

The condition (2.6.40b) requires that q must be an even multiple of π/M:

$$q = 0, \pm 2\pi/M, \pm 4\pi/M, \ldots, \pm(M-2)\pi/M, M. \tag{2.6.44b}$$

We now substitute (2.6.43) into (2.6.41) and carry out the summations over j. It is convenient in writing the result to combine terms with positive and negative values of q. The result is

$$V^{\pm} = (2 \sinh 2K_1)^{M/2} \prod_{0 \leq q \leq \pi} V_q \tag{2.6.45}$$

in which

$$V_q = (V_{2q})^{1/2} V_{1q} (V_{2q})^{1/2}. \tag{2.6.46}$$

These quantities are ($q \neq 0, \pi$)

$$V_{1q} = \exp[-2k_1(\eta_q^+ \eta_q + \eta^+_{-q}\eta_{-q} - 1)] \tag{2.6.47}$$

and

$$V_{2q} = \exp\{2K_2[\cos q\ (\eta_q^+\eta_q + \eta^+_{-q}\eta_{-q}) + \sin q\ (\eta_q\eta_{-q} + \eta^+_{-q}\eta_q^+)]\}. \tag{2.6.48}$$

Note that we treat q and $-q$ as independent variables, and also that

$$\sum_q \cos q = \sum_q \sin q = 0. \tag{2.6.49}$$

The matrices for $q = 0$ and $q = \pi$ are given separately. They may be written as

$$V_0 = \exp[-2(k_1 - K_2)(\eta_0^+\eta_0 - \tfrac{1}{2})], \tag{2.6.50a}$$

$$V_\pi = \exp[-2(k_1 + K_2)(\eta_\pi^+\eta_\pi - \tfrac{1}{2})]. \tag{2.6.50b}$$

The principal result is that it is possible to diagonalize the operators V_q *separately* for each value of q. This occurs because each V_q can be expressed (when the exponentials are expanded) in terms of bilinear products of anticommuting operators. Thus, for instance, $[\eta_q^+\eta_{-q}, \eta_{q'}^+\eta_{-q'}] = 0$ for $q \neq q'$. This implies

$$[V_q, V_{q'}] = 0 \qquad \text{for} \quad q \neq q'. \tag{2.6.51}$$

The operators V_0, V_π are already diagonal. Only positive values of q remain to be considered.

To diagonalize V_q in the general case, we define vectors ϕ_0, ϕ_q, ϕ_{-q}, ϕ_{-qq} by

$$\eta_q\phi_0 = 0; \qquad \eta_{-q}\phi_0 = 0; \qquad \phi_q = \eta_q^+\phi_0;$$
$$\phi_{-q} = \eta^+_{-q}\phi_0; \qquad \phi_{-qq} = \eta^+_{-q}\eta_q^+\phi_0. \tag{2.6.52}$$

These four vectors are a complete set in the space in which V_q acts. It is easy to see that ϕ_q and ϕ_{-q} are already eigenvectors of V_q, since, for example,

$$\eta_q\eta_{-q}\phi_q = 0, \qquad \eta^+_{-q}\eta_q^+\phi_q = 0, \qquad \eta_q^+\eta_q\phi_q = \phi_q. \tag{2.6.53}$$

Then we have

$$V_q\phi_q = \exp(2K_2 \cos q)\phi_q. \tag{2.6.54}$$

It remains to find two combinations of ϕ_0 and ϕ_{-qq} which are eigenfunctions of V_q. We arrange the states in the order ϕ_{-qq}, ϕ_0: so that V_{1q} is diagonal with elements

$$V_{1q} = \begin{pmatrix} \exp(-2k_1) & 0 \\ 0 & \exp(2k_1) \end{pmatrix}. \tag{2.6.55}$$

In the restricted subspace of Hilbert space spanned by the vectors ϕ_0 and ϕ_{-qq}, it is possible to represent the products of η operators in which we are interested by Pauli matrices.

Thus, put

$$\eta^+_{-q}\eta_q^+ = \tfrac{1}{2}(b_q^x + ib_q^y), \qquad \eta_q\eta_{-q} = \tfrac{1}{2}(b_q^x - ib_q^y), \tag{2.6.56}$$

where the b's are the usual Pauli matrices. Further, since

$$\eta_q^+\eta_q\phi_0 = \eta^+_{-q}\eta_{-q}\phi_0 = 0 \qquad \text{and} \qquad \eta^+_q\eta_q\phi_{-qq} = \eta^+_{-q}\eta_{-q}\phi_{-qq} = \phi_{-qq},$$

we can put

$$\eta_q^+\eta_q + \eta^+_{-q}\eta_{-q} = b_q^z + 1. \tag{2.6.57}$$

The utility of the representation in terms of Pauli matrices is that it enables us to construct the exponential in V_{2q}. Thus

$$\begin{aligned} V_{2q}^{1/2} &= \exp\{K_2[\cos q\,(b_q^z + 1) + \sin q\, b_q^x]\} \\ &= \exp(K_2 \cos q) \exp K_2\,(b_q^z \cos q + b_q^x \sin q). \end{aligned} \tag{2.6.58}$$

Explicitly,

$$b_q^z \cos q + b_q^x \sin q = \begin{pmatrix} \cos q & \sin q \\ \sin q & -\cos q \end{pmatrix}$$

and

$$[b_q^z \cos q + b_q^x \sin q]^2 = \begin{pmatrix} 1 & 0 \\ 0 & 1 \end{pmatrix}.$$

A series expansion of the exponential yields

$$\begin{aligned} &\exp K_2\,[b_q^z \cos q + b_q^x \sin q] \\ &\quad = \cosh K_2 I + \sinh K_2\,(b_q^z \cos q + b_q^x \sin q) \\ &\quad = \begin{pmatrix} \cosh K_2 + \sinh K_2 \cos q & \sinh K_2 \sin q \\ \sinh K_2 \sin q & \cosh K_2 - \sinh K_2 \cos q \end{pmatrix}. \end{aligned} \tag{2.6.59}$$

Equations (2.6.59), (2.6.58), and (2.6.55) are substituted into (2.6.46), which gives V_q. The result can be written as

$$V_q = \exp(2K_2 \cos q) \begin{pmatrix} A_q & C_q \\ C_q & B_q \end{pmatrix} \tag{2.6.60}$$

where

$$A_q = e^{-2k_1}(\cosh K_2 + \sinh K_2 \cos q)^2 + e^{2k_1}(\sinh K_2 \sin q)^2, \tag{2.6.61a}$$

$$B_q = e^{-2k_1}(\sinh K_2 \sin q)^2 + e^{2k_1}(\cosh K_2 - \sinh K_2 \cos q)^2, \tag{2.6.61b}$$

$$C_q = 2 \sinh K_2 \sin q\ (\cosh 2k_1 \cosh K_2 - \sinh 2k_1 \sinh K_2 \cos q). \tag{2.6.61c}$$

The eigenvalues of V_q can now be determined. They are

$$\tfrac{1}{2} \exp(2K_2 \cos q)\{(A_q + B_q) \pm [(A_q - B_q)^2 + 4C_q^2]^{1/2}\}.$$

It is convenient to define a quantity ϵ_q by

$$\cosh \epsilon_q = \cosh 2K_2 \cosh 2k_1 - \sinh 2K_2 \sinh 2k_1 \cos q \tag{2.6.62}$$

where it is understood that ϵ_q is positive. Then one finds after considerable algebra that

$$\tfrac{1}{2}\{(A_q + B_q) \pm [(A_q - B_q)^2 + 4C_q^2]^{1/2}\}$$
$$= \cosh \epsilon_q \pm [\cosh^2 \epsilon_q - 1]^{1/2} = \exp(\pm\epsilon_q).$$

Thus the eigenvalues of V_q have the form

$$\exp(2K_2 \cos q) \exp(\pm\epsilon_q) \tag{2.6.63}$$

where ϵ_q is given by (2.6.62).

It should be noted that the cases $q = 0, \pi$ for which the eigenvalues of V are apparent from (2.6.50), are consistent with (2.6.63), so that we may consider these equations as applying for all q.

We now have all the eigenvalues of V_q. To express the operator V_q itself we introduce the transformation

$$\chi_q = \cos \Theta_q \eta_q + \sin \Theta_q \eta^+_{-q}, \qquad \chi_{-q} = -\sin \Theta_q \eta_q^+ + \cos \Theta_q \eta_{-q}, \tag{2.6.64}$$

$$\tan 2\Theta_q = 2C_q/(B_q - A_q). \tag{2.6.65}$$

The eigenvectors of V_q are denoted by ψ_0, ψ_q, ψ_{-q}, and ψ_{-qq}. They can be expressed in terms of these operators by

$$\chi_q\psi_0 = \chi_{-q}\psi_0 = 0, \qquad \psi_{-q} = \phi_{-q} = \chi^+_{-q}\psi_0,$$
$$\psi_q = \phi_q = \chi_q^+\psi_0, \qquad \psi_{-qq} = \chi^+_{-q}\chi_q^+\psi_0. \tag{2.6.66}$$

The operator V_q is given by

$$V_q = \exp(2K_2 \cos q) \exp[-\epsilon_q(\chi_q{}^+\chi_q + \chi^+{}_{-q}\chi_{-q} - 1)]. \quad (2.6.67)$$

It will be seen that this expression reproduces the eigenvalues that haye been obtained.

It is now necessary to construct the matrices $V^{\pm}$ from (2.6.45) and determine the largest eigenvalue. The problem should be considered in two temperature ranges, the division coming when $k_1 = K_2$, at which point $\epsilon(q = 0) = 0$. Note that this temperature, which turns out to be the Curie temperature, is given through the relation

$$\sinh 2k_1 = \sinh 2K_2$$

or more explicitly

$$\sinh(2J_1/K_B T_C) \sinh(2J_2/K_B T_C) = 1. \quad (2.6.68)$$

This result for T_C was first obtained by Kramers and Wannier (1941), prior to the complete solution of the model by Onsager.

Above T_C, $k_1 > K_2$ [see (2.6.21)], and all the eigenvalues are positive. The largest eigenvalue occurs in the state with no "particles" at all (use V^+). In this case, the eigenvalue of V^+ is called $\Lambda_0{}^+$:

$$\Lambda_0{}^+ = (2 \sinh 2K_1)^{M/2} \exp(\tfrac{1}{2} \sum_q \epsilon_q). \quad (2.6.69)$$

However, for $T \leq T_C$, $\epsilon_0 \leq 0$, and the same eigenvalue will be obtained either from V^+ with no particles, or from V^- with $\chi_0{}^+\chi_0 = 1$. (The eigenvalues are actually equal only in the limit $M \to \infty$.)

The effect of this double degeneracy on the free energy per spin is $N^{-1} \ln 2$, which we can neglect.

The sum in (2.6.69) can now be converted to an integral by

$$\sum_q \to (M/2\pi) \int_{-\pi}^{\pi} dq$$

so that (for all T)

$$\Lambda_0 = (2 \sinh 2K_1)^{M/2} \exp(M/4\pi) \int_{-\pi}^{\pi} \epsilon_q \, dq. \quad (2.6.70)$$

2.6.6 *The Phase Transition*

The free energy per spin, (2.6.13), is

$$F = -K_B T[\ln(2 \sinh 2K_1)^{1/2} + (1/4\pi) \int_{-\pi}^{\pi} \epsilon_q \, dq]. \quad (2.6.71)$$

This expression does not appear to show the necessary symmetry between K_1 and K_2; however, it does exist, and can be made explicitly apparent by a transformation due to Onsager. The form above, (2.6.71), is, however, more convenient for the determination of the thermodynamic functions.

The results simplify considerably in the case $K_1 = K_2 = K$ which will be considered from this point on. Equation (2.6.62) becomes

$$\cosh \epsilon_q = \cosh 2K \operatorname{ctnh} 2K - \cos q. \tag{2.6.72}$$

The internal energy U is determined from the formula $U = F + TS$ where $S = -(\partial F/\partial T)$. Thus

$$U = -T^2\, \partial/\partial T(F/T). \tag{2.6.73}$$

It is convenient to carry out the differentiation with respect to K $(=J/K_B T)$:

$$U = J/K_B\, d/dK(F/T)$$

$$= -J\, d/dK[\ln(2 \sinh 2K)^{1/2} + (1/4\pi) \int_{-\pi}^{\pi} d\epsilon_q/dK\, dq]. \tag{2.6.74}$$

When the differentiation is performed, the result is

$$U = -J \operatorname{ctnh} 2K[1 + (1/2\pi)(2 \tanh^2 2K - 1) \int_{-\pi}^{\pi} dq$$

$$\times [1-m \cos q - (m^2/4) \sin^2 q]^{-1/2} \tag{2.6.75}$$

where $m = (2 \sinh 2K)/\cosh^2 2K$.

The integral can be transformed into an elliptic integral. The final result is

$$U = -J \operatorname{ctnh} 2K[1 + (2/\pi)(2 \tanh^2 2K - 1)K_1(m)] \tag{2.6.76}$$

in which K_1 is the complete elliptic integral of the first kind:

$$K_1(m) = \int_0^{\pi/2} d\phi/(1 - m^2 \sin^2\phi)^{1/2}. \tag{2.6.77}$$

The elliptic integral $K_1(m)$ has a logarithmic singularity when $m = 1$. This occurs when

$$\sinh(2J/K_B T_C) = 1 \qquad (K_B T_C/J = 2.27) \tag{2.6.78}$$

which is in accord with (2.6.68). The coefficient of K_1 in the expression for U vanishes at this point, so that the internal energy is continuous through the transition, and there is no latent heat. The specific heat shows, however, a logarithmic singularity. The specific heat is given by

$$C = dU/dT = 2K_B(J/K_BT)^2 \operatorname{csch} 2K \{\operatorname{csch} 2K \times [(2/\pi)(2\tanh^2 2K + 1)K_1(m) - 1] - (4/\pi)(2\tanh^2 2K - 1)^2\, dK_1/dm\}. \qquad (2.6.79)$$

It is the first term in the brackets which is singular at $m = 1$ [note the change of sign with respect to (2.6.76)]. Near this singularity

$$C \approx \ln(|T - T_C|).$$

It is instructive to compare the present exact results with those obtained by the application of molecular field theory to the Ising model. Since molecular field theory ignores the details of the dynamics of the spins, the results may be taken over to the Ising model without change. We merely set $Z = 4$ and $S = \frac{1}{2}$ in the equations of Section 2.3. For spin $\frac{1}{2}$, the Brillouin function becomes the hyperbolic tangent, and we have a reduced magnetization $\sigma = M/M_0$,

$$\sigma = \tanh(g\beta H_T/2K_BT)$$

where $H_T = H_{ext} + \gamma M$ with H_{ext} the external field and γ the molecular field constant. A factor of 2 arises because the J used in this section is half that used in Section 2.3, and $\gamma = 4ZJ/Ng^2\beta^2$.

The Curie temperature can be found from Eq. (2.3.17), taking account of the factor of 2 mentioned above:

$$KT_C/J = 4. \qquad (2.6.80)$$

This is to be compared with (2.6.78). We see that molecular field theory overestimates the Curie temperature rather badly.

The specific heat in the molecular field approach is found from Eq. (2.3.26). We recall that it exhibits a finite discontinuity at the Curie temperature, rather than the logarithmic singularity found in the exact calculation.

The exact results for the internal energy and for the specific heat are shown in Fig. 2.6.1 where they are compared with those obtained from molecular field theory.

Onsager (1944) has also obtained results for the specific heat in the more general case of $J_1 \neq J_2$. Suppose that $J_2 = \alpha J_1$. Without loss of generality, we may suppose $\alpha \leq 1$. It is easy to see by differentiating (2.6.68) that $dT_C/d\alpha > 0$, that is, the Curie temperature decreases with decreasing α. Then T_C goes to zero if α does. This must hold since if $\alpha = 0$, we have just the one-dimensional model. It can be shown that the width of the specific heat peak also decreases with α.

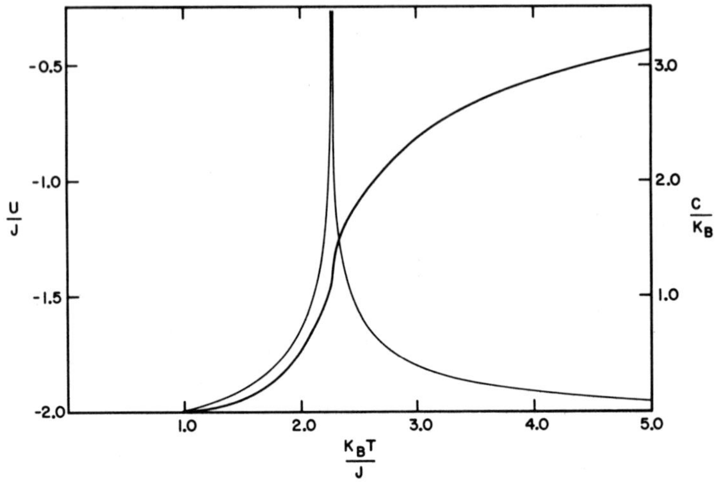

Fig. 2.6.1. Internal energy U and specific heat C as functions of temperature for the Ising model.

It is also possible to solve the problem for certain other two-dimensional lattices. The results are usually not qualitatively different from the case considered here (Newell and Montroll, 1953); however, in the case of a triangular lattice with antiferromagnetic coupling, no transition is found (Wannier, 1950).

Since it is not yet possible to solve the two-dimensional Ising model for any lattice in the presence of an external field the magnetic susceptibility cannot be calculated exactly. However, Fisher (1963) has found by numerical methods that, close to T_C but slightly above it,

$$\chi \propto (T - T_C)^{-7/4}. \tag{2.6.81}$$

Below the Curie temperature, the spontaneous magnetization is of interest. Peierls (1936) showed that the two-dimensional Ising model does have a spontaneous magnetization at low temperatures as expected. It is possible to obtain an exact result for the temperature dependence of the spontaneous magnetization. This was first stated by Onsager (1949) who did not present a proof. Yang (1952) published the first derivation. A simpler procedure was given by Montroll *et al.* (1963). We will give only the result here:

$$\begin{aligned} \sigma = M/M_0 &= (1 - \operatorname{csch}^2 2K_1 \operatorname{csch}^2 2K_2)^{1/8} \quad (T < T_C) \\ &= 0 \qquad\qquad\qquad\qquad\qquad\qquad\quad (T > T_C). \end{aligned} \tag{2.6.82}$$

The magnetization does go to zero with infinite slope as T approaches T_C from below, however, the exponent is different from that obtained in the molecular field theory. It was shown in Section 2.3 that molecular field theory predicts $\sigma \approx (T - T_C)^{1/2}$ for T close to T_C. In contrast, (2.6.82) gives $\sigma \approx (T - T_C)^{1/8}$.

2.7 The Magnetic Phase Transition

In this section, we shall study phase transitions in more detail with emphasis on ferromagnets. Phase transitions have been a topic of increasing interest in recent years. Because of the experimental convenience, the paramagnetic to ferromagnetic transition has become one of the most thoroughly investigated types of phase transition. We are interested in the transition not only for its own sake, but for the information which can be obtained concerning phase transitions in general.

Theoretical understanding of phase transitions is still quite incomplete. This problem is one of the most fascinating, most basic, and most difficult in theoretical physics.

The phase transition is characterized by the appearance, as the temperature is lowered through the critical temperature T_c, of long range order in the system. Above T_c no such order is present; below T_c order increases rapidly (usually with an infinite derivative at T_c). The ordering is not completely determined in the absence of external fields (for instance, the magnetization may point in different directions). Mathematically, the transition is characterized by the divergence of certain thermodynamic quantities: for example, the magnetic susceptibility and the specific heat. Useful general references include Stanley (1971), Kadanoff *et al.* (1967), Heller (1967), Fisher (1967), and Brout (1965).

2.7.1 *The Order Parameter*

Since the phase transition corresponds to a sudden appearance of order, it is convenient to introduce a function to describe this order. We call this the *order parameter*, and will denote this by σ. In the case of a magnetic system σ is the reduced magnetization defined in Eq. (2.3.14):

$$\sigma = M_z/M_0 \tag{2.7.1}$$

where M_z is the magnetization (this defines the z axis), and $M_0 = Ng\beta S$. If the system is uniform, σ is constant; however, it is desirable to be more general, and allow σ to be a function of position. Dynamical properties will not be considered here, so σ will not depend on time:

$$\sigma = \sigma(\mathbf{r}). \tag{2.7.2}$$

The order parameter is expected to have the following properties:

(a) σ is zero above T_c.
(b) σ is not zero below T_c.
(c) σ can approach zero continuously as T approaches T_c from below (in the case of a transition which is not first order).
(d) Below T_c the order parameter is not completely determined in the absence of external fields.

2.7.2 *The Landau Theory of Second-Order Phase Transitions*

The simplest theory of phase transitions which can be constructed is the Landau theory (Landau and Lifshitz, 1958). This is essentially a generalization and systematization of the molecular field theory presented in Section 2.3, and has the same advantages and disadvantages: It is simple, qualitatively useful, but incorrect in detail near the transition temperature.

The essential idea of the Landau theory is that the free energy is to be expressed as a function of the order parameter. The free energy so defined is denoted by G:

$$G = G(T, \sigma). \tag{2.7.3}$$

The entropy S is given by

$$S = -(\partial G/\partial T)_\sigma. \tag{2.7.4}$$

It is convenient to introduce a quantity h related to the external field H_0 by

$$h = M_0 H_0. \tag{2.7.5}$$

We express G as a volume integral of a free energy density $g(\mathbf{r})$:

$$G = \int g(\mathbf{r})\, d^3r. \tag{2.7.6}$$

The fundamental assumption of the Landau theory is that g can be expanded as a power series in σ in the form

$$g(\mathbf{r}) = g_0(T) - h(\mathbf{r})\sigma(\mathbf{r}) + a(T)[\sigma(\mathbf{r})]^2 + b(T)[\sigma(\mathbf{r})]^4 + c(T)\,|\nabla\sigma(\mathbf{r})|^2 + \cdots. \tag{2.7.7}$$

Here g_0, a, b, and c are functions of temperature only. The term involving $|\nabla\sigma|^2$ will be shown to allow for fluctuations. The most probable value of

σ is to be determined by minimization of G. We require

$$\delta \int g(\mathbf{r})\, d^3r = 0. \tag{2.7.8}$$

This gives

$$\int \delta\sigma \left[-h + 2a\sigma + 4b\sigma^3 - 2c\, \nabla^2\sigma(\mathbf{r})\right] d^3r = 0.$$

The following equation for $\sigma(\mathbf{r})$ is obtained:

$$-2c\, \nabla^2\sigma(\mathbf{r}) + [2a + 4b\sigma^2(\mathbf{r})]\sigma(\mathbf{r}) = h(\mathbf{r}). \tag{2.7.9}$$

For the moment, we suppose σ and h to be independent of $\mathbf{r}$. Then

$$[2a + 4b\sigma^2]\sigma = h. \tag{2.7.10}$$

If $h = 0$, two solutions are possible:

$$\sigma = 0, \tag{2.7.11a}$$

$$\sigma = \pm(-a/2b)^{1/2}. \tag{2.7.11b}$$

Solution (2.7.11a) minimizes the free energy if $a > 0$; (2.7.11b) if $a < 0$. We want the free energy to describe a system with nonzero magnetization in the absence of external fields for $T < T_c$. This suggests the assumption

$$a(T) = a'(T - T_c) \tag{2.7.12}$$

where a' is a constant. This assumption leads immediately to the result

$$\sigma(T) \approx (T_c - T)^{1/2} \qquad (T < T_c). \tag{2.7.13}$$

This result is consistent with the molecular field theory of Section 2.3 and motivates the assumption (2.7.12).

To determine the susceptibility, we must consider $h \neq 0$. Then the susceptibility is proportional to $\partial\sigma/\partial h$; we will define this quantity as the reduced susceptibility χ_r. It differs by a constant factor from the usual magnetic susceptibility of Section 2.3:

$$\chi_r = (\partial\sigma/\partial h)_T. \tag{2.7.14}$$

We differentiate (2.7.10) to obtain an equation for χ_r:

$$2a\chi_r + 12b\sigma^2\chi_r = 1. \tag{2.7.15}$$

For $T > T_c$, σ^2 may be considered small, so that the quadratic term in (2.7.15) can be neglected. This gives

$$\chi_r = 1/2a = 1/2a'(T - T_c) \qquad (T > T_c). \tag{2.7.16}$$

This is the Curie–Weiss law. Below T_c, we must retain the term proportional to σ^2 in (2.7.15), which is evaluated using (2.7.11b). Then

$$\chi_r = -1/4a = 1/4a'(T_c - T) \qquad (T < T_c). \tag{2.7.17}$$

The temperature dependence is of the same form; but we note that the proportionality factor is different below and above T_c.

The specific heat (at constant external field) may be found from

$$C_M = -T\,(\partial^2 G/\partial T^2)_h. \tag{2.7.18}$$

We will consider the case of zero external field ($h = 0$). Then, above the transition temperature, $\sigma = 0$; so

$$C = C_0 \qquad (T > T_0) \tag{2.7.19}$$

where

$$C_0 = -T\,d^2/dT^2 \int g_0(\mathbf{r})\,d^3r \qquad (T > T_c).$$

Below T_c, for a uniform system,

$$G = \int d^3r\,[g_0 - (a^2/4b)]. \tag{2.7.20}$$

The specific heat is

$$C = C_0 + T \int \{(a'^2/2b) - [a'^2(T - T_c)^2/4b^2]\,db/dT\}\,d^3r. \tag{2.7.21}$$

Thus there is a finite contribution from the magnetization to the specific heat just below T_c. Thus C has a finite discontinuity at T_c. This is also in accord with Section 2.3.

It is interesting to consider the molecular field ferromagnet more explicitly in the case of spin $\frac{1}{2}$ as an illustration of these ideas since simple, closed results can be written in that case: From (2.3.10), the partition function for $S = \frac{1}{2}$ can be simply expressed as

$$Z = 2\cosh \beta g H_T/2KT \tag{2.7.22}$$

where

$$H_T = H_0 + \gamma M = H_0 + H_e,$$

and γ is defined by Eq. (2.3.5b).

The free energy F, expressed as a function of the total field H_T and temperature, is

$$F = -NKT\ln Z = -NKT\ln 2 - NKT\ln\cosh(g\beta H_T/2KT). \tag{2.7.23}$$

This expression can be used to determine the entropy and the magnetization according to

$$dF = -S\,dT - M\,dH_T, \tag{2.7.24}$$

or

$$M = -(\partial F/\partial H_T)_T; \qquad S = -(\partial F/\partial T)_{H_T}.$$

This gives, for example,

$$M/M_0 = \sigma = \tanh(g\beta H_T/2KT). \tag{2.7.25}$$

The Landau theory is obtained by expressing the free energy in terms of T and M. This can be accomplished by a transformation of the form

$$G = F + \Delta F \tag{2.7.26a}$$

where

$$\Delta F = \int M\,dH_e. \tag{2.7.26b}$$

We make the substitutions and find after some algebra

$$G = NKT\{-\ln 2 + \tfrac{1}{2}\ln(1-\sigma^2)$$
$$+ \tfrac{1}{2}\sigma\ln[(1+\sigma)/(1-\sigma)] - (N/4)ZJ\sigma^2\} - \sigma h \tag{2.7.27}$$

where G possesses a power series expansion consistent with (2.7.7). For example, we obtain from (2.7.27)

$$G = -NKT\ln 2 + (N/2)[KT - (ZJ/2)]\sigma^2 + (NKT/12)\sigma^4 - \sigma h + \cdots. \tag{2.7.28}$$

From Section 2.3, the molecular field Curie temperature is

$$T_c = \tfrac{2}{3}ZJ[S(S+1)/K] \rightarrow ZJ/2K \qquad (S = \tfrac{1}{2}).$$

Thus the coefficient of σ^2 is just $(NK/2)(T - T_c)$. This is consistent with (2.7.12).

The free energy is shown as a function of σ for several different temperatures (zero external field) in Fig. 2.7.1. Note that $t = T/T_c$. Above T_c, $G(\sigma)$ has only a single minimum at $\sigma = 0$. For $T = T_c$ $(t = 1)$, $G(\sigma)$ is quite flat near $\sigma = 0$ since the second-order terms vanish. Below T_c, there are two minima for positive and negative values of σ, corresponding to the two possible equilibrium values of the saturation magnetization. The locus of these minima is shown by the dashed curve. The region of the free energy curve between the two minima represents an unstable situation.

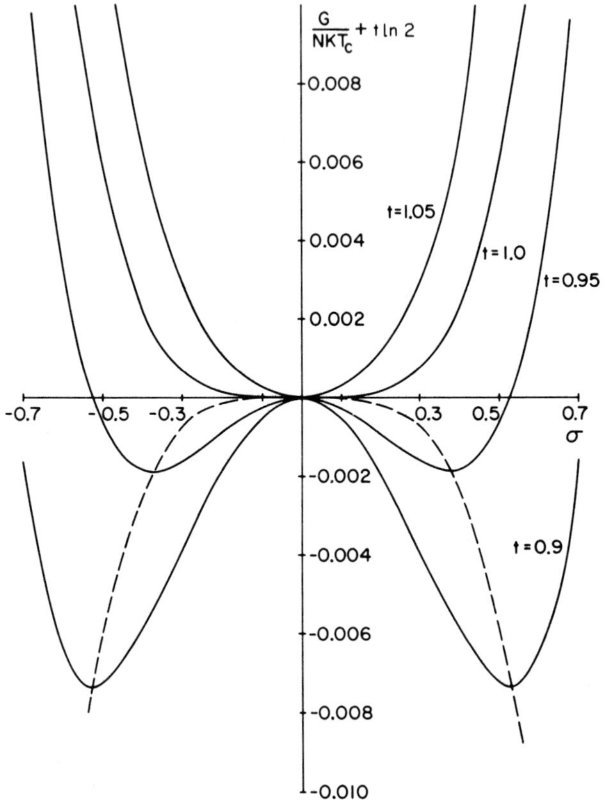

Fig. 2.7.1. Free energy as a function of magnetization according to the Landau mean field theory at several temperatures and $h = 0$. The dashed line shows the equilibrium magnetization.

2.7.3 *Fluctuations of the Order Parameter*

The relative flatness of $G(\sigma)$ near $\sigma = 0$ for $t = T/T_c = 1$, indicates that the magnetization can change substantially with only a small cost in terms of the free energy. It is suggested that large fluctuations may be observed near the phase transition. This does occur as we will see now. Our investigation is based on the Landau theory. The term $c \mid \nabla\sigma \mid^2$ in (2.7.7), which was previously ignored, is now to be retained.

In order to exhibit fluctuations, the magnetization must be locally non-uniform. Let angular brackets $\langle\ \rangle$ denote a thermodynamic average. We define a correlation function $\mathcal{G}(\mathbf{r}, \mathbf{r}')$ by

$$\mathcal{G}(\mathbf{r}, \mathbf{r}') = \langle[\sigma(\mathbf{r}') - \sigma_0][\sigma(\mathbf{r}) - \sigma_0]\rangle = \langle\sigma(\mathbf{r})\sigma(\mathbf{r}')\rangle - \sigma_0^2 \quad (2.7.29)$$

where σ_0 is the average magnetization,

$$\sigma_0 = \langle \sigma(\mathbf{r}) \rangle, \tag{2.7.30}$$

which is assumed to be uniform, that is, independent of $\mathbf{r}$. The function $\mathcal{G}$ is defined in such a way so as to describe the manner in which a fluctuation or departure of the magnetization from its average value at some point $\mathbf{r}$ is related to a fluctuation at a point $\mathbf{r}'$.

It is possible to devise a procedure for the calculation of $\mathcal{G}$. Assume the system is governed by a Hamiltonian containing σ and h of the form

$$H = H_0 - \int d^3r\, \sigma(\mathbf{r})h(\mathbf{r}). \tag{2.7.31}$$

The average value of σ, $\langle\sigma\rangle$ is given by

$$\langle\sigma\rangle = \mathrm{tr}[\sigma(\mathbf{r})e^{-\beta H}]/\mathrm{tr}[e^{-\beta H}] \tag{2.7.32}$$

where, as usual, $\beta = 1/K_BT$. Suppose now that the field $h(\mathbf{r})$ is changed by a small amount δh:

$$h = h_0 + \delta h(\mathbf{r}). \tag{2.7.33}$$

A corresponding small change $\delta\sigma(\mathbf{r})$ is produced in the magnetization. We expand (2.7.32), retaining all terms of first order in δh, and obtain

$$\delta\sigma(\mathbf{r}) = \beta \int d^3r'\, \mathcal{G}(\mathbf{r}, \mathbf{r}')\, \delta h(\mathbf{r}') \tag{2.7.34}$$

where $\mathcal{G}$ is given by (2.7.29).

Next, replace σ by $\sigma + \delta\sigma$ and h by $h + \delta h$ in (2.7.9); then linearize the equation by retaining terms of first order in $\delta\sigma$. The result is

$$-2c\nabla^2\, \delta\sigma(\mathbf{r}) + [2a + 12b\sigma_0^2]\, \delta\sigma(\mathbf{r}) = \delta h(\mathbf{r}). \tag{2.7.35}$$

Substitute (2.7.34) into (2.7.35):

$$\int d^3r'\, \{[2a + 12b\sigma_0^2 - 2c\, \nabla_r^2]\mathcal{G}(\mathbf{r}, \mathbf{r}') - K_BT\, \delta(\mathbf{r} - \mathbf{r}')\}\, \delta h(\mathbf{r}') = 0.$$

This equation must be satisfied for any arbitrary variation in h. Thus

$$[2a + 12b\sigma_0^2 - 2c\, \nabla_r^2]\mathcal{G}(\mathbf{r}, \mathbf{r}') = K_BT\, \delta(\mathbf{r} - \mathbf{r}'). \tag{2.7.36}$$

This equation implies that $\mathcal{G}$ is a Green's function. We can rewrite (2.7.36):

$$[\nabla_r^2 - \lambda^2]\mathcal{G}(\mathbf{r}, \mathbf{r}') = -A\, \delta(\mathbf{r} - \mathbf{r}') \tag{2.7.37}$$

where

$$A = K_BT/2c \tag{2.7.38a}$$

and

$$\lambda^2 = (a + 6b\sigma_0^2)/c \tag{2.7.38b}$$

or

$$\lambda = [a'(T - T_C)/c]^{1/2} \qquad (T > T_c), \tag{2.7.38c}$$

$$= [2a'(T_c - T)/c]^{1/2} \qquad (T < T_c). \tag{2.7.38d}$$

The solution of (2.7.37) is easily seen to be

$$\mathcal{G}(\mathbf{r}, \mathbf{r}') = (A/4\pi)[\exp(-\lambda \,|\, \mathbf{r} - \mathbf{r}' \,|\,)/\,|\, \mathbf{r} - \mathbf{r}' \,|\,]. \tag{2.7.39}$$

This is the Ornstein–Zernike (1914, 1918) form of correlation function; originally obtained for a liquid. The quantity $\xi = 1/\lambda$ is called the correlation length. It is particularly to be noted that λ goes to zero and ξ diverges, at the transition temperature. The quantity ξ can be interpreted as a measure of the size of a region in which a coherent fluctuation in the magnetization will occur. The increase in ξ as $T \to T_c$ indicates an increase in the spatial extent of a typical flutuation. Finally, at the critical point, this range becomes effectively infinite and $\mathcal{G}$ decreases with distance as slowly as $|\, \mathbf{r} - \mathbf{r}' \,|^{-1}$.

Since the reduced magnetic susceptibility $\chi_r = \delta\sigma/\delta h$, it follows from (2.7.36) that, in the case δh is independent of $\mathbf{r}$,

$$\chi_r = \beta \int d^3r' \, \mathcal{G}(\mathbf{r}, \mathbf{r}'). \tag{2.7.40}$$

The reader will easily verify that if (2.7.39) is substituted into (2.7.40), the result agrees with (2.7.16) and (2.7.17) as it should.

It is convenient to introduce Fourier transforms of $\delta\sigma(\mathbf{r})$ and $\delta h(\mathbf{r}')$:

$$\delta h(\mathbf{r}) = (2\pi)^{-3} \int \delta h(\mathbf{q}) \exp(i\mathbf{q}\cdot\mathbf{r}) \, d^3q,$$

$$\delta\sigma(\mathbf{r}) = (2\pi)^{-3} \int \delta\sigma(\mathbf{k}) \exp(i\mathbf{k}\cdot\mathbf{r}) \, d^3k, \tag{2.7.41}$$

$$\delta\sigma(\mathbf{k}) = \int \delta\sigma(\mathbf{r}) \exp(i\mathbf{k}\cdot\mathbf{r}) \, d^3r.$$

These expressions are substituted into (2.7.34). To simplify the result, note that $\mathcal{G}(\mathbf{r}, \mathbf{r}')$ is a function of $\mathbf{r} - \mathbf{r}'$ only. This is a consequence of translational invariance of the system, independent of the validity of specific assumptions made in the Landau theory concerning the free energy. We obtain

$$\delta\sigma(\mathbf{q}) = \beta\mathcal{G}(\mathbf{q}) \, \delta h(\mathbf{q}) \tag{2.7.42}$$

where $\mathcal{G}(\mathbf{q})$ is the Fourier transform of $\mathcal{G}(\mathbf{r} - \mathbf{r}')$:

$$\mathcal{G}(\mathbf{q}) = \int \mathcal{G}(\mathbf{r}) \exp(-i\mathbf{q}\cdot\mathbf{r})\, d^3r. \tag{2.7.43}$$

It is convenient to define a wave-number-dependent susceptibility by

$$\chi(\mathbf{q}) = \delta\sigma(\mathbf{q})/\delta h(\mathbf{q}) \tag{2.7.44}$$

such that $\chi(0) = \chi_r$, the static susceptibility. Thus

$$\chi(\mathbf{q}) = \beta\mathcal{G}(\mathbf{q}). \tag{2.7.45}$$

The specific result of the Landau theory, using (2.7.39), is that

$$\chi(q) = 1/2c(q^2 + \lambda^2) = \chi_r/(1 + q^2\xi^2). \tag{2.7.46}$$

2.7.4 *Critical Exponents*

Molecular field theory has given a simple description of the phase transition and of phenomena near the transition temperature, but there are serious quantitative discrepancies between theory and experiment. For example, ferromagnets do not show a finite discontinuity in the specific heat at T_c, but rather a sharp peak. In addition, the exponents predicted to govern the divergences at the critical temperature are incorrect. In a sense, molecular field theory indicates the reason for its own failure: the fluctuations become too large. A necessary condition for the validity of a molecular field theory is that fluctuations in the order parameter over distances of the order of ξ should be small. This ceases to be the case sufficiently near T_c, as is indicated by the divergence of the susceptibility. Unfortunately, there is at present no quantitative theory of phase transitions, but it has been possible to clarify relations between phenomena near the initial point by means of so-called *scaling laws*.

Let ϵ denote the reduced temperature, defined by $\epsilon = (T/T_c) - 1$. It is convenient to introduce a set of critical exponents which describe the behavior of the system near the phase transition.

For zero external field, we assume that the order parameter σ varies as $|\epsilon|^\beta$ as $T \to T_c$ $(\epsilon \to 0)$ from below. Precisely, we mean that

$$\lim_{\epsilon\to 0^-} [\ln \sigma(\epsilon)/\ln|\epsilon|] = \beta. \tag{2.7.47}$$

This is not quite the same as saying that σ is proportional to $|\epsilon|^\beta$; since we must always expect that correction terms of higher order will be present. Thus one might hope to write

$$\sigma(\epsilon) = A|\epsilon|^\beta(1 + a|\epsilon|^\nu + \cdots) \qquad (\epsilon < 0) \tag{2.7.48}$$

where A and a are constants; however we cannot even guarantee that ν is an integer, or that logarithmic terms are not present. In molecular field theory, $\beta = \frac{1}{2}$; whereas the two-dimensional Ising model gives $\beta = \frac{1}{8}$, and numerical calculations for the three-dimensional Ising model yield $\beta = 0.313 \pm 0.004$.

If, instead of the above situation, we have an external field h, but stay at the critical temperature $\epsilon = 0$, then if h is allowed to tend to zero, σ will tend to zero as well, and we suppose that σ varies as $|h|^{1/\delta}$. This defines another critical index δ. Additional critical indices are associated with the susceptibility χ, the specific heat at constant h, C_h, the correlation function $\mathcal{G}$, and the correlation length ξ. In general, the critical exponents may differ above and below the transition temperature. For example, suppose that the susceptibility χ varies as $\epsilon^{-\gamma}$ for $\epsilon > 0$; for $\epsilon < 0$, we may have $\chi \approx |\epsilon|^{-\gamma'}$ where $\gamma' \neq \gamma$. The critical indices are listed in Table I together with their values in the Landau theory and in the two-dimensional and three-dimensional Ising models, according to Stanley (1971). Experimental values are included for three ferromagnets: Ni, EuS, and $CrBr_3$. Results for other types of systems (fluids) are similar but will not be discussed here.

The specific heat C_h requires additional comment. In this case, let the critical exponents be α, α':

$$C_h \approx \epsilon^{-\alpha} \quad (\epsilon > 0), \qquad C_h \approx |\epsilon|^{-\alpha'} \quad (\epsilon < 0).$$

One frequently encounters small (or zero) values of α and α'. This does not mean that no singularity exists. In fact, we may have a logarithmic singularity or a cusp. The logarithmic behavior was already encountered in the two-dimensional Ising model. In the case of a cusp C_h has an infinite derivative in some finite order. Let k be the smallest integer such that $\partial^k C_h / \partial \epsilon^k$ diverges as $\epsilon \to 0$. We define a new exponent α_s such that

$$\alpha_s = k + \lim_{\epsilon \to 0} [\ln(\partial^k C_h / \partial \epsilon) / \ln \epsilon].$$

2.7.5 *Scaling Laws*

Rigorous inequalities involving these exponents can be deduced from thermodynamics (Fisher, 1967). We shall instead consider in detail certain equalities between critical exponents which are less rigorous and are derived from plausibility arguments. Our treatment follows Kadanoff *et al* (1967). Direct calculations of critical exponents have recently been reported by Wilson and Fisher (1972), Wilson (1972), and Brezin *et al.* (1973).

For simplicity, consider a cubic crystal. Let us divide the crystal into

TABLE I

Physical quality	$\epsilon = \left(\frac{T}{T_c}\right) - 1$	h	Behavior of quantity	Critical index	Landau theory	Ising model 2–d[a]	Ising model 3–d[a]	Experiment Ni	Experiment EuS	Experiment $CrBr_3$
σ	>0	0	$\sigma = 0$							
	<0	0	$\sigma \sim \mid \epsilon \mid^{\beta}$	β	$\frac{1}{2}$	$\frac{1}{8}$	$\sim\frac{5}{16}$	0.42	0.33	0.368
	0	$\neq 0$	$\sigma \sim \mid h \mid^{1/\delta}$	δ	3	15	~ 5	4.22		4.3
χ	>0	0	$\sim \epsilon^{-\gamma}$	γ	1	$\frac{7}{4}$	$\sim\frac{5}{4}$	1.35		1.215
	<0	0	$\sim \mid \epsilon \mid^{-\gamma'}$	γ'	1	$\frac{7}{4}$	$\sim\frac{5}{4}$ or $\sim\frac{21}{16}$			
$\mathcal{G}(\mathbf{r} - \mathbf{r}')$	0	0	$\mid \mathbf{r} - \mathbf{r}' \mid^{-d+2-\eta}$ [a]	η	0	$\frac{1}{4}$	~ 0.041			
ξ	>0	0	$\sim \epsilon^{-\nu}$	ν	$\frac{1}{2}$	1	~ 0.638			
	<0	0	$\mid \epsilon \mid^{-\nu'}$	ν'	$\frac{1}{2}$	1				
C_h	>0	0	$a\epsilon^{-\alpha} + b$	α	0		$\sim\frac{1}{8}$	0		
	<0	0	$a \mid \epsilon \mid^{-\alpha'} + b'$	α'	Finite discontinuity		$\sim\frac{1}{8}$ or $\sim\frac{1}{16}$	$\alpha_s' = -0.3$	$\alpha_s' = -0.15$	
	or >0	0	$A \log \epsilon^{-1} + B$	$\alpha = 0$		Logarithmic singularity				
	<0	0	$A' \log \mid \epsilon \mid^{-1} + B$	$\alpha' = 0$						

[a] d stands for dimensionality.

large cubes of aribtrary, but macroscopic, size. The edge of a cube is La_0 where a_0 is the lattice constant and L is a large number. The essence of the following argument is that physical phenomena should not depend on the choice of L. We label each large cube by an index a, each microscopic atomic cell by an index r, and we suppose there is a variable μ_a which plays the same role for the large cube as does the order parameter σ_r in an atomic cell.

Assume that the external field is altered by an infinitesimal amount and that this field is slowly varying, such that it may be regarded as constant in each large cube. The change in the free energy produced by this effect will involve a term

$$\delta G = -\sum_r \sigma_r \, \delta h = -\sum \mu_a \, \delta \tilde{h} \tag{2.7.49}$$

where h refers to the external field in an atomic cell, and $\tilde{h}$ to the external field as it enters the free energy calculation for the large cube. Further, let $\epsilon = (T/T_c) - 1$ as before in a microscopic theory and correspondingly, let $\tilde{\epsilon}$ denote the similar quantity as related to the large cube. Since the properties of the system near the transition point may depend on interactions between cells, and between these cells and the external field, the way these quantities enter into the free energy may depend on the size of the cell considered. Hence, $\tilde{h}$ and h, $\tilde{\epsilon}$ and ϵ may be different, but we do suppose that they are proportional with a factor involving L, namely,

$$\tilde{h} = L^x h, \qquad \tilde{\epsilon} = L^y \epsilon \tag{2.7.50}$$

where x and y are unknown exponents. Note that $\epsilon = 0$ implies $\tilde{\epsilon} = 0$, so that the transition temperatures will be the same.

The critical exponents can be related to x and y. Since the order parameter and the external fields are slowly varying, we replace the sum over atomic cells in a large cube by the product of the number of cells in each cube and the value of σ in a cell. The number of cells per cube is just L^d where d is the dimensionality of the system. Thus

$$L^d \sigma_r \, \delta h = \mu_a \, \delta \tilde{h}.$$

We use (2.7.48), and obtain

$$\sigma_r = L^{x-d} \mu_a. \tag{2.7.51}$$

Let us now consider the magnetization in a situation in which the field is uniform. The magnetization will depend on temperature and field:

$$\sigma_r = F(\epsilon, h). \tag{2.7.52}$$

However, μ_a is supposed to describe the same problem in terms of new variables $\tilde{\epsilon}$ and $\tilde{h}$; thus μ_a should be specified by the same function F:

$$\mu_a = F(\tilde{\epsilon}, \tilde{h}). \tag{2.7.53}$$

Comparison of (2.7.51)–(2.7.53) shows that

$$\sigma = F(\epsilon, h) = L^{x-d}F(L^y\epsilon, L^x h). \tag{2.7.54}$$

The length L we have introduced is a purely artificial device, and must disappear from expressions which have physical significance. This requirement imposes stringent restrictions on the function F. It is easy to see that this will happen if F has the form

$$\sigma = F(\epsilon, h) = (h/\mid h \mid) \mid \epsilon \mid^{(d-x)/y} f(\epsilon/\mid h \mid^{y/x}). \tag{2.7.55}$$

The unspecified function f depends on the single variable $\epsilon/\mid h \mid^{y/x}$. The multiplying factor of $h/\mid h \mid$ ensures that σ changes sign if h does. Equation (2.7.55) is said to be a scaling law: its significance will be explained below. The function f remains to be calculated by an accurate microscopic theory which does not exist at present. However, we can deduce some of its properties, and find relations between the exponents x, y above and the critical exponents given in Table I by some simple arguments.

We require that $\sigma \approx \mid \epsilon \mid^{\beta}$ for small negative ϵ and $h = 0$. This will occur if f approaches a constant as its argument goes to $-\infty$, and if

$$\beta = (d - x)/y. \tag{2.7.56}$$

Consideration of the susceptibility leads to additional information. Suppose $\epsilon > 0$. It is necessary to determine the temperature dependence of $\partial\sigma/\partial h$ as $h \to 0$. To do this we assume that for large positive values of the argument, f possesses an expansion

$$f(\epsilon/\mid h \mid^{y/x}) = f(\infty) + \text{const}(h/\mid \epsilon \mid^{x/y}). \tag{2.7.57}$$

Then if $\chi \approx \epsilon^{-\gamma}$,

$$\gamma = (2x - d)/y. \tag{2.7.58}$$

We can combine (2.7.56) and (2.7.58) to obtain

$$2\beta + \gamma = d/y. \tag{2.7.59}$$

The basic relation (2.7.55) is supposed to hold both above and below the transition temperature. We suppose that the same expansion (2.7.57) applies when the argument of f approaches $-\infty$; this leads to the conclusion that $\beta' = \beta$. Similar arguments indicate that all the other critical exponents should be the same above and below the transition. From the point of view of comparison with experiment, this is one of the most questionable features of the results.

Next, allow ϵ to vanish, but retain a finite h. We require that σ stay finite. In order for this to happen, we must have

$$f(\epsilon/\mid h\mid^{y/x}) \approx (\epsilon/\mid h\mid^{y/x})^{(x-d)/y}$$

for small ϵ. The dependence on h is now $\sigma \approx h^{[(d-x)/x]}$. This is of the form $h^{1/\delta}$ where

$$\delta = x/(d - x). \tag{2.7.60}$$

From (2.7.56), (2.7.60), and (2.7.59) we obtain

$$\beta(\delta + 1) = d/y = 2\beta + \gamma. \tag{2.7.61}$$

Equation (2.7.61) gives a nontrivial relation between the critical exponents β, γ, and δ which is subject to test. The values given in Table I for Ni and $CrBr_3$ satisfy (2.7.61) satisfactorily. Consideration of the specific heat leads to additional relations.

For this purpose, we must investigate the behavior of the free energy. We assume that, near the transitions, this is determined by a term of the form $\int h\, d\sigma$. We use (2.7.55) and suppose that $h > 0$ so that the factor $h/\mid h\mid$ may be neglected. We have

$$\int h\, d\sigma = \epsilon^{(d-x)/y} \int h\, df(\epsilon/h^{y-x}).$$

Put $z = \epsilon/h^{y/x}$ and $h = (\epsilon/z)^{x/y}$. Then

$$\int h\, d\sigma = \epsilon^{d/y} \int (1/z^{x/y})\, df/dz\, dz. \tag{2.7.62}$$

The integral is to be performed at constant temperature. The magnetization varies from zero, if we are above the critical temperature, or from σ_0, the magnetization in zero field below T_c, to a large value σ_1 involving a strong external field where the free energy is quite smooth. Then z will vary from $\pm\infty$ to close to zero. The integral may or may not converge as it stands.

Suppose (2.7.62) converges. Then there is a term in the free energy proportional to $\epsilon^{d/y}$, and differentiating twice to obtain the specific heat, we find

$$C \approx \epsilon^{d/y-2}. \tag{2.7.63a}$$

Thus

$$\alpha = \alpha' = 2 - d/y. \tag{2.7.63b}$$

This gives, in conjunction with (2.7.61), another relation among the critical exponents,

$$\beta(\delta + 1) = 2\beta + \gamma = 2 - \alpha. \tag{2.7.64}$$

Suppose, however, that the integral in (2.7.62) is singular as $z \to \infty$. The integral must be cut off at a finite limit z_m, say. We consider specifically only the possibility that the integral is logarithmically divergent. In this case,

$$\int h \, d\sigma \approx \epsilon^{d/y} \log \epsilon.$$

The specific heat is found to be

$$C \approx (d/y)[(d/y) - 1]\epsilon^{d/y-2} \log \epsilon + [(d/y) - 1]\epsilon^{d/y-2}.$$

The first term will dominate for small ϵ. It is seen that if $d/y = 2$ (or $\alpha = 0$), a logarithmically singular specific heat will be found.

Arguments similar to those used to establish relations between the exponents governing the behavior of σ, χ, and C can also be applied to the correlation function $\mathcal{G}(\mathbf{r}, \mathbf{r}')$. This function will depend on $R = |\mathbf{r} - \mathbf{r}'|$, ϵ, and h. We will now denote it as $\mathcal{G}(R, \epsilon, h)$.

Thus $\mathcal{G}$ is defined by Eq. (2.7.29). Substitute into (2.7.29) the relation (2.7.51) between the local magnetization σ and the corresponding variable μ_a for the problem when described in terms of a large cube.

$$\mathcal{G}(R, \epsilon, h) = L^{2(x-d)}\langle[\mu_a - \langle\mu_a\rangle][\mu_a' - \langle\mu\rangle]\rangle = L^{2(x-d)}\mathcal{G}(R/L, \epsilon L^y, hL^x). \tag{2.7.65}$$

We now require that the function $\mathcal{G}$ be such that the artificial length L will disappear from (2.7.65). This will occur if

$$\mathcal{G}(R, \epsilon, h) = |\epsilon|^{2(d-x)/y} g(R|\epsilon|^{1/y}, \epsilon|h|^{-y/x}). \tag{2.7.66}$$

The reader will verify that if G has the form (2.7.66), Eq. (2.7.65) is satisfied. Let us now suppose that the molecular field expression (2.7.39) for G is essentially correct, and ask how the correlation length ξ depends on ϵ. From (2.7.39) and (2.7.66), it follows that $\xi \approx |\epsilon|^{-1/y}$. So [from Table I and (2.7.64)]

$$\nu = \nu' = 1/y = (2 - \alpha)/d. \tag{2.7.67}$$

The last critical exponent η is defined by

$$G(R, \epsilon = 0, h) \approx 1/R^{d-2+\eta}.$$

Then from (2.7.65),

$$\eta = d - 2x + 2 \tag{2.7.68a}$$

or

$$d\gamma/(2 - \eta) = 2 - \alpha. \tag{2.7.68b}$$

We see that given any two critical exponents, all the remaining ones can be determined from them.

These considerations which apply to the static correlation function, have been extended to the time- (or frequency-) dependent correlation function by Halperin and Hohenberg (1969).

The two-dimensional Ising model yields results which are in accord with the scaling laws, with $y = 1$, $x = 1.875$. The situation is not so clear cut in the case of the three-dimensional Ising model since exact results are not known; but indications from numerical calculations appear to agree reasonably well with the scaling laws if we take $y \approx 1.55$, $x = 2.5$. There appear to be some discrepancies which may be outside the existing uncertainties in the critical exponents, and it is possible that the critical indices above and below the transition temperature are not, in fact, equal.

The scaling law, Eq. (2.7.55) can be subjected to experimental check. The insulating ferromagnet $CrBr_3$ was carefully studied by Ho and Litster (1969) who determined the critical exponents for that compound presented in Table I. To test (2.7.55) directly, accurate measurements of σ as a function of temperature are made, leading to a determination of T_c and β. Likewise, measurements of $\sigma(h)$ at $T = T_c$ determine δ. Then we may form the so-called "scaled magnetization"

$$m = \sigma \mid \epsilon \mid^{-\beta} \tag{2.7.69}$$

and the scaled magnetic field

$$h_s = h \mid \epsilon \mid^{-\beta\delta}. \tag{2.7.70}$$

Then (2.7.55) implies that m is a function of the single variable h_s, or vice versa,

$$m = m(h_s) \qquad \text{or} \qquad h_s = h_s(m). \tag{2.7.71}$$

Measured values of σ at different fields and temperatures are used to specify the functions m and h_s. If Eq. (2.7.55) is valid, a plot of (2.7.71) should show the experimental points as lying on a single, definite curve. This was, in fact, found to be the case for $CrBr_3$; except that different $h_s(m)$ curves were found for $T > T_c$ and $T < T_c$.

Experimental information concerning the applicability of scaling laws to the correlation function can be derived from neutron scattering measurements. It was shown in Section 2.5 [see Eq. (2.5.28)] that such measurements can be interpreted in terms of the correlation function $\langle \mathbf{S}_\mu(t)\mathbf{S}_0(0)\rangle_T$ where $\mathbf{S}_\mu(t)$ is the spin operator for lattice site $\mathbf{R}_\mu$ at time t. Near the critical temperature, the large fluctuations in the magnetization produce a large scattering cross section. We will consider temperatures close to, but above, the critical temperature, where it is possible to describe the time dependence of $\mathbf{S}_\mu(t)$ in terms of the solution of a diffusion equation with decay governed by a parameter Λ (Van Hove, 1954). As a result, the function $\Gamma_{\alpha\beta}(\mathbf{K}, t)$, (2.5.28), behaves as

$$\sum_{\mu} \exp(i\mathbf{K}\cdot\mathbf{R}_\mu - \Lambda\mathbf{K}^2 t)\langle S_\mu{}^\alpha S_0{}^\beta\rangle_T$$

where $\langle S_\mu S_0\rangle_T$ is a static correlation function. For $\mathbf{R}_\mu \neq 0$, $\langle S_\mu{}^z S_0{}^z\rangle_T$ is the function $\mathcal{G}(\mathbf{R}_\mu)$ which we have discussed previously. Let us assume

$$\langle S_\mu{}^z S_0{}^z\rangle_T = A[\exp(-R_\mu/\xi)/R_\mu{}^{1+\eta}] \qquad (2.7.72)$$

where ξ is the correlation length. If $\eta = 0$, this is the Ornstein–Zernike correlation function. The scattering function $S(\mathbf{K}, \omega)$ [see Eqs. (2.5.20) and (2.5.22)] will be given by

$$S(\mathbf{K}, \omega) = (A/2\pi)\int_{-\infty}^{\infty} dt\, e^{-i\omega t} \sum_{\mu} \exp[i(\mathbf{K}\cdot\mathbf{R}_\mu - \Lambda K^2 t)] \times [\exp(-R_\mu/\xi)/R^{1+\eta}]. \qquad (2.7.73)$$

We replace the sum over μ by an integral over R, and obtain

$$S(\mathbf{K}, \omega) = (\Lambda K/\pi)[\Gamma(1-\eta)/(\omega^2 + \Lambda^2 K^4)][1/(\xi^{-2} + K^2)^{1/2(1-\eta)}] \times \sin[(1-\eta)\tan^{-1}(K\xi)]. \qquad (2.7.74)$$

If η is small, as appears to be the actual case, we may approximate

$$\sin[(1-\eta)\tan^{-1}K\xi)] \approx K/(\xi^{-2} + K^2)^{1/2}.$$

So

$$S(K, \omega) \approx (\Lambda K^2/\pi)[1/(\omega^2 + \Lambda^2 K^4)][1/(\xi^2 + K^2)^{(1-\eta/2)}]. \qquad (2.7.75)$$

If the result is integrated over energy transfers ω, we obtain a differential cross section proportional to

$$d\sigma/d\Omega \approx \int S(K, \omega)\, d\omega \approx 1/(\xi^{-2} + K^2)^{1-\eta/2}. \qquad (2.7.76)$$

Observations of Passell *et al.* (1965) concerning neutron scattering from iron indicate that η is small, of the order of 0.15.

Problems

1 Let u_A and u_B be ls wave functions for electrons bound to two hydrogen atoms, A and B. Consider a properly antisymmetrized wave function corresponding to double occupancy of the symmetric and antisymmetric molecular orbitals. Relate these functions to the molecular wave functions

which appear in the Heitler–London approximation. Explain the significance of the extra terms. Obtain a formal expression for the energy of the ground state in both the molecular orbital and Heitler–London approximations. Which method leads to lower energy?

2 Consider a two-sublattice ferrimagnet in which J_{11}, J_{12}, and J_{22} ($\neq J_{11}$) are the exchange interactions between spins on the sublattices. (a) Find a critical condition for the occurrence of ferrimagnetism ($T_N > 0$) in terms of the J. (b) Determine the magnetic susceptibility above the ferrimagnetic Néel temperature.

3 Use the Heisenberg Hamiltonian including an external magnetic field to find the quantum equation of motion for an atomic spin operator $\mathbf{S}_i$. Obtain the classical limit of this equation in which the spins are treated as ordinary vector functions of position. Show that the equation of motion becomes, for a simple cubic lattice in the long wavelength limit

$$\hbar \frac{d\mathbf{S}}{dt} = \frac{2}{3} JZ\, a^2 \mathbf{S} \times \nabla^2 \mathbf{S} + g\beta \mathbf{S} \times \mathbf{H}.$$

4 Show, by computing the temperature dependence of an assumed spontaneous magnetization, that spin wave theory predicts that ferromagnetic order can not be stable for a two-dimensional lattice with the Heisenberg Hamiltonian.

5 Verify Eqs. (2.4.60), (2.4.65), and (2.4.70).

6 Show that the specific heat of a ferromagnet in spin wave theory is

$$C/N = K[aT^{3/2} + bT^{5/2} + \cdots]$$

in which a is determined by Eq. (2.4.48). Evaluate the coefficient b for a simple cubic ferromagnet with nearest neighbor interactions.

7 Obtain the fourth-order terms in the expansion of the Heisenberg Hamiltonian in spin wave operators using the Holstein–Primakoff substitution [see Eqs. (2.4.29) and (2.4.30)]. Only nearest neighbor interactions need be considered. Show that there is a coupling between spin wave states of the form

$$V_4 = J/2N \sum_{\mathbf{kqps}} a_{\mathbf{k}}^{\dagger} a_{\mathbf{q}}^{\dagger} a_{\mathbf{p}} a_{\mathbf{s}} \delta_{\mathbf{k+q},\mathbf{p+s}} [\gamma(\mathbf{k}) + \gamma(\mathbf{p}) - 2\gamma(\mathbf{k}-\mathbf{p})]$$

where $\gamma(\mathbf{k}) = \sum_{\Delta} \exp(i\mathbf{k}\cdot\boldsymbol{\Delta})$, and $\boldsymbol{\Delta}$ is a nearest neighbor lattice vector.

8 Use the result of Problem 7 to show that the energy of a spin wave in a cubic lattice in the long wavelength limit varies with temperature as

$$\epsilon(\mathbf{k}, T) = \epsilon(\mathbf{k}, 0)(1 - \lambda T^{5/2}).$$

Find λ.

9 Determine the leading contribution from spin wave interactions to the low temperature specific heat of a Heisenberg ferromagnet using the results of Problem 8.

10 In a face centered cubic antiferromagnet of type 2, spins are strongly confined to (1, 1, 1)-type planes, and only weakly aligned along a single axis in such a plane (for example, 1, 1, −2). The Hamiltonian to be considered is

$$H = -2 \sum J_{ij}\mathbf{S}_i \cdot \mathbf{S}_j + \sum_i (D_1 S_{ix}^2 + D_2 S_{iy}^2)$$

The z axis, which is the direction of spin alignment is in a (1, 1, 1) plane, while the x and y axes are perpendicular and parallel to this plane. The quantities D_1 and D_2 are anisotropy constants ($D_1 \gg D_2$). Determine the spin wave energies in such a system. Consider first and second neighbor interactions. Compare your answer with that of Lines and Jones (1965).

11 Verify Eq. (2.6.63).

References

Akhiezer, A. I., Bar'yakhtar, V. G., and Peletminskii, S. V. (1968). "Spin Waves." North Holland, Amsterdam.
Allen, S. J., Loudon, R., and Richards, P. L. (1966). *Phys. Rev. Lett.* **16,** 463.
Anderson, P. W. (1952). *Phys. Rev.* **86,** 694.
Arfken, G. (1970). "Mathematical Methods for Physicists," 2nd. ed. Academic Press, New York.
Boyd, R. G., and Callaway, J. (1965). *Phys. Rev.* **138,** A1621.
Brezin, E., Wallace, D. J., and Wilson, K. G. (1973). *Phys. Rev.* **B7,** 232.
Brockhouse, B. N. (1957). *Phys. Rev.* **106,** 859.
Brout, R. (1965). "Phase Transitions." Benjamin, New York.
Charap, S. H., and Boyd, E. L. (1964). *Phys. Rev.* **133,** A811.
Condon, E. U., and Shortley, G. H. (1964). "The Theory of Atomic Spectra," p. 60. Cambridge Univ. Press, London and New York.
DeGennes, P. G. (1963). Theory of Neutron Scattering by Magnetic Crystals. *In* "Magnetism" (G. Rado and H. Suhl, eds.), Vol. III, p. 115. Academic Press, New York.
Dicke, R. H., and Wittke, J. P. (1960). "Introduction to Quantum Mechanics." Addison-Wesley, Reading, Mass.
Dirac, P. A. M. (1929). *Proc. Roy. Soc.* (*London*) **A123,** 714.
Domb, C. (1960). *Advan. Phys.* **9,** 151.
Dyson, F. J. (1956a). *Phys. Rev.* **102,** 1217.
Dyson, F. J. (1956b). *Phys. Rev.* **102,** 1230.
Fisher, M. E. (1963). *J. Math. Phys.* **4,** 278.
Fisher, M. E. (1967). *Rept. Prog. Phys.* (**GB**)**30,** 615.
Fleury, P. A., and Porto, S. P. S. (1968). *J. Appl. Phys.* **39,** 1035.
Glasser, M. L., and Milford, F. J. (1963). *Phys. Rev.* **130,** 1783.

Goldstone, J., Salam, A., and Weinberg, S. (1962). *Phys. Rev.* **127,** 965.
Halperin, B. I., and Hohenberg, P. C. (1969). *Phys. Rev.* **177,** 952.
Halpern, O., and Johnson, M. H. (1939). *Phys. Rev.* **55,** 898.
Hanus, J. (1963). *Phys. Rev. Lett.* **11,** 336.
Harris, A. B., Kumar, D., Halperin, B. I., and Hohenberg, P. C. (1971). *Phys. Rev.* **B3,** 961.
Heller, P. (1967). *Rep. Progr. Phys.* **30,** 731.
Herring, C. (1966). Direct Exchange between Well Separated Atoms. *In* "Magnetsim" (G. Rado and H. Suhl, eds.), Vol. 2B, p. 1. Academic Press, New York.
Herring, C., and Flicker, M. (1964). *Phys. Rev.* **134,** A362.
Ho, J. T., and Litster, J. D. (1969). *Phys. Rev. Lett.* **22,** 603.
Holstein. T., and Primakoff, H. (1940). *Phys. Rev.* **58,** 1098.
Hove, L. Van (1954). *Phys. Rev.* **95,** 1374.
Huberman, B. A., Burstein, E., and Ito, R. (1972). *Phys. Rev.* **B5,** 168.
Ising, E. (1925). *Z. Phys.* **31,** 253.
Izuyama, T., Kim, D., and Kubo, R. (1963). *J. Phys. Soc. Japan* **18,** 1025.
Izyumov, Yu. A. (1963). *Usp. Fiz. Nauk* **80,** 41 (*English transl.: Sov. Phys. Usp.* **16,** 359).
Kadanoff, L. P. *et al.* (1967). *Rev. Mod. Phys.* **39,** 395.
Keffer, F. (1966). Spin Waves. *In* "Encyclopedia of Physics" (S. Flügge, ed.), Vol. 18/2, p. 1. Springer-Verlag, Berlin.
Kramers, H. A., and Wannier, G. H. (1941). *Phys. Rev.* **60,** 252.
Landau, L., and Lifshitz, E. M. (1958). "Statistical Physics," p. 482. Addison-Wesley, Reading, Massachusetts.
Lee, T. D., and Yang, C. N. (1952). *Phys. Rev.* **87,** 410.
Lines, M. E., and Jones, E. D. (1965). *Phys. Rev.* **139,** A1313.
Loudon, R. (1968). *Advan. Phys.* **17,** 243.
Maleev, S. V. (1958). *J. Exp. Theoret. Phys. USSR* **33,** 1010 (*English transl.: Sov. Phys. JETP* **6,** 776.)
McCollum, D. C., and Callaway, J. (1962). *Phys. Rev. Lett.* **9,** 376.
Minkiewicz, V. J., Collins, M. F., and Shirane, G. (1969). *Phys. Rev.* **182,** 624.
Montroll, E. W., Potts, R. B., and Ward, J. C. (1963). *J. Math. Phys.* **4,** 308.
Moriya, T. (1968). *J. Appl. Phys.* **39,** 1042.
Newell, G. F., and Montroll, E. W. (1953). *Rev. Mod. Phys.* **25,** 353.
Onsager, L. (1944). *Phys. Rev.* **65,** 117.
Onsager, L. (1949). *Nuovo Cimento. Suppl.* **6,** 261.
Ornstein, L. S., and Zernike, F. (1914). *Proc. Acad. Sci. Amsterdam* **17,** 793.
Ornstein, L. S., and Zernike, F. (1918). *Phys. Z.* **19,** 134.
Passel, L., Blinowski, K., Brun, T., and Nielson, P. L. (1965). *Phys. Rev.* **139,** A1866.
Passenheim, B. C., McCollum, D. C., and Callaway, J. (1966). *Phys. Lett.* **23,** 634.
Peierls, R. (1936). *Proc. Cambridge Phil. Soc.* **32,** 477.
Schultz, T. D., Mattis, D. C., and Lieb, E. H. (1964). *Rev. Mod. Phys.* **36,** 856.
Scott, P. D., and Wolf, W. P. (1969). *J. Appl. Phys.* **40,** 1031.
Sell, D. D. (1968). *J. Appl. Phys.* **39,** 1030.
Silberglitt, R., and Harris, A. B. (1968). *Phys. Rev.* **174,** 640.
Smart, J. S. (1966). "Effective Field Theories of Magnetism." Saunders, Philadelphia, Pennsylvania.
Stanley, H. E. (1971). "Introduction to Phase Transactions and Critical Phenomena." Oxford Univ. Press, London and New York.
Sugiura, Y. (1927). *Z. Phys.* **45,** 484.

Tanabe, Y., Moriya, T., and Sugano, S. (1965). *Phys. Rev. Lett.* **15,** 1023.
Turov, E. A. (1965). "Physical Properties of Magnetically Ordered Crystals." Academic Press, New York.
van Vleck, J. H. (1932). "The Theory of Electric and Magnetic Susceptibilities." Oxford Univ. Press, London and New York.
Wannier, G. H. (1950). *Phys. Rev.* **79,** 357.
Weiss, P. (1907). *J. Phys.* **6,** 661.
Wigner, E. P. (1959). "Group Theory and Its Application to the Quantum Mechanics of Atomic Spectra" (J. J. Griffin, transl.). Academic Press, New York.
Wilson, K. G. (1972). *Phys. Rev. Lett.* **28,** 548.
Wilson, K. G., and Fisher, M. E. (1972). *Phys. Rev. Lett.* **28,** 240.
Wortis, M. (1963). *Phys. Rev.* **132,** 85.
Wortis, M. (1965). *Phys. Rev.* **138,** A1126.
Yang, C. N. (1952). *Phys. Rev.* **85,** 808.

CHAPTER 3

Symmetry and Its Consequences

The study of the symmetries of crystals furnishes an important means of describing states and analyzing transitions. Some results of translational symmetry, in particular the existence of the Brillouin zone, were presented in Section 1.2. It is of great significance that solids possess additional symmetries involving rotations and reflections. Some of these will be investigated in this chapter. The fundamental principle is that the quantum mechanical operators representing a symmetry operation of the crystal must commute with the Hamiltonian for the crystal: this enables the derivation of restrictions limiting possible Hamiltonians, the classification of eigenstates according to symmetry, and the determination of selection rules for transitions between states. The proper utilization of symmetry considerations frequently requires group theory. It is therefore desirable to begin with some discussion of the principles of group theory as applied to solids.

We will not attempt to give a detailed introduction to the principles of group theory. There are many books which give an adequate treatment of groups from the physicist's point of view. Useful general references include Eyring *et al.* (1944), Wigner (1959), Lomont (1959), Heine (1960), Lyubarski (1960), Hamermesh (1962), Knox and Gold (1964), Tinkham (1964), Slater (1965), Falicov (1966), Cornwell (1969), and Bradley and Cracknell (1972). We will state the definitions of certain mathematical quantities relevant to our discussion and quote theorems without proof. The reader who finds the general concepts unfamiliar is strongly urged to consult references listed above.

3.1 Space Groups and Point Groups

The elements of the symmetry groups relevant to solids include rotations (both proper and improper) and translations. Such groups are known as *space groups*. A precise definition will be given subsequently.

3.1.1 *Space Group Operations*

The formal properties of space groups were described in detail by Seitz (1936), whose notation is extensively used. An important review is that by Koster (1957). An operator of a space group contains a part which is either a proper, or improper rotation $\boldsymbol{\alpha}$ and a translation part $\mathbf{t}$, and is denoted by the symbol $\{\boldsymbol{\alpha} \mid \mathbf{t}\}$. This operator corresponds to the coordinate transformation

$$\mathbf{x}' = \boldsymbol{\alpha}\mathbf{x} + \mathbf{t}. \tag{3.1.1}$$

In (3.1.1), $\boldsymbol{\alpha}$ can be represented by a 3×3 orthogonal matrix. Two such operators $\{\boldsymbol{\alpha} \mid \mathbf{t}\}$ and $\{\boldsymbol{\beta} \mid \mathbf{t}'\}$ multiply in the following way:

$$\{\boldsymbol{\beta} \mid \mathbf{t}'\}\{\boldsymbol{\alpha} \mid \mathbf{t}\} = \{\boldsymbol{\beta}\boldsymbol{\alpha} \mid \boldsymbol{\beta}\mathbf{t} + \mathbf{t}'\}. \tag{3.1.2}$$

The unit operator is $\{\boldsymbol{\varepsilon} \mid 0\}$. It may be verified, using the multiplication rule (3.1.2), that the inverse of the operator $\{\boldsymbol{\alpha} \mid \mathbf{t}\}$ is

$$\{\boldsymbol{\alpha} \mid \mathbf{t}\}^{-1} = \{\boldsymbol{\alpha}^{-1} \mid -\boldsymbol{\alpha}^{-1}\mathbf{t}\}. \tag{3.1.3}$$

The following matrix representation is useful as an aid to an understanding of the properties of the space group operators: Denote a position vector $\mathbf{x}$ by

$$\begin{pmatrix} 1 \\ x_1 \\ x_2 \\ x_3 \end{pmatrix}$$

Equation (3.1.1) is written as

$$\begin{pmatrix} 1 \\ x_1' \\ x_2' \\ x_3' \end{pmatrix} = \begin{pmatrix} 1 & 0 & 0 & 0 \\ t_1 & \alpha_{11} & \alpha_{12} & \alpha_{13} \\ t_2 & \alpha_{21} & \alpha_{22} & \alpha_{23} \\ t_3 & \alpha_{31} & \alpha_{32} & \alpha_{33} \end{pmatrix} \begin{pmatrix} 1 \\ x_1 \\ x_2 \\ x_3 \end{pmatrix}, \tag{3.1.4}$$

or, in an obvious block notation, as

$$\begin{pmatrix} 1 \\ \mathbf{x}' \end{pmatrix} = \begin{pmatrix} 1 & 0 \\ \mathbf{t} & \boldsymbol{\alpha} \end{pmatrix} \begin{pmatrix} 1 \\ \mathbf{x} \end{pmatrix}. \tag{3.1.5}$$

The multiplication rule (3.1.2) becomes

$$\begin{pmatrix} 1 & 0 \\ \mathbf{t}' & \boldsymbol{\beta} \end{pmatrix}\begin{pmatrix} 1 & 0 \\ \mathbf{t} & \boldsymbol{\alpha} \end{pmatrix} = \begin{pmatrix} 1 & 0 \\ \mathbf{t}' + \boldsymbol{\beta}\mathbf{t} & \boldsymbol{\beta}\boldsymbol{\alpha} \end{pmatrix} \tag{3.1.6}$$

and the inverse is

$$\begin{pmatrix} 1 & 0 \\ \mathbf{t} & \boldsymbol{\alpha} \end{pmatrix}^{-1} = \begin{pmatrix} 1 & 0 \\ -\boldsymbol{\alpha}^{-1}\mathbf{t} & \boldsymbol{\alpha}^{-1} \end{pmatrix}. \tag{3.1.7}$$

The operator representing a lattice translation through $\mathbf{R}_i$ is denoted by $\{\boldsymbol{\varepsilon} \mid \mathbf{R}_i\}$. Since the sum of any two lattice vectors $\mathbf{R}_i$, $\mathbf{R}_j$ is another lattice vector, and since each $\{\boldsymbol{\varepsilon} \mid \mathbf{R}_i\}$ has an inverse $\{\boldsymbol{\varepsilon} \mid -\mathbf{R}_i\}$, it is apparent that the lattice translations form a group. The group of lattice translations is Abelian, which means that its members commute with each other. The more general operators of the full space group containing rotations as well as translations do not possess this property.

The group of lattice translations is evidently a subgroup of full space group. It is also an invariant subgroup. Recall that a subgroup is said to be invariant if the following condition is satisfied: Let A be any member of the subgroup and X any member of the full group. Form the product $B = XAX^{-1}$. If B is in the subgroup concerned for all A and X, then the subgroup is said to be *invariant.* To see that the lattice translation subgroup is invariant, note that if $\boldsymbol{\alpha}$ is the rotational part of any space group operation and $\mathbf{R}_i$ is any direct lattice vector, $\boldsymbol{\alpha}\mathbf{R}_i$ is also a direct lattice vector. Now form the product for arbirtary $\{\boldsymbol{\alpha} \mid \mathbf{t}\}$,

$$\{\boldsymbol{\alpha} \mid \mathbf{t}\}\{\boldsymbol{\varepsilon} \mid \mathbf{R}_i\}\{\boldsymbol{\alpha} \mid \mathbf{t}\}^{-1} = \{\boldsymbol{\alpha} \mid \boldsymbol{\alpha}\mathbf{R}_i + \mathbf{t}\}\{\boldsymbol{\alpha}^{-1} \mid -\boldsymbol{\alpha}^{-1}\mathbf{t}\} = \{\boldsymbol{\varepsilon} \mid \boldsymbol{\alpha}\mathbf{R}_i\}. \tag{3.1.8}$$

Since $\{\boldsymbol{\varepsilon} \mid \boldsymbol{\alpha}\mathbf{R}_i\}$ is a lattice translation if $\{\boldsymbol{\varepsilon} \mid \mathbf{R}_i\}$ is, (3.1.8) shows that the subgroup is invariant.

This property is used as the defining condition for a group to be a space group: A space group is defined to be a group of operators of the form $\{\boldsymbol{\alpha} \mid \mathbf{t}\}$ which possesses an invariant subgroup of pure translations. There are only a finite number of possible space groups in a space of finite dimensions: 230 in three-dimensional space.

We may inquire whether the translation parts of the operators $\{\mathbf{a} \mid \mathbf{t}\}$ are lattice translations, that is, can any $\mathbf{t}$ be written as $l\mathbf{a}_1 + m\mathbf{a}_2 + n\mathbf{a}_3$, where $\mathbf{a}_1$, $\mathbf{a}_2$, $\mathbf{a}_3$ are the primitive lattice vectors and l, m, and n are integers? This is so for certain space groups (called *simple,* or *symmorphic*), but does not hold in general. For a simple space group, we may consider the elements

$\{\boldsymbol{\alpha} \mid \mathbf{0}\}$, which do not involve any translational part. These operators form a subgroup of the space group known as the *point group*. In the case of a nonsymmorphic space group, the point group may be constructed in the same way but it will not be a subgroup of the space group. We may think of a point group as containing all the symmetry operations of a crystal which may be performed with one point held fixed.

3.1.2 *Point Group Operations*

The rotations $\boldsymbol{\alpha}$ which make up a point group (and are incorporated in a space group) are not arbitrary: only rotations through 60°, 90°, or multiples of these are permitted. The particular operations which are permitted in a specific case will vary from example to example; the contention here is that no point group can contain rotations other than through 60°, 90°, and (or) multiples of these. This forbids fivefold rotational symmetry, for example.

The proof is actually simple. It is based on the fact that $\boldsymbol{\alpha}\mathbf{R}$ must be a direct lattice vector if $\mathbf{R}$ is. However,

$$\mathbf{R} = n_1\mathbf{a}_1 + n_2\mathbf{a}_2 + n_3\mathbf{a}_3 \tag{3.1.9}$$

where the n's are integers. A similar equation must hold for $\boldsymbol{\alpha}\mathbf{R}$, that is, $\boldsymbol{\alpha}\mathbf{R}$ must be expressible in terms of the primitive lattice vectors $\mathbf{a}_j$ with integer coefficients. Now let $\boldsymbol{\alpha}$ be a rotation through an angle ϕ about some axis. In considering the operation of $\boldsymbol{\alpha}$ on $\mathbf{R}$, we must allow for the fact that the $\mathbf{a}_j$ may not be orthogonal. Let the $\mathbf{a}_j$ be related to a Cartesian system by a nonsingular matrix $\mathbf{A}$. Denote the vector whose components are n_j in (3.1.9) by $\mathbf{n}$ and the same object for the rotated vector by $\mathbf{n}'$. We must have

$$\mathbf{nA}\boldsymbol{\alpha} = \mathbf{n}'\mathbf{A} \qquad \text{or} \qquad \mathbf{nA}\boldsymbol{\alpha}\mathbf{A}^{-1} = \mathbf{n}'. \tag{3.1.10}$$

Since the elements of $\mathbf{n}$ and $\mathbf{n}'$ are arbitrary integers, the elements of $\mathbf{A}\boldsymbol{\alpha}\mathbf{A}^{-1}$ must be integers. This must also hold for the trace of $\mathbf{A}\boldsymbol{\alpha}\mathbf{A}^{-1}$. However, the trace of a matrix is invariant under a similarity transformation, so

$$\mathrm{tr}(\mathbf{A}\boldsymbol{\alpha}\mathbf{A}^{-1}) = \mathrm{tr}(\boldsymbol{\alpha}) = 1 + 2\cos\phi = \text{integer}.$$

This implies that ϕ must be 60°, 90°, or an integral multiple of these angles.

3.1.3 *Description of Point Groups*

There are only 32 possible crystal point groups. Their properties are described in detail by Koster *et al.* (1963). We will discuss their constitution briefly. Unfortunately, there exist two widely used but different systems of notation for point groups: the international system and the Schoenflies system; there are both "short" and "full" forms for the international notation. We will use the Schoenflies notation. The relation between the different notations is given in Tables I and II.

(1) C_n: These are the point groups which contain only a single axis of symmetry, around which rotations through angles of $2\pi/n$ are permitted. From the preceding discussion, we see that the possible values for n are 1, 2, 3, 4, 6.

Next, we must consider groups involving reflections.

(2) C_{nv}: In addition to the operation C_n or rotation through $2\pi/n$, these groups include reflection in a plane containing the axis of highest symmetry (which is defined as a vertical plane). The reflection symmetry must be consistent with the rotational symmetry, and n-fold rotational symmetry about some axis demands the existence of n reflection planes at angles of π/n if there are any reflection plane containing the axis. The allowed of n are 2, 3, 4, 6.

(3) C_{nh}: These groups contain a "horizontal" reflection plane (operation σ_h), that is, a reflection in a plane through the origin perpendicular to the axis of highest symmetry. If n is even, the group contains the inversion operation i as well. Here $n = 1, 2, 3, 4, 6$.

(4) S_n: These groups have an n-fold improper rotation (rotation combined with reflection in a plane perpendicular to the axis of rotation).

TABLE I

SYMMETRY OPERATIONS

Operation	Shoenflies symbol
Identity	E
Rotation through $2\pi/n$	C_n
Inversion	i
Improper rotation through $2\pi/n$	S_n
Reflection in a plane	σ
Reflection in a plane perpendicular to highest symmetry axis	σ_h
Reflection in a plane containing highest symmetry axis	σ_v
Reflection in a plane containing the highest symmetry axis and bisecting the angle between twofold axes perpendicular to symmetry axis	σ_d

TABLE II

INTERNATIONAL POINT GROUP SYMBOLS

Axis	Symbol[a]
n-Fold rotation axis	n
Improper n-fold rotation axis	$\bar{n}$
Rotation axis with reflection plane perpendicular to it	$\dfrac{n}{m}$
n-Fold rotation axis with twofold axis perpendicular to it	$n2$
n-Fold rotation axis with reflection plane containing the axis	nm
Improper rotation axis with twofold axis perpendicular to it	$\bar{n}2$
Improper rotation axis with reflection planes containing the axis	$\bar{n}m$
Rotation axis with a perpendicular reflection and a set of reflection planes containing it	$\dfrac{n}{mm}$ or $\dfrac{n}{m}m$

[a] n is an integer 1, 2, 4, 6.

The group S_3 is identical with C_{3h} and is therefore not counted. The distinct groups are S_2 (simple inversion), S_4, and S_6.

(5) D_n: In these cases, we have n twofold axes perpendicular to the highest symmetry (C_n) axis. In the case of D_2, there are three twofold axes. We may have $n = 2, 3, 4, 6$.

(6) D_{nd}: Here we have in addition to the operations of D_n, reflections in a "diagonal" plane (σ_d) containing the symmetry axis and bisecting the angle between the twofold axis. The diagonal planes are special cases of vertical planes. There are two cases: D_{2d} and D_{3d}.

(7) D_{nh}: Such groups contain the horizontal reflection plane (σ_h) in addition to the symmetries of D_n. Therefore these are twice as many elements in D_{nh} as in D_n.

The remaining groups are of higher symmetry:

(8) T: Contains the 12 proper rotations which carry a regular tetrahedron into itself.

(9) T_d: Contains the 24 symmetry operations of the regular tetrahedron, including improper rotations.

(10) T_h: This is the 24-element group consisting of the operations of T, and those combined with the inversion i.

TABLE III

THE 32 CRYSTAL POINT GROUPS

Crystal system	Unit cell axes and angles	Groups: Shoenflies symbol	International: Full	International: Short	Number of elements
Triclinic	$a \neq b \neq c$	C_1	1	1	1
	$\alpha \neq \beta \neq \gamma$	S_2	$\bar{1}$	$\bar{1}$	2
Monoclinic	$a \neq b \neq c$	C_{1h}	m	m	2
	$\alpha = \gamma = 90°$	C_2	2	2	2
	$\beta \neq \alpha$	C_{2h}	$2/m$	$2/m$	4
Orthohombic	$a \neq b \neq c$	C_{2v}	$mm2$	$mm2$	4
	$\alpha = \beta = \gamma = 90°$	D_2	222	222	4
		D_{2h}	$(2/m)(2/m)(2/m)$	mmm	8
Tetragonal	$a = b \neq c$	C_4	4	4	4
	$\alpha = \beta = \gamma = 90°$	S_4	$\bar{4}$	$\bar{4}$	4
		C_{4h}	$4/m$	$4/m$	8
		D_{2d}	$\bar{4}2m$	$\bar{4}2m$	8
		C_{4v}	$4mm$	$4mm$	8
		D_4	422	422	8
		D_{4h}	$(4/m)(2/m)(2/m)$	$4/mmm$	16
Rhombohedral (trigonal)	$a = b = c$	C_3	3	3	3
	$\alpha = \beta = \gamma < 120°$	S_6	$\bar{3}$	$\bar{3}$	6
	$\alpha \neq 90°$	C_{3v}	$3m$	$3m$	6
		D_3	32	32	6
		D_{3d}	$\bar{3}(2/m)$	$\bar{3}m$	12
Hexagonal	$a = b \neq c$	C_6	6	6	6
	$\alpha = \beta = 90°$	C_{3h}	$\bar{6}$	$\bar{6}$	6
	$\gamma = 120°$	C_{6h}	$6/m$	$6/m$	12
		D_{3h}	$\bar{6}m2$	$\bar{6}m2$	12
		C_{6v}	$6mm$	$6mm$	12
		D_6	622	622	12
		D_{6h}	$(6/m)(2/m)(2/m)$	$6/mmm$	24
Cubic	$a = b = c$	T	23	23	12
	$\alpha = \beta = \gamma = 90°$	T_h	$(2/m)\bar{3}$	$m\bar{3}$	24
		T_d	$\bar{4}3m$	$\bar{4}3m$	24
		O	432	432	24
		O_h	$(4/m)\bar{3}(2/m)$	$m3m$	48

(11) O: This is the octahedral group, containing the 24 proper rotations which carry a cube into itself.

(12) O_h: This is the largest point group, containing all 48 elements (proper and improper rotations) which carry a cube into itself.

This information is summarized in Tables I–III. Table III also contains a brief description of the crystal systems in which the particular point system may be found.

It is possible and conventional to represent the point groups by a stereogram, which indicates by projection on to a plane, the motion of a representative point on the surface of a sphere under the action of the operators of the group. We will not exhibit the stereograms here (see Tinkham, 1964; or Kittel, 1956).

3.1.4 *The Cubic Group O_h*

In order to have a specific example to which further analysis may be applied, and because of its own intrinsic importance as the point group of highest symmetry, the full cubic group O_h will be described in more detail. The operations of this group may be described as follows (the operations are arranged in ten classes‡):

I. The identity E.

II. Rotation by $\pm 90°$ about a coordinate (fourfold) axis. Class C_4, six operations.

III. Rotations by 180° about the same axis. There are six such axes. Class C_2, six operations.

V. Rotations by $\pm 120°$ about a threefold axis (body diagonal of the cube). There are four such axes. Class C_3, eight operations.

VI. The inversion with respect to the origin. Class J, one operation.

VII. -X Classes JC_4, JC_4^2, JC_2, and JC_3: Combinations of the above operations with the inversion. These four classes contain 23 operations. An alternative designation (σ_h, S_4, σ_d, S_6) is sometimes used for these classes.

The cubic group may be conveniently described in another manner. The operations of the cubic group on a position vector **r** with components x, y, z can be specified as the possible rearrangements or permutations of

‡ A class is a set of elements of a group which are conjugate to each other. If X and A are members of a group, the element $B = XAX^{-1}$ is conjugate to A. If two elements A, C are conjugate to a third element D, they are conjugate to each other. In less formal terminology, two transformations A and B are in the same class if it is possible to find a new coordinate system in which the transformation B has the same effect that transformation A had in the previous system.

TABLE IV

THE CUBIC GROUP O_h

Class	Operation			Class	Operation		
E	x	y	z	J	$-x$	$-y$	$-z$
C_4^2	$-x$	$-y$	z	JC_4^2	x	y	$-z$
	x	$-y$	$-z$		$-x$	y	z
	$-x$	y	$-z$		x	$-y$	z
C_4	$-y$	x	z	JC_4	y	$-x$	$-z$
	y	$-x$	z		$-y$	x	$-z$
	x	$-z$	y		$-x$	z	$-y$
	x	z	$-y$		$-x$	$-z$	y
	z	y	$-x$		$-z$	$-y$	x
	$-z$	y	x		z	$-y$	$-x$
C_2	y	x	$-z$	JC_2	$-y$	$-x$	z
	z	$-y$	x		$-z$	y	$-z$
	$-x$	z	y		x	$-z$	$-x$
	$-y$	$-x$	$-z$		y	x	z
	$-z$	$-y$	$-x$		z	y	x
	$-x$	$-z$	$-y$		x	z	y
C_3	z	x	y	JC_3	$-z$	$-x$	$-y$
	y	z	x		$-y$	$-z$	$-x$
	z	$-x$	$-y$		$-z$	x	y
	$-y$	$-z$	x		y	z	$-x$
	$-z$	$-x$	y		z	x	$-y$
	$-y$	z	$-x$		y	$-z$	x
	$-z$	x	$-y$		z	$-x$	y
	y	$-z$	$-x$		$-y$	z	x

x, y, z including changes of sign. This description is used in Table IV where the operations are presented.

The operations contained in the classes E, C_4^2, C_4, C_2, C_3 form the subgroup O. Those in classes E, C_4^2, JC_4, JC_2, and C_3 comprise the subgroup T_d.

The point groups listed in Table III are found in certain crystal lattices which are also listed in that table. It should be apparent that the requirement that the rotational and the translational symmetry must be compatible limits the possible translations associated with a given rotational symmetry. As a result, the number of space groups is finite, and equal to 230 for three-dimensional space. Since there are only 32 point groups, it is clear that some point groups will occur in many different space groups. For example, ten different space groups are associated with the full cubic point group. These include the ordinary cubic O_h^1 [cesium chloride and perovskite ($CaTiO_3$) structures]; body-centered cubic O_h^9, face-centered cubic O_h^5 [including sodium chloride and fluorite (CaF_2) structures], the diamond lattice O_h^7 and others. One notation just adds a numerical superscript to the Schoenflies point group symbol to designate the space group. We will not give details of the various structures here. The interested reader will find a very useful account in Slater (1965).

3.2 Irreducible Representations: Point Groups

The usefulness of group theory is based on a general principle of quantum mechanics: The wave functions for a quantum system must form bases for irreducible representations of the group of operators which commute with the Hamiltonian of the system (see Wigner, 1959, Chap. 11).

3.2.1 *General Properties of Irreducible Representations*

Let us briefly review what is meant by an irreducible representation.

A group is said to be *represented* by a set of matrices B_i if to each element in the group, there corresponds a matrix such that products correspond to products, etc.

A matrix M is said to be the *direct sum* of matrices $m_1, m_2, \ldots$ if every element of M is zero except for square blocks (the submatrices $m_1, m_2, \ldots$) along the diagonal. If each matrix of a representation can be expressed as a direct sum in the same way, and if the dimensions of the corresponding submatrices are the same in every case, then the submatrices themselves provide a representation of the group, and the original representation has been *reduced*. In order to carry out this reduction, it is necessary to find a unitary transformation U which will bring the matrices to the required form $M_i = U^{-1}B_iU$ (the same matrix U for all the B_i).

In terms of the vectors of the space on which the B_i operate, the reduction separates out the subspaces which are carried into themselves by all the B_i (*invariant subspaces*). If no transformation exists by means of which the matrices of a representation may be expressed as direct sums, the representation is said to be *irreducible*. The wave functions for a quantum system which possesses a symmetry group which commutes with the Hamiltonian are required to be basis functions for such matrix representations.

Perhaps the most familiar example is furnished by the treatment of angular momentum in elementary quantum mechanics. The group in question is the rotation group. It has an infinite number of irreducible representations. For spinless particles, the representations are characterized by the eigenvalue of $\mathbf{L}^2$, which is $l(l + 1)$. Each representation has dimension $(2l + 1)$, and the $2l + 1$ spherical harmonics $Y_{lm}(\theta, \phi)$ are the basis functions for the representation. Any rotation can be represented by a matrix which carries the basis functions for a representation into each other, but does not mix states of different angular momentum.

In some respects the situation in solid state physics is a little simpler since many of the groups of interest are finite. This is the case for point groups. There are two general theorems pertaining to the irreducible representations of point groups which are important in our applications. These are stated without proof (consult the texts listed previously).

(1) The number of irreducible representations equals the number of classes.

(2) The sum of the squares of the dimensions of the irreducible representations equals the number of operations in the group.

As an example of these theorems, consider the cubic group O_h of Table IV. Since there are ten classes, there are ten irreducible representations. Four are one-dimensional, two are two-dimensional, and four are three-dimensional representations. Observe that $4 \times (1)^2 + 2 \times (2)^2 + 4 \times (3)^2 = 48$.

The irreducible representations of a finite group are frequently described by presenting a character table. The character is the trace of a representation matrix. From the invarance of the trace under a similarity transformation, it follows that (1) the character is the same for all operations in a given class, and (2) it is independent of the choice of basis functions for the representation. Specific procedures exist which enable construction of character tables. However, we will not discuss these here (see Hamermesh, 1962). Character tables for all of the 32 point groups are given in several references, one of which is Koster (1957).

3.2.2 *Basis Functions*

For applications in solid state physics it is frequently necessary to find linear combinations of a set of basis functions (spherical harmonics, plane waves, etc.) which transform according to a given irreducible representation. Such functions are basis functions for the given representation. This construction can be made using a projection technique, which we will now describe (Wigner, 1959; Hamermesh, 1962).

It is necessary to describe the effect of a group operator on a function. Suppose $\boldsymbol{\alpha}$ is some orthogonal transformation of coordinates, such that $\boldsymbol{\alpha}$

acting on $\mathbf{r}$ sends it into $\mathbf{r}'$: $\mathbf{r}' = \boldsymbol{\alpha}\mathbf{r}$. We associate with $\boldsymbol{\alpha}$ an operator O_α which acts on functions $\phi(\mathbf{r})$. For any $\phi(\mathbf{r})$, the effect of the operator O_α on ϕ is to change it to a new function $\phi' = O_\alpha\phi$ where

$$\phi'(\mathbf{r}') = O_\alpha\phi(\mathbf{r}') = \phi(\mathbf{r}). \tag{3.2.1}$$

Equation (3.2.1) implies that the value of the transformed function at the rotated or transformed point is the same as the value of the original function at the original point. This may be understood if we interpret $\boldsymbol{\alpha}$ as simply a change of the coordinate system. The "physical" point $\mathbf{r}$ and the function $\phi(\mathbf{r})$ are left exactly as they were, but we describe them now as $\mathbf{r}'$ and $\phi'(\mathbf{r}')$. Thus

$$O_\alpha\phi(\boldsymbol{\alpha}\mathbf{r}) = \phi(\mathbf{r}), \tag{3.2.2}$$

or (after putting $\mathbf{r} = \boldsymbol{\alpha}^{-1}\mathbf{r}'$, then dropping the prime),

$$O_\alpha\phi(\mathbf{r}) = \phi(\boldsymbol{\alpha}^{-1}\mathbf{r}). \tag{3.2.3}$$

It is easy to see that the O_α are in one-to-one correspondence with the $\boldsymbol{\alpha}$, such that products correspond to products, etc. For this reason, we usually may drop the symbol O_α and replace it simply by $\boldsymbol{\alpha}$ without causing misunderstandings. We write

$$\boldsymbol{\alpha}\phi(\mathbf{r}) = \phi(\boldsymbol{\alpha}^{-1}\mathbf{r}). \tag{3.2.4}$$

Suppose that ϕ transforms according to the jth row of the ith irreducible representation of a group of operators of which $\boldsymbol{\alpha}$ is a member. Let the dimension of the representation be $d(i)$. This means that there are $d(i)$ functions, one of which is ϕ, which are carried into linear combinations of each other by the operations of the group. These are called "partner" functions, and are denoted by $\phi_i{}^j(\mathbf{r})$ where j designates the row of the representation. The transformation properties of these functions are expressed by

$$\boldsymbol{\alpha}\phi_i{}^j(\mathbf{r}) = \sum_{m=1}^{d(i)} [\boldsymbol{\alpha}]_{i,mj}\phi_i{}^m(\mathbf{r}). \tag{3.2.5}$$

The symbol $[\boldsymbol{\alpha}]_{i,mj}$ denotes the mjth element of the matrix representing the operation $\boldsymbol{\alpha}$ in the ith representation.

The construction of functions transforming according to a specific representation proceeds as follows. Equation (3.2.5) is multiplied by $[\boldsymbol{\alpha}]^*{}_{i,mj}$ and the result is summed over all the g operations of the group

$$\sum_\alpha [\boldsymbol{\alpha}]^*{}_{i',m'j'}\,\boldsymbol{\alpha}\phi_i{}^j = \sum_\alpha \sum_m [\boldsymbol{\alpha}]^*{}_{i',m'j'}[\boldsymbol{\alpha}]_{i,mj}\phi_i{}^m.$$

(An asterisk denotes complex conjugate.)

The right side of this equation can be evaluated through use of the general orthogonality theorem (for proof, see Wigner, 1959, p. 79)

$$\sum_{\alpha} [\alpha]^*_{i',m'j'}[\alpha]_{i,mj} = (g/d(i))\ \delta_{ii'}\ \delta_{jj'}\ \delta_{mm'} \tag{3.2.6}$$

(where g is the order of the group). Thus we have

$$\sum_{\alpha} [\alpha]^*_{i',m'j'}\ \alpha\phi_i{}^j = (g/d(i))\ \delta_{ii'}\ \delta_{jj'}\ \phi_i{}^{m'}. \tag{3.2.7a}$$

In the particular case $j = j' = m'$, $i = i'$, this reduces to

$$\phi_i{}^j = (d(i)/g) \sum_{\alpha} [\alpha]^*_{i,jj}\ \alpha\phi_i{}^j. \tag{3.2.7b}$$

This result can be used in the following way: Let F be an arbitrary function which can be expressed as a linear combination of functions $\phi_i{}^j$ belonging to the various rows of the irreducible representations of the group

$$F(\mathbf{r}) = \sum_{i,j} a_i{}^j \phi_i{}^j(\mathbf{r}) \tag{3.2.8}$$

where the $a_i{}^j$ are coefficients. Form the sum

$$\sum_{\alpha} [\alpha]^*_{i,jj}\ \alpha F = \sum_{i'j'} a_{i'}{}^{j'} \sum_{\alpha} [\alpha]^*_{i,jj}\ \alpha\phi_{i'}{}^{j'}$$

$$= (g/d(i)) \sum_{i'j'} a_{i'}{}^{j'}\ \delta_{ii'}\ \delta_{jj'}\ \phi_{i'}{}^{j'} = (g/d(i)) a_i{}^j \phi_i{}^j. \tag{3.2.9}$$

Equation (3.2.9) expresses the essential result. If we take an arbitrary function F, and form the sum over all the operations of the group of $[\alpha]^*_{i,jj}\ \alpha F$, the result, if not zero, is a function transforming according to the jth row of the ith irreducible representation. Equation (3.2.9) describes a projection technique, since the operator $\sum [\alpha]^*_{i,jj'}\ \alpha$ projects out of F a function of specified symmetry. In the particular case of a one-dimensional representation, the matrix element $[\alpha]^*_{i,jj}$ is just the complex conjugate of the character of the operation in that representation.

3.2.3 *Decomposition of a Reducible Representation*

One additional item of general theory is required. It frequently happens that we wish to decompose a reducible representation of a group into a (direct) sum of irreducible representations. Let Γ denote the reducible representation and Γ_i the ith irreducible representation. We write

$$\Gamma = \sum a_i \Gamma_i \tag{3.2.10}$$

where the a_i are coefficients which specify how many times the ith irreducible representation occurs in the decomposition of the reducible representation. The a_i are determined as follows: Let $\chi(\boldsymbol{\alpha})$ be the character of operation $\boldsymbol{\alpha}$ in the reducible representation and $\chi_i(\boldsymbol{\alpha})$ the character of the same operation in the ith irreducible representation. It is a consequence of the definition of the direct sum that (3.2.10) must hold for the characters of each operation:

$$\chi(\boldsymbol{\alpha}) = \sum_i a_i \chi_i(\boldsymbol{\alpha}). \tag{3.2.11}$$

We now require a theorem which is the specialization of (3.2.6) to the characters: Take (3.2.6); set $m = j$, $m' = j'$. The character χ_i is given by

$$\chi_i(\boldsymbol{\alpha}) = \sum_j [\boldsymbol{\alpha}]_{i,jj}. \tag{3.2.12}$$

We have

$$\sum_\alpha [\boldsymbol{\alpha}]^*_{i',j'j'}[\boldsymbol{\alpha}]_{i,jj} = (g/d(i))\,\delta_{ii'}\,\delta_{jj'}.$$

Now sum over j and j'

$$\sum_\alpha \chi_{i'}^*(\boldsymbol{\alpha})\chi_i(\boldsymbol{\alpha}) = (g/d(i))\,\delta_{ii'} \sum_{jj'} \delta_{jj'} = g\,\delta_{ii'}. \tag{3.2.13}$$

Multiply (3.2.11) by $\chi_j(\boldsymbol{\alpha})$, sum over $\boldsymbol{\alpha}$, and use (3.2.13). The result is

$$a_j = (1/g) \sum_\alpha \chi_j^*(\boldsymbol{\alpha})\chi(\boldsymbol{\alpha}). \tag{3.2.14}$$

Since the character of an operation depends only on the class, (3.2.14) can be rewritten as follows: Let the classes of the group be denoted by an index r, such that $\chi(r)$ is the character of the representation for class r, and let there be N_r elements in class r. Then (3.2.14) is equivalent to

$$a_j = (1/g) \sum_r N_r \chi_j^*(r)\chi(r). \tag{3.2.15}$$

3.2.4 C_{4v}: *An Example*

To illustrate these considerations, it is desirable to study a simple example where many of the results can be deduced by inspection. We consider the group C_{4v}, which is the group of the square. We can label the operations just as was done for the cubic group O_h of which C_{4v} is a subgroup. The operations are listed in Table V, which also presents the multiplication table for this eight-element group. It will be noted that we have described the operations in such a way that x is unchanged by any operation, that is, we can imagine the square to lie in the yz plane perpendicular to the x axis which is the axis of rotation. The entries in the multiplication table are

TABLE V

GROUP C_{4v}

Class	Operation	Designation	E	α	β	γ	δ	ϵ	ζ	η
E	$x \quad y \quad z$	E	E	α	β	γ	δ	ϵ	ζ	η
$(C_2)C_4^2$	$x \quad -y \quad -z$	α	α	E	γ	β	ϵ	δ	η	ζ
C_4	$x \quad -z \quad y$	β	β	γ	α	E	ζ	η	ϵ	δ
	$x \quad z \quad -y$	γ	γ	β	E	α	η	ζ	δ	ϵ
$(\sigma_v)JC_4^2$	$x \quad -y \quad z$	δ	δ	ϵ	η	ζ	E	α	γ	β
	$x \quad y \quad -z$	ϵ	ϵ	δ	ζ	η	α	E	β	γ
$(\sigma_d)JC^2$	$x \quad -z \quad -y$	ζ	ζ	η	δ	ϵ	β	γ	E	α
	$x \quad z \quad y$	η	η	ζ	ϵ	δ	γ	β	α	E

worked out in the following manner:

$$\zeta\delta = (x, -z, -y)(x, -y, z) = (x, -z, y) = \beta.$$

The table is constructed so that the operator which designates a row appears on the left in multiplication. Multiplication is not commutative; for instance $\delta\zeta = \gamma$ while $\zeta\delta = \beta$. Since there are five classes, there are five irreducible representations, four of which are one-dimensional, and the other two-dimensional. This is all, since $[4 \times (1)^2 + 1 \times (2)^2 = 8]$. The one-dimensional representations have as bases functions which are carried into themselves (or their negatives) by the operations of the group. In these cases, the group operations are represented by numbers: either $+1$ or -1. The two-dimensional representation involves 2×2 square matrices, and acts in a two-dimensional vector space.

For most purposes, the simplest basis functions are constructed from elements of the form $x^l y^m z^n$ and their linear combinations. In this example, a possible set of basis functions can be chosen very simply; there are, however, an infinite set of possible basis functions for each representation. Unfortunately, there is no generally accepted system for the labeling of these representations. Since most of our applications will be in the area of energy bands, we will adopt the notation which is most commonly employed in that area, due to Bouckaert *et al.* (BSW) (1936).

(1) There is a symmetric representation denoted Δ_1, in which each operation is represented by the number $+1$. A representation of this type always exists for any group. Acceptable basis functions for this representation are 1, x, x^2, $y^2 + z^2$, etc.

(2) In the representation Δ_2, the operators E, α, δ, and ϵ are represented by $+1$; the operators β, γ, ρ, and η are represented by -1. The reader can

easily verify that this assignment is consistent with the multiplication table. A basis function for this representation is $y^2 - z^2$.

(3) In the representation $\Delta_{2'}$, the operators E, α, ζ, and η are represented by $+1$; the operators β, γ, and ϵ are represented by -1. A basis function for this representation is yz.

(4) In the representation of $\Delta_{1'}$, the operators E, α, β, and γ are represented by $+1$; the operators δ, ϵ, ζ, and η by -1. A basis function for this representation is $yz(y^2 - z^2)$.

(5) For the two-dimensional representation Δ_5, the following assignment of matrices to operations may be made:

$$E = \begin{pmatrix} 1 & 0 \\ 0 & 1 \end{pmatrix}; \quad \alpha = \begin{pmatrix} -1 & 1 \\ 0 & -1 \end{pmatrix}; \quad \beta = \begin{pmatrix} 0 & -1 \\ 1 & 0 \end{pmatrix}; \quad \gamma = \begin{pmatrix} 0 & 1 \\ -1 & 0 \end{pmatrix};$$

$$\delta = \begin{pmatrix} -1 & 0 \\ 0 & 1 \end{pmatrix}; \quad \epsilon = \begin{pmatrix} 1 & 0 \\ 0 & -1 \end{pmatrix}; \quad \zeta = \begin{pmatrix} 0 & -1 \\ -1 & 0 \end{pmatrix}; \quad \eta = \begin{pmatrix} 0 & 1 \\ 1 & 0 \end{pmatrix}. \tag{3.2.16}$$

The functions y, z are suitable basis functions, as are also the functions xy, xz. The character table for C_{4v} is presented in Table VI.

3.2.5 *Irreducible Representations of O_h*

Special interest attaches to the full cubic group O_h not only because it is the point group of highest symmetry, but because it describes the point symmetry in many materials and applies in an approximate fashion in others. The character table for O_h is given in Table VII. The representations are labeled according to three sets of notation in common use. Throughout this work, we will use the BSW notation. The polynomial basis functions

TABLE VI

CHARACTER TABLE: GROUP C_{4v}

Representation	Basis	E	$C_4^2(C_2)$	C_4	JC_4^2 (σ_v)	JC_2 (σ_d)
Δ_1	$1, x, 2x^2 - y^2 - z^2$	1	1	1	1	1
Δ_2	$y^2 - z^2$	1	1	-1	1	-1
$\Delta_{2'}$	yz	1	1	-1	-1	1
$\Delta_{1'}$	$yz(y^2 - z^2)$	1	1	1	-1	-1
Δ_5	$y, z; xy, xz$	2	-2	0	0	0

TABLE VII

CHARACTER TABLE FOR THE CUBIC GROUP O_h

Representation														
BSW[a]	EWK[b]	Bethe[c]	Basis	BSW[a] : E / Koster[d] : E	$3C_4^2$ / $3C_2$	$6C_4$ / $6C_4$	$6C_2$ / $6C_{2'}$	$8C_3$ / $8C_3$	J / I	$3JC_4^2$ / $3\sigma_h$	$6JC_4$ / $6S_4$	$6JC_2$ / $6\sigma_d$	$6JC_3$ / $8S_6$
Γ_1	A_{1g}	Γ_1	1	1	1	1	1	1	1	1	1	1	1
Γ_2	A_{2g}	Γ_2	$x^4(y^2 - z^2) + y^4(z^2 - x^2) + z^4(x^2 - y^2)$	1	1	−1	−1	1	1	1	−1	−1	1
Γ_{12}	E_g	Γ_3	$x^2 - y^2$										
			$2z^2 - x^2 - y^2$	2	2	0	0	−1	2	2	0	0	−1
Γ_{15}	T_{1u}	$\Gamma_{4'}$	x, y, z	3	−1	1	−1	0	−3	1	−1	1	0
Γ_{25}	T_{2u}	$\Gamma_{5'}$	$z(x^2 - y^2)$	3	−1	−1	1	0	−3	1	1	−1	0
$\Gamma_{1'}$	A_{1u}	$\Gamma_{1'}$	$xyz[x^4(y^2 - z^2) + y^4(z^2 - x^2) + z^4(x^2 - y^2)]$	1	1	1	1	1	−1	−1	−1	−1	−1
$\Gamma_{2'}$	A_{2u}	$\Gamma_{2'}$	xyz	1	1	−1	−1	1	−1	−1	1	1	−1
$\Gamma_{12'}$	E_u	$\Gamma_{3'}$	$xyz(x^2 - y^2)$	2	2	0	0	−1	−2	−2	0	0	1
$\Gamma_{15'}$	T_{1g}	Γ_4	$xy(x^2 - y^2)$	3	−1	1	−1	0	3	−1	1	−1	0
$\Gamma_{25'}$	T_{2g}	Γ_5	xy, yz, zx	3	−1	−1	1	0	3	−1	−1	1	0

[a] Bouckaert *et al.* (1936).
[b] Eyring *et al.* (1944).
[c] Bethe (1929).
[d] Koster (1957).

were presented by von der Lage and Bethe (1947). These functions are frequently called *Kubic harmonics.* In the case of a representation of dimension greater than one, only one basis function is given in most cases; others may be determined by interchange of x, y, and z.

The determination of basis functions makes possible an approximate correspondence with atomic central field wave functions, conventionally designated s, p, d, f, ... This can be accomplished if one observes that a term $x^m y^n z^p$ is proportional to r^{m+n+p} times a linear combination of spherical harmonics whose total angular momentum quantum number $l = m + n + p$.

This correspondence is useful in considering the states in cubic solids likely to be occupied by electrons deriving from atomic levels of a given symmetry. We see that $\Gamma_1 \to$ s; $\Gamma_{15} \to$ p; Γ_{12}, $\Gamma_{25'} \to$ d; $\Gamma_{2'}$, $\Gamma_{25'}$, $\Gamma_{15} \to$ f. . . . There are two features of this correspondence which require comment.

(1) There are only ten irreducible representations in the cubic group, and the maximum dimension of a representation is three. This contrasts with the full rotation group which applies to atoms in free space and has an infinite number of irreducible representations; the dimension of a representation being $(2l + 1)$ where l is any integer.

Since the spherical harmonics of order l form a basis for the group of all rotations, they certainly are a basis for a representation of the cubic group, or any other point group. However, this representation will in general be reducible, that is, the spherical harmonics of order l will in general belong to different representations of cubic group. This decomposition can be worked out using the formulas (3.2.14) or (3.2.15) for the reduction of a reducible representation. In the simpler cases of small l, the results can be determined by inspection of Table VII, and from the comments above: s and p states are not split; but d states are divided into two kinds: those of Γ_{12} and $\Gamma_{25'}$ symmetry; doubly and triply degenerate, and f states are split into three: $\Gamma_{2'}$, Γ_{25}, and Γ_{15} with degeneracies 1, 3, and 3, respectively.

(2) The correspondence between angular momentum wave functions and those transforming according to O_h is only approximate in another way. A function belonging to $\Gamma_{25'}$ for instance need not be a pure d function; it may contain components of $l = 4$ and $l = 6$ as well. A function belonging to Γ_1 may contain $l = 4, 6, \ldots$ in addition to $l = 0$. The amount of mixing of higher angular momentum states is determined by the details of the electron dynamics, and cannot be predicted from considerations of symmetry.

The Kubic harmonics which are bases for the Γ_1 (completely symmetric) representation will be of considerable importance in the following section. They may be obtained by the projection technique described above, Eq.

(3.2.9); however, the following procedure is quite simple. We require a combination of monomials $x^m y^n z^p$ which is unchanged by any operation of the cubic group. We see immediately that $m + n + p$ must be an even integer. If this integer is zero, we have the simplest such function. In second order, we have the combination $x^2 + y^2 + z^2$; however, this is just $r^2 \times 1$ and so does not give anything new. In fourth order, we can form combinations like $x^4 + y^4 + z^4$ and $x^2y^2 + x^2z^2 + y^2z^2$. These can be orthogonalized to the leading zero-order harmonic; when this is done the results are not independent and we adopt, apart from a normalizing factor, the function

$$[(x^4 + y^4 + z^4)/r^4] - \tfrac{3}{5}.$$

In sixth order we obtain again a single harmonic after orthogonalization to the zero- and fourth-order terms. In Table VIII, we list the harmonics through eighth order, normalized to unity after integration over angles. The notation is $K_{l,i}$; this is the Kubic harmonic of order l belonging to the ith irreducible representation of O_h. The table is taken from von der Lage and Bethe (1947). A more complete listing of Kubic harmonics has been given by Altmann and Cracknell (1965).

3.2.6 *Selection Rules*

In addition to furnishing a classification of states according to symmetry, group theory enables the derivation of selection rules for matrix elements. The mathematical basis is the following: Consider the scalar product of two functions ϕ_i^j belonging to the jth row of the ith irreducible representation, and ϕ_k^l belonging to the lth row of the kth representation. The scalar product will be shown to be zero unless $i = k$ and $j = l$. First, observe that it must be unchanged by a rotation of coordinates

$$(\phi_i^j, \phi_k^l) = (\alpha\phi_i^j, \alpha\phi_k^l) = (1/g) \sum_{\alpha} (\alpha\phi_i^j, \alpha\phi_k^l).$$

TABLE VIII

KUBIC[a] HARMONICS FOR Γ_1

l	$K_{l,1}$
0	$(4\pi)^{-1/2}$
4	$\frac{5}{4}(21/4\pi)^{1/2}\{[(x^4 + y^4 + z^4)/r^4] - \frac{3}{5}\}$
6	$\frac{231}{8}(26/4\pi)^{1/2}\{(x^2y^2z^2/r^6) + \frac{1}{22}K_{4,1}^{u} - \frac{1}{105}\}$
8	$\frac{65}{16}(561/4\pi)^{1/2}\{[(x^8 + y^8 + z^8)/r^8] - \frac{28}{5}K_{6,1}^{u} - \frac{210}{143}K_{4,1}^{u} - \frac{1}{3}\}$

[a] The normalized Kubic harmonic is indicated by $K_{l,1}^{u}$.

Now use (3.2.6):

$$(\phi_i^j, \phi_k^l) = (1/g) \sum_{m,n} \sum_{\alpha} [\alpha]^*_{i,mj} [\alpha]_{k,nl} (\phi_i^m, \phi_k^n).$$

We may now apply (3.2.6) to obtain

$$(\phi_i^j, \phi_k^l) = (1/d(i))\, \delta_{ik}\, \delta_{jl} \sum_n (\phi_i^n, \phi_i^n). \tag{3.2.17}$$

This is the desired result: Two functions belonging to different rows of the same irreducible representation of a symmetry group, or to different irreducible representations, are orthogonal.

These results may be extended to matrix elements of quantum operators. Let $\mathcal{O}$ be some operator, and consider

$$\mathcal{O}_{ik}^{jl} = (\phi_i^j, \mathcal{O}\phi_k^l). \tag{3.2.18}$$

We begin by considering important special case in which $\mathcal{O}$ is unchanged by any operation of the group

$$[\alpha, \mathcal{O}] = 0 \qquad (\text{all} \quad \alpha). \tag{3.2.19}$$

Here, the preceding analysis may be applied without change, and we see that

$$\mathcal{O}_{ik}^{jl} = (1/d(i))\, \delta_{ik}\, \delta_{jl} \sum_n \mathcal{O}_{ii}^{nn}. \tag{3.2.20}$$

Moreover, if we set $j = l$ (and $i = k$), we see that the value of the matrix element $\mathcal{O}_{ii}^{jj}$ is actually independent of the row j. It will depend only on the representation i.

The most important operator to which these considerations may be applied is the Hamiltonian of the system. The Hamiltonian of a system will not have any matrix elements connecting functions of different symmetries, and the value of the matrix elements between functions of the same symmetry is independent of the row.

Next, consider the case of an operator which is not unchanged by the symmetry operations α. We denote such an operator by Q in order to distinguish it from a $\mathcal{O}$ which commutes with all α. In this situation we must determine the transformation properties of the function $Q\phi_k^l$. We suppose that Q itself transforms according to some irreducible representation of the group; to be precise, let it be the pth row of the qth irreducible representation. Then the product $Q\phi_k^l$ transforms according to the plth row of the direct product representation. This representation, which we will denote $\Gamma_q \times \Gamma_k$ is frequently a reducible one, and must be decomposed according to (3.2.10) into its irreducible components. Our conclusion concerning the matrix element is that $Q_{ik}{}^{jl} = 0$ unless a function transforming according to the jth row of the ith irreducible representation appears when the function $Q\phi_k{}^l$ is expressed as a sum of functions belonging to irreducible representations. A necessary condition may be expressed as follows. Write

$$\Gamma_q \times \Gamma_k = \sum a_s \Gamma_s \tag{3.2.21}$$

where the sum is over the irreducible representations. The coefficient a_i must be different from zero in order to obtain a nonzero matrix element.

These results may be expressed in a slightly different way. The essence of our reasoning is that the integrand in a matrix element calculation, say, $\phi_i{}^j Q \phi_k{}^l$ must be invariant under the group operations; or less restrictively, must contain a portion which is invariant. Otherwise, the result is zero. Invariance implies a function transforming according to the scalar representation Γ_1. Consequently, we may consider the triple direct product

$$\Gamma_i{}^* \times \Gamma_q \times \Gamma_k = \sum_s a_s \Gamma_s. \tag{3.2.22}$$

The matrix element will be zero unless Γ_1 appears on the right side of (3.2.21) with a nonzero coefficient.

3.3 Symmetry with Spin

The preceding considerations have not included electron spin in a specific way. Some unavoidable complications arise when this is attempted. This occurs because an electron is described by a two-component wave function, or Pauli spinor, in nonrelativistic quantum theory. Spin–orbit coupling, which is significant in solids as well as in atomic spectra, can remove some of the degeneracies present in a theory in which spin is neglected.

3.3.1 *Transformation Properties of Spinors*

There are fundamental differences in the description of the effect of rotations on a spinor (as compared to a scalar or an ordinary vector). Such an object transforms under rotations according to the $j = \frac{1}{2}$ representation of the rotation group. Two different quantum mechanical operators exist which correspond to the same physical transformation of points in space. It is these operators, rather than the physical transformations themselves, which form the group whose representations are required. This double valuedness has far-reaching consequences. We begin by examining its origin.

Let u be a constant spinor in some coordinate system (a subscript ν

indicates components: $\nu = 1, 2$), and let $\mathbf{R}$ be a rotation characterized by Euler angles α, β, γ.‡ In the new coordinate system, the spinor is (Tinkham, 1964, p. 109)

$$u_\nu = (S_{\mathbf{R}} u)_\nu = \sum D_{\lambda\nu}{}^{(1/2)}(\mathbf{R}) u_\lambda. \tag{3.3.1}$$

The operator $S_{\mathbf{R}}$ induces the rotation Δ on the spinor Δ. The result is expressed in terms of the matrix $D^{(1/2)}$, which belongs to the $j = \frac{1}{2}$ representation of the rotation group. It is given by

$$\mathbf{D}^{1/2}(\mathbf{R}) = \begin{pmatrix} \exp[-i(\alpha + \gamma)/2] \cos \beta/2 & -\exp[-i(\alpha - \gamma)/2] \sin \beta/2 \\ \exp[i(\alpha - \gamma)/2] \sin \beta/2 & \exp[i(\alpha + \gamma)] \cos \beta/2 \end{pmatrix}. \tag{3.3.2}$$

The appearance of half angles in (3.3.2) is particularly to be noted, as this gives rise to the "double valuedness" mentioned above. If one of the Euler angles is increased by 2π, the physical transformation is unchanged, but the representation matrix changes sign. In particular, the matrices

$$\begin{pmatrix} 1 & 0 \\ 0 & 1 \end{pmatrix} \quad \text{and} \quad \begin{pmatrix} -1 & 0 \\ 0 & -1 \end{pmatrix}$$

both correspond to the identity.

The wave functions of interest to us are not constant spinors, but contain functions of position as well. Let us consider a function

$$\psi_{l,i,\nu} = {}^\nu\phi_l{}^i u_\nu \tag{3.3.3a}$$

in which u_ν is a component of a Pauli spinor and ${}^\nu\phi_l{}^i$ is an ordinary function of position which is a basis function for the ith row of the lth irreducible representation of some point group. The index ν as applied to ϕ indicates that a different function ϕ may be associated with each of the two components of the spinor; written as a column vector we have

$$\psi = \begin{pmatrix} {}^1\phi_l{}^i \\ {}^2\phi_l{}^i \end{pmatrix}. \tag{3.3.3b}$$

‡ Euler angles are defined here as follows. Begin with a rectangular coordinate system (x, y, z). First, rotate through an angle α about the z axis; call the resulting system (x', y', z'). Then rotate through β about the y' axis, leading to the system (x'', y'', z''). The final rotation is through γ about the z'' axis, the resulting system being labeled (x''', y''', z''').

The transformation of ϕ under the rotation **R** is

$$P_{\mathrm{R}}\, {}^{\nu}\phi_l{}^i(\mathbf{r}) = \sum_{j=1}^{n} \Gamma_{ji}{}^{(l)}\, {}^{\nu}\phi_l{}^j(\mathbf{r}). \tag{3.3.4}$$

In this equation P_{R} is the operator which induces the rotation **R**. We depart slightly from our previous notation at this point to distinguish specifically the operators P_{R} from the rotation **R** and we write $\Gamma_{ji}{}^{(l)}$ to denote the jith element of the matrix representing the operation **R** in the representation $\Gamma^{(l)}$: $\Gamma_{ji}{}^{(l)}(\mathbf{R}) = [\mathbf{R}]_{l,ji}$. The sum over j includes the n functions which form a basis for the lth representation. To determine the transformation properties of $\psi_{l,i\nu}$, both space and spin variables must be considered. Then let the operator Q_{R} induce the transformation **R** on $\psi_{l,i\nu}$ where Q_{R} may be regarded as the product of two matrices—P_{R} which acts on the space function in accord with (3.3.4), and S_{R} which acts on the spinor:

$$\begin{aligned} Q_{\mathrm{R}}\psi_{l,i,\nu} &= \sum_{j=1}^{n} \sum_{\lambda=1}^{2} \Gamma_{ji}{}^{(l)}(\mathbf{R}) D_{\lambda\nu}{}^{(1/2)}(\mathbf{R})\, {}^{\lambda}\phi_l{}^j(\mathbf{r})\, u_\lambda \\ &= \sum_{j\lambda} [\Gamma^{(l)}(\mathbf{R}) \times D^{(1/2)}(\mathbf{R})]_{j\lambda,i\nu}\psi_{l,j,\lambda}. \end{aligned} \tag{3.3.5}$$

We infer from (3.3.5) that $\psi_{l,i,\nu}$ transforms according to the direct product representation: the representation whose matrices are direct products of those belonging to $\Gamma^{(l)}$ and those belonging to $D^{(1/2)}$. The double valuedness of $D^{(1/2)}$ implies that the direct product representation is also double valued.

3.3.2 *Double Groups*

Let us consider the group G formed by the operators Q_{R} which correspond to the transformations **R** which belong to a point group g. The group G contains twice as many elements as g does: hence the commonly used term *double group*. Also G is homomorphic rather than isomorphic to g. The general principles of quantum theory require that the one-electron wave functions form bases for the irreducible representations of G. The irreducible representations of g may be trivially extended to form representations of G merely by assigning the same matrix to represent both operations of G which correspond to the same operation **R** of g. However, because G contains more operations than g, it will contain more classes and thus must possess additional irreducible representations. These have the property that the two matrices E and $\bar{E}$ which correspond to the identity ϵ of g have characters which differ in sign. One-electron wave functions which include spin transform according to these additional representations. This follows from (3.3.5), since the elements of the direct product representation $\Gamma^{(l)}(\mathbf{R}) \times D^{(1/2)}(\mathbf{R})$ contain the half angle functions of $D^{(1/2)}(\mathbf{R})$ so that the elements of $\bar{E}$ differ in sign from those of E.

Each element in g has two corresponding elements in G. Let $\mathbf{R}$ be an element of g, and the two corresponding elements of G be $\mathbf{R}'$ and $\bar{\mathbf{R}}'$. We have $\bar{\mathbf{R}}' = \bar{\mathbf{E}}\mathbf{R}'$, where $\bar{\mathbf{E}}$ is the *negative* of a unit matrix. Thus the character of $\bar{\mathbf{R}}'$ is opposite in sign to that of $\mathbf{R}'$.

If a set of elements $\{\mathbf{R}\}$ is a class in g, then the elements $\{\mathbf{R}'\}$ form a class in G and usually the $\{\bar{\mathbf{R}}'\}$ form a distinct class. An exception occurs if $\mathbf{R}$ is a twofold rotation in g, and g contains either another rotation about a twofold axis perpendicular to the axis of $\mathbf{R}$ or a reflection in a plane containing the axis. In this case the elements $\{\mathbf{R}', \bar{\mathbf{R}}'\}$ constitute a single class in G. Similar conclusions apply to improper rotations. The difference between the number of classes of G and g is the number of additional irreducible representations of G. It is these additional irreducible representations which must be used to describe the transformation of election wave functions when spin is included.

We wish to determine the behavior of an electron state which belonged to the representation $\Gamma^{(l)}$ before spin was considered. To this end, we form the direct product $\Gamma^{(l)} \times D^{(1/2)}$. The representation so obtained may be reducible as a representation of G. As in (3.2.10) we write

$$\Gamma^{(l)} \times D^{(1/2)} = \sum_i c^{(i)}\nu^{(i)}, \tag{3.3.6}$$

in which the $c^{(i)}$ are coefficients, the $\nu^{(i)}$ are the irreducible representations of G, and the sum is a direct sum as described in Section 3.2. Equation (3.3.6) has a physical interpretation: it expresses the splitting of a degenerate state l by spin orbit coupling into states of symmetry i. Only the additional representations of G occur on the right of (3.3.6).

Up to this point, proper rotations have been emphasized. The extension of these results to point groups which contain improper rotations is easily accomplished when one observes that such a group is either isomorphic to a group containing proper rotations only, or else is formed by supplementing a group containing proper rotations only by a product of its operations with the inversion J. Further, a Pauli spinor is invariant under inversion: the representation matrix in $D^{(1/2)}$ for the inversion is just the unit matrix. Hence, for any operation $\mathbf{R}$, we may set $D^{(1/2)}(\mathbf{R}) = D^{(1/2)}(J\mathbf{R})$. It is then possible to employ (3.3.5) to obtain the direct product representation of a group containing improper rotations.

These considerations will be applied to the point groups C_{4v} and O_h discussed previously. In the case of C_{4v}, there is a reflection plane containing the axis so C_4^2 and $\bar{C}_4^2$ are in the same class. Likewise JC_4^2 and $\bar{J}\bar{C}_4^2$ are in

TABLE IX

CHARACTERS OF THE ADDITIONAL REPRESENTATIONS OF THE DOUBLE GROUP C_{4v}

Represen-tation	E	$\bar{E}$	C_4	$\bar{C}_4$	$(C_4^2, \bar{C}_4^2)$	$(JC_4^2, \bar{J}\bar{C}_4^2)$	$(JC_2, \bar{J}C_2)$
Δ_6	2	-2	$2^{1/2}$	$-2^{1/2}$	0	0	0
Δ_7	2	-2	$-2^{1/2}$	$2^{1/2}$	0	0	0

the same class, as are JC_2 and $\bar{J}\bar{C}_2$. There are thus only two new classes: $\bar{E}$ and $\bar{C}_4$. Consequently there are two additional irreducible representations (Δ_6 and Δ_7). Since these are eight additional operations, these representations must be two-dimensional. The characters of the additional representations are given in Table IX. By comparison of the character tables,‡ we see that $\Delta_1 \times D^{(1/2)} = \Delta_6$; $\Delta_1' \times D^{(1/2)} = \Delta_6$; $\Delta_2 \times D^{(1/2)} = \Delta_7$, $\Delta_2' \times D^{(1/2)} = \Delta_7$, $\Delta_5 \times D^{(1/2)} = \Delta_6 + \Delta_7$.

In the case of O_h, there are six additional classes. The character table is presented in Table X. It is to be noted that the sixfold degenerate states (or threefold if spin is neglected) are split into a fourfold and a twofold degenerate state, but the fourfold states (doubly degenerate when spin is neglected) are not split.

The character tables for the additional representations of all the double groups formed from the 32 point groups are given by Koster (1957). The number of representations added is usually smaller than the number of representations in the single point group. Hence the classification of states is, in a sense, less detailed when spin is included. Fortunately, in many cases, spin–orbit coupling can be regarded as a perturbation on a level structure calculated ignoring the spin.

3.4 IONS IN CRYSTALS

In this section, we wish to consider some of the problems associated with the description of the states of a single atom or ion in a crystal environment. The theory began with the basic work of Bethe (1929), Penney and Schlapp (1932), and Van Vleck (1932). Our attention is focused on transition metal

‡ A character table may be considered to be a matrix whose rows designate classes and whose columns designate irreducible representations. The character table appropriate to a direct product group is the direct product of the matrix character tables of the groups which are factors of the product group (Murnaghan, 1938).

TABLE X

CHARACTERS OF THE ADDITIONAL REPRESENTATIONS OF THE DOUBLE GROUP O_h

Representation	Class							
	E	$\bar{E}$	$(C_4^2, \bar{C}_4^2)$	C_4	$\bar{C}_4$	$(\bar{C}_2, C_2)$	C_3	$\bar{C}_3$
Γ_6^+	2	−2	0	$2^{1/2}$	$-2^{1/2}$	0	1	−1
Γ_7^+	2	−2	0	$-2^{1/2}$	$2^{1/2}$	0	1	−1
Γ_8^+	4	−4	0	0	0	0	−1	1
Γ_6^-	2	−2	0	$2^{1/2}$	$-2^{1/2}$	0	1	−1
Γ_7^-	2	−2	0	$-2^{1/2}$	$2^{1/2}$	0	1	−1
Γ_8^-	4	−4	0	0	0	0	−1	1
	J	$\bar{J}$	$(JC_4^2, \bar{J}\bar{C}_4^2)$	JC_4	$\bar{J}\bar{C}_4$	$(JC_2, \bar{J}\bar{C}_2)$	JC_3	$\bar{J}\bar{C}_3$
Γ_6^+	2	−2	0	$2^{1/2}$	$-2^{1/2}$	0	1	−1
Γ_7^+	2	−2	0	$-2^{1/2}$	$2^{1/2}$	0	1	−1
Γ_8^+	4	−4	0	0	0	0	−1	1
Γ_6^-	−2	2	0	$-2^{1/2}$	$2^{1/2}$	0	−1	1
Γ_7^-	−2	2	0	$2^{1/2}$	$-2^{1/2}$	0	−1	1
Γ_8^-	−4	4	0	0	0	0	1	−1

$$\Gamma_1 \times D^{(1/2)} = \Gamma_6^+ \qquad \Gamma_{1'} \times D^{(1/2)} = \Gamma_6^-$$
$$\Gamma_2 \times D^{(1/2)} = \Gamma_7^+ \qquad \Gamma_{2'} \times D^{(1/2)} = \Gamma_7^-$$
$$\Gamma_{12} \times D^{(1/2)} = \Gamma_8^+ \qquad \Gamma_{12'} \times D^{(1/2)} = \Gamma_8^-$$
$$\Gamma_{15} \times D^{(1/2)} = \Gamma_6^- + \Gamma_8^- \qquad \Gamma_{15'} \times D^{(1/2)} = \Gamma_6^+ + \Gamma_8^+$$
$$\Gamma_{25} \times D^{(1/2)} = \Gamma_7^- + \Gamma_8^- \qquad \Gamma_{25'} \times D^{(1/2)} = \Gamma_7^+ + \Gamma_8^+$$

and rare earth atoms, as these have been the subject of intense study. The atom considered may be either an impurity, or part of the regular lattice structure. The fundamental approximation is that the atomic energy levels are sharp; that is, that band formation which broadens energy levels through interaction with neighboring atoms is negligible. The extent of validity of this approximation is not known at present, and is the subject of considerable controversy. However, systems exist in which it is certainly correct. Useful general references include Griffith (1961), Ballhausen (1962), Di Bartolo (1968), White (1970); reviews include Bleaney and Stevens (1953), Low (1960), Anderson (1963), McClure (1959).

3.4.1 *Crystal Fields*

The first problem that we shall investigate is the splitting of atomic energy levels by crystal field effects. In a crystal, the effective potential at an atomic site has the symmetry of the crystal structure, rather than the spherical symmetry appropriate to atoms in free space. As a result, some of the degeneracy of free atom energy levels will be removed in the crystal. The treatment of this effect depends on how the strength of the effective potential compares with (1) the multiplet splitting between atomic energy levels, and (2) the spin–orbit coupling of electrons in the atom. We can distinguish three cases.

(1) Crystal field weak compared to spin–orbit coupling: This occurs in rare earth atoms in crystal environments. The crystal field is treated as a perturbation on atomic states characterized by definite values of the total angular momentum $\mathbf{J}^2$ ($\mathbf{J} = \mathbf{L} + \mathbf{S}$).

(2) Intermediate fields: In transition metal atoms, the crystal field is frequently found to be strong compared to atomic spin–orbit coupling, but small compared to the separation between multiplets. This case is the one we will stress in our treatment, which considers the crystal potential as a perturbation on states of definite L and S.

(3) Strong fields: In this case, the crystal field is large compared to the multiplet splitting, and it is not possible to consider the atoms to be, in first approximation, in states of definite L and S. The multiplet splitting must be regarded as a perturbation on a level structure determined by the crystal potential.

Let us begin by considering a model which indicates the principal features of the crystal potential of interest here although, quantitatively, it is unrealistic. Specifically, we consider a cubic lattice of ions which are point charges. The lattice positions are denoted by double indices; $\mu\lambda(\mathbf{R}_{\mu\lambda})$ in which μ designates a shell [thus in simple cubic, $\mu = 1$ indicates the $(1, 0, 0)$ atoms, $\mu = 2$ designates those at $(1, 1, 0)$ positions, etc.], and λ designates a particular atom in the shell. The charge on an atom is written as $z_\mu e$, where z_μ can have positive or negative values depending on the shell. The electrostatic potential energy of an electron at $\mathbf{r}$ may be written as

$$V(\mathbf{r}) = -e^2 \sum_{\mu\lambda} z_\mu \,|\, \mathbf{r} - \mathbf{R}_{\mu\lambda} \,|^{-1}. \tag{3.4.1}$$

According to the multipole expansion

$$\begin{aligned} |\, \mathbf{r} - \mathbf{R}_{\mu\nu} \,|^{-1} &= \sum_{l=0}^{\infty} (r_<^l / r_>^{l+1}) P_l(\cos \theta_{\mathrm{rR}}) \\ &= \sum_{l=0}^{\infty} \sum_{m=-l}^{l} [4\pi/(2l+1)](r_<^l/r_>^{l+1}) Y^*_{lm}(\theta, \phi) Y_{lm}(\theta_R, \phi_R). \end{aligned} \tag{3.4.2}$$

Here $r_<$ ($r_>$) is the lesser (greater) of r and $R_{\mu\lambda}$; r, θ, ϕ are the spherical coordinates of $\mathbf{r}$ with respect to a fixed set of axes in the crystal, and R, θ_R, ϕ_R are similar quantities referring $\mathbf{R}_{\mu\lambda}$ to the fixed axes. The distance $R_{\mu\lambda}$ is the same for all atoms in a shell and thus is independent of λ: $R_\mu = |\mathbf{R}_{\mu\lambda}|$. It is convenient to express $V(\mathbf{r})$ as a sum of terms from each order l of spherical harmonic

$$V(\mathbf{r}) = \sum_l V_l(\mathbf{r}) \tag{3.4.3}$$

where, after substituting (3.4.2) in (3.4.1), we have

$$V_l(\mathbf{r}) = [-4\pi e^2/(2l+1)] \sum_\mu z_\mu (r_<^l/r_>^{l+1}) \sum_{m,\lambda} [Y^*_{lm}(\theta, \phi) Y_{lm}(\theta_R, \phi_R)].$$

We have used the fact that the only dependence on the index λ occurs in the $Y_{lm}(\theta_R, \phi_R)$.

It is useful to express $V_l(\mathbf{r})$ in terms of angular functions with lattice symmetry. The Kubic harmonics $K_{l,s}$, which were introduced in Section 3.2.5, are related to the spherical harmonics Y_{lm} by a unitary transformation

$$K_{l,s}(\theta, \phi) = \sum_m U_{sm}^{(l)} Y_{lm}(\theta, \phi). \tag{3.4.4}$$

The index s is used to denote collectively the irreducible representation, row, and number of the function should a given representation occur more than once for fixed l. The elements of U satisfy

$$\sum_m U_{s,m}^{(l)*} U_{t,m}^{(l)} = \delta_{st}.$$

It follows from these relations that

$$\sum_m Y^*_{lm}(\theta, \phi) Y_{lm}(\theta_R, \phi_R) = \sum_s K^*_{l,s}(\theta, \phi) K_{l,s}(\theta_R, \phi_R). \tag{3.4.5}$$

This expression is to be inserted in (3.4.3). Summation over λ is equivalent to summation over group operators, apart from a numerical factor. We recall that the character of the completely symmetric representation is unity for all group operations. It follows from (3.2.7a) that only the contribution from the symmetric representation Γ_1 survives after the sum on λ has been performed. Thus we can write‡

$$V_l(\mathbf{r}) = v_l(r) K_{l,1}(\theta, \phi). \tag{3.4.6a}$$

This expression is essential to crystal field theory, and to a substantial extent its form is independent of the specific model employed. The coeffi-

‡ For $l \geq 12$, there will be more than one Kubic harmonic belonging to Γ_1 for fixed (even) l. This complication is not significant here.

cients $v_l(r)$ do depend on the model and, in the case we are considering, have the value

$$v_l(r) = [-4\pi e^2/(2l+1)] \sum_\mu z_\mu (r_<^l / r_>^{l+1}) B_{l,\mu} \qquad (3.4.6b)$$

where $B_{l,\mu}$ is a numerical coefficient containing the sum on λ. We investigate in detail the important case of $l = 4$. The fourth-order Kubic harmonic $K_{4,1}$ (see Table X) is related to spherical harmonics by

$$K_{4,1}(\theta, \phi) = \tfrac{5}{4}(21/4\pi)^{1/2}\{[(x^4 + y^4 + z^4)/r^4] - \tfrac{3}{5}\}$$
$$= (\tfrac{7}{12})^{1/2}[Y_{40} + (\tfrac{5}{14})^{1/2}(Y_{44} + Y_{4-4})]. \qquad (3.4.7)$$

An expression for $B_{4,\mu}$ is obtained,

$$B_{4,\mu} = \sum_\lambda K_{4,1}(\theta_R, \phi_R) = \tfrac{5}{4}(21/4\pi)^{1/2} N_\mu \{[x_\mu^4 + y_\mu^4 + z_\mu^4 - \tfrac{3}{5}]/R_\mu^4\}, \qquad (3.4.8)$$

in which x_μ, y_μ, z_μ are the components of any one of the $\mathbf{R}_{\mu\lambda}$ with respect to the crystal axes, and N_μ is the number of atoms in the shell.

In the case in which r is smaller than the nearest neighbor distance, we can write (3.4.6) in the form

$$v_4(r) = (-4\pi e^2/9a)(r/a)^4 \sum_\mu z_\mu (a/R_\mu)^5 B_{4,\mu} \qquad (3.4.9)$$

where a is the lattice constant.

In the rare earth series, where 4f electrons are involved, it is necessary to consider the sixth-order term $v_6(r)$ as well as v_4.

3.4.2 *Splitting of* d *Levels*

Now let us consider the effect of this potential on an ion with d electrons. For simplicity, the first case studied will be that of a single d electron. A system with this configuration would be the trivalent titanium ion Ti^{3+}. The analysis can also be applied to an ion with a single hole in the 3d shell (example: Cu^{2+}) provided that the sign of the effective interaction is changed. The potential is assumed to be weak enough so that first-order perturbation theory can be used.

It follows from the discussion of Section 3.2.5 that, in the presence of the potential (3.4.3), which has cubic rather than spherical symmetry, the fivefold degeneracy of d states will be split into a threefold degenerate state $\Gamma_{25'}$ (also called T_{2g}) and a doubly degenerate state Γ_{12} (also called E_g). We may proceed in either of two ways to calculate the amount of this splitting: (1) We can consider the fivefold degenerate set of d electron wave functions of the form $R_d(r)Y_{2m}(\theta, \phi)$ $(m = -2, \ldots, 2)$, set up the 5×5 matrix of the potential (3.4.3) on the basis of these states, and diagonalize it. (2) An equivalent method is to construct functions with the required angular symmetry of $\Gamma_{25'}$ and Γ_{12}; and calculate the average of V with these functions. There will be no matrix elements connecting the functions of these two symmetries.

We will follow the second procedure. The wave functions are

$$\psi(\Gamma_{25'}) = R_d(r)K_{2,25'}(\theta, \phi), \qquad \psi(\Gamma_{12}) = R_d(r)K_{2,12}(\theta, \phi) \quad (3.4.10)$$

where R_d is a common radial function and the $K_{2,i}$ are angular functions of the appropriate symmetry

$$K_{2,25'} = (15/4\pi)^{1/2}(xy/r^2); \qquad K_{2,12} = \tfrac{1}{2}(15/4\pi)^{1/2}[(x^2 - y^2)/r^2]. \quad (3.4.11)$$

The energy difference between these states ΔE is

$$\Delta E = E(\Gamma_{25'}) - E(\Gamma_{12}) = \int R_d^2(r)[K^2_{2,25'}(\theta, \phi) - K^2_{2,12}(\theta, \phi)]V(\mathbf{r})\, d^3r. \quad (3.4.12)$$

We substitute (3.4.11) into (3.4.12) and find after some algebra, which makes use of the symmetric nature of V,

$$\Delta E = [-5/2(21\pi)^{1/2}] \int R_d^2(r)K_{4,1}(\theta, \phi)V(\mathbf{r})\, d^3r. \quad (3.4.13)$$

Then we substitute (3.4.3) and (3.4.6a) and make use of the orthonormality properties of the Kubic harmonics:

$$\int K_{l,i,j}(\theta, \phi)K_{l',i',j'}(\theta, \phi)\, d\Omega = \delta_{ll'}\,\delta_{ii'}\,\delta_{jj'} \quad (3.4.14)$$

(where l designates angular momentum, i denotes the irreducible representation, and j is the row of a degenerate representation). We obtain

$$\Delta E = [-5/2(21\pi)^{1/2}] \int R_d^2(r)v_4(r)r^2\, dr. \quad (3.4.15)$$

The degenerate d levels are split by the fourth-order nonspherical portion of the crystal potential. Conventionally this energy difference is denoted $-10Dq$, or

$$E(\Gamma_{12}) - E(\Gamma_{25'}) = 10Dq = -\Delta E. \quad (3.4.16)$$

If one recalls that

$$\sum_{m=-l}^{l} | Y_{lm} |^2 = (2l + 1)/4\pi, \tag{3.4.17}$$

it is easy to see that the trace of the perturbation matrix (v_4K_4 being the perturbation) vanishes. Thus the "center of gravity" of the originally degenerate d levels is unchanged, so that

$$4E(\Gamma_{12}) + 6E(\Gamma_{25'}) = 0. \tag{3.4.18}$$

The energy of the unperturbed state has been set equal to zero as a convention. The solution of (3.4.16) and (3.4.18) is

$$E(\Gamma_{12}) = 6Dq, \qquad E(\Gamma_{25'}) = -4Dq. \tag{3.4.19}$$

Thus the Γ_{12} level is raised and the $\Gamma_{25'}$ state is lowered.

The splitting $10Dq$ is usually regarded as an empirical parameter to be determined by experiment. The model we have chosen permits an estimation of that parameter if a radial wave function is specified. A simple hydrogenic function will be assumed for this purpose.

$$R_d = (\alpha^7/6!)^{1/2} r^2 \exp(-\tfrac{1}{2}\alpha r). \tag{3.4.20}$$

Equation (3.4.20) is not particularly accurate quantatively, but actual d electron wave functions in transition metal atoms and ions can be precisely represented as sums of such terms. Substitution of (3.4.20) and (3.4.9) into (3.4.15) gives

$$\begin{aligned} 10Dq &= 11200(\pi/21)^{1/2}(e^2/a)(\alpha a)^{-4} \sum_{\mu} z_\mu (a/R_\mu)^5 B_{4,\mu} \\ &= 7000(e^2/a)(\alpha a)^{-4} \sum_{\mu} N_\mu z_\mu (a/R_\mu)^5 \{[(x_\mu{}^4 + y_\mu{}^4 + z_\mu{}^4)/R_\mu{}^4] - \tfrac{3}{5}\}. \end{aligned} \tag{3.4.21}$$

The lattice sum remains to be evaluated. It converges readily enough so that special techniques are not required. It should be noted that the result will differ certainly in magnitude and possibly in sign for different types of atomic arrangements (simple cubic, body-centered, face-centered, tetrahedral, etc.). Note that Dq varies as the inverse fifth power of the lattice constant. Numerical results will not be given.

3.4.3 *Inclusion of Spin–Orbit Coupling*

If there were no crystal field effects, spin–orbit coupling would split the degenerate d states into a fourfold $j = \frac{3}{2}$ level and a sixfold $j = \frac{5}{2}$ state.

In the presence of a cubic field, we see from Table X that the Γ_{12} level is not split, going directly into Γ_8, while the $\Gamma_{25'}$ state splits into a doubly degenerate Γ_7 and a fourfold Γ_8. If the spin–orbit coupling is small compared to the crystal field, it may be included by first-order degenerate perturbation theory using the $\Gamma_{25'}$ states (supplemented by spin functions) as a basis. It is sufficient to consider a perturbation of the form

$$H_{\text{soc}} = \xi_d \mathbf{L}\cdot\mathbf{S}. \tag{3.4.22}$$

where ξ_d is a constant. This is valid under the assumption that the potential is spherically symmetric (this is true to an adequate extent since we are interested in the potential near the nucleus), and that all the d states have the same radial function. Then

$$\xi_d = (\hbar^2/2m^2c^2) \int R_d^2(r)\,(1/r\,dV/dr)r^2\,dr. \tag{3.4.23}$$

The result of the perturbation procedure in which the $\Gamma_{25'}$ states alone are considered is that a splitting of $\frac{3}{2}\xi_d$ is obtained, the Γ_7 state being raised by ξ_d, and the Γ_8 being lowered by $\frac{1}{2}\xi_d$.

It is not difficult to set up the complete problem of crystal field splitting plus spin–orbit coupling for a single d electron on the basis of the ten d states, including spin. We use as basis states the spherical harmonics Y_{2m} times spin functions; α representing spin up and β representing spin down. Let us arrange the states in the order $Y_{22}\alpha$; $Y_{22}\beta$, $Y_{21}\alpha, \ldots, Y_{2-2}\beta$. We will measure the crystal field effect in units of $10Dq$, where $10Dq$ is given by (3.4.16). After a straightforward calculation, the matrix (3.4.24) (page 202) is found.

The eigenvalue equation factors into two identical quadratic equations and one sixth-order equation. However, we may use our knowledge of the degeneracies to extract the eigenvalues from the 6×6. The solutions are

$$E(\Gamma_8) = Dq - \tfrac{1}{4}\xi_d \pm \tfrac{1}{2}[\tfrac{25}{4}\xi_d^2 + 10Dq\xi_d + 100(Dq)^2]^{1/2}, \tag{3.4.25a}$$

$$E(\Gamma_7) = -4Dq + \xi_d. \tag{3.4.25b}$$

For small values of the ratio ξ_d/Dq, the energy of the Γ_8 levels can be expanded to give

$$E(\Gamma_8) = 6Dq + \tfrac{3}{20}(\xi_d^2/Dq), \tag{3.4.26a}$$

$$E(\Gamma_8) = -4Dq - (\xi_d/2) - \tfrac{3}{20}(\xi_d^2/Dq). \tag{3.4.26b}$$

The first solution (3.4.26a) refers to the Γ_8 level which derives from Γ_{12}; the other (3.4.26b) to that which derives from $\Gamma_{25'}$.

In the opposite limit of large spin–orbit coupling, an expansion of

	2α	2β	1α	1β	0α	0β	-1α	-1β	-2α	-2β
2α	$Dq+\xi_d$	0	0	0	0	0	0	0	5Dq	0
2β	0	$Dq-\xi_d$	ξ_d	0	0	0	0	0	0	5Dq
1α	0	ξ_d	$-4Dq+\xi_d/2$	0	0	0	0	0	0	0
1β	0	0	0	$-4Dq-\xi_d/2$	$\xi_d(\frac{3}{2})^{1/2}$	0	0	0	0	0
0α	0	0	0	$\xi_d(\frac{3}{2})^{1/2}$	$6Dq$	0	0	0	0	0
0β	0	0	0	0	0	$6Dq$	$\xi_d(\frac{3}{2})^{1/2}$	0	0	0
-1α	0	0	0	0	0	$\xi_d(\frac{3}{2})^{1/2}$	$-4Dq-\xi_d/2$	0	0	0
-1β	0	0	0	0	0	0	0	$-4Dq+\xi_d/2$	ξ_d	0
-2α	$5Dq$	0	0	0	0	0	0	ξ_d	$Dq-\xi_d$	0
-2β	0	$5Dq$	0	0	0	0	0	0	0	$Dq+\xi_d$

(3.4.24)

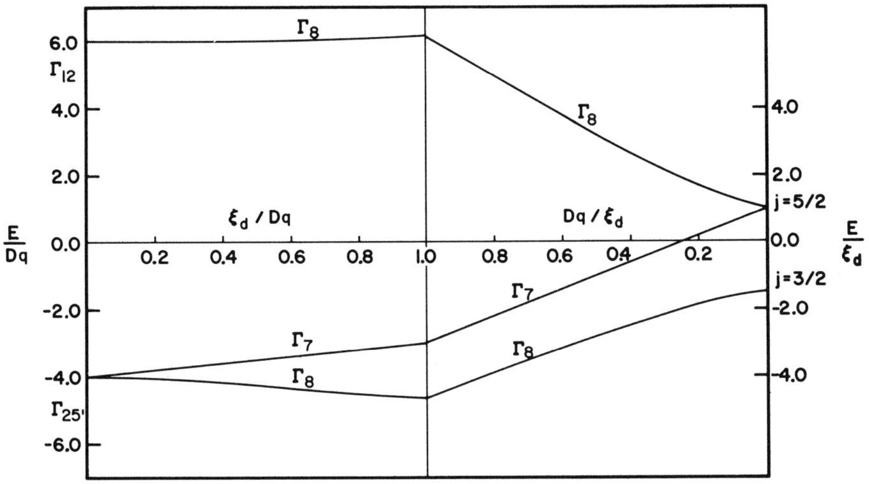

Fig. 3.4.1. Crystal field and spin–orbit splitting of d levels.

(3.4.25a) gives (to first order)

$$E(\Gamma_8) = \xi_d + 2Dq, \tag{3.4.27a}$$

$$E(\Gamma_8) = -\tfrac{3}{2}\xi_d. \tag{3.4.27b}$$

The upper solution (3.4.27a) and the Γ_7 state derive from the $j = \frac{5}{2}$ level; these two states are split by $6Dq$. The lower solution (3.4.27b) gives the $j = \frac{3}{2}$ level in the absence of crystal field effects, which change its energy only in second order.

The behavior of the levels is shown in Fig. 3.4.1. In the left half of the figure, we plot E/Dq as a function of ξ_d/Dq when this is less than one. In the right half, we show E/ξ_d as a function of Dq/ξ_d when this is less than one.

3.4.4 *Ions with More Than One* d *Electron*

In the study of crystal field effects on more complex ions, we will keep to the intermediate field case in which the crystal field is small compared to the multiplet splittings of the free atom, but is large compared to spin–orbit coupling. For zero crystal field, the atom is in a state of definite L and S. We consider how a state of given L and S is split by the crystal field, and defer study of the possible interaction between states of different values of these quantities.

To determine the splitting in a qualitative sense, we need only to determine how the $(2L + 1)$ dimensional representation of the operators of the point group which is afforded by the free atom states is decomposed into irreducible representations. If we have two d electrons, or two d holes, the ionic ground state is likely to be ^{3}F. From Section 3.2, we see that this splits as

$$^3\mathrm{F} \rightarrow {}^3\Gamma_{2'} + {}^3\Gamma_{15} + {}^3\Gamma_{25}. \tag{3.4.28}$$

Only this case will be considered in detail. A quantitative determination of the splitting requires a set of two-electron wave functions for the particular atomic state. Let us denote these as $\psi(L, S; M_L, M_S)$. These states are taken to be linear combinations of Slater determinants. It will be sufficient to consider a 2×2 determinantal wave function for the d^2 configuration, since the core electrons do not play any role in the calculation. Let us denote such a determinant as $(m_1{}^\alpha, m_2{}^\beta)$ where m_1 and m_2 denote eigenvalues of l_z for the two d electrons considered, and α and β take the values $+$ or $-$ corresponding to the values of the z component of an individual electron spin. Thus $(2^+, 1^-)$ represents a determinantal wave function

$$(2)^{-1/2} \begin{vmatrix} R_d(r_1)Y_{22}(\theta_1, \phi_1)\alpha(1) & R_d(r_1)Y_{21}(\theta_1, \phi_1)\beta(1) \\ R_d(r_2)Y_{22}(\theta_2, \phi_2)\alpha(2) & R_d(r_2)Y_{21}(\theta_2, \phi_2)\beta(2) \end{vmatrix}, \tag{3.4.29}$$

$\alpha(i)$ and $\beta(i)$ being "up" or "down" spin functions for electron i. Such a determinantal wave function has a definite value of M_L and M_S; in the example given, these are 3 and 0, respectively. However, a single such determinant is usually not an eigenfunction of total angular momentum and total spin; instead it is necessary to find linear combinations of different determinants with the same values of M_L and M_S. These combinations may be found as described in standard texts on atomic theory (see, for instance, Slater, 1960, or Ballhausen 1962). In the present example, we will ignore spin–orbit coupling. Then it is necessary to consider only a single value of M_S. In the case of ^{3}F, we have $S = 1$ and choose $M_S = 1$. There are seven wave functions to consider. Since L, S, and M_S are now fixed, the notation can be simplified by replacing $\psi(L, S; M_L, M_S)$ by $\psi(M_L)$:

$$\begin{aligned}
\psi(3) &= (2^+, 1^+), \\
\psi(2) &= (2^+, 0^+), \\
\psi(1) &= (\tfrac{3}{5})^{1/2}(2^+, -1^+) + (\tfrac{2}{5})^{1/2}(1^+, 0^+), \\
\psi(0) &= (\tfrac{1}{5})^{1/2}(2^+, -2^+) + (\tfrac{4}{5})^{1/2}(1^+, -1^+), \\
\psi(-1) &= (\tfrac{3}{5})^{1/2}(1^+, -2^+) + (\tfrac{2}{5})^{1/2}(0^+, -1^+), \\
\psi(-2) &= (0^+, -2^+), \\
\psi(-3) &= (-1^+, -2^+).
\end{aligned} \tag{3.4.30}$$

Next, it is necessary to set up the matrix elements of the operator

$$\sum_{i=1}^{2} v_4(r_i) K_{4,1}(\theta_i, \phi_i) \tag{3.4.31}$$

between these functions. This may be done in a straightforward way by expanding the determinants, and then using (3.4.24) with $\xi_d = 0$. The eigenvalues of this matrix give the energies we want. However, the calculation can be shortened by use of symmetry considerations to construct combinations of the $\psi(M_L)$ which transform according to the irreducible representations which appear in (3.4.28). These combinations may be found in a straightforward way using the projection techniques; or more simply by comparing the basis functions for O_h given in Table VII with a table giving explicit spherical harmonics for $l = 3$, treating the wave functions for L, M_L as if they were single-particle functions of the corresponding values of l and m. In this way, we find that one wave function belonging to Γ_{15} is

$$\psi(\Gamma_{15}) = \psi(0), \tag{3.4.32a}$$

and one belonging to Γ_{25} is

$$\psi(\Gamma_{25}) = 2^{-1/2}[\psi(2) + \psi(-2)], \tag{3.4.32b}$$

whereas for $\Gamma_{2'}$ we have

$$\psi(\Gamma_{2'}) = 2^{-1/2}[\psi(2) - \psi(-2)]. \tag{3.4.32c}$$

It is only necessary to calculate the expectation of (3.4.31) with these functions. We sketch this calculation for the Γ_{25} state. Let

$$V = \sum V(i) = \sum v_4(r_i) K_4(\theta_i, \phi_i).$$

Then

$$\int \psi(\Gamma_{25}) V \psi(\Gamma_{25})\, d\tau$$

$$= \tfrac{1}{2}\Big\{\int (2^+, 0^+) V (2^+, 0^+)\, d\tau + \int (2^+, 0^+) V (0^+, -2^+)\, d\tau$$

$$+ \int (0^+, -2^+) V (2^+, 0^+)\, d\tau + \int (0^+, -2^+) V (0^+, -2^+)\, d\tau\Big\}.$$

Next, we expand the determinantal wave function and obtain the matrix elements from (3.4.24). The matrix elements of $v_4 K_{4,1}$ between single particle states of $l = 2$ and definite m are denoted by $(m \mid v \mid m')$,

$$E(\Gamma_{25}) = \int \psi(\Gamma_{25}) V \psi(\Gamma_{25})\, d\tau$$

$$= (0 \mid v \mid 0) + (2 \mid v \mid 2) - (2 \mid v \mid -2) = 2Dq. \quad (3.4.33a)$$

By similar procedures,

$$E(\Gamma_{2'}) = 12Dq \quad (3.4.33b)$$

and

$$E(\Gamma_{15}) = -6Dq. \quad (3.4.33c)$$

The sum of the products of the degeneracies and the energies for the three states is zero.

These states cannot be described in terms of an integral occupation of single-particle states of $\Gamma_{25'}$ or Γ_{12} symmetry. Consider Γ_{15} above. Suppose we try to construct a single Γ_{15} state by supposing that n electrons are in $\Gamma_{25'}$ with energy $-4Dq$ and $2 - n$ are in Γ_{12} with energies $6Dq$. Then we must have

$$n(-4Dq) + (2 - n)(6Dq) = -6Dq$$

which gives $n = \frac{9}{5}$. Electron correlation, which forces the ion to be in a state of definite L, requires that the higher energy Γ_{12} be partially occupied.

These calculations can be extended to determine the spin–orbit splittings of the states for d^2. From Table VI we see that Γ_{15} and Γ_{25} are both split into a fourfold and a twofold state, while $\Gamma_{2'}$ is not split. We will not, however, discuss the computation of the splittings.

In the strong crystal field situation mentioned previously (Case III), the results we have just obtained would not be applicable. Instead, we would simply add electrons into the states of lowest crystal field energy ($\Gamma_{25'}$) until these were filled; only afterwards would there by any occupation of the Γ_{12} levels. Thus, in lowest approximation, the energy of the ground state of d^2 would be just $-8Dq$. States must be classified according to the representations of O_h (example, ${}^3\Gamma_{25'}$) in first approximation. It is then necessary to consider the repulsion of the d electrons. Such a calculation can be made by the methods used in multiplet theory for free atoms. The matrices which must be diagonalized are listed by McClure (1959).

Experimentally, typical values of $10Dq$ for transition metal ions are of the order of 1 eV, compared to typical multiplet splittings of several electron volts. Spin–orbit coupling contributes splitting in the range of a hundredth to a tenth of an electron volt. Thus it is legitimate to consider spin–orbit coupling as a perturbation on the crystal field splitting. However, multiplet splittings are typically of the order of a few volts, so that one must frequently contend with a situation intermediate between case (2) and the case (3) limits (Tanabe and Sugano, 1954).

In order to avoid a deep involvement with the theory of multiplet structure of atoms, we will quote some results where needed for the d^2 configuration. The reader is referred to McClure (1959) or Ballhausen (1962) for more details.

In the case of d^2 we have seen that the case (2) result has the degenerate state ${}^3\Gamma_{15}$ lowest in energy. This comes from the ${}^3F$ atomic state. However, there is a ${}^3P$ state of slightly higher energy, which in a cubic environment goes over to ${}^3\Gamma_{15}$. The difference in energy between these states is, in the limit of free atoms,

$$E({}^3\mathrm{P}) - E({}^3\mathrm{F}) = 15F_2 - 75F_4 \tag{3.4.34}$$

where F_2 and F_4 are integrals defined by Condon and Shortley (1951, p. 177):

$$F_2 = (e^2/49) \int_0^\infty \int_0^\infty (r_<^2/r_>^3) R_d^2(r_1) R_d^2(r_2) r_1^2 r_2^2 \, dr_1 \, dr_2,$$

$$F_4 = (e^2/441) \int_0^\infty \int_0^\infty (r_<^4/r_>^5) R_d^2(r_1) R_d^2(r_2) r_1^2 r_2^2 \, dr_1 \, dr_2. \tag{3.4.35}$$

In V^{3+}, $E({}^3\mathrm{P}) - E({}^3\mathrm{F}) = 1.6$ eV.

The two ${}^3\Gamma_{15}$ states have the same symmetry, and can be connected by a matrix element of the crystal field. The wave function for one of these states is, in the $\psi(L, S; M_L, M_S)$ notation,

$$\psi(1, 1; 0, 1) = (\tfrac{4}{5})^{1/2}(2^+, -2^+) - (\tfrac{1}{5})^{1/2}(1^+, -1^+). \tag{3.4.36}$$

[This is derived simply, as it is the only other state of the same M_L and M_S as $\psi(3, 1; 0, 1)$ which can be formed from $(2^+, -2^+)$ and $(1^+, -1^+)$, and therefore determine the coefficients by requiring orthogonality to $\psi(3, 1; 0, 1)$ and normalization.] The cubic field has no diagonal matrix element in this state, but it connects $\psi(3, 1; 0, 1)$ and $\psi(1, 1; 0, 1)$ by a matrix element we can easily work out to be $4Dq$. As a result we have the 2×2 matrix of crystal field plus electrostatic interaction

$$\begin{vmatrix} -6Dq & 4Dq \\ 4Dq & 15F_2 - 75F_4 \end{vmatrix}. \tag{3.4.37}$$

The energies are

$$E(\Gamma_{15}) = \tfrac{1}{2}\{15F_2 - 75F_4 - 6Dq \pm [(15F_2 - 75F_4)^2 + 12Dq(15F_2 - 75F_4) + 100(Dq)^2]^{1/2}\}. \tag{3.4.38}$$

In a case approximating (2) (small Dq) we obtain the results previously stated, whereas in case (3) limit, the energies $-8Dq$ and $2Dq$ result.

3.4.5 *Symmetries Lower Than Cubic. The Jahn–Teller Effect*

Our discussion of crystal fields has concerned only potentials of cubic symmetry. Quite frequently, one encounters situations of lower symmetry. In such cases, additional splittings of the remaining degeneracies are to be expected. For example, if the system has hexagonal, trigonal, tetragonal symmetry, etc., there can be a term in the potential proportional to Y_{20} which will have a nonzero expectation value in any state belonging to $l = 2$. To be specific, consider the case of tetragonal symmetry described by the point group D_{4h}. The character table for this group is given in Appendix C, Table IV. We see from this table that the d states of Γ_{12} symmetry are split by a tetragonal field

$$\Gamma_{12} \rightarrow X_1 + X_2 \tag{3.4.39a}$$

(notation is that of Appendix C, Table IV), both of which are nondegenerate in the absence of spin, and the $\Gamma_{25'}$ states also split

$$\Gamma_{25'} \rightarrow X_3 + X_5 \tag{3.4.39b}$$

where the X_5 states are doubly degenerate.

In such a splitting, we expect one of the states to be lowered in energy, the other raised. This possibility raises the question whether a system, assumed to be in a degenerate state, could lower its energy by distorting so as to split the degeneracy, and then occupying the lower state (or states whose energy is lower in total). That such a phenomena would be expected to occur quite frequently was pointed out by Jahn and Teller (1937).

The basic point is that in any assumed equilibrium atomic configuration, the energy will be a minimum with respect to small displacements. Suppose that the system is in a degenerate electronic state. If the system distorts by a small amount so as to split the degeneracy, and is able to occupy the lower state, the energy should decrease in proportion to the distortion. On the other hand, the gain in energy due to elastic restoring forces will only be quadratic in the distortion. Thus equilibrium is attained for finite distortion. This phenomenon is known as the *Jahn–Teller effect.*

A schematic theory follows. Let ζ be a parameter measuring the distortion. The energy lost by splitting the assumed degeneracy is $-a\zeta$. The gain in energy produced by elastic forces is $b\zeta^2$. The energy depends on ζ as

$$E(\zeta) = -a\zeta + b\zeta^2. \tag{3.4.40}$$

This is a minimum for a nonzero distortion ζ_0,

$$\zeta_0 = a/2b, \tag{3.4.41}$$

and the energy is

$$E(\zeta_0) = E_{JT} = -a^2/4b. \tag{3.4.42}$$

These considerations do not apply to the tightly bound electrons in inner atomic shells, which are very weakly coupled to the crystal environment. In this case, ζ is effectively zero. Also, static distortion does not occur unless the reduction of the energy E_{JT} so produced is greater than the zero-point energy of associated lattice vibrational modes. If it is not, then even at 0°K, the vibrational motion of the relevant complex is not localized about a single stable configuration, and the vibrations can carry the system between the neighborhoods of different distorted configurations. Such a situation is termed "the dynamical Jahn–Teller effect" (Liehr, 1963; Ham, 1965).

3.4.6 *Operator Equivalents. The Spin Hamiltonian*

The calculation of crystal field effects can become rather cumbersome if there are many electrons and spin–orbit coupling is included. In order to alleviate this difficulty, it is desirable to use the operator equivalents introduced by Bleaney and Stevens (1953) and Stevens (1954).

The potentials of interest can be represented as a sum over the coordinates of a number of electrons of the form

$$V = \sum_i f(\mathbf{r}_i)$$

where the $f(\mathbf{r}_i)$ can be expressed as some simple polynomial in the Cartesian coordinates of electron i. It is useful to observe that within a subspace of Hilbert space in which J is constant, there are simple relations between matrix elements of the coordinates and momentum. In fact, matrix elements of two operators having the same transformation properties are proportional as long as we stay within a subspace of definite J. Thus $\sum_i z_i^2 \approx J_z^2$. However, we must be cautious in the case of angular momentum components which do not commute, so that

$$\sum_i x_i y_i \rightarrow \tfrac{1}{2}(J_x J_y + J_y J_x).$$

Thus we have

$$\sum_i (3z_i^2 - r_i^2) \rightarrow \alpha(3J_z^2 - J^2), \tag{3.4.43}$$

for example. The fourth-order Kubic harmonic becomes

$$\sum_i K_{4,1}(\theta_i, \phi_i) \to \beta[J_z^4 + J_y^4 + J_x^4 - \tfrac{1}{5}J(J+1)(3J^2 + 3J - 1)]. \tag{3.4.44}$$

Here, α and β are constants of proportionality. The last term of (3.4.44) has the particular form it does since the operator equivalent to r^4 is not $\mathbf{J}^4$ but $\mathbf{J}^2(\mathbf{J}^2 - \frac{1}{3})$. Other equivalents are given by Stevens in the references cited. Moreover, since the potentials of interest are independent of spin, similar operator equivalents may be used in a subspace of constant L [here we use L_x, L_y, L_z in (3.4.43) and (3.4.44) instead of J_x, J_y, and J_z]. The constants of proportionality can be determined by explicit comparison of matrix elements of the related expressions for a few specific states.

Let us consider the ground state of the system. If this has spin degeneracy, it can be split by the application of an external magnetic field, and transitions can be observed in the microwave range between the separated levels. This is the phenomenon of paramagnetic resonance, and an extensive literature exists concerning this effect (Bleaney and Stevens, 1953; Bowers and Owen, 1955; Low, 1960; Pake, 1962).

Suppose the ion in question has a ground state which is orbitally nondegenerate, that is, it belongs to a nondegenerate representation of the appropriate crystal (single) point group. However, let us suppose there is a nonzero total spin, and therefore some spin degeneracy in the absence of spin–orbit coupling and external magnetic fields. This degeneracy will be split in the presence of these effects. It is convenient to describe those low lying energy states by an operator which is a polynomial in the spin vector $\mathbf{S}$. This operator is known as the *spin Hamiltonian* (Pryce, 1950; Abragam and Pryce, 1951).

Let us consider a perturbation of the form

$$\mathcal{H}' = \xi \mathbf{L}\cdot\mathbf{S} + \beta \mathbf{H}\cdot(\mathbf{L} + 2\mathbf{S}) \tag{3.4.45}$$

where the first term represents spin–orbit coupling and the second represents the interaction with the external field. In (3.4.45), β is the Bohr magneton, and explicit account is taken of the g factor 2 associated with spin. The effect of this operator will be examined using second-order perturbation theory; however, in making this calculation, we will calculate the scalar products which are the matrix elements as space coordinate integrals only. The spin is left as an operator.

The assumption of an orbitally nondegenerate level implies that the ground state expectation value of $\mathbf{L}$ vanishes. This may be seen as follows: Since the level is nondegenerate, the ground state wave function may be chosen to be real. The operator **L** is purely imaginary; however, it is Hermitian, and as a consequence the average value of **L** in any state must be a vector with real components. This is impossible with an imaginary operator and real wave functions, consequently the average of **L** is zero.

Let us call the ground state $|\,0\rangle$. The first-order perturbation result is

$$\langle 0 \,|\, \mathcal{H}' \,|\, 0\rangle = 2\beta \mathbf{H}\cdot\mathbf{S}. \tag{3.4.46}$$

The second-order perturbation can be expressed in terms of a complete set of states $|\,n\rangle$:

$$\begin{aligned} E_2 &= \sum_{n\neq 0} |\,\langle 0 \,|\, \xi \mathbf{L}\cdot\mathbf{S} + \beta\mathbf{H}\cdot(\mathbf{L} + 2\mathbf{S}) \,|\, n\rangle\,|^2/(E_0 - E_n) \\ &= -\sum_{n\neq 0} |\,\langle 0 \,|\, \xi\mathbf{L}\cdot\mathbf{S} + \beta\mathbf{H}\cdot\mathbf{L} \,|\, n\rangle\,|^2/(E_n - E_0). \end{aligned} \tag{3.4.47}$$

The combination of (3.4.46) and (3.4.47) can be expressed as

$$E = \sum_{i,j} \{2\beta(\delta_{ij} - \xi\Lambda_{ij})S_iH_j - \xi^2\Lambda_{ij}S_iS_j - \beta^2\Lambda_{ij}H_iH_j\} \tag{3.4.48}$$

where

$$\Lambda_{ij} = \sum_{n\neq 0} \langle 0 \,|\, L_i \,|\, n\rangle\langle n \,|\, L_j \,|\, 0\rangle/(E_n - E_0). \tag{3.4.49}$$

The summation is over Cartesian vector components. Since the components of L are Hermitian operators, Λ is a real symmetric tensor.

The last term of (3.4.48) gives rise to a uniform displacement of the levels in an external field. This is of little interest and will be discarded. The first term of (3.4.48) is of the form

$$\beta\mathbf{S}\cdot\mathbf{g}\cdot\mathbf{H} \tag{3.4.50}$$

where g is a tensor "g" factor

$$g_{ij} = 2(\delta_{ij} - \xi\Lambda_{ij}). \tag{3.4.51}$$

The second term of (3.4.48) gives a small splitting of the spin states in zero field, induced by spin–orbit coupling.

To make these considerations more concrete, suppose we have a system with tetragonal symmetry described by D_{4h}. Then Λ has only two independent components: $\Lambda_\perp = \Lambda_{xx} = \Lambda_{yy}$ and $\Lambda_{||} = \Lambda_{zz}$, all others being zero. In this case (3.4.48) becomes (neglecting all terms constant within the multiplet)

$$E = \beta[g_\perp(S_xH_x + S_yH_y) + g_{||}S_zH_z] - DS_z^2 \tag{3.4.52}$$

where $D = \xi^2(\Lambda_\perp - \Lambda_{||})$: $g_{\perp(||)} = 2(1 - \xi\Lambda_{\perp(||)})$ and we have used the identity

$$S_x^2 + S_y^2 = \mathbf{S}^2 - S_z^2 = S(S+1) - S_z^2. \tag{3.4.53}$$

The last step holds for all states of a given S.

Equations (3.4.48) and (3.4.53) define the spin Hamiltonian. The energy levels associated with the spin orientation are to be found by diagonalizing these expressions which, as we recall, are operators on spin coordinates.

In a cubic system, $\Lambda_{\|} = \Lambda_{\perp}$, so the g factor is isotropic and D vanishes. Higher-order effects which we will not discuss can give rise to a term with cubic symmetry containing $(S_x^4 + S_y^4 + S_z^4)$.

In general, the spin Hamiltonian will include terms such that the order of any polynomial appearing is less than or equal to $2S$. The matrix representing a component of the operator S has dimension $(2S + 1) \times (2S + 1)$. Consequently any polynomial in S of order $2S + 1$ or higher can be expressed in terms of lower-order polynomials. Thus, for an ion in a $^6S_{5/2}$ state in a cubic environment, the spin Hamiltonian is of the form

$$a(S_x^4 + S_y^4 + S_z^4) + g\beta\mathbf{H}\cdot\mathbf{S}. \tag{3.4.54}$$

A more general treatment of the effective Hamiltonian for ions in crystals free of the restrictions of perturbation theory has been given by Koster and Statz (1959); Statz and Koster (1959).

3.4.7 *Computation of Dq. Molecular Orbitals*

The preceding discussion has been largely phenomenological. Except in the illustrative calculation based on the point charge model at the beginning of this section, the results have been based mainly on symmetry considerations combined with some parameterized matrix elements.

Serious problems arise when a calculation of the parameters is attempted. Consider in particular, the fundamental quantity Dq. This has frequently been regarded simply as an empirical constant to be determined by fitting experimental data. It should, however, be calculable from first principles. Early attempts to do this (Van Vleck, 1939; Polder, 1942) used a point ion model and simple Slater orbital wave functions, and obtained reasonably good agreement with experimental results. These authors considered transition metal ions surrounded by six water molecules, with the O^{2-} ions pointing toward the metal. Unfortunately, the agreement disappeared when Kleiner (1952) used an extended charge distribution and consequently a more realistic potential for the oxygen ions in to calculate Dq for Cr^{3+} in this environment. As calculated by Kleiner, Dq had the wrong sign compared with experiment. This occurred because the attractive potential of the oxygen nuclei overpowered the repulsion of the extra two electrons bound to the ion in its effect on the chromium d wave functions.

The origin of the difficulty is that two important physical effects have been neglected. These are (1) the d wave functions on the metal ions must be orthogonalized to the 2s and 2p wave functions of the oxygen ligands, and (2) the formation of chemical bonds (the covalency effect) must be considered. The latter effect is generally described by forming molecular orbitals, which in effect, allow the metal d electron wave functions to mix with those of the ligands.

We will consider in outline how a more correct calculation can be performed. This will be done with reference to the specific example of $KNiF_3$, which has been extensively studied (Sugano and Shulman, 1963; Simanek and Sroubek, 1964; Watson and Freeman, 1964; Ellis *et al.*, 1968).

$KNiF_3$ has the perovskite structure. A cubic cell for this compound (see Fig. 3.4.2) has the nickel atoms at the body center, fluorine at the face centers, and potassium at the corners. The arrangement is cubic, with a lattice constant of 4.014 Å ($7.585a_0$). It is antiferromagnetic below 275°K. Attention focuses on the cluster consisting of the Ni^{2+} ion surrounded by six neighboring F^- ions. In the simplest approximation, these F^- ions would be considered to create an electrostatic field of cubic symmetry, which would split the d states of the nickel atom. It is this approximation which the calculations of Kleiner showed to be inadequate in the case he studied. Sugano and Shulman obtained a similar result in this instance. The nickel d wave functions must be allowed to combine with s and p functions on the fluorine atoms to form molecular orbitals. In this way, the d electrons from nickel are allowed to be distributed in part throughout the cluster. Consider the 2s and 2p wave functions on the six fluorine atoms, located as shown in Fig. 3.4.3. We can form linear combinations of these wave functions which

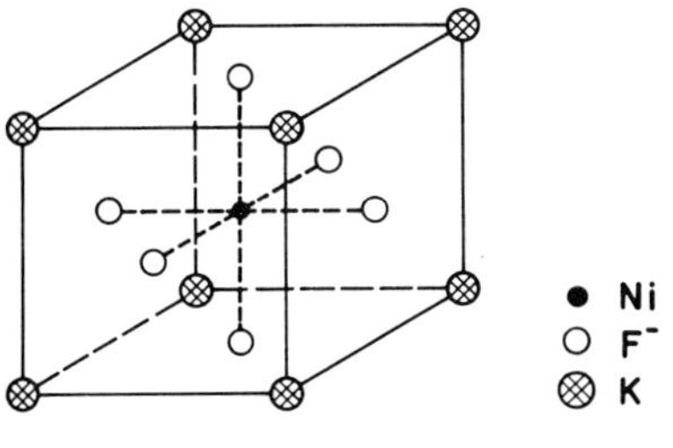

Fig. 3.4.2

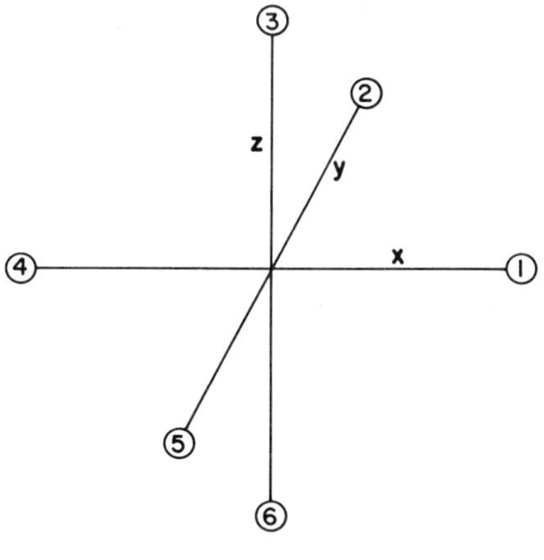

Fig. 3.4.3

Fig. 3.4.2. Structure of $KNiF_3$ (perovskite).
Fig. 3.4.3. Coordinate system used to describe arrangement of F^- ions in $KNiF_3$.

transform according to the $\Gamma_{25'}$ and Γ_{12} representations. These combinations can mix with the d functions of the same symmetry on the Ni ion, giving a wave function with s, p, and d components which spreads through the cluster. Direct evidence that a picture of this sort is necessary (in contrast to the more localized picture of the purely ionic model) is provided by the nuclear magnetic resonance measurements of Shulman and Sugano (1963). These observations reveal that there are large internal magnetic fields at the fluorine nuclei. These fields arise from hyperfine interactions with electrons of unpaired spin. The unpaired electrons are based on the Ni^{2+} ion, and the hyperfine interaction mentioned indicates these have an effect at fluorine sites, rather than being restricted to the nickel.

In order to construct molecular orbitals, it is conventional to divide the 2p functions of F^- into two sets denoted σ and π. The σ functions are the atomic p functions which have no angular momentum about the axis connecting the F and Ni nuclei; the π states are the remaining p states which have one unit of angular momentum about this axis. We must also consider s functions on F^-: for most purposes only the 2s is important. Let us denote a fluorine wave function of symmetry i ($i = s, x, y, z$) on atom μ ($\mu = 1, \ldots, 6$) by $\phi_{\mu,i}$. We follow the procedure of Section 3.2 to form molecular orbitals as combinations of these functions. Let $\boldsymbol{\alpha}$ be an operation of O_h. Construct the sum

$$\sum_{\alpha} [\boldsymbol{\alpha}]^*_{l,nn} \,\alpha\phi_{\mu,i} \tag{3.4.55}$$

where l and n denote the irreducible representation and row desired, respectively. We can write $\phi_{\mu,i}$ as $\phi_i(\mathbf{r} - \mathbf{R}_\mu)$. The action of $\boldsymbol{\alpha}$ on this function is

$$\begin{aligned} \alpha\phi_i(\mathbf{r} - \mathbf{R}_\mu) &= \phi_i(\boldsymbol{\alpha}^{-1}\mathbf{r} - \mathbf{R}_\mu) = \phi_i[\boldsymbol{\alpha}^{-1}(\mathbf{r} - \boldsymbol{\alpha}\mathbf{R}_\mu)] \\ &= \sum_j [\boldsymbol{\alpha}]_{k,ji}\phi_j(\mathbf{r} - \boldsymbol{\alpha}\mathbf{R}_\mu). \end{aligned} \tag{3.4.56}$$

In this equation k denotes the irreducible representation (Γ_1 or Γ_{15}) that is based on s or p functions. Thus if k refers to Γ_1, $[\alpha] = 1$, the character of operation α in Γ_1; however, if k refers to Γ_{15}, we have a 3×3 matrix. The elements $[\boldsymbol{\alpha}]_{l,nn}$ are determined by considering the cubic basis functions listed in Table VII. Let us denote the molecular orbitals which are constructed as $\chi_{i,l,n}$ where l and n are as defined above and $i = s, \sigma$, or π denotes the set of functions used in construction.

By convention the σ functions have their "z" axis in coincidence with the line joining the fluorine and nickel atoms. This implies that these functions effectively transform like s functions. We simply replace $[\boldsymbol{\alpha}]_\sigma$ (and $[\boldsymbol{\alpha}]_s$) by 1. Then from (3.4.55) and (3.4.56) we can construct an s-like (Γ_1) combination, three p-like combinations (transforming according to Γ_{15}), and two d-like combinations (Γ_{12}). These are

$$\begin{aligned}
\chi_{i,1,1} &= 6^{-1/2}(\phi_{1,i} + \phi_{2,i} + \phi_{3,i} + \phi_{4,i} + \phi_{5,i} + \phi_{6,i}),\\
\chi_{i,15,1} &= 2^{-1/2}(\phi_{1,i} - \phi_{4,i}),\\
\chi_{i,15,2} &= 2^{-1/2}(\phi_{2,i} - \phi_{5,i}),\\
\chi_{i,15,3} &= 2^{-1/2}(\phi_{3,i} - \phi_{6,i}),\\
\chi_{i,12,1} &= \tfrac{1}{2}\cdot 3^{-1/2}(2\phi_{3,i} + 2\phi_{6,i} - \phi_{1,i} - \phi_{2,i} - \phi_{4,i} - \phi_{5,i}),\\
\chi_{i,12,2} &= \tfrac{1}{2}(\phi_{1,i} - \phi_{2,i} + \phi_{4,i} - \phi_{5,i}).
\end{aligned} \tag{3.4.57}$$

In the above, $i = s$ or σ.

There are twelve π functions. These can be used to construct combinations belonging to the Γ_{15}, $\Gamma_{25'}$, $\Gamma_{25'}$, and Γ_{25} representation. Expressions for the $\chi_{\pi,l,n}$ are given by Ballhausen (1962).

The wave functions for the cluster are formed as combinations of the nickel d wave functions, which will be denoted by d_j (where $j = xy, yz, zx$ for $\Gamma_{25'}$; $x^2 - y^2$ and $3z^2 - r^2$ for Γ_{12}). Thus, to obtain the cluster wave function of $3z^2 - r^2$ symmetry, we write

$$\Phi_{3z^2-r^2} = N_{3z^2-r^2}(d_{3z^2-r^2} - \lambda_s\chi_{s,12,1} - \lambda_\sigma\chi_{\sigma,12,1}) \tag{3.4.58}$$

where λ_s and λ_σ are mixing coefficients and N is the normalization factor given by

$$N = (1 - 2\lambda_s S_s - 2\lambda_\sigma S_\sigma + \lambda_s^2 + \lambda_\sigma^2 + 2\lambda_s\lambda_\sigma S_{s\sigma})^{1/2}. \tag{3.4.59}$$

It is assumed here that the orbitals of (3.4.57) have been normalized; this has not been shown explicitly in (3.4.57). The S are overlap integrals:

$$S_s = \int d_{3z^2-r^2}\chi_{s,12,1}\,d\tau, \qquad S_\sigma = \int d_{3z^2-r^2}\chi_{\sigma,12,1}\,d\tau, \qquad S_{s\sigma} = \int \chi_{s,12,1}\chi_{\sigma,12,1}\,d\tau. \tag{3.4.60}$$

The coefficients λ_s and λ_σ cannot be determined by symmetry considerations. To do this, we must set up the matrix of the Hamiltonian on the basis of the three wave functions $d_{3z^2-r^2}$, $\chi_{s,12,1}$, and $\chi_{\sigma,12,1}$ and solve the resulting secular equations. Three independent solutions will be obtained [if fluorine 1s functions are included there will be another term in (3.4.58) and four solutions]. The lowest of these is called a *bonding orbital*; the upper are *antibonding orbitals*. The wave function for the entire cluster can be regarded as a Slater determinant of molecular orbitals.

We proceed in a similar way for the $\Gamma_{25'}$ wave functions. For example,

$$\Psi_{xy} = N_{xy}(d_{xy} - \lambda_\pi\chi_{\pi,25',xy}) \tag{3.4.61}$$

in which

$$N_{xy} = (1 - 2\lambda_\pi S_\pi + \lambda_\pi^2), \qquad (3.4.62a)$$

$$S_\pi = \int d_{xy}\chi_{\pi,25',xy}\, d\tau. \qquad (3.4.62b)$$

The states of the molecular cluster can be classified according to the representations of the cubic group. This applies to both the single-particle levels obtained by solution of the matrix diagonalization problem previously described, and to the energy states of the entire system, including all 86 electrons. The arrangement of the molecular orbital levels is shown in Fig. 3.4.4 according to Ellis *et al.* (EFR) (1968). Spin splitting has been neglected. The lower case Greek letter γ is used here for the representations of the single-particle energy levels. The notation used by EFR is given in parentheses. The relation of the molecular orbital levels to the atomic

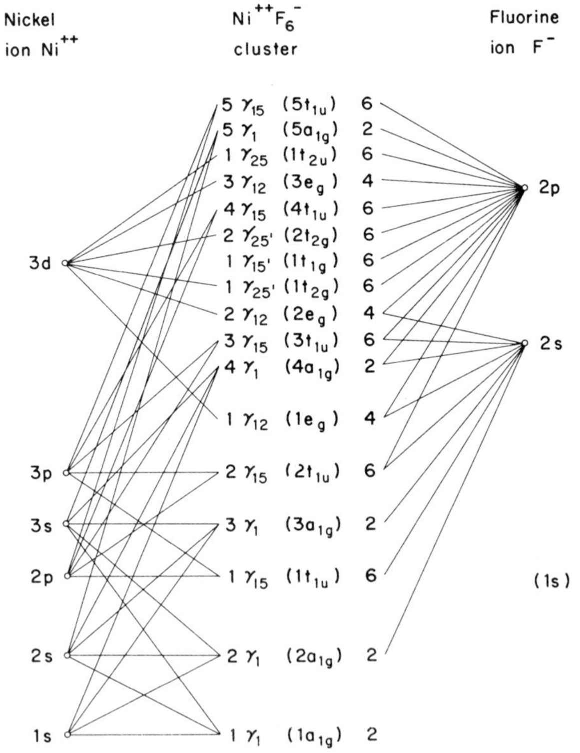

Fig. 3.4.4. Energy levels for a $Ni^{2+}\ F_6^-$ cluster according to molecular orbital theory.

energy levels is shown by lines. Levels arising from the 1s states of the fluorine ions have not been shown. The number of electrons which can be contained in the molecular orbital is also shown.

The quantity corresponding to the crystal field splitting of our previous discussion is determined from the energy difference between the ground state of the cluster ($^3\Gamma_{2'}$) and the various excited states ($^3\Gamma_{15}$), ($^3\Gamma_{25}$), etc. Note that the energy difference between $^3\Gamma_{2'}$ and $^3\Gamma_{25}$ is $10Dq$ according to (3.4.33). Experimental results (Knox *et al.*, 1963) can be interpreted to give a value for $10Dq$ of 0.90 eV. It would be pleasing to be able to account for this quantity simply as the difference in one-electron energies between γ_{12} and $\gamma_{25'}$ orbitals. In the present case, we may think of $^3\Gamma_{25}$ as being formed from the ground $^3\Gamma_{2'}$ by transferring an electron from an occupied γ_{25} orbital (in $^3\Gamma_{2'}$) to an unoccupied γ_{12} orbital. Unfortunately, this procedure is not accurate quantitatively since the transfer of electrons between orbitals is accompanied by relaxation, that is, distortion of the orbitals themselves. Thus, it is not possible to write a simple formula for the effective $10Dq$; one must instead calculate the total energies of the $^3\Gamma_{2'}$ and the $^3\Gamma_{25}$ states. This has been done by EFR, who obtained 1.34 eV for this quantity. The earlier calculation of Sugano and Shulman (1963) gave 0.79 eV. It is clear that quantitative success has not yet been obtained.

An alternative to the molecular orbital approach to the problem of covalency in transition metal salts should be mentioned. Hubbard *et al.* (1966) developed an analog of the Heitler–London treatment of the hydrogen molecule. This method will not be discussed in detail; however, it has the merit of allowing for configuration interaction (electron correlation) in a natural way.

3.4.8 *Magnetic Order*

We will conclude this section with a discussion of the coupling of ionic spins. A cursory survey of the magnetic properties of insulating crystals containing a single species of transition metal atom reveals that the overwhelming majority of these materials order antiferromagnetically. The number of true ferromagnets in this category is quite small (For a survey of magnetic properties of such systems, see Goodenough, 1963). In this account, we will focus on the origin of antiferromagnetism, and ignore the relatively rare situations in which ferromagnetism can occur. Our approach follows that of Anderson (1963).

The wave functions on the transition metal ions are mixed with those of the ligands by covalency effects as previously described. This serves in effect to connect the transition metal ions even when they are so far apart (4.0 Å in $KNiF_3$) that direct overlap of 3d wave functions on different atoms would be negligible. The role of the ligand is to establish this connection between metal ions. Beyond this, they need not be considered further in a first approximation. Even though the transition metal ions are connected, it is energetically quite unfavorable to have the wrong number of electrons on any ion. The extra electron experiences a strong coulomb repulsion produced by its interaction with the other electrons on the ion. While the numerical magnitude of this effect is not known precisely because one would have to obtain accurate wave functions, and allow for relaxation effects; it will be sufficient to suppose that it is large compared to the "transfer integral" to be defined below. Should this fail, our argument collapses, and we may have a paramagnetic (or even ferromagnetic) metal instead of an antiferromagnetic insulator.

Let us suppose that the individual ions in question have an effective spin of $\frac{1}{2}$, that is, that the lowest state after all large effects have been included is a doublet. For reasons of conceptual simplicity, it is assumed to be an orbital s state. There can be only two electrons in this state: if one is present, we have the doublet; if a second is added, a singlet. The energy of the singlet contains a large term U. Let the spatial wave function for ion a be denoted by $a(i)$ where i is an electron coordinate. Then U is given by

$$U = \int |a(1)|^2 |a(2)|^2 (e^2/r_{12}) \, dr_1^3 \, d^3r_2 \qquad (3.4.63)$$

and can be estimated with typical 3d atomic wave functions to have a value of roughly 25 eV. Now consider a second transition metal ion b, also with a single electron [whose wave function is $b(j)$] outside of an inert core. Intervening ligands are ignored.

The Hamiltonians for the two-electron, two-atom system can be written as

$$H = H_A(1) + V_B(1) + H_B(2) + V_A(2) + V(1, 2) \qquad (3.4.64)$$

where $H_A(1)$ contains the kinetic energy of electron 1 plus its interaction with the nucleus and core electrons of A; $V_B(1)$ contains the interaction of electron 1 with the nucleus and core electrons of atom B; and $V(1, 2)$ is the interaction of the two external electrons with each other, $V(1, 2) = e^2/r_{12}$. Since the wave function on atoms A and B overlap, $a(1)$ and $b(1)$ are not orthogonal. We define

$$(a \mid b) = S. \qquad (3.4.65)$$

It is convenient to introduce a set of orthonormal eigenfunctions

$$A(1) = N[a(1) - \gamma b(1)], \qquad B(1) = N[b(1) - \gamma a(1)]. \qquad (3.4.66)$$

From the requirement of orthogonality

$$(A \mid B) = 0, \tag{3.4.67}$$

we deduce that

$$S - 2\gamma + \gamma^2 S = 0$$

or

$$\gamma = S^{-1}[1 \pm (1 - S^2)^{1/2}]. \tag{3.4.68}$$

The negative sign is chosen since we require that $\gamma = 0$ if $S = 0$. Then for small S, $\gamma \approx S/2$. The normalization constant is given by the relation

$$N^{-2} = 1 + \gamma^2 - 2\gamma S. \tag{3.4.69}$$

The Hamiltonian can be expressed in the notation of second quantization in the following way (see Appendix D or Landau and Lifshitz, 1965, Chap. IX):

$$H = \sum_{\sigma} [(H^{(1)}{}_{A\sigma,A\sigma} c^{\dagger}{}_{A\sigma} c_{A\sigma} + H^{(1)}{}_{B\sigma,B\sigma} c^{\dagger}{}_{B\sigma} c_{B\sigma} + H^{(1)}{}_{A\sigma,B\sigma} (c^{\dagger}{}_{A\sigma} c_{B\sigma} + c^{\dagger}{}_{B\sigma} c_{A\sigma})]$$

$$+ \tfrac{1}{2} \sum_{\mu\nu\rho\tau,\sigma_1\sigma_2\sigma_3\sigma_4} H^{(2)}{}_{\mu\sigma_1,\nu\sigma_2;\,\rho\sigma_3,\tau\sigma_4} c^{\dagger}{}_{\mu\sigma_1} c^{\dagger}{}_{\nu\sigma_2} c_{\tau\sigma_4} c_{\rho\sigma_3}. \tag{3.4.70}$$

In the last term μ, ν, ρ, τ take the values A, B. The spin variables represent $\uparrow$ or $\downarrow$ states. The operators $c^{\dagger}{}_{A\sigma}$ ($c_{A\sigma}$) create (destroy) an electron in a state of spin σ on atom A. The elements of the Hamiltonian have superscripts (1) or (2) according as they refer to the one- or two-electron portions of H. The single-particle elements are independent of spin and will be denoted $H_{A,A}$ and $H_{A,B}$:

$$H_{A,A} = H^{(1)}{}_{A\sigma,A\sigma} = (A(1) \mid H_A(1) + V_B(1) \mid A(1)),$$

$$H_{A,B} = H^{(1)}{}_{A\sigma,B\sigma} = (A(1) \mid H_A(1) + V_B(1) \mid B(1)). \tag{3.4.71}$$

The following definitions are introduced in order to simplify (3.4.71):

$$K = (a(1) \mid V_B(1) \mid a(1)), \qquad t = (a(1) \mid V_A(1) \mid b(1)). \tag{3.4.72}$$

The original states from which we started are eigenstates of H_A and H_B:

$$H_A a(1) = \epsilon_0 a(1); \qquad H_B b(1) = \epsilon_0 b(1). \tag{3.4.73}$$

The quantity t is referred to as a *transfer integral*; K will be called the *potential integral*. After a little algebra, $H_{A,A}$ and $H_{A,B}$ can be expressed as

$$H_{A,A} = \epsilon_0 + [(K - St)/(1 - S^2)], \qquad H_{A,B} = (t - SK)/(1 - S^2). \tag{3.4.74}$$

We will consider in detail only the case in which S, t, K are small. Then, to first order,

$$H_{A,A} = \epsilon_0 + K, \qquad H_{A,B} = t. \tag{3.4.75}$$

Only the leading portion of the two-particle term, which is $Un\uparrow n\downarrow$, will be retained. These approximations lead to the following expression for H:

$$H = \sum_{\sigma} [(\epsilon_0 + K)(n_{A\sigma} + n_{B\sigma}) + t(c^\dagger{}_{A\sigma}c_{B\sigma} + c^\dagger{}_{B\sigma}c_{A\sigma})] + U[n_{A\uparrow}n_{A\downarrow} + n_{B\uparrow}n_{B\downarrow}]. \tag{3.4.76}$$

This Hamiltonian appears quite frequently in the theory of magnetic order in solids. It is easily generalized to the case of an arbitrary number of interacting atoms, in the form

$$H = \sum_{\mu,\nu,\sigma} \epsilon_{\mu\nu}c^\dagger{}_{\mu\sigma}c_{\nu\sigma} + U\sum_{\mu} n_{\mu\uparrow}n_{\mu\downarrow}. \tag{3.4.77}$$

Here the atoms are labeled by an index μ; the diagonal elements of $\epsilon_{\mu\nu}$ are just $\frac{1}{2}(\epsilon_0 + K)$, etc. This Hamiltonian has come to be called the *Hubbard Hamiltonian* (Hubbard, 1963). It will be considered in another context in Part B, Section 8.4.

The Hamiltonian of (3.4.76) can be diagonalized in a straightforward way (but only when only two atoms are considered). We determine the matrix representing H on the basis of states $c^\dagger{}_{\mu\sigma}c^{\dagger\prime}{}_{\nu\sigma}\,|\,0\rangle$ (μ, ν = A, B; $\sigma, \sigma' = \uparrow, \downarrow$). There are only six possible states since if $\mu = \nu$, we cannot have $\sigma = \sigma'$. Further, the triplet states $c^\dagger{}_{A\sigma}c^\dagger{}_{B\sigma}\,|\,0\rangle$ are not coupled to the rest (or to each other), so we have only a 4 × 4 matrix to diagonalize. This matrix acts on the four states which have the z component of total spin equal to zero. These are (1) $c^\dagger{}_{A\uparrow}c^\dagger{}_{A\downarrow}\,|\,0\rangle$; (2) $c^\dagger{}_{B\uparrow}c^\dagger{}_{B\downarrow}\,|\,0\rangle$; (3) $c^\dagger{}_{A\uparrow}c^\dagger{}_{B\downarrow}\,|\,0\rangle$, and (4) $c^\dagger{}_{B\uparrow}c^\dagger{}_{A\downarrow}\,|\,0\rangle$. The matrix is

$$\begin{bmatrix} 2(\epsilon_0 + K) + U & 0 & t & t \\ 0 & 2(\epsilon_0 + K) + U & t & t \\ t & t & 2(\epsilon_0 + K) & 0 \\ t & t & 0 & 2(\epsilon_0 + K) \end{bmatrix} \tag{3.4.78}$$

The eigenvalues are

$$E_1 = 2(\epsilon_0 + K) + \tfrac{1}{2}U[1 - (1 + 16t^2/U^2)^{1/2}], \tag{3.4.79a}$$

$$E_2 = 2(\epsilon_0 + K), \tag{3.4.79b}$$

$$E_3 = 2(\epsilon_0 + K) + U, \tag{3.4.79c}$$

$$E_4 = 2(\epsilon_0 + K) + \tfrac{1}{2}U[1 + (1 + 16t^2/U^2)^{1/2}]. \tag{3.4.79d}$$

The lowest eigenvalue is (3.4.79a), belonging to a singlet state. The next eigenvalue E_2 is the energy of the triplet state which has z component of total spin equal to zero. The two higher singlets do not concern us. The energy of the lowest singlet differs from that of the triplets by an amount δ,

$$\delta = E_1 - E_2 = \tfrac{1}{2}U[1 - (1 + 16t^2/U^2)^{1/2}]. \tag{3.4.80}$$

Physical systems have $t \ll U$. In this case, we may expand (3.4.80a) and obtain the leading term

$$\delta = -4t^2/U. \tag{3.4.81}$$

That a singlet state is lower in energy than the triplet states is in accord with the prevalence of antiferromagnetism in contrast to ferromagnetism in systems to which similar analysis could be applied. Exact solution for a system of many atoms is not possible; however, application of perturbation theory leads to results consistent with those presented above (Mattheiss, 1961). An interaction of this type which tends to produce an antiferromagnetic ground state of a system of coupled spin is frequently called "superexchange."

3.5 Irreducible Representations: Translation Groups and Bloch's Theorem

The subject of irreducible representations of space groups is more complicated than that of point groups since the former are not finite. Space groups contain an infinite, invariant Abelian subgroup of pure translations. In this section, we will study the representations of this subgroup, which are of vital importance for our future work.

Wave functions for electrons in crystals must transform under translations according to the irreducible representations of this group. Since the translation group is Abelian, the irreducible representations are one-dimensional. This leads to a result of fundamental importance for the theory of solids, known as "Bloch's theorem." It is important to consider the argument carefully.

It is assumed that the state of the solid can be specified by giving the wave function of each electron. This is not rigorous, since in principle only the full many-electron wave function is well defined, but the concept of an individual electron wave functions is an extremely useful approximation. The approximations involved in a one-electron approach to the full many-body problem will be discussed in Section 4.9 and Part B, Section 8.1. Further, it is supposed that these wave functions can be determined by the solution of some effective Schrödinger equation. We will not try to write this equation explicitly at this point. It is only necessary to assume that the one-electron Hamiltonian which appears in this equation has the periodicity of the crystal lattice. Let us denote this Hamiltonian by H. It is assumed to commute with all the operators of the space group. If $\{\boldsymbol{\alpha} \mid \mathbf{t}_\alpha\}$ is such an operator,

$$[H, \{\boldsymbol{\alpha} \mid \mathbf{t}_\alpha\}] = 0. \tag{3.5.1}$$

To simplify the notation for the translation operators somewhat, we define

$$T(\mathbf{R}_i) = \{\epsilon \mid -\mathbf{R}_i\}. \tag{3.5.2}$$

The minus sign is chosen in (3.5.2) so that the effect of T on a function $\phi(\mathbf{r})$ [see Eq. (3.2.4)] may be expressed by

$$T(\mathbf{R}_i)\phi(\mathbf{r}) = \phi(\mathbf{r} + \mathbf{R}_i). \tag{3.5.3}$$

These operators commute with each other and with the Hamiltonian

$$[T(\mathbf{R}_i), T(\mathbf{R}_j)] = 0, \tag{3.5.4}$$

$$[T(\mathbf{R}_i), H] = 0. \tag{3.5.5}$$

Equations (3.5.4) and (3.5.5) imply that the one-electron wave functions may be chosen to be simultaneously eigenfunctions of the Hamiltonian and of all the translations. Let ψ be such an eigenfunction:

$$T(\mathbf{R}_i)\psi(\mathbf{r}) = \psi(\mathbf{r} + \mathbf{R}_i) = \lambda_i\psi(\mathbf{r}). \tag{3.5.6}$$

The eigenvalue λ_i describes the effect of the operator $T(\mathbf{R}_i)$ on the function ψ. We require that the representation be unitary; in other words that the norm of ψ is unchanged by translation. In order for this to be true, λ_i must be a complex number of modulus unity, and so can be written as

$$\lambda_i = \exp(i\theta_i) \tag{3.5.7}$$

where θ_i is real.

Equation (3.5.7) is really based on the following physical reasoning. The electron distribution will determine, in part, the potential energy function which appears in the effective single-particle Hamiltonian H. Even if we cannot write a simple expression for this function, we may suppose that H will be unchanged by a lattice translation if $|\psi|^2$ is, that is, if

$$|\psi(\mathbf{r} + \mathbf{R}_i)|^2 = |\psi(\mathbf{r})|^2. \tag{3.5.8}$$

Equation (3.5.8) implies that $|\lambda_i|^2 = 1$, and so leads to (3.5.7).

Now let two translation operators, say $T(\mathbf{R}_j)$ and $T(\mathbf{R}_i)$, act in succession. The result is equivalent to a single translation $T(\mathbf{R}_j + \mathbf{R}_i)$. Thus

$$T(\mathbf{R}_j)T(\mathbf{R}_i)\psi(\mathbf{r}) = \psi(\mathbf{r} + \mathbf{R}_i + \mathbf{R}_j) = \lambda_i\lambda_j\psi(\mathbf{r}) = \lambda_{i+j}\psi(\mathbf{r}). \quad (3.5.9)$$

Evidently, the product of eigenvalues corresponding to different displacements must be equal to the eigenvalue of the combined translation. This can only be satisfied if the phase θ_i is linearly related to the displacement $\mathbf{R}_i$. If $\mathbf{R}_i = 0$, there is no translation and it is required that $\lambda_0 = 1$. Thus we must have

$$\theta_i = \mathbf{k}\cdot\mathbf{R}_i \quad (3.5.10)$$

where $\mathbf{k}$ is an arbitrary vector which is the same for each of the operations.

The wave function of an electron in a crystal is characterized by the wave vector $\mathbf{k}$ which appears in the eigenvalue of each translation operation; and we will generally incorporate $\mathbf{k}$ into the argument of the function ψ. We have now established Bloch's theorem (Bloch, 1928).

$$\psi(\mathbf{k}, \mathbf{r} + \mathbf{R}_i) = \exp(i\mathbf{k}\cdot\mathbf{R}_i)\psi(\mathbf{k}, \mathbf{r}). \quad (3.5.11)$$

This theorem may be interpreted as a boundary condition on the solution of the Schrödinger equation with a periodic potential.

It should be evident from the argument employed to justify (3.5.7) that this theorem is not absolute. We may encounter disturbances of the electron distribution in a crystal which are not periodic, both as a result of external fields (including imperfections which might destroy the periodicity), and as a result of the interactions of electrons with each other. When this happens, Bloch's theorem will not hold.

It is desirable to define a function $u(\mathbf{k}, \mathbf{r})$ by the relation

$$\psi(\mathbf{k}, \mathbf{r}) = \exp(i\mathbf{k}\cdot\mathbf{r})u(\mathbf{k}, \mathbf{r}). \quad (3.5.12)$$

Then, from Bloch's theorem,

$$\begin{aligned}\psi(\mathbf{k}, \mathbf{r} + \mathbf{R}_i) &= \exp[i\mathbf{k}\cdot(\mathbf{r} + \mathbf{R}_i)]u(\mathbf{k}, \mathbf{r} + \mathbf{R}_i)\\ &= \exp(i\mathbf{k}\cdot\mathbf{R}_i)[\exp(i\mathbf{k}\cdot\mathbf{r})u(\mathbf{k}, \mathbf{r})].\end{aligned}$$

Consequently, $u(\mathbf{k}, \mathbf{r})$ must be unchanged by translation by any lattice vector:

$$u(\mathbf{k}, \mathbf{r} + \mathbf{R}_i) = u(\mathbf{k}, \mathbf{r}). \quad (3.5.13)$$

This function is referred to as the *cell periodic function*, and is said to have the full periodicity of the potential.

Now consider two vectors in $\mathbf{k}$ space $\mathbf{k}$, $\mathbf{k}'$ which satisfy

$$\mathbf{k} = \mathbf{k}' + \mathbf{K}_s \quad (3.5.14)$$

where $\mathbf{K}_s$ is an arbitrary reciprocal lattice vector. Two such vectors are said to be equivalent. Evidently

$$\exp(i\mathbf{k}\cdot\mathbf{R}_i) = \exp(i\mathbf{k}'\cdot\mathbf{R}_i) \tag{3.5.15}$$

for all lattice vectors $\mathbf{R}_i$. Consequently, the wave functions $\psi(\mathbf{k}, \mathbf{r})$ and $\psi(\mathbf{k}', \mathbf{r})$ satisfy the same boundary conditions, that is, they possess the same eigenvalue λ for all lattice translations. We adopt the convention that these functions must be the same,

$$\psi(\mathbf{k}, \mathbf{r}) = \psi(\mathbf{k} + \mathbf{K}_s, \mathbf{r}), \tag{3.5.16}$$

for any $\mathbf{K}_s$. This convention implies that the domain of $\mathbf{k}$ is restricted to the first Brillouin zone, as described in Section 1.2.

When the convention (3.5.16) is adopted, we have the possibility, or more realistically speaking, the certainty, that there will be more than one state, or wave function with a given wave vector $\mathbf{k}$. These different solutions are labeled by an index n; frequently (but not necessarily) in order of increasing energy. The energy of a state of wave vector $\mathbf{k}$ is denoted by $E_n(\mathbf{k})$; the equation satisfied by $\psi_n(\mathbf{k}, \mathbf{r})$ is

$$H\psi_n(\mathbf{k}, \mathbf{r}) = E_n(\mathbf{k})\psi_n(\mathbf{k}, \mathbf{r}) \tag{3.5.17}$$

where H is the effective Hamiltonian. The quantity $E_n(\mathbf{k})$ is referred to as the *energy band function*; $\psi_n(\mathbf{k}, \mathbf{r})$ is called a *Bloch function*. The function $E_n(\mathbf{k})$ is a continuous and differentiable function of the wave vector $\mathbf{k}$ (see Section 4.1). In accord with (3.5.16) we require that

$$E_n(\mathbf{k}) = E_n(\mathbf{k} + \mathbf{K}_s). \tag{3.5.18}$$

We may also require that $\psi_n(\mathbf{k}, \mathbf{r})$ is a continuous and differentiable function of $\mathbf{k}$, except for possible discontinuities associated with existence of degeneracies.

3.6 Irreducible Representations: Space Groups

In this section we are concerned essentially with the transformation properties of the one-electron wave function or Bloch function $\psi_n(\mathbf{k}, \mathbf{r})$ under operations of a space group. From general principles, it is required that $\psi_n(\mathbf{k}, \mathbf{r})$ must transform according to one of the irreducible representations of the space group of the crystal. The behavior under translations was studied in Section 3.5. We must now consider the behavior with respect to more general space group operations which include a rotational part as well.

3.6.1 *Transformation Properties of Bloch Functions*

It is convenient to introduce an expansion of the Bloch wave function in terms of plane waves. To do this, observe that functions of the form

$\exp(i\mathbf{K}_s \cdot \mathbf{r})$ satisfy the periodicity conditions obeyed by the cell periodic function $u_n(\mathbf{k}, \mathbf{r})$, namely,

$$\exp[i\mathbf{K}_s \cdot (\mathbf{r} + \mathbf{R}_i)] = \exp(i\mathbf{K}_s \cdot \mathbf{r}) \tag{3.6.1a}$$

for all reciprocal lattice vectors $\mathbf{K}_s$ and all direct lattice vectors $\mathbf{R}_i$. Consequently, functions of the type $\exp[i(\mathbf{k} + \mathbf{K}_s) \cdot \mathbf{r}]$ obey Bloch's theorem for a state of wave vector $\mathbf{k}$:

$$\exp[i(\mathbf{k} + \mathbf{K}_s) \cdot (\mathbf{r} + \mathbf{R}_i)] = \exp(i\mathbf{k} \cdot \mathbf{R}_i) \exp[i(\mathbf{k} + \mathbf{K}_s) \cdot \mathbf{r}]. \tag{3.6.1b}$$

These functions are therefore suitable elements for the expansion of the actual Bloch functions, and we can write

$$\psi_n(\mathbf{k}, \mathbf{r}) = \sum_s b_n(\mathbf{k}, \mathbf{K}_s) \exp[i(\mathbf{k} + \mathbf{K}_s) \cdot \mathbf{r}]. \tag{3.6.2}$$

Now consider the effect of the space group operation $\{\boldsymbol{\alpha} \mid \mathbf{t}_\alpha\}$ on (3.6.2). Here $\boldsymbol{\alpha}$ is the rotational part of the operation and $\mathbf{t}_\alpha$ is the translational part. We will suppose for the moment that no lattice translation is involved, so that $\mathbf{t}_\alpha$ is a nonprimitive translation which is present only for a nonsymmorphic space group. If $\phi(\mathbf{r})$ is an arbitrary function of position, we have from (3.2.4) and (3.1.3) that

$$\{\boldsymbol{\alpha} \mid \mathbf{t}_\alpha\}\phi(\mathbf{r}) = \phi[\{\boldsymbol{\alpha} \mid \mathbf{t}_\alpha\}^{-1}\mathbf{r}] = \phi(\boldsymbol{\alpha}^{-1}\mathbf{r} - \boldsymbol{\alpha}^{-1}\mathbf{t}_\alpha). \tag{3.6.3}$$

Thus

$$\begin{aligned}\{\boldsymbol{\alpha} \mid \mathbf{t}_\alpha\} \exp[i(\mathbf{k} + \mathbf{K}_s) \cdot \mathbf{r}] &= \exp[i(\mathbf{k} + \mathbf{K}_s) \cdot (\boldsymbol{\alpha}^{-1}\mathbf{r} - \boldsymbol{\alpha}^{-1}\mathbf{t}_\alpha)] \\ &= \exp[i\boldsymbol{\alpha}(\mathbf{k} + \mathbf{K}_s) \cdot (\mathbf{r} - \mathbf{t}_\alpha)].\end{aligned}$$

If we now apply $\{\boldsymbol{\alpha} \mid \mathbf{t}_\alpha\}$ to $\psi_n(\mathbf{k}, \mathbf{r})$, we have

$$\begin{aligned}\{\boldsymbol{\alpha} \mid \mathbf{t}_\alpha\}\psi_n(\mathbf{k}, \mathbf{r}) = \exp(-i\boldsymbol{\alpha}\mathbf{k} \cdot \mathbf{t}_\alpha) &\sum_s b_n(\mathbf{k}, \mathbf{K}_s) \exp(i\boldsymbol{\alpha}\mathbf{K}_s \cdot \mathbf{t}_\alpha) \\ &\times \exp[i\boldsymbol{\alpha}(\mathbf{k} + \mathbf{K}_s) \cdot \mathbf{r}].\end{aligned} \tag{3.6.4}$$

This transformed function has a different wave vector from the original one. However, since all space group operations commute with the Hamiltonian, the energy of the transformed state must be the same as that of the original one. Unless we are at a point in $\mathbf{k}$ space where different bands have the same energy (this possibility is ignored for the moment), the transformed function must belong to the same energy band as the original one. Thus let us write

$$\{\boldsymbol{\alpha} \mid \mathbf{t}_\alpha\}\psi_n(\mathbf{k}, \mathbf{r}) = C\psi_n(\mathbf{k}', \mathbf{r})$$

where C is a complex number of unit modulus, which still has to be specified.

To determine $\mathbf{k}'$ we must ask how $\psi_n(\mathbf{k}', \mathbf{r})$ transforms with respect to a pure translation. Therefore, we apply a translation operator $T(\mathbf{R}_i) = \{\boldsymbol{\varepsilon} \mid -\mathbf{R}_i\}$ to $\psi_n(\mathbf{k}', \mathbf{r})$. From (3.6.4) and (3.5.3), we see that

$$T(\mathbf{R}_i)\{\boldsymbol{\alpha} \mid \mathbf{t}_\alpha\}\psi_n(\mathbf{k}, \mathbf{r}) = \exp(i\boldsymbol{\alpha}\mathbf{k}\cdot\mathbf{R}_i)\{\boldsymbol{\alpha} \mid \mathbf{t}_\alpha\}\psi_n(\mathbf{k}, \mathbf{r}).$$

Comparison with Bloch's theorem (3.5.11) indicates that the wave vector of the transformed function $\mathbf{k}' = \boldsymbol{\alpha}\mathbf{k}$. Thus

$$\{\boldsymbol{\alpha} \mid \mathbf{t}_\alpha\}\psi_n(\mathbf{k}, \mathbf{r}) = \exp[i\phi_n(\boldsymbol{\alpha}, \mathbf{k})]\psi_n(\boldsymbol{\alpha}\mathbf{k}, \mathbf{r}) \tag{3.6.5}$$

where $\phi_n(\mathbf{k})$ is a real function of $\mathbf{k}$, which we will want to be at least piecewise continuous and differentiable for fixed α.

Equation (3.6.5) is a fundamental result. For many purposes, the specification of the phase $\phi_n(\mathbf{k})$ is of no importance. However, it will be significant when we (later on) wish to construct localized functions from the running wave Bloch functions.

3.6.2 *The Group of the Wave Vector*

The different vectors $\boldsymbol{\alpha}\mathbf{k}$ which may appear in (3.6.5) when the operations of the space group are applied to $\psi_n(\mathbf{k}, \mathbf{r})$ form a geometric figure called the "star" of $\mathbf{k}$. It is conventional to include $\mathbf{k}$ itself in the star, so that we may say that the star of $\mathbf{k}$ is generated by applying to $\mathbf{k}$ all the rotations $\boldsymbol{\alpha}$ in the point group of the crystal. Corresponding states belonging to different vectors in the star have the same energy; so that the energy band function is invariant under all the operations in the point group of the crystal:

$$E_n(\boldsymbol{\alpha}\mathbf{k}) = E_n(\mathbf{k}). \tag{3.6.6}$$

However, even if the point group of the crystal does not contain the inversion, we must have

$$E_n(-\mathbf{k}) = E_n(\mathbf{k}) \tag{3.6.7}$$

as a consequence of time reversal symmetry which will be discussed subsequently.

For certain values of $\mathbf{k}$, the point group of the crystal will contain operations which carry $\mathbf{k}$ either into itself or into an equivalent $\mathbf{k}$. Call such an operation $\boldsymbol{\beta}$:

$$\boldsymbol{\beta}\mathbf{k} = \mathbf{k} + \mathbf{K}_s \tag{3.6.8}$$

where $\mathbf{K}_s$ is a reciprocal lattice vector (including zero). The operators $\{\boldsymbol{\beta} \mid 0\}$ form a point group, which is a subgroup of the crystal point group which we denote by $\mathcal{K}_p$. This group is called the "group" of the wave vector. We may also define a space group $\mathcal{K}_s$ consisting of those operations

$\{\beta \mid \mathbf{t}_\beta\}$ of the space group such that $\{\beta \mid 0\}$ is in $\mathcal{K}_p$, and containing the pure translations as a subgroup. The irreducible representations of $\mathcal{K}_p$ are of great importance in the classification of states in the crystal. Note that if $\mathbf{k} = 0$, $\mathcal{K}_p$ is the entire crystal point group. If there are n operations in the crystal point group and g operations in $\mathcal{K}_p$, the ratio $n/g = q$ is an integer which gives the number of vectors in the star of $\mathbf{k}$.

At a general point of the Brillouin zone, the (point) group of the wave vector contains only the identity, and there are n different vectors in the star. The Bloch functions belonging to these $\mathbf{k}$ vectors are seen from (3.6.5) to be basis functions for an n-dimensional irreducible representation of the space group. For symmetry points in the zone (by definition: those points where the point group of the wave vector contains more than the identity), the irreducible representations of the space group are smaller in dimension. Let us first consider the irreducible representations of $\mathcal{K}_p$. Suppose there are s irreducible representations of $\mathcal{K}_p$ whose dimensions are denoted d_i $(i = 1, \ldots, s)$. This means that there are d_i orthogonal Bloch functions $\psi_n(\mathbf{k}, \mathbf{r})$ $(n = 1, \ldots, d_i)$ which transform among themselves under the operations of $\mathcal{K}_p$, and form basis functions for the ith irreducible representation of $\mathcal{K}_p$. If $d_i > 1$, different energy bands must touch or "stick together" at that point. These bands are said to be *degenerate*. We may now select q different operations $\{\gamma \mid \mathbf{t}_\gamma\}$ such that the vectors $\gamma\mathbf{k}$ are all different and generate the star of $\mathbf{k}$. These operators enable us to express the full space group as the sum of its left cosets with respect to $\mathcal{K}_s$ (for proofs, see Koster, 1957):

$$G = \mathcal{K}_s + \{\gamma_2 \mid \mathbf{t}_{\gamma 2}\}\mathcal{K}_s + \cdots + \{\gamma_q \mid \mathbf{t}_{\gamma q}\}\mathcal{K}_s. \tag{3.6.9}$$

The set of $D_i = qd_i$ functions

$$\{\gamma \mid \mathbf{t}_\gamma\}\psi_n(\mathbf{k}, \mathbf{r}) \qquad (n = 1, \ldots, d_i; \quad \gamma = 1, \ldots, q)$$

form bases for a D_i-dimensional irreducible representation of the full space group.

Since the irreducible representations of the space group can be generated once the point group of the wave vector $\mathcal{K}_p$ is known, the essential problem in the classification of electron states is the determination of these groups whenever they are nontrivial, and the construction of their irreducible representations.

Character tables for the point groups of the wave vector have been given for a number of space groups, beginning with Bouckaert *et al.* (1936). Useful references include Koster (1957), Jones (1960), Koster *et al.* (1963), and Slater (1965). We will not attempt an extensive compilation of these results. Results for simple, body-centered, and face-centered cubic crystals are tabulated in Appendix C.

Symmetry considerations lead to a determination of certain required degeneracies. It is of interest to inquire whether there may be others, that is, whether energy bands can cross or touch when they are not compelled by symmetry considerations to do so. These intersections of bands are called *accidental degeneracies*, and have been studied by Herring (1937b). We will not examine Herring's arguments in detail: the essential results are as follows:

I. It is possible for energy bands belonging to different representations to cross. Such a crossing, which may occur along a symmetry axis or in a symmetry plane, does produce an accidental degeneracy.

II. It is unlikely that there will be accidental degeneracies between bands of the same spin. It is meant by this that (except for certain rather specialized possibilities discussed in detail by Herring) if such contact occurs for some specific crystal potential, it will be removed by almost any small change in the potential. This remark will be made plausible if we consider that since the crystal potential has full lattice symmetry, a change in it will have a nonvanishing matrix element between degenerate states. However, such matrix elements vanish between states of different symmetry mentioned in I above. Since all states at general points of the Brillouin zone have the same symmetry, it follows that isolated accidental degeneracies of bands at general points of the zone are (for a crystal with a center of inversion) vanishingly improbable.

3.6.3 *Phases of Bloch Functions*

We will now consider the choice of phase factor $\phi_n(\alpha, \mathbf{k})$ in (3.6.5). Inspection of (3.6.4) suggests it might be desirable to separate a factor $-\alpha\mathbf{k}\cdot\mathbf{t}_\alpha$ which arises naturally from the plane wave expansion. Let us write

$$\phi_n(\alpha, \mathbf{k}) = -\alpha\mathbf{k}\cdot\mathbf{t}_\alpha + \theta_n(\alpha, \mathbf{k}) \tag{3.6.10}$$

where θ is the residual part of the factor. It might be considered desirable to set θ_n equal to zero for all α, $\mathbf{k}$. However, this leads to difficulties, since at a symmetry point where $\alpha\mathbf{k} = \mathbf{k} + \mathbf{K}_s$, ψ_n must transform according to one of the irreducible representations of the group of the wave vector. Suppose that ψ_n is not degenerate at the symmetry point. Then ϕ_n must be determined by the fact that the complete phase factor $\exp(i\phi_n)$ must be equal to the character of $\{\alpha \mid \mathbf{t}_\alpha\}$ in the group of the wave vector.

To determine the phase factor away from a symmetry point, we require that the wave function should vary smoothly with $\mathbf{k}$ throughout the zone. Smoothness of the wave function may be examined by considering the relation between the cell periodic part of the Bloch function u_n at neighboring points of the zone. It is a consequence of $\mathbf{k}\cdot\mathbf{p}$ perturbation theory (which will be developed subsequently in Section 4.1) that for small $\delta\mathbf{k}$,

$$[(2\pi)^3/\Omega] \int_{\text{cell}} u_n^*(\mathbf{k}, \mathbf{r}) u_n(\mathbf{k} + \delta\mathbf{k}, \mathbf{r})\, d^3r = 1 - O(\delta\mathbf{k})^2 \quad (3.6.11)$$

where Ω is the volume of the unit cell, which is the volume through which the integral is performed. For justification, see the discussion following Eq. (4.1.15). The integral differs from unity by terms of second order in $\delta\mathbf{k}$. Consider now a similar integral defined for rotated wave vectors $\boldsymbol{\alpha}\mathbf{k}$, $\boldsymbol{\alpha}(\mathbf{k} + \delta\mathbf{k})$. Equations (3.6.5) and (3.6.10) imply the following transformation rule for the cell periodic function:

$$\{\boldsymbol{\alpha} \mid \mathbf{t}_\alpha\} u_n(\mathbf{k}, \mathbf{r}) = \exp[i\theta_n(\boldsymbol{\alpha}, \mathbf{k})] u_n(\boldsymbol{\alpha}\mathbf{k}, \mathbf{r}). \quad (3.6.12)$$

Since the operator $\{\boldsymbol{\alpha} \mid \mathbf{t}_\alpha\}$ is unitary on the basis of the functions considered, it may be applied to each of the functions u_n in (3.6.11) without changing the scalar product. Thus

$$\begin{aligned}
&[(2\pi)^3/\Omega] \int_{\text{cell}} [\{\boldsymbol{\alpha} \mid \mathbf{t}_\alpha\} u_n(\mathbf{k}, \mathbf{r})]^* [\{\boldsymbol{\alpha} \mid \mathbf{t}_\alpha\} u_n(\mathbf{k} + \delta\mathbf{k}, \mathbf{r})]\, d^3r \\
&\quad = [(2\pi)^3/\Omega] \exp\{i[\theta_n(\boldsymbol{\alpha}, \mathbf{k} + \delta\mathbf{k}) - \theta_n(\boldsymbol{\alpha}, \mathbf{k})]\} \\
&\qquad \times \int_{\text{cell}} u_n^*(\boldsymbol{\alpha}\mathbf{k}, \mathbf{r}) u_n[\boldsymbol{\alpha}(\mathbf{k} + \delta\mathbf{k})]\, d^3r \\
&\quad = 1 - O(\delta\mathbf{k})^2. \qquad (3.6.13)
\end{aligned}$$

However, the vectors $\boldsymbol{\alpha}\mathbf{k}$ and $\boldsymbol{\alpha}(\mathbf{k} + \delta\mathbf{k})$ are close if $\delta\mathbf{k}$ is small, so that we naturally require that the integral

$$[(2\pi)^3/\Omega] \int_{\text{cell}} u_n^*(\boldsymbol{\alpha}\mathbf{k}, r) u_n[\boldsymbol{\alpha}(\mathbf{k} + \delta\mathbf{k}), \mathbf{r}]\, d^3r = 1 - O(\delta\mathbf{k})^2.$$

This implies that the difference of the phase factors on the left side of (3.6.13) should vanish to first order, or that

$$\nabla_{\mathbf{k}}\theta_n(\boldsymbol{\alpha}, \mathbf{k}) = 0. \quad (3.6.14a)$$

Thus, for reasons of continuity, we should choose

$$\theta_n(\boldsymbol{\alpha}, \mathbf{k}) = C_{n,\alpha} \quad (3.6.14b)$$

where $C_{n,\alpha}$ is a constant depending on α, but independent of $\mathbf{k}$.

Can this argument fail? It is a differential argument based on the behavior of the wave function at neighboring points of the zone. It breaks down only if (3.6.5) fails, which can happen only at a point at which two bands are degenerate.

In the case of an isolated energy band, which has no degeneracies at any point, the phase factor will be constant for all $\mathbf{k}$. The phase factor is determined from the behavior of the wave function at the point of highest symmetry $\mathbf{k} = 0$. At this point, the wave function must transform according to one of the one-dimensional representations of the crystal point group; and this determines the phase factor. We then have

$$\exp(i\theta_n) = \chi_n^{(j)}(\boldsymbol{\alpha}) \tag{3.6.15}$$

where $\chi_n^{(j)}(\boldsymbol{\alpha})$ is the character of the jth (one-dimensional) irreducible representation of the crystal point group. Our final statement is that, for an isolated energy band,

$$\{\boldsymbol{\alpha} \mid \mathbf{t}_\alpha\}\psi_n(\mathbf{k}, \mathbf{r}) = \chi_n^{(j)}(\boldsymbol{\alpha}) \exp(-i\boldsymbol{\alpha}\mathbf{k}\cdot\mathbf{t}_\alpha)\psi_n(\boldsymbol{\alpha}\mathbf{k}, \mathbf{r}). \tag{3.6.16}$$

We now must consider the more complex case of degenerate bands. Unfortunately, it is the usual situation. If $\mathbf{k}$ is not in the neighborhood of a point of degeneracy, (3.6.5) will still apply and the residual phase θ will have to be constant. However, suppose $\mathbf{k}$ is a point of degeneracy. Then $\psi_n(\mathbf{k}, \mathbf{r})$ must be a basis function for one of the degenerate irreducible representations of the group of $\mathbf{k}$. Let $[\boldsymbol{\alpha}]_{l,mn}$ be the mnth element of the matrix representing the operation $\boldsymbol{\alpha}$ in the lth irreducible representation of the point group of the wave vector $\mathbf{k}$, K_p. Then we should have

$$\{\boldsymbol{\alpha} \mid \mathbf{t}_\alpha\}\psi_n(\mathbf{k}, \mathbf{r}) = \exp(-i\boldsymbol{\alpha}\mathbf{k}\cdot\mathbf{t}_\alpha) \sum_m [\boldsymbol{\alpha}]_{l,mn}\psi_m(\boldsymbol{\alpha}\mathbf{k}, \mathbf{r}) \qquad (\boldsymbol{\alpha} \in \mathcal{K}_p). \tag{3.6.17}$$

In the case of operations $\boldsymbol{\beta}$ not in the group of $\mathbf{k}$, $\boldsymbol{\beta}\mathbf{k}$ is some other vector in the star of $\mathbf{k}$. The number of operations $\boldsymbol{\beta}$ which produces a given member of the star of $\mathbf{k}$ is the same as the order of the group of $\mathbf{k}$. We use the coset decomposition (3.6.9) to express the operators generating the particular member of the star in the form of a product $\{\boldsymbol{\beta} \mid \mathbf{t}_\beta\}\ \{\boldsymbol{\alpha} \mid \mathbf{t}_\alpha\}$ where $\boldsymbol{\alpha} \in \mathcal{K}_p$ and $\{\boldsymbol{\beta} \mid \mathbf{t}_\beta\}$ is one of the operators $\{\boldsymbol{\gamma}_i \mid \mathbf{t}_{\gamma i}\}$ generating the coset decomposition (3.6.9). Thus a typical member of this coset $\{\mathbf{n} \mid \mathbf{t}_\eta\}$, say, can be expressed as

$$\{\mathbf{n} \mid \mathbf{t}_\eta\} = \{\boldsymbol{\beta} \mid \mathbf{t}_\beta\}\{\boldsymbol{\alpha} \mid \mathbf{t}_\alpha\} = \{\boldsymbol{\beta}\boldsymbol{\alpha} \mid \boldsymbol{\beta}\mathbf{t}_\alpha + \mathbf{t}_\beta\}.$$

We require the following transformation rule to hold:

$$\begin{aligned}\{\mathbf{n} \mid \mathbf{t}_\eta\}\psi_n(\mathbf{k}, \mathbf{r}) &= \chi_n(\boldsymbol{\beta}) \exp[-i(\boldsymbol{\alpha}\mathbf{k}\cdot\mathbf{t}_\alpha + \boldsymbol{\beta}\boldsymbol{\alpha}\mathbf{k}\cdot\mathbf{t}_\beta)] \sum_m [\boldsymbol{\alpha}]_{l,mn}\psi_m(\boldsymbol{\beta}\boldsymbol{\alpha}\mathbf{k}, \mathbf{r}) \\ &= \chi_n(\boldsymbol{\beta}) \exp[-i\mathbf{n}\mathbf{k}\cdot\mathbf{t}_\eta] \sum_m [\boldsymbol{\alpha}]_{l,mn}\psi_m(\boldsymbol{\beta}\mathbf{k}, \mathbf{r}).\end{aligned} \tag{3.6.18}$$

The appropriate factor $\chi_n(\boldsymbol{\beta})$ is identified by continuity considerations to be the same as that which occurs in (3.6.16) for the relevant band when **k** is not a point of degeneracy [and $\boldsymbol{\alpha}$ in (3.6.16) is replaced by $\boldsymbol{\beta}$].

3.6.4 *Double Groups*

The preceding considerations have considered the "single" groups only, that is, we have not studied the double groups which arise when spin is included in the wave function. It is evident from Section 3.3 that we must associate two quantum operators with each space group operation. The description of the Brillouin zone is a consequence of the translational periodicity, and is unaffected by inclusion of spin. Bloch spinors of the form $\exp(i\mathbf{k}\cdot\mathbf{r})u(\mathbf{k}, \mathbf{r})$ where $u(\mathbf{k}, \mathbf{r})$ is a two-component spinor, are basis functions for irreducible representations of the "double" space group. The point group of the wave vector is a double group, as described in Section 3.3, and the representations of this group may be used to generate representations of the double space group.

It is the additional representations of the double groups which are of interest. Character tables for these representations for the simple, body-centered, and face-centered cubic structures, the diamond, and the hexagonal close packed lattices have been given by Elliott (1954). Of most interest is the splitting of degeneracies at symmetry points and along lines of symmetry. With reference to Appendix C, we note that the sixfold degenerate states at Γ, H, $P(\Gamma_{15}, \Gamma_{25'}$, etc.) are each split into a fourfold and a twofold degenerate state but the originally fourfold degenerate states at these points (Γ_{12}, H_{12}, P_3) are not split. The fourfold degenerate states at X, W, and L are split into two doubly degenerate states each. The (1, 0, 0) axis Δ is the only symmetry axis for which there are two distinct irreducible representations. Both fourfold states, Δ_5 and Λ_3 are split. As a consequence, many of the accidental degeneracies which might be predicted along symmetry axes in a calculation without spin-orbit coupling will be removed when spin is included.

3.7 Time Reversal Symmetry

3.7.1 *The Time Inversion Operator*

In addition to the spatial translations, rotations, and reflections contained in space groups, it is also necessary to consider time inversion. This operation is of a different nature from those previously considered in that it is not described by a linear, unitary operator, but by one which is instead antilinear and antiunitary. We denote this operator by $\mathfrak{J}$. There are considerable mathematical complications, so complete proofs will not be given for the properties asserted; the reader is referred to Wigner (1959) or Falicov (1966) for further information. The antilinearity property means that

$$\mathfrak{I}(a\phi + b\psi) = a^*\mathfrak{I}\phi + b^*\mathfrak{I}\psi \tag{3.7.1}$$

for arbitrary functions ψ, ϕ (a^* is the complex conjugate of a). The operator is antiunitary since, in addition to (3.7.1),

$$(\mathfrak{I}\psi, \mathfrak{I}\phi) = (\psi, \phi)^* \tag{3.7.2}$$

where the parentheses denote the scalar product.

Application of the time reversal operator twice to a wave function must produce a state physically indistinguishable from the original state. The only permissible effect of the double application would be the introduction of a phase factor $e^{i\theta}$. It can further be shown that this phase factor can be only ± 1 (Wigner, 1959). Thus

$$\mathfrak{I}^2 = \pm I \tag{3.7.3}$$

where I is the identity.

The simplest example of an antilinear, antiunitary operator is the operator which produces complex conjugation. This operator is denoted by K. This is the time reversal operator in a theory in which spin is neglected

$$\mathfrak{I} = K, \qquad \mathfrak{I}\phi = \phi^*. \tag{3.7.4}$$

The connection between time reversal and complex conjugation is plausible if we consider a simple stationary state wave function of the form $u(\mathbf{r})\exp(-iEt/\hbar)$ in which u is real. The complex conjugate of such a function is the same as a function in which t has been replaced by $-t$. Further, we have

$$K\mathbf{p}K^{-1} = -\mathbf{p} \tag{3.7.5}$$

where p is the momentum operator ($\mathbf{p} = \hbar/i\nabla$ in the Schrödinger picture). Thus the time reversal operator reverses the momentum as it should

When spin is included, we require in addition to (3.7.5) that

$$\mathfrak{I}\mathbf{s}\mathfrak{I}^{-1} = -\mathbf{s} \tag{3.7.6}$$

where $\mathbf{s}$ is a spin operator. The choice

$$\mathfrak{I} = \sigma_y K \tag{3.7.7a}$$

will be seen to satisfy (3.7.6). The operator σ_y is the usual Pauli spinor

$$\sigma_y = \begin{pmatrix} 0 & -i \\ i & 0 \end{pmatrix}.$$

In this case, we see that (3.7.3) holds with the negative sign. If we have a system of many (N say) spin $\frac{1}{2}$ particles, then

$$\mathfrak{I} = \left(\prod_{i}^{N} \sigma_{y,i}\right) K \tag{3.7.7b}$$

where $\sigma_{y,i}$ operates on the spin coordinate of the ith particle and

$$\mathfrak{I}^2 = (-1)^N. \tag{3.7.8}$$

Suppose that the Hamiltonian of a physical system is a function of coordinates, momenta, and spins. It is a consequence of (3.7.5) and (3.7.6) that

$$\mathfrak{I} H(\mathbf{r}, \mathbf{p}, \mathbf{s}) \mathfrak{I}^{-1} = H(\mathbf{r}, -\mathbf{p}, -\mathbf{s}). \tag{3.7.9}$$

Except when an external magnetic field is present, the Hamiltonian will be invariant under the simultaneous transformation $\mathbf{p} \to -\mathbf{p}$, $\mathbf{s} \to -\mathbf{s}$. A simple example for which this is true is a single-particle Hamiltonian containing spin–orbit coupling:

$$H = (\mathbf{p}^2/2m) + V(\mathbf{r}) + (\hbar/4m^2c^2)\boldsymbol{\sigma}\cdot(\nabla V \times \mathbf{p}). \tag{3.7.10}$$

This is obviously invariant. However, the addition of a term proportional to $\boldsymbol{\sigma}\cdot\mathbf{B}$ where $\mathbf{B}$ is a fixed external magnetic field, destroys the invariance.

3.7.2 *Kramers' Theorem*

We will examine some of the consequences of time reversal symmetry. Specifically, it is shown that time reversal symmetry requires that $E(\mathbf{k}) = E(-\mathbf{k})$ [where $E(\mathbf{k})$ describes an energy band] regardless of the spatial symmetry of the system (Kramers' theorem) and that a double degeneracy of the band system throughout the Brillouin zone must exist if the potential has a center of inversion. It is supposed in this proof that the electron energies and wave functions are eigenvalues and eigenfunctions of a one-particle Hamiltonian of the form (3.7.10). These eigenfunctions have the Bloch form

$$H\psi(\mathbf{k}, \mathbf{r}) = E(\mathbf{k})\psi(\mathbf{k}, \mathbf{r}), \qquad \psi(\mathbf{k}, \mathbf{r}) = \exp(i\mathbf{k}\cdot\mathbf{r})u(\mathbf{k}, \mathbf{r})$$

where $u(\mathbf{k}, \mathbf{r})$ is a Pauli spinor. Application of the time reversal operator gives (since $\mathfrak{I}$ commutes with H)

$$\mathfrak{I} H\psi(\mathbf{k}, \mathbf{r}) = H\mathfrak{I}\psi(\mathbf{k}, \mathbf{r}) = E(\mathbf{k})\sigma_y\psi^*(\mathbf{k}, \mathbf{r}). \tag{3.7.11}$$

Hence $\sigma_y\psi^*(\mathbf{k}, \mathbf{r})$ is an eigenfunction of H with eigenvalue $E(\mathbf{k})$. Also

$$\sigma_y\psi^*(\mathbf{k}, \mathbf{r}) = \exp(-i\mathbf{k}\cdot\mathbf{r})\omega(\mathbf{k}, \mathbf{r})$$

where $\omega(\mathbf{k}, \mathbf{r})$ is a Pauli spinor with the full periodicity of the potential. Evidently $\sigma_y\psi^*(\mathbf{k}, \mathbf{r})$ is a Bloch function for a state of wave vector $-\mathbf{k}$, and we have

$$E(\mathbf{k}) = E(-\mathbf{k}). \tag{3.7.12}$$

It requires only a slight change in the wording of the proof to establish that, for a single-particle localized state described by the usual quantum numbers (n, j, l, s, m_j) with j the total angular momentum [or alternately (n, l, s, m_l, m_s)] that

$$E(m_j) = E(-m_j).$$

The levels are at least doubly degenerate. This result actually holds for any system with an odd number of electrons. The degeneracy here brought into evidence is called the *Kramers degeneracy.*

Next, consider the equation satisfied by $u(\mathbf{k}, \mathbf{r})$ for the Hamiltonian (3.7.10). This is easily worked out to be

$$[(\mathbf{p}^2/2m) + V(\mathbf{r}) + (\hbar/m)\mathbf{k}\cdot\mathbf{p} + (\hbar/4m^2c^2)\sigma\cdot(\nabla V \times \mathbf{p}) \\ + (\hbar^2/4m^2c^2)\mathbf{k}\cdot(\sigma \times \nabla V)]u(\mathbf{k}, \mathbf{r}) = [E(\mathbf{k}) - \hbar^2\mathbf{k}^2/2m]u(\mathbf{k}, \mathbf{r}). \tag{3.7.13}$$

Let us apply the inversion operator J to this equation. Then $\mathbf{r}$ changes into $-\mathbf{r}$ and $\mathbf{p}$ into $-\mathbf{p}$. We obtain

$$[(\mathbf{p}^2/2m) + V(-\mathbf{r}) - (\hbar/m)\mathbf{k}\cdot\mathbf{p} + (\hbar/4m^2c^2)\boldsymbol{\sigma}\cdot(\nabla V(-\mathbf{r}) \times \mathbf{p}) \\ - (\hbar^2/4m^2c^2)\mathbf{k}\cdot(\boldsymbol{\sigma} \times \nabla V(-\mathbf{r}))]u(\mathbf{k}, -\mathbf{r}) \\ = [E(\mathbf{k}) - \hbar^2\mathbf{k}^2/2m]u(\mathbf{k}, -\mathbf{r}). \tag{3.7.14}$$

If $V(\mathbf{r}) = V(-\mathbf{r})$, we see that this equation is the same as (3.7.13) with $\mathbf{k}$ changed into $-\mathbf{k}$. Hence, if the potential has inversion symmetry,

$$u(-\mathbf{k}, \mathbf{r}) = u(\mathbf{k}, -\mathbf{r}) \tag{3.7.15}$$

apart from a constant factor.

The solution of (3.7.13) for wave vector $-\mathbf{k}$ is required by (3.7.12) to have the same energy as the solution for $\mathbf{k}$. However, at a general point in the zone, the wave function is not an eigenfunction of the inversion, so $u(\mathbf{k}, -\mathbf{r})$ is a different function from $u(\mathbf{k}, \mathbf{r})$. Hence there are two different wave functions corresponding to the same wave vector that have the same energy. Thus we conclude that if the crystal potential has inversion symmetry, there is a double degeneracy of the band structure throughout the zone.

3.7.3 *Additional Degeneracies Produced by Time Reversal Symmetry*

Inclusion of time reversal symmetry augments the group of transformations which commute with the Hamiltonian. No additional quantum numbers are introduced, but additional degeneracies may be required. The preceding arguments furnish one example of this. Now we examine the effect on representations at symmetry points of the zone. The basic analysis is due to Herring (1937a) and Elliott (1954).

The problem can be formulated as follows: Suppose we have an eigenfunction of the crystal Hamiltonian or, more comprehensively, a set of degenerate eigenfunctions. Let us operate on these with the time reversal operator $\mathfrak{J}$. The result is another eigenfunction, or a set of them, which are degenerate with each other and with the first set. The new set of states can be shown to transform according to the complex conjugate representation of the relevant group rather than the original representation. It must be determined whether the new set of functions is linearly independent of the old set or not. If the answer is yes, a new degeneracy has been introduced into the problem; otherwise not.

Results concerning this question will be stated without proof (see Falicov, 1966). Let the original representation matrices be denoted by Γ; the matrices describing the transformation of the "time reversed functions" are denoted by Γ^*. There are three cases to be considered.

(1) Γ and Γ^* are equivalent, and can be chosen to be real and identical;
(2) Γ and Γ^* are inequivalent;
(3) Γ and Γ^* are equivalent but cannot be transformed to real form.

The consequences of the three cases are as follows. Unfortunately, the results are different depending on whether spin is considered

I. Spinless particles
  - Case 1. No additional degeneracy.
  - Case 2. Degeneracy is doubled: Γ and Γ^* correspond to the same eigenvalue.
  - Case 3. Degeneracy is doubled.

II. Particles of spin $\frac{1}{2}$
  - Case 1. Degeneracy is doubled.
  - Case 2. Degeneracy is doubled, and the two representations Γ and Γ^* occur together.
  - Case 3. No additional degeneracy.

The following test may be used to determine which case applies. Let $\mathbf{R}$ designate an operation; $\chi(\mathbf{R})$ is the character of $\mathbf{R}$ in the irreducible representation considered; and g is the order of the group. We form the sum over all operations of $\chi(\mathbf{R}^2)$. Then if

$$\begin{aligned}\sum_{\mathbf{R}} \chi(\mathbf{R}^2) &= g, \qquad \text{we have case 1,}\\ &= 0, \qquad \text{we have case 2,}\\ &= -g, \qquad \text{we have case 3.}\end{aligned} \tag{3.7.16}$$

These results are directly applicable to point groups but not to space groups, since the latter are of infinite order. However, Herring (1937a) showed that it is sufficient to consider operations Q_0 in the space group which send a given $\mathbf{k}$ into its negative $-\mathbf{k}$. For such operations, $Q_0{}^2$ is in the group of $\mathbf{k}$. In this case, one forms the sum over the operations in the group of $\mathbf{k}$, of $\chi(Q_0)^2$:

$$\begin{aligned}\sum_{Q_0} \chi(Q_0{}^2) &= g_{\mathrm{k}} \qquad \text{(case 1),}\\ &= 0 \qquad \text{(case 2),}\\ &= -g_{\mathrm{k}} \qquad \text{(case 3),}\end{aligned} \tag{3.7.17}$$

where g_{k} is the order of the group of $\mathbf{k}$. This test is applicable to either the single or the double group; however, the conclusions drawn from the results are different, as described previously.

Degeneracies imposed by time reversal occur fairly frequently in the double groups of interest in band theory. For one example, consider the group of the point L (fcc—center of a hexagonal face) (see Appendix C). The single group contains six classes and twelve operations; the double group has six additional classes. The six additional representations and their degeneracies are $L_4{}^{\pm}(1)$, $L_5{}^{\pm}(1)$, $L_6{}^{\pm}(2)$. The fourfold degenerate L_3 splits into $L_4{}^+ + L_5{}^+ + L_6{}^+$; however, $L_4{}^+$ and $L_5{}^+$ are degenerate by time reversal symmetry, so that two doublets are produced. Likewise $L_4{}^-$ and $L_5{}^-$ are degenerate by time reversal so that $L_3{}^-$ likewise splits into two doublets: $(L_4{}^-, L_5{}^-)$ and $L_6{}^-$. Degeneracies imposed by time reversal are listed in the character tables given by Elliott (1954).

3.7.4 *Magnetic Structures*

Special problems arise when magnetic structures are investigated. In antiferromagnetic and ferromagnetic systems, the periodicity of the crystal may be different when the magnetic order is considered than if only the chemical structure is observed. It is conventional to describe the system in terms of a magnetic unit cell (instead of a chemical unit cell) and a magnetic Brillouin zone.

The problem was first studied by Shubnikov (1951), who introduced an additional operation: that of interchanging black and white colors, in addition to the usual geometric operations. More recent investigations are based on considerations of time reversal invariance because, as seen from (3.7.5), time reversal leads to the reversal of spins. In the case of a crystal without magnetic moments (or loosely speaking, in the case of a paramagnetic substance in which the randomness of local moments is ignored) the crystal Hamiltonian will be invariant under the time reversal operation. In the case of an antiferromagnet, the Hamiltonian will not be invariant under time reversal by itself, but there will be symmetry operations in which time reversal is combined with operations from the usual space group.

An introduction to the theory of magnetic crystal groups has been given by Dimmock and Wheeler (1962) and by Tinkham (1964). An extensive compilation of representations has been presented by Miller and Love (1967).

Let us denote the operations of the ordinary space group which refers to the chemical structure of a crystal by $\mathbf{H}$. In the paramagnetic state, the crystal will also be invariant under $\mathfrak{J}$ and under operations in which $\mathfrak{J}$ is combined with a member of $\mathbf{H}$. Then the full space group, including time reversal, of the paramagnetic crystal is denoted by $\mathbf{G}$, such that $\mathbf{G} = \mathbf{H} + \mathbf{H}\mathfrak{J}$. Below the ordering temperature, the crystal is not invariant under all the operations of G since $\mathfrak{J}$ changes the magnetic structure. There will be some set of spatial operations which carry not only the chemical lattice but the magnetic structure as well into itself. Let this group be denoted by $\mathfrak{H}$. There may also exist some spatial operations which carry the chemical lattice into itself but reverse the sign of the distribution of magnetic moments. Let such an operation be denoted by $\mathbf{v}_0$. Then the magnetic crystal will be invariant under the operation $\mathbf{v}_0\mathfrak{J}$. The magnetic space group G will consist of $\mathfrak{H} + \mathfrak{H}\mathbf{a}_0$ where $\mathbf{a}_0 = \mathbf{v}_0\mathfrak{J}$. There are 1191 different magnetic space groups (as compared to 230 ordinary space groups).

It is necessary to investigate groups which contain antiunitary elements since $\mathfrak{J}$ is antiunitary. Representation theory becomes more complex, since the matrices which describe the transformation of a set of functions under the operations of such a group no longer form a representation of that group. However, these matrices are required: they are said to form a "corepresentation."

Let $\mathbf{u}, \mathbf{u}', \ldots$ be unitary operators belonging to the group in question, and let $\mathbf{a}, \mathbf{a}', \ldots$ be antiunitary operators. The matrices $\mathbf{D}(\mathbf{u})$, $\mathbf{D}(\mathbf{a})$ which form a corepresentation have the property that (Wigner, 1959)

$$\mathbf{D}(\mathbf{u})\mathbf{D}(\mathbf{u}') = \mathbf{D}(\mathbf{uu}'), \qquad \mathbf{D}(\mathbf{u})\mathbf{D}(\mathbf{a}) = \mathbf{D}(\mathbf{ua}),$$
$$\mathbf{D}(\mathbf{a})\mathbf{D}(\mathbf{u})^* = \mathbf{D}(\mathbf{au}), \qquad \mathbf{D}(\mathbf{a})\mathbf{D}(\mathbf{a}')^* = \mathbf{D}(\mathbf{aa}'). \qquad (3.7.18)$$

The presence of the complex conjugation sign in the last two equations is to be noted. The symmetry properties of the eigenstates of the magnetic crystal are to be described in terms of these corepresentations. The construction of corepresentations is described by Miller and Love (1967).

The effects of ferromagnetic order on the symmetry properties of Bloch functions have been studied by Falicov and Ruvalds (1968).

Problems

1 Derive the character table and basis functions for the irreducible representations of C_{3v}.

2 The ion Ce^{3+} has a ground state $^2F_{5/2}$ and an excited state $^2F_{7/2}$. Determine the splitting of these states in a crystal field of cubic symmetry, assuming that the crystal field is weak compared to spin orbit coupling.

3 Determine the crystal field splitting of the levels of a single d electron in point charge lattices of the zinc blend structure and the rocksalt structure by performing the lattice sum of Eq. (3.4.21) to 10% accuracy.

4 Show that the operator equivalent of $x^4 + y^4 - 6x^2y^2$ within a set of states of definite L is proportional to

$$\tfrac{1}{2}[(L_+)^4 + (L_-)^4].$$

5 Suppose a Cu^{++} ion is situated in an environment in which a large cubic field splitting is supplemented by a weak tetragonal field. Assume that the energy level scheme is

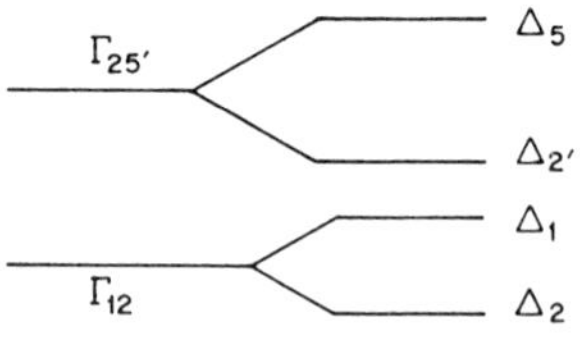

The notation is that of Table C.4. Obtain an explicit expression for the g factor in terms of the strength parameter ζ for spin orbit coupling, and the energy differences between states.

6 Consider a spin Hamiltonian of the form

$$H = \beta \mathbf{S}\cdot\mathbf{g}\cdot\mathbf{H} + D[S_z^2 - \tfrac{1}{3}S(S+1)] + E(S_x^2 - S_y^2).$$

The g tensor will be assumed to have two independent components $g_\perp$ and $g_{\shortparallel}$, as is specified above Eq. (3.4.52). Suppose **H** is parallel to the z axis. Find the energy levels if $S = \frac{3}{2}$.

7 Suppose the spin Hamiltonian for a crystal includes hyperfine interactions between the electron spin S and nuclear spin I and is of the form

$$H = \beta \mathbf{S}\cdot\mathbf{g}\cdot\mathbf{H} + AS_zI_z + B(S_xI_x + S_yI_y).$$

The g tensor has the same form as in Problem 6, and the field is in the z direction. Assume $S = 1$, $I = \frac{1}{2}$. Find the energy levels. Is it important to include in H a term coupling the nucleus and the magnetic field?

8 Determine basis functions for the irreducible representations of the group of G (bcc lattice) (see Appendix C).

9 Consider plane waves of the type exp $(i\mathbf{k}\cdot\mathbf{r})$ in which $k = (p, q, s)$ with $p \neq q \neq s$ and none of the quantities p, q, s is zero. Construct two orthogonal linear combinations of functions of this type which transform according to the $x^2 - y^2$ row of the Γ_{12} representation of O_h.

10 Integrals of the type $V_{ij}(\mathbf{R}) = \int u_i^*(\mathbf{r})V(r)u_j(\mathbf{r} - \mathbf{R})\,d^3r$ are of importance in certain methods of energy band calculations. Suppose u_i transforms according to the xz row of $\Gamma_{25'}$ and u_j according to the $x^2 - y^2$ row of Γ_{12}. The angular dependence of u_i and u_j is expressed in terms of linear combinations of orthonormal spherical harmonics. Let $R = (a, b, c)$, $a \neq b \neq c$. None of the components of R vanish, and the crystal is cubic, so that $V(r)$ is invariant under all the operations of O_h. Prove that
$V_{xz,x^2-y^2}(a, b, c) = -(1/2)V_{xz,x^2-y^2}(b, c, a) - [(3^{1/2}/2]V_{xz,3z^2-r^2}(b, c, a)$.

References

Abragam, A., and Pryce, M. H. L. (1951). *Proc. Roy. Soc. (London)* **A205,** 135.

Altmann, S. L., and Cracknell, A. P. (1965). *Rev. Mod. Phys.* **37,** 19.

Anderson, P. W. (1963). Exchange In Insulators: Superexchange, Direct Exchange, and Double Exchange. *In* "Magnetism" (G. T. Rado and H. Suhl, eds.), Vol. 1, p. 25. Academic Press, New York.

Ballhausen, C. J. (1962). "Introduction to Ligand Field Theory." McGraw-Hill, New York.

Bethe, H. A. (1929). *Ann. Phys.* **3,** 133; primes added by A. W. Overhauser, *Phys. Rev.* **101,** 1702.

Bleaney, B., and Stevens, K. W. H. (1953). *Rep. Progr. Phys.* **16,** 108.

Bloch, F. (1928). *Z. Phys.* **52,** 555.

Bowers, K. D., and Owen, J. (1955). *Rep. Progr. Phys.* **18,** 304.

Bouckaert, L. P., Smoluchowski, R., and Wigner, E. (1936). *Phys. Rev.* **50,** 58.

Condon, E. U., and Shortley, G. H. (1951). "The Theory of Atomic Spectra." Cambridge Univ. Press, London and New York.

Di Bartolo, B. (1968). "Optical Interactions in Solids." Wiley, New York.

Dimmock, J. O., and Wheeler, R. G. (1962). *Phys. Rev.* **177,** 391.

Elliott, R. J. (1954). *Phys. Rev.* **96,** 280.

Ellis, D. E., Freeman, A. J., and Ros, P. (1968). *Phys. Rev.* **176,** 688.

Eyring, H., Walter, J., and Kimball, G. E. (1944). "Quantum Chemistry." Wiley, New York.

Falicov, L. M. (1966). "Group Theory and Its Physical Applications." Univ. of Chicago Press, Chicago, Illinois.

Falicov, L. M., and Ruvalds, J. (1968). *Phys. Rev.* **172,** 499.

Goodenough, J. B. (1963). "Magnetism and the Chemical Bond." Wiley (Interscience), New York.

Griffith, J. S. (1961). "The Theory of Transition Metal Ions." Cambridge Univ. Press, London and New York.

Ham, F. S. (1965). *Phys. Rev.* **138,** A1727.

Hamermesh, M. (1962). "Group Theory." Addison-Wesley, Reading, Massachusetts.

Heine, V. (1960). "Group Theory in Quantum Mechanics." Pergamon, Oxford.

Herring, C. (1937a). *Phys. Rev.* **52,** 361.

Herring, C. (1937b). *Phys. Rev.* **52,** 365.

Hubbard, J. (1963). *Proc. Roy. Soc. (London)* **276A,** 238.

Hubbard, J., Rimer, D. E., and Hopgood, F. R. A. (1966). *Proc. Phys. Soc. (London)* **88,** 13.

Jahn, H. A., and Teller, E. (1937). *Proc. Roy. Soc. (London)* **A161,** 220.

Jones, H. (1960). "The Theory of Brillouin Zones and Electronic States in Crystals." North-Holland Publ., Amsterdam.

Kittel, C. (1956). "Introduction to Solid State Physics," 2nd ed. Wiley, New York.

Kleiner, W. H. (1952). *J. Chem. Phys.* **20,** 1784.

Knox, K., Shulman, R. G., and Sugano, S. (1963). *Phys. Rev.* **130,** 512.

Knox, R. S., and Gold, A. (1964)." Symmetry in the Solid State." Benjamin, New York.

Koster, G. (1957). *Solid State Phys.* **5,** 173.

Koster, G. F., and Statz, H. (1959). *Phys. Rev.* **113,** 445.

Koster, G. F., Dimmock, J. O., Wheeler, R. G., and Statz, H. (1963). "Properties of the Thirty-Two Point Groups." MIT Press, Cambridge, Massachusetts.

von der Lage, F. C., and Bethe, H. A. (1947). *Phys. Rev.* **71,** 612.

Landau, L. P., and Lifshitz, E. M. (1965). "Quantum Mechanics, Non-Relativistic Theory." Addison-Wesley, Reading, Massachusetts.

Liehr, A. D. (1963). *J. Phys. Chem.* **67,** 389.

Lomont, J. S. (1959). "Applications of Finite Groups." Academic Press, New York.

Low, W. (1960). "Paramagnetic Resonance in Solids," *Solid State Phys.* Suppl. *2*. Academic Press, New York.

Lyubarski, G. Ya. (1960). "The Application of Group Theory in Physics." Pergamon, Oxford.

Mattheiss, L. F. (1961). *Phys. Rev.* **123,** 1219.

McClure, D. S. (1959). *Solid State Phys.* **9,** 399.

Miller, S. C., and Love, W. F. (1967). "Tables of Irreducible Representations of Space Groups and Corepresentations of Magnetic Space Groups." Pruett Press, Boulder, Colorado.

Murnaghan, F. D. (1938). "Theory of Group Representations." Johns Hopkins Press, Baltimore, Maryland.

Pake, G. E. (1962). "Paramagnetic Resonance." Benjamin, New York.

Penney, W. G., and Schlapp, R. (1932). *Phys. Rev.* **41,** 194.

Polder, D. (1942). *Physica* **9,** 709.

Pryce, M. H. L. (1950). *Proc. Phys. Soc.* (*London*) **A63,** 25.

Seitz, F. (1936). *Ann. Math.* **37,** 17.

Shubnikov, A. V. (1951). "Symmetry and Anti-Symmetry of Finite Figures." USSR Press, Moscow.

Shulman, R. G., and Sugano, S. (1963). *Phys. Rev.* **130,** 506.

Simanek, E., and Sroubek, Z. (1964). *Phys. Status Solidi* **4,** 251.

Slater, J. C. (1960). "Quantum Theory of Atomic Structure." McGraw-Hill, New York.

Slater, J. C. (1965). "Quantum Theory of Molecules and Solids," Vol. III, "Symmetry and Energy Bands in Crystals." McGraw-Hill, New York.

Statz, H., and Koster, G. F. (1959). *Phys. Rev.* **115,** 1568.

Stevens, K. W. H. (1954). *Proc. Phys. Soc.* (*London*) **A65,** 311.

Sugano, S., and Shulman, R. G. (1963). *Phys. Rev.* **130,** 517.

Tanabe, Y., and Sugano, S. (1954). *J. Phys. Soc. Japan* **9,** 753, 766.

Tinkham, M. (1964). "Group Theory and Quantum Mechanics." McGraw-Hill, New York.

Van Vleck. J. H. (1932). *Phys. Rev.* **41,** 208.

Van Vleck, J. H. (1939). *J. Chem. Phys.* **7,** 72.

Watson, R. E., and Freeman, A. J. (1964). *Phys. Rev.* **134,** A1526.

White, R. M. (1970). "Quantum Theory of Magnetism." McGraw-Hill, New York.

Wigner, E. P. (1959). "Group Theory and Its Application to the Quantum Mechanics of Atomic Spectra" (J. J. Griffin, transl.). Academic Press, New York.

CHAPTER 4

Energy Bands

The calculation of the energy levels of electrons in solids, that is, the determination of energy bands, is a central theoretical problem of solid state physics. Knowledge of these energies and of electron wave functions is required, in principle, for any calculation of more directly observable properties including vibrational spectra, magnetic order, electrical and thermal conductivities, optical dielectric function, and so on. In practice, phenomenological models are often employed which apparently do not require such specific information, as in the treatment of lattice dynamics in Chapter 1; however, it remains a task for fundamental theory to account for the values obtained for the parameters of such a model. The discussion of lattice vibrations assumed the existence of interatomic forces specified by certain force constants which are second derivatives of a crystal potential function. This potential function can, in principle, be determined from the results of a sufficiently complete energy band calculation.

It is already implied in speaking of energy levels of electrons in solids that it is possible substantially to separate the descriptions of electronic and nuclear motions, and that the influence of the nuclear motions on the electrons is small. This is usually true; the occasional failure, as in the case of superconductivity, can produce spectacular results. A fundamental idealization is also present in the theory of energy bands: that one may consider the states of a single electron in a rigid, infinite, periodic lattice. The assembly of many electrons is regarded as a collection of particles occupying one-electron states in accord with the rules of Fermi statistics. It is possible to describe many of the properties of real materials on this basis, without a detailed account of the interaction of electrons with each other. However, in some interesting situations, the interaction of electrons is of great significance, and procedures must be developed for its inclusion.

These methods rest on assumed single-particle levels, even though certain classes of perturbations may be summed to all orders. Some of these techniques are developed in Chapter 8.

Our discussion of energy bands in this chapter is divided into three distinct parts. In the first (Section 4.1), the nature of the energy levels as functions of wave vectors is described in a general way through the use of $\mathbf{k}\cdot\mathbf{p}$ perturbation theory. Subsequently (Sections 4.2–4.8) we consider some of the methods for performing the calculation of energy bands by solution of an effective Schrödinger equation which includes a specified potential energy function. Finally (Sections 4.9–4.10) the determination of the potential energy is discussed.

There is an extensive literature concerned with methods of band calculations. Specific references will be found in the sections devoted to methods. General reviews have been presented by Ziman (1971), Fletcher (1971), and Callaway (1964). The reader who wishes to survey the state of computational procedures should consult Marcus *et al.* (1971) and Alder *et al.* (1968).

4.1 General Properties of Energy Bands

4.1.1 *Properties of Bloch Functions*

It is necessary in much of the subsequent considerations to make use of certain general properties of one-electron crystal wave functions, or Bloch functions. Such functions will be assumed to be solutions of an effective Schrödinger equation

$$[(\mathbf{p}^2/2m) + V(\mathbf{r})]\psi_n(\mathbf{k}, \mathbf{r}) = E_n(\mathbf{k})\psi_n(\mathbf{k}, \mathbf{r}) \tag{4.1.1}$$

in which $V(\mathbf{r})$ is a periodic potential function which does not need to be specified in detail at this time (see Sections 4.8 and 4.10). The index n denotes the energy band and $\mathbf{k}$ is the wave vector. The set of Bloch functions will be shown to be orthonormal and complete.

In order to establish properties of Bloch functions, it is convenient to expand these functions in plane waves. This is not always useful in actual calculations; however, the existence of a convergent plane wave expansion is of great assistance in proving theorems. We have, from Eq. (3.6.2),

$$\psi_n(\mathbf{k}, \mathbf{r}) = (2\pi)^{-3/2} \sum_s b_n(\mathbf{k} + \mathbf{K}_s) \exp[i(\mathbf{k} + \mathbf{K}_s)\cdot\mathbf{r}] \tag{4.1.2}$$

where $\mathbf{K}_s$ is a reciprocal lattice vector. The normalizing factor $(2\pi)^{-3/2}$ has been chosen to give simple orthogonality relations. The coefficients b_n are functions of the single (vector) variable $\mathbf{k} + \mathbf{K}_s$. This serves to guarantee that the Bloch function will satisfy

$$\psi_n(\mathbf{k} + \mathbf{K}_t, \mathbf{r}) = \psi_n(\mathbf{k}, \mathbf{r})$$

as the reader may verify for himself. It follows from (4.1.1) that the coefficients $b_n(\mathbf{k} + \mathbf{K}_s)$ for a particular value of $\mathbf{k}$ are the orthonormal eigenvectors of a Hermitian matrix. They may therefore be regarded as elements of a unitary transformation which diagonalizes H_0. As such they satisfy

$$\sum_s b_n^*(\mathbf{k} + \mathbf{K}_s) b_l(\mathbf{k} + \mathbf{K}_s) = \delta_{nl} \tag{4.1.3}$$

and

$$\sum_n b_n^*(\mathbf{k} + \mathbf{K}_s) b_n(\mathbf{k} + \mathbf{K}_t) = \delta_{st}. \tag{4.1.4}$$

It is convenient to make use of the plane wave expansion to establish the orthonormality relation for Bloch functions. Substitute (4.1.2) into the orthogonality integral

$$\begin{aligned}\int \psi_n^*(\mathbf{k}, \mathbf{r}) \psi_l(\mathbf{q}, \mathbf{r}) \, d^3r \\ &= (2\pi)^{-3} \sum_{s,t} b_n^*(\mathbf{k} + \mathbf{K}_s) b_l(\mathbf{q} + \mathbf{K}_t) \\ &\quad \times \int \exp[i(\mathbf{q} + \mathbf{K}_t - \mathbf{k} - \mathbf{K}_s) \cdot \mathbf{r}] \, d^3r \\ &= \sum_{s,t} b_n^*(\mathbf{k} + \mathbf{K}_s) b_l(\mathbf{q} + \mathbf{K}_t) \, \delta(\mathbf{q} + \mathbf{K}_t - \mathbf{k} - \mathbf{K}_s).\end{aligned} \tag{4.1.5}$$

If $\mathbf{k}$ and $\mathbf{q}$ are confined to the interior of the Brillouin zone, they cannot differ by a reciprocal lattice vector. Therefore the only contribution to (4.1.5) comes from the term with $\mathbf{K}_s - \mathbf{K}_t = 0$. If $\mathbf{q}$ and $\mathbf{k}$ are on the surface of the Brillouin zone, the argument still applies unless $\mathbf{q}$ and $\mathbf{k}$ are equivalent points. In such a case $\mathbf{q} = \mathbf{k} + \mathbf{K}_1$, say, but we still must have $(\mathbf{q} + \mathbf{K}_t) = \mathbf{k} + \mathbf{K}_s$, and the problem is the same as if $\mathbf{q}$ and $\mathbf{k}$ were the same point. We now have

$$\begin{aligned}\int \psi^*_n(\mathbf{k}, \mathbf{r}) \psi_l(\mathbf{q}, \mathbf{r}) \, d^3r &= \delta(\mathbf{q} - \mathbf{k}) \sum_s b_n^*(\mathbf{k} + \mathbf{K}_s) b_l(\mathbf{k} + \mathbf{K}_s) \\ &= \delta_{nl} \, \delta(\mathbf{q} - \mathbf{k}).\end{aligned} \tag{4.1.6a}$$

The last step follows through the use of (4.1.3). It is understood in the use of (4.1.6a) that equivalent points are treated as identical.

It is also permissible to use discrete normalization for the Bloch functions. In this case the factor $(2\pi)^{-3/2}$ in (4.1.2) is replaced by $(N\Omega)^{-1/2}$ where N is the number of unit cells in the normalization volume and each cell has volume Ω. Equation (4.1.6a) is then replaced by

$$\int \psi_n^*(\mathbf{k}, \mathbf{r})\psi_l(\mathbf{q}, \mathbf{r})\, d^3r = \delta_{nl}\, \delta_{\mathbf{kq}}. \tag{4.1.6b}$$

Discrete and continuous normalizations will be used in the following according to the requirements of convenience.

An important corollary of (4.1.6a) is the orthonormality relation obeyed by the cell periodic function $u_n(\mathbf{k}, \mathbf{r})$ (see Sections 3.5 and 3.6)

$$\psi_n(\mathbf{k}, \mathbf{r}) = \exp(i\mathbf{k}\cdot\mathbf{r})u_n(\mathbf{k}, \mathbf{r}). \tag{4.1.7}$$

This turns out to be

$$\int_{\text{cell}} u_n^*(\mathbf{k}, \mathbf{r})u_l(\mathbf{k}, \mathbf{r})\, d^3r = [\Omega/(2\pi)^3]\, \delta_{nl} \tag{4.1.8}$$

where the integral includes only a single unit cell, whose volume is Ω. The cell periodic functions are not orthogonal for different wave vectors.

Equation (4.1.8) may be established as follows: Substitute (4.1.7) into the left side of (4.1.6a). We have

$$\int \exp[i(\mathbf{q} - \mathbf{k})\cdot\mathbf{r}]u_n^*(\mathbf{k}, \mathbf{r})u_l(\mathbf{q}, \mathbf{r})\, d^3r.$$

The product $u_n^*u_l$ is the same in each unit cell in the crystal. The integral over all space may therefore be expressed as the product of an integral over a single cell and a sum over all cells in the following way:

$$\begin{aligned}
&\sum_\nu \left\{\exp[i(\mathbf{q} - \mathbf{k})\cdot\mathbf{R}_\nu] \int_{\text{cell}} \exp[i(\mathbf{q} - \mathbf{k})\cdot(\mathbf{r} - \mathbf{R}_\nu)]\right.\\
&\qquad \left.\times\, u_n^*(\mathbf{k}, \mathbf{r} - \mathbf{R}_\nu)u_l(\mathbf{q}, \mathbf{r} - \mathbf{R}_\nu)\, d^3r\right\}\\
&= \left\{\int_{\text{cell } 0} \exp[i(\mathbf{q} - \mathbf{k})\cdot\mathbf{r}]u_n(\mathbf{k}, \mathbf{r})u_l(\mathbf{q}, \mathbf{r})\, d^3r\right\} \sum_\nu \exp[i(\mathbf{q} - \mathbf{k})\cdot\mathbf{R}_\nu]\\
&= [(2\pi)^3/\Omega]\, \delta(\mathbf{k} - \mathbf{q}) \int_{\text{cell}} \exp[i(\mathbf{q} - \mathbf{k})\cdot\mathbf{r}]u_n^*(\mathbf{k}, \mathbf{r})u_l(\mathbf{q}, \mathbf{r})\, d^3r\\
&= \delta_{nl}\, \delta(\mathbf{k} - \mathbf{q}).
\end{aligned}$$

We obtain (4.1.8) by equating the coefficients of the delta function. When we do this, we must set $\mathbf{k} = \mathbf{q}$ in the coefficient. A direct proof using (4.1.2) is also easy to construct.

To prove the completeness of Bloch functions, we consider

$$\sum_n \int \psi_n^*(\mathbf{k}, \mathbf{r})\psi_n(\mathbf{k}, \mathbf{r}')\, d^3k$$

$$= (2\pi)^{-3} \sum_{s,t} \int \sum_n b_n^*(\mathbf{k} + \mathbf{K}_s) b_n(\mathbf{k} + \mathbf{K}_t) \times \exp[-i(\mathbf{k} + \mathbf{K}_s)\cdot\mathbf{r}] \exp[i(\mathbf{k} + \mathbf{K}_t)\cdot\mathbf{r}']\, d^3k. \quad (4.1.9)$$

This integral (and all other $\mathbf{k}$ integrals, unless specified differently) includes the Brillouin zone. Now we use (4.1.4) to simplify our expression to

$$(2\pi)^{-3} \sum_s \int \exp[-i(\mathbf{k} + \mathbf{K}_s)\cdot(\mathbf{r} - \mathbf{r}')]\, d^3k.$$

Integration over the Brillouin zone plus summation over all reciprocal lattice vectors is equivalent to integration over all $\mathbf{k}$ space. Thus we obtain

$$\sum_n \int \psi_n^*(\mathbf{k}, \mathbf{r})\psi_n(\mathbf{k}, \mathbf{r}')\, d^3k$$

$$= (2\pi)^{-3} \int_\infty \exp[-i\mathbf{k}\cdot(\mathbf{r} - \mathbf{r}')]\, d^3k = \delta(\mathbf{r} - \mathbf{r}'). \quad (4.1.10)$$

Equation (4.1.10) is the desired result.

4.1.2 *The* $\mathbf{k}\cdot\mathbf{p}$ *Method*

In Sections 3.5 and 3.6 some of the properties of energy bands and wave functions which could be determined by general considerations of symmetry and continuity were examined. In the present section the relation of energies and wave functions at closely spaced $\mathbf{k}$ values will be discussed in a more detailed fashion which will enable us to obtain quantitative results. A review of the procedure has been given by Kane (1966).

More precisely, let us suppose that energy bands and wave functions have been determined by some process for a reference point in the Brillouin zone $\mathbf{k}_0$. Then we ask for these quantities at a neighboring point $\mathbf{k}$. Thus let $\psi_n(\mathbf{k}, \mathbf{r})$ be the wave function for a state in the nth band at position $\mathbf{k}$ in the zone, and let $\psi_j(\mathbf{k}_0, \mathbf{r})$ refer to the jth band at $\mathbf{k}_0$. A perturbation procedure which relates these functions will be developed.

To this end, it is convenient to define a set of functions

$$\chi_j(\mathbf{k}, \mathbf{r}) = \exp[i(\mathbf{k} - \mathbf{k}_0)\cdot\mathbf{r}]\psi_j(\mathbf{k}_0, \mathbf{r}). \tag{4.1.11}$$

These functions, which were originally introduced by Luttinger and Kohn (1955) in another context, are a suitable set for the expansion of the wave function for a state of wave vector **k**. The procedure used to establish the orthonormality and completeness of Bloch functions in Section 4.1.1 can be adopted in a straightforward way to prove corresponding relations for the χ. Specifically,

$$\int \chi_n^*(\mathbf{q}, \mathbf{r})\chi_l(\mathbf{k}, \mathbf{r})\, d^3r = \delta_{nl}\,\delta(\mathbf{k} - \mathbf{q}) \tag{4.1.12}$$

and

$$\sum_n \int \chi_n^*(\mathbf{k}, \mathbf{r})\chi_n(\mathbf{k}, \mathbf{r}')\, d^3k = \delta(\mathbf{r} - \mathbf{r}'). \tag{4.1.13}$$

The proof is left to the reader. It is easy to see that $\chi_n(\mathbf{k}, \mathbf{r})$ obeys Bloch's theorem for a state of wave vector **k**, since (4.1.11) can be written in the alternative form

$$\chi_n(\mathbf{k}, \mathbf{r}) = \exp(i\mathbf{k}\cdot\mathbf{r})u_n(\mathbf{k}_0, \mathbf{r}) \tag{4.1.14}$$

in which u_n is the cell periodic function introduced in Section 3.5.

The unknown $\psi_n(\mathbf{k}, \mathbf{r})$ is expanded in the known functions χ_j:

$$\psi_n(\mathbf{k}, \mathbf{r}) = \sum_j A_{nj}(\mathbf{k})\chi_j(\mathbf{k}, \mathbf{r}). \tag{4.1.15}$$

Note that we cannot expand $\psi_n(\mathbf{k}, \mathbf{r})$ directly in terms of the $\psi_j(\mathbf{k}_0, \mathbf{r})$ since functions belonging to different wave vectors are orthogonal.

The Bloch functions are solutions of the effective Schrödinger equation (4.1.1).

An equation for the A_{nj} is determined by substituting (4.1.15) and (4.1.11) into (4.1.1):

$$\begin{aligned}
&[(\mathbf{p}^2/2m) + V]\psi_n(\mathbf{k}, \mathbf{r}) \\
&\quad = \exp[i(\mathbf{k} - \mathbf{k}_0)\cdot\mathbf{r}] \sum_j A_{nj}(\mathbf{k})[E_j(\mathbf{k}_0) \\
&\qquad + (\hbar/m)(\mathbf{k} - \mathbf{k}_0)\cdot\mathbf{p} + (\hbar/2m)(\mathbf{k} - \mathbf{k}_0)^2]\psi_j(\mathbf{k}_0, \mathbf{r}) \\
&\quad = E_n(\mathbf{k}) \exp[i(\mathbf{k} - \mathbf{k}_0)\cdot\mathbf{r}] \sum_j A_{nj}(\mathbf{k})\psi_j(\mathbf{k}_0, \mathbf{r}).
\end{aligned}$$

It is desirable to replace ψ_j by u_j. We find

$$\sum_j A_{nj}(\mathbf{k})[E_j(\mathbf{k}_0) + (\hbar/m)(\mathbf{k} - \mathbf{k}_0)\cdot\mathbf{p} + (\hbar^2/2m)(\mathbf{k}^2 - \mathbf{k}_0^2)]u_j(\mathbf{k}_0, \mathbf{r})$$
$$= E_n(\mathbf{k}) \sum_j A_{nj}(\mathbf{k})u_j(\mathbf{k}_0, \mathbf{r}). \tag{4.1.16}$$

Next we multiply (4.1.16) by $u_l^*(\mathbf{k}_0, \mathbf{r})$ and integrate over a unit cell of the crystal. The orthogonality relation for the u's [Eq. (4.1.8)] is employed. The result is

$$\sum_j \{[E_j(\mathbf{k}_0) - E_n(\mathbf{k}) + (\hbar^2/2m)(\mathbf{k}^2 - \mathbf{k}_0^2)]\,\delta_{jl}$$
$$+ (\hbar/m)(\mathbf{k} - \mathbf{k}_0)\cdot\mathbf{p}_{lj}\}A_{nj}(\mathbf{k}) = 0 \tag{4.1.17}$$

in which

$$\mathbf{p}_{lj} = [(2\pi)^3/\Omega]\int_\Omega d^3r\ u_l^*(\mathbf{k}_0, \mathbf{r})\,\mathbf{p}u_j(\mathbf{k}_0, \mathbf{r})\ d^3r. \tag{4.1.18}$$

There is one equation for each value of the band index l. The index n (on A_{nj}) refers to the nth solution of the equations, and may be suppressed without leading to misunderstanding. The condition for this infinite set of simultaneous, linear, homogeneous equations to have a nontrivial solution is that the determinant of the coefficients should vanish. A general element of this determinant has the form

$$H_{jl} - E(\mathbf{k})\,\delta_{jl} \tag{4.1.19}$$

with

$$H_{jl} = [E_j(\mathbf{k}_0) + (\hbar^2/2m)(\mathbf{k}^2 - \mathbf{k}_0^2)]\,\delta_{jl} + (\hbar/m)(\mathbf{k}-\mathbf{k}_0)\cdot\mathbf{p}_{lj}. \tag{4.1.20}$$

The problem is to diagonalize the effective Hamiltonian matrix, whose elements are given by (4.1.20).

Although this procedure has been developed with the intention that it will be used to relate energies of states at neighboring points of the zone, there is nothing in the derivation of the effective Hamiltonian which restricts us to this situation. If enough states are included in the Hamiltonian and if the eigenvalues are obtained accurately, there is no reason why the bands cannot be obtained throughout the zone. This supposes that there is enough experimental or theoretical information available from other sources to fix the values of the momentum matrix elements on which the calculation depends, and which may be regarded as adjustable parameters. This procedure has been carried through by Cardona and collaborators for some semiconductors (see e.g., Cardona and Pollak, 1966).

However, the simplest and most common use of this Hamiltonian is in a calculation of effective masses using ordinary perturbation theory. This will be applicable provided that the off-diagonal terms are small, and that we are not concerned with degenerate levels (at $\mathbf{k}_0$). Let $\mathbf{k} - \mathbf{k}_0 = \mathbf{s}$. The energy of a state in the nth band at $\mathbf{k}$ is related to that at $\mathbf{k}_0$ in second order by

$$E_n(\mathbf{k}) = E_n(\mathbf{k}_0) + (\hbar/m)\mathbf{s}\cdot\mathbf{p}_{nn} + (\hbar^2/2m)(\mathbf{k}^2 - \mathbf{k}_0{}^2) + (\hbar^2/m^2) \sum_{j(j\neq n)} \{(\mathbf{s}\cdot\mathbf{p}_{nj})(\mathbf{s}\cdot\mathbf{p}_{jn})/[E_n(\mathbf{k}_0) - E_j(\mathbf{k}_0)]\}. \quad (4.1.21)$$

Perturbation theory gives a Taylor series expansion of the energy as a function of $\mathbf{k}$. The convergence of the series is governed by the energy denominators which appear in (4.1.21). If these are small, the second-order term in the energy will be large, and higher terms in the series will be important. Under such circumstances, it is desirable to diagonalize the portion of the effective Hamiltonian (4.1.20) which connects the nearly degenerate states.

The wave function may also be obtained in perturbation theory. To first order, the expansion coefficients A_j are

$$A_{nj} = (\hbar/m)\{\mathbf{s}\cdot\mathbf{p}_{jn}/[E_n(\mathbf{k}_0) - E_j(\mathbf{k}_0)]\} \qquad (n \neq j). \quad (4.1.22)$$

An expression for the cell periodic part of the Bloch function is obtained from (4.1.14), (4.1.15), and (4.1.22):

$$u_n(\mathbf{k}, \mathbf{r}) = u_n(\mathbf{k}_0, \mathbf{r}) + (\hbar/m)\mathbf{s}\cdot \sum_{j\neq n} \{\mathbf{p}_{jn}u_j(\mathbf{k}_0, \mathbf{r})/[E_n(\mathbf{k}_0) - E_j(\mathbf{k}_0)]\}. \quad (4.1.23)$$

It is of interest to compute the overlap integral between cell periodic functions for different values of $\mathbf{k}$. This quantity was required in Eq. (3.6.11):

$$[(2\pi)^3/\Omega] \int_\Omega u_n{}^*(\mathbf{k}, \mathbf{r})u_l(\mathbf{k}, \mathbf{r})\, d^3r = \delta_{nl} + 2(\hbar/m)\{\mathbf{s}\cdot\mathbf{p}_{nl}/[E_n(\mathbf{k}_0) - E_l(\mathbf{k}_0)]\}(1 - \delta_{nl}). \quad (4.1.24)$$

The overlap between the cell periodic functions in different bands at neighboring points $\mathbf{k}_0$ and $\mathbf{k}_0 + \delta\mathbf{k}$ is thus of order $\delta\mathbf{k}$. However, if the bands are the same, the first-order term vanishes. A contribution will be obtained in second order. By Schwarz's inequality, the sign of the second-order term must be negative, as was asserted in (3.6.11).

Let us consider a situation in which band n has an extremum at $\mathbf{k}_0$ so that the term linear in $\mathbf{s}$ in (4.1.21) vanishes. This implies $\mathbf{p}_{nn} + \hbar\mathbf{k}_0 = 0$. Then the second-order terms dominate. Let s_α and s_β be rectangular components of $\mathbf{s}$ with respect to some fixed axes, and $p^{\alpha}{}_{nj}$, etc. be corresponding components of the matrix element. We differentiate (4.1.21) twice, and obtain

$$m/\hbar^2\, \partial^2 E_n/\partial s_\alpha\, \partial s_\beta$$
$$= m/\hbar^2\, \partial^2 E_n/\partial k_\alpha\, \partial k_\beta$$
$$= \delta_{\alpha\beta} + (1/m) \sum_{j(j\neq n)} \{(p^{\alpha}{}_{nj} p^{\beta}{}_{jn} + p^{\beta}{}_{nj} p^{\alpha}{}_{jn})/[E_n(\mathbf{k}_0) - E_j(\mathbf{k}_0)]\}. \tag{4.1.25}$$

It is convenient to define a reciprocal effective mass tensor by

$$(m/m_n^*)_{\alpha\beta} = m/\hbar^2\, \partial^2 E_n/\partial k_\alpha\, \partial k_\beta. \tag{4.1.26}$$

The elements are

$$(m/m_n^*)_{\alpha\beta} = \delta_{\alpha\beta} + (1/m) \sum_{j(j\neq n)} \{(p^{\alpha}{}_{nj} p^{\beta}{}_{jn} + p^{\beta}{}_{nj} p^{\alpha}{}_{jn})/[E_n(\mathbf{k}_0) - E_j(\mathbf{k}_0)]\}. \tag{4.1.27}$$

In these equations, m is the free electron mass.

It is always possible to diagonalize the reciprocal effective mass tensor by a proper choice of coordinate axes. If the extremum point $\mathbf{k}_0$ is a general point in the zone, the choice of the axes depends on the details of the dynamics, that is, on the crystal potential. However, if the extremum occurs at a symmetry point or along an axis of symmetry, the axes may be partially determined by symmetry. For example, if $\mathbf{k}_0$ is along the 100, 111, or 110 axes in a cubic crystal, the symmetry axis must be a principal axis. In the case of the 100 and 111 axes, the surfaces of constant energy in the effective mass approximation must be ellipsoids of revolution, characterized by a longitudinal effective mass referring to displacements along the symmetry axis, and a transverse mass referring to perpendicular displacements. If $\mathbf{k} = 0$ is the extremum, the surfaces of constant energy in a cubic crystal must be spherical.

In the principal axis system, the reciprocal effective masses are

$$(m/m_n^*)_{\alpha\alpha} = 1 + (2/m) \sum_{j(j\neq n)} \{|\, p^{\alpha}{}_{nj}\,|^2/[E_n(\mathbf{k}_0) - E_j(\mathbf{k}_0)]\} \tag{4.1.28}$$

where α now refers to one of the principal axes.

The interaction of a given level with lower lying levels or core states $(E_j < E_n)$ tends to decrease an effective mass, while interaction with higher states tends to increase it.

There are two terms in (4.1.21) which tend to give rise to a linear dependence of energy on wave vector going away from $\mathbf{k}_0$. This will be seen if we rewrite (4.1.21) in the form

$$E_n(\mathbf{k}) = E_n(\mathbf{k}_0) + (\hbar/m)\mathbf{s}\cdot(\mathbf{p}_{nn} + \hbar\mathbf{k}_0) + (\hbar^2/2m)\mathbf{s}^2 + (\hbar^2/m^2) \sum_{j(j\neq n)} \{(\mathbf{s}\cdot\mathbf{p}_{nj})(\mathbf{s}\cdot\mathbf{p}_{jn})/[E_n(\mathbf{k}_0) - E_j(\mathbf{k}_0)]\}. \quad (4.1.29)$$

The linear term must vanish, however, for certain states at symmetry points of the Brillouin zone. In this case, the energy is quadratic in $\mathbf{s}$ as we have discussed. To determine whether the linear term vanishes by reason of symmetry, we observe that

$$\mathbf{p}_{nn} + \hbar\mathbf{k}_0 = [(2\pi)^3/\Omega] \int_\Omega \psi_n^*(\mathbf{k}_0, \mathbf{r})\,\mathbf{p}\psi_n(\mathbf{k}_0, \mathbf{r})\, d^3r \quad (4.1.30)$$

where on the right we have the full Bloch function instead of the cell periodic function.

Consider for the moment a general matrix element of the form

$$\int \psi_i^*(\mathbf{k}_0, \mathbf{r})\, \mathcal{O}\psi_j(\mathbf{k}_0, \mathbf{r})\, d^3r$$

where $\mathcal{O}$ is some operator. In order that this integral not be zero, it is necessary that the integral contain a scalar component. Suppose that the functions $\psi_i(\mathbf{k}_0, \mathbf{r})$, $\psi_j(\mathbf{k}_0, \mathbf{r})$ and the operator $\mathcal{O}$ transform according to irreducible representations of the group of $\mathbf{k}_0$, say $\Gamma(i)$, $\Gamma(j)$, $\Gamma(\mathcal{O})$. The integral then transforms according to the direct product representation $\Gamma^*(i) \times \Gamma(\mathcal{O}) \times \Gamma(j)$. It follows from the discussion in Section 3.2 that the integral will be zero unless the direct product representation (usually a reducible one) contains $\Gamma(1)$ where $\Gamma(1)$ is the symmetric representation of the group of $\mathbf{k}_0$. In the case of effective mass theory, we are concerned with $\Gamma(\mathcal{O}) = \Gamma(\mathbf{V})$, where $\Gamma(\mathbf{V})$ is the representation (or one of the representations) appropriate for quantities transforming as a vector.

At a general point of the zone, all the representations are the same, and a linear dependence will usually exist. At symmetry points where there are a number of representations, the linear dependence will often vanish. In particular, if the group of $\mathbf{k}_0$ contains the inversion, the wave functions will be eigenfunctions of definite parity; the product $\psi_i^* \nabla\psi_i$ will be an odd function whether ψ_i is odd or even, and the integral must vanish. For instance, $\nabla_\mathbf{k} E = 0$ for all representations at Γ, X, and L in the face-centered cubic lattice.

4.1.3 *A Two-Band Model*

Some insight into the nature of the results to be expected in more general cases can be obtained by considering a two-band model. We will obtain the eigenvalues of the Hamiltonian (4.1.20) in the case of two bands, denoted 0 and 1. We set $\mathbf{p}_{01} = \mathbf{p}^*_{10} = \mathbf{p}$, and for additional simplicity, take $k_0 = 0$, set $E_0(\mathbf{k}_0) = 0$, $E_1(\mathbf{k}_0) = E_g$, the band gap, and assume that $\mathbf{p}$ is isotropic. The effective Hamiltonian is

$$H = \begin{pmatrix} \hbar^2k^2/2m & (\hbar/m)kp \\ (\hbar/m)kp & E_g + \hbar^2k^2/2m \end{pmatrix}. \tag{4.1.31}$$

The eigenvalues are

$$E_{0,1}(k) = (E_g/2) + (\hbar^2k^2/2m) \pm [(E_g^2/4) + (\hbar^2/m^2)k^2p^2]^{1/2}. \tag{4.1.32}$$

The square root may be expanded for small k:

$$E_0(k) = (\hbar^2k^2/2m)[1 - (2p^2/mE_g)], \tag{4.1.33a}$$

$$E_1(k) = E_g + (\hbar^2k^2/2m)[1 + (2p^2/mE_g)]. \tag{4.1.33b}$$

The reciprocal effective masses are scalars in consequence of the assumption of an isotropic matrix element. Equation (4.1.26) yields

$$(m/m^*)_0 = 1 - (2p^2/mE_g), \qquad (m/m^*)_1 = 1 + (2p^2/mE_g). \tag{4.1.34}$$

Particular interest attaches to the case in which $E_g \ll 2p^2/m$. This occurs in several common semiconductors; notably GaAs, InAs, and InSb. We may then neglect the 1 and find two bands whose masses are equal in magnitude but are opposite in sign. The magnitudes of the masses, in a two-band model with a small energy gap, are directly proportional to the gap. It is found that this proportionality is roughly obeyed in the comparison of the conduction band masses at $\mathbf{k} = 0$ in Ge and GaAs, indicating that with similar electronic structures the momentum matrix element does not vary much.

Specifically, in reference to the conduction band near $\mathbf{k} = 0$, let us examine the assumed proportionality

$$m^*(\mathrm{GaAs})/m^*(\mathrm{Ge}) = E_g(\mathrm{GaAs})/E_g(\mathrm{Ge}).$$

Use of experimental values, $E_g(\mathrm{GaAs}) = 1.52$ eV, $E_g(\mathrm{Ge}) = 0.889$ eV, and $m^*(\mathrm{Ge}) = 0.041$, leads to $m^*(\mathrm{GaAs}) = 0.070$ in fair agreement with the measured value $m^* = 0.0665$. The calculation should be improved by taking into account the spin–orbit splitting of the valence band (Moss and Walton, 1959; Ehrenreich, 1961).

It is to be noted also that the range of validity of the expanded forms (4.1.33) will be quite restricted when $2p^2/mE_g \gg 1$. The bands begin to depart from parabolic form when k is still relatively small. As an example, in an n-type semiconductor doped to degeneracy, the effective mass at the Fermi energy will increase with carrier concentration. This is a result of the higher-order terms of opposite sign which appear in (4.1.33b) when the expansion is carried further.

4.1.4 *Band Degeneracies*

It frequently occurs that the state of interest at $\mathbf{k}_0$ is degenerate. In this case the perturbation will remove the degeneracy, at least in some directions. This means that in going from a point of higher symmetry to one of lower symmetry the energy bands split. It is then necessary to use degenerate perturbation theory. If the momentum operator has matrix elements connecting the members of the degenerate set, the degeneracy will be removed in first order, and the split bands will go away from the symmetry point with a nonzero slope. This can occur only in the vicinity of symmetry points whose group does not contain the inversion. In many interesting cases, however, the momentum operator has nonvanishing matrix elements only between the degenerate subset and states of different energy. The degeneracy is then removed in second order, and the perturbation theory appropriate for this case must be employed (Shockley, 1950; Dresselhaus *et al.*, 1955; Kane, 1956).

Often it is desirable to treat a situation in which some bands, although not quite degenerate, approach each other so closely at $\mathbf{k}_0$ that the range of usefulness of the expansion (4.1.21) is quite small. This case, as well as that of actual degeneracy, are treated by means of a form of perturbation theory introduced by Lowdin (1951). The essential idea of this procedure is to separate the states considered in the perturbation calculation into two sets: one of which involves a small number of strongly coupled states whose interaction is treated exactly; and the other, more numerous, contains those which are well removed in energy from the first set, and interact with it only weakly. We will sketch the derivation in the case in which there are only two states in the first category. These are labeled 0 and 1. The generalization to more closely coupled states is immediate. We consider Eq. (4.1.17) in which the following abbreviations are introduced:

$$\epsilon = E(\mathbf{k}) - (\hbar^2/2m)(\mathbf{k}^2 - \mathbf{k}_0^2), \qquad h_{lj} = (\hbar/m)(\mathbf{k} - \mathbf{k}_0)\cdot\mathbf{p}_{lj}. \tag{4.1.35}$$

Equation (4.1.17) is broken into two sets. The first set consists of $l = 0, 1$:

$$\begin{aligned} [E_0(\mathbf{k}_0) - \epsilon + h_{00}]A_0 + h_{01}A_1 + \sum_{j>1} h_{0j}A_j &= 0, \\ h_{10}A_0 + [E_1(\mathbf{k}_0) - \epsilon + h_{11}]A_1 + \sum_{j>1} h_{1j}A_j &= 0. \end{aligned} \tag{4.1.36}$$

The second set includes all values of l greater than 1. In this case only those terms connecting state l to states 0, 1 are retained. This approximation gives

$$[E_l(\mathbf{k}_0) - \epsilon + h_{ll}]A_l + h_{l0}A_0 + h_{l1}A_1 = 0. \tag{4.1.37}$$

Equation (4.1.37) for A_l is solved and the result is substituted into (4.1.36). The resulting equations now involve only A_0 and A_1. They have the form

$$\begin{aligned}\{E_0(\mathbf{k}_0) + h_{00} + \sum_{j\neq 0,1} (h_{0j})^2/[\epsilon - E_j(\mathbf{k}_0) - h_{jj}] - \epsilon\}A_0 \\ + \{h_{01} + \sum_{j\neq 0,1} h_{0j}h_{j1}/[\epsilon - E_j(\mathbf{k}_0) - h_{jj}]\}A_1 = 0,\\ \{h_{10} + \sum_{j\neq 0,1} h_{1j}h_{j0}/[\epsilon - E_j(\mathbf{k}_0) - h_{jj}]\}A_0 \\ + \{E_1(\mathbf{k}_0) + h_{11} + \sum_{j\neq 0,1} | h_{1j} |^2/[\epsilon - E_j(\mathbf{k}_0) - h_{jj}] - \epsilon\}A_1 = 0.\end{aligned} \tag{4.1.38}$$

The 2×2 determinant of the coefficients of A_0 and A_1 in (4.1.38) must vanish. The resulting equation determines the values of the energy. We note that the equation involved is somewhat complicated in form since the unknown energy ϵ appears in the denominator of the sums. This feature is a common one in perturbation procedures of the Brillouin–Wigner type. Solutions usually must be obtained by a self-consistent numerical procedure.

The generalization to any number of strongly interacting states is immediate. The problem is equivalent to the diagonalization of an effective Hamiltonian whose general element is of the form

$$\begin{aligned}H_{ln}^{(\mathrm{eff})} = [E_l(\mathbf{k}_0) + (\hbar^2/2m)(\mathbf{k}^2 - \mathbf{k}_0^2)]\,\delta_{ln} + (\hbar/m)\mathbf{s}\cdot\mathbf{p}_{ln} \\ + (\hbar^2/m^2) \sum_j \{(\mathbf{s}\cdot\mathbf{p}_{lj})(\mathbf{s}\cdot\mathbf{p}_{jn})/[\epsilon - E_j(\mathbf{k}_0) - h_{jj}]\}.\end{aligned} \tag{4.1.39}$$

If the separation of the strongly interacting levels at $\mathbf{k}_0$ is small (or if they are actually degenerate), we may replace the $E(\mathbf{k})$ appearing in ϵ as a further approximation, by the average energy of these levels at $\mathbf{k}_0$. The perturbation procedure can be expected to converge well if the matrix elements connecting states in the two groups are small compared to the separation in energy between these groups. As long as this condition is satisfied, that is, the bands in group 1 must not approach those in group 2 closely, the procedure may be extended to large values of $(\mathbf{k} - \mathbf{k}_0)$.

4.1.5 *Extensions*

These procedures may be extended to include spin–orbit coupling. Suppose we add to (4.1.1) a term H_{soc} representing this effect

$$H_{\mathrm{soc}} = (\hbar/4m^2c^2)\,\boldsymbol{\sigma}\cdot(\nabla V \times \mathbf{p}) \tag{4.1.40}$$

where $\boldsymbol{\sigma}$ is the Pauli spin operator. At this point, we have two choices: (1) we may take a set of basis states at $\mathbf{k}_0$ which include spin–orbit coupling. In this case, Eq. (4.1.17) is unchanged, except for the replacement of $\mathbf{p}_{lj}$ by $\mathbf{\Pi}_{lj}$ where

$$\mathbf{\Pi}_{lj} = \mathbf{p}_{lj} + [(2\pi)^3/\Omega](\hbar/4m^2c^2) \int_\Omega u_l^*(\mathbf{k}_0, \mathbf{r})(\sigma \times \nabla V) u_j(\mathbf{k}_0, \mathbf{r})\, d^3r. \tag{4.1.41}$$

Specifically, we obtain

$$\sum_j \{[E_j(\mathbf{k}_0) - E(\mathbf{k}) + (\hbar^2/2m)(\mathbf{k}^2 - \mathbf{k}_0^2)]\,\delta_{lj} + (\hbar/m)(\mathbf{k} - \mathbf{k}_0)\cdot\mathbf{\Pi}_{lj}\}A_j = 0. \tag{4.1.42}$$

Alternately, spin–orbit coupling may not be included in the zero-order Hamiltonian which defines the basis sets at $\mathbf{k}_0$. In this case we treat H_{soc} by perturbation theory, and degenerate states at $\mathbf{k}_0$ may be split by this perturbation.

For applications to heavy elements it may be desirable to base a band calculation on the Dirac equation for a single particle in a periodic potential, V. This equation can be written as (see Schiff, 1968, p. 472)

$$[-c\boldsymbol{\alpha}\cdot\mathbf{p} - \beta mc^2 + V]\psi_n(\mathbf{k}, \mathbf{r}) = E_n(\mathbf{k})\psi_n(\mathbf{k}, \mathbf{r}). \tag{4.1.43}$$

The argument of Section 3.5, establishing Bloch's theorem, is still valid since it depends only on the translation invariance of the Hamiltonian. Hence, we may write

$$\psi_n(\mathbf{k}, \mathbf{r}) = \exp(i\mathbf{k}\cdot\mathbf{r})u_n(\mathbf{k}, \mathbf{r}) \tag{4.1.44}$$

where $u_n(\mathbf{k}, \mathbf{r})$ is a four-component spinor. Our general procedure for the calculation of effective masses is still valid; however, we must replace Eq. (4.1.17) by

$$\sum_j \{[E_l(\mathbf{k}_0) - E(\mathbf{k})]\,\delta_{lj} - \hbar c\mathbf{s}\cdot\boldsymbol{\alpha}_{lj}\}A_j = 0 \tag{4.1.45}$$

where

$$\boldsymbol{\alpha}_{lj} = [(2\pi)^3/\Omega] \int u_l^*(\mathbf{k}_0, \mathbf{r})\,\boldsymbol{\alpha} u_j(\mathbf{k}_0, \mathbf{r})\, d^3r, \tag{4.1.46}$$

and the integration includes summation over spinor indices.

The energy bands which are derived from the Dirac equation (4.1.43) differ qualitatively from those obtained from the Schrödinger equation in that some degeneracies present in the latter case are removed. The splitting of these degeneracies is given correctly qualitatively and with substantial accuracy quantitatively if the spin–orbit coupling Hamiltonian (4.1.40) is treated as a perturbation on a band structure calculated with the Schrödinger equation.

For some purposes, particularly in the theory of tunneling, it is desirable to determine the properties of the energy as a function of a complex wave vector. The mathematical analysis is based on the perturbation procedure we have discussed. The problem has been treated extensively by Blount (1962), to whose article the reader is referred for more detail. Some of the essential results are: (1) Bands in complex **k** space may be determined as solutions of an effective Hamiltonian [Eq. (4.1.20) in nonrelativistic theory] whose matrix elements are simple analytic functions (in fact just quadratic functions) of complex **k**. (2) Consider a band which is not degenerate at a (real) point **k**. Then the energy is an analytic function of complex **k** in a region about $\mathbf{k}_0$. The energy may be determined as a function of **k** from the Taylor expansion, Eq. (4.1.21), which is convergent in this region. The region of analyticity is terminated by surfaces of branch points which must exist when the imaginary components of **k** are large. (3) In the case of a group of bands which are degenerate at $\mathbf{k}_0$, the behavior of the energy in complex **k** space is considerably more complicated. It suffices here to state that $E(\mathbf{k})$ will not usually be analytic near $\mathbf{k}_0$; instead $\mathbf{k}_0$ will be a branch point. This follows from consideration of the determinantal equation resulting from (4.1.39).

Let us now summarize the general consequences of these considerations with reference to our discussion of group theory in the previous chapter. Our investigations show that if energy levels and wave functions have been determined at a point $\mathbf{k}_0$, the energies of states at $\mathbf{k}_0 + \mathbf{s}$, where $\mathbf{s}$ is small, may be determined by considering the effect of the perturbation $(\hbar/m)\mathbf{s}\cdot\mathbf{p}$. The solution of this perturbation problem establishes that the energy is a continuous and differentiable function of (real) **k** throughout the Brillouin zone. The same remarks evidently apply to the Bloch function so long as no arbitrary phase factors are allowed to enter.

If $\mathbf{k}_0$ is a general point of the zone, symmetry considerations do not add anything to the analysis. If $\mathbf{k}_0$ is a symmetry point and $\mathbf{k}_0 + \mathbf{s}$ is a general point, all the degeneracy which may be present at $\mathbf{k}_0$ is removed. If the group of $\mathbf{k}_0 + \mathbf{s}$ is a subgroup of the group of **k**, but still contains more than the identity, as occurs on going away from $\mathbf{k} = 0$ along a symmetry axis, the wave functions at $\mathbf{k}_0 + \mathbf{s}$ transform according to the subgroup. If the appropriate representation of the group of $\mathbf{k}_0$ is reducible as a representation of the subgroup at $\mathbf{k}_0 + \mathbf{s}$, the degeneracy at $\mathbf{k}_0$ will be removed, at least in part. Of course, if the groups at $\mathbf{k}_0$ and $\mathbf{k}_0 + \mathbf{s}$ are the same, the degeneracy will also be the same.

Information concerning the connections of bands and the splittings of degeneracies can be obtained by determining how the representations of the group at $\mathbf{k}_0$ are reducible (are expressed as direct sums) in terms of the representations at $\mathbf{k}_0 + \mathbf{s}$. This may be formally determined according to the procedures of Section 3.2 concerning the reduction of a reducible representation. These results are summarized in compatibility tables, two of which are presented in Appendix C.

The nature of this procedure may be appreciated from a simple argument which relates the representations at Γ to those along the 100 axis Δ in a cubic lattice. We examine Tables VI and VI, in Chapter 3, which list basis functions for the representations for Γ and Δ. From these tables, we see that a function which has Γ_1 symmetry will go into one with Δ_1 symmetry. The triply degenerate state Γ_{15} splits into Δ_1 and Δ_5, with functions of symmetry y and z going with Δ_5, while those of x symmetry go into Δ_1. Similarly the doubly degenerate Γ_{12} splits into Δ_1 and Δ_2, while the triply degenerate $\Gamma_{25'}$ splits into $\Delta_{2'}$ and Δ_5. The splitting may be discovered by comparing the basis functions for the representations. The subscript notation for the state at Γ gives the appropriate compatibility relations for this axis. Unfortunately, the compatibility results cannot always be stated so concisely when other axes are considered (see Appendix C). Note that in going from a point of high symmetry to an axis of lower symmetry, the classification of states becomes less detailed, simply because fewer representations exist, and states may acquire significant admixtures of other angular momentum components. (A Δ_1 state deriving from an s like Γ_1 will pick up p, d, etc., components, for example). The amount of this mixing depends in a detailed way on the crystal potential.

4.1.6 *The Electronic Density of States and Specific Heat*

A density of states function is defined for energy bands precisely as was done for lattice vibrations in Section 1.5. Specifically, let $G_n(E)$ be the fractional number of states (per unit cell) belonging to band n and lying in the range between E and $E + dE$. This function is defined according to Eq. (1.5.13),

$$G_n(E) = [\Omega/(2\pi)^3] \int d^3k\, \delta[E - E_n(\mathbf{k})], \tag{4.1.47}$$

and is normalized so that

$$\int G_n(E)\, dE = 1. \tag{4.1.48}$$

In the usual case in which one considers a band which has a double degeneracy throughout the Brillouin zone, as is the case whenever spin–orbit coupling or exchange interactions in ferromagnetic materials are neglected (see Section 3.7), the number of states in the band should be normalized to two per atom per band, and it is conventional to multiply (4.1.47) by a factor of 2. The total density of states at energy E, $G(E)$, is the sum of (4.1.47) over all bands

$$G(E) = \sum_n G_n(E). \tag{4.1.49}$$

The calculation of the electronic density of states is subject to all of the complications present in the lattice vibration problem (Section 1.5). See Gilat (1972), Cooke and Wood (1972), and Mueller *et al.* (1971) for a discussion of computational methods.

At $T = 0°\text{K}$, and neglecting the interaction of electrons other than as may be described by an average field, states are occupied up to the Fermi energy (or chemical potential) to be denoted by μ. This energy is determined by the requirement that

$$\int_{-\infty}^{\mu} G(E)\, dE = N \tag{4.1.50}$$

where N is the number of electrons (per unit cell) which must be accommodated. At a nonzero temperature T, the probability that a state of energy E is occupied is specified according to Fermi statistics by the function

$$f(E - \mu) = \{\exp[(E - \mu)/KT] + 1\}^{-1}. \tag{4.1.51}$$

The Fermi energy is a function of temperature and must now be determined by a generalization of (4.1.50):

$$\int_{-\infty}^{\infty} G(E)f(E - \mu)\, dE = N. \tag{4.1.52}$$

One important way in which experimental information is determined concerning the density of states is through measurement of the low-temperature specific heat. To this end, we wish to compute the total single-particle energy

$$U(T) = \int_{-\infty}^{\infty} EG(E)f(E - \mu)\, dE \tag{4.1.53}$$

and differentiate with respect to temperature.

It is necessary to do integrals with the Fermi function. We follow an operator procedure due to Blankenbecler (1957). Consider the integral

$$I = \int_{-\infty}^{\infty} \phi(E) f(E - \mu)\, dE \tag{4.1.54}$$

where $\phi(E)$ is a function which vanishes at both limits and possesses derivatives of all orders at $E = \mu$. Further, let $\Phi(E)$ be the integral of ϕ:

$$\Phi(E) = \int_{-\infty}^{E} \phi(E')\, dE'. \tag{4.1.55}$$

Integrate by parts:

$$I = [\Phi(E) f(E - \mu) |_{-\infty}^{\infty} - \int_{-\infty}^{\infty} \Phi(E)\, df(E - \mu)/dE\, dE.$$

Then $\Phi(E)$ must vanish at the lower limit and $f(E)$ vanishes at the upper limit. Introduce a new variable $z = E - \mu$. Then we have

$$I = -\int_{-\infty}^{\infty} \Phi(\mu + z)\, df(z)/dz\, dz.$$

The essential point is that df/dz is nearly a delta function so that Φ is only required for small z. Hence it is useful to expand Φ, using the identity

$$\Phi(\mu + z) = e^{z(\partial/\partial\mu)} \Phi(\mu).$$

Thus

$$I = \left[-\int_{-\infty}^{\infty} e^{z(\partial/\partial\mu)} df/dz\, dz \right] \Phi(\mu).$$

However,

$$df/dz = (-1/KT)[e^{z/KT}/(e^{z/KT} + 1)^2].$$

Finally, put $x = e^{-z/KT}$, $D = KT\, \partial/\partial\mu$. The integral becomes

$$\begin{aligned} I &= \int_{0}^{\infty} [x^{-D}/(x + 1)^2\, dx] \Phi(\mu) \\ &= \Gamma(1 - D)\Gamma(1 + D)\Phi(\mu)/\Gamma(2) \\ &= [\pi D \csc \pi D]\Phi(\mu). \end{aligned}$$

The trigonometric function is now expanded. It is necessary to retain only terms through second order

$$I = [1 + (\pi^2/6)(KT)^2\, d^2/d\mu^2 + \cdots]\Phi(\mu). \tag{4.1.56}$$

This procedure must first be applied to (4.1.52) with $\phi(E)$ set equal to the density of states in order to determine $\mu(T)$. Let μ_0 be the value of μ at $T = 0$. We expand the integral $G(E)$ in terms of $\mu - \mu_0$:

$$\int_0^{\mu} G(E)\, dE = N + (\mu - \mu_0)G(\mu_0) + \tfrac{1}{2}(\mu - \mu_0)^2\, (\partial G/\partial E)_{\mu_0} + \cdots.$$

We may now apply (4.1.52) and (4.1.56). The result is

$$\mu = \mu_0[1 - (\pi^2/6)(KT/\mu_0)^2(\mu_0/G(\mu_0))(\partial G/\partial E)_{\mu_0}]. \tag{4.1.57}$$

The Fermi energy decreases slightly with increasing temperature.

The same procedure can be applied to the integral (4.1.53) for the internal energy

$$U = U_0 + (\pi^2/6)(KT)^2 G(\mu_0) \tag{4.1.58}$$

in which U_0 is the energy at $T = 0$. The specific heat per atom at constant volume is

$$C_v = (\partial U/\partial T)_v = \gamma T \tag{4.1.59a}$$

where

$$\gamma = (\pi^2/3)K^2 G(\mu_0). \tag{4.1.59b}$$

The specific heat is a linear function of temperature at low temperatures. The coefficient of the temperature dependence is the density of states at the Fermi energy. Hence we can determine the density of states at μ_0 from measurements of the linear term in the specific heat at low temperatures.

Results for the density of states according to this method are frequently described in terms of a so-called "thermal" effective mass. Suppose, for simplicity, that the Fermi surface lies in a single band. Equation (4.1.47) can be rewritten, after introducing a factor of 2 for spin:

$$G(\mu) = \Omega/4\pi^3 \int dS_F/|\nabla_k E(\mathbf{k})| \tag{4.1.60}$$

where the integral runs over the Fermi surface. If the energy band were parabolic, characterized by an effective mass m^*, we would have

$$G(\mu) = \Omega m^* k_F/\pi^2\hbar^2 \qquad \text{or} \qquad m^* = \pi^2\hbar^2 G(\mu)/\Omega k_F \tag{4.1.61}$$

where k_F is the radius of the (spherical) Fermi surface. It is convenient to define a thermal effective mass m_t through (4.1.61) for arbitrary band structures, k_F still being given by the usual expression for free electrons

$$k_F = k_F^{(0)} = (3\pi^2 N)^{1/3}$$

where N is the number density of electrons. Then

$$m_t = 3\hbar^2\gamma/\Omega k_F K^2. \tag{4.1.62}$$

It is also possible to write the thermal mass as

$$m_t = \hbar k_F^{(0)}/S_F^{(0)} \int dS_F/|\,\mathbf{v}(\mathbf{k})\,| \tag{4.1.63}$$

where $S_F^{(0)} = 4\pi k_F^{(0)2}$ is the area of a spherical Fermi surface which would enclose the proper number of states and $v(\mathbf{k})$ is the group velocity on the real Fermi surface,

$$\mathbf{v}(\mathbf{k}) = (1/\hbar)\ \nabla_{\mathbf{k}} E(\mathbf{k}). \tag{4.1.64}$$

4.2 Plane Wave Expansions

In this section, we are concerned with the solution of an effective Schrödinger equation for the one-particle Bloch function by means of an expansion in plane waves. Other sets of expansion functions will be discussed subsequently. For convenience we will adopt atomic units: set $\hbar = 1$. The unit of length is the Bohr radius of hydrogen $a_0 = \hbar/me^2$ (5.2917×10^{-9} cm). The unit of energy is the rydberg $e^2/2a_0$ (13.6049 eV), which is the ionization energy of hydrogen. In order for our units of energy and length to have the numerical value 1, we must choose $e^2 = 2$ and $m = \frac{1}{2}$. The speed of light is determined from the dimensionless relation $\hbar c/e^2 = 137.037$ to have the numerical value $c = 274.074$. The reader should be aware that another set of atomic units in which $\hbar$, m, and e all have the numerical value 1 exist; in this case the unit of energy is the hartree, or double rydberg ($e^2/a_0 = 27.2098$ eV).

4.2.1 *General Principles*

In atomic units, the Schrödinger equation has the form

$$[-\nabla^2 + V(\mathbf{r})]\psi_n(\mathbf{k}, \mathbf{r}) = E_n(\mathbf{k})\psi_n(\mathbf{k}, \mathbf{r}) \tag{4.2.1}$$

in which we refer to a state of wave vector $\mathbf{k}$ in band n. If (4.2.1) is regarded as a form of the Hartree–Fock equation, $V(\mathbf{r})$ is actually nonlocal and furthermore depends on the state considered. We will ignore such complications at the moment, this being the usual practice in band calculations. Thus $V(\mathbf{r})$ is considered to be an effective local potential which is the same for all states. Although this restriction on V is undoubtedly an oversimplification, it permits us to focus attention on the problem of finding a solution of (4.2.1) in the solid.

The problem of solving (4.2.1) is somewhat complicated. In the first place, the effective potential V is required to have the symmetry of the crystal, and therefore need not have spherical symmetry around any atomic site. The site symmetry will be that of the symmetric representation of the crystal point group. Consequently, separation of variables in (4.2.1) using spherical coordinates is not possible, or in other terms, we may say that $V(\mathbf{r})$ contains components coupling functions of different angular momentum, and different values of the "z" component of angular momentum, as was discussed in Section 3.4. However, in metals with crystal structures of high symmetry, it is likely that the departures of the potential around an atomic site from spherical symmetry are relatively small, so that a spherically symmetric atomic potential is a legitimate first approximation. Second, the determination of the solution of a partial differential equation requires specification of boundary conditions. In the case of an infinite crystal the boundary conditions are specified by Bloch's theorem (Section 3.5) in the form

$$\psi_n(\mathbf{k}, \mathbf{r} + \mathbf{R}_i) = \exp(i\mathbf{k}\cdot\mathbf{R}_i)\psi_n(\mathbf{k}, \mathbf{r}) \tag{4.2.2}$$

where $\mathbf{R}_i$ is any direct lattice vector. In addition, $\psi_n(\mathbf{k}, \mathbf{r})$ must have the rotational and reflection symmetry implied by one of the irreducible representations of the group of the wave vector $\mathbf{k}$. The conditions on the solution of (4.2.1) implied by (4.2.2) are not generally consistent with the separation of variables in (4.2.1). That is, if we approximate $V(\mathbf{r})$ in (4.2.1) by a spherically symmetric function, we still have difficulties since surfaces of constant r, spheres, are not naturally adapted to the imposition of (4.2.2). [A reasonably good approximation may be obtained by replacing the atomic cell by a sphere in the *very special* case of $\mathbf{k} = 0$, a cubic lattice, and the Γ_1 representation.]

In view of the complications caused by (4.2.2), a natural approach to the solution of (4.2.1) is to expand ψ in a set of functions which satisfy (4.2.2) initially. The expansion coefficients are then chosen so that the Schrödinger equation becomes satisfied to an adequate approximation. It is actually not difficult to construct such functions, and the use of different sets of such functions characterizes the different methods of band calculation. We will begin by consideration of the simplest set of such functions: plane waves.

It was shown in Section 3.6 that functions $\exp[i(\mathbf{k} + \mathbf{K}_s)\cdot\mathbf{r}]$ in which $\mathbf{K}_s$ is any reciprocal lattice vector satisfy (4.2.2). It is convenient to use the expansion (3.6.2):

$$\psi_n(\mathbf{k}, \mathbf{r}) = \sum_s b_n(\mathbf{k}, \mathbf{K}_s)\{\exp[i(\mathbf{k} + \mathbf{K}_s)\cdot\mathbf{r}]/(N\Omega)^{1/2}\}. \tag{4.2.3}$$

Discrete normalization will be employed here [in contrast to the continuum normalization employed in (4.1.2)]. The normalization volume contains N

unit cells of volume Ω. The factor $(N\Omega)^{-1/2}$ ensures that individual plane waves are normalized to unity in this volume.

To determine an equation satisfied by the b_n, we substitute (4.2.3) into (4.2.1), multiply by $\exp[-i(\mathbf{k} + \mathbf{K}_t)\cdot\mathbf{r}]$, and integrate over the crystal. The results are

$$\sum_s \Big\{[(\mathbf{k} + \mathbf{K}_s)^2 - E_n(\mathbf{k})]\,\delta_{st} + (N\Omega)^{-1} \int \exp[i(\mathbf{K}_s - \mathbf{K}_t)\cdot\mathbf{r}]V(\mathbf{r})\, d^3r\Big\}\, b_n(\mathbf{k}, \mathbf{K}_s) = 0. \tag{4.2.4}$$

The potential $V(\mathbf{r})$ can be expressed as a sum of N identical terms centered on each unit cell in the crystal, which we denote as V_c,

$$V(\mathbf{r}) = \sum_\mu V_c(\mathbf{r} - \mathbf{R}_\mu). \tag{4.2.5}$$

Then we interchange the sum and the integral and use the fact that

$$(\mathbf{K}_s - \mathbf{K}_t)\cdot\mathbf{R}_\mu = 2\pi \times (\text{integer})$$

to shift the origin of integration to $\mathbf{R}_\mu$. The terms in (4.2.5) contribute an equal amount. Equation (4.2.4) becomes

$$\sum_s \{[(\mathbf{k} + \mathbf{K}_s)^2 - E_n(\mathbf{k})]\,\delta_{st} + V(\mathbf{K}_t - \mathbf{K}_s)\}\, b_n(\mathbf{k}, \mathbf{K}_s) = 0 \tag{4.2.6}$$

in which

$$V(\mathbf{K}_t - \mathbf{K}_s) = (1/\Omega) \int \exp[i(\mathbf{K}_t - \mathbf{K}_s)\cdot\mathbf{r}]V_c(\mathbf{r})\, d^3r. \tag{4.2.7}$$

The quantity $V(\mathbf{K})$ is a Fourier coefficient of the crystal potential.

There is one equation of the form (4.2.7) for each reciprocal lattice vector $\mathbf{K}_s$. The condition for this set of homogeneous linear equations to have a nontrivial solution is that the (infinite) determinant of the coefficients must vanish

$$\det |\, [(\mathbf{k} + \mathbf{K}_s)^2 - E_n(\mathbf{k})]\,\delta_{st} + V(\mathbf{K}_t - \mathbf{K}_s)\, | = 0. \tag{4.2.8}$$

The roots of this equation give the allowed energies. Once these have been determined, the coefficients $b_n(\mathbf{k}, \mathbf{K}_s)$ can be found except for a normalization factor.

This problem can be stated in equivalent way: we set up the matrix of the Hamiltonian on the basis of plane waves and diagonalize it. Since the Hamiltonian is Hermitian, the eigenvectors corresponding to different eigenvalues are orthogonal, and we have upon normalization the result expressed previously by Eq. (4.1.3).

4.2.2 *Symmetrized Combinations of Plane Waves*

If **k** is a symmetry point in the zone, it is possible to reduce the number of independent terms in the expansion through the use of symmetry conditions. At such a point, the wave function belongs to a specific irreducible representation of the group of **k**. If combinations of plane waves are constructed which transform according to particular irreducible representations (and particular rows in the case of a degenerate representation), the principles of group theory ensure that the crystal Hamiltonian will not have matrix elements between combinations of different symmetry (see Section 3.2).

This process is equivalent to the determination of a unitary transformation which partially diagonalizes the Hamiltonian matrix into blocks coming from specific irreducible representations. The construction is frequently referred to as "leading to a factorization of the secular equation."

The coefficients in the combination can be determined by applying the procedure of Eq. (3.2.9). We will consider explicitly here only the case of a symmorphic group. The extension to more general cases is easily accomplished with the aid of Eq. (3.6.3). It was shown in the argument leading to Eq. (3.2.9), that if F is an arbitrary function and $[\boldsymbol{\alpha}]_{i,jj}$ is the jjth element of the matrix representing the operation $\boldsymbol{\alpha}$ in the ith irreducible representation, the sum over all operations in the group $\sum_\alpha [\boldsymbol{\alpha}]^*_{i,jj}\boldsymbol{\alpha}F$ (if not zero) is a function transforming according to the jth row of the ith representation. In the present instance, the operators $\boldsymbol{\alpha}$ are those of the group of the wave vector, and it is usually convenient to choose F to be a single plane wave $\exp[i(\mathbf{k}+\mathbf{K}_s)\cdot\mathbf{r}]$. Let us define the combination

$$\Phi_s^{(i,j)}(\mathbf{k},\mathbf{r}) = N_s(N\Omega)^{-1/2}\sum_\alpha [\boldsymbol{\alpha}]^*_{i,jj}\exp[i(\mathbf{k}+\boldsymbol{\alpha}\mathbf{K}_s)\cdot\mathbf{r}] \tag{4.2.9}$$

where N_s is a factor chosen so that the combination is normalized to unity,

$$\int |\Phi_s^{(i,j)}(\mathbf{k},\mathbf{r})|^2\,d^3r = 1, \tag{4.2.10}$$

and we have used the fact that

$$\boldsymbol{\alpha}\exp(i\mathbf{k}\cdot\mathbf{r}) = \exp(i\mathbf{k}\cdot\boldsymbol{\alpha}^{-1}\mathbf{r}) = \exp(i\boldsymbol{\alpha}\mathbf{k}\cdot\mathbf{r}) = \exp(i\mathbf{k}\cdot\mathbf{r}). \tag{4.2.11}$$

The last step follows since $\boldsymbol{\alpha}$ is in the group of **k**. If **k** is on the surface of the Brillouin zone, we may have $\boldsymbol{\alpha}\mathbf{k} = \mathbf{k}+\mathbf{K}_u$, in which case a term $i\mathbf{K}_u\cdot\mathbf{r}$ is added to the exponent.

The function $\Phi_s^{(i,j)}(\mathbf{k}, \mathbf{r})$ belongs to the jth row of the ith irreducible representation of wave vector $\mathbf{k}$ and is therefore a suitable basis function for the expansion of a Bloch function with the same symmetry. Discussions of the construction of such functions have been given by Schlosser (1962) and Mariot (1962). Let $\psi_n(\mathbf{k}, \mathbf{r})$ be a Bloch function of the same symmetry as $\Phi_s^{(i,j)}$, so that it can be expanded in terms of such Φ:

$$\psi_n(\mathbf{k}, \mathbf{r}) = \sum_s c_{n,s}(\mathbf{k})\Phi_s^{(i,j)}(\mathbf{k}, \mathbf{r}). \tag{4.2.12}$$

The sum must include all distinct combinations $\Phi_s^{(i,j)}$. The number of such distinct combinations for fixed $|\mathbf{k} + \mathbf{K}_s|$ is equal to the number of times the representation i occurs in the reduction of the reducible representation of the group of $\mathbf{k}$ formed by plane waves of wave vector $\mathbf{k} + \mathbf{K}_s$. Only those plane waves are included whose wave vectors may be obtained by applying the operators of the group of $\mathbf{k}$ to any one of them. The number of the independent functions can be determined by application of Eq. (3.2.15). For general vectors $\mathbf{k} + \mathbf{K}_s$ such that the vectors $[\boldsymbol{\alpha}(\mathbf{k} + \mathbf{K}_s)]$ are all distinct, the representation formed will be the regular representation of the group of $\mathbf{k}$, and each irreducible representation will occur a number of times equal to its degeneracy.

We may repeat the procedure which led to (4.2.6) in order to obtain an equation satisfied by the coefficients C. A secular equation similar to (4.2.8) is obtained. This is

$$\det |[(\mathbf{k} + \mathbf{K}_s)^2 - E_n(\mathbf{k})]\,\delta_{st} + U_{s,t}| = 0 \tag{4.2.13}$$

where we have used the fact that $|\boldsymbol{\alpha}(\mathbf{k} + \mathbf{K}_s)|^2 = |(\mathbf{k} + \mathbf{K}_s)|^2$, and $U_{s,t}$ is a combination of Fourier coefficients of potential

$$U_{s,t} = N_s N_t \sum_{\alpha,\beta} [\boldsymbol{\beta}]_{i,jj}[\boldsymbol{\alpha}]^*_{i,jj} V(\boldsymbol{\beta}\mathbf{K}_t - \boldsymbol{\alpha}\mathbf{K}_s). \tag{4.2.14}$$

4.2.3 *The Point Charge Model*

Considerable insight into the nature of plane wave expansions is afforded by a simple model. We consider a system resembling that studied in Section 3.4, a lattice of point charges Ze. Here e is the magnitude of the electronic charge and Z is positive. There is only one kind of atom, and the lattice structure is chosen to be body-centered cubic. In order to preserve electrical neutrality and increase the resemblance between the model and a real metal, we will suppose that the system contains a uniform distribution of negative charge $-Ze$ per unit cell. The potential energy function for a single electron in this system $V(\mathbf{r})$ satisfies Poisson's equation, which is, in atomic units

$$\nabla^2 V(\mathbf{r}) = 8\pi\rho \tag{4.2.15}$$

where ρ is the charge density.

Fourier expansions are introduced for these functions:

$$V = \sum_s V(\mathbf{K}_s) \exp(i\mathbf{K}_s \cdot \mathbf{r}); \qquad \rho = \sum \rho(\mathbf{K}_s) \exp(i\mathbf{K}_s \cdot \mathbf{r}) \quad (4.2.16)$$

in which $V(\mathbf{K}_s)$ is given by (4.2.7). When (4.2.16) is substituted into (4.2.15), a relation is obtained between $V(\mathbf{K}_s)$ and $\rho(\mathbf{K}_s)$,

$$V(\mathbf{K}_s) = (-8\pi/\mathbf{K}_s^2)\rho(\mathbf{K}_s). \quad (4.2.17)$$

The case $\mathbf{K}_s = 0$ must be considered separately. In spite of appearances, a finite limit can be obtained.

The validity of Eq. (4.2.17) is not restricted to the model described above. A particular model enters only when the function $\rho(\mathbf{r})$ is specified. The charge density can be written as a sum of terms from the atomic cells:

$$\rho(\mathbf{r}) = \sum_\mu \rho_c(\mathbf{r} - \mathbf{R}_\mu). \quad (4.2.18)$$

The charge density within a single cell is assumed to be

$$\rho_c(\mathbf{r}) = Z[\delta(\mathbf{r}) - 1/\Omega]. \quad (4.2.19)$$

The Fourier coefficient of charge density $\rho(\mathbf{K}_s)$ is given by

$$\rho(\mathbf{K}_s) = (1/\Omega) \int \exp(i\mathbf{K}_s \cdot \mathbf{r}) \rho_c(\mathbf{r}) \, d^3r \quad (4.8.20)$$

in which the integral includes the contribution from the charge in a single cell only. Hence $\rho(\mathbf{K}_s) = Z/\Omega$, and

$$V(\mathbf{K}_s) = -8\pi Z/\mathbf{K}_s^2\Omega. \quad (4.2.21)$$

For the body-centered cubic lattice, $\Omega = a^3/2$ where a is the lattice constant, and $\mathbf{K}_s^2 = (4\pi^2/a^2)\mathbf{S}^2$ with $\mathbf{S}^2 = s_1^2 + s_2^2 + s_3^2$. Then

$$V(\mathbf{K}_s) = -4Z/\pi a \mathbf{S}^2. \quad (4.2.22)$$

We define $V(0)$ as the limit of $V(\mathbf{K}_s)$, in which $\mathbf{K}_s$ is treated as a continuous variable, and allowed to approach zero:

$$V(0) = -8\pi \lim_{\mathbf{K}\to 0} \rho(\mathbf{K})/\mathbf{K}^2. \quad (4.2.23)$$

To evaluate (4.2.23), (4.2.20) is expanded:

$$\rho(\mathbf{K}) = (1/\Omega)\left[\int \rho_c(\mathbf{r}) \, d^3r + i \int \rho_c(\mathbf{r})(\mathbf{K}\cdot\mathbf{r}) \, d^3r \right.$$

$$\left. - (\mathbf{K}^2/2) \int r^2 \rho_c(\mathbf{r}) \cos^2\theta \, d^3r \right]$$

(where θ is the angle between $\mathbf{K}$ and $\mathbf{r}$). Since the charge distribution is neutral, the first term in this expansion vanishes, and since it has inversion symmetry, the second term also vanishes. In the third term, put $\cos^2\theta = \frac{1}{3} + \frac{2}{3}P_2(\cos\theta)$ (P_2 is the second Legendre polynomial). Then P_2 is a basis function for the Γ_{12} representation in a cubic crystal; hence if the charge density has cubic symmetry Γ_1, the integral of this function must vanish. Thus we have the general result

$$V(0) = (4\pi/3\Omega) \int \rho_c(\mathbf{r}) r^2 \, d^3r. \tag{4.2.24}$$

For the specific model under consideration, the integral becomes

$$V(0) = (-4\pi Z/3\Omega^2) \int r^2 \, d^3r. \tag{4.2.25}$$

The integral can be worked out exactly for cubic lattices. In the case of the body-centered cubic lattice

$$\Omega^{-1} \int r^2 \, d^3r = [(221)^{1/2}/100] a^2$$

and so

$$V(0) = -1.2454Z/a. \tag{4.2.26}$$

For arbitrary states in the zone and arbitrary values of Z and a, the eigenvalues of the Hamiltonian with this V must be determined numerically. However, considerable insight can be gained if we look only at states at $\mathbf{k} = 0$ and employ ordinary perturbation theory on the supposition that the matrix elements of potential are small compared to the kinetic energies. Actually, the kinetic energy terms will be of order K_s^2, and this is proportional to $1/a^2$. The matrix elements of potential are proportional to Z/a. The second-order term will be of order $(V)^2/K^2$, and thus will be proportional to Z^2. Each successive term in the expansion introduces a matrix element of order Z/a and another energy denominator of order $1/a^2$, and thus gives an additional factor of Za. The nth-order term, counting the kinetic energy as of order zero, is proportional to $(1/a^2)(Za)^n$. Evidently the quantity aE/Z is a function of the single variable Za,

$$aE/Z = f(Za). \tag{4.2.27}$$

Perturbation theory yields a power series expansion of the function $f(Za)$ in which the lowest term is of order $1/Za$. The function f is different for each state.

As a specific example, consider the state $\Gamma_{25'}$. The terms of the perturbation series are computed using symmetrized linear combinations of plane waves for this representation. The leading plane waves are those of type $(2\pi/a)(1, 1, 0)$. Evaluation of the coefficients through second order gives

$$E = (8\pi^2/a^2) - 0.7679Z/a - 0.00509Z^2 + \cdots. \qquad (4.2.28)$$

The rapid decrease of the coefficients with increasing order should be noted. A numerical calculation of the energy of this state by straightforward diagonalization of the Hamiltonian matrix shows that the first three terms of the series given in (4.2.28) are a good approximation to the energy up to about $Za \approx 25$. There are two reasons for the apparently good convergence of this series. First, the minimum energy of the denominator in perturbation theory is $8\pi^2/a^2$. Second, there is a considerable amount of cancellation in the calculation of matrix elements. The results may be quite different for a state of different symmetry, such as the s-like Γ_1. In this case, all the signs are positive in the calculation of the second-order perturbation, so that this term is much larger, and the convergence of the series is restricted to a smaller range of Za.

4.3 Orthogonalized Plane Waves

The plane wave expansion method does not converge rapidly enough to be of much practical use with realistic crystal potentials. Plane waves of large kinetic energy may make a substantial contribution to the wave function. In addition the plane wave expansion for a state of wave vector $\mathbf{k}$ will converge to the state of lowest energy for that wave vector, and this will frequently be an uninteresting state. For example, consider sodium. The states which are important to the electronic properties of the solid are those related to the 3s and 3p free atom wave functions. A plane wave expansion for a general $\mathbf{k}$ will converge to a 1s state. This problem can be avoided, in principle, by considering higher eigenvalues of the Hamiltonian matrix, but to obtain any degree of accuracy, it would be necessary to employ such large matrices that extensive calculations would be impractical even with the aid of modern, high speed computers.

4.3.1 *The Orthogonalization Procedure*

A way out of this difficulty was proposed by Herring (1940). In order to present this method in a relatively simple way, we consider a crystal with only one kind of atom. Suppose that the wave functions for the core states (1s, 2s, and 2p in the case of sodium) are known (perhaps from a self-consistent field calculation for the free atom, the assumption being made that these are unaltered in the solid) or that they can be calculated by other means. Let the wave function for a state of the core be designated $u_j(\mathbf{r})$. In a description appropriate to free atoms, j stands for the three usual quantum numbers n, l, and m, with n the principal quantum number, l the angular momentum quantum number, and m the z component of angular momentum. These functions are assumed to be orthonormal. It is possible to combine these "free atom" functions to form combinations possessing the periodicity required by Bloch's theorem. Let $\mathbf{R}_\mu$ be a direct lattice vector; then the required functions are

$$\phi_j(\mathbf{k}, \mathbf{r}) = N^{-1/2} \sum_\mu \exp(i\mathbf{k}\cdot\mathbf{R}_\mu) u_j(\mathbf{r} - \mathbf{R}_\mu). \tag{4.3.1}$$

Observe that

$$\begin{aligned}\phi_j(\mathbf{k}, \mathbf{r} + \mathbf{R}_\nu) &= N^{-1/2} \sum_\mu \exp(i\mathbf{k}\cdot\mathbf{R}_\mu) u_j(\mathbf{r} + \mathbf{R}_\nu - \mathbf{R}_\mu) \\ &= N^{-1/2} \exp(i\mathbf{k}\cdot\mathbf{R}_\nu) \sum_{\mu-\nu} \exp[i\mathbf{k}\cdot(\mathbf{R}_\mu - \mathbf{R}_\nu)] u_j[\mathbf{r} - (\mathbf{R}_\mu - \mathbf{R}_\nu)] \\ &= \exp(i\mathbf{k}\cdot\mathbf{R}_\nu)\phi_j(\mathbf{k}, \mathbf{r}). \end{aligned} \tag{4.3.2}$$

The last step follows since the sum over $\mathbf{R}_\mu$ may be replaced by one over $(\mathbf{R}_\mu - \mathbf{R}_\nu)$, which is identical to that in (4.3.1). As in Section 4.2, N represents the number of atoms in the crystal, here assumed to be large but finite, and the factor $N^{-1/2}$ ensures that the ϕ are normalized when the atoms are far enough apart so that functions u_j on different atoms do not overlap appreciably. These combinations are frequently called *tight binding functions*, and will be discussed more extensively in Section 4.5. The functions $\phi_j(\mathbf{k}, \mathbf{r})$ are solutions of the Schrödinger equation only if overlap can be ignored; however, this important condition will generally be satisfied by core wave functions. In fact, what we mean by a core state is one which does not overlap appreciably similar states on other atoms.

A plane wave of wave vector $\mathbf{k}$ can be made orthogonal by the Schmidt process to tight binding core wave functions of an equivalent $\mathbf{k}$. Orthogonality to functions of inequivalent $\mathbf{k}$ is guaranteed by general principles, since such functions transform according to inequivalent irreducible representations of the crystal space group. Let us denote a function constructed in this way by $\chi(\mathbf{k}, \mathbf{r})$; we call it an *orthogonalized plane wave*:

$$\chi(\mathbf{k}, \mathbf{r}) = (N\Omega)^{-1/2} \exp(i\mathbf{k}\cdot\mathbf{r}) - \sum_j \mu_j(\mathbf{k})\phi_j(\mathbf{k}, \mathbf{r}). \tag{4.3.3}$$

The sum includes all core states. It is required that

$$\int \phi_j^*(\mathbf{k}, \mathbf{r})\chi(\mathbf{k}, \mathbf{r})\, d^3r = 0. \tag{4.3.4}$$

This condition determines the "orthogonality coefficient" $\mu_j(\mathbf{k})$ as

$$\int \phi_j^* (\mathbf{k}, \mathbf{r})\chi(\mathbf{k}, \mathbf{r})\, d^3r = N^{-1}\Omega^{-1/2} \sum_\nu \int \exp[i\mathbf{k}\cdot(\mathbf{r} - \mathbf{R}_\nu)]u_j^*(\mathbf{r} - \mathbf{R}_\nu)\, d^3r$$
$$- N^{-1} \sum_l \mu_l(\mathbf{k}) \sum_{\nu\rho} \exp[i\mathbf{k}\cdot(\mathbf{R}_\rho - \mathbf{R}_\nu)]$$
$$\times \int u_j^*(\mathbf{r} - \mathbf{R}_\nu)u_l(\mathbf{r} - \mathbf{R}_\rho)\, d^3r. \tag{4.3.5}$$

This expression may be greatly simplified if use is made of the assumed orthonormality (and lack of overlap) of the atomic functions u:

$$\int u_j^*(\mathbf{r} - \mathbf{R}_\nu)u_l(\mathbf{r} - \mathbf{R}_\rho)\, d^3r = \delta_{jl}\,\delta_{\nu\rho}. \tag{4.3.6}$$

Also, the integral in the first term on the right of (4.3.5) is the same for each cell. Hence (4.3.3) yields

$$\mu_j(\mathbf{k}) = \Omega^{-1/2} \int_\infty \exp(i\mathbf{k}\cdot\mathbf{r})u_j^*(\mathbf{r})\, d^3r. \tag{4.3.7}$$

An orthogonalized plane wave (OPW) is a function which behaves as a plane wave at large distances from an atom, but possesses the rapidly varying character of an atomic wave function near any nucleus. Since both of these features characterize electron wave functions in solids, orthogonalized plane waves are useful functions for the construction of wave function expansions. The properties of the OPW method have been reviewed by Woodruff (1957).

As an example, let us consider a situation in which there is only a single type of core function, which we assume to be a 1s function (as would be the case in lithium), and which is approximated by a hydrogenic function

$$u_{1s}(\mathbf{r}) = (\alpha^3/\pi)^{1/2}e^{-\alpha r}. \tag{4.3.8}$$

Then

$$\mu_{1s}(\mathbf{k}) = \Omega^{-1/2}(\alpha^3/\pi)^{1/2} \int \exp(i\mathbf{k}\cdot\mathbf{r} - \alpha r)\, d^3r$$
$$= 8(\alpha\pi/\Omega)^{1/2}[\alpha^2/(\alpha^2 + \mathbf{k}^2)^2]. \tag{4.3.9}$$

Then, in the cell centered at $\mathbf{R}_\nu = 0$,

$$\chi(\mathbf{k}, \mathbf{r}) = (N\Omega)^{-1/2}[\exp(i\mathbf{k}\cdot\mathbf{r}) - [8\alpha^4/(\alpha^2 + \mathbf{k}^2)^2]e^{-\alpha r}]. \tag{4.3.10}$$

The effect of orthogonalization is apparent most simply if we consider the special case of $\mathbf{k} = 0$, for which we have

$$\chi(0, \mathbf{r}) = (N\Omega)^{-1/2}(1 - 8e^{-\alpha r}). \tag{4.3.11}$$

This function possesses the single radial node characteristic of 2s functions (here the node occurs at $\alpha r = 2.07$, while in the 2s hydrogenic function similar to (4.3.8), the node occurs for $\alpha r = 2$), but is flat for large values of αr where a hydrogenic function would decay exponentially. This behavior is characteristic of s-like (Γ_1) states in metals at the center of the Brillouin zone.

Symmetrized linear combinations of orthogonalized plane waves which transform according to a particular irreducible representation can be formed with exactly the same coefficients as in the case of ordinary plane waves previously discussed. The Bloch function which is to be determined may then be expanded in a series of symmetrized linear combinations of orthogonalized plane waves. Some degree of complication arises because the orthogonalized plane waves are not mutually orthogonal. It is necessary to diagonalize simultaneously the Hamiltonian matrix and the overlap matrix on the basis of these functions.

4.3.2 *OPW Matrix Elements*

Now we consider the determination of the basic matrix elements. To simplify the appearance of the equations, we use Dirac notation with $|\chi_{\mathbf{k}}\rangle$ representing an orthogonalized plane wave and $|\mathbf{k}\rangle$ an ordinary plane wave, etc. or more precisely,

$$\langle \mathbf{r} | \chi_{\mathbf{k}} \rangle = \chi(\mathbf{k}, \mathbf{r}), \quad \langle \mathbf{r} | \mathbf{k} \rangle = (N\Omega)^{-1/2} \exp(i\mathbf{k}\cdot\mathbf{r}), \quad \langle \mathbf{r} | \phi_{j\mathbf{k}} \rangle = \phi_j(\mathbf{k}, \mathbf{r}). \tag{4.3.12}$$

Then

$$\langle \chi_{\mathbf{k}'} | H | \chi_{\mathbf{k}} \rangle = \langle \mathbf{k}' | H | \mathbf{k} \rangle - \sum_j [\mu_j^*(\mathbf{k}') \langle \phi_{j\mathbf{k}'} | H | \mathbf{k} \rangle + \mu_j(\mathbf{k}) \langle \mathbf{k}' | H | \phi_{j\mathbf{k}} \rangle]$$
$$+ \sum_{jj'} \mu_{j'}^*(\mathbf{k}') \mu_j(\mathbf{k}) \langle \phi_{j'\mathbf{k}'} | H | \phi_{j\mathbf{k}} \rangle. \tag{4.3.13}$$

The $\phi_j(\mathbf{k}, \mathbf{r})$ are assumed to be eigenfunctions of the Hamiltonian H which is used for the energy band calculation:

$$H\phi_j(\mathbf{k}, \mathbf{r}) = E_j\phi_j(\mathbf{k}, \mathbf{r}). \tag{4.3.14}$$

This supposition makes possible a major simplification of the matrix element

$$\begin{aligned}\langle \chi_{\mathbf{k}'} \mid H \mid \chi_{\mathbf{k}} \rangle &= \langle \mathbf{k}' \mid H \mid \mathbf{k} \rangle - \sum \mu_j^*(\mathbf{k}')\mu_j(\mathbf{k})E_j \\ &= \mathbf{k}^2 \delta_{\mathbf{k}\mathbf{k}'} + V(\mathbf{k} - \mathbf{k}') - \sum_j \mu_j^*(\mathbf{k}')\mu_j(\mathbf{k})E_j \qquad (4.3.15)\end{aligned}$$

in which $V(\mathbf{k} - \mathbf{k}')$ is given by (4.2.7).

The overlap matrix elements are given by

$$\langle \chi_{\mathbf{k}'} \mid \chi_{\mathbf{k}} \rangle = \delta_{\mathbf{k}\mathbf{k}'} - \sum_j \mu_j^*(\mathbf{k}')\mu_j(\mathbf{k}). \qquad (4.3.16)$$

The Fourier coefficients of potential $V(\mathbf{k} - \mathbf{k}')$ which appear in (4.3.15) will generally be negative in sign since they pertain to an attractive potential. Likewise, the energies of the core states will be negative. Experience has shown that the product $\mu_j^*(\mathbf{k}')\mu_j(\mathbf{k})$ is effectively positive, at least when $\mathbf{k}$ and $\mathbf{k}'$ are small. The result is that the orthogonality term tends to cancel the potential term, so that the combination of the two is much weaker than either term individually. This makes it possible to obtain a reasonable estimate of the energy of a state on the basis of only a few orthogonalized plane waves in circumstances where many ordinary plane waves would be required.

The orthogonality coefficients are actually rather complicated expressions. Suppose that wave functions for states of the same n, l, but different m have the same radial part,

$$u_j(\mathbf{r}) = R_{nl}(r)Y_{lm}(\theta, \phi) \qquad (4.3.17)$$

where Y_{lm} is a normalized spherical harmonic and R is normalized so that

$$\int | R_n |^2 r^2 \, dr = 1. \qquad (4.3.18)$$

Then the sum on m implied in (4.3.15) and (4.3.16) can be carried out with the aid of the additional theorem for spherical harmonics:

$$\sum_j \mu_j^*(\mathbf{k}')\mu_j(\mathbf{k}) = (4\pi/\Omega) \sum_{nl} (2l + 1)P_l(\cos\theta_{\mathbf{k}\mathbf{k}'})I_{k'nl}I_{knl} \qquad (4.3.19)$$

where P_l is a Legendre polynomial, $\theta_{\mathbf{k}\mathbf{k}'}$ is the angle between $\mathbf{k}$ and $\mathbf{k}'$,

$$I_{knl} = \int_0^\infty j_l(kr)R_{nl}(r)r^2 \, dr, \qquad (4.3.20)$$

and j_l is a spherical Bessel function.

The OPW method has been extensively applied to Group IV and II–V semiconductors. As an example, portions of the calculated band structure of silicon are shown in Fig. 4.3.1 according to Stuckel and Euwema (1970).

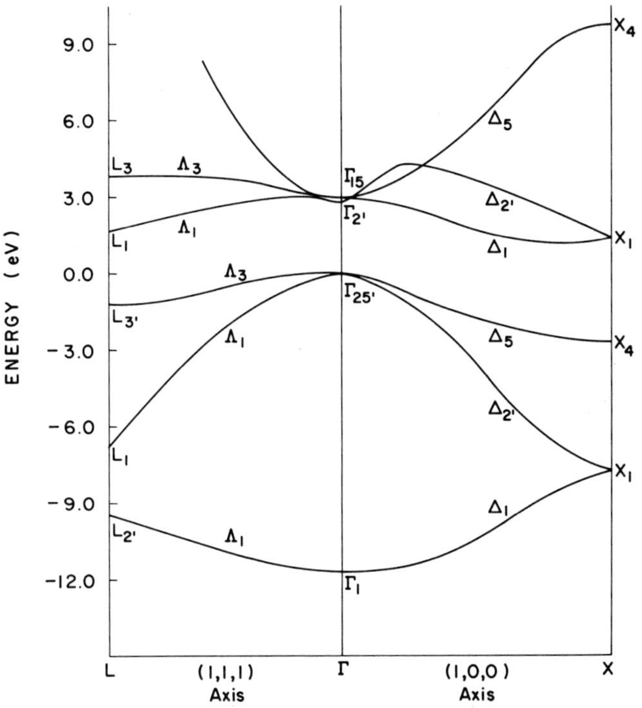

Fig. 4.3.1. Energy bands in silicon according to the OPW calculation of Stuckel and Euwema (1970).

4.3.3 *Limitations of the OPW Procedure. Extensions and Modifications*

There are, however, difficulties in connection with (4.3.14), particularly if it is thought to obtain the u_j and hence the ϕ_j from an atomic self-consistent field calculation. The problem is that the Hamiltonian assumed for the solid will usually not be the same as that used in the atomic problem because the charge densities of the valence electrons associated with atoms in different cells overlap. As a result the potential field in which a core electron moves is different; and this problem becomes particularly acute if exchange is represented by a local potential in the solid (see Section 4.10), while Hartree–Fock functions pertaining to the free atom are used for the core states.

In order to take this effect into account, we may write instead of (4.3.14)

$$H\phi_j(\mathbf{k}, r) = [H_0 + U(\mathbf{r})]\phi_j(\mathbf{k}, \mathbf{r}) = [E_j(\mathbf{k}) + U(\mathbf{r})]\phi_j(\mathbf{k}, \mathbf{r}) \quad (4.3.21a)$$

in which H_0 is the Hamiltonian of which the ϕ_j are eigenstates and U is the difference between the crystal and atomic Hamiltonians. Then we have, instead of (4.3.15),

$$\langle \chi_{\mathbf{k}'} \mid H \mid \chi_{\mathbf{k}} \rangle = k^2 \,\delta_{\mathbf{k}\mathbf{k}'} + V(\mathbf{k} - \mathbf{k}') - \sum_j \mu_j{}^*(\mathbf{k}')\mu_j(\mathbf{k})E_j$$
$$- \sum_j [\mu_j{}^*(\mathbf{k}')\langle \phi_{j\mathbf{k}'} \mid U \mid \mathbf{k}\rangle + \mu_j(\mathbf{k})\langle \mathbf{k}' \mid U \mid \phi_{j\mathbf{k}}\rangle]$$
$$+ \sum_{j'} \mu_{j'}{}^*(\mathbf{k}')\mu_j(\mathbf{k})\langle \phi_{j\mathbf{k}'} \mid U \mid \phi_{j\mathbf{k}}\rangle + U(\mathbf{k} - \mathbf{k}'). \qquad (4.3.21b)$$

As a practical matter, the terms involving U may be numerically quite significant; and their effect within the core cannot usually be represented in the simple manner one might expect from a glance at (4.3.15), that of correcting the energy E_j in perturbation theory for the effect of U. An exception occurs if U is constant over the core orbitals, in which case it suffices to increase E_j by this constant U. The same zero of energy must be used for both the crystal potential and the core states. If this is done, addition of a constant potential cannot effect the relative position of levels.

Even if the more elaborate calculation based on (4.3.21) rather than (4.3.15) is carried out, the energies resulting from solutions of the secular equation

$$\det[\langle \chi_{\mathbf{k}'} \mid H \mid \chi_{\mathbf{k}} \rangle - E(\mathbf{k})\langle \chi_{\mathbf{k}'} \mid \chi_{\mathbf{k}} \rangle] = 0 \qquad (4.3.22)$$

will not necessarily be correct. Since the states $\mid \phi_{j\mathbf{k}} \rangle$ were not eigenstates of H, one has orthogonalized to the wrong states. The error in the energy from this cause can be estimated from an analysis given by Herring (1940). Let the exact eigenstates of the full Hamiltonian be denoted $\mid \psi_{s\mathbf{k}} \rangle$ with energies $E_s(\mathbf{k})$. The subscript j will continue to be used to designate a core state. A particular approximate valence state in the lowest band above the core will be denoted $\mid \psi'_{0\mathbf{k}} \rangle$; this is obtained from the solution of the eigenvalue problem (4.3.19). The state $\mid \psi'_{0\mathbf{k}} \rangle$ may be expanded in the $\mid \psi_{s\mathbf{k}} \rangle$:

$$\mid \psi'_{0\mathbf{k}} \rangle = \sum_s \langle \psi_{s\mathbf{k}} \mid \psi'_{0\mathbf{k}} \rangle \mid \psi_{s\mathbf{k}} \rangle. \qquad (4.3.23)$$

The energy of the approximate function $E_0'(\mathbf{k})$ is the expectation value of the Hamiltonian in the state $\mid \psi'_{0\mathbf{k}} \rangle$. Thus

$$E_0'(\mathbf{k}) = \langle \psi'_{0\mathbf{k}} \mid H \mid \psi'_{0\mathbf{k}} \rangle = \sum_s \mid \langle \psi_{s\mathbf{k}} \mid \psi'_{0\mathbf{k}} \rangle \mid^2 E_s(\mathbf{k}). \qquad (4.3.24)$$

The difference between the exact energy $E_0(\mathbf{k})$ and the approximate value $E_0'(\mathbf{k})$ is

$$E_0(\mathbf{k}) - E_0'(\mathbf{k}) = \sum_s \mid \langle \psi_{s\mathbf{k}} \mid \psi'_{0\mathbf{k}} \rangle \mid^2 [E_0(\mathbf{k}) - E_s(\mathbf{k})]. \qquad (4.3.25)$$

Only the core states have energies less than $E_0(\mathbf{k})$ so the error will be overestimated algebraically by including just the exact core states $|\psi_{j\mathbf{k}}\rangle$ in (4.3.22). Thus we have a bound

$$E_0(\mathbf{k}) - E_0'(\mathbf{k}) \leq \sum_j |\langle\psi_{j\mathbf{k}}|\psi'_{0\mathbf{k}}\rangle|^2[E_0(\mathbf{k}) - E_j(\mathbf{k})]. \quad (4.3.26)$$

The approximate function $|\psi'_{0\mathbf{k}}\rangle$ will be orthogonal to the approximate core states, which we denote $|\phi_{j\mathbf{k}}\rangle$. We may use a form of Bessel's inequality

$$|\langle\psi_{j\mathbf{k}}|\psi_{j\mathbf{k}}\rangle|^2 = 1 \leq |\langle\psi_{jk}|\psi'_{0\mathbf{k}}\rangle|^2 + |\langle\psi_{j\mathbf{k}}|\phi_{j\mathbf{k}}\rangle|^2. \quad (4.3.27)$$

Thus we also have

$$E_0(\mathbf{k}) - E_0'(\mathbf{k}) \leq \sum_j [1 - |\langle\psi_{j\mathbf{k}}|\phi_{j\mathbf{k}}\rangle|^2][E_0(\mathbf{k}) - E_j(\mathbf{k})]. \quad (4.3.28)$$

The approximation to the energy obtained by orthogonalization to incorrect states is too low, that is, the approximate wave function for the valence state incorporates some components from the true core states, and so lies below the energy it would have if it were properly orthogonal to them. Since the scalar product $\langle\psi_{j\mathbf{k}}|\phi_{j\mathbf{k}}\rangle$ is analogous to the cosine of the angle between vectors, the error produced by orthogonalization to the wrong functions is a second-order effect. Unfortunately, in practical cases the energy differences which are important in (4.3.28) may be quite large (say 10–50 Ry) so that it does not take large differences between the $|\psi_{j\mathbf{k}}\rangle$ and the $|\phi_{j\mathbf{k}}\rangle$ to produce significant effects.

This problem limits the utility of the OPW method. There are two possible solutions:

(1) Find the actual core states for the potential employed, and use these functions in (4.3.15). Such core states may be mathematical artifacts, which have no physical significance.

(2) Modify the OPW method itself by adding the core functions to the basis set. In this case, there is no necessity to orthogonalize the plane waves to the core states, since this will be in effect taken care of by the matrix diagonalization process. We will discuss this possibility subsequently. In this case, use of approximate core functions does not introduce large errors, as the core functions can correct themselves.

Symmetrized linear combinations of orthogonalized plane waves, which are constructed to transform according to particular irreducible representations, will be automatically orthogonal to functions belonging to other representations. For example, in a cubic lattice, a combination belonging to Γ_{15} will be orthogonal to all s-like core functions, which give Γ_1; likewise $\Gamma_{15'}$ and Γ_{12} combinations appropriate to d bands will be orthogonal by reason of symmetry to s and p core states. It is to be emphasized that this orthogonality occurs without the necessity for the introduction of any orthogonality coefficients. In these circumstances, the expansion may reduce to one involving only ordinary plane waves.

Such an expansion may not converge rapidly enough to be useful. In this case, as in case (2), what is required is a modification of the basis set by the addition of functions which are sufficiently rapidly varying near a nucleus (Herring, 1940; Callaway, 1955; Brown and Krumhansl, 1958; Gray and Brown, 1967; Deegan and Twose, 1967). In both cases, we form linear combinations of a set of localized functions u_j according to (4.3.1); however we are not concerned as to whether u_j is the solution to some atomic problem. It may be an arbitrary function chosen to simulate the behavior expected of a 3d function near the nucleus. It remains convenient to choose the u_j so that they do not overlap from one atomic site to another. We will assume this has been done, although it is not absolutely necessary. The expansion for a state of wave vector $\mathbf{k}$ is now

$$\psi_n(\mathbf{k}, \mathbf{r}) = \sum_j a_{nj}(\mathbf{k})\phi_j(\mathbf{k}, \mathbf{r}) + \sum_s b_n(\mathbf{k}, \mathbf{K}_s)\{\exp[i(\mathbf{k} + \mathbf{K}_s)\cdot\mathbf{r}]/(N\Omega)^{1/2}\}. \quad (4.3.29)$$

There is a formal problem associated with such an expansion: Plane waves are themselves a complete set of states, and if a plane wave expansion is summed to convergence, there is nothing to add. Therefore addition of the functions ϕ is redundant. One could avoid this problem by orthogonalizing the plane waves to the functions ϕ_j as in the standard OPW procedure; however, as a practical matter, redundancy in (4.3.29) is not a serious problem, since only a relatively small number of plane waves are to be included.

The Hamiltonian and overlap matrices are easily constructed on the basis of the functions appearing in (4.3.26). These matrices are conveniently expressed in a block form

$$\begin{pmatrix} \text{Loc–Loc} & \text{Loc–PW} \\ \text{PW–Loc} & \text{PW–PW} \end{pmatrix} \quad (4.3.30)$$

where Loc stands for localized function and PW stands for plane wave. The formal expressions for the matrix elements of the Hamiltonian are

$$\begin{aligned} &\text{Loc–Loc:} \quad \langle \phi_{j\mathbf{k}} \mid H \mid \phi_{j'\mathbf{k}'} \rangle = \langle u_j \mid H \mid u_{j'} \rangle \\ &\text{Loc–PW:} \quad \langle \phi_{j\mathbf{k}} \mid H \mid \mathbf{k}' \rangle = \mathbf{k}'^2 \mu_j(\mathbf{k}') + \langle u_j \mid V \mid \mathbf{k}' \rangle \quad (4.3.31) \\ &\text{PW–PW:} \quad \langle \mathbf{k} \mid H \mid \mathbf{k}' \rangle = k^2\,\delta_{\mathbf{k}\mathbf{k}'} + V(\mathbf{k} - \mathbf{k}'). \end{aligned}$$

In these expressions

$$\langle u_j \mid H \mid u_{j'} \rangle = \int u_j^*(\mathbf{r}) H u_{j'}(\mathbf{r}) \, d^3r$$

$$\langle u_j \mid V \mid \mathbf{k} \rangle = \Omega^{-1/2} \int u_j^*(\mathbf{r}) V(\mathbf{r}) \exp(i\mathbf{k}\cdot\mathbf{r}) \, d^3r. \qquad (4.3.32)$$

The quantities $u_j(\mathbf{k}')$ and $V(\mathbf{k} - \mathbf{k}')$ are given by (4.3.7) and (4.2.8), respectively. It is necessary that $\mathbf{k}$ and $\mathbf{k}'$ should be equivalent, that is, $\mathbf{k}$ and $\mathbf{k}$ can differ only by a reciprocal lattice vector in order to obtain a nonzero value for any of these matrix elements. The assumption of no overlap of localized functions on different sites has been used. We also note that the overlap matrix has the form (under these assumptions)

$$\begin{pmatrix} I & \text{Loc–PW} \\ \text{PW–Loc} & I \end{pmatrix} \qquad (4.3.33)$$

where I represents a unit matrix and

$$\text{Loc–PW:} \quad \langle \phi_{j\mathbf{k}} \mid \mathbf{k}' \rangle = \mu_j(\mathbf{k}'). \qquad (4.3.34)$$

Convergence difficulties also appear to occur in the OPW method in diatomic crystals when one atom is much bigger than the other, and especially if the anion is much larger than the cation (example ZnO) (Euwema and Stuckel, 1970). Likewise, difficulties will occur if the valence electron charge density is quite compact (in comparison with the lattice spacing) as in a crystal of a solidified rare gas. The essential point is that if $k_{\max}$ is the maximum wave vector contained in the expansion, a charge density smaller in extent than $d_{\min} = 2\pi/k_{\max}$ cannot be adequately described by the expansion. The solution would appear to be the addition of some suitably localized function or functions to the expansion, as described above.

In considering energy bands in materials containing atoms of large atomic number, it may be desirable to base the calculation on the Dirac equation (for a single particle in a potential) rather than the Schrödinger equation. There is no essential difficulty connected with this process (Soven, 1965). The ordinary plane waves of the usual formulation must be replaced by four-component plane wave spinors, and the core functions must likewise be considered to be spinors, obtained as solutions of the Dirac equation for the appropriate potential.

4.4 Pseudopotential Methods

It was observed in the previous section that the orthogonalization terms which arise in the OPW method tend to cancel the potential terms and thus give rise to an effective potential which is much weaker than the true crystal potential. It is possible to develop an approach to band structure calculations which is based on this cancellation; this is the pseudopotential method (Phillips and Kleinman, 1959; Harrison, 1966). We will examine this procedure, following the original method of Phillips and Kleinman. Recent reviews of different aspects of the pseudopotential have been presented by Heine (1970), Cohen and Heine (1970), and Heine and Weaire (1970).

4.4.1 *Derivation of the Pseudopotential*

Let $\psi_n(\mathbf{k}, \mathbf{r})$ be the wave function for a state of wave vector $\mathbf{k}$ in band n. In a broad-band system, this function will tend to behave as a plane wave at large distances from an atomic nucleus; but will vary rapidly near a nucleus. As in Section 4.3 we represent the tight binding wave functions for core state j by $\phi_j(\mathbf{k}, \mathbf{r})$. The function ψ_n can be regarded as the sum of a smooth, plane-wave-like function we denote as $f_n(\mathbf{k}, \mathbf{r})$ and a part proportional to core functions. Thus we write

$$\psi_n(\mathbf{k}, \mathbf{r}) = f_n(\mathbf{k}, \mathbf{r}) - \sum_j a_{nj}(\mathbf{k})\phi_j(\mathbf{k}, \mathbf{r}). \tag{4.4.1}$$

The coefficients $a_{nj}(\mathbf{k})$ are determined by the requirement that the ψ_n and the ϕ_j should be orthogonal:

$$\int \phi_j^*(\mathbf{k}, \mathbf{r})\psi_n(\mathbf{k}, \mathbf{r})\, d^3r = 0. \tag{4.4.2}$$

The calculation here is parallel to (4.3.5) and makes use of the fact that $f_n(\mathbf{k}, \mathbf{r})$ must obey Bloch's theorem:

$$f_n(\mathbf{k}, \mathbf{r} + \mathbf{R}_\nu) = \exp(i\mathbf{k}\cdot\mathbf{R}_\nu)f_n(\mathbf{k}, \mathbf{r}). \tag{4.4.3}$$

We find

$$a_{nj}(\mathbf{k}) = N^{1/2} \int u_j^*(\mathbf{r})f_n(\mathbf{k}, \mathbf{r})\, d^3r. \tag{4.4.4}$$

If $f_n(\mathbf{k}, \mathbf{r})$ is just a single plane wave $\exp(i\mathbf{k}\cdot\mathbf{r})/(N\Omega)^{1/2}$, $a_{nj}(\mathbf{k})$ reduces to the orthogonality coefficient $\mu_j(\mathbf{k})$ given by (4.3.7).

An equation satisfied by f_n can be determined by substituting (4.4.1) into the Schrödinger equation: $H\psi_n(\mathbf{k}, \mathbf{r}) = E_n(\mathbf{k})\psi_n(\mathbf{k}, \mathbf{r})$. It is assumed that the core functions ϕ are eigenfunctions of this Hamiltonian. The result is

$$Hf_n(\mathbf{k}, \mathbf{r}) - \sum_j a_{nj}(\mathbf{k})[E_j - E_n(\mathbf{k})]\phi_j(\mathbf{k}, \mathbf{r}) = E_n(\mathbf{k})f_n(\mathbf{k}, \mathbf{r}). \quad (4.4.5)$$

Equation (4.4.5) is in fact an integrodifferential equation for $f_n(\mathbf{k}, \mathbf{r})$. We may use (4.4.4) and (4.3.1) to write (4.4.5) in the form

$$Hf_n(\mathbf{k}, \mathbf{r}) + \int V_{\mathrm{R}}(\mathbf{r}, \mathbf{r}')f_n(\mathbf{k}, \mathbf{r}')\, d^3r' = E_n(\mathbf{k})f_n(\mathbf{k}, \mathbf{r}) \quad (4.4.6\mathrm{a})$$

where

$$V_{\mathrm{R}}(\mathbf{r}, \mathbf{r}') = \sum_\nu \sum_j u_j^*(\mathbf{r}' - \mathbf{R}_\nu)[E_n(\mathbf{k}) - E_j]u_j(\mathbf{r} - \mathbf{R}_\nu). \quad (4.4.7)$$

Just as in Section 4.3 we have considered here only a monatomic crystal.

Suppose that the ordinary crystal potential is $V_{\mathrm{c}}(\mathbf{r})$:

$$H = T + V_{\mathrm{c}}(\mathbf{r}) \quad (4.4.8)$$

(in which T is the kinetic energy operator). The combination of V_{c} and V_{R} gives the pseudopotential V_{p},

$$V_{\mathrm{p}}(\mathbf{r}, \mathbf{r}') = V_{\mathrm{c}}(\mathbf{r})\, \delta(\mathbf{r} - \mathbf{r}') + V_{\mathrm{R}}(\mathbf{r}, \mathbf{r}'). \quad (4.4.9)$$

The effective Schrödinger equation (4.4.6) is

$$Tf_n(\mathbf{k}, \mathbf{r}) + \int V_{\mathrm{p}}(\mathbf{r}, \mathbf{r}')f_n(\mathbf{k}, \mathbf{r}')\, d^3r' = E_n(\mathbf{k})f_n(\mathbf{k}, \mathbf{r}). \quad (4.4.6\mathrm{b})$$

The pseudopotential is nonlocal, that is, it depends on the two coordinates $\mathbf{r}$ and $\mathbf{r}'$ separately and it is energy-dependent through the presence of the energy eigenvalues $E_n(\mathbf{k})$. It can be expressed as a sum of contributions from each atom (atomic pseudopotentials).

The function $V_{\mathrm{R}}(\mathbf{r}, \mathbf{r}')$ is generally repulsive, since the dominant contributions to (4.4.7) can be expected to come from the outer region of both the (usually real) core orbitals, so that the product of these functions will be positive, as also is the energy difference. The repulsive contribution distorts the actual crystal potential so that the pseudopotential, which incorporates both V_{R} and the true crystal potential, is effectively weak. The wave function of lowest energy for the modified Schrödinger equations (4.4.6b) is the smooth function $f_n(\mathbf{k}, \mathbf{r})$.

4.4.2 *Approximate and Empirical Forms*

The relatively complicated form of the actual pseudopotential suggests that it would be desirable to approximate V_{R} by a function which is local and energy independent. To see how this may be done, we insert (4.4.7) in (4.4.6a), integrate, and multiply and divide the second term by $f_n(\mathbf{k}, \mathbf{r})$, so that we have

$$[H + V_{\mathrm{R}}'(\mathbf{r})]f_n(\mathbf{k}, \mathbf{r}) = E_n(\mathbf{k})f_n(\mathbf{k}, \mathbf{r}) \tag{4.4.10}$$

where

$$V_{\mathrm{R}}'(\mathbf{r}) = \sum_{\nu j} [E_n(\mathbf{k}) - E_j] \int u_j^*(\mathbf{r}' - \mathbf{R}_\nu)f_n(\mathbf{k}, \mathbf{r}')\, d^3r'\, u_j(\mathbf{r} - \mathbf{R}_\nu)/f_n(\mathbf{k}, \mathbf{r}). \tag{4.4.11}$$

Since the u_j are wave functions for tightly bound core states, they will be appreciable on only a single site. Thus, with the use of (4.4.3), the repulsive potential V_{R}' can be expressed as the sum of individual atomic pseudopotentials:

$$V_{\mathrm{R}}'(\mathbf{r}) = \sum_{\nu} V_{\mathrm{R}}^{(\mathrm{A})}(\mathbf{r} - \mathbf{R}_\nu). \tag{4.4.12}$$

A similar decomposition is possible for the real crystal potential V_{c}:

$$V_{\mathrm{c}}(\mathbf{r}) = \sum_{\nu} V_{\mathrm{c}}^{(\mathrm{A})}(\mathbf{r} - \mathbf{R}_\nu). \tag{4.4.13}$$

An atomic pseudopotential may now be defined as the sum of $V_{\mathrm{c}}^{(\mathrm{A})}$ and $V_{\mathrm{R}}^{(\mathrm{A})}$. It is given by

$$V_{\mathrm{p}}^{(\mathrm{A})}(\mathbf{r}) = V_{\mathrm{c}}^{(\mathrm{A})}(\mathbf{r}) + \sum_{j} [E_n(\mathbf{k}) - E_j] \times \int d^3r'\, u_j^*(\mathbf{r}')f_n(\mathbf{k}, \mathbf{r}')[u_j(\mathbf{r})/f_n(\mathbf{k}, \mathbf{r})]. \tag{4.4.14}$$

Usually, the energy difference $E_n(\mathbf{k}) - E_j$ is reasonably large (of the order of several rydbergs). Thus, if interest centers on states in a small range of energies, the variation of $E_n(\mathbf{k}) - E_j$ can be neglected, and the term can be replaced by an average. An approximation expression for the dependence on coordinates can be obtained as follows: The most important contribution to the pseudopotential comes from the outermost lobe of the core orbital u_j, which may be represented with moderate accuracy by a Slater-type orbital, proportional to $r^p e^{-sr}$. It will also usually be a reasonable approximation to assume that in this region the smooth function f_n is proportional to a Slater-type orbital $r^q e^{-tr}$ with a larger power q and smaller exponent t. It is sufficient for these qualitative purposes to neglect all but the outermost orbital of the same angular symmetry as the dominant symmetry component of f_n. Then, $q \approx p + 1$, and we have

$$V_{\mathrm{p}}^{(\mathrm{A})}(\mathbf{r}) \approx Ae^{-\beta r}/r \tag{4.4.15}$$

where A and β are constants. This form of pseudopotential was first introduced by Hellman (1934) (however, without the "derivation"). If A and β are regarded as empirical, adjustable constants to be determined from fits to the spectra of the free atom, rather good results can be obtained, for energy bands (and other properties) of alkali metals (Hughes and Callaway, 1964).

The wave function of an electron in the pseudopotential approach (the pseudo wave function) differs from the true wave function close to a nucleus because the inner nodes have been subtracted. In Fig. 4.4.1 we show the comparison between the pseudo wave function and the "actual" wave function for an electron in the 3s state of sodium. The pseudofunction is the solution (Callaway and Laghos, 1969) of the radial Schrödinger equation for the observed energy (−0.3778 Ry) and a potential of the Hellman type

$$V(r) = (-2 + Ae^{-\beta r})/r. \tag{4.4.16}$$

The parameters of this potential, A and β, can be determined by empirically fitting spectroscopic date for the free atom. In the case shown A and β were chosen to fit the 3s and 3p energy levels. Reasonable agreement is obtained for the others as well. For comparison, the radial 3s function of Bierman and Lubeck (1948) is also shown.

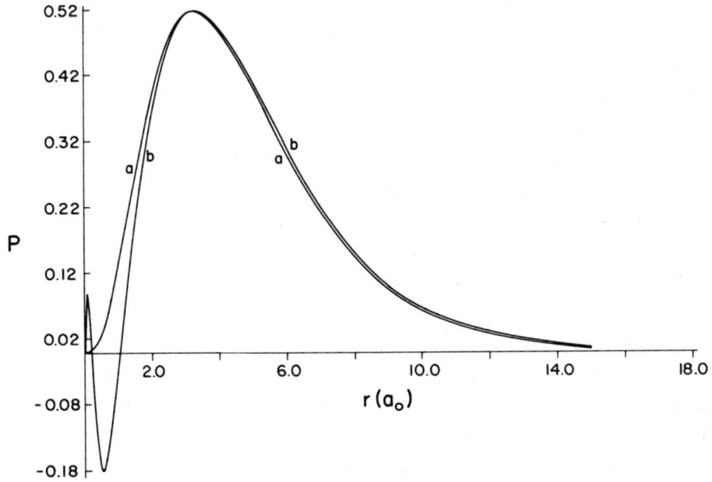

Fig. 4.4.1. Radial wave functions for sodium ($P = rR_{3s}$) are shown as a function of distance in atomic units, as calculated by solution of (a) the Schrödinger equation with the pseudopotential [Eq. (4.4.16), with $A = 20.43$, $\beta = 2.0475$], and (b) according to the calculation of Bierman and Lubeck (1948).

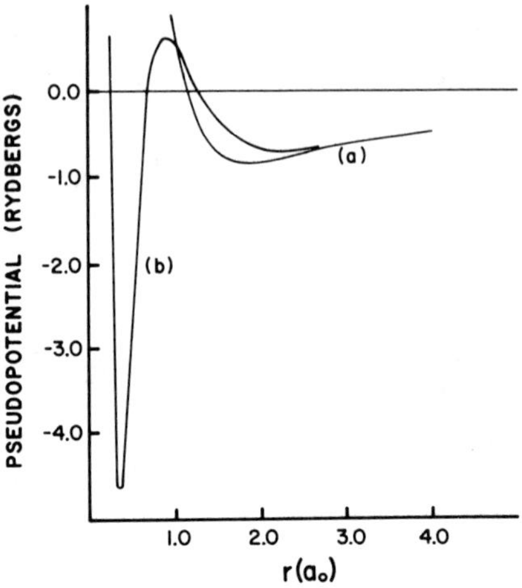

Fig. 4.4.2. Pseudopotentials for sodium. (a) The pseudopotential of Eq. (4.4.13) (with $A = 20.43$, $\beta = 2.0475$) is compared with (b) a calculation of the Phillips–Kleinman pseudopotential due to Szasz and McGinn (1967).

The simple pseudopotential of the form (4.4.15) is not a close approximation to the actual function specified by the Phillips–Kleinman formula (4.4.14) over the whole range of values of r. The Phillips–Kleinman function must be more complicated since $u_j(r)$ will have radial nodes. As a result, (4.4.14) actually gives a rapidly varying pseudopotential inside the atom. This is illustrated in Fig. 4.4.2 where we compare the Phillips–Kleinman pseudopotential for sodium (taken from Szasz and McGinn, 1967).

Other approximate pseudopotentials have been proposed (Abarenkov and Heine, 1965; Animalu and Heine, 1965; Ashcroft, 1968). The former authors suggest the simple form for a single atom,

$$V(r) = -2Z/r \quad (r > r_0), \qquad V(r) = \sum_l A_l(E)P_l \quad (r < r_0)$$

in which Z is the valence, P_l is a projection operator for states of angular momentum l, and $A_l(E)$ is constant in space for functions of the same angular momentum, but depends on energy. The transition radius r_0 is also an adjustable parameter. Ashcroft suggests that it is satisfactory to set $A_l = 0$. The parameters of these potentials are to be determined from spectroscopic data for the free atom. One notes that pseudopotentials of this type are not continuous. This may cause complications.

In many of the applications of pseudopotentials in solid state problems, the individual atomic pseudopotentials are screened by the introduction of the dielectric function (see Section 7.1.4). Such screening describes the response of the electrons to the pseudopotentials of the ions (Cohen and Phillips, 1961). Tables of values of r_0 and A_l for a number of elements are given by Abarenkov and Heine.

4.4.3 *The General Pseudopotential*

As the reader has probably inferred from the preceding discussion the pseudopotential is a rather complicated construction. The main problem is the lack of uniqueness to which we have referred. Let us see how far this extends (Cohen and Heine 1961; Austin *et al.*, 1962; Pendry, 1968). Consider a trial nonlocal $V_{\mathbf{R}}$ defined with respect to an arbitrary function $\chi_n(\mathbf{k}, \mathbf{r})$:

$$V_{\mathbf{R}}\chi_n(\mathbf{k}, \mathbf{r}) = \sum_j \int F_j(\mathbf{r}')\chi_n(\mathbf{k}, \mathbf{r}')\, d^3r'\, \phi_j(\mathbf{k}, \mathbf{r}) \tag{4.4.17}$$

where the sum runs over core states as before, but the $F_j(\mathbf{r}')$ are entirely arbitrary objects. We now suppose χ_n to be an eigenfunction of the Hamiltonian $H + V_{\mathbf{R}}$, and its energy eigenvalue to be E_n': we ask what relation exists between the eigenvalue E_n' and the "true" energy value $E_n(\mathbf{k})$, which is the energy of the true Bloch function $\psi_n(\mathbf{k}, \mathbf{r})$. Thus suppose that

$$[H + V_{\mathbf{R}}]\chi_n(\mathbf{k}, \mathbf{r}) = E_n'(\mathbf{k})\chi_n(\mathbf{k}, \mathbf{r}). \tag{4.4.18}$$

Substitute (4.4.17) in (4.4.18), multiply by $\psi_l^*(\mathbf{k}, \mathbf{r})$, and integrate. Since the true Bloch function for any band above the core is orthogonal to all core states,

$$\int \psi_l^*(\mathbf{k}, \mathbf{r})\phi_j(\mathbf{k}, \mathbf{r})\, d^3r = 0;$$

$V_{\mathbf{R}}$ does not contribute to the integral. Hence we find

$$[E_l(\mathbf{k}) - E_n'(\mathbf{k})] \int \psi_l^*(\mathbf{k}, \mathbf{r})\chi_n(\mathbf{k}, \mathbf{r})\, d^3r = 0. \tag{4.4.19}$$

This indicates that unless the energy of χ_n coincides with the energy of some Bloch state, χ_n will be orthogonal to all these states. If it is not orthogonal, its energy is the same as that of a Bloch state. Hence, we may write

$$\chi_n(\mathbf{k}, \mathbf{r}) = \psi_n(\mathbf{k}, \mathbf{r}) + \sum_j b_j(\mathbf{k})\phi_j(\mathbf{k}, \mathbf{r}) \tag{4.4.20}$$

where the sum includes core states only. The energy of such a χ_n coincides with the true energy $E_n(\mathbf{k})$, indicating that $V_{\mathbf{R}}$ may be used to construct an acceptable pseudopotential. The reason is that V_R acts only in the subspace of Hilbert space spanned by the core states ϕ_j. Since the true Bloch functions are orthogonal to all functions in this subspace, the energy is not altered.

Next, we substitute (4.4.20) into (4.4.18), multiply by $\phi_c^*(\mathbf{k}, \mathbf{r})$ (c is a core state), and integrate. Since the ϕ are an orthonormal set, we have

$$[E_c - E_n(\mathbf{k})]b_c(\mathbf{k}) + \int F_c(\mathbf{r}')\chi_n(\mathbf{k}, \mathbf{r}')\, d^3r' = 0.$$

We can substitute (4.4.20) again into this equation and obtain

$$\sum_j \left\{[E_j - E_n(\mathbf{k})]\,\delta_{cj} + \int F_c(\mathbf{r}')\phi_j(\mathbf{k}, \mathbf{r}')\, d^3r'\right\} b_j(\mathbf{k})$$

$$= -\int F_c(\mathbf{r}')\psi_n(\mathbf{k}, \mathbf{r}')\, d^3r'. \qquad (4.4.21)$$

Equation (4.4.21) is one of a set of inhomogeneous equations for the coefficients b in (4.4.20). These equations possess a solution unless the determinant of the coefficients vanishes.

It is also interesting to set up the matrix of the pseudo Hamiltonian $H + V_R$ in the subspace of the core states. These states are orthogonal to the true Bloch states for higher bands, and so are not mixed with them. Consequently, the eigenvalues of this matrix determine linear combinations of the core states which are also solutions of the pseudopotential problem. The determinantal equation of the matrix diagonalization problem is

$$\det\left[(E_j - \mathcal{E})\,\delta_{jc} + \int F_j(\mathbf{r}')\phi_c(\mathbf{r}')\, d^3r'\right] = 0 \qquad (4.4.22)$$

where j and c label core states, and $\mathcal{E}$ is the desired eigenvalue. Comparison of (4.4.21) and (4.4.22) indicates that (4.4.21) possesses a nontrivial solution unless $E_n(\mathbf{k})$ coincides with one of the eigenvalues $\mathcal{E}$ of (4.4.22).

The general pseudopotential V_R of (4.4.17) is not only nonlocal, it is not Hermitian. For most purposes, this gives an unpleasantly large amount of arbitrariness. However, this characteristic of the pseudopotential procedure is of interest just because it does suggest that there will be quite a variety of useful approximations.

4.4.4 *Description of Simple Metals*

The significance of the pseudopotential approach is due to the fact that the band structures of many materials can be described in terms of plane wave states with only small modifications produced by a weak effective lattice potential. Such materials include the simple metals: those metals in which d states do not contribute to the occupied bands. These are the nontransition metals, including the alkali metals, magnesium, aluminum, etc. For a more complete account, see Harrison (1966, 1970).

We will consider some of the applications of pseudopotential theory to the calculation of the properties of simple metals. One of the simplest applications, and perhaps the most striking success has been the determination of Fermi surfaces.

At the absolute zero of temperature, the occupied one-electron states fill some bounded region of **k** space; all states whose energy is less than the Fermi energy are full. The surface bounding the occupied region is called the *Fermi surface*. It can be shown that the existence of a Fermi surface across which the occupation number of one-electron states is discontinuous is an almost exact result of the quantum theory of a many-Fermion system (Luttinger and Ward, 1960, see also Part B, Section 8.2). This remark applies to normal metals (those not in a superconducting state) which are characterized by the existence of vacant states immediately adjacent to occupied states. In insulators to which the one-electron approximation may be applied, a set of energy bands is full. In such a case (and also in a superconductor) there is no Fermi surface since there are no allowed vacant states adjacent in energy to the occupied states.

There are several types of experiments, some of which will be discussed in Part B, Chapters 6 and 7 which are able to determine the dimensions of the Fermi surface with considerable precision. It is of considerable importance that it is possible to describe quantitatively the Fermi surfaces of simple metals using the pseudopotential approach.

In a first approximation, the pseudopotential can actually be set equal to zero. Then the electron states are simply represented by plane waves, and the Fermi surface can be found by a geometric construction. Next, the effects of a weak pseudopotential can be considered. There are usually two principal effects: (1) sharp edges which occur on the Fermi surface in the free-electron approximation are smoothed off, and (2) small pockets of electrons (or holes) in higher (or lower) bands predicted on account of degeneracies which occur in the free-electron approximation may disappear as states are forced to move apart by a potential which splits the degeneracies.

It might seem on first consideration that the free-electron model would require the Fermi surface to be spherical. This, however, is true only for monovalent materials. When there is more than one electron per atom, the Brillouin zone structure must be taken into account. If we imagine gradually adding electrons to the "empty lattice," the Fermi surface is at first a sphere which expands until it is in contact with the zone boundary. This occurs at the point N for the body-centered cubic lattice; and at L for the face-centered cubic lattice. The Fermi surface is, in a sense, reflected inward from the boundary. The first band continues to fill while some states in higher bands are occupied. The Fermi surface is said to have portions in the higher bands. If there are enough electrons to fill states at a point where a band degeneracy occurs, states in several bands begin to fill at the same time. All pieces of Fermi surface are portions of spheres.

The energy bands in this approximation are described algebraically by

$$E_s(\mathbf{k}) = (\hbar^2/2m) \mid \mathbf{k} + \mathbf{K}_s \mid^2 \tag{4.4.23}$$

where $\mathbf{K}_s$ is any reciprocal lattice vector. The band structure resulting from this formula is surprisingly complex; examples are shown in Fig. 4.4.3. The construction of the Fermi surface has been described by Harrison (1960), who has considered the body-centered cubic, face-centered cubic, and hexagonal close packed structures.

The actual construction of the surfaces is facilitated by use of an extended zone scheme. We can draw about each reciprocal lattice point a

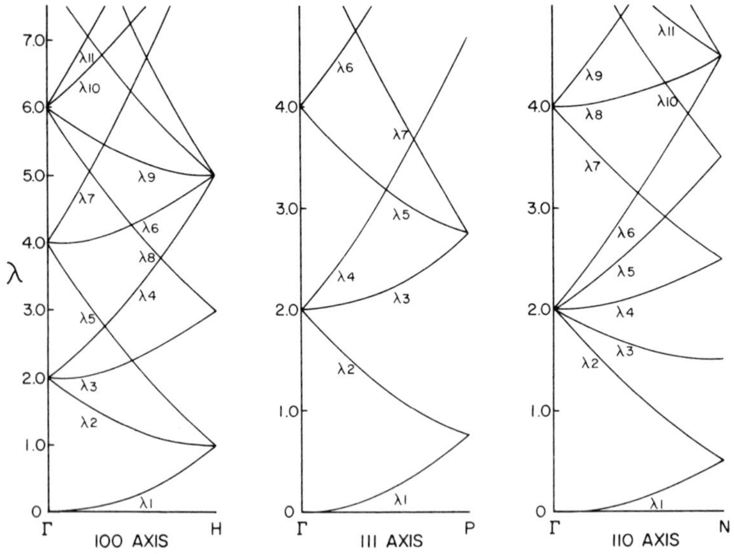

Fig. 4.4.3. Energy bands in the free-electron limit. The dimensionless quantity $\lambda = ma^2E^2/\hbar^2\pi^2$ is plotted as a function of wave vector for points along the 100, 111, and 110 axes in a body-centered cubic crystal.

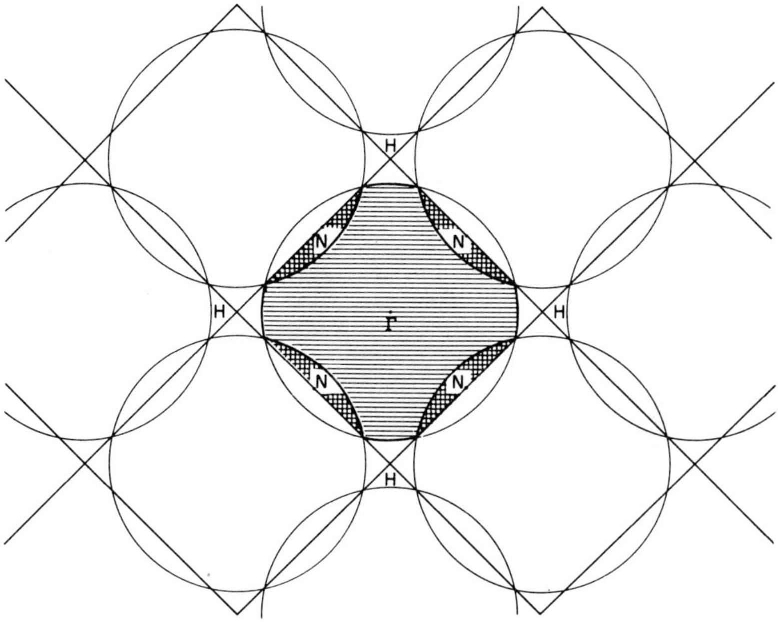

Fig. 4.4.4. Construction of the Fermi surface in the free-electron approximation. The $k_x k_y$ plane in the reciprocal lattice of a body-centered cubic crystal is considered. The circles are drawn for a radius corresponding to three electrons per atom. The singly shaded area in the central zone represents a region in which states in the first band only are occupied, the doubly shaded area represents occupation of both first and second bands, and the unshaded area represents an unoccupied region.

sphere whose volume corresponds to the appropriate number of electrons per atom. These spheres will intersect when the number of electrons is sufficiently large. These intersections can occur only on the surface of a Brillouin zone. The constant energy surface passes into a higher band on the surface of the zone.

Consider a point **k**, lying within the central Brillouin zone. If it is contained within a single sphere, a free-electron state at that point in the first band is occupied. If it belongs to two spheres, states belonging to the first and second bands are occupied, and so on. If it does not lie within a sphere, the state is unoccupied. As an illustration, Fig. 4.4.4 shows the construction for the $k_x k_y$ plane for the body-centered cubic lattice for the case of a hypothetical metal with three electrons per atom. Small regions around the corner H are unoccupied, and in the second band, regions around the face centers are occupied. In Fig. 4.4.5 two of the Fermi surfaces constructed by Harrison are shown.

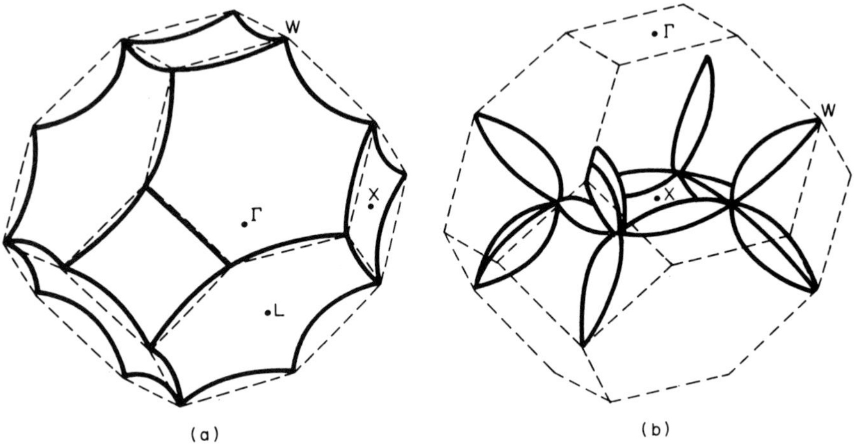

Fig. 4.4.5. Fermi surfaces in the free-electron approximation. The Fermi surface is shown for a face-centered cubic crystal with three electrons per atoms (example: aluminum). The first band is full: (a) The portion of the Fermi surface in the second band; (b) the portion in the third band. The free-electron approximation also predicts the existence of small pockets of occupied states around the corners W in the fourth band. Note that part (b) is centered on X.

To go beyond the free-electron approximation, a nonzero pseudopotential is introduced. This may be done empirically, that is, the Fourier coefficients of potential are regarded as disposable parameters to be determined in order to fit an experimentally determined Fermi surface. Energy bands are obtained from (4.2.8) or (4.2.13). It is possible to include spin–orbit coupling and also to allow the pseudopotential to contain a nonlocal part (see, for example, Stark and Falicov, 1967).

It has been assumed that the pseudopotential is weak enough so that its effect is a small perturbation on a band structure which is basically free electron-like. This assumption facilitates calculation of the total energy of the crystal. In fact, it is not necessary to assume that we have a crystal at all: if perturbation theory is valid, we can treat a disordered system. We will consider below that part of the total energy (at $T = 0$) which can be represented as the sum of the one-electron energies over occupied states. Although this does not give the total energy completely, it sometimes furnishes a good approximation to it. Thus in the alkali metals, if we assume that each electron moves in the field of a singly charged ion, the sum of the differences between the one-electron energies for the band electrons and the eigenvalue of the corresponding free-atom state (for example, 3s in sodium) is known from the classic work of Wigner and Seitz (1933) to furnish an excellent approximation to the cohesive energy of the metal.

In perturbation theory, the energy of a state of wave vector $\mathbf{k}$ may be written in second order as

$$E(\mathbf{k}) = \frac{\hbar^2\mathbf{k}^2}{2m} + \langle \mathbf{k} \mid V_{\mathrm{p}} \mid \mathbf{k} \rangle + \sum_{q\neq 0} \frac{\mid \langle \mathbf{k} \mid V_{\mathrm{p}} \mid \mathbf{k}+\mathbf{q} \rangle \mid^2}{\hbar^2/2m[\mathbf{k}^2 - (\mathbf{k}+\mathbf{q})^2]} \tag{4.4.24}$$

where V_{p} is the total pseudopotential (real potential plus repulsive part as in Eq. (4.4.9). It is to be remembered that V_{p} is a nonlocal operator because it contains V_{R}. However, it may be written as a sum of contributions from individual atomic pseudopotentials, $V_{\mathrm{p}}^{(\mathrm{A})}$ [see (4.4.7) and (4.4.14)]. The matrix elements of V_{p} are defined by

$$\begin{aligned}\langle \mathbf{k} \mid V_{\mathrm{p}} \mid \mathbf{k}+\mathbf{q} \rangle &= (N\Omega)^{-1} \int \exp(-i\mathbf{k}\cdot\mathbf{r}) V_{\mathrm{p}}(\mathbf{r}, \mathbf{r}') \\ &\qquad \times \exp[i(\mathbf{k}+\mathbf{q})\cdot\mathbf{r}']\, d^3r\, d^3r' \\ &= [N^{-1} \sum_{\nu} \exp(i\mathbf{q}\cdot\mathbf{R}_\nu)]\Omega^{-1} \int \exp(-i\mathbf{k}\cdot\mathbf{r}) V_{\mathrm{p}}^{(\mathrm{A})}(\mathbf{r}, \mathbf{r}') \\ &\qquad \times \exp[i(\mathbf{k}+\mathbf{q})\cdot\mathbf{r}']\, d^3r\, d^3r' \\ &= S(\mathbf{q}) \langle \mathbf{k} \mid V_{\mathrm{p}}^{(\mathrm{A})} \mid \mathbf{k}+\mathbf{q} \rangle \end{aligned} \tag{4.4.25}$$

where $S(\mathbf{q})$ is the structure factor,

$$S(\mathbf{q}) = N^{-1} \sum_{\nu} \exp(i\mathbf{q}\cdot\mathbf{R}_\nu), \tag{4.4.26}$$

and

$$\langle \mathbf{k} \mid V_{\mathrm{p}}^{(\mathrm{A})} \mid \mathbf{k}+\mathbf{q} \rangle = \Omega^{-1} \int \exp(-i\mathbf{k}\cdot\mathbf{r}) V_{\mathrm{p}}^{(\mathrm{A})}(\mathbf{r}, \mathbf{r}') \times \exp[i(\mathbf{k}+\mathbf{q})\cdot\mathbf{r}']\, d^3r\, d^3r'. \tag{4.4.27}$$

If a regular crystal lattice is considered, $S(\mathbf{q})$ is zero unless $\mathbf{q}$ is a reciprocal lattice vector. Since the matrix elements of the atomic pseudopotential $\langle \mathbf{k} \mid V_{\mathrm{p}}^{(\mathrm{A})} \mid \mathbf{k}+\mathbf{q} \rangle$ (which are called the *pseudopotential form factor*, or sometimes the *OPW form factor*) do not depend on the position of the atoms, the entire dependence of the theory on these positions is contained in the structure factor $S(\mathbf{q})$. In the case of a liquid metal, in which the atoms are disordered, an estimate of $S(\mathbf{q})$ may be obtained from neutron diffraction (Ziman, 1961).

The calculation of the total energy, approximated by the sum of the energies of the occupied one-electron states $E(\mathbf{k})$, is performed by integrating (4.4.24) over the Fermi distribution. At $T = 0$, and in the absence of magnetic order, each $\mathbf{k}$ state lying below the Fermi energy will be doubly occupied. Since perturbation theory is being used for the energies, we may determine the Fermi wave vector in the free-electron approximation. Let Z represent the number of band electrons per atom (the effective valence). Then

$$[2\Omega/(2\pi)^3] \int_{\mathbf{k}<\mathbf{k}_F} d^3k = Z$$

or

$$k_F = (3\pi^2 Z/\Omega)^{1/3}. \tag{4.4.28}$$

The total energy per atom is

$$E_T = 2\Omega/(2\pi)^3 \int_{\mathbf{k}<\mathbf{k}_F} E(\mathbf{k})\, d^3k. \tag{4.4.29}$$

The result may be expressed as

$$E_T = \tfrac{3}{5}E_F + V_{Av} + \sum_{q\neq 0} |S(\mathbf{q})|^2 F(\mathbf{q}) \tag{4.4.30}$$

in which E_F is the free-electron approximation to the Fermi energy

$$E_F = \hbar^2 k_F^2/2m, \tag{4.4.31}$$

V_{Av} is the average of the diagonal pseudopotential matrix element,

$$V_{Av} = [2\Omega/(2\pi)^3] \int d^3k \langle \mathbf{k} \mid V_p^{(A)} \mid \mathbf{k} \rangle, \tag{4.4.32}$$

and

$$F(\mathbf{q}) = \frac{2\Omega}{(2\pi)^3} \int d^3k \frac{|\langle \mathbf{k} \mid V_p^{(A)} \mid \mathbf{k}+\mathbf{q} \rangle|^2}{\hbar^2/2m(\mathbf{k}^2 - |\mathbf{k}+\mathbf{q}|^2)}\,. \tag{4.4.33}$$

The second-order term $\sum_q |S(\mathbf{q})|^2 F(\mathbf{q})$ is called the *band structure energy*. The contribution of this term to the total energy can be rewritten as

$$\begin{aligned} \sum_{q\neq 0} |S(\mathbf{q})|^2 F(\mathbf{q}) &= N^{-2} \sum_{q,\mu,\nu} F(\mathbf{q}) \exp[i\mathbf{q}\cdot(\mathbf{R}_\nu - \mathbf{R}_\mu)] \\ &= (2N)^{-1} \sum_{\mu\neq\nu} V(\mathbf{R}_\mu - \mathbf{R}_\nu) + N^{-1} \sum_{q\neq 0} F(\mathbf{q}) \end{aligned} \tag{4.4.34}$$

where

$$V(\mathbf{R}) = 2N^{-1} \sum_{q\neq 0} F(\mathbf{q}) \exp(-i\mathbf{q}\cdot\mathbf{R}). \tag{4.4.35}$$

The result is that the band structure energy has been expressed as a contribution from pairs of ions which interact through a potential $V(\mathbf{R})$. If, in (4.4.33), the only dependence of F on the orientation of $\mathbf{q}$ occurs through terms proportional to $\mathbf{k}\cdot\mathbf{q}$ (which is the case for the denominator), F will depend on the magnitude of q only, and this indicates that the effective interaction between ions will be central, that is, will depend on $|\mathbf{R}|$ only, to a good approximation. These observations suggest that the pseudopotential method should make possible a reasonable account of the elastic and vibrational properties of metals. This is indeed possible (see, for example, Wallace, 1969).

4.5 The Tight Binding Method

The methods of band calculations just described are most applicable to simple, i.e., nontransition, metals and to some wide-band semiconductors. Plane wave, or OPW expansions, may converge too slowly to be of much use in other circumstances. However, another set of functions is available for expansion of the wave function: these are the tight binding functions mentioned briefly in Section 4.3. Once regarded as suitable only for a very restricted class of materials, the tight binding procedure has been shown to give satisfactory results even for nearly free electron systems such as the alkali metals (Stern, 1959; Lafon and Lin, 1966). It possesses the advantage with respect to methods based on plane waves, that relatively smaller basis sets may be employed; and a partially compensating disadvantage, that extensive computation of complicated integrals is required for quantitative results.

4.5.1 *Formulation*

The method supposes the existence of a set of functions localized on the atoms of the crystal. The ith such function at lattice site $\mathbf{R}_\mu$ is $u_i(\mathbf{r} - \mathbf{R}_\mu)$. For convenience we restrict our attention to a crystal with one atom per unit cell. The extension to more complex systems is quite straightforward. The functions u_i are frequently considered to be wave functions for the free atom, possibly as calculated from solution of the Hartree–Fock equations. However, assumptions of this type are not necessary in the application of the method. It will be supposed that the u_i have the characteristic angular dependence of atomic wave functions

$$u_i(\mathbf{r}) = R_{il}(r) Y_{lm}(\theta, \phi) \tag{4.5.1}$$

in which the Y_{lm} are spherical harmonics (or real linear combinations of them). The radial functions $R_{il}(r)$ may be either atomic wave functions, or simply members of a discrete set (such as Slater-type orbitals $r^n e^{-\alpha r}$, Gaussian orbitals, etc).

Linear combinations of these functions can be constructed which obey Bloch's theorem for a state of wave vector $\mathbf{k}$. As in Section 4.3, we have

$$\phi_i(\mathbf{k}, \mathbf{r}) = N^{-1/2} \sum_{\mu} \exp(i\mathbf{k}\cdot\mathbf{R}_\mu) u_i(\mathbf{r} - \mathbf{R}_\mu). \tag{4.5.2}$$

These functions are neither orthogonal nor normalized when the overlaps of functions based on different atoms are included. They are still usable. The exact Bloch wave function for band n and wave vector $\mathbf{k}$ is expressed as a linear combination of such functions:

$$\psi_n(\mathbf{k}, \mathbf{r}) = \sum_i C_{ni}(\mathbf{k}) \phi_i(\mathbf{k}, \mathbf{r}). \tag{4.5.3}$$

The coefficients $C_{ni}(\mathbf{k})$ and the energies $E_n(\mathbf{k})$ are to be found by simultaneous diagonalization of the Hamiltonian and overlap matrices on the basis of the functions ϕ_i.

Let us consider a general matrix element of the Hamiltonian, which is denoted by $H_{ij}(\mathbf{k})$,

$$H_{ij}(\mathbf{k}) = \int \phi_i^*(\mathbf{k}, \mathbf{r}) H \phi_j(\mathbf{k}, \mathbf{r})\, d^3r. \tag{4.5.4}$$

This becomes, with the use of (4.5.2),

$$\begin{aligned} H_{ij}(\mathbf{k}) &= N^{-1} \sum_{\mu\nu} \exp[i\mathbf{k}\cdot(\mathbf{R}_\mu - \mathbf{R}_\nu)] \int u_i^*(\mathbf{r} - \mathbf{R}_\nu) H u_j(\mathbf{r} - \mathbf{R}_\mu)\, d^3r \\ &= \sum_\rho \exp(i\mathbf{k}\cdot\mathbf{R}_\rho) \int u_i^*(\mathbf{r}) H u_j(\mathbf{r} - \mathbf{R}_\rho)\, d^3r \\ &= \sum_\rho \exp(i\mathbf{k}\cdot\mathbf{R}_\rho) E_{ij}(\mathbf{R}_\rho) \end{aligned} \tag{4.5.5}$$

in which

$$E_{ij}(\mathbf{R}_\rho) = \int u_i^*(\mathbf{r}) H u_j(\mathbf{r} - \mathbf{R}_\rho)\, d^3r. \tag{4.5.6}$$

We have made use of the invariance of the Hamiltonian with respect to displacement of coordinates

$$H(\mathbf{r} + \mathbf{R}_\nu) = H(\mathbf{r}). \tag{4.5.7}$$

The integrals E_{ij} in (4.5.5) are the fundamental parameters of the tight binding method. Since the functions ϕ_i are not orthogonal, the elements of the overlap matrix must be computed explicitly. We find by an analogous calculation

$$S_{ij}(\mathbf{k}) = \int \phi_i^*(\mathbf{k}, \mathbf{r})\phi_j(\mathbf{k}, \mathbf{r})\, d^3r = \sum_{\rho} \exp(i\mathbf{k}\cdot\mathbf{R}_\rho) S_{ij}(\mathbf{R}_\rho) \qquad (4.5.8)$$

where

$$S_{ij}(\mathbf{R}_\rho) = \int u_i^*(\mathbf{r}) u_j(\mathbf{r} - \mathbf{R}_\rho)\, d^3r. \qquad (4.5.9)$$

The energies and the expansion coefficients of the wave function $\psi_n(\mathbf{k}, \mathbf{r})$ are determined from the solution of the equations

$$\sum_j H_{ij}(\mathbf{k}) C_{nj}(\mathbf{k}) = E_n(\mathbf{k}) \sum_j S_{ij}(\mathbf{k}) C_{nj}(\mathbf{k}). \qquad (4.5.10)$$

The expression of the matrix elements as sums over lattice sites with a specified $\mathbf{k}$ dependence, as in (4.5.5) and (4.5.8) is characteristic of the tight binding method. These sums can be simplified by symmetry considerations as follows: There is a relation between the parameters E_{ij} for values of $\mathbf{R}$ which are connected by an operation in the point group of the crystal. Let $\boldsymbol{\alpha}$ be such an operation and consider

$$E_{ij}(\boldsymbol{\alpha}\mathbf{R}_\rho) = \int u_i^*(\mathbf{r}) H u_j(\mathbf{r} - \boldsymbol{\alpha}\mathbf{R}_\rho)\, d^3r = \int u_i^*(\boldsymbol{\alpha}\mathbf{r}') H u_j[\boldsymbol{\alpha}(\mathbf{r}' - \mathbf{R}_\rho)]\, d^3r'.$$

We have used the invariance of the Hamiltonian (and the volume element) under the transformation $\mathbf{r} = \boldsymbol{\alpha}\mathbf{r}'$. The atomic functions will transform in some specified way under the rotation α. This may be stated as

$$u_i(\boldsymbol{\alpha}\mathbf{r}) = \sum_l D_{li}(\boldsymbol{\alpha}^{-1}) u_l(\mathbf{r}) \qquad (4.5.11)$$

where the D_{li} are the representation matrices which are relevant to the atomic functions. We then have

$$E_{ij}(\boldsymbol{\alpha}\mathbf{R}_\rho) = \sum_{l,n} D^*_{li}(\boldsymbol{\alpha}^{-1}) D_{nj}(\boldsymbol{\alpha}^{-1}) E_{ln}(\mathbf{R}_\rho). \qquad (4.5.12)$$

Also, as a result of the Hermitian nature of the Hamiltonian, we have

$$E_{ij}(\mathbf{R}_\rho) = E^*_{ji}(-\mathbf{R}_\rho). \qquad (4.5.13)$$

This result can be used to reduce the number of independent parameters which must be determined in order to specify a matrix element. For example, if u_i and u_j are both s-like functions and therefore are unchanged by rotation, $E_{ij}(\boldsymbol{\alpha}\mathbf{R}_\rho)$ is the same for all α. A more complete analysis of the symmetry properties of the E_{ij} is given by Egorov *et al.* (1968). Let us consider a hypothetical simple cubic lattice of lattice parameter a. We find that

$$\sum_{100 \text{ pos}} \exp(i\mathbf{k}\cdot\mathbf{R}_\rho) = 2(\cos k_x a + \cos k_y a + \cos k_z a) \tag{4.5.14a}$$

$$\sum_{110 \text{ pos}} \exp(i\mathbf{k}\cdot\mathbf{R}_\rho) = 4(\cos k_x a \cos k_y a + \cos k_x a \cos k_z a + \cos k_y a \cos k_z a) \tag{4.5.14b}$$

$$\sum_{111 \text{ pos}} \exp(i\mathbf{k}\cdot\mathbf{R}_\rho) = 8 \cos k_x a \cos k_y a \cos k_z a. \tag{4.5.14c}$$

Thus, if atoms at (000), (100), (110), and (111) positions are considered, Eq. (4.5.8) takes the form, in the case i and j refer to s functions,

$$\begin{aligned} H_{ij}(\mathbf{k}) = E_{ij}(0) &+ 2E_{ij}(a)(\cos k_x a + \cos k_y a + \cos k_z a) \\ &+ 4E_{ij}(2^{1/2}a)(\cos k_x a \cos k_y a + \cos k_x a \cos k_z a \\ &+ \cos k_y a \cos k_z a) + 8E_{ij}(3^{1/2}a) \cos k_x a \cos k_y a \cos k_z a. \end{aligned} \tag{4.5.15}$$

Tables of similar expressions for matrix elements involving s-, p-, and d-like functions have been presented by Slater and Koster (1954) for cubic lattices, and by Miasek (1956) for the hexagonal close-packed lattice.

4.5.2 *Two- and Three-Center Integrals*

The calculation of the parameters E_{ij} and S_{ij} is a central problem in the tight binding method. It is instructive to express these parameters in terms of two- and three-center integrals, although this decomposition is not particularly useful for computational purposes. Thus, in (4.5.5) it is convenient to express the Hamiltonian as

$$H = T + \sum_\mu V(\mathbf{r} - \mathbf{R}_\mu) = T + V(\mathbf{r}) + V(\mathbf{r} - \mathbf{R}_\rho) + \sum_{\mu\neq 0,\rho} V(\mathbf{r} - \mathbf{R}_\mu) \tag{4.5.16}$$

in which T is the kinetic energy and $V(\mathbf{r} - \mathbf{R}_\mu)$ represents the contribution to the total crystal potential from the atom located at site μ. Let us also suppose that $u_i(\mathbf{r})$ is an eigenfunction of a single atom Hamiltonian

$$[T + V(\mathbf{r})]u_i(\mathbf{r}) = E_i^{(0)} u_i(\mathbf{r}). \tag{4.5.17}$$

Then

$$E_{ij}(\mathbf{R}_\rho) = E_i^{(0)} S_{ij}(\mathbf{R}_\rho) + \int u_i^*(\mathbf{r}) V(\mathbf{r} - \mathbf{R}_\rho) u_j(\mathbf{r} - \mathbf{R}_\rho)\, d^3r$$

$$+ \sum_{\mu \neq 0, \rho} \int u_i^*(\mathbf{r}) V(\mathbf{r} - \mathbf{R}_\mu) u_j(\mathbf{r} - \mathbf{R}_\rho)\, d^3r. \qquad (4.5.18)$$

This expression contains two-center and three-center terms; the computation of the latter has been such a difficult problem that they have often been arbitrarily neglected.

The simplest possible case is that in which only a single band is to be considered; all interactions between bands being disregarded. In this case, the expression for the energy can be put in the form

$$E_i(\mathbf{k}) = H_{ii}(\mathbf{k})/S_{ii}(\mathbf{k})$$

$$= E_i^{(0)} + \frac{[K_i + \sum_{\rho \neq 0} J_i(\mathbf{R}_\rho) \exp(i\mathbf{k}\cdot\mathbf{R}_\rho)]}{[1 + \sum_{\rho \neq 0} S_i(\mathbf{R}_\rho) \exp(i\mathbf{k}\cdot\mathbf{R}_\rho)]} \qquad (4.5.19)$$

in which

$$K_i = \int |u_i(\mathbf{r})|^2 [\sum_{\mu \neq 0} V(\mathbf{r} - \mathbf{R}_\mu)]\, d^3r \qquad (4.5.20a)$$

$$J_i(\mathbf{R}_\rho) = \int u_i^*(\mathbf{r}) [\sum_{\mu \neq 0} V(\mathbf{r} - \mathbf{R}_\mu)] u_i(\mathbf{r} - \mathbf{R}_\rho)\, d^3r \qquad (4.5.20b)$$

and S_i is obtained from (4.5.9). The quantity K_i contains the effect of all distant potentials on the function in the central cell. We may think of this as containing crystal field effects. The interaction integrals $J(\mathbf{R}_\rho)$ are smaller than the crystal field terms for atoms which are far apart, since $J(\mathbf{R}_\rho)$ [and $S(\mathbf{R}_\rho)$] will decrease exponentially while K falls off only as a power of $|\mathbf{R}_\rho|$. We may expand the denominator in (4.5.19),

$$E_i(\mathbf{k}) \approx E_i^{(0)} + K_i + \sum_{\rho \neq 0} [J_i(\mathbf{R}_\rho) - K_i S_i(\mathbf{R}_\rho)] \exp(i\mathbf{k}\cdot\mathbf{R}_\rho). \qquad (4.5.21)$$

This shows that in the limit of large atomic separations, the overlap and crystal field terms can be regarded as modifying the values of the interaction integral without changing the functional $\mathbf{k}$ dependence of the energy band. The band is centered not at the position of the free atom level, but at the displaced energy $E_0 + K_i$.

Formulas of the type (4.5.18) are not convenient for practical calculations, since the number of three-center integrals which must be computed in order to obtain accurate expressions for matrix elements is very large. It is desirable, as suggested by Lafon and Lin (1966) to consider instead a Fourier expansion for the potential energy of the form

$$V(\mathbf{r}) = \sum_s V(\mathbf{K}_s) \exp(i\mathbf{K}_s \cdot \mathbf{r}). \tag{4.5.22}$$

In a crystal with inversion symmetry, the exponential is replaced by a cosine. We now have

$$E_{ij}(\mathbf{R}_\rho) = T_{ij}(\mathbf{R}_\rho) + V_{ij}(\mathbf{R}_\rho) \tag{4.5.23}$$

where

$$T_{ij}(\mathbf{R}_\rho) = \int u_i^*(\mathbf{r})[(-\hbar^2/2m)\ \nabla^2]u_j(\mathbf{r} - \mathbf{R}_\rho)\ d^3r \tag{4.5.24}$$

and

$$V_{ij}(\mathbf{R}_\rho) = \sum_s V(\mathbf{K}_s) \int u_i^*(\mathbf{r}) \exp(i\mathbf{K}_s \cdot \mathbf{r}) u_j(\mathbf{r} - \mathbf{R}_\rho)\ d^3r. \tag{4.5.25}$$

The integrals involved here are two-center integrals. Further, given a fixed set of basis functions, the integrals in (4.5.25) need be computed only once no matter how often the potential is changed (as in an attempt to calculate a self-consistent band structure by iteration). If Slater-type orbitals are used as basis functions, the integrals in (4.5.25) can be reduced to a one-dimensional numerical integration, while if Gaussian orbitals are used, the integrals may be performed analytically. Useful formulas for this computation have been presented by Lafon and Lin (1966), and Langlinais and Callaway (1972).

4.5.3 *Determination of the Fourier Coefficients*

The Fourier coefficient of potentials can be found by the procedure described in Section 4.2, in which they are related to the Fourier coefficients of change density $\rho(\mathbf{K}_s)$. For $\mathbf{K}_s \neq 0$, we have, in atomic units

$$V(\mathbf{K}_s) = (-8\pi/\mathbf{K}_s^2)\rho(\mathbf{K}_s).$$

Let us suppose that the charge density is to be obtained from a set of wave functions $\psi_n(\mathbf{k}, \mathbf{r})$, which are expressed in terms of the basis set $\phi_i(\mathbf{k}, \mathbf{r})$ according to (4.5.3). Then

$$\rho(\mathbf{r}) = \sum_{n\mathbf{k}ij} C^*_{nj}(\mathbf{k}) C_{ni}(\mathbf{k}) \phi_j^*(\mathbf{k}, \mathbf{r}) \phi_i(\mathbf{k}, \mathbf{r}) \tag{4.5.26}$$

in which the sum on n and $\mathbf{k}$ includes only those states which are occupied.

The Fourier coefficients of charge density are given by

$$\rho(\mathbf{K}) = (N\Omega)^{-1} \int \rho(\mathbf{r}) \exp(i\mathbf{K}\cdot\mathbf{r})\, d^3r. \qquad (4.5.27)$$

We substitute (4.5.26) and (4.5.2) into (4.5.27). The result may be expressed as

$$\rho(\mathbf{K}) = (N\Omega)^{-1} \sum_{n\mathbf{k}ij} C^*_{ni}(\mathbf{k}) C_{nj}(\mathbf{k}) S_{ij}(\mathbf{k}, \mathbf{K}) \qquad (4.5.28)$$

where

$$S_{ij}(\mathbf{k}, \mathbf{K}) = \sum_{\rho} \exp(i\mathbf{k}\cdot\mathbf{R}_\rho) \int u_i^*(\mathbf{r}) \exp(i\mathbf{K}\cdot\mathbf{r}) u_j(\mathbf{r} - \mathbf{R}_\rho)\, d^3r. \quad (4.5.29)$$

The integrals required in (4.5.29) are the same as those used in (4.5.25). In the case of $\mathbf{K} = 0$, we follow the limiting procedure of Section 4.2:

$$V(\mathbf{K} = 0) = \lim_{\mathbf{K}\to 0} [\rho(\mathbf{K})/\mathbf{K}^2]. \qquad (4.5.30)$$

This gives, for a cubic crystal,

$$V(0) = (-4\pi/3N\Omega) \sum_{n\mathbf{k}ij} C^*_{ni}(\mathbf{k}) C_{nj}(\mathbf{k}) S_{ij}^{(2)}(\mathbf{k}) \qquad (4.5.31)$$

with

$$S_{ij}^{(2)}(\mathbf{k}) = \sum_{\rho} \exp(i\mathbf{k}\cdot\mathbf{R}_\rho) \int u_i^*(\mathbf{r}) r^2 u_j(\mathbf{r} - \mathbf{R}_\rho)\, d^3r. \qquad (4.5.32)$$

These expressions suffice to indicate that the tight binding method may be, in fact, a powerful tool for energy band calculations in a wide variety of materials. Two points deserve comment in the connection.

(1) A problem can arise in connection with core states, as in the OPW method. Unless the basis functions u_i are explicitly orthogonalized to core functions on neighboring atoms, the energy obtained will be too low. Since such orthogonalization is usually undesirable computationally, it is better to add core functions (or reasonable approximations to them) to the basis set.

(2) Bound atomic wave functions do not form a complete set of states, since functions exist (continuum or scattering functions) which are orthogonal to all bound functions. This objection does not, however, appear to be of much practical importance as far as is known at this time. In case of difficulty, it is possible to use, instead of atomic wave functions for highly excited states, discrete localized functions (for example, Sturmian func-

tions) which do form a complete set. Alternately, plane waves may be added to the basis, as discussed in Section 4.3.

4.5.4 *Interpolation Schemes*

A different utilization of the tight binding method deserves mention. The elements of the Hamiltonian and overlap matrices have relatively simple expressions, as far as the **k** dependence is concerned. The Hamiltonian matrix may be of relatively small dimension. However, the $E_{ij}(\mathbf{R})$ parameters are difficult to evaluate. If the energy band structure is known at some points in the zone, usually those of highest symmetry, either as a result of calculations by other methods or from experiment, it is possible to use the tight binding method as an interpolation scheme (Slater and Koster, 1954, Dresselhaus and Dresselhaus, 1967). One determines the E_{ij} and S_{ij} parameters by fitting known energy levels at certain points. Energy bands can then be found at general points of the zone.

In application to the transition metals, one considers a situation in which a narrow set of d bands mix with a wide s-p band. Several authors have proposed to combine the tight binding and pseudopotential method to describe this situation (Hodges *et al.*, 1966; Mueller, 1967; Heine, 1967; Ehrenreich and Hodges, 1968). In this approach, one considers an effective Hamiltonian of the form of (4.3.30) involving both tight binding and plane wave parts (combined interpolation scheme). Schematically

$$H = \begin{pmatrix} \text{TB–TB} & \text{TB–PW} \\ \\ \text{PW–TB} & \text{PW–PW} \end{pmatrix} \tag{4.5.33}$$

where the elements of the tight binding part would be given by (4.5.8). The remaining elements are as specified in (4.3.31). It is desirable to include at least as much of the overlap matrix as connects the tight binding functions with the plane waves. This matrix would then be given by (4.3.33). A purely tight binding part of the overlap matrix with elements given by (4.5.8) can also be included. The Fourier coefficients of potential which appear in the plane wave part of (4.5.33) can be regarded as referring to a pseudopotential, and determined empirically. In this way, it is sufficient to include only a small number of plane waves in (4.5.33). Most calculations for face-centered cubic transition metals (Ni and Cu) employ, when spin–orbit coupling is neglected, a 9×9 matrix in which the tight binding part is based on five d functions. Four (unsymmetrized) plane waves are included. The tight binding parameters are determined either from a fit to other calculations, or directly from estimates of the integrals. Such methods seem to be able to describe what is known about the band structure and Fermi surface of transition metals with considerable efficiency.

4.6 The Cellular Method

We will now consider a set of methods for band calculation which are based on the direct solution of the Schrödinger equation within an atomic cell. These are the cellular method, the Green's function method, and the augmented plane wave method. The cellular method is the oldest and most straightforward of these (Wigner and Seitz, 1933; Slater, 1934; Shockley, 1937). In outline, the method is extremely simple: The Schrödinger equation is solved in a single atomic cell subject to the boundary conditions for the wave function and its derivatives on the surface of an atomic cell which are implied by Bloch's Theorem. For some states in simple crystals, the boundary conditions are sufficiently simple so that a solution can be obtained with very little labor. (This applies specifically to states at the bottom of the lowest valence electron band in cubic metals.) For many states of interest, however, the complexity of the boundary conditions cause serious difficulties, so much so, that the cellular method is seldom employed for quantitative calculations at the present time.

4.6.1 *Boundary Conditions for the Bloch Function*

Our discussion of the cellular method will follow the work of von der Lage and Bethe (1947). A symmetric atomic cell (also known as the "polyhedral cell" or the "Wigner–Seitz cell") is constructed in the direct lattice exactly as the Brillouin zone was constructed in Section 1.2 in the reciprocal lattice. Lines are drawn which connect an atom with its neighbors; then the planes which are the perpendicular bisectors of these lines are constructed. The smallest, central, solid figure so formed is the desired cell.

It is conventionally assumed in the cellular method that the crystal potential is spherically symmetric within any given atomic cell. This assumption is made so that it will be possible to separate variables in the Schrödinger equation, and obtain a single differential equation for the radial wave function associated with a given angular momentum. If the potential contains nonspherical terms, and is expanded in spherical harmonics, a set of coupled differential equations will be obtained in which functions belonging to different l are connected. Unfortunately, the assumption of a spherically symmetric potential, which is common to the cellular method, the Green's function method, and the augmented plane wave method, is never exactly true, although in simple metals, it does not cause appreciable error. This assumption is not required in methods based on plane wave expansions, or in the tight binding method.

In the cellular method, the one-electron wave function for a state in band n belonging to wave vector $\mathbf{k}$ is expressed as a sum of products of radial wave functions $R_l(E, r)$ and spherical harmonics $Y_{lm}(\theta, \phi)$:

$$\psi_n(\mathbf{k}, \mathbf{r}) = \sum_l A_{nl}(\mathbf{k})[\sum_m C_{nlm}(\mathbf{k}) Y_{lm}(\theta, \phi)] R_l(E, r). \quad (4.6.1)$$

The radial functions satisfy the differential equation (atomic units)

$$\frac{1}{r^2}\frac{d}{dr}\left(\frac{r^2\, dR_l}{dr}\right) + \left(E - V(r) - \frac{l(l+1)}{r^2}\right) R_l = 0 \quad (4.6.2)$$

in which V is the spherically symmetric potential existing in a single cell. The combination of spherical harmonics of different m values for a given l is chosen to produce a wave function with appropriate transformation properties. These combinations are determined from group theory (see Sections 3.2 and 3.6) so that $\psi_n(\mathbf{k}, \mathbf{r})$ belongs to a particular irreducible representation of the group of wave vector $\mathbf{k}$, and in the case of cubic lattices, are frequently called *Kubic harmonics*. Some of these functions were listed in Table VII, Chapter 3, for the Γ_1 representation. More extensive compilations have been given by Bell (1953), Altmann and Cracknell (1965), and Altmann and Bradley (1965a).

The boundary conditions on the wave function are obtained as follows. Let $\mathbf{A}$ be a point on some face of the polyhedral cell and let $\mathbf{B}$ be a point perpendicularly opposite on a parallel face. These points are separated by a lattice translation vector $\mathbf{T}$. Since Bloch's theorem asserts that

$$\psi(\mathbf{k}, \mathbf{r} + \mathbf{T}) = \exp(i\mathbf{k}\cdot\mathbf{T})\psi(\mathbf{k}, \mathbf{r}),$$

it follows that

$$\psi(\mathbf{k}, \mathbf{A}) = \exp(i\mathbf{k}\cdot\mathbf{T})\psi(\mathbf{k}, \mathbf{B}). \quad (4.6.3)$$

In this form, the boundary conditions relate the wave function at different points. It is desirable to restate these conditions to refer to values of the wave function, or its derivatives at a single point.

To see how this can be done, consider the body-centered cubic lattice. The atomic cell for this lattice is shown in Fig. 4.6.1. The parallel square faces are located on planes x, y, or $z = \pm(a/2)$ (a is the lattice parameter) while the hexagonal faces have normal vectors $(a/4)(\pm 1, \pm 1, \pm 1)$. Hence, the translation $\mathbf{T}$ is of the type $(a, 0, 0)$ for points on the square faces and $(a/2)(1, 1, 1)$ for the hexagonal faces. To have a specific example, consider states belonging to $\mathbf{k} = \mathbf{0}$. For these states (4.6.3) implies

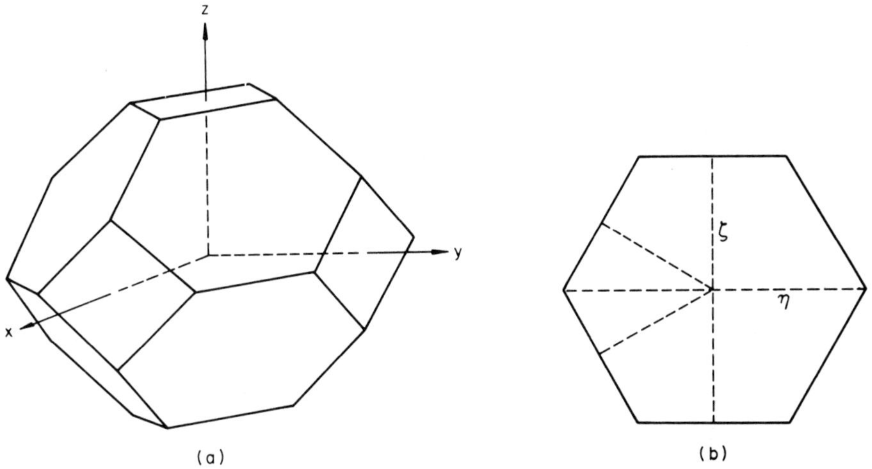

Fig. 4.6.1. (a) Atomic cell for body-centered cubic lattice. (b) Coordinate system in a hexagonal face of the cell.

$\psi(0, A) = \psi(0, B)$; such functions are said to be *periodic* with the periods of all pairs of parallel faces of the cell. Suppose ψ belongs to the Γ_1 representation of O_h and so is unchanged by any operation of the full cubic group. Consider the function at a point close to a square face of the cell, say, $(\frac{1}{2}a + \epsilon, y, z)$. Let this be in the neighborhood of A; then a corresponding point near B is $(-\frac{1}{2}a + \epsilon, y, z)$. Hence

$$\psi(\tfrac{1}{2}a + \epsilon, y, z) = \psi(-\tfrac{1}{2}a + \epsilon, y, z).$$

However, the operation $(-x, y, z)$ leaves ψ unchanged, so that

$$\psi(\tfrac{1}{2}a + \epsilon, y, z) = \psi(\tfrac{1}{2}a - \epsilon, y, z).$$

We conclude that the normal derivative of this function vanishes on the square faces. A similar argument shows that the normal derivative also vanishes at the center of a hexagonal face and along certain lines in this face (but not everywhere): Let (x, y, z) be the coordinates of a point on such a face. On applying the lattice translation $(-\frac{1}{2}a, -\frac{1}{2}a, -\frac{1}{2}a)$ followed by the operation $(-y, -x, -z)$, we find

$$\psi(x, y, z) = \psi(x - \tfrac{1}{2}a, y - \tfrac{1}{2}a, z - \tfrac{1}{2}a) = \psi(\tfrac{1}{2}a - y, \tfrac{1}{2}a - x, \tfrac{1}{2}a - z).$$

However, $(\frac{1}{2}a - y, \frac{1}{2}a - x, \frac{1}{2}a - z)$ are the coordinates of a point in the same hexagonal face which is symmetrically situated across the line $\zeta = 0$ illustrated in Fig. 4.6.1. Again, considering a point close to (x, y, z), but displaced along the normal to this face by an infinitesimal amount, we have

$$\psi(x + \epsilon, y + \epsilon, z + \epsilon) = \psi(\tfrac{1}{2}a - y - \epsilon, \tfrac{1}{2}a - x - \epsilon, \tfrac{1}{2}a - z - \epsilon).$$

We conclude that the function ψ is symmetric about the line $\zeta = 0$ and the normal derivative is antisymmetric about this line.

Boundary conditions may be deduced in a similar manner for functions with other wave vectors. In the case of the point $H = (2\pi/a)(1, 0, 0)$ in the Brillouin zone for the body-centered cubic lattice, the wave function is periodic with the periods of pairs of pairs of square faces, but is anti-periodic, or changes sign, when pairs of hexagonal faces are considered. In the case of the symmetric representation H_1, this implies that, while the normal derivative still vanishes on the square faces, the wave function itself must vanish at the center of a hexagonal face, and on the line $\zeta = 0$. The normal derivative on the hexagonal face is symmetric about the line $\zeta = 0$.

The complexity of the boundary conditions is evident. One may attempt to satisfy these conditions in a step by step manner by requiring that they be satisfied at a small number of selected points; one for each radial function in the expansion (4.6.1). The utility of a given approximation to the satisfaction of boundary conditions may be partially judged by the empty lattice test of Shockley (1937). If the potential $V = 0$, the possible energies of a state of wave vector $\mathbf{k}$ are $|\mathbf{k} + \mathbf{K}_s|^2$ where $\mathbf{K}_s$ is any reciprocal lattice vector. To obtain this result in the cellular method is by no means trivial, and requires that the boundary conditions be properly obeyed. It appears from the work of Howarth and Jones (1952) that success in the empty lattice test is more of a necessary than sufficient condition with regard to application to real metals, as various choices of approximations which give quite good results in the empty lattice still give a not inconsiderable spread of energies for a state when V is not zero. More recently, a least squares approach to the fitting of boundary conditions has been developed which yields results in good agreement with those obtained by other methods in significant test cases, so that the cellular method may still be regarded as competitive with other procedures for band calculations (Altmann and Bradley, 1965b; Altmann *et al.*, 1968).

In the early calculations (and in some later work as well), a simple approximation to the boundary conditions was employed. The polyhedral cell is replaced by the sphere of equal volume (the spherical approximation). Under these circumstances, only a single term is required in (4.6.1) for states at $\mathbf{k} = \mathbf{0}$. Functions of even l are then periodic in the crystal and even under inversion; hence, these must have vanishing radial derivative on the sphere. Functions of odd l must vanish on the sphere.

In this approximation, the detailed crystal structure is irrelevant; results depend only on the volume of the cell. The spherical approximation leads to accurate energies only for states near the bottom of a simple s band. In other cases, the results may not only be poor quantitatively; they may be qualitatively misleading. As one example of this, we see that the fivefold degeneracy of the d states is not split at the center of the zone.

4.6.2 *Calculation of the Effective Mass*

An interesting and important application of the cellular method is involved in the determination of effective masses and wave functions for states near the bottom of the lowest valence electrons band in a cubic crystal (Bardeen, 1938). The procedure used here is that of Silverman (1952), and differs from that of Section 4.1 in that here the differential equation of first-order perturbation theory is solved subject to the boundary conditions imposed by the spherical approximation to the cellular method. A variational principal for the determination of the components of the effective mass tensor at any point in the Brillouin zone was given by Cohen and Ham (1960).

The bottom of the band in question is assumed to occur at $\mathbf{k} = 0$. We consider a state of small $\mathbf{k}$, and study the differential equation satisfied by the cell periodic part of the Bloch function $u(\mathbf{k}, \mathbf{r})$ where $\psi(\mathbf{k}, \mathbf{r}) = \exp(i\mathbf{k}\cdot\mathbf{r})u(\mathbf{k}, \mathbf{r})$. Spin–orbit coupling is neglected. It follows from Eq. (3.7.13) that the equation satisfied by $u(\mathbf{k}, \mathbf{r})$ is, in atomic units,

$$(-\nabla^2 + V)u(\mathbf{k}, \mathbf{r}) - 2i\mathbf{k}\cdot\nabla u(\mathbf{k}, \mathbf{r}) = [E(\mathbf{k}) - \mathbf{k}^2]u(\mathbf{k}, \mathbf{r}). \quad (4.6.4)$$

The function $u(\mathbf{k}, \mathbf{r})$ is expanded as a power series in k with coefficients which are functions of position, and are to be determined from (4.6.4).

$$u(\mathbf{k}, \mathbf{r}) = \sum_{n=0}^{\infty} u_n(\mathbf{r})k^n. \quad (4.6.5)$$

Since $u(\mathbf{k}, \mathbf{r})$ is periodic in the atomic cell, it must satisfy the same boundary conditions as a wave function with $k = 0$.

The energy is also expanded in powers of k:

$$E(\mathbf{k}) = \sum_{n=0}^{\infty} E_{2n}k^{2n}. \quad (4.6.6)$$

Odd powers of k are absent because the energy has inversion symmetry and, from Section 4.1, may be required to have continuous derivatives of all orders at $k = 0$. Because spherical boundary conditions are used, terms in the energy possessing cubic rather than spherical symmetry [such as a possible term proportional to $K_{4,1}(\theta_k, \phi_k)$] will not appear. When (4.6.5) and (4.6.6) are substituted into (4.6.4), and coefficients of like powers of k are equated, a set of equations is obtained for the functions u_n. These are

$$(-\nabla^2 + V - E_0)u_0(\mathbf{r}) = 0, \tag{4.6.7a}$$

$$(-\nabla^2 + V - E_0)u_1(\mathbf{r}) = 2i\hat{\mathbf{k}}\cdot\nabla u_0(\mathbf{r}), \tag{4.6.7b}$$

$$(-\nabla^2 + V - E_0)u_2(\mathbf{r}) = 2i\hat{\mathbf{k}}\cdot\nabla u_1(\mathbf{r}) + (E_2 - 1)u_0(\mathbf{r}), \tag{4.6.7c}$$

$$(-\nabla^2 + V - E_0)u_n(\mathbf{r}) = 2i\hat{\mathbf{k}}\cdot\nabla u_{n-1}(\mathbf{r}) + (E_2 - 1)u_{n-2}(\mathbf{r}) + E_4 u_{n-4}(\mathbf{r}) + \cdots. \tag{4.6.7d}$$

In these equations $\hat{\mathbf{k}}$ is a unit vector in the direction of $\mathbf{k}$. It is assumed that the solution of (4.6.7a) is known [subject to the boundary condition $(du_0/dr)_{r_s} = 0$, where r_s is the radius of the sphere of volume equal to that of the atomic cell].

Solutions of the remaining equations can be determined. The particular integral of (4.6.7b) can be verified to be $-i\hat{\mathbf{k}}\cdot\mathbf{r}u_0(\mathbf{r})$. Evidently, $u_1(\mathbf{r})$ is an odd function; and thus must satisfy the boundary conditions $u_1(r_s) = 0$. This may be accomplished by adding a multiple of a p ($l = 1$) solution of the homogeneous equation [obtained by dropping the right side of (4.6.7b)]. Let the radial part of this function be denoted by $f_p(r)$. Then

$$u_1(\mathbf{r}) = i\hat{\mathbf{k}}\cdot\mathbf{r}[(f_p/r) - u_0]. \tag{4.6.8}$$

Similarly, it can be verified that a particular solution of (4.6.7c) is

$$-i\hat{\mathbf{k}}\cdot\mathbf{r}u_1(\mathbf{r}) + \tfrac{1}{2}(\hat{\mathbf{k}}\cdot\mathbf{r})^2 u_0(\mathbf{r}) + E_2 g(\mathbf{r}) \tag{4.6.8a}$$

where g satisfies

$$(-\nabla^2 + V - E_0)g(\mathbf{r}) = u_0(\mathbf{r}). \tag{4.6.8b}$$

A function g which satisfies (4.6.8b) may be determined by differentiating Eq. (4.6.7a) with respect to E and setting $E = E_0$:

$$(-\nabla^2 + V - E_0)\ (\partial u_0/\partial E)_{E_0} = u_0. \tag{4.6.8c}$$

The function u_2 is even under inversion. Consequently, it must satisfy the condition $(\partial u_2/\partial r)_{r_s} = 0$. In order to satisfy this condition it is first necessary to separate it into s and d functions. This may be done by use of the identity $\cos^2\theta = \frac{2}{3}P_2 + \frac{1}{3}$, where P_2 is the second Legendre polynomial. Then

$$u_2 = (\tfrac{2}{3}rf_p - \tfrac{1}{3}r^2u_0 + c_d f_d)P_2 + (\tfrac{1}{3}rf_p - \tfrac{1}{6}r^2u_0 + E_2\ (\partial u_0/\partial E)_{E_0} + c_s u_0). \tag{4.6.9}$$

The function f_d is the radial part of the d solution of the homogeneous equation. The coefficient c_d is determined by the boundary condition. On differentiating the multiplier of P_2 with respect to r, we obtain

$$c_d = \tfrac{2}{3} r_s \,(df_p/dr)_{r_s}\, (df_d/dr)_{r_s}^{-1}. \tag{4.6.10}$$

However, the boundary condition on the s part of u_2 cannot be satisfied by a choice of c_s, since $(du_0/dr)_{r_s}$ is already zero. The function $\partial u_0/\partial E$ can be uniquely determined by the condition that $u_0(E, r)$ be normalized for all energies. If

$$\int u_0^2(E, r)\, d^3r = \text{constant},$$

then

$$\int u_0(r)\, \partial u_0/\partial E\, d^3r = 0. \tag{4.6.11}$$

Hence u_0 and $\partial u_0/\partial E$ are orthogonal. It is apparent that the additional term $c_s u_0(r)$ in (4.6.9) is superfluous, and may be dropped. The boundary condition on the s-like part of u_2 actually determines E_2:

$$E_2 = -r_s/3\, (\partial f_p/dr)_{r_s}\, (\partial^2 u_0/\partial r\, \partial E)^{-1}{}_{r_s, E_0}. \tag{4.6.12}$$

This expression is inconvenient because of the presence of the second derivative. However, this term can be replaced with something simpler with the aid of Green's theorem. We multiply both sides of (4.6.8c) by u_0, integrate over the all volume, and apply Green's theorem:

$$\begin{aligned}
&\int_\Omega u_0[-\nabla^2 + V - E_0]\,(\partial u_0/\partial E)\, d^3r \\
&\quad = \int (\partial u_0/\partial E)\,[-\nabla^2 + V - E_0] u_0\, d^3r \\
&\qquad + \int_s [(\partial u_0/\partial r)_{r_s}\, (\partial u_0/\partial E)_{r_s} - u_0\, (\partial^2 u_0/\partial r\, \partial E)_{r_s, E_0}]\, ds \\
&\quad = \int_n u_0^2(r)\, d^3r = 1.
\end{aligned} \tag{4.6.13}$$

In (4.6.13) ds is the element of area, and the surface integral includes the surface of the atomic sphere. It is convenient (and customary) to choose the normalization so that the volume integral of u_0^2 over the atomic cell is unity (this contrasts with the usage in previous sections). Equation (4.6.13) reduces to

$$-4\pi r_s^2 u_0(r_s)\ (\partial^2 u_0/\partial r\ \partial E)_{r_s,E_0} = 1. \qquad (4.6.14)$$

Then Eq. (4.6.12) may be reexpressed as

$$E_2 = [4\pi r_s^3 u_0(r_s)/3][r_s/f_p(r_s)]\ (df_p/dr)_{r_s}. \qquad (4.6.15)$$

[Here we have $r_s u_0(r_s) = f_p(r_s)$, as follows from the vanishing of u_1 on the atomic sphere.] If we introduce the functions $R_0 = ru_0/(4\pi)^{1/2}$, $P = rf_p$, which are the functions usually calculated, we have

$$E_2 = (r_s R_0^2/3)[r/P\ dP/dr - 1]_{r_s}. \qquad (4.6.16)$$

The expression for c_d can be converted into a more convenient form

$$c_d = [-2E_2/r_s R_0(r_s)](df_d/dr)_{r_s}^{-1}. \qquad (4.6.17)$$

Silverman has continued this procedure through the calculation of E_4. His result is

$$E_4 = \tfrac{2}{5} r_s^2 E_2 - \tfrac{4}{15}(E_2^2/\gamma) r_s^2 (r/f_d\ df_d/dr)_{r_s}^{-1} + [\gamma E_2/u_0(r_s)]$$
$$\times \left\{ E_2/\gamma\ (\partial u_0/\partial E)_{r_s,E_0} - r_s u_0(r_s) \left[\left[\int_0^{r_s} P^2(E_0, r)\ dr/P^2(E_0, r_s) \right] \right] \right\} \qquad (4.6.18)$$

where

$$\gamma = (4\pi/3) r_s^3 u_0^2(r_s) = -r_s[u_0(r_s)/3](\partial^2 u_0/\partial r\ \partial E)^{-1}_{r_s,E_0} \qquad (4.6.19)$$

and

$$\int_0^{r_s} P^2(E_0, r)\ dr/P^2(E_0, r_s) = -(d/dE)[1/P\ dP/dr]_{E_0,r_s}. \qquad (4.6.20)$$

An important property of these formulas is that, in spite of first appearances, they are independent of the normalization of the wavefunction and depend in fact only on logarithmic derivatives of the functions on the surface of the sphere. The significance of this fact is that computations may then be performed using a pseudopotential or quantum defect approach (Section 4.11) which determines the wavefunction of the valence electron correctly only in the region outside the atomic core. This is easy to see in the case of E_2. Equation (4.6.12) can be rewritten in a form explicitly independent of normalization:

$$E_2 = -[\tfrac{1}{3} r_s^2/f_p(r_s)](df_p/dr)_{r_s} u_0(r_s)\ (\partial^2 u_0/\partial r\ \partial E)^{-1}_{r_s,E_0}. \qquad (4.6.21)$$

It is convenient to introduce the notation

$$\phi_l = r_s/f_l(r_s)\ (df_l/dr)_{r_s}; \qquad \phi_l' = -(\partial\phi_l/\partial E)_{E_0}; \qquad \phi_l'' = (\partial^2\phi_l/\partial E^2)_{E_0}. \tag{4.6.22}$$

Then (4.6.21) becomes

$$E_2 = \tfrac{1}{3}r_s^2(\phi_1/\phi_0'). \tag{4.6.23}$$

Brooks (1958) has shown that E_4 can be expressed in the normalization-independnet form

$$E_4 = (2r_s^4/15)(\phi_1/\phi_0')[1 - \tfrac{2}{3}(\phi_1/\phi_2) - \tfrac{5}{6}(\phi_1/\phi_0') + \tfrac{5}{12}(\phi_0''\phi_1/(\phi_0')^2)]. \tag{4.6.24}$$

This procedure can be carried through for the Dirac equation describing the motion of a single electron in a periodic potential. Relativistic effects can be of some importance in solid state problems concerning heavy atoms, since the valence electron penetrates the core into a region in which the electrostatic potential is quite strong. In the relativistic case, the effective mass is determined from an expression similar to (4.6.16) which contains in place of the single function P the large components of the radial wave functions for the p states of $j = \frac{3}{2}(P_{3/2})$ and $j = \frac{1}{2}(P_{1/2})$. The function R_0 goes over to the similar large component of the $j = \frac{1}{2}$, s-like function $R_{1/2}$ (Callaway *et al.*, 1957):

$$E_2 = [r_sR^2_{1/2}(r_s)/3][\tfrac{2}{3}(r_s/P_{3/2})\ (dP_{3/2}/dr) + \tfrac{1}{3}(r_s/P_{1/2})\ (dP_{1/2}/dr) - 1]_{r_s}. \tag{4.6.25}$$

4.7 The Green's Function Method

4.7.1 *The Variational Principle*

The difficulties experienced with point matching of boundary conditions in the cellular method provide a motivation for the development of a variational approach. One of the most interesting of these is the Green's function or Korringa–Kohn–Rostoker (KKR) method (Korringa, 1947; Kohn and Rostoker, 1954; Ham and Segall, 1961; Ziman, 1965; Slater, 1966).

Consider the functional

$$I = \int \psi^*(H - E)\psi\ d^3r \tag{4.7.1}$$

where H is some Hamiltonian. The demand that

$$\delta I = 0 \tag{4.7.2}$$

when a small variation is made in ψ leads to the usual Schrödinger equation

$$H\psi = E\psi \tag{4.7.3}$$

provided that suitable boundary conditions are satisfied. Normally, these conditions amount to specified behavior of ψ and its derivatives on the boundaries of the region of integration. In the case of an isolated system (a free atom for instance), it is sufficient to require that ψ vanish sufficiently rapidly at infinity.

Care is required in a solid. Let us suppose in (4.7.1) that the region of integration has been chosen to be the atomic (Wigner–Seitz) cell of a monatomic crystal. A variation of ψ is made by replacing ψ in (4.7.1) by ψ' where $\psi' = \psi + \delta\psi$, and ψ satisfies both the Schrödinger equation and the boundary conditions. We find, with the use of Green's theorem,

$$\delta I = \int (\delta\psi \, \nabla\psi^* - \psi^* \, \nabla\delta\psi) \cdot d\mathbf{s}. \tag{4.7.4}$$

The integral goes over the surface of the atomic cell. The quantity I will vanish in general only if $\delta\psi$ obeys the correct boundary conditions prescribed by Bloch's theorem, for then the contribution from conjugate points on the cell boundary (points separated by a lattice translation vector) will cancel. It is possible, however, to find a different functional which does vanish for arbitrary variations of ψ (Kohn, 1952).

The approach of Kohn and Rostoker is based on an integral equation which is equivalent to the usual Schrödinger equation. It is first necessary to find the Green's function $\mathcal{G}(\mathbf{k}, \mathbf{r} - \mathbf{r}')$ which satisfies the equation

$$(\nabla^2 + E)\mathcal{G}(\mathbf{k}, \mathbf{r} - \mathbf{r}') = \delta(\mathbf{r} - \mathbf{r}') \tag{4.7.5}$$

subject to the boundary conditions required by Bloch's theorem

$$\mathcal{G}(\mathbf{k}, \mathbf{r} + \mathbf{R}_\mu) = \exp(i\mathbf{k}\cdot\mathbf{R}_\mu)\mathcal{G}(\mathbf{k}, \mathbf{r}). \tag{4.7.6}$$

As usual, $\mathbf{R}_\mu$ is any lattice translation vector.

The standard technique for determining a Green's function relies on use of eigenfunctions of the homogenous equation which satisfy the appropriate boundary conditions. If these eigenfunctions (of ∇^2) are distinguished by an index j (eigenvalues are E_j) we have (Goertzel and Tralli, 1960)

$$\mathcal{G}(\mathbf{r}, \mathbf{r}') = \sum_j \psi_j^*(\mathbf{r}')(E - E_j)^{-1}\psi_j(\mathbf{r}). \tag{4.7.7}$$

In the present case, the eigenfunctions are plane waves, which we choose to be normalized in the unit cell (volume Ω)

$$\psi_j(\mathbf{r}) = \Omega^{-1/2} \exp[i(\mathbf{k} + \mathbf{K}_s)\cdot\mathbf{r}] \tag{4.7.8}$$

where $\mathbf{K}_s$ is any reciprocal lattice vector. The Green's function is

$$\mathcal{G}(\mathbf{k}, \mathbf{r} - \mathbf{r}') = \Omega^{-1} \sum_s \exp[i(\mathbf{k} + \mathbf{K}_s) \cdot (\mathbf{r} - \mathbf{r}')]/[E - (\mathbf{k} + \mathbf{K}_s)^2]. \tag{4.7.9}$$

If we operate on this $\mathcal{G}$ with $(\nabla^2 + E)$, it will be verified that (4.7.5) is satisfied since, according to the closure property of the eigenfunctions

$$\sum_j \psi_j^*(\mathbf{r}')\psi_j(\mathbf{r}) = \Omega^{-1} \sum_s \exp[i(\mathbf{k} + \mathbf{K}_s) \cdot (\mathbf{r} - \mathbf{r}')] = \delta(\mathbf{r} - \mathbf{r}'). \tag{4.7.10}$$

The normalization chosen for the eigenfunctions in Eq. (4.7.8) is appropriate for (4.7.1). It will be seen from (4.7.9) that $\mathcal{G}$ has the properties

$$\mathcal{G}(\mathbf{r} - \mathbf{r}') = \mathcal{G}^*(\mathbf{r}' - \mathbf{r}) \tag{4.7.11a}$$

and

$$\mathcal{G}(\mathbf{r} + \mathbf{R}_\mu - \mathbf{r}') = \exp(i\mathbf{k} \cdot \mathbf{R}_\mu)\mathcal{G}(\mathbf{r} - \mathbf{r}'). \tag{4.7.11b}$$

The Schrödinger equation can be written in the form

$$(\nabla^2 + E)\psi(\mathbf{k}, \mathbf{r}) = V\psi(\mathbf{k}, \mathbf{r}). \tag{4.7.12}$$

This equation can be converted into an integral equation

$$\psi(\mathbf{k}, \mathbf{r}) = \int_\Omega \mathcal{G}(\mathbf{k}, \mathbf{r} - \mathbf{r}') V(\mathbf{r}')\psi(\mathbf{k}, \mathbf{r}')\, d^3r' \tag{4.7.13}$$

in which $\mathcal{G}$ is given by (4.7.9). Equation (4.7.13) may be verified by direct substitution. Unlike the differential equation (4.7.12), the integral equation contains the boundary conditions: (4.7.11b) guarantees that a solution of (4.7.13) satisfies the boundary conditions. Since (4.7.13) is a homogeneous integral equation, it possesses a solution only in exceptional cases, that is, for certain values of the energy which are the desired eigenvalues. The energy enters the integral equation through the Green's function.

The integral equation (4.7.13) can be derived from the variational principle

$$\delta\Lambda = 0 \tag{4.7.14a}$$

where

$$\Lambda = \int_\Omega \psi^*(\mathbf{r}) V(\mathbf{r})\psi(\mathbf{r})\, d^3r - \int_\Omega \int_{\Omega'} \psi^*(\mathbf{r}) V(\mathbf{r})\mathcal{G}(\mathbf{k}, \mathbf{r} - \mathbf{r}') V(\mathbf{r}')\psi(\mathbf{r}')\, d^3r\, d^3r'. \tag{4.7.14b}$$

This variational principle has the important property that $\delta\Lambda$ vanishes in first order for all variations from the solution of (4.7.13) regardless of whether the variations satisfy the boundary conditions. Further, for the exact solution of the problem,

$$\Lambda(\psi, \mathbf{k}, E) = 0. \tag{4.7.15}$$

If an approximate Λ is computed from some trial function ψ_t, and (4.7.15) is solved for the energy, the error in the energy is of second order compared to that of the trial function.

4.7.2 *The KKR Expansion*

A convenient method of employing the variational principle is to choose a trial function which is a linear combination of a finite number N of basis functions with undetermined coefficients

$$\psi(\mathbf{k}, \mathbf{r}) = \sum_{n}^{N} C_n(\mathbf{k})\phi_n(\mathbf{k}, \mathbf{r}). \tag{4.7.16}$$

This expansion is used to determine an approximate Λ, which has the form

$$\Lambda = \sum_{n,l}^{N} C_n^*(\mathbf{k})\Lambda_{nl}(\mathbf{k})C_l(\mathbf{k}) \tag{4.7.17}$$

in which the quantities Λ_{nl} are given by

$$\Lambda_{nl} = \int \phi_n^*(\mathbf{k}, \mathbf{r})V(\mathbf{r})\phi_l(\mathbf{k}, \mathbf{r})\, d^3r$$

$$- \iint \phi_n^*(\mathbf{k}, \mathbf{r})V(\mathbf{r})\mathcal{G}(\mathbf{k}, \mathbf{r} - \mathbf{r}')V(\mathbf{r}')\phi_l(\mathbf{k}, \mathbf{r}')\, d^3r\, d^3r'. \tag{4.7.18}$$

The requirement that Λ be stationary means that the partial derivatives of Λ with respect to the coefficients C must vanish or that, for each n,

$$\sum_{l}^{N} \Lambda_{nl}(\mathbf{k})C_l(\mathbf{k}) = 0. \tag{4.7.19}$$

A set of N linear, homogenous, equations is obtained. The condition for a nontrivial solution is that

$$\det(\Lambda_{nl}) = 0. \tag{4.7.20}$$

This equation may be used to determine the energy for a state of given $\mathbf{k}$ (or the wave vectors $\mathbf{k}$ for which the wave function will have a prescribed energy). If the functions $\phi_n(\mathbf{k}, \mathbf{r})$ are members of a complete set, the energy obtained will approach the correct energy as the number of functions increases.

One possible choice of functions $\phi_n(\mathbf{k}, \mathbf{r})$ is plane waves:

$$\phi_n(\mathbf{k}, \mathbf{r}) = \Omega^{-1/2} \exp[i(\mathbf{k} + \mathbf{K}_n)\cdot\mathbf{r}]. \tag{4.7.21}$$

In this case, the Λ_{nl} are given by

$$\Lambda_{nl} = V(\mathbf{K}_l - \mathbf{K}_n) + \sum_j V(\mathbf{K}_j - \mathbf{K}_n)V(\mathbf{K}_l - \mathbf{K}_j)/[(\mathbf{k} + \mathbf{K}_j)^2 - E]. \tag{4.7.22}$$

This result could be used in (4.7.20) in place of the plane wave method previously discussed; however, the sum in (4.7.22) would not be easy to evaluate. Use of such a Λ would involve a sort of complicated energy-dependent effective potential, whose form is quite similar to that obtained in the Lowdin form of perturbation theory discussed in Section 4.1.

The possibilities of practical application of the KKR method are limited by the difficulties which arise in evaluating (4.7.14b). To simplify the problem, it has been customary to restrict consideration to a particular class of potentials. Specifically, it is assumed that the potential in an atomic cell is constant outside of the largest sphere which can be inscribed inside the atomic (Wigner–Seitz) cell. The radius of this sphere is denoted r_i. It is also assumed that the potential is spherically symmetric inside the sphere. By a proper choice of the zero of energy, the potential may be made to vanish outside the inscribed sphere

$$V(\mathbf{r}) = V(|\mathbf{r}|) \quad \text{for} \quad r < r_i, \qquad V(\mathbf{r}) = 0 \quad \text{for} \quad r > r_i. \tag{4.7.23}$$

Potentials of this type are frequently called "muffin-tin" potentials, and have been widely used in recent years. Of course actual crystal potentials will not be of this form, and it must be assumed (or hoped) that not too much violence is done to whatever real problem is being studied by this approximation. Experience seems to show that this is the case in simple metals; however, in the case of transition metals, semiconductors, and insulators, the situation may be less pleasant.

The trial wave function to be used in the variational principle is taken to be

$$\psi(\mathbf{k}, \mathbf{r}) = \sum_{lm} C_{lm}(\mathbf{k}) R_l(E, r) Y_{lm}(\theta, \phi) \tag{4.7.24}$$

in which the radial function R_l is a solution of the radial Schrödinger equation (4.6.2) for the potential $V(r)$ inside the muffin tin. As in the cellular method, if $\mathbf{k}$ is a symmetry point, linear combinations of spherical harmonics may be chosen so that ψ has specific symmetry properties. Since the variational principle involves $V(r)$, and since this is zero outside the muffin tin, we do not have to write an explicit trial function in the exterior region.

It is necessary to determine the "matrix elements" of Λ on the basis of the functions $R_l(r) Y_{lm}(\theta, \phi)$. The derivation is quite lengthy and complete details will not be given here. The interested reader should consult the paper of Kohn and Rostoker (1954) for a more complete treatment.

The restriction to potentials which vanish outside the inscribed sphere makes possible the conversion of the expression (4.7.14b) for Λ to a form which does not depend explicitly on the potential but involves surface integrals over the inscribed sphere. It is essential to the argument that the expansion (4.7.24) will be a solution of the Schrödinger equation (4.7.12) within the inscribed sphere. Also, some care has to be employed in the derivation since the Green's function is singular when its argument is zero. Where necessary, we will consider a sphere whose radius is slightly smaller than that of the inscribed sphere $r_i - \epsilon$, and subsequently let ϵ tend to zero. First, let us observe that as a consequence of (4.7.12),

$$\int \mathcal{G}(\mathbf{r} - \mathbf{r}') V(\mathbf{r}') \psi(\mathbf{r}') \, d^3r' = \int \mathcal{G}(\mathbf{r} - \mathbf{r}') (\nabla'^2 + E) \psi(\mathbf{r}') \, d^3r'.$$

The operator ∇' differentiates functions with respect to coordinate $\mathbf{r}'$. Green's theorem enables us to write

$$\begin{aligned} \int \mathcal{G}(\mathbf{r} - \mathbf{r}') (\nabla'^2 + E) \psi(\mathbf{r}') \, d^3r' &= \int (\nabla'^2 + E) \mathcal{G}(\mathbf{r} - \mathbf{r}') \psi(\mathbf{r}') \, d^3r' \\ &\quad + \int [\mathcal{G}(\mathbf{r} - \mathbf{r}') \, \nabla' \psi(\mathbf{r}') - \psi(\mathbf{r}') \, \nabla' \mathcal{G}(\mathbf{r} - \mathbf{r}')] \cdot d\mathbf{s}' \\ &= \psi(\mathbf{r}) + \int_{r_i - \epsilon} [\mathcal{G}(\mathbf{r} - \mathbf{r}') \, \partial\psi/\partial r' - \psi(\mathbf{r}') \, \partial \mathcal{G}(\mathbf{r} - \mathbf{r}')/\partial r'] \, ds'. \end{aligned} \tag{4.7.25}$$

Thus

$$\begin{aligned} \psi(\mathbf{r}) - \int \mathcal{G}(\mathbf{r} - \mathbf{r}') V(\mathbf{r}') \psi(\mathbf{r}') \, d^3r' \\ = - \int_{r_i - \epsilon} [\mathcal{G}(\mathbf{r} - \mathbf{r}') \, \partial\psi(\mathbf{r}')/\partial r' - \psi(\mathbf{r}') \, \partial \mathcal{G}(\mathbf{r} - \mathbf{r}')/\partial r'] \, ds'. \end{aligned} \tag{4.7.26}$$

The volume integrals include the interior of the sphere of radius $r_i - \epsilon$; the surface integrals pertain to the surface of this sphere. If we let $\epsilon \to 0$, we find that for potentials which vanish beyond the inscribed sphere and for wave functions which satisfy the Schrödinger equation inside the sphere, the basic integral equation (4.7.13) reduces to

$$\int_{r_i} [\mathcal{G}(\mathbf{r}-\mathbf{r}')\,\partial\psi(\mathbf{r}')/\partial r' - \psi(\mathbf{r}')\,\partial\mathcal{G}(\mathbf{r}-\mathbf{r}')/\partial r']\,ds' = 0. \tag{4.7.27}$$

We return to the determination of Λ from (4.7.14b). Since there are two integrations, a second sphere of slightly smaller radius, which is taken to be $r_i - 2\epsilon$, is also required. Then by an argument similar to that leading to (4.7.26), we have, for $r' < r$

$$\begin{aligned}\int \psi^*(\mathbf{r})V(\mathbf{r})\mathcal{G}(\mathbf{r}-\mathbf{r}')\,d^3r &= \int_\Omega (\nabla^2 + E)\psi^*(\mathbf{r})\mathcal{G}(\mathbf{r}-\mathbf{r}')\,d^3r \\ &= \int_{r_i-2\epsilon} [\partial\psi^*/\partial r\,\mathcal{G}(\mathbf{r}-\mathbf{r}') - \psi^*\,\partial\mathcal{G}(\mathbf{r}-\mathbf{r}')/\partial r]\,ds. \end{aligned}\tag{4.7.28}$$

Substitute (4.7.26) and (4.7.28) into (4.7.14b)

$$\begin{aligned}\Lambda &= \int_\Omega \psi^*(\mathbf{r})V(\mathbf{r})\left[\psi(\mathbf{r}) - \int \mathcal{G}(\mathbf{r}-\mathbf{r}')V(\mathbf{r}')\psi(\mathbf{r}')\,d^3r'\right]d^3r \\ &= -\int \psi^*(\mathbf{r})V(\mathbf{r})\left[\int_{r_i-\epsilon} \{\mathcal{G}(\mathbf{r}-\mathbf{r}')\,\partial\psi(\mathbf{r}')/\partial r' \right. \\ &\quad \left. - \psi(\mathbf{r}')\,\partial\mathcal{G}(\mathbf{r}-\mathbf{r}')/\partial r'\}\,ds'\right]d^3r \\ &= \int_{r_i-2\epsilon} ds \int_{r_i-\epsilon} ds'[\partial\psi^*(\mathbf{r})/\partial r - \psi^*(\mathbf{r})\partial/\partial r][\psi(\mathbf{r}')\,\partial\mathcal{G}(\mathbf{r}-\mathbf{r}')/\partial r' \\ &\quad - \mathcal{G}(\mathbf{r}-\mathbf{r}')\,\partial\psi/\partial r'].\end{aligned}\tag{4.7.29}$$

In order to compute the "matrix elements" of Λ, we replace ψ^* in (4.7.29) by $R_{l'}Y^*_{l'm'}(\theta, \phi)$ and ψ by $R_l Y_{lm}(\theta, \phi)$. It is then necessary to express the Green's function (4.7.9) in spherical waves. This may be done with the use of the expansion for plane waves

$$\exp(i\mathbf{k}\cdot\mathbf{r}) = 4\pi \sum_{lm} i^l j_l(kr)\,Y_{lm}(\theta, \phi)\,Y^*_{lm}(\theta_k\phi_k). \tag{4.7.30}$$

The angles θ, ϕ and θ_k, ϕ_k specify the orientation of $\mathbf{r}$ and $\mathbf{k}$ with respect to a fixed set of axes. Then Green's function is

$$\mathcal{G}(\mathbf{k}, \mathbf{r} - \mathbf{r}') = \frac{(4\pi)^2}{\Omega_0} \sum_{lm,l'm'} \sum_n i^{l-l'} \frac{j_l(|\mathbf{K}_n + \mathbf{k}|r) j_{l'}(|\mathbf{K}_n + \mathbf{k}|r)}{E - (\mathbf{K}_n + \mathbf{k})^2}$$
$$\times Y_{lm}(\theta, \phi) Y^*_{l'm'}(\theta', \phi') Y^*_{lm}(\theta_K, \phi_K) Y_{l'm'}(\theta_K, \phi_K). \tag{4.7.31}$$

The angles θ', ϕ' and θ_K, ϕ_K now refer to the vectors $\mathbf{r}'$ and $\mathbf{K}_{n+}\mathbf{k}$, respectively.

It is possible to rewrite this expression in several different ways. One of these is (valid for $r < r' < r_i$)

$$\mathcal{G}(\mathbf{k}, \mathbf{r} - \mathbf{r}') = \sum_{lm,l'm'} [A_{lm,l'm'}(\mathbf{k}, E) j_l(\kappa r) j_{l'}(\kappa r')$$
$$+ \kappa\, \delta_{ll'}\, \delta_{mm'}\, j_l(\kappa r) n_l(\kappa r')] Y_{lm}(\theta\phi) Y^*_{l'm'}(\theta', \phi') \tag{4.7.32}$$

where $\kappa^2 = E$ and

$$n_l(x) = (\pi/2x)^{1/2} J_{-l-1/2}(x). \tag{4.7.33}$$

The quantities $A_{lm,l'm'}(\mathbf{k}, E)$ are given by fairly complex summations, which will be stated subsequently.

Equation (4.7.32) can be derived as follows: Within an atomic cell, it must be possible to express the Green's function as the sum of two parts, one of which is a particular integral of (4.7.5) and hence singular when $r = r'$, the other a solution of the homogeneous equation, regular for $r = r'$.

It can be verified (see Goertzel and Tralli, 1960, Chap. XIII) that a singular solution of (4.7.5), valid within the inscribed sphere is

$$-(1/4\pi)[\cos(\kappa|\mathbf{r} - \mathbf{r}'|)/|r - r'|]$$
$$= \kappa \sum_{lm} j_l(\kappa r_<) n_l(\kappa r_>) Y_{lm}(\theta, \phi) Y^*_{lm}(\theta', \phi') \tag{4.7.34}$$

in which $r_<$ ($r_>$) is the lesser (greater) of r and r'. Similarly, the regular part of $\mathcal{G}$ must have the form

$$\sum_{lm,l'm'} A_{lm,l'm'} j_l(\kappa r) j_{l'}(\kappa r') Y_{lm}(\theta\phi) Y^*_{l'm'}(\theta', \phi') \tag{4.7.35}$$

in which the $A_{lm,l'm'}$ are constants to be determined so that the boundary conditions (4.7.11b) are satisfied. (This function can be seen to be regular, and is easily found to be a solution of the homogenous equation.) Equation (4.7.32) is simply the sum of (4.7.34) and (4.7.35); and comparison of

(4.7.32) and (4.7.31) shows that

$$A_{lm,l'm'}(\mathbf{k}, E)$$

$$= [(4\pi)^2/\Omega] i^{(l-l')} [j_l(\kappa r) j_{l'}(\kappa r')]^{-1}$$

$$\times \sum_n \frac{j_l(|\mathbf{K}_n + \mathbf{k}| r) j_{l'}(|\mathbf{K}_n + \mathbf{k}| r') Y^*_{lm}(\theta_K, \phi_K) Y_{l'm'}(\theta_K, \phi_K)}{E - (\mathbf{K}_n + \mathbf{k})^2}$$

$$- \frac{\kappa\, \delta_{ll'}\, \delta_{mm'}\, n_l(\kappa r')}{j_l(\kappa r')}. \qquad (4.7.36)$$

In spite of the appearance of r and r' in (4.7.36), the preceding general argument implies that the $A_{lm,l'm'}$ must actually be independent of these quantities. Kohn and Rostoker have given other expressions for the $A_{lm,l'm'}$ which are simpler to evaluate, but we will not discuss these here.

We now return to (4.7.29) to evaluate the quantities $\Lambda_{lm,l'm'}$ which are the "matrix elements of Λ" computed with the basis functions $R_l Y_{lm}(\theta, \phi)$. Equation (4.7.32) is substituted, and the integrals over the spheres are performed using the orthogonality properties of spherical harmonics. The result is, in the limit $\epsilon \to 0$,

$$\begin{aligned} \Lambda_{lm,l'm'} &= R_l R_{l'} [L_l j_l - j_l'][(A_{lm,l'm'}\, j'_{l'} + \kappa\, \delta_{mm}\, \delta_{ll'}\, n'_{l'}) \\ &\qquad - (A_{lm,l'm'}\, j_{l'} + \kappa\, \delta_{ll'}\, \delta_{mm'}\, n_{l'}) L_l] \\ &= -R_l R_{l'} [L_l j_l - j_l'][L_{l'} j_{l'} - j'_{l'}][A_{lm,l'm'} \\ &\qquad + \kappa\, \delta_{ll'}\, \delta_{mm'} (n_l L_l - n_l')/(j_l L_l - j_l')]. \end{aligned} \qquad (4.7.37)$$

In (4.7.37),

$$L_l = (1/R_l)\, dR_l/dr\,|_{r=r_i}; \qquad j_l' = dj_l(\kappa r)/dr\,|_{r=r_i}, \qquad (4.7.38)$$

and all functions are evaluated at r_i.

The determinantal equation (4.7.20) may now be constructed, and simplified by division of the common factors $-R_l[L_l j_l - j_l']$ from each row and $R_{l'}[L_{l'} j_{l'} - j'_{l'}]$ from each column. The resulting equation

$$\det |A_{lm,l'm'} + \kappa\, \delta_{ll'}\, \delta_{mm'}\, (n_l' - n_l L_l)/(j_l' - j_l L_l)| = 0 \qquad (4.7.39)$$

gives the required connection between E and $\mathbf{k}$. If L is the maximum value of the angular momentum l included in (4.7.39), the determinant contains $(L + 1)^2$ rows and columns. This equation can be written in an interesting form if we recall the relation between the scattering phase shifts δ_l for the lth partial wave and the logarithmic derivative of the radial wave function in the case of a spherical potential which is zero for $r > r_i$ (Schiff, 1968, page 121):

$$\tan \delta_l = (j_l' - L_l j_l)/(n_l' - L_l n_l). \qquad (4.7.40)$$

Our notation differs from that of Schiff in that the prime here indicates derivative with respect to r rather than with respect to κr. Equation (4.7.39) becomes

$$\det | A_{lm,l'm'} + \delta_{ll'} \delta_{mm'} \kappa \operatorname{ctn} \delta_l | = 0. \tag{4.7.41}$$

This result is quite interesting. It indicates that the band structure produced by a periodic array of muffin-tin potentials depends on the potential only through the scattering phase shifts. Thus, two potentials which give the same phase shifts in a given energy region will produce the same band structure for those energies.

The KKR method has been applied fairly extensively to band structure calculations in metals. If the muffin-tin potential approximation is assumed to be acceptable, the utility of the method depends on the availability of the structure constants $A_{lm,l'm'}$. These quantities can be computed once and for all for a given lattice. The calculation is not too difficult with modern computers. It might appear that some difficulty would occur in these formulas for energies ($E = \kappa^2 < 0$) below the zero of the muffin-tin potential. This is not the case as one may introduce $\kappa = i | \kappa |$ and revise the formulas slightly to keep everything real. Relativistic effects have been included by Onodera and Okazaki (1966), who have derived an expression similar to (4.7.39) based on the Dirac equation.

The convergence of the KKR method is determined by the scattering phase shifts δ_l in (4.7.41). These become less important as the angular momentum increases. The convergence appears to be rather rapid; four phase shifts [$l = 0, 1, 2, 3$; the determinant in (4.7.41) is 16×16] being sufficient to give quite good results for the Fermi surface of noble metals (Shaw *et al.*, 1972). In analogy with the pseudopotential method where Fourier coefficients of the potential could be considered to be disposable parameters, the scattering phase shifts δ_l in (4.7.41) can be considered as parameters to be determined empirically so as to obtain a band structure or Fermi surface in agreement with experiment. Such an approach can determine phase shifts only up to an arbitrary multiple of π. Neglect of such factors of π is related to the removal of core states from the potential by the pseudopotential approach, since at $E = 0$, $\delta_l = n_{Bl}\pi$ where n_{Bl} is the number of bound states of angular momentum l in the particular potential. (This is Levinson's theorem; see Newton, 1966, Chap. XII.)

4.7.3 *Transformation to a Plane Wave Basis*

It is instructive to derive a determinental equation equivalent to (4.7.41) in which the elements are labeled by reciprocal lattice vectors instead of angular momentum states. This plane wave representation was obtained by Ziman (1965), and we follow his procedure.

A modified phase shift η_l is defined by

$$\operatorname{ctn} \eta_l = \operatorname{ctn} \delta_l - [n_l(\kappa r_i)/j_l(\kappa r_i)]. \tag{4.7.42}$$

Equation (4.7.39) can be rewritten with the aid of (4.7.36) in the form

$$\det \left| \sum_n [B_{lm,n}B^*_{l'm',n}/\mathcal{E}(n)] - C_l\, \delta_{ll'}\, \delta_{mm'} \right| = 0 \tag{4.7.43}$$

where

$$B_{lm,n} = (4\pi i^l/\Omega^{1/2})[j_l(|\mathbf{K}_n + \mathbf{k})|r_i)/j_l(\kappa r_i)]Y^*_{lm}(\theta_K, \phi_K) \tag{4.7.44}$$

$$\mathcal{E}(n) = E - (\mathbf{K}_n + \mathbf{k})^2 \tag{4.7.45}$$

and

$$C_l = \kappa \operatorname{ctn} \eta_l. \tag{4.7.46}$$

We have used the fact that $A_{lm,l'm'}$ is independent of the values of r and r' to make the convenient choice of $r = r' = r_i$. Next, we divide the rows and columns of the determinant (4.7.43) by $C_l^{1/2}$, and define a matrix

$$\Phi_{lm,n} = B_{lm,n}/[C_l\mathcal{E}(n)]^{1/2}. \tag{4.7.47}$$

This matrix need not be square. Its rows are labeled by angular momentum quantum numbers, and its columns by reciprocal lattice vectors. Let the complex conjugate of the transposed matrix be denoted by $\tilde{\Phi}_{n,lm}$. In terms of these quantities (4.7.43) becomes

$$\det \left| \sum_n \Phi_{lm,n}\tilde{\Phi}_{n,l'm'} - \delta_{ll'}\, \delta_{mm'} \right| = 0. \tag{4.7.48}$$

This equation implies that there exists some vector $\mathbf{u}$ (components u_{lm}) such that

$$\sum_{l'm'} \sum_n \Phi_{lm,n}\tilde{\Phi}_{n,l'm'}u_{l'm'} = u_{lm}. \tag{4.7.49}$$

Define a vector $\mathbf{v}$ (components v_n) by

$$\sum_{l'm'} \tilde{\Phi}_{n,l'm'}u_{l'm'} = v_n. \tag{4.7.50}$$

Then v_n must satisfy

$$\sum_{n'} \Phi_{lm,n'}v_{n'} = u_{lm}. \tag{4.7.51}$$

Multiply (4.7.51) by $\tilde{\Phi}_{n,lm}$, sum on lm, and use (4.7.50):

$$\sum_{lmn'} \tilde{\Phi}_{n,lm}\Phi_{lm,n'}v_{n'} = v_n. \tag{4.7.52}$$

Equations (4.7.52) are a set of simultaneous, homogenous, linear equations. The condition for a nontrivial solution is then that

$$\det \left| \sum_{lm} \tilde{\Phi}_{n,lm}\Phi_{lm,n'} - \delta_{nn'} \right| = 0. \tag{4.7.53}$$

We replace the Φ by the B, with the use of (4.7.47), and then multiply the rows and columns by $\mathcal{E}^{1/2}(n)$. The result is

$$\det \left| \sum_{lm} (B_{lm,n}B^*_{lm,n'}/C_l) - \mathcal{E}(n)\,\delta_{nn'} \right| = 0 \tag{4.7.54}$$

or, equivalently,

$$\det \left| [(\mathbf{k} + \mathbf{K}_n)^2 - E]\delta_{nn'} + \mathcal{V}_{nn'} \right| = 0 \tag{4.7.55}$$

where

$$\mathcal{V}_{nn'} = -\sum_{lm} (B_{lm,n}B^*_{lm,n'}/C_l). \tag{4.7.56}$$

The addition theorem for spherical harmonics asserts that

$$\sum_m Y^*_{lm}(\theta_{\mathrm{K}}, \phi_{\mathrm{K}})Y_{lm}(\theta_{\mathrm{K}'}, \phi_{\mathrm{K}'}) = [(2l+1)/4\pi]P_l(\cos\theta_{nn'}) \tag{4.7.57}$$

where $\theta_{nn'}$ is the angle between the vectors $\mathbf{k} + \mathbf{K}_n$ and $\mathbf{k} + \mathbf{K}_{n'}$. We can write $\mathcal{V}_{nn'}$ as

$$\mathcal{V}_{nn'} = \frac{-4\pi}{\kappa\Omega} \sum_l (2l+1) \tan\eta_l \frac{j_l(|\mathbf{k}+\mathbf{K}_n|r_{\mathrm{i}})j_l(|\mathbf{k}+\mathbf{K}_{n'}|r_{\mathrm{i}})}{j_l^2(\kappa r_{\mathrm{i}})} P_l(\cos\theta_{nn'}). \tag{4.7.58}$$

The secular equation of the KKR method, based on functions of definite angular momentum, has been transformed to a secular equation based on plane waves. The effective Hamiltonian is quite different in appearance from that considered in Sections 4.2 and 4.3. The interpretation of (4.7.58) has been extensively discussed by Slater (1966). An approach based on (4.7.58) has been of considerable interest in connection with d bands, and forms the basis for several attempts to justify the combined interpolation of Section 4.5 in a more fundamental way (Heine, 1967; Hubbard, 1967; Hubbard and Dalton, 1968).

The essential idea of these approaches is that the interaction between the d ($l = 2$) portion of a plane wave and the atomic potentials gives rise to a scattering resonance; so that the phase shift has the form

$$\delta_2 = \tan^{-1}[\tfrac{1}{2}W/(E_d - E)] \tag{4.7.59a}$$

where W is the width of the resonance and E_d is its position. The modified phase shift η_2 also has a similar form with a displaced resonant position E_d'

$$\eta_2 = \tan^{-1}[\tfrac{1}{2}W/(E_d' - E)] \tag{4.7.59b}$$

with

$$E_d' = E_d - \tfrac{1}{2}W[n_2(\kappa r_i)/j_2(\kappa r_i)]. \quad (4.7.60)$$

Other partial waves may be expected to be nonresonant, that is, to have slowly varying phase shifts. Use of (4.7.59b) in the secular equation (4.7.55) appears to give the sort of band structure expected for transition metals where a broad free-electron-like band overlaps, and hybridizes strongly with a set of d electrons. To see that this can happen, it is sufficient to consider the secular equation in the greatly oversimplified case that only terms with $n = 0$ are considered. Then we find

$$k^2 = E + (20\pi/\kappa\Omega)[W/2(E_d' - E)][j_2^2(kr_i)/j_2^2(\kappa r_i)]. \quad (4.7.61)$$

This transcendental equation is to be solved for real values of $k(E)$. For quantitative purposes, a more elaborate treatment is required. The interested reader should consult the references cited for further detail.

One computational feature of the KKR method requires comment. Even when the determinantal equation (4.7.41) is rewritten, as in (4.7.55), to have the desired energy value $\mathcal{E}$ appear on the diagonal, the matrix elements $\mathcal{V}_{nn'}$, Eq. (4.7.58), are functions of this energy. This means that (4.7.55) cannot be solved by matrix diagonalization to find all the energies for a given $\mathbf{k}$ at once, as is possible in the OPW and tight binding methods. Instead, it is necessary to obtain the matrix elements for specific values of $\mathcal{E}$, evaluate the determinant at that energy, and repeat the process as required to find zeros. This procedure must also be applied if (4.7.41) is used. In the case of a double degeneracy, the determinant, regarded as a function of energy, will not change sign as the energy passes through a root. This may cause computational difficulties in locating roots.

4.8 The Augmented Plane Wave Method

The fundamental reason for the difficulties encountered in energy band calculations is that the simplest functions which satisfy the boundary conditions imposed by Bloch's theorem are plane waves, but plane wave expansions do not converge readily in the interior of an atomic cell. One way to avoid this difficulty was proposed by Slater (1937). The wave function is expanded in a set of functions composed of plane waves in the outer regions of the atomic cell, and a sum of spherical waves in the interior. Such a function is called an *augmented plane wave*; and the use of such expansions characterizes the augmented plane wave method (APW). The method has been developed by several authors from different points of view: Slater (1953), Saffern and Slater (1953), Leigh (1956), Schlosser and Marcus (1963), Kleinman and Shurtleff (1969). The method has been reviewed by Dimmock (1971), and by Loucks (1967), whose description also presents a set of computer programs.

4.8.1 *Properties of Augmented Plane Waves*

A single augmented plane wave is defined to be the function (within a single unit cell)

$$\Phi(\mathbf{k}, \mathbf{r}) = \epsilon(r - r_{\mathrm{i}}) \exp(i\mathbf{k}\cdot\mathbf{r}) + \sum_{lm} a_{lm}\epsilon(r_{\mathrm{i}} - r) Y_{lm}(\theta, \phi) R_l(E, r). \tag{4.8.1}$$

The function ϵ is a unit step function

$$\epsilon(x) = 1 \quad \text{for} \quad x \geq 0, \qquad \epsilon(x) = 0 \quad \text{for} \quad x < 0. \tag{4.8.2}$$

As in our discussions of other methods, we consider only crystals containing a single atom in each unit cell. The plane wave is joined to the spherical waves on a sphere whose radius is r_{i} (usually, this sphere is the largest sphere which can be inscribed in the atomic cell). It is convenient to choose the coefficients a_{lm} so that the function $\Phi(\mathbf{k}, \mathbf{r})$ is continuous across the sphere. To do this, we expand the plane wave in spherical harmonics according to (4.7.30). Then we find that

$$a_{lm} = 4\pi i^l Y^*_{lm}(\theta_{\mathrm{k}}, \phi_{\mathrm{k}})[j_l(kr_{\mathrm{i}})/R_l(E, r_{\mathrm{i}})]. \tag{4.8.3}$$

Equation (4.8.3) implies that $\Phi(\mathbf{k}, \mathbf{r})$ would be continuous at r_{i} if all spherical harmonics were included. In any actual calculation only a finite number of terms (typically up to $l \approx 12$) can be incorporated. Then Φ actually has a small discontinuity across the sphere. Moreover, even if we consider Φ to be continuous, its normal derivative is discontinuous. A plane wave cannot in general be joined smoothly onto spherical waves in the interior of some region: There must be scattered waves as well. However, in spite of these seemingly unpleasant properties of the individual augmented plane waves, they remain as useful functions in which to expand the actual solid state wave function which is smooth across the sphere.

The function $R_l(E, r)$ is a solution of the radial wave equation (4.6.2) for some energy E. There are two ways in which this energy may be chosen. In the original paper of Slater (1937), E represents the energy of the wave function $\psi_n(\mathbf{k}, \mathbf{r})$ which we are trying to determine. In the subsequent work of Saffern and Slater (1953), the energy is set equal to the expectation value of the Hamiltonian with a single augmented plane wave. The former procedure is more commonly used, and will be adopted here.

It is not actually necessary to introduce the composite function Φ.

Schlosser and Marcus (1963) expanded the wave function in two different set of functions: plane waves in the region external to the spheres, and spherical functions in the interior. We will, however, follow the more conventional approach.

The APW method is most frequently used in conjunction with the muffin-tin approximation to the crystal potential discussed in the previous section. Thus it is assumed that the potential is constant in the region outside the inscribed sphere, and usually this constant is chosen to be zero by proper choice of the zero of energy. However, the muffin-tin approximation is less intimately embedded in the structure of the APW method than in the case of the KKR procedure and considerable attention has been given to removing this restriction (Rudge, 1969; Kleinman and Shurtleff, 1969; Koelling *et al.*, 1970). In order not to become too deeply involved in the details of the APW procedures, we will confine our attention in this discussion to muffin-tin potentials.

The Bloch function $\psi_n(\mathbf{k}, \mathbf{r})$ is expanded in the $\Phi(\mathbf{k}, \mathbf{r})$ as

$$\psi_n(\mathbf{k}, \mathbf{r}) = \sum_s c_n(\mathbf{k}, \mathbf{K}_s)\Phi(\mathbf{k} + \mathbf{K}_s, \mathbf{r}) \tag{4.8.4}$$

where the sum includes reciprocal lattice vectors. The coefficients c_n and the energy of the state are to be determined variationally. The augmented plane wave Φ already obeys the proper boundary conditions on the surface of the atomic cell; however, the variational calculation is not as simple as might be desired because of the discontinuities which exist inside the cell.

4.8.2 *Variational Procedure*

A variational expression for the energy valid in the case of functions which are discontinuous in any cell, but otherwise satisfy the correct boundary conditions was given by Schlosser and Marcus (1963). Let $u_\mathrm{i}(\mathbf{r})$ be a trial wave function which is continuous and differentiable in the interior of the cell, and let $u_\mathrm{o}(\mathbf{r})$ be a similar function in the exterior. The surface of discontinuity S has an element of area denoted dS, and the unit normal vector is $\hat{\mathbf{n}}$, which points out of S. For convenience, the inner and outer portions of the cell volume will be denoted i and o, respectively. The expression which is stationary is

$$E \int_{\mathrm{i+o}} u^* u \, d^3r = \int_{\mathrm{i+o}} u^* H u \, d^3r + \tfrac{1}{2} \int \{(u_\mathrm{o} - u_\mathrm{i})[\nabla(u_\mathrm{o}^* + u_\mathrm{i}^*)] - (u_\mathrm{o}^* + u_\mathrm{i}^*)[\nabla(u_\mathrm{o} - u_\mathrm{i})]\} \cdot d\mathbf{S}. \tag{4.8.5}$$

This equation may be shown to be the proper variational expression by the following argument. We write

$$E = E_t + \delta E, \qquad u_i = u_t + \delta u_i, \qquad u_o = u_t + \delta u_o \tag{4.8.6}$$

where E_t and u_t are the true energy and wave function; the latter being smooth inside the cell. The variation δu_o vanishes on the cell boundary and δu_i vanishes at the origin. Equation (4.8.5) will be verified if we substitute (4.8.6), collect all terms which are of first order in the variations, and show that E vanishes to this order.

We find, on substitution,

$$\begin{aligned}
&[E_t + \delta E]\int_{i+o} u_t^* u_t \, d^3r + E_t \int_i [\delta u_i^* u_t + u_t^* \delta u_i]\, d^3r \\
&\quad + E_t \int_o [\delta u_o^* \, u_t + u_i^* \, \delta u_o]\, d^3r \\
&= E_t\left[\int_{i+o} u_t^* u_t \, d^3r + \int_o \delta u_o^* u_t \, d^3r + \int_i \delta u_i^* u_t \, d^3r\right] + \int_i u_t^* H \, \delta u_i \, d^3r \\
&\quad + \int_0 u_t^* H \, \delta u_o \, d^3r + \int (\delta u_o - \delta u_i)\, \nabla u_t^* \cdot d\mathbf{S} - \int u_t^* \, \nabla(\delta u_o - \delta u_i) \cdot d\mathbf{S}.
\end{aligned} \tag{4.8.7}$$

This expression can be simplified with the aid of Green's theorem applied to the kinetic energy part of the Hamiltonian. In this way, we find that

$$\int_i u_t^* H \, \delta u_i \, d^3r = E_t \int_i u_t^* \, \delta u_i \, d^3r - \int [u_t^* \, \nabla \, \delta u_i - \delta u_i \, \nabla u_t^*] \cdot d\mathbf{S} \tag{4.8.8}$$

$$\int_o u_t^* H \, \delta u_o \, d^3r = E_t \int_o u_t^* \, \delta u_o \, d^3r + \int [u_t^* \, \nabla \, \delta u_o - \delta u_o \, \nabla u_t^*] \cdot d\mathbf{S}. \tag{4.8.9}$$

The opposite sign of the second term in (4.8.9) as compared to that in (4.8.8) results from the fact that $d\mathbf{S}$ points out of the inner region but into the outer region. We substitute (4.8.8) and (4.8.9) into (4.8.7), and find finally

$$\delta E \int u_t^* u_t \, d^3r = 0 \tag{4.8.10}$$

as required.

In our applications, we will suppose that the surface separating the regions is spherical, and that $u_o = u_i$ on this surface, although this is not exactly true. Then (4.8.5) simplifies to

$$E \int_{i+o} u^* u \, d^3r = \int_{i+o} u^* H u \, d^3r - \int u_i^* \, \partial(u_o - u_i)/\partial r \, dS. \qquad (4.8.11)$$

4.8.3 *APW Matrix Elements*

We now follow a procedure similar to that described in the previous section to determine the expansion coefficients for the wave function. Substitute (4.8.4) into (4.8.11), and differentiate the resulting expression with respect to the coefficients. A set of equations are obtained which have the form

$$\sum_t \{\langle s \mid H \mid t \rangle + \langle s \mid \mathcal{S} \mid t \rangle - E_n(\mathbf{k}) \langle s \mid t \rangle\} c_n(\mathbf{k}, \mathbf{K}_t) = 0. \qquad (4.8.12a)$$

In these equations, the matrix elements are

$$\langle s \mid H \mid t \rangle = \int_{i+o} \Phi^*(\mathbf{k} + \mathbf{K}_s, \mathbf{r}) H \Phi(\mathbf{k} + \mathbf{K}_t, \mathbf{r}) \, d^3r \qquad (4.8.12b)$$

$$\langle s \mid \mathcal{S} \mid t \rangle = - \int \Phi_o^*(\mathbf{k} + \mathbf{K}_s, \mathbf{r}_i) \, (\partial/\partial r)[\Phi_o(\mathbf{k} + \mathbf{K}_t, \mathbf{r}) - \Phi_i(\mathbf{k} + \mathbf{K}_t, \mathbf{r})]_{r_i} \, dS \qquad (4.8.13)$$

$$\langle s \mid t \rangle = \int_{i+o} \Phi^*(\mathbf{k} + \mathbf{K}_s, \mathbf{r}) \Phi(\mathbf{k} + \mathbf{K}_t, \mathbf{r}) \, d^3r. \qquad (4.8.14)$$

Here, Φ_o is the outer part of Φ while Φ_i is the inner part.

We will consider the evaluation of these matrix elements in more detail beginning with (4.8.12) and (4.8.14) in the combination $\langle s \mid H \mid t \rangle - E_n(\mathbf{k}) \langle s \mid t \rangle$. In the outer region V is zero according to the muffin-tin approximation, and Φ is a plane wave. The contribution from the inner region vanishes since Φ satisfies the Schrödinger equation there

$$\int_{i+o} \Phi^*(\mathbf{k} + \mathbf{K}_s, \mathbf{r}) (H - E) \Phi(\mathbf{k} + \mathbf{K}_t, \mathbf{r}) \, d^3r$$

$$= [(\mathbf{k} + \mathbf{K}_t)^2 - E] \int_o \exp[i(\mathbf{K}_t - \mathbf{K}_s) \cdot \mathbf{r}] \, d^3r. \qquad (4.8.15)$$

The integral is conveniently performed by extending it over the entire cell, then subtracting the portion resulting from the inscribed sphere. Thus

(4.8.15) becomes

$$[(\mathbf{k}+\mathbf{K}_t)^2 - E]\left[\Omega\,\delta_{st} - \int_{\mathrm{i}} \exp[i(\mathbf{K}_t - \mathbf{K}_s)\cdot\mathbf{r}]\,d^3r\right].$$

However,

$$\int_{\mathrm{i}} \exp(i\mathbf{K}\cdot\mathbf{r})\,d^3r = 4\pi\int_0^{r_{\mathrm{i}}} [(\sin Kr)/Kr]r^2\,dr = 4\pi r_{\mathrm{i}}^2 j_1(Kr_i)/K$$

in which j_1 is a spherical Bessel function. Consequently (4.8.15) becomes

$$\int_{\mathrm{i+o}} \Phi(\mathbf{k}+\mathbf{K}_s,\mathbf{r})(H-E)\Phi(\mathbf{k}+\mathbf{K}_t,\mathbf{r})\,d^3r$$
$$= [(\mathbf{k}+\mathbf{K}_t)^2 - E][\Omega\,\delta_{st} - 4\pi r_{\mathrm{i}}^2 j_1[(|\,\mathbf{K}_s - \mathbf{K}_t\,|\,r_{\mathrm{i}})]/|\,\mathbf{K}_s - \mathbf{K}_t\,|]. \tag{4.8.16}$$

We now consider the surface integral term (4.8.13). This can be done with the aid of the expansion of a plane wave in spherical harmonics, the orthogonality properties of the spherical harmonics, and the addition theorem. The result is

$$\langle s\,|\,\mathcal{S}\,|\,t\rangle = -4\pi r_{\mathrm{i}}^2 \sum_l (2l+1)P_l(\cos\theta_{st})j_l(|\,\mathbf{k}+\mathbf{K}_s\,|\,r_{\mathrm{i}})j_l(|\,\mathbf{k}+\mathbf{K}_t\,|\,r_{\mathrm{i}})$$
$$\times\left(\frac{1}{j_l(|\,\mathbf{k}+\mathbf{K}_t\,|\,r)}\frac{dj_l(|\,\mathbf{k}+\mathbf{K}_t\,|\,r)}{dr} - \frac{1}{R_l(E,r)}\frac{dR_l(E,r)}{dr}\right)_{r_{\mathrm{i}}} \tag{4.8.17}$$

in which θ_{st} is the angle between the vectors $\mathbf{k}+\mathbf{K}_s$ and $\mathbf{k}+\mathbf{K}_t$. A general element of (4.8.12) may be expressed as the sum of (4.8.17) and (4.8.18):

$$\langle s\,|\,M\,|\,t\rangle = \langle s\,|\,H\,|\,t\rangle + \langle s\,|\,\mathcal{S}\,|\,t\rangle - E\langle s\,|\,t\rangle$$
$$= [(\mathbf{k}+\mathbf{K}_t)^2 - E]\left[\Omega\,\delta_{st} - 4\pi r_{\mathrm{i}}^2\frac{j_1(|\,\mathbf{K}_s - \mathbf{K}_t\,|\,r_{\mathrm{i}})}{|\,\mathbf{K}_s - \mathbf{K}_t\,|}\right]$$
$$- 4\pi r_{\mathrm{i}}^2 \sum_l (2l+1)P_l(\cos\theta_{st})j_l(|\,\mathbf{k}+\mathbf{K}_s\,|\,r_{\mathrm{i}})j_l(|\,\mathbf{k}+\mathbf{K}_t\,|\,r_{\mathrm{i}})$$
$$\times\left(\frac{1}{j_l(|\,\mathbf{k}+\mathbf{K}_t\,|\,r)}\frac{dj_l(|\,\mathbf{k}+\mathbf{K}_t\,|\,r)}{dr} - \frac{1}{R_l(E,r)}\frac{dR_l(E,r)}{dr}\right)_{r_{\mathrm{i}}}. \tag{4.8.18}$$

This expression can be simplified through the use of an identity involving spherical Bessel functions. If R, r, ρ, and θ are related by

$$R^2 = r^2 + \rho^2 - 2r\rho \cos \theta,$$

one has, for arbitrary λ,

$$j_0(\lambda R) = \sum_0^\infty (2l + 1) j_l(\lambda r) j_l(\lambda \rho) P_l(\cos \theta). \tag{4.8.19}$$

Differentiate (4.8.19) with respect to ρ. We obtain

$$[(\rho - r \cos \theta)/R] j_1(\lambda R) = -\sum_0^\infty (2l + 1) j_l(\lambda r)\, d/d\rho\, j_l(\lambda \rho) P_l(\cos \theta). \tag{4.8.20}$$

This can be used to simplify (4.8.18), if we make the correspondence $\boldsymbol{\varrho} = \mathbf{k} + \mathbf{K}_t$, $\mathbf{r} = \mathbf{k} + \mathbf{K}_s$, $\lambda = r_\mathrm{i}$:

$$\begin{aligned}\langle s \mid M \mid t \rangle = {}& [(\mathbf{k} + \mathbf{K}_t)^2 - E]\Omega\, \delta_{st} - 4\pi r_\mathrm{i}^2 \\ & \times \Bigg\{ [(\mathbf{k} + \mathbf{K}_s) \cdot (\mathbf{k} + \mathbf{K}_t) - E] \frac{j_1(|\mathbf{K}_s - \mathbf{K}_t| r_\mathrm{i})}{|\mathbf{K}_s - \mathbf{K}_t|} \\ & - \sum_l (2l + 1) P_l(\cos \theta_{st}) j_l(|\mathbf{k} + \mathbf{K}_s| r_\mathrm{i}) \\ & \times \frac{j_l(|\mathbf{k} + \mathbf{K}_t| r_\mathrm{i})}{R_l(E, r_\mathrm{i})} \left(\frac{dR_l}{dr}\right)_{r_\mathrm{i}} \Bigg\}. \end{aligned} \tag{4.8.21}$$

The crystal potential enters the matrix elements through the logarithmic derivative $R_l^{-1}\, dR_l/dr$, which is to be evaluated on the inscribed sphere. If the potential were zero inside the sphere, we would have $R_l(E, r) = j_l(\kappa r)$, where $\kappa^2 = E$, and a free-electron band structure would be obtained. The logarithmic derivatives can be replaced by an expression involving scattering phase shifts through the relation (valid for a muffin-tin potential):

$$\left(\frac{1}{R_l} \frac{dR_l}{dr}\right) = k \frac{[j_l'(kr_\mathrm{i}) - \tan \delta_l n_l'(kr_\mathrm{i})]}{[j_l(kr_\mathrm{i}) - \tan \delta_l n_l(k, r_\mathrm{i})]} \tag{4.8.22}$$

in which the prime indicates derivative with respect to the argument of the function. Equations (4.8.21) and (4.8.22) furnish the basis for an empirical form of the APW method in which the scattering phase shifts are chosen to fit experimental data (Lee, 1969). A considerable similarity to the KKR method exists in that the matrix elements are expressible in terms of phase shifts; but the equations for the elements are not identical (Slater, 1966).

Energy levels are obtained from the zeros of the equation

$$\det[\langle s \mid M \mid t \rangle] = 0. \qquad (4.8.23)$$

It is seen from (4.8.21) that the APW matrix elements are functions of the energy which is to be obtained from (4.8.23). This feature, which we also observed in the KKR method, implies that the roots of (4.8.22) must be obtained individually: the Hamiltonian cannot be diagonalized to obtain all the energy levels for a given **k** at one time.

The APW method has been quite popular and successful in the calculation of energy bands in metals. Band structures and Fermi surfaces in good agreement with experiment have been obtained. A portion of the band structure of copper is shown in Fig. 4.8.1, according to the APW calculation of Snow (1968). For applications to heavier metals, relativistic effects

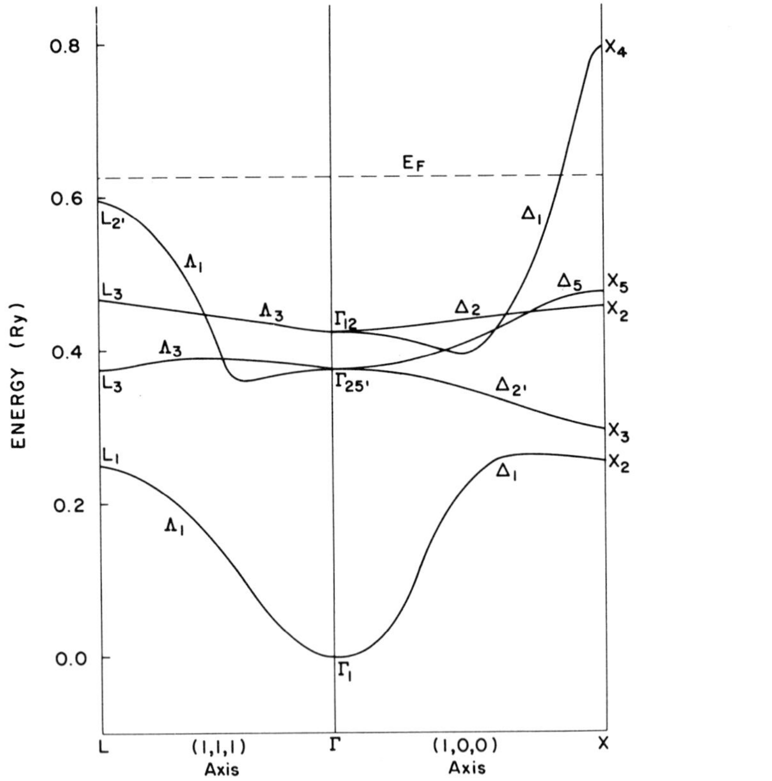

Fig. 4.8.1. Energy bands in copper from the APW calculations of Snow (1968). The dashed line denotes the Fermi energy.

should be included through a formalism based on the Dirac equation, as has been described by Loucks (1967) and Koelling (1969). Applications to covalently bonded semiconductors and semimetals, and to ionic crystals have been less successful, presumably because the muffin-tin approximation to the crystal potential ceases to be valid. In addition, it should be mentioned that the convergence of the APW method appears to be generally poorer than that of the KKR method previously described. Calculations frequently must be taken through $l = 12$, which implies that the determinant (4.8.23) is of order 169×169 at a general point of the zone. In the case of symmetry points, the size of the determinant can be substantially reduced by the introduction of symmetrized combinations of augmented plane waves, as was described in Section 4.2 for plane wave expansions. The size of the determinant that must be considered at a general point is so large that it frequently becomes desirable to avoid such calculations, and use an interpolation scheme instead.

4.9 The Hartree–Fock Method

Our discussion of the methods of calculating energy levels of electrons in crystals has been based on a single-particle Schrödinger equation with a periodic potential. It is now desirable to inquire more closely into the basis of this approach and to consider how this potential may be determined.

4.9.1 *General Properties of the Many-Body Wave Function*

The discussion begins from a many-electron point of view. The electrons of the crystal interact with the atomic nuclei and with each other. The Hamiltonian for the N-electron system can be expressed in atomic units as

$$H = \sum_{i=1}^{N} \left[-\nabla_i^2 + V(\mathbf{r}_i) + \sum_{j(j>i)}^{N} 2/r_{ij} \right]. \tag{4.9.1}$$

The first term represents the kinetic energy of the electrons; the second, their potential energy in the field of all the nuclei; the third, the electrostatic interaction of the electrons with each other ($\mathbf{r}_i$ is the position of electron i and r_{ij} is the distance between electrons i and j). The Schrödinger equation with this Hamiltonian is not separable on account of the interaction term, so that approximate methods of solution must be employed.

The principal approximation method for reducing the complexities of the many-body problem is the Hartree–Fock method. Most energy band calculations involve this approximation although it is usually not possible to solve the Hartree–Fock equations directly. These equations will be derived in this section, and some of the limitations of the procedure will be mentioned. Chapter 8, Part B, contains further discussion of many-body techniques.

It is desirable to begin by considering some of the general properties of the wave function of a many-electron system which is governed by the Hamiltonian (4.9.1). A total spin operator may be defined

$$\mathbf{S} = \sum_i \mathbf{s}_i. \tag{4.9.2}$$

The following commutation rules hold for the Hamiltonian (4.9.1):

$$[H, \mathbf{S}^2] = 0, \qquad [H, S_z] = 0, \qquad [H, \mathbf{s}_i^2] = 0. \tag{4.9.3}$$

Since the operators $\mathbf{S}^2$, S_z, $\mathbf{s}_i^2$ also commute with each other, the stationary states of the system will be eigenfunctions of these operators as well as of the energy. In the case of a single atom, H commutes with $\boldsymbol{L}^2$ ($\boldsymbol{L}$ is the total orbital angular momentum), L_z, and the angular momentum for each electron $\boldsymbol{l}_i^2$, but in the solid these results do not apply since the system is not spherically symmetric.

Also, if the potential $V(\mathbf{r}_i)$, in (4.9.1) is invariant with respect to some space group, we must have

$$[H, \{\boldsymbol{\alpha} \mid \mathbf{t}\}] = 0 \tag{4.9.4}$$

where we understand that the operator $\{\boldsymbol{\alpha} \mid \mathbf{t}\}$ now acts on the coordinates of *each* electron. These considerations are not altered by the inclusion of the electrostatic interaction since the distance between any two electrons is unaffected by an operation which moves all of them equally. In consequence, the total wave function of the system must belong to one of the irreducible representations of the relevant space group. A special case of importance is that (4.9.4) obviously holds for lattice translations; thus the wave function of the entire system satisfies Bloch's theorem and can be characterized by a total wavevector $\mathbf{K}$:

$$\Psi(\mathbf{K}, \mathbf{r}_1 + \mathbf{R}_\mu, \ldots, \mathbf{r}_N + \mathbf{R}_\mu) = \exp(i\mathbf{K}\cdot\mathbf{R}_\mu)\Psi(\mathbf{K}, \mathbf{r}_1, \ldots, \mathbf{r}_N). \tag{4.9.5}$$

As a result, the total wave function can be expanded in plane waves in which the set of wave vectors is restricted only by the condition that

$$\sum_{j=1}^{N} \mathbf{k}_j = \mathbf{K} + \mathbf{K}_s \tag{4.9.6}$$

in which $\mathbf{K}_s$ is any reciprocal lattice vector.

A further general requirement is antisymmetry. The wave function of the system must change sign if the coordinates and spins of any two electrons are interchanged.

The energies of the states of the system are functions of the total wave vector **K** (and the other relevant quantum numbers). The arguments presented in Sections 3.6 and 3.7 can be applied immediately so that

$$E(\alpha\mathbf{K}) = E(\mathbf{K}) \tag{4.9.7}$$

and

$$E(\mathbf{K}) = E(-\mathbf{K}) \tag{4.9.8}$$

in which α is the rotational part of any operation in the space group, and (4.9.8) holds even if the inversion is not in the space group. The ordinary Brillouin zone structure can be used to describe the properties of the total energy as a function of the total wave vector.

4.9.2 *Approximations to the Wave Function*

The Hartree–Fock method furnishes a prescription for the construction of an approximate wave function for the N-electron system. The fundamental idea is to associate a specific wave function with each electron. These N independent functions of the coordinates and spin of a single particles are used in forming the wave function for the entire system. The construction is to be made in such a way that the approximate total wave function is antisymmetric and possesses other relevant space and spin symmetries, thus being characterized by the appropriate (conserved) quantum numbers.

It is to be emphasized that the essential feature is the use of N functions $\Psi_j(\mathbf{r}, \mathbf{s})$ for an N electron system. Improvements beyond the Hartree–Fock method require the use of a greater number of functions (infinite in principle) for the same number of electrons, as in the method of configuration interaction.

The relevant functions Ψ_j are "spin orbitals," that is, each $\Psi_j(\mathbf{r}, \mathbf{s})$ may be written as

$$\Psi_j(\mathbf{r}, \mathbf{s}) = \psi_{j1}(\mathbf{r})\alpha + \psi_{j2}(\mathbf{r})\beta \tag{4.9.9}$$

in which α and β represent the usual two-component spinors for up and down spin: $\binom{1}{0}$ and $\binom{0}{1}$, respectively. The functions of position which appear in (4.9.9), ψ_{j1} and ψ_{j2}, may be different, that is, we allow the possibility of having different spatial wave functions for different spins. Further, since the Ψ_j are independent by hypothesis, they may be chosen without loss of generality to be orthonormal

$$\langle \Psi_j \mid \Psi_l \rangle = \delta_{jl}. \tag{4.9.10}$$

The scalar product includes both an integration over space coordinates and a summation over spin components.

The simplest way to construct an antisymmetric wave function for the N electron system is to arrange the Ψ_j into an $N \times N$ determinant

$$\Psi(1, \ldots, N) = (N!)^{-1/2} \begin{vmatrix} \Psi_1(\mathbf{r}_1, \mathbf{s}_1) & \cdots & \Psi_N(\mathbf{r}_1, \mathbf{s}_1) \\ \cdots & \Psi_j(\mathbf{r}_i, \mathbf{s}_i) & \cdots \\ \Psi_1(\mathbf{r}_N, \mathbf{s}_N) & \cdots & \Psi_N(\mathbf{r}_N, \mathbf{s}_N) \end{vmatrix} . \tag{4.9.11}$$

This function can be seen to be antisymmetric, since if rows (or columns) of the determinant are interchanged, the sign of the function is changed.

In general, however, this determinant will not be an eigenfunction of the relevant symmetry operators. It is necessary to impose some restrictions on the spin orbitals. For example, if we require the N-electron function (4.9.11) to be an eigenfunction of S_z, it is sufficient to impose the condition that the one-electron functions Ψ_j should all be eigenfunctions of the relevant S_{iz}. Usually a single determinant which is an eigenfunction of S_z will not be an eigenfunction of $\mathbf{S}^2$. For this purpose it is necessary to form linear combinations of determinantal functions with definite S_z. Two important exceptions in which a single determinant is an eigenfunction of both S^2 and S_z include (1) the state of complete spin alignment, in which all spins are "up" along a "z" axis and (2) the singlet state in which the space part of each Ψ_j is the same for two states with opposite spins. The formation of the required linear combinations in other cases has been discussed by Lowdin (1955).

4.9.3 *The Hartree–Fock Equations*

Assume that some N-particle wave function has been constructed in a formal sense from N one-particle functions. The optimum single-particle wave functions for this purpose are those which lead to the lowest total energy of the system. This leads to a variational calculation. One sets up the expression for the expectation value of the energy for the system in terms of the Ψ_j and requires that this expectation value be a minimum, subject possibly to certain constraints. The normalization and orthogonality of the one-electron functions are to be preserved. We will illustrate this briefly for the single-determinant wave function (4.9.11). The process is more complicated in detail but not in principle when a linear combination of determinants is employed.

The expectation value of the Hamiltonian (4.9.1) with the wave function

(4.9.11) can be expressed as

$$E = \sum_{j=1}^{N} \left\{ \int \Psi_j^*(\mathbf{r}_1)[-\nabla_1^2 + V(\mathbf{r}_1)]\Psi_j(\mathbf{r}_1)\, d^3r_1 \right.$$

$$+ \tfrac{1}{2} \sum_{l=1}^{N} \iint |\Psi_j(\mathbf{r}_1)|^2 |\Psi_l(\mathbf{r}_2)|^2 (2/r_{12})\, d^3r_1\, d^3r_2$$

$$\left. - \tfrac{1}{2} \sum_{l=1}^{N} \iint \Psi_j^*(\mathbf{r}_1)\Psi_l^*(\mathbf{r}_2)(2/r_{12})\Psi_j(\mathbf{r}_2)\Psi_l(\mathbf{r}_1)\, d^3r_1\, d^3r_2 \right\}. \quad (4.9.12)$$

The integrations include summation over the spinor indices implicit in the one-electron wave function. It is required that the expression (4.9.12) for E be stationary when the Ψ_j are varied, subject to preservation of the orthonormality of these functions. The equation for the one-electron functions which is obtained is given below. It is convenient in writing this equation, to exhibit the spinor indices explicitly, which we do as in (4.9.9) through the notation $\psi_{j\alpha}$ ($\alpha = 1, 2$). The summation convention is adopted with respect to spinor indices: a repeated spin index implies a sum over both possible values.

$$\left[-\nabla_1^2 + V(\mathbf{r}_1) + \sum_l \int \psi^*_{l\alpha}(\mathbf{r}_2)\psi_{l\alpha}(\mathbf{r}_2)(2/r_{12})\, d^3r_2 \right] \psi_{j\beta}(\mathbf{r}_1)$$

$$- \sum_l \int \psi^*_{l\alpha}(\mathbf{r}_2)\psi_{j\alpha}(\mathbf{r}_2)(2/r_{12})\, d^3r_2\, \psi_{l\beta}(\mathbf{r}_1) = \epsilon_j \psi_{j\beta}(\mathbf{r}_1). \quad (4.9.13)$$

These are the Hartree–Fock equations (Fock, 1930a, b) in the form given by Thompson (1960). For a system of N electrons, one evidently has a set of $2N$ coupled integrodifferential equations.

These equations may be rewritten in a form with the spin indices suppressed, and with the Pauli operators appearing explicitly

$$\left[-\nabla_1^2 + V(\mathbf{r}_1) + \sum_l \int |\Psi_l(\mathbf{r}_2)|^2 (2/r_{12})\, d^3r_2 \right] \Psi_j(\mathbf{r}_1)$$

$$- \sum_l \int \Psi_l^*(\mathbf{r}_2)[(I + \boldsymbol{\sigma}_1 \cdot \boldsymbol{\sigma}_2)/2]\Psi_l(\mathbf{r}_1)(2/r_{12})\Psi_j(\mathbf{r}_2)\, d^3r_2 = \epsilon_j \Psi_j(\mathbf{r}_1).$$

(4.9.14)

In the exchange term of this equation (second integral), σ_1 acts on the spinor Ψ_l, σ_2 acts on Ψ_j, and I is a unit operator on both Ψ_l and Ψ_j. The equality of the exchange terms in (4.9.13) and (4.9.14) may be verified by explicit comparison of the components.

The Hartree–Fock equations can be interpreted in the following way (Slater, 1951). The third term on the left evidently represents the electrostatic potential energy of an electron in state j in the field of other electrons. The exchange term reduces the electrostatic interaction; the reduction being greatest when the spins are parallel. This occurs because the probability of close encounters between electrons of parallel spin is decreased by the Pauli principle. Contributions to (4.9.12) and (4.9.14) from terms with $l = j$ cancel: the electron does not act on itself.

It must be noted that the exchange term does not have the form of an ordinary potential: an integral operator is involved. An exchange operator may be defined as follows: The exchange term in (4.9.14) is written as

$$\int A(\mathbf{r}_1, \mathbf{r}_2)\Psi_j(\mathbf{r}_2)\, d^3r_2 \tag{4.9.15}$$

where

$$A(\mathbf{r}_1, \mathbf{r}_2) = \sum_l \Psi_l^*(\mathbf{r}_2)[(I + \boldsymbol{\sigma}_1 \cdot \boldsymbol{\sigma}_2)/2](2/r_{12})\Psi_l(\mathbf{r}_1). \tag{4.9.16}$$

We can also express A in the form

$$A(1, 2) = \tfrac{1}{2}[u(\mathbf{r}_1, \mathbf{r}_2) + \mathbf{M}(\mathbf{r}_1, \mathbf{r}_2) \cdot \boldsymbol{\sigma}_2] \tag{4.9.17}$$

where

$$u(\mathbf{r}_1, \mathbf{r}_2) = \tfrac{1}{2} \sum_l \Psi_l^*(\mathbf{r}_2)\,(2/r_{12})\Psi_l(\mathbf{r}_1) \tag{4.9.18}$$

and

$$\mathbf{M}(\mathbf{r}_1, \mathbf{r}_2) = \tfrac{1}{2} \sum_l \Psi_l^*(\mathbf{r}_2)\,(2/r_{12})\,\boldsymbol{\sigma}_1\Psi_l(\mathbf{r}_1). \tag{4.9.19}$$

Summation over spin components is understood in (4.9.18) and (4.9.19). In spite of the fact that (4.9.16) is not equivalent to an ordinary potential, it is frequently approximated by one. Two such approximations have been proposed by Slater (1951), and an extensive literature has developed on this problem. This will be discussed subsequently.

If wave functions satisfying the Hartree–Fock equations are used in the evaluation of the energy of the N-electron system according to (4.9.12), the result is

$$E = \sum_j [\epsilon_j - \tfrac{1}{2} \sum_l \iint |\Psi_l(\mathbf{r}_2)|^2 \, |\Psi_j(\mathbf{r}_1)|^2 (2/r_{12})\, d^3r_1\, d^3r_2$$

$$+ \tfrac{1}{2} \sum_l \iint \Psi_j^*(\mathbf{r}_1)\Psi_l^*(\mathbf{r}_2)\,(2/r_{12})\Psi_j(\mathbf{r}_2)\Psi_l(\mathbf{r}_1)\, d^3r_1\, d^3r_2]. \tag{4.9.20}$$

The total energy is not just a sum of "one-particle energies" ϵ_j. Such a sum counts each interaction twice, and thus must be corrected as in (4.9.17). It is a general characteristic of the theory of interacting particles that the total energy is not simply the sum of one-particle energies, but rather depends on the distribution of the particles.

4.9.4 *Koopmans' Theorem*

The justification for the designation "one-particle energy" comes from Koopmans' theorem (Koopmans, 1934): Let us consider the difference in energy between two systems containing N and $N - 1$ particles, respectively, but otherwise identical. In particular, we suppose that the one-electron wave functions Ψ_l are the same in each case; however, in one, the state Ψ_j is unoccupied. The approximation of unchanged wave functions is likely to be reasonably valid in a solid, at least for states belonging to a wide band, since the electrons are not bound to any particular atom but spread throughout the solid. From (4.9.12), we find that the difference in energy ΔE is given by

$$\Delta E = \int \Psi_j^*[-\nabla^2 + V(\mathbf{r})]\Psi_j \, d^3r + \sum_l \iint |\Psi_l(\mathbf{r}_1)|^2 |\Psi_j(\mathbf{r}_2)|^2 \times (2/r_{12}) \, d^3r_1 \, d^3r_2 - \sum_l \iint \Psi_j^*(\mathbf{r}_1)\Psi_l^*(\mathbf{r}_2)(2/r_{12})\Psi_j(\mathbf{r}_1)\Psi_l(\mathbf{r}_1) \, d^3r_1 \, d^3r_2. \tag{4.9.21}$$

The factors of $\frac{1}{2}$ in (4.9.12) disappear because the deleted state Ψ_j occurs in both summations. If the wave functions satisfy (4.9.13), we have

$$\Delta E = \epsilon_j. \tag{4.9.22}$$

The energy parameter ϵ_j specifies the energy required to remove an electron from the state j, leaving the other electrons of the system undisturbed. As a corollary, the energy which has to be added to the system to remove an electron from state j and place it in state i is $\epsilon_i - \epsilon_j$. Since the energy bands in a solid are usually determined, at least in principle, from the Hartree–Fock equations, the physical interpretation of the energy bands is that afforded by Koopmans' theorem.

The essential nature of the hypothesis that the wave functions are unchanged should be noted. An extreme example of the error involved is furnished by the case of the free helium atom in its ground state. According to the calculations of Roothaan *et al.* (1960) and Clementi, (1965), the one-electron energy for a 1s electron is -1.8359 Ry. However, if an electron is removed, leaving a He^+ ion, the wave function of the remaining electron becomes hydrogenic with $Z = 2$ and an energy -4 Ry. The difference between the total energy of the He atom in the same Hartree–Fock calculation and that of the ion is 1.7234 Ry; this is the removal energy in the Hartree–Fock approximation. The difference between the one-electron energy and the ionization energy computed as the difference in total energies is accounted for in terms of rearrangement, or readjustment, of the atomic wave functions. Obviously, helium is a very severe case; in contrast, for alkali metal atoms, Koopmans' theorem is an excellent approximation. For example, in the case of the ionization of the valence electron of sodium, the rearrangement energy in the Hartree–Fock approximation is only 4×10^{-5} Ry. However, in other atoms, the error may be quite appreciable. The possible errors in solid state problems in this connection have been discussed by Phillips (1961), and Doniach (1971).

4.9.5 *Symmetry Restrictions*

We will now return to the questions of symmetry discussed at the beginning of this section. There we saw that the exact wave function of the system must be an eigenfunction of certain operators. In particular, if Θ is an operator which commutes with the Hamiltonian, $H\Theta = \Theta H$, every eigenfunction Ψ of H is either automatically an eigenfunction of Θ, or in the case of a set of degenerate levels, may be constructed to be one:

$$H\Psi = E\Psi; \qquad \Theta\Psi = \theta\Psi.$$

It is natural to demand that an assumed approximate wave function for the total system should also be an eigenfunction of Θ. However, this condition refers to the assumed N particle wave function only, and not to the one-electron functions themselves. There is a delicate point here. Suppose we construct a determinantal wave function from single-particle functions which have proper symmetry, that is, form a basis for an irreducible representation of the group of operators which commutes with the Hamiltonian. These one-particle functions may be required to be solutions of the Hartree–Fock equations. It has been shown (Roothaan, 1960; Lowdin, 1962) that the energy computed from such functions is extreme in the variational sense: first-order changes in the wave functions produce a second-order change in the energy. Unfortunately, it does not follow that the value is the lowest possible; and in fact, it may easily happen that a lower extreme value is obtained by relaxing the symmetry requirements. The absolute minimum of the energy of the system may not have the required symmetry—and if the determinantal function is forced to have the proper symmetry, a considerably higher energy may be obtained. This is the symmetry dilemma of the Hartree–Fock method (Lowdin, 1966). However, in this case it can be shown that the determinant is then a sum of terms transforming according to irreducible representations of the correct group, and that at least one of these components has an energy lower than that of the determinantal function. It is, obviously, not a single determinant.

As a simple example, consider the application of the Hartree–Fock method to systems in a singlet state. It is a common procedure to assume that the space part of the wave function is the same for electrons of ↑ and ↓ spins. This is the concept of doubly occupied orbitals. In this case, the determinantal wave function is actually a singlet. However, the restriction to a single space wave function for each pair of electrons is not favorable energetically, as may be seen in the case of the helium atom. The assumption of a doubly occupied orbital leads to a ground state total energy $E = -5.7234$ Ry (Roothaan *et al.*, 1960). A substantial improvement can be obtained by removing the restriction of a single space function. When two functions are used, as we have described, the ground state energy is lowered to -5.7560 Ry (Froese, 1966). For comparison, the exact energy, according to an elaborate variational calculation by Pekeris (1962) is -5.8074 Ry.

4.10 Determination of the Crystal Potential

4.10.1 *Approximate Exchange Potentials*

Although the Hartree–Fock equations are only approximations to a complex physical situation, they are generally too difficult to solve for solids. Numerical solutions are available for a large number of single atoms (Clementi, 1965) and some molecules. A few Hartree–Fock calculations have been attempted for solids (Harris and Monkhorst, 1969; Kunz and Lipari, 1971; Euwema *et al.*, 1973; Harris *et al.*, 1973). It is customary to replace the exchange operator by an ordinary potential. Instead of (4.9.14) we write

$$[-\nabla^2 + V_{\mathrm{N}}(\mathbf{r}) + V_{\mathrm{o}}(\mathbf{r}) + V_{\mathrm{ex}}(\mathbf{r})]\Psi_j(\mathbf{r}) = E_j\Psi_j(\mathbf{r}). \quad (4.10.1)$$

In this equation, $V_{\mathrm{N}}(\mathbf{r})$ is the potential energy of an electron in the field of the nuclei of the system; $V_{\mathrm{o}}(\mathbf{r})$ is the ordinary average electrostatic potential of an electron in the field of all the charges of the system, and V_{ex} is an average exchange potential, which may depend on the spin associated with the state $\psi_j(\mathbf{k}, \mathbf{r})$.

The exchange potential, which is spin-dependent, is formally defined by

$$\int A(1, 2)\Psi_j(\mathbf{r}_2)\, d^3r_2 = V_{\mathrm{ex}}(\mathbf{r}_1)\Psi_j(\mathbf{r}_1) \quad (4.10.2)$$

where A is given by (4.9.16). In a situation in which Ψ_j is a general two-component spinor, V_{ex} can be written in the form

$$V_{ex}(\mathbf{r}_1) = \tfrac{1}{2}[U(\mathbf{r}_1) + \mathbf{M}(\mathbf{r}_1)\cdot\sigma] \tag{4.10.3}$$

where $U(\mathbf{r}_1)$ and $\mathbf{M}$ are suitably defined in accord with (4.9.18) and (4.9.19). The explicit and exact forms of V_{ex}, U, and $\mathbf{M}$ are not very useful as these quantities will actually depend both on the particular state and the particular spin component considered. The object of the procedure is to develop an approximation which replaces these complicated quantities by suitable state-independent average values.

The most widely used approximation of this type was proposed by Slater (1951): The exchange potential in an electron system with a given charge density $\rho = \sum_l \Psi_l{}^*\Psi_l$ should be the same as in a free-electron gas of the same density. This leads to an expression of the form

$$V_{ex}(\sigma, \mathbf{r}) = -6\alpha[(3/4\pi)\rho_\sigma(\mathbf{r})]^{1/3}. \tag{4.10.4}$$

It is assumed here that the wave functions considered are eigenfunctions of the z component of spin. The quantity ρ_σ is the density of electrons of spin σ ($\sigma = \uparrow, \downarrow$). Thus V_{ex} is different for different spin states if the density of these electrons is different. The quantity α is a numerical factor which takes values of 1 (Slater, 1951), $\frac{2}{3}$ (Gaspar, 1954; Kohn and Sham, 1965; Rajagopal and Callaway, 1973), or may be varied between these limits to improve the results obtained (Slater *et al.*, 1969a, b).

We will indicate how these results are obtained. To this end, consider the exchange energy of the free electron gas. Suppose that the wave functions are eigenfunctions of S_z and, for their spatial dependence, are plane waves:

$$\psi(\mathbf{k}, \mathbf{r}) = \Omega^{-1/2}\exp(i\mathbf{k}\cdot\mathbf{r}) \tag{4.10.5}$$

where Ω is the volume in which the wave functions are normalized.

Plane waves are, in fact, eigenfunctions of the Hartree–Fock equations under suitable circumstances. The system described is the neutralized free electron gas, that is, a collection of electrons in which the expected periodic potential has been replaced by a uniform distribution of positive charge. This positive charge serves to keep the system electrically neutral, and its uniformity ensures that the average electrostatic potential in the system is zero. The energy of the state (4.10.5) is, in atomic units,

$$E(\mathbf{k}) = \mathbf{k}^2 + \epsilon_1(\mathbf{k}) \tag{4.10.6}$$

where $-\epsilon_1$ is the $\mathbf{k}$-dependent exchange energy, which is the eigenvalue of the exchange operator

$$\int A(1,2)\psi(\mathbf{k},\mathbf{r}_2)\,d^3r_2 = -\epsilon_1(\mathbf{k})\psi(\mathbf{k},\mathbf{r}_1). \tag{4.10.7}$$

Substitution of the plane wave functions shows that ϵ_1 is given by

$$\begin{aligned}\epsilon_1 &= (-1/\Omega)\sum_{\mathbf{q}}\int \exp[i(\mathbf{k}-\mathbf{q})\cdot(\mathbf{r}_1-\mathbf{r}_2)](2/r_{12})\,d^3r_2\\ &= -\sum_{q} 8\pi/\Omega\,|\,\mathbf{k}-\mathbf{q}\,|^2.\end{aligned} \tag{4.10.8}$$

Although this result was derived for the case of plane waves, the singularity as $\mathbf{k}\to\mathbf{q}$ is typical of solid state problems (Herring, 1966, p. 149). In the summation over $\mathbf{q}$, we include only occupied states of the same spin as ψ. Thus we replace $\sum_q$ by $[\Omega/(2\pi)^3]\int d^3q$. We suppose that the electrons occupy all states with wave vectors $q \le k_{\mathrm{F}}$, where k_{F} is the radius of the Fermi sphere

$$\begin{aligned}\epsilon_1(k) &= (-2/\pi)\int_0^{k_{\mathrm{F}}} q^2\,dq\int_0^{\pi}\sin\theta\,d\theta\,(q^2+k^2-2kq\cos\theta)^{-1}\\ &= (-2k_{\mathrm{F}}/\pi)\{1+[(k^2-k_{\mathrm{F}}^2)/2kk_{\mathrm{F}}]\ln|\,(k-k_{\mathrm{F}})/(k+k_{\mathrm{F}})\,|\}\\ &= -2k_{\mathrm{F}}/\pi F(k/k_{\mathrm{F}})\end{aligned} \tag{4.10.9}$$

where F is given by

$$F(x) = 1+[(1-x^2)/2x]\ln|\,(1+x)/(1-x)\,|. \tag{4.10.10}$$

A graph of $F(x)$ is shown in Fig. 4.10.1.

There are several ways in which (4.10.9) can be used to determine an effective exchange potential. All involve the assumption that k_{F} is to be related to the local density of particles. In the case that an equal number of electrons of each spin are present in an electron gas of (constant) density ρ, this relation is

$$k_{\mathrm{F}} = (3\pi^2\rho)^{1/3}. \tag{4.10.11}$$

It is now assumed that (4.10.11) may be used in the case of a nonuniform system, by regarding it as valid locally, that is, one has a function $k_{\mathrm{F}}(\mathbf{r})$ connected to a density function $\rho(\mathbf{r})$ through (4.10.11). This approximation is fundamental in the Thomas–Fermi method. From this point, different paths are possible. Liberman (1968) replaces k by a function of $\mathbf{r}$ as well through

$$k(\mathbf{r}) = [E - V(\mathbf{r})]^{1/2} \tag{4.10.12}$$

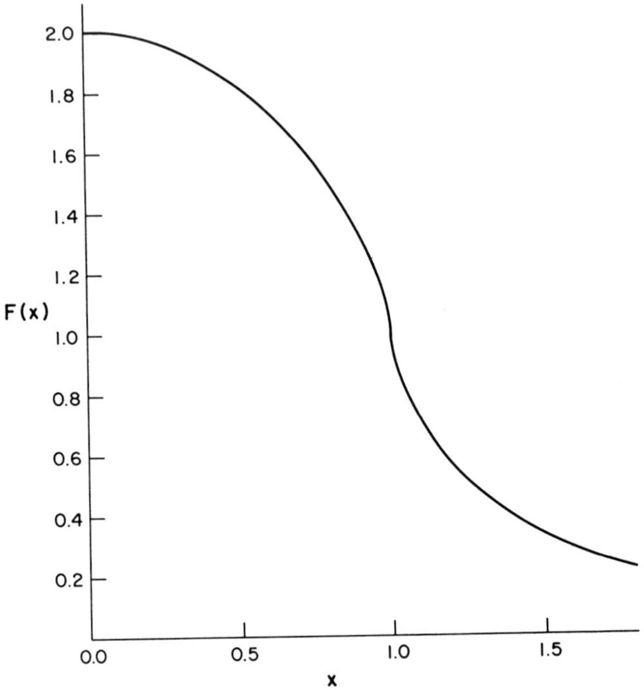

Fig. 4.10.1. The function $F(x)$ given by Eq. (4.10.10) is shown. It has a vertical tangent at $x = 1.0$.

where $V(\mathbf{r}) = V_N(\mathbf{r}) + V_o(\mathbf{r})$. Thus $V(\mathbf{r})$ is the total "ordinary potential." Equation (4.10.12) follows the spirit of the WKB approximation. The approximate exchange potential is then just

$$V_{ex,1} = \epsilon_1(k_F(\mathbf{r}), k(\mathbf{r})). \tag{4.10.13}$$

Alternately, one may replace $F(k/k_F)$ by some suitable constant. Slater chooses to average F over the Fermi distribution of occupied states. This average is

$$F_{Av} = (3/4\pi k_F^3) \int F(k/k_F)\, d^3k = 3 \int_0^1 x^2 F(x)\, dx = \tfrac{3}{2}. \tag{4.10.14}$$

Slater's average exchange potential is then

$$V_{ex,s} = -6(3\rho/8\pi)^{1/3}. \tag{4.10.15}$$

In order to obtain agreement with (4.10.4), we observe that in the present case, both spins are included in ρ, and we may simply substitute $\rho = 2\rho_\sigma$.

It should be also noted that Eq. (4.10.15), with constant ρ ($= N/V$), gives *twice* the exchange energy per electron in the neutralized electron gas:

$$E_x/N = (1/2N) \sum_{\mathbf{k}} \epsilon_1(\mathbf{k}) = -3(3\rho/8\pi)^{1/3}, \qquad (4.10.16)$$

in which the summation includes both spins and the factor of $\frac{1}{2}$ is included so that the interactions between pairs are not counted twice.

The approach of Kohn and Sham derives from the theory of the inhomogenous electron gas (Hohenberg and Kohn, 1964). This method is based on a theorem establishing that the energy of the ground state of any system of interacting electrons is uniquely determined in principle if the charge density is specified. Formally, the ground state energy is a functional of the charge density: $E = E[\rho]$. If the charge density is varied, the energy attains its minimum value for the correct charge density function $\rho(\mathbf{r})$. The variational principle for the density can be shown to lead to a set of single-particle effective Schrödinger equations of the general form of Eq. (4.10.1) for functions ψ_j which are related to the density by the usual formula

$$\rho(\mathbf{r}) = \sum_{j=1}^{N} |\psi_j(\mathbf{r})|^2. \qquad (4.10.17)$$

The quantity V_{ex} in Eq. (4.10.1) is now an object $\mathrm{V}_{\mathrm{xc}}(r)$ which is to be found as the functional derivative of the exchange and correlation contribution to $E[\rho]$.

The preceding remarks are exact. Unfortunately, the functional $E[\rho]$ is not known in general. When the density $\rho(\mathbf{r})$ is slowly varying, it is permissible to use expressions derived for a free electron gas (constant density) with the density regarded as a function of position. If the Hartree–Fock expression for the energy of a free electron gas is used, an effective exchange potential can be found:

$$V_{\mathrm{ex},K} = (\partial/\partial\rho)[(1/V)E_x] = (\partial/\partial\rho)\{-3(3/8\pi)^{1/3}\rho^{4/3}\} = -4(3\rho/8\pi)^{1/3} \qquad (4.10.18)$$

where V is the volume of the system. It will be observed that this exchange potential differs from that proposed by Slater (4.10.18) by a factor of $\frac{2}{3}$. This amounts to replacing $F(k/k_{\mathrm{F}})$, not by its average value, but by its value at the Fermi energy $F(1) = 1$.

The Kohn–Sham–Gaspar exchange potential (4.10.18) leads, through (4.10.1) and (4.10.17) to a charge density ρ, such that if this charge density is used to calculate the total energy of the system via expressions valid for an electron gas of slowly varying density, the resulting total energy is a minimum. It does not follow that if the wave functions which result from (4.10.1) in this case are substituted into (4.9.12), that a minimum value would be obtained, nor does it follow that the one-electron eigenvalues are necessarily good approximations to Hartree–Fock eigenvalues or to experimental single particle energy levels. Slater *et al.* (1969a) have shown that these eigenvalues do give a good approximation to the derivative of the energy with respect to the occupation number of the orbitals. This is what is required in an energy band calculation in order to determine the occupancy of states and the Fermi surface (but not the energies of transitions).

It is possible to take an empirical view of the exchange potential problem by considering an expression for V_{ex} which contains an undetermined parameter, as in (4.10.4) (the "$X\alpha$" method of Slater *et al.* (1969b). This parameter can be varied so as to minimize the energy of the ground state of the system. However, since this is a variation of the Hamiltonian of the system, there is no guarantee that minimization of the ground state energy gives a solution which is optimum in other respects. The "best" value of α will vary from atom to atom.

The preceding discussion involves approximate exchange potentials which are simply related to the exchange energy for a uniform electron gas. In a physical system, where the density is not constant, it must be expected that there will be corrections involving derivatives of the density. In order to take these corrections into account, Herman *et al.* (1969) have proposed to add to (4.10.4), a term involving derivatives of ρ. Their expression for V_{ex} is

$$V_{ex,HVO} = -6(3\rho/8\pi)^{1/3}\{\alpha + \beta\rho^{-2/3}[\tfrac{4}{3}(\nabla\rho/\rho)^2 - (2\,\nabla^2\rho/\rho)]\} \quad (4.10.19)$$

in which both α and β are adjustable constants. (However, it appears to be satisfactory to fix α at the value $\frac{2}{3}$ appropriate to the Kohn–Sham–Gaspar treatment, and simply vary β.) This form gives difficulties, however, at both small and large values of r in an atom. The best total energies obtained using (4.10.19) are somewhat closer to the Hartree–Fock results than those obtained with (4.10.4).

Part of the confusion surrounding the question of an effective exchange potential results from the fact that the Hartree–Fock eigenvalues have undesirable properties. This may be seen in the case of a free electron gas, by differentiating ϵ_1, as given by (4.10.9), with respect to k:

$$d\epsilon_1(k)/dk = -(1/\pi)[(1 + (k_F^2/k^2))\ln|(k - k_F)/(k + k_F)| + (2k_F/k)]. \quad (4.10.20)$$

The derivative contains a term which is logarithmically infinite at $k = k_F$.

From this, we see that the density of states, which depends inversely on dE/dk, would then vanish on the Fermi surface. This result would have a great influence on the calculation of the properties of metals, which depend critically on the state of affairs at the Fermi surface, and is in violent conflict with experiment.

It is evident that something is fundamentally wrong with the Hartree–Fock theory. In fact, the influence of the electron interaction when calculated by a more correct theory (see Section 8.2) turns out to be weaker than is predicted by (4.10.20). It would be undesirable to include the full exchange interaction between nearly free electrons in a band calculation.

4.10.2 *The Wigner–Seitz Approximation and the Quantum Defect Method*

Even if one ignores the difficulties associated with the choice of exchange potential, and decides to use one of the available expressions described in the preceding subsection, the task of a band calculation is not easy. The electrostatic potential produced by the electron distribution must be consistent with that used in its calculation. This can only be accomplished by an iteration procedure in which the charge density calculated at one stage is used to determine an input potential for the next stage. The exchange potential also must be iterated. This process requires sampling the calculated charge distribution throughout the zone, and in general can be quite time consuming.

The Wigner–Seitz approximation offers one way to avoid these difficulties (Wigner and Seitz, 1933). It was introduced at a time when there was no possibility of a direct attack on the problem of self-consistency, but remains a relevant and useful approximation at present. For simplicity, consider an alkali metal in which each atom possesses a single valence electron. According to the discussion of Sections 4.9.3 and 4.10.1, the charge density of the valence electron distribution should be included in determining the crystal potential; instead Wigner and Seitz propose that each electron should be considered to move in the field of a singly charged ion. This is based on the qualitative argument than in a monovalent metal, the coulomb repulsion of the electrons (and also exchange effects) makes it unlikely that two valence electrons will be found on the same lattice site. Consequently, the potential energy of a single electron will be that due to its presence in the field of a single ion, the rest of the lattice being neutral. The polyhedral cells approximate spheres reasonably closely so that multipole components in the potential may be neglected as long as one is concerned with s and p bands. The potential within the cell does not differ appreciably from what it is in the free atom at the same position.

The Wigner–Seitz approximation is not just a simplification of the Hartree–Fock method. It moves beyond Hartree–Fock by imposing the requirement that each electron moves in the field of an ion. The Hartree–Fock procedure is known to give a substantial probability that too many electrons will be found on some atoms and too few on others and thus leads, especially at large interatomic distances, to an energy which is too high. The Wigner–Seitz approximation enables one to avoid this difficulty, and appears to lead to a very substantial measurement of agreement between calculated and observed values of the cohesive energies for the alkali metals. It is probable, however, that the Wigner–Seitz approximation tends to overcorrelate the electrons in that the exclusion of two electrons from the same cell should not be absolute; and moreover, that the relative importance of this exclusion should depend on the state considered.

If the Wigner–Seitz approximation is accepted as reasonable in regard to the crystal potential, it is possible to use spectroscopic data from the free atom as a guide in the construction of this potential. However, as early as 1929 Prokofjew was able to construct an empirical potential for sodium which was used by Wigner and Seitz (1933) in their calculation of the cohesive energy of sodium. Similarly, Seitz (1935) determined the potential to be used in a calculation of the cohesive energy of lithium from the spectrum of the free atom.

The construction of an empirical potential in the fashion of Prokofjew and Seitz presents considerable problems. It is complicated by questions of uniqueness and by the fact that in the Hartree–Fock approximation, the effective potential acting on an electron is nonlocal and state-dependent. The practical difficulties are apparently not serious for lithium and sodium; for substantially heavier atoms, such as potassium, construction of a single local potential for all states may not be possible (Gorin, 1936). It was suggested by Kuhn and Van Vleck (1950) that spectroscopic energy level information could be used almost directly in the energy band problem without the necessity of constructing a potential explicitly. This procedure, which has come to be known as the "quantum defect method" has been significantly refined and extended by Brooks (1953, 1958; Brooks and Ham, 1958), and by Ham (1954, 1955). It may also be used in atomic scattering problems (Seaton, 1966).

The essential idea of the procedure is the following. In a free atom of an alkali metal, the valence electron is loosely bound to a compact spherical core which contains the electrons in closed shells. These closed shells will not be significantly modified in the solid where they occupy only a small portion of the volume of an atomic cell. For this reason, the electrostatic field which acts on the valence electron is nearly a coulomb field throughout most of the cell. Hence, its wave function must be a (negative energy) coulomb wave function in the outer portions of the cell. A similar situation exists in the free atom.

It is not implied that the wave function is hydrogenic. For any energy, there are two linearly independent solutions of Schrödinger's equation with a coulomb potential. These are confluent hypergeometric functions. In the case of the free atom, the wave function must vanish exponentially at infinity. At a given energy there is a unique "coupling constant" which determines, except for a multiplicative factor, the combination of the two functions which does vanish properly at infinity. Thus, the coupling constant is determined for energies corresponding to eigenvalues of the free atom. If it is assumed that the coupling constant depends smoothly on energy, it may be interpolated or extrapolated as a function of energy, and thus determined for any energy. The coupling constant may also be determined by solving the wave equation with the actual potential energy inside the core; however, it is possible to determine this constant empirically from the observed eigenvalues without taking explicit account of interactions in the core.

In the solid, the wave function must satisfy certain specified boundary conditions (see Section 4.6). These conditions may be stated in a fashion independent of the normalization of the wave function. Thus, in the case of the lowest state Γ_1, the logarithmic derivative of the wave function must vanish on the atomic sphere (in the spherical approximation). These boundary conditions can be satisfied only for certain specified values of the coupling constant. Since the coupling constant is now known as a function of energy, the energies of interest in the solid state problem are also determined. The calculation is reasonably simple only for states which are predominantly of a single angular momentum (such as Γ_1). Also, the parameters E_2 and E_4 may be calculated from the normalization independent expressions given in Section 4.6. Calculations for other states are beset by the same difficulties previously mentioned in the discussion of the cellular method. It is possible, however, to use either the Green's function method or the APW method for the band calculation, since in these procedures, the information concerning the potential inside the muffin-tin sphere can be summarized by presenting the scattering phase shifts. These are, in effect, determined from the coupling constant. Information can also be obtained about the behavior of the wave function in the core region (Brooks and Ham, 1958).

We have described the method in a fashion appropriate to the alkali metals, where it has been quite successful (Ham, 1962a, b). Applications to more complicated materials are difficult because the basic assumption that the potential is coulombic in the outer portions of the atomic cell is much less justified.

To formalize this discussion, we consider the radial wave equation of the cellular method (4.6.2) and make the following substitutions: $U_l = rR_l$; $E = -1/n^2$ (n is an integer in the hydrogenic problem, but is not so restricted here), and $V = -2/r$ (since we are concerned with the outer portion of the cell). Then we have

$$(d^2U_l/dr^2) + \{-(1/n^2) + (2/r) - [l(l+1)/r^2]\}U_l = 0. \quad (4.10.21)$$

Only negative energies will be considered. Two linearly independent solutions of (4.10.21) are required. It is convenient to choose them so that one, $U_{l,0}$, vanishes at the origin, and the other, $U_{l,1}$, is singular there. Consequently, $U_{l,1}$ could not appear in the usual hydrogenic problem. These functions may be expressed in terms of Bessel functions (Wannier, 1943; Kuhn, 1951; Ham, 1957). To make this connection we introduce the substitutions

$$z = (8r)^{1/2} \quad (4.10.22)$$

and

$$U_l = (z/2)V_l. \quad (4.10.23)$$

We find from (4.10.21)

$$z^2 (dV_l/dz^2) + z (dV_l/dz) + [z^2 - (2l+1)^2]V_l = (1/n^2)(z/2)^4V_l. \quad (4.10.24)$$

If $n \to \infty$ ($E \to 0$), V_l is evidently a Bessel function of order $(2l+1)$. Further, for interesting states in the alkali metals, one has $|E| < 1$, so that $n^2 > 1$. The regular and irregular solutions are denoted by

$$U_{l,0} = (z/2)V_{l,0} = (z/2)J^n{}_{2l+1}(z), \qquad U_{l,1} = (z/2)V_{l,1} = (z/2)N^n{}_{2l+1}(z). \quad (4.10.25)$$

These functions are determined by expansion with respect to $1/n^2$:

$$V_{l,i}(z) = \sum_k (1/n^{2k})V_{l,i}{}^{(k)}(z). \quad (4.10.26)$$

The functions $V_{l,i}{}^{(k)}$ satisfy

$$\nabla_l V_{l,i}{}^{(0)} = 0, \qquad \nabla_l V_{l,i}{}^{(k)} = (z/2)^4 V_{l,i}{}^{(k-1)}(z) \quad (4.10.27)$$

in which

$$\nabla_l = z^2 (d^2/dz^2) + z (d/dz) + z^2 - (2l+1)^2. \quad (4.10.28)$$

Kuhn (1951) has shown how the functions $V_{l,i}{}^{(k)}$ may be generated from cylindrical functions. Let C_{2l+1} be an arbitrary linear combination of the ordinary Bessel function $J_{2l+1}(z)$ and the Weber function $Y_{2l+1}(z)$

$$C_{2l+1}(z) = AJ_{2l+1}(z) + BY_{2l+1}(z). \tag{4.10.29}$$

(To construct $J^n{}_{2l+1}$, set $A = 1$, $B = 0$.) It follows from the recurrence relations obeyed by the cylindrical functions that

$$\nabla_l \left\{ \frac{2l+2+q}{4(2+q)} \left(\frac{z}{2}\right)^{q+2} C_{2l+3+q}(z) - \frac{1}{4(3+q)} \left(\frac{z}{2}\right)^{q+3} C_{2l+4+q}(z) \right\}$$
$$= \left(\frac{c}{2}\right)^{q+4} C_{2l+1+q}(z). \tag{4.10.30}$$

It is evident that $V_{l,i}{}^{(0)} = C_{2l+1}$. Next, put $q = 0$ in (4.10.30). The equation is then identical with that satisfied by $V_{l,i}{}^{(1)}$ provided that $V_{l,i}{}^{(1)}$ is given by the bracket on the left of (4.10.30), namely

$$V_{l,i}{}^{(1)} = \tfrac{1}{4}(l+1)(z/2)^2 C_{2l+3} - \tfrac{1}{12}(z/2)^3 C_{2l+4}. \tag{4.10.31}$$

Higher terms of the series may be generated. The details of the series for the regular function $J^n{}_{2l+1}$ and the irregular function $N^n{}_{2l+1}$ are given by Kuhn (1951) and, more completely, by Ham (1957). The functions $U_{l,0}$ and $U_{l,1}$ may be obtained from tables computed by Blume *et al.* (1959).

The general solution of (4.10.21) can be expressed as

$$U_l(r) = \alpha(n) U_{l,0}(n, r) + \gamma(n) U_{l,1}(n, r). \tag{4.10.32}$$

Both $U_{l,0}$ and $U_{l,1}$ are present because the potential in the core region of the atom is different from the $(-2/r)$ assumed in (4.10.21).

The problem is to determine the ratio $\alpha(n)/\gamma(n)$. At an eigenvalue of the free atom, the wave function vanishes exponentially at large r and consequently, must be represented by the function $W_{n,l+1/2}(2r/n)$ which has this property (Wittaker and Watson, 1952, Chapter 16). Wannier (1943) has given the relation between the quasi-Bessel functions $J^n{}_{2l+1}$, $N^n{}_{2l+1}$ and the Whittaker function $W_{n,l+1/2}$. It is

$$W_{n,l+1/2}(2r/n) = \Gamma(n+l+1) n^{-l-1}(z/2) J^n{}_{2l+1}(z) \cos[\pi(n-l-1)]$$
$$+ \Gamma(n-l) n^l (z/2) N^n{}_{2l+1}(z) \sin[\pi(n-l-1)]. \tag{4.10.33}$$

Hence, at eigenvalues of the free atom, the ratio $\alpha(n)/\gamma(n)$ is determined to be

$$\alpha(n)/\gamma(n) = \Gamma(n+l+1)/[n^{2l+1}\Gamma(n-l) \tan \pi(n-l-1)]. \tag{4.10.34}$$

The right side of (4.10.34) can be computed from the energy of an atomic state. In order to obtain the wave function in the solid state problem, we require an interpolation or extrapolation of the ratio α/γ to other energies. It is convenient to define this ratio to be, for an arbitrary energy,

$$\alpha(n)/\gamma(n) = -\Gamma(n+l+1)/[n^{2l+1}\Gamma(n-l) \tan \pi\nu(n)]. \qquad (4.10.35)$$

Thus at an eigenvalue,

$$\nu(n) = l + m' + 1 - n \qquad (4.10.36)$$

where m' is an arbitrary integer. The energy at an eigenvalue is

$$E = -1/n^2 = -1/(m-\nu)^2 \qquad (4.10.37)$$

in which $m = l + m' + 1$ is an integer. Consequently, ν differs from the experimental quantum defect by an integer at most. The procedure of the quantum defect method is to set ν equal to the observed quantum defect at energies corresponding to eigenvalues of the free atom, and to determine it at other energies by putting a smooth curve through the experimental points. The justification for this procedure has been given by Ham (1955) and Brooks and Ham (1958) in terms of the WKB approximation. As a practical matter it appears to be a better procedure to extrapolate or interpolate the quantity $\eta(n)$ defined by

$$\eta(n) = -(1/\pi) \tan^{-1}[\gamma(n)/\alpha(n)] \qquad (4.10.38)$$

because the quantity ν in (4.10.35) must be an integer whenever n is an integer, since for integer values of n, the wave function is hydrogenic and $\gamma = 0$. Once η (or ν) is determined, the radial wave function is specified as a function of energy in the outer part of the atomic cell, except for a normalization factor. This information is sufficient to permit band calculations by the Green's function, APW, or cellular methods, which require knowledge only of logarithmic derivatives. The WKB approximation can be used to obtain information about the wave function in the interior of the atom.

It is possible also to extend these procedures to include relativistic effects if one takes account of the spin–orbit splitting of atomic energy levels of nonzero angular momentum (Callaway *et al.*, 1957). In many respects the quantum defect method is extremely attractive because it enables one to bypass many of the difficulties of the Hartree–Fock procedure. It is not exact; even in the alkali metals the crystal potential in the outer portion of the atomic cell differs from the simple coulomb form assumed through the influence of core polarization, which adds a term proportional to $-\alpha/r^4$, where α is the polarizability of the ion (Callaway, 1957). However, these effects are not large, except possibly in the heavier alkali metals, and in any case are not included in conventional band calculations.

PROBLEMS

1 Derive an equation giving the energy bands for a one-dimensional periodic potential of the form shown. Find the width of the lowest band and the smallest gap between the first and second bands if $V_o = 0.4$ Ry, $a = 4$, $b - a = 2$ (atomic units).

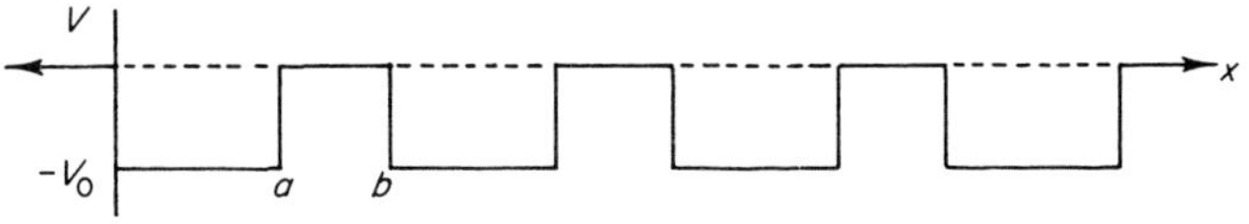

2 Use the simplest form of the tight binding method to determine the lowest energy band in the one-dimensional potential $V = V_o \cos 2\pi x/a$. Assume that the individual site wave functions are of the form $u(x) = e^{-\alpha^2 x^2}$, and determine α by minimizing the energy. Include only first neighbor interactions. For numerical calculations, assume $V_o = 3.0$ Ry, $a = 4.0$ (atomic units).

3 Determine the energies of the lowest and highest states in the first band for the potential of problem 2 by expansion in plane waves.

4 Use the following pseudopotential parameters:

$$\begin{aligned} V(\mathbf{K}_s) &= -0.23 \quad \text{for} \quad \mathbf{K}_s = (2\pi/a) \quad (1, 1, 1) \\ &\ \ 0.0 \qquad\qquad\qquad\qquad\qquad\ \ (2, 2, 0) \\ & +0.08 \qquad\qquad\qquad\qquad\qquad (3, 1, 1) \end{aligned}$$

(in rydbergs) to calculate the energy difference between the valence band maximum ($\Gamma_{25'}$) and the local conduction band minimum ($\Gamma_{2'}$) at $k = 0$ in germanium. Use the 15 plane waves of lowest kinetic energy. The lattice constant is 5.65 Å. You will have to form appropriate symmetrized combinations of plane waves for the germanium lattice. It is important to remember that there are two atoms in each unit cell, so that there is a structure factor to be included in the Fourier coefficients of potential. The space group is not simple, but contains a nonprimitive translation.

5 Use the wave functions and energy levels obtained in Problem 4 to calculate the effective mass in the $\Gamma_{2'}$ conduction band near $k = 0$ for germanium. Consider the interaction between the $\Gamma_{2'}$ and $\Gamma_{25'}$ levels. What would be the effect of including other bands in the calculation?

6 Use the Seitz empirical potential for lithium [given by Kohn and Rostoker (1954)] to calculate the energy and wave function of the lowest Γ_1 state (2s band) by the cellular method. Assume $r_s = 3.21$ in atomic units. This calculation is not too difficult to do on a desk calculator, but you may prefer to use a computer.

7 Find the pressure exerted by a gas of electrons, using Fermi statistics. Evaluate the result (in nt/m^2) for sodium for which $r_s = 3.96$ in atomic units.

8 Show that in germanium, $\mathbf{k}\cdot\mathbf{p}$ perturbation theory leads to the following effective Hamiltonian for states near the of the $\Gamma_{25'}$ valence. band.

$$\begin{pmatrix} Ak_x^2 + B(k_y^2 + k_z^2) & Ck_xk_y & Ck_xk_z \\ Ck_xk_y & Ak_y^2 + B(k_x^2 + k_z^2) & Ck_yk_z \\ Ck_xk_z & & Ak_z^2 + B(k_x^2 + k_y^2) \end{pmatrix}$$

Obtain expressions for A, B, and C in terms of specific matrix elements and energy denominators. The level structure at $k = 0$ may be assumed to be (schematically)

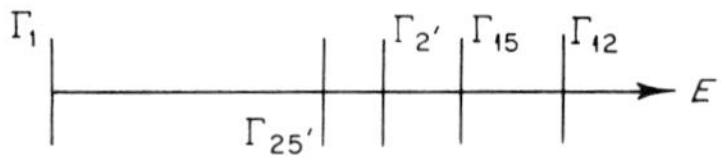

9 Find the Fermi energy for an intrinsic semiconductor if Boltzmann statistics can be applied to both conduction and valence bands.

10 Set up the expectation value of the Hamiltonian using determinantal wave functions, and verify Eqs. (4.9.12) and (4.9.13).

References

Abarenkov, I. V., and Heine, V. (1965). *Phil. Mag.* **12,** 529.

Alder, B., Fernbach, S., and Rotenberg, M. (eds.) (1968). "Methods in Computational Physics," Vol. 8, Energy Bands of Solids. Academic Press, New York.

Altmann, S. L., and Bradley, C. J. (1965a). *Rev. Mod. Phys.* **37,** 33.

Altmann, S. L., and Bradley, C. J. (1965b). *Proc. Phys. Soc.* (*London*) **86,** 915.

Altmann, S. L., and Cracknell, A. P. (1965). *Rev. Mod. Phys.* **37,** 19.

Altmann, S. L., Davies, B. L., and Harford, A. R. (1968). *J. Phys.* **C1,** 1633.

Animalu, A. O. E., and Heine, V. (1965). *Phil. Mag.* **12,** 1249.

Ashcroft, N. W. (1968). *J. Phys.* **C1,** 232.

Austin, B. J., Heine, V., and Sham, L. J. (1962). *Phys. Rev.* **127,** 276.

Bardeen, J. (1938). *J. Chem. Phys.* **6,** 367.

Bell, D. G. (1953). *Rev. Mod. Phys.* **26,** 311.

Bierman, L., and Lubeck, K. (1948). *Z. Astrophys.* **25,** 325.

Blankenbecler, R. (1957). *Amer. J. Phys.* **25,** 279.

Blount, E. I. (1962). *Solid State Phys.* **13,** 305.

Blume, M., Briggs, N., and Brooks, H. (1959). Tables of Coulomb Wave Functions. Cruft Lab., Harvard Univ., Cambridge, Massachusetts.

Brooks, H. (1953). *Phys. Rev.* **91,** 1027.

Brooks, H. (1958). *Nuovo Cimento Suppl.* **7,** 165.
Brooks, H., and Ham, F. S. (1958). *Phys. Rev.* **112,** 344.
Brown, E., and Krumhansl, J. A. (1958). *Phys. Rev.* **109,** 30.
Callaway, J. (1955). *Phys. Rev.* **97,** 933.
Callaway, J. (1957). *Phys. Rev.* **106,** 868.
Callaway, J. (1964). "Energy Band Theory." Academic Press, New York.
Callaway, J., and Laghos, P. S. (1969). *Phys. Rev.* **137,** 192.
Callaway, J., Woods, R. D., and Sirounian, V. (1957). *Phys. Rev.* **107,** 934.
Cardona, M., and Pollak, F. H. (1966). *Phys. Rev.* **142,** 530.
Clementi, E. (1965). Tables of Atomic Functions. IBM Corp., Yorktown Heights, New York.
Cohen, M. H., and Ham, F. S. (1960). *J. Phys. Chem. Solids* **16,** 177.
Cohen, M. H., and Heine, V. (1961). *Phys. Rev.* **122,** 1821.
Cohen, M. H., and Heine, V. (1970). *Solid State Phys.* **24,** 37.
Cohen, M. H., and Phillips, J. C. (1961). *Phys. Rev.* **124,** 1818.
Cooke, J. F., and Wood, R. F. (1972). *Phys. Rev.* **B5,** 1276.
Deegan, R. A., and Twose, W. D. (1967). *Phys. Rev.* **164,** 993.
Dimmock, J. O. (1971). *Solid State Phys.* **26,** 103.
Doniach, S. (1971). "Single Particle States in Many Body Systems in Computational Methods in Band Theory" (P. M. Marcus, J. F. Janak, and A. F. Williams, eds.), p. 500. Plenum Press, New York.
Dresselhaus, G., and Dresselhaus, M. S. (1967). *Phys. Rev.* **160,** 649.
Dresselhaus, G., Kip, A. F., and Kittel, C. (1955). *Phys. Rev.* **98,** 368.
Egorov, R. F., Reser, B. I., and Shirokovskii, V. P. (1968). *Phys. Status Solidi.* **26,** 391.
Ehrenreich, H. (1961). *J. Appl. Phys.* **32,** 2155.
Ehrenreich, H., and Hodges, L. (1968). *Methods Computat. Phys.* **8,** 149.
Euwema, R. N., and Stuckel, D. J. (1970). *Phys. Rev.* **B1,** 4692.
Euwema, R. N., Wilhite, D. L., and Surratt, G. T. (1973). *Phys. Rev.* **B7,** 818.
Fletcher, G. C. (1971). "The Electron Band Theory of Solids." North Holland, Amsterdam.
Fock, V. (1930a). *Z. Phys.* **61,** 126.
Fock, V. (1930b). *Z. Phys.* **2,** 795.
Froese, C. (1966). *Phys. Rev.* **150,** 1.
Gaspar, R. (1954). *Acta Phys. Acad. Sci. Hung.* **3,** 263.
Gilat, G. (1972). *J. Comp. Phys.* **10,** 432.
Goertzel, G., and Tralli, N. (1960). "Some Mathematical Methods of Physics." McGraw Hill, New York.
Gorin, E. (1936). *Phys. Z. Sowjetunion* **9,** 328.
Gray, D., and Brown, E. (1967). *Phys. Rev.* **160,** 567.
Ham, F. S. (1954). Electronic Energy Bands in Metals. Ph.D. Thesis, Harvard Univ. (unpublished).
Ham, F. S. (1955). *Solid State Phys.* **1,** 127.
Ham, F. S. (1957). *Quart. Appl. Math.* **15,** 31.
Ham, F. S. (1962a). *Phys. Rev.* **128,** 82.
Ham, F. S. (1962b). *Phys. Rev.* **128,** 2524.
Ham, F. S., and Segall, B. (1961). *Phys. Rev.* **124,** 1786.
Harris, F. E., and Monkhorst, H. J. (1969). *Phys. Rev. Lett.* **23,** 1026.
Harris, F. E., Kumar, L., and Monkhorst, H. J. (1973). *Phys. Rev.* **B7,** 2850.
Harrison, W. (1960). *Phys. Rev.* **118,** 1190.
Harrison, W. A. (1966). "Pseudopotentials in the Theory of Metals." Benjamin, New York.

Harrison, W. A. (1970). "Solid State Theory." McGraw-Hill, New York.
Heine, V. (1967). *Phys. Rev.* **153,** 673.
Heine, V. (1970). *Solid State Phys.* **24,** 1.
Heine, V., and Weaire, D. (1970). *Solid State Phys.* **24,** 249.
Hellman, H. (1934). *Acta Physicochim. URSS* **1,** 913.
Herman, F., Van Dyke, J. P., and Ortenburger, I. B. (1969). *Phys. Rev. Lett.* **22,** 807.
Herring, C. (1940). *Phys. Rev.* **57,** 1169.
Herring, C. (1966). Exchange Interactions among Itinerant Electrons. *In* "Magnetism" (G. T. Rado and H. Suhl, eds.), Vol. 4. Academic Press, New York.
Hodges, L., Ehrenreich, H., and Lang, N. D. (1966). *Phys. Rev.* **152,** 505.
Hohenberg, P., and Kohn, W. (1964). *Phys. Rev.* **136,** B864.
Howarth, D. J., and Jones, H. (1952). *Proc. Phys. Sod. (London)* **A65,** 355.
Hubbard, J. (1967). *Proc. Phys. Soc.* **92,** 921.
Hubbard, J., and Dalton, N. W. (1968). *J. Phys.* **C2,** 1637.
Hughes, A. J., and Callaway, J. (1964). *Phys. Rev.* **136,** A1390.
Kane, E. O. (1956). *J. Phys. Chem. Solids* **1,** 83.
Kane, E. O. (1966). The $\mathbf{k} \cdot \mathbf{p}$ Method. *In* "Semiconductors and Semimetals" (R. K. Willardson and A. C. Beer, eds.), Vol. 1, p. 75. Academic Press, New York.
Kleinman, L., and Shurtleff, R. (1969). *Phys. Rev.* **188,** 1111.
Koelling, D. D. (1969). *Phys. Rev.* **188,** 1049.
Koelling, D. D., Freeman, A. J., and Mueller, F. M. (1970). *Phys. Rev.* **Bl,** 1318.
Kohn, W. (1952). *Phys. Rev.* **87,** 472.
Kohn, W., and Rostoker, N. (1954). *Phys. Rev.* **94,** 1111.
Kohn, W., and Sham, L. J. (1965). *Phys. Rev.* **140,** A1133.
Koopmans, T. (1934). *Physica* **1,** 104.
Korringa, J. (1947). *Physica* **13,** 392.
Kuhn, T. S. (1951). *Quart. Appl. Math.* **9,** 1.
Kuhn, T. S., and Van Vleck, J. H. (1950). *Phys. Rev.* **79,** 382.
Kunz, A. B., and Lipari, N. O. (1971). *Phys. Rev.* **B4,** 1374.
Lafon, E., and Lin, C. C. (1966). *Phys. Rev.* **152,** 579.
von der Lage, F. C., and Bethe, H. A. (1947). *Phys. Rev.* **71,** 612.
Langlinais, J., and Callaway, J. (1972). *Phys. Rev.* **B5,** 124.
Lee, M. J. G. (1969). *Phys. Rev.* **178,** 953.
Leigh, R. S. (1956). *Proc. Phys. Soc. (London)* **A69,** 388.
Liberman, D. A. (1968). *Phys. Rev.* **171,** 1.
Loucks, T. L. (1967). "Augmented Plane Wave Method." Benjamin, New York.
Lowdin, P. O. (1951). *J. Chem. Phys.* **19,** 1396.
Lowdin, P. (1955). *Phys. Rev.* **97,** 1509.
Lowdin, P. O. (1962). *J. Appl. Phys.* **33,** 251.
Lowdin, P. O. (1966). *In* "Quantum Theory of Atoms, Molecules, and the Solid State" (P. O. Lowdin, ed.), p. 601. Academic Press, New York.
Luttinger, J. M., and Kohn, W. (1955). *Phys. Rev.* **97,** 869.
Luttinger, J. M., and Ward, J. C. (1960). *Phys. Rev.* **118,** 1417.
Marcus, P. M., Janak, J. F., and Williams, A. R. (eds.) (1971). "Computational Methods in Band Theory." Plenum, New York.
Mariot, L. (1962). "Group Theory and Solid State Physics." Prentice-Hall, Englewood Cliffs, New Jersey.
Miasek, M. (1956). *Bull. Acad. Pol. Sci. Cl. III,* **4,** 805.
Moss, T. S., and Walton, A. K. (1959). *Physica* **25,** 1142.
Mueller, F. M. (1967). *Phys. Rev.* **153,** 659.

Mueller, F. M., Garland, J. W., Cohen, M. H., and Benneman, K. H. (1971). *Ann. Phys.* (New York) **67,** 19.

Newton, R. G. (1966). "Scattering Theory of Waves and Particles." McGraw-Hill, New York.

Onodera, Y., and Okazaki, M. (1966). *J. Phys. Soc. Japan* **21,** 1273.

Pekeris, C. L. (1962). *Phys. Rev.* **126,** 1470.

Pendry, J. B. (1968). *J. Phys.* **C1,** 1065.

Phillips, J. C. (1961). *Phys. Rev.* **123,** 420.

Phillips, J. C., and Kleinman, C. (1959). *Phys. Rev.* **116,** 287.

Prokofjew, W. K. (1929). *Z. Phys.* **58,** 255.

Rajagopal, A. K., and Callaway, J. (1973). *Phys. Rev.* **B7,** 1912.

Roothaan, C. C. J. (1960). *Rev. Mod. Phys.* **32,** 179.

Roothaan, C. C. J., Sachs, L. M., and Weiss, A. W. (1960). *Rev. Mod. Phys.* **32,** 186.

Rudge, W. E. (1969). *Phys. Rev.* **181,** 1024.

Saffern, M. M., and Slater, J. C. (1953). *Phys. Rev.* **92,** 1126.

Schiff, L. I. (1968). "Quantum Mechanics," 3rd ed. McGraw-Hill, New York.

Schlosser, H. (1962). *J. Phys. Chem. Solids* **23,** 963.

Schlosser, H., and Marcus, P. M. (1963). *Phys. Rev.* **131,** 2529.

Seaton, M. J. (1966). *Proc. Phys. Soc. (London)* **88,** 801.

Seitz, F. (1935). *Phys. Rev.* **47,** 400.

Shaw, J. C., Ketterson, J. B., and Windmiller, L. R. (1972). *Phys. Rev.* **B5,** 3894.

Shockley, W. (1937). *Phys. Rev.* **52,** 866.

Shockley, W. (1950). *Phys. Rev.* **78,** 173.

Silverman, R. A. (1952). *Phys. Rev.* **85,** 227.

Slater, J. C. (1934). *Phys. Rev.* **45,** 794.

Slater, J. C. (1937). *Phys. Rev.* **51,** 151.

Slater, J. C. (1951). *Phys. Rev.* **81,** 385.

Slater, J. C. (1953). *Phys. Rev.* **92,** 603.

Slater, J. C. (1966). *Phys. Rev.* **145,** 599.

Slater, J. C., and Koster, G. F. (1954). *Phys. Rev.* **94,** 1498.

Slater, J. C., Mann, J. B., Wilson, T. M., and Wood, J. H. (1969a). *Phys. Rev.* **184,** 672.

Slater, J. C., Wilson, T. M., and Wood, J. H. (1969b). *Phys. Rev.* **179,** 28.

Soven, P. (1965). *Phys. Rev.* **137,** A1706.

Snow, E. C. (1968). *Phys. Rev.* **171,** 785.

Stark, R. W., and Falicov, L. M. (1967). *Phys. Rev. Lett.* **19,** 795.

Stern, F. (1959). *Phys. Rev.* **116,** 1399.

Stuckel, D. J., and Euwema, R. N. (1970). *Phys. Rev.* **B1,** 1635.

Szasz, L., and McGinn, G. (1967). *J. Chem. Phys.* **47,** 3495.

Thompson, E. D. (1960). Low Lying Energy Levels in Ferromagnetic Metals. Ph.D. Thesis, Mass. Inst. Technol. (unpublished).

Wallace, D. C. (1969). *Phys. Rev.* **187,** 991.

Wannier, G. (1943). *Phys. Rev.* **64,** 358.

Whittaker, E. T., and Watson, G. N. (1952). "A Course in Modern Analysis," 4th ed. Cambridge Univ. Press, London and New York.

Wigner, E., and Seitz, F. (1933). *Phys. Rev.* **43,** 804.

Woodruff, T. O. (1957). *Solid State Phys.* **4,** 367.

Ziman, J. M. (1961). *Phil. Mag.* **6,** 1013.

Ziman, J. M. (1965). *Proc. Phys. Soc. (London)* **86,** 337.

Ziman, J. M. (1971). *Solid State Phys.* **26,** 1.

Appendices to Part A

APPENDIX A

Summation Relations

Certain useful summation and closure relations are contained in this appendix. These concern two expressions which occur frequently in the theory:

$$\sum_{\nu} \exp[i(\mathbf{k} - \mathbf{k}') \cdot \mathbf{R}_\nu] \qquad \text{and} \qquad \sum_{\mathbf{k}} \exp[i\mathbf{k} \cdot (\mathbf{R}_\nu - \mathbf{R}_{\nu'})].$$

The summations include all direct lattice vectors and all $\mathbf{k}$ vectors in the Brillouin zone, respectively. It is useful to consider first a crystal on which periodic boundary conditions have been imposed, and subsequently to pass to the limit as the periodic volume becomes infinite. Periodic boundary conditions imply that we suppose the crystal repeats itself in all respects after translation through vectors $\mathbf{R}_1 = (2N_1 + 1)\mathbf{a}$, $\mathbf{R}_2 = (2N_2 + 1)\mathbf{b}$, $\mathbf{R}_3 = (2N_3 + 1)\mathbf{c}$ where $\mathbf{a}$, $\mathbf{b}$, and $\mathbf{c}$ are the three primitive translation vectors of the lattice. The use of an odd number is just a mathematical convenience. The concept is made more precise by employing the translation operators $T(\mathbf{R})$. It is then required

$$T[(2N_1 + 1)\mathbf{a}] = T(\mathbf{a})^{(2N_1+1)} = \mathbf{1}, \tag{A.1}$$

etc., where $\mathbf{1}$ designates the unit operator.

The derivation of the rules is given below for the case of a simple cubic lattice: $|\mathbf{a}| = |\mathbf{b}| = |\mathbf{c}| = a$. An extension to other cubic structures can be made quite simply by considering a cubic cell, which may contain several atoms.

Let $\mathbf{k} - \mathbf{k}' = \mathbf{s}$; $\mathbf{s} = (s_1, s_2, s_3)$; $\mathbf{R}_\nu = a(n_1, n_2, n_3)$. Then consider the summation

$$S_1 = (1/\mathfrak{N}) \sum_{n_1=-N_1}^{N_1} \sum_{n_2=-N_2}^{N_2} \sum_{n_3=-N_3}^{N_3} \exp[i(s_1n_1 + s_2n_2 + s_3n_3)a] \tag{A.2}$$

where

$$\mathfrak{N} = (2N_1 + 1)(2N_2 + 1)(2N_3 + 1).$$

The summations are evidently independent. A single one is just a geometric series:

$$\frac{1}{2N_1+1}\sum_{-N_1}^{N_1} \exp(is_1n_1a) = \frac{1}{2N_1+1}\exp(-is_1N_1a)\sum_{0}^{2N_1}\exp(is_1n_1a)$$

$$= \frac{\exp(-is_1N_1a)}{(2N_1+1)}\left[\frac{1-\exp[is_1(2N_1+1)a]}{1-\exp(is_1a)}\right]$$

$$= \frac{\sin[s_1(N_1+\frac{1}{2})a]}{(2N_1+1)\sin(s_1a/2)}. \tag{A.3}$$

In the limit that N_1 is large, the sum is appreciable only when both the numerator and denominator are zero. When this happens, the result is unity. The condition under which it will occur is $s_1a/2 = l\pi$ or $s_1 = 2\pi l/a$, where l is any integer. Identical results are obtained in the other summations and one observes that for the condition above to hold for all three components, **s** must be a reciprocal lattice vector. Further, the quantity $\mathfrak{N}$ equals the number of atoms in the block of atoms on which we have imposed periodic boundary conditions. Hence, we have in the limit of large $\mathfrak{N}$

$$S_1 = (1/\mathfrak{N})\sum_{\nu}\exp[i(\mathbf{k}-\mathbf{k}')\cdot\mathbf{R}_\nu] = 1 \qquad \text{if}\quad \mathbf{k}-\mathbf{k}' = \mathbf{K}_s,$$

$$= 0 \qquad \text{otherwise,} \tag{A.4}$$

where $\mathbf{K}_s$ is a reciprocal lattice vector. If **k** is chosen in accord with the rules (1.2.7) implied by periodic boundary conditions, the zero of the second line of Eq. (A.4) is correct; otherwise we have a quantity of order $\mathfrak{N}^{-1/3}$.

In the second summation, it is convenient to put $\mathbf{R}_\nu - \mathbf{R}_{\nu'} = \mathbf{R}$, $\mathbf{R} = a(m_1, m_2, m_3)$, and

$$\mathbf{k} = t[n_1/(2N_1+1),\ n_2/(2N_2+1),\ n_3/(2N_3+1)],$$

$$\mathbf{k}\cdot\mathbf{R} = at[n_1m_1/(2N_1+1) + n_2m_2/(2N_2+1) + n_3m_3/(2N_3+1)],$$

in which n_1, m_1, etc., are integers. Since the sum includes only values of **k** lying inside or on the surface of a single Brillouin zone, we have $-N_1 \le n_1 \le N_1$, etc. So consider

$$S_2 = (1/\mathfrak{N})\sum_{n_1=-N_1}^{N_1}\sum_{n_2=-N_2}^{N_2}\sum_{n_3=-N_3}^{N_3}\exp\{iat[n_1m_1/(2N_1+1) + n_2m_2/(2N_2+1) + n_3m_3/(2N_3+1)]\}. \tag{A.5}$$

As before, a single one of the independent sums gives

$$\sum_{-N_1}^{N_1} \exp[ian_1m_1t/(2N_1+1)] = \sin(am_1t/2)/\sin\{am_1t/[2(2N_1+1)]\}. \quad \text{(A.6)}$$

The quantity $am_1t/2(2N_1+1)$ will not be an integral multiple of 2π, so there will be a contribution of order $(2N_1+1)$ only when $m_1 = 0$. Hence, for large N_1, N_2, N_3

$$S_2 = (1/\mathfrak{N}) \sum_{\mathbf{k}} \exp[i\mathbf{k}\cdot(\mathbf{R}_\nu - \mathbf{R}_{\nu'})] = \delta_{\nu,\nu'}. \quad \text{(A.7)}$$

Equations (A.4) and (A.7) are the fundamental summation rules. It is useful to examine these sums in the limit $\mathfrak{N}$ becomes infinite. To do this, we write (A.3) in the form

$$\sum_{-N_1}^{N_1} \exp(ias_1n_1) = (2\pi/a)[(as_1/2)/\sin as_1/2][\sin as_1(N_1+\tfrac{1}{2})/(\pi s_1)].$$

Suppose first that $s_1a/2$ is not an integral multiple of π. The Dirac delta function can be defined through the relation

$$\delta(x) = (1/2\pi) \lim_{b\to\infty} \int_{-b}^{b} \exp(ikx)\, dk = \lim_{b\to\infty} (\sin xb)/(\pi x).$$

Hence

$$\lim_{N_1\to\infty} \sum_{-N_1}^{N_1} \exp(ias_1n_1) = (2\pi/a)\,\delta(s_1). \quad \text{(A.8)}$$

If $s_1a = 2l\pi$, this situation is identical with the case $s_1 = 0$, since the exponentials on the left are unity. Hence we have

$$\sum_{\nu} \exp[i(\mathbf{k}-\mathbf{k}')\cdot\mathbf{R}_\nu] = [(2\pi)^3/\Omega] \sum_{l} \delta(\mathbf{k}-\mathbf{k}'-\mathbf{K}_l) \quad \text{(A.9a)}$$

where the sum over $\mathbf{K}_l$ includes all reciprocal lattice vectors and Ω is the volume of the unit cell. If, however, $\mathbf{k}$ and $\mathbf{k}'$ are constrained to lie inside a single zone, then only the zero reciprocal lattice vector contributes, so that we obtain

$$\sum_{\nu} \exp[i(\mathbf{k}-\mathbf{k}')\cdot\mathbf{R}_\nu] = [(2\pi)^3/\Omega]\,\delta(\mathbf{k}-\mathbf{k}'). \quad \text{(A.9b)}$$

The continuous $\mathbf{k}$ limit of (A.7) is obvious. Since the volume of $\mathbf{k}$ space which is included in the integral is finite and equal to $(2\pi)^3/\Omega$, we have immediately

$$\int \exp[i\mathbf{k}\cdot(\mathbf{R}_\nu - \mathbf{R}_{\nu'})]\, d^3k = [(2\pi)^3/\Omega]\,\delta_{\nu,\nu'}. \quad \text{(A.10)}$$

APPENDIX B

Quantization of the Free Electromagnetic Field

This appendix contains a brief description of the nonrelativistic quantization of a free electromagnetic field in the Schrödinger picture. The starting point is an expression for the energy of an electromagnetic field in the absence of charges and currents

$$U = \int d^3r[\tfrac{1}{2}\epsilon_0\mathbf{E}^2 + \tfrac{1}{2}(\mathbf{B}^2/\mu_0)] \tag{B.1}$$

in which $\mathbf{E}$ is the electric field strength and $\mathbf{B}$ is the magnetic flux density. MKS units are used, with ϵ_0 and μ_0 the permittivity and permeability of free space, respectively. It is convenient to choose a gauge in which the scalar potential is zero so that both $\mathbf{E}$ and $\mathbf{B}$ may be derived from a vector potential A

$$\mathbf{E} = -\partial\mathbf{A}/\partial t, \qquad B = \nabla \times A. \tag{B.2}$$

We write $\mathbf{A}(\mathbf{r}, t)$ as a Fourier transform. The procedure is quite similar to that followed in quantizing lattice vibrations in Section 1.5

$$\mathbf{A} = c(\mu_0^{1/2}/8\pi^3) \sum_p \int \boldsymbol{\varepsilon}_p(\mathbf{k})q_p(\mathbf{k}, t) \exp(i\mathbf{k}\cdot\mathbf{r})\, d^3k. \tag{B.3}$$

It is to be noted that $\mathbf{k}$ is not restricted to a Brillouin zone. The quantity $\boldsymbol{\varepsilon}_p$ is a polarization vector, and q_p is to become the normal coordinate. Since $\nabla\cdot\mathbf{A} = 0$, we have

$$\mathbf{k}\cdot\boldsymbol{\varepsilon}_p(\mathbf{k}) = 0.$$

Then

$$\mathbf{B} = ic(\mu_0^{1/2}/8\pi^3) \sum_p \int (\mathbf{k} \times \boldsymbol{\varepsilon}_p)q_p(\mathbf{k}, t) \exp(i\mathbf{k}\cdot\mathbf{r})\, d^3k. \tag{B.4}$$

Since the vector potential must be real (and $\boldsymbol{\varepsilon}_\rho$ is real), we have

$$q_\rho^*(\mathbf{k}, t) = q_\rho(-\mathbf{k}, t).$$

Also

$$E = -c(\mu_0^{1/2}/8\pi^3) \sum_\rho \int \epsilon_\rho \dot{q}(\mathbf{k}, t) \exp(i\mathbf{k}\cdot\mathbf{r})\, d^3k \tag{B.5}$$

where the overdot indicates time derivative.

The contribution to the energy from the magnetic field is

$$(-c^2/128\pi^6) \sum_{\rho\rho'} \int d^3r\, d^3k\, d^3k'\, (\mathbf{k} \times \boldsymbol{\varepsilon}_\rho)\cdot(\mathbf{k}' \times \boldsymbol{\varepsilon}_{\rho'}) q_\rho(\mathbf{k}, t) q_{\rho'}(\mathbf{k}'t)$$
$$\times \exp[i(\mathbf{k} + \mathbf{k}')\cdot\mathbf{r}]$$
$$= (2\pi)^{-3}(c^2/2) \sum_{\rho\rho'} \int d^3k (\mathbf{k} \times \boldsymbol{\varepsilon}_\rho)\cdot(\mathbf{k} \times \boldsymbol{\varepsilon}_{\rho'}) q_\rho(\mathbf{k}, t) q_{\rho'}(-\mathbf{k}, t).$$

Now

$$(\mathbf{k} \times \boldsymbol{\varepsilon}_\rho)\cdot(\mathbf{k} \times \boldsymbol{\varepsilon}_{\rho'}) = \mathbf{k}^2(\boldsymbol{\varepsilon}_\rho\cdot\boldsymbol{\varepsilon}_{\rho'}) = \mathbf{k}^2\, \delta_{\rho\rho'}.$$

Hence the magnetic energy is (with $c^2\mathbf{k}^2 = \omega^2$)

$$U_\mathrm{M} = \tfrac{1}{2}(2\pi)^{-3} \sum_\rho \int d^3k\, \omega^2 \mid q_\rho(\mathbf{k}, t) \mid^2. \tag{B.6}$$

The energy in the electric field is similarly found to be

$$U_\mathrm{E} = \tfrac{1}{2}(2\pi)^{-3} \sum_\rho \int d^3k \mid \dot{q}_\rho(\mathbf{k}, t) \mid^2. \tag{B.7}$$

The system may be considered to be equivalent to a collection of uncoupled simple harmonic oscillators (radiation oscillators). We proceed as in Section 1.4 by defining Lagrangian and cannonical momenta $p_\rho(\mathbf{k}, t) = \dot{q}_\rho(\mathbf{k}, t)$. The field energy (B.1) becomes the Hamiltonian H,

$$H = \tfrac{1}{2}(2\pi)^{-3} \sum_\rho \int d^3k\, [\mid p_\rho(\mathbf{k}, t) \mid^2 + \omega^2(\mathbf{k}) \mid q_\rho(\mathbf{k}, t) \mid^2]. \tag{B.8}$$

To quantize the theory we may now introduce the usual commutation relations

$$[q_\rho^*(\mathbf{k}), p_{\rho'}(\mathbf{k}')] = i\hbar\, \delta_{\rho\rho'}\, \delta(\mathbf{k} - \mathbf{k}')$$
$$[q_\rho(\mathbf{k}), q_{\rho'}(\mathbf{k}')] = [p_\rho(\mathbf{k}), p_{\rho'}(\mathbf{k}')] = 0, \quad \text{etc.} \tag{B.9}$$

Creation and annihilation operators $b_\rho^\dagger$, b_ρ are introduced exactly in parallel with Section 1.4:

$$b_\rho(\mathbf{k}) = (\omega(\mathbf{k})/2\hbar)^{1/2}q_\rho(\mathbf{k}) + \{i/[2\hbar\omega(\mathbf{k})]^{1/2}\}p_\rho(\mathbf{k})$$

$$b_\rho^\dagger(\mathbf{k}) = (\omega(\mathbf{k})/2\hbar)^{1/2}q_\rho^*(\mathbf{k}) - \{i/[2\hbar\omega(\mathbf{k})]^{1/2}\}p_\rho(\mathbf{k}). \qquad \text{(B.10)}$$

These relations may be inverted to yield

$$q_\rho(\mathbf{k}) = [\hbar/2\omega(\mathbf{k})]^{1/2}[b_\rho(\mathbf{k}) + b_\rho^\dagger(-\mathbf{k})]$$

$$p_\rho(\mathbf{k}) = (1/i)[\hbar\omega(\mathbf{k})/2]^{1/2}[b_\rho(\mathbf{k}) - b_\rho^\dagger(-\mathbf{k})]. \qquad \text{(B.11)}$$

The commutation relations for the b_ρ are

$$[b_\rho(\mathbf{k})b_{\rho'}^\dagger(\mathbf{k}')] = \delta_{\rho\rho'}\,\delta(\mathbf{k} - \mathbf{k}') \qquad \text{(B.12)}$$

and

$$[b_\rho(\mathbf{k}), b_\rho'(\mathbf{k}')] = [b_\rho^\dagger(\mathbf{k}), b_{\rho'}^\dagger(\mathbf{k}')] = 0.$$

The substitution of (B.11) enables us to write the total energy H of (B.8) in a simple form

$$H = (2\pi)^{-3}\sum_\rho \int d^3k\; \hbar\omega_\rho(\mathbf{k})[b_\rho^\dagger(\mathbf{k})b_\rho(\mathbf{k}) + \tfrac{1}{2}]. \qquad \text{(B.13)}$$

The first term in (B.13) evidently represents the number of photons of wave vector $\mathbf{k}$ and polarization ρ, and is denoted by $n_\rho(\mathbf{k})$:

$$n_\rho(\mathbf{k}) = b_\rho^\dagger(\mathbf{k})b_\rho(\mathbf{k}). \qquad \text{(B.14)}$$

For use in the main text, we require an expression for the vector potential in terms of creation and annihilation operators. This is obtained from (B.3) and (B.11):

$$\mathbf{A} = c[(\hbar\mu_0)^{1/2}/8\pi^3] \sum_\rho \int [\boldsymbol{\varepsilon}_\rho/(2\omega(\mathbf{k}))^{1/2}][b_\rho(\mathbf{k}) + b_\rho^\dagger(-\mathbf{k})]\exp(i\mathbf{k}\cdot\mathbf{r})\, d^3k$$

$$= c[(\hbar\mu_0)^{1/2}/8\pi^3] \sum_\rho \int [\boldsymbol{\varepsilon}_\rho/(2\omega(\mathbf{k}))^{1/2}][b_\rho(\mathbf{k}) \exp(i\mathbf{k}\cdot\mathbf{r})$$

$$+ b_\rho^\dagger(\mathbf{k}) \exp(-i\mathbf{k}\cdot\mathbf{r})]\, d^3k. \qquad \text{(B.15)}$$

For use in Section 1.9 we wish to modify Eq. (B.15) by inserting the high frequency dielectric constant κ_∞. This is done simply by replacing c by $c/\kappa_\infty^{1/2}$.

APPENDIX C

Character Tables and Compatibility Tables

In this appendix, we present character tables and compatibility tables for the group of the wave vector at symmetry points of the Brillouin zone for the simple cubic, body-centered cubic, and face-centered cubic lattices.

The group of the center of the zone Γ is the full cubic group O_h. The character table for this group has been given in the main text (Chapter 3, Table VII). This has also been done for the group C_{4v} which is the group of the 100 axis (Δ) (Chapter 3, Table VI). Diagrams of the Brillouin zone for these lattices have also been given (Figs. 1.2.1–1.2.3). This material is not repeated here.

The points $R[(2\pi/a)(1, 1, 1)]$ in the Brillouin zone for the simple cubic lattice and $H[(2\pi/a)(1, 0, 0)]$ in the body-centered cubic also have full cubic symmetry. The character tables for these points is the same as that for Γ.

The group of $P[(2\pi/a)(\frac{1}{2}, \frac{1}{2}, \frac{1}{2})]$ in the body-centered cubic BZ is the tetrahedral group T_d. This has five irreducible representations—half the number for O_h. The representation of P are related to those at Γ in the following way: P_1 contains functions belonging to Γ_1 and Γ_2'; P_2 contains those belonging to Γ_2 and $\Gamma_{1'}$; P_3 contains Γ_{12} and $\Gamma_{12'}$; P_4, Γ_{15} and $\Gamma_{25'}$; P_5, Γ_{25} and $\Gamma_{15'}$.

Basis functions for N(bcc) have been chosen which are appropriate to the point $(2\pi/a)(\frac{1}{2}, \frac{1}{2}, 0)$. The operation $C_{2\parallel}$ is a rotation about an axis parallel to the wave vector, while $C_{2\perp}$ is a rotation about an axis perpendicular to the wave vector. These operations are not equivalent. The representations are all nondegenerate. Note that the representation N_1 contains not only *s* functions, but two different d-like functions as well.

The points on the surface of the Brillouin zone of the face-centered cubic lattice do not have as high symmetry as do the corresponding points pertaining to the body-centered cubic lattice. There are three inequivalent points X (instead of the single H), so that the group of X contains 16 operations. It is necessary to distinguish rotations about axis parallel and perpendicular to the wave vector. Note that the representation X_1, has a d function in the basis, as well as an s function. There are four inequivalent points L compared to two inequivalent points P. In addition to the basis functions listed for the doubly degenerate representation L_3, there are two additional linear combinations of spherical harmonics with $l = 2$ which belong to L_3; thus all the d-like functions are contained in L_1 and L_3. There are six inequivalent points W. Since the group of W does not contain the inversion, the representations may contain both even and odd basis functions. For instance W_3 contains both p- and d-like functions.

Information concerning the connection of bands and the splitting of degeneracies as one moves away from a symmetry point along lines of symmetry is summarized in compatibility tables. The construction of these tables was discussed briefly in Section 4.1. The most important compatibility relations are summarized in Tables C.IX and C.X. Other sets of relations, i.e., between P and Λ and N and Σ, may be easily deduced from the character tables, C.I–C.VIII.‡

‡ Tables C.I–C.X taken from J. Callaway (1964). "Energy Band Theory." Academic Press, New York.

TABLE C.I

CHARACTER TABLE, GROUP OF $P = (2\pi/a)(\frac{1}{2}, \frac{1}{2}, \frac{1}{2})$ (T_d)

Representation	Basis	E	$3C_4^2$	$8C_3$	$6JC_4$	$6JC_2$
P_1	$1, xyz$	1	1	1	1	1
P_2	$x^4(y^2 - z^2) + y^4(z^2 - x^2) + z^4(x^2 - y^2)$	1	1	1	−1	−1
P_3	$x^2 - y^2, xyz(x^2 - y^2)$	2	2	−1	0	0
P_4	$x, y, z; xy; yz; zx$	3	−1	0	−1	1
P_5	$z(x^2 - y^2)$	3	−1	0	1	−1

TABLE C.II

CHARACTER TABLE, GROUP OF $N = (2\pi/a)(\frac{1}{2}, \frac{1}{2}, 0)$ (D_{2h})

Representation	Basis	E	C_4^2	$C_2\|\|$	$C_2\perp$	J	JC_4^2	$JC_2\perp$	$JC_2\|\|$
N_1	$1, xy, 3z^2 - r^2$	1	1	1	1	1	1	1	1
N_2	$z(x - y)$	1	−1	1	−1	1	−1	−1	1
N_3	$z(x + y)$	1	−1	−1	1	1	−1	1	−1
N_4	$x^2 - y^2$	1	1	−1	−1	1	1	−1	−1
N_1'	$x + y$	1	−1	1	−1	−1	1	1	−1
N_2'	$z(x^2 - y^2)$	1	1	1	1	−1	−1	−1	−1
N_3'	z	1	1	−1	−1	−1	−1	1	1
N_4'	$x - y$	1	−1	−1	1	−1	1	−1	1

TABLE C.III

CHARACTER TABLE, GROUP OF Λ, F[a] (C_{3v})

Representation	Basis	E	$2C_3$	$3JC_2$
Λ_1	$1, x + y + z$	1	1	1
Λ_2	$x(y^2 - z^2) + y(z^2 - x^2) + z(x^2 - y^2)$	1	1	−1
Λ_3	$2x - y - z, y - z$	2	−1	0

[a] $F = (2\pi/a)(\frac{1}{2} + x, \frac{1}{2} - x, \frac{1}{2} - x)$; $0 \leqslant x \leqslant \frac{1}{2}$.

TABLE C.IV

CHARACTER TABLE, GROUP OF $X^a = (2\pi/a)(1, 0, 0)$ (D_{4h})

Representation	Basis	E	$2C_4^2\perp$	$C_4^2\|\|$	$2C_4\|\|$	$2C_2$	J	$2JC_4^2\perp$	$JC_4^2\|\|$	$2JC_4\|\|$	$2JC_2$
X_1	$1, 2x^2 - y^2 - z^2$	1	1	1	1	1	1	1	1	1	1
X_2	$y^2 - z^2$	1	1	1	−1	−1	1	1	1	−1	−1
X_3	yz	1	−1	1	−1	1	1	−1	1	−1	1
X_4	$yz(y^2 - z^2)$	1	−1	1	1	−1	1	−1	1	1	−1
X_5	xy, xz	2	0	−2	0	0	2	0	−2	0	0
X_1'	$xyz(y^2 - z^2)$	1	1	1	1	1	−1	−1	−1	−1	−1
X_2'	xyz	1	1	1	−1	−1	−1	−1	−1	1	1
X_3'	$x(y^2 - z^2)$	1	−1	1	−1	1	−1	1	−1	1	−1
X_4'	x	1	−1	1	1	−1	−1	1	−1	−1	1
X_5'	y, z	2	0	−2	0	0	−2	0	2	0	0

[a] Also applies to $M = (2\pi/a)(1, 1, 0)$.

TABLE C.V

CHARACTER TABLE, GROUP OF $L = (2\pi/a)(\frac{1}{2}, \frac{1}{2}, \frac{1}{2})$ (D_{3d})

Representation	Basis	E	$2C_3$	$3C_2$	J	$2JC_3$	$3JC_2$
L_1	$1, xy + yz + xz$	1	1	1	1	1	1
L_2	$yz(y^2 - z^2) + xy(x^2 - y^2) + xz(z^2 - x^2)$	1	1	-1	1	1	-1
L_3	$2x^2 - y^2 - z^2;\ y^2 - z^2$	2	-1	0	2	-1	0
L_1'	$x(y^2 - z^2) + y(z^2 - x^2) + z(x^2 - y^2)$	1	1	1	-1	-1	-1
L_2'	$x + y + z$	1	1	-1	-1	-1	1
L_3'	$y - z;\ 2x - y - z$	2	-1	0	-2	1	0

TABLE C.VI

CHARACTER TABLE, GROUP OF $Wi = (2\pi/a)(1, \frac{1}{2}, 0)$ (D_{2d})

Representation	Basis	E	C_4^2	$2C_2$	$2JC_4$	$2JC_4^2$
W_1	$1, 2y^2 - x^2 - z^2$	1	1	1	1	1
W_1'	xz	1	1	1	-1	-1
W_2	xyz	1	1	-1	1	-1
W_2'	$y, z^2 - x^2$	1	1	-1	-1	1
W_3	$xy, yz; x, z$	2	-2	0	0	0

TABLE C.VII

CHARACTER TABLE, GROUP OF $\Sigma = (2\pi/a)(x, x, 0)$ (C_{2v})

Representation	Basis	E	C_2	JC_4^2	JC_2
Σ_1	$1, x + y$	1	1	1	1
Σ_2	$z(x - y);\ z(x^2 - y^2)$	1	1	-1	-1
Σ_3	$z;\ z(x + y)$	1	-1	-1	1
Σ_4	$x - y;\ x^2 - y^2$	1	-1	1	-1

TABLE C.VIII

CHARACTER TABLES OF G, K, U, D, Z, S (C_{2v})

Representation	Z	E	C_4^2	JC_4^2	$JC_4^2 \perp$
	G, K, U, S	E	C_2	JC_4^2	JC_2
	D	E	C_4^2	JC_2	$JC_2 \perp$
K_1	$1, x+y$	1	1	1	1
K_2	$z(x-y), \quad z(x^2-y^2)$	1	1	-1	-1
K_3	$z, \quad z(x+y)$	1	-1	-1	1
K_4	$x-y; \quad x^2-y^2$	1	-1	1	-1

$G = \frac{2\pi}{a}(\frac{1}{2}+x, \frac{1}{2}-x, 0)$ (bcc); $K = \frac{2\pi}{a}(\frac{3}{4}, \frac{3}{4}, 0)$ (fcc)

$U = \frac{2\pi}{a}(1, \frac{1}{4}, \frac{1}{4})$ (fcc); $D = \frac{2\pi}{a}(\frac{1}{2}, \frac{1}{2}, x)$ (bcc)

$Z = \frac{2\pi}{a}(1, x, 0)$ (fcc); $S = \frac{2\pi}{a}(1, x, x)$ (fcc)

TABLE C.IX

COMPATIBILITY RELATIONS BETWEEN Γ AND Δ, Λ, Σ

Γ_1	Γ_2	Γ_{12}	Γ_{15}	Γ_{25}'	Γ_1'	Γ_2'	Γ_{12}'	Γ_{15}'	Γ_{25}
Δ_1	Δ_2	$\Delta_1\,\Delta_2$	$\Delta_1\,\Delta_5$	$\Delta_2'\,\Delta_5$	Δ_1'	Δ_2'	$\Delta_1'\,\Delta_2'$	$\Delta_1'\,\Delta_5$	$\Delta_2\,\Delta_5$
Λ_1	Λ_2	Λ_3	$\Lambda_1\,\Lambda_3$	$\Lambda_1\,\Lambda_3$	Λ_2	Λ_1	Λ_3	$\Lambda_2\,\Lambda_3$	$\Lambda_2\,\Lambda_3$
Σ_1	Σ_4	$\Sigma_1\,\Sigma_4$	$\Sigma_1\,\Sigma_3\,\Sigma_4$	$\Sigma_1\,\Sigma_2\Sigma_3$	Σ_2	Σ_3	$\Sigma_2\,\Sigma_3$	$\Sigma_2\,\Sigma_3\,\Sigma_4$	$\Sigma_1\,\Sigma_2\,\Sigma_4$

TABLE C.X

COMPATIBILITY RELATIONS BETWEEN X AND Δ, Z, S

X_1	X_2	X_3	X_4	X_5	X_1'	X_2'	X_3'	X_4'	X_5'
Δ_1	Δ_2	Δ_2'	Δ_1'	Δ_5	Δ_1'	Δ_2'	Δ_2	Δ_1	Δ_5
Z_1	Z_1	Z_4	Z_4	$Z_3\,Z_2$	Z_2	Z_2	Z_3	Z_3	$Z_1\,Z_4$
S_1	S_4	S_1	S_4	$S_2\,S_3$	S_2	S_3	S_2	S_3	$S_1\,S_4$

APPENDIX D

Second Quantization for a System of Fermions

The second quantization method is quite useful as a systematic approach to the many fermion problem. Such a procedure is needed because it is difficult to work directly with a many-electron wave function. A brief survey of second quantization for a system of fermions is given here. The corresponding procedures for phonons and photons were discussed in Section 1.4 and Appendix B, respectively. For a more complete account, consult Schweber (1961, Chap. 6).

We begin with a set of single-particle wave functions, including spin, which will be denoted here by $u_k(\mathbf{x})$. These functions will be assumed to be orthonormal, but they are otherwise arbitrary. Frequently, the space part of u_k is taken to be a plane wave. Alternately, Bloch functions may be employed. The index k designates the single-particle state, while $\mathbf{x}$ represents both space and spin coordinates. Now consider a set of N noninteracting particles, such that each particle will be in one of the states u_k. Let n_k be the number of particles in k: This can be either 1 or 5. The quantity n_k is called the *occupation number*. Such quantities, rather than the particle coordinates will be the fundamental variables of the theory.

A wave function for the system of noninteracting particles can be constructed as an antisymmetrized product of single-particle functions. Such antisymmetrized functions can be expressed as determinants. In order not to have an ambiguity of algebraic sign, we agree to arrange the indices k of the single-particle states in a definite order, such that k_1 occurs before k_2, etc. There must be N states having nonzero occupation numbers: The wave function for the system is

$$\Psi(\mathbf{x}_1 \cdots \mathbf{x}_N) = (N!)^{-1/2} \begin{vmatrix} u_{k_1}(\mathbf{x}_1) & u_{k_1}(\mathbf{x}_2) \cdots & u_{k_1}(\mathbf{x}_N) \\ u_{k_2}(\mathbf{x}_1) & u_{k_2}(\mathbf{x}_2) \cdots & u_{k_2}(\mathbf{x}_N) \\ u_{k_N}(\mathbf{x}_1) & u_{k_N}(\mathbf{x}_2) \cdots & \cdot u_{k_N}(\mathbf{x}_N) \end{vmatrix} . \tag{D.1}$$

Note that no two states can have the same quantum numbers (no two k_i are equal). Likewise, the coordinates of two particles cannot be equal; otherwise the wave function vanishes. If the basis functions u_k form a complete set, the set of all possible independent determinantal functions (D.1) is a complete set for the expansion of the many-body function for interacting particles.

It can be verified easily that (D.1) is normalized. This is accomplished by the factor $(N!)^{-1/2}$. Likewise, functions with any differences in the occupied single-particle states are orthogonal to (D.1).

It is necessary to determine the effect of operators on determinantal functions. The operators to be considered are of two types: single-particle and two-particle operators. A single-particle operator is one which depends on the coordinates of a single particle, e.g., $f(\mathbf{x}_i)$; however, in considering a system of identical particles, we always encounter sums of such operators, identical except for the coordinate, over all the particles of the system. Such a sum will be referred to as a single-particle operator; the sum being implied. Thus

$$F = \sum_i f(\mathbf{x}_i). \tag{D.2}$$

Similarily a two-particle operator G will refer to a sum of operators $g(\mathbf{x}_i, \mathbf{x}_j)$ which depend on the coordinates of a pair of particles. All distinct pairs are included in the sum. However, a particle does not interact with itself, so there is no term of the form $g(\mathbf{x}_i, \mathbf{x}_i)$ in G

$$G = \sum_{i, j>i} g(\mathbf{x}_i, \mathbf{x}_j). \tag{D.3}$$

Consider first a single-particle operator F. Its matrix elements will be zero unless: (1) all occupation numbers are unchanged (diagonal element) or (2) the states differ in regard to a single occupation number. In the second case, consider the element $\langle B \mid F \mid A \rangle$, $\mid A \rangle$ and $\mid B \rangle$ being represented by wave functions of the form (D.1). These states are specified by the occupation numbers of the single-particle states. Suppose that in B the occupation number of state l is 1 and that of n is zero, while in A that of l is zero and n is 1. It is necessary to introduce f_{ln}, the matrix element between single-particle states

$$f_{ln} = \int u_l^*(\mathbf{x}) f(\mathbf{x}) u_n(\mathbf{x})\, d^3x. \tag{D.4}$$

Integrations include summations over spin coordinates. It is easy to see that the N particle matrix element is proportional to f_{ln}; however, the sign is difficult to specify. Some experimentation, which will be left to the reader shows that $(l < n)$

$$\langle B \mid F \mid A \rangle = \langle \cdots 1_l \cdots 0_n \cdots \mid F \mid \cdots 0_l \cdots 1_n \cdots \rangle = f_{ln}(-1)^{S(l+1,n-1)} \tag{D.5}$$

in which S is the sum of the occupation numbers from (and including) $l + 1$ through $n - 1$. In general,

$$S(i, j) = \sum_{a=i}^{j} n_a. \tag{D.6}$$

If $l > n$, the exponent in (D.5) is $S(n + 1, l - 1)$: we always count the numbers of occupied states between the one of smaller index in our conventional order and the one of larger index. If $i = j \pm 1$, $S = 0$. The quantity S specifies the number of permutations of the $\mathbf{k}_i$ required to make the determinants line up with all rows identical except for the one substituted.

The diagonal matrix element of F (between identical determinants) $\langle A \mid F \mid A \rangle$ is simply

$$\langle A \mid F \mid A \rangle = \sum_i f_{ii} n_i. \tag{D.7}$$

In order to keep the complicated business of algebraic signs in order, we introduce operators c_n, etc. defined so that c_n removes row n from the determinant. This operator connects an N particle state with one containing $N - 1$ particles: c_n is said to be an annihilation operator for state n. We define

$$\langle \cdots 0_n \cdots \mid c_n \mid \cdots 1_n \cdots \rangle = (-1)^{S(1,n-1)}. \tag{D.8}$$

Other elements of c_n are zero. The operator c_n is not Hermitian. Its adjoint $c_n^\dagger$ evidently adds an electron in state n if none was present initially. It is therefore a creation operator.

$$\langle \cdots 1_n \cdots \mid c_n^\dagger \mid \cdots 0_n \cdots \rangle = (-1)^{S(1,n-1)}. \tag{D.9}$$

Equations (D.8) and (D.9) suffice to enable us to work out the commutation rules algebraically. In particuler (for $n > l$),

$$\langle \cdots 1_l \cdots 0_n \mid c_l^\dagger c_n \mid \cdots 0_l \cdots 1_n \cdots \rangle = (-1)^{S(1,l-1)}(-1)^{S(1,l-1)+S(l+1,n-1)}$$
$$= (-1)^{S(l+1,n-1)}. \tag{D.10a}$$

Note the omission of the contribution from the state l (which is empty initially) to the algebraic sign. If we consider the operators in opposite order $c_n c_l^\dagger$, the state l is present when c_n acts and an additional negative sign is obtained.

$$\langle \cdots 1_l \cdots 0_n \cdots \mid c_n c_l^\dagger \mid \cdots 0_l \cdots 1_n \cdots \rangle = -(-1)^{S(l+1,n-1)}. \tag{D.10b}$$

The reader should investigate what happens if $l > n$. In the case $l = n$, the operator $c_n^\dagger c_n$ is diagonal with unit matrix element if n is occupied initially, and is zero otherwise. In the case of $c_n c_n^\dagger$ we obtain unity if n is initially empty.

These results can be combined as

$$c_l^\dagger c_n + c_n c_l^\dagger = \delta_{nl}. \tag{D.11}$$

We can also obtain

$$c_l c_n + c_n c_l = c_l^\dagger c_n^\dagger + c_n^\dagger c_l^\dagger = 0. \tag{D.12}$$

Equations (D.11) and (D.12) are the fundamental anticommutation relations for a system of fermions. The operator $c_l^\dagger c_l$ is called the *number operator*

$$n_l = c_l^\dagger c_l. \tag{D.13}$$

The basis states of interest are eigenstates of the number operator. Recall that in the case of a boson system, such as the electromagnetic field, the anticommutation rules become commutation rules.

We can now see by comparison of (D.5) and (D.10a) that the single-particle operator F can be expressed as

$$F = \sum_{ln} f_{ln} c_l^\dagger c_n. \tag{D.14}$$

The results for the two-body operator, (D.3), are similar. It turns out that

$$G = \tfrac{1}{2} \sum_{ikln} g_{ik,ln} c_i^\dagger c_k^\dagger c_n c_l, \tag{D.15}$$

in which the matrix element $g_{ik,ln}$ is given by

$$g_{ik,ln} = \int u_i^*(\mathbf{x}_1) u_k^*(\mathbf{x}_2) g(\mathbf{x}_1, \mathbf{x}_2) u_l(\mathbf{x}_1) u_n(\mathbf{x}_2)\, d^3x_1\, d^3x_2. \tag{D.16}$$

It is important to note that the order of the operators in (D.15) (i, k, n, l) is different from that appearing in the matrix element (i, k, l, n).

It is frequently convenient to introduce the electron field operators $\psi(\mathbf{x}), \psi^\dagger(\mathbf{x})$ through the definitions

$$\psi(\mathbf{x}) = \sum_k u_k(\mathbf{x})c_k; \qquad \psi^\dagger(\mathbf{x}) = \sum_k u_k^*(\mathbf{x})c_k^\dagger. \tag{D.17}$$

We will suppose that the single-particle functions u are complete in the sense that

$$\sum_k u_k(\mathbf{x})u_k^*(\mathbf{x}') = \delta(\mathbf{x} - \mathbf{x}'). \tag{D.18}$$

It is then easy to obtain the anticommutation rules for $\psi, \psi^\dagger$

$$\psi(\mathbf{x})\psi^\dagger(\mathbf{x}') + \psi^\dagger(\mathbf{x}')\psi(\mathbf{x}) = \delta(\mathbf{x} - \mathbf{x}') \tag{D.19}$$

and

$$\psi(\mathbf{x})\psi(\mathbf{x}') + \psi(\mathbf{x}')\psi(\mathbf{x}) = \psi^\dagger(\mathbf{x})\psi^\dagger(\mathbf{x}') + \psi^\dagger(\mathbf{x}')\psi^\dagger(\mathbf{x}) = 0 \tag{D.20}$$

The operator $\psi(\mathbf{x})$ destroys a particle at the point $\mathbf{x}$, while $\psi^\dagger$ creates a particle at $\mathbf{x}$.

Let us suppose the Hamiltonian for the system of interacting particles has the form

$$H = H_0 + H_1 \tag{D.21a}$$

with

$$H_0 = \sum_i [T(\mathbf{x}_i) + V(\mathbf{x}_i)] \tag{D.21b}$$

and

$$H_1 = \sum_{i,j>i} V(\mathbf{x}_i, \mathbf{x}_j). \tag{D.21c}$$

In Eq. (D.21b), $T(\mathbf{x}_i)$ and $V(\mathbf{x}_i)$ are the single-particle kinetic and potential energies, respectively, while $V(\mathbf{x}_i, \mathbf{x}_j)$ in (D.21c) is the two-body interaction potential, for example, the Coulomb interaction. It is usually convenient to choose the basis functions u_k to be eigenfunctions of the single particle Hamiltonian

$$[T(\mathbf{x}) + V(\mathbf{x})]u_k(\mathbf{x}) = E_k u_k(\mathbf{x}). \tag{D.22}$$

If this is done, the Hamiltonian is expressed in second quantized form as:

$$H = \sum_k E_k c_k^\dagger c_k + \sum_{ijkl} V_{ij,kl} c_i^\dagger c_j^\dagger c_l c_k. \tag{D.23}$$

An equivalent expression can be obtained in terms of the field operators $\psi, \psi^\dagger$

$$H = \int \psi^\dagger(\mathbf{x})[T(\mathbf{x}) + V(\mathbf{x})]\psi(\mathbf{x})\, d^3x$$

$$+ \tfrac{1}{2} \int \psi^\dagger(\mathbf{x})\psi^\dagger(\mathbf{x}')\, V(\mathbf{x}, \mathbf{x}')\psi(\mathbf{x}')\psi(\mathbf{x})\, d^3x\, d^3x'. \qquad \text{(D.24)}$$

Reference

Schweber, S. (1961). "An Introduction to Relativistic Quantum Field Theory." Harper, New York.

Part B

CHAPTER 5

Impurities and Alloys

5.1 Representation Theory

Throughout this chapter and the subsequent one we are concerned with solutions of a Schrödinger equation of the form

$$(H_0 + U)\Psi = i\hbar\, \partial\Psi/\partial t \tag{5.1.1}$$

where H_0 is a Hamiltonian which contains a periodic potential and U represents an external potential. We will consider three different situations: (1) (this chapter) point impurities and alloys; (2) (Chapter 6) uniform electric and magnetic fields in crystals; and (3) (also Chapter 6) the response of electrons to electromagnetic radiation. The method of treatment of these diverse problems has a unifying feature, the expansion of the wave function of an electron in a complete set of functions originally defined in the perfect crystal.

5.1.1 *The Crystal Momentum Representation*

One rather obvious procedure is to expand quantities in terms of the Bloch functions $\psi_n(\mathbf{k}, \mathbf{r})$ which are eigenfunctions of the perfect crystal Hamiltonian

$$H_0\psi_n(\mathbf{k}, \mathbf{r}) = E_n(\mathbf{k})\psi_n(\mathbf{k}, \mathbf{r}) \tag{5.1.2}$$

where $\mathbf{k}$ is the wave vector and n is the band index. This procedure is known as the *crystal momentum representation* (CMR) (Adams, 1952, 1953) since it is based on states of definite $\mathbf{k}$, the "crystal momentum." It was shown in Section 4.1.1, Part A, that the Bloch functions are orthonormal,

$$\int \psi_n^*(\mathbf{k}, \mathbf{r})\psi_l(\mathbf{q}, \mathbf{r})\, d^3r = \delta_{nl}\, \delta(\mathbf{k} - \mathbf{q}), \tag{5.1.3}$$

and in this equation and everywhere else, unless explicitly specified otherwise, the integral on $\mathbf{r}$ includes all space. They were also shown to be complete:

$$\sum_n \int \psi_n^*(\mathbf{k}, \mathbf{r})\psi_n(\mathbf{k}, \mathbf{r}')\, d^3k = \delta(\mathbf{r} - \mathbf{r}'). \tag{5.1.4}$$

The integral includes a single Brillouin zone.

Since the Bloch functions are complete, any one-electron function may be expanded in Bloch functions. Let $\Psi(\mathbf{r}, t)$ be a general wave function. We write

$$\Psi(\mathbf{r}, t) = \sum_n \int \phi_n(\mathbf{k}, t)\psi_n(\mathbf{k}, \mathbf{r})\, d^3k. \tag{5.1.5}$$

The expansion coefficients $\phi_n(\mathbf{k}, t)$ describe the wave function in the crystal momentum representation. They satisfy an effective Schrödinger equation which can be obtained by substituting (5.1.5) into (5.1.1), multiplying by $\psi_l^*(\mathbf{q}, \mathbf{r})$, and integrating

$$\sum_n \iint \psi_l^*(\mathbf{q}, \mathbf{r})\{[E_n(\mathbf{k}) + U]\phi_n(\mathbf{k}, t) - i\hbar\, \partial\phi_n(\mathbf{k}, t)/\partial t\}\psi_n(\mathbf{k}, \mathbf{r})\, d^3k\, d^3r$$

$$= [E_l(\mathbf{q}) - i\hbar\, \partial/\partial t]\phi_l(\mathbf{q}, t) + \sum_n \int d^3k \langle l\mathbf{q} \mid U \mid n\mathbf{k}\rangle \phi_n(\mathbf{k}, t) = 0. \tag{5.1.6}$$

We have introduced the abbreviation

$$\langle l\mathbf{q} \mid U \mid n\mathbf{k}\rangle = \int \psi_l^*(\mathbf{q}, \mathbf{r})U(\mathbf{r})\psi_n(\mathbf{k}, \mathbf{r})\, d^3r. \tag{5.1.7}$$

Note that if we are concerned with an energy eigenstate, we have

$$(H_0 + U)\Psi = E\Psi \tag{5.1.8}$$

which implies

$$i\hbar\, \partial\phi_n(\mathbf{k}, t)/\partial t = E\phi_n(\mathbf{k}, t), \tag{5.1.9}$$

so that the time-independent form of (5.1.6) is

$$[E_l(\mathbf{q}) - E]\phi_l(\mathbf{q}) + \sum_n \int d^3k \langle l\mathbf{q} \mid U \mid n\mathbf{k}\rangle \phi_n(\mathbf{k}) = 0. \tag{5.1.10}$$

It is convenient in some of the following to use Dirac notations and work directly with the states $\mid n\mathbf{k}\rangle$. The Bloch wave function $\psi_n(\mathbf{k}, \mathbf{r})$ may be interpreted as a transformation coefficient

$$\psi_n(\mathbf{k}, \mathbf{r}) = \langle \mathbf{r} \mid n\mathbf{k}\rangle. \tag{5.1.11}$$

5.1.2 *The Effective Mass Representation*

Use of the crystal momentum representation may require knowledge of crystal wave functions throughout the Brillouin zone. Although it is by no means impossible with modern computing equipment to obtain wave functions at a representative sample of general points in the zone, it is sometimes desirable to expand in terms of functions simply related to wave functions at a single point of a band. Such a representation was introduced by Luttinger and Kohn (1955); we shall call it the *effective mass representation.*

The basis functions of the Luttinger–Kohn representation are denoted by $\chi_n(\mathbf{k}, \mathbf{r})$. They are defined in terms of the Bloch functions at some conveniently chosen reference point in the Brillouin zone $\mathbf{k}_0$ by

$$\chi_n(\mathbf{k}, \mathbf{r}) = \exp[i(\mathbf{k} - \mathbf{k}_0)\cdot\mathbf{r}]\psi_n(\mathbf{k}_0, \mathbf{r}) = \exp(i\mathbf{k}\cdot\mathbf{r})u_n(\mathbf{k}_0, \mathbf{r}). \quad (5.1.12)$$

These functions were introduced in Section 4.1.2, Part A, in connection with $\mathbf{k}\cdot\mathbf{p}$ perturbation theory. We will find them particularly convenient in problems involving a small number of particles–electrons (or holes) in a nearly empty (or nearly full) band. They may be shown to obey the same orthogonality and completeness relations as the Bloch functions.

If $\mathbf{k}_0$ is not invariant under all the operations of the crystal point group, there will be other points (those in its star) which are equivalent to it. In such cases, it will usually be desirable to perform parallel expansions about all such points. However, for the present we will ignore this problem and consider only a single point $\mathbf{k}_0$.

Consider the expansion of an arbitrary wave function in these functions:

$$\Psi(\mathbf{r}, t) = \sum_n \int A_n(\mathbf{k}, t)\chi_n(\mathbf{k}, \mathbf{r})\, d^3k. \quad (5.1.13)$$

The $A_n(\mathbf{k}, t)$ are the expansion coefficients. The representation of the Hamiltonian on the basis of the χ_n is somewhat more complicated than in the case of the ψ_n since the χ's are not eigenfunctions of H_0. The matrix elements of H_0 are determined in the following manner:

$$\begin{aligned}[n\mathbf{k} \mid H_0 \mid l\mathbf{q}] &= \int \chi_n^*(\mathbf{k}, \mathbf{r})H_0\chi_l(\mathbf{q}, \mathbf{r})\, d^3r \\ &= \int \exp(i(\mathbf{q} - \mathbf{k})\cdot\mathbf{r})u_n^*(\mathbf{k}_0, \mathbf{r}) \\ &\quad \times [E_l(\mathbf{k}_0) + (\hbar(\mathbf{q} - \mathbf{k}_0)/m)\cdot\mathbf{p} + (\hbar^2/2m)(\mathbf{q}^2 - \mathbf{k}_0^2)] \\ &\quad \times u_l(\mathbf{k}_0, \mathbf{r})\, d^3r. \end{aligned} \quad (5.1.14)$$

We will use square brackets $[\cdots]$ to indicate matrix elements in the effective mass formalism. We reserve angular brackets $\langle\cdots\rangle$ for the CMR and parentheses for the Wannier functions to be discussed in Section 5.1.3.

Let us examine the term involving $\mathbf{p}$ in Eq. (5.1.14). We observe that $\mathbf{p}u_l(\mathbf{k}_0, \mathbf{r})$ is periodic if u_l is. Therefore we can break up the integral over all space into an integral over cells:

$$\begin{aligned}
&\int \exp(i(\mathbf{q}-\mathbf{k})\cdot\mathbf{r})u_n^*(\mathbf{k}_0,\mathbf{r})\mathbf{p}u_l(\mathbf{k}_0,\mathbf{r})\,d^3r \\
&\qquad = \sum_\nu \exp(i(\mathbf{q}-\mathbf{k})\cdot\mathbf{R}_\nu)\int \exp(i(\mathbf{q}-\mathbf{k})\cdot(\mathbf{r}-\mathbf{R}_\nu)) \\
&\qquad\quad \times u_n^*(\mathbf{k}_0,\mathbf{r}-\mathbf{R}_\nu)\mathbf{p}u_l(\mathbf{k}_0,\mathbf{r}-\mathbf{R}_\nu)\,d^3r \\
&\qquad = \int_{\text{cell }0} \exp(i(\mathbf{q}-\mathbf{k})\cdot\mathbf{r})u_n^*(\mathbf{k}_0,\mathbf{r})\mathbf{p}u_l(\mathbf{k}_0,\mathbf{r})\,d^3r \sum_\nu \exp(i(\mathbf{q}-\mathbf{k})\cdot\mathbf{R}_\nu) \\
&\qquad = \delta(\mathbf{q}-\mathbf{k})\mathbf{p}_{nl}(\mathbf{k}_0) \qquad (5.1.15)
\end{aligned}$$

where

$$\mathbf{p}_{nl}(\mathbf{k}_0) = [(2\pi)^3/\Omega]\int_{\text{cell}} u_n^*(\mathbf{k}_0,\mathbf{r})\mathbf{p}u_l(\mathbf{k}_0,\mathbf{r})\,d^3r. \qquad (5.1.16)$$

Thus Eq. (5.1.14) yields

$$\begin{aligned}
[n\mathbf{k}\mid H_0\mid l\mathbf{q}] = \delta(\mathbf{q}-\mathbf{k})\{&\delta_{nl}[E_l + (\hbar^2/2m)(\mathbf{q}^2-\mathbf{k}_0^2)] \\
&+ (\hbar/m)(\mathbf{q}-\mathbf{k}_0)\cdot\mathbf{p}_{nl}(\mathbf{k}_0)\}. \qquad (5.1.17)
\end{aligned}$$

Consequently, the equation satisfied by the expansion coefficients $A_n(\mathbf{k}, t)$ of Eq. (5.1.13) is

$$\begin{aligned}
&[E_n + (\hbar^2/2m)(\mathbf{k}^2-\mathbf{k}_0^2) - i\hbar\,\partial/\partial t]A_n(\mathbf{k},t) \\
&\quad + [\hbar(\mathbf{k}-\mathbf{k}_0)/m]\cdot\sum_l \mathbf{p}_{nl}(\mathbf{k}_0)A_l(\mathbf{k},t) \\
&\quad + \sum_l\int d^3q\,[n\mathbf{k}\mid U\mid l\mathbf{q}]A_l(\mathbf{q},t) = 0. \qquad (5.1.18)
\end{aligned}$$

This is the Schrödinger equation in the effective mass representation. We use the term "effective mass representation" because (5.1.18) contains the $\mathbf{k}\cdot\mathbf{p}$ Hamiltonian of effective mass theory (see Section 4.1, Part A). In addition, the usual procedure for solving (5.1.18), which we will examine in Section 5.3, employs a unitary transformation which exhibits the effective mass explicitly.

5.1.3 *Wannier Functions*

We will now consider expanding quantities in terms of orthogonal localized functions (Wannier functions). Use of atomic wave functions along the lines of the tight binding method discussed in Section 4.5, Part A, is also possible (for an example, see Gauthier and Lenglart, 1965). Because some controversy has, on occasion, attended use of Wannier functions, we will examine their properties in some detail.

The Wannier functions (Wannier, 1937) are characterized by a band index and a lattice site vector $\mathbf{R}_\mu$. They are denoted by $a_n(\mathbf{r} - \mathbf{R}_\mu)$ and are defined in terms of the basic Bloch functions by

$$a_n(\mathbf{r} - \mathbf{R}_\mu) = (\Omega^{1/2}/(2\pi)^{3/2}) \int \exp(-i\mathbf{k}\cdot\mathbf{R}_\mu)\psi_n(\mathbf{k}, \mathbf{r})\, d^3k. \quad (5.1.19)$$

The integral includes the Brillouin zone. We note that a single Wannier function is defined for each band and each unit cell. If the unit cell contains more than one atom, the Wannier function so defined may not be localized around a single atom. In such an event, it will be necessary to combine Wannier functions for two (or more) bands in order to obtain single atom Wannier functions.

It is not evident from the definition that a_n is in fact a function only of the difference of $\mathbf{r}$ and $\mathbf{R}_\mu$. This may, however, easily be established if we rewrite (5.1.19) in the form

$$\begin{aligned} a_n(\mathbf{r} - \mathbf{R}_\mu) &= (\Omega^{1/2}/(2\pi)^{3/2}) \int \exp(i\mathbf{k}\cdot(\mathbf{r} - \mathbf{R}_\mu))u_n(\mathbf{k}, \mathbf{r})\, d^3k \\ &= (\Omega^{1/2}/(2\pi)^{3/2}) \int \exp(i\mathbf{k}\cdot(\mathbf{r} - \mathbf{R}_\mu))u_n(\mathbf{k}, \mathbf{r} - \mathbf{R}_\mu)\, d^3k. \end{aligned} \quad (5.1.20)$$

In the last step, we have used the fact that $u_n(\mathbf{k}, \mathbf{r}) = u_n(\mathbf{k}, \mathbf{r} - \mathbf{R}_\mu)$. It is also apparent from this argument that in order to obtain this result, the argument of the exponential in (5.1.19) must be $\mathbf{k}\cdot$(direct lattice vector). Thus, if we wish to displace $\mathbf{r}$ by $\mathbf{x}$, we must write

$$a_n(\mathbf{r} + \mathbf{x} - \mathbf{R}_\mu) = (\Omega^{1/2}/(2\pi)^{3/2}) \int \exp(-i\mathbf{k}\cdot\mathbf{R}_\mu)\psi_n(\mathbf{k}, \mathbf{r} + \mathbf{x})\, d^3k.$$

The apparent alternate form

$$(\Omega^{1/2}/(2\pi)^{3/2}) \int \exp(i\mathbf{k}\cdot(\mathbf{x} - \mathbf{R}_\mu))\psi_n(\mathbf{k}, \mathbf{r})\, d^3k$$

is not acceptable unless $\mathbf{x}$ is a direct lattice vector.

Now we obtain the orthonormality and completeness relations for the Wannier functions. Consider

$$\int a_n^*(\mathbf{r} - \mathbf{R}_\mu) a_l(\mathbf{r} - \mathbf{R}_\nu)\, d^3r$$

$$= (\Omega/(2\pi)^3) \int \exp(i\mathbf{k}\cdot\mathbf{R}_\mu - i\mathbf{q}\cdot\mathbf{R}_\nu)\psi_n^*(\mathbf{k}, \mathbf{r})\psi_l(\mathbf{q}, \mathbf{r})\, d^3r\, d^3k\, d^3q$$

$$= (\Omega/(2\pi)^3)\, \delta_{nl} \int \exp(i\mathbf{k}\cdot(\mathbf{R}_\mu - \mathbf{R}_\nu))\, d^3k = \delta_{nl}\, \delta_{\mu\nu}. \tag{5.1.21}$$

We see that Wannier functions satisfy orthonormality relations characteristic of localized (normalizable) functions in contrast to the continuum relations (5.1.3) obeyed by Bloch functions. Evidently (5.1.21) implies at least some degree of localization of Wannier functions. In particular, the Wannier function must decay more rapidly than $|\mathbf{r} - \mathbf{R}_\mu|^{-3/2}$ at infinity.

To establish completeness, we must sum over all bands and all lattice sites

$$\sum_{n\mu} a_n^*(\mathbf{r} - \mathbf{R}_\mu) a_n(\mathbf{r}' - \mathbf{R}_\mu)$$

$$= (\Omega/(2\pi)^3) \int \sum_n \psi_n^*(\mathbf{k}, \mathbf{r}')\psi_n(\mathbf{q}, \mathbf{r}) \sum_\mu \exp(i(\mathbf{k} - \mathbf{q})\cdot\mathbf{R}_\mu)\, d^3k\, d^3q$$

$$= \int \sum_n \psi_n^*(\mathbf{k}, \mathbf{r}')\psi_n(\mathbf{q}, \mathbf{r})\, \delta(\mathbf{k} - \mathbf{q})\, d^3k\, d^3q$$

$$= \sum_n \int \psi_n^*(\mathbf{k}, \mathbf{r}')\psi_n(\mathbf{k}, \mathbf{r})\, d^3k = \delta(\mathbf{r} - \mathbf{r}'). \tag{5.1.22}$$

It is possible to invert Eq. (5.1.19) to express Bloch functions in terms of Wannier functions. To do this, we multiply both sides of that equation by $(\Omega^{1/2}/(2\pi)^{3/2}) \exp(i\mathbf{q}\cdot\mathbf{R}_\mu)$ and then sum over μ.

$$(\Omega^{1/2}/(2\pi)^{3/2}) \sum_\mu \exp(i\mathbf{q}\cdot\mathbf{R}_\mu) a_n(\mathbf{r} - \mathbf{R}_\mu)$$

$$= (\Omega/(2\pi)^3) \sum_\mu \int \exp(i(\mathbf{q} - \mathbf{k})\cdot\mathbf{R}_\mu)\psi_n(\mathbf{k}, \mathbf{r})\, d^3k$$

$$= \int \delta(\mathbf{q} - \mathbf{k})\psi_n(\mathbf{k}, \mathbf{r})\, d^3k.$$

Thus we have

$$\psi_n(\mathbf{q}, \mathbf{r}) = (\Omega^{1/2}/(2\pi)^{3/2}) \sum_{\mu} \exp(i\mathbf{q}\cdot\mathbf{R}_\mu) a_n(\mathbf{r} - \mathbf{R}_\mu). \quad (5.1.23)$$

Equation (5.1.23) appears quite similar to the expression for wave functions in the tight binding approximation. On closer observation significant differences will be found. In particular, Wannier functions form a complete orthonormal set, whereas bound atomic functions do not. Furthermore, the Hamiltonian of a (perfect) crystal does not have matrix elements between Wannier functions belonging to different energy bands, as will be seen below.

If we wish to use discrete normalization for the Bloch functions, we write instead of (5.1.19) and (5.1.23)

$$a_n(\mathbf{r} - \mathbf{R}_\mu) = N^{-1/2} \sum_{\mathbf{k}} \exp(-i\mathbf{k}\cdot\mathbf{R}_\mu)\psi_n(\mathbf{k}, \mathbf{r}) \quad (5.1.24a)$$

and

$$\psi_n(\mathbf{q}, \mathbf{r}) = N^{-1/2} \sum_{\mu} \exp(i\mathbf{k}\cdot\mathbf{R}_\mu) a_n(\mathbf{r} - \mathbf{R}_\mu), \quad (5.1.24b)$$

in which N is the number of unit cells considered.

An important theorem connects a determinantal wave function (see Section 4.9, Part A) whose elements are Bloch functions, and one whose elements are Wannier functions. Suppose that the $\mathfrak{M}$ electrons present just suffice to fill states in a Brillouin zone at the absolute zero of temperature. We suppose that $\mathfrak{M} = nN$, that is, that the number of electrons is an integral multiple of the number of lattice sites. We may consider either the case of complete ferromagnetic alignment (odd n) or a singlet state in which n is even, and there are an equal number of electrons of up and down spins. Consider a wave function for the system which is a determinant of Bloch functions

$$\Psi(1, \ldots, \mathfrak{M}) = (\mathfrak{M}!)^{-1/2} \det | \psi_j(\mathbf{k}, \mathbf{r}) |. \quad (5.1.25)$$

We do not write the spin explicitly. Now substitute (5.1.24b) in (5.1.25):

$$\begin{aligned} \Psi(1, \ldots, \mathfrak{M}) &= (\mathfrak{M}!)^{-1/2} \det | N^{-1/2} \sum_{\mu} \exp(i\mathbf{k}\cdot\mathbf{R}_\mu) a_j(\mathbf{r} - \mathbf{R}_\mu) | \\ &= (\mathfrak{M}!)^{-1/2} \det | \exp(i\mathbf{k}\cdot\mathbf{R}_\mu)/N^{-1/2} | \det | a_j(\mathbf{r} - \mathbf{R}_\mu) | \\ &= e^{i\theta}/(\mathfrak{M}!)^{-1/2} \det | a_j(\mathbf{r} - \mathbf{R}_\mu) |. \end{aligned} \quad (5.1.26)$$

In the second line of (5.1.26) we have made use of the rule for multiplying determinants, and then observed that since the transformation between Bloch and Wannier functions is unitary, the determinant of the coefficients is a complex number of modulus unity. The argument requires the existence of a one-to-one correspondence between Bloch functions for occupied states and Wannier functions, which can exist only when all bands are full. Since the phase factor can be discarded we see that the many-electron wave functions can be equally well represented as a determinant of Bloch functions or as a determinant of Wannier functions. Conversely, in the case of a system of atoms with filled atomic shells, a determinant of localized atomic functions may be transformed to a determinant of tight binding functions (Seitz, 1940).

There do not appear to be any simple examples in which Wannier functions can be expressed in closed form in terms of more usual functions. Hence, we are forced to consider a rather unrealistic case to give an example of Wannier function. Let us consider a simple cubic lattice of lattice constant a and replace the Bloch functions by plane waves

$$\psi(\mathbf{k}, \mathbf{r}) = (2\pi)^{-3/2} \exp(i\mathbf{k}\cdot\mathbf{r}).$$

The Brillouin zone is a cube: $-\pi/a \leq k_x, k_y, k_z \leq \pi/a$. We put $\mathbf{r} - \mathbf{R}_\mu = \hat{\mathbf{i}}X + \hat{\mathbf{j}}Y + \hat{\mathbf{k}}Z$ where $\hat{\mathbf{i}}$, $\hat{\mathbf{j}}$, $\hat{\mathbf{k}}$ are the usual unit vectors along the x, y, and z axes. We obtain

$$a(\mathbf{r} - \mathbf{R}_\mu) = ((2\pi)^{3/2}/\Omega^{1/2}) \frac{(\sin \pi X/a \sin \pi Y/a \sin \pi Z/a)}{\pi XYZ/a^3}. \quad (5.1.27)$$

This function oscillates and decreases rather slowly with distance. A more complete examination of the properties of Wannier functions formed from free-electron wave functions has been given by Winston (1954). This paper is of some interest because of the discussion of symmetry properties. Slater (1952) has discussed the Wannier functions associated with a periodic potential of the form $(\cos x + \cos y + \cos z)$.

In general, we expect that Wannier functions which are derived from Bloch functions which are solutions of a periodic potential problem will decay exponentially at large distances. Kohn (1959) has shown that this is so for one-dimensional periodic systems. In such cases there are no band degeneracies. Blount (1962) has shown that exponential decay will be obtained in three dimensions for nondegenerate bands. If the band of interest is degenerate with another band, substantial complications may arise.

One other case is simple enough to be of some interest at this point. Suppose the simplest tight binding approximation is valid in a monatomic crystal so that

$$\psi(\mathbf{k}, \mathbf{r}) = (\Omega^{1/2}/(2\pi)^{3/2})(1/N(\mathbf{k})) \sum \exp(i\mathbf{k}\cdot\mathbf{R}_\nu) u(\mathbf{r} - \mathbf{R}_\nu) \quad (5.1.28)$$

where u is an atomic function and $N(\mathbf{k})$ is a normalization constant obtained by requiring that

$$\int \psi^*(\mathbf{k}, \mathbf{r})\psi(\mathbf{q}, \mathbf{r})\, d^3r = \delta(\mathbf{k} - \mathbf{q}).$$

Then $N(\mathbf{k})$ can be found from Eq. (4.5.8), Part A,

$$N^2(\mathbf{k}) = \sum_\rho \exp(-i\mathbf{k}\cdot\mathbf{R}_\rho) S(\mathbf{R}_\rho) \tag{5.1.29}$$

with

$$S(\mathbf{R}_\rho) = \int u^*(\mathbf{r} - \mathbf{R}_\rho)u(\mathbf{r})\, d^3r. \tag{5.1.30}$$

Then we have in this case

$$a(\mathbf{r} - \mathbf{R}_\mu) = \sum_\nu F(\mathbf{R}_\nu - \mathbf{R}_\mu)u(\mathbf{r} - \mathbf{R}_\nu) \tag{5.1.31}$$

where

$$F(\mathbf{R}_\nu - \mathbf{R}_\mu) = (\Omega/(2\pi)^3) \int \{\exp(i\mathbf{k}\cdot(\mathbf{R}_\mu - \mathbf{R}_\nu))/N(\mathbf{k})\}\, d^3k. \tag{5.1.32}$$

If $N(\mathbf{k})$ could be replaced by 1, F would be $\delta_{\mu\nu}$, and the Wannier function would reduce to the atomic function. When atomic functions overlap, the requirement that the Wannier functions on different sites be orthogonal causes atomic functions on different lattice sites to be included. This procedure for constructing approximate Wannier functions from atomic wave functions is equivalent to the symmetric orthogonalization procedure of Lowdin (1962).

If the overlap integrals $S(\mathbf{R}_\rho)$ [Eq. (4.5.9), Part A], are small for $\rho \neq 0$, so that terms of order S^2 may be neglected, and fall off sufficiently rapidly with distance so that they may be neglected beyond first neighbors, we have

$$N^2(\mathbf{k}) = 1 + S(\Delta) \sum_\Delta \exp(i\mathbf{k}\cdot\boldsymbol{\Delta}). \tag{5.1.33}$$

It has been assumed that $S(\mathbf{R}_\rho)$ depends on $|R_\rho|$ only and we denote by $\boldsymbol{\Delta}$ any vector connecting an atom to a nearest neighbor. Then we expand $N^{-1}(\mathbf{k})$ and perform the integration in (5.1.32). The result is

$$F(\mathbf{R}_\mu) = \delta_{\mu,0} - \tfrac{1}{2}S(\Delta)\, \delta_{\mu,\Delta} \tag{5.1.34}$$

so that

$$a(\mathbf{r} - \mathbf{R}_\mu) = u(\mathbf{r} - \mathbf{R}_\mu) - \tfrac{1}{2}S(\Delta) \sum_\Delta u(\mathbf{r} - \mathbf{R}_\mu - \boldsymbol{\Delta}). \tag{5.1.35}$$

In this limit, the Wannier function is simply a linear combination of atomic functions on nearest neighbor atoms.

There is a problem of some importance concerning the localization of Wannier functions. Any given Bloch function $\psi_n(\mathbf{k}, \mathbf{r})$ may be multiplied by a phase factor $\exp[i\phi_n(\mathbf{k})]$. The result is still a Bloch function insofar as it is still a solution of the Schrödinger equation for the same energy. However, we see from (5.1.19) that the Wannier functions formed from phase-modified Bloch functions will be altered. In particular, their localization can be changed since one understands intuitively that rapid variation in $\mathbf{k}$ in the Bloch function will produce unpleasant behavior of the Wannier function for large arguments. If expansions in terms of Wannier functions are carried to covergence, no physical result will be altered by a change of phase; however, the rate of convergence may be affected. In a calculation in which Wannier functions are used for more than formal arguments, it is usually desirable to take advantage of this freedom to choose a phase factor in the Bloch functions to make the $\mathbf{k}$ dependence as smooth as possible, and to keep the Wannier functions as localized as possible. One attempt to do this is described by Callaway and Hughes (1967).

All of the preceding discussion has assumed implicitly that the band in question is nondegenerate; specifically, that there are no points or lines of contact with other bands. Such an assumption is always valid for one-dimensional systems, but only rarely so for three dimensions. The problem for degenerate bands is to avoid rapid change in the Bloch function as a function of $\mathbf{k}$ as $\mathbf{k}$ goes through degeneracy. In this case, the construction of Wannier functions is facilitated by use of the symmetry analysis of Section 3.6.3, Part A (see also Blount, 1962 and des Cloizeaux, 1964a, b). We assume that degeneracy may be ignored in the remainder of this discussion.

Wannier functions are not energy eigenfunctions, but are, instead, combinations of Bloch functions which have different wave vectors and therefore different energies. We will now obtain the matrix elements of the (perfect) crystal Hamiltonian between Wannier functions:

$$\begin{aligned}\int a_l^*(\mathbf{r} - \mathbf{R}_\nu) H_0 a_n(\mathbf{r} - \mathbf{R}_\mu)\, d^3r &\\ = (\Omega/(2\pi)^3) \int \exp(i(\mathbf{q}\cdot\mathbf{R}_\nu - \mathbf{k}\cdot\mathbf{R}_\mu))\psi_l(\mathbf{q}, \mathbf{r}) H_0 \psi_n(\mathbf{k}, \mathbf{r})\, d^3k\, d^3q\, d^3r &\\ = \delta_{nl}\, \mathcal{E}_n(\mathbf{R}_\nu - \mathbf{R}_\mu) & \qquad (5.1.36)\end{aligned}$$

where

$$\mathcal{E}_n(\mathbf{R}_\sigma) = (\Omega/(2\pi)^3) \int \exp(i\mathbf{k}\cdot\mathbf{R}_\sigma) \mathrm{E}_n(\mathbf{k})\, d^3k. \qquad (5.1.37)$$

The quantity $\mathcal{E}_n(\mathbf{R}_\sigma)$ appears as a Fourier component of the energy band function. Equation (5.1.37) may be inverted by the same technique used to express the Bloch functions in terms of Wannier functions. The result is

$$E_n(\mathbf{k}) = \sum_\sigma \exp(-i\mathbf{k}\cdot\mathbf{R}_\sigma)\mathcal{E}_n(\mathbf{R}_\sigma). \tag{5.1.38}$$

Equation (5.1.38) indicates that a Fourier expansion of an energy band function in terms of direct lattice vectors is always possible. Of course, if $E_n(\mathbf{k})$ varies rapidly, as it will if two bands are degenerate or nearly so at some points, the expansion will converge only slowly. We know from (3.6.6), Part A, that $E_n(\alpha\mathbf{k}) = E_n(\mathbf{k})$ for any $\boldsymbol{\alpha}$ in the point group. We can easily show that $\mathcal{E}_n(\mathbf{R})$ has a similar property. Consider

$$\mathcal{E}_n(\boldsymbol{\alpha}\mathbf{R}_\sigma) = (\Omega/(2\pi)^3)\int \exp(i\mathbf{k}\cdot\boldsymbol{\alpha}\mathbf{R}_\sigma)E_n(\mathbf{k})\,d^3k$$

$$= (\Omega/(2\pi)^3)\int \exp(i\boldsymbol{\alpha}^{-1}\mathbf{k}\cdot\mathbf{R}_\sigma)E_n(\mathbf{k})\,d^3k.$$

We can now make the change of variables $\mathbf{k}' = \boldsymbol{\alpha}^{-1}\mathbf{k}$; $E_n(\mathbf{k}') = E_n(\mathbf{k})$. The Jacobian of this transformation is unity. Therefore we have

$$\mathcal{E}_n(\alpha\mathbf{R}_\sigma) = \mathcal{E}_n(\mathbf{R}_\sigma). \tag{5.1.39}$$

Thus, there is only one independent coefficient $\mathcal{E}_n(\mathbf{R}_\sigma)$ for all the lattice sites which can be obtained from any one of them by any operation in the point group.

It has occasionally been proposed to use Wannier functions directly in the calculation of energy bands through a variational method (Koster, 1953; Parzen, 1953; Kohn, 1973). While such a procedure is possible in principle, it has not as yet found wide application. One quantitative attempt reported by Wainwright and Parzen (1953) contains a serious numerical error.

5.1.4 *The Local Representation*

Wannier functions are, however, quite useful in considering the behavior of electrons in external fields; particularly when the potential U is of short range in space. We now expand the solution Ψ of (5.1.1) in the form

$$\Psi(\mathbf{r}, t) = \sum_{n\mu} B_n(\mathbf{R}_\mu, t)a_n(\mathbf{r} - \mathbf{R}_\mu). \tag{5.1.40}$$

The expansion in terms of Wannier functions will be described as the "local representation" (LR). Note that we must sum over all bands and all lattice sites. We wish to determine the equation satisfied by B_n. As usual, we substitute (5.1.40) into (5.1.1), multiply by $a_l^*(\mathbf{r} - \mathbf{R}_\nu)$, and integrate. We obtain, with the use of (5.1.36),

$$\sum_\nu [\mathcal{E}_n(\mathbf{R}_\nu - \mathbf{R}_\mu) - E\,\delta_{\mu\nu}]B_l(\mathbf{R}_\mu) + \sum_{n\mu} (l\nu \mid U \mid n\mu)B_n(\mathbf{R}_\mu) = 0. \tag{5.1.41}$$

We use parentheses to denote matrix elements of quantities on the Wannier function basis. Specifically, we have

$$(l\nu \mid U \mid n\mu) = \int a_l^*(\mathbf{r} - \mathbf{R}_\nu)Ua_n(\mathbf{r} - \mathbf{R}_\mu)\, d^3r. \tag{5.1.42}$$

We will refer to Eq. (5.1.41) as the Schrödinger equation in the local representation. Since Wannier functions form a discrete basis, we have obtained a matrix equation in contrast to the integral equation derived in the crystal momentum representation.

It must be possible to express a quantity given originally in the CMR in terms of the LR and vice versa. In the case of the wave function, we equate Eqs. (5.1.5) and (5.1.40). Then one multiplies by either $a_l^*(\mathbf{r} - \mathbf{R}_\nu)$ or $\psi_l^*(\mathbf{q}, \mathbf{r})$, integrates, and uses (5.1.19). The results are

$$B_l(\mathbf{R}_\nu) = (\Omega^{1/2}/(2\pi)^{3/2}) \int \exp(i\mathbf{k}\cdot\mathbf{R}_\nu)\phi_l(\mathbf{k})\, d^3k \tag{5.1.43}$$

and

$$\phi_l(\mathbf{q}) = (\Omega^{1/2}/(2\pi)^{3/2}) \sum_\mu \exp(-i\mathbf{q}\cdot\mathbf{R}_\mu)B_n(\mathbf{R}_\mu). \tag{5.1.44}$$

The localized representation may offer considerable advantages when potentials which have only a small number of nonnegligible matrix elements on the Wannier function basis are considered. In this case, it is possible to solve Eq. (5.1.41) in a straightforward way, as will be seen in Section 5.2. On the other hand, if U is of long range, or has a $\mathbf{k}$ conservation selection rule, the CMR will probably be more appropriate.

Occasionally we will use Dirac notation in which we introduce states $\mid n\mu)$ such that

$$a_n(r - \mathbf{R}_\mu) = \langle \mathbf{r} \mid n\mu). \tag{5.1.45}$$

Use of the states $\mid n\mu)$ enables us to describe other types of excitations (phonons or spin waves) within a similar formalism.

5.1.5 *The* **kq** *Representation*

Another representation of some interest was introduced by Zak (1968, 1969, 1972). This is called the **kq** *representation*. Previously, in the crystal momentum and effective mass representations we specified precisely the **k** vector and the band index of our basis states. The localized representation, based on Wannier functions, did not involve **k**, but described states in terms of a cell index and a band index. In the present case, we attempt to specify both **k** and position. Of course, this cannot be done precisely because of the limitations imposed by the uncertainty principle, but we are able to specify position within an atomic cell, and wave vector within the Brillouin zone. We lose knowledge of which cell is involved and the wave vector is uncertain in the amount of a reciprocal lattice vector (if we consider the extended zone scheme); or in the reduced zone, we lose information concerning the particular energy band.

The representation is precisely defined by determining the eigenfunctions of a complete set of commuting operators. The operators chosen are

$$T(\mathbf{R}_\mu) = \exp(i\mathbf{p}\cdot\mathbf{R}_\mu) \tag{5.1.46}$$

and

$$T(\mathbf{K}_s) = \exp(i\mathbf{r}\cdot\mathbf{K}_s) \tag{5.1.47}$$

in which $\mathbf{R}_\mu$ and $\mathbf{K}_s$ are direct and reciprocal lattice vectors while **p** and **r** are the ordinary momentum and position operators, respectively. The reader can verify that these operators do commute, as advertised. It is essential to the argument that $\mathbf{K}_s\cdot\mathbf{R}_\mu = 2\pi \times$ integer for all μ and s. Zak has shown that this set of operators is complete in the sense that any operator which is a function of **r** and **p** can be expressed as a sum of the T.

It will be shown that the functions

$$\Xi(\mathbf{k}, \mathbf{q}, \mathbf{r}) = (\Omega^{1/2}/(2\pi)^{3/2}) \sum_\nu \delta(\mathbf{r} - \mathbf{q} - \mathbf{R}_\nu) \exp(i\mathbf{k}\cdot\mathbf{R}_\nu) \tag{5.1.48}$$

are the required eigenfunctions. In the case of $T(\mathbf{R}_\mu)$ we simply use the fact that

$$\exp(i\mathbf{p}\cdot\mathbf{R}_\mu)f(\mathbf{r}) = f(\mathbf{r} + \mathbf{R}_\mu)$$

to enable us to write (with C replacing the normalizing constant)

$$\begin{aligned}
T(\mathbf{R}_\mu)\Xi(\mathbf{k}, \mathbf{q}, \mathbf{r}) \\
&= C \sum_\nu \delta(\mathbf{r} - \mathbf{q} + \mathbf{R}_\mu - \mathbf{R}_\nu) \exp(i\mathbf{k}\cdot\mathbf{R}_\nu) \\
&= \exp(i\mathbf{k}\cdot\mathbf{R}_\mu) C \sum_\nu \delta[\mathbf{r} - \mathbf{q} - (\mathbf{R}_\nu - \mathbf{R}_\mu)] \exp(i\mathbf{k}\cdot(\mathbf{R}_\nu - \mathbf{R}_\mu)) \\
&= \exp(i\mathbf{k}\cdot\mathbf{R}_\mu)\Xi(\mathbf{k}, \mathbf{q}).
\end{aligned} \tag{5.1.49}$$

Similarly, we have

$$T(\mathbf{K}_s)\Xi(\mathbf{k}, \mathbf{q}, \mathbf{r}) = \exp(i\mathbf{q}\cdot\mathbf{K}_s)C \sum_\nu \delta(\mathbf{r} - \mathbf{q} - \mathbf{R}_\nu) \times \exp(i(\mathbf{r} - \mathbf{q})\cdot\mathbf{K}_s) \exp(i\mathbf{k}\cdot\mathbf{R}_\nu) = \exp(i\mathbf{q}\cdot\mathbf{K}_s)\Xi(\mathbf{k}, \mathbf{q}, \mathbf{r}) \tag{5.1.50}$$

since by the properties of the delta function, $\mathbf{r} - \mathbf{q}$ must be a direct lattice vector.

These functions are found to be orthonormal in that

$$\int \Xi^*(\mathbf{k}', \mathbf{q}', \mathbf{r})\Xi(\mathbf{k}, \mathbf{q}, \mathbf{r})\, d^3r = \sum_s \delta(\mathbf{k}' - \mathbf{k} - \mathbf{K}_s) \sum_\mu \delta(\mathbf{q} - \mathbf{q}' - \mathbf{R}_\mu). \tag{5.1.51}$$

This relation indicates that we should restrict $\mathbf{k}'$ and $\mathbf{k}$ to be within a single Brillouin zone while $\mathbf{q}$ and $\mathbf{q}'$ are in a single Wigner–Seitz cell. If this is done, we have

$$\int \Xi^*(\mathbf{k}', \mathbf{q}', \mathbf{r})\Xi(\mathbf{k}, \mathbf{q}, \mathbf{r})\, d^3r = \delta(\mathbf{k}' - \mathbf{k})\, \delta(\mathbf{q}' - \mathbf{q}). \tag{5.1.52}$$

The form of the Schrödinger equation (5.1.1) in the $\mathbf{kq}$ representation is quite different from that encountered previously since the basis states are not related in any simple way to the eigenfunctions of the perfect crystal Hamiltonian. We must consider the matrix elements of $\mathbf{p}$ and $\mathbf{r}$ on the basis of the functions Ξ. When this is done, we see that we should replace an electron coordinate $\mathbf{r}$ whenever it appears in the Hamiltonian by

$$\mathbf{r} = i\,\nabla_\mathbf{k} + \mathbf{q} \tag{5.1.53}$$

and the momentum operator $\mathbf{p}$ by

$$\mathbf{p} = -i\hbar\,\nabla_\mathbf{q}. \tag{5.1.54}$$

However, in the case of the periodic potential $V(\mathbf{r})$ it suffices to replace $V(\mathbf{r})$ by $V(\mathbf{q})$. This follows since the periodicity of $V(\mathbf{r})$ ensures that it will connect only those states whose wave vectors differ by a reciprocal lattice vector. However, according to the restriction imposed in order to derive (5.1.52), such states are not included in the basis. We write the solution of (5.1.1) in the form

$$\Psi(\mathbf{r}, t) = \iint d^3k\, d^3q\, F(\mathbf{k}, \mathbf{q}, t)\Xi(\mathbf{k}, \mathbf{q}, \mathbf{r}). \tag{5.1.55}$$

Consider a situation in which $H = \mathbf{p}^2/2m + V(\mathbf{r}) + U(\mathbf{r}, t)$, V being periodic and U an external perturbation depending on position and time. We find

$$[-(\hbar^2/2m)\nabla_{\mathbf{q}}^2 + V(\mathbf{q}) + U(i\nabla_{\mathbf{k}} + \mathbf{q}, t)]F(\mathbf{k}, \mathbf{q}, t) = i\hbar\,\partial F(\mathbf{k}, \mathbf{q}, t)/\partial t. \tag{5.1.56}$$

Usually one obtains a differential equation in six variables (plus time).

5.2 Localized Impurity States

Theoretical study of the effect of impurities and defects on the electronic structure of solids is motivated by the observation that many electrical, magnetic, and optical properties are quite sensitive to the presence of a small quantity of impurities. For a single example, transistors would not exist were it not possible to control the conductivity type (n or p) by suitable doping. Techniques used in studying impurities and defects fall naturally into two categories according to whether the change in the crystal potential produced by the presence of the impurity or defect is of short or long range.

We will first consider the short range case. This obtains naturally in metals, where any charge unbalance will be screened by a redistribution of electrons. Electrically neutral imperfections in semiconductors may also be studied by these methods. There are also other problems which may be handled by the same methods including the effect of a magnetic impurity on the spin wave spectrum of a Heisenberg ferromagnet, and the effect of a substituted atom of different mass on the lattice vibrational system. We will discuss the problems in terms of the CMR.

5.2.1 *The t Matrix*

Let us begin by considering the scattering of an electron between two Bloch states caused by a static potential U. For such potentials, which do not possess any internal degrees of freedom, the scattering may be adequately described in terms of a one-electron theory. No complications arise from the exclusion principle. Some difficulties which occur when the scattering center has its own degrees of freedom are discussed in Section 5.6. Suppose the initial state has an electron in band n with wave vector $\mathbf{k}$. We require the probability per unit time of transitions to a state in band l, with wave vector $\mathbf{q}$. A formal expression for this quantity can be derived from the work of Gell-Mann and Goldberger (1953). It is

$$W(n\mathbf{k} \to l\mathbf{q}) = (2\pi)^7 \,|\langle l\mathbf{q} \mid U \mid +n\mathbf{k}\rangle|^2\, \delta[E_l(\mathbf{q}) - E_n(\mathbf{k})]. \tag{5.2.1}$$

This expression must be integrated over a range of final states. The peculiar multiplying factor of $(2\pi)^7$ is a consequence of using Bloch functions normalized according to (5.1.3). The state $| +n\mathbf{k}\rangle$ is one which evolves from the original Bloch state $| n\mathbf{k}\rangle$ through the scattering process. Thus $\langle \mathbf{r} | l\mathbf{q}\rangle = \psi_l(\mathbf{q}, \mathbf{r})$ (a Bloch function), and $\langle \mathbf{r} | +n\mathbf{k}\rangle$ is a solution of the complete Schrödinger equation which would go smoothly into $\psi_n(\mathbf{k}, \mathbf{r})$ if the interaction were smoothly turned off. The + indicates that certain boundary conditions have been imposed which will be described later.

An exact solution of the Schrödinger equation $| +n\mathbf{k}\rangle$ appears in (5.2.1). It is convenient to eliminate this formally by introducing an operator t (the t matrix) which satisfies

$$U | +n\mathbf{k}\rangle = t | n\mathbf{k}\rangle \tag{5.2.2}$$

where a free Bloch state appears on the right side. We can determine t in principle as follows. The state $| +n\mathbf{k}\rangle$ satisfies

$$(E - H_0) | +n\mathbf{k}\rangle = U | +n\mathbf{k}\rangle. \tag{5.2.3}$$

Since the scattering process may be assumed not to change the energy of individual continuum states, $E = E_n(\mathbf{k})$. Equation (5.2.3) may be rewritten as

$$| +n\mathbf{k}\rangle = | n\mathbf{k}\rangle + [1/(E - H_0)]U | +n\mathbf{k}\rangle. \tag{5.2.4}$$

The state $| n\mathbf{k}\rangle$ appears on the right since it satisfies

$$(E - H_0) | n\mathbf{k}\rangle = 0.$$

Equation (5.2.4) is not quite right as it stands. Since E is in the continuous spectrum of H_0 the operator $E - H_0$ is singular; it does not have an inverse. However, formal scattering theory (Gell-Mann and Goldberger, 1953) shows us how to avoid this difficulty. We are supposed to replace E by $E^+ = E + i\epsilon$, where ϵ is positive and infinitesimal, and is allowed to vanish after certain integrals are performed. This device corresponds to the imposition of an outgoing wave boundary condition on the wave function $\psi^+ = \langle \mathbf{r} | +n\mathbf{k}\rangle$. We will use the notation $\mathcal{G}$ for the operator $1/(E^+ - H_0)$,

$$| +n\mathbf{k}\rangle = | n\mathbf{k}\rangle + [1/(E^+ - H_0)]U | +n\mathbf{k}\rangle = | n\mathbf{k}\rangle + \mathcal{G}U | +n\mathbf{k}\rangle. \tag{5.2.5}$$

This is the Lippmann–Schwinger equation.

Now we multiply both sides by U and use the definition of t, Eq. (5.2.2). We obtain

$$U | +n\mathbf{k}\rangle = U | n\mathbf{k}\rangle + U\mathcal{G}U | +n\mathbf{k}\rangle \quad \text{or} \quad t | n\mathbf{k}\rangle = U | n\mathbf{k}\rangle + U\mathcal{G}t | n\mathbf{k}\rangle.$$

Since the state $| n\mathbf{k}\rangle$ is an arbitrary member of a complete set, we have the operator equation

$$t = U + U\mathcal{G}t, \tag{5.2.6}$$

and Eq. (5.2.1) takes the form

$$W(n\mathbf{k} \to l\mathbf{q}) = (2\pi)^7 | \langle l\mathbf{q} | t | n\mathbf{k} \rangle |^2 \delta[E_l(\mathbf{q}) - E_n(\mathbf{k})]. \tag{5.2.7}$$

Equation (5.2.6) has the formal solution

$$t = U[1/(1 - \mathcal{G}U)]. \tag{5.2.8}$$

This may be verified by comparing the binomial expansion of (5.2.8) with the iterated solution of (5.2.6), or more simply by multiplying on the left by the inverse of the operator $\{U[1/(1 - \mathcal{G}U)]$ which is $[(1 - \mathcal{G}U)(1/U)]$ (followed by multiplication on the left by U).

The utility of Eq. (5.2.8) rests on the possibility of actually constructing the inverse operator $1/(1 - \mathcal{G}U)$. This can actually be done exactly for a class of potentials we define as potentials of finite range, and approximately in other cases. For this purpose it is convenient to work in the local representation. The transformation between Bloch functions and Wannier functions relates t matrix elements in the two bases

$$\langle l\mathbf{q} | t | n\mathbf{k} \rangle = [\Omega/(2\pi)^3] \sum_{\mu\nu} \exp[i(\mathbf{k}\cdot\mathbf{R}_\mu - \mathbf{q}\cdot\mathbf{R}_\nu)](l\nu | t | n\mu). \tag{5.2.9}$$

If the t matrix is calculated on the Wannier function basis, it may be immediately transformed to the crystal momentum representation. We will obtain the matrix elements of t in the local representation directly from (5.2.8); the quantities which appear in that equation being regarded as matrices. This procedure is equivalent to the solution of the scattering problem directly from Eq. (5.1.41).

Let us look for a moment at this equation, which is the effective Schrödinger equation in the local representation. We can write it as

$$\sum_{\mu} (l\nu | E - H_0 | l\mu) B_l(\mathbf{R}_\mu) = \sum_{n\mu} (l\nu | U | n\mu) B_n(\mathbf{R}_\mu) \tag{5.2.10}$$

since H_0 is diagonal in the band index. It is convenient to introduce the Green's function $\mathcal{G}$ in this representation

$$\mathcal{G}_n(\mathbf{R}_\mu - \mathbf{R}_\nu) = (n\mu | 1/(E^+ - H_0) | n\nu). \tag{5.2.11}$$

Thus $\mathcal{G}$ is diagonal in the bands since H_0 is. From the definition it is apparent that $\mathcal{G}$ satisfies the equation

$$\sum_{\mu} (l\nu | E - H_0 | l\mu) \mathcal{G}_l(\mathbf{R}_\mu - \mathbf{R}_\rho)$$

$$= \sum_{\mu} [E\,\delta_{\mu\nu} - \mathcal{E}_l(\mathbf{R}_\nu - \mathbf{R}_\mu)] \mathcal{G}_l(\mathbf{R}_\mu - \mathbf{R}_\rho) = \delta_{\nu\rho}. \tag{5.2.12}$$

We may now rewrite (5.2.10) in a form exactly analogous to (5.2.5)

$$B_l(\mathbf{R}_\mu) = B_l^{(0)}(\mathbf{R}_\mu) + \sum_{\nu n \rho} \mathcal{G}_l(\mathbf{R}_\mu - \mathbf{R}_\nu)(l\nu \mid U \mid n\rho) B_n(\mathbf{R}_\rho) \quad (5.2.13)$$

where $B_l^{(0)}(\mathbf{R}_\mu)$ is a solution of the homogeneous equation

$$\sum_\mu [E\,\delta_{\mu\nu} - \mathcal{E}_l(\mathbf{R}_\nu - \mathbf{R}_\mu)] B_l^{(0)}(\mathbf{R}_\mu) = 0. \quad (5.2.14)$$

We can show from Eq. (5.1.37) that an explicit solution of (5.2.14) is

$$B_l^{(0)}(\mathbf{R}_\mu) = [\Omega^{1/2}/(2\pi)^{3/2}] \exp(i\mathbf{k}\cdot\mathbf{R}_\mu) \quad (5.2.15)$$

provided $E_l(\mathbf{k}) = E$. The normalization factor is chosen so that

$$\sum_\mu B_l^{(0)}(\mathbf{R}_\mu) a_l(\mathbf{r} - \mathbf{R}_\mu) = \psi_l(\mathbf{k}, \mathbf{r}). \quad (5.2.16)$$

We will return to (5.2.13) below.

The central problem in the determination of t is the construction of the inverse operator $[1 - \mathcal{G}U]^{-1}$. This inverse may always be written in the form

$$[1 - \mathcal{G}U]^{-1} = P'/D \quad (5.2.17)$$

where

$$D = \det[1 - \mathcal{G}U]. \quad (5.2.18)$$

A necessary and sufficient condition for the existence of an inverse is that the determinant D must not vanish. The matrix P' is known as the *adjugate* or *adjoint matrix*. The determinant can vanish only if there is some vector which is annihilated by the operator $(1 - \mathcal{G}U)$. Let such a vector be denoted by B. Then B must satisfy the equation

$$B = \mathcal{G}UB, \quad (5.2.19)$$

or, with indices inserted,

$$B_l(\mathbf{R}_\nu) = \sum_{\mu n \rho} \mathcal{G}_l(\mathbf{R}_\nu - \mathbf{R}_\mu)(l\mu \mid U \mid n\rho) B_n(\mathbf{R}_\rho). \quad (5.2.20)$$

This is just Eq. (5.2.13) without a homogenous term. A solution of this equation can be found only in a region outside the band (below or above) where there is no possibility of satisfying the condition $E_l(\mathbf{k}) = E$. This is characteristic of bound states, and we will show subsequently that the $B(\mathbf{R}_\nu)$ which are solutions of (5.2.20) do decrease exponentially at large distances from the impurity site.

The determinant D depends on energy implicitly through the dependence of the Green's function matrix on energy [see Eq. (5.2.11)]. Thus we can write

$$D = D(E).$$

The determinant will vanish only for some specific values of the energy which are the energies of the localized states. Consequently, in order to find the energy E_0 of a localized state, we look for a root of $D(E)$: the equation

$$D(E_0) = 0 \tag{5.2.21}$$

determines the energy of such a state.

We will consider explicitly potentials of finite range. This means that if U is expressed in the Wannier function basis, only a finite number of matrix elements are different from zero. Such an assumption will be exactly true only in certain models; but it can be an adequate approximation in realistic circumstances. (Then one must investigate the question of convergence: do the results approach a limit as the number of matrix elements of the potential is increased?)

The significance of the restriction to finite range potentials is that, in such cases, the determinant D and the entire inverse matrix $[1 - \mathcal{G}U]^{-1}$ can be constructed from finite matrices (whereas the matrix $\mathcal{G}$ will have, in general, an infinite number of nonvanishing elements). To see how this comes about, let us arrange the matrix U in block form, so that the portion of U containing nonzero elements forms an $n \times n$ submatrix in the upper left. Then $\mathcal{G}$ is arranged in a corresponding manner. Indices a, b are used on the submatrices as shown:

$$\mathbf{U} = \begin{pmatrix} \mathbf{U}_{aa} & 0 \\ 0 & 0 \end{pmatrix}, \qquad \mathcal{G} = \begin{pmatrix} \mathbf{g}_{aa} & \mathbf{g}_{ab} \\ \mathbf{g}_{ba} & \mathbf{g}_{bb} \end{pmatrix}. \tag{5.2.22a}$$

Then the matrix $1 - \mathcal{G}U$ becomes

$$1 - \mathcal{G}U = \begin{pmatrix} 1 - \mathbf{g}_{aa}\mathbf{U}_{aa} & 0 \\ -\mathbf{g}_{ba}\mathbf{U}_{aa} & 1 \end{pmatrix} \tag{5.2.22b}$$

in which the right-hand box contains a unit matrix. One easily verifies (for instance, by expanding according to minors, beginning at the lower right) that the determinant of the entire matrix is just the determinant of the upper $n \times n$ part:

$$D = \det(1 - \mathbf{g}_{aa}\mathbf{U}_{aa}). \tag{5.2.23}$$

Further, if we construct the inverse of the upper left matrix $1 - \mathbf{g}_{aa}\mathbf{U}$, the inverse of the full matrix can be found immediately. Let us denote this inverse as $\mathbf{P}_{aa}/D$,

$$(1 - \mathbf{g}_{aa}\mathbf{U}_{aa})^{-1} = \mathbf{P}_{aa}/D.$$

Then

$$[1 - \mathcal{G}\mathbf{U}]^{-1} = \begin{pmatrix} \mathbf{P}_{aa}/D & 0 \\ \mathbf{g}_{ba}\mathbf{U}_{aa}(\mathbf{P}_{aa}/D) & 1 \end{pmatrix}. \tag{5.2.24}$$

To construct t, we multiply on the left by U,

$$\mathbf{t} = \begin{pmatrix} \mathbf{U}_{aa}\mathbf{P}_{aa}/D & 0 \\ 0 & 0 \end{pmatrix}. \tag{5.2.25}$$

The nonzero portion of t has the same dimensionality as the nonzero portion of the potential. We can now write an expression for the t matrix on the Bloch function basis. From (5.2.9) and (5.2.25)

$$\langle l\mathbf{q} \mid t \mid n\mathbf{k} \rangle = [\Omega/(2\pi)^3 D] \sum \exp[i(\mathbf{k}\cdot\mathbf{R}_\mu - \mathbf{q}\cdot\mathbf{R}_\nu)](l\nu \mid U \mid j\sigma)(j\sigma \mid P \mid n\mu). \tag{5.2.26}$$

5.2.2 *The Scattering Amplitude*

Similar procedures enable us to solve (5.2.13). We represent B, B^0 as column vectors in accord with submatrix description of U and $\mathcal{G}$. Then Eq. (5.1.13) takes the form

$$\begin{pmatrix} \mathbf{B}_a \\ \mathbf{B}_b \end{pmatrix} = \begin{pmatrix} \mathbf{B}_a^0 \\ \mathbf{B}_b^0 \end{pmatrix} + \begin{pmatrix} \mathbf{g}_{aa}\mathbf{U}_{aa}\mathbf{B}_a \\ \mathbf{g}_{ba}\mathbf{U}_{aa}\mathbf{B}_a \end{pmatrix} \tag{5.2.27}$$

where $\mathbf{B}_b$ is determined after $\mathbf{B}_a$ has been found, and $\mathbf{B}_a$ is given by

$$\mathbf{B}_a = (1 - \mathbf{g}_{aa}\mathbf{U}_{aa})^{-1}\mathbf{B}_a^{(0)} = (\mathbf{P}_{aa}/D)\mathbf{B}_a^{(0)}. \tag{5.2.28}$$

Then for $\mathbf{B}_b$ we obtain

$$\mathbf{B}_b = \mathbf{B}_b^{(0)} + \mathbf{g}_{ba}\mathbf{U}_{aa}(\mathbf{P}_{aa}/D)\mathbf{B}_a^{(0)}. \tag{5.2.29}$$

We will require the expansion coefficients B for positions well beyond the range of the potential. This may be obtained from (5.2.29), which gives with indices restored

$$B_l(\mathbf{R}_\mu) = B_l^{(0)}(\mathbf{R}_\mu) + (1/D) \sum_{\rho n \nu} \mathcal{G}_l(\mathbf{R}_\mu - \mathbf{R}_\rho)(l\rho \mid U \mid n\nu) \times (n\nu \mid P \mid m\sigma) B_m^{(0)}(\mathbf{R}_\sigma). \tag{5.2.30}$$

All the sums in Eq. (5.2.30) involve only a finite number of terms.

At this point, it is desirable to examine the properties of the Greens' function. From (5.2.11) and (5.1.19), we have

$$\mathcal{G}_l(\mathbf{R}_\mu - \mathbf{R}_\rho)$$

$$= [\Omega/(2\pi)^3] \iiint d^3k\, d^3q\, d^3r \exp(i\mathbf{k}\cdot\mathbf{R}_\mu)\psi_l^*(\mathbf{k},\mathbf{r})$$

$$\times [1/(E^+ - H_0)] \exp(-i\mathbf{q}\cdot\mathbf{R}_\rho)\psi_l(\mathbf{q},\mathbf{r})$$

$$= [\Omega/(2\pi)^3] \iiint d^3k\, d^3q\, d^3r \exp(i\mathbf{k}\cdot\mathbf{R}_\mu)\psi_l^*(\mathbf{k},\mathbf{r})$$

$$\times [1/(E^+ - E_l(\mathbf{q})] \exp(-i\mathbf{q}\cdot\mathbf{R}_\rho)\psi_l(\mathbf{q},\mathbf{r})$$

$$= [\Omega/(2\pi)^3] \iint d^3k\, d^3q [\exp(i\mathbf{k}\cdot\mathbf{R}_\mu - i\mathbf{q}\cdot\mathbf{R}_\rho)/(E^+ - E_l(\mathbf{q}))]\, \delta(\mathbf{k}-\mathbf{q})$$

$$= [\Omega/(2\pi)^3] \int d^3k [\exp(i\mathbf{k}\cdot(\mathbf{R}_\mu - \mathbf{R}_\rho))/(E^+ - E_l(\mathbf{k}))]. \qquad (5.2.31)$$

Equation (5.2.31) furnishes us with a useful representation of the Green's function which is independent of the specific form of the Bloch functions.

For use in Eq. (5.2.30), we require $\mathcal{G}_l(\mathbf{R}_\mu - \mathbf{R}_\rho)$ for R outside of the range of the potential. Let us therefore examine the properties of $\mathcal{G}_l(\mathbf{R})$ in the asymptotic region of large $|\mathbf{R}|$. The asymptotic form of $\mathcal{G}$ may be worked out using the method of stationary phase. This calculation has been described by Lifshitz (1948), Koster (1954), Callaway (1964), and Preziosi (1971). There are some fairly subtle points for a general band structure; complications are particularly apt to arise if the surfaces of constant energy contain concave portions. We will restrict attention here to the case of a parabolic band

$$E_l(\mathbf{k}) = \gamma_l k^2$$

for which an approximate calculation is simple. The reader is referred to the references cited above for a discussion of more complex cases. We replace the Brillouin zone by a sphere of radius k_m,

$$\mathcal{G}_l(R_\mu) = (\Omega/2\pi^2 R) \lim_{\epsilon\to 0} \int_0^{km} (q \sin q R_\mu)/(E + i\epsilon - \gamma_l q^2)\, dq$$

$$\approx (\Omega/2\pi^2 R) \lim_{\epsilon\to 0} \int_0^{\infty} (q \sin q R_\mu)/(E + i\epsilon - \gamma_l q^2)\, dq$$

$$= -(\Omega/4\pi\gamma_l)(e^{ikR_\mu}/R_\mu) \qquad (5.2.32)$$

where

$$k = (E/\gamma_l)^{1/2}. \tag{5.2.33}$$

In the second line the upper limit has been made infinite. This approximation is consistent with the stationary phase method provided that $E < \gamma k_m^2$, which we assume to hold.

Equation (5.2.32) is actually valid for a general band provided that the surface of energy E defined by $E_l(\mathbf{q}) = E$ is convex if some reinterpretations are made (Callaway, 1964). The function $\exp(ikR_\mu)$ must be replaced by $\exp(i\mathbf{k}\cdot\mathbf{R}_\mu)$ where $\mathbf{k}$ is a vector whose length and direction are determined by the conditions that $E_l(\mathbf{k}) = E$ and that $[\nabla E_l(\mathbf{q})]_\mathrm{k}$ must be a vector parallel to $\mathbf{R}_\mu$. The quantity γ_l becomes a more complicated constant depending on the first and second derivatives of the energy for $\mathbf{q} = \mathbf{k}$, and which reduces to the reciprocal effective mass in the case of a parabolic band.

We are interested for the moment in $\mathcal{G}_n(\mathbf{R}_\mu - \mathbf{R}_\rho)$ in the limit of $R_\mu \gg R_\rho$. In this case, we have

$$\mathcal{G}(\mathbf{R}_\mu - \mathbf{R}_\rho) = (-\Omega/4\pi\gamma_l R) \exp(ikR_\mu) \exp(i\mathbf{k}'\cdot\mathbf{R}_\rho) \tag{5.2.34}$$

where $\mathbf{k}'$ is a vector of length k in the direction of $\mathbf{R}_\mu$

$$\mathbf{k}' = k\mathbf{R}_\mu/|\,\mathbf{R}_\mu\,|.$$

With the use of Eq. (5.2.15), we can rewrite Eq. (5.2.30). Let us assume an incoming Bloch wave in band j characterized by wave vector $\mathbf{k}_0$.

Asymptotically, $B_l(\mathbf{R}_\mu)$ has the form

$$B_l(\mathbf{R}_\mu) \to [\Omega^{1/2}/(2\pi)^{3/2}][\exp(i\mathbf{k}_0\cdot\mathbf{R}_\mu)\,\delta_{lj} + f_{lj}(\mathbf{k}, \mathbf{k}_0)\exp(ikR_\mu)/R_\mu] \tag{5.2.35}$$

where

$$f_{lj}(\mathbf{k}, \mathbf{k}_0) = -(\Omega/4\pi\gamma_l D) \sum_{\rho\nu n} \exp[i(\mathbf{k}_0\cdot\mathbf{R}_\sigma - \mathbf{k}\cdot\mathbf{R}_\rho)](l\rho\,|\,U\,|\,n\nu)(n\nu\,|\,P\,|\,j\sigma). \tag{5.2.36}$$

We have used the symbol f in Eq. (5.1.35) by analogy with conventional free space potential scattering theory in which the asymptotic form of the wave function is

$$\psi^+(\mathbf{r}) \to e^{ikz} + fe^{ikr}/r$$

and f is the scattering amplitude. In solids, the scattering amplitude is a matrix in the bands. If a wave enters in band j, an outgoing wave will occur in band l if the conservation conditions are satisfied.

What happens if the energy E in (5.2.32) does not coincide with the energy of any Bloch state (that is, E is either below or above the band considered)? Suppose that $E < 0$, and let us define κ through

$$E_l = -\gamma\kappa^2. \tag{5.2.37}$$

Then instead of (5.2.32), we have

$$\mathcal{G}_l(R_\mu) = (-\Omega/2\pi^2\gamma R_\mu) \int_0^{km} q \sin qR_\mu/(q^2 + \kappa^2)\, dq. \tag{5.5.38}$$

If R_μ is large and κ reasonably small, we may approximate this integral by allowing the upper limit to be infinite. Then

$$\mathcal{G}_l(R_\mu) = (-\Omega/4\pi\gamma R_\mu) \exp(-\kappa R_\mu) \tag{5.2.39}$$

which is just what would have been obtained by making the replacement $k \to i\kappa$ in (5.2.32). It follows from Eq. (5.2.20) that in the limit of R_ν large compared to the range of the potential,

$$B_l(R_\nu) \approx \exp(-\kappa R_\nu)/R_\nu. \tag{5.2.40}$$

This is just what would be expected for a bound state wave function. (Compare the result of elementary quantum mechanics for the wave function at large distances from a potential well

$$\psi \approx h_l^{(0)}(i\kappa r) \to e^{-\kappa r}/r \qquad \text{for any angular momentum } l.)$$

If $E > E_m$, where E_m is the highest energy in the band, we are concerned with a bound state above the band. A similar analysis applies if we now put $E_l(k) = E_m - \gamma k^2$ where γ is the reciprocal effective mass at the top of the band, γ being positive, and k is measured from the band maximum. In order to obtain the same result (5.2.40), all we have to do is to set $E - E_m = \gamma\kappa^2$. Hence the term "bound state" may be just as legitimately applied to states above the top of the band.

Now let us return to the scattering problem. Comparison of (5.2.36) and (5.2.26) shows that

$$f_{lj}(\mathbf{q}, \mathbf{k}) = (-2\pi^2/\gamma_l)\langle l\mathbf{q} \mid t \mid j\mathbf{k}\rangle. \tag{5.2.41}$$

Thus the scattering amplitude is simply proportional to the t matrix. This relation is just that obtained in the ordinary potential scattering theory (see Goldberger and Watson, 1964, p. 235), in the case of a single parabolic band.

The close analogy between the relations obtained in this discussion, particularly (5.2.36) and (5.2.41) and those of conventional approaches, suggest that we should look for a phase shift or partial wave expansion for the scattering amplitude. In ordinary scattering theory, one writes

$$f(\mathbf{q}, \mathbf{k}) = (1/k) \sum_l (2l+1) \exp(i\delta_l) \sin \delta_l P_l(\cos \theta_{\mathbf{qk}}) \quad (5.2.42)$$

where $\theta_{\mathbf{qk}}$ is the angle between $\mathbf{q}$ and $\mathbf{k}$, and $|\mathbf{q}| = |\mathbf{k}|$. We would expect to have some relation of this type in the present case. However, Eq. (5.2.42) will not hold exactly in a solid even in the limit of a parabolic energy band because we do not have spherical symmetry. The change in crystal potential which produces the scattering cannot be expected to have spherical symmetry, as is required for the validity of the usual partial wave expansion: in general, the scattering potential will be invariant under the crystal point group or some subgroup of it. Use of a more general $E(\mathbf{k})$ leads to a similar conclusion. It follows from Eq. (5.2.31) by the same argument used in deriving Eq. (5.1.39) that

$$\mathcal{G}_l(\alpha\mathbf{R}_\mu) = \mathcal{G}_l(\mathbf{R}_\mu) \quad (5.2.43)$$

for any α in the point group. Thus $\mathcal{G}$ is invariant under the point group but does not have spherical symmetry.

What remains of the usual partial wave analysis may be uncovered by constructing symmetrized combinations of the coefficients $B_m(\mathbf{R}_\mu)$ which transform according to the irreducible representations of the symmetry group under which the defect potential is invariant. Let us denote the irreducible representations of this group by indices s, t; the operation of the group by α. In the case of degenerate representations, we distinguish the rows by a subscript η; thus s_η designates the ηth row of representation s. The construction of symmetrized combinations may be accomplished by introducing a unitary transformation S, whose elements may be denoted by $S(s_\eta, \mathbf{R}_\mu)$. Let us consider quantities $B_n(\mathbf{R}_\mu)$ such that all the $\mathbf{R}_\mu$ involved may be obtained from any one by a rotation by one of the operations in the group of the defect potential. Such vectors will be called a *type*. The symmetrized functions will be denoted by $C_{s\eta}(R_\mu)$. If irreducible representation s occurs only once in the reduction of the representation formed by the B the elements of U may be constructed as described in Section 3.2.2, Part A. If s occurs more than once, additional orthogonalization may be required. Explicit expressions for the elements of U will not be required. We have

$$C_{s\eta}{}^{(n)}(R_\mu) = \sum_\mu S(s\eta, \mathbf{R}_\mu) B_n(\mathbf{R}_\mu). \quad (5.2.44)$$

The sum runs over all the vectors of the particular type; the result still depends on the type. For this reason, we retain the R_μ in the designation of the functions.

If the unitary transformation S is applied to the quantities $B_n{}^{(0)}(\mathbf{k}, \mathbf{R}_\mu)$

where

$$B_n^{(0)}(\mathbf{k}, \mathbf{R}_\mu) = [\Omega^{1/2}/(2\pi)^{3/2}] \exp(i\mathbf{k}\cdot\mathbf{R}_\mu),$$

the result is a symmetrized combination of such functions which we denote by $C_{s\eta}^{(0)}(k, R_\mu)$, in accord with (5.2.34). Such functions are in fact nothing more than the symmetrized linear combinations of plane waves previously introduced in Section 4.2, Part A, which appear here in another context.

One property of the $C_{s\eta}^{(0)}$ which will be important to us later is the following. Suppose the function $C_{s\eta}^{(0)}$ is expanded in powers of $\mathbf{k}$. Then the terms of the expansion are proportional to functions of k times Kubic harmonics for the particular representation and row. If we retain only the leading term in the expansion, we have

$$C_{s\eta}^{(0)}(k, R_\mu) = \lambda_s k^\sigma K_{s\eta}(\theta, \phi) \tag{5.2.45}$$

in which λ_s is a constant depending on R_μ, the exponent σ depends on the representation s, $K_{s\eta}$ is the relevant Kubic harmonic (von der Lage and Bethe, 1947) for row η of representation s, and the angles θ, ϕ specify the orientation of $\mathbf{k}$ with respect to the crystal axes. Note that λ_s and σ are independent of η.

Introduction of the symmetrized functions causes the determinant D to factor into a product of subdeterminants pertaining to different representations. This is a result of the general theorem (Section 3.2.6, Part A) that the potential will not have (nonzero) matrix elements between functions belonging to different irreducible representations or to different rows of the same representation. This applies specifically to the defect potential, but also to the unperturbed Hamiltonian, and consequently to the Green's function as well. Thus we can write

$$D = \prod_s (D_s)^{g_s} \tag{5.2.46}$$

where g_s is the degeneracy of representations, and D_s is the subdeterminant derived from functions belonging to any single row of representation s. Since such subdeterminants are independent of the row of the representation, the contribution from a representation is the g_sth power of the contribution from a row.

The matrices $\mathcal{G}$ and U are block diagonal. For example,

$$\sum_{\mu\rho} S(s_\eta, \mathbf{R}_\mu)(n\mu \mid U \mid l\rho) S^+(\mathbf{R}_\rho, t_\zeta) = \delta_{st}\,\delta_{\zeta\eta}\, U^{(s)}{}_{n\mu,l\rho}. \tag{5.2.47}$$

The matrix $U^{(s)}$ does not depend on the row of the representation considered. The rows and columns will be designated by the band index and lattice vector type. The matrix P must also be block diagonal. After some thought, it will be realized that we must have

$$\sum_{\mu\rho} S(s_\eta, \mathbf{R}_\mu)(n\mu \mid P \mid l\rho) S^+(\mathbf{R}_\rho, t_\zeta) = \delta_{st}\, \delta_{\zeta\eta}\, (D/D_s) P^{(s)}{}_{n\mu, l\rho} \tag{5.2.48}$$

where the submatrix $P^{(s)}$ satisfies

$$[I - \mathcal{G}^{(s)} U^{(s)}]^{-1} = P^{(s)}/D_s. \tag{5.2.49}$$

Thus, if we construct $\mathcal{G}^s$ and U^s on the basis of symmetrized functions, as can be done using the techniques of Section 3.2.2 and invert the symmetrized matrix, we obtain just $P^{(s)}$ and D_s; the contributions from other rows and other representations do not enter.

In matrix notation, the expression for the t matrix we have given previously [Eq. (5.2.26)] may be written as

$$t = (1/D)\mathbf{B}^{0*}\mathbf{UPB}^0.$$

We introduce the symmetrizing transformation S as

$$t = (1/D)\mathbf{B}^{0*}(S^+S)\mathbf{U}(S^+S)\mathbf{P}(S^+S)\mathbf{B}^0.$$

The result is that the t matrix becomes expressed as a sum of contributions from the irreducible representations. We have, on restoring the indices,

$$\langle l\mathbf{q} \mid t \mid n\mathbf{k}\rangle = \sum_s \langle l\mathbf{q} \mid t_s \mid n\mathbf{k}\rangle \tag{5.2.50a}$$

in which the partial wave t matrices are given by

$$\langle l\mathbf{q} \mid t_s \mid n\mathbf{k}\rangle = (1/D_s) \sum_{\substack{\nu\sigma\mu \\ j}} U^{(s)}{}_{l\nu, j\sigma} P^{(s)}{}_{j\sigma, n\mu} \sum_\eta C_{s\eta}{}^{(0)*}(q, R_\nu) C_{s\eta}{}^{(0)}(k, R_\mu). \tag{5.2.50b}$$

Equation (5.2.50b) is the general partial wave formula of solid state scattering theory.

We will begin the analysis of this result by considering the quantities D_s. These subdeterminants are complex functions of energy. It is useful to define a phase shift δ_s by

$$\tan \delta_s = -\mathrm{Im}\, D_s/\mathrm{Re}\, D_s \tag{5.2.51}$$

where Im and Re refer to imaginary part and real part, respectively.

This phase shift, which is defined without explicit reference to a wave function or a Schrödinger equation, is actually the same phase shift that appears in the usual partial wave expansion for the scattering amplitude (or t matrix) when the appropriate limiting processes are performed. We can prove this without explicit reference to an ordinary Schrödinger equation involving ∇^2. This demonstration is quite instructive and will be presented subsequently in Section 5.2.5.

5.2.3 *The Change in the Density of States*

It is interesting to examine another use of the phase shift: the computation of the change in density of states produced by the impurity. Some discussion of the general ideas of this calculation may be helpful. In scattering theory, we are concerned with processes occurring in the continuous spectrum. The continuous energy spectrum itself is not changed, although one or more states may be displaced below (or in solids, above) the continuum to form localized or bound states. However, although the spectrum is the same, the density of states in the spectrum may be altered. Physically, this effect may be traced to the retardation or acceleration of a wave packet passing by the scattering center (Wigner, 1955). The calculation is quite straightforward.

The density of states at energy E may be simply defined as (where $\mathfrak{N}$ is number of cells in the crystal)

$$N(E) = (1/\mathfrak{N})\ \mathrm{Tr}[\delta(E - H)]. \tag{5.2.52}$$

In a representation in which H is diagonal, this simply counts the number of states which have energy E. Thus, in the perfect crystal (no impurity) we have

$$N(E) \to G(E) = (1/\mathfrak{N})\ \mathrm{Tr}[\delta(E - H_0)] = [\Omega/(2\pi)^3] \int d^3k\ \delta(E - E(\mathbf{k})). \tag{5.2.53}$$

We will henceforth use $G(E)$ to denote the density of states in the perfect crystal and $N(E)$ to denote that function in the real, imperfect crystal. The trace of an operator is unchanged by a unitary transformation, so that if (5.2.52) is correct in the representation in which H is diagonal, it holds in any representation. This property, which is shared by the determinant of an operator, will be of considerable use. Thus, even if we do not know the diagonal representation, the trace can be calculated in any representation which is handy, in particular, one for the perfect crystal.

Let us also recall that for any operator $\mathcal{O}$,

$$\mathrm{Tr}[\ln \mathcal{O}] = \ln \det \theta. \tag{5.2.54}$$

This relation is obvious for $\mathcal{O}$ diagonal, since the determinant is equal to the product of the eigenvalues and the logarithm converts the product into a sum. Since the relation involves the trace and the determinant, it holds in any representation. We also use the identity

$$\lim_{\epsilon \to 0} 1/(x + i\epsilon) = \mathrm{P}(1/x) - i\pi\ \delta(x) \tag{5.2.55}$$

where P stands for principal part. Thus we may write

$$N(E) = (-1/\pi\mathfrak{N}) \text{ Im Tr } 1/(E^+ - H) \tag{5.2.56}$$

where $E^+ = E + i\epsilon$ and the limit is understood. This may be rewritten as

$$N(E) = (-1/\pi\mathfrak{N}) \text{ Im } d/dE \text{ Tr ln}(E^+ - H)$$

which becomes, through (5.2.54),

$$N(E) = (-1/\pi\mathfrak{N}) \text{ Im } d/dE \ln \det(E^+ - H).$$

The contribution from impurities may be separated by writing

$$E^+ - H = E^+ - H_0 - V = (E^+ - H_0)[1 - (E^+ - H_0)^{-1} V].$$

The determinant of the product of two operators is the product of the determinants. Also,

$$G(E) = (-1/\pi\mathfrak{N}) \text{ Im } d/dE \ln \det(E^+ - H_0).$$

Thus the change in density of states is given by

$$\begin{aligned} \Delta N(E) &= N(E) - G(E) \\ &= (-1/\pi\mathfrak{N}) \text{ Im } d/dE\{\ln \det[1 - (E^+ - H_0)^{-1} V]\}. \end{aligned} \tag{5.2.57}$$

Equation (5.2.57) must be true for any representation so that it holds. in particular, if the basis functions are Wannier functions for the perfect crystal. It is also important to note, and will be useful later, that nowhere in the argument leading to this equation has any assumption been made concerning the number of impurities present in the crystal. Therefore Eq. (5.2.57) is an exact formal result for any defect concentration, and will form the basis of our examination of alloys in Section 5.5. It also may be noted that we have not made any essential assumption about the one-electron (or many-electron) character of the problem either.

We return to the problem of a single scattering center. The operator inside square brackets will be recognized as $I - \mathcal{G}V$, so that we have simply

$$\Delta N(E) = (-1/\pi\mathfrak{N}) \text{ Im } d \ln D(E)/dE. \tag{5.2.58}$$

Now we use Eqs. (5.2.46) and (5.2.51) to bring in the scattering phase shifts. Since

$$\ln D(E) = \sum_s g_s \ln D_s = \sum g_s[\ln |D_s| - i\,\delta_s + 2in\pi]$$

where n is arbitrary, we have at once

$$\Delta N(E) = (1/\pi\mathfrak{N}) \sum_s g_s\, d\delta_s/dE. \tag{5.2.59a}$$

If only a small number of impurities are present (say n in a monatomic crystal, with $n/\mathfrak{N} = c$, $c \ll 1$), the contribution of each impurity to the change in density of states may be added. This is obvious physically, and will be justified formally in Section 5.5. Thus, for small but finite concentrations, we have

$$\Delta N(E) = c/\pi \sum_s g_s \, d\delta_s/dE. \tag{5.2.59b}$$

It is interesting, at this point, to consider what happens in the vicinity of a scattering resonance. A scattering resonance is said to occur if one of the partial wave phase shifts δ_s increases rapidly through an odd integral multiple of $\pi/2$. This causes the scattering cross section to have a maximum. In free space, scattering resonances are directly observable. In a solid we cannot measure cross sections directly: we have to look for properties determined by the t matrix. A scattering resonance produces a maximum in the (change in the) density of states $\Delta N(E)$.

Suppose δ_s passes through an odd multiple of $\pi/2$ when $E = E_s$. From (5.2.51) we see that this happens if

$$\operatorname{Re} D_s(E_s) = 0. \tag{5.2.60}$$

In order to study the behavior of the scattering phase shift near a resonance, we expand $\operatorname{Re} D_s(E)$

$$\operatorname{Re} D_s(E) = \operatorname{Re} D_s(E_s) + (E - E_s)\,(d/dE \operatorname{Re} D_s)_{E_s}.$$

In general, $\operatorname{Im} D_s(E_s)$ will not vanish. Therefore, we may replace $\operatorname{Im} D_s(E)$ by its value at $E = E_s$. Let us define Γ_s by

$$\Gamma_s = 2 \operatorname{Im} D_s(E_s)/[d(\operatorname{Re} D_s)/dE]_{E_s}. \tag{5.2.61}$$

Then we have, close to E_s,

$$\tan \delta_s = \Gamma_s/2(E_s - E). \tag{5.2.62}$$

Equation (5.2.62) shows that for positive Γ, the phase shift increases through $\pi/2$ as the energy goes through E_s from below. We now obtain $d\delta_s/dE$. Let the contribution to $\Delta N(E)$ in (5.2.59b) be $\Delta N_s(E)$

$$\Delta N_s(E) \equiv (cg_s/\pi)\, d\delta_s(E)/dE = (cg_s\Gamma_s/2\pi)[1/((E - E_s)^2 + \Gamma_s^2/4)]. \tag{5.2.63}$$

This expression has the characteristic Breit–Wigner form (see Newton, 1966, Section 11.2) and indicates that Γ_s is the width of the resonance.

Let us now suppose that the defects introduced into the crystal are such that the total number of states is unchanged, and that a single band (or group of bands) is isolated from all others. The lowest energy in the band is E_0 and the highest is E_m. Then if we integrate Eq. (5.2.59b) with respect to energy from E_0 to E_m, the result must be minus the number of states forced out of the band by the perturbation, that is, the number of bound states in representation s lying either above or below the band. This quantity is $cg_s n_s$, where n_s is the number of states in any row of representation s forced out of the band by a single impurity atom. We obtain

$$\delta_s(E_0) - \delta_s(E_m) = \pi n_s. \tag{5.2.64}$$

This result is analogous to Levinson's theorem in ordinary scattering theory (see Newton, 1966, Section 11.2).

5.2.4 *The Optical Theorem*

There is an important relation in ordinary scattering theory known as the *optical theorem* (Newton, 1966, Section 7.2). In its simplest form this theorem states that the imaginary part of the forward scattering amplitude is proportional to the total cross section. This theorem is actually a consequence of the unitarity of the S matrix and so must hold in all systems if the relevant quantities are defined appropriately. The form we require can be obtained as follows.

An alternative formal expression for the t matrix is

$$t = U + U[1/(E^+ - H)]U \tag{5.2.65}$$

where the full Hamiltonian $H = H_0 + U$ appears. This expression may be seen to be equivalent to our previous expression, Eq. (5.2.6), by employing the identity

$$(E^+ - H_0 - U)^{-1} = (E^+ - H_0)^{-1} + (E^+ - H_0)^{-1}U(E_0^+ - H)^{-1}$$

and comparing the iterated expressions.

The t matrix is not Hermitian even though U is. The adjoint matrix is

$$t^+ = U + U(E^- - H)^{-1}U. \tag{5.2.66}$$

The difference between the operators is

$$t - t^+ = U[(E^+ - H)^{-1} - (E^- - H)^{-1}]U = -2\pi i U\,\delta(E - H)U. \tag{5.2.67}$$

We consider the matrix elements of Eq. (5.2.67) in the crystal momentum representation, and introduce a complete set of eigenstates of H, $| +l\mathbf{q}\rangle$

$$\begin{aligned}\langle n\mathbf{k} \mid t \mid n'\mathbf{k}'\rangle &- \langle n\mathbf{k} \mid t^+ \mid n'\mathbf{k}'\rangle \\ &= -2\pi i\langle n\mathbf{k} \mid U\,\delta(E - H)U \mid n'\mathbf{k}'\rangle \\ &= -2\pi i \sum_l \int d^3q\langle n\mathbf{k} \mid U \mid +l\mathbf{q}\rangle\,\delta(E - E_l(\mathbf{q}))\langle +l\mathbf{q} \mid U \mid n'\mathbf{k}'\rangle.\end{aligned} \tag{5.2.68}$$

However,

$$\langle n\mathbf{k} \mid U \mid +l\mathbf{q}\rangle = \langle n\mathbf{k} \mid t \mid l\mathbf{q}\rangle \quad \text{and} \quad \langle +l\mathbf{q} \mid U \mid n'\mathbf{k}'\rangle = \langle l\mathbf{q} \mid t^+ \mid n'\mathbf{k}'\rangle.$$

Also, we have

$$\int d^3q = \int dS_{\mathbf{q}}(E_l)\, dE/| \nabla_{\mathbf{q}}E_l |$$

where $dS_q(E_l)$ is an element of area on a surface of constant energy in band l. Formally, bound states produced by the scattering potential must be included in the complete set of states; however, since we are concerned with an energy E in the continuum, the delta function ensures that they make no contribution. Equation (5.2.68) reduces to

$$\begin{aligned}\langle n\mathbf{k} \mid t \mid n'\mathbf{k}'\rangle - \langle n\mathbf{k} \mid t^+ \mid n'\mathbf{k}'\rangle = -2\pi i \sum \int dS_{\mathbf{q}}(E_l)/| \nabla_{\mathbf{q}}(E_l) | \\ \times \langle n\mathbf{k} \mid t \mid l\mathbf{q}\rangle\langle l\mathbf{q} \mid t^+ \mid n'\mathbf{k}'\rangle.\end{aligned} \tag{5.2.69}$$

In this equation, and subsequently, the integration over a surface of constant energy refers to a surface in band l on which the energy is E; the summation is restricted [by the delta function which appeared in (5.2.66)] to those bands in which there are states of energy E. We now specialize to the case $| n'\mathbf{k}'\rangle = | n\mathbf{k}\rangle$ for which

$$\langle n\mathbf{k} \mid t^+ \mid n\mathbf{k}\rangle = \langle n\mathbf{k} \mid t \mid n\mathbf{k}\rangle^*$$

(asterisk indicates complex conjugate). Then we have

$$\operatorname{Im}\langle n\mathbf{k} \mid t \mid n\mathbf{k}\rangle = -\pi \sum_l \int dS_{\mathbf{q}}(E_l)/| \nabla_{\mathbf{q}}E_l |\, | \langle n\mathbf{k} \mid t \mid l\mathbf{q}\rangle |^2. \tag{5.2.70a}$$

This is the optical theorem. It may be expressed in terms of the scattering amplitude through the use of (5.2.41)

$$\operatorname{Im} f_{nn}(k) = (\gamma_n/2\pi) \sum_l \int dS_{\mathbf{q}}(E_l)/| \nabla_{\mathbf{q}}(E_l) |\, | f_{nl}(\mathbf{k}, \mathbf{q}) |^2. \tag{5.2.70b}$$

The reader may convince himself that this is really the optical theorem by considering the case of a single parabolic band for which $E(\mathbf{k}) = \gamma k^2$, $|\mathbf{k}| = |\mathbf{q}|$, $dS \to k^2\, d\Omega$ ($d\Omega$ being an element of solid angle). In this case $f \equiv f(\mathbf{k} - \mathbf{q})$, and

$$\int |f|^2\, d\Omega = \sigma_{\mathrm{T}}$$

the total cross section. We have

$$\sigma_{\mathrm{T}} = (4\pi/k)\, \mathrm{Im}\, f(0), \tag{5.2.70c}$$

$f(0)$ being the forward scattering amplitude. This is the familiar result.

Equation (5.2.70a) differs from the usual result in two nontrivial aspects: (1) a sum over band indices appears, and (2) the surfaces of constant energy may be anisotropic.

5.2.5 *Phase Shifts and the Scattering Amplitude*

Use of the optical theorem allows us to show that under certain circumstances a relation analogous to that usually obtained in the theory of scattering of free particles by spherically symmetric potentials connecting the phase shift and the t matrix also holds in solids.

In the more familiar case, we have for the scattering amplitude

$$f(\theta) = (1/k) \sum_l (2l + 1) \exp(i\delta_l) \sin \delta_l P_l(\cos\theta) \tag{5.2.71a}$$

or for the t matrix

$$t(\mathbf{k}, \mathbf{q}) = (-\hbar^2/4\pi^2 mk) \sum_l (2l + 1) \exp(i\delta_l) \sin \delta_l P_l(\cos\theta) \tag{5.2.71b}$$

where θ is the angle between $\mathbf{k}$ and $\mathbf{q}$, $|\mathbf{k}| = |\mathbf{q}|$, l is the angular momentum of the lth partial wave, and P_l is a Legendre polynomial. These equations have quite a different appearance from (5.2.50). A connection however, should exist.

In order to reduce Eq. (5.2.50) to (5.2.71b), certain restrictions must be imposed: (1) Elements of the t matrix connecting different bands are neglected. We will therefore drop the band index in the specification of t and consider only a single band. (2) The energy band function has the simple form $E = \gamma k^2$. Surfaces of constant energy are spherical. (3) The symmetrized functions $C_{s\eta}^{(0)}(k, R)$ which appear in Eq. (5.1.50) are to be expanded in powers of k. We retain only the leading term, which is given by Eq. (5.2.46).

These restrictions have the effect of making the scattering process resemble that for which (5.2.71) was derived. However, we will not assume that the potential is spherically symmetric; nor that the B's satisfy a differential equation of the effective mass type.

From the definition of the phase shift, we have

$$1/D_s = -\sin \delta_s \exp(i\delta_s)/\mathrm{Im}\, D_s. \tag{5.2.72}$$

Then we introduce this plus Eq. (5.2.56) into (5.2.50b) and have

$$\langle \mathbf{k} \mid t_s \mid \mathbf{q} \rangle = (-\sin \delta_s \exp(i\delta_s)/\mathrm{Im}\, D_s)\, \Lambda_s k^{2\sigma} \sum_{\eta} K^*_{s\eta}(\theta_{\mathrm{q}}, \phi_{\mathrm{q}}) K_{s\eta}(\theta_{\mathrm{k}}, \phi_{\mathrm{k}}). \tag{5.2.73}$$

The angles ϕ_{k}, ϕ_{q}, etc. specify the orientation of the vectors $\mathbf{k}$ and $\mathbf{q}$ with respect to the crystal axes. The quantity Λ_s contains sums of elements of the U and P matrices and factors of λ_s from Eq. (5.2.45). It will be of some importance below that Λ_s depends only on the representation and not on the row. In addition, Λ may be complex; although in the simplest cases it turns out to bé real. We write Λ_s in the form

$$\Lambda_s = \mid \Lambda_s \mid \exp(i\rho_s).$$

We substitute (5.2.73) into (5.2.70a). In order to simplify some of the subsequent formulas, we use real Kubic harmonics normalized to 4π. Moreover, Kubic harmonics belonging to different representations or to different rows of the same representation are orthogonal.

$$\int K_{s\nu}(\theta, \phi) K_{s'\nu'}(\theta, \phi)\, d\Omega = 4\pi\, \delta_{ss'}\, \delta_{\nu\nu'}. \tag{5.2.74}$$

As a result of Eq. (5.2.74) there are no cross terms in (5.2.70a) connecting different representations or different rows of the same. representation. Equation (5.2.70a) holds for each component t_s separately, and this result is general, not merely a low energy approximation. After the substitution the left-hand side of Eq. (5.2.70a) becomes

$$\mathrm{Im}\langle \mathbf{k} \mid t_s \mid \mathbf{k} \rangle = [-\mid \Lambda_s \mid \sin \delta_s \sin(\delta_s + \rho_s)/\mathrm{Im}\, D_s] k^{2\sigma} \sum_{\eta} K^2_{s\eta}(\theta_{\mathrm{k}}, \phi_{\mathrm{k}}). \tag{5.2.75}$$

The right side is

$$\begin{aligned}(-\pi q/2\gamma)[\sin^2 \delta_s \mid \Lambda_s \mid^2/(\mathrm{Im}\, D_s)^2] k^{2\sigma} q^{2\sigma} \int d\Omega_q \sum_{\nu\eta} K_{s\eta}(\theta_{\mathrm{k}}, \phi_{\mathrm{k}}) K_{s\eta}(\theta_{\mathrm{q}}, \phi_{\mathrm{q}}) \\ \times K_{s\nu}(\theta_{\mathrm{k}}, \phi_{\mathrm{k}}) K_{s\nu}(\theta_{\mathrm{q}}, \phi_{\mathrm{q}}) \\ = (-2\pi^2/\gamma) k^{4\sigma+1} [\sin^2 \delta_s/(\mathrm{Im}\, D_s)^2] \mid \Lambda_s \mid^2 \sum_{\eta} K^2_{s\eta}(\theta_{\mathrm{k}}, \phi_{\mathrm{k}}).\end{aligned} \tag{5.2.76}$$

We have used the fact that for spherical energy surfaces, $|\mathbf{k}| = |\mathbf{q}|$. We now equate (5.2.75) and (5.2.76), and solve for $\mathrm{Im}\, D_s$

$$\mathrm{Im}\, D_s = (2\pi^2/\gamma)[\sin \delta_s\, k^{2\sigma+1}/\sin(\delta_s + \rho_s)]\,|\,\Lambda_s\,|. \qquad (5.2.77)$$

This result is substituted back into Eq. (5.2.73). The result is

$$\langle \mathbf{k} \mid t_s \mid \mathbf{q} \rangle = (-\gamma/2\pi^2 k) \exp[i(\delta_s + \rho_s)] \times \sin(\delta_s + \rho_s) \sum_\eta K_{s\eta}(\theta_{\mathrm{k}}, \phi_{\mathrm{k}}) K_{s\eta}(\theta_{\mathrm{q}}, \phi_{\mathrm{q}}). \qquad (5.2.78a)$$

This may be compared with (5.2.71b). The scattering amplitude is now found from (5.2.41) to be

$$f_s(\mathbf{q}, \mathbf{k}) = (1/k) \exp[i(\delta_s + \rho_s)] \sin(\delta_s + \rho_s) \sum_\eta K_{s\eta}(\theta_{\mathrm{k}}, \phi_{\mathrm{k}}) K_{s\eta}(\theta_{\mathrm{q}}, \phi_{\mathrm{q}}) \qquad (5.2.78b)$$

which may be compared with (5.2.71a). Some important similarities and differences appear:

There are only a finite number of representations whose contributions need to be added to give the full t matrix. This contrasts with the infinite number of angular momentum states which contribute to (5.2.71). The presence of a finite number of terms in the solid state analog of the partial wave expansion for t is a general feature of solid state scattering theory which results from the fact that the scattering potential and the energy bands are invariant under only a finite group. Each phase shift δ_s must contain contributions from an infinite number of spherical partial waves in the limit of complete spherical symmetry.

If we imagine a limiting process in which the lattice constant becomes infinite and the scattering potential acquires spherical symmetry, the δ_s must become combinations of spherical wave phase shifts δ_l. In this limit, the determinant D is still well defined although our methods of calculating it would not be useful. The subdeterminants D_s would then factor still further into a product of terms coming from each of the partial waves which in the solid form possible basis functions for representation s. From this we see that each δ_s would become in this limit equal to the sum of the phase shifts for all those waves which go into representation s. One must note that in the actual solid state scattering problem only the δ_s are well defined, not the δ_l.

The angular dependence of the t matrix elements given by (5.2.78) is also different from the one of (5.2.71). Both sets of angles θ_{k}, ϕ_{k} and θ_{q}, ϕ_{q} appear in (5.2.78), so that t does not depend only on the angle between $\mathbf{k}$ and $\mathbf{q}$. However, in some special cases, further reduction is possible. In particular, if the representation s has as basis function all of the spherical harmonics of a given l, then the sum over η can be performed. In cubic crystals, this happens only for the s-like representation Γ_1 with $l = 0$ and for the p-like representation Γ_{15} if $l = 1$. In the latter case, we may choose a basis in the representation such that

$$K_{p1}(\theta_k, \phi_k) = 3^{1/2}k_x/k = 3^{1/2} \sin\theta_k \cos\phi_k,$$

etc. The sum over η gives immediately

$$3\mathbf{k}\cdot\mathbf{q}/kq = 3\cos\theta$$

in agreement with Eqs. (5.2.71). These are the only two situations in which reduction to the form of Eqs. (5.2.71) is always possible. In the case of $l = 2$, for instance, the five spherical harmonics are divided between Γ_{12} and $\Gamma_{25'}$ representations. If $\delta_{12} + \rho_{12} = \delta_{25'} + \rho_{25'}$, the kubic harmonics from both representations can be combined to yield $5P_2(\cos\theta)$. Obviously, the condition $\delta_{12} + \rho_{12} = \delta_{25'} + \rho_{25'}$ will be satisfied only for restricted potentials.

Next we should comment on the presence of an additional contribution to the phase ρ_s in Eqs. (5.2.78). This quantity arose from the possibility that Λ_s might be complex. We could have defined our phases so that ρ did not appear explicitly in (5.2.78), but then the relations between the change in the density of states and the phase shift such as Eq. (5.2.59) would not hold. Although no theorem seems to exist which would require Λ_s to be real, these are some important cases in which this happens. For example, if the matrix U_s is one-dimensional, $P_s = 1$, and Λ_s is real. A less trivial example concerns spin wave impurity scattering in the Heisenberg model of a ferromagnet (Callaway and Boyd, 1964; Callaway, 1967).

5.2.6 *The Koster–Slater Model*

After this long and rather formal discussion we will turn to the simplest example, which is furnished by the Koster–Slater (1954a, b) model. In this, we consider only a single band (so we will drop the band index) and a single nonzero matrix element of the impurity potential.

$$(\mu \mid U \mid \nu) = V_0\, \delta_{\mu 0}\, \delta_{\nu 0}. \tag{5.2.79}$$

This is an oversimplification of realistic impurity potentials but it does furnish a very instructive mathematical model. In this case, only one element of the Green's function is to be considered: that with $R_\mu = R_\nu = 0$. From Eq. (5.2.31), this is

$$\mathcal{G}(0) = [\Omega/(2\pi)^3] \int d^3k/[E^+ - E(\mathbf{k})]. \tag{5.2.80}$$

This may be expressed in an interesting form if we introduce the density of states $G(E)$ for the perfect crystal. This is [see Eq. (5.2.53)]

$$G(E) = [\Omega/(2\pi)^3] \int d^3k\, \delta[E - E(\mathbf{k})]. \tag{5.2.81}$$

Equation (5.2.81) is substituted into (5.2.80), which becomes

$$\mathcal{G}(0) = \int dE'\, G(E')/(E^+ - E'). \tag{5.2.82}$$

This can be simplified with the aid of the identity (5.2.55)

$$\mathcal{G}(0) = I(E) - i\pi G(E) \tag{5.2.83}$$

where

$$I(E) = P \int_{-\infty}^{\infty} dE'\, G(E')/(E - E') \tag{5.2.84}$$

and $I(E)$ and $G(E)$ are difficult functions to compute for realistic bands. However, their general behavior can be described without a detailed calculation. Subsequently we will show numerical results for a specific assumed $E(\mathbf{k})$. The density of states will be finite and nonzero outside of a finite interval on the energy axis, from E_0 to E_m, say. Only when E in (5.2.80) is in this region ($E_0 < E < E_m$) will $\mathcal{G}$ have a nonvanishing imaginary part. However, $I(E)$ exists and is continuous for all values of E. For $E < E_0$, $I(E)$ is negative, as may be seen immediately from (5.2.84), and decreases in magnitude as the energy decreases. Conversely, for $E > E_m$, $I(E)$ is positive and decreases with increasing energy. Since I must be continuous it will have to have a minimum, a zero, and a maximum in the region $E_0 \leq E \leq E_m$. This behavior is shown in Fig. 5.2.1 for the case of a tight binding s band in a simple cubic lattice,

$$E(\mathbf{k}) = E_0 + 2E_1(\cos k_xa + \cos k_ya + \cos k_za). \tag{5.2.85}$$

It was shown by Koster and Slater (1954b) that it is possible to write for such a band

$$I(E') = \int_0^{\infty} J_0^3(t) \sin E't\, dt, \qquad G(E') = (1/\pi) \int_0^{\infty} J_0^3(t) \cos E't\, dt \tag{5.2.86}$$

where $E' = (E_0 - E)/2\,|E_1|$ and J_0 is an ordinary Bessel function of zero order.

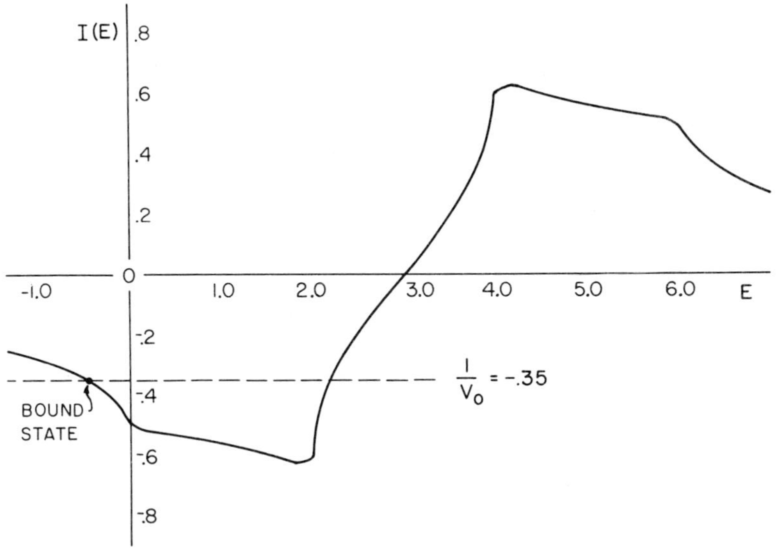

Fig. 5.2.1. The function $I(E)$ defined according to Eq. (5.2.84) is shown for the band described by Eq. (5.2.85). The parameters of Eq. (5.2.85) have been chosen to be $E_0 = -3.0$, $E_1 = -0.5$. The intersection of the dashed line with $I(E)$ illustrates the determination of the energy of a bound state below the band according to Eq. (5.2.91) in the case $1/V_0 = -0.35$.

With the use of (5.2.84), we have

$$D(E) = 1 - V_0 I(E) + i\pi V_0 G(E). \tag{5.2.87}$$

Thus from (5.2.51),

$$\tan \delta = -\pi V_0 G(E)/[1 - V_0 I(E)]. \tag{5.2.88}$$

Since $P = 1$, and we have to consider only $R_\mu = R_\nu = 0$, we obtain a t matrix

$$t(E) = V_0[1 - \mathcal{G}(0) V_0] = [-1/\pi G(E)]e^{i\delta} \sin \delta. \tag{5.2.89}$$

Near the bottom of the band, this gives a scattering amplitude

$$f = (1/k)e^{i\delta} \sin \delta. \tag{5.2.90}$$

Only the completely symmetric (Γ_1) representation contributes to the scattering. The scattering amplitude for this case, which is essentially s wave scattering, is independent of angle and depends only on energy.

The condition for the occurence of a bound state of energy E_b, outside the band, is

$$I(E_b) = 1/V_0. \tag{5.2.91}$$

If this equation is satisfied for an energy E_0 inside the band, a resonance may occur. The width of the resonance is

$$\Gamma = -2\pi G(E_0)/I'(E_0). \tag{5.2.92}$$

A more accurate treatment will yield a shift of the resonance away from E_0, and a slightly more complicated expression for the width (see Callaway, 1964).

Let us consider the behavior of the phase shift as a function of the potential strength V_0. We suppose that V_0 is negative, corresponding to an attractive potential. (In the case of positive V_0, the same remarks apply if we refer to the top instead of the bottom of the band.)

(A) Suppose V_0 is sufficiently small in magnitude so that $1/V_0 \neq I(E)$ for any E. This is the weak potential limit, and there is neither a bound state nor a resonance. The phase shift δ starts at 0° at the bottom of the band, rises to a maximum value (but less than 90°) and decays to 0° again at the top of the band. If V_0 is sufficiently small so that only first-order terms in V_0 need be considered, we can expand (also assume a parabolic band)

$$\tan \delta_0 \approx \delta_0 = -\pi V_0 G(E) = -V_0 \Omega k/4\pi. \tag{5.2.93}$$

In this limit

$$\langle \mathbf{k} \mid t \mid \mathbf{q} \rangle = V_0(\Omega/8\pi^3) \qquad \text{and} \qquad f(\mathbf{k}, \mathbf{q}) = V_0\Omega/4\pi\gamma = a \tag{5.2.94}$$

where a is the scattering length. These are the Born approximation results, which could have been obtained directly by considering just the first term in the Born series for the t matrix, that is, $t = U$.

(B) Suppose a resonance occurs. Note that the equation $1/V_0 = I(E_R)$ must have two solutions or none [unless we have a band of such a pathological shape that $I(E)$ has more than one minimum. This possibility will be neglected]. If the solution of lowest energy occurs within the band, there may be a resonance close by in energy. The higher energy solution corresponds to a depletion in the density of states [since $I'(E)$ must be positive there] and usually will not correspond to a resonance. The phase shift starts at zero at the bottom of the band, passes through $\pi/2$ at the resonance, reaches a maximum (less than π), and decreases. Close to the resonance, Eq. (5.2.62) will hold and we obtain a cross section

$$d\sigma/d\Omega = |f|^2 = (1/4k^2)\{\Gamma^2/[(E - E_R)^2 + \Gamma^2/4]\}. \tag{5.2.95}$$

This is of standard Breit–Wigner form. Such resonant behavior is usually significant only if Γ is small, which generally means at low enough energy so that $G(E)$ is small. Whether or not the cross section actually has a maximum within the band depends on the size of Γ, which will (for *s* wave scattering at low energies) generally be proportional to k itself. The phase shift falls through $\pi/2$ at the second intersection of $1/V_0$ with $I(E)$ and goes to zero at the top of the band. Usually, the cross section does not show important structure where the phase shift falls through $\pi/2$.

(C) Suppose a bound state exists below the band. Then the first intersection of $1/V_0$ with $I(E)$ has already occurred before the band begins. The quantity $1 - V_0 I$ is negative at the bottom of the band. This situation is illustrated in Fig. 5.2.1. The phase shift starts out at π and decreases as a function of energy. It falls through $\pi/2$ and goes to zero at the top of the band. This behavior is consistent with Eq. (5.2.64). Only one bound state can exist for this potential since dI/dE is always negative below the band. Also, if a bound state exists in this model, a resonance is not to be expected for usual band shapes, since this would require $I(E)$ to have two minima.

The transition between these forms of behavior as $|V_0|$ increases appears to be a smooth one; that is, for small $|V_0|$, we have a gentle, nonresonant maximum in the cross section—then for larger $|V_0|$ a wide resonance. As $|V_0|$ increases still further, the point of first intersection moves to lower energies, giving a narrower resonance because of the decrease of $G(E)$ with decreasing energy. Finally, the intersection passes out of the band, giving a bound state, i.e., a resonance of zero width. The idea of the smoothness of the transition at which a bound state is formed with increasing $|V_0|$ has been given precise meaning by Kohn and Majumdar (1965) who have shown that if we replace V_0 by λV_0, where λ is varied continuously, the density matrix of a system of noninteracting Fermions is an analytic function of λ even at the value of λ at which a bound state appears.

5.3 Impurities with Long Range Potentials

If the potential produced by the impurity is of long range, the techniques of the preceding section are not convenient. This situation is best treated in the effective mass representation described in Section 5.1. The case of a long range potential is of particular interest with reference to semiconductors. Consider, for example, an arsenic impurity atom in germanium. The additional electron furnished by the arsenic atom experiences a long range Coulomb potential produced by an As^+ ion, and screened by the dielectric constant κ of the host crystal. A spectrum of localized energy levels exists within the band gap which resembles that of a hydrogen atom; the energy scale being reduced by the factor (m^*/κ^2) (m^* being the effective mass). In the case of a bound hole, as would be produced by a Ga^- impurity in germanium, the situation is qualitatively similar in many respect, but complications are produced by the degenerate nature of the valence band structure.

5.3.1 *The Effective Mass Equation*

The theory of such states will be presented in this section according to the development of Luttinger and Kohn (1955). This is a one-particle approach, but it can be extended and justified from a many-body point of view, to which we will return in Section 8.5 (Kohn, 1957, 1958; Sham, 1966). Since the Coulomb potential is long range, its Fourier transform connects mainly states which are close in **k** space. This argument provides the principal motivation for the success of the effective mass representation which is most useful when attention can be restricted to states which are in the neighborhood of some reference point $\mathbf{k}_0$ in the Brillouin zone.

We begin with Eq. (5.1.18) which gives an equation for the expansion coefficients $A_n(\mathbf{k})$ for the wave function in the effective mass representation. For a stationary state of energy E, this is

$$[E_n + (\hbar^2/2m)(\mathbf{k}^2 - \mathbf{k}_0^2) - E]A_n(\mathbf{k}) + (\hbar/m)(\mathbf{k} - \mathbf{k}_0)\cdot\sum_l \mathbf{p}_{nl}(\mathbf{k}_0)A_l(\mathbf{k})$$

$$+ \sum_l \int d^3q[n\mathbf{k} \mid U \mid l\mathbf{q}]A_l(\mathbf{q}) = 0. \tag{5.3.1}$$

The quantities which appear in this equation are defined in Section 5.1.2. Consider the impurity potential U. The matrix element is

$$[n\mathbf{k} \mid U \mid l\mathbf{q}] = \int \exp[i(\mathbf{q} - \mathbf{k})\cdot\mathbf{r}]U(\mathbf{r})u_n^*(\mathbf{k}_0, \mathbf{r})u_l(\mathbf{k}_0, \mathbf{r})\, d^3r. \tag{5.3.2}$$

The product of the cell periodic functions, u_n^* and u_l, is periodic in the crystal, and thus can be expanded in plane waves

$$u_n^*(\mathbf{k}_0, \mathbf{r})u_l(\mathbf{k}_0, \mathbf{r}) = \sum_s B_{nl}(\mathbf{K}_s)\exp(i\mathbf{K}_s\cdot\mathbf{r}) \tag{5.3.3}$$

where

$$B_{nl}(\mathbf{K}_s) = (1/\Omega)\int_{\text{cell}} u_n^*(\mathbf{k}_0, \mathbf{r})u_l(\mathbf{k}_0, \mathbf{r})\exp(-i\mathbf{K}_s\cdot\mathbf{r})\, d^3r. \tag{5.3.4}$$

In particular, we note that on account of the orthonormality of the u_n,

Eq. (4.1.8),

$$B_{nl}(0) = [1/(2\pi)^3]\,\delta_{nl}. \tag{5.3.5}$$

Equation (5.3.2) becomes

$$[n\mathbf{k} \mid U \mid l\mathbf{q}] = \sum_s B_{nl}(\mathbf{K}_s) \int \exp[i(\mathbf{q} - \mathbf{k} + \mathbf{K}_s)\cdot\mathbf{r}]U(\mathbf{r})\, d^3r. \tag{5.3.6}$$

The fundamental approximation of the present approach is that $U(\mathbf{r})$ is sufficiently slowly varying so that terms in (5.3.6) with $\mathbf{K}_s \neq 0$ can be neglected. To the extent this is true, we may write

$$[n\mathbf{k} \mid U \mid l\mathbf{q}] = \delta_{nl}\, U(\mathbf{k} - \mathbf{q}) \tag{5.3.7}$$

with

$$U(\mathbf{k} - \mathbf{q}) = [1/(2\pi)^3] \int \exp[i(\mathbf{q} - \mathbf{k})\cdot\mathbf{r}]U(\mathbf{r})\, d^3r. \tag{5.3.8}$$

Equation (5.3.1) now has the form

$$[E_n + (\hbar^2/2m)(\mathbf{k}^2 - \mathbf{k}_0^2) - E]A_n(\mathbf{k}) + (\hbar/m)(\mathbf{k} - \mathbf{k}_0)\cdot \sum_l \mathbf{p}_{nl}(\mathbf{k}_0)A_l(\mathbf{k})$$

$$+ \int d^3q\, U(\mathbf{k} - \mathbf{q})A_n(\mathbf{q}) = 0. \tag{5.3.9}$$

The only terms coupling bands in (5.3.9) are those involving the momentum matrix element $\mathbf{p}_{nl}$.

These terms can be removed by a unitary transformation. The success of this procedure depends on the assumption that $\mathbf{k} - \mathbf{k}_0$ is small, so that we may proceed order by order in this quantity. For practical purposes, we must attain reasonable accuracy in low order.

In general, if we have an eigenvalue equation

$$H\psi = E\psi \tag{5.3.10}$$

and make the unitary transformation

$$\psi = e^{iS}\phi \tag{5.3.11}$$

in which S is Hermitian, a transformed eigenvalue equation is obtained,

$$\bar{H}\phi = E\phi, \tag{5.3.12}$$

in which the transformed Hamiltonian $\bar{H}$ is given by

$$\bar{H} = e^{-iS}He^{iS}. \tag{5.3.13}$$

If S is in some sense "small," the exponential functions in (5.3.13) can be expanded, as is done below. Terms of third and higher order in S are neglected:

$$\bar{H} = (1 - iS - \tfrac{1}{2}S^2)H(1 + iS - \tfrac{1}{2}S^2) = H + i[H, S] - \tfrac{1}{2}[[H, S], S] \tag{5.3.14}$$

in which the square bracket denotes the commutator

$$[H, S] = HS - SH.$$

In order to apply this general procedure to (5.3.9), we introduce the abbreviations

$$[n\mathbf{k} \mid H_0 \mid l\mathbf{q}] = [E_n + (\hbar^2/2m)(\mathbf{k}^2 - \mathbf{k}_0^2)]\,\delta_{nl}\,\delta(\mathbf{k} - \mathbf{q}) \tag{5.3.15}$$

$$[n\mathbf{k} \mid H_1 \mid l\mathbf{q}] = (\hbar/m)(\mathbf{k} - \mathbf{k}_0)\cdot\mathbf{p}_{nl}(\mathbf{k}_0)\,\delta(\mathbf{k} - \mathbf{q}). \tag{5.3.16}$$

Equation (5.3.9) has the form (5.3.10) in which

$$H = H_0 + H_1 + U. \tag{5.3.17}$$

The off-diagonal terms of (5.3.9) can be eliminated to first order by a unitary transformation provided S is given by

$$i[H_0, S] = -H_1. \tag{5.3.18}$$

The matrix elements of S in the representation employed are determined to be

$$\begin{aligned} [n\mathbf{k} \mid S \mid l\mathbf{q}] &= i[n\mathbf{k} \mid H_1 \mid l\mathbf{q}]/(E_n - E_l) \\ &= i[(\mathbf{k} - \mathbf{k}_0)\cdot\mathbf{p}_{nl}/m\omega_{nl}]\,\delta(\mathbf{k} - \mathbf{q}) \qquad \text{for} \quad n \neq l \qquad (5.3.19) \\ &= 0 \qquad \text{for} \quad n = l \end{aligned}$$

where $\hbar\omega_{nl} = E_n - E_l$.

The transformed Hamiltonian, including terms of second order in S, can now be formed. Note that

$$i[H_1, S] = [[H_0, S], S]. \tag{5.3.20}$$

We obtain

$$\bar{H} = H_0 + U + i[U, S] - \tfrac{1}{2}[[U, S], S] + \tfrac{1}{2}i[H_1, S]. \tag{5.3.21}$$

The matrix elements of the last term in (5.3.21) can be evaluated readily from (5.3.16) and (5.3.19) since both H_1 and S are diagonal with respect to $\mathbf{k}$ and $\mathbf{q}$.

$$\begin{aligned} &[n\mathbf{k} \mid [H_1, S] \mid l\mathbf{q}] \\ &\quad = \sum_{j\mathbf{p}} \{[n\mathbf{k} \mid H_1 \mid j\mathbf{p}][j\mathbf{p} \mid S \mid l\mathbf{q}] - [n\mathbf{k} \mid S \mid j\mathbf{p}][j\mathbf{p} \mid H_1 \mid l\mathbf{q}]\} \\ &\quad = (i\hbar/m^2)\,\delta(\mathbf{k} - \mathbf{q}) \sum_j (\delta\mathbf{k}\cdot\mathbf{p}_{nj})[(1/\omega_{jl}) + (1/\omega_{jn})]. \qquad (5.3.22) \end{aligned}$$

We have introduced the notation $\delta\mathbf{k} = \mathbf{k} - \mathbf{k}_0$.

The matrix elements of $[U, S]$ are found by a similar procedure (note that U is diagonal in the band index but not in $\mathbf{k}$):

$$[n\mathbf{k} \mid [U, S] \mid l\mathbf{q}]$$

$$= \int d^3p \{[n\mathbf{k} \mid U \mid n\mathbf{p}][n\mathbf{p} \mid S \mid l\mathbf{q}] - [n\mathbf{k} \mid S \mid l\mathbf{p}][l\mathbf{p} \mid U \mid l\mathbf{q}]\}$$

$$= i(\mathbf{q} - \mathbf{k}) \cdot (\mathbf{p}_{nl}(\mathbf{k}_0)/m\omega_{nl}) U(\mathbf{k} - \mathbf{q}) \qquad \text{if} \quad n \neq l \tag{5.3.23}$$

$$= 0 \qquad \text{for} \quad n = l.$$

This term contains interband matrix elements and is of first order in the wave vectors. However, the momentum matrix elements are reduced by a factor which is the ratio of a Fourier coefficient of potential to an interband energy difference. Such terms can contribute only in second order since there are no diagonal matrix elements. In second order, the factor U/ω is effectively squared. We will suppose that such contributions are negligible. This implies we will also neglect the second-order term $[[U, S], S]$.

Our treatment of the second-order terms in (5.3.22) depends on whether the bands of interest at $\mathbf{k}_0$ are degenerate or not. To obtain results valid to second order in $\delta\mathbf{k}$, we can neglect the terms in (5.3.22) which are not diagonal in the band index when we are interested in a nondegenerate band. In the case of a degenerate set, we must retain all terms coupling the degenerate bands.

Consider the simpler, nondegenerate case first. We write $C = e^{-iS}A$ to represent the transformed wave function. We have, instead of (5.3.1),

$$[E_n + (\hbar^2/2m)(\mathbf{k}^2 - \mathbf{k}_0^2) + \hbar \sum_j ((\delta\mathbf{k} \cdot \mathbf{p}_{nj})(\delta\mathbf{k} \cdot \mathbf{p}_{jn})/m^2\omega_{nj}) - E]C_n(\mathbf{k})$$

$$+ \int d^3q \, U(\mathbf{k} - \mathbf{q}) C_n(\mathbf{q}) = 0. \tag{5.3.24}$$

The first three terms in (5.3.24) are equivalent to the expression given in (4.1.21) for the energy as a function of wave vector to second order in $\delta\mathbf{k}$, so that we replace those terms by $E_n(\mathbf{k})$:

$$(E_n(\mathbf{k}) - E)C_n(\mathbf{k}) + \int d^3q \, U(\mathbf{k} - \mathbf{q}) C_n(\mathbf{q}) = 0. \tag{5.3.25}$$

This equation resembles the Schrödinger equation (in momentum space) for one particle in the potential U. There is a significant difference in that the effective mass tensor m^* is involved, rather than the free electron mass m_0. All of the effects of the periodic potential are incorporated in the effective mass.

It is desirable to transform (5.3.25) to a differential equation in ordinary space. This process encounters a formal difficulty because the range of the **k** variable is not infinite, but is limited to the Brillouin zone. We define a function $F_n(\mathbf{r})$ by

$$F_n(\mathbf{r}) = \int \exp(i\,\delta\mathbf{k}\cdot\mathbf{r})C_n(\mathbf{k})\,d^3k. \tag{5.3.26}$$

The integration in (5.3.26) includes only the Brillouin zone. Next, multiply (5.3.25) by $\exp(i\,\delta\mathbf{k}\cdot\mathbf{r})$ and integrate over the zone. Let us consider the term

$$\int E_n(\mathbf{k})C_n(\mathbf{k})\exp(i\,\delta\mathbf{k}\cdot\mathbf{r})\,d^3k \tag{5.3.27}$$

where $E_n(\mathbf{k})$ can be expressed as

$$E_n(\mathbf{k}) = E_n + \sum \alpha_{ij}\,\delta k_i\,\delta k_j \tag{5.3.28}$$

where α_{ij} is the reciprocal effective mass tensor, given implicitly in (5.3.24). We obtain, on substitution into (5.3.27),

$$\begin{aligned} E_nF_n(\mathbf{r}) &+ \sum_{ij} \alpha_{ij} \int \delta k_i\,\delta k_j\,C_n(\mathbf{k})\exp(i\,\delta\mathbf{k}\cdot\mathbf{r})\,d^3k \\ &= E_nF_n(\mathbf{r}) + \sum_{ij} \alpha_{ij}(1/i\;\partial/\partial x_i)(1/i\;\partial/\partial x_j)F_n(\mathbf{r}) \\ &= E_n(1/i\;\nabla)F_n(\mathbf{r}). \end{aligned} \tag{5.3.29}$$

The expression $E_n(1/i\;\nabla)$ means that we are to substitute $1/i\;\partial/\partial x_j$ for δk_j, whenever δk_j appears in the expansion of the energy as a function of **k**.

The last term of (5.3.25) is transformed as

$$\begin{aligned} &\iint d^3q\,d^3k\,U(\mathbf{k}-\mathbf{q})\exp(i\,\delta\mathbf{k}\cdot\mathbf{r})C_n(\mathbf{q}) \\ &\quad = [1/(2\pi)^3] \iiint d^3q\,d^3k\,d^3r'\,U(\mathbf{r}')\exp[i(\mathbf{q}-\mathbf{k})\cdot\mathbf{r}'] \\ &\qquad \times \exp[i(\mathbf{k}-\mathbf{k}_0)\cdot\mathbf{r}]C_n(\mathbf{q}). \end{aligned} \tag{5.3.30}$$

Let us consider the object

$$\Delta(\mathbf{r}-\mathbf{r}') = [1/(2\pi)^3] \int \exp[i\mathbf{k}\cdot(\mathbf{r}-\mathbf{r}')]\,d^3k. \tag{5.3.31}$$

If the integration included all k space we would have simply $\Delta(\mathbf{r}-\mathbf{r}') =$

$\delta(\mathbf{r} - \mathbf{r}')$. When the volume of integration is finite, we still have

$$\int \Delta(\mathbf{r})\, d^3r = 1. \tag{5.3.32}$$

For $|\mathbf{r}|$ large compared to a lattice spacing, $\Delta(\mathbf{r})$ oscillates and decreases as r^{-3}. When the product of Δ and a function $f(\mathbf{r})$, which is slowly varying and extends over many cells, is integrated it is a good approximation to treat Δ as a δ function:

$$\int \Delta(\mathbf{r} - \mathbf{r}')f(\mathbf{r}')\, d^3r' \approx f(\mathbf{r}). \tag{5.3.33}$$

Qualitatively, one can estimate that the error involved in (5.3.33) is of the order of $(a/a_f)^2$ where a is the lattice constant and a_f is some measure of the extent of $f(\mathbf{r})$. This is comparable to the error introduced by the neglect of higher Fourier coefficients of potential and of the neglect of the commutator $[U, S]$. These errors are not important for slowly varying impurity potentials.

These arguments enable us to simplify (5.3.30) as

$$\iint d^3q\, d^3r'\, \Delta(\mathbf{r} - \mathbf{r}') \exp(i\,\delta\mathbf{q}\cdot\mathbf{r})U(\mathbf{r}')C_n(\mathbf{q}) = U(\mathbf{r})F_n(\mathbf{r}). \tag{5.3.34}$$

We now have, in place of (5.3.25),

$$[E_n(1/i\,\nabla) - E]F_n(\mathbf{r}) + U(\mathbf{r})F_n(\mathbf{r}) = 0. \tag{5.3.35}$$

This is the transformed effective mass equation.

In order to investigate the significance of the wave function we return to Eq. (5.1.23). The transformation (5.3.11) implies that

$$\begin{aligned} A_n(\mathbf{k}) &= \sum_l \int [n\mathbf{k} \mid e^{iS} \mid l\mathbf{q}]C_l(\mathbf{q})\, d^3q \\ &= C_n(\mathbf{k}) - (\delta\mathbf{k}/m) \sum_l (\mathbf{p}_{nl}/\omega_{nl})C_l(\mathbf{k}). \end{aligned} \tag{5.3.36}$$

The leading term in A_n is just C_n, and we neglect the first-order correction. To this order

$$\begin{aligned} \Psi(\mathbf{r}) &= \sum_n \int C_n(\mathbf{k}) \exp[i(\mathbf{k} - \mathbf{k}_0)\cdot\mathbf{r}]\psi_n(\mathbf{k}_0, \mathbf{r})\, d^3k \\ &= \sum_n \psi_n(\mathbf{k}_0, \mathbf{r})F_n(\mathbf{r}). \end{aligned} \tag{5.3.37}$$

Equation (5.3.35) does not contain any terms coupling different bands.

Thus if we are interested in the wave function associated with a particular impurity level under the conduction band, for instance $n = c$, we have finally

$$\Psi = \psi_c(\mathbf{k}_0, \mathbf{r})F(\mathbf{r}). \tag{5.3.38}$$

The impurity wave function is then an oscillatory band wave function modulated by a slowly varying, but exponentially decreasing, envelop function $F(\mathbf{r})$.

5.3.2 *Properties of Donor States in Semiconductors*

In the case mentioned at the beginning of this section, that of a single electron loosely bound to a donor ion, the potential $U(\mathbf{r})$ at large distances from the impurity will be that of a single point charge, screened by the dielectric constant κ of the crystal. Of course, the potential will be different near the impurity, and in this region effective mass theory may break down, but we will ignore this effect. Consider first the case of a parabolic band of effective mass m^*. We set the zero of energy to be that of the band minimum. Equation (5.3.35) becomes

$$(-\hbar^2/2m^*)\,\nabla^2 F - (e^2/\kappa r)F = EF. \tag{5.3.39}$$

This is just a simple hydrogenic problem for which the energies are

$$E_n = -m^* e^4/2\kappa^2 n^2 \hbar^2. \tag{5.3.40}$$

If we take $\kappa = 12.5$ and an effective mass $m^* = 0.0665$, as are roughly appropriate for gallium arsenide, we find

$$E_n = -(0.00579/n^2) \quad \text{eV}. \tag{5.3.41a}$$

The effective Bohr radius for the lowest orbit is $\kappa\hbar^2/m^*e^2$, which is about 100 A in the present case. The orbit includes thousands of cells. The experimental ionization energies for Ge, Si, Se, and S donors in this material are in the range of 0.0061 to 0.0058 eV and the agreement with effective mass theory appears to be quite good (Summers *et al.*, 1970).

In the case of gallium arsenide, the lowest conduction band minimum is at the center of the zone, and the constant energy surfaces are spherical, as is assumed in deriving (5.3.40). The situation in germanium is different. If we try to apply (5.3.40) with $\kappa = 16$, $m^* = 0.12$, as are roughly appropriate, we find

$$E_n = -(0.0064/n^2) \quad \text{eV}. \tag{5.3.41b}$$

Experimental ionization energies for groups V donors in germanium are slightly less than twice as large (in magnitude) about 0.010 eV.

In order to improve our description we must take into account the anisotropy of the conduction band. When the band extremum of interest is not located at $k = 0$, there will generally be more than one extremum of the same energy—as many as there are vectors in the star of $\mathbf{k}$. In the effective mass approximation there will be a set of impurity states associated with each minimum, and corresponding states associated with different minima will be degenerate. For example, in silicon, there are six equivalent conduction band minima along the 100 axis, near the face centers X. Thus the lowest impurity state would be sixfold degenerate in the effective mass approximation.

This degeneracy is removed in part when we go beyond the effective mass approximation. As was pointed out in the previous section, the states associated with an impurity must be classified according to the irreducible representations of the point group, if the potential has point group symmetry. Just as in the crystal field problem examined in Chapter 3, Part A, a splitting of degenerate states is to be expected. In the case of silicon, the relevant point group is T_d, the tetrahedral group. The six degenerate effective mass wave functions form the basis for a reducible representation of T_d. This can be decomposed into a nondegenerate state, a doubly degenerate state, and a triply degenerate state (Kohn and Luttinger, 1955). The wave function for these states is, in first approximation, a linear combination of functions of the form (5.3.38) associated with the degenerate minima; these combinations can be constructed according to the procedures discussed in Chapter 3, Part A. It is natural to expect the symmetric (nondegenerate) combination to have the lowest energy; and this is found to be the case. Since this state has a wave function of large amplitude at the donor site, the energy of this state should be depressed somewhat below the position predicted in the effective mass approximation. The separation of this state from its position according to the effective mass procedure is known as the *valley-orbit splitting* or the *chemical splitting* (Aggarwal and Ramdas, 1965). The triply degenerate state has the next lowest energy, and the doubly degenerate state is highest (in the case of P, As, and Sb donors in Si).

Let us consider the effective mass equation for shallow donors in germanium. Valley-orbit splitting effects will be ignored from this point. The lowest conduction band minima in germanium are located at the centers of hexagonal faces (L point in the Brillouin zone; refer to Fig. 5.3.1). There are four inequivalent points of this type, and thus four minima. The surfaces of constant energy for energies slightly above the minimum are needle shaped ellipsoids of revolution with the long axis being that (1, 1, 1) type axis which goes through the minimum. The band structure can be specified in terms of the transverse and longditudinal effective mass ratios: $m_t = 0.08152 \pm 0.00008$; $m_l = 1.588 \pm 0.005$ (Levinger and Frankl, 1961).

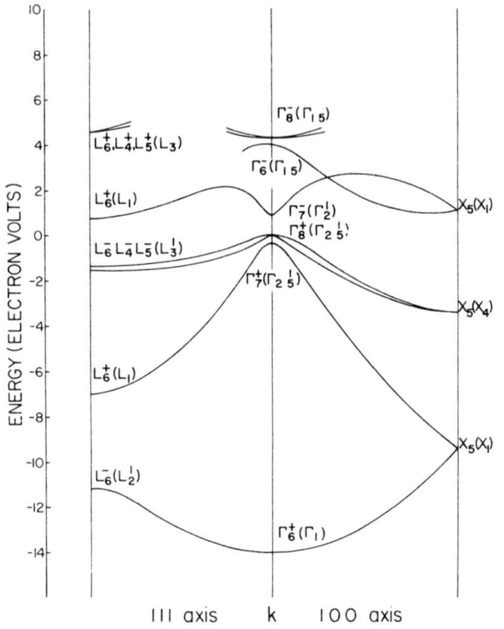

Fig. 5.3.1. Energy bands in germanium are shown along the (100) and (111) axes with spin–orbit coupling included. Representations are labeled in the notation appropriate to the double groups; the labels in parentheses indicate the single group designation which would apply if spin–orbit coupling were neglected. This band structure is a composite of experimental and theoretical results.

Let the x axis be along the relevant (1, 1, 1) axis. We obtain in place of (5.3.39)

$$[(-\hbar^2/2m_l)\,\partial^2/\partial x^2 - (\hbar^2/2m_t)(\partial^2/\partial y^2 + \partial^2/\partial z^2)]F - (e^2/\kappa r)F = EF. \tag{5.3.42}$$

An exact solution to this equation is not known, but a variational calculation can be performed using a trial function for the lowest state of the form

$$F = (a^2b/\pi)^{1/2} \exp\{-[a(y^2 + z^2) + bx^2]^{1/2}\}. \tag{5.3.43}$$

This appears to be a reasonable generalization of the usual 1s hydrogenic function, and would be exact in the limit of a parabolic band. The parameters a and b are to be varied. The variational calculation with this function has been performed by several authors (Kittel and Mitchell, 1954; Lampert,

1955; Luttinger and Kohn, 1955). The ground state energy is -0.0090 eV, which is in rather better agreement with experiment.

5.3.3 *Degenerate Bands: Acceptor States*

Let us now consider what happens if the band of interest is degenerate at $\mathbf{k}_0$. This situation is relevant to the discussion of acceptors in germanium or silicon, since in these materials, the valence band maximum at $k = 0$ is the fourfold degenerate Γ_8^+ state (including spin–orbit coupling). Slightly below this state is the Γ_7^+ state, split from the original sixfold degenerate $\Gamma_{25'}$ by spin–orbit effects (see Fig. 5.3.1).

Equation (5.3.9) is still valid in this case. We may suppose that there are no matrix elements of momentum between members of the degenerate set (the degeneracy is removed in second order) and that the reference point $\mathbf{k}_0$ is the origin. These correspond to the actual situation. We reserve the indices l and l' to indicate members of the degenerate set of bands, and denote the remaining bands by an index m. Equation (5.3.9) becomes

$$[E_l + (\hbar^2\mathbf{k}^2/2m) - E]A_l(\mathbf{k}) + (\hbar\mathbf{k}/m)\cdot\sum_m \mathbf{p}_{lm}A_m(\mathbf{k}) + \int d^3q\, U(\mathbf{k}-\mathbf{q})A_l(\mathbf{q}) = 0.$$

The momentum matrix elements are eliminated to first order in $\mathbf{k}$ by a unitary transformation as before. The elements of S are still given by (5.3.19). The difference occurs from the fact that the A_l are all of the same order of magnitude so that we must retain the terms in (5.3.22) which connect them. We have, instead of (5.3.24), a set of equations of the form

$$[E_l + (\hbar^2\mathbf{k}^2/2m) - E]C_l + \hbar\sum_{l'}[\sum_m (\mathbf{k}\cdot\mathbf{p}_{lm})(\mathbf{k}\cdot\mathbf{p}_{ml'})/m^2\omega_{lm}]C_{l'} + \int d^3q\, U(\mathbf{k}-\mathbf{q})C_l(\mathbf{q}) = 0. \quad (5.3.44)$$

There is one such equation for each value of l. If the perturbing potential U were zero, the same determinantal equation would be obtained for the energies as in Section 4.1.4, Part A. If we introduce rectangular components of $\mathbf{k}$: k_α and k_β, and take the zero of energy at E_l $(= E_{l'})$, the equations may be abbreviated as

$$\sum_{l'}\sum_{\alpha\beta}[D_{ll',\alpha\beta}k_\alpha k_\beta - E\,\delta_{ll'}]C_{l'} + \int d^3q\, U(\mathbf{k}-\mathbf{q})C_l(\mathbf{q}) = 0. \quad (5.3.45)$$

In (5.3.45), the coefficient $D_{ll',\alpha\beta}$ stands for

$$D_{ll',\alpha\beta} = (\hbar^2/2m)\,\delta_{ll'}\,\delta_{\alpha\beta} + \hbar \sum_m (p^\alpha{}_{lm}p^\beta{}_{ml'}/m^2\omega_{lm}) \qquad (5.3.46)$$

and $p^\alpha{}_{lm}$ is the αth rectangular component of the vector matrix element $\mathbf{p}_{lm}$. The quantities $D_{ll'\alpha\beta}$ may be determined from a band calculation or they may be inferred from experiments, such as cyclotron resonance, which determine the band shape near $\mathbf{k} = 0$. The transformation back to $\mathbf{r}$ space proceeds as in (5.3.26). A set of coupled differential equations (as many as there are degenerate bands) results, which replace the single equation (5.3.35):

$$\sum_{l'} [\sum_{\alpha\beta} D_{ll',\alpha\beta}(1/i\,\partial/\partial x_\alpha)\,(1/i\,\partial/\partial_\beta) + U(\mathbf{r})\,\delta_{ll'}]F_{l'}(\mathbf{r}) = EF_l(\mathbf{r}). \qquad (5.3.47)$$

These equations generalize effective mass theory to degenerate bands. The leading term in the wave function is, in analogy with (5.3.38),

$$\Psi = \sum_l \psi_l(0, \mathbf{r})F_l(\mathbf{r}). \qquad (5.3.48)$$

The solution of Eq. (5.3.47) is quite difficult, even when approximation techniques are used. Schechter (1962) applied a variational procedure to obtain the ionization energy and low lying excited states of acceptors in germanium and silicon. He found, in the case of germanium, an ionization energy of 0.0089 eV in fair agreement with the experimental values which range from 0.0102 to 0.0112 eV, depending on the impurity.

Effective mass theory predicts that the energy levels of the localized electron or hole should be independent of the specific donor or acceptor, depending only on the valence. This is not observed. Discrepancies of this type, and the generally lower values of observed ionization energies may be attributed to the departure of the potential near an impurity from the simple form $(-e^2/\kappa r)$ previously used. Obviously, screening of the impurity potential by the dielectric constant of the host crystal cannot be an accurate approximation close to the impurity. The corrections of this sort are smaller for excited states of the excess electron or hole (for p states, the region where the perturbing potential is strong will be avoided) than for the ground state. The energies of the excited p states of group V donors in germanium and silicon are nearly independent of the specific donor. Agreement between effective mass theory and experimental (Reuszer and Fisher, 1964; Faulkner, 1969) is quite good for such states. The wave function of this theory can be used to study optical transitions, and with the inclusion of certain corrections, the hyperfine structure observed in spin resonance experiments (Kohn and Luttinger, 1955).

5.4 Localized Moments

The discussion of impurity problems in the preceding two sections has considered the impurity as the source of a fixed scattering potential. In the case of a substitutional atom, this potential can be considered in a first approximation to be the difference between the electrostatic potential of the impurity atom and that of an atom of the perfect crystal host: There are complications which have been ignored up to this point due to exchange interactions and the readjustment of the charge distribution due to electron scattering by the impurity. Such effects can lead to observable phenomena of great interest.

Suppose that a transition metal atom (iron, say) is present as an impurity in copper. The impurity may (and in this example does) acquire a local magnetic moment. In some cases, the local moments may be quite large; in other, apparently similar cases (Ni in Cu), no local moment is produced. The explanation of the occurrence of such local moments has turned out to be quite interesting, and requires extension of our previous discussions. Connections have been made with other peculiar phenomena observed in transition metal alloys, such as the low-temperature resistance minimum (see Section 5.6).

Theoretical investigation of the local moment problem is based on the fundamental work of Anderson (1961) and Wolff (1961). The subject has been reviewed by Bailyn (1966), Abrikosov (1969), Kondo (1969), and Heeger (1969). Anderson considers the impurity as supplying a localized extra orbital to the Bloch functions of the metal. The eigenfunctions of the one-electron Hamiltonian include both the band states of the metal and the localized state associated with the impurity. Coupling is provided by a (one-electron) s–d mixing interaction which has the effect of broadening the previously sharp localized state. An approximation to the Coulomb interaction between electrons in the localized state of the form $Un\uparrow n\downarrow$ (where $n\uparrow$ is the number of electrons of $\uparrow$ spin in that state) is also incorporated (see Section 3.4, Part A). This term favors the occupancy of the localized state by electrons of a single spin, and thus provides a tendency toward the formation of a local moment.

Wolff considers the interaction between the electron and the impurity atom as a scattering problem of the sort described in Section 5.2. The localized state appears as a scattering resonance. The scattering potential is allowed to depend on the electron distribution at the impurity through Hartree–Fock theory and the exchange interactions so included lead to local moment formation. The most important matrix elements of this interaction are of the form $Un\uparrow n\downarrow$ considered by Anderson. Our discussion here will follow that that of Wolff, since this makes use of the scattering theory we have developed.

5.4.1 *The Wolff Model*

We consider a single-band Slater–Koster problem, in which the impurity is located at the origin. The impurity potential, which is localized in the Wannier function basis, is given by (5.2.79) except that we allow the constant V_0, the potential strength, to depend on spin. It was shown in Section 5.2 that the condition for the existence of a resonance in the band is

$$\mathrm{Re}[D(E)] = 0 \tag{5.2.60'}$$

where

$$D(E) = 1 - V_0 I(E) + i\pi V_0 G(E). \tag{5.2.87'}$$

In these equations $G(E)$ is the density of states in the band and

$$I(E) = \mathrm{P}\int G(E')/(E - E')\, dE'. \tag{5.2.84'}$$

The Bloch functions of the system are modified by the scattering. The scattered wave functions for a state of spin σ are expanded in Wannier functions associated with that particular spin

$$\Psi_\sigma(\mathbf{q}, \mathbf{r}) = \sum_\mu B_\sigma(\mathbf{q}, \mathbf{R}_\mu) a_\sigma(\mathbf{r} - \mathbf{R}_\mu) \tag{5.4.1}$$

where $\Psi_\sigma(\mathbf{q}, \mathbf{r})$ is the state which evolves from the perfect crystal Bloch state $\psi_\sigma(\mathbf{q}, \mathbf{r})$ through the scattering process. The coefficients $B_\sigma(\mathbf{q}, \mathbf{R}_\mu)$ are given by (5.2.30), which is in this case

$$B_\sigma(\mathbf{q}, \mathbf{R}_\mu) = [\Omega^{1/2}/(2\pi)^{3/2}][\exp(i\mathbf{q}\cdot\mathbf{R}_\mu) + [1/D_\sigma(E)]V_{0,\sigma}\mathcal{G}(\mathbf{R}_\mu)] \tag{5.4.2}$$

in which $\mathcal{G}(\mathbf{R}_\mu)$ is the scattering Green's function given by (5.2.31), and $V_{0,\sigma}$ is the potential strength for particles of spin σ:

$$\mathcal{G}(\mathbf{R}_\mu) = [\Omega/(2\pi)^3]\int d^3q\, [\exp(i\mathbf{q}\cdot\mathbf{R}_\mu)/(E^+ - E(\mathbf{q}))]. \tag{5.2.31'}$$

We note that since $E^+ = E + i\epsilon$

$$\mathcal{G}(0) = I(E) - i\pi G(E).$$

We will separate $V_{0,\sigma}$ into two parts: an ordinary potential V_0 which is independent of spin and a specifically spin-dependent "exchange potential" δV_σ

$$V_{0,\sigma} = V_0 + \delta V_\sigma. \tag{5.4.3}$$

These quantities are to be computed using the Hartree–Fock method. The action of the nonlocal Hartree–Fock "potential" on a Wannier function associated with spin σ is defined by

$$V_{\mathrm{HF}} a_\sigma(\mathbf{r} - \mathbf{R}_\mu) = \sum_{\mathbf{q}s\,(\mathrm{occ})} \{ \int | \Psi_s(\mathbf{q}, \mathbf{r}') |^2 (e^2/ | \mathbf{r} - \mathbf{r}' |)\, d^3r'\, a_\sigma(\mathbf{r} - \mathbf{R}_\mu)$$

$$- \delta_{\sigma s} \int \Psi_s^*(\mathbf{q}, \mathbf{r}') (e^2/ | \mathbf{r} - \mathbf{r}' |)$$

$$\times a_\sigma(\mathbf{r}' - \mathbf{R}_\mu)\, d^3r'\, \Psi_s(\mathbf{q}, \mathbf{r}) \}; \tag{5.4.4}$$

$$(\nu\sigma' \mid V_{\mathrm{HF}} \mid \mu\sigma) = \int a_{\sigma'}^*(\mathbf{r} - \mathbf{R}_\nu) V_{\mathrm{HF}} a_\sigma(\mathbf{r} - \mathbf{R}_\mu)\, d^3r. \tag{5.4.5}$$

It is easily verified that the matrix element is diagonal in the spin index

$$(\nu\sigma' \mid V_{\mathrm{HF}} \mid \mu\sigma) = \delta_{\sigma\sigma'} (\nu\sigma \mid V_{\mathrm{HF}} \mid \mu\sigma)$$

$$= \delta_{\sigma\sigma'} \sum_{\mathbf{q}s\,(\mathrm{occ})} \sum_{\rho\tau} B_s^*(\mathbf{q}, \mathbf{R}_\rho) B_s(\mathbf{q}, \mathbf{R}_\tau)$$

$$\times \{ \int a_\sigma^*(\mathbf{r} - \mathbf{R}_\nu) a_s^*(\mathbf{r}' - \mathbf{R}_\rho) (e^2/ | \mathbf{r} - \mathbf{r}' |)$$

$$\times a_s(\mathbf{r}' - \mathbf{R}_\tau) a_\sigma(\mathbf{r} - \mathbf{R}_\mu)\, d^3r\, d^3r'$$

$$- \delta_{s\sigma} \int a_\sigma^*(\mathbf{r} - \mathbf{R}_\nu) a_s^*(\mathbf{r}' - \mathbf{R}_\rho) (e^2/ | \mathbf{r} - \mathbf{r}' |)$$

$$\times a_\sigma(\mathbf{r}' - \mathbf{R}_\mu) a_s(\mathbf{r} - \mathbf{R}_\tau)\, d^3r\, d^3r' \}. \tag{5.4.6}$$

The largest matrix element will be that in which all the Wannier functions are on the same site, and this site should be the origin (the impurity site) since the B are expected to be largest there. The result is a contribution to $V_{0,\sigma}$. If all the Wannier functions are restricted to the same site, the direct and exchange integrals are the same, and are given by

$$U = \iint | a(\mathbf{r}) |^2 | a(\mathbf{r}') |^2 (e^2/| \mathbf{r} - \mathbf{r}' |)\, d^3r\, d^3r'. \tag{5.4.7}$$

This is the same sort of matrix element that appears in Anderson's model. Equation (5.4.6) simplifies to

$$(0\sigma \mid V_{\mathrm{HF}} \mid 0\sigma) = U \sum_{\mathbf{q}s\,(\mathrm{occ})} | B_s(\mathbf{q}, 0) |^2 [1 - \delta_{s\sigma}]. \tag{5.4.8}$$

The contribution of electrons of spin s parallel to σ vanishes, and we have

$$(0\uparrow \mid V_{\text{HF}} \mid 0\uparrow) = U \sum_{\mathbf{q}\,(\text{occ})} \mid B_{\downarrow}(\mathbf{q}, 0) \mid^2. \tag{5.4.9}$$

The quantity $\mid B_\sigma(\mathbf{q}, 0) \mid^2$ is given by

$$\mid B_\sigma(\mathbf{q}, 0) \mid^2 = [\Omega/(2\pi)^3][(1 - V_{0,\sigma}I(E))^2 + \pi^2 V^2_{0,\sigma} G^2(E)]^{-1}. \tag{5.4.10}$$

If there is no localized moment, $B_\uparrow = B_\downarrow = B$ and we may choose $\delta V_\uparrow = \delta V_\downarrow = 0$. Should a state with a spin excess exist, there will be a difference between the values of these quantities for $\uparrow$ and $\downarrow$ spins. A consistency condition must be imposed. This is

$$\delta V_\uparrow = U \sum_{\mathbf{q}\,(\text{occ})} [\mid B_\downarrow(\mathbf{q}, 0) \mid^2 - \mid B(\mathbf{q}, 0) \mid^2] \tag{5.4.11}$$

and similarly for $\delta V_\downarrow$. The normalization of the functions is such that we may replace $\sum_{\mathbf{q}}$ by $\int dE\, G(E)$, and drop the factor $\Omega/(2\pi)^3$ in (5.4.10). Then we have

$$\begin{aligned}\delta V_\uparrow = U \int_{-\infty}^{E_{\mathrm{F}}} & dE\, G(E)\{[(1 - (V_0 + \delta V_\downarrow)I(E))^2 \\ & + \pi^2(V_0 + \delta V_\downarrow)^2 G^2(E)]^{-1} \\ & - [(1 - V_0 I(E))^2 + \pi^2 V_0^2 G^2(E)]^{-1}\}\end{aligned} \tag{5.4.12}$$

$$\begin{aligned}\delta V_\downarrow = U \int_{-\infty}^{E_{\mathrm{F}}} & dE\, G(E)\{[(1 - (V_0 + \delta V_\uparrow)I(E))^2 \\ & + \pi^2(V_0 + \delta V_\uparrow)^2 G^2(E)]^{-1} \\ & - [(1 - V_0 I(E))^2 + \pi^2 V_0^2 G^2(E)]^{-1}\}.\end{aligned} \tag{5.4.13}$$

5.4.2 *The Scattering Resonance*

In general these expressions cannot be evaluated without detailed information concerning the band structure. An exception exists in the case of a narrow resonance, which we will proceed to consider. The condition for a resonance at energy E_0 in this model is given by (5.2.90)

$$I - V_0 I(E_0) = 0.$$

We assume that the "ordinary" part of the scattering potential produces the scattering resonance, and that $\delta V \ll V_0$, which implies that the location of the resonance is not substantially changed. Then $I(E)$ may be expanded near E_0

$$I(E) = 1/V_0 + (E - E_0)I'(E_0) \tag{5.4.14}$$

where I' is the derivative of I. We suppose that the density of states is slowly varying near the resonance, so that $G(E)$ can be replaced by $G(E_0)$. The integrals are elementary when these approximations are made. To simplify the results, we use the assumption that $\delta V/V_0 \ll 1$, as mentioned above. Equation (5.4.12) becomes

$$\delta V_\uparrow = (-U/\pi V_0^2 I')\{\tan^{-1}[2(E_0 - E_F)/\Gamma - \delta V_\downarrow/\pi G V_0^2] - \tan^{-1}[2(E_0 - E_F)/\Gamma]\}, \tag{5.4.15}$$

where Γ is the width of the resonance, which is given by (5.2.92) to be

$$\Gamma = -2\pi G(E_0)/I'(E_0). \tag{5.2.92}$$

A similar result holds for $\delta V_\downarrow$:

$$\delta V_\downarrow = (-U/\pi V_0^2 I')\{\tan^{-1}[2(E_0 - E_F)/\Gamma - \delta V_\uparrow/\pi G V_0^2] - \tan^{-1}[2(E_F - E_0)/\Gamma]\}. \tag{5.4.16}$$

Equations (5.4.15) and (5.4.16) must be solved simultaneously. These equations appear to be quite complicated, but they can be stated in terms of a small number of parameters if we introduce the following abbreviations:

$$x = \delta V_\downarrow/\pi G V_0^2, \qquad y = \delta V_\uparrow/\pi G V_0^2,$$
$$\mathcal{E} = 2(E_F - E_0)/\Gamma, \qquad \xi = U/(\pi^2 V_0^4 G I'). \tag{5.4.17}$$

We will assume that I' is negative, as is necessary to obtain a positive width from (5.2.92). Then

$$x = \xi[\tan^{-1}(\mathcal{E} - y) - \tan^{-1}\mathcal{E}] \tag{5.4.18a}$$

$$y = \xi[\tan^{-1}(\mathcal{E} - x) - \tan^{-1}\mathcal{E}]. \tag{5.4.18b}$$

It is informative to examine these equations graphically. An illustrative example is shown in Fig. 5.4.1, where Eqs. (5.4.18a) and (5.4.18b) are plotted for the case $\mathcal{E} = 0.5$, $\xi = 2.0$. There are three solutions in this case, one of which ($x = y = 0$) occurs for any values of the parameters $\mathcal{E}$ and ξ.

The dependence of these curves on the parameters $\mathcal{E}$ and ξ can be understood from the derivatives dy/dx. Equation (5.4.18a) gives

$$\text{curve (a):} \quad dy/dx = -[1 + (\mathcal{E} - y)^2]/\xi, \tag{5.4.19a}$$

while (5.4.18b) yields

$$\text{curve (b):} \quad dy/dx = -\xi/[1 + (\mathcal{E} - x)^2]. \tag{5.4.19b}$$

Suppose, with reference to Fig. 5.4.1, that the parameter ξ is decreased. Curve (a) acquires a steeper negative slope, while the slope of curve (b)

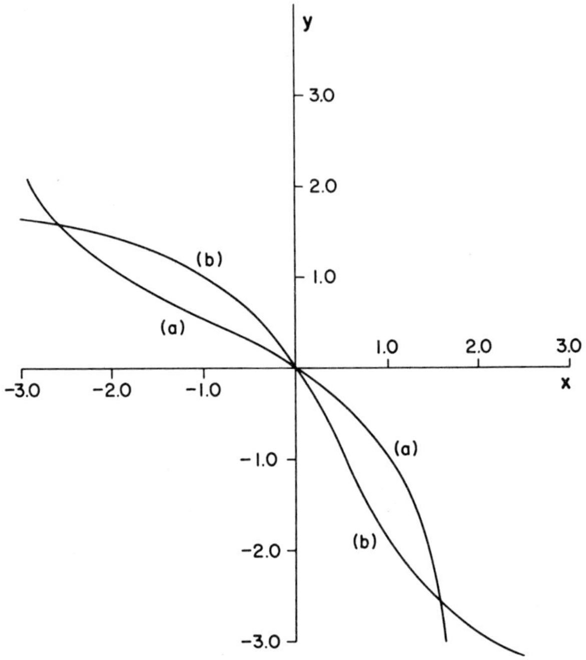

Fig. 5.4.1. Graph of functions defined by Eqs. (5.4.18a) [curve (a)] and (5.4.18b) [curve (b)] with $\xi = 2$ and $\mathcal{E} = 0.5$.

becomes less steep. In other words, the curves tend to uncross, and the points of intersection move toward the origin. It is seen from the figure that in order to obtain a solution (other than at the origin) the slope of (b) must be greater than that of (a) at the origin. Consequently, the critical condition for the existence of solutions is that the derivatives are equal at the origin, or

$$\xi = 1 + \mathcal{E}^2. \tag{5.4.20}$$

Solutions are obtained for values of ξ larger than that given by (5.4.20). The variation with $\mathcal{E}$ may be analyzed in the same way: increasing $\mathcal{E}$ has the same general effect as decreasing ξ, for fixed ξ, $\mathcal{E}$ must be smaller than the critical value determined from (5.4.20).

In the case that ξ is larger (or $\mathcal{E}$ is smaller) than the critical condition given by (5.4.20), three solutions of Eq. (5.4.18b) exist. The two solutions which have $x \neq y$ ($\delta V_\uparrow \neq \delta V_\downarrow$) imply, through (5.4.11) that the density of electrons of spin $\uparrow$ is different from that referring to spin $\downarrow$. This excess spin density is built up in the neighborhood of the impurity, and thus corresponds to a localized moment. The spin polarization does have a long tail since the scattered part of the wave function falls off rather slowly.

The Hartree–Fock equations are derived from the variational principle, as discussed in Section 4.9, Part A. This means that the solutions of these equations correspond to an extreme value of the energy, which may be—when one parameter is considered—either a maximum or a minimum. Let us consider the difference between ↑ and ↓ charge densities as the relevant parameter. It is not plausible that values of this quantity at the ends of its range should correspond to minimum values of the energy, and the solutions are clearly symmetric between ↑ and ↓. We conclude that if three solutions exist, we should have two minima and one maximum, and the maximum should be the solution with $x = y = 0$, or no spin polarization. This solution is therefore unstable, and the solutions with the local moments will occur. Of course, if only the one solution for $x = y = 0$ is allowed by the parameters, this will correspond to the minimum energy.

The properties of localized moments have been studied extensively, most of the investigation being based on the Anderson model. The reader is referred to the references cited at the beginning of this section for further information. Some of the consequences of the existence of these moments in related problems will be considered in Section 5.6. We will see that much more must be said about the subject: the simple Hartree–Fock treatment of the interaction of the electrons omits several essential features.

5.5 Alloys

Considerable effort has been devoted in recent years to the study of the energy levels of disordered alloys and amorphous systems. An introduction is given by Landsberg *et al.* (1969); for reviews see Mott (1967), Mott and Davis (1971). Much of our treatment is based on the work of Lifshitz (1964).

The fundamental problem which arises in the study of such systems is that the crystal potential is not periodic. As a result, the conditions for the validity of Bloch's theorem, Section 3.5, Part A, are not satisfied and it is no longer possible to characterize the electron states by a wave vector **k**. An effect of this type was already encountered in Section 5.2, where we saw that the presence of a defect would cause scattering of an electron from a state of one wave vector to another. Thus the fundamental concepts we have previously employed—the energy band and the Brillouin zone—lose some of their meaning. However, if the disorder is small, if the concentration of impurities is small, and if the scattering from one is weak, we may continue to use our usual notions. The effect of disorder will just be to produce occasional transitions between states of definite **k**. This is the usual point of view in calculations of the electrical resistivity of nearly perfect crystals (see Section 7.2).

We want to consider the problem in a more general way here. Our attention will be focused on the calculation of the density of states, which remains meaningful and important even in systems with random potentials.

5.5.1 *Dilute Alloys*

The case of a dilute alloy can be treated in a rather straightforward way. We will examine this situation first. As a convention, the minority constituent will be called the "impurity." The meaning of the concept "dilute" is that the density of states is expanded in powers of the impurity concentration and only the leading terms are retained. Our considerations will be restricted to two-component systems. To begin, we recall the exact formal result of Section 5.2 [Eq. (5.2.58)], that the change in density of states of a crystal due to the presence of some impurity atoms is given by

$$\Delta N(E) = (-1/\pi\mathfrak{N}) \operatorname{Im} d \ln D(E)/dE \tag{5.2.58'}$$

where there are $\mathfrak{N}$ sites in the crystal and

$$D(E) = \det[I - \mathcal{G}V]. \tag{5.5.1}$$

In the present case I is a unit matrix, V is the matrix representing the defect potential—here this means the difference between the potential energy matrix of the entire crystal containing all impurity atoms and that of the perfect host, and $\mathcal{G}$ is the Greens' function matrix which is formally expressed as

$$\mathcal{G} = 1/(E^+ - H_0) \tag{5.5.2}$$

($E^+ = E + i\epsilon$, ϵ being a positive infinitesimal, and H_0 is the Hamiltonian of the perfect crystal). These equations are valid for any set of basis functions; however, for most purposes, it will be convenient to use the localized orthonormal basis of Wannier functions. In this case, it follows from Section 5.2 that only those sites which are connected by a nonzero matrix element of the defect potential are included in D. (Example: suppose we are concerned with a single band system containing n impurity atoms, which yield a nonzero matrix element contribution to V only for diagonal elements involving the site containing the impurity. Then only an $n \times n$ portion of the matrix $I - \mathcal{G}V$ need be considered.)

For simplicity, let us suppose that the potentials from different atoms do not overlap. Then the terms coupling different defect atoms are of the form

$$-\mathcal{G}(\mathbf{R}_a - \mathbf{R}_b) V(\mathbf{R}_b) \tag{5.5.3}$$

where $\mathbf{R}_a$ and $\mathbf{R}_b$ are defect sites. However, if $|\mathbf{R}_a - \mathbf{R}_b|$ is large, it follows from (5.2.32) that $\mathcal{G}$ falls off like $|\mathbf{R}_a - \mathbf{R}_b|^{-1}$. If the distances between impurity atoms is large, these coupling terms can be ignored. Specifically, if the concentration of impurities is c, then the average distance between impurities is proportional to $c^{-1/3}$, while the probability of close encounters involving pairs of impurities is proportional to c^2; thus for small c, close pairs are highly unlikely.

It follows from these considerations that for low concentrations, the determinantal function for the whole crystal effectively contains just diagonal blocks $D_a(E)$ where $D_a(E)$ is the determinant of $(I - \mathcal{G}V)$ associated with a single atom. Thus

$$D(E) = [D_a(E)]^n \tag{5.5.4}$$

and from (5.2.58), we see that

$$\Delta N(E) = (-c/\pi)\,\mathrm{Im}\,d \ln D_a(E)/dE \equiv \Delta N_1(E) \tag{5.5.5}$$

where $c = n/\mathfrak{N}$. The contributions of single impurities add in this limit. This result was asserted without proof in Section 5.2. We call the result ΔN_1 as it will represent the first-order term in the expansion of the full $\Delta N(E)$ in powers of the concentration.

Let us consider the corrections to this result. It helps to fix our ideas to adopt a more specific model; and to this end we consider a collection of Koster–Slater-type impurities, which were studied in Section 5.2. The determinant $D_a(E)$ is simply

$$D_a(E) = 1 - \mathcal{G}(0)V_0 \tag{5.5.6}$$

where V_0 is the potential strength associated with a single impurity and

$$\mathcal{G}(0) = I(E) - i\pi G(E)$$

with $G(E)$ being the density of states of the perfect crystal, and

$$I(E) = P\int dE'\, G(E')/(E - E'). \tag{5.2.84$'$}$$

We consider expanding the determinantal function for the entire crystal, the leading term being the product of diagonal elements given by (5.5.4). The corrections to this must involve at least two off-diagonal terms. Thus

$$\begin{aligned} D(E) = [D_a(E)]^n &+ [D_a(E)]^{n-2}V_0^2 \sum_{\text{pairs}} \mathcal{G}(\mathbf{R}_a - \mathbf{R}_b)\mathcal{G}(\mathbf{R}_b - \mathbf{R}_a) \\ &+ [D_a(E)]^{n-3}V_0^3 \sum_{\text{triads}} \mathcal{G}(\mathbf{R}_a - \mathbf{R}_b)\mathcal{G}(\mathbf{R}_b - \mathbf{R}_c)\mathcal{G}(\mathbf{R}_c - \mathbf{R}_a) \\ &+ \cdots \end{aligned} \tag{5.5.7}$$

The summation includes only different pairs, triads, etc. Thus, in the second term, there is no contribution from $\mathbf{R}_b = \mathbf{R}_a$. Let us recall that the t matrix for this problem is given by (5.2.89)

$$t = V_0/[1 - \mathcal{G}(0)V_0] = V_0/D_a(E). \tag{5.5.8}$$

Thus we may rewrite (5.5.7) as

$$D(E) = [D_a(E)]^n[1 + t^2 \sum_{\text{pairs}} \mathcal{G}(\mathbf{R}_a - \mathbf{R}_b)\mathcal{G}(\mathbf{R}_b - \mathbf{R}_a) + t^3 \sum_{\text{triads}} \mathcal{G}(\mathbf{R}_a - \mathbf{R}_b)\mathcal{G}(\mathbf{R}_b - \mathbf{R}_c)\mathcal{G}(\mathbf{R}_c - \mathbf{R}_a) + \cdots]. \tag{5.5.9}$$

Only distinct pairs, triads, etc. are included in (5.5.9), and the sites a, b, c, . . . , etc. must all be different. We may now apply Eq. (5.2.58) which yields

$$\Delta N(E) = \Delta N_1(E) - (1/\pi\mathfrak{N}) \operatorname{Im} d \ln[1 + t^2 \sum_{\text{pairs}} \mathcal{G}(\mathbf{R}_a - \mathbf{R}_b) \times \mathcal{G}(\mathbf{R}_b - \mathbf{R}_a) + t^3 \sum_{\text{triads}} \cdots]/dE, \tag{5.5.10}$$

where ΔN_1 is the first-order term we have already obtained in (5.5.5). The second-order term can be estimated if it is supposed that the concentration of impurities is small enough so that the logarithm may be expanded. We then have

$$\begin{aligned}\Delta N(E) &= \Delta N_1(E) + \Delta N_2(E) \\ &= \Delta N_1(E) - (1/\pi\mathfrak{N}) \operatorname{Im} (d/dE) \\ &\quad \times [t^2 \sum_{\text{pairs}} \mathcal{G}(\mathbf{R}_a - \mathbf{R}_b)\mathcal{G}(\mathbf{R}_b - \mathbf{R}_a)].\end{aligned} \tag{5.5.11}$$

Let us consider the sum over pairs in (5.5.11). The expression apparently depends on the actual impurity positions. These are unknown. However, if there is no correlation between impurities, we may replace the actual sum by one in which impurities are uniformly distributed with a concentration c. We proceed as follows, using (5.2.31),

$$\begin{aligned}&(1/\mathfrak{N}) \sum_{\text{pairs}} \mathcal{G}(\mathbf{R}_a - \mathbf{R}_b)\mathcal{G}(\mathbf{R}_b - \mathbf{R}_a) \\ &= [(\Omega^2/(2\pi)^6] \int d^3k\, d^3q\, (1/\mathfrak{N}) \sum_{\text{pairs}} \\ &\quad \times \{\exp[i(\mathbf{k} - \mathbf{q})\cdot(\mathbf{R}_a - \mathbf{R}_b)]/[E^+ - E(\mathbf{k})][E^+ - E(\mathbf{q})]\}.\end{aligned}$$

However,

$$(1/\mathfrak{N}) \sum_{\text{pairs}} \exp[i(\mathbf{k} - \mathbf{q}) \cdot (\mathbf{R}_a - \mathbf{R}_b)] \approx (c^2/2) \sum_{\mu} \exp[i(\mathbf{k} - \mathbf{q}) \cdot \mathbf{R}_\mu]$$

$$= (c^2/2)[(2\pi)^3/\Omega]\, \delta(\mathbf{k} - \mathbf{q}). \tag{5.5.12}$$

We have dropped a term of order $1/N$. A correction should be applied since $\mathbf{R}_\mu = 0$ must be excluded from the sum; however, this will not be considered here. Then we have

$$(1/\mathfrak{N}) \sum_{\text{pairs}} \mathcal{G}(\mathbf{R}_a - \mathbf{R}_b)\mathcal{G}(\mathbf{R}_b - \mathbf{R}_a)$$

$$= (c^2/2)[\Omega/(2\pi)^3] \int d^3k/[E^+ - E(\mathbf{k})]^2$$

$$= (c^2/2) \int G(E')/(E^+ - E')^2\, dE'. \tag{5.5.13}$$

We have introduced the density of states just as in (5.2.81).

Since

$$[E - E' + i\epsilon]^{-1} = P(1/E - E') - i\pi\, \delta(E - E'),$$

we have, on differentiation with respect to E,

$$[E - E' + i\epsilon]^{-2} = -d/dE\, P(1/(E - E')) + i\pi\, d\, \delta(E - E')/dE.$$

Thus, (5.5.13) becomes

$$(-c^2/2)[dI(E)/dE - i\pi\, dG(E)/dE]. \tag{5.5.14}$$

We substitute (5.5.14) into (5.5.11), and find with the use of (5.2.89), that the second-order change in the density of states is given by

$$\begin{aligned} \Delta N_2(E) &= (c^2/2\pi)\, \mathrm{Im}\, (d/dE)\{t^2[dI/dE - i\pi\, dG/dE]\} \\ &= (c^2/2\pi^3)\, (d/dE)\{\sin^2 \delta(E)/G^2(E) \\ &\qquad \times [\sin 2\delta(E)\, dI/dE - \pi \cos 2\delta(E)\, dG/dE]\} \end{aligned} \tag{5.5.15}$$

where δ is the scattering phase shift given through (5.2.88),

$$\tan \delta = -\pi V_0 G(E)/[1 - V_0 I(E)]. \tag{5.2.88}$$

Equation (5.5.15) enables the computation of the second-order correction to the density of states provided that the density of states of the perfect crystal is known. To appreciate the sort of result that is obtained, let us consider the simplest possible sort of density of states

$$G(E) \approx E^{1/2} \quad (0 \le E \le E_m), \qquad G(E) = 0 \quad \text{outside this range.} \tag{5.5.16}$$

We will confine our attention to energies near the bottom of the band. In this region

$$\delta(E) \approx E^{1/2}.$$

A straightforward calculation shows that, for $E > 0$,

$$I(E) \approx -\{2E_m^{1/2} + E^{1/2} \ln[(1 - (E/E_m)^{1/2})/(1 + (E/E_m)^{1/2})]\}. \tag{5.5.17}$$

Thus dI/dE approaches a constant limit as $E \to 0$. The term involving dG/dE dominates, and we find

$$\Delta N(E) \approx E^{-3/2}. \tag{5.5.18}$$

In this limit (5.5.5) yields $\Delta N_1 \approx E^{-1/2}$. Evidently the series diverges at $E \to 0$. This appears to be connected with a shift of the bottom of the band to lower energies. Restrictions can be placed on such shifts in the case of certain potential models, including the one we have been using as will be discussed subsequently.

5.5.2 *Friedel Oscillations*

Let us return to the dilute situation in which corrections to the change in density of states of order c^2 and higher are ignored, and consider the behavior of the charge density around an impurity atom in a metal. We wish to show that this perturbation of the charge density in the system possesses long-range oscillations (Friedel, 1952, 1954, 1958; Kohn and Vosko, 1960). The change in charge density is $\Delta\rho(\mathbf{r})$,

$$\Delta\rho(\mathbf{r}) = \int d^3k \, \Delta\rho(\mathbf{k}, \mathbf{r}) \, \eta[\epsilon_F - \epsilon(\mathbf{k})] \tag{5.5.19}$$

where $\Delta\rho(\mathbf{k}, \mathbf{r})$ is the change in charge density produced by scattering of electrons of wavevector $\mathbf{k}$, and η is a unit step function

$$\eta(x) = 0 \quad \text{if} \quad x < 0$$
$$= 1 \quad \text{if} \quad x > 0,$$

which ensures that we count only occupied states ϵ_F as the Fermi energy, and our discussion refers only to $T = 0°\text{K}$. The quantity $\Delta\rho(\mathbf{k}, \mathbf{r})$ may be written as

$$\Delta\rho(\mathbf{k}, \mathbf{r}) = |\Psi^+(\mathbf{k}, \mathbf{r})|^2 - |\psi(\mathbf{k}, \mathbf{r})|^2 \tag{5.5.20}$$

where Ψ^+ is the scattered wave function and ψ is the Bloch function for the unperturbed system. To avoid unessential complications, we will consider here only a single band, so that no band index appears in (5.5.20). We expand the wavefunctions in (5.5.20) in terms of Wannier functions according to the procedures used in Sections 5.1 and 5.2.

$$\Psi^+(\mathbf{k}, \mathbf{r}) = \sum_\mu B(\mathbf{k}, \mathbf{R}_\mu) a(\mathbf{r} - \mathbf{R}_\mu) \tag{5.5.21a}$$

$$\psi(\mathbf{k}, \mathbf{r}) = \sum_\mu B^{(0)}(\mathbf{k}, \mathbf{R}_\mu) a(\mathbf{r} - \mathbf{R}_\mu), \tag{5.5.21b}$$

with

$$B^{(0)}(\mathbf{k}, \mathbf{R}_\mu) = [\Omega^{1/2}/(2\pi)^{3/2}] \exp(i\mathbf{k}\cdot\mathbf{R}_\mu). \tag{5.5.21c}$$

Thus we must consider

$$\Delta\rho(\mathbf{r}) = \int d^3k\, \eta[\epsilon_F - \epsilon(\mathbf{k})] \sum_{\mu\nu} [B^*(\mathbf{k}, \mathbf{R}_\nu) B(\mathbf{k}, \mathbf{R}_\mu) - B^{(0)*}(\mathbf{k}, \mathbf{R}_\nu) B^{(0)}(\mathbf{k}, \mathbf{R}_\mu)] a^*(\mathbf{r} - \mathbf{R}_\nu) a(\mathbf{r} - \mathbf{R}_\mu). \tag{5.5.22}$$

It is convenient to introduce some additional approximations before proceeding. It is assumed that the Wannier functions are sufficiently localized so that we may neglect products of such functions centered on different sites. Moreover, we will consider only large values of $\mathbf{r}$ (and therefore $\mathbf{R}_\mu$, since the functions are supposed to be well localized) for which the asymptotic form for the B, Eq. (5.2.35) is valid. Finally, we will consider a metal with spherical energy surfaces. Then we have

$$B(\mathbf{k}, \mathbf{R}_\mu) = [\Omega^{1/2}/(2\pi)^{3/2}][\exp(i\mathbf{k}\cdot\mathbf{R}_\mu) + f(\theta) \exp(ikR_\mu)/R_\mu] \tag{5.5.23}$$

where $f(\theta)$ is the scattering amplitude.

We substitute (5.5.23) and (5.5.21c) into (5.5.22) and obtain, in accord with the preceding argument

$$\Delta\rho(r) = [\Omega/(2\pi)^3] \sum_\mu \{| a(\mathbf{r} - \mathbf{R}_\mu) |^2 \int d^3k\, \eta[\epsilon_F - \epsilon(\mathbf{k})] \times [\exp(-i\mathbf{k}\cdot\mathbf{R}_\mu)(\exp(ikR_\mu)/R_\mu) f(\theta) + \text{c.c.} + (1/R_\mu^2) | f(\theta) |^2]\}. \tag{5.5.24}$$

Here, c.c. indicates the complex conjugate of the preceding expression. The integral is considered in spherical coordinates. The angular part contains a term

$$\int \exp(-i\mathbf{k}\cdot\mathbf{R}_\mu) f(\theta)\, d\Omega = 2\pi \int_0^\pi \exp(-ikR\cos\theta) f(\theta) \sin\theta\, d\theta.$$

For large values of R, this integral may be attacked by repeated integration by parts which yields a series in powers of $(1/R)$. In the asymptotic region, we require only the first term of this series which is

$$(2\pi/ikR_\mu)[f(\pi)\exp(ikR_\mu) - f(0)\exp(-ikR_\mu)].$$

The complex conjugate of this term also appears. The resulting expression can be simplified using the optical theorem (5.2.70c)

$$\int |f(\theta)|^2\, d\Omega = (4\pi/k)\,\mathrm{Im}\, f(0).$$

It is also useful to write

$$f(\pi) = |f(\pi)|\, e^{i\tau}$$

where τ is a quantity which can be determined from the phase shifts. Equation (5.5.24) now becomes

$$\Delta\rho(\mathbf{r}) = \Omega/2\pi^2 \sum_\mu \{[|a(\mathbf{r} - \mathbf{R}_\mu)|^2/R_\mu{}^2] \int_0^{k_\mathrm{F}} k\, dk\, |f(\pi)| \sin(2kR_\mu + \tau)\}. \tag{5.5.25}$$

Once again, it is legitimate to approximate the integral by the leading term in an asymptotic expansion obtained by integration by parts. The result is written as

$$\Delta\rho(\mathbf{r}) = C \sum_\mu [|a(\mathbf{r} - \mathbf{R}_\mu)|^2/R_\mu{}^3] \cos(2k_\mathrm{F}R_\mu + \tau). \tag{5.5.26}$$

Here C is a constant which is easily determined from (5.5.25). When the Wannier functions are strongly localized, as has been assumed previously, the only term in the sum which will be appreciable is that for which $\mathbf{R}_\mu$ is as close as possible to $\mathbf{r}$. Thus we have, approximately,

$$\Delta\rho(\mathbf{r}) = C' \cos(2k_\mathrm{F}r + \tau)/r^3. \tag{5.5.27}$$

The perturbed charge density oscillates and decreases only slowly with distance. These oscillations are a consequence of the discontinuous drop in the occupation number of the electron states as the Fermi energy, which leads to a specific upper limit to the integral in (5.5.25). If the occupation probability were a smooth function of k, no long-range oscillations would be found.

The existence of the long range tail of the perturbed charge density is supported by experimental measurements of nuclear magnetic resonance in dilute alloys (Bloembergen and Rowland, 1953; Rowland, 1960). Specifically, the addition of about 4 (atomic) % antimony to copper causes a drop in the intensity of the nuclear magnetic resonance signal (due to copper at 4 MHz) and similar behavior is produced by other elements. The decrease in intensity is attributed to the presence of strong electric field gradients associated with the displaced charge distribution. (No field gradients occur at nuclear sites in pure copper as they are forbidden in a system with cubic symmetry.) The field gradients interact with quadrupole moments of the Cu nuclei and broaden the resonance line.

5.5.3 *The Friedel Sum Rule*

Another characteristic of scattering in metallic systems is the existence of a sum rule for scattering phase shifts derived by Friedel (1952, 1958). Suppose an impurity atom is introduced into a metal. Let the nucleus of the impurity atom have a charge Z units greater than that of the host. The charge distribution of the host will distort so as to screen the long range Coulomb potential at large distances. Charge will accumulate around the impurity in a sufficient amount to balance the excess ionic charge and to produce a system which is electrically neutral. It must therefore be required that the total displaced charge in $\Delta\rho$ equal Z:

$$\int \Delta\rho(\mathbf{r})\, d^3r = Z. \tag{5.5.28}$$

It is clear from the meaning of the quantities involved that (5.5.28) must be equivalent to

$$\int_0^{E_F} \Delta N(E)\, dE = Z \tag{5.5.29}$$

where ΔN is the change in density of states previously discussed and E_F is the Fermi energy. The possible change in the Fermi energy of the system is of order $1/N$ for a single impurity. It is negligible in this case and can still be neglected when the impurity concentration is small. The representation (5.2.59a) of ΔN in terms of scattering phase shifts is used

$$\Delta N(E) = (2/\pi) \sum_s g_s\, d\delta_s/dE$$

where δ_s is the phase shift for representation s and g_s is the degeneracy of that representation. A factor of 2 is included to count both directions of electron spin. Thus Eq. (5.5.29) yields [assuming $\delta_s(0) = 0$]

$$\sum_s g_s\, \delta_s(\epsilon_F) = \pi Z/2. \tag{5.5.30a}$$

If spherical symmetry is a reasonable approximation, $g_s = (2l + 1)$, l being the usual angular momentum, and we have

$$\sum_{s} (2l + 1)\ \delta_l(\epsilon_F) = \pi Z/2. \tag{5.5.30b}$$

These equations, (5.5.30a) or (5.5.30b), state the Friedel sum rule. This sum rule is an important condition which must be satisfied by certain impurity potentials and may be imposed as a self-consistency requirement on such potentials.

5.5.4 *Localized States*

Another feature of the single-impurity problem is the formation of localized states. These result from the zeros of $D(E)$ outside the continuous spectrum. As a first approximation in the many impurity situation we find that if $D_a(E)$ has a zero at $E = E_0$, n states will be obtained at that energy. Interactions between impurities will tend to remove this degeneracy, and cause the distribution of these levels to broaden.

The simplest case to consider is that of two impurities. We will keep to the Slater–Koster model for the potentials. We set up the function $D(E)$ for the case of two impurities a distance R apart. The energies are determined by

$$D(E) = \begin{vmatrix} 1 - V_0 I(E) & -V_0 \mathcal{G}(R) \\ -V_0 \mathcal{G}(R) & 1 - V_0 I(E) \end{vmatrix} = 0. \tag{5.5.31}$$

The Green's function $\mathcal{G}(R)$ is real, since the energies of interest are negative, $E < 0$. We will suppose that the band of interest is parabolic: $E(\mathbf{k}) = \gamma \mathbf{k}^2$. The Green's function can be obtained by setting $k = (E/\gamma)^{1/2} = i\alpha$ in (5.2.32)

$$G(R) = (-\Omega/4\pi\gamma)(e^{-\alpha R}/R). \tag{5.5.32}$$

In addition, let us suppose that the single impurity would have a localized state for $E = E_0$,

$$I - V_0 I(E_0) = 0.$$

The energies will be only slightly displaced from E_0 if the separation is large, so we may expand $I(E)$ near E_0: $I(E) = I(E_0) + (E - E_0)I'(E_0)$; I' being the derivative of I. Thus

$$1 - V_0 I(E) = -V_0(E - E_0)I'(E_0). \tag{5.5.33}$$

Equation (5.5.31) can be rewritten with the use of (5.5.33) in the form

$$\begin{vmatrix} \epsilon & -(\lambda/R)e^{-\alpha R} \\ -(\lambda/R)e^{-\alpha R} & \epsilon \end{vmatrix} = 0. \tag{5.5.34}$$

We have defined $\epsilon = E - E_0$, and

$$\lambda = \Omega/4\pi\gamma I'(E_0). \tag{5.5.35}$$

The solutions of (5.5.34) are evidently

$$\epsilon = \pm\lambda e^{-\alpha R}/R. \tag{5.5.36}$$

Thus we see that the original doubly degenerate localized state associated with a pair of impurities is split by an amount which decreases exponentially with the separation of the pair.

These results can be qualitatively applied to the many-impurity problem. Because the interaction between impurities decreases exponentially with distance, as is indicated by (5.5.36), we may, in first approximation, consider only the interaction of an impurity with the neighboring impurity closest to it. In this case, we will evidently obtain a set of impurity states whose energies are split by varying amounts according to the distribution of distances.

The distribution function for nearest neighbor distances can be worked out quite simply if there is no correlation between impurities. A Poisson distribution is assumed. The probability that *NO* (other) impurity is present in a sphere of volume Ω (which we may take to be centered about an impurity is proportional to $\exp[-(\Omega/\Omega_0)]$ where Ω_0 is the average volume occupied by an impurity,

$$\Omega_0 = V/n = 4\pi R_s^3/3 \tag{5.5.37}$$

(where V is the volume of the system). Equation (5.5.37) defines an average impurity radius R_s. The probability of finding a first neighbor impurity at R [we call this $P(R)$] is proportional to the derivative of the exponential function with respect to R. When this is properly normalized, we find

$$P(R) = (3R^2/R_s^3) \exp[-(R/R_s)^3]. \tag{5.5.38}$$

The density of states will be found by considering R in (5.5.38) to be a function of ϵ, which is to be determined by solving (5.5.36). Since $1/R$ varies slowly compared to the exponential in (5.5.36), an approximate solution is

$$R = (-1/\alpha) \ln(R_s |\epsilon|/\lambda). \tag{5.5.39}$$

The density of states is then

$$N(\epsilon) = P[R(\epsilon)] \, dR/dE \tag{5.5.40}$$

(normalized to unity). The final result is

$$N(\epsilon) = [3/(\alpha R_s)^3]\{[\ln(R_s |\epsilon|/\lambda)]^2/|\epsilon|\} \exp\{[\ln(R_s |\epsilon|/\lambda)]^3/(\alpha R_s)^3\}. \tag{5.5.41}$$

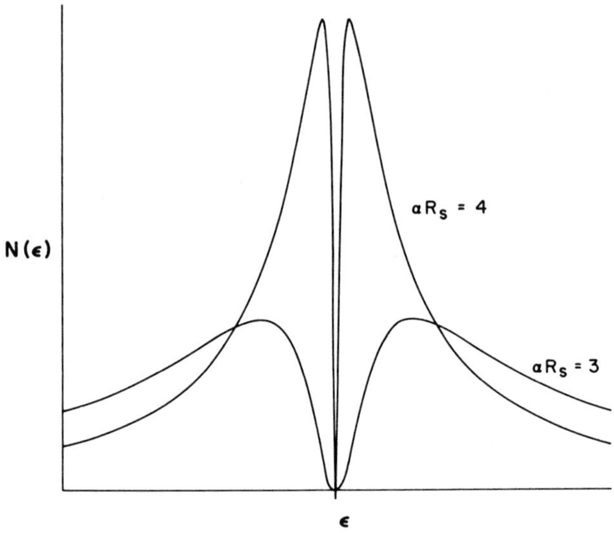

Fig. 5.5.1. Energy distribution of impurity states according to Eq. (5.5.41). Note that the curve with $\alpha R_s = 4$ corresponds to a smaller concentration of impurities than that with $\alpha R_s = 3$.

A graph of $N(\epsilon)$ according to (5.5.41) is shown in Fig. 5.5.1. The function vanishes at $\epsilon = 0$, as a consequence of the fact that all impurities have been considered to be arranged in pairs and all pairs are split, and it is symmetrical about $\epsilon = 0$ in consequence of the fact that the states associated with a pair are split symmetrically. This result must not be taken too literally: It is not valid for pairs which are very close, where (5.5.32) and (5.5.39) fail; and it is not valid for very distant pairs either, since the approximation of considering only pairs is likely to break down if an atom has two (say), neighbors at comparable distances.

Since $R_s \propto c^{-1/3}$, decreasing R_s corresponds to increasing impurity concentration. As would be expected, the effect of increasing the concentration is to decrease the height of the peak while broadening the wings of the distribution. This broadening results from an increase in the number of pair states with large splittings. A more complete description of the density of states in the region of localized states is given by Lifshitz (1964).

The density of states we have calculated for the splitting of localized impurity levels differs in one important aspect from the density of states obtained in an energy band calculation. The states here remain localized, though split; they are not extended throughout the crystal and thus cannot be associated with a band-type mobility or conductivity (Anderson, 1958).

However, if the concentration of impurities is sufficiently large, band formation in the usual sense must take place, that is, states will be extended throughout the entire crystal. [This apparently does not occur in a one dimension, in which the states of a disordered system are always localized (Mott and Twose, 1961; Mott, 1967).]

The nature of the transition between localized and conducting states has been of considerable interest. It might be expected that the transition would be gradual, as the effective distance between impurity atoms is decreased by an increase in the concentration. Mott has, however, presented convincing arguments that the metal–insulator transition should be sharp at $T = 0$ (Mott, 1949, 1956, 1961, 1968). To understand why this should be the case requires some consideration of the effects of electron–electron interactions. Consider specifically shallow donor impurities in a semiconductor. At very low concentrations, a theory quite similar to that presented above will be applicable. In order to make electrical conduction possible it is necessary to ionize an electron from one of the donor states. This will require an energy I which is the binding energy as calculated in Section 5.3 with some small correction of the sort we have discussed above. As the concentration of impurities increases, we might expect the activation energy for conduction to decrease gradually as the broadening of the distribution of impurity states in energy becomes more pronounced. When this distribution becomes broad enough to overlap the conduction band the activation energy would be zero. However, something has been omitted: If there are an appreciable number of free electrons present, dielectric screening of a long-range Coulomb potential will occur (This is discussed in detail subsequently in Section 7.1.) The effective interaction potential between an electron and an ion then has the approximate form

$$V(r) = (-e^2/\kappa r)e^{-\mu r} \qquad (5.5.42)$$

in which κ is the background dielectric constant and μ is related to the electron density. In the Thomas–Fermi approximation [see Eq. (7.1.57)]

$$\mu^2 = (4m^*e^2/\kappa\hbar^2)(3N_e/\pi)^{1/3} \qquad (5.5.43)$$

where m^* is the effective mass in the conduction band and N_e is the electron density.

In contrast to the ordinary Coulomb potential, the screened Coulomb (or Yukawa) potential does not always have bound states. For this potential, the condition that a bound state exist is (Hulthēn and Laurikainen, 1951)

$$2m^*e^2/\kappa\mu\hbar^2 \geq 1.68. \qquad (5.5.44)$$

This means that no bound state can exist if N_e is sufficiently large. The condition can be stated in a simple way if we assume that each impurity atom could contribute a single electron, thus $N_e = 3/4\pi R_s^3$. We combine (5.5.43) and (5.5.44), introduce the Bohr radius for the impurity state $a = \kappa\hbar^2/m^*e^2$, and obtain

$$(R_s/a) > (1.68)^2(9/4\pi^2)^{1/3} = 1.72. \tag{5.5.45}$$

Mott argues that localized states will not exist if R_s is smaller than the value on the right side of (5.5.45). A small number of free carriers is not possible in the ground state of the system; they will fall into bound states associated with the donors. On the other hand, if the number of free electrons is sufficiently large, bound states cannot exist, and a metallic type of conductivity should be observed. (However, the argument of exact sharpness probably applies only to ordered systems, the transition may be somewhat smeared in disordered materials.) The numerical value of (R_s/a) at which the transition should occur will not be given precisely by (5.5.45) because our considerations have been quite schematic (we have ignored anisotropy of the conduction band, have used too simple a function for $V(r)$ in (5.5.42), etc.), but it does seem to be reasonably satisfactory.

The transition between the metallic and insulating state so described has come to be called the *Mott transition*. It is believed that this would occur in any atomic system were it possible to alter the atomic separation: the material always being an insulator at large separations. However, it is only in disordered systems, of which doped semiconductors are an example, that it is possible to change experimentally the effective atomic distance as required by the argument.

5.5.5 *The Coherent Potential Approximation*

Much of the preceding discussion has concerned systems with small disorder. Thus we have considered a power series expansion of the density of states in terms of the concentration of the minority constituent as a small parameter. These techniques are not useful for alloys in which the concentrations of the constituents are nearly equal, nor do they apply to liquid metals. We will now examine the so-called coherent potential theory (Soven, 1967; Velicky *et al.*, 1968). The essential physical idea of this approach is to describe the system by an appropriate effective Hamiltonian. This Hamiltonian is defined by the condition that if it is employed, there is no further scattering from the individual sites on the average. It incorporates a potential such that with its use, a wave will propagate coherently through the material. It is a necessary complication of the approach that the effective Hamiltonian turns out to be non-Hermitian and energy-dependent.

Let H be the true Hamiltonian, containing all the effects of randomness. The system would be adequately described if we knew the true "Green's function" or resolvent operator

$$\mathcal{G}(z) = 1/(z - H) \tag{5.5.46}$$

where z is a complex variable. It is clear that $\mathcal{G}(z)$ has poles at the eigenvalues of H, and we have an expression for the density of states

$$N(E) = (-1/\pi\mathfrak{N}) \operatorname{Im} \operatorname{Tr} \mathcal{G}(E + i\epsilon),$$

ϵ being an infinitesimal positive quantity, which is allowed to go to zero appropriately (see Section 5.2).

It is usually inconvenient to work with the full operator $\mathcal{G}(z)$ since it involves the random positions of the various kinds of atoms. We consider instead an average of $\mathcal{G}(z)$ over all possible configurations of atoms; this operator is denoted by $\langle \mathcal{G}(z) \rangle$. The notation $\langle \cdots \rangle$ will be used below to indicate an average over configurations. The result of the averaging process is an operator with the symmetry of the lattice. We write

$$\langle \mathcal{G}(z) \rangle = 1/(z - H_{\text{eff}}) \tag{5.5.47}$$

where H_{eff} is to be determined.

Let us suppose some starting approximation to H_{eff} is available with the same analytic properties. We denote this by K. The operator identity

$$(A - B)^{-1} = A^{-1} + A^{-1}B(A - B)^{-1} \tag{5.5.48}$$

can be applied (with $A = z - K, B = H_{\text{eff}} - K$) to give

$$\langle \mathcal{G}(z) \rangle = R(z) + R(z)(H_{\text{eff}} - K)\langle \mathcal{G}(z) \rangle \tag{5.5.49}$$

where

$$R(z) = [z - K]^{-1}. \tag{5.5.50}$$

We also have

$$\mathcal{G}(z) = R(z) + R(z)[H - K]\mathcal{G}(z). \tag{5.5.51}$$

Let us now introduce the T matrix for the entire system, relative to the operator K. This operator satisfies

$$T[K] = (H - K) + (H - K)R(z)T[K]. \tag{5.5.52a}$$

This equation follows from (5.2.6) with $U = H - K$. It also follows from (5.2.65) that

$$T[K] = (H - K) + (H - K)(1/(z - H))(H - K). \tag{5.5.52b}$$

The T matrix is related to $\mathcal{G}$ by the identity

$$\mathcal{G}(z) = R(z) + R(z)T[K]R(z). \tag{5.5.53}$$

Let us consider an average of (5.5.53) over impurity configurations. The operator R is unaffected by this process, so that we may write

$$\langle \mathcal{G}(z) \rangle = R(z) + R(z)\langle T[K]\rangle R(z). \tag{5.5.54}$$

We can now combine (5.5.49) and (5.5.54) to obtain

$$H_{\text{eff}} = K + \langle T\rangle(1/(1 + R\langle T\rangle)). \tag{5.5.55}$$

There are two possible approaches to the use of (5.5.54): We can make some assumptions about the system in order to permit us to define K, calculate $\langle T\rangle$ using this K, and then employ (5.5.55) to deduce an effective Hamiltonian; (2) Alternately, we can require as mentioned previously

$$\langle T[K]\rangle = 0 \tag{5.5.56}$$

in which case,

$$H_{\text{eff}} = K. \tag{5.5.57}$$

Equation (5.5.56) is a transcendental equation which may be difficult to solve.

Any attack on (5.5.55) or (5.5.56) requires use of the full T matrix. This is not known, and it is necessary to develop a relation between the complete operator T for the entire system, and the scattering matrix for single potentials which can be determined by the procedures of Section 5.2. To this end, let us suppose that $(H - K)$ can be expressed as a sum of random potentials V_n associated with the individual sites

$$H - K = \sum_n V_n. \tag{5.5.58}$$

We have from (5.5.53)

$$\mathcal{G}^{-1} = R^{-1}[1 + RT]^{-1}, \tag{5.5.59}$$

while use of (5.5.51) and (5.5.53) gives

$$(H - K)\mathcal{G} = TR. \tag{5.5.60}$$

We combine (5.5.58)–(5.5.60) to give

$$T = \sum_n V_n(1 + RT) = \sum_n Q_n. \tag{5.5.61}$$

In (5.5.61), we have attempted to write the total T matrix as the sum of contributions from each of the sites. These contributions are denoted Q_n. It might at first be expected that Q_n should be the single site T matrix, but the relation is in fact more complex because of the possibility that an electron may scatter on site 1, then on site 2, say, then back again to 1, and so on. The single site T matrix (we will henceforth use a lower case letter for this) t_n, say, is given according to (5.2.8) by

$$t_n = V_n(1 - RV_n)^{-1} = (1 - V_nR)^{-1}V_n. \tag{5.5.62}$$

Equations (5.5.61) and (5.5.62) can be combined to give an expression for Q_n:

$$Q_n = t_n[1 + R \sum_{m\neq n} Q_m]. \tag{5.5.63}$$

The factor in brackets in (5.5.63) contains the rescattering correction mentioned above. Equations (5.5.61) and (5.5.63) can now be combined to give a series relating T to t (Edwards, 1961):

$$T = \sum_n t_n + \sum_n t_nR \sum_{m\neq n} t_m + \cdots. \tag{5.5.64}$$

Equations (5.5.61)–(5.5.64) apply to the operators themselves. In order to obtain the configurational average $\langle \mathcal{G} \rangle$ we require $\langle T \rangle$. The average may be applied to (5.5.61)

$$\begin{aligned}\langle T \rangle = \sum_n \langle Q_n \rangle &= \langle t_n(1 + R \sum_{m\neq n} Q_m) \rangle \\ &= \langle t_n \rangle[1 + R \sum_{m\neq n} \langle Q_m \rangle + \langle t_nR \sum_{m\neq n} (Q_m - \langle Q_m \rangle) \rangle].\end{aligned} \tag{5.5.65}$$

We will make the approximation of neglecting the last term. It is possible to eliminate the sum over $\langle Q_m \rangle$ using the fact that $\sum_{m\neq n} \langle Q_m \rangle = T - Q_n$

$$\langle Q_n \rangle = [1 + \langle t_n \rangle R]^{-1}\langle t_n \rangle[1 + R\langle T \rangle]. \tag{5.5.66}$$

This result may be substituted into (5.5.55), and the combination into (5.5.65) to obtain

$$H_{\text{eff}} = K + \sum_n \langle t_n \rangle(1 + R\langle t_n \rangle)^{-1}. \tag{5.5.67}$$

The last term in (5.5.67) represents the effect of the average scattering. If we choose to proceed from this point according to (5.5.56), we have to consider the equation

$$\langle t_n[K] \rangle = 0. \tag{5.5.68}$$

At this point, it is desirable to introduce a model. We will proceed by making a slight generalization of the Koster–Slater problem that has frequently been discussed in this chapter. The assumed Hamiltonian for the model is

$$H = H_{\text{p}} + H_{\text{r}} \tag{5.5.69}$$

where H_{p} is periodic and H_{r} contains the random elements. It is supposed here that H_r has matrix elements which are diagonal with respect to the band and site indices in a representation based on Wannier functions. The bands [energy $\epsilon(\mathbf{k})$] and the Wannier functions mentioned are those deriving from the periodic portion of the Hamiltonian H_p

$$H_p\psi(\mathbf{k}, \mathbf{r}) = \epsilon(\mathbf{k})\psi(\mathbf{k}, \mathbf{r}). \tag{5.5.70}$$

Two kinds of sites are assumed to exist: those of type A, for which H_r contributes a matrix element V_A, and those of type B for which the matrix element is V_B. The location of these A and B sites is the random element of the problem. The B atoms are present with concentration c; thus the concentration of the A atoms is $(1 - c)$. We will find it convenient to define a quantity V_0 which is the average of H_r over all sites. Thus,

$$cV_B + (1 - c)V_A = V_0. \tag{5.5.71}$$

It is possible to adjust H_p as a function of concentration to make $V_0 = 0$. In this case H_p contains the average potential in the system. Such an H_p is called the "virtual crystal Hamiltonian."

Some characteristics of the energy levels of the system can be determined by physical arguments. Suppose that, as a fluctuation, there exists a substantial region of the crystal which consists of atoms of type A. These atoms will in effect define a small crystal. The lowest energy of states based on this region will be $\epsilon_{\min} + V_A$, where $\epsilon_{\min}$ is the minimum value of $\epsilon(\mathbf{k})$; and the highest energy will be $\epsilon_{\max} + V_A$, $\epsilon_{\max}$ being the maximum of $\epsilon(\mathbf{k})$. Likewise, if we consider a region composed of type B atoms, the lowest energy will be $\epsilon_{\min} + V_B$, and the highest $\epsilon_{\max} + V_B$. It is therefore to be expected that the spectrum of the full Hamiltonian H is contained in the union of the two regions $(\epsilon_{\min} + V_B, \epsilon_{\max} + V_B)$ and $(\epsilon_{\min} + V_A, \epsilon_{\max} + V_A)$. If these regions do not intersect, the band should be split into B and A subbands, containing Nc and $N(1 - c)$ levels per spin, respectively.

These results may be established as follows (Kirkpatrick *et al.*, 1970; see also Messiah, 1962, p. 712): Consider the resolvent operator

$$G(\lambda, z) = (z - H_r - \lambda H_p)^{-1} \tag{5.5.72}$$

This operator has poles at the eigenvalues of $H_r + \lambda H_p$, and is analytic everywhere else. We may write $G(\lambda, z)$ as a power series in terms of the operator λH_p by repeated iteration of the identity (5.5.48)

$$\begin{aligned} G(\lambda, z) &= (z - H_r)^{-1} + (z - H_r)^{-1}\lambda H_p(z - H_r)^{-1} \\ &\quad + (z - H_r)^{-1}\lambda H_p(z - H_r)^{-1}\lambda H_p(z - H_r)^{-1} + \cdots. \end{aligned} \tag{5.5.73}$$

We inquire whether the series (5.5.73) converges. If it converges for some value of z, that value of z is *not* an eigenvalue of $H_r + \lambda H_p$.

The operators H_r and H_p are (for the model considered) both bounded operators in Hilbert space. [An operator Q is bounded if there exists a finite number such that for any $| u \rangle$

$$\langle u | Q^+Q | u \rangle / \langle u | u \rangle \leq M^2.$$

The smallest value of M is by definition the upper bound or norm of Q, and is denoted by $|| Q ||$. We may take M to be the largest eigenvalue of Q.]

The series (5.5.73) is known to converge absolutely for values of z such that

$$\lambda \, || (z - H_r)^{-1} H_p || \leq \lambda \, || (z - H_r)^{-1} || \, || H_p || < 1. \qquad (5.5.74)$$

In the model considered, H_r is diagonal on the Wannier function basis. The eigenvalues are simply V_A and V_B. The norm of $(z - H_r)^{-1}$ is simply either $1/| z - V_A |$ or $1/| z - V_B |$, whichever is larger. The eigenvalues of H_p lie on the real energy axis between ϵ_{min} and ϵ_{max}; for simplicity we suppose that the zero of energy has been chosen so that $\epsilon_{min} = -\epsilon_{max}$. Then

$$|| H_p || = \epsilon_{max}.$$

As a result, (5.5.74) implies that (5.5.73) converges for values of z such that (both)

$$\lambda\epsilon_{max}/| z - V_A | < 1 \qquad \text{and} \qquad \lambda\epsilon_{max}/| z - V_B | < 1. \qquad (5.5.75)$$

In other words, (5.5.73) converges outside of two circles of radius $\lambda\epsilon_{max}$, centered about the points V_A and V_B. If we now let $\lambda \to 1$, we obtain the Hamiltonian (5.5.69), and see that the spectrum of this operator is confined to the union of the two intervals $(V_A + \epsilon_{min}, V_A + \epsilon_{max})$ and $(V_B + \epsilon_{min}, V_B + \epsilon_{max})$.

Further, if $\lambda = 0$, there are cN states with energies V_B and $(1 - c)N$ with energies V_A. The eigenvalues are continuous functions of λ. Therefore there will be the same number of eigenstates in the intervals $(V_B + \lambda\epsilon_{min}, V_B + \lambda\epsilon_{max})$ and $(V_A + \lambda\epsilon_{min}, V_A + \lambda\epsilon_{max})$ as at $\lambda = 0$ until these intervals overlap. If the bands are still separate at $\lambda = 1$, they will contain cN and $(1 - c)N$ states each.

We will now consider the construction of the spectrum in the coherent potential approximation. It will be assumed that the operator $R = [z - K]^{-1}$ can be written in the following way on the basis of eigenstates of H_p:

$$R = 1/[z - \epsilon(\mathbf{k}) - \Sigma(z)] \qquad (5.5.76)$$

where Σ depends on z but not on $\mathbf{k}$. We will see whether a solution for the quantity Σ, conventionally called the *self-energy* (see Section 8.1), can be found. The assumed form of R is simple enough so that we can obtain an explicit expression for the single site to matrix t_n. The effect of Σ merely corresponds to a displacement of the energy in the considerations of Section 5.2

$$t_n = [V_n - \Sigma(z)]/\{1 - F(z)[V_n - \Sigma(z)]\}, \tag{5.5.77}$$

where

$$F(z) = [\Omega/(2\pi)^3] \int d^3k/[z - \epsilon(\mathbf{k}) - \Sigma(z)]. \tag{5.5.78}$$

It is to be remembered that z and Σ are complex. The averaged t matrix is

$$\langle t_n \rangle = ct_{\mathrm{B}} + (1 - c)t_{\mathrm{A}}. \tag{5.5.79}$$

When (5.5.77) is substituted into (5.5.79), the following equation is obtained

$$\Sigma(z) = V_0 - [V_{\mathrm{A}} - \Sigma(z)]F(z)[V_{\mathrm{B}} - \Sigma(z)]. \tag{5.5.80}$$

This equation must be solved for Σ. It is highly implicit; however, if V_{A} and V_{B} are not much different, the solution may be found by iteration

$$\begin{aligned}\Sigma(z) &= V_0 - [V_{\mathrm{A}} - V_0]F_0(z)[V_{\mathrm{B}} - V_0] + \cdots \\ &= V_0 + c(1 - c)(V_{\mathrm{A}} - V_{\mathrm{B}})^2F_0(z) + \cdots \end{aligned} \tag{5.5.81}$$

in which

$$F_0(z) = [\Omega/(2\pi)^3] \int d^3k/[z - \epsilon(\mathbf{k}) - V_0]. \tag{5.5.82}$$

In (5.5.81), the virtual crystal result $\Sigma(z) = V_0$ is explicitly exhibited as a first approximation with a correction which depends on the square of the difference between the A and the B potentials.

Once Σ has been determined, the energies and the densities of states must be determined from the poles of the operator R for energies E. From (5.5.76), it is seen that this requires solution of another implicit equation

$$E = \epsilon(\mathbf{k}) + \Sigma(E). \tag{5.5.83}$$

The wavevector $\mathbf{k}$ has its usual significance. The function $F(z)$ defined by (5.5.78) is analytic in the complex z plane except for a cut along the real axis spanning the region of the eigenvalues $\epsilon(\mathbf{k})$. If $G_0(E)$ is the density of states corresponding to $\epsilon(\mathbf{k})$, (5.5.78) can be written in the simpler form

$$F(z) = \int dE'\, G_0(E')/[z - \Sigma(z) - E']. \tag{5.5.84}$$

The function F will be complex for all z except those lying on the real axis on either side of the cut. As a result, the self-energy Σ will be complex, and (5.5.83) furnishes the possibility of obtaining complex energies for real values of $\mathbf{k}$. In this case, the imaginary part (which must be negative) corresponds to a width associated with a state of energy $\mathrm{Re}(E)$.

The possibility of splitting the energy band as mentioned earlier can come about in the following way: It is possible for the self-energy to develop a pole. Let us see what happens in this case.

It is convenient to separate V_0 from Σ by the definition

$$\Sigma(z) = V_0 + S(z). \tag{5.5.85}$$

The function $S(z)$ satisfies the equation

$$S = [c\Delta + S]F[(1-c)\Delta - S] \tag{5.5.86}$$

where $\Delta = V_B - V_A$. Now suppose that $S(z)$ and hence $\Sigma(z)$ has a pole at some value of z. In the vicinity of this point, $|z - \Sigma| \to \infty$, so that (5.5.84) becomes

$$F(z) \approx [z - \Sigma(z)]^{-1} \int dE'\, G_0(E') = [z - \Sigma(z)]^{-1}. \tag{5.5.87}$$

In this approximation, we may solve (5.5.86) for $S(z)$

$$S(z) = c(1-c)\Delta^2/\{z - [V_0 + (1-2c)\Delta]\}. \tag{5.5.88}$$

This shows that a pole exists on the real axis: thus that the assumption of a pole is self-consistent.

We may use (5.5.83) to determine the band structure of the alloy in this approximation. The allowed energies are now determined as solutions of the equation

$$E - \epsilon(\mathbf{k}) - V_0 - [A/(E - E_0)] = 0 \tag{5.5.89}$$

where

$$A = c(1-c)\Delta^2, \qquad E_0 = V_0 + (1-2c)\Delta = (1-c)V_B + cV_A. \tag{5.5.90}$$

The solutions of (5.5.89) are

$$E = \tfrac{1}{2}\{(E_0 + \epsilon(\mathbf{k}) + V_0) \pm [(\epsilon(\mathbf{k}) + V_0 - E_0)^2 + 4A]^{1/2}\}. \tag{5.5.91}$$

If A is small compared to $|\epsilon(\mathbf{k}) + V_0 - E_0|$, we have by expansion of (5.5.91), the two roots

$$E_1 = \epsilon(\mathbf{k}) + V_0 + \{c(1-c)\Delta^2/[\epsilon(\mathbf{k}) + (2c-1)\Delta]\} \tag{5.5.92a}$$

and

$$E_2 = E_0 - \{c(1-c)\Delta^2/[\epsilon(\mathbf{k}) + (2c-1)\Delta]\}. \tag{5.5.92b}$$

The principal significance of these energies is that they may be used in a calculation of the density of states by standard techniques.

The present approximation always leads to a pole in Σ and consequently to a split band. However, this splitting is not an inevitable consequence of the coherent potential method. For small Δ, S and Σ possess power series expansions in Δ, in which the first term (beyond V_0 in the case of Σ) is of order Δ^2, as is indicated by (5.5.81). In the case of a symmetrical band [such that $\int \epsilon G_0(\epsilon)\, d\epsilon = 0$], it can be shown that a necessary condition for the existence of a pole is

$$c(1-c)\Delta^2 \geq \int \epsilon^2 G_0(\epsilon)\, d\epsilon. \tag{5.5.93}$$

Applications of the coherent potential method to real systems are reasonably practical. For example, nickel–copper alloys have been studied by Kirkpatrick *et al.* (1970) and by Stocks *et al.* (1971).

The idea was introduced in our discussion concerning the location of the spectrum of the model Hamiltonian previously considered that the extremities of the band structure would be determined by considering regions in which a fluctuation has produced all A- or all B-type atoms. This argument leads to bounds on the spectrum, but does not directly produce a quantitative description of it. Butler and Kohn (1971) have significantly generalized and extended this point of view by developing a procedure which makes possible, in principle, an accurate calculation of the density of states.

Their approach may be qualitatively described as follows. Consider, in a disordered material, a region of a specific size (for example, a cube of side L). In this region, determine all possible relevant arrangements of atoms. Each arrangement is characterized by a statistical probability of occurrence. For a given arrangement within this region, define a hypothetical crystal in which this particular atomic arrangement is repeated periodically. A band structure can be calculated for this crystal. Roughly speaking, the density of states for the disordered material can be described as an average of the densities of states for the various periodic crystals that can be constructed in this way, each weighted by the statistical probability mentioned above. It turns out that such a procedure will give results which increase exponentially in accuracy with the size of the basic region considered. More precisely, one may calculate in this way the density of states density $\langle \mathbf{r} \mid G(z) \mid \mathbf{r}' \rangle$ whose trace gives, in the proper limit, the density of states of the material.

5.6 THE KONDO EFFECT

In Section 5.4, we saw that the exchange interaction between electrons could give rise to a localized moment, or spin polarization, associated with a narrow scattering resonance. The consequences of the existence of such moments are fascinating. One such which has received extensive attention in recent years is the *Kondo effect*. Study of this phenomenon will be important here in that it indicates clearly the nature of the limitations on our single-particle treatment of scattering processes which are imposed by the many-electron characteristics of real metals.

5.6.1 *The Resistance Minimum*

It has been known for many years that certain dilute alloys involving transition metals exhibit a minimum in the electrical resistivity at low temperature. For example, an alloy containing 0.1% of iron in copper exhibits a shallow minimum in the resistivity close to 30°K. Figure 5.6.1 illustrates this effect for dilute alloys of manganese in zinc. This is quite contrary to elementary theoretical expectations which show that the electron–phonon scattering contribution should increase with temperature while that due to impurity scattering should become constant as $T \to 0$. This resistance minimum phenomenon is quite different from superconductivity, in which a sharp drop of resistance to zero is observed (in other metals). Other physical quantities show unexpected behavior as well. The specific heat of such a material (0.1% Fe in Cu) is larger than that obtained for pure Cu by an amount which shows a broad maximum in the neighborhood of 5°K. Measurements on samples of different impurity concentrations show that the entropy associated with the excess specific heat is of the order of $NK \ln 2$ where N is the number of impurity atoms. An excess magnetic susceptibility is also observed at low temperatures and a peak appears in the thermoelectric power as well.

The theoretical and experimental situations are summarized in reviews mentioned in Section 5.4 (Bailyn, 1966; Kondo, 1969; Heeger, 1969; Abrikosov, 1969, see also Daybell and Steyert, 1968). The explanation of the effects is based on the assumption that a local moment exists, and that the exchange coupling between the conduction electron and a single local moment can be written as

$$U = -Jf(\mathbf{r})\,\mathbf{S}\cdot\mathbf{s} \tag{5.6.1}$$

where J is an exchange parameter measuring the strength of the interaction, $\mathbf{S}$ is the spin associated with the moment, $\mathbf{s}$ is the spin operator of a conduction electron, and $f(\mathbf{r})$ is a "form factor" describing the position dependence of the interaction. It is assumed that $f(\mathbf{r})$ is positive and is

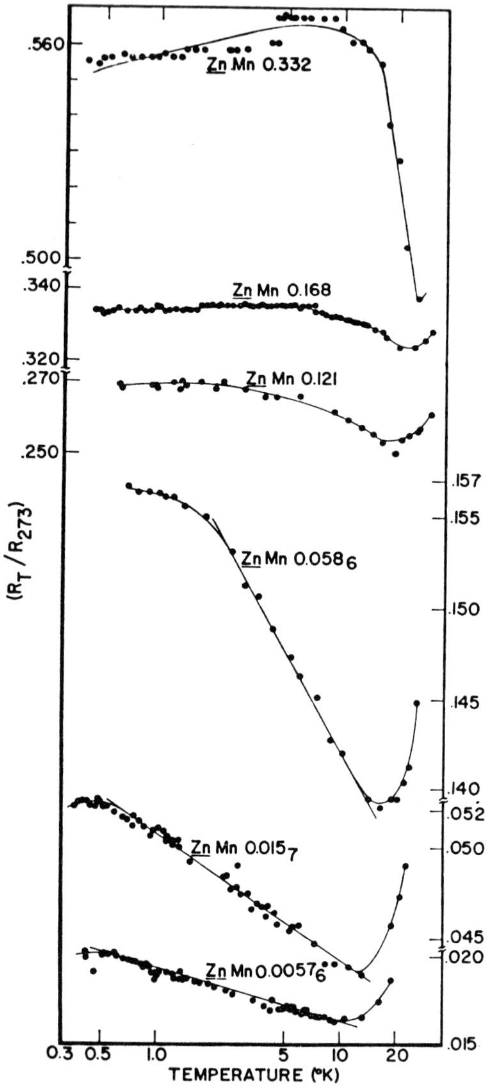

Fig. 5.6.1. Low temperature electrical resistance of dilute alloys of manganese in zinc. [From Hedgcock and Rizutto (1967).]

normalized to unity. The physically interesting case is that in which J is negative, that is, the interaction tends to produce antiferromagnetic spin alignment. An interaction of this type can be derived from the Anderson model mentioned in Section 5.4 (Kondo, 1962; Schrieffer and Wolff, 1966; Bailyn, 1966). The interaction is produced by a second-order treatment of the **k**–d mixing term of this model. Let us start with an interaction connecting a plane wave state of wavevector **k** and spin s with a localized "atomic state" $d\sigma$. The strength of the interaction is $V_{\mathbf{k}d}$ This coupling causes transitions between the localized state $d\sigma$ and the band states **k**s. A second-order treatment leads to a spin–spin interaction corresponding to processes in which an initial state $\mathbf{k}\downarrow d\uparrow$ goes through an intermediate state (either $\mathbf{k}\downarrow \mathbf{k}'\uparrow$ or $d\downarrow d\uparrow$) to a final state ($d\downarrow \mathbf{k}'\uparrow$) in which the spins of the band and the bound electrons have been reversed. The energy denominators $[E(\text{init}) - E(\text{inter})]^{-1}$ are negative because the intermediate state is of higher energy than the initial state; however, an additional minus sign comes about at the end. This is a consequence of the fact that the order of states in a determinantal wavefunction must be preserved, with $d\downarrow$ replacing $d\uparrow$ and $\mathbf{k}'\uparrow$ replacing $\mathbf{k}\downarrow$. However, as we have seen above, we have gone from ($\mathbf{k}\downarrow d\uparrow$) to ($d\downarrow, \mathbf{k}'\uparrow$); an interchange must now be performed which contributes a minus sign to the interaction. To this effect we must add the contribution of direct exchange between the band and bound electrons (Kasuya, 1956). This is always ferromagnetic. Thus some compensation will occur; however, it is possible for antiferromagnetism to dominate. Other spin orientations may be treated similarly.

Kondo (1964) assumed the existence of such an interaction with negative (antiferromagnetic) J, and calculated its contribution to the electrical resistivity of a metal containing localized moments using the second Born approximation. He found that the resistivity should show a logarithmic temperature dependence at low temperatures, which results from a divergence in the scattering cross section at the Fermi energy due to exchange effects. This logarithmic term, which increases as $T \to 0$, must be combined with the normal resistivity, which approaches a constant in this limit, and increases for $T > 0$. The resulting combined resistivity shows the observed minimum.

Further investigations by several authors (see the reviews mentioned above for detailed references) showed that below a characteristic temperature T_K, a system containing a localized moment plus free electrons develops a new type of ground state. This new state, which is a collective, many-electron state, is a singlet (no magnetic moment) in which electrons are strongly coupled to the impurity. The energy of this state at $T = 0°$K is lower than that obtained for the system if coupling is neglected by approximately KT_K per impurity atom. The Kondo temperature T_K itself is approximately given by

$$KT_K \approx \epsilon_F \exp[-1/|J| G(\epsilon_F)]$$

where ϵ_F is the Fermi energy and $G(\epsilon_F)$ is the density of states for a single spin (per atom) at the Fermi energy. The transition to the "bound" state is not sharp, as in a phase transition, but is broadened by thermal fluctuations. Thus it turns out that the simple Hartree–Fock treatment of the magnetic impurity problem may not be adequate at low temperatures if an antiferromagnetic interaction exists between the quasi-bound and band electrons. At $T = 0$ K, the moment is quenched by compensating spin polarization. Only at temperatures above T_K does the local moment "break free," and establish a regime close to the Hartree–Fock approximation.

We will begin our detailed discussion with a consideration of the resistance minimum. It is convenient to express the Hamiltonian (5.6.1) in second quantized form. Let $c_s^+(\mathbf{k})$, $c_s(\mathbf{k})$ be creation and annihilation operators for free electrons in a state of wavevector $\mathbf{k}$ and spin s. Only a single band is considered so that no band index is included. We adopt discrete normalization for our states so that the anticommutation relations are

$$c_s^+(\mathbf{k})c_{s'}(\mathbf{q}) + c_{s'}(\mathbf{q})c_s^+(\mathbf{k}) = \delta_{\mathbf{kq}}\,\delta_{ss'} \tag{5.6.2}$$

$$c_s(\mathbf{k})c_{s'}(\mathbf{q}) + c_{s'}(\mathbf{q})c_s(\mathbf{k}) = 0, \qquad \text{etc.} \tag{5.6.3}$$

The conversion to second quantized form can be carried out by standard procedures (see, for example, Landau and Lifshitz, 1965, Chapter 9 or Appendix D). We find

$$U = (-1/2N) \sum_{\mathbf{kq}} J(\mathbf{k}, \mathbf{q}) \{ S_z[c_\uparrow^+(\mathbf{k}+\mathbf{q})c_\uparrow(\mathbf{k}) - c_\downarrow^+(\mathbf{k}+\mathbf{q})c_\downarrow(\mathbf{k})]$$
$$+ S_+ c_\downarrow^+(\mathbf{k}+\mathbf{q})c_\uparrow(\mathbf{k}) + S_- c_\uparrow^+(\mathbf{k}+\mathbf{q})c_\downarrow(\mathbf{k})]\}, \tag{5.6.4}$$

in which $J(\mathbf{k}, \mathbf{q})$ is given by

$$J(\mathbf{k}, \mathbf{q}) = J \int \psi^*(\mathbf{k}, \mathbf{r}) f(\mathbf{r}) \psi(\mathbf{q}, \mathbf{r})\, d^3r. \tag{5.6.5}$$

Our previous discussion indicates that the determination of $Jf(\mathbf{r})$ would be a difficult task. It is probably reasonable to assume that $f(\mathbf{r})$ is of relatively short range, which implies that $J(\mathbf{k}, \mathbf{q})$ should be slowly varying. We will make the simple approximation

$$J(\mathbf{k}, \mathbf{q}) = J \qquad \text{for} \quad |\epsilon(\mathbf{q}) - \epsilon_F| < D \quad \text{and} \quad |\epsilon(\mathbf{k}) - \epsilon_F| < D$$
$$= 0 \qquad \text{otherwise} \tag{5.6.6}$$

where $\epsilon(\mathbf{k})$ is the band energy and ϵ_F is the Fermi energy (at $T = 0$). Then J is a negative number and D is an arbitrary constant which is introduced to simulate the effect of the finite range of the form factor $f(\mathbf{r})$.

We will calculate the scattering for an electron by the interaction Hamiltonian (5.6.4) using the second Born approximation. It is easily seen by iterating (5.2.6) once that we must evaluate (to this order)

$$t = U + U\mathcal{G}U \tag{5.6.7}$$

where $\mathcal{G}$ is the free electron Greens' function for the perfect crystal. For reasons which will become clear later, it is not practical to use the formal solution (5.2.8) of the integral equation for the t matrix. The matrix elements of t are required on the basis of Slater determinants of single-particle states $|\,\mathbf{k}, s\rangle$ (wave vector $\mathbf{k}$ and spin s). These states are eigenstates of the perfect crystal Hamiltonian with energies $\epsilon(\mathbf{k})$ which are assumed independent of spin. For simplicity, we will study scattering with no change of spin ($\mathbf{k}\uparrow \to \mathbf{p}\uparrow$). Since we are concerned with a real scattering process, $\epsilon(\mathbf{k}) = \epsilon(\mathbf{p})$. In first order, we have simply

$$\langle \mathbf{p}\uparrow |\, t^{(1)}\,| \mathbf{k}\uparrow \rangle = \langle \mathbf{p}\uparrow |\, U\,| \mathbf{k}\uparrow \rangle = (-J/2N)S_z. \tag{5.6.8}$$

The transition rate is then

$$W^{(1)}(\mathbf{k}\uparrow \to \mathbf{p}\uparrow) = (2\pi/\hbar)(J/2N)^2 S_z^2\, \delta[\epsilon(\mathbf{k}) - \epsilon(\mathbf{p})]. \tag{5.6.9}$$

This transition rate may be used in connection with the Boltzmann equation to calculate a contribution to the electrical resistivity (see Section 7.2).

Complications arise in second order. Both direct and exchange processes are possible. In the direct process we simply transfer an electron from $\mathbf{k}$ to $\mathbf{q}$. The energy denominator (factor $\mathcal{G}$) is $[\epsilon(\mathbf{k}) - \epsilon(\mathbf{q})]^{-1}$ and a factor $(1 - f[\epsilon(\mathbf{q})])$ must be inserted. Here f is the Fermi function which gives the probability of occupation of a state of wave vector $\mathbf{q}$,

$$f[\epsilon(\mathbf{q})] = \{\exp[\epsilon(\mathbf{q}) - \epsilon_F]/KT + 1\}^{-1}$$

and ϵ_F is the Fermi energy (at temperature T). In the exchange process, the intermediate state contains an electron in the final state $\mathbf{p}$, a hole in some state $\mathbf{q}$, and the electron in the initial state $\mathbf{k}$. The energy denominator is $[\epsilon(\mathbf{q}) - \epsilon(\mathbf{p})]^{-1}$. A factor of $f[\epsilon(\mathbf{q})]$ must be inserted since the process can occur only if $\mathbf{q}$ is occupied in the initial state, and an overall minus sign is required because of the presence of the intermediate hole. This process may be represented by diagrams shown in Fig. 5.6.2. The general concepts and methods of diagrammatic perturbation theory are discussed in Section 8.1. The second-order contribution to t is

$$\begin{aligned}\langle \mathbf{p}\uparrow |\, t^{(2)}\,| \mathbf{k}\uparrow \rangle = &\sum_{\mathbf{q}s} \{\langle \mathbf{p}\uparrow |\, U\,| \mathbf{q}s\rangle\langle \mathbf{q}s\,|\, U\,| \mathbf{k}\uparrow \rangle/[\epsilon(\mathbf{k}) - \epsilon(\mathbf{q})]\}\\ &\times (1 - f[\epsilon(\mathbf{q})])\\ &+ \sum_{\mathbf{q}s} \{\langle \mathbf{p}\uparrow |\, U\,| \mathbf{q}s\rangle\langle \mathbf{q}s\,|\, U\,| \mathbf{k}\uparrow \rangle/[\epsilon(\mathbf{k}) - \epsilon(\mathbf{q})]\}\\ &\times f[\epsilon(\mathbf{q})].\end{aligned} \tag{5.6.10}$$

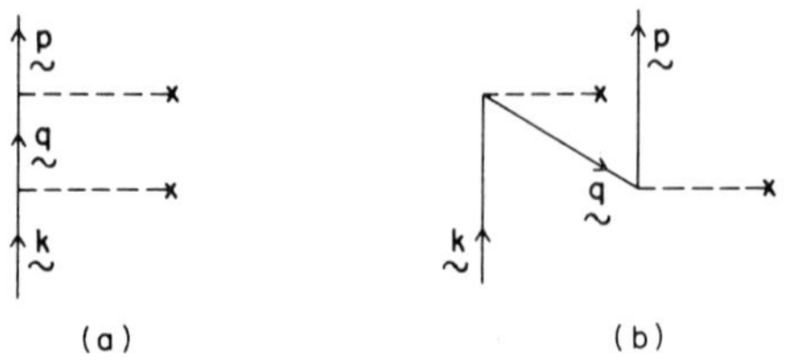

Fig. 5.6.2. Diagrammatic representation of (a) direct and (b) exchange scattering processes. The vertical direction represents the time order and ---× indicates interaction with the local spin.

The contribution of the part of U containing S_z is simply

$$\sum_{\mathbf{q}} (J/2N)^2 S_z^2/[\epsilon(\mathbf{k}) - \epsilon(\mathbf{q})] \tag{5.6.11}$$

since the Fermi function $f[\epsilon(\mathbf{q})]$ cancels on addition of the two terms of (5.6.10). There are no cross terms of the form $S_z S_+$ since we require that the final state have the same spin as the initial state. However, there are terms involving two changes of spin. In writing these matrix elements we must take the operators according to the time order of Fig. 5.6.2. We obtain from the portion of U involving S_+ and S_-

$$\sum_{\mathbf{q}} \{(J/2N)^2 S_- S_+/[\epsilon(\mathbf{k}) - \epsilon(\mathbf{q})]\}(1 - f[\epsilon(\mathbf{q})]) + \sum_{\mathbf{q}} \{(J/2N)^2 S_+ S_- f[\epsilon(\mathbf{q})]/[\epsilon(\mathbf{k}) - \epsilon(\mathbf{q})]\}. \tag{5.6.12}$$

These expressions still contain the spin operators for the local moment. Since S_- and S_+ do not commute, the Fermi function $f[\epsilon(\mathbf{q})]$ does not cancel as it did in (5.6.11). Instead, we have

$$S_+ S_- = S_- S_+ + 2S_z, \qquad S_- S_+ = S^2 - S_z^2 - S_z.$$

The combination of (5.6.11) and (5.6.12) gives

$$\langle \mathbf{p}\uparrow | t^{(2)} | \mathbf{k}\uparrow \rangle = (J/2N)^2 (S^2 - S_z) \sum_{\mathbf{q}} 1/[\epsilon(\mathbf{k}) - \epsilon(\mathbf{q})] + 2(J/2N)^2 S_z \sum_{\mathbf{q}} f[\epsilon(\mathbf{q})]/[\epsilon(\mathbf{k}) - \epsilon(\mathbf{q})]. \tag{5.6.13}$$

The two terms in (5.6.13) have quite different behavior. The first term, which has no Fermi factor, has the same form as would be obtained in ordinary second-order perturbation theory with no allowance for the fact that electrons in the metal occupy a Fermi distribution. The case of ordinary potential scattering, as considered in Section 5.2, is of this type. The apparent singularity which occurs for $\epsilon(\mathbf{k}) = \epsilon(\mathbf{q})$ is to be handled by the addition of a small imaginary part of $\epsilon(\mathbf{k})$, as discussed in Section 5.2. We will not consider this term further, as it is unimportant.

Interest centers on the second term of (5.6.13). It will be sufficiently accurate to evaluate it using the low-temperature approximation to the Fermi function

$$f(\epsilon) = 1 \quad \text{for} \quad \epsilon \leq \epsilon_F$$
$$= 0 \quad \text{for} \quad \epsilon > \epsilon_F.$$

In this limit, we must consider the integral

$$1/N \sum_{\mathbf{q}} f[\epsilon(\mathbf{q})]/[\epsilon(\mathbf{k}) - \epsilon(\mathbf{q})] \approx \int_{\epsilon_F - D}^{\epsilon_F} G(E)/[\epsilon(\mathbf{k}) - E]\, dE \qquad (5.6.14)$$

in which $G(E)$ is the density of states of a single spin and the lower limit of integration is determined by (5.6.6). We will suppose that the region of integration does not cover any singularities of the density of states, and that $G(E)$ is slowly varying compared to the rest of the integrand. Then $G(E)$ can be approximated by $G(\epsilon_F)$, and we have

$$G(\epsilon_F) \int_{\epsilon_F - D}^{\epsilon} dE/[\epsilon(\mathbf{k}) - E] = G(\epsilon_F) \ln([\epsilon(\mathbf{k}) + D - \epsilon_F]/[\epsilon(\mathbf{k}) - \epsilon_F])$$
$$\approx G(\epsilon_F) \ln(D/[\epsilon(\mathbf{k}) - \epsilon_F]). \qquad (5.6.15)$$

The expression diverges logarithmically as $\epsilon(\mathbf{k})$ approaches the Fermi energy. Since the states of interest for electrical conduction are within energy KT of the Fermi energy, there is a marked effect on the resistivity. We now combine (5.6.15), (5.6.13), and (5.6.8) [and continue to ignore the first term of (5.6.13)] to obtain

$$\langle \mathbf{p} \uparrow | t | \mathbf{k} \uparrow \rangle = (-J/2N) S_z\{1 + JG(\epsilon_F) \ln([\epsilon(\mathbf{k}) - \epsilon_F]/D)\} \qquad (5.6.16)$$

and

$$W(\mathbf{k} \uparrow \rightarrow \mathbf{p} \uparrow) = (2\pi/\hbar)(J/2N)^2 S_z^2\{1 + 2JG(\epsilon_F)$$
$$\times \ln([\epsilon(\mathbf{k}) - \epsilon_F]/D)\}\, \delta[\epsilon(\mathbf{k}) - \epsilon(\mathbf{p})]. \qquad (5.6.17)$$

It will be shown subsequently that in order to calculate the electrical conductivity σ (Section 7.2 or Wilson, 1953, Chapter 9) we have to evaluate

$$\sigma = (e^2/12\pi^3) \int (-df(\epsilon)/d\epsilon)\, \tau(\mathbf{k}) v^2(\mathbf{k})\, d^3k \qquad (5.6.18)$$

where $\tau(\mathbf{k})$ is the relaxation time and cubic symmetry has been assumed.

It will be assumed that τ is a function of energy only.

$$W(\epsilon)\,\delta[\epsilon(\mathbf{k}) - \epsilon(\mathbf{p})] = \tau(\epsilon)^{-1}\,\delta[\epsilon(\mathbf{k}) - \epsilon(\mathbf{p})].$$

We will suppose that the band structure is isotropic so that (5.6.18) can be converted to an integral over energy.

$$\sigma = (e^2/12\pi^3)v^2(\epsilon_F)G(\epsilon_F)\int (-\partial f/\partial\epsilon)\,\tau(\epsilon)\,d\epsilon. \tag{5.6.19}$$

Slowly varying quantities have been taken outside the integral. We expand the bracket in (5.6.17). Thus we must consider the integral

$$\int (-df/d\epsilon)\,\{1 - 2JG(\epsilon_F)\ln[(\epsilon - \epsilon_F)/D]\}\,d\epsilon. \tag{5.6.20}$$

The integral may be evaluated as follows. Introduce the variable $x = (\epsilon - \epsilon_F)/KT$. Then (5.6.20) may be written as

$$\int (-df/dx)\,\{1 - 2JG(\epsilon_F)[\ln x - \ln D/KT]\}\,dx$$

$$\approx 1 + 2JG(\epsilon_F)[\ln(D/KT) + \text{const}]. \tag{5.6.21}$$

The conductivity of the system is proportional to (5.6.21). Thus the resistivity, which is the reciprocal of the conductivity, contains a term

$$2JG(\epsilon_F)\,\ln(KT/D). \tag{5.6.22}$$

This result indicates that for negative (antiferromagnetic) J, the resistivity contains a term which increases as $T \to 0$. In an actual material, which contains a finite concentration c, of impurities, the resistivity will be of the form

$$\rho = A - Bc\ln(T/T_0)$$

in which B and T_0 are positive constants: A approaches a constant limit as $T \to 0$, and increases with increasing T ($A \approx \alpha + \beta T^5$), the T^5 term being the electron–phonon contribution discussed in Section 7.6. Thus the resistivity has a minimum at a temperature T^D which is proportional to $c^{1/5}$. Experimental measurements of the resistivity for dilute alloys do show the expected logarithmic behavior, and the $c^{1/5}$ dependence of the position of the minimum has been observed.

A backward look at the calculation just performed indicates that the essential feature is the presence of the Fermi function $f[\epsilon(\mathbf{q})]$ in the second term of (5.6.13). This means that states above the Fermi energy do not contribute in the integration over $\mathbf{q}$, and thus leads to a logarithmic singularity in the t matrix at the Fermi energy. This singularity is sufficiently weak for the electrical resistivity to be finite; however, since electrons in the range of energies $-KT$ about the Fermi energy contribute to conduction, the resistivity contains a term proportional to ln T. The logarithmic temperature dependence of the resistivity cannot persist, however, for extremely low temperatures. The preceding analysis is incomplete in that it does not indicate what happens to this term as $T \rightarrow 0$. We will not investigate this in detail, but remark that the condition that the S matrix be unitary imposes restrictions on the t matrix. In particular the maximum possible scattering occurs when a phase shift passes through $\pi/2$. Suhl and Wong (1967) showed that the resistivity tends to a finite limit at $T = 0$ as the preceding argument indicates that it must. This result has been confirmed by Hepp (1970) who shows rigorously that the scattering matrix is not singular at $T = 0°$K.

5.6.2 *The Ground State*

We will now consider the ground state energy of the system. Mattis (1967) showed rigorously that the ground state of a magnetic atom of spin S interacting through exchange with a nonmagnetic host metal is $S \mp \frac{1}{2}$; where the upper sign pertains to antiferromagnetic coupling and the lower to ferromagnetic coupling. Thus, in particular, if $S = \frac{1}{2}$, the ground state will be singlet or triplet according to the sign of the interaction. However, the argument of Mattis does not permit us to estimate the energies involved. The exact ground state energy is not known. In order to make estimates, we will employ an approximate but revealing variational approach, due to Yosida (1966).

Let us consider the case in which the localized spin has magnitude $\frac{1}{2}$. Our calculation is restricted to $T = 0$. The variational method assumes a trial wave function for the system of localized spin plus free electrons. This function is assumed to be a combination of states in which an electron is excited above the Fermi surface, and the spin of the impurity is adjusted to give a state of the appropriate total spin. This function is, in the notation of second quantization,

$$\Psi = \sum_{\epsilon(k) \geq \epsilon_F} a(\mathbf{k})[c_{\downarrow}^{+}(\mathbf{k})\alpha \mp c_{\uparrow}^{+}(\mathbf{k})\beta) \mid F\rangle \tag{5.6.23}$$

in which α and β represent up and down spin states for the localized spin and the state vector $| F\rangle$ designates the Fermi sea of noninteracting particles. The negative sign is to be used in the case of antiferromagnetic (negative) J and the positive sign in the case of positive J. The choice of sign ensures that the system is in a total spin singlet or triplet state, respectively. The assumed Hamiltonian is

$$H = \sum_{\epsilon(k)<\epsilon_F} \epsilon(\mathbf{k}) c_s^+(\mathbf{k}) c_s(\mathbf{k}) + U \tag{5.6.24}$$

in which U is given by (5.6.4), J being restricted by (5.6.6). A further approximation is made: only those values of $\mathbf{k}$ and $\mathbf{q}$ are retained in (5.6.4) for which both $\epsilon(\mathbf{k})$ and $\epsilon(\mathbf{k} + \mathbf{q})$ are above the Fermi energy.

Equation (5.6.23) is to be substituted into the Schrödinger equation

$$H\Psi = E\Psi. \tag{5.6.25}$$

It is desirable to obtain an equation satisfied by the coefficients $a(\mathbf{k})$. To this end, we form the scalar product of (5.6.25) with the conjugate vector

$$\langle F \mid [c_\downarrow(\mathbf{k}')\alpha^+ + c_\uparrow(\mathbf{k}')\beta^+] \qquad [\epsilon(\mathbf{k}') > \epsilon_F].$$

The resulting expression is evaluated through the use of the properties of the spin operators and the anticommutation rules (5.6.3) for the operators $c_s(\mathbf{k})$. It is useful to recall that

$$c_s^+(\mathbf{k}) \mid F\rangle = 0 \qquad \text{if} \quad \epsilon(\mathbf{k}) < \epsilon_F \tag{5.6.26a}$$

and

$$c_s(\mathbf{k}) \mid F\rangle = 0 \qquad \text{if} \quad \epsilon(\mathbf{k}) > \epsilon_F \tag{5.6.26b}$$

since we are considering only $T = 0$. It is also useful to change the summation from $\mathbf{q}$ to $\mathbf{k} + \mathbf{q}$ in (5.6.4).

The result of this calculation is

$$[\epsilon(\mathbf{k}') - E]a(\mathbf{k}') + (3J/4N) \sum_{\mathbf{q}} a(\mathbf{q}) = 0 \tag{5.6.27a}$$

in the singlet case and

$$[\epsilon(\mathbf{k}') - E]a(\mathbf{k}') - (J/4N) \sum_{\mathbf{q}} a(\mathbf{q}) = 0 \tag{5.6.27b}$$

for the triplet. An equation can be readily obtained for the eigenvalues. Divide (5.6.26) and (5.6.27) by $[\epsilon(\mathbf{k}') - E]$ and sum over $\mathbf{k}'$. We then use the fact that

$$\sum_{\mathbf{k}'} a(\mathbf{k}') = \sum_{\mathbf{q}} a(\mathbf{q}) = \text{const} \tag{5.6.28}$$

and obtain

$$1 = -(3J/4N) \sum_{\mathbf{k}'} 1/[\epsilon(\mathbf{k}') - E] \tag{5.6.29}$$

for the singlet state and

$$1 = (J/4N) \sum_{\mathbf{k}'} 1/[\epsilon(\mathbf{k}') - E] \tag{5.6.30}$$

for the triplet. The sum can be converted to an integral over the density of states

$$J/N \sum 1/[\epsilon(\mathbf{k}') - E] = J \int G(\epsilon)\, d\epsilon/(\epsilon - E)$$

$$\approx JG(\epsilon_F) \int_{\epsilon_F}^{\epsilon_F+D} d\epsilon/(\epsilon - E). \quad (5.6.31)$$

In the last step we have used (5.6.6), and the fact that our summations have been restricted to states above the Fermi energy. Equation (5.6.29) then gives

$$1 = (-3J/4)G(\epsilon_F)\ln[(\epsilon_F + D - E)/(\epsilon_F - E)]. \quad (5.6.32)$$

This equation is readily solved to give

$$\epsilon_F - E = D\{\exp[-4/(3JG(\epsilon_F))] - 1\}^{-1}$$

$$\approx D\exp[4/(3JG(\epsilon_F))] = D\exp\{-4/[3\,|\,J\,|\,G(\epsilon_F)]\}. \quad (5.6.33)$$

The approximation in the last line of (5.6.33) is valid in the (usual) case where $|\,J\,|\,G(\epsilon_F)$ is small in magnitude. Similarly it follows from (5.6.30) that in the ferromagnetic case (positive J) that

$$\epsilon_F - E = D\{\exp[4/JG(\epsilon_F)] - 1\}^{-1} \approx D\exp[-4/JG(\epsilon_F)]. \quad (5.6.34)$$

Thus in both cases a bound state is obtained with energy below the Fermi energy. The form of these expressions is such that the energies do not possess an expansion in powers of J, even though the binding is very small for small J. In fact, the binding energy vanishes faster as $|\,J\,| \to 0$ than any power of J. Since no expansion in J is possible, we see that these energies could not have been obtained through the application of Raleigh–Schrödinger perturbation theory, which would yield such an expansion. In fact, it is apparent that the perturbation series does not converge, even though the interaction seems to be weak.

The variational treatment does not yield the exact energy of the ground state (which is not known at present) because the assumed wave function is not an eigenfunction of the full Hamiltonian. Study of the details of the calculations leading to (5.6.26) and (5.6.27) shows, however, that the assumed function (5.6.23) is an eigenfunction of the truncated Hamiltonian, including only terms involving states with energies above the Fermi energy. Our answers to this modified problem are exact. This occurs because the truncation of the Hamiltonian implies $U\,|\,F\rangle = 0$. As a result, we see that the deficiency of the present approach lies in the neglect of scattering across the Fermi surface (or electron–hole interactions).

Let us consider the nature of the assumed wave function in more detail, with the object of determining the asymptotic spatial dependence of the spin polarization produced in the electron system by interaction with the local spin. For simplicity, we consider a parabolic band $\epsilon(\mathbf{k}) = \gamma\mathbf{k}^2$. The energy of the bound state will be written

$$E = \gamma k_F^2(1 - \Delta) \tag{5.6.35}$$

where k_F is the radius of the Fermi sphere and

$$\Delta = (D/\epsilon_F) \exp[-4/(3\,|\,J\,|\,G(\epsilon_F))] \tag{5.6.36}$$

in the case of an antiferromagnetic interaction. The quantity of interest is the Fourier transform of the eigenvector $a(\mathbf{k})$ which will be denoted $a(\mathbf{r})$:

$$a(\mathbf{r}) = (1/N) \sum_{k>k_F} a(\mathbf{k}) \exp(i\mathbf{k}\cdot\mathbf{r}) \propto \text{const} \int_{k_F}^{\infty} \exp(i\mathbf{k}\cdot\mathbf{r})/[\epsilon(\mathbf{k}) - E]\, d^3k. \tag{5.6.37}$$

The integral becomes

$$\int_{k_F}^{\infty} \exp(i\mathbf{k}\cdot\mathbf{r})/[\epsilon(\mathbf{k}) - E]\, d^3k$$

$$= 4\pi/\gamma r \int_{k_F}^{\infty} k \sin kr/[k^2 - k_F^2(1 - \Delta)]\, dk.$$

We are interested in the result in the limit of large r. In this case, we can obtain a series in terms of $(1/r)$ by repeated integration by parts. The leading term is the one we want, and this yields

$$a(\mathbf{r}) \propto (\cos k_F r)/\Delta r^2. \tag{5.6.38}$$

This function oscillates, and falls off quite slowly with distance. The spin disturbance in the system has a rather long range. This result is a consequence of the sharp cutoff of the integration on the Fermi sphere, as was seen to be the case for the charge density oscillations discussed in Section 5.5.

The results presented above pertain to the truncated Hamiltonian described at the beginning of this subsection. The situation for the full Hamiltonian containing the interaction (5.6.4) is still somewhat obscure. Hepp (1970) has shown that the time development operator (see Section 8.1) is analytic in the strength of the potential. This implies that the energy of the system is also analytic in the potential strength in contrast to Eqs.

(5.6.33) and (5.6.34). Furthermore, the disturbance of the spin density in the system is apparently milder than that described by (5.6.38).

PROBLEMS

1 The impurity levels associated with an isolated donor atom are spin degenerate. However, the repulsion between electrons is strong enough so that only one electron can occupy such a state. Show that, in order to describe this situation, the usual Fermi distribution function must be replaced by

$$f(E) = \tfrac{1}{2}\{1 + \tfrac{1}{2}\exp[(E - \mu)/KT]\}^{-1}.$$

2 A semiconductor contains a concentration n_a of acceptor states (degeneracy 4) and n_d donor states (degeneracy 2). Find the position of the Fermi level at low temperatures if (a) $n_a \gg n_d$ or (b) $n_a \gg n_a$.

3 Show that the change in the total energy of a semiconductor, approximated by the sum of one-electron energies, which contains a small concentration c of impurities producing a localized potential too weak to create bound states is

$$\Delta E = -(2c/\pi) \sum_s g_s \int \delta_s(E)\, dE$$

in which the integral includes all the occupied states.

4 Suppose that a scattering potential is localized and is weak enough to be treated in first-order Born approximation. Prove that

$$\delta_T(E) = \sum g_s\, \delta_s(E) = -(2m^*k/\hbar^2) \int V(r) r^2\, dr.$$

Assume that the band is parabolic and that the scattering potential has spherical symmetry.

5 Show that, if the results of Problem 4 are applied, and the Friedel sum rule holds, the change in the total energy of a monovalent metal caused by the introduction of a vacancy is $\tfrac{2}{3}E_F$. Suppose further that the volume of the system is increased (the displaced atom is placed on the surface). Then show that the change in the total energy of the system is $4E_F/15$. Treat the electrons as free in the final computation.

6 A vibrating lattice contains an atom whose mass is different from that of all the other atoms by an amount δM. Under what conditions will a localized vibrational mode or a scattering resonance be found?

7 Determine the scattering cross section for phonons in a simple cubic lattice in the presence of a mass defect (see Problem 6). Assume nearest neighbor harmonic forces.

8 Determine the scattering cross section for spin waves in a Heisenberg model of a simple cubic ferromagnet with nearest neighbor interactions if one atom has a different spin and a different exchange coupling.

9 Evaluate $I(E)$ as given by Eq. (5.2.84) for a rectangular density of states: $G(E) = 1/W$ $(0 \leq E \leq W)$, and G is zero outside this range. Determine the positions of bound states and construct the scattering phase shifts in a Koster–Slater model.

10 Obtain the solutions of Eq. (5.4.18a) for $\epsilon = 1.0$, $\zeta = 3.0$.

References

Abrikosov, A. A. (1969). *Usp. Phys. Nauk* **97,** 403 [*English transl.: Sov. Phys. Usp.* **12,** 168].
Adams, E. N. (1952). *Phys. Rev.* **85,** 41.
Adams, E. N. (1953). *J. Chem. Phys.* **21,** 2013.
Aggarwal, R. L., and Ramdas, A. K. (1965). *Phys. Rev.* **140,** A1246.
Anderson, P. W. (1958). *Phys. Rev.* **109,** 1492.
Anderson, P. W. (1961). *Phys. Rev.* **124,** 41.
Bailyn, M. (1966). *Advan. Phys.* **15,** 179.
Bloembergen, N., and Rowland, T. J. (1953). *Acta. Met.* **1,** 731.
Blount, E. I. (1962). *Solid State Phys.* **13,** 305.
Butler, W. H., and Kohn, W. (1971). "Electronic Density of States" (L. H. Bennett, ed.), p. 465. NBS Special Publ. Washington, D.C.
Callaway, J. (1964). *J. Math. Phys.* **5,** 783.
Callaway, J. (1967). *Phys. Rev.* **154,** 515.
Callaway, J., and Boyd, R. (1964). *Phys. Rev.* **134,** A1655.
Callaway, J., and Hughes, A. J. (1967). *Phys. Rev.* **156,** 860.
des Cloizeaux, J. (1964a). *Phys. Rev.* **135,** A685.
des Cloizeaux, J. (1964b). *Phys. Rev.* **135,** A698.
Daybell, M. D., and Steyert, W. A. (1968). *Rev. Mod. Phys.* **40,** 238.
Edwards, S. F. (1961). *Phil. Mag.* **6** (Ser. 8), 617.
Faulkner, R. A. (1969). *Phys. Rev.* **184,** 713.
Friedel, J. (1952). *Phil. Mag.* [7], 43, 153.
Friedel, J. (1954). *Advan. Phys.* **3,** 446.
Friedel, J. (1958). *Nuovo Cimento* [10], **7,** Suppl. 287.
Gauthier, F., and Lenglart, P. (1965). *Phys. Rev.* **139,** A705.
Gell-Mann, M., and Goldberger, M. L. (1953). *Phys. Rev.* **91,** 398.
Goldberger, M. L., and Watson, K. M. (1964). "Collison Theory," p. 236. Wiley, New York.
Hedgcock, F. T., and Rizzuto, C. (1967). *Phys. Rev.* **163,** 517.
Heeger, A. J. (1969). *Solid State Phys.* **23,** 283.
Hepp, K. (1970). *Solid State Commun.* **8,** 2087.

Hulthén, L., and Laurikainen, T. (1951). *Rev. Mod. Phys.* **23,** 1.
Kasuya, T. (1956). *Progr. Theor. Phys.* **16,** 45.
Kirkpatrick, S., Velicky, B., and Ehrenreich, H. (1970). *Phys. Rev.* **B1,** 3250.
Kittel, C., and Mitchell, A. H. (1954). *Phys. Rev.* **96,** 1488.
Kohn, W. (1957). *Phys. Rev.* **105,** 509.
Kohn, W. (1958). *Phys. Rev.* **10,** 857.
Kohn, W. (1959). *Phys. Rev.* **115,** 809.
Kohn, W. (1973). *Phys. Rev.* **B7,** 4388.
Kohn, W., and Luttinger, J. M. (1955). *Phys. Rev.* **98,** 915.
Kohn, W., and Majumdar, C. (1965). *Phys. Rev.* **138,** A1617.
Kohn, W., and Vosko, S. H. (1960). *Phys. Rev.* **119,** 912.
Kondo, J. (1962). *Progr. Theor. Phys.* **28,** 846.
Kondo, J. (1964). *Progr. Theor. Phys.* **32,** 37.
Kondo, J. (1969). *Solid State Phys.* **23,** 183.
Koster, G. F. (1953). *Phys. Rev.* **89,** 67.
Koster, G. F. (1954). *Phys. Rev.* **95,** 1436.
Koster, G. F., and Slater, J. C. (1954a). *Phys. Rev.* **95,** 1167.
Koster, G. F., and Slater, J. C. (1954b). *Phys. Rev.* **96,** 1208.
von der Lage, F. C., and Bethe, H. A. (1947). *Phys. Rev.* **71,** 612.
Lampert, M. (1955). *Phys. Rev.* **97,** 352.
Landau, L. D., and Lifshitz, E. M. (1965). "Quantum Mechanics." Addison Wesley, Reading, Massachusetts.
Landsberg, P. T. (ed.) (1969). "Solid State Theory: Methods and Applications." Wiley (Interscience), New York.
Levinger, B., and Frankl, D. (1961). *J. Phys. Chem. Solids* **20,** 281.
Lifshitz, I. M. (1948). *Zh. Ekspim. Theoret. Fiz.* **18,** 293.
Lifshitz, I. M. (1964). *Usp. Fiz. Nauk* **83,** 617. [*English transl.: Sov. Phys. Usp.* **7,** 549.]
Lowdin, P. O. (1962). *J. Appl. Phys.* **33,** 251.
Luttinger, J. M., and Kohn, W. (1955). *Phys. Rev.* **97,** 869.
Mattis, D. C. (1967). *Phys. Rev. Lett.* **19,** 479.
Messiah, A. (1962) "Quantum Mechanics." North Holland, Amsterdam.
Mott, N. F. (1949). *Proc. Phys. Soc.* **A62,** 416.
Mott, N. F. (1956). *Can. J. Phys.* **34,** 1336.
Mott, N. F. (1961). *Phil. Mag.* **6,** 287.
Mott, N. F. (1967). *Advan. Phys.* **16,** 49.
Mott, N. F. (1968). *Rev. Mod. Phys.* **40,** 677.
Mott, N. F., and Davis, E. A. (1971). "Electronic Processes in Non Crystalline Materials." Clarendon Press, Oxford.
Mott, N. F., and Twose, W. D. (1961). *Advan. Phys.* **10,** 107.
Newton, R. G. (1966). "Scattering Theory of Particles and Waves." McGraw–Hill, New York.
Parzen, G. (1953). *Phys. Rev.* **89,** 237.
Preziosi, B. (1971). *Nuovo Cimento* **6B,** 131.
Reuszer, J. H., and Fisher, P. (1964). *Phys. Rev.* **135,** A1125.
Rowland, T. J. (1960). *Phys. Rev.* **119,** 900.
Schechter, D. (1962). *J. Phys. Chem. Solids* **23,** 237.
Schrieffer, J. R., and Wolff, P. A. (1966). *Phys. Rev.* **149,** 491.
Seitz, F. (1940). "The Modern Theory of Solids." McGraw–Hill, New York.
Sham, L. J. (1966). *Phys. Rev.* **150,** 720.
Slater, J. C. (1952). *Phys. Rev.* **87,** 807.

Soven, P. (1967). *Phys. Rev.* **156,** 809.
Stocks, G. M., Williams, R. W., and Faulkner, J. S. (1971). *Phys. Rev.* **B4,** 4390.
Suhl, H., and Wong, D. (1967). *Physics* **3,** 17.
Summers, C. D., Dingle, R., and Hill, D. E. (1970). *Phys. Rev.* **B1,** 1603.
Velicky, B., Kirkpatrick, S., and Ehrenreich, H. (1968). *Phys. Rev.* **175,** 747.
Wainwright, T., and Parzen, G. (1953). *Phys. Rev.* **92,** 1129.
Wannier, G. (1937). *Phys. Rev.* **52,** 191.
Wigner, E. P. (1955). *Phys. Rev.* **98,** 145.
Wilson, A. H. (1953). "The Theory of Metals." Cambridge Univ. Press, London and New York.
Winston, H. (1954). *Phys. Rev.* **94,** 328.
Wolff, P. A. (1961). *Phys. Rev.* **124,** 1030.
Yosida, K. (1966). *Phys. Rev.* **147,** 223.
Zak, J. (1968). *Phys. Rev.* **168,** 687.
Zak, J. (1969). *Phys. Rev.* **177,** 1151.
Zak, J. (1972). *Solid State Phys.* **27,** 1.

CHAPTER 6

External Fields

In this chapter, we will investigate the effects of external electric and magnetic fields on the energy levels and wave functions of solids. We will also consider the transitions between these levels which may be produced by time-dependent electromagnetic fields. Some of the more important experimental techniques by which information is obtained about energy bands will be analyzed. Our treatment will be based, in the main, on the use of the crystal momentum representation described in Section 5.1.

6.1 The Steady Electric Field

In this section we will consider solids subjected to steady and uniform electric fields. The study of the motion of electrons in solids in such fields has turned out, perhaps rather surprisingly, to be quite complicated. The field is considered to contribute to the Hamiltonian (for a single electron) a term

$$U = -\mathbf{F}\cdot\mathbf{r} \tag{6.1.1}$$

where the force $\mathbf{F} = e\boldsymbol{\mathcal{E}}$ and $\boldsymbol{\mathcal{E}}$ is the electric field strength. An alternative approach to the electric field problem which uses, in the case of a uniform and time-independent field, a vector potential proportional to time, is possible, but will not be employed here. These approaches are connected by a gauge transformation.

The source of the difficulties connected with the electric field problem is that no matter how small the field strength is, for sufficiently large distances, the potential (6.1.1) becomes arbitrarily strong. In more mathematical terms, it is said that U (and thus the Hamiltonian, when U is included) is not a bounded operator. In consequence, straightforward

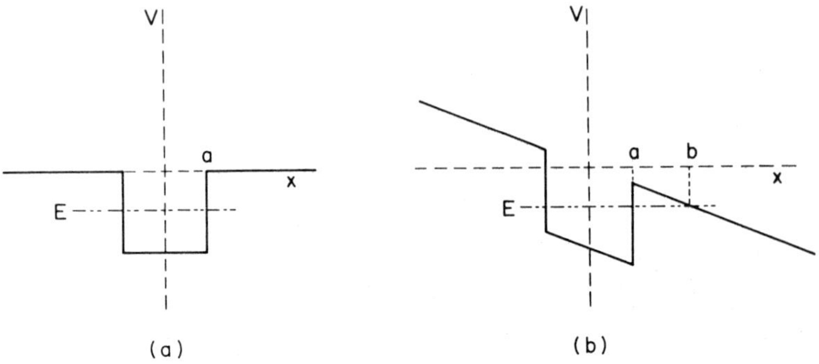

Fig. 6.1.1. Effect of an electric field on a square well in one dimension.

application of perturbation theory is dangerous, and important physical quantities may not possess a power series expansion in the field strength.

An additional complication is that there are no ordinary localized bound states when the Hamiltonian contains a term of the form (6.1.1). To see this, consider the potential energy diagrams of Fig. 6.1.1, which show the effect of an electric field on a square well potential in one dimension. It is seen that, for any energy E which, in the absence of the field, produces a bound state in the well, there is no bound state when the field is present because the electron has a finite probability of tunneling out to the right. On the right, the wave function will be in (in the WKB approximation)

$$[\text{const}/(E + Fx)^{1/4}] \exp[i(2m)^{1/2}\{(-2/3F\hbar)(E+Fx)^{3/2}\}]. \quad (6.1.2)$$

This is not normalizable. The transmission coefficient through the barrier, which is the ratio of the square magnitude of the function outside the well to that in the well is, in the WKB approximation, approximately (Bohm, 1951)

$$\exp[-\tfrac{4}{3}(2m)^{1/2}(E^{3/2}/\hbar \mid F \mid)]. \quad (6.1.3)$$

Although this expression is very small for small F, it does not possess a power series in F, and thus could not have been obtained by a straightforward perturbation treatment of (6.1.1) which would generate a power series. Of course, in very many practical situations, the field is sufficiently weak so that the lifetime of a bound state is very long indeed, and the energy is, for all reasonable purposes, quite sharp. In the case of strong fields, tunneling is quite important as we shall see in more detail later.

6.1.1 *Motion of a Wave Packet*

Let us consider the time-dependent Schrödinger equation for an electron in an electric field. For convenience, we will assume that the electric field is in the x direction. The crystal momentum representation is employed, which means that the electron wave function $\Psi(\mathbf{r}, t)$ is expanded in Bloch functions $\psi_n(\mathbf{k}, \mathbf{r})$

$$\Psi(\mathbf{r}, t) = \sum_n \int d^3k\, \phi_n(\mathbf{k}, t)\psi_n(\mathbf{k}, \mathbf{r}). \tag{6.1.4}$$

The function $\phi_n(\mathbf{k}, t)$ is the wave function in the crystal momentum representation. It satisfies the time-dependent Schrödinger equation in the crystal momentum representation, Eq. (5.1.6)

$$[E_n(\mathbf{k}) - i\hbar\, \partial/\partial t]\phi_n(\mathbf{k}) - F \sum_l \int d^3q \langle n\mathbf{k} \mid x \mid l\mathbf{q}\rangle \phi_l(\mathbf{q}) = 0. \tag{6.1.5}$$

It is necessary to determine the matrix elements of the coordinate x

$$\begin{aligned}\langle n\mathbf{k} \mid x \mid l\mathbf{q}\rangle &= \int \exp[i(\mathbf{q} - \mathbf{k})\cdot\mathbf{r}]u_n^*(\mathbf{k}, \mathbf{r})\, x u_l(\mathbf{q}, \mathbf{r})\, d^3r \\ &= i^{-1}[(\partial/\partial q_x) \int \exp[i(\mathbf{q} - \mathbf{k})\cdot\mathbf{r}]u_n^*(\mathbf{k}, \mathbf{r}) u_l(\mathbf{q}, \mathbf{r})\, d^3r \\ &\quad - \int \exp[i(\mathbf{q} - \mathbf{k})\cdot\mathbf{r}]u_n^*(\mathbf{k}, \mathbf{r})\, (\partial/\partial q_x) u_l(\mathbf{q}, \mathbf{r})\, d^3r].\end{aligned}$$

The quantity $u_n^*(\partial/\partial q_x) u_l$ is a periodic function of position. As a result, the second integral is proportional to $\delta(\mathbf{q} - \mathbf{k})$. We obtain

$$\langle n\mathbf{k} \mid x \mid l\mathbf{q}\rangle = \delta_{nl}\, i\, (\partial/\partial k_x)\, \delta(\mathbf{k} - \mathbf{q}) + X_{nl}\, \delta(\mathbf{k} - \mathbf{q}) \tag{6.1.6}$$

where

$$X_{nl} = [(2\pi)^3 i/\Omega] \int u_n^*(\mathbf{k}, \mathbf{r})\, (\partial/\partial k_x)\, u_l(\mathbf{k}, \mathbf{r})\, d^3r, \tag{6.1.7}$$

and the integral in (6.1.7) includes a single cell. We have calculated the matrix element in a manner which apparently ignores questions of convergence. For a more careful discussion, see Blount (1962). The effective Schrödinger equation now has the form

$$[E_n(\mathbf{k}) - iF\, (\partial/\partial k_x) - i\hbar\, (\partial/\partial t)]\phi_n(\mathbf{k}) - F \sum_l X_{nl}(\mathbf{k})\phi_l(\mathbf{k}) = 0. \tag{6.1.8}$$

Our further discussion is based on this equation.

An important general result can be obtained immediately. If we multiply (6.1.8) by $\phi_n^*(\mathbf{k})$, subtract from the resulting equation its complex conjugate, and then sum on the band index n, the term involving the X_{nl} disappears on summation. This follows because $X_{nl} = X^*_{ln}$. We then obtain

$$[F\,(\partial/\partial k_x) + \hbar\,(\partial/\partial t)] \sum_n |\,\phi_n(\mathbf{k})\,|^2 = 0. \tag{6.1.9}$$

The general solution of this equation is

$$\sum_n |\,\phi_n(\mathbf{k})\,|^2 = G(k_x - Ft/\hbar,\, k_y,\, k_z) \tag{6.1.10}$$

where G is an arbitrary function of its arguments. We will consider a normalizable wave function in (6.1.4). This implies that

$$\sum_n \int |\,\phi_n(\mathbf{k})\,|^2\, d^3k = \int G(k_x - Ft/\hbar,\, k_y,\, k_z)\, d^3k = 1 \tag{6.1.11}$$

for all times. Consider now the average $\mathbf{k}$ contained in Ψ, that is, the centroid of the wave packet. This is defined to be

$$\langle\mathbf{k}\rangle = \sum_n \int |\,\phi_n(\mathbf{k})\,|^2 \mathbf{k}\, d^3k. \tag{6.1.12}$$

Only the x component is of interest as only it is affected by the field. We find

$$\langle k_x\rangle = \langle k_x\rangle_0 + Ft/\hbar \tag{6.1.13a}$$

where $\langle k_x\rangle_0$ is the location of the centroid at $t = 0$. This equation, which is just what would be obtained in classical mechanics for a particle of charge e and momentum $\hbar\mathbf{k}$, is valid for small times, and requires a rather obvious modification if a boundary of the Brillouin zone is crossed (see Section 6.1.3). One often finds in the literature the alternative form

$$dk_x/dt = F/\hbar \tag{6.1.13b}$$

which must be understood as referring to the centroid of the packet.

6.1.2 *Velocity and Acceleration Theorems*

We will now examine (6.1.8) in more detail. The diagonal matrix element X_{nn} will not vanish at a general point of the Brillouin zone, nor need it vanish at symmetry points unless the group of the wave vector contains the inversion. This matrix element gives rise to a displacement of the band structure which is linear in the applied field. We can incorporate this shift into the first term of (6.1.8) through the definition

$$E_n^{(1)}(\mathbf{k}) = E_n(\mathbf{k}) - FX_{nn}(\mathbf{k}). \tag{6.1.14}$$

Then Eq. (6.1.8) becomes

$$[E_n^{(1)}(\mathbf{k}) - iF\,\partial/\partial k_x - i\hbar\,\partial/\partial t]\phi_n(\mathbf{k}) - F\sum_{l,\,l\neq n} X_{nl}\phi_l(\mathbf{k}) = 0. \qquad (6.1.15)$$

The motion of an electron in a given band is described by the terms in the square bracket; the off-diagonal elements produce transitions between bands. These transitions comprise the phenomena of tunneling.

Tunneling will be seen to be important only for strong fields. We will neglect the off-diagonal terms for the present and consider the resulting reduced equation for functions we now call $\phi_n^{(1)}$:

$$[E_n^{(1)}(\mathbf{k}) - iF\,\partial/\partial k_x - i\hbar\,\partial/\partial t]\phi_n^{(1)}(\mathbf{k}) = 0. \qquad (6.1.16)$$

Our first step will be to derive some results concerning the motion of electrons in very weak fields. Let us suppose we are concerned with a wave packet, formed from states in a single band:

$$\Psi(\mathbf{r}, t) = \int \phi(\mathbf{k}, t)\psi(\mathbf{k}, \mathbf{r})\, d^3k. \qquad (6.1.17)$$

Since only a single band is considered, we suppress the band index. The reason for discussing a wave packet is that we wish to describe a situation in which the electron is roughly localized in both $\mathbf{r}$ and $\mathbf{k}$; such a situation requires a superposition of Bloch states. Let us consider the velocity of an electron in such a state. Initially (but obviously not for long times), we may begin by neglecting the electric field in (6.1.16) in which case

$$\phi(\mathbf{k}, t) = f(\mathbf{k}) \exp[-(it/\hbar)E(\mathbf{k})] \qquad (6.1.18)$$

where $f(\mathbf{k})$ is independent of time. We assume that $f(\mathbf{k})$ is strongly peaked at some $\mathbf{k}_0$. Then $E(\mathbf{k})$ is expanded about this point

$$E(\mathbf{k}) = E(\mathbf{k}_0) + (\mathbf{k} - \mathbf{k}_0)\cdot\nabla_{\mathbf{k}}E(\mathbf{k}) + \cdots. \qquad (6.1.19)$$

We write the Bloch functions in the form $\exp(i\mathbf{k}\cdot\mathbf{r})u(\mathbf{k}, \mathbf{r})$, and suppose that the variation of $u(\mathbf{k}, \mathbf{r})$ with respect to $\mathbf{k}$ is much slower than that of $f(\mathbf{k})$, so that it may be taken out of the integral in the form $u(\mathbf{k}_0, \mathbf{r})$. Then

$$\Psi(\mathbf{r}, t) = \psi(\mathbf{k}_0, \mathbf{r}) \exp[-iE(\mathbf{k}_0)t/\hbar]$$
$$\times \int f(\mathbf{k}) \exp i(\mathbf{k} - \mathbf{k}_0)\cdot[\mathbf{r} - \nabla_{\mathbf{k}}E(\mathbf{k}_0)t/\hbar]\, d^3k. \qquad (6.1.20)$$

The wave packet has constant amplitude for positions for which

$$\mathbf{r} - \nabla_{\mathbf{k}}E(\mathbf{k}_0)t/\hbar = \text{const.} \qquad (6.1.21)$$

It follows from this that the group velocity of the packet is

$$\mathbf{v}_g = (1/\hbar)\ \nabla_k E(\mathbf{k}). \tag{6.1.22}$$

A more formal argument can be based on considerations of the velocity operator

$$\mathbf{v} = (i/\hbar)[H\mathbf{r} - \mathbf{r}H]. \tag{6.1.23}$$

Let us consider this operator in the crystal momentum representation. The matrix elements in the crystal momentum representation are highly singular; however, a meaningful result is obtained if we consider a state of the form (6.1.17) and define

$$\begin{aligned}\langle \mathbf{v} \rangle &= \int \Psi^*(\mathbf{r}, t)\mathbf{v}\Psi(\mathbf{r}, t)\ d^3r \\ &= (i/\hbar) \int \phi^*(\mathbf{q}, t)\{E(\mathbf{q})\langle \mathbf{q} \mid \mathbf{r} \mid \mathbf{k}\rangle - \langle \mathbf{q} \mid \mathbf{r} \mid \mathbf{k}\rangle E(\mathbf{k})\}\phi(\mathbf{k}, t)\ d^3k\ d^3q.\end{aligned} \tag{6.1.24}$$

The matrix element $\langle \mathbf{q} \mid \mathbf{r} \mid \mathbf{k}\rangle$ is obtained from (6.1.6). The following result is obtained after an integration by parts:

$$\langle \mathbf{v} \rangle = (1/\hbar) \int \phi^*(\mathbf{k}, t)\ \nabla_k E(\mathbf{k})\ \phi(\mathbf{k}, t)\ d^3k. \tag{6.1.25}$$

It is useful to obtain an expression for the time rate of change of the average velocity as given by (6.1.25). An acceleration is defined by

$$d\langle \mathbf{v} \rangle/dt = 1/\hbar\ (d/dt) \int \phi^*(\mathbf{k}, t)\ \nabla_k E(\mathbf{k})\ \phi(\mathbf{k}, t)\ d^3k. \tag{6.1.26}$$

This expression may be reduced to an interesting form with the use of (6.1.16):

$$\begin{aligned}d\langle \mathbf{v} \rangle/dt &= (1/\hbar) \int [\partial\phi^*(\mathbf{k}, t)/\partial t\ \nabla_k E(\mathbf{k})\ \phi(\mathbf{k}, t) + \phi^*(\mathbf{k}, t) \\ &\qquad \times \nabla_k E(\mathbf{k})\ \partial\phi(\mathbf{k}, t)/\partial t]\ d^3k \\ &= (-1/\hbar)(F/\hbar) \int [\partial\phi^*/\partial k_x\ \nabla_k E\phi + \phi^*\ \nabla_k E\ \partial\phi/\partial k_x]\ d^3k \\ &= (F/\hbar^2) \int \phi^*\ [\partial(\nabla_k E)/\partial k_x]\phi\ d^3k.\end{aligned} \tag{6.1.27}$$

The third line of (6.1.27) is obtained from the second through an integration by parts. The result may be interpreted with the aid of the following definition of the acceleration associated with a state of wave vector **k**, $\boldsymbol{\alpha}(\mathbf{k})$:

$$d\langle\mathbf{v}\rangle/dt = \int \phi^* \boldsymbol{\alpha}(\mathbf{k})\phi \; d^3k. \tag{6.1.28}$$

The acceleration is

$$\boldsymbol{\alpha}(\mathbf{k}) = F/\hbar^2 \, \partial(\nabla_{\mathbf{k}} E(\mathbf{k}))/\partial k_x \tag{6.1.29a}$$

or, in the case of an arbitrarily directed field

$$\boldsymbol{\alpha}(\mathbf{k}) = (1/\hbar^2)(\mathbf{F}\cdot\nabla_{\mathbf{k}}) \, \nabla_{\mathbf{k}} E(\mathbf{k}). \tag{6.1.29b}$$

This expression can be simplified with the aid of the reciprocal effective mass tensor defined by Eq. (4.1.11):

$$\alpha_j(\mathbf{k}) = \sum_i (1/m^*)_{ji} F_i \tag{6.1.30a}$$

(where α_α is the jth rectangular component of the acceleration and F_i the ith component of the force exerted by the field). An alternate form is

$$\sum_j m^*_{ij}\alpha_j = F_i. \tag{6.1.30b}$$

These results are just what would be obtained in classical mechanics for a particle characterized by a tensor effective mass.

6.1.3 *Wannier Levels*

We will now obtain a stationary solution of (6.1.16) which shows the existence of a set of discrete levels (Kane, 1959; Argyres, 1962). Assume that the functions $\phi_n^{(1)}$ have the time dependence $\exp[(-i/\hbar)Et]$. Then we obtain

$$[E_n^{(1)}(\mathbf{k}) - iF \, \partial/\partial k_x]\phi_n^{(1)}(\mathbf{k}) = E\phi_n^{(1)}(\mathbf{k}). \tag{6.1.31}$$

A solution of this equation can be obtained immediately:

$$\phi^{(1)}(\mathbf{k}) = f(k_y, k_z) \exp\left\{i/F \int_0^{k_x} [E - E^{(1)}(\mathbf{k}')] \, dk_x'\right\}. \tag{6.1.32}$$

The function $f(k_y, k_z)$ is arbitrary. It is natural to impose the condition that the wave packet (6.1.17) should be characterized by definite values of the components of **k** perpendicular to the direction of the field. Let these components be k_{0y} and k_{0z}. Thus we choose

$$\phi^{(1)}(\mathbf{k}) = \kappa^{-1/2}\,\delta(k_y - k_{0y})\delta(k_z - k_{0z})\exp\left\{(i/F)\int_0^{k_x}[E - E^{(1)}(\mathbf{k}')]\,dk_x'\right\}. \tag{6.1.33}$$

The function has been normalized by the introduction of the constant κ which is the length of the line (k_x, k_{0y}, k_{0z}) lying within the Brillouin zone. We assume, for simplicity, that this line is parallel to a reciprocal lattice vector; or in other words, that the direction of the electric field coincides with one of the vectors of the reciprocal lattice. The energy eigenvalue E is then determined by the condition that the wave function must be the same at the end points of this line, since these are equivalent points. Then the change of the phase of the wave function in a distance κ must be an integral multiple of 2π:

$$(1/F)\int_{-\kappa/2}^{\kappa/2}[E - E^{(1)}(\mathbf{k}')]\,dk_x' = 2\nu\pi$$

where ν is a positive or negative integer. The eigenvalues in the presence of the field can be designated by an index ν,

$$E_\nu = \nu 2\pi F/\kappa + W(k_y, k_z) \tag{6.1.34a}$$

where

$$W = (1/\kappa)\int_{-\kappa/2}^{\kappa/2} E^{(1)}(k_x', k_y, k_z)\,dk_x'. \tag{6.1.34b}$$

The wave function will be denoted $\phi_\nu^{(1)}$. The energy spectrum in the presence of the electric field contains a series of discrete levels, separated by $2\pi F/\kappa$. The existence of these levels was first pointed out by Wannier (1959, 1960, 1962), and they will be designated as Wannier levels. The term "Stark ladder" is also applied. If the lattice parameter is d, their separation is of the order Fd, but there will be a constant of proportionality which will depend on the field direction. (However, the separation is undoubtedly small for attainable fields; being of the order of 0.005 eV in the case of fields of strength $\sim 10^7$ V/m.) In the simple case of a parabolic band $E = \gamma\mathbf{k}^2$ we obtain (neglecting a possible contribution from the X_{nn} term)

$$E_\nu = \nu(2\pi F/\kappa) + (\gamma\kappa^2/12) + \gamma(k_y^2 + k_z^2). \tag{6.1.35}$$

In this case, the band structure is unaltered in directions perpendicular to the electric field, but has been destroyed, and replaced by a set of discrete steps in the case of motion parallel to the field.

The existence of these levels, which have no counterpart in nonperiodic systems, can be made plausible by the following argument: According to (6.1.13), the centroid of the electron wave packet moves through **k** space steadily. The time required to cross the Brillouin zone in some direction (assumed to be of length κ) is

$$T = \hbar\kappa/F. \tag{6.1.36}$$

When the electron wave vector reaches the Brillouin zone boundary, an umklapp process will occur, that is, the electron wave vector is changed by a reciprocal lattice vector. If the field direction is parallel to the relevant reciprocal lattice vector, the wave vector will appear at an opposite point of the Brillouin zone, and the process will be repeated. As long as the electron does not tunnel into another band at higher energy, it must oscillate within the single band. Equation (6.1.36) gives the period of such oscillations, and we know that an oscillatory motion always corresponds to a series of equally spaced energy levels whose separation is

$$\Delta E = \hbar\omega = \hbar 2\pi/T = 2\pi F/\kappa. \tag{6.1.37}$$

The existence and nature of these levels have been quite controversial (Zak, 1968, 1969; Wannier, 1969, Shockley, 1972). Experimental observations supporting their existence have been reported by Koss and Lambert (1972). The question is, in fact, rather intricate. The present treatment suggests that the Stark levels should be considered as approximate eigenstates. Our calculation has neglected tunneling, which allows an electron to escape from one band into another. If only one energy band is considered, the existence of the Stark ladder appears to be rigorously established (Hacker and Obermair, 1970). Our discussion suggests that, for a many-band system, tunneling should be regarded as broadening the Stark levels; the amount of broadening being

$$\delta_{\mathrm{T}} \approx \hbar/\tau_{\mathrm{T}} \tag{6.1.38a}$$

where τ_{T} is the lifetime for tunneling. Additional broadening would be caused by scattering of the electron (lifetime τ_{s}) by phonons or by impurities and by misalignment of the applied electric field (which would allow the electron to escape from a periodic motion between plane parallel but finite-sized portions of the zone), giving a lifetime τ_{m}. We may add the reciprocal lifetimes for these processes

$$\tau^{-1} = \tau_{\mathrm{T}}^{-1} + \tau_{\mathrm{s}}^{-1} + \tau_{\mathrm{m}}^{-1} \tag{6.1.38b}$$

and compute a total width

$$\delta = \hbar/\tau. \tag{6.1.38c}$$

Observability of the levels requires that δ should be smaller than the level spacing ΔE. In more pictorial language, the electron must be able to make at least one oscillation through the zone before being deflected. This condition is, however, difficult to satisfy experimentally (the lifetime must be of the order of 10^{-12} sec or longer in a field of 10^7 V/m).

The above argument neglects the fact, that even in a molecular crystal in which the probability of motion of an electron from one site to another is small, a uniform electric field $\boldsymbol{\mathcal{E}}$ will cause an energy difference $\sim e\boldsymbol{\mathcal{E}}\cdot\mathbf{a}$ between two otherwise equivalent energy states of neighboring molecules separated by $\mathbf{a}$. This situation may not be adequately described by band theory.

In contrast to the point of view presented here, Wannier and Fredkin (1962) have shown that closed bands can still be found in the presence of an electric field. These bands are not, however, necessarily related in any simple way to the bands which exist in zero field (Wannier and Van Dyke, 1968).

Let us consider the nature of the wave function formed by superposing the functions $\phi_\nu^{(1)}$ given by (6.1.33) to form the wave packet $\Psi(\mathbf{r}, t)$ by (6.1.17). Let us construct the superposition in such a way that states of different ν have equal amplitude. Then

$$\begin{aligned}\Psi(\mathbf{r}, t) &= \int \phi^{(1)}(\mathbf{k}, t)\psi(\mathbf{k}, \mathbf{r})\, d^3k \\ &= \kappa^{-1/2} \sum_\nu \int \exp\left\{(i/F) \int_0^{k_x} [E_\nu - E^{(1)}(\mathbf{k}')]\, dk_x' - i(E_\nu t/\hbar)\right\} \\ &\quad \times \psi(\mathbf{k}_x, \mathbf{k}_{0y}, \mathbf{k}_{0z}, \mathbf{r})\, dk_x. \qquad (6.1.39)\end{aligned}$$

We insert (6.1.34) and interchange the sum and the integral. The sum on ν is of the form

$$\sum_\nu \exp\{(i)[(2\pi\nu/\kappa)(k_x - (Ft/\hbar))]\} = \kappa \sum_m \delta[(k_x - (Ft/\hbar)) - m\kappa]. \qquad (6.1.40)$$

Since k_x is inside the zone, $k_x < \kappa$, and thus for a given value of t, only a single value of m can contribute. Let this be $m = 0$, for simplicity, corresponding to small times. Then we obtain

$$\Psi(\mathbf{r}, t) = \kappa^{1/2}\psi(Ft/\hbar, k_{0y}, k_{0z}) \exp\left[-i/F \int_0^{Ft/\hbar} E^{(1)}(\mathbf{k}')\, dk_x'\right]. \qquad (6.1.41)$$

This is known as a *Houston function* (Houston, 1940). It may be extended to large times using the summation on m in (6.1.40) and this describes, in the correct quantum fashion, the oscillation through the zone mentioned previously.

6.1.4 *Tunneling*

We will now consider the process of tunneling of an electron between bands according to the method of Argyres (1962). The interband matrix element FX_{nl} which appears in (6.1.15) is to be treated as a perturbation which causes transitions between states $\phi_{n,\nu}^{(1)}(\mathbf{k}_\perp)$ given by (6.1.33), which are solutions of (6.1.16). It is necessary to use these states, rather than the original Bloch states, as bases for the calculation in order to obtain the correct dependence of the tunneling probability on the electric field strength. Here we use $\mathbf{k}_\perp$ to indicate the components of the wave vector perpendicular to the field; the additional quantum numbers involved are the band number n and the Wannier level index ν. The energy of such a state is denoted by $E_{n,\nu}(\mathbf{k}_\perp)$; this quantity being given by (6.1.34). We may apply standard first-order time-dependent perturbation theory to determine the probability of transition between a state in band n, level ν to one in band l, level μ. The transverse wave vector $\mathbf{k}_\perp$ must be conserved in such a process. The transition rate is

$$w_{l\mu,n\nu} = (2\pi/\hbar) \mid M_{l\mu,n\nu}(\mathbf{k}_\perp) \mid^2 \delta[E_{l,\mu}(\mathbf{k}_\perp) - E_{n,\nu}(\mathbf{k}_\perp)]. \quad (6.1.42)$$

The matrix element $M_{l\mu,n\nu}$ is given by

$$M_{l\mu,n\nu}(\mathbf{k}_\perp) = F\int_{-\kappa/2}^{\kappa/2} \phi_{l,\mu}^{(1)*}(\mathbf{k}) X_{ln}(\mathbf{k}) \phi_{n,\nu}^{(1)}(\mathbf{k})\, dk_x. \quad (6.1.43)$$

The number of electrons per unit volume of sample which tunnel from band n to band l per unit time is found by summing (6.1.42) with respect to $k_\perp$, ν, and μ. We call this quantity W_{ln}

$$W_{ln} = (2/L_x)(2\pi/\hbar) \sum_{\nu,\mu} \int [d\mathbf{k}_\perp/(2\pi)^2] \mid M_{l\mu,n\nu}(\mathbf{k}_\perp) \mid^2 \times \delta[E_{l,\mu}(\mathbf{k}_\perp) - E_{n,\nu}(\mathbf{k}_\perp)]. \quad (6.1.44)$$

Here, L_x is the length of the specimen in the x direction and the factor of 2 allows for both directions of electron spin.

The matrix element (6.1.43) can be rewritten, with the use of (6.1.33). It turns out to be independent of μ and ν, so we drop these indices

$$M_{l,n}(\mathbf{k}_\perp) = (F/\kappa) \int_{-\kappa/2}^{\kappa/2} X_{ln}(\mathbf{k}) \exp\left\{(i/F) \int_0^{k_x} E_{ln}(k_x', \mathbf{k}_\perp)\, dk_x'\right\} dk_x. \quad (6.1.45)$$

We have introduced the definition

$$E_{ln}(\mathbf{k}) = E_l^{(1)}(\mathbf{k}) - E_n^{(1)}(\mathbf{k}) \tag{6.1.46}$$

and have taken account of the presence of the delta function in (6.1.42) to eliminate factors $E_{l,\mu}(\mathbf{k}_\perp)$, $E_{n\nu}(\mathbf{k}_\perp)$ which would otherwise occur.

The sum over ν and μ in (6.1.44) can be simplified if we observe that the expression depends on these quantities only through the difference $(\mu - \nu)$ which occurs in the argument of the delta function. Thus we may replace $(\mu - \nu)$ by ρ (ρ is an integer) and consider the sum as referring to μ (say) and ρ. The expression to be summed is now independent of μ, and the number of terms is $L_x\kappa/2\pi$. The remaining sum over ρ can be transformed using the Poisson summation formula (Morse and Feshbach, 1953, p. 466). This states that

$$\sum_{\rho=-\infty}^{\infty} f(\alpha\rho) = (1/\alpha) \sum_{m=-\infty}^{\infty} \int f(x) \exp(2\pi imx/\alpha)\, dx \tag{6.1.47}$$

[α being an arbitrary constant in (6.1.47); in the present case, $\alpha = 2\pi F/\kappa$]. Thus

$$W_{ln} = (2\kappa/\hbar)(\kappa/2\pi F) \sum_{m=-\infty}^{\infty} \int [d\mathbf{k}_\perp/(2\pi)^2]\, |M_{l,n}(\mathbf{k}_\perp)|^2 \int dx$$

$$\times \exp(imx\kappa/F)\, \delta[x - \Delta_{ln}(\mathbf{k}_\perp)]$$

where

$$\Delta_{ln}(\mathbf{k}_\perp) = (1/\kappa) \int_{-\kappa/2}^{\kappa/2} E_{nl}(k_x', \mathbf{k}_\perp)\, dk_x'. \tag{6.1.48}$$

The integral over x can be performed, and we obtain

$$W_{ln} = (\kappa^2/\pi\hbar F) \Bigg[\int [d\mathbf{k}_\perp/(2\pi)^2]\, |M_{l,n}(\mathbf{k}_\perp)|^2$$

$$+ 2 \sum_{m=1}^{\infty} \int [d\mathbf{k}_\perp/(2\pi)^2]\, |M_{ln}(\mathbf{k}_\perp)|^2 \cos(m\, \Delta_{ln}(\mathbf{k}_\perp)\kappa/F) \Bigg]. \tag{6.1.49a}$$

The term with $m = 0$ has been separated. This term, which is the most important, describes what may be called steady tunneling: the second term produces structure which is an oscillatory function of the field strength and is due to the discreteness of the Wannier levels. In most circumstances of practical interest, the contribution from the terms with $m \neq 0$ is small enough to be neglected. We then have

$$W_{ln} \approx (\kappa^2/4\pi^3\hbar F) \int d\mathbf{k}_\perp \mid M_{l,n}(\mathbf{k}_\perp) \mid^2. \tag{6.1.49b}$$

This is the best we can do without considering details of the band structure. In order to continue, a model must be introduced. We will use the isotropic two-band model described in Section 4.1.3, Part A [see Eqs. (4.1.31)–(4.1.34)]. Let the two bands be called c (conduction) and v (valence), respectively; they are separated by a gap E_g at $\mathbf{k} = 0$. For $\mathbf{k} \neq 0$, the energies are determined as the eigenvalues of the 2×2 matrix

$$H = \begin{pmatrix} \hbar^2k^2/2m & (\hbar/m)kp \\ (\hbar/m)kp & E_g + \hbar^2k^2/2m \end{pmatrix} \tag{4.1.31}$$

in which p is the momentum matrix element connecting the bands. In general, the difference in energy between the bands is denoted by η:

$$\eta = E_v - E_c = [E_g^2 + (4\hbar^2k^2p^2/m^2)]^{1/2} = [E_g^2 + (\hbar^2k^2E_g/\mu)]^{1/2} \tag{6.1.50}$$

where μ is the reduced effective mass for the pair of bands, defined by

$$\mu^{-1} = m_c^{*-1} + \mid m_v^* \mid^{-1}. \tag{6.1.51}$$

The individual masses are given by Eq. (4.1.34), Part A.

The matrix element M_{cv} is to be calculated according to (6.1.45). The quantity X_{cv} is given by (6.1.7). In order to evaluate X_{cv} we must express $u_n(\mathbf{k}, \mathbf{r})$ in terms of the assumed basis functions. Let these be $u_v(0, \mathbf{r})$ and $u_c(0, \mathbf{r})$. A straightforward calculation of the eigenvectors of H gives

$$\begin{aligned} u_v(\mathbf{k}, \mathbf{r}) &= (2\eta)^{-1/2}[(\eta - E_g)^{1/2}u_c(0, \mathbf{r}) - (\eta + E_g)^{1/2}u_v(0, \mathbf{r})] \\ u_c(\mathbf{k}, \mathbf{r}) &= (2\eta)^{-1/2}[(\eta + E_g)^{1/2}u_c(0, \mathbf{r}) + (\eta - E_g)^{1/2}u_v(0, \mathbf{r})]. \end{aligned} \tag{6.1.52}$$

It is now possible to obtain X_{cv}. The k dependence of the functions is entirely contained in η; and the functions $u_c(0, \mathbf{r})$, $u_v(0, \mathbf{r})$ satisfy the orthonormality relations (4.1.8), Part A. We obtain

$$X_{cv} = iE_g^2\hbar^2k_x/2\mu\eta^2(\eta^2 - E_g^2)^{1/2} = [iE_g^{3/2}\hbar/2\eta^2(\mu)^{1/2}](k_x/k). \tag{6.1.53}$$

We substitute (6.1.53) into (6.1.45) and obtain

$$M_{cv} = (iF/2\kappa)[\hbar E_g^{3/2}/(\mu)^{1/2}] \int_{-\kappa/2}^{\kappa/2} (k_x\, dk_x/k\eta^2) \exp\left[-i/F \int_0^{k_x} \eta(k_x', \mathbf{k}_\perp)\, dk_x'\right]. \tag{6.1.54}$$

It turns out that only small values of the perpendicular components of **k** are of any significance in the final evaluation of (6.1.49). It will therefore be sufficiently accurate to replace k_x by k in (6.1.54). Further, it is a reasonable approximation to treat the integral (6.1.54) as if the only singularity in the integrand were a simple pole at $\eta = 0$. In fact, this is not strictly valid as this is actually a branch point since the behavior η is more complicated but we will obtain a result that is almost correct from the simpler approach. We see from (6.1.50) that $\eta = 0$ when $k_x = \pm iq$, q being given by

$$q^2 = \mathbf{k}_\perp^2 + \beta^2; \qquad \beta^2 = \mu E_g/\hbar^2. \tag{6.1.55}$$

The present approximation gives for the integral

$$\begin{aligned} M_{cv} &= (iF/\kappa)(\pi/2q)(\mu E_g/\hbar^2)^{1/2} \exp\left[(-i/F)\int_0^{iq} \eta(k_x', \mathbf{k}_\perp)\,dk_x'\right] \\ &= (iF/\kappa)(\pi/2)(\beta/q)\exp[(-\pi/4\beta)(E_g q^2/F)]. \end{aligned} \tag{6.1.56}$$

We now substitute (6.1.56) into (6.1.49b). Because the integral is rapidly convergent, the upper limit may be made infinite

$$\begin{aligned} W_{cv} &= (F/8)(\beta^2/\hbar)\int_0^\infty [k_\perp\,dk_\perp/(k_\perp^2 + \beta^2)] \\ &\quad \times \exp\{-(\pi/2)[E_g(\beta^2 + k_\perp^2)/\beta F]\} \\ &\approx (F^2\beta/8\pi\hbar E_g)\exp(-\pi\beta E_g/2F) \\ &= (F^2/8\pi\hbar^2)(\mu/E_g)^{1/2}\exp[(-\pi E_g/2F)(\mu E_g/\hbar^2)^{1/2}]. \end{aligned} \tag{6.1.57}$$

A more exact treatment of the singularity leads to replacement of the numerical coefficient 8 in the denominator of (6.1.57) by 18 (Kane, 1959). It is to be noted that the tunneling rate depends exponentially on $1/F$: although this is extremely small for small F, it could not have been obtained by straightforward perturbation theory starting from a basis of Bloch functions. The introduction of eigenfunctions for the single band in the presence of the field was a necessary step in the calculation.

6.2 The Steady Magnetic Field

The problem of the behavior of electrons in solids in the presence of an external magnetic field is both important and difficult. Many of the experimental techniques for determining information about the band structures of solids employ magnetic fields, and it is vital to be able to interpret correctly the data obtained. The difficulties of the topic arise from the same source as in the case of the electric field: the Hamiltonian contains the electron coordinate and therefore contains a term which can become arbitrarily large.

6.2.1 *The Hamiltonian: Gauge Transformations*

The Hamiltonian for a single electron in the presence of a magnetic field can be obtained by replacing the momentum operator $\mathbf{p}$ wherever it appears in the Hamiltonian for zero field by $\mathbf{P} = \mathbf{p} + e\mathbf{A}$, where $\mathbf{A}$ is the vector potential and by including a term $(e\hbar/m)\mathbf{s}\cdot\mathbf{B}$ to represent the energy of the spin. The electron charge e is taken to be positive in these expressions. We use MKS units in which the vector potential $\mathbf{A}$ is related to the flux density $\mathbf{B}$ (in units of $\mathrm{Wb/m^2}$) by

$$\mathbf{B} = \nabla \times \mathbf{A} \tag{6.2.1}$$

$$H = (1/2m)(\mathbf{p} + e\mathbf{A})^2 + V(\mathbf{r}) + (e\hbar/m)\mathbf{s}\cdot\mathbf{B}. \tag{6.2.2}$$

The vector potential $\mathbf{A}$ is not uniquely determined by the field, since the gradient of any scalar function of position may be added to $\mathbf{A}$ without changing $\mathbf{B}$. Such a change in $\mathbf{A}$ is referred to as a *gauge transformation.* A change in the phase of the wave function must accompany a gauge transformation of the vector potential. Thus, suppose we replace $\mathbf{A}$ by $\mathbf{A}'$ where

$$\mathbf{A}' = \mathbf{A} + \nabla f(\mathbf{r}). \tag{6.2.3}$$

The wave function of the electron changes to ψ'

$$\psi' = U\psi \tag{6.2.4a}$$

where

$$U = \exp[-ief(\mathbf{r})/\hbar]. \tag{6.2.4b}$$

We may also consider a unitary transformation of the Hamiltonian

$$H' = UHU^{+} \tag{6.2.5a}$$

which yields

$$H' = (1/2m)(\mathbf{p} + e\mathbf{A}')^2 + V(\mathbf{r}) + (e\hbar/m)\mathbf{s}\cdot\mathbf{B}. \tag{6.2.5b}$$

The new Hamiltonian has the same form in terms of the new potentials as the first one did in terms of the original potentials. Since all physical properties of a quantum system are preserved by a unitary transformation, it follows that the physical results of quantum theory are unaffected by a change of gauge.

The Hamiltonian (6.2.2) is considerably more complicated than in the case of $\mathbf{A} = 0$. Note that the components of the canonical momentum $\mathbf{P}$

do not commute with each other. Let P_α and P_β be rectangular components of **P**. Their commutator is

$$P_\alpha P_\beta - P_\beta P_\alpha = (e\hbar/i)(\partial A_\beta/\partial x_\alpha - \partial A_\alpha/\partial x_\beta) = (e\hbar/i) \sum_\gamma B_\gamma \epsilon_{\gamma\alpha\beta} \tag{6.2.6}$$

in which B_γ is the γ component of the flux density and $\epsilon_{\gamma\alpha\beta}$ is the antisymmetric Levi–Civitta symbol (which has the value +1 or −1 according as the arrangement of the numbers $\gamma\alpha\beta$ is an even or odd permutation of 1, 2, 3).

6.2.2 *Magnetic Translations*

The presence of a magnetic field implies that simple translation invariance has been lost: Consider the effect of the operator $T(\mathbf{R}_i) = \{\epsilon \mid -\mathbf{R}_i\}$ on H (see Section 3.5, Part A):

$$T(\mathbf{R}_i)H = \{[\mathbf{p} + e\mathbf{A}(\mathbf{r} + \mathbf{R}_i)]^2/2m\} + V(\mathbf{r}) + (e\hbar/m)\mathbf{s}\cdot\mathbf{B}. \tag{6.2.7}$$

The translation has shifted the vector potential by a lattice vector. However, the physical situation is unchanged, since a shift in the origin of the vector potential does not alter the magnetic field. It should, therefore, be possible to find some new translation operators which commute with the Hamiltonian.

We consider the case of a uniform magnetic field and adopt the symmetric gauge in which

$$A = \tfrac{1}{2}\mathbf{B} \times \mathbf{r}. \tag{6.2.8}$$

Let us consider the operators (Peterson, 1960; Brown, 1964, 1968; Zak, 1964)

$$\begin{aligned}\mathfrak{J}(\mathbf{R}_j) &= \exp\{(i/\hbar)\mathbf{R}_j\cdot[\mathbf{p} + e(\mathbf{r} \times \mathbf{B})/2]\} \\ &= \exp[(i/\hbar)\mathbf{R}_j\cdot\mathbf{p}] \exp[i(\boldsymbol{\beta} \times \mathbf{R}_j)\cdot\mathbf{r}/2]\end{aligned} \tag{6.2.9}$$

where $\boldsymbol{\beta} = e\mathbf{B}/\hbar$. The exponential can be written as a product of two terms since the terms commute (they involve perpendicular components of **p** and **r**). The first factor in (6.2.9) is a representation of the usual translation operator

$$T(\mathbf{R}_j) = \exp[(i/\hbar)\mathbf{R}_j\cdot\mathbf{p}]. \tag{6.2.10}$$

Furthermore, the operators $\mathfrak{J}(\mathbf{R}_j)$ commute with the Hamiltonian. This may be shown as follows. Let

$$\boldsymbol{\Pi} = \mathbf{p} + e(\mathbf{r} \times \mathbf{B})/2. \tag{6.2.11}$$

We find by a straightforward calculation that

$$[\mathbf{\Pi}, \mathbf{P}] = 0 \tag{6.2.12a}$$

and thus that

$$[\mathbf{\Pi}, \mathbf{P}^2] = 0 \tag{6.2.12b}$$

(where $\mathbf{P}^2 = \sum_\alpha P_\alpha^2$). Since $\mathfrak{J}(\mathbf{R}_j) = \exp(i\mathbf{R}_j \cdot \mathbf{\Pi}/2)$, it follows that

$$[\mathfrak{J}(\mathbf{R}_j), \mathbf{P}^2] = 0. \tag{6.2.13}$$

Also, we have from (6.2.9), that if $f(\mathbf{r})$ is an arbitrary function of position

$$\mathfrak{J}(\mathbf{R}_j)f(\mathbf{r}) = \exp[i(\boldsymbol{\beta} \times \mathbf{R}_j)\cdot\mathbf{r}/2]f(\mathbf{r} + \mathbf{R}_j). \tag{6.2.14}$$

Thus we see that

$$[\mathfrak{J}(\mathbf{R}_j), H] = 0 \tag{6.2.15}$$

for Hamiltonians of the form (6.2.2) which contain a periodic potential. We also find

$$\mathfrak{J}(\mathbf{R}_j)\mathfrak{J}(\mathbf{R}_i) = \exp[i\boldsymbol{\beta}\cdot(\mathbf{R}_i \times \mathbf{R}_j)/2]\mathfrak{J}(\mathbf{R}_j + \mathbf{R}_i) \tag{6.2.16}$$

and consequently that

$$\mathfrak{J}(\mathbf{R}_j)\mathfrak{J}(\mathbf{R}_i) = \exp[i\boldsymbol{\beta}\cdot(\mathbf{R}_i \times \mathbf{R}_j)]\mathfrak{J}(\mathbf{R}_i)\mathfrak{J}(\mathbf{R}_j). \tag{6.2.17}$$

The operators $\mathfrak{J}(\mathbf{R}_i)$ are called *magnetic translation operators* [Readers consulting other works on this subject should note that our $\mathfrak{J}(\mathbf{R}_i)$ is the same as $\mathbf{T}(-\mathbf{R}_i)$ defined by Brown, 1968]. They commute with the Hamiltonian for the solid; however, we note from (6.2.16) that the product of two translation operators differs from a translation operator by a phase factor, and from (6.2.17), that the operators do not commute. In consequence of these facts, the group theoretical procedures discussed in Sections 3.5 and 3.6, Part A, must be generalized: we are here concerned with ray representations of the translation group, rather than the more usually discussed vector representations (Hamermesh, 1962; Brown, 1968).

The phase factor which appears in (6.2.16) has a simple physical interpretation. The vectors $\mathbf{R}_j$, $\mathbf{R}_i$, $-(\mathbf{R}_j + \mathbf{R}_i)$ may be connected to form a triangle. The magnetic flux passing through this triangle is

$$\Phi = \tfrac{1}{2}\mathbf{B}\cdot(\mathbf{R}_j \times \mathbf{R}_i). \tag{6.2.18}$$

Thus (6.2.16) becomes

$$\mathfrak{J}(\mathbf{R}_j)\mathfrak{J}(\mathbf{R}_i) = e^{-ie\Phi/\hbar}\mathfrak{J}(\mathbf{R}_j + \mathbf{R}_i). \tag{6.2.19}$$

Consequently, if the flux Φ through the triangle is an integral multiple of the flux quantum

$$\Phi_0 = 2\pi\hbar/e, \tag{6.2.20}$$

the multiplication relation becomes that for ordinary translation operators. The numerical value of Φ_0 is 4.14×10^{-15} Wb (or 4.14×10^{-7} G-cm^2). This is twice the size of the flux quantum which is relevant in the theory of superconductivity, in which case the relevant particles are electron pairs with charge $2e$.

An interesting situation arises if it is possible to construct a superlattice within the ordinary lattice such that the magnetic flux through any face of the unit cell of the superlattice is an integral multiple of Φ_0. This amounts to supposing that there exists a subset of the lattice vectors to be denoted by $\mathbf{R}_i'$, which themselves form a lattice, and have the following properties involving the magnetic field: Let $\mathbf{a}_1'$, $\mathbf{a}_2'$, and $\mathbf{a}_3'$ be the primitive vectors of the superlattice. Consider a primitive cell of the superlattice in the form of a parallelepiped. The fluxes through the faces are of the form $\mathbf{B}\cdot(\mathbf{a}_i' \times \mathbf{a}_j')$; and we require that the three independent quantities of this type $\mathbf{B}\cdot(\mathbf{a}_1' \times \mathbf{a}_2')$, $\mathbf{B}\cdot(\mathbf{a}_3' \times \mathbf{a}_1')$, $\mathbf{B}\cdot(\mathbf{a}_2' \times \mathbf{a}_3')$ should be all integral multiples of Φ_0; thus that their ratios are the ratios of integers

$$\mathbf{B}\cdot(\mathbf{a}_2' \times \mathbf{a}_3'):\mathbf{B}\cdot(\mathbf{a}_3' \times \mathbf{a}_1'):\mathbf{B}\cdot(\mathbf{a}_1' \times \mathbf{a}_2') = l:m:n. \quad (6.2.21)$$

We see from the definition of the reciprocal lattice, Eq. (1.2.3), Volume A, that the vectors $(\mathbf{a}_i' \times \mathbf{a}_j')$ determine the directions of primitive reciprocal lattice vector for the magnetic superlattice. Consequently $\mathbf{B}$ itself must be in the direction of a lattice vector of the (direct) superlattice

$$\mathbf{B} = C(l\mathbf{a}_1' + m\mathbf{a}_2' + n\mathbf{a}_3'). \quad (6.2.22)$$

Here C is a constant and l, m, and n are the integers appearing in (6.2.21). Each face of the unit cell of the superlattice contains an integral number of parallelogram faces of the unit cell in the direct lattice. Hence, the flux through any face of a cell in the direct lattice must be a rational number. If we choose (or redefine) a primitive lattice vector of the ordinary direct lattice to be in the direction of $\mathbf{B}$, and call it $\mathbf{A}_3$ ($\mathbf{B}$ = const $\mathbf{A}_3$) while two other primitive vectors are $\mathbf{A}_1$ and $\mathbf{A}_2$, then the flux through a parallelogram defined by $\mathbf{A}_1$ and $\mathbf{A}_2$ must be a rational multiple of Φ_0. Since $\mathbf{A}_3\cdot(\mathbf{A}_1 \times \mathbf{A}_2)$ is the volume Ω of the unit cell of the direct lattice, we finally have

$$\mathbf{B} = (L/N\Omega)\phi_0\mathbf{A}_3 \quad (6.2.23)$$

where L and N are integers with no common factor.

Magnetic fields satisfying this condition will be called *rational fields*. The essential result is that if $\mathbf{B}$ is a rational field, we can find a subset of magnetic translation operators which combine among themselves as an ordinary (vector) translation group: The general principles of band theory, including Bloch's theorem apply with respect to the superlattice constructed as described above. We have therefore proved that some sort of band structure must exist in the case of rational fields. The magnetic Brillouin zone may, however, be quite small: The volume of the magnetic zone is of the order of $1/N$ of the volume of the real zone. From (6.2.20) and (6.2.23) we can make an order of magnitude estimate of the number N involved. For this purpose, assume $\Omega \approx a^3$, with $a \approx 4 \times 10^{-10}$ m. Then

$$B \approx 2.5L/N \times 10^4 \text{ Wb/m}^2 = 2.5L/N \times 10^8 \text{ G}. \qquad (6.2.24)$$

A strong field for laboratory purposes is of the order of 10 Wb/m^2; so we see that attainable fields correspond to values of $N > 1000$. The magnetic Brillouin zone can then be visualized as a thin strip.

An arbitrary magnetic field can be approximated by some rational field. However, if we imagine varying the strength of the field, we see that N can change very rapidly, and may become very large (as an example we observe that fields characterized by the pairs of integers (L, N) given by (1, 1000), (10, 10001), (100, 100001), etc., differ by extremely small amounts. A simple Bloch band would split into the order of 1000 subbands if $N \approx 1000$, and correspondingly greater numbers for larger N. These bands would be expected to cluster in such a way that the density of states of the system would not be radically altered by a small change in the field. For further consideration of the nature of the energy levels, see Blount (1962), Brown (1968), Butler and Brown (1968). The last mentioned work reports the calculation of the magnetic band structure for a model. It is, however, quite unlikely that experimental measurements of magnetic band structures will be possible until fields an order of magnitude larger than now available can be attained.

6.2.3 *The One-Band Effective Hamiltonian to Second Order*

The object of much work on the problem of a solid in an external magnetic field has been the derivation of an effective Hamiltonian. It was conjectured by Peierls (1933) that the energy levels of an electron in a magnetic field could be found in the following way: Let the energy band function for some band in the absence of a field be $E(\mathbf{k})$. In the presence of an external magnetic field, we form the operator $E(\mathbf{P}/\hbar) = E[(\mathbf{p} + e\mathbf{A})/\hbar]$ and solve the effective Schrödinger equation

$$E[(\mathbf{p} + e\mathbf{A})/\hbar]\phi = E\phi. \qquad (6.2.25)$$

This result turns out not to be strictly correct, although it is valid for a single nondegenerate band to second order in $\mathbf{P}$. There exists, however, an effective Hamiltonian H_{eff} which does not couple bands and yields the energy levels in a magnetic field. Two effects occur: (1) the parameters of the energy band are gradually transformed by the field, and (2) the band breaks up into a set of more or less discrete states.

Our further discussion will first consider the approach of Luttinger and Kohn (1955) (see also Kjeldaas and Kohn, 1957), which establishes the form of the effective Hamiltonian to second order in P and is very useful in studying the properties of semiconductors in which only a small number of electrons and holes need be considered. We will then establish the existence of an effective Hamiltonian in a more complete fashion following Wannier and Fredkin (1962). In a subsequent section we will use this result to calculate the steady diamagnetic susceptibility of a metal. Finally, we will discuss effects which depend on the existence of discrete levels, specifically the de Haas–van Alphen oscillations of the magnetic susceptibility which appear at high field strengths.

We begin by considering the Hamiltonian (6.2.2) without the spin term (which will be added later). The effective mass representation (see Section 5.1) will be employed. As a preliminary to the body of the calculation, it is desirable to obtain the representation of the (rectangular) coordinate $\mathbf{r}$ which enters the Hamiltonian through the vector potential

$$
\begin{aligned}
[n\mathbf{k} \mid \mathbf{r} \mid l\mathbf{q}] &= \int \exp[i(\mathbf{q} - \mathbf{k}) \cdot \mathbf{r}] u_n^*(\mathbf{k}_0, \mathbf{r}) \mathbf{r} u_l(\mathbf{k}_0, \mathbf{r})\, d^3r \\
&= i\, \nabla_{\mathbf{k}} \int \exp[i(\mathbf{q} - \mathbf{k}) \cdot \mathbf{r}] u_n^*(\mathbf{k}_0, \mathbf{r}) u_l(\mathbf{k}_0, \mathbf{r})\, d^3r \\
&= i\, \nabla_{\mathbf{k}}\, \delta(\mathbf{q} - \mathbf{k})\, \delta_{nl}.
\end{aligned} \tag{6.2.26}
$$

Here, $\mathbf{k}_0$ is the reference point for the basis functions. Rectangular brackets denote matrix elements in the effective mass representation. It is convenient to introduce the symbol $(\mathbf{k} \mid Q \mid \mathbf{q})$ to designate the matrix elements of an operator Q on the basis of plane waves

$$
(\mathbf{k} \mid Q \mid \mathbf{q}) = [1/(2\pi)^3] \int \exp(-i\mathbf{k} \cdot \mathbf{r}) Q \exp(i\mathbf{q} \cdot \mathbf{r})\, d^3r. \tag{6.2.27}
$$

Then we have for $\mathbf{r}$

$$
[n\mathbf{k} \mid \mathbf{r} \mid l\mathbf{q}] = (\mathbf{k} \mid \mathbf{r} \mid \mathbf{q})\, \delta_{nl}. \tag{6.2.28}
$$

Similarly, we obtain for the momentum $\mathbf{p}$

$$
\begin{aligned}
[n\mathbf{k} \mid \mathbf{p} \mid l\mathbf{q}] &= \hbar\mathbf{k}\, \delta_{nl}\, \delta(\mathbf{k} - \mathbf{q}) + \mathbf{p}_{nl}\, \delta(\mathbf{k} - \mathbf{q}) \\
&= \delta_{nl}(\mathbf{k} \mid \mathbf{p} \mid \mathbf{q}) + \mathbf{p}_{nl}\, \delta(\mathbf{k} - \mathbf{q})
\end{aligned} \tag{6.2.29}
$$

in which

$$\mathbf{p}_{nl} = [(2\pi)^3/\Omega] \int_{\text{cell}} u_n^*(\mathbf{k}_0, \mathbf{r}) \mathbf{p} u_l(\mathbf{k}_0, \mathbf{r})\, d^3r. \tag{6.2.30}$$

We can now calculate the matrix elements of $\mathbf{P}$

$$[n\mathbf{k} \mid \mathbf{P} \mid l\mathbf{q}] = (\mathbf{k} \mid \mathbf{P} \mid \mathbf{q})\, \delta_{nl} + \mathbf{p}_{nl}\, \delta(\mathbf{k} - \mathbf{q}) \tag{6.2.31}$$

since the vector potential depends on coordinates only and thus is diagonal in the bands in this representation. The singular character of $(\mathbf{k} \mid \mathbf{P} \mid \mathbf{q})$ comes about through (6.2.26). The matrix elements of the kinetic energy are

$$\begin{aligned} [n\mathbf{k} \mid \mathbf{P}^2/2m \mid l\mathbf{q}] &= (1/2m) \sum_j \int d^3k'\, [n\mathbf{k} \mid \mathbf{P} \mid j\mathbf{k}'][j\mathbf{k}' \mid \mathbf{P} \mid l\mathbf{q}] \\ &= (\mathbf{k} \mid \mathbf{P}^2/2m \mid \mathbf{q})\, \delta_{nl} + (\mathbf{p}_{nl}/m) \cdot (\mathbf{k} \mid \mathbf{P} \mid \mathbf{q}) \\ &\quad + \sum_j (\mathbf{p}_{nj}\mathbf{p}_{jl}/2m)\, \delta(\mathbf{k} - \mathbf{q}). \end{aligned} \tag{6.2.32}$$

The potential energy V is periodic in the crystal and thus has no matrix elements between states characterized by inequivalent wavevectors. Therefore we can write

$$[n\mathbf{k} \mid V \mid l\mathbf{q}] = V_{nl}\, \delta(\mathbf{k} - \mathbf{q}). \tag{6.2.33}$$

The matrix elements of the Hamiltonian are found by combining (6.2.32) and (6.2.33):

$$\begin{aligned} [n\mathbf{k} \mid H \mid l\mathbf{q}] &= [(1/2m)(\mathbf{p}^2)_{nl} + V_{nl}]\, \delta(\mathbf{k} - \mathbf{q}) \\ &\quad + (\mathbf{p}_{nl}/m) \cdot (\mathbf{k} \mid \mathbf{P} \mid \mathbf{q}) + \delta_{nl}(\mathbf{k} \mid \mathbf{P}^2/2m \mid \mathbf{q}). \end{aligned} \tag{6.2.34}$$

The energy of a state at the reference point $\mathbf{k}_0$ is given through the expression

$$[(1/2m)(\mathbf{p}^2)_{nl} + (1/m)\hbar\mathbf{k}_0 \cdot \mathbf{p}_{nl} + V_{nl}] = [E_n(\mathbf{k}_0) - (\hbar^2\mathbf{k}_0^2/2m)]\, \delta_{nl}. \tag{6.2.35}$$

Thus we have

$$\begin{aligned} [n\mathbf{k} \mid H \mid l\mathbf{q}] &= \{[E_n(\mathbf{k}_0) - (\hbar^2\mathbf{k}_0^2/2m)]\, \delta(\mathbf{k} - \mathbf{q}) + (\mathbf{k} \mid \mathbf{P}^2/2m \mid \mathbf{q})\}\, \delta_{nl} \\ &\quad + (\mathbf{p}_{nl}/m) \cdot (\mathbf{k} \mid \mathbf{P} - \hbar\mathbf{k}_0 \mid \mathbf{q}). \end{aligned} \tag{6.2.36}$$

The terms in (6.2.36) which are not diagonal in the band index can be removed by the unitary transformation procedure employed in Section 5.3. The Hamiltonian consists of three parts:

$$H = H_0 + H_1 + H_2 \tag{6.2.37a}$$

where

$$[n\mathbf{k} \mid H_0 \mid l\mathbf{q}] = [E_n(\mathbf{k}_0) - (\hbar^2\mathbf{k}_0^2/2m)]\,\delta(\mathbf{k} - \mathbf{q})\,\delta_{nl} \quad (6.2.37\text{b})$$

$$[n\mathbf{k} \mid H_1 \mid l\mathbf{q}] = (\mathbf{p}_{nl}/m)\cdot(\mathbf{k} \mid \mathbf{P} - \hbar\mathbf{k}_0 \mid \mathbf{q}) \quad (6.2.37\text{c})$$

$$[n\mathbf{k} \mid H_2 \mid l\mathbf{q}] = (\mathbf{k} \mid \mathbf{P}^2/2m \mid \mathbf{q})\,\delta_{nl}. \quad (6.2.37\text{d})$$

The transformation for which we are looking is defined by a Hermitian operator S_1 which produces a transformed Hamiltonian $\bar{H}$ [see Eq. (5.3.13)]

$$\bar{H} = \exp(-iS_1)\,\mathrm{H}\,\exp(iS_1)$$

where S_1 is chosen so that $\bar{H}$ will have no off-diagonal elements to first order. It can be determined from (5.3.18) and (5.3.19) that

$$[n\mathbf{k} \mid S_1 \mid l\mathbf{q}] = i([n\mathbf{k} \mid H_1 \mid l\mathbf{q}]/\hbar\omega_{nl}) = (i\mathbf{p}_{nl}/m\hbar\omega_{nl})\cdot(\mathbf{k} \mid \mathbf{P} - \hbar\mathbf{k}_0 \mid \mathbf{q}) \quad (6.2.38)$$

in which $\hbar\omega_{nl} = E_n(\mathbf{k}_0) - E_l(\mathbf{k}_0)$ as before.

The transformed Hamiltonian is given by an expression similar to (5.3.21), including terms of second order

$$\bar{H} = H_0 + H_2 + \tfrac{1}{2}i[H_1, S_1] + \cdots. \quad (6.2.39)$$

The matrix elements of the commutator in (6.2.39) are

$$[n\mathbf{k} \mid [H_1, S_1] \mid l\mathbf{q}] = (i/m^2\hbar) \sum_j \{\mathbf{p}_{nj}\cdot(\mathbf{k} \mid [\mathbf{P} - \hbar\mathbf{k}_0][\mathbf{P} - \hbar\mathbf{k}_0] \mid \mathbf{q})\cdot\mathbf{p}_{jl}\}\,(\omega_{jl}^{-1} + \omega_{jn}^{-1}). \quad (6.2.40)$$

This is not diagonal in the band index, but the off-diagonal elements are evidently of second order. These may be removed by a further unitary transformation using a Hermitian matrix S_2 in which the term in (6.2.39), $\frac{1}{2}i[H_1, S_1]$, is treated in the same manner previously applied to H_1, namely

$$i[H_0, S_2] = -\tfrac{1}{2}i[H_1, S_1]. \quad (6.2.41)$$

This leads to

$$\begin{aligned}[n\mathbf{k} \mid S_2 \mid l\mathbf{q}] &= -i(2m^2\hbar^2\omega_{nl})^{-1} \\ &\quad \times \sum_j \{\mathbf{p}_{nj}\cdot(\mathbf{k} \mid [\mathbf{P} - \hbar\mathbf{k}_0][\mathbf{P} - \hbar\mathbf{k}_0] \mid \mathbf{q})\cdot\mathbf{p}_{jl}\} \\ &\quad \times (\omega_{jl}^{-1} + \omega_{jn}^{-1}) \qquad \text{if} \quad n \neq l \\ &= 0 \qquad\qquad\qquad\qquad \text{if} \quad n = l. \end{aligned} \quad (6.2.42)$$

A series of successive unitary transformations can be carried out to eliminate the off-diagonal elements to arbitrary order. (Note that if we wish to go beyond the second order, it is necessary to include higher-order terms from S_1, such as $[H_2, S_1]$, etc.) To obtain the portion of the effective Hamiltonian matrix which is diagonal in the band index, including all second-order terms it is necessary to retain only the diagonal part of (6.2.40). This is substituted into (6.2.39), and the following result is obtained:

$$[n\mathbf{k} \mid \bar{H} \mid l\mathbf{q}] = \delta_{nl}(\mathbf{k} \mid H_n \mid \mathbf{q}) \tag{6.2.43}$$

where

$$\begin{aligned}(\mathbf{k} \mid H_n \mid \mathbf{q}) = {} & E_n(\mathbf{k}_0)\, \delta(\mathbf{k} - \mathbf{q}) + (1/m)\mathbf{p}_{nn}\cdot(\mathbf{k} \mid \mathbf{P} - \hbar\mathbf{k}_0 \mid \mathbf{q}) \\ & + (1/2m)(\mathbf{k} \mid \mathbf{P}^2 - \hbar^2\mathbf{k}_0{}^2 \mid \mathbf{q}) \\ & + (1/m^2) \sum_j \mathbf{p}_{nj}\cdot\{(\mathbf{k} \mid [\mathbf{P} - \hbar\mathbf{k}_0][\mathbf{P} - \hbar\mathbf{k}_0] \mid \mathbf{q})/\hbar\omega_{nj}\}\cdot\mathbf{p}_{jn}.\end{aligned} \tag{6.2.44}$$

This expression should be compared with Eq. (4.1.21), Part A, which gives the expansion of $E_n(\mathbf{k})$ for a single nondegenerate band through second order in $(\mathbf{k} - \mathbf{k}_0)$. The terms correspond exactly if we replace $\mathbf{k}$ in (4.1.21), Part A, by $\mathbf{P}/\hbar$. The conclusion is that through second order, the effective Hamiltonian in the presence of a magnetic field is simply obtained by replacing $\mathbf{k}$ (but not $\mathbf{k}_0$, which is a fixed, reference point) by $\mathbf{P}/\hbar$. We show below that (6.2.25) is valid to this order.

Kjeldaas and Kohn (1957) considered higher-order terms in the expansion whose initial terms are given by (6.2.44) for the case of a cubic crystal with $k_0 = 0$. In this case, the linear (and cubic) terms in P vanish. An explicit expression was obtained for the coefficients $E_n{}^{\alpha\beta\gamma\delta}$, in the fourth-order term of the effective Hamiltonian

$$\sum_{\alpha\beta\gamma\delta} E_n{}^{\alpha\beta\gamma\delta}(\mathbf{k} \mid P_\alpha P_\beta P_\gamma P_\delta \mid \mathbf{q}) \tag{6.2.45}$$

(α, β, γ, δ represent Cartesian coordinate indices). They found that more complicated quantities are involved in (6.2.45) than the coefficients of the fourth-order term in the expansion of $E(\mathbf{k})$ (The latter determine only certain sums of the $E_n{}^{\alpha\beta\gamma\delta}$).

Before the transformation e^{iS} was applied, the equation satisfied by the expansion coefficients $A_l(\mathbf{q})$ of the wave function in the effective mass representation was [see Eq. (5.1.18)]

$$\sum_l \int \{[n\mathbf{k} \mid H \mid l\mathbf{q}] - E\, \delta_{nl}\, \delta(\mathbf{k} - \mathbf{q})\}A_l(\mathbf{q})\, d^3q = 0.$$

After the transformation the new expansion coefficients $C_n(\mathbf{k})$, which are related to the A by $C = e^{-iS}A$, satisfy an equation which we can write as

$$\int (\mathbf{k} \mid E_n(\mathbf{P}/\hbar) \mid \mathbf{q}) C_n(\mathbf{q}) \, d^3q = EC_n(\mathbf{k}) \tag{6.2.46}$$

in which it is understood that we retain only terms through second order in the expansion of $E_n(\mathbf{P})$. The transformation back to real space is accomplished through the procedure used in Section 5.3: Multiply (6.2.46) by $\exp[i(\mathbf{k} - \mathbf{k}_0)\cdot\mathbf{r}]$ and integrate over the Brillouin zone. A function $F_n(\mathbf{r})$ is defined as in (5.3.26):

$$F_n(\mathbf{r}) = \int \exp[i(\mathbf{k} - \mathbf{k}_0)\cdot\mathbf{r}] C_n(\mathbf{k}) \, d^3k. \tag{6.2.47}$$

Consider the contribution

$$\iint \exp[i(\mathbf{k} - \mathbf{k}_0)\cdot\mathbf{r}](\mathbf{k} \mid E_n(\mathbf{P}/\hbar) \mid \mathbf{q}) C_n(\mathbf{q}) \, d^3q \, d^3k$$

$$= (2\pi)^{-3} \iiint \exp[i(\mathbf{k} - \mathbf{k}_0)\cdot\mathbf{r}] \exp(-i\mathbf{k}\cdot\mathbf{r}') E_n(\mathbf{P}'/\hbar)$$

$$\times \exp(i\mathbf{q}\cdot\mathbf{r}') C_n(\mathbf{q}) \, d^3q \, d^3k \, d^3r'$$

where we have used (6.2.27). The notation $\mathbf{P}'$ has been employed to indicate that $\mathbf{P}$ is a differential operator with respect to the coordinates $\mathbf{r}'$. We expand E_n, and carry out the indicated differentiations. The technique of Eqs. (5.3.29)–(5.3.34) may now be employed to reduce this to

$$\iint \Delta(\mathbf{r} - \mathbf{r}') \exp(-i\mathbf{k}_0\cdot\mathbf{r}) E_n[(\mathbf{q} - \mathbf{k}_0) + e\mathbf{A}(\mathbf{r}')/\hbar]$$

$$\times \exp(i\mathbf{q}\cdot\mathbf{r}') C_n(\mathbf{q}) \, d^3q \, d^3r'.$$

We treat the function Δ as an ordinary delta function and obtain

$$\int E_n[(\mathbf{q} - \mathbf{k}_0) + e\mathbf{A}(\mathbf{r})/\hbar] \exp[i(\mathbf{q} - \mathbf{k}_0)\cdot\mathbf{r}] C_n(\mathbf{q}) \, d^3q$$

$$= E_n(\mathbf{P}/\hbar) \int \exp[i(\mathbf{q} - \mathbf{k}_0)\cdot\mathbf{r}] C_n(\mathbf{q}) \, d^3q$$

$$= E_n(\mathbf{P}/\hbar) F_n(\mathbf{r}) \tag{6.2.48}$$

where

$$\mathbf{P} = (\hbar\nabla/i) + eA(\mathbf{r}). \tag{6.2.49}$$

This establishes (6.2.25) in the form

$$E_n(\mathbf{P}/\hbar)F_n(\mathbf{r}) = EF_n(\mathbf{r}) \qquad (6.2.50)$$

where (we repeat) it is understood that only terms through second order are to be retained in the expansion of $E_n(\mathbf{P}/\hbar)$ in terms of $(\mathbf{P}/\hbar) - \mathbf{k}_0$.

6.2.4 *Landau Levels*

To proceed further, we will suppose that the reference point $\mathbf{k}_0$ is taken as the origin and that $\mathbf{p}_{nn} = 0$. We may then write the expansion of $1 \circ \mathbf{P}/\hbar)$ in the form

$$E_n(\mathbf{P}/\hbar) = E_n(0) + \sum_{\alpha\beta} E_n{}^{\alpha\beta} P_\alpha P_\beta \qquad (6.2.51)$$

where

$$E_n{}^{\alpha\beta} = (1/2m_0)\, \delta_{\alpha\beta} + \sum_j (p^\alpha{}_{nj} p^\beta{}_{jn} / \hbar m_0{}^2 \omega_{nj}). \qquad (6.2.52)$$

Equation (6.2.50) now becomes

$$[E_n(0) - E]F_n(\mathbf{r}) + \sum_{\alpha\beta} E_n{}^{\alpha\beta} P_\alpha P_\beta F_n(\mathbf{r}) = 0. \qquad (6.2.53)$$

Let us determine the solutions of (6.2.53) in the simple special case of a spherical band with effective mass m^*: thus, we suppose

$$E_n{}^{\alpha\beta} = (1/2m^*)\, \delta_{\alpha\beta}. \qquad (6.2.54)$$

We will put $E_n(0) = 0$ and drop the index n. It is convenient to take the magnetic field B to be along the z direction and to choose a gauge slightly different from (6.2.8):

$$A_x = -By, \qquad A_y = A_z = 0. \qquad (6.2.55)$$

Then

$$P_1 = \hbar/i\, \partial/\partial x - eBy, \qquad P_2 = \hbar/i\, \partial/\partial y, \qquad P_3 = h/i\, \partial/\partial z. \qquad (6.2.56)$$

With these simplifications (6.2.53) takes the form

$$-\hbar^2/2m^* \,[(\partial/\partial x - ieBy/\hbar)^2 + \partial^2/\partial y^2 + \partial^2/\partial z^2]F(\mathbf{r}) = EF(\mathbf{r}). \qquad (6.2.57)$$

This equation is separable in rectangular coordinates. Let us put

$$F(\mathbf{r}) = \exp[i(k_x x + k_z z)]g(y). \qquad (6.2.58)$$

This leads to

$$(d^2g/dy^2) + (2m^*/\hbar^2)[E - (\hbar^2 k_z{}^2/2m^*) - (1/2m^*)(\hbar k_x - eBy)^2]g = 0. \qquad (6.2.59)$$

It is convenient to define

$$y_0 = \hbar k_x/eB. \tag{6.2.60}$$

We may rewrite (6.2.59) as

$$(d^2g/dy^2) + (2m^*/\hbar^2)[\epsilon - \tfrac{1}{2}m^*\omega_c^2(y - y_0)^2]g(y) = 0 \tag{6.2.61}$$

in which

$$\omega_c = eB/m^*, \qquad \epsilon = E - (\hbar^2 k_z^2/2m^*). \tag{6.2.62}$$

Equation (6.2.61) is the equation for a simple harmonic oscillator of frequency ω_c, with the equilibrium point located at y_0. The eigenvalue ϵ_l is given by

$$\epsilon_l = (l + \tfrac{1}{2})\hbar\omega_c \tag{6.2.63}$$

where l is any positive integer, including zero. The quantity ω_c is usually called the *cyclotron frequency*. The energy of an electron in the magnetic field (remember, spin has been neglected) is

$$E = (\hbar^2 k_z^2/2m^*) + (l + \tfrac{1}{2})\hbar\omega_c. \tag{6.2.64}$$

The result (6.2.64) is quite remarkable. The continuous, three-dimensional, parabolic band structure from which we started has been split up into a series of lines (the oscillator levels) which we can associate with the classical circular motion of the electron in a plane perpendicular to the magnteic field, plus a one-dimensional parabolic term coming from the free electron behavior in a direction parallel to the field. The discrete levels are known as *Landau levels* (Landau, 1930). The energy of the lowest state is no longer zero, but has been raised to $\frac{1}{2}\hbar\omega_c$.

The Landau levels are highly degenerate. To determine the degeneracy, let the system be contained in a large rectangular box with sides of length L_x, L_y, and L_z. The number of possible values of k_α (where $\alpha = x, y, z$) in a small interval Δk_α is $L_\alpha\, \Delta k_\alpha/2\pi$. All values of k_x are permissible provided that the "orbit center" y_0 lies within the box: $-L_y/2 \le y_0 \le L_y/2$. Here, we neglect the extent of the orbit relative to the size of the container. From this, we can determine the range of allowed values of k_x

$$-eBL_y/2\hbar \le k_x \le eBL_y/2\hbar. \tag{6.2.65}$$

The number of ordinary states (with fixed k_z) in a single Landau level is

$$L_x\, \Delta k_x/2\pi = eBL_xL_y/2\pi\hbar. \tag{6.2.66}$$

If we next consider an interval Δk_z in k_z, the number of states in a Landau level with k_z lying in the range Δk_z is

$$(eBV/4\pi^2\hbar)\, \Delta k_z \tag{6.2.67}$$

in which V is the volume of the box. Note that the degeneracy is proportional to B.

The function $g(y)$ which is a solution of (6.2.61) is a harmonic oscillator function

$$g_l(y) = (2\pi^{1/2}\alpha/\Omega 2^l l!)^{1/2} H_l[\alpha(y - y_0)] \exp[-\tfrac{1}{2}\alpha^2(y - y_0)^2] \quad (6.2.68)$$

in which $\alpha^2 = m^*\omega_c/\hbar = eB/\hbar$, H_l is a Hermite polynomial, and l is the oscillator quantum number which appears in (6.2.64). We can obtain a measure of the radius of the orbit by computing the root mean square value of y. After a simple calculation, we obtain

$$y_{\text{rms}} = [\langle(y - y_0)^2\rangle]^{1/2} = (l + \tfrac{1}{2})^{1/2}/\alpha = (\hbar/eB)^{1/2}(l + \tfrac{1}{2})^{1/2}. \quad (6.2.69)$$

In a magnetic field of 1 Wb/m^2 (10^4 G), we find y_{rms} is about 180 Å for the lowest state. This is certainly small enough to justify our neglect of the orbit radius compared to L_y in calculating the density of states; it is also large enough to justify replacement of $\Delta(\mathbf{r} - \mathbf{r}')$ by $\delta(\mathbf{r} - \mathbf{r}')$ in (6.2.48).

The Landau levels are separated by an energy

$$\hbar eB/m^* = 1.1577 \times 10^{-4}\ \text{eV}\ (B)/(m^*/m_0) \quad (6.2.70)$$

where B is in webers per square meter and m_0 is the free electron mass. For low fields, and for materials with effective mass ratios of the order of unity, this energy is small compared to thermal energies except at the very lowest temperatures; the quantization of levels can usually be ignored. However, if we consider materials with effective mass ratios of the order of 10^{-2} (InSb, Bi, etc.) in strong fields, the formation of the Landau levels becomes quite significant.

The structure of the spectrum obtained in the present approximation differs in some respects from what may be inferred from the rigorous discussion at the beginning of this section. In fact, for rational fields, a band structure will be formed from the Landau levels. This requires a better approximation to the effective Hamiltonian than we have obtained. The degeneracy of the Landau levels is split by band formation. One source of interaction leading to a broadening of the Landau levels is (according to Blount, 1962) that the orbit centers y_0 will have differing positions in the unit cell and thus will have different energies. Numerical calculations for models of the effective Hamiltonian show that only a few low lying Landau levels show appreciable broadening and that the simple picture with discrete levels is substantially accurate (Butler and Brown, 1968).

6.2.5 *The g Factor*

We will now consider the effects of adding spin to the picture. A term representing the interaction of the electron spin with the magnetic field is added to (6.2.53) which becomes

$$[E_n(0) - E]F_n(\mathbf{r}) + \sum_{\alpha\beta} E_n{}^{\alpha\beta}P_\alpha P_\beta F_n(\mathbf{r}) + (e\hbar/m_0)\,\mathbf{s}\cdot\mathbf{B}F_n(\mathbf{r}) = 0. \tag{6.2.71}$$

The function $F_n(\mathbf{r})$ must now be interpreted as a two-component spinor.

First, suppose that we are concerned with a spherical band as described by (6.2.54). Then the energies are, evidently,

$$E = (\hbar^2 k_z{}^2/2m^*) + (l + \tfrac{1}{2})\hbar\omega_c + (e\hbar/m_0)Bm_s \tag{6.2.72}$$

where m_s is the spin quantum number in $m_s = \pm\frac{1}{2}$. Here we note that the free electron mass m_0 enters into the spin term, whereas the effective mass m^* enters into the cyclotron frequency. Also note that as a result of the spin term, we find that if $m^* = m_0$, the energy of the lowest state is $E = 0$, that is, the lowering of energy caused by alignment of the spin cancels the increase produced by confinement of an electron to an orbit. However, in general $m^* \neq m_0$, and one effect or the other may predominate.

The preceding paragraph presents a somewhat oversimplified picture of the spectrum particularly in heavier materials where spin–orbit coupling is significant. The effective g factor of the electron may depart significantly from 2, as we shall now proceed to discuss. Our approach follows that of Roth (1960). The calculation of g factors has been reviewed by Yafet (1963), and simplified recently by de Graaf and Overhauser (1969).

The considerations are based on (6.2.71), which is more complex than our previous considerations would suggest if the quantities $E_n{}^{\alpha\beta}$ are not symmetric ($E_n{}^{\alpha\beta} \neq E_n{}^{\beta\alpha}$). We noted in (6.2.6) that the different components of $\mathbf{P}$ do not commute with each other. We can write

$$\sum_{\alpha\beta} E_n{}^{\alpha\beta}P_\alpha P_\beta = \tfrac{1}{4}\sum_{\alpha\beta}\Big[(E_n{}^{\alpha\beta} + E_n{}^{\beta\alpha})(P_\alpha P_\beta + P_\beta P_\alpha) + (e\hbar/i)(E_n{}^{\alpha\beta} - E_n{}^{\beta\alpha})\sum_\gamma \epsilon_{\gamma\alpha\beta}B_\gamma\Big]. \tag{6.2.73}$$

The second term on the right side of (6.2.73) can be expressed as

$$(e\hbar/2m_0)\mathbf{M}\cdot\mathbf{B} \tag{6.2.74}$$

where we have defined the vector $\mathbf{M}$, dual to the antisymmetric tensor whose components are $(E_n{}^{\alpha\beta} - E_n{}^{\beta\alpha})$, by

$$M_\gamma = (m_0/2i)\sum_{\alpha\beta}(E_n{}^{\alpha\beta} - E_n{}^{\beta\alpha})\epsilon_{\gamma\alpha\beta}. \tag{6.2.75}$$

When this term is included, (6.2.71) becomes (with $\mathbf{S} = \frac{1}{2}\boldsymbol{\sigma}$ and using the fact that α, β are dummy labels)

$$[E_n(0) - E]F_n(\mathbf{r}) + \tfrac{1}{2}\sum_{\alpha\beta}(E_n^{\alpha\beta} + E_n^{\beta\alpha})P_\alpha P_\beta F_n(\mathbf{r})$$

$$+ (e\hbar/2m_0)(\boldsymbol{\sigma} + \mathbf{M})\cdot\mathbf{B}F_n(\mathbf{r}) = 0. \quad (6.2.76)$$

Now we consider the evaluation of **M**. Equation (6.2.52) is substituted into (6.2.75) yielding

$$M_\gamma = (1/2\hbar i m_0)\sum_j\sum_{\alpha\beta}\epsilon_{\gamma\alpha\beta}(p^\alpha{}_{nj}p^\beta{}_{jn} - p^\beta{}_{nj}p^\alpha{}_{jn})/\omega_{nj} \quad (6.2.77a)$$

or

$$\mathbf{M} = (1/\hbar i m_0)\sum_j \mathbf{p}_{nj}\times\mathbf{p}_{jn}/\omega_{nj}. \quad (6.2.77b)$$

If spin–orbit coupling is neglected, M_γ must vanish. This may be seen as follows. Since **p** is an Hermitian operator, $p^\alpha{}_{nj} = p_{jn}{}^{\alpha*}$. For wave functions which do not include spin, the time reversal operator which changes **k** into $-\mathbf{k}$ was shown in Section 3.7, Part A, to be equivalent to complex conjugation. Thus in the absence of spin–orbit coupling, the functions $u_n(0, \mathbf{r})$ may always be chosen to be real, leading to a pure imaginary momentum matrix element $\mathbf{p}_{nj}$. Then (6.2.77) vanishes. Therefore we must consider the effect of spin–orbit coupling, which adds to the Hamiltonian a term

$$H_{\text{so}} = (\hbar/4m^2c^2)\boldsymbol{\sigma}\cdot(\nabla V\times\mathbf{p}) = \boldsymbol{\sigma}\cdot\mathbf{h}. \quad (6.2.78)$$

Relativistic effects other than spin–orbit coupling will be neglected here [although this means we ignore an additional correction, which appears to be unimportant except for lithium (Overhauser and de Graaf, 1969)]. In the presence of a magnetic field we replace **p** by $\mathbf{p} + e\mathbf{A}$ as before

$$H_{\text{so}} = (\hbar/4m^2c^2)\boldsymbol{\sigma}\cdot[\nabla V\times(\mathbf{p} + e\mathbf{A})]. \quad (6.2.79)$$

If this term is added to the Hamiltonian (6.2.2), one can easily show that the formal theory remains unchanged except that the matrix elements $\mathbf{p}_{nj}$ must be replaced by

$$\pi_{nj} = [(2\pi)^3/\Omega]\int_{\text{cell}} u_n^*(0, \mathbf{r})[\mathbf{p} + (\hbar/4m^2c^2)\boldsymbol{\sigma}\times\nabla V]u_j(0, \mathbf{r})\, d^3r. \quad (6.2.80)$$

The contribution of the $\boldsymbol{\sigma}\times\nabla V$ term π_{nj} is usually negligible. The modification of the wave function at $k = 0$, u_n, by spin–orbit coupling is of more significance for the present problem. We will take this into account through first-order perturbation theory, in which we treat H_{so} [as given by (6.2.78)] as a perturbation. Let $u_n^{(0)}$ be the unperturbed wave function. Then we have

$$u_n = u_n^{(0)} + \boldsymbol{\sigma}\cdot\sum_l (\mathbf{h}_{ln}/\hbar\omega_{nl})\,u_l^{(0)} \tag{6.2.81}$$

where

$$\mathbf{h}_{ln} = [(2\pi)^3/\Omega]\int_{\text{cell}} u_l^{(0)*}\mathbf{h}u_n^{(0)}\,d^3r. \tag{6.2.82}$$

We substitute (6.2.82) into the definition (6.2.30) of the matrix element P_{nj}. This gives, in first order

$$\mathbf{p}_{nj} = \mathbf{p}_{nj}^{(0)} + \boldsymbol{\sigma}\cdot\sum_l [(\mathbf{h}_{lj}\mathbf{p}_{nl}^{(0)}/\hbar\omega_{jl}) + (\mathbf{h}_{nl}\mathbf{p}_{lj}^{(0)}/\hbar\omega_{nl})]. \tag{6.2.83}$$

The superscript (0) indicates that the matrix element is obtained without considering spin–orbit coupling. A similar expression holds for $\mathbf{p}_{lj}$. These results are substituted into (6.2.77b). The result can be simplified by noting that $\mathbf{p}_{nj}^{(0)} \times \mathbf{p}_{jn}^{(0)} = 0$, and that $\mathbf{p}_{nj}^{(0)} = -\mathbf{p}_{jn}^{(0)}$, etc. We obtain

$$\mathbf{M} = (1/i\hbar^2m_0)\sum_{jl}(1/\omega_{nj})\{(1/\omega_{jl})\,\boldsymbol{\sigma}\cdot(\mathbf{h}_{jl} - \mathbf{h}_{lj})(\mathbf{p}_{nj}\times\mathbf{p}_{ln})$$
$$+ (1/\omega_{nl})\,\boldsymbol{\sigma}\cdot(\mathbf{h}_{ln} - \mathbf{h}_{nl})(\mathbf{p}_{nj}\times\mathbf{p}_{jl})\}. \tag{6.2.84}$$

We have dropped the superscript on the $\mathbf{p}$'s since it is superfluous from this point on.

Let us write the spin-dependent term which appears in (6.2.76) in the (conventional) form

$$(e\hbar/2m_0)(\boldsymbol{\sigma} + \mathbf{M})\cdot\mathbf{B} = (e\hbar/4m_0)\,\boldsymbol{\sigma}\cdot\mathbf{g}\cdot\mathbf{B} \tag{6.2.85}$$

where $\mathbf{g}$ is a dyadic (tensor) which may be expressed with the aid of (6.2.84) as

$$\mathbf{g} = 2\mathbf{1} + (2/i\hbar^2m_0)\sum_{jl}(1/\omega_{nj})\{(1/\omega_{jl})(\mathbf{h}_{jl} - \mathbf{h}_{lj})(\mathbf{p}_{nj}\times\mathbf{p}_{ln})$$
$$+ (1/\omega_{nl})(\mathbf{h}_{ln} - \mathbf{h}_{nl})(\mathbf{p}_{nj}\times\mathbf{p}_{jl})\}. \tag{6.2.86}$$

Here $\mathbf{1}$ is the unit dyadic and $\mathbf{g}$ is the g tensor. Use of (6.2.85) converts (6.2.76) to

$$[E_n(0) - E] + \tfrac{1}{2}\sum_{\alpha\beta}(E_n^{\alpha\beta} + E_n^{\beta\alpha})P_\alpha P_\beta F_n(\mathbf{r})$$
$$+ (e\hbar/4m_0)\,\boldsymbol{\sigma}\cdot\mathbf{g}\cdot\mathbf{B}F_n(\mathbf{r}) = 0. \tag{6.2.87}$$

The g tensor is essential in order to describe the effect of spin on the energy levels of an electron in a magnetic field. In the absence of spin–orbit coupling, $\mathbf{g} = 2\mathbf{1}$; however, the deviations from this value may be quite large. The calculation of $\mathbf{g}$ may be dominated by the contribution from a single band if one of the energy denominators in (6.2.86) is small, as is the case in some semiconductors. When the effective mass is small, a large effect can be expected if there is possible spin–orbit coupling. For example, in indium antimonide, a large, negative g factor is obtained.

6.2.6 *Degenerate Bands*

The preceding discussion has been limited to the case of a single band. In the case of bands which are degenerate at the reference point, the theory becomes more complex. This situation is actually of considerable practical importance, since it applies to the valence band in many semiconductors. The theory follows the procedures of the previous discussion except that we must not ignore the off-diagonal elements of $[H_1, S_1]$ in (6.2.40) which connect members of the degenerate set. This means that the effective Hamiltonian of (6.2.43) is not diagonal in the band index, but retains terms coupling the bands. Suppose, as is true in most cases of practical interest, that $\mathbf{k}_0 = 0$, and that $\mathbf{p}_{nn} = 0$. Then

$$[n\mathbf{k} \mid \bar{H} \mid l\mathbf{q}] = (\mathbf{k} \mid H_{nl} \mid \mathbf{q}) \tag{6.2.88}$$

where n and l refer to the degenerate set only, and

$$(\mathbf{k} \mid H_{nl} \mid \mathbf{q}) = E_n(0)\, \delta_{nl}\, \delta(\mathbf{k} - \mathbf{q}) + \sum_{\alpha\beta} D_{nl,\alpha\beta}(\mathbf{k} \mid P_\alpha P_\beta \mid \mathbf{q}) \tag{6.2.89}$$

with [see Eqs. (5.3.44)–(5.3.46) for the analogous impurity case]

$$D_{nl,\alpha\beta} = (1/2m)\, \delta_{nl}\, \delta_{\alpha\beta} + (1/m^2) \sum (p^\alpha{}_{nj} p^\beta{}_{jl}/\hbar\omega_{nj}). \tag{6.2.90}$$

The remainder of the calculation proceeds as before. We obtain, instead of (6.2.53), a set of coupled equations

$$\sum_{l,\alpha,\beta} D_{nl,\alpha\beta}[(1/i)\, \partial/\partial x_\alpha + eA_\alpha][(1/i)\, \partial/\partial x_\beta + eA_\beta]F_l(\mathbf{r})$$
$$= [E - E_n(0)]F_n(\mathbf{r}). \tag{6.2.91}$$

Calculations with this equation are quite difficult. The level spectrum is quite different from the simple structure of Landau levels which is obtained for a nondegenerate band. For details of this structure, see for example, Hensel and Suzuki (1970).

6.2.7 *The General Effective Hamiltonian*

We will now consider the construction of the full effective Hamiltonian for a single band following the procedure of Wannier and Fredkin (1962).

This procedure does not lead to an explicit expression for this object, but does succeed in confirming its existence. As was pointed out earlier, the explicit construction is apparently quite cumbersome, and must be undertaken by some procedure similar to the successive application of the cannonical transformation procedure previously described. We consider the Hamiltonian of (6.2.2), neglect spin, and use the gauge of Eq. (6.2.8)

$$H = (1/2m)(\mathbf{p} + \tfrac{1}{2}e\mathbf{B} \times \mathbf{r})^2 + V(\mathbf{r}). \qquad (6.2.92)$$

A set of functions $b_n(\mathbf{k}, \mathbf{r})$ are introduced which satisfy the equation

$$\{(1/2m)[\mathbf{p} + \tfrac{1}{2}e\mathbf{B} \times (\mathbf{r} + i\,\nabla_k)]^2 + V(\mathbf{r})\}b_n(\mathbf{k}, \mathbf{r})$$
$$= \sum_\mu \mathcal{W}_n(\mathbf{R}_\mu) \exp(i\mathbf{k}\cdot\mathbf{R}_\mu)$$
$$\times \exp[i(e/2\hbar)\mathbf{B} \times \mathbf{R}_\mu\cdot\mathbf{r}]b_n(\mathbf{k} - (e/2\hbar)\mathbf{B} \times \mathbf{R}_\mu, \mathbf{r}) \qquad (6.2.93)$$

where the $\mathcal{W}_n(\mathbf{R}_\mu)$ are coefficients analogous to the Fourier coefficients of the usual energy band function, defined through (5.1.47). However, $\mathcal{W}_n \neq \mathcal{E}_n$, since $\mathcal{W}_n$ depends on the external magnetic field B. If B is set equal to zero, $\mathcal{W}_n \to \mathcal{E}_n$ and the summation on the right side of (6.2.93) becomes

$$\sum_\mu \mathcal{E}_n(\mathbf{R}_\mu) \exp(i\mathbf{k}\cdot\mathbf{R}_\mu) = E_n(\mathbf{k})$$

so that (6.2.93) becomes the ordinary Schrödinger equation. Thus

$$\lim_{B\to 0} b_n(\mathbf{k}, \mathbf{r}) = \psi_n(\mathbf{k}, \mathbf{r}). \qquad (6.2.94)$$

The functions b_n defined by (6.2.93) have the property [in common with the ordinary Bloch functions $\psi_n(\mathbf{k}, \mathbf{r})$] that they may be written as

$$b_n(\mathbf{k}, \mathbf{r}) = \exp(i\mathbf{k}\cdot\mathbf{r})g_n(\mathbf{k}, \mathbf{r}) \qquad (6.2.95)$$

where g_n is periodic in the unit cell. The equation obeyed by g_n is found by substituting (6.2.95) into (6.2.93):

$$\{(1/2m)(\mathbf{p} + \hbar\mathbf{k} + \tfrac{1}{2}ei\mathbf{B} \times \nabla_k)^2 + V(\mathbf{r})\}g_n(\mathbf{k}, \mathbf{r})$$
$$= \sum_\mu \mathcal{W}_n(\mathbf{R}_\mu) \exp(i\mathbf{k}\cdot\mathbf{R}_\mu)g_n(\mathbf{k} - (e/2\hbar)\mathbf{B} \times \mathbf{R}_\mu, \mathbf{r}). \qquad (6.2.96)$$

This equation is simpler in appearance than (6.2.93) since some of the exponential factors on the right have canceled. Equation (6.2.96) can be solved, in principle, by expansion in powers of $\mathbf{B}$. If $\mathbf{B} = 0$, $g_n(\mathbf{k}, \mathbf{r}) = u_n(\mathbf{k}, \mathbf{r})$, the usual cell periodic part of the Bloch function. A solution could then be obtained to first order in $\mathbf{B}$, and so on. It will be assumed that this process converges, at which point the functions $b_n(\mathbf{k}, \mathbf{r})$ and $\mathcal{W}_n(\mathbf{R}_\mu)$ are known. A magnetic energy band function $W_n(\mathbf{k})$ can now be constructed:

$$W_n(\mathbf{k}) = \sum_\mu \exp(i\mathbf{k}\cdot\mathbf{R}_\mu)\mathcal{W}_n(\mathbf{R}_\mu). \tag{6.2.97}$$

Next, we define functions $C_n(\mathbf{k}, \mathbf{r})$ by

$$C_n(\mathbf{k}, \mathbf{r}) = b_n(\mathbf{k} + (e/2\hbar)\mathbf{B} \times \mathbf{r}, \mathbf{r}). \tag{6.2.98}$$

These functions are introduced in order to simplify (6.2.93). This is possible, since one can show by differentiation, that the following identity holds:

$$(\mathbf{p} + \tfrac{1}{2}e\mathbf{B} \times \mathbf{r})C_n(\mathbf{k}, \mathbf{r}) = \{[\mathbf{p} + (e/2\hbar)\mathbf{B} \times (\mathbf{r} + i\nabla_{\mathbf{k}})]b_n(\mathbf{k}, \mathbf{r})\}_{\mathbf{k}\to k+(e/2\hbar)\mathbf{B}\times\mathbf{r}}. \tag{6.2.99}$$

Use of (6.2.98) also simplifies the right side of (6.2.93). Since one of the exponentials cancels, Eq. (6.2.93) then becomes

$$HC_n(\mathbf{k}, \mathbf{r}) = \sum_\mu \mathcal{W}_n(\mathbf{R}_\mu) \exp(i\mathbf{k}\cdot\mathbf{R}_\mu)C_n(\mathbf{k} - (e/2\hbar)\mathbf{B} \times \mathbf{R}_\mu, \mathbf{r}). \tag{6.2.100}$$

The functions C_n are coupled with each other by the Hamiltonian with coefficients which are independent of $\mathbf{r}$ [in contrast with (6.2.93)]. Unfortunately the definition (6.2.98) of the C_n implies that the C_n will not satisfy Bloch's theorem. We may, however, proceed by defining a set of functions resembling Wannier functions (we use discrete normalization and consider a sum on $\mathbf{k}$, rather than an integral, for convenience)

$$A_n(\mathbf{r}, \mathbf{R}_\nu) = N^{-1/2} \sum_{\mathbf{k}} \exp(-i\mathbf{k}\cdot\mathbf{R}_\nu)C_n(\mathbf{k}, \mathbf{r}). \tag{6.2.101}$$

These functions differ from the more usual Wannier functions in that the argument is not simply $\mathbf{r} - \mathbf{R}_\nu$. This would be the argument only if the C_n satisfied Bloch's theorem. The functions A_n are supposed to be linearly independent, but need not be orthogonal. However, in order to have linear independence, it is necessary that the index n specify something more general than a single magnetic band referred to at the beginning of this section. The A_n must, in fact, combine contributions from many such bands. This situation arises since there are many more unit cells in the crystal than there are eigenstates in a magnetic band.

An equation satisfied by the A_n may be determined from (6.2.100) and (6.2.101) to be

$$HA_n(\mathbf{r}, \mathbf{R}_\nu) = \sum_\mu \exp[(-ie/2\hbar)\mathbf{B}\cdot(\mathbf{R}_\mu \times \mathbf{R}_\nu)]\mathcal{W}_n(\mathbf{R}_\mu)A_n(\mathbf{r}, \mathbf{R}_\nu - \mathbf{R}_\mu) \tag{6.2.102a}$$

or, with the definition $\mathbf{R}_\rho = \mathbf{R}_\nu - \mathbf{R}_\mu$

$$HA_n(\mathbf{r}, \mathbf{R}_\nu) = \sum_\rho \exp[(ie/2\hbar)\mathbf{B}\cdot(\mathbf{R}_\rho \times \mathbf{R}_\nu)]\mathcal{W}_n(\mathbf{R}_\nu - \mathbf{R}_\rho)A_n(\mathbf{r}, \mathbf{R}_\rho). \tag{6.2.102b}$$

None of the functions (b_n, C_n, A_n, etc.) are eigenfunctions of the Hamiltonian (6.2.92). It is supposed, however, that the actual eigenfunction $\Psi(\mathbf{r})$ can be expanded in terms of these functions. Let us write (still using discrete notation for convenience)

$$\Psi(\mathbf{r}) = \sum_{n,\mathbf{k}} f_n(\mathbf{k})C_n(\mathbf{k}, \mathbf{r}). \tag{6.2.103}$$

The eigenvalue equation

$$H\Psi(\mathbf{r}) = E\Psi(\mathbf{r})$$

becomes, using (6.2.100),

$$\begin{aligned}\sum_{n\mathbf{k}\mu} f_n(\mathbf{k})\mathcal{W}_n(\mathbf{R}_\mu) \exp(i\mathbf{k}\cdot\mathbf{R}_\mu)C_n(\mathbf{k} - (e/2\hbar)\mathbf{B} \times \mathbf{R}_\mu, \mathbf{r}) \\ = \sum_{n\mathbf{k}\mu} \mathcal{W}_n(\mathbf{R}_\mu) \exp(i\mathbf{k}\cdot\mathbf{R}_\mu)f_n(\mathbf{k} + (e/2\hbar)\mathbf{B} \times \mathbf{R}_\mu)C_n(\mathbf{k}, \mathbf{r}) \\ = E\sum_{\mathbf{k}} f_n(\mathbf{k})C_n(\mathbf{k}, \mathbf{r}).\end{aligned} \tag{6.2.104}$$

In the second line, we have redefined the summation index $\mathbf{k}$. It is now convenient to write

$$f_n(\mathbf{k} + (e/2\hbar)\mathbf{B} \times \mathbf{R}_\mu) = \exp[(e/2\hbar)\mathbf{B} \times \mathbf{R}_\mu\cdot\nabla_{\mathbf{k}}]f_n(\mathbf{k}). \tag{6.2.105}$$

Equation (6.2.105) is substituted into (6.2.104). It is possible to combine exponentials since their arguments commute [see (6.2.10)]. Then (6.2.104) can be rewritten in the form

$$\sum_{n\mathbf{k}\mu} \{\mathcal{W}_n(\mathbf{R}_\mu) \exp[i(\mathbf{k} + i(e/2\hbar)\mathbf{B} \times \nabla_{\mathbf{k}})\cdot\mathbf{R}_\mu] - E\}f_n(\mathbf{k})C_n(\mathbf{k}, \mathbf{r}) = 0. \tag{6.2.106}$$

Since the functions $C_n(\mathbf{k})$ are linearly independent, we must have

$$\sum_\mu \{\mathcal{W}_n(\mathbf{R}_\mu) \exp[i(\mathbf{k} + i(e/2\hbar)\mathbf{B} \times \nabla_{\mathbf{k}})\cdot\mathbf{R}_\mu] - E\}f_n(\mathbf{k}) = 0. \tag{6.2.107}$$

An alternate form is obtained with the aid of (6.2.97) to be

$$W_n(\mathbf{k} + i(e/2\hbar)\mathbf{B} \times \nabla_{\mathbf{k}})f_n(\mathbf{k}) = Ef_n(\mathbf{k}). \tag{6.2.108}$$

This establishes the existence of an effective Hamiltonian. It is of the expected form (6.2.25) except that we did not—and cannot at the present—given an explicit expression for the function W_n. We do know that, to second order in its argument, W is just the usual energy band function. Equation (6.2.108) may be transformed to position space if desired, to give a sort of one-band Schrödinger equation.

6.3 The Low Field Diamagnetic Susceptibility

The problem of the low field susceptibility of electrons in crystals is old, interesting, and, unfortunately, quite complicated. In classical mechanics, it can be shown quite generally that the magnetic susceptibility of a system of charges is zero (the Bohr–van Leeuwen theorem). It can easily be seen that this must be correct, since the energy of a charged particle is, in classical theory, unaffected by a magnetic field. Hence the free energy is independent of the field, and the susceptibility is zero. In quantum theory the energies are, however, changed by the field, and it is possible to calculate a nonzero susceptibility.

Such a calculation begins with a determination of the free energy, which is given for the case of Fermi statistics by the expression (Wilson, 1953)

$$\mathfrak{F} = N\mu - KT \sum_l \ln\{1 + \exp[(\mu - E_l)/KT]\} \qquad (6.3.1)$$

where the index l is used to designate some complete set of electron states. The dependence on magnetic field in the present problem is contained in the energies E_l, which should, in principle, be found as solutions of (6.2.106). The quantity μ is the chemical potential (or Fermi energy); it is determined from the requirement that N, the number of electrons, be given by

$$N = \sum_l n_l. \qquad (6.3.2)$$

This condition is evidently equivalent to

$$\partial\mathfrak{F}/\partial\mu = 0. \qquad (6.3.3)$$

We shall here require the determination of the free energy to second order in B. Unfortunately, we do not have an explicit expression for the energy E_l. However, since we require a sum over all states, we may use the fact that the trace of an operator is unchanged by a unitary transformation of the basis functions. Thus we have

$$\sum_l \ln(1 + \exp[(\mu - E_l)/KT]) = \operatorname{Tr} \ln(1 + \exp[(\mu - H)/KT]) \tag{6.3.4}$$

where H can be taken as the single electron Hamiltonian (6.2.92).

Equation (6.3.4) can be evaluated by expansion. If $F(H)$ is any function of the Hamiltonian which possess a power series expansion in H, we have

$$\operatorname{Tr} F(H) = \sum_n (1/n!) F^{(n)}(0) \operatorname{Tr}(H^n) \tag{6.3.5}$$

where

$$F^{(n)}(0) = (d^n F(x)/dx^n)_{x=0}. \tag{6.3.6}$$

The problem is, in effect, reduced to the calculation of the trace of the nth power of the Hamiltonian correctly through second order in B. In spite of the fact that we have only an implicit equation which describes the effect of H, the problem can be solved. We follow here the method of Wannier and Upadhyaya (1964). Other calculations of the susceptibility are due to Hebborn and Sondheimer (1960), Enz (1960), Roth (1962), Blount (1962), Misra and Roth (1969), and Misra and Kleinman (1972).

The procedure for calculating traces is simple in appearance. We consider the function $A_n(\mathbf{r}, \mathbf{R}_\nu)$ defined in (6.2.101). Let us suppose that there is an operator θ which acts on these functions in accord with

$$\theta A_n(\mathbf{r}, \mathbf{R}_\nu) = \sum_{m,\sigma} \theta_{n\nu,m\sigma} A_m(\mathbf{r}, \mathbf{R}_\sigma). \tag{6.3.7}$$

The functions A_n are assumed to be linearly independent, but need not be orthogonal. We have

$$\operatorname{Tr}(\theta) = \sum_{n,\nu} \theta_{n\nu,n\nu}. \tag{6.3.8}$$

In the present problem only one "band" is considered, so there is no sum on n. We are concerned with the Hamiltonian, which satisfies (6.2.102b)

$$HA_n(\mathbf{r}, \mathbf{R}_\nu) = \sum_\sigma \exp[(i/2\hbar) e\mathbf{B} \cdot (\mathbf{R}_\sigma \times \mathbf{R}_\nu)] \mathcal{W}_n(\mathbf{R}_\nu - \mathbf{R}_\sigma) A_n(\mathbf{r}, \mathbf{R}_\sigma).$$

From (6.3.7) and (6.3.8) we have

$$H_{n\nu,n\sigma} = \exp[-(ie/2\hbar)\mathbf{B} \cdot (\mathbf{R}_\nu \times \mathbf{R}_\sigma)] \mathcal{W}_n(\mathbf{R}_\nu - \mathbf{R}_\sigma) \tag{6.3.9}$$

so that

$$\operatorname{Tr}(H) = \sum_\nu \mathcal{W}_n(0). \tag{6.3.10}$$

Equation (6.2.97) can be inverted to give

$$\mathcal{W}_n(\mathbf{R}_\mu) = (1/N) \sum_{\mathbf{k}} \exp(i\mathbf{k}\cdot\mathbf{R}_\mu) W_n(\mathbf{k}). \tag{6.3.11}$$

Equation (6.3.10) can be rewritten in the form

$$\mathrm{Tr}(H) = \sum_{\mathbf{k}} W_n(\mathbf{k}). \tag{6.3.12}$$

It is possible to calculate matrix elements of powers of H using (6.3.9)

$$(H^2)_{n\nu,n\sigma} = \sum_{\rho} \exp[-i(e/2\hbar)\mathbf{B}\cdot(\mathbf{R}_\nu \times \mathbf{R}_\rho)] \exp[-i(e/2\hbar)\mathbf{B}\cdot(\mathbf{R}_\rho \times \mathbf{R}_\sigma)]$$
$$\times \mathcal{W}_n(\mathbf{R}_\nu - \mathbf{R}_\rho)\mathcal{W}_n(\mathbf{R}_\rho - \mathbf{R}_\sigma).$$

Thus

$$\mathrm{Tr}(H^2) = \sum_{\nu\rho} \mathcal{W}_n(\mathbf{R}_\nu - \mathbf{R}_\rho)\mathcal{W}_n(\mathbf{R}_\rho - \mathbf{R}_\nu) = \sum_{\mathbf{k}} [W_n(\mathbf{k})]^2. \tag{6.3.13}$$

The calculation becomes more complicated for third and higher powers since the exponentials involving B no longer drop out on taking the trace

$$(H)^3{}_{n\nu,n\sigma} = \sum_{\rho\tau} \exp[-i(e/2\hbar)\mathbf{B}\cdot(\mathbf{R}_\nu \times \mathbf{R}_\rho)] \exp[-i(e/2\hbar)\mathbf{B}\cdot(\mathbf{R}_\rho \times \mathbf{R}_\tau)]$$
$$\times \exp[-i(e/2\hbar)B\cdot(\mathbf{R}_\tau \times \mathbf{R}_\sigma)]$$
$$\times \mathcal{W}_n(\mathbf{R}_\nu - \mathbf{R}_\rho)\mathcal{W}_n(\mathbf{R}_\rho - \mathbf{R}_\tau)\mathcal{W}_n(\mathbf{R}_\tau - \mathbf{R}_\sigma).$$

The expression for the trace which results from this can be simplified by introducing

$$\mathbf{R}_\rho = \mathbf{R}_\nu + \mathbf{R}_\mu, \qquad \mathbf{R}_\tau = \mathbf{R}_\nu + \mathbf{R}_\kappa.$$

We then obtain

$$\mathrm{Tr}(H^3) = N \sum_{\mu\kappa} \exp[-i(e/2\hbar)\mathbf{B}\cdot(\mathbf{R}_\mu \times \mathbf{R}_\kappa)]\mathcal{W}_n(-\mathbf{R}_\mu)\mathcal{W}_n(\mathbf{R}_\mu - \mathbf{R}_\kappa)\mathcal{W}_n(\mathbf{R}_\kappa). \tag{6.3.14}$$

It is not necessary to evaluate this and higher traces exactly, instead it is sufficient to expand in powers of B and retain only terms through second order. In order to simplify the algebra, let us suppose that $\mathbf{B}$ is along the z axis of a rectangular coordinate system. Further, let us replace $\mathbf{R}_\mu$ by $\mathbf{r}$ (components x, y, z) and $\mathbf{R}_\kappa$ by $\mathbf{r}'$. We obtain, after introducing (6.3.11),

$$\mathrm{Tr}(H^3) = (1/N^2) \sum_{\mathbf{k},\mathbf{q},\mathbf{p}} \sum_{\mathbf{r},\mathbf{r}'} [1 - (1/2\hbar)ieB(xy' - yx')$$
$$- (1/8\hbar^2)e^2B^2(xy' - yx')^2]$$
$$\times \exp\{i[(\mathbf{q} - \mathbf{k})\cdot\mathbf{r} + (\mathbf{p} - \mathbf{q})\cdot\mathbf{r}']\} W_n(\mathbf{k}) W_n(\mathbf{q}) W_n(\mathbf{p}). \tag{6.3.15}$$

It is now possible to replace x by $i\,\partial/\partial k_x$ and y' by $-i\,\partial/\partial p_y$, operating on the exponential function. An integration by parts then allows the derivatives to act on the functions W. The integrated part vanishes because all terms are periodic in reciprocal space. Thus

$$\begin{aligned}\mathrm{Tr}(H^3) = (1/N^2) \sum_{\mathbf{k},\mathbf{q},\mathbf{p}} \sum_{\mathbf{r},\mathbf{r}'} \exp\{i[(\mathbf{q}-\mathbf{k})\cdot\mathbf{r} + (\mathbf{p}-\mathbf{q})\cdot\mathbf{r}']\} \\ \times\left[1 - \frac{1}{2\hbar}\, ieB\left(\frac{\partial^2}{\partial k_x\,\partial p_y} - \frac{\partial^2}{\partial k_y\,\partial p_x}\right)\right. \\ \left. - \frac{1}{8\hbar^2}\, e^2B^2\left(\frac{\partial^2}{\partial k_x\,\partial p_y} - \frac{\partial^2}{\partial k_y\,\partial p_x}\right)^2\right] W_n(\mathbf{k})\, W_n(\mathbf{q})\, W_n(\mathbf{p}).\end{aligned} \tag{6.3.16}$$

The derivatives act on the W functions. Then the sum over $\mathbf{r}$ and $\mathbf{r}'$ is performed to yield $N^2\,\delta_{\mathbf{kq}}\,\delta_{\mathbf{pq}'}$, so that ultimately, all the wave vectors become equal. The linear term vanishes, and we find

$$\begin{aligned}\mathrm{Tr}(H^3) = \sum_{\mathbf{k}} [W_n(\mathbf{k})]^3 - \frac{1}{4\hbar^2}\, e^2B^2 \\ \times \sum_{\mathbf{k}} W_n(\mathbf{k}) \left[\frac{\partial^2 W_n}{\partial k_x^2}\,\frac{\partial^2 W_n}{\partial k_y^2} - \left(\frac{\partial^2 W_n}{\partial k_x\,\partial k_y}\right)^2\right] + O(B^3).\end{aligned} \tag{6.3.17}$$

The traces of higher power terms do not yield additional complications to second order in B. Wannier and Upadhyaya find that

$$\begin{aligned}\mathrm{Tr}(H^m) = \sum_{\mathbf{k}} [W(\mathbf{k})]^m - \frac{m(m-1)}{24}\,\frac{e^2B^2}{\hbar^2} \\ \times \sum_{\mathbf{k}} W_n^{m-2}(\mathbf{k}) \left[\frac{\partial^2 W_n}{\partial k_x^2}\,\frac{\partial^2 W_n}{\partial k_y^2} - \left(\frac{\partial^2 W_n}{\partial k_x\,\partial k_y}\right)^2\right] + O(B^3).\end{aligned} \tag{6.3.18}$$

We may now employ (6.3.5) to calculate the trace of an arbitrary function of H: This gives

$$\begin{aligned}\mathrm{Tr}[F(H)] = \sum_{\mathbf{k}} F[W_n(\mathbf{k})] - \tfrac{1}{24}(e^2B^2/\hbar^2) \sum_{\mathbf{k}} d^2F[W_n(\mathbf{k})]/dW_n^2 \\ \times [\partial^2 W_n/\partial k_x^2\, \partial^2 W_n/\partial k_y^2 - (\partial^2 W_n/\partial k_x\,\partial k_y)^2] + O(B^3) + \cdots.\end{aligned} \tag{6.3.19}$$

In the present case

$$F(H) = \ln(1 + \exp[(\mu - H)KT]).$$

Thus

$$\mathrm{Tr}[F(H)] = \sum_{\mathbf{k}} \ln(1 + \exp[(\mu - W_n)/KT]) + \tfrac{1}{24}(e^2B^2/KT\hbar^2) \sum_{\mathbf{k}} df(W_n - \mu)/dW_n \times [\partial^2W_n/\partial k_x^2\, \partial^2W_n/\partial k_y^2 - (\partial^2W_n/\partial k_x\, \partial k_y)^2] \quad (6.3.20)$$

where $f(\epsilon)$ is the Fermi function

$$f(\epsilon) = (e^{\epsilon/KT} + 1)^{-1}. \quad (6.3.21)$$

In order to obtain the free energy to second order in B we must expand W in powers of $eB/\hbar$. We write

$$W_n(\mathbf{k}) = E_n(\mathbf{k}) + (eB/\hbar)W_n^{(1)}(\mathbf{k}) + (e^2B^2/\hbar^2)W_n^{(2)}(\mathbf{k}) \quad (6.3.22)$$

in which $E_n(\mathbf{k})$ is the usual energy band function. Since the second term in (6.3.20) already contains an explicit factor of B^2, we may replace W_n by E_n in it. On the other hand, second-order quantities must be retained in the expansion of the first term:

$$\ln(1 + \exp[(\mu - W_n)/KT]) = \ln(1 + \exp[(\mu - E_n)/KT]) - W_n^{(1)}(eB/\hbar KT)f(E_n - \mu) - W_n^{(2)}(e^2B^2/\hbar^2KT)f(E_n - \mu) - (e^2B^2/2KT\hbar^2)[W_n^{(1)}]^2 f'(E_n - \mu) \quad (6.3.23)$$

where f' is the derivative of the Fermi function (6.3.21) with respect to E_n. Equation (6.3.20) must be multiplied by a factor of 2 before insertion into (6.3.1) in order to account for directions of the electron spin. Our result for the free energy to second order in B is

$$\mathfrak{F} = N\mu - 2KT \sum_{\mathbf{k}} \ln(1 + \exp[(\mu - E_n)/KT]) + (2eB/\hbar) \sum_{\mathbf{k}} W_n^{(1)} f(E_n - \mu) + (2e^2B^2/\hbar^2) \sum_{\mathbf{k}} \{W_n^{(2)}(\mathbf{k})f(E_n - \mu) + \tfrac{1}{2}[W_n^{(1)}(\mathbf{k})]^2 f'(E_n - \mu) - \tfrac{1}{24}f'(E_n - \mu) \times [\partial^2E_n/\partial k_x^2\, \partial^2E_n/\partial k_y^2 - (\partial^2E_n/\partial k_x\, \partial k_y)^2]\}. \quad (6.3.24)$$

We must now consider the determination of the magnetic susceptibility. The electrons interact with the magnetic flux density B (rather than with the field intensity H). However, the induced magnetization is the negative of the first derivative of $\mathfrak{F}$ with respect to H. In MKS units, we have

$$\mu_0 M = -\partial\mathfrak{F}/\partial H \quad (6.3.25a)$$

where μ_0 is the permeability of free space $(4 \times 10^{-7}\ \mathrm{H/m})$. Also since

$$M = \chi_m H$$

$$B = \mu_0(1 + \chi_m)H \tag{6.3.25b}$$

where χ_m is the dimensionless magnetic susceptibility. Thus if $\mathfrak{F} = \text{const} - CB^2$, where C is a constant, we find

$$\chi_m/(1 + \chi_m)^2 = 2\mu_0 C. \tag{6.3.25c}$$

The preceding discussion may require modification if the interaction of electrons is considered. Pippard (1963) and Condon (1966) argue that it is then desirable to use the $\mathfrak{F}(B)$ computed above for noninteracting electrons but to change (6.3.25a) to $M = -\partial\mathfrak{F}/\partial B$. This is equivalent to the addition of a term $\frac{1}{2}\mu_0 M^2$ to $\mathfrak{F}$ as previously computed with (6.3.25a) retained, in partial account of the effect of electron interactions on the free energy. We will restrict our attention here to situations in which the magnetic susceptibility is small ($\chi_m \ll 1$) so that

$$\chi_m \approx 2\mu_0 C. \tag{6.3.25d}$$

It turns out that $W_n^{(1)}(\mathbf{k}) = -W_n^{(1)}(-\mathbf{k})$. As a result, the term linear in B in (6.3.24) vanishes. The combination of (6.3.25d) and (6.3.24) gives us (with the summation being replaced by an integral)

$$\begin{aligned}\chi_m = -[4e^2\mu_0/(2\pi)^3\hbar^2] \int d^3k \, \{&W_n^{(2)}(\mathbf{k}) f(E_n - \mu) \\ &+ \tfrac{1}{2}[W_n^{(1)}(\mathbf{k})]^2 f'(E_n - \mu) \\ &- \tfrac{1}{24} f'(E_n - \mu)\,[\partial^2 E_n/\partial k_x^2 \, \partial^2 E_n/\partial k_y^2 - (\partial^2 E_n/\partial k_x \, \partial k_y)^2]\}. \end{aligned} \tag{6.3.26}$$

This result contains three terms. The third term, which is known as the Landau–Peierls susceptibility is given by

$$\begin{aligned}\chi_m^{\text{L-P}} = \tfrac{1}{48}(e^2\mu_0/\pi^3\hbar^2) \int d^3k [&\partial^2 E_n/\partial k_x^2 \, \partial^2 E_n/\partial k_y^2 \\ &- (\partial^2 E_n/\partial k_x \, \partial k_y)^2] f'(E_n - \mu). \end{aligned} \tag{6.3.27}$$

This term turns out to be the only one to survive in the case of free electrons. If we assume a spherical band characterized by an effective mass m^*, (6.3.27) is easily integrated at $T = 0$ [where $f'(\epsilon) = -\delta(\epsilon)$] and yields

$$\chi_m^{\text{L-P}} = -e^2\mu_0 k_F \hbar^2/12\pi^2 m_0^2 (m^*/m_0)^2 \tag{6.3.28}$$

in which m_0 is the free electron mass. The diamagnetic susceptibility is quite small. Suppose we consider a cubic material of lattice constant 4 Å = 4×10^{-10} m. Then $\chi_m^{\text{L-P}}$ becomes

$$|\chi_m^{\text{L-P}}| = 2.3 \times 10^{-6}/(m^*/m_0)^2. \tag{6.3.29}$$

Note the presence of the effective mass in the denominator, indicating that the diamagnetic effect becomes large when the effective mass is small. The

Landau–Peierls susceptibility results from the breakup of an energy band into discrete levels. The separation between these levels is inversely proportional to m^*, and this feature carries over into the susceptibility.

The other terms in (6.3.26) arise from modification of the band parameters by the field. The quantity $W^{(1)}$ enters the expression for the free energy like a magnetic moment and gives a contribution to (6.3.26) of the sign appropriate to paramagnetism. This term is called "crystalline paramagnetism" by Wannier and Upadhyaya. It vanishes for crystals having inversion symmetry. The first term in χ, which involves $W^{(2)}$, contains the atomic diamagnetism which must assert itself in the tight binding limit. The expression for W_2 is extremely complicated, and will not be given here. [For further details, see Wannier and Upadhaya (1964), or Hebborn and Sondheimer (1960)]. The physical interpretation of these terms is not properly understood at present, and no numerical evaluation for an actual material has been reported. However, simplifications are found to occur if a pseudopotential approximation is employed (Glasser, 1964; Misra and Roth, 1969), and some calculations have been performed for simple metals.

6.4 The de Haas–van Alphen Effect

In this section we will study some phenomena which are observed in metals in the presence of strong magnetic fields at low temperatures. Under these circumstances, the formation of discrete Landau levels from a band structure becomes significant. Many properties of the material show an oscillatory behavior as a function of $1/B$ in this region. A specific, and typical, example is the magnetic susceptibility. The high field oscillations of this quantity comprise the de Haas–van Alphen effect, which furnishes one of the most important techniques for the experimental investigation of Fermi surfaces. Another example, which we will not discuss here, is the high field magnetoresistance, which shows similar oscillations (Shubnikov–de Haas effect). Gold (1968) has reviewed theoretical and experimental studies of the de Haas–van Alphen effect.

6.4.1 *Nearly Free Electrons*

A complete and practically applicable theory of oscillatory magnetic effects in real metals does not yet exist. The foundation for such a theory is the effective Hamiltonian discussed in Section 6.2. Unfortunately, we do not have explicit expressions for the functions involved, and the power series expansion method used to calculate the low field susceptibility no longer is useful. In the absence of a complete theory, we will consider first a simple model for which calculations are possible: that of free electrons with effective mass m^*. The interaction of the electron spin with the field is included, and spin–orbit coupling is approximately incorporated through an effective (scalar) g factor. The results will be qualitatively extended to more complicated band structures through the WKB approximation. Our treatment of the parabolic band problem will follow, to a large extent, the work of Sondheimer and Wilson (1951) and Wilson (1953).

We must evaluate the free energy, Eq. (6.3.1), for all fields, rather than just to second order in B. The calculation is made possible by the use of a simple expression for the energy. Equation (6.3.1) states that

$$\mathfrak{F} = N\mu - KT \sum_l \ln(1 + \exp[(\mu - E_l)/KT]) \tag{6.3.1}$$

in which the index l designates a member of a complete set of states. Here we combine (6.2.64) with (6.2.85) and obtain

$$E_l = (\hbar^2 k_z^2/2m^*) + (l + \tfrac{1}{2})\hbar\omega_c \pm (g\hbar\omega_0/4) \tag{6.4.1}$$

in which $\omega_c = eB/m^*$ is the cyclotron frequency for the band, $\omega_0 = eB/m_0$, with m_0 being the free electron mass, and g is the effective g factor, here assumed to be a scalar. The ($\pm$) sign results from different directions of spins.

The calculation proceeds as follows. The classical partition function $Z(\beta)$ is introduced.

$$Z(\beta) = \sum_l \exp(-\beta E_l) \tag{6.4.2}$$

(with $\beta = 1/KT$). We also define an auxiliary function

$$g(E) = \ln(1 + \exp[\beta(\mu - E)]). \tag{6.4.3}$$

It is convenient to introduce the Laplace transform of $g(E)$, called $\phi(s)$,

$$\phi(s) = \int_0^\infty g(E)e^{-sE}\,dE \tag{6.4.4}$$

and to express $Z(\beta)/\beta^2$ as the Laplace transform of a function $z(E)$

$$Z(\beta)/\beta^2 = \int_0^\infty z(E)e^{-\beta E}\,dE. \tag{6.4.5}$$

These transforms may be inverted. According to the theory of the Laplace transform, we have

$$z(E) = (1/2\pi i)\int_{c-i\infty}^{c+i\infty} e^{Es}s^{-2}\,Z(s)\,ds \tag{6.4.6}$$

and

$$g(E) = (1/2\pi i) \int_{c-i\infty}^{c+i\infty} \phi(s) e^{Es}\, ds. \tag{6.4.7}$$

The path of integration is parallel to the imaginary axis. The constant c must be chosen so that all of the singularities of the integral are on the left, but it is otherwise arbitrary. These functions are now introduced into the free energy:

$$\mathfrak{F} = N\mu - KT \sum_l g(E_l) = N\mu - (KT/2\pi i) \int_{c-i\infty}^{c+i\infty} \sum_l e^{E_l s} \phi(s)\, ds$$

$$= N\mu - (KT/2\pi i) \int_{c-i\infty}^{c+i\infty} [Z(-s)/s^2] s^2 \phi(s)\, ds.$$

Since $s^2\phi(s)$ is the Laplace transform of $\partial^2 g/\partial E^2$, the final integral can be expressed as

$$\mathfrak{F} = N\mu - KT \int_0^\infty z(E)\, \partial^2 g/\partial E^2\, dE. \tag{6.4.8}$$

Now

$$\partial^2 g/\partial E^2 = -1/KT\, \partial f/\partial E \tag{6.4.9}$$

where $f(E - \mu)$ is the Fermi function

$$f(E - \mu) = \{\exp[(E - \mu)/KT] + 1\}^{-1} \tag{6.4.10}$$

so that

$$\mathfrak{F} = N\mu + \int_0^\infty z(E)\, df/dE\, dE. \tag{6.4.11}$$

We can calculate $Z(\beta)$ directly from (6.4.1) and (6.4.2). The number of states in an interval Δk_z is given by Eq. (6.2.67):

$$Z(\beta) = (eBV/4\pi^2\hbar) \sum_{\text{spin}} \sum_{l=0}^{\infty} \exp[-\beta(l + \tfrac{1}{2})\hbar\omega_c] \exp[\pm\beta g\hbar(\omega_0/4)]$$

$$\times \int_{-\infty}^{\infty} dk_z \exp(-\beta\hbar^2 k_z^2/2m^*)$$

$$= (eBV/2\pi\hbar^2)(m^*/2\beta\pi)^{1/2}[(\cosh \beta g\hbar\, \omega_0/4)/(\sinh \beta\hbar\, \omega_c/2)] \tag{6.4.12}$$

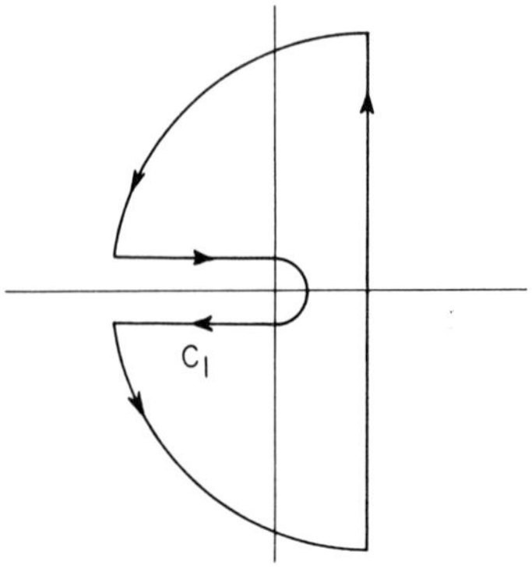

Fig. 6.4.1. Contour of integration for Eq. (6.4.13).

This result is to be substituted into (6.4.6):

$$z(E) = (1/2\pi i) \int e^{E\beta}(Z(\beta)/\beta^2)\, d\beta$$

$$= \hbar\omega_c V(m^*/2\pi\hbar^2)^{3/2}(1/2\pi i)$$

$$\times \int_{c-i\infty}^{c+i\infty} [(e^{E\beta} \cosh \beta g\hbar\, \omega_0/4)/(\beta^{5/2} \sinh \beta\hbar\, \omega_c/2)]\, d\beta. \quad (6.4.13)$$

The path of integration is parallel to the imaginary axis. It is desirable to close the contour to the left, so that Cauchy's theorem can be used. The integrand has poles on the imaginary axis at points $\beta\hbar\omega_c/2 = n\pi i$ (n is any integer, positive or negative, but not zero); and a branch point at the origin. Because the branch point is present, we finally arrive at the contour shown in Fig. 6.4.1. Since the contribution from the large arcs tends to zero as they are indefinitely enlarged, we need consider only the contribution from the residues and from the portion of the path parallel to the negative real axis. The contribution from a single residue is

$$(-1)^{n+1}2\pi i(\hbar\omega_c/2)^{3/2}[\exp[i(2n\pi E/\hbar\omega_c - \pi/4)]/(n\pi)^{5/2}] \cos(n\pi gm^*/2m_0).$$

The sum of the residues is, evidently,

$$-2\pi i(\hbar\omega_c/2)^{3/2}\, 2\sum_{n=1}^{\infty} (-1)^n[\cos(2n\pi E/\hbar\omega_c - \pi/4)/(n\pi)^{5/2}] \cos(n\pi gm^*/2m_0). \quad (6.4.14)$$

It is this portion of $z(E)$ which comes from poles of the integrand and produces the oscillatory behavior of the magnetic susceptibility observed in the de Haas–van Alphen effect.

The steady low field susceptibility discussed in Section 6.3 results from the integral along that part of the contour parallel to the negative real axis (C_1). This is considered in the limit that the radius of the semicircle at the origin tends to zero. Near $\beta = 0$, we may use the small argument expansion of the hyperbolic functions in (6.4.13). For small z, we have

$$\operatorname{csch} z = (1/z)[1 - (z^2/6) + (7z^4/360) + \cdots].$$

Now put $z = \beta\hbar\omega_c/2$. We obtain a contribution

$$[1/(2\pi i)](\hbar\omega_c/2)^{3/2} \int_{c1} \exp(2zE/\hbar\omega_c) \times [z^{-7/2} + z^{-3/2}(m^{*2}g^2/8m_0^2 - \tfrac{1}{6}) + \cdots]\, dz. \tag{6.4.15}$$

This integral can be evaluated through the use of the identity (Jeffreys and Jeffreys, 1950, p. 401)

$$(1/2\pi i) \int_{c1} z^{-\nu} e^{zt}\, dz = t^{\nu-1}/\Gamma(\nu).$$

Then (6.4.15) becomes

$$(16E^{5/2}/15\hbar\omega_c\pi^{1/2}) + (\hbar\omega_c E^{1/2}/\pi^{1/2})(m^{*2}g^2/8m_0^2 - \tfrac{1}{6}) + \cdots. \tag{6.4.16}$$

Higher terms in the series can be neglected, provided (as we shall be able to infer subsequently) $\mu \gg \hbar\omega_c$. The neglected terms would give rise to a field dependence of the normal diamagnetism and paramagnetism. We will now combine (6.4.13), (6.4.14), and (6.4.16) to find $z(E)$

$$\begin{aligned} z(E) = V(m^*/2\pi\hbar^2)^{3/2}\Big[&(16E^{5/2}/15\pi^{1/2}) \\ &+ ((\hbar\omega_c)^2/\pi^{1/2})E^{1/2}(m^{*2}g^2/8m_0^2 - \tfrac{1}{6}) + \cdots \\ &+ [-(\hbar\omega_c)^{5/2}/2^{1/2}] \sum_{n=1}^{\infty} (-1)^n[\cos(2\pi nE/\hbar\omega_c - \pi/4)/(n\pi)^{5/2}] \\ &\times \cos(n\pi g m^*/2m_0)\Big]. \end{aligned} \tag{6.4.17}$$

This result is to be substituted into (6.4.11) to obtain the free energy. It is tempting simply to write $\partial f/\partial E = -\delta(E - \mu)$, which enables the integral to be performed trivially. This is in fact an adequate approximation as far as the first two terms are concerned, since they would exhibit only a weak temperature dependence if the calculation were done more adequately. The oscillatory term does depend strongly on temperature, and greater care is desirable.

It follows from Eq. (6.4.10) that

$$\partial f/\partial E = -(4KT\cosh^2[(E-\mu)/2KT])^{-1}.$$

Then consider the integral

$$\begin{aligned}\int_0^\infty &\cos(2\pi nE/\hbar\omega_c - \pi/4)\ \partial f/\partial E\ dE \\ &= -\tfrac{1}{4}\operatorname{Re}\exp[i(2\pi n\mu/\hbar\omega_c - \pi/4)] \\ &\quad\times\int_{-\infty}^{\infty}[\exp(2\pi inKTy/\hbar\omega_c)/\cosh^2(y/2)]\,dy. \qquad (6.4.18)\end{aligned}$$

We have defined $y = (E-\mu)/KT$ and extended the lower limit of the integration on y from $-\mu/KT$ to $-\infty$. The error arising from this replacement is of the order $e^{-\mu/KT}$, which is negligible. The integral is a tabulated one (Erdelyi *et al.*, 1954, p. 31) and gives

$$-(2\pi^2 nKT/\hbar\omega_c)[\cos(2\pi n\mu/\hbar\omega_c - \pi/4)/\sinh(2\pi^2 nKT/\hbar\omega_c)].$$

The free energy is found from (6.4.11) to be

$$\begin{aligned}\mathfrak{F} = N\mu - V\frac{16\mu^{5/2}}{15\pi^{1/2}}\left(\frac{m^*}{2\pi\hbar^2}\right)^{3/2} &\\ \times\Bigg[1 + \frac{15}{16}\left(\frac{\hbar\omega_c}{\mu}\right)^2\left(\frac{m^{*2}g^2}{8m_0^2} - \frac{1}{6}\right) - \frac{15}{8(2)^{1/2}}\left(\frac{KT}{\mu}\right)\left(\frac{\hbar\omega_c}{\mu}\right)^{3/2} &\\ \times\sum_{n=1}^{\infty}\frac{(-1)^n}{n^{3/2}}\frac{\cos(2n\pi\mu/\hbar\omega_c - \pi/4)\cos(n\pi gm^*/2m_0)}{\sinh(2\pi^2 nKT/\hbar\omega_c)}\Bigg]. &\qquad (6.4.19)\end{aligned}$$

The magnetization is found by differentiating (6.4.19) with respect to H according to (6.3.25a–d). Further, in differentiating the oscillatory part of the free energy, we neglect the contribution from all except the cosine term, which should be dominant under the conditions of interest. We divide by the volume V in order to obtain the magnetization per unit volume:

$$\begin{aligned}M = M_0\Bigg[\left(\frac{3m^{*2}g^2}{4m_0^2} - 1\right) + \frac{6\pi KT}{\hbar\omega_c}\left(\frac{2\mu}{\hbar\omega_c}\right)^{\tau/2}\sum_n\frac{(-1)^n}{n^{1/2}} &\\ \times\frac{\sin(2\pi n\mu/\hbar\omega_c - \pi/4)\cos(n\pi gm^*/2m_0)}{\sinh(2\pi^2 nKT/\hbar\omega_c)}\Bigg] &\qquad (6.4.20)\end{aligned}$$

in which

$$M_0 = (e/6\pi^2\hbar)(m^*\mu/2)^{1/2}\omega_c = (e^2/12\pi^2\hbar)(2\mu/m^*)^{1/2}B. \qquad (6.4.21)$$

The magnetic susceptibility is found by dividing M by H. The leading term is the paramagnetic susceptibility due to electron spins; the second term is the steady diamagnetic term which is one third as large when the effective mass ratio is unity and $g = 2$: these terms are independent of temperature. The oscillatory term is significant at low temperatures and high fields; it produces the de Haas–van Alphen effect.

The chemical potential μ must now be determined in accord with (6.3.3) by differentiating $\mathfrak{F}$ with respect to μ

$$N = \frac{V}{3\pi^2}\left(\frac{2m^*\mu}{\hbar^2}\right)^{3/2}\left[1 + \frac{3}{16}\left(\frac{\hbar\omega_c}{\mu}\right)^2\left(\frac{m^{*2}g^2}{8m_0^2} - \frac{1}{6}\right) + \frac{3\pi KT}{\hbar\omega_c}\left(\frac{\hbar\omega_c}{2\mu}\right)^{3/2}\right.$$
$$\left.\times \sum_{n=1}^{\infty} \frac{(-1)^n}{n^{1/2}} \frac{\sin(2\pi n\mu/\hbar\omega_c - \pi/4)\cos(n\pi g m^*/2m_0)}{\sinh(2\pi^2 nKT/\hbar\omega_c)}\right]. \qquad (6.4.22)$$

It is necessary to solve (6.4.22) to determine μ in terms of N. However, it is usually sufficient to include only the first term: this is the case as long as $(\hbar\omega_c) \ll \mu$. Then the relation between N and μ is the same as for a free electron gas. The chemical potential differs from the free electron value by terms proportional to B^2 when B is small. We set $\mu = \hbar^2k_F^2/2m^*$, in which k_F is the wavevector on the Fermi surface. We obtain for M_0 from (6.4.21)

$$M_0 = (e^2k_F/12\pi^2m^*)B. \qquad (6.4.23)$$

This agrees with (6.3.28).

It is interesting to consider the origin of the oscillatory term in the magnetization. The application of the external field causes the band structure to break up (for directions of **k** perpendicular to the field) into a set of Landau levels. For fields which are not too strong, the Fermi level can be regarded as a constant. The lowest energy level now has energy $\hbar\omega_c/2 - g\hbar\omega_0/4$. The increase in energy of the electron states due to formation of the Landau level structure gives rise to a steady diamagnetic contribution to the susceptibility which will dominate the spin paramagnetism if the effective mass is small. As the field increases, the spacing between the Landau levels increases, so that the levels are gradually forced through the Fermi energy and depopulated. At the same time, the degeneracy of the Landau levels increases. The average energy of the system fluctuates as this "forcing through" occurs. The amplitude of the oscillations increases with increasing field, and the period is proportional to B^{-1}. The amplitude is a rapidly decreasing function of $KT/\hbar\omega_c$, which implies that observation of the oscillations requires low temperature.

6.4.2 *General Fermi Surfaces*

The extension of these results to the case of a band with anisotropic effective masses is straightforward (Blackman, 1938), but for arbitrary, nonparabolic band structures, the problem is much more difficult. A fairly extensive literature exists on this subject. A few references are given: Onsager, 1952; Lifshitz and Kosevich, 1955; Lifshitz and Kaganov 1959, 1962; Pippard 1962, 1964; Chambers, 1965; Roth, 1966. The problem is not altogether in a satisfactory condition at the present time since, as we have noted several times before, an explicit expression for the effective Hamiltonian does not exist. The problem can be resolved into two parts: (1) the computation of the separation of the Landau levels (the cyclotron frequency) for an arbitrary Fermi surface, and (2) the subsequent calculation of the free energy.

In the absence of an adequate effective Hamiltonian formalism for nonparabolic bands, the determination of the level structure in the presence of a magnetic field is based on semiclassical arguments employing the WKB approximation. We begin by considering the commutation rule (6.2.6) for the components of the cannonical momentum and let the direction of B define the z axis

$$[P_x, P_y] = (e\hbar/i)B, \qquad [P_y, P_z] = [P_x, P_z] = 0. \tag{6.4.24}$$

An analogy can be drawn between the first of Eqs. (6.4.24) and the ordinary commutation rule between coordinates and momenta

$$[p_k, q_j] = (\hbar/i)\,\delta_{kj} \tag{6.4.25}$$

by making the identification

$$p_x = p_x; \qquad q_x = p_y/eB. \tag{6.4.26}$$

The semiclassical quantum condition

$$(1/2\pi)\oint p_x\,dq_x = (n + \gamma)\hbar, \tag{6.4.27}$$

in which n is a positive integer and γ is some constant "phase factor" ($\gamma = \frac{1}{2}$ for free electrons), can be applied. We substitute (6.4.26) into (6.4.27) and find

$$\oint P_x\,dP_y = 2\pi(n + \gamma)\hbar eB = \mathcal{A}. \tag{6.4.28}$$

The integral in (6.4.28) runs along the curve bounding a cross section of a surface of constant energy in momentum space and has a value equal to the area $\mathcal{A}$ of this cross section [note that $\mathcal{A} = \hbar^2\mathcal{A}'(\mathbf{k})$, where $\mathcal{A}'$ is the area in $\mathbf{k}$ space). The area of the cross section depends on the energy of the state whose quantum number is n. We will suppose that n is sufficiently large that it may be in effect regarded as a continuous variable. Then the separation of the quantum states in energy, which is defined in terms of the cyclotron frequency ω_c, is

$$dE/dn = \hbar\omega_c. \tag{6.4.29}$$

Differentiation of (6.4.28) with respect to energy yields

$$\omega_c = 2\pi eB(\partial\mathcal{A}/\partial\epsilon)^{-1}. \tag{6.4.30}$$

Note that for a parabolic band with effective mass m^*

$$\mathcal{A} = \pi p^2 = \pi\hbar^2k^2 = 2\pi m^*\epsilon.$$

Then (6.4.30) yields the usual result $\omega_c = eB/m^*$, and comparison with (6.4.28) shows that ϵ is given correctly if $\gamma = \frac{1}{2}$, as mentioned above.

We can easily see that ω_c is the circular frequency of rotation of a classical orbit on a surface of constant energy. The orbit defines a cross section of this surface perpendicular to the magnetic field. The Lorentz force on a particle is evB, so that the time required for the momentum to change by dp is dp/evB. The time required to complete a revolution is thus

$$T = (1/eB)\int dp/v. \tag{6.4.31}$$

Now consider similar cross sections of surfaces of constant energy E and $E + dE$. The separation between these surfaces is $dE/|\nabla_p E| = dE/\text{v}$. Thus the area of the annular ring between E and $E + dE$ is

$$d\mathcal{A} = dE\oint dp/v \tag{6.4.32}$$

where dp is a momentum increment along the ring. Evidently

$$T = (1/eB)\, d\mathcal{A}/dE \qquad \text{and} \qquad \omega_c = 2\pi/T = 2\pi eB(d\mathcal{A}/d\epsilon)^{-1} \tag{6.4.33}$$

as required.

A more complete treatment (Roth, 1966) shows that γ in (6.4.28) is a slowly varying function of energy, which differs from $\frac{1}{2}$ by terms which depend on the detailed shape of the cross section (vanishing for a sphere) and on the first- and second-order (in B) terms in the effective Hamiltonian. A quantitative investigation of this correction for a model (Butler and Brown, 1968) yielded results for γ which differed from $\frac{1}{2}$ by a correction which was, in fact, small and slowly varying.

The existence and periodicity of oscillations in the magnetic susceptibility can be inferred from these considerations. Since $\hbar\omega_c/\mu$ is quite small in usual circumstances, the Fermi energy will be essentially independent of the field. Consider a particular level of oscillator quantum number n. This corresponds to an electron orbit in momentum space in a plane perpendicular to the magnetic field. For small fields, orbits with this n are small and lie well inside the Fermi surface. Now increase the field. The area $\mathcal{A}$ of an electron orbit in momentum space increases with increasing field according to (6.4.28). As the area expands, the energy of an electron in such an orbit increases. When the area equals that of a cross section of the Fermi surface $\mathcal{A}_0$ the oscillator level will be half depopulated. This occurs when

$$2\pi(n + \gamma) = \mathcal{A}_0(p_z)/\hbar eB. \tag{6.4.34}$$

The magnetization will vary with the state of depletion of the level with energy. A periodic dependence of magnetization on field is to be expected as successive levels come up to the Fermi energy and depopulate. Thus we expect that the susceptibility contains, at least as lowest harmonic, a term proportional to

$$\sin[(\mathcal{A}_0(p_z)/\hbar eB + \phi] \tag{6.4.35}$$

where ϕ is some phase factor. So far we have considered a cross section corresponding to a definite value of p_z. The contribution of all cross sections must be summed. Since the trigonometric function oscillates rapidly as a function of $\mathcal{A}_0$ (usually $\hbar eB \ll \mathcal{A}_0$), the contribution from different cross sections will tend to cancel. The resultant will be governed by the region of Fermi surface for which the area is an extremum with respect of p_z. This is what is observed experimentally: susceptibility measurements show periodic behavior as a function of $\mathcal{A}_0/\hbar eB$, where $\mathcal{A}_0$ is the area of an extremal cross section of the Fermi surface perpendicular to the field. In the scalar effective mass approximation

$$\mathcal{A}_0/\hbar eB = 2\pi m^*\mu/\hbar eB = 2\pi\mu/\hbar\omega_c. \tag{6.4.36}$$

This is consistent with Eq. (6.4.20) as it should be.

Equation (6.4.20) indicates that harmonics of the basic period $\mathcal{A}_0/\hbar eB$ are present in the magnetization for the case of a spherical Fermi surface. The calculation of Roth (1966) confirms their presence in the more general case just discussed. However, their amplitude decreases quite rapidly with order because of the presence of the hyperbolic sine in the denominator of (6.4.20) if, as is usual $2\pi^2KT > \hbar\omega_c$. Ordinarily, only a single harmonic need be considered. Figure 6.4.2 shows as an example de Haas–van Alphen oscillations of the magnetic susceptibility observed in zirconium.

The theory presented above is obviously applicable to a closed convex

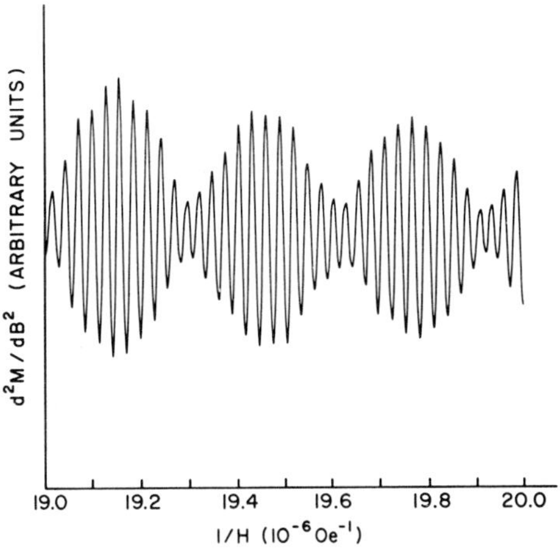

Fig. 6.4.2. De Haas–van Alphen oscillations in zirconium, according to Everett (1972). Two different periods are present, indicating that there are contributions from two different portions of the Fermi surface.

Fermi surface or to such portions of a Fermi surface which consists of several pieces. More complicated Fermi surfaces are possible which permit qualitatively different orbits. In particular, if the Fermi surface contains a saddle point, a self-intersecting (figure eight) orbit is possible. Also, if the Fermi surface is in contact with the Brillouin zone, the intersection of a plane perpendicular to the field with the Fermi surface may not give a closed curve. This situation may be visualized with reference to Fig. 6.4.3, which shows orbits of different types. If an extended zone scheme were adopted,

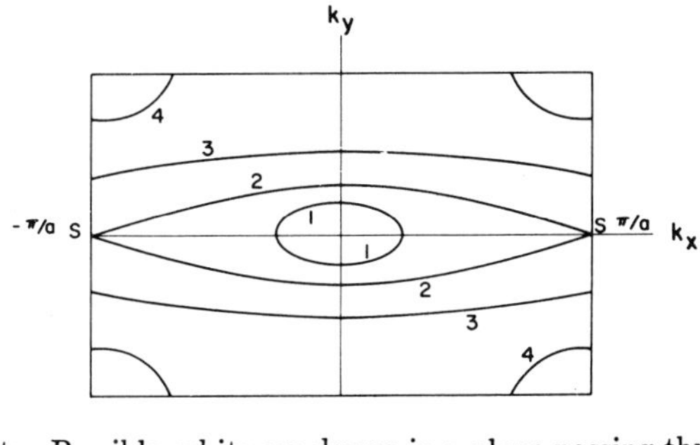

Fig. 6.4.3. Orbits. Possible orbits are drawn in a plane passing through the Brillouin zone of a hypothetical crystal. Each orbit corresponds to a Fermi surface of different size. 1, a closed electron orbit; 2, a self-intersecting orbit passing through the saddle point S; 3, an open orbit; 4, a hole orbit.

the "nonclosed" orbit mentioned above would continue throughout many cells in reciprocal space. If the magnetic field is a rational field, in the sense of Section 6.2, the orbit will be periodic with some long period, but for an irrational field, the orbit would continue indefinitely. The effect of such orbits on the susceptibility has been analyzed by Roth (1966) and others. We will not present details.

In sufficiently strong magnetic fields, the theory presented above fails for another reason: The assumption that it is sufficient to consider a single isolated energy band breaks down. The physical situation is somewhat analogous to the tunneling problem for an electric field described in Section 6.1: Under the influence of the Lorentz force, an electron describes a periodic orbit on a surface of constant energy within a band. This motion is described by an effective single-band Hamiltonian. However, there remains a probability that the electron may jump into another band. Pictorially, we may think of an electron orbit being continued through a Brillouin zone boundary where a gap would normally be present. This effect was named *magnetic breakdown* by Cohen and Falicov (1961), and has been studied by several authors (see particularly Blount, 1962; Pippard 1962, 1964; Brown, 1968). The analogy with the electric tunneling problem suggests that the probability of magnetic breakdown should be proportional to $\exp(-\text{const}/B)$. It can further be shown that magnetic breakdown will be a significant effect if $\hbar\omega_c E_F > E_g^2$ where E_g is the gap across a zone face.

6.5 Optical Properties

In this section the optical properties of solids will be considered from the point of view of energy band theory. This involves electronic transitions between band states. These transitions occur in different energy ranges in different materials; however, we are mostly concerned with processes involving photons of energies between 0.1 and 20 eV. At longer wavelengths, lattice vibrations are important (the influence of lattice vibrations on the optical properties of crystals was discussed in Sections 1.3 and 1.9, Part A). Excitation of levels in the cores of atoms becomes significant at energies higher than those we will consider here. The wavelength of a 20-eV photon is 600 Å; this is still large compared to a typical lattice constant (6 Å, say), thus the spatial variation of the electric field over a unit cell can be neglected to a reasonable approximation. The wave vector of the light is effectively zero, and we describe the properties of materials by conductivity and dielectric functions of frequency (rather than of frequency and wave vector). A useful introduction to the theory of optical properties of solids has been given by Stern (1963). MKS units will be used in our discussion.

6.5.1 *The Kramers–Kronig Relations*

We will begin by deriving some relations between the relevant physical quantities which are consequences of general considerations, such as causality. Let a time-dependent electric field $\mathbf{E}(t)$ be applied to a solid. A current density $\mathbf{J}(t)$ will be produced. These quantities are connected linearly by the time-dependent conductivity $\mathbf{\Sigma}$. As stated above, we will neglect any spatial variation of these quantities. In general, $\mathbf{E}$ and $\mathbf{J}$ are vectors, and $\mathbf{\Sigma}$ is a tensor; however, consideration will be limited to cubic materials for which $\mathbf{E}$ and $\mathbf{J}$ are in the same direction. Their vector nature can then be neglected. We may thus take Σ to be a scalar. The relation between E and J is assumed to be linear, that is, we consider only weak fields. Nonlinear effects are presently of considerable importance, but we will postpone consideration of these to Section 6.5.5.

A general linear relation between J and E can be written as

$$J(t) = (2\pi)^{-1/2} \int \Sigma(t - t')E(t')\, dt'. \tag{6.5.1}$$

The function Σ will be called the *time-dependent conductivity*. The factor $(2\pi)^{-1/2}$ is a convention introduced so that the usual Ohm's law will be obtained for the Fourier transforms. Since the origin of the time scale should not be of physical significance, $\Sigma(t, t')$ is a function of the difference of its arguments, and this has been used above.

Equation (6.5.1) takes a simpler form if we introduce the Fourier transforms of the functions. Let $\mathcal{J}(\omega)$, $\mathcal{E}(\omega)$, and $\sigma(\omega)$ be the transforms of J, E, and Σ:

$$J(t) = (2\pi)^{-1/2} \int_{-\infty}^{\infty} \mathcal{J}(\omega)e^{-i\omega t}\, d\omega \tag{6.5.2a}$$

$$\mathcal{J}(\omega) = (2\pi)^{-1/2} \int_{-\infty}^{\infty} J(t)e^{i\omega t}\, dt \tag{6.5.2b}$$

$$E(t) = (2\pi)^{-1/2} \int_{-\infty}^{\infty} \mathcal{E}(\omega)e^{-i\omega t}\, d\omega \tag{6.5.2c}$$

$$\Sigma(t) = (2\pi)^{-1/2} \int_{-\infty}^{\infty} \sigma(\omega)e^{-i\omega t}\, d\omega. \tag{6.5.2d}$$

Then we find

$$\mathcal{J}(\omega) = \sigma(\omega)\mathcal{E}(\omega). \tag{6.5.3}$$

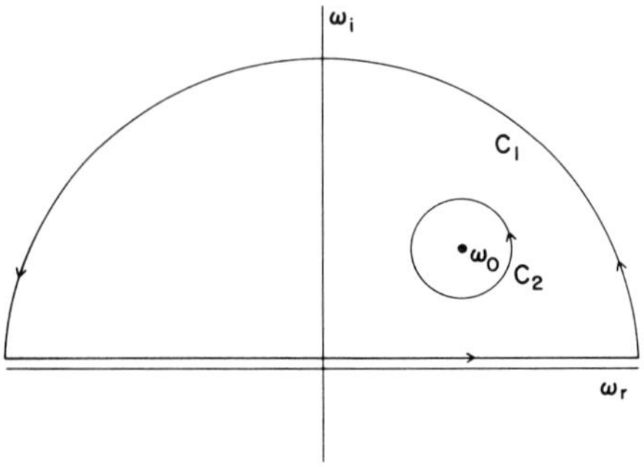

Fig. 6.5.1. Contours of integration for Eq. (6.5.2d) (C_1), if $t < 0$, and for (6.5.4) (C_2).

The function $\sigma(\omega)$ is the frequency-dependent conductivity. It will be a major subject of interest, and its properties will be investigated in detail.

We will assume that $\Sigma(t)$ is zero for negative values of its argument. This is the assumption of causality: the current cannot start before the electric field arrives. Thus if t_0 is the earliest value of t' for which $E(t')$ is nonzero, (6.5.1) implies, with this assumption, that $J(t) = 0$ if $t < t_0$. The assumption that Σ is zero for negative values of its argument implies that its Fourier transform $\sigma(\omega)$ is an analytic function in the upper half of the complex ω plane. This is necessary since, for negative t, the contour of integration for (6.5.2d) should be closed by a large semicircle enclosing the upper half complex plane (Fig. 6.5.1). By Cauchy's theorem, the integral will vanish if $\sigma(\omega)$ is analytic everywhere in the upper half plane (and goes to zero properly at ∞, etc.).

Now let ω_0 be a frequency with a positive imaginary part. We can write

$$\sigma(\omega_0) = (1/2\pi i) \oint_{C_2} \sigma(\omega)/(\omega - \omega_0)\, d\omega. \tag{6.5.4}$$

Initially, the contour may be chosen as a small circle about ω_0 (C_2 in Fig. 6.5.1), but it may be expanded to C_1 without changing the result. Now suppose that $\sigma(\omega) \to 0$ if $|\omega| \to \infty$. Then the integral in (6.5.4), after expansion, has a contribution only from the portion along the real axis. Let $\nu = \omega_r$. Then

$$\sigma(\omega_0) = (1/2\pi i) \int_{-\infty}^{\infty} \sigma(\nu)/(\nu - \omega_0)\, d\nu. \tag{6.5.5}$$

Let ω_0 approach the real axis from above, and come within some small distance ϵ of it: $\omega_0 = \omega + i\epsilon$, ω being real. Use the identity

$$1/(\nu - \omega - i\epsilon) = P(1/\nu - \omega) + i\pi\, \delta(\nu - \omega) \tag{6.5.6}$$

(where P indicates principal value). Substitute (6.5.6) in (6.5.5). We obtain

$$\sigma(\omega) = (1/\pi i)P\int_{-\infty}^{\infty} \sigma(\nu)/(\nu - \omega)\, d\nu. \tag{6.5.7}$$

Equation (6.5.7) can be separated into equations involving the real and imaginary parts of the function σ. Let $\sigma(\omega) = \sigma_1(\omega) + i\sigma_2(\omega)$. Then

$$\sigma_1(\omega) = (1/\pi)P\int_{-\infty}^{\infty} \sigma_2(\nu)/(\nu - \omega)\, d\nu \tag{6.5.8a}$$

$$\sigma_2(\omega) = -(1/\pi)P\int_{-\infty}^{\infty} \sigma_1(\nu)/(\nu - \omega)\, d\nu. \tag{6.5.8b}$$

Thus the real and imaginary parts of σ are Hilbert transforms of each other. It will be assumed that σ_1 and σ_2 vanish sufficiently rapidly as $|\nu| \to \infty$ so that the integrals exist. This will be satisfied if σ_1 and σ_2 are square integrable.

Next, we observe that the time-dependent conductivity function Σ must be real, since any real electric field $E(t)$ must produce a real current $J(t)$. This can be seen to require that

$$\sigma^*(\nu) = \sigma(-\nu). \tag{6.5.9}$$

Consequently,

$$\sigma_1(\nu) = \sigma_1(-\nu) \tag{6.5.10a}$$

$$\sigma_2(\nu) = -\sigma_2(-\nu). \tag{6.5.10b}$$

The real part of σ is an even function and the imaginary part an odd function of frequency.

Equations (6.5.10) can be used to eliminate the negative frequency portion of (6.5.8). Thus consider (6.5.8a):

$$\sigma_1(\omega) = (1/\pi)P\left[\int_{-\infty}^{0} \sigma_2(\nu)/(\nu - \omega)\, d\nu + \int_{0}^{\infty} \sigma_2(\nu)/(\nu - \omega)\, d\nu\right]$$

$$= (1/\pi)P\int_{0}^{\infty} \sigma_2(\nu)[1/(\nu - \omega) + 1/(\nu + \omega)]\, d\nu$$

$$= (2P/\pi)\int_{0}^{\infty} \nu\sigma_2(\nu)/(\nu^2 - \omega^2)\, d\nu. \tag{6.5.11a}$$

In the second step, we have replaced ν by $-\nu$, and used (6.5.10b). Also,

we have

$$\sigma_2(\omega) = -(2/\pi)\omega P \int_0^\infty \sigma_1(\nu)/(\nu^2 - \omega^2)\, d\nu. \qquad (6.5.11b)$$

Equations (6.5.11) comprise the Kramers–Kronig relations (Kramers, 1927; Kronig, 1926, 1942) as applied to the conductivity. It is also possible to obtain equivalent relations involving the frequency-dependent dielectric function of the system:

$$\kappa(\omega) = \epsilon(\omega)/\epsilon_0 = \kappa_1(\omega) + i\kappa_2(\omega) \qquad (6.5.12)$$

where ϵ is the permittivity of the material, ϵ_0 that of free space, and κ the dielectric function. We have from electromagnetic theory for an electric field proportional to $e^{-i\omega t}$, $\epsilon(\omega) = \epsilon_0 + i\sigma/\omega$. Thus

$$\kappa_1(\omega) = 1 - [\sigma_2(\omega)/\epsilon_0\omega] \qquad (6.5.13a)$$

$$\kappa_2(\omega) = \sigma_1(\omega)/\epsilon_0\omega \qquad (6.5.13b)$$

and

$$\kappa_1(\omega) = 1 + (2/\pi)P \int_0^\infty \nu\kappa_2(\nu)/(\nu^2 - \omega^2)\, d\nu \qquad (6.5.14a)$$

$$\kappa_2(\omega) = (2/\pi\omega)P \int_0^\infty \nu^2[1 - \kappa_1(\nu)]/(\nu^2 - \omega^2)\, d\nu. \qquad (6.5.14b)$$

Equation (6.5.14b) can be simplified if we note that from (6.5.11a) and (6.5.13a) that the dc conductivity $\sigma_1(0)$ is given by

$$\sigma_1(0) = 2(\epsilon_0/\pi) \int_0^\infty [1 - \kappa_1(\nu)]\, d\nu. \qquad (6.5.15)$$

Also,

$$P \int_0^\infty [\nu^2 - \omega^2]^{-1}\, d\nu = 0.$$

Then we have

$$\kappa_2(\omega) - [\sigma_1(0)/\epsilon_0\omega] = -2P\omega/\pi \int_0^\infty \kappa_1(\nu)/(\nu^2 - \omega^2)\, d\nu. \qquad (6.5.14c)$$

The quantities κ_1 and κ_2 (or σ_1 and σ_2) are, in principle, independently measurable. Thus these relations are subject to experimental verification. However, the assumptions which enter their derivation are so general that it has become customary to accept the Kramers–Kronig relations as valid, and to use them instead as an aid in the analysis of experimental data. Thus if $\kappa_1(\omega)$ is measured over a sufficiently wide range of frequencies, κ_2 may be extracted by integration, using (6.5.14c). Various forms of these relations have been constructed to facilitate such analysis.

Of particular interest in this respect is the reflectivity of a solid at normal incidence, which is a directly measurable experimental quantity. The measured reflectivity $R(\omega)$ is the ratio of reflected to incident intensities. This is the absolute square of the (complex) ratio of reflected and incident electric or magnetic field strengths $r(\omega)$:

$$R(\omega) = |r(\omega)|^2, \qquad r(\omega) = |r(\omega)|\, e^{i\theta(\omega)}. \tag{6.5.16}$$

Given the amplitude and phase of the reflectivity, it is possible to deduce the real part of the index of refraction n and the extinction coefficient k, and from these one can work back to the real and imaginary parts of the dielectric function. We quote the relations here without proof (for further details, see Stern, 1963). For a material with the magnetic properties of free space

$$[n(\omega) + ik(\omega)]^2 = \kappa(\omega) \tag{6.5.17}$$

where $\kappa(\omega)$ is the dielectric function given by (6.5.12). Also

$$\begin{aligned} n &= (1 - R)/(1 + R - 2R^{1/2}\cos\theta) \\ k &= 2R^{1/2}\sin\theta/(1 + R - 2R^{1/2}\cos\theta) \end{aligned} \tag{6.5.18}$$

where θ is the phase given by (6.5.16).

The Kramers–Kronig relations can be applied to determine θ from R. The derivation of this relation is not so straightforward as is the argument leading to Eqs. (6.5.11) since it is apparent from (6.5.16) that we could multiply $r(\omega)$ by any function which has unit absolute value for all real ω and which has zeros but no poles in the upper half plane without changing the observed R. However, the possible contributions from such a function can ultimately be ruled out by physical arguments, and one finds (Toll, 1956; Stern, 1963)

$$\theta(\omega) = (\omega/\pi)P\int_0^\infty \ln[R(\nu)/R(\omega)]/(\omega^2 - \nu^2)\, d\nu. \tag{6.5.19}$$

6.5.2 *Simple Theory of Optical Absorption: Direct Transitions*

We will next consider a simple theory of optical absorption in band to band transitions. Subsequently, we will develop a more complete approach, leading to the calculation of the dielectric function. The absorption of light by a solid excites an electron but must, at the same time, leave behind an unoccupied state or hole. The present approach is applicable in circumstances where the interaction between the excited electron and the residual hole can be neglected. The electromagnetic field is treated classically. A more detailed treatment in which theory and experiment are compared has been given by Johnson (1967).

The Hamiltonian for an electron in an external electromagnetic field characterized by a vector potential $\mathbf{A}$ has been given previously,

$$H = [(\mathbf{p} + e\mathbf{A})^2/2m] + V(\mathbf{r}).$$

In the standard semiclassical theory of radiation, it is customary to neglect the portion of the Hamiltonian quadratic in $\mathbf{A}$ and to choose a gauge in which $\nabla \cdot \mathbf{A} = 0$. Then $[\mathbf{p}, \mathbf{A}] = 0$, and we may consider a perturbation

$$H' = (e/m)\mathbf{A} \cdot \mathbf{p} = (-ie\hbar/m)\mathbf{A} \cdot \nabla. \tag{6.5.20}$$

Let us consider a matrix element of the perturbation between a final state $\psi_n(\mathbf{k}, \mathbf{r})$ and an initial state in a different band $\psi_l(\mathbf{q}, \mathbf{r})$. We choose for the (real) vector potential a monochromatic plane wave $A_0 \boldsymbol{\varepsilon}_s \cos(\mathbf{s} \cdot \mathbf{r} - \omega t)$ in which $\boldsymbol{\varepsilon}_s$ is a polarization vector. Then we may consider the matrix element of H' between initial and final states in different bands:

$$H'_{fi} = (e/m) \langle n\mathbf{k} | \mathbf{A} \cdot \mathbf{p} | l\mathbf{q} \rangle = (-ie\hbar A_0/2m)\, \boldsymbol{\varepsilon}_s \cdot \{ \langle n\mathbf{k} | \exp(i\mathbf{s} \cdot \mathbf{r}) \nabla | l\mathbf{q} \rangle e^{-i\omega t}$$

$$+ \langle n\mathbf{k} \mid \exp(-i\mathbf{s} \cdot \mathbf{r}) \nabla \mid l\mathbf{q} \rangle e^{i\omega t} \}. \tag{6.5.21}$$

Let us consider one matrix element from the right side of (6.5.21):

$$\langle n\mathbf{k} \mid \exp(i\mathbf{s} \cdot \mathbf{r}) \nabla \mid l\mathbf{q} \rangle$$

$$= \int \exp(-i\mathbf{k} \cdot \mathbf{r}) u_n^*(\mathbf{k}, \mathbf{r}) \exp(i\mathbf{s} \cdot \mathbf{r}) \nabla[\exp(i\mathbf{q} \cdot \mathbf{r}) u_l(\mathbf{q}, \mathbf{r})]\, d^3r$$

$$= \int \exp[i(\mathbf{q} + \mathbf{s} - \mathbf{k}) \cdot \mathbf{r}] u_n^*(\mathbf{k}, \mathbf{r})\, \nabla u_l(\mathbf{q}, \mathbf{r})\, d^3r$$

$$+ i\mathbf{q} \int \exp[i(\mathbf{q} + \mathbf{s} - \mathbf{k}) \cdot \mathbf{r}] u_n^*(\mathbf{k}, \mathbf{r}) u_l(\mathbf{q}, \mathbf{r})\, d^3r.$$

The functions $u_n^* u_l$ and $u_n^*\ \nabla u_l$ are periodic in the crystal. The integral over all space can be separated into a sum over unit cells times an integral over a single unit cell, as was done in Section 4.1, Part A [below Eq. (4.1.8)]. It is assumed that the wavelength of the electromagnetic wave is sufficiently long so that if **k** and **q** are within the zone, $\mathbf{q} + \mathbf{s} - \mathbf{k}$ is not a reciprocal lattice vector.

$$\langle n\mathbf{k} \mid \exp(i\mathbf{s}\cdot\mathbf{r})\nabla \mid l\mathbf{q}\rangle$$

$$= [(2\pi)^3/\Omega]\,\delta(\mathbf{q} + \mathbf{s} - \mathbf{k}) \int_{\text{cell}} u_n^*(\mathbf{k}, \mathbf{r})\,\nabla u_l(\mathbf{q}, \mathbf{r})\,d^3r. \quad (6.5.22)$$

The integral includes a single unit cell whose volume is Ω. The delta function expresses the conservation of wave vector in the absorption. As was implied in the introduction to this section, s is small compared to the dimensions of a typical Brillouin zone: in a simple cubic lattice of lattice parameter a, s/K_1 (where K_1 is the length of the first reciprocal lattice vector) $= a/\lambda$. Since this ratio is smaller than 10^{-2} over the range of interest, it is a good approximation to neglect **s** and set $\mathbf{q} = \mathbf{k}$. With reference to an energy band diagram, first-order optical transitions are referred to as vertical. We now have

$$H'_{\text{fi}} = (e/2m)A_0\boldsymbol{\varepsilon}_s\cdot\mathbf{p}_{nl}(\mathbf{k})\,\delta(\mathbf{q} - \mathbf{k})\,(e^{-i\omega t} + e^{i\omega t}) \quad (6.5.23)$$

where $\mathbf{p}_{nl}(\mathbf{k})$ is the momentum matrix element given by Eq. (4.1.18), Part A.

It is necessary to compute the transition rate using ordinary first-order time-dependent perturbation theory. This is slightly more subtle than might seem to be the case. Suppose we consider an absorption process $[E_n(\mathbf{k}) = E_l(\mathbf{k}) + \hbar\omega]$: Then we must include only the positive frequency portion of the cosine. This is the part proportional to $e^{-i\omega t}$; the negative frequency part $e^{i\omega t}$ would contribute to stimulated emission. Second, one must square the matrix element and multiply by $2\pi/\hbar$ times a delta function which expresses conservation of energy. Some difficulty might be expected in connection with the square of the delta function in (6.5.23). However, one can show that the transition probability per unit volume per unit time is well defined and contains only one delta function of momentum conservation (Bethe *et al.*, 1955). Further, since one must consider in a solid not transitions between discrete states, but rather transitions between groups of states, it is necessary to integrate over ranges of states in **k** and **q**. When this is done, the delta function of momentum conservation renders one integration trivial, and the following expression for the transition rate is obtained:

$$W = 2(2\pi/\hbar)[1/(2\pi)^3] \int d^3k\,|\bar{H}_{\text{fi}}|^2\,\delta[E_n(\mathbf{k}) - \hbar\omega - E_l(\mathbf{k})] \quad (6.5.24)$$

(we count both directions of spin) in which $\bar{H}_{\text{fi}}$ is a reduced matrix element

$$\bar{H}_{\text{fi}} = (e/2m)A_0\boldsymbol{\varepsilon}_s\cdot\mathbf{p}_{nl}(\mathbf{k}). \tag{6.5.25}$$

Equation (6.5.24) can be converted to an integration over a surface of constant energy dS_ω where $\hbar\omega = E_n(\mathbf{k}) - E_l(\mathbf{k})$, by the technique described in Section 1.5, Part A:

$$W = 2(2\pi/\hbar)(2\pi)^{-3}\int dS_\omega\,|\bar{H}_{\text{fi}}|^2/|\nabla_{\mathbf{k}}[E_n(\mathbf{k}) - E_l(\mathbf{k})]|. \tag{6.5.26}$$

As an approximation, we may suppose that $\bar{H}_{\text{fi}}$ does not vary strongly with angle on a surface of constant energy so that it may be taken outside the integral. It is then convenient to introduce the joint density of states for bands n and l through the expression [see Eqs. (1.5.13) and (1.5.15), Part A]

$$G_{nl}(E) = [\Omega/(2\pi)^3]\int d^3k\,\delta[E_n(\mathbf{k}) - E_l(\mathbf{k}) - E]$$

$$= [\Omega/(2\pi)^3]\int dS_E/|\nabla_{\mathbf{k}}[E_n(\mathbf{k}) - E_l(\mathbf{k})]|. \tag{6.5.27}$$

The joint density of states has all of the singularity structure ascribed to the ordinary density of states described in Section 1.5, Part A. We see that

$$W = (4\pi/\hbar\Omega)\,|\bar{H}_{\text{fi}}|^2 G_{nl}(\hbar\omega). \tag{6.5.28}$$

The transition rate is thus seen to be proportional to the joint density of states, provided that $\bar{H}_{\text{fi}}$ is reasonably constant on a surface of constant energy.

The absorption constant is of greater experimental interest than the transition rate. It is defined as the ratio of the energy removed from the incident beam per unit time and per unit volume to the incident flux [see Eq. (2.4.89), Part A]. The energy flux is interpreted as the product of the energy density and the speed of flow. The energy density in the medium is ϵE^2, which, when averaged over a cycle, gives $\epsilon\omega^2 A_0^2/2$ in MKS units (with ϵ the permittivity, A_0 the amplitude of the vector potential, and ω the circular frequency of the wave). The speed of propagation is c/n where n is the index of refraction at frequency ω. Note that $\epsilon/n = n\epsilon_0$ where ϵ_0 is the permittivity of free space. Hence, from (6.5.28), (6.5.25) and (2.4.89), Part A,

$$\alpha = 2\hbar W/\omega n\epsilon_0 c A_0^2 = (8\pi/\Omega)(|H_{\text{fi}}|^2/\omega_n\epsilon_0 c A_0^2)G_{nl}(\hbar\omega)$$

$$= (2\pi/\Omega)(e^2\,|\boldsymbol{\varepsilon}_s\cdot\mathbf{p}_{nl}|^2/m^2 n\epsilon_0 c\omega)G_{nl}(\hbar\omega). \tag{6.5.29}$$

The absorption constant is also proportional to the joint density of states, and will thus exhibit the same structure, including the sharp corners arising from van Hove singularities. An important special case is that of transitions between two parabolic bands with extrema at $\mathbf{k} = 0$. We put

$$E_n(\mathbf{k}) = E_g + (\hbar^2\mathbf{k}^2/2m_c^*), \qquad E_l(\mathbf{k}) = -\hbar^2\mathbf{k}^2/2m_v^*. \quad (6.5.30)$$

The bands, henceforth referred to as *conduction* and *valence bands,* are separated by a gap E_g at $\mathbf{k} = 0$, and have effective masses m_c^* and m_v^*, respectively. Then

$$E_n - \hbar\omega - E_l = E_g + (\hbar^2\mathbf{k}^2/2\mu) - \hbar\omega \quad (6.5.31)$$

in which the reduced mass $\mu = m_c^*m_v^*/(m_c^* + m_v^*)$. The joint density of states is found by direct integration using (6.5.27) to be

$$G_{nl}(\hbar\omega) = (\Omega/4\pi^2)(2\mu/\hbar^2)^{3/2}(\hbar\omega - E_g)^{1/2}. \quad (6.5.32)$$

The final expression for the absorption constant is

$$\alpha = K[(2\mu)^{3/2}/(\hbar\omega)^{1/2}](1 - E_g/\hbar\omega)^{1/2} \text{ with } K = e^2 \,|\, \boldsymbol{\varepsilon}_s\cdot\mathbf{p}_{nl} \,|^2/2\pi m^2\hbar^2 n\epsilon_0 c. \quad (6.5.33)$$

This applies to allowed direct transitions between valence and conduction bands. The onset of absorption is quite sharp. As an example, the experimental absorption edge of InSb is shown in Fig. 6.5.2 according to Gobeli and Fan (1956).

If the transition between valence and conduction bands is forbidden by a selection rule at the band extremum, it may occur in the vicinity of the extremum as components of different symmetry are included in the wavefunction. In this case, the matrix element $\bar{H}_{fi}$ would be expected to be proportional to $|\mathbf{k} - \mathbf{k}_0|$ (where $\mathbf{k}_0$ locates the extremum and $\mathbf{k}$ is the wave vector at which the transition occurs). If this is used in (6.5.26), an absorption coefficient proportional to $(\hbar\omega - E_g)^{3/2}$ results. Information concerning the symmetry of valence and conduction band wave functions can thus be determined by observing the energy dependence of the absorption coefficient near threshold.

6.5.3 *Indirect Transitions*

It frequently happens, however, that the valence and conduction band extrema are located at different points in $\mathbf{k}$ space. An optical transition between them usually requires, in this case, the assistance of a phonon to supply the additional momentum. Such processes, which are called *indirect transitions* (Bardeen *et al.*, 1957), may occur in two ways: (1) An electron

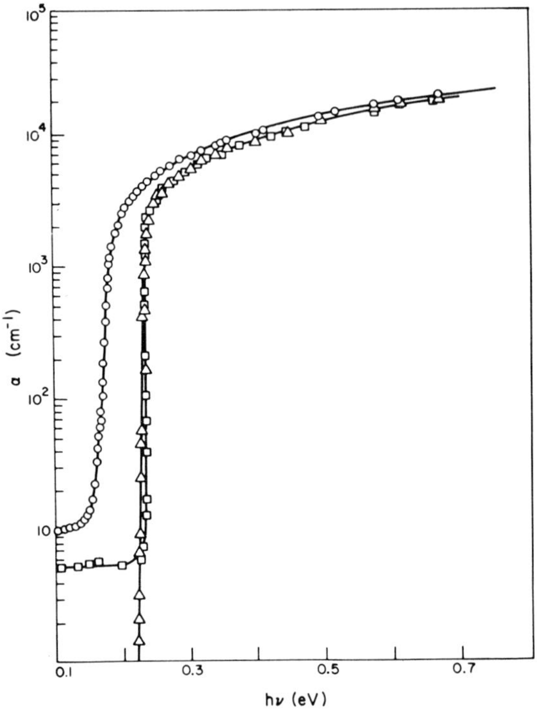

Fig. 6.5.2. Absorption edge of pure InSb. The values shown are ○ $T = 298°K$, △ $T \simeq 90°K$, □ $T \sim 5°K$. [From Gobeli and Fan (1956).]

in the valence band may absorb a photon and make a transition to an intermediate state in the conduction band of essentially the same wave vector, and then a phonon may be emitted or absorbed to yield the final state. (2) Alternately, the photon may excite an electron from a valence band state directly below the conduction band minimum, with the hole being transferred to the valence band maximum by phonon emission or absorption. The final state is the same in both cases.

Indirect transitions can be studied using second-order time-dependent perturbation theory. It is shown in elementary quantum mechanics (Landau and Lifshitz, 1965, p. 143) that if there is no first-order contribution, the matrix element $\bar{H}_{\mathrm{fi}}$ which enters the usual transition probability formula must be replaced by

$$\sum_m \bar{H}_{\mathrm{f}m}\bar{H}_{m\mathrm{i}}/(E_\mathrm{i} - E_m) \tag{6.5.34}$$

where the index m refers to the intermediate states. In an indirect transition, one of the matrix elements has the form (6.5.25), in which the electron interacts with the electromagnetic field, and the other involves the electron–phonon coupling. The matrix element for the absorption of a phonon of wave vector $\mathbf{k}$ is proportional to $[n(\mathbf{k})]^{1/2}$, where $n(\mathbf{k})$ is the number of phonons already present with wave vector $\mathbf{k}$, while the matrix element for emission is proportional to $[n(\mathbf{k}) + 1]^{1/2}$. Since, in thermal equilibrium at temperature T, $n(\mathbf{k})$ is proportional to $(e^{\Theta/T} - 1)^{-1}$ (where the energy of the phonon has been written as $K\Theta$, K being Boltzmann's constant), there is a characteristic temperature dependence of the absorption constant for these transitions.

We usually cannot evaluate the sum in (6.5.34) since we do not have sufficiently detailed knowledge of the intermediate states and of the electron–phonon interaction. It is possible, however, to determine the dependence of the absorption coefficient on photon energy under the assumption that the energy dependence of the effective matrix element (6.5.34) can be neglected. This assumption is probably valid as long as we consider photon energies close to threshold. This calculation requires an integration of the delta function of energy conservation over a range of initial and final states. In the present case, the requirement of wave vector conservation does not make one integration trivial, because the phonon can take up what momentum is required.

To have a more specific example, we will study a simple model of a semiconductor. Consider a transition in which a phonon is absorbed. Let us choose the zero of energy at the conduction band minimum, and let E_f, E_i, and E_g represent the energy of states in the conduction and valence bands, and the energy gap, respectively. Energy conservation demands that

$$E_f = E_i + \hbar\omega + K\Theta. \qquad (6.5.35)$$

We will assume that the densities of states in the conduction and valence band are proportional to $E_f^{1/2}$ and $(-E_g - E_i)^{1/2}$, respectively $[(-)(E_g + E_i)$ is a positive number$]$. One has to evaluate

$$\iint E_f^{1/2}(-E_g - E_i)^{1/2}\,\delta(E_i + \hbar\omega + K\theta - E_f)\,dE_i\,dE_f. \qquad (6.5.36)$$

The integral over E_i may be done immediately. The remaining integral has limits 0 and $\hbar\omega + K\Theta - E_g$, and is

$$\int_0^{\hbar\omega+K\Theta-E_g} dE_f\,[E_f(\hbar\omega + K\Theta - E_g - E_f)]^{1/2} = \tfrac{1}{8}\pi(\hbar\omega + K\Theta - E_g)^2. \qquad (6.5.37)$$

It suffices to change the sign of $K\Theta$ in order to obtain the result for phonon emission. This expression may be combined with factors involving the number of phonons to put the absorption coefficient in the form

$$\alpha = (C/\hbar\omega)[(1/1 - e^{-\Theta/T})(\hbar\omega - E_g - K\Theta)^2\eta(\hbar\omega - E_g - K\Theta) + (1/e^{\Theta/T} - 1)(\hbar\omega - E_g + K\Theta)^2\eta(\hbar\omega - E_g + K\Theta)] \tag{6.5.38}$$

in which C contains the unknown factors and η is a unit step function. This formula was first given by Macfarlane and Roberts (1955). When $(\hbar\omega\alpha)^{1/2}$ is plotted against $\hbar\omega$, the curve lies close to one straight line for $E_g - K\Theta < \hbar\omega < E_g + K\Theta$, and (at low temperatures where phonon absorption is considerably less likely than phonon emission) close to another straight line for $\hbar\omega > E_g + K\Theta$. From analysis of the absorption in indirect transitions, it is possible to deduce the dependence of the energy upon temperature, and by comparison with the vibrational spectrum, if the latter is known, to determine the separation in **k** space of the valence and conduction band extrema. It may occur that more than one type of phonon is involved.

As an example of the effect of indirect transitions, the absorption constant of germanium is shown in Fig. 6.5.3 according to McFarlane *et al.* (1957).

6.5.4 *A General Formula for the Conductivity*

We will now obtain a general expression for the conductivity (the Kubo formula, see Kubo, 1957). This is an important example of so-called linear response theory. The specific treatment follows Viswanathan (1965).

The problem is to determine the current which flows when an external time-dependent electromagnetic field is applied to a solid. A sample of unit (macroscopic) volume is considered. A many-body approach will be employed. The interaction Hamiltonian between a single electron and a weak external field is given by (6.5.20). This must be summed over the coordinates of all electrons

$$H' = (e/m) \sum_i \mathbf{A}(\mathbf{r}_i) \cdot \mathbf{p}_i \tag{6.5.39}$$

where $\mathbf{p}_i$ is the momentum operator for the ith electron and $\mathbf{r}_i$ is its coordinate. This Hamiltonian can be expressed in second quantized form (see Appendix D) as

$$H' = (e\hbar/2im) \int d^3r\, \mathbf{A}(\mathbf{r}) \cdot [\psi^+(\mathbf{r})(\nabla\psi(\mathbf{r})) - (\nabla\psi^+(\mathbf{r}))\psi(\mathbf{r})]. \tag{6.5.40}$$

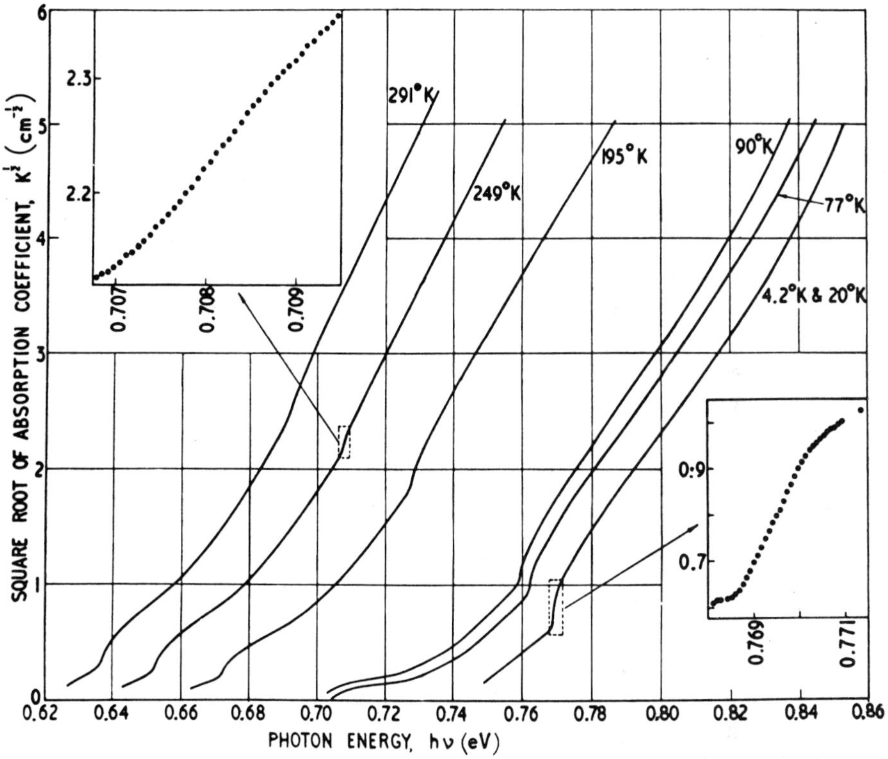

Fig. 6.5.3. Absorption spectrum of germanium in the indirect transition region. The inserts show the definition of the drawn lines in terms of the experimental points. [From Macfarlane *et al.* (1957).]

Here $\psi(\mathbf{r})$ is the electron field operator (Schweber, 1961, Chapter 6). It is also useful to define the electron current operator (Landau and Lifshitz, 1965, p. 435)

$$\mathbf{j}(\mathbf{r}) = -(e\hbar/2im)[\psi^+(\mathbf{r})\,\nabla\psi(\mathbf{r}) - \nabla\psi^+(\mathbf{r})\psi(\mathbf{r})] - (e^2/m)A\psi^+(\mathbf{r})\psi(\mathbf{r}). \tag{6.5.41}$$

A term representing the spin contribution has been dropped. Evidently we can write, correct to first order in $\mathbf{A}$,

$$H' = -\int d^3r\,\mathbf{A}(\mathbf{r})\cdot\mathbf{j}(\mathbf{r}). \tag{6.5.42}$$

Our objective will be to compute the average of the current operator $\mathbf{j}$ in the presence of a specified external field. Averages are to be computed using a density matrix $\rho(\mathbf{r}, t)$:

$$\mathbf{J}(\mathbf{r}, \mathbf{t}) = \mathrm{Tr}(\rho\mathbf{j}). \tag{6.5.43}$$

Further, let H_0 be the full Hamiltonian for the crystal including all interactions *except* those involving the external field which are specified by H'. The density matrix is to be found as a solution of the Liouville equation

$$i\hbar\, \partial\rho/\partial t = [H_0 + H', \rho]. \tag{6.5.44}$$

The bracket denotes the commutator.

It is convenient to transform to the interaction picture according to

$$\rho_{\mathrm{I}} = \exp(iH_0t/\hbar)\rho \exp(-iH_0t/\hbar). \tag{6.5.45}$$

Differentiate (6.5.45) with respect to time and use (6.5.44). The density matrix in the interaction picture is found to satisfy

$$i\hbar\, \partial\rho_{\mathrm{I}}/\partial t = [H_{\mathrm{I}}', \rho_{\mathrm{I}}] \tag{6.5.46}$$

where

$$H_{\mathrm{I}}' = \exp(iH_0t/\hbar)H' \exp(-iH_0t/\hbar). \tag{6.5.47}$$

It will be assumed that the interaction H' vanished in the remote past. Then we can write an integral equation equivalent to (6.5.46):

$$\rho_{\mathrm{I}}(t) = \rho_0 + (1/i\hbar) \int_{-\infty}^{t} dt' \, [H_{\mathrm{I}}'(t'), \rho_{\mathrm{I}}(t')] \tag{6.5.48}$$

in which ρ_0 is the equilibrium density matrix which is a function of H_0 only.

The essential feature of this procedure is that, if we are only interested in the linear response of the system, that is, terms in the current which are linear in the external field, it suffices to obtain $\rho_{\mathrm{I}}(t)$ to first order in the interaction. To this order, we can replace ρ_{I} on the right side of (6.5.48) by ρ_0,

$$\rho_{\mathrm{I}}(t) = \rho_0 + (1/i\hbar) \int_{-\infty}^{t} dt' \, [H_{\mathrm{I}}'(t'), \rho_0]. \tag{6.5.49}$$

This expression for the density matrix in the presence of the external field is used to compute the average current according to (6.5.43),

$$\begin{aligned} \mathbf{J}(\mathbf{r}, \mathbf{t}) &= \mathrm{Tr}(\rho_0\, \mathbf{j}) - (i/\hbar) \int_{-\infty}^{t} dt' \, \mathrm{Tr}\{[H_{\mathrm{I}}(t'), \rho_0]\, \mathbf{j}_{\mathrm{I}}\} \\ &= -(Ne^2/m)\mathbf{A}(\mathbf{r}, t) - (i/\hbar) \int_{-\infty}^{t'} dt' \, \langle [\, \mathbf{j}_{\mathrm{I}}, H_{\mathrm{I}}] \rangle_0. \end{aligned} \tag{6.5.50}$$

In order to obtain the second line of (6.5.50) we have used the fact that only the second term in (6.5.41) will have a nonzero average in equilibrium, since

$$\mathrm{Tr}(\rho_0 \psi^+ \psi) = N$$

where N is the number of electrons per unit volume. Also, we have been able to rearrange the commutator since operators may be permuted cyclically within the trace operation. Finally, $\langle \cdots \rangle_0$ indicates that the average of the quantity inside the angular brackets is to be computed with the equilibrium density matrix ρ_0.

Let us now substitute (6.5.42) into (6.5.50) and introduce Cartesian indices α, β:

$$\begin{aligned} J_\alpha(\mathbf{r}, t) &= \langle j_\alpha(\mathbf{r}, t) \rangle \\ &= -(Ne^2/m) A_\alpha(\mathbf{r}, t) \\ &\quad + (i/\hbar) \int_{-\infty}^{t} dt' \int d^3r \langle [j_\alpha(\mathbf{r}, t), j_\beta(\mathbf{r}', t')] \rangle_0 A_\beta(\mathbf{r}', t'). \end{aligned} \tag{6.5.51}$$

The subscript I has been dropped; however, we note that all the operators in (6.5.51) are in the interaction picture. Next, we introduce a Fourier representation for A_β and j_β under the integral:

$$A_\beta(\mathbf{r}', t) = (2\pi)^{-3/2} \int \exp(i\mathbf{q}'\cdot\mathbf{r}') \mathcal{A}_\beta(\mathbf{q}', t)\, d^3q'$$

$$\mathrm{j}_\beta(\mathbf{q}', t) = (2\pi)^{-3/2} \int j_\beta(\mathbf{r}', t) \exp(-i\mathbf{q}'\cdot\mathbf{r}')\, d^3r. \tag{6.5.52}$$

Then we multiply both sides of the result by $\exp(-i\mathbf{q}\cdot\mathbf{r})/(2\pi)^{-3/2}$ and integrate. The result gives the qth Fourier component of the average current,

$$\begin{aligned} \mathcal{J}_\alpha(\mathbf{q}, t) &= \langle \mathrm{j}_\alpha(\mathbf{q}, t) \rangle \\ &= -(Ne^2/m) \mathcal{A}_\alpha(\mathbf{q}, t) \\ &\quad + (i/\hbar) \int_{-\infty}^{t} dt' \int d^3q' \langle [\mathrm{j}_\alpha(\mathbf{q}, t), \mathrm{j}_\beta(-\mathbf{q}', t')] \rangle_0 \mathcal{A}_\beta(\mathbf{q}', t'). \end{aligned} \tag{6.5.53}$$

The average of the commutator means that we sum the product of its diagonal elements weighted by their probability of occurrence. It can be shown by a straightforward but somewhat lengthy calculation that the commutator has no diagonal elements unless $\mathbf{q}' = \mathbf{q}$. We obtain

$$\mathcal{J}_\alpha(\mathbf{q}, t) = -(Ne^2/m)\,\mathcal{A}_\alpha(\mathbf{q}, t) + (i/\hbar)\int_{-\infty}^{t} dt'\,\langle[\,\mathrm{j}_\alpha(\mathbf{q}, t), \mathrm{j}_\beta(-\mathbf{q}, t')\,]\rangle_0 \mathcal{A}_\beta(\mathbf{q}, t'). \quad (6.5.54)$$

The electric field is related to the vector potential by

$$\mathbf{E}(\mathbf{r}, t) = -\partial\mathbf{A}/\partial t\,(\mathbf{r}, t). \quad (6.5.55)$$

We may now introduce the Fourier transform of the fields and currents with respect to time according to (6.5.2). We also use the cyclic property of the trace operation

$$\mathrm{Tr}[ABC] = \mathrm{Tr}[BCA] = \mathrm{Tr}[CAB], \quad (6.5.56)$$

to write

$$\langle[\,\mathrm{j}_\alpha(\mathbf{q}, t), \mathrm{j}_\beta(-\mathbf{q}, t')\,]\rangle = \langle[\,\mathrm{j}_\alpha(\mathbf{q}, 0), \mathrm{j}_\beta(-\mathbf{q}, t'-t)\,]\rangle. \quad (6.5.57)$$

Then, since $\mathcal{A}(\mathbf{q}, \omega) = \mathcal{E}(\mathbf{q}, \omega)/(i\omega)$,

$$\begin{aligned}\mathcal{J}_\alpha(\mathbf{q}, \omega) &= (2\pi)^{-1/2}\int_{-\infty}^{\infty} \mathcal{J}_\alpha(\mathbf{q}, t)e^{i\omega t}\,dt \\ &= -(Ne^2/im\omega)\,\mathcal{E}_\alpha(\mathbf{q}, \omega) \\ &\quad + (1/\hbar\omega)\int_{-\infty}^{0} dt\,\langle[\,\mathrm{j}_\alpha(\mathbf{q}, 0), \mathrm{j}_\beta(-\mathbf{q}, t)\,]\rangle e^{-i\omega t}\mathcal{E}_\beta(\mathbf{q}, \omega).\end{aligned}$$

We use the relation between current and field

$$\mathcal{J}_\alpha(\mathbf{q}, \omega) = \sigma_{\alpha\beta}\mathcal{E}_\beta(\mathbf{q}, \omega)$$

to obtain the conductivity tensor

$$\sigma_{\alpha\beta}(\mathbf{q}, \omega) = -(Ne^2/im\omega)\,\delta_{\alpha\beta} + (1/\hbar\omega)\int_{-\infty}^{0} dt\,\langle[\,\mathrm{j}_\alpha(\mathbf{q}, 0), \mathrm{j}_\beta(-\mathbf{q}, t)\,]\rangle e^{-i\omega t}. \quad (6.5.58)$$

This is the general result which has come to be known as the *Kubo formula*. Our interest in optical properties requires that we consider the $q = 0$ limit of this expression. Then (6.5.13) can be used to obtain an expression for the dielectric tensor

$$\begin{aligned}\kappa_{\alpha\beta}(\omega) &= \delta_{\alpha\beta} + [i\sigma_{\alpha\beta}(0, \omega)/\epsilon_0\omega] \\ &= \delta_{\alpha\beta}[1 - (Ne^2/m\epsilon_0\omega^2)] + (i/\hbar\omega^2\epsilon_0)\int_{-\infty}^{0} dt\,\langle[\,\mathrm{j}_\alpha(0), \mathrm{j}_\beta(t)\,]\rangle e^{-i\omega t}\end{aligned} \quad (6.5.59)$$

in which $\mathrm{j}_\beta(t) = \mathrm{j}_\beta(0, t)$, etc.

Suppose the system has finite conductivity ($\sigma_{\alpha\beta} < \infty$) as $\omega \to 0$. It is seen that (6.5.58) apparently diverges in this limit because of the presence of the ω^{-1} factor. It is necessary to obtain a finite conductivity that the coefficients of ω^{-1} cancel. This leads us to the relation

$$(Ne^2/m)\,\delta_{\alpha\beta} = i/\hbar \int_{-\infty}^{0} \langle[\mathrm{j}_\alpha(0), \mathrm{j}_\beta(t)]\rangle\, dt. \tag{6.5.60}$$

The current operator j is Hermitian. Thus the commutator $[\mathrm{j}_\alpha(0), \mathrm{j}_\beta(t)]$ must be an anti-Hermitian operator. The diagonal elements of such an operator must be imaginary, so that

$$\langle[\mathrm{j}_\alpha(0), \mathrm{j}_\beta(t)]\rangle = 2i\,\mathrm{Im}\langle \mathrm{j}_\alpha(0)\mathrm{j}_\beta(t)\rangle. \tag{6.5.61}$$

Thus (6.5.59) becomes

$$\kappa_{\alpha\beta}(\omega) = \delta_{\alpha\beta}[1 - (Ne^2/m\epsilon_0\omega^2)] - (2/\hbar\omega^2\epsilon_0)\int_{-\infty}^{0} dt\, e^{-i\omega t}\,\mathrm{Im}\langle \mathrm{j}_\alpha(0)\mathrm{j}_\beta(t)\rangle. \tag{6.5.62}$$

It is convenient at this point to introduce a set of states, formally designated as $|l\rangle$, $|n\rangle$, etc. which are eigenstates of the Hamiltonian of the system in the absence of the time-dependent external field:

$$H_0|l\rangle = E_l|l\rangle. \tag{6.5.63}$$

The density matrix ρ_0 is (Kittel, 1958, p. 109)

$$\rho_0 = Z^{-1}e^{-\beta H} \tag{6.5.64}$$

where $\beta = 1/KT$ and $Z = \mathrm{Tr}\, e^{-\beta H}$. The quantity Z is the usual partition function. The average can be expressed as

$$\langle \mathrm{j}_\alpha(0)\mathrm{j}_\beta(t)\rangle = (1/Z)\sum_{n,l}\exp(-\beta E_l)\langle l|\mathrm{j}_\alpha|n\rangle\langle n|\mathrm{j}_\beta|l\rangle\exp(i\omega_{nl}t) \tag{6.5.65}$$

where $\omega_{nl} = \hbar^{-1}(E_n - E_l)$.

From this point, we will restrict our attention to a simple cubic crystal in which the dielectric tensor reduces to a scalar $\kappa_{\alpha\beta} = \kappa\,\delta_{\alpha\beta}$. Then we may replace $\langle \mathrm{j}_\alpha(0)\mathrm{j}_\beta(t)\rangle$ by $\frac{1}{3}\langle \mathbf{j}(0)\cdot\mathbf{j}(t)\rangle$ and obtain

$$\mathrm{Im}\,\tfrac{1}{3}\langle \mathbf{j}(0)\cdot\mathbf{j}(t)\rangle = \tfrac{1}{3}\sum_{n,l}\exp[-\beta(E_l - \mu N)]\,|\langle l|\mathbf{j}|n\rangle|^2\sin\omega_{nl}t.$$

This result is inserted in (6.5.62). We consider the frequency ω which appears in the Fourier transform to have a small positive imaginary part

$i\epsilon$ so that the integral will converge at the lower limit:

$$\int_{-\infty}^{0} \exp[-i(\omega + i\epsilon)t] \sin \omega_{nl}t \, dt = \tfrac{1}{2}[(\omega - \omega_{nl} + i\epsilon)^{-1} - (\omega + \omega_{nl} + i\epsilon)^{-1}].$$

Now ϵ may be allowed to vanish. The usual identity $(x + i\epsilon)^{-1} = P(1/x) + i\pi\,\delta(x)$ is employed. Equation (6.5.62) becomes

$$\kappa(\omega) = 1 - (Ne^2/m\epsilon_0\omega^2) - (1/3Z\hbar\omega^2\epsilon_0) \sum_{nl} \exp(-\beta E_l) \mid \langle l \mid \mathbf{j} \mid n \rangle \mid^2$$
$$\times [(2\omega_{nl}/(\omega^2 - \omega_{nl}^2)) - i\pi\,\delta(\omega - \omega_{nl}) + i\pi\,\delta(\omega + \omega_{nl})]. \quad (6.5.66)$$

The sum involving the first term in the bracket gives rise to a principal value integral.

Equation (6.5.66) yields the dielectric function at any temperature. The separation into real and imaginary parts is obvious, and the generalization to materials of lower symmetry is easy. At zero temperature, further simplification is possible, since only the state of lowest energy, $l = 0$, can contribute. In this case,

$$\kappa(\omega) = 1 - (Ne^2/m\epsilon_0\omega^2) - (1/3\hbar\omega^2\epsilon_0) \sum_{n} \mid \langle 0 \mid \mathbf{j} \mid n \rangle \mid^2$$
$$\times [2\omega_{n0}/(\omega^2 - \omega_{n0}^2) - i\pi\,\delta(\omega - \omega_{n0}) + i\pi\,\delta(\omega + \omega_{n0})]. \quad (6.5.67)$$

We see from (6.5.67) that for very high frequencies the real part of the dielectric function is simply

$$\kappa_1(\omega) = 1 - (Ne^2/m\epsilon_0\omega^2) + O(\omega^{-4}). \quad (6.5.68)$$

Also at high frequencies, we have from (6.5.14a) that

$$\kappa_1(\omega) = 1 - (2/\pi\omega^2) \int_0^\infty \nu\kappa_2(\nu)\, d\nu. \quad (6.5.69)$$

Comparison of (6.5.68) and (6.5.69) enables us to derive a sum rule for the imaginary part of the dielectric function

$$\int_0^\infty \nu\kappa_2(\nu)\, d\nu = (\pi/2)(Ne^2/m\epsilon_0). \quad (6.5.70)$$

This may be stated in an alternative form as a sum rule for the real part of the conductivity $\sigma_1 = \epsilon_0\omega\kappa_2$,

$$\int_0^\infty \sigma_1(\nu)\, d\nu = (\pi/2)(Ne^2/m). \quad (6.5.71)$$

Unfortunately, Eqs. (6.5.66) and (6.5.67) are not of much direct use for practical calculations since they involve the wave functions and energies of the exact, many-body states. More explicit results are possible within the Hartree–Fock approximation. In this case, each wave function for the system is considered to be a Slater determinant of single-particle Bloch functions. Since the current operator **j** for the system is the sum of one-particle operators, the possible matrix elements $\langle 0 \mid \mathbf{j} \mid n \rangle$ involve the excitation of a single particle from a state which is occupied in the ground state of the system to some unoccupied state. Thus if a state $\psi_l(\mathbf{q}, \mathbf{r})$ is occupied and $\psi_n(\mathbf{k}, \mathbf{r})$ is unoccupied, matrix elements $\langle l\mathbf{q} \mid \mathbf{j}_s \mid n\mathbf{k} \rangle$ must be considered, where $\mathbf{j}_s$ is the current operator for a single particle, and the notation now refers to the crystal momentum representation. In fact

$$\langle l\mathbf{q} \mid \mathbf{j}_s \mid n\mathbf{k} \rangle = -(e/m) \langle l\mathbf{q} \mid \mathbf{p} \mid n\mathbf{k} \rangle = -(e/m) \mathbf{p}_{ln}(\mathbf{k})\, \delta_{\mathbf{qk}}$$

where **p** is the single-particle momentum operator, Eq. (6.5.22) has been used, and $\mathbf{p}_{nl}(\mathbf{k})$ is given by (4.1.18), Part A. For convenience, we employ discrete normalization. Thus (6.5.67) reduces to

$$\kappa(\omega) = 1 - (Ne^2/m\epsilon_0\omega^2) - (e^2/3\hbar m^2\epsilon_0\omega^2) \sum_{l\mathbf{k},\mathrm{o}} \sum_{n\mathbf{k},\mathrm{u}} | \mathbf{p}_{ln}(\mathbf{k}) |^2$$

$$\times [2\omega_{nl}/(\omega^2 - \omega_{nl}^2) - i\pi\, \delta(\omega - \omega_{nl}) + i\pi\, \delta(\omega + \omega_{nl})]. \quad (6.5.72)$$

The subscript o under a summation symbol indicates that only occupied states are to be included, while u indicates that unoccupied states are involved. Let us examine the imaginary and real parts of (6.5.72) separately. The imaginary part is

$$\kappa_2(\omega) = (\pi e^2/3\hbar m^2\epsilon_0\omega^2) \sum_{l\mathbf{k},\mathrm{o}} \sum_{n\mathbf{k},\mathrm{u}} | \mathbf{p}_{ln}(\mathbf{k}) |^2 [\delta(\omega - \omega_{nl}) - \delta(\omega + \omega_{nl})]. \quad (6.5.73)$$

Equation (6.5.73) is not quite correct since the Hartree–Fock functions are not exact eigenstates. A formal modification is required in the limit $\omega \to 0$. We will not discuss this in detail here (however, see Problem 6.6). In the case of the simple band model examined in subsection 6.5.2, this expression reduces to that already obtained for the absorption constant. The absorption constant is related to κ_2 by

$$\alpha(\omega) = \omega\kappa_2(\omega)/nc \quad (6.5.74)$$

where n is the index of refraction (This relation is obtained by considering the propagation of an electromagnetic wave in an absorbing medium). Thus if we consider the two bands n, l (and for convenience, suppose l to be full and n to be empty), we see from (6.5.27) that

$$\sum_{\mathbf{k}} \delta(\omega - \omega_{nl}(\mathbf{k})) = (\hbar/\Omega) G_{nl}(\hbar\omega) \quad (6.5.75)$$

where $G_{nl}(\mathbf{k})$ is the joint density of states. Summation over $\mathbf{k}$ here implicitly includes spin directions and this produces a factor of 2. Thus

$$\alpha(\omega) = (2\pi/\Omega)(e^2/m^2 n\epsilon_0 c\omega)\tfrac{1}{3} \mid \mathbf{p}_{nl} \mid^2 G_{nl}(\hbar\omega) \tag{6.5.76}$$

where, in comparison with (6.5.29), $\mid \boldsymbol{\varepsilon}\cdot\mathbf{p}_{nl} \mid^2$ has been replaced by its average $\tfrac{1}{3} \mid \mathbf{p}_{nl} \mid^2$.

Next, we consider the real part. Note that

$$\sum_{l\mathbf{k},\mathrm{o}} (1) = N.$$

Also, we may formally extend the sum on n to include all bands, since the factor ω_{nl} in the numerator ensures that the contribution from occupied states will cancel. We have

$$\kappa_1(\omega) = 1 - (e^2/m\epsilon_0\omega^2) \sum_{l\mathbf{k},\mathrm{o}} \{1 + (2/3\hbar m) \sum_n [\mid \mathbf{p}_{ln}(\mathbf{k}) \mid^2/(\omega^2 - \omega_{nl}^2)]\omega_{nl}\}. \tag{6.5.77}$$

The second term can be simplified by the introduction of an effective mass. We introduce $m_l^*(\mathbf{k})$ [see Eq. (4.1.28), Part A]

$$m/m_l^*(\mathbf{k}) = 1 + (2/3\hbar m) \sum_{n\neq l} \mid \mathbf{p}_{ln}(\mathbf{k}) \mid^2/\omega_{ln}(\mathbf{k}). \tag{6.5.78}$$

The effective mass defined by (6.5.78) is summed over occupied states in (6.5.77). It is convenient to define an optical effective mass averaged over occupied states in band l. Let

$$(m_l^*)_{\mathrm{op}}^{-1} = N_l^{-1} \sum_{\mathbf{k}\epsilon l,\mathrm{o}} [m_l^*(\mathbf{k})]^{-1} \tag{6.5.79}$$

in which N_l is the number of electrons occupying states in band l. Equation (6.5.77) becomes

$$\kappa_1(\omega) = 1 - \sum_l [N_l e^2/(m_l^*)_{\mathrm{op}}\epsilon_0\omega^2]$$
$$- (2e^2/3m^2\hbar\epsilon_0) \sum_{l\mathbf{k}} \sum_n \mid \mathbf{p}_{ln}(\mathbf{k}) \mid^2/\omega_{nl}(\omega^2 - \omega_{nl}^2). \tag{6.5.80}$$

Electrons in full bands make no contribution to the second term in (6.5.80). This may be seen as follows: The effective mass defined in (6.5.78) can be seen from Eq. (4.1.25), Part A, to be simply related to the Laplacian of the energy band function

$$1/m_l^*(\mathbf{k}) = (1/3\hbar^2)\, \nabla_{\mathbf{k}}^2 E_l(\mathbf{k}). \tag{6.5.81}$$

The sum on $\mathbf{k}$ in (6.5.79) can be converted into an integral in the usual way, and we obtain

$$(m_l^*)_{\mathrm{op}}^{-1} = [2/3N_l(2\pi)^3\hbar^2] \int d^3k \, \nabla^2 E_l(\mathbf{k}). \qquad (6.5.82)$$

The integral may be transformed by Gauss's theorem into a surface integral of the group velocity in band l

$$\mathbf{v}_l(\mathbf{k}) = (1/\hbar) \, \nabla_{\mathbf{k}} E_l(\mathbf{k})$$

over the Fermi surface

$$(m_l^*)_{\mathrm{op}}^{-1} = 1/12\pi^3 N_l \hbar \int d\mathbf{S}_{\mathrm{F}} \cdot \mathbf{v}_l(\mathbf{k}). \qquad (6.5.83)$$

If band l is full, the Fermi surface has no component in that band. Hence the integral vanishes.

Consider for a moment the special case of a monovalent metal, in which the Fermi surface is contained within a single band. It is useful to introduce $k_{\mathrm{F}}^{(0)}$ the radius, and $S_{\mathrm{F}}^{(0)}$ the area of a spherical free electron Fermi surface containing N_l electrons, $k_{\mathrm{F}}^{(0)3}/3\pi^2 = N_l$. Equation (6.5.83) can be rewritten as (Cohen and Heine, 1958)

$$m_{\mathrm{op}}^{*-1} = [\hbar k_{\mathrm{F}}^{(0)} S_{\mathrm{F}}^{(0)}]^{-1} \int d\mathbf{S}_{\mathrm{F}} \cdot \mathbf{v}(\mathbf{k}). \qquad (6.5.84)$$

The optical effective mass determined here is, in general, different from that determined from the electronic density of states. The density of states, or thermal effective mass m_{t}^*, was given by Eq. (4.1.63), Part A, for the case of a single band:

$$m_{\mathrm{t}}^* = \hbar k_{\mathrm{F}}/S_{\mathrm{F}}^{(0)} \int dS_{\mathrm{F}}/v(\mathbf{k}).$$

Hence

$$m_{\mathrm{t}}^*/m^*_{\mathrm{op}} = 1/(S_{\mathrm{F}}^{(0)})^2 \left[\int dS_{\mathrm{F}}/v(\mathbf{k}) \right] \left[\int dS_{\mathrm{F}} v(\mathbf{k}) \right]. \qquad (6.5.85)$$

Averages of v and $1/v$ over the Fermi surface are involved. The Schwartz inequality, which applies to any two arbitrary functions f and g, states that

$$\left| \int dS_{\mathrm{F}} \, fg \right|^2 \leq \left| \int dS_{\mathrm{F}} \, f^2 \right| \, \left| \int dS_{\mathrm{F}} \, g^2 \right|. \qquad (6.5.86)$$

Hence

$$\left[\int dS_{\mathrm{F}}\, v(\mathbf{k})\right]\left[\int dS_{\mathrm{F}}/v(\mathbf{k})\right] \geq S_{\mathrm{F}}^{2} \tag{6.5.87}$$

in which S_{F} is the area of the actual Fermi surface. Thus we have the inequality

$$m_{\mathrm{t}}^*/m^*_{\mathrm{op}} \geq (S_{\mathrm{F}}/S_{\mathrm{F}}^{(0)})^2. \tag{6.5.88}$$

Equality holds only if the Fermi surface is spherical which implies that $v(\mathbf{k})$ is constant over the surface.

In a monovalent metal, we must have $S_{\mathrm{F}} \geq S_{\mathrm{F}}^{(0)}$ and hence $m_{\mathrm{t}}^* \geq m_{\mathrm{op}}$, if the Fermi surface does not touch the Brillouin zone. However, if contact occurs, it is possible that $S_{\mathrm{F}} < S_{\mathrm{F}}^{(0)}$ since the area of contact is not actually part of the Fermi surface.

After this digression, we return to a discussion of Eq. (6.5.80). The second term in this equation is analagous to the contribution to the dielectric constant from the free electrons of a classical plasma. We define an an effective plasma frequently ω_{p} for the material by

$$\omega_{\mathrm{p}}^2 = (e^2/\epsilon_0) \sum_l [N_l/(m_l^*)_{\mathrm{op}}].$$

Equation (6.5.80) becomes

$$\kappa_1(\omega) = 1 - (\omega_{\mathrm{p}}^2/\omega^2) - (2e^2/3m^2\hbar\epsilon_0) \sum_{l\mathbf{k}} \sum_n |\, \mathbf{p}_{ln}(\mathbf{k})\, |^2/\omega_{nl}(\omega^2 - \omega^2{}_{nl}). \tag{6.5.89}$$

Let us consider the behavior of κ_1 for frequencies substantially smaller than that corresponding to the smallest direct band gap. In this case, ω^2 can be neglected compared to $\omega^2{}_{nl}$. Then

$$\kappa_1(\omega) = 1 + \kappa_{\mathrm{A}} - (\omega_{\mathrm{p}}^2/\omega^2) \tag{6.5.90}$$

where

$$\kappa_{\mathrm{A}} = 2e^2/3m^2\hbar\epsilon_0 \sum_{l\mathbf{k}} \sum_n |\, \mathbf{p}_{ln}(\mathbf{k})\, |^2/\omega^3{}_{nl}(\mathbf{k}). \tag{6.5.91}$$

In the limit of tight binding in which the wave functions and energies occurring in (6.5.91) can be replaced by the corresponding atomic (or ionic) quantities which are independent of $\mathbf{k}$, κ_{A} reduces to the atomic (or ionic) polarizability. We will refer to κ_{A} as the *atomic polarizability* although it must be understood that solid state effects can modify its numerical value.

The dielectric function of a metal is negative for frequencies ω below a critical frequency ω_{c} given by

$$\omega_{\mathrm{c}}^2 = \omega_{\mathrm{p}}^2/(1 + \kappa_{\mathrm{A}}). \tag{6.5.92}$$

An electromagnetic wave will not propagate in the metal for frequencies smaller than ω_c; the metal is then totally reflecting. For $\omega > \omega_c$, κ_1 becomes positive and the metal becomes transparent, provided that ω is not large enough for real interband transitions to occur in sufficient strength to produce a large attenuation by absorption. It is found, in fact, that the alkali metals become transparent in the near ultraviolet.

In order to develop understanding of the properties of κ_1, we will evaluate the dielectric function for a simple model. We will consider a hypothetical semiconductor with two bands: band 0 is full and band 1 is empty. The interband matrix element $\mathbf{p}_{01} = \mathbf{p}$ will be supposed to be independent of $\mathbf{k}$. The band energies will be given by (6.5.30), so that

$$\hbar\omega_{nl}(\mathbf{k}) = E_g + \gamma\mathbf{k}^2 \tag{6.5.93}$$

where $\gamma = \hbar^2/2\mu$, μ being the reduced effective mass. Since we are considering a situation in which bands are either entirely full or entirely empty, the second term of (6.5.80) vanishes. Then

$$\kappa_1 - 1 = e^2\hbar^2 \mid \mathbf{p}^2 \mid/6\pi^3 m^2\epsilon_0 \int d^3k/\{[(E_g + \gamma k^2)^2 - \hbar^2\omega^2](E_g + \gamma k^2)\}. \tag{6.5.94}$$

The integral must be interpreted as a principal value. We will ignore the detailed shape of the Brillouin zone, and replace it by a sphere of equal volume whose radius is k_m. It is convenient to introduce dimensionless quantities

$$\mathcal{E} = \hbar\omega/E_g, \qquad x = \gamma k^2/E_g, \qquad x_m = \gamma k_m^2/E_g. \tag{6.5.95}$$

Thus

$$\kappa_1 - 1 = e^2\hbar^2 \mid \mathbf{p} \mid^2/[3\pi^2 m^2\epsilon_0(\gamma E_g)^{3/2}] \int_0^{x_m} x^{1/2}\,dx/\{[(1+x)^2 - \mathcal{E}^2](1+x)\}. \tag{6.5.96}$$

The integral is straightforward, but somewhat tedious. We will simplify the result by introducing a restriction to energies $\mathcal{E} \ll x_m$ (and also the assumption that the band gap E_g is much smaller than the combined band width, measured by x_m). Then

$$\kappa_1 - 1 = (e^2 \mid \mathbf{p} \mid^2/6\pi m^2\epsilon_0\hbar\mathcal{E}^2)(2\mu/E_g)^{3/2}$$
$$\times [2 - (1+\mathcal{E})^{1/2} - (1-\mathcal{E})^{1/2}\theta(1-\mathcal{E})] \tag{6.5.97}$$

in which θ is a unit step function: $\theta(x) = 1$ if $x > 0$, and $\theta(x) = 0$ if $x < 0$.

The frequency dependence of the dielectric function is described by the function

$$f(\mathcal{E}) = \mathcal{E}^{-2}[2 - (1 + \mathcal{E})^{1/2} - (1 - \mathcal{E})^{1/2}\theta(1 - \mathcal{E})]. \quad (6.5.98)$$

This function is shown in Fig. 6.5.4. It is seen that $f(\mathcal{E})$ exhibits a cusp at $\mathcal{E} = 1$, which marks the onset of allowed interband transition, and thereafter decreases, ultimately becoming negative. As $\mathcal{E} \to 0$, $f(\mathcal{E}) \approx \mathcal{E}^2/4$. Thus the two-band contribution to the zero-frequency dielectric function is

$$\kappa_1(0) - 1 = (e^2 \mid \mathbf{p} \mid^2/24\pi m^2\epsilon_0\hbar)(2\mu/E_g)^{3/2}. \quad (6.5.99)$$

The two-band model can be carried one step further: We may express $2\mu E_g$ in terms of the momentum matrix element through (4.1.34), Part A. The final result is

$$\kappa_1(0) - 1 = (48\pi\sqrt{2})^{-1}(me^2/\hbar\epsilon_0 \mid \mathbf{p} \mid). \quad (6.5.100)$$

Caution: Eq. (6.5.100) is valid only if the bands contributing to the optical transition also control each other's effective mass: specifically (6.5.100) cannot be applied for small $\mid \mathbf{p} \mid$.

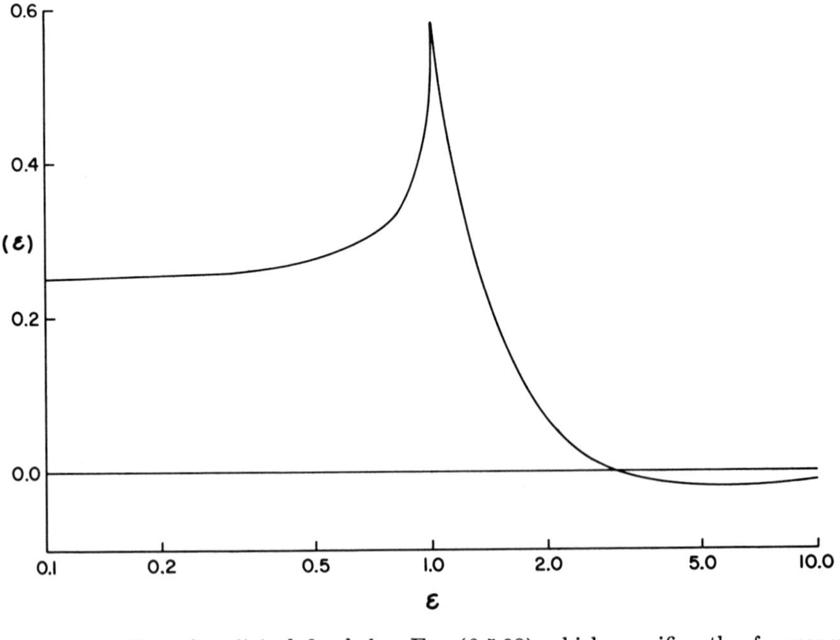

Fig. 6.5.4. Function $f(\mathcal{E})$ defined by Eq. (6.5.98) which specifies the frequency dependence of the dielectric function for a two-band model.

6.5.5 *Nonlinear Effects*

The availability of intense electric fields at optical frequencies, as provided by lasers, has stimulated interest in the nonlinear optical properties of solids. A typical example of such an effect is second harmonic generation in which two photons are annihilated and a third one, of twice the energy, is created. Two-photon absorption also occurs in which an electron must take up the energy from two light quanta in order to cross a band gap. There is an extensive literature on the subject of nonlinear optics (for an early review, see Bloembergen, 1965). We will not attempt a detailed discussion of this topic, and will confine our attention to a single aspect: the calculation of the second-order response function (Butcher and McLean, 1963; Kelley, 1963; Cheng and Miller, 1964).

Our previous discussion of optical properties took as its premise the existence of a general linear relation between current and field; Eq. (6.5.1). This form is now too restrictive. Let us suppose that

$$\mathbf{J}(t) = \sum_n \mathbf{J}^{(n)}(t)$$

where $\mathbf{J}^{(n)}$ involves nth order effects. The discussion of the preceding section has concerned only $\mathbf{J}^{(1)}(t)$. The second-order term $\mathbf{J}^{(2)}$ will be related to two electric fields $\boldsymbol{\mathcal{E}}_1(t_1)$ and $\boldsymbol{\mathcal{E}}_2(t_2)$ by

$$J_\alpha^{(2)}(t) = (2\pi)^{-1/2} \int \Sigma_{\alpha\beta\gamma}(t - t_1, t - t_2)\mathcal{E}_{\beta 1}(t_1)\mathcal{E}_{\gamma 2}(t_2)\, dt_1\, dt_2. \qquad (6.5.101)$$

The indices α, β, γ denote Cartesian components, and a summation over repeated indices is understood. The second-order frequency dependent conductivity is the Fourier transform of Σ:

$$\sigma_{\alpha\beta\gamma}^{(2)}(\omega_1, \omega_2) = (1/2\pi) \int \Sigma_{\alpha\beta\gamma}(t, t') \exp(i\omega_1 t) \exp(i\omega_2 t')\, dt\, dt'. \qquad (6.5.102)$$

The relation between the Fourier transforms of the current and the fields is

$$\mathcal{J}_\alpha(\omega) = \int \sigma_{\alpha\beta\gamma}^{(2)}(\omega', \omega - \omega')\mathcal{E}_\beta(\omega')\mathcal{E}_\gamma(\omega - \omega')\, d\omega'. \qquad (6.5.103)$$

Evidently, the second-order conductivity is a third rank tensor. As such, it must be odd under inversion. Consequently, the tensor must vanish in crystals whose point groups contain the inversion.

The calculation of $J^{(2)}$ and $\sigma^{(2)}$ can be performed in a rather straightforward way as an extension of the procedure in Section 6.5.4; it is, however, rather tedious. Complete details will not be given. The essential point is that we must solve the integral equation for the density matrix (6.5.48) to second order in the interaction. This can be done by iteration. Substitution of (6.5.49) into (6.5.48) yields

$$\rho_{\mathrm{I}}(t) = \rho_0 + \rho_1(t) + [1/(i\hbar)^2] \int_{-\infty}^{t} dt_1 \int_{-\infty}^{t_1} dt_2 \, [H_{\mathrm{I}}(t_1), [H_{\mathrm{I}}(t_2), \rho_0]] \tag{6.5.104}$$

where ρ_1 is the first-order term previously considered and is given by the second term in (6.5.49). This expression is to be substituted into (6.5.43). The resulting equation for the second-order current $J^{(2)}(t)$ is then simplified through the introduction of the Fourier transform of the relevant operators and then through the use of a determinantal wave function for the states of the system. The resulting equation is still quite lengthy. We will quote the form given by Bloembergen (1965) (cgs units):

$$\sigma_{\alpha\beta\gamma}{}^{(2)}(\omega_1, \omega_2) = (e^3/\hbar^2 m^3 \omega_1 \omega_2 c^2) \sum \mathrm{Perm}(\beta, \omega_1; \gamma, \omega_2) \sum_{n,n',n''} (2\pi)^{-3}$$

$$\times \int f[E_n(\mathbf{k})] \, d^3k [\cdots] \tag{6.5.105}$$

where

$$[\cdots] = \frac{p^{\alpha}{}_{nn'} p^{\beta}{}_{n'n''} p^{\gamma}{}_{n''n}}{[\omega_1 + \omega_2 - \omega_{n'n}(\mathbf{k})][\omega_2 - \omega_{n'n''}(\mathbf{k})]}$$

$$+ \frac{p^{\beta}{}_{nn'} p^{\gamma}{}_{n'n''} p^{\alpha}{}_{n''n}}{[\omega_1 + \omega_2 + \omega_{n''n}(\mathbf{k})][\omega_1 + \omega_{n'n}(\mathbf{k})]}$$

$$+ \frac{p^{\gamma}{}_{nn'} p^{\alpha}{}_{n'n''} p^{\beta}{}_{n''n}}{[\omega_1 + \omega_2 - \omega_{n''n}(\mathbf{k})][\omega_2 + \omega_{n'n}(\mathbf{k})]}$$

$$+ \frac{p^{\gamma}{}_{nn'} p^{\alpha}{}_{n'n''} p^{\beta}{}_{n''n'}}{[\omega_1 + \omega_2 - \omega_{n''n'}(\mathbf{k})][\omega_1 - \omega_{n''n}(\mathbf{k})]}. \tag{6.5.106}$$

In this equation $f[E_n(\mathbf{k})]$ is the Fermi function, the $p^{\alpha}{}_{nn'}$ are the usual momentum matrix elements, which are understood to be functions of $\mathbf{k}$. The frequencies ω_1 and ω_2 may be regarded as having an infinitesimal imaginary part. Alternately, we may suppose the energies $E_n(\mathbf{k})$ contain an imaginary part representing the finite lifetime of an electron in $|n\mathbf{k}\rangle$. As a result, a term of the form $i\Gamma(\mathbf{k})$, with appropriate signs and indices will be inserted in the energy denominators.

The real and imaginary parts of $\sigma^{(2)}$ can be separated using the identity

$$\lim_{\epsilon\to 0} [(x + i\epsilon)(x + y + i\epsilon)]^{-1}$$
$$= (1/y)[P(1/x) - P(1/x + y) - i\pi\,\delta(x) + i\pi\,\delta(x + y)]. \tag{6.5.107}$$

The calculation of the second-order conductivity according to (6.5.105) and (6.5.106) is quite complex. Multiple sums over intermediate states are present. For this reason, most attempts to calculate $\sigma^{(2)}$, or equivalently $\epsilon^{(2)}$, have been based on quite simplified models of the electronic structure of the solid under consideration.

6.6 Excitons

The theory of optical properties that we have presented has been based on a one-electron picture, and has emphasized the role of band to band transitions. This approach works well in common semiconductors and perhaps also in metals; but it clearly fails in ionic crystals. An examination of the absorption constant of NaCl (Eby, Teegarden, and Dutton, 1959) as one example, shows considerable structure; but it is not easy to relate this structure to band to band transitions. Although the band structure near the direct gap is of a fairly standard type (a p-like valence band and s-like lowest conduction band), an optical absorption edge of the expected form is not observed: indeed, the threshold of band–band transitions must be searched for rather carefully in the midst of more prominent structure.

The essential missing element in the preceding theory is the coulomb interaction between the excited electron and the hole that has been left behind. When this interaction is strong, it is necessary to consider the electron–hole pair as a unit, which is called an *exciton*. We will present here a brief survey of the optical properties of systems where exciton effects are important. The problem was first taken up by Frenkel (1931a, b). Our approach is based on the work of Wannier (1937). Useful reviews have been given by Knox (1963), Dexter and Knox (1965), and Dimmock (1967). A related problem arises in the x-ray emission and absorption spectra of metals when tightly bound core levels are involved (Combescot and Nozières, 1971).

6.6.1 *General Theory*

We will develop the theory of excitons according to the general procedures of Wannier. We consider an insulating crystal at $T = 0°\mathrm{K}$ in which all bands are either completely full or completely empty.

The wave function for the ground state will be approximated by a single determinant. It was shown in Section 5.1 that this wave function may be constructed using either Wannier functions or Bloch functions as single-particle states: the same total wave function is obtained. For convenience, we will use Wannier functions as our basis. Let Ψ_G be the (N-particle) ground state wave function, and let $a_{n\sigma}(\mathbf{r} - \mathbf{R}_\mu)$ denote a Wannier function centered on site $\mathbf{R}_\mu$ belonging to band n. For convenience, we also attach a spin index σ ($\sigma = \uparrow, \downarrow$) to this function. Then

$$\Psi_G = \mathcal{A} a_{n\sigma_1}(\mathbf{r}_1 - \mathbf{R}_1) a_{n'\sigma_2}(\mathbf{r}_2 - \mathbf{R}_2) \cdots \tag{6.6.1}$$

where $\mathcal{A}$ is the antisymmetrizing operator

$$\mathcal{A} = (1/N!) \sum_P (-1)^P P \tag{6.6.2}$$

in which P is an operator which exchanges coordinates and spins. The band indices $n, n', \ldots,$ etc. run over all occupied bands. We suppose that all spin states are doubly occupied, and that the spatial portion of the Wannier function is the same for states of $\uparrow$ and $\downarrow$ spin. The wave function Ψ_G is then a singlet.

It is not a good approximation to consider the excited states of the system to be single determinants. However, we may construct an excited state as a linear combination of determinantal functions. The basis functions of this type are formed by transferring an electron from band n and site μ to (unoccupied) band l and site ν. The state is then characterized by the indices $l, \nu; n, \mu$ and may in fact be thought of as a two-particle state. It is necessary to take account of the spin of the excited state: this will be a singlet or a triplet. A triplet state is formed by reversing the spin of the electron transferred to the conduction band. The singlet excited state is more complicated: it requires a combination of two determinantal functions with $\uparrow\downarrow$ and $\downarrow\uparrow$ spin configurations. The singlet and triplet Wannier states will be denoted by $|\sigma; l\nu; n\mu\rangle$ where $\sigma = 0$ refers to the singlet and $\sigma = 1$ to the triplet state, respectively. We require the matrix elements of the Hamiltonian between these states.

The Hamiltonian is

$$H = \sum_i \left[T_i + U_i + \sum_{j>i} V(\mathbf{r}_i - \mathbf{r}_j)\right] \tag{6.6.3}$$

in which T_i is the kinetic energy for the ith particle, U_i is the single-particle potential energy used in the calculation of the band structure, and $V(\mathbf{r}_i - \mathbf{r}_j)$ contains the electron–electron interaction not included in the average field U_i. The basis states are orthogonal:

$$\langle\sigma'; l'\nu'; n'\mu' \mid \sigma; l\nu; n\mu\rangle = \delta_{\sigma\sigma'}\, \delta_{ll'}\, \delta_{nn'}\, \delta_{\mu\mu'}\, \delta_{\nu\nu'}. \tag{6.6.4}$$

Let us denote an excited state of spin σ and energy E by $|\sigma, E\rangle$. We will determine other quantum numbers necessary subsequently. This state is expanded as

$$|\sigma, E\rangle = \sum_{l\nu, n\mu} C(l\nu, n\mu) \, |\sigma; l\nu; n\mu\rangle. \tag{6.6.5}$$

The coefficients C satisfy the equation

$$\sum \langle\sigma; l'\nu'; n'\mu' \,|\, H \,|\, \sigma; l\nu; n\mu\rangle C(l\nu, n\mu) = EC(l'\nu'; n'\mu'). \tag{6.6.6}$$

We must determine the matrix elements of the Hamiltonian. These can be worked out by a straightforward but somewhat tedious calculation, which we shall not give in detail. The result is

$$\begin{aligned}\langle\sigma; l'\nu'; n'\mu' \,|\, H \,|\, \sigma; l\nu; n\mu\rangle &= E_0\, \delta_{ll'}\, \delta_{nn'}\, \delta_{\mu\mu'}\, \delta_{\nu\nu'} + \delta_{ll'}\, \delta_{\mu\mu'}\, \delta_{nn'}\, \mathcal{E}_l(\mathbf{R}_{\nu'} - \mathbf{R}_\nu) \\ &\quad - \delta_{nn'}\, \delta_{ll'}\, \delta_{\nu\nu'}\, \mathcal{E}_l(\mathbf{R}_\mu' - \mathbf{R}_\mu) + 2(1-\sigma)\langle l'\nu'; n\mu \,|\, V \,|\, n'\mu'; l\nu\rangle \\ &\quad - \langle l'\nu'; n\mu \,|\, V \,|\, l\nu; n'\mu'\rangle \end{aligned} \tag{6.6.7}$$

in which the $\mathcal{E}$'s are given by (5.1.37) and

$$\begin{aligned}&\langle l'\nu'; n\mu \,|\, V \,|\, l\nu; n'\mu'\rangle \\ &\quad = \int a^*_{l'}(\mathbf{r}_1 - \mathbf{R}_\nu')a_n^*(\mathbf{r}_2 - \mathbf{R}_\mu)V(\mathbf{r}_1 - \mathbf{r}_2) \\ &\qquad \times a_l(\mathbf{r}_1 - \mathbf{R}_\nu)a_n'(\mathbf{r}_2 - \mathbf{R}_\mu')\, d^3r_1\, d^3r_2. \end{aligned} \tag{6.6.8a}$$

The quantity E_0 is the average energy of the state $|\sigma; l\nu; n\mu\rangle$ with which we shall not be concerned. The exchange terms in (6.6.7) occur in a different fashion from what one would perhaps expect: they are present in the singlet state but are absent in the triplet. Likewise the direct term has a negative sign which results from the fact that we have here an interaction between an electron and a hole. The hole behaves as a positive charge.

It is an important property of the potential matrix element (6.6.8) that we can displace all sites by a lattice vector without changing its value. This occurs because the interaction depends on $\mathbf{r}_1 - \mathbf{r}_2$. Thus, for example,

$$\langle l'\nu'; n\mu \,|\, V \,|\, l\nu; n'\mu'\rangle = \langle l, \nu' - \nu; n, \mu - \nu \,|\, V \,|\, l, 0; n', \mu' - \nu\rangle. \tag{6.6.8b}$$

(Here $\nu' - \nu$ designates $\mathbf{R}_\nu' - \mathbf{R}_\nu$, etc.)

Equation (6.6.6) can be simplified if we recognize that the system under consideration is effectively one of two particles. The total wave vector of the electron–hole pair should be a constant of the motion. Thus it is desirable to introduce states characterized by a total wave vector $\mathbf{K}$ and a relative coordinate $\mathbf{R}_\rho$. These states are denoted by $|\sigma; \mathbf{K}\rho; ln\rangle$ and are constructed as follows:

$$| \sigma; \mathbf{K}\rho; ln\rangle = \Omega^{1/2}/(2\pi)^{3/2} \sum_{\mu} \exp(i\mathbf{K}\cdot\mathbf{R}_\mu) \mid \sigma; l, \mu + \rho; n\mu\rangle. \quad (6.6.9)$$

In such a state, the electron and the hole are separated by $\mathbf{R}_\rho$, and propagate together through the lattice with wave vector $\mathbf{K}$. It follows from (6.6.4) that

$$\langle\sigma'; \mathbf{Q}\rho'; l'n' \mid \sigma; \mathbf{K}\rho; ln\rangle = \delta_{\sigma\sigma'}\, \delta_{\rho\rho'}\, \delta_{ll'}\, \delta_{nn'}\, \delta(\mathbf{K} - \mathbf{Q}). \quad (6.6.10)$$

We use the new set of functions as a basis set for the expansion of the wave function for the excited state. It is important to observe that the electron interaction does not connect states of different total wave vectors. We must consider the matrix element:

$$\begin{aligned}&\langle\sigma', \mathbf{Q}\rho'; l'n' \mid H \mid \sigma, \mathbf{K}\rho; ln\rangle \\ &\quad = [\Omega/(2\pi)^3] \sum_{\mu\nu} \exp[i(\mathbf{K}\cdot\mathbf{R}_\mu - \mathbf{Q}\cdot\mathbf{R}_\nu)] \\ &\quad\quad \times \langle\sigma'; l', \nu + \rho'; n'\nu \mid H \mid \sigma; l, \mu + \rho; n\mu\rangle.\end{aligned}$$

This can be worked out with the aid of (6.6.7) and (6.6.9). The result is

$$\begin{aligned}&\langle\sigma'; \mathbf{Q}\rho'; l'n' \mid H \mid \sigma; \mathbf{K}\rho; ln\rangle \\ &\quad = \delta_{\sigma\sigma'}\, \delta(\mathbf{K} - \mathbf{Q}) \{\delta_{nn'}\, \delta_{ll'}[E_0\, \delta_{\rho\rho'} + \mathcal{E}_l(\mathbf{R}_{\rho'} - \mathbf{R}_\rho) \\ &\quad\quad - \exp[i\mathbf{K}\cdot(\mathbf{R}_\rho{}' - \mathbf{R}_\rho)]\mathcal{E}_n(\mathbf{R}_\rho{}' - \mathbf{R}_\rho)] + \sum_{\tau} \exp(i\mathbf{K}\cdot\mathbf{R}_\tau) \\ &\quad\quad \times [2(1 - \sigma)\langle l'\rho'; n\tau \mid V \mid n'0; l, \tau + \rho\rangle \\ &\quad\quad - \langle l'\rho'; n\tau \mid V \mid l, \tau + \rho; n'0\rangle]\}. \end{aligned} \quad (6.6.11)$$

The multiplying delta function indicates that the Hamiltonian is diagonal in the total wave vector $\mathbf{K}$ which should evidentally be incorporated in the specification of the excited state as $| \sigma, \mathbf{K}, E\rangle$. Then we have, instead of (6.6.5) and (6.6.6),

$$| \sigma, \mathbf{K}, E\rangle = \sum_{ln\rho} | \sigma, \mathbf{K}\rho; ln\rangle U(\sigma, \mathbf{K}\rho; ln). \quad (6.6.12)$$

The coefficients U satisfy the equation

$$\sum_{ln\rho} [\langle\sigma, \mathbf{K}\rho'; l'n' \mid H \mid \sigma, \mathbf{K}\rho; l, n\rangle U(\sigma, \mathbf{K}\rho; ln) = EU(\sigma, \mathbf{K}\rho'; l'n'). \quad (6.6.13)$$

The delta function is discarded when (6.6.11) is inserted into (6.6.13); however, a special notation will not be introduced to describe this. The explicit equation which determines the amplitudes U is

$$\sum_{\mu} \{\mathcal{E}_l(\mathbf{R}_\rho - \mathbf{R}_\mu) - \mathcal{E}_n(\mathbf{R}_\rho - \mathbf{R}_\mu) \exp[i\mathbf{K}\cdot(\mathbf{R}_\rho - \mathbf{R}_\mu)]\} U(\sigma, \mathbf{K}\mu; ln)$$

$$+ \sum_{\mu\tau, l'n'} \exp(i\mathbf{K}\cdot\mathbf{R}_\tau)[2(1-\sigma)\langle l\rho; n'\tau \mid V \mid n, 0; l', \tau+\mu\rangle$$

$$- \langle l\rho; n'\tau \mid V \mid l', \tau+\mu; n0\rangle] U(\sigma, K\mu; l'n') = (E - E_0) U(\sigma, \mathbf{K}\rho; ln). \tag{6.6.14}$$

The one-electron terms in (6.6.14) can be rewritten in terms of the band energies in the following way [using (5.1.37)]:

$$\mathcal{E}_l(\mathbf{R}_\rho - \mathbf{R}_\mu) - \mathcal{E}_n(\mathbf{R}_\rho - \mathbf{R}_\mu) \exp[i\mathbf{K}\cdot(\mathbf{R}_\rho - \mathbf{R}_\mu)]$$

$$= [\Omega/(2\pi)^3] \int d^3k \exp[i\mathbf{k}\cdot(\mathbf{R}_\rho - \mathbf{R}_\mu)][E_l(\mathbf{k}) - E_n(\mathbf{k} - \mathbf{K})]$$

$$= [\Omega/(2\pi)^3] \exp[i\mathbf{K}\cdot(\mathbf{R}_\rho - \mathbf{R}_\mu)/2] \int d^3k \exp[i\mathbf{k}\cdot(\mathbf{R}_\rho - \mathbf{R}_\mu)]$$

$$\times [E_l(\mathbf{k} + \mathbf{K}/2) - E_n(\mathbf{k} - \mathbf{K}/2)]. \tag{6.6.15}$$

This expression indicates that the excited state, or "exciton" (as it will be called from here on), is formed as a wave packet by transferring an electron from one state to another whose wave vector differs by $\mathbf{K}$.

6.6.2 *The Wannier Exciton*

We have obtained a set of linear equations for the exciton amplitudes. The complete solution of these is seldom attempted. Two limits are of particular interest: (1) the loosely bound (Wannier) exciton, and (2) the tightly bound (Frenkel) exciton.

In the first case, that of the Wannier exciton, it is supposed that the amplitudes U are appreciable for rather large separations of the electron and the hole. We begin by considering the potential matrix elements in (6.6.14). If $|\mathbf{R}_\rho - \mathbf{R}|$ is large, and the Wannier functions are well localized, the electron–hole interaction $V(\mathbf{r}_1 - \mathbf{r}_2)$ should be reasonably approximated by a Coulomb potential screened by the (low-frequency) dielectric function of the system:

$$V(\mathbf{r}_1 - \mathbf{r}_2) = e^2/\kappa \mid \mathbf{r}_1 - \mathbf{r}_2 \mid. \tag{6.6.16}$$

This representation of the electron–hole interaction is adequate for our purposes but is somewhat oversimplified for applications to real materials, in that the wave vector and frequency dependence of the dielectric function, which leads to a position dependence of the effective κ in (6.6.16), has been neglected.

The assumed interaction (6.6.16) is slowly varying. As a result, the orthogonality of the Wannier functions from different bands will tend to reduce matrix elements except the direct case with $n' = n$ and $l' = l$. The dominant term will be the direct term with $\mathbf{R}_\tau = 0$, $\mathbf{R}_\rho = \mathbf{R}_\mu$. This element is

$$\langle l\rho; n0 \mid V \mid l\rho; n0 \rangle = \int \mid a_l(\mathbf{r}_1 - \mathbf{R}_\rho) \mid^2 \mid a_n(\mathbf{r}_2) \mid^2 (e^2/\kappa \mid \mathbf{r}_1 - \mathbf{r}_2 \mid) \, d^3r_1 \, d^3r_2$$

$$\approx e^2/\kappa R_\rho. \tag{6.6.17}$$

This approximation will be valid when R_ρ is large.

We retain the matrix element (6.6.17) and ignore the rest, which are of shorter range. This approximation is quite analogous to the treatment of the shallow impurity states in Section 5.3, and has a similar range of validity. The terms which have been dropped here could, at least in principle, be treated as perturbations on the simplified problem. Equation (6.6.14) now becomes

$$\sum_\mu \{\mathcal{E}_l(\mathbf{R}_\rho - \mathbf{R}_\mu) - \mathcal{E}_n(\mathbf{R}_\rho - \mathbf{R}_\mu) \exp[i\mathbf{K}\cdot(\mathbf{R}_\rho - \mathbf{R}_\mu)]\} U(\mathbf{K}, \mu)$$

$$- (e^2/\kappa \mid \mathbf{R}_\rho \mid) U(\mathbf{K}, \rho) = (E - E_0) U(\mathbf{K}, \rho). \tag{6.6.18}$$

Superfluous indices have been dropped. We now proceed by treating the lattice vectors $\mathbf{R}_\rho$, $\mathbf{R}_\mu$ as continuous variables, and the discrete coefficients $U(\mathbf{K}, \rho)$ as ordinary functions. The coefficients $\mathcal{E}_l$, $\mathcal{E}_n$ become differential operators. The last step is accomplished with the aid of (6.6.15). Substitute (6.6.15) into (6.6.18) and define

$$U'(\mathbf{K}, \mathbf{R}_\mu) = \exp(-i\mathbf{K}\cdot\mathbf{R}_\mu/2) U(\mathbf{K}, \boldsymbol{\mu}).$$

The result is

$$\sum_\mu \left\{[\Omega/(2\pi)^3] \int d^3k \exp[i\mathbf{k}\cdot(\mathbf{R}_\rho - \mathbf{R}_\mu)]\right.$$

$$\left. \times [E_l(\mathbf{k} + \tfrac{1}{2}\mathbf{K}) - E_n(\mathbf{k} - \tfrac{1}{2}\mathbf{K})]\right\} U'(\mathbf{K}, \mathbf{R}_\mu)$$

$$- (e^2/\kappa R_\rho) U'(\mathbf{K}, \mathbf{R}_\rho) = (E - E_0) U'(\mathbf{K}, \mathbf{R}_\rho). \tag{6.6.19}$$

Now consider the integral over $\mathbf{k}$. It is supposed that the band functions E_l, E_n possess power series expansions: In general, let $f(\mathbf{k})$ be an arbitrary function possessing a power series expansion in $\mathbf{k}$ about some arbitrary reference point $\mathbf{k}_0$:

$$f(\mathbf{k}) = \sum_{rst} A_{rst}(k-k_0)_x{}^r (k-k_0)_y{}^s (k-k_0)_z{}^t.$$

Then

$$[\Omega/(2\pi)^3] \int f(\mathbf{k}) \exp[i\mathbf{k}\cdot(\mathbf{R}_\rho - \mathbf{R}_\mu)]\, d^3k$$

$$= [\Omega/(2\pi)^3] \exp[i\mathbf{k}_0\cdot(\mathbf{R}_\rho - \mathbf{R}_\mu)] \sum_{rst} A_{rst}$$

$$\times \int \exp[i(\mathbf{k}-\mathbf{k}_0)\cdot(\mathbf{R}_\rho - \mathbf{R}_\mu)](k-k_0)_x{}^r (k-k_0)_y{}^s (k-k_0)_z{}^t\, d^3k$$

$$= [\Omega/(2\pi)^3] \exp[i\mathbf{k}_0\cdot(\mathbf{R}_\rho - \mathbf{R}_\mu)] \sum A_{rst}(-i\,\partial/\partial R_{\rho x})^r$$

$$\times (-i\,\partial/\partial R_{\rho y})^s (-i\,\partial/\partial R_{\rho z})^t \int \exp[i(\mathbf{k}-\mathbf{k}_0)\cdot(\mathbf{R}_\rho - \mathbf{R}_\mu)]\, d^3k$$

$$= \delta_{\mu\rho} f(-i\nabla). \tag{6.6.20}$$

This result will be valid when one operates on a function of lattice coordinate. Equation (6.6.19) becomes

$$[E_l(-i\nabla + \tfrac{1}{2}\mathbf{K}) - E_n(-i\nabla - \tfrac{1}{2}\mathbf{K}) - (e^2/\kappa R)]U'(\mathbf{K},\mathbf{R})$$

$$= (E - E_0)\,U'(\mathbf{K},\mathbf{R}). \tag{6.6.21}$$

It is not practical to solve this equation directly; instead, one expands the operators, retaining terms through second order. Clearly, there are many possibilities depending on what the valence and conduction band structure actually is. We will consider only the simplest possibility: nondegenerate, parabolic, valence and conduction bands centered at $\mathbf{k} = 0$:

$$E_n(\mathbf{k}) = -\hbar^2k^2/2m_v^*, \qquad E_l = E_g + (\hbar^2\mathbf{k}^2/2m_c^*). \tag{6.6.22}$$

From this point, we refer to the bands as v and c, rather than n and l, and E_g is the band gap at $\mathbf{k} = 0$. Equation (6.6.21) yields

$$[-(\hbar^2/2\mu)\nabla^2 - \tfrac{1}{2}\hbar^2 i(1/m_c^* - 1/m_v^*)\mathbf{K}\cdot\nabla + \hbar^2\mathbf{K}^2/8\mu - (e^2/\kappa R)]U'(\mathbf{K},\mathbf{R})$$

$$= (E - E_0 - E_g)\,U'(\mathbf{K},\mathbf{R}). \tag{6.6.23}$$

in which $\mu^{-1} = m_c^{*-1} + m_v^{*-1}$. The linear term in this equation can be eliminated by a transformation

$$U'(\mathbf{K},\mathbf{R}) = \exp(i\lambda\mathbf{K}\cdot\mathbf{R})F(\mathbf{K},\mathbf{R}). \tag{6.6.24}$$

When (6.6.24) is substituted into (6.6.23), it is found that λ should be chosen to be

$$\lambda = \tfrac{1}{2}(m_c^* - m_v^*)/(m_c^* + m_v^*).$$

Then (6.6.23) becomes

$$[-(\hbar^2/2\mu)\nabla^2 - (e^2/\kappa R)]F(\mathbf{K}, \mathbf{R}) = [E - E_0 - E_g - [\hbar^2\mathbf{K}^2/2(m_c^* + m_v^*)]]F(\mathbf{K}, \mathbf{R}). \quad (6.6.25)$$

We now have a simple problem analogous to the hydrogen atom, or the shallow impurity. The energy levels are

$$E - E_0 = E_g + [\hbar^2K^2/2(m_c^* + m_v^*)] - (\mu e^4/2n^2\hbar^2\kappa^2). \quad (6.6.26)$$

The quantity E_0 is the energy of the ground state, so that $E - E_0$ represents the excitation energy. This differs from the band gap energy E_g by the binding energy of the electron and the hole, and by the kinetic energy associated with motion of the center of mass. As has been discussed in previous sections, only excitons with wave vectors close to zero can be created by absorption of visible light. In some materials, including solidified rare gases and such materials as cuprous oxide, Cu_2O, and cadmium sulfide, CdS, well-defined hydrogenic series of exciton lines are observed in absorption. An example is shown in Fig. 6.6.1.

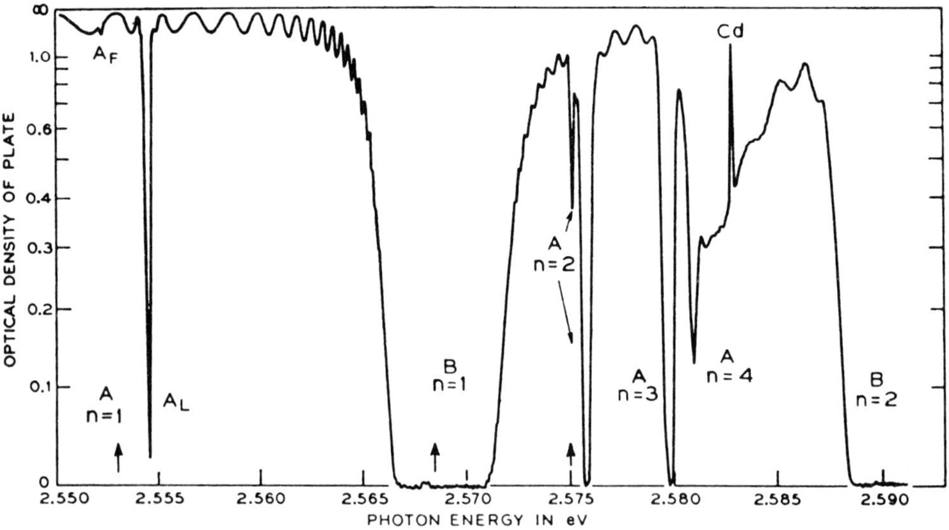

Fig. 6.6.1. A microphotometer trace of a photographic plate showing absorption due to two series (A and B) of excitons in CdS at 1.6°K, $t = 6\mu$. The light is polarized with **E** parallel to the c axis. The arrows indicate the location of transitions seen in the reflection spectrum for $\mathbf{E} \parallel c$; A_F indicates a strongly forbidden transition and A_L a longitudinal exciton. Note the contrast in the widths of the A and B series lines. [From Thomas and Hopfield (1961).]

In addition to the discrete, or bound states, whose energies are given by (6.6.26), continuum states pertaining to unbound excitons must also be considered. A continuum exciton state is characterized by a wave vector **k** such that

$$k^2 = (2\mu/\hbar^2)[E - E_0 - E_g - [\hbar^2 K^2/2(m_c^* + m_v^*)]]. \quad (6.6.27)$$

6.6.3 *The Frenkel Exciton*

In this subsection we will consider the opposite limit of the exciton problem: the Frenkel exciton. In this case, the exciton is strongly localized: the electron and the hole are on the same atomic site. An individual excited atom is involved, as in the crystal field problem of Section 3.4, Part A, except that we do not know which atom is excited.

The relevant matrix elements of the Hamiltonian in the case of the Frenkel exciton may be found from (6.6.11) by taking the relative coordinates $R_\rho = R_\rho' = 0$. For simplicity, we will consider a situation in which all the exciton states have the hole in the same band (or in the same atomic state). This would be the case if the ground atomic state is s-like. Hence, we set $n = n' = i$ in (6.6.11) to indicate this state. We write

$$\langle \sigma, \mathbf{Q}0; l'i \mid H \mid \sigma, \mathbf{K}0; li \rangle = \delta_{\sigma\sigma'}\, \delta(\mathbf{K} - \mathbf{Q})\, H_{l'l}^{(\sigma)}(\mathbf{K}) \quad (6.6.28)$$

in which

$$H_{l'l}^{(\sigma)}(\mathbf{K}) = \delta_{ll'}\, [E_0 + \varepsilon_l(0) - \varepsilon_i(0)] + \sum_\tau \exp(i\mathbf{K}\cdot\mathbf{R}_\tau)[2(1 - \sigma)$$
$$\times \langle l'0, i\tau \mid V \mid i0, l\tau \rangle - \langle l'0, i\tau \mid V \mid l\tau, i0 \rangle]. \quad (6.6.29)$$

The energy of an excited state of the crystal of wave vector **K** can be approximated by diagonalizing the matrix $H_{l'l}^{(\sigma)}$. This process yields energies which are functions of **K** and defines an exciton band. The process resembles quite closely the calculation of energy bands in the tight binding approximation except that we are here concerned with a correlated two particle state.

The **K** dependence of the matrix element (6.6.29) arises from the terms $(\mathbf{R}_\tau \neq 0)$

$$2(1 - \sigma) \int a^*_{l'}(\mathbf{r}_1) a_i^*(\mathbf{r}_2 - \mathbf{R}_\tau) V(\mathbf{r}_1 - \mathbf{r}_2) a_i(\mathbf{r}_1) a_l(\mathbf{r}_2 - \mathbf{R}_\tau)\, d^3r_1\, d^3r_2$$

$$- \int a^*_{l'}(\mathbf{r}_1) a_i^*(\mathbf{r}_2 - \mathbf{R}_\tau) V(\mathbf{r}_1 - \mathbf{r}_2) a_l(\mathbf{r}_1 - \mathbf{R}_\tau) a_i(\mathbf{r}_2)\, d^3r_1\, d^3r_2.$$

(6.6.30)

The components of (6.6.30) may be considered as representing the effective interaction of certain charge distributions: in the first term, the extended charge $a^*_{l'}(\mathbf{r}_1)a_i(\mathbf{r}_1)$ interacts with a similar charge distribution displaced to the lattice site $\mathbf{R}_\tau$: $a_l(\mathbf{r}_2 - \mathbf{R}_\tau)a_i^*(\mathbf{r}_2 - \mathbf{R}_\tau)$. Each charge distribution is effectively neutral since the Wannier functions are orthogonal. The dominant interaction at large R should then be of the dipole–dipole type if this is allowed by symmetry. In contrast, the second term of (6.6.30) involves the interactions between distributions $a^*_{l'}(\mathbf{r}_1)a_l(\mathbf{r}_1 - \mathbf{R}_\tau)$ and $a_i(\mathbf{r}_2)a_i^*(\mathbf{r}_2 - \mathbf{R}_\tau)$. In this case, the distributions themselves vanish exponentially as $|\mathbf{R}_\tau|$ becomes large. Therefore this term is of shorter range than the first term (which is, however, present only in the singlet state). In consequence, we expect substantially different band structures for singlet and triplet excitons.

Let us consider as an example a situation in which the function a_i is s-like and a_l and $a_{l'}$ are p-like. Then the charge distributions a_ia_l and $a_ia_{l'}$ possess dipole moments. The interaction V is taken to be the coulomb potential. The question of screening is somewhat delicate in the present instance since we are dealing with a system which is much more tightly bound than a Wannier exciton. However, we can absorb a possible contribution from the dielectric function into the effective dipole moments and retain the same functional form for the dipole–dipole interaction, which is

$$2\int a^*_{l'}(\mathbf{r}_1)a_i(\mathbf{r}_1)V(\mathbf{r}_1 - \mathbf{r}_2)a_i^*(\mathbf{r}_2 - \mathbf{R})a_l(\mathbf{r}_2 - \mathbf{R})\, d^3r\, d^3r_2$$
$$\approx (\mathbf{p}_{l'i}\cdot\mathbf{p}_{il}/R^3) - 3[(\mathbf{p}_{l'i}\cdot\mathbf{R})(\mathbf{p}_{il}\cdot\mathbf{R})/R^5] \tag{6.6.31}$$

where

$$\mathbf{p}_{l'i} = e\sqrt{2}\int a^*_{l'}(\mathbf{r})\mathbf{r}a_i(\mathbf{r})\, d^3r. \tag{6.6.32}$$

A factor of $\kappa^{-1/2}$ may be included if appropriate.

Since (6.6.31) does not apply for $\mathbf{R} = 0$, where there are in fact no singularities, it is desirable to separate out of (6.6.29) the term with $\mathbf{R} = 0$ and consider the lattice sum

$$S_{l'l}(\mathbf{K}) = \sum_{\tau\neq 0} \exp(i\mathbf{K}\cdot\mathbf{R}_\tau)\{(\mathbf{p}_{l'i}\cdot\mathbf{p}_{il}/|\mathbf{R}_\tau|^3)$$
$$- 3[(\mathbf{p}_{l'i}\cdot\mathbf{R}_\tau)(\mathbf{p}_{il}\cdot\mathbf{R}_\tau)/|\mathbf{R}_\tau|^5]\}. \tag{6.6.33}$$

The evaluation of sums of this type is rather tricky (see, for example, Cohen and Keffer, 1955). We will investigate the behavior of this sum for small

$\mathbf{K}$ by replacing it by an integral over R space, excluding a sphere around the origin. A factor ρ representing the number of lattice points per unit volume must be included. For simplicity, let $\mathbf{p}_{l'i}$ be in the $x_{l'}$ ($l' = 1, 2$, or 3) direction in a rectangular coordinate system. Consider the integral

$$\left[\int_{\infty} - \int_{V}\right] [\exp(i\mathbf{K}\cdot\mathbf{R})(\delta_{ll'}/R^3 - 3x_l x_{l'}/R^5)]\, d^3R. \tag{6.6.34}$$

The excluded region is taken to be a sphere of radius R_s, where $4\pi R_s^3/3 = \Omega$, the central cell volume. The integral is not difficult if we observe that the function $\delta_{ll'} - 3x_l x_{l'}/R^2$ is, for all l and l', a linear combination of spherical harmonics Y_{2m}, and that

$$\int \exp(i\mathbf{K}\cdot\mathbf{R}) Y_{lm}(\Omega)\, d\Omega = 4\pi i^l j_l(KR) Y_{lm}(K_\theta, K_\phi).$$

Then (6.6.34) becomes

$$-4\pi[\delta_{ll'} - (3K_l K_{l'}/K^2)] \int_R^\infty j_2(KR)/R\, dR.$$

The radial integral is easily performed. The result is

$$S_{l'l}(\mathbf{K}) = -(4\pi/3) p_{li} p_{il} \rho [\delta_{l'l} - (3K_l K_{l'}/K^2)][3j_1(KR_s)/KR_s]. \tag{6.6.35}$$

For small x, $3j_1(x)/x = 1 - x^2/10$; thus we have for small K

$$S(K) \approx -(4\pi/3) p_{li} p_{il} \rho [\delta_{ll'} - (3K_l K_{l'}/K^2)][1 - \tfrac{1}{10}(KR_s)^2]. \tag{6.6.36}$$

Examination of the angular integration which occurs in the calculation shows that the terms in (6.6.29) with $\tau = 0$ are already diagonal with respect to l and l' for the basis functions considered. Let

$$V_l^{(\sigma)} = 2(1 - \sigma) \int a^*_l(\mathbf{r}_1) a_i(\mathbf{r}_1) V(\mathbf{r}_1 - \mathbf{r}_2) a_i(\mathbf{r}_2) a_l(\mathbf{r}_2)\, d^3r_1\, d^3r_2$$

$$- \int |a_i(\mathbf{r}_1)|^2 |a_l(\mathbf{r}_2)|^2 V(\mathbf{r}_1 - \mathbf{r}_2)\, d^3r_1\, d^3r_2. \tag{6.6.37}$$

We retain only V_l and the long-range part of the potential. Then (6.6.29) becomes, in the small K limit,

$$H_{l'l}^{(\sigma)} = \delta_{ll'}[E_0 + \epsilon_l(0) - \epsilon_i(0) + V_l^{(\sigma)}] - (4\pi/3)(1 - \sigma) |p_{il}|^2 \rho$$
$$\times [\delta_{ll'} - (3K_l K_{l'}/K^2)][1 - \tfrac{1}{10}(KR_s)^2]. \tag{6.6.38}$$

The energies of exciton states are to be found by diagonalizing this matrix. This is accomplished quite simply, since the only nondiagonal part is of the form $K_l K_l'$. The eigenvalues of a matrix $M_{ll'} = K_l K_{l'}$ are just $\mathbf{K}^2$ (once) and zero (twice). Diagonalization of this matrix is equivalent to a rotation of the coordinates so that one axis is parallel to $\mathbf{K}$ (longitudinal exciton) and two axes are perpendicular to $\mathbf{K}$ (transverse exciton). The energies are then:

longitudinal exciton

$$E_T = E_0 + \epsilon_l(0) - \epsilon_i(0) + V_l^\sigma + (1 - \sigma)(8\pi/3) \mid p_{il} \mid^2 \rho[1 - \tfrac{1}{10}(KR_s)^2]; \tag{6.6.39}$$

transverse exciton

$$E_T = E_0 + \epsilon_l(0) - \epsilon_i(0) + V_l^\sigma - (1 - \sigma)(4\pi/3) \mid p_{il} \mid^2 \rho[1 - \tfrac{1}{10}(KR_s)^2]. \tag{6.6.40}$$

The splitting in energy between longitudinal and transverse excitations was first obtained by Heller and Marcus (1951). Equation (6.6.39) and (6.6.40) can be used for an estimate of the exciton effective masses, but a more accurate calculation would require numerical evaluation of the lattice sum. The necessity for other improvements should also be evident.

6.6.4 *Optical Absorption by Excitons*

We will discuss optical absorption by excitons by generalizing the treatment presented in Section 6.5.2. The essential quantity is the matrix element of the operator

$$H' = (e/m) \sum_i \mathbf{A}(\mathbf{r}_i) \cdot \mathbf{p}_i \tag{6.6.41}$$

between the initial state in which all bands are filled and the final state (6.6.12). To determine this, we first consider the matrix element connecting the ground state to the state $| \sigma, l\nu; n\mu \rangle$. The ground state is assumed to be a singlet. This implies that the excited state must also be a singlet; and this will be assumed. Let the ground state be $| G \rangle$. We have

$$\langle 0, l\nu; n\mu \mid H' \mid G \rangle = (l\nu \mid H' \mid n\mu) \tag{6.6.42}$$

where

$$(l\nu \mid H' \mid n\mu) = (e/m) \int a_l^*(\mathbf{r} - \mathbf{R}_\nu) \mathbf{A} \cdot \mathbf{p} a_n(\mathbf{r} - \mathbf{R}_\mu) \, d^3r. \tag{6.6.43}$$

We may now use (6.6.9) and (6.6.12) to obtain

$$\langle 0, \mathbf{K}, E \mid H' \mid G \rangle = [\Omega^{1/2}/(2\pi)^{3/2}] \sum_{\mu\rho, ln} U^*(0, \mathbf{K}\rho; ln)$$

$$\times \exp(-i\mathbf{K}\cdot\mathbf{R}_\mu)\,(l, \mu + \rho \mid H' \mid n\mu). \quad (6.6.44)$$

Consider a term in the vector potential proportional to $\exp(i\mathbf{s}\cdot\mathbf{r})$. The origin of integration can be shifted so that

$$(l, \mu + \rho \mid \exp(i\mathbf{s}\cdot\mathbf{r})\nabla \mid n\mu) = \exp(i\mathbf{s}\cdot\mathbf{R}_\mu)\,(l\rho \mid \exp(i\mathbf{s}\cdot\mathbf{r})\nabla \mid n0).$$

The only dependence on $\mathbf{R}_\mu$ now occurs in the exponential factors; so that the sum can be performed. Thus

$$\langle 0, \mathbf{K}, E \mid \exp(i\mathbf{s}\cdot\mathbf{r})\nabla \mid G \rangle$$

$$= [(2\pi)^{3/2}/\Omega^{1/2}]\,\delta(\mathbf{s} - \mathbf{K}) \sum_{\rho ln} U^*(0, \mathbf{K}\rho; ln)\,(l, \rho \mid \exp(i\mathbf{s}\cdot\mathbf{r})\nabla \mid n0). \quad (6.6.45)$$

The delta function indicates that the wave vector of the light must match the total momentum of the exciton state. For the reasons indicated in the previous section, this implies that light can excite only excitons of wave vectors $\mathbf{K}$ close to zero. We will make the approximations of setting $\mathbf{K}$ exactly equal to zero from this point, and introduce a two-band model in which only a single valence and conduction band are considered. Our attention will be focused on the Wannier exciton. The delta function disappears from the calculation of the absorption constant as was described previously.

We are therefore concerned with an effective matrix element

$$H'_{\mathrm{cv}} = [(2\pi)^{3/2}/\Omega^{1/2}] \sum_{\rho} F^*(0, \mathbf{R}_\rho)\,(\mathrm{c}\rho \mid H' \mid \mathrm{v}0). \quad (6.6.46)$$

It is useful to express the matrix element on the Wannier function basis in terms of Bloch functions. Note that

$$(\mathrm{c}\rho \mid \mathbf{p} \mid \mathrm{v}0) = [\Omega/(2\pi)^3] \int \exp(i\mathbf{k}\cdot\mathbf{R}_\rho)\,\langle \mathrm{c}\mathbf{k} \mid \mathbf{p} \mid \mathrm{v}\mathbf{q} \rangle\, d^3k\, d^3q$$

$$= [\Omega/(2\pi)^3] \int \exp(i\mathbf{k}\cdot\mathbf{R}_\rho)\,\mathbf{p}_{\mathrm{cv}}(\mathbf{k})\, d^3k. \quad (6.6.47)$$

We have used (6.5.22) to relate the momentum matrix element on the Bloch function basis to that involving the cell periodic functions [see Eq. (4.1.18), Part A]. We now combine (6.6.42) and (6.6.43) and express the result in a form analogous to (6.5.25)

$$\bar{H}_{\text{cv}} = (e/2m)A_0\boldsymbol{\varepsilon}_{\text{s}}\cdot\Big\{[\Omega^{1/2}/(2\pi)^{3/2}]$$

$$\times \int [\sum_{\rho} \exp(i\mathbf{k}\cdot\mathbf{R}_\rho)F^*(0, \mathbf{R}_\rho)]\mathbf{p}_{\text{cv}}(\mathbf{k})\, d^3k\Big\}.$$

Now let

$$f(\mathbf{K}, \mathbf{k}) = [\Omega^{1/2}/(2\pi)^{3/2}] \sum_{\rho} \exp(-i\mathbf{k}\cdot\mathbf{R}_\rho)F(\mathbf{K}, \mathbf{R}_\rho). \tag{6.6.48}$$

The function $f(\mathbf{K}, \mathbf{k})$ is the Fourier representative of the exciton wave function. Thus

$$\bar{H}_{\text{cv}} = (e/2m)A_0\boldsymbol{\varepsilon}_{\text{s}}\cdot \int f^*(0, \mathbf{k})p_{\text{cv}}(\mathbf{k})\, d^3k. \tag{6.6.49}$$

In an optical absorption experiment, we may observe either a line absorption corresponding to the discrete states (with $\mathbf{K} = 0$) whose energies are given approximately by (6.6.26) or a continuum absorption involving transitions to unbound states. Let us examine (6.6.45) in the limit in which the exciton wave function is slowly varying over large distances in the crystal. Then $f(0, \mathbf{k})$ will be well localized in the Brillouin zone. Suppose that $\mathbf{p}_{\text{cv}}(\mathbf{k})$ can be regarded as a constant throughout the region of $\mathbf{k}$ space in which $f(0, \mathbf{k})$ is appreciable. We then have

$$H_{\text{cv}} = (e/2m)a_0(\boldsymbol{\varepsilon}_{\text{s}}\cdot\mathbf{p}_{\text{cv}})[(2\pi)^{3/2}/\Omega^{1/2}]F^*(0, 0). \tag{6.6.50}$$

From this point, the factor $(2\pi)^{3/2}/\Omega^{1/2}$ will be absorbed in the quantity F. Thus the transition matrix element that is obtained in the simple theory of Section 6.5.2 when the electron–hole interaction is neglected, Eq. (6.5.25), is modified through multiplication by a factor which is essentially the amplitude of the hole–electron pair wave function in the central cell.

The remainder of the calculation of the absorption constant proceeds exactly as in Section 6.5 in the case of transitions to unbound exciton states. We have, in place of (6.5.29),

$$\alpha = (2\pi/\Omega)(e^2 \mid \boldsymbol{\varepsilon}_{\text{s}}\cdot\mathbf{p}_{\text{cv}} \mid^2/m^2 n\epsilon_0 c\omega) \mid F(0, 0) \mid^2 G_{\text{cv}}(\hbar\omega) \tag{6.6.51}$$

where G_{cv} is the joint density of states. The factor $\mid F(0, 0) \mid^2$ depends on energy. In the simple case in which (6.6.25) is valid, F will be a positive energy coulomb wave function. The wave function may be obtained from standard sources (Schiff, 1968, p. 142). Clearly only s waves can contribute, and

$$F(0, R) = e^{+\pi\gamma/2}\Gamma(1 - i\gamma)e^{ikR}\Phi(1 - i\gamma, 2, -2ikR) \tag{6.6.52}$$

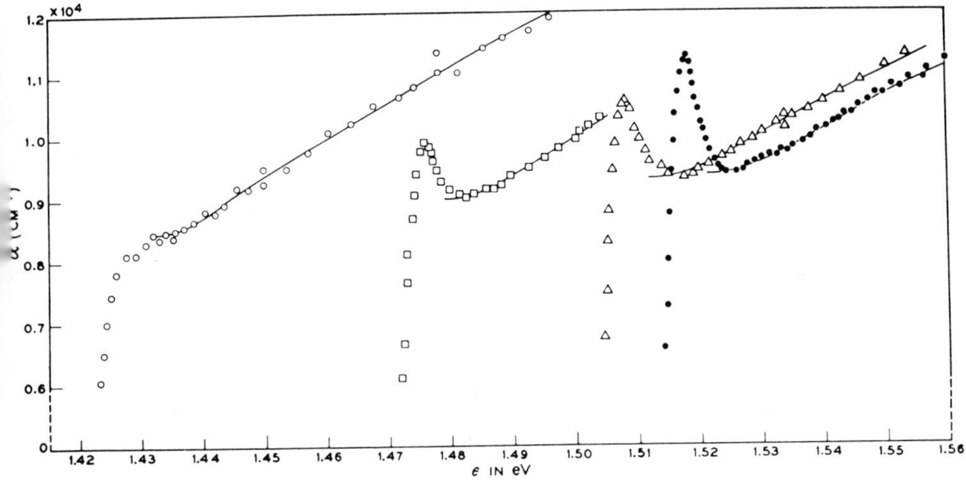

Fig. 6.6.2. The continuum exciton peak at the onset of direct transitions is shown for gallium arsenide. The measurements are for temperatures of 294°K (○), 180°K (□), 90°K (△), and 21°K (●). [From Sturge (1962).]

where

$$k^2 = (2\mu/\hbar^2)(E - E_0 - E_g), \qquad \gamma = \mu e^2/\kappa\hbar^2 k, \tag{6.6.53}$$

and Φ is a confluent hypergeometric function. The enhancement of the absorption coefficient through the Coulomb interaction is expressed by the factor (Elliott, 1957)

$$[F(0,0)]^2 = e^{\pi\gamma}\,|\,\Gamma(1 - i\gamma)\,|^2 = \pi\gamma e^{\pi\gamma}/\sinh \pi\gamma. \tag{6.6.54}$$

Figure 6.6.2 shows the optical absorption of gallium arsenide according to the measurements of Sturge (1962). The peak close to the band edge is attributed to the Coulomb enhancement described by (6.6.54).

The line spectrum must be treated slightly differently. In this case, we can define a total absorption strength of the line. Only s states can contribute in the limit we are considering. For a level of principal quantum number n, this will give an absorption strength proportional to

$$(1/\Omega)\,|\,F(0,0)\,|^2 = 1/\pi a^3 n^3 \tag{6.6.55}$$

where $a = \kappa\hbar^2/\mu e^2$ is the Bohr radius of the exciton.

The theory of optical absorption by excitons can be expressed in a more general form which is free of the limitation of the previous discussion to a loosely bound hydrogenic system (Toyozawa *et al.*, 1967; Hermanson, 1968).

It follows from Eq. (6.6.46) and the discussion of Section 6.5.2 that the absorption coefficient of the crystal is

$$\alpha = (2\pi/\Omega)(e^2/m^2 n_0 c\omega) S(\hbar\omega) \tag{6.6.56}$$

where

$$S(\hbar\omega) = \sum \delta(\hbar\omega - E) \mid \sum_{\rho} F_{\mathrm{E}}(0, \mathbf{R}_\rho)(c\rho \mid \boldsymbol{\varepsilon}\cdot\mathbf{p} \mid \mathrm{v}0) \mid^2. \tag{6.6.57}$$

The first summation includes all exciton (or two-particle) states of energy E and total wave vector zero. Note that a factor $(2\pi)^{3/2}/\Omega^{1/2}$ has been included in F. We will concentrate our attention on the function S. Let us consider $F_{\mathrm{E}}(0, \mathbf{R}_\rho)$ as a transformation coefficient $\langle\rho \mid E\rangle$, where $\mid E\rangle$ is used to represent the exciton state. Also, let $\boldsymbol{\varepsilon}\cdot\mathbf{p} \mid \mathrm{v}0\rangle$ be denoted by a ket $\mid \mu\rangle$. Equation (6.6.57) can be written in a simplified way using matrix notation

$$S(\hbar\omega) = -(1/\pi) \lim_{\epsilon\to 0^+} \mathrm{Im}\langle\mu \mid 1/(\hbar\omega + i\epsilon - H) \mid \mu\rangle. \tag{6.6.58}$$

The Hamiltonian H referred to here is that determining the energy of the electron–hole pair with $\mathbf{K} = 0$, and may be obtained from Eq. (6.6.14). We will also introduce a Hamiltonian H_0 which describes an electron–hole pair when interaction is neglected. Let $\mathcal{G}^{(0)}$ be the Green's function for H_0:

$$\mathcal{G}^{(0)}(z) = (z - H_0)^{-1}. \tag{6.6.59}$$

Also, let $\mathcal{G}(z)$ refer to the full Hamiltonian H:

$$\mathcal{G}(z) = (z - H)^{-1}. \tag{6.6.60}$$

The elements of $\mathcal{G}^{(0)}$ are given on the local basis by

$$\langle\rho \mid \mathcal{G}^{(0)}(z) \mid \tau\rangle = \mathcal{G}^{(0)}(\mathbf{R}_\tau - \mathbf{R}_\rho, z)$$

$$= [\Omega/(2\pi)^3] \int d^3k/\{z - [E_{\mathrm{c}}(\mathbf{k}) - E_{\mathrm{v}}(\mathbf{k})]\}. \tag{6.6.61}$$

We have as an identity

$$\mathcal{G}(z) = \mathcal{G}^{(0)}(z) + \mathcal{G}^{(0)}(z)t(z)\mathcal{G}^{(0)}(z), \tag{6.6.62}$$

in which t is the two-particle t matrix which satisfies

$$t(z) = V + V\mathcal{G}^{(0)}(z)t. \tag{6.6.63}$$

Here V represents the electron–hole interaction. Equation (6.6.61) is substituted into (6.6.58):

$$S(\hbar\omega) = S^{(0)}(\hbar\omega) - (1/\pi) \lim_{\epsilon\to 0^+} \mathrm{Im}\langle\mu \mid \mathcal{G}^{(0)}t\mathcal{G}^{(0)} \mid \mu\rangle. \tag{6.6.64}$$

The first term $S^{(0)}$ evidently determines the absorption for noninteracting particles and the second represents the correction due to their interaction. The t matrix can be found according to the procedures of Section 5.2. Bound exciton states appear as poles of the t matrix. Equation (6.6.63) allows for the possibility of exciton resonances in the continuum as well (unbound excitons). The formal solution of (6.6.62) is

$$t = V[1/(1 - \mathcal{G}^{(0)}V)]. \tag{6.6.65}$$

As was discussed in Section 5.2, resonances appear as solutions of the equation

$$\det[1 - \mathcal{G}^{(0)}(z)V] = 0. \tag{6.6.66}$$

The real and imaginary parts of the complex energy z for which (6.6.66) is satisfied determine, respectively, the position E_0 and the width Γ of the resonance. Near a resonance, the energy dependence of a typical t matrix element is given by

$$t \approx \Gamma/2/[(E - E_0) + i\Gamma/2]. \tag{6.6.67}$$

If the elements of the matrix $\mathcal{G}^{(0)}$ are slowly varying functions of energy near the resonance, and if the resonance width is narrow, Eq. (6.6.64) shows that this shape of the absorption near such a resonance can be represented by

$$S(\hbar) \approx \{A(\hbar\omega)[\hbar\omega - E_0] + B(\hbar\omega)\Gamma(\hbar\omega)\}/\{[\hbar\omega - E_0]^2 + \tfrac{1}{4}[\Gamma(\hbar\omega)]^2\}. \tag{6.6.68}$$

Here A and B are slowly varying functions of energy which must be determined by a detailed calculation. If the energy dependences of A and B can be neglected near a resonance, the line shape can be described as an asymmetric Lorentzian of width Γ with peak location at E_0.

6.7 Effect of External Fields on Optical Properties

In Sections 6.1 and 6.2, we investigated the effects of external electric and magnetic fields on band structure. The changes in the energy levels of a solid can be studied through the resulting alteration of its optical properties. This has become a very fruitful area for experimentation and has lead to a substantial increase in our knowledge of energy bands.

6.7.1 *Optical Magnetoabsorption*

Application of a magnetic field causes an energy band to separate into a series of Landau levels. We will consider here the direct manifestation of the Landau level structure in optical absorption. The theory was developed by Roth *et al.* (1959) and Burstein *et al.* (1959). A review of theory and experiment has been given by Lax and Mavroides (1967). Other effects of a magnetic field, such as the rotation of the plane of polarization of transmitted light (the Faraday effect), will not be investigated here.

We consider nondegenerate "conduction" and "valence" bands, denoted by indices c and v, and characterized by effective masses m_c^* and m_v^*, respectively. Both bands are assumed to have extrema at $\mathbf{k} = 0$. Let an external field B be applied in the z direction. The energies are

$$E_c = E_g + \hbar\omega_c(l + \tfrac{1}{2}) + (\hbar^2 k_z^2/2m_c^*) + (e\hbar/2m_0)g_c B m_s$$
$$E_v = -\hbar\omega_v(l + \tfrac{1}{2}) - (\hbar^2 k_z^2/2m_v^*) + (e\hbar/2m_0)g_v B m_s. \quad (6.7.1)$$

The quantities ω_c and ω_v are the cyclotron frequencies for the conduction and valence bands, respectively: $\omega_c = eB/m_c^*$. The spin quantum number m_s has values $\pm\frac{1}{2}$. Also, E_g is the energy gap in the absence of the field. The valence and conduction band g factors are g_v and g_c. The maximum of the valence band in the absence of the field has been chosen as the zero of energy. When the field is present the minimum separation between the bands is

$$\Delta E = E_g + \tfrac{1}{2}\hbar(\omega_c + \omega_v) - (\hbar e B/4m_0)(g_c + g_v). \quad (6.7.2)$$

The term involving the cyclotron frequencies is frequently dominant so that the effect of the magnetic field is to increase the energy gap and hence to displace the absorption edge.

We will obtain the selection rules and the absorption constant for the allowed transitions. We will follow the procedure of Section 6.5.2, and consider the effects of the magnetic field on both the matrix element and on the joint density of states. Effects due to spin and spin–orbit coupling will be neglected. In the simplest approximation, the wave functions are products of the periodic part of the Bloch wave functions for a given band (index n) and oscillator wave functions for the Landau levels (index l)

$$\psi_n = u_n(\mathbf{r})F_l^{(n)}(\mathbf{r}). \quad (6.7.3)$$

Let $\mathbf{A}_{\text{ext}}$ be the vector potential of the (steady) external field, which is in the z direction, and let $\mathbf{A}_{\text{rad}}$ be the same function for the radiation field. The Hamiltonian is

$$H = (1/2m)(\mathbf{p} + e\mathbf{A}_{\text{ext}} + e\mathbf{A}_{\text{rad}})^2 + V(\mathbf{r}) \quad (6.7.4)$$

in which V is the periodic potential of the crystal. The wave functions (6.7.3) are approximate solutions of the eigenvalue problem

$$(1/2m)(\mathbf{p} + e\mathbf{A}_{\text{ext}})^2\psi_n + V\psi_n = E_n\psi_n,$$

so that, if the radiation field is weak, the perturbing Hamiltonian H' is

$$H' = (e/m)(p + e\mathbf{A}_{\text{ext}})\cdot\mathbf{A}_{\text{rad}}. \tag{6.7.5}$$

The matrix element H'_{fi}, which appears in the calculations of the absorption coefficient, is given by

$$H'_{\text{fi}} = (e/m)\int_{\infty} u_{\text{c}}^*(\mathbf{r})F_l^{(\text{c})*}(\mathbf{r})[(\mathbf{p} + e\mathbf{A}_{\text{ext}})\cdot\mathbf{A}_{\text{rad}}]u_{\text{v}}(\mathbf{r})F_{l'}^{(\text{v})}(\mathbf{r})\,d^3r \tag{6.7.6}$$

in which c and v refer to the conduction and valence bands, respectively. Since the oscillator functions F_l and the field $\mathbf{A}_{\text{ext}}$ are slowly varying compared to the cell periodic functions u_{c}, u_{v}, it is reasonable to approximate the integral in (6.7.6) by splitting it into two parts as

$$H'_{\text{fi}} = (e/m)\Bigg[\int_{\text{cell}} u_{\text{c}}^*(\mathbf{r})\,\mathbf{p}\cdot\mathbf{A}_{\text{rad}}u_{\text{v}}(\mathbf{r})\,d^3r\int_{\infty} F_l^{(\text{c})*}(\mathbf{r})F_{l'}^{(\text{v})}(\mathbf{r})\,d^3r$$
$$+\int_{\text{cell}} u_{\text{c}}^*(\mathbf{r})u_{\text{v}}(\mathbf{r})\,d^3r\int_{\infty} F^{(\text{c})*}(\mathbf{r})(\mathbf{p} + e\mathbf{A}_{\text{ext}})\cdot\mathbf{A}_{\text{rad}}F_{l'}^{(\text{v})}(\mathbf{r})\,d^3r\Bigg]. \tag{6.7.7}$$

The first term of (6.7.7) is responsible for the ordinary optical transitions. It follows from Eq. (6.2.68) that the functions F are independent of the effective masses of the bands. Furthermore, it will be convenient to suppose consistently with our previous conventions about normalization that

$$\int F_l^{(\text{c})*}(\mathbf{r})F_{l'}^{(\text{v})}(\mathbf{r})\,d^3\mathbf{r} = \delta_{ll'}\,\delta(k_z - k_z')\,\delta(k_x - k_{x'})\,(2\pi)^3/\Omega \tag{6.7.8}$$

(see Section 6.2). Since the functions F_l, F_l' are orthonormal, transitions between bands in which there is no change in the quantum number of the Landau levels are obtained. These are the transitions of interest in the optical problem. The $\Delta l = 0$ selection rule replaces the requirement that the y component of $\mathbf{k}$ be conserved. No change of k_x or k_z is permitted.

Since the cell periodic functions u are also orthogonal, the second term of (6.7.7) induces transitions between Landau levels belonging to the same band. The wave functions F_l, which are given by Eqs. (6.2.58) and (6.2.68), are products of exponentials and simple harmonic oscillator functions. It follows from the momentum matrix elements for these wave functions that the selection rules for transitions produced by the second term are

$$\Delta k_x = 0, \qquad \Delta k_z = 0, \qquad \Delta l = \pm 1. \tag{6.7.9}$$

The condition $\Delta l = \pm 1$ indicates that the energy change in the transition is $\pm\hbar\omega_c$ or $\pm\hbar\omega_v$. This is just cyclotron resonance absorption. The $\Delta l = \pm 1$ selection rule applies, however, only when the radiation field, A_{rad} is uniform over the "orbit." If the field varies appreciably in a distance comparable to the extent of the wave function, the $\Delta l = \pm 1$ selection does not apply, and the transition will occur for any integer value of Δl. Cyclotron resonance absorption will then occur for subharmonics of the fundamental frequency.

Introduction of the dipole approximation in the first term of (6.7.7) leads to

$$H'_{fi} = \delta(k_x - k_x')\,\delta(k_z - k_z')\,\bar{H}_{fi}\,\delta_{ll'} + I_{ll'}\,\delta_{nn'} \qquad (6.7.10)$$

in which $\bar{H}_{fi}$ is given by (6.5.25) and

$$I_{ll'}\,\delta(k_x - k_x')\,\delta(k_z - k_z') = [\Omega/(2\pi)^3] \int F_l(\mathbf{r})(\mathbf{p} + e\mathbf{A}_{ext})\cdot\mathbf{A}_{rad}F_{l'}(\mathbf{r})\,d^3r. \qquad (6.7.11)$$

The term $\delta_{nn'}$ in (6.7.10) indicates that the transition occurs only between levels associated with the same band.

Only the optical transitions will be considered further. We would like to apply Eq. (6.5.24), which determines the transition probability, to calculate the absorption coefficient. However, it is necessary to use the density of states appropriate to a Landau level, which may be obtained from Eq. (6.2.67) except for a factor of 2 required by spin degeneracy. We then have, in place of (6.5.24), for a fixed l (and unit volume)

$$W_l = (eB/2\pi^2\hbar)(2\pi/\hbar) \int dk_z\,\bar{H}^2_{fi}\,\delta[E_g + \hbar\omega_l + (\hbar^2k_z^2/2\mu) - \hbar\omega]$$

$$= (2eB/\pi\hbar^3)(\mu/2)^{1/2}\,|\,\bar{H}_{fi}\,|^2/[\hbar\omega - (E_g + \hbar\omega_l)]^{1/2} \qquad (6.7.12)$$

in which

$$\omega_l = (l + \tfrac{1}{2})(\omega_c + \omega_v) = (l + \tfrac{1}{2})(eB/\mu) \qquad (6.7.13)$$

and

$$\mu^{-1} = m_c^{*-1} + m_v^{*-1}. \qquad (6.7.14)$$

The absorption coefficient is obtained from (6.7.12) by summing over all l for which the denominator is positive. The result is

$$\alpha = K(2\mu)^{1/2}(eB/\omega) \sum_l [\hbar\omega - (E_g + \hbar\omega)]^{-1/2} \qquad (6.7.15)$$

in which $K = e^2\,|\,\boldsymbol{\varepsilon}\cdot\mathbf{p}_{cv}\,|^2/2\pi m^2\hbar^2 n\epsilon_0 c$. The constant K previously appeared in the absorption coefficient (6.5.33) in the absence of a magnetic field.

The onset of strong absorption is shifted from the energy $\hbar\omega = E_g$ for which it occurs in the absence of a field to

$$\hbar\omega = E_g + (\hbar eB/2\mu).$$

This result is in agreement with (6.7.2) since spin effects have been neglected. Provided that this neglect is justified (this requires weak spin–orbit coupling), it is possible to determine μ experimentally by observing the displacement of the absorption threshold in a magnetic field.

The expression for the absorption coefficient contains inverse square root singularities. These result from the fact that the joint density of states is that of a one-dimensional band. In reality, however, the predicted infinite peaks are not observed, not only because of natural linewidth considerations and instrumental resolution but also because, as we observed in Section 6.2, the Landau levels are not arbitrarily sharp. The effects of line broadening have been discussed in the papers cited (Burstein *et al.*, 1959; Roth *et al.*, 1959). The result is that one observes a series of absorption peaks centered approximately around the frequencies for which the denominators vanish. Figure 6.7.1 illustrates this effect in the case of germanium, according to Zwerdling *et al.* (1957).

The effect of an external field on the absorption coefficient for indirect

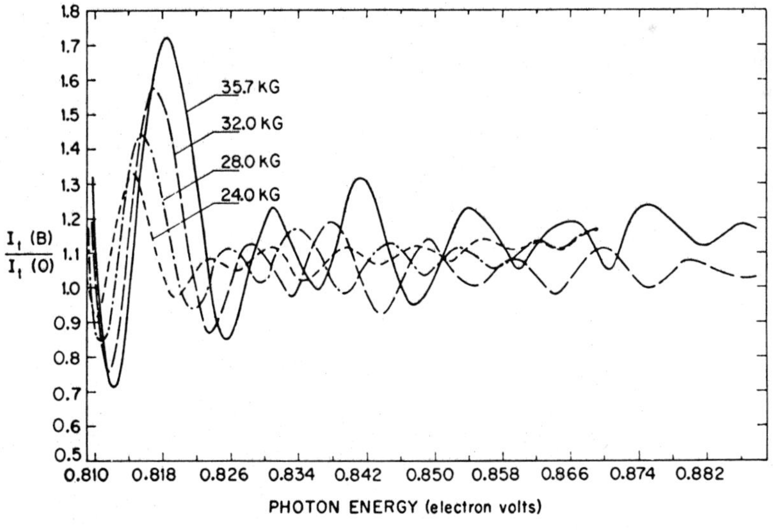

Fig. 6.7.1. Oscillatory magnetoabsorption associated with the direct interband transition in germanium. The ratio of the transmitted light intensity in the presence of the field is shown as a function of photon energy for several different field strengths. The field is parallel to the (1, 1, 1) crystal axis. [From Zwerdling *et al.* (1957).]

transitions has also been determined (Roth *et al.*, 1959). In this case, the absorption does not oscillate but rather exhibits a series of steps. The theory has also been worked out for degenerate bands. Extensive studies of this magnetooptical optical effect have been made on some semiconductors (Burnstein *et al.*, 1959; Zwerdling *et al.*, 1959). There are additional complications caused by spin–orbit coupling and by the complex nature of the valence band structure; however, experiments have been successful in determining accurate values of band parameters (example: the effective mass of the $k = 0$ minimum in germanium).

6.7.2 *Electroabsorption*

The effects of an external electric field on the optical properties of solids are not so striking as in the case of a magnetic field. There are a variety of reasons of both practical and theoretical nature for this: perhaps the most basic is that an electric field destroys the band structure only with respect to a direction parallel to the field, whereas a magnetic field destroys the bands in two dimensions. Our discussion will be confined to semiconductors, since a strong uniform field cannot be maintained over appreciable distances in a metal.

The theory of optical absorption in the presence of an external electric field has been investigated by many authors (Franz, 1958; Keldysh, 1958; Callaway, 1963; Dow and Redfield, 1970; Blossey, 1970). A simple approach due to Tharmalingam (1963); will be followed here. This is based on Eq. (6.6.53), which indicates that in the presence of interactions, the transition probability is multiplied by $|F(0,0)|^2$ [in which $F(0,0)$ is the value of the wave function for a hole–electron pair of zero total wave vector for zero relative coordinate]. Instead of including the electron–hole interaction as was done in the previous section, we here consider the effect of an external electric field. In place of Eq. (6.6.25), we write [with $F(\mathbf{R})$ replacing $F(0, \mathbf{R})$]

$$[-(\hbar^2/2\mu)\nabla^2 - \mathfrak{F}z]F(\mathbf{R}) = \epsilon F(\mathbf{R}). \tag{6.7.16}$$

We have assumed the presence of a uniform force $\mathfrak{F}$ in the z direction. Here μ is the reduced mass of the pair and $\mathcal{E}$ is the difference between the excitation energy and the band gap. Equation (6.7.16) can be solved exactly (see Landau and Lifshitz, 1965, p. 73). The result is

$$F(\mathbf{R}) = A \exp[i(k_x x + k_y y)]\,\mathrm{Ai}(-\xi) \tag{6.7.17}$$

in which Ai is an Airy function, A is a normalization constant,

$$\xi = [z + (\epsilon'/\mathfrak{F})](2\mu\mathfrak{F}/\hbar^2)^{1/2}, \tag{6.7.18}$$

and

$$\epsilon' = \epsilon - (\hbar^2/2\mu)(k_x^2 + k_y^2). \tag{6.7.19}$$

The conventional definition of the Airy function is

$$\mathrm{Ai}(x) = (1/\pi) \int_0^{\infty} \cos(sx + s^3/3)\, ds. \tag{6.7.20}$$

This function decays exponentially for x large and positive

$$\mathrm{Ai}(x) \approx (\tfrac{1}{2}\pi^{-1/2}x^{-1/4}) \exp(-\tfrac{2}{3}x^{3/2}) \tag{6.7.21}$$

and oscillates for large negative values of its argument

$$\mathrm{Ai}(-x) \sim (1/\pi^{-1/2}x^{-1/4}) \sin(2/3 \mid x \mid^{3/2} + \pi/4). \tag{6.7.22}$$

We shall require the normalization constant. This is determined according to the procedure of Landau and Lifshitz. We must note that our definition of Ai differs by a factor of $\pi^{1/2}$ from theirs. In order to be consistent with the normalization used in Section 6.6 we must multiply by a factor of $\Omega^{1/2}(2\pi)^{-3/2}$. This gives

$$A = [\Omega^{1/2}/(2\pi)^{3/2}](2\mu)^{1/3}/\mathfrak{F}^{1/6}\hbar^{2/3}. \tag{6.7.23}$$

Thus we have for $F(0)$

$$F(0) = [\Omega^{1/2}/(2\pi)^{3/2}][(2\mu)^{1/3}/\mathfrak{F}^{1/6}\hbar^{2/3}]\, \mathrm{Ai}(-\sigma\beta^{1/3}) \tag{6.7.24}$$

where

$$\beta = 2\mu/\hbar^2, \qquad \sigma = (1/\mathfrak{F})[(\hbar^2/2\mu)(k_x^2 + k_y^2) - \epsilon]. \tag{6.7.25}$$

The transition matrix element is

$$(e/2m)A_0 \mid \boldsymbol{\varepsilon}_s \cdot \mathbf{p}_{cv} \mid (\beta^{1/3}/\mathfrak{F}^{1/2})\, \mathrm{Ai}(\sigma\beta^{1/3}). \tag{6.7.26}$$

The problem of calculating the transition probability is slightly different in this case, since the band structure is present only with respect to directions of motion perpendicular to the field. To calculate the transition probability, we square (6.7.26), multiply by $2(2\pi/\hbar)$, and integrate over the directions of $\mathbf{k}$ perpendicular to the magnetic field. A factor of $(2\pi)^{-2}$ must accompany this integration. We also must include a delta function of energy conservation $\delta[\epsilon - (\hbar\omega - E_g)]$, with E_g being the band gap. Then we integrate over ϵ. To determine the absorption constant α from the transition probability, we follow the recipe of (6.5.29). The result is

$$\alpha = 2\pi\omega^{-1}K\mu\beta^{1/3} \int_y^{\infty} \mathrm{Ai}^2(t)\, dt \tag{6.7.27}$$

where

$$y = (\beta^{1/3}/\mathfrak{F})(E_g - \hbar\omega). \tag{6.7.28}$$

The Airy function Ai satisfies the differential equation

$$d^2\, \mathrm{Ai}(t)/dt^2 = t\, \mathrm{Ai}(t). \tag{6.7.29}$$

As a result, the integral in (6.7.27) can be performed exactly, and gives

$$\alpha = 2\pi\omega^{-1}K\mu\beta^{1/3}\{d\,\mathrm{Ai}(y)/dy - y\,\mathrm{Ai}^2(y)\}. \qquad (6.7.30)$$

Equation (6.7.30) gives the absorption constant in closed form. There are two interesting limits. Let us first consider large positive values of y, corresponding to photon energies below the band gap. In this case, an explicit result can be obtained using the asymptotic expansion (6.7.21) for the Airy function. This leads to an expression

$$\alpha = (K\mu/4\omega)[\mathfrak{F}/(E_g - \hbar\omega)]\exp\{-\tfrac{4}{3}(2\mu/\hbar^2)^{1/2}[(E_g - \hbar\omega)/\mathfrak{F}]\}. \qquad (6.7.31)$$

No longer is there a sharp cut off in the optical absorption at the band gap. Instead, the absorption exhibits an exponential fall off in the region of the gap, whose steepness increases dramatically with decreasing field strength. This effect has been observed experimentally (for example, see French, 1968), and has come to be known as the Franz–Keldysh effect. Figure 6.7.2a compares theory and experiment in the case of gallium arsenide. Above the gap, y is negative and (6.7.22) must be used. In this case,

$$\alpha = K[(2\mu)^{3/2}/(\hbar\omega)](\hbar\omega - E_g)^{1/2}$$
$$- (K\mu/2\omega)[\mathfrak{F}/(\hbar\omega - E_g)]\cos\{\tfrac{4}{3}(2\mu/\hbar^2)^{1/2}[(E_g - \hbar\omega)/\mathfrak{F}]\}. \qquad (6.7.32)$$

The first term is the zero field absorption, which agrees with (6.5.33). The second term gives rise to oscillatory behavior, which has also been observed.

The absorption coefficient is related to the imaginary part of the dielectric function through Eq. (6.5.74). The real part of this function and hence the index of refraction can be obtained by an analysis using the Kramers–Kronig relations (6.5.14). It follows that the change in the optical absorption produced by an electric field is accompanied by a change in the index of refraction and of the reflectance (Seraphin and Bottka, 1965). An explicit expression can be obtained for this (Aspnes, 1967). This effect was first observed by Seraphin and Hess (1965). It is found that changes in the absorption and the reflectance can be associated with critical points in the joint density of states, not solely with the region around the lowest band gap. This furnishes the basis for a popular and useful method of studying the band structure of semiconductors (Seraphin, 1972; Cardona, 1969; Aspnes and Bottka, 1971).

The preceding treatment neglects the discrete nature of the levels associated with the Stark ladder. When this level structure is included, it is predicted that the optical absorption should show a step-like structure, with the steps being separated by the Wannier level spacing (6.1.37) (Callaway, 1964). Observational evidence for this effect has been provided by Koss and Lambert (1972). Figure 6.7.2b shows the step-like structure observed in galluim arsenide at high fields.

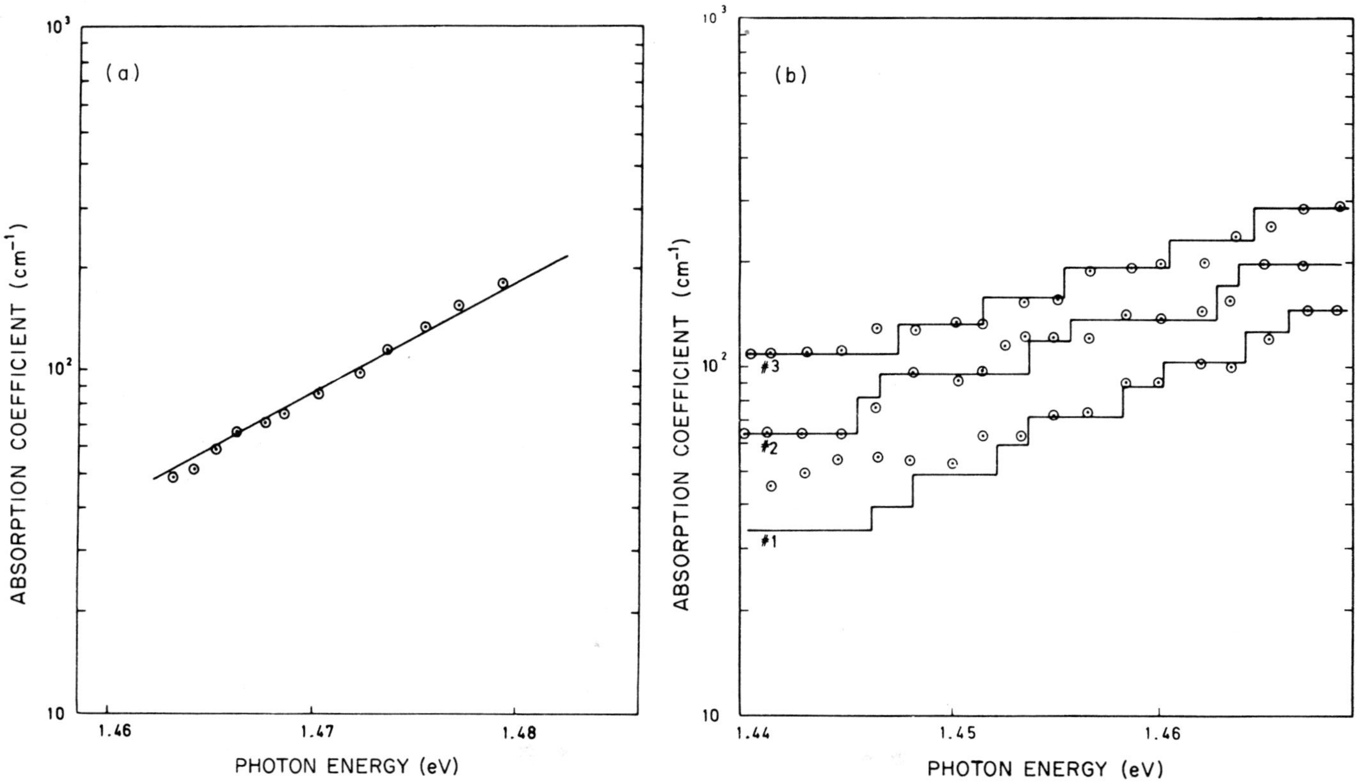

Fig. 6.7.2. Electric field effects on the optical absorption of gallium arsenide. (a) Smearing of the absorption edge at $E = 7.5 \times 10^4$ V/cm for experimental (○) and theoretical (—) values (compare with Fig. 6.6.2); (b) appearance of structure due to Wannier levels (see Section 6.1). #1, $E = 1.1 \times 10^5$ V/cm; #2, $E = 1.45 \times 10^5$ V/cm; #3, $E = 1.6 \times 10^5$ V/cm. [From Koss and Lambert (1972)].

Problems

1 Evaluate the integral (6.1.54) exactly and obtain a corrected form of Eq. (6.1.57).

2 Assume that the vector potential in the Landau level problem is $\mathbf{A} = \frac{1}{2}\mathbf{B} \times \mathbf{r}$ instead of (6.2.5). Solve Eq. (6.2.53) in this case.

3 What is the expected spacing of the Wannier levels in GaAs for a field of 10^5 V/cm if applied in the (a) (1, 0, 0); (b) (1, 1, 0), (c) (1, 1, 1) directions. The lattice constant of GaAs is 5.65 Å.

4 Consider the Hamiltonian for an interacting electron gas in a magnetic field

$$H = \sum_i \mathbf{P}_i^2/2m + \sum_{i>j} V(\mathbf{r}_i - \mathbf{r}_j)$$

in which $\mathbf{P}_i = \mathbf{p}_i + eA(\mathbf{r}_i)$. Construct the total cannonical momentum $\mathbf{P} = \sum_i \mathbf{P}_i$. Define $P_\pm = P_x \pm iP_y$. Show by constructing the commutator $[H, P_\pm]$ that the cyclotron frequency eB/m is not affected by electron–electron interactions (Kohn, 1961).

5 Suppose that a small energy gap E_g separates energy bands in a metal on a face of the Brillouin zone. Show that magnetic breakdown can occur provided that $E_F\hbar\omega_c/E_g^2 > 1$ (Blount, 1962).

6 Show that the real part of the conductivity tensor in the Hartree–Fock approximation should be modified so as to be written

$$\sigma_1 = \pi e^2\{\tfrac{1}{2}\delta(\omega) \sum_l [N_l/(m_l^*)_{\mathrm{op}}] + (1/3m^2\hbar\omega) \sum_{l\mathbf{k},\mathrm{o}} \sum_{n\mathbf{k},\mathrm{u}} |\mathbf{P}_{ln}|^2$$

$$\times [\delta(\omega - \omega_{nl}) - \delta(\omega + \omega_{nl})].$$

Investigate the origin of the term containing $\delta(\omega)$. Show by integration that the sum rule, Eq. (6.5.71), is satisfied.

7 Determine the wavelengths at which sodium, potassium, rubidium, and cesium should become transparent. Assume that the optical effective mass ratios m^*/m_0 are 1.07 (Na), 1.04 (K), 1.03 (Ro), and 1.05 (Cs).

8 Calculate the optical absorption coefficient caused by transitions between hydrogenic donor states and a parabolic conduction band in a semiconductor.

9 Suppose the momentum matrix element for the optical absorption problem is $p_{nl} = c_{nl}k \exp(-\gamma k^2)$ where c_{nl} and γ are constants. Evaluate

the absorption constant for transitions between two bands, $E_l(\mathbf{k}) = -\hbar^2\mathbf{k}^2/2m_l^*$, $E_n(\mathbf{k}) = E_g + \hbar^2\mathbf{k}^2/2m_n^*$; both m_l and m_n^* are positive. Assume that all states in band 1 are filled, all in band n are empty.

10 Consider a crystal with a preferred axis, such that $m^*_\perp \neq m^*_{||}$ for both electrons and holes, and $\kappa_{xx} = \kappa_{yy} = \kappa_\perp$, $\kappa_{zz} = \kappa_{||}$. Determine the energy levels for the 1s, 2s, $2p_0$ and $2p_{\pm1}$ states of a hydrogenic exciton if the departures from isotropy are small (Hopfield and Thomas, 1961).

References

Argyres, P. N. (1962). *Phys. Rev.* **126,** 1386.

Aspnes, D. E. (1967). *Phys. Rev.* **153,** 972.

Aspnes, D. E., and Bottka, N. (1971). Electric field effects on the dielectric functions of semiconductors and insulators. *In* "Semiconductors and Semimetals" (R. K. Willardson and A. Beer, eds.), Vol. 6. Academic Press, New York.

Bardeen, J., Blatt, F. J., and Hall, L. J. (1957). Indirect transitions from the valence to the conduction band. *In* "Photoconductivity Conference" (R. G. Breckenridge *et al.*, eds.), p. 146. Wiley, New York.

Bethe, H., Schweber, S., and de Hoffman, F. (1955). "Mesons and Fields," Vol. 1. Row-Peterson. New York.

Blackman, M. (1938). *Proc. Roy. Soc.* (*London*) **A166,** 1.

Bloembergen, N. (1965). "Nonlinear Optics." Benjamin, New York.

Blossey, D. F. (1970). *Phys. Rev.* **B 2,** 3976.

Blount, E. I. (1962). *Phys. Rev.* **126,** 1636.

Bohm, D. (1951). "Quantum Theory." Prentice-Hall, Englewood Cliffs, New Jersey.

Brown, E. (1964). *Phys. Rev.* **133,** A1038.

Brown, E. (1968). *Solid State Phys.* **22,** 313.

Burstein, E., Picus, G. S., Wallis, R. F., and Blatt, F. (1959). *Phys. Rev.* **112,** 15.

Butcher, P. N., and McLean, T. P. (1963). *Proc. Phys. Soc.* (*London*) **81,** 219.

Butler, F. A., and Brown, E. (1968). *Phys. Rev.* **166,** 630.

Callaway, J. (1963). *Phys. Rev.* **130,** 549.

Callaway, J. (1964). *Phys. Rev.* **134,** A998.

Cardona, M. (1969). "Modulation Spectroscopy" (*Solid State Phys., Suppl.* **11**). Academic Press, New York.

Chambers, W. G. (1965). *Phys. Rev.* **140,** A135.

Cheng, H., and Miller, P. B. (1964). *Phys. Rev.* **134,** A683.

Cohen, M. H., and Falicov, L. M. (1961). *Phys. Rev. Lett.* **7,** 231.

Cohen, M. H., and Heine, V. (1958). *Advan. Phys.* **7,** 395.

Cohen, M. H., and Keffer, F. (1955). *Phys. Rev.* **99,** 1128.

Combescot, M., and Nozières, P. (1971). *J. Phys.* **32,** 913.

Condon, J. H. (1966). *Phys. Rev.* **145,** 526.

Dexter, D. L., and Knox, R. S. (1965). "Excitons." Wiley (Interscience), New York.

Dimmock, J. O. (1967). Introduction to the theory of exciton states. *In* "Semiconductors and Semimetals" (P. K. Willardson and A. C. Beer, eds.), Vol. 3, p. 259. Academic Press, New York.

Dow, J. D., and Redfield, D. (1970). *Phys. Rev.* **B 1,** 3358.

Eby, J. E., Teegarden, K. J., and Dutton, D. B. (1959). *Phys. Rev.* **116,** 1099.

Elliott, R. J. (1957). *Phys. Rev.* **108,** 1384.

Enz, C. P. (1960). *Helv. Phys. Acta* **33,** 89.

Erdelyi, A. (1954). "Table of Integral Transforms," Vol. I. McGraw-Hill, New York.

Everett, P. M. (1972). *Phys. Rev.* **B6,** 3559.

Franz, W. (1958). *Z. Naturforsch.* **13,** 484.

French, B. T. (1968). *Phys. Rev.* **174,** 991.

Frenkel, J. (1931a). *Phys. Rev.* **37,** 17.

Frenkel, J. (1931b). *Phys. Rev.* **37,** 1276.

Glasser, M. L. (1964). *Phys. Rev.* **134,** A1296.

Gobeli, G. W., and Fan, H. Y. (1956). Semiconductor Res. 2nd Quart. Rep., Purdue Univ.

Gold, A. V. (1968). "The De Haas–van Alphen Effect in Solid State Physics" (J. F. Cochran and R. R. Haering, eds.), Vol. 1, p. 39. Gordon and Breach, New York.

de Graaf, A. M., and Overhauser, A. W. (1969). *Phys. Rev.* **180,** 701.

Hacker, K., and Obermair, G. (1970). *Z. Phys.* **234,** 1.

Hamermesh, M. (1962). "Group Theory." Addison Wesley, Reading, Massachusetts.

Hebborn, J. E., and Sondheimer, E. H. (1960). *J. Phys. Chem. Solids* **13,** 105.

Heller, W. R., and Marcus, A. (1951). *Phys. Rev.* **84,** 809.

Hensel, J. C., and Suzuki, J. (1970). "Proceedings of the 10th International Conference on the Physics of Semiconductors." U. S. Atomic Energy Commission, Washington, D.C., p. 541.

Hermanson, J. (1968). *Phys. Rev.* **166,** 893.

Hopfield, J. J., and Thomas, D. G. (1961). *Phys. Rev.* **122,** 35.

Houston, W. V. (1940). *Phys. Rev.* **57,** 184.

Jeffreys, H., and Jeffreys, B. S. (1950). "Methods of Mathematical Physics." Cambridge Univ. Press, London and New York.

Johnson, E. J. (1967). Absorption near the fundamental edge. *In* "Semiconductors and Semimetals" (R. K. Willardson and A. C. Beer, eds.), Vol. 3, p. 154. Academic Press, New York.

Kane, E. O. (1959). *J. Phys. Chem. Solids* **12,** 181.

Keldysh, L. V. (1958). *Zh. Eksperim. Teor. Fiz.* **34,** 1138 (*English transl.: Sov. Phys. JETP* **7,** 788).

Kelley, P. L. (1963). *J. Phys. Chem. Solids* **24,** 607.

Kittel, C. (1958). "Elementary Statistical Physics." Wiley, New York.

Kjeldaas, T., and Kohn, W. (1957). *Phys. Rev.* **105,** 806.

Knox, R. S. (1963). "Theory of Excitons." Academic Press, New York.

Kohn, W. (1961). *Phys. Rev.* **123,** 1242.

Koss, R. W., and Lambert, L. M. (1972). *Phys. Rev.* **B 5,** 1479.

Kramers, H. A. (1927). *Estratto dagli Atti del Congr. Int. Fis.* **2,** 545.

Kronig, R. de L. (1926). *J. Opt. Soc. Amer.* **12,** 547.

Kronig, R. de L. (1942). *Ned. Tijdschn. Natuurk* **9,** 402.

Kubo, R. (1957). *J. Phys. Soc. Japan* **12,** 570.

Landau, L. (1930). *Z. Phys.* **64,** 629.

Landau, L. D., and Lifshitz, E. M. (1965). "Quantum Mechanics," 2nd ed. Addison-Wesley, Reading, Massachusetts.

Lax, B., and Mavroides, J. G. (1967). Interband magnetooptical effects. *In* "Semiconductors and Semimetals" (R. K. Willardson and A. C. Beer, eds.), Vol. 3, p. 321. Academic Press, New York.

Lifshitz, I. M., and Kaganov, M. I. (1959). *Usp. Fiz. Nauk* **69,** 419 (*English Transl.: Sov. Phys. Usp.* **2,** 831).

Lifshitz, I. M., and Kaganov, M. I. (1962). *Usp. Fiz. Nauk* **78,** 411 (*English Transl.: Sov. Phys. Usp.* **5,** 878).
Lifshitz, I. M., and Kosevich, A. M. (1955). *Zh. Ekspeirn. Teoret. Fiz.* **29,** 730 (*English transl.: Sov. Phys. JETP* **2,** 636).
Luttinger, J. M., and Kohn, W. (1955). *Phys. Rev.* **97,** 869.
Macfarlane, G. G., and Roberts, V. (1955). *Phys. Rev.* **94,** 1714.
Macfarlane, G. G., McLean, T. P., Quarrington, J. E., and Roberts, V. (1957). *Phys. Rev.* **108,** 1377.
Misra, P. K., and Kleinman, L. (1972). *Phys. Rev.* **B5,** 4581.
Misra, P. K., and Roth, L. M. (1969). *Phys. Rev.* **177,** 1089.
Morse, P. M., and Feshbach, H. (1953). "Methods of Theoretical Physics." McGraw-Hill, New York.
Onsager, L. (1952). *Phil Mag.* (7) **43,** 1006.
Overhauser, A. W., and de Graaf, A. M. (1969). *Phys. Rev. Lett.* **22,** 127.
Peierls, R. (1933). *Z. Phys.* **80,** 763; **81,** 186.
Peterson, G. A. (1960). Thesis, Cornell Univ., Ithaca, New York (unpublished).
Pippard, A. B. (1962). *Proc. Roy. Soc.* (*London*) **A270,** 7.
Pippard, A. B. (1963). *Proc. Roy. Soc.* (*London*) **A272,** 192.
Pippard, A. B. (1964). *Phil. Trans. Roy. Soc.* (*London*) **A256,** 317.
Roth, L. M. (1960). *Phys. Rev.* **118,** 1534.
Roth, L. M. (1962). *J. Phys. Chem. Solids* **23,** 433.
Roth, L. M. (1966). *Phys. Rev.* **145,** 434.
Roth, L. M., Lax, B., and Zwerdling, S. (1959). *Phys. Rev.* **114,** 90.
Schiff, L. I. (1968). "Quantum Mechanics," 3rd ed. McGraw-Hill, New York.
Schweber, S. S. (1961). "An Introduction to Relativistic Quantum Field Theory." Harper, New York.
Seraphin, B. O. (1972). Electroreflectance. *In* "Semiconductors and Semimetals" (R. K. Willardson and A. C. Beer, eds.), Vol. 9. Academic Press, New York.
Seraphin, B. O., and Bottka, N. (1965). *Phys. Rev.* **139,** A560.
Seraphin, B. O., and Hess, R. B. (1965). *Phys. Rev. Lett.* **14,** 138.
Shockley, W. (1972). *Phys. Rev. Lett.* **28,** 349.
Sondheimer, E. H., and Wilson, A. H. (1951). *Proc. Roy. Soc.* **A210,** 173.
Stern, F. (1963). *Solid State Phys.* **15,** 299.
Sturge, M. D. (1962). *Phys. Rev.* **127,** 768.
Tharmalingan, K. (1963). *Phys. Rev.* **130,** 2204.
Thomas, D. G., and Hopfield, J. J. (1961). *Phys. Rev.* **124,** 657.
Toll, J. S. (1956). *Phys. Rev.* **104,** 1760.
Toyozawa, Y., Inoue, M., Inui, T., Okazaki, M., and Hanamura, E. (1967). *J. Phys. Soc. Japan* **22,** 1337.
Viswanathan, K. S. (1965). Ph. D. Thesis, Univ. of Calif. Riverside (unpublished).
Wannier, G. H. (1937). *Phys. Rev.* **52,** 191.
Wannier, G. (1959). "Elements of Solid State Theory." Cambridge Univ. Press, London and New York.
Wannier, G. (1960). *Phys. Rev.* **117,** 432.
Wannier, G. H. (1962). *Rev. Mod. Phys.* **34,** 645.
Wannier, G. H. (1969). *Phys. Rev.* **181,** 1364.
Wannier, G. H., and Fredkin, D. R. (1962). *Phys. Rev.* **125,** 1910.
Wannier, G. H., and Upadhaya, U. N. (1964). *Phys. Rev.* **136,** A803.
Wannier, G. H., and Van Dyke, J. P. (1968). *J. Math. Phys.* **9,** 899.
Wilson, A. H. (1953). "The Theory of Metals," 2nd ed. Cambridge Univ. Press, London and New York.

Yafet, Y. (1963). *Solid State Phys.* **14,** 1.
Zak, J. (1964). *Phys. Rev.* **134,** A1602.
Zak, J. (1968). *Phys. Rev. Lett.* **20,** 1477.
Zak, J. (1969). *Phys. Rev.* **181,** 1366.
Zwerdling, S., Lax, B., and Roth, L. M (1957). *Phys. Rev.* **108,** 1402.
Zwerdling, S., Lax, B., Roth, L. M., and Button, K. J. (1959). *Phys. Rev.* **114,** 80.

CHAPTER 7

Electrons and Phonons

This chapter will be concerned with some properties of solids which depend on the interaction between electrons and phonons. The most obvious of these are the so-called "transport properties," which involve the flow of electricity and of heat: the electrical and thermal conductivities. We will also consider how the electron–phonon interaction influences the calculation of vibrational frequencies and how the electron states themselves are altered by the interaction. The nature of the electron–phonon interaction depends on whether long range forces are involved. In the first part of this chapter, we will be concerned primarily with metals in which the forces have relatively short range. Subsequently, we will study insulators, where the consequences of long range interactions are dramatic. A discussion of superconductivity concludes the chapter.

An important general reference, especially relevant to the early part of this chapter, is Ziman (1960). A more recent presentation has been given by Abrikosov (1972). Specific references will be listed in the individual sections.

7.1 The Electron–Phonon Interaction

In this section, we will consider how the electron–phonon interaction originates, is described, and may be calculated. This subject clearly forms the basis for our subsequent investigations of transport properties and of superconductivity. Review articles concerned with the electron–phonon interaction have been presented by Chester (1961), Sham and Ziman (1963) and Joshi and Rajagopal (1968).

7.1.1 *The Adiabatic Approximation*

The basis for a discussion of the electron–phonon interaction is the adiabatic (or Born–Oppenheimer) approximation, which is also of vital importance in the physics of molecules (Born and Oppenheimer, 1927). We begin by considering the fundamental question: Why is it that the motion of the electrons and of the nuclei can be considered separately in a first approximation? This has been assumed in all of our discussions up to the present. The answer can be stated quite simply. Because the electrons are very light compared to the nuclei, they move much more rapidly, and can follow the slower motions of the nuclei quite accurately. At the same time, the electron distribution determines the potential in which the nuclei move.

Let us consider a system containing ions of mass M with coordinates $\mathbf{X}$ ($\mathbf{X}_\mu$ designates the position of the μth ion) and electrons of mass m with coordinates $\mathbf{r}_i$. It is convenient to consider the tightly bound electrons of an atomic core as firmly attached to their nucleus. The Hamiltonian for the system is

$$H = \sum_\mu [-(\hbar^2 \nabla_\mu^2/2M) + \sum_{\nu>\mu} V_\mathrm{I}(\mathbf{X}_\mu - \mathbf{X}_\nu)] + \sum_i [-(\hbar^2/2m)\nabla_i^2 + \sum_{j>i} (e^2/|\mathbf{r}_i - \mathbf{r}_j|) + \sum_\mu U_{\mathrm{e-I}}(\mathbf{r}_i - \mathbf{X}_\mu)]. \tag{7.1.1}$$

The term $V_\mathrm{I}(\mathbf{X}_\mu - \mathbf{X}_\nu)$ is the interaction potential of the ions with each other, while $U_{\mathrm{e-I}}(\mathbf{r}_i - \mathbf{X}_\mu)$ represents the interaction between an electron at $\mathbf{r}_i$ and an ion at $\mathbf{X}_\mu$. It is natural to consider this Hamiltonian to be the sum of an ionic and an electronic part

$$H = H_\mathrm{I} + H_\mathrm{e} \tag{7.1.2}$$

in which H_I contains the first two terms of (7.1.1)

$$H_\mathrm{I} = \sum_\mu [-(\hbar^2/2M)\nabla_\mu^2 + \sum_{\nu>\mu} V_\mathrm{I}(\mathbf{X}_\mu - \mathbf{X}_\nu)] \tag{7.1.3}$$

and H_e contains the remainder, including the interaction of the electrons with the ions,

$$H_\mathrm{e} = \sum_i [-(\hbar^2/2m)\nabla_i^2 + \sum_{j>i} (e^2/|\mathbf{r}_i - \mathbf{r}_j|) + \sum_\mu U_{\mathrm{e-I}}(\mathbf{r}_i - \mathbf{X}_\mu)]. \tag{7.1.4}$$

The Schrödinger equation for the electrons in the presence of fixed ions is

$$H_\mathrm{e}\Psi(\mathbf{K}, \mathbf{X}, \mathbf{r}) = E(\mathbf{K}, \mathbf{X})\Psi(\mathbf{K}, \mathbf{X}, \mathbf{r}), \tag{7.1.5}$$

in which $\mathbf{K}$ is the total wave vector of the system; $\mathbf{X}$ and $\mathbf{r}$ denote the set of all electronic and ionic coordinates. The energy of the electronic system and the wave function of the electronic state depend on the ionic positions. In practice, we are unable to solve (7.1.5) exactly, as is obvious, and must resort to approximation procedures. At this point, however, it is desirable to proceed as if a complete set of solutions $\Psi(\mathbf{K}, \mathbf{X}, \mathbf{r})$ could be obtained.

The wave function for the entire system of electrons plus ions is to be expanded with respect to the Ψ as basis functions. Let $\mathbf{Q}$ denote the quantum numbers required to specify the total state of the system. The wave function is

$$\Phi(\mathbf{Q}, \mathbf{X}, \mathbf{r}) = \sum_{\mathbf{K}} \chi(\mathbf{Q}, \mathbf{K}, \mathbf{X})\Psi(\mathbf{K}, \mathbf{X}, \mathbf{r}). \tag{7.1.6}$$

The complete wave function Φ must satisfy a Schrödinger equation with the full Hamiltonian of (7.1.1):

$$H\Phi(\mathbf{Q}, \mathbf{X}, \mathbf{r}) = \mathcal{E}(\mathbf{Q})\Phi(\mathbf{Q}, \mathbf{X}, \mathbf{R}). \tag{7.1.7}$$

For convenience, $\mathbf{K}$ will be considered to be a discrete quantity. The electronic functions are assumed to be normalized for all values of $\mathbf{X}$ and are orthogonal with respect to $\mathbf{K}$ for fixed $\mathbf{X}$:

$$\int \psi^*(\mathbf{K}, \mathbf{X}, \mathbf{r})\Psi(\mathbf{K}', \mathbf{X}, \mathbf{r})\, d\mathbf{r} = \delta_{\mathbf{K}'\mathbf{K}}. \tag{7.1.8}$$

Substitute (7.1.6) into the Schrödinger equation for Φ, use (7.1.5), then multiply on the left by $\Psi^*(\mathbf{K}', \mathbf{X}, \mathbf{r})$, and integrate over $\mathbf{r}$. The result is a set of coupled equations for the functions χ of the form

$$\sum_{\mathbf{K}} \{[H_{\mathrm{I}} + E(\mathbf{K}, \mathbf{X})]\,\delta_{\mathbf{K}'\mathbf{K}} + C(\mathbf{K}', \mathbf{K}, \mathbf{X})\}\chi(\mathbf{Q}, \mathbf{K}, \mathbf{X})$$
$$= \mathcal{E}(\mathbf{Q})\chi(\mathbf{Q}, \mathbf{K}', \mathbf{X}). \tag{7.1.9}$$

The operator $C(\mathbf{K}', \mathbf{K}, \mathbf{X})$ has the form

$$C(\mathbf{K}', \mathbf{K}, \mathbf{X}) = -\int \Psi^*(\mathbf{K}', \mathbf{X}, \mathbf{r})\,(\hbar^2/2m) \sum_{\mu} [\nabla_\mu{}^2\Psi(\mathbf{K}, \mathbf{X}, \mathbf{r})$$
$$+ 2\nabla_\mu\Psi(\mathbf{K}, \mathbf{X}, \mathbf{r}) \cdot \nabla_\mu]\, d\mathbf{r}. \tag{7.1.10}$$

In the lowest (or adiabatic) approximation, the coupling term $C(\mathbf{K}', \mathbf{K}, \mathbf{X})$ is ignored. Then (7.1.9) is diagonal; indicating that the energy levels of the system of ions are determined by solving the Schrödinger equation

$$[H_{\mathrm{I}} + E(\mathbf{K}, \mathbf{X})]\chi(\mathbf{Q}, \mathbf{K}, \mathbf{X}) = \mathcal{E}(\mathbf{Q})\chi(\mathbf{Q}, \mathbf{K}, \mathbf{X}). \tag{7.1.11}$$

The Hamiltonian of (7.1.11) is obtained by adding to H_{I}, as given by (7.1.3), the term $E(\mathbf{K}, \mathbf{X})$, which thus is seen to represent a contribution to the potential energy of the ion system. This implies that the potential energy depends on the state of the electrons. However, this dependence should not be strong under usual circumstances; since normal conductivity processes in solids involve a rearrangement of only a few electrons near the Fermi surface.

The term $C(\mathbf{K}', \mathbf{K}, \mathbf{X})$ couples different states of the ionic lattice. It is not diagonal in the electron wave vector and thus involves transitions between differing electronic states. The diagonal components of C contribute an additional term to the leading approximation (7.1.11), which we will not consider here. The matrix element for a simultaneous transition of the lattice and the electronic system is

$$\mathfrak{M}(\mathbf{Q}', \mathbf{K}'; \mathbf{Q}, \mathbf{K}) = \int \chi^*(\mathbf{Q}', \mathbf{K}', \mathbf{X}) C(\mathbf{K}', \mathbf{K}, \mathbf{X}) \chi(\mathbf{Q}, \mathbf{K}, \mathbf{X}) \, d\mathbf{X} \tag{7.1.12}$$

in which the χ are solutions of (7.1.11).

7.1.2 *Energy Band Formulation of the Electron–Phonon Interaction*

An approach based on (7.1.11) and (7.1.12) suffers from the disadvantage of applications to real systems that the basis functions involved depend on the vibrational coordinates (ionic positions) and the determination of such functions is difficult. It is therefore useful to consider an alternative approach which uses a fixed set of basis states.

We begin by supposing that the ions of the crystal lattice execute only small oscillations around their equilibrium positions. An idealized system in which the ions are fixed in these positions is first considered, and the energy levels and wave functions are determined. This still involves, in principle, a many-electron problem; however, we will consider this in a one-electron approximation. This supposes that the energy bands $E_n(\mathbf{k})$ and $\psi_n(\mathbf{k}, \mathbf{r})$ are known. As a result of the oscillations of the ions, the actual crystal potential differs from that which would exist if the ions were located in the equilibrium positions. This difference will be treated as a perturbation.

Let $V(\mathbf{r} - \mathbf{X}_\mu)$ be the potential energy of an electron at $\mathbf{r}$ in the field of a ion at $\mathbf{X}_\mu$: We write $\mathbf{X}_\mu = \mathbf{R}_\mu + \mathbf{u}_\mu$ where $\mathbf{R}_\mu$ is the equilibrium position and $\mathbf{u}_\mu$ is the atomic displacement. In order not to complicate the notation more than necessary, we restrict attention to a crystal with only one atom in each unit cell. Then

$$V(\mathbf{r} - \mathbf{X}_\mu) = V(\mathbf{r} - \mathbf{R}_\mu) - \mathbf{u}_\mu \cdot \nabla V(\mathbf{r} - \mathbf{R}_\mu) + \cdots. \tag{7.1.13}$$

The perturbation potential, including all atoms in the crystal, is

$$V_{\mathrm{p}} = -\sum_\mu \mathbf{u}_\mu \cdot \nabla V(\mathbf{r} - \mathbf{R}_\mu). \tag{7.1.14}$$

This perturbation will produce transitions between one-electron states. The relevant matrix element is

$$M_{n\mathbf{k},l\mathbf{q}} = \int \psi_n^*(\mathbf{k}, \mathbf{r}) V_{\mathrm{p}} \psi_l(\mathbf{q}, \mathbf{r}) \, d^3r. \tag{7.1.15}$$

In order that the changes in phonon system may be properly included, it is convenient to express $\mathbf{u}_\mu$ as an operator on the phonons according to Eqs. (1.4.22), using discrete normalization

$$u_\mu = (\hbar/2\mathfrak{N}M)^{1/2} \sum_{j\mathbf{p}} [\omega_j^{1/2}(\mathbf{p})]^{-1} \mathbf{e}^{(j)}(\mathbf{p}) [\exp(i\mathbf{p}\cdot\mathbf{R}_\mu) a_j(\mathbf{p}) + \exp(-i\mathbf{p}\cdot\mathbf{R}_\mu) a_j^+(\mathbf{p})] \tag{7.1.16}$$

in which j denotes a branch of the phonon spectrum, $\mathbf{e}_j(\mathbf{p})$ is the eigenvector for a vibrational state of wave vector $\mathbf{p}$ and branch j, and a_j^+ (a_j) is a phonon creation (annihilation) operator. We substitute (7.1.14) and (7.1.16) into (7.1.15). Terms of the following form are encountered:

$$\begin{aligned} &\sum_\mu \exp(i\mathbf{p}\cdot\mathbf{R}_\mu) \int_\infty \psi_n^*(\mathbf{k}, \mathbf{r}) \; \nabla V(\mathbf{r} - \mathbf{R}_\mu) \psi_l(\mathbf{q}, \mathbf{r}) \, d^3r \\ &= \sum_\mu \exp[i(\mathbf{q} + \mathbf{p} - \mathbf{k})\cdot\mathbf{R}_\mu] \int \psi_n^*(\mathbf{k}, \mathbf{r}) \; \nabla V(\mathbf{r}) \; \psi_l(\mathbf{q}, \mathbf{r}) \, d^3r \\ &= \sum_s \delta_{\mathbf{q}+\mathbf{p},\mathbf{K}_s+\mathbf{k}} \, \mathbf{C}_{nl}(\mathbf{k}, \mathbf{q}), \end{aligned} \tag{7.1.17}$$

in which

$$\mathbf{C}_{nl}(\mathbf{k}, \mathbf{q}) = \mathfrak{N} \int \psi_n^*(\mathbf{k}, \mathbf{r}) \; \nabla V(\mathbf{r}) \; \psi_l(\mathbf{q}, \mathbf{r}) \, d^3r. \tag{7.1.18}$$

The matrix element becomes

$$\begin{aligned} M_{n\mathbf{k},l\mathbf{q}} = &-(\hbar/\mathfrak{N}M)^{1/2} \sum \{\mathbf{e}^{(j)}(\mathbf{k} - \mathbf{q}) \cdot \mathbf{C}_{nl}(\mathbf{k}, \mathbf{q}) [\omega_j(\mathbf{k} - \mathbf{q})]^{-1/2} \\ &\times [a_j(\mathbf{k} - \mathbf{q}) + a_j^+(\mathbf{q} - \mathbf{k})]\}. \end{aligned} \tag{7.1.19}$$

It is understood in (7.1.19) that if $\mathbf{k} - \mathbf{q}$ lies outside the Brillouin zone, it is to be brought back in by the addition of an appropriate reciprocal lattice vector. As in the discussion of lattice thermal conductivity in Section 1.8, Part A, we may introduce the distinction between normal processes (in which $\mathbf{k} - \mathbf{q}$ is inside the Brillouin zone) and Umklapp processes (in which $\mathbf{k} - \mathbf{q}$ must be brought back into the zone by addition of a reciprocal lattice vector). A graphical description of Umklapp processes is found in Section 1.8.

In the very simple limit in which the Bloch functions are replaced by plane waves

$$\psi_n(\mathbf{k}, \mathbf{r}) = \mathcal{V}^{-1/2} \exp(i\mathbf{k}\cdot\mathbf{r})$$

in which $\mathcal{V}$ is the volume of the system, Eq. (7.1.18) can be simplified. We integrate by parts, and obtain [since $V(\mathbf{r})$ vanishes at large distances]

$$\mathbf{C}_{nl}(\mathbf{k}, \mathbf{q}) = (i/\Omega)(\mathbf{k} - \mathbf{q}) \int V(\mathbf{r}) \exp[i(\mathbf{q} - \mathbf{k})\cdot\mathbf{r}]\, d^3r$$

$$= i(\mathbf{k} - \mathbf{q}) V(\mathbf{k} - \mathbf{q}) \tag{7.1.20}$$

in which Ω is the cell volume and $V(\mathbf{k} - \mathbf{q})$ is a Fourier component of the ionic potential $V(\mathbf{r})$. In this limit, we see that the matrix element M is proportional to $\mathbf{e}^{(j)}(\mathbf{k} - \mathbf{q}) \cdot (\mathbf{k} - \mathbf{q})$. This indicates that only components of the eigenvector parallel to the wave vector difference contribute. This approximation will not be valid for more realistic wave functions.

Inspection of (7.1.19) leads to an expression for the effective electron–phonon interaction in the language of second quantization (see Appendix D). Let $c_n^+(\mathbf{k})$, $c_n(\mathbf{k})$ be creation and annihilation operators for an electron of wave vector $\mathbf{k}$ in band n. These operators obey the anticommutation rules appropriate to fermions; for example (using discrete normalization),

$$c_n^+(\mathbf{k})c_l(\mathbf{q}) + c_l(\mathbf{q})c_n^+(\mathbf{k}) = \{c_n^+(\mathbf{k}), c_l(\mathbf{q})\} = \delta_{nl}\,\delta_{\mathbf{kq}}. \tag{7.1.21}$$

Further, we introduce the field operators $\psi(\mathbf{r})$, $\psi^+(\mathbf{r})$:

$$\psi(\mathbf{r}) = \sum_{n\mathbf{k}} \psi_n(\mathbf{k}, \mathbf{r}) c_n(\mathbf{k}), \quad \text{etc.} \tag{7.1.22}$$

We also define a set of quantities $D_{nl,j}$:

$$D_{nl,j}(\mathbf{k}, \mathbf{q}) = -(\hbar/2M\omega_j(\mathbf{k} - \mathbf{q}))^{1/2}\mathbf{e}^{(j)}(\mathbf{k} - \mathbf{q}) \cdot \mathbf{C}_{nl}(\mathbf{k}, \mathbf{q}). \tag{7.1.23}$$

Then we can write an interaction Hamiltonian for the electron–phonon

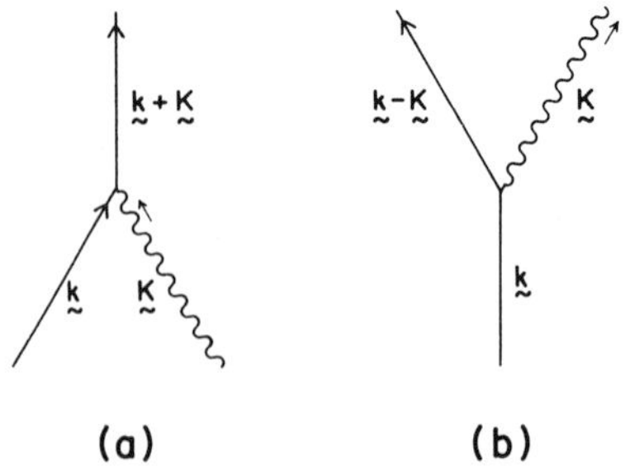

Fig. 7.1.1. Diagrams for phonon (a) absorption and (b) emission.

system in the form

$$H_{\text{el-ph}} = \int \psi^+(\mathbf{r}) V_{\text{p}} \psi(\mathbf{r})\, d^3r \tag{7.1.24}$$

where V_{p} is given by (7.1.14). This can be expressed as

$$\begin{aligned} H_{\text{el-ph}} &= (\mathfrak{N})^{-1/2} \sum_{jnl\mathbf{kq}} D_{nl,j}(\mathbf{k}, \mathbf{q})[c_n^+(\mathbf{k}) c_l(\mathbf{q}) a_j(\mathbf{k} - \mathbf{q}) \\ &\quad + c_n^+(\mathbf{k}) c_l(\mathbf{q}) a_j^+(\mathbf{q} - \mathbf{k})] \\ &= (\mathfrak{N})^{-1/2} \sum_{jnl\mathbf{kK}} [D_{nl,j}(\mathbf{k} + \mathbf{K}, \mathbf{k}) c_n^+(\mathbf{k} + \mathbf{K}) c_l(\mathbf{k}) a_j(\mathbf{K}) \\ &\quad + D_{nl,j}(\mathbf{k} - \mathbf{K}, \mathbf{k}) c_n^+(\mathbf{k} - \mathbf{K}) c_l(\mathbf{k}) a_j^+(\mathbf{K})]. \end{aligned} \tag{7.1.25}$$

In the last portion of (7.1.25), **K** is the wave vector supplied by (or given to) the phonon.

The processes of phonon absorption or emission by an electron can be conveniently represented by the diagrams shown in Fig. 7.1.1 where the wavy line represents the phonon.

The derivation of Eq. (7.1.25) has, however, hidden some significant points. Most importantly, the nature of the ionic potential introduced in (7.1.13) is not clear. The simplest interpretation is that the ions vibrate rigidly (the *rigid ion approximation*), carrying their potentials with them as they move. This is certainly oversimplified, as an important effect is neglected: the electron distribution will distort around the moving ion. This screening effect must be incorporated in the theory. Nonetheless, Eq. (7.1.25) has a phenomenological validity that extends beyond the approximations introduced in its derivation. We will use (7.1.25) freely, regarding $D_{nl,j}$ as a disposable parameter in most of our subsequent work.

7.1.3 *The Deformation Potential*

The problem of determining an effective potential to use in a calculation of the electron–phonon matrix element is quite complicated. In the limit that only phonons whose wavelength is long compared to a lattice spacing are considered, accurate results can be obtained. In this situation, the vibrations of the solid resemble those of an elastic continuum, and their effects can be described in a manner related to the theory of elasticity. The essential concept is that if the solid is subject to a (tensor) strain, $Y_{\alpha\beta}$ say, which is a slowly varying function of position, there will be a change in energy of each electronic state which is proportional to the strain. This leads us to the so-called *deformation potential* (Bardeen and Shockley, 1950). We may write for the change in energy of a state of wave vector **k** in band n

$$\delta E_n(\mathbf{k}) = \sum_{\alpha\beta} \epsilon_{\alpha\beta}(n, \mathbf{k}) Y_{\alpha\beta} \tag{7.1.26}$$

in which $\epsilon_{\alpha\beta}$ is the deformation potential tensor.

The derivation of Eq. (7.1.26) is quite straightforward in the case of a strain which is independent of position. One simply considers a calculation of energy bands in the distorted crystal, either by perturbation theory or directly. As long as the strain is small, the change in an energy level will be proportional to the strain, unless some unlikely analytic catastrophe occurred; hence (7.1.26). The strains of greatest interest are those which preserve the symmetry of the crystal. These are dilitations: expansions and contractions of the volume. In a simple cellular method calculation of energy bands, for example, only the volume of the unit cell is important. Furthermore, we will neglect the possible dependence of $\epsilon_{\alpha\beta}$ on the wave vector **k**. This procedure is valid in a situation in which electrons are restricted to a small portion of the Brillouin zone, as in a ideal semiconductor, and appears to be a reasonable approximation in other cases. Thus (7.1.26) is simplified to

$$\delta E_n(\mathbf{k}) = C_n \Delta \tag{7.1.27}$$

where Δ is the dilitation.

This simple expression may also be applied when Δ is a slowly varying function of position. In this case, δE is to be interpreted as a potential energy function which appears in an effective mass equation of the sort derived in Section 5.3, and simply expresses the fact that the potential energy of an electron will depend on the average atomic volume in the region in which it is located. It must be emphasized that only slowly varying, i.e., long wavelength dililations can be considered in this way.

Let us estimate C for electrons on the Fermi surface in a simple free electron-like metal. In such a system, the Fermi energy ϵ_F is

$$\epsilon_F = (\hbar^2/2m^*)(3\pi^2 N/V)^{2/3}$$

in which N/V is the electron density. Consider now a change in the volume of the region containing N electrons by an amount δV. The change in Fermi energy is

$$\delta\epsilon_F = \partial\epsilon_F/\partial V\, \delta V = -\tfrac{2}{3}(\epsilon_F/V)\, \delta V = -\tfrac{2}{3}\epsilon_F \Delta$$

where $\Delta = \delta V/V$ is the dilitation. Hence

$$C = -\tfrac{2}{3}\epsilon_F \tag{7.1.28}$$

in this simple model.

The deformation potential interaction can be expressed in terms of phonon operators in a manner consistent with (7.1.25). Let $\mathbf{u}_\mu$ be the displacement vector for an element of volume located at $\mathbf{R}_\mu$. The displacement $\mathbf{u}_\mu$ is assumed to be a continuous function of position, and the dilitation is simply

$$\Delta = \nabla \cdot \mathbf{u}_\mu. \tag{7.1.29}$$

We may also interpret $\mathbf{u}_\mu$ as an atomic displacement vector and employ (7.1.16) treating $\mathbf{R}_\mu = \mathbf{r}$ as a continuous variable. Thus

$$\Delta(\mathbf{r}) = i(\hbar/2\mathfrak{N}M)^{1/2} \sum_{j\mathbf{p}} [\mathbf{p}\cdot\mathbf{e}^{(j)}(\mathbf{p})/\omega_j^{1/2}(\mathbf{p})] \times [\exp(i\mathbf{p}\cdot\mathbf{r})a_j(\mathbf{p}) - \exp(-i\mathbf{p}\cdot\mathbf{r})a_j^+(\mathbf{p})]. \tag{7.1.30}$$

We follow the procedure leading to (7.1.25). This time the perturbation V_p to be substituted into (7.1.24) is

$$V_p = C_n\, \Delta(\mathbf{r}).$$

The matrix elements which arise in the process are of the form

$$\langle n\mathbf{k} \mid \exp(\pm i\mathbf{p}\cdot\mathbf{r}) \mid l\mathbf{q}\rangle = \int_\infty \exp[i(\mathbf{q} \pm \mathbf{p} - \mathbf{k})\cdot\mathbf{r}]u_n^*(\mathbf{k}, \mathbf{r})u_l(\mathbf{q}, \mathbf{r})\, d^3r$$

in which $u_n(\mathbf{k}, \mathbf{r})$ is periodic in the unit cell. This can be reduced to

$$\langle n\mathbf{k} \mid \exp(\pm i\mathbf{p}\cdot\mathbf{r}) \mid l\mathbf{q}\rangle = \mathfrak{N} \sum_s \delta_{\mathbf{q}\pm\mathbf{p},\mathbf{k}+\mathbf{K}_s} \int_\Omega \exp(i\mathbf{K}_s\cdot\mathbf{r})u_n^*(\mathbf{k}, \mathbf{r})u_l(\mathbf{q}, \mathbf{r})\, d^3r. \tag{7.1.31}$$

The usual separation into normal and Umklapp processes can be made at this point. However, since the deformation potential theory is valid only for long wavelength phonons, Umklapp processes will be ignored. For the same reason, we will suppose that no transitions between bands are produced. This will be correct except in the neighborhood of a point of degeneracy. With these simplifications, we may observe that, since $\mathbf{q}$ and $\mathbf{k}$ are close together,

$$\mathfrak{N} \int_\Omega u_n^*(\mathbf{k}, \mathbf{r})u_l(\mathbf{q}, \mathbf{r})\, d^3r \approx 1. \tag{7.1.32}$$

The error in (7.1.32) is of order $|\mathbf{k} - \mathbf{q}|^2$, which we will regard as negligible. We now have, for $H_{\text{el-ph}}$,

$$H_{\text{el-ph}} = -\mathfrak{N}^{-1/2} \sum_{nk\mathbf{K}j} [\bar{D}_{nj}(\mathbf{K}) c_n^+(\mathbf{k} + \mathbf{K}) c_n(\mathbf{k}) a_j(\mathbf{K})$$
$$+ \bar{D}_{nj}(-\mathbf{K}) c_n^+(\mathbf{k} - \mathbf{K}) c_n(\mathbf{k}) a_j^+(\mathbf{K})] \qquad (7.1.33)$$

where

$$\bar{D}_{nj}(\mathbf{K}) = i(\hbar/2\omega_j M)^{1/2} \mathbf{K} \cdot \mathbf{e}_j(\mathbf{K}) C_n. \qquad (7.1.34)$$

The general form of these results is consistent with those obtained in subsection 7.1.2; however, the expression for $\bar{D}_n$ is simpler as it contains only a single disposable constant, which can at least be estimated, rather than an unknown function of two variables. The simplification has been purchased at the price, however, of a restriction to long wavelengths.

7.1.4 *Dielectric Screening*

We will now consider a more comprehensive calculation of the electron–phonon interaction in simple metals, which includes the principal effect we have not discussed: the response of the free electrons to the oscillations of the ions. The basic principles of such a calculation were formulated by Bardeen (1937).

The fundamental point is that the electrons readjust themselves to reduce their effective interaction with the vibrating ions. We can begin by considering a perturbation $\delta V_{\text{I}}(\mathbf{r})$ caused by the displacement of the ions alone (the "bare" ions). This causes a change in the electron distribution which in turn produces a change in the potential $\delta V_{\text{e}}(\mathbf{r})$. The interaction V_{p} which must be incorporated in (7.1.24) is the sum of δV_{I} and δV_{e}; and it is this total interaction to which the electrons must respond. Since V_{p} must be used to determine δV_{e} and is partially determined by it, a self-consistent problem results.

To formulate this description in a more precise way, we consider the change in the wave function $\delta\psi_n(\mathbf{k}, \mathbf{r})$ produced by the potential $\delta V_{\text{I}} + \delta V_{\text{e}}$. Our treatment here follows Sham and Ziman (1963). The quantity $\delta\psi_n(\mathbf{k}, \mathbf{r})$ is to be determined using first-order perturbation theory, which gives (using discrete characterization of the states)

$$\delta\psi_n(\mathbf{k}, \mathbf{r}) = \sum_{l\mathbf{q}} \{\langle l\mathbf{q} \mid \delta V_{\text{I}} + \delta V_{\text{e}} \mid n\mathbf{k}\rangle / [E_n(\mathbf{k}) - E_l(\mathbf{q})]\} \psi_l(\mathbf{q}, \mathbf{r}). \qquad (7.1.35)$$

The distortion of the electron wave functions gives rise to a change in the charge density of the electrons, which we denote by $\delta\rho$,

$$\delta\rho(\mathbf{r}) = \sum_{n\mathbf{k}} N_n(\mathbf{k})[\,|\psi_n(\mathbf{k},\mathbf{r}) + \delta\psi_n(\mathbf{k},\mathbf{r})|^2 - |\psi_n(\mathbf{k},\mathbf{r})|^2] \tag{7.1.36}$$

in which $N_n(\mathbf{k})$ is the occupation number of the state $|n\mathbf{k}\rangle$. To first order in $\delta\psi$, we have

$$\begin{aligned}\delta\rho(\mathbf{r}) &= \sum_{n\mathbf{k}} N_n(\mathbf{k})[\psi_n^*\delta\psi_n + \delta\psi_n^*\psi_n] \\ &= \sum_{n\mathbf{k}} N_n(\mathbf{k}) \sum_{l\mathbf{q}} (\{\langle l\mathbf{q}|\,\delta V_{\mathrm{I}} + \delta V_{\mathrm{e}}\,|n\mathbf{k}\rangle/[E_n(\mathbf{k}) - E_l(\mathbf{q})]\}\psi_n^*\psi_l \\ &\quad + \{\langle n\mathbf{k}|\,\delta V_{\mathrm{I}} + \delta V_{\mathrm{e}}\,|l\mathbf{q}\rangle/[E_n(\mathbf{k}) - E_l(\mathbf{q})]\}\psi_l^*\psi_n).\end{aligned} \tag{7.1.37}$$

This equation can be simplified by the interchange of the dummy labels $n\mathbf{k}$ with $l\mathbf{q}$ in the last term

$$\begin{aligned}\delta\rho(\mathbf{r}) &= \sum_{n\mathbf{k},l\mathbf{q}} \{[N_n(\mathbf{k}) - N_l(\mathbf{q})]/[E_n(\mathbf{k}) - E_l(\mathbf{q})]\} \\ &\quad \times \langle l\mathbf{q}|\,\delta V_{\mathrm{I}} + \delta V_{\mathrm{e}}\,|n\mathbf{k}\rangle\psi_n^*\psi_l.\end{aligned} \tag{7.1.38}$$

We may now use Poisson's equation to compute $\delta V_{\mathrm{e}}(\mathbf{r})$ from $\delta\rho(\mathbf{r})$. In mks units

$$\nabla^2\,\delta V_{\mathrm{e}} = -(e^2/\epsilon_0)\,\delta\rho, \tag{7.1.39}$$

in which $\delta\rho$ is given by (7.1.38) and ϵ_0 is the permittivity of free space. To obtain results in cgs units, replace ϵ_0 by $(4\pi)^{-1}$. Since Eqs. (7.1.38) and (7.1.39) are linear, it is possible to solve the combination by introducing a Fourier representation. Furthermore, it is sufficient to consider a perturbation resulting from a single mode of wave vector $\mathbf{p}$. The response can contain components with wave vectors $\mathbf{p} + \mathbf{K}_s$ where $\mathbf{K}_s$ is a reciprocal lattice vector. Thus we write

$$\delta V_{\mathrm{e}}(\mathbf{r}) = \sum_s \delta V_{\mathrm{e}}(\mathbf{p} + \mathbf{K}_s)\exp[i(\mathbf{p} + \mathbf{K}_s)\cdot\mathbf{r}]. \tag{7.1.40}$$

Equation (7.1.40) is substituted into (7.1.39). A similar expansion may be applied to the matrix element

$$\langle l\mathbf{q}|\,\delta V_{\mathrm{I}}\,|n\mathbf{k}\rangle = \sum_s \langle l\mathbf{q}|\exp[i(\mathbf{p} + \mathbf{K}_s)\cdot\mathbf{r}]|n\mathbf{k}\rangle\,\delta V_{\mathrm{I}}(\mathbf{p} + \mathbf{K}_s). \tag{7.1.41}$$

We substitute (7.1.38), (7.1.40), and (7.1.41) into (7.1.39), multiply the result by $\exp[-i(\mathbf{p} + \mathbf{K}_t)\cdot\mathbf{r}]$, and integrate over a volume containing $\mathfrak{N}$

cells of volume Ω. The result is

$$\begin{aligned}\delta V_e(\mathbf{p}+\mathbf{K}_t) &= [e^2/\mathfrak{N}\Omega(\mathbf{p}+\mathbf{K}_t)^2\epsilon_0] \sum_{n\mathbf{k},l\mathbf{q}} \{[N_n(\mathbf{k}) - N_l(\mathbf{q})]/[E_n(\mathbf{k}) - E_l(\mathbf{q})]\} \\ &\quad \times \sum_s \langle l\mathbf{q} \mid \exp[i(\mathbf{p}+\mathbf{K}_s)\cdot\mathbf{r}] \mid n\mathbf{k}\rangle\langle n\mathbf{k} \mid \exp[-i(\mathbf{p}+\mathbf{K}_t)\cdot\mathbf{r}] \mid l\mathbf{q}\rangle \\ &\quad \times [\delta V_{\mathrm{I}}(\mathbf{p}+\mathbf{K}_s) + \delta V_e(\mathbf{p}+\mathbf{K}_s)]. \end{aligned} \tag{7.1.42}$$

We have here a set of simultaneous, inhomogeneous linear equations which determine δV_e in terms of δV_{I}. It is convenient to write these equations in a matrix form as

$$[\mathbf{I} - \mathbf{F}]\,\delta\mathbf{V}_e = \mathbf{F}\,\delta\mathbf{V}_{\mathrm{I}} \tag{7.1.43}$$

in which the rows and columns are labeled by the reciprocal lattice vectors $\mathbf{K}_s$, $\mathbf{K}_t$, and I is a unit matrix. Equation (7.1.43) has the formal solution

$$\delta\mathbf{V}_e = [\mathbf{F}/(I - \mathbf{F})]\,\delta\mathbf{V}_{\mathrm{I}}. \tag{7.1.44}$$

We are interested in the sum of δV_e and δV_{I}:

$$\mathbf{V}_{\mathrm{p}} = \delta\mathbf{V}_{\mathrm{I}} + \delta\mathbf{V}_e = [I/(I - \mathbf{F})]\,\delta\mathbf{V}_{\mathrm{I}}. \tag{7.1.45}$$

The matrix $I - \mathbf{F}$ will be referred to as the *dielectric function matrix* κ:

$$\begin{aligned}\kappa_{ts}(\mathbf{p}) &= \kappa(\mathbf{p}+\mathbf{K}_s, \mathbf{p}+\mathbf{K}_t) \\ &= \delta_{st} - [e^2/\mathfrak{N}\Omega\epsilon_0(\mathbf{p}+\mathbf{K}_t)^2] \sum_{n\mathbf{k},l\mathbf{q}} \{[N_n(\mathbf{k}) - N_l(\mathbf{q})]/[E_n(\mathbf{k}) - E_l(\mathbf{q})]\} \\ &\quad \times \langle l\mathbf{q} \mid \exp[i(\mathbf{p}+\mathbf{K}_s)\cdot\mathbf{r}] \mid n\mathbf{k}\rangle\langle n\mathbf{k} \mid \exp[-i(\mathbf{p}+\mathbf{K}_t)\cdot\mathbf{r}] \mid l\mathbf{q}\rangle. \end{aligned} \tag{7.1.46}$$

The result (7.1.45) can be expressed as

$$V_{\mathrm{p}}(\mathbf{p}+\mathbf{K}_t) = \sum_s [\kappa^{-1}(\mathbf{p})]_{ts}\,\delta V_{\mathrm{I}}(\mathbf{p}+\mathbf{K}_s). \tag{7.1.47}$$

The role of κ may be qualitatively interpreted as follows: Given a perturbation δV_{I}, the electron system responds, producing a change in the potential δV_e which is proportional to δV_{I}. The total potential V_{p} is proportional to δV_{I}, but is reduced by a factor κ^{-1}. This process is quite analogous to polarization in a dielectric which leads to similar screening. For this reason the term "dielectric function" is applied. The considerations here differ from those of Section 6.5 in that we are presently considering a perturbation which varies in space but not in time. In our discussion of optical properties the opposite situation was investigated. We also have here a longitudinal field, rather than a transverse one. We should note that the present considerations may be applied to the response of a solid to any position-dependent perturbation; specifically to the screening of a charge by the system.

Equation (7.1.47) may be used to reconstitute the perturbation potential in $\mathbf{r}$ space, which is to be substituted into (7.1.24):

$$\begin{aligned} V_{\mathrm{P}}(\mathbf{r}) &= \sum_{\mathbf{p},t} V_{\mathrm{P}}(\mathbf{p}+\mathbf{K}_t) \exp[i(\mathbf{p}+\mathbf{K}_t)\cdot\mathbf{r}] \\ &= \sum_{\mathbf{p},s,w} [\kappa^{-1}(\mathbf{p})]_{w+s,s} \exp(i\mathbf{K}_w\cdot\mathbf{r}) \\ &\qquad \times \delta V_{\mathrm{I}}(\mathbf{p}+\mathbf{K}_s) \exp[i(\mathbf{p}+\mathbf{K}_s)\cdot\mathbf{r}]. \end{aligned} \tag{7.1.48}$$

We have introduced $\mathbf{K}_w = \mathbf{K}_t - \mathbf{K}_s$. The presence of the function $[\kappa^{-1}(\mathbf{p})]$ alters the position dependence of the effective potential.

In order to understand what is happening here, it is useful to examine a limiting case in which κ may be computed explicitly. To this end, we consider the limit in which the Bloch functions $\psi_n(\mathbf{k}, \mathbf{r})$ are replaced by plane waves. Since we are using discrete normalization at this point,

$$\langle l\mathbf{q} \mid \exp[i(\mathbf{p}+\mathbf{K}_s)\cdot\mathbf{r}] \mid n\mathbf{k}\rangle = \delta_{\mathbf{q},\mathbf{K}_s+\mathbf{p}+\mathbf{k}}, \tag{7.1.49}$$

and the dielectric function matrix becomes a multiple of the unit matrix:

$$\kappa_{st}(\mathbf{p}) = \kappa(\mathbf{p}+\mathbf{K}_t)\,\delta_{st} \tag{7.1.50}$$

in which (we also abandon the band index)

$$\begin{aligned} \kappa(\mathbf{K}) &= 1 - (e^2/\mathfrak{N}\Omega\mathbf{K}^2\epsilon_0) \\ &\qquad \times \sum_{\mathbf{k}} \{[N(\mathbf{k}) - N(\mathbf{k}+\mathbf{K})]/[E(\mathbf{k}) - E(\mathbf{k}+\mathbf{K})]\}. \end{aligned} \tag{7.1.51}$$

This formula for the dielectric function is frequently attributed to Lindhard (1954); however, it was first obtained by Bardeen (1937). Use of (7.1.50) in (7.1.47) leads to the simplification

$$V_{\mathrm{P}}(\mathbf{p}+\mathbf{K}_t) = \delta V_{\mathrm{I}}(\mathbf{p}+\mathbf{K}_t)/\kappa(\mathbf{p}+\mathbf{K}_t) \tag{7.1.52}$$

so that the Fourier coefficients of V_{P} are simply obtained from those of δV_{I} by dividing by the dielectric function.

It is useful to note at this point an essential difference between semiconductors and metals in regard to the long wavelength behavior of the dielectric function. In a semiconductor, the occupied and the empty states are separated by an energy gap. It will be seen from (7.1.46) that the energy denominator never vanishes. Further, the off-diagonal matrix elements (with respect to the band index) vanish if $\mathbf{p} + \mathbf{K}_s$ approaches zero. In fact they vanish linearly so that κ is finite as $p \to 0$. In a metal, however, empty states adjoin filled states and caution is required. In this case, we have to consider the diagonal (in bands) elements, which are close to 1 (error of order p^2) as $p \to 0$. Thus we may use (7.1.51) for small K regardless of the validity of the single-plane wave model for the wave functions. In the small K limit we have

$$E(\mathbf{k}) - E(\mathbf{k} + \mathbf{K}) = -\mathbf{K} \cdot \nabla_{\mathbf{k}} E(\mathbf{k}),$$

and

$$N(\mathbf{k}) - N(\mathbf{k} + \mathbf{K}) = -\mathbf{K} \cdot \nabla_{\mathbf{k}} N(\mathbf{k}) = -dN/dE\, \mathbf{K} \cdot \nabla_{\mathbf{k}} E(\mathbf{k})$$

since N is a function of energy only. At $T = 0°\mathrm{K}$, dN/dE becomes a delta function, so that we have

$$\kappa(\mathbf{K}) = 1 + [e^2/\Omega\epsilon_0 \mathbf{K}^2] G(E_{\mathrm{F}}) \tag{7.1.53}$$

where $G(E_{\mathrm{F}})$ is the density of states at the Fermi energy. It is indicated that $\kappa(\mathbf{K})$ must become infinite as $K \to 0$. Consider then the screening of a point charge qe which produces a potential

$$\delta V(\mathbf{K}) = qe^2/\mathbf{K}^2\epsilon_0. \tag{7.1.54}$$

The system modifies this to

$$V_{\mathrm{p}} = qe^2/[\epsilon_0 \mathbf{K}^2 + e^2 G(E_{\mathrm{F}})\Omega]. \tag{7.1.55}$$

This expression is valid for small K. There is no longer an infinity at $K = 0$. Equation (7.1.55) can be transformed to position space:

$$V_{\mathrm{p}}(r) = (qe^2/4\pi\epsilon_0 r)e^{-\lambda r} \tag{7.1.56}$$

where

$$\lambda^2 = e^2 G(E_{\mathrm{F}})/\epsilon_0 \Omega. \tag{7.1.57}$$

As K increases, the results depend more strongly on the model (for semiconductors see Penn, 1962; Fry, 1969). Calculations are now possible using realistic band structures (see Walter and Cohen, 1970). We will consider here just a parabolic band structure $E = \gamma \mathbf{k}^2$. Define

$$\begin{aligned} S(K) &= (-1/\mathfrak{N}) \sum_{\mathbf{k}} \{[N(\mathbf{k}) - N(\mathbf{k} + \mathbf{K})]/[E(\mathbf{k}) - E(\mathbf{k} + \mathbf{K})]\} \\ &= (2/\mathfrak{N}) \sum_{\mathbf{k}} \{N(\mathbf{k})/[E(\mathbf{k} + \mathbf{K}) - E(\mathbf{k})]\} \\ &= [2\Omega/(2\pi)^3] \int \{d^3k\, N(\mathbf{k})/[E(\mathbf{k} + \mathbf{K}) - E(\mathbf{k})]\} \end{aligned} \tag{7.1.58}$$

in which Ω is the volume of a unit cell.

For the assumed parabolic band structure, the integral can be done in spherical coordinates with $\mathbf{K}$ as the polar axis. Let $x = \cos\phi_{\mathbf{Kk}}$, and include a factor of 2 for spin:

$$S(K) = [8\pi\Omega/(2\pi)^3\gamma]\int_0^{k_F} k^2\,dk \int_{-1}^{1}[dx/(K^2+2Kkx)]$$

$$= \Omega/2\pi^2\gamma K \int_0^{k_F} k\,dk \ln|(K+2k)/(K-2k)|$$

$$= (3z/4E_F)\{1+[(4k_F^2-K^2)/4k_FK]\ln|(2k_F+K)/(2k_F-K)|\}. \qquad (7.1.59)$$

We have used a relation valid for a free electron metal when there is one atom per unit cell: $\Omega k_F^3 = 3\pi^2 z$ (where z is the valence). The dielectric function is

$$\kappa(\mathbf{K}) = 1 + (e^2/\epsilon_0\mathbf{K}^2\Omega)S(K). \qquad (7.1.60)$$

As $K \to 0$, $S(K)$ approaches $3z/2E_F$. This is consistent with (7.1.53) since the density of states at the Fermi energy for a free electron gas with z electrons per unit cell is just this. We also note that if we take for $\delta V_I(K)$, the Kth Fourier component of the potential energy of an electron in the field of an ion of charge ze, namely $-ze^2/K^2\epsilon_0$, it follows from (7.1.55) that, in the small K limit,

$$\lim_{K\to 0} V_p(K) = -z\Omega/G(E_F) = -\tfrac{2}{3}E_F \qquad (7.1.61)$$

(the last step requiring a free electron model). This is exactly the same as the result obtained in (7.1.28) using the dilitation potential method.

This result for the small K limit of the effective interaction may be extended to larger values of K if we add to the coulomb potential a pseudopotential representing the effect of the ion core. Let the $\mathbf{K}$th Fourier component of a single bare ionic pseudopotential be $U_{psp}(\mathbf{K})$:

$$\delta V_I = (-ze^2/\mathbf{K}^2\epsilon_0) + U_{psp}(\mathbf{K}). \qquad (7.1.62)$$

The Coulomb part of (7.1.62) already includes the contribution from a uniform neutralizing distribution of negative charge (see Section 4.2.3, Part A. Then if the dielectric screening is treated in the free electron model,

$$V_p(\mathbf{K}) = [-ze^2/\epsilon_0\mathbf{K}^2 + U_{psp}(\mathbf{K})]/[1+(e^2/\epsilon_0\mathbf{K}^2\Omega)S(K)]. \qquad (7.1.63)$$

Let the dielectric screening be treated in the small K limit. Then

$$V_p(\mathbf{K}) = [-\tfrac{2}{3}E_F + \lambda^2\mathbf{K}^2U_{psp}(\mathbf{K})]/(1+\lambda^2\mathbf{K}^2) \qquad (7.1.64)$$

where λ^2 is given by (7.1.57). Since $U_{psp}(\mathbf{K})$ should be finite as $K \to 0$, (7.1.61) remains valid. The effective scattering matrix element may now be obtained from (7.1.18) or more crudely in the free electron limit, from (7.1.20), using (7.1.64) to describe the effective potential.

The preceding calculations require improvement in several respects. We have treated the electron–electron interaction in a simple way, amounting to a Hartree approximation. Each electron has been assumed to move independently except for an average field produced by the other electrons. Exchange has been ignored, as have more detailed correlation effects. Joshi and Rajagopal (1968) describe various attempts to improve the model. For more recent results, see Toigo and Woodruff (1970, 1971) and Vashishta and Singwi (1972).

7.1.5 *Calculation of Lattice Vibrational Frequencies*

The preceding investigation of the electron–phonon interaction enables us to compute lattice vibration frequencies. We saw in subsection 7.1.1 that the potential in which the ions move is determined as the sum of the direct ion–ion interaction and the energy of the electronic system, regarded as a function of the ionic positions. Since the dynamical matrix is a function of the second derivatives of the effective ion–ion interaction, it is necessary to obtain the terms in the electronic energy which are of second order in the displacement of the ions from their equilibrium positions.

Our discussion will be based on the simple approach of Woll and Kohn (1962). More inclusive treatments have been given by Vosko *et al.* (1965) and Joshi and Rajagopal (1968).

Let us consider a simple model consisting of an array of ions of charge ze immersed in a uniform electron gas (sometimes called the "jellium model"). Our object is to determine the interaction energy between a pair of ions. Let one of these ions be located at the origin and the other be at $\mathbf{R}$. The energy of interaction of this pair is denoted by $\Phi(\mathbf{R})$. It is convenient to represent this function as a Fourier sum

$$\Phi(\mathbf{R}) = (1/\mathcal{V}) \sum_{\mathbf{p}} \phi(\mathbf{p}) \exp(i\mathbf{p}\cdot\mathbf{R}). \tag{7.1.65}$$

Let the potential energy of an electron in the field of a "bare" ion at the origin be denoted by $V_{\mathrm{I}}(\mathbf{r})$. The quantity $\mathcal{V}$ is the volume of the system. We introduce the Fourier transform of $V_{\mathrm{I}}(\mathbf{r})$ by

$$V_{\mathrm{I}}(\mathbf{p}) = \int V_{\mathrm{I}}(\mathbf{r}) \exp(-i\mathbf{p}\cdot\mathbf{r})\, d^3r. \tag{7.1.66}$$

We may also express the change in the electron (number) density which builds up around the ion at the origin as

$$\delta\rho(\mathbf{r}) = (1/\mathcal{V}) \sum_{\mathbf{p}} \delta\rho(\mathbf{p}) \exp(i\mathbf{p}\cdot\mathbf{r}). \tag{7.1.67}$$

From (7.1.52), (7.1.44), and (7.1.39), the charge density $\delta\rho$ is related to V_{I} by

$$\delta\rho(\mathbf{p}) = \{\epsilon_0\mathbf{p}^2[1 - \kappa(\mathbf{p})]/e^2\kappa(\mathbf{p})\} V_{\mathrm{I}}(\mathbf{p}) \tag{7.1.68}$$

in which $\kappa(\mathbf{p})$ is the dielectric function we have constructed. Use of free electron gas model reduces all quantities in (7.1.44) to scalars. We can obtain the interaction energy of the ion located at $\mathbf{R}$ with this displaced charge density as

$$\int V_{\mathrm{I}}(\mathbf{R} - \mathbf{r})\, \delta\rho(\mathbf{r})\, d^3r. \tag{7.1.69}$$

This gives a contribution to $\Phi(\mathbf{R})$, whose Fourier representation is

$$\{\epsilon_0\mathbf{p}^2[1 - \kappa(\mathbf{p})]/e^2\kappa(\mathbf{p})\} V_{\mathrm{I}}^2(\mathbf{p}). \tag{7.1.70}$$

The direct interaction between the ions can be found from an expression similar to (7.1.69) in which $\delta\rho$ is the charge distribution of the ion itself. This leads to a contribution to $\phi(\mathbf{p})$ of the form

$$\epsilon_0\mathbf{p}^2 V_{\mathrm{I}}^2(\mathbf{p})/e^2. \tag{7.1.71}$$

The sum of (7.1.70) and (7.1.71) gives

$$\begin{aligned}\phi(\mathbf{p}) &= [\epsilon_0\mathbf{p}^2 V_{\mathrm{I}}^2(\mathbf{p})/e^2]\{1 + [1 - \kappa(\mathbf{p})]/\kappa(\mathbf{p})\} \\ &= \epsilon_0\mathbf{p}^2 V_{\mathrm{I}}^2(\mathbf{p})/e^2\kappa(\mathbf{p}).\end{aligned} \tag{7.1.72a}$$

Equation (7.1.72a) expresses the screening of the interaction of the ions by the dielectric function of the free electron gas in which they are immersed. An alternate form of this expression is

$$\phi(\mathbf{p}) = [\epsilon_0\mathbf{p}^2 V_{\mathrm{I}}^2(\mathbf{p})/e^2]\{1 - [e^2S(\mathbf{p})/(\epsilon_0\mathbf{p}^2 + e^2S(\mathbf{p}))]\} \tag{7.1.72b}$$

in which $S(\mathbf{p})$ is given by (7.1.59).

We can now calculate the vibrational frequencies of the ionic lattice. Let the equilibrium position of the σth ion be denoted by $\mathbf{R}_\sigma$ and its deviation from the equilibrium position by $\mathbf{u}_\sigma$. The force on this ion is

$$\begin{aligned}F_\sigma &= -\nabla_\sigma \sum_\tau \Phi(\mathbf{R}_\sigma + \mathbf{u}_\sigma - \mathbf{R}_\tau - \mathbf{u}_\tau) \\ &= (1/\mathcal{V}) \sum_\tau \sum_{\mathbf{p}} \exp[i\mathbf{p}\cdot(\mathbf{R}_\sigma - \mathbf{R}_\tau)]\mathbf{p}[\mathbf{p}\cdot(\mathbf{u}_\sigma - \mathbf{u}_\tau)]\phi(\mathbf{p}).\end{aligned} \tag{7.1.73}$$

We have used the fact that the net force on an atom in its equilibrium position must vanish and have retained only the leading term in the atomic displacements. We now consider the $\mathbf{u}_\sigma$ to be functions of time which obey the equations of motion

$$M\, d^2\mathbf{u}_\sigma/dt^2 = \mathbf{F}_\sigma. \tag{7.1.74}$$

Solutions corresponding to normal modes are obtained if $\mathbf{u}_\sigma$ is expressed as

$$\mathbf{u}_\sigma = \boldsymbol{\varepsilon} \exp[i(\mathbf{q}\cdot\mathbf{R}_\sigma - \omega t)] \tag{7.1.75}$$

in which $\boldsymbol{\varepsilon}$ is a unit vector describing the polarization of the vibrational wave and $\mathbf{q}$ is its wave vector. Equations (7.1.73) and (7.1.75) are substituted into (7.1.74). After some algebra, we find that ω is given by

$$\omega^2 = (N_\mathrm{I}/M) \sum_s [A(\mathbf{K}_s + \mathbf{q}) - A(\mathbf{K}_s)]. \tag{7.1.76}$$

In this equation, N_I is the number of ions per unit volume, $\mathbf{K}_s$ is a reciprocal lattice vector,

$$A(\mathbf{q}) = (\boldsymbol{\varepsilon}\cdot\mathbf{q})^2\phi(\mathbf{q}) = [\epsilon_0\mathbf{q}^4V_\mathrm{I}^2(\mathbf{q})/e^2]B(\mathbf{q}), \tag{7.1.77}$$

and

$$B(\mathbf{q}) = (\mathbf{e}\cdot\mathbf{q}/\mathbf{q})^2\{1 - [e^2S(\mathbf{q})/\epsilon_0\mathbf{q}^2 + e^2S(\mathbf{q})]\}. \tag{7.1.78}$$

Suppose that the ions are effectively point charges. From (7.1.66), we have

$$V_\mathrm{I}(\mathbf{q}) = -ze^2/\epsilon_0\mathbf{q}^2. \tag{7.1.79}$$

We may also introduce the ionic plasma frequency

$$\omega_\mathrm{p}^2 = N_\mathrm{I}(ze)^2/M\epsilon_0. \tag{7.1.80}$$

Then (7.1.76) becomes

$$\omega^2 = \omega_\mathrm{p}^2\{B(\mathbf{q}) + \sum_{s\neq 0} [B(\mathbf{q} + \mathbf{K}_s) - B(\mathbf{K}_s)]\}. \tag{7.1.81}$$

There are two important features of this result. We see from (7.1.81) that were it not for the dielectric screening, the ions would vibrate at their plasma frequency ω_p. As a consequence of the infinity in the dielectric function at $q = 0$, ω goes to zero (proportional to q) as q vanishes. This is as required by the general principles described in Section 1.1, Part A.

As an example, consider longitudinal vibrations in the limit $q \to 0$, and ignore contributions from terms with $K_s \neq 0$. We find

$$\omega = cq \tag{7.1.82}$$

where c, the speed of sound, is given, in the case of a monovalent metal (Bohm and Staver, 1951), by

$$c = 2E_\mathrm{F}/3M = (m/3M)v_\mathrm{F}^2 \tag{7.1.83}$$

where v_F is the Fermi velocity, $E_F = mv_F^2/2$. This formula gives reasonable results. For example, for sodium with $E_F = 3.2$ eV, Eq. (7.1.83) gives $c_s \approx 3.0 \times 10^5$ cm/sec. The experimental values for longitudinal phonons are not greatly different from this (2.9×10^5 cm/sec for the 100 axis, 3.6×10^5 for the 110 axis).

Second, it should be noted that the derivative of the function $S(\mathbf{q})$, given by (7.1.59), has a logarithmic singularity at $q = 2k_F$. The vertical slope of $S(\mathbf{q})$, which is a consequence of the discontinuity in the occupation number of single-particle states at the Fermi energy, is transmitted to the phonon dispersion curves. Thus the phonon spectrum is expected to show a rapid, "anomalous" change in slope, in the vicinity of wavevectors $\mathbf{q}$ such that for some reciprocal lattice vector $\mathbf{K}_s$

$$|\mathbf{q} + \mathbf{K}_s| = 2k_F. \tag{7.1.84}$$

This effect was predicated by Kohn (1959) and has been observed in certain metals [for example, see Brockhouse *et al.* (1962), in regard to Pb and Stedman and Nilsson (1965) concerning Al].

The singular derivative of $S(q)$ has other manifestations as well. The rapid change of S cannot be represented accurately by Fourier series containing only a small number of terms. Consequently, we expect that the Fourier transform of $\phi(\mathbf{p})$, which is $\Phi(\mathbf{r})$, will not fall off rapidly at large r. This implies the existence of a long-range interaction between atoms. For point ions

$$\phi(\mathbf{p}) = (ze)^2/[\epsilon_0\mathbf{p}^2 + e^2S(p)]. \tag{7.1.85}$$

Thus on replacing the sum in (7.1.65) by an integral

$$\begin{aligned}\Phi(\mathbf{r}) &= (2\pi)^{-3}\int \exp(i\mathbf{p}\cdot\mathbf{r})\phi(\mathbf{p})\, d^3p \\ &= [(ze^2)^2/2\pi^2r]\int_0^\infty dp\, p \sin pr/[\epsilon_0 p^2 + e^2S(p)]. \end{aligned} \tag{7.1.86}$$

If only the small p behavior of S is considered, the result is a Yukawa-type potential as discussed previously. The actual integral cannot be done analytically in closed form. An asymptotic expression can, however, be obtained (Langer and Vosko, 1959), according to the methods of Lighthill, (1958). A similar calculation is described in detail in Section 8.4.1. The result is

$$\Phi(\mathbf{r}) = \text{const}\ (\cos 2k_F r)/r^3. \tag{7.1.87}$$

Long range, slowly damped oscillations are present. This long-range interaction precludes accurate application of the force constant techniques of Chapter 1, Part A, to actual metals.

The theory presented here requires substantial improvement for quantitative purposes. In essence what one wishes to do is to calculate the energy of the system to second order in the electron–phonon interaction. Our simple model takes no account of the band structure and, further, does not consider the phonon energies in the calculation of screening by the dielectric function. The essential features are, however, correct even though the quantitative results of this simple approach are not to be trusted.

7.1.6 *Mass Renormalization*

One of the interesting consequences of the existence of the electron–phonon interaction is a change in the effective mass of electrons. This effect is particularly significant near the Fermi surface, and apparently accounts, in large part, for the disagreement between results of energy band calculations of effective masses on the Fermi surface and the results of cyclotron resonance and other similar experiments. The theory of this effect was given by Migdal (1958), and extensions have been made by many authors (see, for example, Joshi and Rajagopal, 1968).

We will use perturbation theory here to calculate the change in energy of a Bloch state $E_n(\mathbf{k})$ to second order in the electron–phonon interaction. We will use the Hamiltonian (7.1.25) applied to a single band.

There is no effect in first order. In second order, we must consider the virtual emission and absorption of a phonon by an electron. We shall furthermore consider only very low temperatures, for which all phonon occupation numbers are zero so that emission must precede absorption. The processes which are considered are represented diagrammatically in Fig. 7.1.2. The general principles of diagrammatic perturbation theory are described in Section 8.1. In (a), a phonon of wave vector $\mathbf{q}$ is emitted by an electron in a state of wave vector $\mathbf{k}$ and is subsequently reabsorbed. The

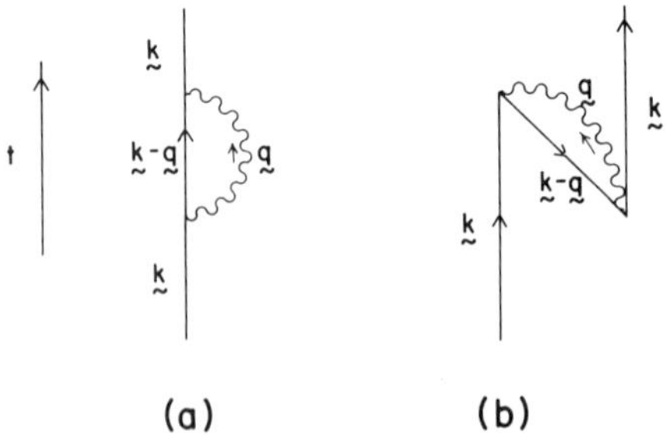

Fig. 7.1.2. Processes contributing to the electron self-energy: (a) direct; (b) exchange.

second process (b) involves emission by an electron in a state of wave vector **k**–**q** leaving a hole in that state, which subsequently annihilates the initial electron with reabsorption of the phonon. The contributions from these diagrams are

$$E_2(\mathbf{k}) = \sum_{j\mathbf{q}} |D_j(\mathbf{q})|^2 \left\{ \frac{1 - f[E(\mathbf{k}-\mathbf{q})]}{E(\mathbf{k}) - E(\mathbf{k}-\mathbf{q}) - \hbar\omega_j(\mathbf{q})} - \frac{f[E(\mathbf{k}-\mathbf{q})]}{E(\mathbf{k}-\mathbf{q}) - \hbar\omega_j(\mathbf{q}) - E(\mathbf{k})} \right\} \tag{7.1.88}$$

in which $f[E]$ is the Fermi function

$$f[E] = [e^{(E-\mu)/KT} + 1]^{-1} \tag{7.1.89}$$

with μ being the chemical potential. The energy of an electron is then $\mathcal{E}(\mathbf{k})$ where

$$\mathcal{E}(\mathbf{k}) = E(\mathbf{k}) + \sum_{\mathbf{q},j} |D_j(\mathbf{q})|^2 \left\{ \frac{1}{E(\mathbf{k}) - E(\mathbf{k}+\mathbf{q}) - \hbar\omega_j(\mathbf{q})} - \frac{2\hbar\omega_j(\mathbf{q}) f[E(\mathbf{k}-\mathbf{q})]}{[E(\mathbf{k}) - E(\mathbf{k}-\mathbf{q})]^2 - [\hbar\omega_j(\mathbf{q})]^2} \right\}. \tag{7.1.90}$$

Equation (7.1.90) gives the energy of an electron in a state of wave-vector **k** in second order in the electron–phonon interaction. The integrals are singular as they stand; however, they are to be interpreted by adding an infinitesimal, positive, imaginary quantity to $E(\mathbf{k})$ [replace $E(\mathbf{k})$ by $E(\mathbf{k}) + i\delta$]. The real part of E_2 is then found by interpreting the sum in (7.1.89) or (7.1.90) as a principal value integral. The imaginary part is proportional to the lifetime. We will consider only the real part at this time.

The first term of (7.1.90) gives rise to a change in the energy which is a slowly varying function of **k**. We are interested in energies close to the Fermi energy; however, there will be a substantial tendency for contributions from regions of **q** above and below the Fermi surface to cancel. The second term is more interesting since there is a sharp cutoff at the Fermi energy. The situation here is quite parallel to that we encountered in discussing the Kondo problem. A schematic diagram of this effect is shown in Fig. 7.1.3.

We will proceed by ignoring the first term (it will be considered later in the polaron problem) and concentrate on the second term. Let us look at the density of states which, in an isotropic situation, is simply propor-

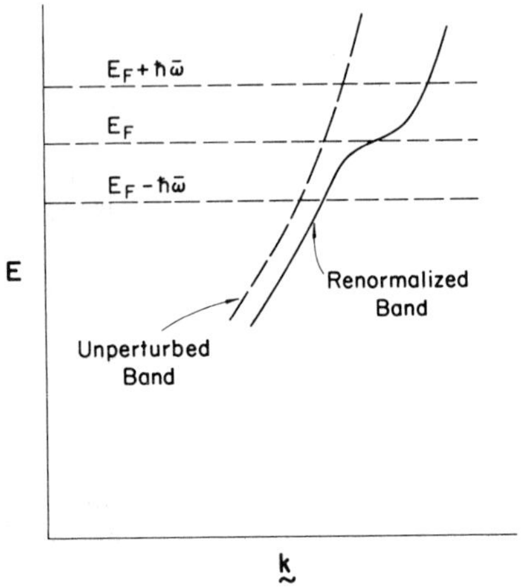

Fig. 7.1.3. Effect of electron–phonon interaction on an energy band (schematic). $\hbar\bar{\omega}$ is an average phonon energy.

tional to $|\nabla_k \mathcal{E}|^{-1}$. The dominant contribution to the derivative comes from the Fermi function present in the second term. The derivative of this function is nearly a delta function. Thus if we are concerned with electron energies $\mathcal{E}(\mathbf{k})$ quite close to the Fermi surface, it is a good approximation to put

$$\nabla_k \mathcal{E}(\mathbf{k}) \approx \nabla_k E(\mathbf{k}) - \sum_{jq} \frac{|D_j(\mathbf{q})|^2 2\hbar\omega_j(\mathbf{q}) \nabla_k f[E(\mathbf{k}-\mathbf{q})]}{[E(\mathbf{k}) - E(\mathbf{k}-\mathbf{q})]^2 - [\hbar\omega_j(\mathbf{q})]^2}$$

$$= \nabla_k E(\mathbf{k})[1 - \alpha(\mathbf{k})]. \tag{7.1.91}$$

A simple form for $\alpha(\mathbf{k})$ is obtained if we introduce the further approximation of replacing $\nabla_k E(\mathbf{k}-\mathbf{q})$ by $\nabla_k E(\mathbf{k})$. Then

$$\alpha(\mathbf{k}) = [2\Omega/(2\pi)^3] \sum_j \int [dS_{k'}/|\nabla E(\mathbf{k}')|]$$

$$\times [|D_j(\mathbf{k}-\mathbf{k}')|^2/\hbar\omega_j(\mathbf{k}-\mathbf{k}')]. \tag{7.1.92}$$

Thus, to first order in α, the modified density of states $G_R(E)$ is related to the usual density G_0 for E close to E_F by

$$G_R(E) = G_0[1 + \bar{\alpha}] \tag{7.1.93}$$

in which $\bar{\alpha}$ is the average of α over a surface of constant energy. This increase in the density of states can be interpreted as a change in the effective mass by the same factor,

$$m_R^* = m_0^*[1 + \bar{\alpha}] \tag{7.1.94}$$

in which m_0^* is the ordinary band mass.

Equation (7.1.92) is suitable for numerical calculation; however, the nature of the approximations which are involved should be remembered. To estimate the orders of magnitude involved, we may express $\bar{\alpha}$ as

$$\bar{\alpha} = \bar{D}^2 G_0(E_F)/\hbar\bar{\omega} \tag{7.1.95}$$

where $\bar{D}$ is an average electron–phonon matrix element and $\hbar\bar{\omega}$ is an average phonon energy. A rough approximation employing the deformation potential approach gives

$$D = \tfrac{2}{3}(\hbar/2M\bar{\omega})^{1/2}(\bar{\omega}/v_s)E_F \tag{7.1.96}$$

where v_s is the velocity of sound. Equation (7.1.95) becomes

$$\bar{\alpha} = \tfrac{2}{9}(E_F/Mv_s^2)E_F G_0(E_F) \approx (z/3)E_F/mv_s^2 \tag{7.1.97}$$

for a free electron metal. For sodium, we evaluate (7.1.97) to give $\bar{\alpha} \approx 0.5$. This is larger by about a factor of 2 than what is inferred from experiment. Typical experimental values of α are in the range of 10 to 30% (see, for example, Lee, 1970, concerning copper).

7.1.7 *Electron–Phonon Interaction in Polar Crystals*

The longitudinal optical modes of an ionic crystal involve opposing displacements of oppositely charged atoms (see Section 1.1, Part A). Such oscillations produce long range dipole fields to which an electron will couple strongly. A slow, itinerant electron wandering through the crystal will disturb the charge balance and will be accompanied by a moving wave of lattice deformation. The coupled system of an electron plus longitudinal optical phonons is referred to as a *polaron*. In this section, we will set up the Hamiltonian describing the electron–phonon interaction under these circumstances. Subsequently, in Section 7.7, we will consider some of the consequences of this interaction.

The electron–phonon interaction as applied to the polaron problem is described by the Fröhlich Hamiltonian (Fröhlich, 1954). Our approach to its derivation will be based on Kittel (1963). Let us consider a longitudinal optical phonon mode of wave vector $\mathbf{q}$, described by creation and annihilation operators $a_\mathbf{q}^+$, $a_\mathbf{q}$. The energy of this mode is $\hbar\omega(\mathbf{q})$. It is customary, however, to note that the frequencies of such phonons do not usually depend strongly on $\mathbf{q}$, and to ignore this dependence, setting $\omega(\mathbf{q}) = \omega$, the energy of an LO phonon at $\mathbf{q} = 0$. The phonon part of the Hamiltonian is then

$$H_{\text{ph}} = \hbar\omega \sum_{\mathbf{q}} a_{\mathbf{q}}^{+} a_{\mathbf{q}}. \tag{7.1.98}$$

We consider the polarization $\mathbf{P}$ in the material. The polarization is produced by the displacement of the atoms from their equilibrium positions. We expand this in phonon operators in a fashion analogous to the atomic displacements. The polarization at a point $\mathbf{r}$ is given by

$$\mathbf{P}(\mathbf{r}) = F\mathfrak{N}^{-1/2} \sum_{\mathbf{q}} \boldsymbol{\varepsilon}_{\mathbf{q}}[a_{\mathbf{q}} \exp(i\mathbf{q}\cdot\mathbf{r}) + a_{\mathbf{q}}^{+} \exp(-i\mathbf{q}\cdot\mathbf{r})]. \tag{7.1.99}$$

Here $\boldsymbol{\varepsilon}_{\mathbf{q}}$ is a unit vector in the direction of $\mathbf{q}$ and F is a constant which will be determined subsequently. Use of (7.1.99) for all values of $\mathbf{r}$ amounts to considering the material to be an elastic continuum. The displacement $\mathbf{D}$ is related to the electric field $\boldsymbol{\mathcal{E}}$ and the polarization $\mathbf{P}$ by (in MKS units)

$$\mathbf{D} = \epsilon_0 \boldsymbol{\mathcal{E}} + \mathbf{P}. \tag{7.1.100}$$

The material contains no free charges, so that

$$\nabla\cdot\mathbf{D} = 0. \tag{7.1.101}$$

Now let the electric field $\boldsymbol{\mathcal{E}}$ be derived from a potential

$$\boldsymbol{\mathcal{E}} = -\nabla\phi \tag{7.1.102}$$

and let the operator ϕ be represented by

$$\phi(\mathbf{r}) = \mathfrak{N}^{-1/2} \sum_{\mathbf{q}} [\phi_{\mathbf{q}} \exp(i\mathbf{q}\cdot\mathbf{r}) + \phi_{\mathbf{q}}^{+} \exp(-i\mathbf{q}\cdot\mathbf{r})]. \tag{7.1.103}$$

To determine $\phi_{\mathbf{q}}$, we substitute (7.1.99), (7.1.103), (7.1.102) into (7.1.100) and then (7.1.101). We obtain

$$-i\epsilon_0 \mathbf{q}\phi_{\mathbf{q}} + F\boldsymbol{\varepsilon}_{\mathbf{q}} a_{\mathbf{q}} = 0. \tag{7.1.104}$$

Since $\boldsymbol{\varepsilon}_{\mathbf{q}}$ is parallel to $\mathbf{q}$, we have

$$\phi_{\mathbf{q}} = -iFa_{\mathbf{q}}/q\epsilon_0, \qquad \phi_{\mathbf{q}}^{+} = iFa_{\mathbf{q}}^{+}/q\epsilon_0. \tag{7.1.105}$$

The interaction energy between an electron at $\mathbf{r}$ and this field is given by

$$\begin{aligned} H'(\mathbf{r}) &= e\phi(\mathbf{r}) \\ &= (-ieF/\mathfrak{N}^{1/2}\epsilon_0) \sum_{\mathbf{q}} (1/q)[a_{\mathbf{q}} \exp(i\mathbf{q}\cdot\mathbf{r}) - a_{\mathbf{q}}^{+} \exp(-i\mathbf{q}\cdot\mathbf{r})]. \end{aligned} \tag{7.1.106}$$

We will now consider the determination of F. To this end we work out the contribution of the electron–phonon interaction to the interaction of a pair of charges located at $\mathbf{r}_1$ and $\mathbf{r}_2$, using second-order perturbation theory (for $T = 0°\mathrm{K}$). Let the particles have charge Q. We consider an interaction Hamiltonian

$$H'' = Q[\phi(\mathbf{r}_1) + \phi(\mathbf{r}_2)]. \tag{7.1.107}$$

In second-order perturbation theory, the change in energy of the system is

$$E_2 = \sum_n \langle 0 \mid H'' \mid n\rangle\langle n \mid H'' \mid 0\rangle/(E_0 - E_n) \tag{7.1.108}$$

in which the excited states of the system are denoted by $\mid n\rangle$. Substituting (7.1.107) in (7.1.108)

$$E_2 = Q^2 \sum_n 1/(E_0 - E_n)\Big[\sum_{i,j=1}^{2} \langle 0 \mid \phi(\mathbf{r}_i) \mid n\rangle\langle n \mid \phi(\mathbf{r}_j) \mid 0\rangle\Big]. \tag{7.1.109}$$

We will neglect the terms in (7.1.109) which correspond to a self-interaction [such as $\mid \langle 0 \mid \phi(\mathbf{r}_1) \mid n\rangle \mid^2$]. Since the ground state at $T = 0$ contains no real phonons, we can obtain only terms involving emission of a single phonon followed by absorption. The energy difference between $\mid n\rangle$ and $\mid 0\rangle$ is $\hbar\omega$. Thus

$$\Delta E_2 = (-2Q^2/\hbar\omega) \sum_{\mathbf{q}} \langle 0 \mid \phi(\mathbf{r}_1) \mid \mathbf{q}\rangle\langle \mathbf{q} \mid \phi(\mathbf{r}_2) \mid 0\rangle. \tag{7.1.110}$$

However,

$$\langle 0 \mid \phi(\mathbf{r}_1) \mid \mathbf{q}\rangle = (-iF/q\epsilon_0\mathfrak{N}^{1/2}) \exp(i\mathbf{q}\cdot\mathbf{r}_1). \tag{7.1.111}$$

Then we have

$$\Delta E_2 = (-2Q^2F^2/\mathfrak{N}\hbar\omega\epsilon_0) \sum_{\mathbf{q}} \exp[i\mathbf{q}\cdot(\mathbf{r}_1 - \mathbf{r}_2)]/q^2. \tag{7.1.112}$$

The sum is converted to an integral

$$\mathfrak{N}^{-1}\sum_{\mathbf{q}} \exp(i\mathbf{q}\cdot\mathbf{r})/q^2 = (2\pi)^{-3} \int d^3q \exp(i\mathbf{q}\cdot\mathbf{r})/q^2 = 1/4\pi r. \tag{7.1.113}$$

Thus (7.1.112) becomes

$$\Delta E_2 = (-F^2/2\pi\epsilon_0{}^2\hbar\omega)(Q^2/\mid \mathbf{r}_1 - \mathbf{r}_2 \mid). \tag{7.1.114}$$

Equation (7.1.114) gives the contribution from polarization of the system to the effective interaction between two fixed test charges. Let ϵ_T be the permittivity of the system, including both the contribution from electronic transitions and the ionic contribution obtained here, and let ϵ_e be the high frequency part of the permittivity, containing the only electronic contribution (see Section 1.3, Part A). The quantity ϵ_T determines the response of the system to a low frequency disturbance, whereas ϵ_e is used if the frequency is much higher than that of the LO phonons. Since the interaction between two charges in MKS units is $Q^2/4\pi\epsilon r$, we may relate ϵ_0 and ϵ_e by

$$1/\epsilon_T = (1/\epsilon_e) - (2F^2/\epsilon_0^2\hbar\omega). \qquad (7.1.115)$$

Let $\epsilon_T = \kappa_0\epsilon_0$ and $\epsilon_e = \kappa_\infty\epsilon_0$, where κ_0 and κ_∞ are the low frequency and high frequency dielectric constants, respectively. Thus

$$F^2 = \tfrac{1}{2}\epsilon_0\hbar\omega[(1/\kappa_\infty) - (1/\kappa_0)]. \qquad (7.1.116)$$

We are now in a position to write the Hamiltonian for an electron interacting with the phonon system. An effective mass approximation is used to remove the periodic potential. Let m^* be the electron effective mass as calculated in band theory, and let $\mathbf{p}$ be its momentum. The Hamiltonian is

$$H = (p^2/2m^*) + \hbar\omega \sum_q a_q^+ a_q + i\,\mathfrak{N}^{-1/2} \sum_q V(\mathbf{q}) \times [a_q^+ \exp(-i\mathbf{q}\cdot\mathbf{r}) - a_q \exp(i\mathbf{q}\cdot\mathbf{r})] \qquad (7.1.117)$$

in which

$$V(\mathbf{q}) = eF/q\epsilon_0 = (e/q)[(\hbar\omega/2\epsilon_0)(1/\kappa_\infty - 1/\kappa_0)]^{1/2}. \qquad (7.1.118)$$

It has become conventional to introduce a dimensionless coupling constant α which is defined by

$$\alpha = (e^2/8\pi\epsilon_0)(1/\hbar\omega)(2m^*\omega/\hbar)^{1/2}[(1/\kappa_\infty) - (1/\kappa_0)]. \qquad (7.1.119)$$

Then $V(\mathbf{q})$ can be written in terms of α as

$$V(\mathbf{q}) = (4\pi\alpha)^{1/2}[(\hbar\omega/q)(\hbar/2m^*\omega)^{1/4}]. \qquad (7.1.120)$$

This rather complicated appearing rearrangement is motivated by the fact that the results of calculations in polaron theory are expressible in a simple form in terms of α. We will see this in Section 7.7 below. Equation (7.1.117) defines the Fröhlich Hamiltonian. A table of values of α for a representative sample of compounds is given subsequently (Section 7.7.1).

7.2 Transport Phenomena

In this section we will study the conduction of heat and electricity by electrons in crystals. Previously we have discussed the conduction of heat by phonons in insulators (Section 1.8, Part A). Important general references include Wilson (1953) Jones (1956), Ziman (1960), Smith *et al.* (1967), Blatt (1968).

7.2.1 *The Boltzmann Equation*

We shall approach transport theory from the point of view of the Boltzmann equation, starting from an individual electron model. We assume that the system can be described in terms of single-particle Bloch functions. The effects of the exclusion principle are contained in the distribution function, which specifies the number of electrons in a given state.

Let us consider an element of volume of the system ΔV which is small compared to the size of the crystal, but large enough to contain many electrons. This element is centered at $\mathbf{r}$. The number of electrons in this volume of spin σ, belonging to band n and having wave vectors in a range d^3k centered about $\mathbf{k}$ is specified by the distribution function $f_{\sigma n}(\mathbf{k}, \mathbf{r})$. It is

$$(\Delta V/8\pi^3) f_{\sigma n}(\mathbf{k}, \mathbf{r})\, d^3k. \tag{7.2.1}$$

It will be sufficient for most of this work to consider only a single band and to restrict attention to systems without magnetization, so that both the spin and the band indices may be suppressed.

In the presence of external fields, the distribution function will change as electrons are accelerated by the field. Further, the distribution function may depend on position, as occurs when a temperature gradient is present, so that there will be a tendency for electrons to diffuse. A third effect which alters the distribution function is scattering: the electrons interact with impurities, phonons, and with each other. We write

$$(\partial f/\partial t)_{\text{tot}} = (\partial f/\partial t)_{\text{fld}} + (\partial f/\partial t)_{\text{diff}} + (\partial f/\partial t)_{\text{scat}} \tag{7.2.2}$$

Under steady state conditions, there is no change in f with time, and (7.2.2) vanishes. Scattering processes counteract the effects of diffusion and of external fields:

$$(\partial f/\partial t)_{\text{fld}} + (\partial f/\partial t)_{\text{diff}} = -(\partial f/\partial t)_{\text{scat}}. \tag{7.2.3}$$

This is the Boltzmann equation in symbolic form. Explicit representations of the terms are constructed as follows: External electric and magnetic fields ($\boldsymbol{\mathcal{E}}$ and $\mathbf{B}$) cause a change in the wave vector of an electron (whose charge is $-e$)‡

$$(\partial f/\partial t)_{\text{fld}} = -d\mathbf{k}/dt \cdot \nabla_{\mathbf{k}} f = (e/\hbar)(\boldsymbol{\mathcal{E}} + \mathbf{v} \times \mathbf{B}) \cdot \nabla_{\mathbf{k}} f. \tag{7.2.4}$$

‡ It was shown in Section 6.1.1 that the rate of change of the average $\mathbf{k}$ of a wave packet in the presence of an electric field $\boldsymbol{\mathcal{E}}$ is, as expected, $\hbar d(\mathbf{k})/dt = -e\boldsymbol{\mathcal{E}}/\hbar$. The extension of this result to the case where a magnetic field is present is possible but is more complicated and will not be described here. The proof is given by Jones and Zener (1934a). It is apparent from the transformation rules for electromagnetic fields as given by special relativity that the desired result must be correct, since a suitable change of the frame of reference can be used to reduce the magnetic force to zero.

The diffusion term produced by a temperature gradient is written as

$$(\partial f/\partial t)_{\text{diff}} = -\mathbf{v}(\mathbf{k}) \cdot \nabla_{\mathbf{r}} f = -(\mathbf{v}(\mathbf{k}) \cdot \nabla T)\, \partial f/\partial T, \qquad (7.2.5)$$

in which $\mathbf{v}(\mathbf{k})$ is the velocity of an electron wave packet centered at $\mathbf{k}$. We now have from (7.2.3)

$$-(e/\hbar)(\boldsymbol{\mathcal{E}} + \mathbf{v} \times B) \cdot \nabla_{\mathbf{k}} f + \mathbf{v} \cdot \nabla T\, df/dT = (\partial f/\partial t)_{\text{scat}}. \qquad (7.2.6)$$

Since the electron and phonon systems are coupled, as was discussed in Section 7.1, we should add to (7.2.6) an equation for the phonon distribution function $n_j(\mathbf{q}, \mathbf{r})$ which similarly specifies the number of phonons in branch j of wave vector $\mathbf{q}$ which are in ΔV. This equation was presented previously as

$$\mathbf{v}_j(\mathbf{q}) \cdot \nabla T\, \partial n_j(\mathbf{q})/\partial T = (\partial n_j/\partial t)_{\text{scat}}. \qquad (1.8.8)$$

Fortunately it is sufficient for most calculations of electronic properties to assume that the phonon distribution is in thermal equilibrium. We shall do this unless the contrary has been stated explicitly.

Assume for the moment that the scattering term in (7.2.6) has been specified and that the Boltzmann equation has been solved. The electric current density $\mathbf{j}$ is found as

$$\mathbf{j} = -(e/4\pi^3) \int \mathbf{v}(\mathbf{k}) f(\mathbf{k}, \mathbf{r})\, d^3k. \qquad (7.2.7)$$

We have included a factor of 2 for spin directions. Likewise, the heat current $\mathbf{Q}$ is expressed as

$$\mathbf{Q} = (1/4\pi^3) \int E(\mathbf{k}) \mathbf{v}(\mathbf{k}) f(\mathbf{k}, \mathbf{r})\, d^3k. \qquad (7.2.8)$$

The central problem connected with the use of the Boltzmann equation is that of obtaining an adequate representation of the collision term $(\partial f/\partial t)_{\text{scat}}$ in (7.2.6). To see what is involved, suppose that the function $W(\mathbf{q}, \mathbf{k})$ specifies the rate at which scattering processes transfer electrons from state $|\,\mathbf{q}\rangle$ to state $|\,\mathbf{k}\rangle$. The rate of change of $f(\mathbf{k})$ due to scattering will be the difference of the rates at which electrons are (1) scattered into $|\,\mathbf{k}\rangle$ from other states and the rate at which (2) they are scattered out of $|\,\mathbf{k}\rangle$ into other states. Process (1) requires that $|\,\mathbf{k}\rangle$ be empty and $|\,\mathbf{q}\rangle$ be occupied, while the reverse is true for (2). Thus

$$[\partial f(\mathbf{k})/\partial t]_{\text{scat}} = \sum_{\mathbf{q}} \{W(\mathbf{q}, \mathbf{k}) f(\mathbf{q}) \times [1 - f(\mathbf{k})] - W(\mathbf{k}, \mathbf{q}) f(\mathbf{k})[1 - f(\mathbf{q})]\}. \qquad (7.2.9)$$

This expression is to be inserted into (7.2.6). The result is a nonlinear integral equation for the distribution function.

The usual circumstances in which we require a solution of the Boltzmann equation involve small external fields and small temperature gradients. In such situations, the distribution function differs by only a small amount from the usual Fermi function, which depends on **k** only through the energy and will be denoted by f_0:

$$f_0(\mathbf{k}) = f_0[E(\mathbf{k})] = \{\exp[E(\mathbf{k}) - \mu]/KT + 1\}^{-1}.$$

It is convenient to define auxiliary functions f_1 and Φ through the relations

$$\begin{aligned} f(\mathbf{k}, \mathbf{r}) = f_0(\mathbf{k}) + f_1(\mathbf{k}, \mathbf{r}) &= f_0(\mathbf{k}) + \Phi(\mathbf{k}, \mathbf{r})\, df_0(\mathbf{k})/dE \\ &= f_0(\mathbf{k}) - (KT)^{-1}\Phi(\mathbf{k}, \mathbf{r}) f_0(\mathbf{k})[1 - f_0(\mathbf{k})]. \end{aligned} \tag{7.2.10}$$

A simple identity involving the Fermi function has been used in the last step of (7.2.10). The function df_0/dE is a delta function at $T = 0°\text{K}$ and varies rapidly at ordinary temperatures. In contrast, the function Φ can be expected to be smooth.

It is natural to treat f_1 and Φ as small quantities. This means that we may try to simplify (7.2.6) by replacing f by f_0 on the left side. However, since

$$\nabla_{\mathbf{k}} f_0(\mathbf{k}) = \hbar \mathbf{v}(\mathbf{k})\, df_0(\mathbf{k})/dE, \tag{7.2.11}$$

the lowest term involving the magnetic field contains f_1. Further, it is evident from physical reasoning that collisions cannot change f_0. This implies that

$$\sum_{\mathbf{q}} \{W(\mathbf{q}, \mathbf{k}) f_0(\mathbf{q})[1 - f_0(\mathbf{k})] - W(\mathbf{k}, \mathbf{q}) f_0(\mathbf{k})[1 - f_0(\mathbf{q})]\} = 0. \tag{7.2.12}$$

This equation will be satisfied if the function

$$P(\mathbf{q}, \mathbf{k}) = W(\mathbf{q}, \mathbf{k}) f_0(\mathbf{q})[1 - f_0(\mathbf{k})] \tag{7.2.13a}$$

is symmetric:

$$P(\mathbf{q}, \mathbf{k}) = P(\mathbf{k}, \mathbf{q}). \tag{7.2.13b}$$

It follows that

$$W(\mathbf{q}, \mathbf{k}) \exp[-E(\mathbf{k})/KT] = W(\mathbf{k}, \mathbf{q}) \exp[-E(\mathbf{q})/KT]. \tag{7.2.13c}$$

Equation (7.2.6) becomes

$$\mathbf{v}(\mathbf{k}) \cdot (\nabla T\, \partial f_0/\partial T - e\mathcal{E}\, df_0/dE) - (e/\hbar)\mathbf{v}(\mathbf{k}) \times \mathbf{B} \cdot \nabla_{\mathbf{k}} f_1 = (\partial f_1/\partial t)_{\text{scat}}. \tag{7.2.14}$$

The term involving the magnetic field in (7.1.14) enters in higher order than is the case for the electric field and the thermal gradient. We can write

$$\partial f_0/\partial T = -[T\, \partial(\mu/T)/\partial T + (E/T)]\, df_0/dE. \tag{7.2.15}$$

Equation (7.1.14) can now be put in the form

$$\mathbf{v}(\mathbf{k}) \cdot [e\mathcal{E} + T\, \nabla_{\mathrm{r}}(\mu/T) + (E/T)\, \nabla T]\, df_0/dE = [-(\partial/\partial t)_{\text{scat}} - (e/\hbar)\mathbf{v}(\mathbf{k}) \times \mathbf{B} \cdot \nabla_{\mathbf{k}}] f_1. \tag{7.2.16}$$

In the determination of the scattering term, corresponding simplifications suggest the deletion of all terms involving two factors of f_1 or Φ:

$$(\partial f/\partial t)_{\text{scat}} = (\partial f_1/\partial t)_{\text{scat}} = (1/KT) \sum_{\mathbf{q}} P(\mathbf{k}, \mathbf{q})[\Phi(\mathbf{k}) - \Phi(\mathbf{q})]. \tag{7.2.17}$$

Equation (7.2.17) is inserted into (7.2.16). The result is a linear integro-differential equation which determines f_1 as a linear function of $\mathcal{E}$ and ∇T with coefficients which are functions of $\mathbf{B}$. Thus the electrical and thermal currents which are calculated from (7.2.7) and (7.2.8) can be expressed in a general way as

$$j_\alpha = \sum_{\beta=1}^{3} \{S_{\alpha\beta}^{(1)}(\mathbf{B})[\mathcal{E}_\beta + (T/e)\, \partial(\mu/T)/\partial x_\beta] + S_{\alpha\beta}^{(2)}(\mathbf{B})\,(1/T)\, \partial T/\partial x_\beta\} \tag{7.2.18a}$$

$$Q_\alpha = \sum_{\beta=1}^{3} \{S_{\alpha\beta}^{(3)}(\mathbf{B})[\mathcal{E}_\beta + (T/e)\, \partial(\mu/T)/\partial x_\beta] + S_{\alpha\beta}^{(4)}(\mathbf{B})\,(1/T)\, \partial T/\partial x_\beta\}. \tag{7.2.18b}$$

Tensor notation has been introduced, in which α and β represent Cartesian components. The object of transport theory is to calculate the tensors $S_{\alpha\beta}^{(i)}$ $(i = 1, \ldots, 4)$.

7.2.2 *Impurity Scattering and Electrical Conductivity*

Since the distribution function satisfies a linear integrodifferential equation, a variety of techniques are available for its solution. For example, variational methods may be employed (Kohler, 1948; Wilson, 1953, Chap. 10). We will, however emphasize applications in which exact solutions can be obtained, at least in limiting cases. One of the most important of these involves elastic scattering by impurities with no internal degrees of freedom.

We will consider a system in which the impurity concentration c is sufficiently small so that the impurities scatter independently. The transition rate between states $|\,\mathbf{k}\rangle$ and $|\,\mathbf{q}\rangle$ is obtained from Eq. (5.2.7). Since only a single band is being considered, the band index is suppressed:

$$W(\mathbf{q},\mathbf{k}) = c[(2\pi)^7/\hbar]\,|\,\langle\mathbf{q}\,|\,t\,|\,\mathbf{k}\rangle\,|^2\,\delta[E(\mathbf{q}) - E(\mathbf{k})], \tag{7.2.19}$$

in which t is the t matrix which satisfies Eq. (5.2.6). In this case, $f_0(\mathbf{k}) = f_0(\mathbf{q})$ for all states which are connected by scattering, and $W(\mathbf{q},\mathbf{k})$ is symmetric [as well as $P(\mathbf{q},\mathbf{k})$]. It is then possible to remove the function df_0/dE from the combination of Eqs. (7.2.16) and (7.2.17). The result is

$$\begin{aligned}\mathbf{v}(\mathbf{k})[e\boldsymbol{\mathcal{E}} + T\,\nabla_{\mathbf{r}}(\mu/T) + (E/T)\,\nabla_{\mathbf{r}}T]\\ = -\sum_{\mathbf{q}} W(\mathbf{q},\mathbf{k})[\Phi(\mathbf{q}) - \Phi(\mathbf{k})] - (e/\hbar)\mathbf{v}(\mathbf{k}) \times \mathbf{B}\cdot\nabla_{\mathbf{k}}\Phi(\mathbf{k})\end{aligned} \tag{7.2.20}$$

since

$$\mathbf{v}(\mathbf{k})\cdot\mathbf{B} \times \nabla_{\mathbf{k}}\,\partial f_0/\partial E = 0.$$

The simplest case to consider is one in which the magnetic field $\mathbf{B}$ vanishes. It will also be assumed that the band is parabolic: $E(\mathbf{k}) = \gamma\mathbf{k}^2$. Surfaces of constant energy are then spheres and $\mathbf{v}(\mathbf{k}) = 2\gamma\mathbf{k}/\hbar$. The t matrix element will be assumed to depend on the angle $\theta_{\mathbf{kq}}$ between $\mathbf{k}$ and $\mathbf{q}$ and not on the orientations of the vectors individually. This will be true for a parabolic band provided that we do not have to consider scattering in partial waves of high angular momentum (see Section 5.2). Denote temporarily the vector in brackets on the left side of (7.2.20) by $-\mathbf{A}$ and convert the sum on $\mathbf{q}$ in the collision term to an integral in the usual way. We have

$$\begin{aligned}\mathbf{v}(\mathbf{k})\cdot\mathbf{A} &= (2\gamma k/\hbar)A\cos\theta_{\mathbf{k}A}\\ &= [1/(2\pi)^3]c[(2\pi)^7/\hbar]\int|\,\langle\mathbf{q}\,|\,t\,|\,\mathbf{k}\rangle\,|^2\,\delta[\gamma\mathbf{q}^2 - \gamma\mathbf{k}^2]\\ &\quad\times[\Phi(\mathbf{q}) - \Phi(\mathbf{k})]\,d^3q.\end{aligned} \tag{7.2.21}$$

The t matrix can be replaced by the usual scattering amplitude (5.2.41). Equation (7.2.21) can be simplified to

$$\cos\theta_{\mathbf{k}A} = c/A\int d\Omega_{\mathbf{q}}\,d\sigma/d\Omega\,(\theta_{\mathbf{kq}})\,[\Phi(k,\Omega_{\mathbf{q}}) - \Phi(k,\Omega_{\mathbf{k}})], \tag{7.2.22}$$

in which $d\sigma/d\Omega$ is the differential scattering cross section. The integration is reduced to one involving only angles. Equation (7.2.21) can now be solved if we make the substitution

$$\Phi(k, \Omega_{\mathbf{k}}) = \lambda(k) \cos \theta_{\mathbf{k}A}, \qquad \Phi(k, \Omega_{\mathbf{q}}) = \lambda(k) \cos \theta_{\mathbf{q},A}. \tag{7.2.23}$$

We have

$$\cos \theta_{\mathbf{q},A} = \cos \theta_{\mathbf{kq}} \cos \theta_{\mathbf{k}A} + \sin \theta_{\mathbf{kq}} \sin \theta_{\mathbf{k}A} \cos(\phi_A - \phi_{\mathbf{q}}).$$

Since $d\sigma/d\Omega$ depends only on $\theta_{\mathbf{kq}}$, the integration over ϕ vanishes and we obtain finally an expression for λ. It is convenient to recognize in the result the so-called "momentum transfer cross section"

$$\sigma_{\mathrm{m}} = 2\pi \int_0^{\pi} d\theta \sin\theta \,(d\sigma/d\Omega)\,(1 - \cos\theta) \tag{7.2.24}$$

and to define a quantity with the units of length which is effectively a mean free path for scattering

$$l(\mathbf{k}) = 1/[c\sigma_{\mathrm{m}}(\mathbf{k})]. \tag{7.2.25}$$

Then we have $\lambda = -Al(\mathbf{k})$. A relaxation time can also be defined by

$$\tau(\mathbf{k}) = l(\mathbf{k})/v(\mathbf{k}) = \hbar l(\mathbf{k})/2\gamma k. \tag{7.2.26}$$

We may rewrite (7.2.23) in the form

$$\Phi(\mathbf{k}) = -\tau(\mathbf{k})\mathbf{v}(\mathbf{k})\cdot\mathbf{A}. \tag{7.2.27}$$

Equation (7.2.27) is an exact solution of the Boltzmann equation. It is established that, under the somewhat restrictive assumptions stated previously, a well-defined relaxation time exists. In other words, if we had made the substitution in (7.2.16) (with $B = 0$),

$$(\partial f^{(1)}/\partial t)_{\mathrm{scat}} = -f^{(1)}/\tau(\mathbf{k}), \tag{7.2.28}$$

the result would have been the same as (7.2.27). Equation (7.2.28) is known as the *relaxation time approximation*. It is exact in the following circumstances: (1) $B = 0$, and (2) elastic scattering. The assumptions of spherical energy surfaces and dependence of the t matrix on $\mathbf{k}-\mathbf{q}$ only can be removed in principle if we are willing to cope with a more complicated integral equation in following the path (7.2.20) → (7.2.21) → (7.2.22).

To determine the electrical conductivity, we make the temperature uniform. Then $\mathbf{A} = -e\boldsymbol{\mathcal{E}}$ and the current density $\mathbf{j}$ is

$$\mathbf{j} = (-e^2/4\pi^3) \int \tau(\mathbf{k})\mathbf{v}(\mathbf{k})\mathbf{v}(\mathbf{k})\cdot\boldsymbol{\mathcal{E}}\, df^{(0)}/dE\, d^3k. \tag{7.2.29a}$$

At low temperatures, we replace $df^{(0)}/dE$ by $-\delta(E-\mu)$, and find

$$\mathbf{j} = (e^2/4\pi^3) \int \tau(\mathbf{k})\mathbf{v}(\mathbf{k})\mathbf{v}(\mathbf{k}) \cdot \boldsymbol{\mathcal{E}}\, dS_\mu/|\nabla E(\mathbf{k})|. \tag{7.2.29b}$$

The integration runs over the Fermi surface. In a system with cubic symmetry, the current must be in the direction of the field, and the conductivity reduces to a scalar $\sigma = S_{\alpha\alpha}^{(1)}$:

$$\sigma = (e^2/12\pi^3) \int \tau(\mathbf{k}) v^2(\mathbf{k})\, dS_\mu/|\nabla E(\mathbf{k})|. \tag{7.2.30}$$

To simplify this further, we suppose that $|\mathbf{v}(\mathbf{k})|$ and $\tau(\mathbf{k})$ are functions of energy only. Then the conductivity is given by

$$\sigma = (e^2/12\pi^3\hbar) l(\mu) S(\mu), \tag{7.2.31a}$$

in which $S(\mu)$ is the area of the Fermi surface. If this is spherical, we have

$$\sigma = (e^2/3\pi^2\hbar) l(k_F) k_F^2 = e^2\tau(k_F) k_F^3/3\pi^2 m^* = ne^2\tau(k_F)/m^*, \tag{7.2.31b}$$

in which n is the electron density and γ has been replaced by $\hbar^2/2m^*$.

Equation (7.2.31b) is exactly the result of elementary kinetic arguments. Specifically, we may consider electrons of mass m^* moving in an external electric field, obeying the equation of motion

$$m^* [d\mathbf{v}/dt + \mathbf{v}/\tau] = -e\boldsymbol{\mathcal{E}}. \tag{7.2.32}$$

The equilibrium velocity is $\mathbf{v} = -e\boldsymbol{\mathcal{E}}\tau/m^*$. The current density is

$$\mathbf{j} = \sigma\boldsymbol{\mathcal{E}} = -ne\mathbf{v} = (ne^2\tau/m^*)\,\boldsymbol{\mathcal{E}},$$

which gives σ in agreement with (31b).

7.2.3 *Thermal Conductivity*

The thermal conductivity of metals is dominated by the contribution from electrons near the Fermi surface. Lattice conductivity is secondary. The problem of calculating the thermal conductivity of a metal in a regime dominated by impurity scattering is considerably simpler than is the case for a phonon system. The essential difference is a consequence of the exclusion principle; while long wave phonons tend to dominate lattice thermal conductivity because they are not scattered efficiently by point defects, long wavelength electrons do not participate. Only electrons near the top of the Fermi sea can contribute, and these have short enough wavelengths to scatter effectively. For a more complete discussion than is presented here, see Klemens (1956) and Mendelssohn and Rosenberg (1961).

For simplicity, consider a cubic crystal in which the thermal conductivity, like the electrical conductivity, is a scalar. Let the thermal gradient be in the z direction. However, in contrast to the situation involving electrical conductivity, we cannot simply set the electric field equal to zero and calculate the heat current. Experimentally, thermal conductivity is measured under circumstances in which no electrical current flows. We see from (7.2.18a) that a thermal gradient acting alone would tend to produce an electric current. This tendency is countered by a small electric field which builds up to cancel the flow of current, and this field must be in turn included in the calculation of the heat current. Since the field will be in the same direction as the thermal gradient, the vector nature of these quantities can be disregarded, and we may proceed by treating both $\mathcal{E}$ and ∇T as scalars.

From (7.2.10) and (7.2.27) we have

$$f^{(1)} = \tau(\mathbf{k})\mathbf{v}(\mathbf{k})\cdot[e\boldsymbol{\mathcal{E}} + T\ \nabla_{\mathbf{r}}(\mu/T) + (E/T)\ \nabla_{\mathbf{r}}T]\ \partial f^{(0)}/\partial E. \qquad (7.2.33)$$

We then find from (7.2.7) and (7.2.8),

$$j = (-e/4\pi^3)\int \mathbf{v}(\mathbf{k})\mathbf{v}(\mathbf{k})\cdot[e\boldsymbol{\mathcal{E}} + T\ \nabla(\mu/T) + (E/T)\ \nabla T]$$
$$\times\ \tau(\mathbf{k})\ \partial f^{(0)}/\partial E\ d^3k$$
$$\mathbf{Q} = (1/4\pi^3)\int E(\mathbf{k})\mathbf{v}(\mathbf{k})\mathbf{v}(\mathbf{k})\cdot[e\boldsymbol{\mathcal{E}} + T\ \nabla(\mu/T) + (E/T)\ \nabla T]$$
$$\times\ \tau(\mathbf{k})\ \partial f^{(0)}/\partial E\ d^3k. \qquad (7.2.34)$$

We now make the simplifications mentioned previously concerning the vector quantities, and replace $\mathbf{vv}$ by $\frac{1}{3}v^2$. It is also convenient to define integrals of the following class:

$$\mathfrak{F}_n = (-1/12\pi^3)\int \tau(\mathbf{k})v^2(\mathbf{k})E^{n-1}(\mathbf{k})\ df^{(0)}/dE\ d^3k. \qquad (7.2.35)$$

Then

$$j = e\{\mathfrak{F}_1[e\mathcal{E} + T\ \nabla(\mu/T)] + (\mathfrak{F}_2/T)\ \nabla T\}$$
$$Q = -\mathfrak{F}_2[e\mathcal{E} + T\ \nabla(\mu/T)] - (\mathfrak{F}_3/T)\ \nabla T. \qquad (7.2.36)$$

Set $j = 0$ to determine $\mathcal{E}$: $e\mathcal{E} = -T\ \nabla(\mu/T) - (\mathfrak{F}_2/\mathfrak{F}_1)\ \nabla T/T$. This is substituted into (7.2.36), which gives

$$Q = [(\mathfrak{F}_2^2/\mathfrak{F}_1) - \mathfrak{F}_3]\ \nabla T/T. \qquad (7.2.37)$$

The thermal conductivity κ is given through $Q = -\kappa\nabla T$ to be

$$\kappa = (\mathfrak{F}_1\mathfrak{F}_3 - \mathfrak{F}_2^2)/\mathfrak{F}_1 T. \qquad (7.2.38)$$

It remains to evaluate the integrals $\mathfrak{F}_n$. Unfortunately, the simple approximation in which $df^{(0)}/dE$ is replaced by a delta function is not adequate here since it would lead to a vanishing thermal conductivity. We must evaluate $\mathfrak{F}_n$ through second order in T. This can be done readily with the aid of the procedure developed in Section 4.1.6, Part A. According to Eq. (4.1.56), we have for a smooth function $\phi(E)$

$$\int \phi(E)\, \partial f^{(0)}/\partial E\, dE = -[1 + (\pi^2/6)(KT)^2\, d^2/d\mu^2]\phi(\mu). \tag{7.2.39}$$

We can use (7.2.39) to express $\mathfrak{F}_n(\mu, T)$ in terms of $\mathfrak{F}_1(\mu, 0)$ [where $\mathfrak{F}_1(\mu, 0)$ is the value of $\mathfrak{F}_1$ at $T = 0°\text{K}$]. This relation is

$$\begin{aligned}\mathfrak{F}_n(\mu, T) &= \mu^{n-1}\mathfrak{F}_1(\mu, 0) + \xi^2[(n-1)(n-2)\mu^{n-3}\mathfrak{F}_1(\mu, 0) \\ &\quad +2(n-1)\mu^{n-2}\, d\mathfrak{F}_1(\mu, 0)/d\mu + \mu^{n-1}\, d^2\mathfrak{F}_1(\mu, 0)/d\mu^2]\end{aligned} \tag{7.2.40}$$

in which

$$\xi^2 = (\pi^2/6)K^2T^2 \tag{7.2.41}$$

and

$$\mathfrak{F}_1(\mu, 0) = (1/12\pi^3) \int v^2(\mathbf{k})\tau(\mathbf{k})\, dS_{\text{F}}/|\nabla E(\mathbf{k})|. \tag{7.2.42}$$

These expressions are substituted into (7.2.38). The temperature-independent terms cancel. The term of lowest order in temperature is

$$\kappa = (\pi^2/3)K^2T\mathfrak{F}_1(\mu, 0) = (K^2T/36\pi) \int \tau(\mathbf{k})v^2(\mathbf{k})\, dS_{\text{F}}/|\nabla E(\mathbf{k})|. \tag{7.2.43}$$

We can compare this expression with that given in (7.2.29) for the electrical conductivity. A remarkable result is obtained. The ratio $L = \kappa/\sigma T$ is predicted to be a constant, independent of specific material properties:

$$L = \kappa/\sigma T = (\pi^2/3)(K/e)^2 = 2.45 \times 10^{-8} \quad \text{W-}\Omega/\text{deg}^2. \tag{7.2.44}$$

The qualitative result, that $\kappa/\sigma T$ is a constant for a particular metal (the Wiedemann–Franz law), holds quite well. The numerical value of the Lorenz number L given by (7.2.45) is close to the experimental results for many metals.

If we assume a parabolic band structure with τ constant on the Fermi surface, κ simplifies to

$$\kappa = \tau K^2Tk_{\text{F}}^3/9m^* = (\pi^2K^2T/3)(n\tau/m^*). \tag{7.2.45}$$

We saw earlier, in the discussion of electrical conductivity that simple kinetic arguments give a value of σ in agreement with that obtained from the Boltzmann equation. In the present case, we may use the simple formula of kinetic theory

$$\kappa = \tfrac{1}{3}Cvl = \tfrac{1}{3}Cv^2\tau \qquad (7.2.46)$$

in conjunction with the electron specific heat as given by Eq. (4.1.58), Part A, to obtain exactly (7.2.46).

The essential assumption underlying the derivation of the Wiedemann–Franz law is that the relaxation time approximation is valid. This holds when electrons are scattered elastically. In fact, at sufficiently low temperatures, impurity scattering dominates in normal metals. Since this scattering is elastic, the Wiedemann–Franz law should be obeyed at low temperatures.

7.2.4 *Thermoelectric Effects*

An electric field and a thermal gradient enter into the Boltzmann equation is similar ways. This fact gives rise to a number of thermoelectric effects [for a survey, see MacDonald (1962)]. We consider a metal in which a current is flowing and a thermal gradient is also present. The rate of accumulation of heat $\mathfrak{Q}$ in a volume ΔV of metal is

$$d\mathfrak{Q}/dt = \Delta V[\mathbf{j}\cdot\boldsymbol{\varepsilon} - \nabla\cdot\mathbf{Q}]. \qquad (7.2.47)$$

The first term represents the Joule heating and the second term is the net flow of heat into the volume due to the presence of the heat current. We will evaluate this using the expressions (7.2.36) for $\mathbf{j}$ and $\mathbf{Q}$. We solve for $\boldsymbol{\varepsilon}$ and obtain

$$\boldsymbol{\varepsilon} = (\mathbf{j}/e^2\mathfrak{F}_1) - (T/e)\,[d(\mu/T)/dT]\,\nabla T - (\mathfrak{F}_2/T\mathfrak{F}_1 e)\,\nabla T. \qquad (7.2.48)$$

This is to be substituted into the equation for $\mathbf{Q}$,

$$\mathbf{Q} = -(\mathfrak{F}_2/e\mathfrak{F}_1)\mathbf{j} + (\mathfrak{F}_2{}^2 - \mathfrak{F}_3\mathfrak{F}_1/T\mathfrak{F}_1)\,\nabla T. \qquad (7.2.49)$$

We substitute (7.2.46) and (7.2.47) into (7.2.44). The electrical conductivity is $\sigma = e^2\mathfrak{F}_1$. Equation (7.2.38) is used for the thermal conductivity κ. Note that $\nabla\cdot\mathbf{j} = 0$:

$$d\mathfrak{Q}/dt = (\mathbf{j}^2/\sigma) + (T/e)\mathbf{j}\cdot\nabla_{\mathbf{r}}(\mathfrak{F}_2/T\mathfrak{F}_1 - \mu/T) + \nabla\cdot(\kappa\,\nabla T). \qquad (7.2.50)$$

The first term in (7.2.50) represents the Joule heat; the third term is the ordinary diffusion of heat; the middle term contains the effects in which we are interested. It depends on both the direction of current flow and of the thermal gradient.

Let us consider a sample in which the thermal gradient is uniform and the thermal conductivity is independent of position. The contribution of the second term, which is linear in **j**, dominates the joule heating when the current density is small. This is the Thompson effect. Its contribution can be written as

$$dQ_T/dt = -\sigma_T \mathbf{j} \cdot \nabla T \tag{7.2.51}$$

where σ_T is the Thompson coefficient. We see that

$$\sigma_T = -(T/e)(d/dT)(\mathfrak{F}_2/T\mathfrak{F}_1 - \mu/T). \tag{7.2.52}$$

It is also useful to define the absolute thermoelectric power $\mathcal{S}$ in terms of which the various thermoelectric effects can be described by

$$\mathcal{S} = \int_0^T \sigma_T/T \, dT. \tag{7.2.53}$$

From (7.2.40) we see that $\sigma_T(T) \approx T$ as $T \to 0$, so that the integral exists

$$\mathcal{S} = (1/eT)[\mu - (\mathfrak{F}_2/\mathfrak{F}_1)]. \tag{7.2.54}$$

A more explicit form can be obtained with the use of the expansions (7.2.40) for the functions $\mathfrak{F}_\mu$. The result is

$$\mathcal{S} = -(\pi^2/3e)K^2T \, d/d\mu \ln \sigma(\mu) \tag{7.2.55}$$

in which σ is given by (7.2.30)

An accurate evaluation of (7.2.55) requires knowledge of the energy dependence of the relaxation time τ. Use of the approximation (7.2.31b) for σ leads to

$$\mathcal{S} = (-\pi^2/2e)(K^2T/\mu)[1 + \tfrac{2}{3}(\mu/\tau)\, d\tau/d\mu]. \tag{7.2.56}$$

Unfortunately, Eq. (7.2.56) may not be a useful approximation in real metals, since the relaxation time may be quite anisotropic, even in simple systems.

7.3 The Hall Effect and Magnetoresistance

We will now investigate the effects of an external magnetic field on the conductivity of a metal. As might be surmised from Chapter 6, this turns out to be a very complex problem. We will begin with the simplest case—that of a parabolic band structure characterized by an effective mass m^*. It is assumed that no thermal gradient is present, and thus possible thermomagnetic effects will be neglected. For reviews see Chambers (1960) and Fawcett (1964).

7.3.1 *Parabolic Bands*

We begin with the Boltzmann equation in the form (7.2.20) with $\nabla T = 0$. The relaxation time approximation is introduced.

$$e\mathbf{v}(\mathbf{k})\cdot\boldsymbol{\mathcal{E}} = [\Phi(\mathbf{k})/\tau(\mathbf{k})] - (e/\hbar)\mathbf{v}(\mathbf{k}) \times \mathbf{B}\cdot\nabla_{\mathbf{k}}\Phi(\mathbf{k}). \tag{7.3.1}$$

We try as a solution

$$\Phi^{(1)}(\mathbf{k}) = \mathbf{k}\cdot\mathbf{C}(E) \tag{7.3.2}$$

where $\mathbf{C}$ is a vector to be determined, which is, however, assumed to be independent of the orientation of $\mathbf{k}$. We have

$$(\mathbf{v} \times \mathbf{B})\cdot\nabla_{\mathbf{k}}(\mathbf{k}\cdot\mathbf{C}) = (\mathbf{v} \times \mathbf{B})\cdot\mathbf{C} = \mathbf{v}\cdot(\mathbf{B} \times \mathbf{C}).$$

Thus (7.3.1) becomes

$$e\mathbf{v}\cdot\boldsymbol{\mathcal{E}} = (\mathbf{k}\cdot\mathbf{C}/\tau) - (e/\hbar)\mathbf{v}\cdot(\mathbf{B} \times \mathbf{C}). \tag{7.3.3}$$

Since $\mathbf{v} = \hbar\mathbf{k}/m^*$, we have

$$(m^*/\hbar\tau)\mathbf{C} - (e/\hbar)(\mathbf{B} \times \mathbf{C}) = e\boldsymbol{\mathcal{E}}. \tag{7.3.4}$$

This equation can be solved if we express $\mathbf{C}$ as

$$\mathbf{C} = \alpha\boldsymbol{\mathcal{E}} + \beta(\mathbf{B} \times \boldsymbol{\mathcal{E}}) + \gamma\mathbf{B}(\mathbf{B}\cdot\boldsymbol{\mathcal{E}}). \tag{7.3.5}$$

The coefficients α, β, γ are determined by substituting (7.3.5) into (7.3.4). The result is

$$\mathbf{C} = (e\hbar\tau/m^*)(1 + \omega_c^2\tau^2)^{-1}[\boldsymbol{\mathcal{E}} + (e\tau/m^*)\mathbf{B} \times \boldsymbol{\mathcal{E}} + (e\tau/m^*)^2\mathbf{B}(\mathbf{B}\cdot\boldsymbol{\mathcal{E}})] \tag{7.3.6}$$

where $\omega_c = eB/m^*$ is the cyclotron frequency. In order to include a thermal gradient, all that is necessary at this point is to replace $e\boldsymbol{\mathcal{E}}$ by $e\boldsymbol{\mathcal{E}} + T\,\nabla(\mu/T) + E\,\nabla T/T$ in accord with (7.2.20). The electric current $\mathbf{j}$, which is calculated by substituting (7.3.6) and (7.3.2) into (7.2.7) is, in the limit in which $\partial f^{(0)}/\partial E$ is a delta function,

$$\mathbf{j} = (e^2/4\pi^3)\int [\tau(\mathbf{k})/(1 + \omega_c^2\tau^2(\mathbf{k})]\mathbf{v}(\mathbf{k})\mathbf{v}(\mathbf{k})$$

$$\cdot[\boldsymbol{\mathcal{E}} + (e\tau/m^*)\mathbf{B} \times \boldsymbol{\mathcal{E}} + (e\tau/m)^2\mathbf{B}(\mathbf{B}\cdot\boldsymbol{\mathcal{E}})]\,dS_F/|\nabla E(\mathbf{k})|. \tag{7.3.7}$$

This is to be compared with (7.2.29b). Equation (7.3.7) makes allowance for the deflection of the electrons by the magnetic field. To make matters more explicit, let us suppose **B** is in the z direction, and write out the components of (7.3.7):

$$j_x = (e^2/4\pi^3) \int [\tau v_x^2/(1 + \omega_c^2\tau^2)](\mathcal{E}_x - \omega_c\tau\mathcal{E}_y)\, dS_F/|\nabla_k E| \qquad (7.3.8a)$$

$$j_y = (e^2/4\pi^3) \int [\tau v_y^2/(1 + \omega_c^2\tau^2)](\mathcal{E}_y + \omega_c\tau\mathcal{E}_x)\, dS_F/|\nabla_k E| \qquad (7.3.8b)$$

and

$$j_z = (e^2/4\pi^3) \int \tau v_z^2\, dS_F/|\nabla_k E|\, \mathcal{E}_z. \qquad (7.3.8c)$$

The system can now evidently be described by a conductivity tensor

$$\sigma = \begin{pmatrix} \sigma_d & -\sigma_n & 0 \\ \sigma_n & \sigma_d & 0 \\ 0 & 0 & \sigma_0 \end{pmatrix} \qquad (7.3.9)$$

in which

$$\sigma_0 = (e^2/12\pi^3) \int \tau v^2(\mathbf{k})\, dS_F/|\nabla_k E| \qquad (7.3.10)$$

is the conductivity in the absence of magnetic effects, and

$$\begin{aligned} \sigma_d &= (e^2/12\pi^3) \int [\tau(\mathbf{k}) v^2(\mathbf{k})/(1 + \omega_c^2\tau^2(\mathbf{k}))]\, dS_F/|\nabla_k E| \\ &\approx \sigma_0/(1 + \omega_c^2\tau^2) \end{aligned} \qquad (7.3.11)$$

$$\begin{aligned} \sigma_n &= (e^2/12\pi^3) \int [\omega_c\tau^2(\mathbf{k}) v^2(\mathbf{k})/(1 + \omega_c^2\tau(\mathbf{k}))]\, dS_F/|\nabla_k E| \\ &\approx \omega_c\tau\sigma_0/(1 + \omega_c^2\tau^2). \end{aligned} \qquad (7.3.12)$$

The last step in (7.3.11) and (7.3.12) follows in the approximation that τ is a constant on the Fermi surface. The off-diagonal components $\sigma_{\alpha\beta}$ satisfy the condition

$$\sigma_{\alpha\beta}(\mathbf{B}) = -\sigma_{\beta\alpha}(\mathbf{B}) \qquad (7.3.13)$$

which is required by the thermodynamics of irreversible processes.

Let us consider a situation in which current is allowed by the construction of the specimen to flow only in the x direction. Then $j_y = 0$, and the field $\mathcal{E}_y$ must be given by

$$\mathcal{E}_y = -(\sigma_n/\sigma_d)\mathcal{E}_x \approx -\omega_c\tau\mathcal{E}_x. \tag{7.3.14}$$

We must also have $\mathcal{E}_z = 0$. The field $\mathcal{E}_y$ which cancels the transverse flow of current is known as the *Hall field*. With this field, the current in the x direction is simply

$$j_x = \sigma_0\mathcal{E}_x. \tag{7.3.15}$$

In these circumstances, we see that the magnetic field disappears from the calculation of the current. The Hall constant R is defined by

$$R = \mathcal{E}_y/Bj_x = -\sigma_n/(\sigma_D\sigma_0 B) \approx -1/ne. \tag{7.3.16}$$

The approximation (7.2.30b) has been used in the last step.

Equation (7.3.15) indicates that, within the approximations employed, no transverse magnetoresistance is found. The Hall field cancels the tendency of the Lorentz force to deflect electrons. However, it should be apparent that (7.3.14) is valid only if the relaxation time is constant, independent of $\mathbf{k}$ on the Fermi surface. Observation of a transverse magnetoresistance in a material with a simple parabolic band structure is evidence for anisotropy of the relaxation time.

7.3.2 *Two Carriers*

In the presence of a magnetic field, simple kinetic arguments lead to the same results, (7.3.10)–(7.3.12), as are obtained from the Boltzmann equation, provided again that a constant relaxation time is employed. All that is necessary is to add a Lorentz force term $e\mathbf{v} \times \mathbf{B}$ to the equation of motion for an electron. Such an approach can easily be extended to include contributions from two or more sets of carriers (for example, set 1 of number density n_1, effective mass m_1^*, relaxation time τ_1; set 2 with n_2, m_2^*, τ_2). The contribution from different sets of carriers to the current is simply additive provided that if the second set consists of holes, proper account is taken of the sign of the effective charge. In such a case (with $\mathbf{B}$ in the z direction)

$$j_x = \left[\frac{\sigma_{10}}{1+\omega_{c1}^2\tau_1^2} + \frac{\sigma_{20}}{1+\omega_{c2}^2\tau_2^2}\right]E_x - \left[\frac{\omega_{c1}\tau_1\sigma_{10}}{1+\omega_{c1}^2\tau_1^2} - \frac{\omega_{c2}\tau_2\sigma_{20}}{1+\omega_{c2}^2\tau_2^2}\right]E_y$$

$$\tag{7.3.17}$$

$$j_y = \left[\frac{\omega_{c1}\tau_1\sigma_{10}}{1+\omega_{c1}^2\tau_1^2} - \frac{\omega_{c2}\tau_2\sigma_{20}}{1+\omega_{c2}^2\tau_2^2}\right]E_x + \left[\frac{\sigma_{10}}{1+\omega_{c1}^2\tau_1^2} + \frac{\sigma_{20}}{1+\omega_{c2}^2\tau_2^2}\right]E_y$$

in which the subscript 1 refers to electrons and 2 to holes;

$$\sigma_{10} = n_1 e^2 \tau_1 / m_1^*,$$

etc. It is straightforward to follow the previous analysis by imposing the condition that $j_y = 0$. An expression for j_x is then obtained. The results are unfortunately rather cumbersome for arbitrary field strengths. Let us first consider weak fields. In this case, the Hall constant can be expressed as

$$R = (\sigma^2_{10} R_1 - \sigma^2_{20} R_2)/(\sigma_{10} + \sigma_{20})^2 \tag{7.3.18}$$

in which $R_1 = -1/n_1 e$, etc. The electric field which develops in the y direction cannot eliminate the deflections of both sets of carriers. Therefore the conductivity does show a dependence on magnetic field:

$$\begin{aligned}\Delta\sigma/\sigma_0 &= [\sigma - (\sigma_{10} + \sigma_{20})]/(\sigma_{10} + \sigma_{20}) \\ &= -[\sigma_{10}\sigma_{20}/(\sigma_{10} + \sigma_{20})^2](\omega_{c1}\tau_1 + \omega_{c2}\tau_2)^2 \\ &= -(eB)^2(n_1\tau_1/m_1^*)(n_2\tau_2/m_2^*)(\tau_1/m_1^* + \tau_2/m_2^*)^2 \\ &\quad \times (n_1\tau_1/m_1^* + n_2\tau_2/m_2^*)^2.\end{aligned} \tag{7.3.19}$$

The conductivity decreases with field in a manner proportional to B^2. For an arbitrary value of the field, we have

$$R = -\frac{1}{e}\frac{(\sigma^2_{10}/n_1 - \sigma^2_{20}/n_2) + (B/e)^2[(n_1 - n_2)/n_1^2 n_2^2]\sigma^2_{10}\sigma^2_{20}}{(\sigma_{10} + \sigma_{20})^2 + (B/e)^2[(n_1 - n_2)^2/n_1^2 n_2^2]\sigma^2_{10}\sigma^2_{20}}. \tag{7.3.20a}$$

In the high field limit, $\omega_c \tau \gg 1$, and

$$R \approx -1/e(n_1 - n_2) \qquad \text{unless} \quad n_1 \approx n_2, \tag{7.3.20b}$$

$$\approx -(1/ne)(\tau_1/m_1^* - \tau_2/m_2^*)/(\tau_1/m_1^* + \tau_2/m_2^*) \qquad \text{if} \quad n_1 = n_2 = n. \tag{7.3.20c}$$

In the high field limit, the Hall coefficient is much smaller when $n_1 = n_2$. In this case the field $\mathcal{E}_y$, which tries to counter the deflection produced by the magnetic field, is small. The general formula for the conductivity is

$$\sigma = \frac{(\sigma_{10} + \sigma_{20})^2 + (\omega_{c1}\tau_1)(\omega_{c2}\tau_2)e^4\tau_1\tau_2(n_1 - n_2)^2/m_1^* m_2^*}{\sigma_{10} + \sigma_{20} + (\omega_{c1}\tau_1)(\omega_{c2}\tau_2)e^2(n_1\tau_2/m_2^* + n_2\tau_1/m_1^*)}. \tag{7.3.21a}$$

In a high field limit, we see that

$$\sigma \approx e^2\tau_1\tau_2(n_1 - n_2)^2/(n_1\tau_2 m_1^* + n_2\tau_1 m_2^*) \qquad \text{unless} \quad n_1 \approx n_2. \tag{7.3.21b}$$

This expression is independent of the magnetic field. Equations (7.3.19) and (7.3.21b) imply that the conductivity tends to decrease with increasing field, but ultimately the resistance saturates, that is, it approaches a limit independent of the value of the field. The situation is quite different, however, if n_1 is close to n_2. Then the high field limit is

$$\sigma \approx (\sigma_{10} + \sigma_{20})^2/[(\omega_{c1}\tau_1)(\omega_{c2}\tau_2)e^2(n_1\tau_2/m_2^* + n_2\tau_1/m_1^*)]$$
$$= (n_1\tau_1 m_2^* + n_2\tau_2 m_1^*)^2/[B^2\tau_1\tau_2(n_1 m_1\tau_2 + n_2 m_2\tau_1)] \qquad (n_1 \approx n_2). \qquad (7.3.21c)$$

If equal numbers of electrons and holes are present, the conductivity continues to decrease as B increases. The magnetoresistance does not saturate, but increases as B^2. This result is a consequence of the requirement that j_y vanish. The conclusion turns out to be independent of our initial assumptions of isotropy of the band structure and relaxation time, and holds for a general Fermi surface in the high field limit. Wannier (1972) has shown quite generally that as long as scattering processes are independent of the magnetic field, the conductivity is a monotonically nonincreasing function of the magnetic field strength.

It should be noted here that the meaning of high field limit is simply $\omega_c\tau \gg 1$, and does not extend to fields high enough so that the discrete structure of the Landau levels is significant. Under such circumstances, oscillations are observed in the resistivity (Schubnikov–de Haas effect) and in other transport properties, just as was found to be the case for the magnetic susceptibility discussed in Section 6.4. Quantum oscillations will not be considered in detail here.

The behavior of the magnetoresistance as actually observed in metals falls into three categories, two of which are in accord with the theory as presented above.

(1) The resistance may saturate, i.e., become independent of field. This occurs for materials with simple, closed, convex Fermi surfaces or with unequal numbers of electrons and holes.

(2) The resistance may increase in proportion to B^2 for all crystal directions. This is the case if equal numbers of electrons and holes are present (a "compensated" metal), and is observed in Bi, Sb, W.

(3) The resistance may saturate for some crystal orientations but increases indefinitely in other directions. A very striking anisotropy of the resistance is then observed. This situation cannot be explained by our previous arguments, and is due to the existence of open orbits previously mentioned in Section 6.4. An extension of transport theory to this situation is required.

7.3.3 *Chambers' Solution of the Boltzmann Equation*

The simple solutions of the Boltzmann equation which have been discussed up to this point become useless when there are pronounced anisotropies in the band structure or in the relaxation time. Chambers (1952) obtained a solution of the Boltzmann equation (assuming that a relaxation time exists) which is both appealing and generally valid. We give here Chambers' semiclassical argument:

Let us consider a volume element ΔV in the conductor, and focus attention on electrons in it having wave vectors in d^3k. It is assumed that collisions are described by a relaxation time τ, and that after a collision, electrons are in the equilibrium distribution ($f = f_0; f_1 = 0$). Electrons which pass through ΔV with wave vectors in a certain range at time t_0 have followed some trajectory since their last collision. The value of f is obtained by adding the contributions from all electrons scattered into the trajectory at a time t prior to t_0, which had energy $E - \Delta E(t)$, where $\Delta E(t)$ is the energy gained from the applied field prior to reaching ΔV. The probability that an electron scattered into the proper trajectory at time t will remain until t_0 without an additional collision is, according to the Poisson distribution, proportional to $\exp[-(t_0 - t)/\tau]$. Thus

$$f = \int_{-\infty}^{t_0} dt/\tau\, f_0(E - \Delta E) \exp[-(t_0 - t)/\tau]. \tag{7.3.22}$$

This result is correct if the relaxation time is independent of $\mathbf{k}$. To include the more general case, we write the probability that scattering has not occurred between t and t_0 as

$$\exp\left[-\int_t^{t_0} dt'/\tau(t')\right]$$

in which τ depends on t' through the dependence of $\mathbf{k}$ (or $\mathbf{v}$) on t'. Thus

$$f(t_0) = \int_{-\infty}^{t_0} \left\{dt/\tau(\mathbf{k}(t))\, f_0[E - \Delta E(t)] \exp\left[-\int_t^{t_0} dt'/\tau(\mathbf{k}(t'))\right]\right\}. \tag{7.3.23}$$

In these equations, ΔE is given by

$$\Delta E(t) = \int_t^{t_0} \mathbf{F}(t') \cdot \mathbf{v}(t')\, dt' \tag{7.3.24}$$

in which $\mathbf{F}$ is the force acting on the electron. Although the derivation is, in many respects, intuitive, Eq. (7.3.23) can be shown to be an exact solution of the Boltzmann equation in the relaxation time approximation (Budd, 1962).

In order to obtain the conductivity, a term in f linear in the applied field is required. To obtain this $f_0[E - \Delta E]$ is expanded to first order in ΔE. The result may be integrated by parts once to give

$$f = f_0 - df_0/dE \int_{-\infty}^{t_0} \mathbf{F}(t) \cdot \mathbf{v}(t) \exp[- \int_t^{t_0} dt'/\tau(\mathbf{k}(t'))]\, dt. \qquad (7.3.25)$$

In application, we set $t_0 = 0$. The current is found by using (7.2.7) with $\mathbf{F} = -e(\boldsymbol{\mathcal{E}} + \mathbf{v} \times \mathbf{B})$

$$\mathbf{j} = (-e^2/4\pi^3) \int d^3k\, df_0/dE\, \mathbf{v}(\mathbf{k}) \int_{-\infty}^{0} \boldsymbol{\mathcal{E}} \cdot \mathbf{v}(\mathbf{k}, t) \exp[- \int_t^0 dt'/\tau(\mathbf{k}(t'))]\, dt. \qquad (7.3.26)$$

The elements of the conductivity tensor $\sigma_{\alpha\beta}$ are given by

$$\sigma_{\alpha\beta} = (-e^2/4\pi^3) \int d^3k\, df_0/dE\, v_\alpha(\mathbf{k}) \int_{-\infty}^{0} dt \times \{v_\beta(\mathbf{k}, t) \exp[- \int_t^0 dt'/\tau(\mathbf{k}(t'))]\}. \qquad (7.3.27)$$

In order to see how this expression can be employed, let us consider the small field limit. Define a quantity s by

$$s(t) = - \int_t^0 dt'/\tau(\mathbf{k}(t')).$$

Then we have

$$\int_{-\infty}^{0} dt\, v_\beta(\mathbf{k}, t) \exp[- \int_t^0 dt'/\tau(\mathbf{k}(t'))] = \int_{-\infty}^{0} ds\, v_\beta(\mathbf{k}, s)\tau(\mathbf{k}(s))e^s.$$

The product $v_\beta\tau$ is expanded in a power series in s:

$$v_\beta(\mathbf{k}, s)\tau(\mathbf{k}(s)) = v_\beta(\mathbf{k}, 0)\tau(\mathbf{k}, 0) + s\,[d(v\tau)/ds]_0 + \tfrac{1}{2}s^2\,[d^2(v\tau)/ds^2]_0 + \cdots.$$

The integral is performed, giving

$$v_\beta(\mathbf{k})\tau(\mathbf{k}) - [d(v\tau)/ds]_{s=0} + [d^2(v\tau)/ds^2]_0 + \cdots.$$

Thus

$$\sigma_{\alpha\beta} = (-e^2/4\pi^3) \int d^3k\, df_0/dE\, v_\alpha(\mathbf{k}) \times \{v_\beta(\mathbf{k})\tau(\mathbf{k}) + [d(v_\beta\tau)/ds]_0 + [d^2(v_\beta\tau)/ds^2]_0 + \cdots\}. \qquad (7.3.28)$$

The leading term agrees with (7.2.30) in the low temperature limit. Higher terms contain the effects of a magnetic field. To see how this occurs, return to the case of a constant relaxation time

$$d(\mathbf{v}\tau)/ds = \tau^2\, d\mathbf{v}/dt = \tau^2\, d\mathbf{k}/dt \cdot \nabla_k \mathbf{v} = -\,\tau^2 (e/\hbar)\mathbf{v}(\mathbf{k}) \times \mathbf{B} \cdot \nabla_k \mathbf{v}(\mathbf{k}). \tag{7.3.29}$$

We substitute this into (7.3.28) and also replace df_0/dE by $-\delta(E - \mu)$. Evidently, an expansion of the elements of the conductivity tensor in powers of B is generated. This is the Jones–Zener (1934b) expansion. For example, consider a cubic crystal with $\mathbf{B}$ in the z direction. We evaluate σ_{xy} in this case. The leading term vanishes on account of symmetry. The first nonvanishing term in σ_{xy} is

$$\sigma_{xy} = \frac{e^2}{4\pi^3}\frac{eB\tau^2}{\hbar^4}\int \frac{dS_F}{|\nabla_k E|}\frac{\partial E}{\partial k_x}\left[\frac{\partial E}{\partial k_y}\frac{\partial^2 E}{\partial k_x\,\partial k_y} - \frac{\partial E}{\partial k_x}\frac{\partial^2 E}{\partial k_y{}^2}\right]. \tag{7.3.30}$$

The reader may verify that in the case of a parabolic band, this result agrees with the leading term in the expansion of (7.3.12) for small fields.

7.3.4 *Open Orbits*

We now use Chambers' result (7.3.27) to discuss the nature of the high field magnetoresistance. The results depend in an essential way on the nature of the orbits; and this in turn depends on the properties of the Fermi surface. In the high magnetic field limit, the motion of an electron is mainly determined by B. Since $\hbar\, d\mathbf{k}/dt = -\,e\mathbf{v} \times \mathbf{B}$, the projection of the orbit in real space in a plane perpendicular to the magnetic field (the xy plane) is similar to the orbit in $\mathbf{k}$ space but rotated by 90° about the field. Moreover, in high fields, several tranversals of a closed orbit will occur, or the electron will move a substantial distance along an open orbit between collisions. We can then replace $1/\tau$ by its average along the orbit, which we denote by $1/\bar{\tau}$. Then (7.3.27) becomes

$$\sigma_{\alpha\beta} = (-e^2/4\pi^3)\int d^3k\, df_0/dE\, v_\alpha(\mathbf{k})\int_{-\infty}^{0} v_\beta(\mathbf{k}, t)\exp(t/\bar{\tau})\, dt. \tag{7.3.31}$$

At this point, the distinction between closed and open orbits becomes important. Suppose there is an open orbit in the k_x direction. Then v_y will have a finite time average $\bar{v}_y$. In this case, we may replace $v_y(\mathbf{k}, t)$ by $\bar{v}_y$ in (7.3.31). We will then obtain a contribution to σ_{yy} which is finite and independent of B in leading order. In contrast, if all orbits are closed, the time averages of v_x and v_y will vanish. This will also be true for v_x in the open orbit case above. For such cases, there will be no contribution to σ which is independent of the field. The high field limit of these elements will remain consistent with (7.3.11) and (7.3.12): σ_{xx} (and σ_{yy} if there are no open orbits) $\approx 1/(\omega_c\bar{\tau})^2$; $\sigma_{xy} \approx 1/\omega_c\bar{\tau}$. The average velocity in parallel to the field direction will remain finite, so that

$$\sigma_{zz} = (-e^2/4\pi^3) \int d^3k \, df_0/dE \, \bar{\tau} v_z \bar{v}_z. \tag{7.3.32}$$

This approaches a finite limit independent of B which should be smaller than the zero field value (where $\bar{v}_z = v_z$) due to the anisotropy of v_z around the orbit. Hence a longitudinal magnetoresistance results, which must saturate.

We may now proceed to determine the transverse magnetoresistance

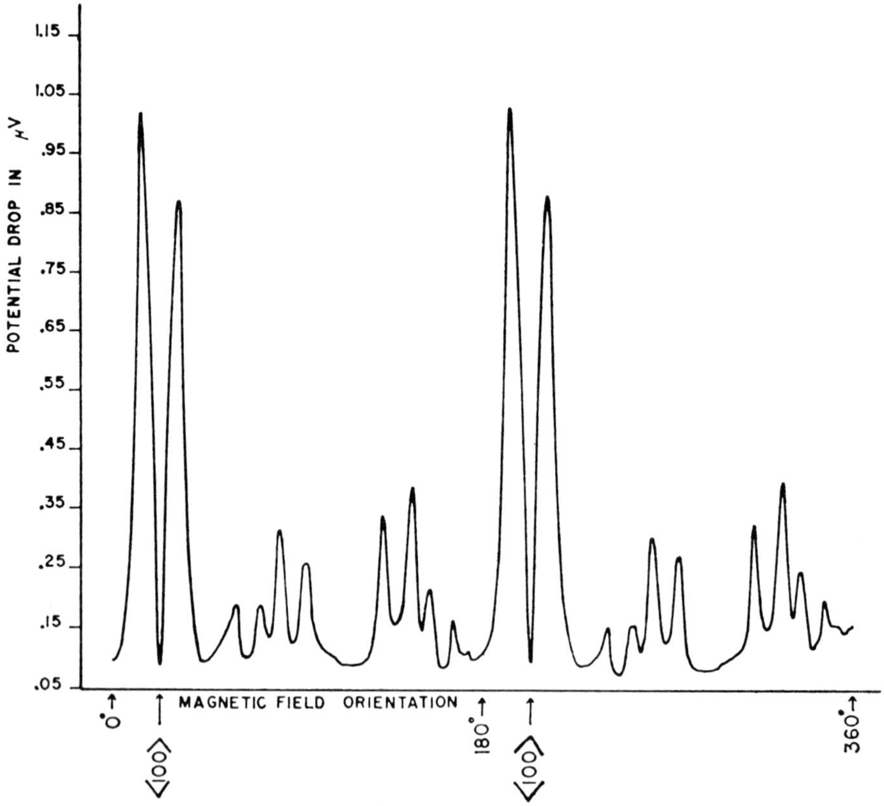

Fig. 7.3.1. Transverse magnetoresistance rotation curve for copper. [From Coleman *et al.* (1964).]

from the conductivity as before. The condition that $j_y = 0$ implies that

$$\sigma_{yx}\mathcal{E}_x + \sigma_{yy}\mathcal{E}_y = 0.$$

Hence

$$\begin{aligned} \mathcal{E}_y = -(\sigma_{yx}/\sigma_{yy})\mathcal{E}_x &\approx -(1/\omega_c\bar{\tau})\mathcal{E}_x && \text{(open orbit)} \\ &\approx \omega_c\bar{\tau}\mathcal{E}_x && \text{(no open orbit).} \end{aligned} \tag{7.3.33}$$

Then

$$\begin{aligned} j_x = (\sigma_{xx}\mathcal{E}_x - \sigma_{yx}\mathcal{E}_y) &\approx \text{const}/(\omega_c\bar{\tau})^2\mathcal{E}_x && \text{(open orbit)} \\ &\approx \text{const}'\mathcal{E}_x && \text{(no open orbit).} \end{aligned} \tag{7.3.34}$$

Therefore the magnetoresistance saturates if no open orbits are present, but increases as B^2 when there is an open orbit. If the open orbit lies in a general direction in the xy plane, both σ_{xx} and σ_{yy} will have finite limits independent of B. In this case, the magnetoresistance will saturate, unless the open orbit runs very close to the y direction (or x direction in $\mathbf{k}$ space).

Magnetoresistance measurements are generally made on oriented single-crystal specimens in the form of fine wires with the magnetic field perpendicular to the axis of the wire. In high fields, the magnetoresistance will be observed to saturate except in those directions in which open orbits exist (or unless the material has an equal concentration of electrons and holes). The anisotropy effects, if they exist, are frequently quite pronounced. An example of this effect (in copper) is shown in Fig. 7.3.1, taken from Coleman *et al.* (1964).

7.4 Electromagnetic Properties of Metals

In our discussion of transport theory in Sections 7.2 and 7.3 we have considered only steady fields. Previously (Section 6.5) we have discussed the optical properties of metals. In this section we will investigate phenomena which occur in an intermediate range of frequencies between dc and optical. This has proved to be an important area for study and has yielded some surprising results. Reviews relevant to this topic have been presented by Kaner and Skobov (1968), Walsh (1968), Baynham and Boardman (1970), and Platzman and Wolff (1973). The reader should bear in mind that MKS units will be used throughout this discussion.

7.4.1 *Skin Effects*

In the frequency range of interest, $\sigma \gg \epsilon\omega$, and the conduction current dominates the displacement current. A typical value for the room temperature conductivity of a metal is 10^7 mhos/meter; if we take $\epsilon \approx \epsilon_0 = 8.85 \times 10^{-12}$, the condition above will certainly be satisfied for frequencies ω less than 10^{17} sec (unless the dielectric function is extremely large). In this region we may neglect the displacement current $\partial D/\partial t$ in Maxwell's equations, which become

$$\nabla\cdot\mathbf{D} = 0, \quad \nabla\cdot\mathbf{B} = 0, \qquad \nabla\times\boldsymbol{\mathcal{E}} = -\partial\mathbf{B}/\partial t, \quad \nabla\times\mathbf{H} = \mathbf{j}. \tag{7.4.1}$$

The usual relations between $\mathbf{j}$ and $\boldsymbol{\mathcal{E}}$, $\boldsymbol{\mathcal{E}}$ and $\mathbf{D}$, $\mathbf{B}$ and $\mathbf{H}$ are

$$\mathbf{D}(\mathbf{r}, \omega) = \epsilon(\omega)\,\boldsymbol{\mathcal{E}}(\mathbf{r}, \omega), \qquad \mathbf{j}(\mathbf{r}, \omega) = \sigma(\omega)\,\boldsymbol{\mathcal{E}}(\mathbf{r}, \omega),$$

$$\mathbf{B}(\mathbf{r}, \omega) = \mu(\omega)\mathbf{H}(\mathbf{r}, \omega) \tag{7.4.2}$$

in which $\boldsymbol{\mathcal{E}}(\mathbf{r}, \omega)$ is the Fourier transform of $\boldsymbol{\mathcal{E}}(\mathbf{r}, t)$ with respect to time. It is supposed in writing (7.4.2) that the relations between $\mathbf{j}(\mathbf{r}, t)$ and $\boldsymbol{\mathcal{E}}(\mathbf{r}, t)$, for example, are local with respect to the space coordinate; which which means that the value of $\mathbf{j}$ at a given point is determined by the value of $\boldsymbol{\mathcal{E}}$ at the same point, perhaps, however, integrated over time. It would be expected from the discussion at the beginning of Section 6.5 that this would be a valid approximation as long as the wavelength of the electromagnetic field is large compared to an atomic spacing. However, in the present context, there is another important length to be considered: the electron mean free path. Departures from local relations of the type (7.4.2) will be found to occur at longer wavelengths than would have been previously anticipated.

Equations (7.4.1) and (7.4.2) lead to the following equation for a monochromatic field: $H(\mathbf{r}, t) \propto \exp[i(\mathbf{k}\cdot\mathbf{r} - \omega t)$,

$$k^2\mathbf{H} = i\mu\sigma\omega\mathbf{H}. \tag{7.4.3}$$

For low frequencies, a simple band, and a constant relaxation time

$$\sigma = \sigma_0 = ne^2\tau/m^* \tag{7.2.31b}$$

in which n is the electron concentration and m^* is the effective mass. We then find from (7.4.3)

$$\mathbf{k} = (1 + i)(\mu\sigma\omega/2)^{1/2}\hat{\mathbf{n}} \tag{7.4.4}$$

in which $\hat{\mathbf{n}}$ is a unit vector in the direction of propagation. The wave is thus damped exponentially in a characteristic distance, called the *skin depth* and denoted δ_0,

$$\delta_0 = (2/\omega\mu\sigma)^{1/2}. \tag{7.4.5}$$

For an electromagnetic field with a free space wavelength of 1 cm, δ at room temperature is of the order of magnitude of 10^{-4} cm. At liquid helium temperature a very pure metal might have $\delta_0 \approx 10^{-6}$ cm. It is clear that a low frequency field is attenuated in a small fraction of a free space wavelength.

The situation is different for higher frequencies for which $\omega\tau > 1$. The conductivity can be obtained by a simple kinetic argument based on (7.2.32). If we do not neglect $d\mathbf{v}/dt$ in that equation, and put $v \propto e^{-i\omega t}$, we find a conductivity

$$\sigma = (ne^2\tau/m^*)/(1 - i\omega\tau) = \sigma_0/(1 - i\omega\tau). \tag{7.4.6}$$

Then if $\omega\tau \gg 1$,

$$\sigma \equiv ine^2/m^*\omega. \tag{7.4.7}$$

The relaxation time does not appear. Suppose that the material has no striking magnetic properties, so that $\mu = \mu_0$. Substituting (7.4.7) in (7.4.3) yields

$$k^2 = -\mu_0 ne^2/m^* = -\omega_p{}^2/c^2 \tag{7.4.8}$$

in which ω_p is the electron plasma frequency. The metal is totally reflecting. The skin depth δ is just

$$\delta = c/\omega_p. \tag{7.4.9}$$

This in the range of 10^{-5} to 10^{-6} cm. The situation of total reflectance persists until the frequency becomes large enough for band structure effects to become important, as is discussed in more detail in Section 6.5.

The situation becomes more interesting and more complex if we observe that in very pure metals at liquid helium temperatures, the mean free path l may be of the order 10^{-2} cm and may therefore exceed the skin depth by a substantial factor. When $l \gg \delta_0$, the local relation between current and field assumed in (7.4.2) (Ohm's law) is no longer valid. The skin effect is anomalous under these circumstances. The current density is determined by the electric field in a region with dimensions of order l. A rigorous approach must use a nonlocal relation between current and field. However, the essential features of the anomalous skin effect can be understood in terms of the "ineffectiveness" concept of Pippard (1947). This involves the observation that not all the electrons near the Fermi surface in the metal are of the same importance in determining the current. Most of the electrons move rapidly out of the range of the electric field, which must alternate rapidly in space. Only those electrons whose velocities make a small angle of order δ_s/l with the surface (where δ_s is the actual skin depth, to be determined) can contribute. Thus we replace the electron concentration in (7.2.31b) by n_{eff} where

$$n_{\text{eff}} = an\delta_s/l \tag{7.4.10}$$

and a is a constant which should be of order unity. The conductivity is now

$$\sigma = a(ne^2\tau/m^*)(\delta_s/l) = a\sigma_0\delta_s/l. \tag{7.4.11}$$

This must be substituted into (7.4.5) to determine the skin depth, which is found to be

$$\delta_s = (2l/\omega\mu\sigma_0 a)^{1/3} = \delta_0(l/a\delta_0)^{1/3} \tag{7.4.12a}$$

in which δ_0 is the ordinary skin depth, determined from (7.4.5) with $\sigma = \sigma_0$. We can rewrite this in the form

$$\delta_s = (2m^* v_F/a\omega ne^2\mu_0)^{1/3} \tag{7.4.12b}$$

where v_F is the speed of an electron at the Fermi surface. The skin depth is seen to be independent of τ, and falls off as $\omega^{-1/3}$.

The electromagnetic properties of a metallic surface are described by the surface impedance function $Z(\omega)$:

$$Z_s(\omega) = \mathcal{E}_t(\omega)/H_t(\omega) \tag{7.4.13}$$

in which $\mathcal{E}_t$ and H_t are the tangential components of the fields at the surface. Z is complex:

$$Z(\omega) = R(\omega) - iX(\omega) \tag{7.4.14}$$

in which $R(\omega)$ is the surface resistance which describes the absorption of power and $X(\omega)$ is the surface reactance. Let the normal vector pointing into the surface under consideration define the **z** direction. We find from Maxwell's equation that

$$i\omega\mu H_t = \partial\mathcal{E}_t/\partial z \tag{7.4.15}$$

where E_t and H_t are continuous across the surface. Hence

$$Z = i\omega\mu\mathcal{E}_t/(\partial\mathcal{E}_t/\partial z)_{z=0} = \omega\mu/k. \tag{7.4.16}$$

The absorption of energy can be determined from the Poynting vector **S**:

$$\mathbf{S} = \boldsymbol{\mathcal{E}} \times \mathbf{H}. \tag{7.4.17}$$

For simplicity, let us suppose that the wave is incident perpendicularly to the surface, and let $\boldsymbol{\mathcal{E}}$ be in the x direction, **H** in the y direction. Then **S** is in the z direction

$$S_z = \mathcal{E}_x Hy = ZH_y^2. \tag{7.4.18}$$

The time average of the real part of S is $\frac{1}{2}R\bar{H}_y^2$ in which $\bar{H}_y$ is the amplitude of H_y at the surface. Evidently, Z can be expressed in terms of the skin depth δ. In the region of frequencies in which (7.4.4) is valid, we have

$$Z = \tfrac{1}{2}(1 - i)\omega\mu\delta. \tag{7.4.19}$$

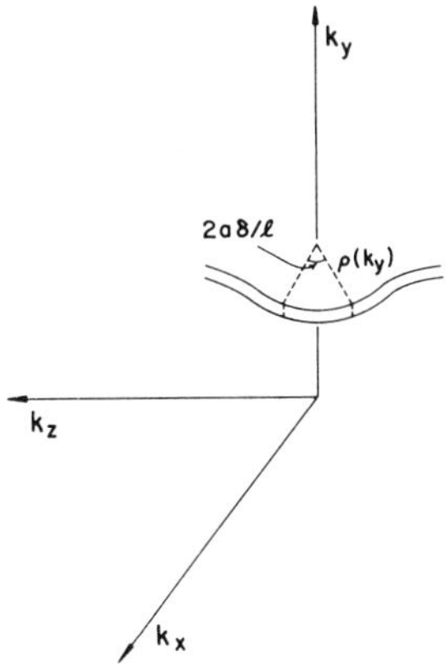

Fig. 7.4.1. Effective portion of the Fermi surface for the anomalous skin effect.

The surface resistance is thus

$$R = (\omega\mu_0/2)(2m^* v_F/a\omega n e^2 \mu_0)^{1/3} = (\tfrac{3}{4}\pi^2\omega^2\mu_0{}^2\hbar/ae^2k_F{}^2)^{1/3}. \qquad (7.4.20)$$

This expression involves the wave vector k_F of the Fermi sphere and could, in principle, be used to determine this quantity. However, the presence of the constant a in (7.4.20) limits the usefulness of this approach. A proper determination of a from a more rigorous theory is required. However, the simple ineffectiveness approach is capable of accounting for the angular variation of the surface resistance when the Fermi surface is not spherical.

Suppose, as before, that the electric field is in the x direction and the normal to the surface is in the z direction. Only those electrons with velocities nearly parallel to the surface can contribute. This means that if we consider a cross section of the Fermi surface perpendicular to the k_y axis, those electrons in a sector making an angle $2a\delta/l$ about the x axis can contribute. For k_y in the range dk_y, the number of electrons in the sector is (see Fig. 7.4.1) $(1/4\pi^3) \mid \rho(k_y) \mid (2a\delta/l)\, dk_y$ in which $\rho(k_y)$ is the radius of the portion of Fermi surface considered. Such electrons are displaced in $\mathbf{k}$ space by the electric field by an amount $e\mathcal{E}_x\tau$. Their contribution to the x component of the current is obtained by multiplying by ev. We may then integrate over k_y:

$$j_x = (\mathcal{E}_x/4\pi^3)(2a\delta/l)e^2\tau v \int \mid \rho(k_y) \mid dk_y. \qquad (7.4.21)$$

Hence, since $v\tau = l$,

$$\sigma_{xx} = (e^2/4\pi^3)\cdot 2a\delta \int \mid \rho(k_y) \mid dk_y. \qquad (7.4.22)$$

This may be used in (7.4.5) in place of the isotropic conductivity σ. Evidently, the x–x component of the surface resistivity R_{xx} is

$$R_{xx} \approx \text{const} \times \left(\omega^2 \Big/ \int |\rho(k_y)|\, dk_y\right)^{1/3}. \tag{7.4.23}$$

A similar result is obtained by R_{yy}. These quantities can vary dramatically with orientation of the electric field.

A more rigorous theory of the anomalous skin effect has been given by Reuter and Sondheimer (1948) (see also Kittel, 1963). The mathematical techniques involved are of some interest, and we will outline the calculation. Consider a semiinfinite metal surface lying in the xy plane, with the positive z axis directed into the metal. The electric field is taken to be in the x direction, $\mathcal{E}_x(z)e^{-i\omega t}$, and the magnetic field is in the y direction. Our treatment is based on the Boltzmann equation. An alternative, purely quantum, approach is due to Mattis and Bardeen (1958).

We calculate first the wave vector and frequency dependence of the conductivity σ_{xx}. To this end, we will consider a system with a single parabolic band $E = \gamma k^2$ and assume the validity of the relaxation time approximation. Terms must be added to the Boltzmann equation to allow for the explicit time and position dependence introduced by the varying external field. Instead of (7.2.6), we have

$$(\partial f/\partial t) - (e/\hbar)(\mathbf{E} + \mathbf{v} \times \mathbf{B}) \cdot \nabla_{\mathbf{k}} f + \mathbf{v}(\mathbf{k}) \cdot \nabla_{\mathbf{r}} f = -(f - f^{(0)})/\tau. \tag{7.4.24}$$

We expect $f^{(1)} = f - f^{(0)}$ to have the same frequency dependence $e^{-i\omega t}$ as the external field and to depend on position only through the distance z into the metal. Magnetic effects are dropped, and the equation is linearized in the usual way. Equation (7.4.24) simplifies to

$$[(1 - i\omega\tau)/\tau] f^{(1)}(z) + v_z\, \partial f^{(1)}/\partial z = e\mathcal{E}_x(z) v_x(\mathbf{k})\, \partial f^{(0)}/\partial E. \tag{7.4.25}$$

It is convenient to solve this by a Fourier transform method. Define

$$f(p) = (2\pi)^{-1/2} \int_{-\infty}^{\infty} f^{(1)}(z) e^{ipz}\, dz, \qquad \mathcal{E}_x(p) = (2\pi)^{-1/2} \int_{-\infty}^{\infty} \mathcal{E}_x(z) e^{ipz}\, dz. \tag{7.4.26}$$

Equation (7.4.25) yields an equation for the Fourier transform,

$$(1 - i\omega\tau + ipv_z\tau) f(p) = e\mathcal{E}_x(p) v_x\, \partial f^{(0)}/\partial E. \tag{7.4.27}$$

This can be solved immediately to yield

$$f(p) = \{[ev_x\tau\, \partial f^{(0)}/\partial E]/(1 - i\omega\tau + ipv_z\tau)\}\mathcal{E}_x(p). \tag{7.4.28}$$

It will be sufficient to use the low temperature approximation $\partial f^{(0)}/\partial E = -\delta(E - \mu)$. The electric current density is determined from Eq. (7.2.7). This relation can be expressed in terms of the Fourier transform of the current

$$j_x(p) = (2\pi)^{-1/2} \int j_x(z) e^{ipz}\, dz.$$

We obtain

$$\begin{aligned} j_x(p) &= (-e/4\pi^3) \int v_x(\mathbf{k}) f(\mathbf{k}, p)\, d^3k \\ &= (-e^2/4\pi^3) \int (v_x^2 \tau\, \partial f^{(0)}/\partial E)/(1 - i\omega\tau + ipv_z\tau)\, d^3k\, \mathcal{E}_x(p). \end{aligned} \tag{7.4.29}$$

This equation may be used immediately to define the Fourier transform of the conductivity $\sigma_{xx}(p)$ by the relation

$$\begin{aligned} j_x(p) &= \sigma_{xx}(p) \mathcal{E}_x(p) \\ \sigma_{xx}(p) &= (-e^2/4\pi^3) \int (v_x^2 \tau\, \partial f^{(0)}/\partial E)/(1 - i\omega\tau + ipv_z\tau)\, d^3k \\ &= (e^2/4\pi^3) \int dS_{\mathrm{F}}/|\nabla_{\mathbf{k}} E|\, v_x^2 \tau/(1 - i\omega\tau + ipv_z\tau). \end{aligned} \tag{7.4.30}$$

This may be compared with (7.2.30). In the usual limit of a constant relaxation time and isotropic velocity, Eq. (7.4.30) reduces to

$$\begin{aligned} \sigma_{xx} &= \sigma_0 \left[(3/4\pi) \int (d\Omega \sin^2\theta \cos^2\phi)/(1 - i\omega\tau + pv\tau \cos\theta) \right] \\ &= \sigma_0 \left[\tfrac{3}{4} \int_0^\pi (\sin^3\theta\, d\theta)/(1 - i\omega\tau + ipv\tau\cos\theta) \right] \end{aligned} \tag{7.4.31}$$

in which $\sigma_0 = ne^2\tau/m^*$ and we have chosen the k_z axis as the polar axis for a spherical coordinate system. The integral is straightforward. Define $\zeta = pv\tau[1 - i\omega\tau]^{-1}$. We have, with $\cos\theta = x$,

$$\int_{-1}^{1} (1 - x^2)\, dx/(1 + i\zeta x) = (2/\zeta^3)[(1 + \zeta^2) \tan^{-1}\zeta - \zeta].$$

Thus

$$\sigma_{xx} = [3\sigma_0/2(1 - i\omega\tau)\zeta^3][(1 + \zeta^2)\tan^{-1}\zeta - \zeta]. \tag{7.4.32}$$

In the limit $p = 0$, (7.4.32) reduces to (7.4.6), the drift velocity result. We are, however, primarily concerned with a different situation. The important values of p will be of order $1/\delta$; thus $pv\tau$ is of order l/δ, which will be assumed to be large compared to $\omega\tau$. In this limit, (7.4.32) becomes

$$\sigma_{xx} = (3\pi/4)(\sigma_0/|p|v\tau). \tag{7.4.33}$$

This result is to be used in conjunction with the Maxwell equations to obtain the electric field near the surface. As before, we suppose that the displacement current may be neglected. We then obtain from Maxwell's equations

$$d^2\mathcal{E}_x/dz^2 = i\omega\mu\, dH_y/dz = -i\omega\mu j_x. \tag{7.4.34}$$

This equation must be solved subject to the proper boundary conditions at the surface. In general, this is difficult because the behavior of the current is affected by the manner in which electrons are reflected when striking the surface. One idealized case can be treated simply. We assume the electrons are reflected specularly at the surface (the z component of velocity is reversed, while components in the plane of the surface are unaltered). This contrasts with the (more likely) possibility of diffuse reflection in which the z component is randomized. It turns out that there is not much difference in the final result between specular and diffuse reflection.

Specular reflection is simpler because we may then conceptually extend the metal to fill all space. Each trajectory leaving the original metal is exactly matched by one entering it with the same transverse component of velocity but opposite z component. It is necessary only to provide a cusp for the electric field at $z = 0$ so that the field decays on both sides of the xy plane. This (artificial) cusp can be provided by the addition of a term $2(d\mathcal{E}/dz)_{0^+}\delta(z)$ to (7.4.34). The notation 0^+ indicates that the derivative is taken as $z \rightarrow 0$ on the positive z axis:

$$d^2\mathcal{E}_x/dz^2 = -i\omega\mu j_x(z) + 2(d\mathcal{E}_x/dz)_{0^+}\delta(z). \tag{7.4.35}$$

The additional term in (7.4.35) indicates that there will be a discontinuity in the first derivative equal to $2(d\mathcal{E}_x/dz)_{0^+}$ on passing through the xy plane; thus the derivative on the negative side will be the negative of the derivative on the positive side. It appears that an additional complication has been introduced, since the value of this quantity is not specified; however, it will disappear in the calculation of the surface impedance. The Fourier transform of (7.4.35) is

$$-p^2\mathcal{E}_x(p) = i\omega\mu\sigma_{xx}(p)\mathcal{E}_x(p) + (2/\pi)^{1/2}(d\mathcal{E}_x/dz)_{0^+}.$$

Thus

$$\mathcal{E}_x(z) = (2\pi)^{-1/2} \int \mathcal{E}_x(p) e^{-ipz}\, dp$$

$$= -(1/\pi)(d\mathcal{E}_x/dz)_{0^+} \int_{-\infty}^{\infty} e^{-ipz}/[p^2 - i\mu\omega\sigma_{xx}(p)]\, dp. \qquad (7.4.36)$$

The surface impedance can now be obtained from (7.4.16)

$$Z = i\mu\omega\mathcal{E}(z)/(d\mathcal{E}/dz)_{0^+}$$

$$= -i\mu\omega/\pi \int_{-\infty}^{\infty} dp/[p^2 - i\mu\omega\sigma_{xx}(p)] \approx -2i\omega\mu/\pi \int_0^{\infty} p\, dp/(p^3 - i\lambda)$$

$$(7.4.37)$$

in which $\lambda = \frac{3}{4}(\pi\omega\mu\sigma_0/v\tau)$ and (7.4.33) has been used. The integral can be found in standard tables

$$Z = [2\omega\mu/3\sqrt{3}\lambda^{1/3}](1 - i\sqrt{3}) = \tfrac{4}{9}(\sqrt{3}\omega^2\mu^2{}_0 l/2\pi\sigma_0)^{1/3}(1 - i\sqrt{3}).$$

$$(7.4.38)$$

The surface resistance R, which is the real part of Z, is of the same form as that obtained from (7.4.20), and agrees numerically if

$$a = (4\pi/\sqrt{3})(\tfrac{9}{8})^3 = 10.3. \qquad (7.4.39)$$

The relatively large value of a is somewhat unexpected. The phase of Z as given by (7.4.38) does not coincide with (7.4.20). The agreement with respect to the functional dependence of R on the parameter $\omega^2 l/\sigma_0$ confirms the qualitative validity of the ineffectiveness concept. The results do depend somewhat on the nature of the reflection of electrons at the surface. In the case of diffuse reflection, Reuter and Sondheimer obtained a result which is larger than (7.4.38) by a factor of $\frac{9}{8}$.

7.4.2 *Helicons*

The electromagnetic properties of a metal are dramatically altered by the presence of a strong magnetic field. In particular, it is possible to propagate electromagnetic waves under certain conditions.

We will consider metals with (nearly) spherical Fermi surfaces as occur in the alkalis. In this discussion it will be sufficient to neglect any frequency and wave vector dependence of the conductivity, which can then be treated as a constant tensor. From Maxwell's equations we have

$$\partial\mathbf{j}/\partial t = \boldsymbol{\sigma}\, \partial\boldsymbol{\mathcal{E}}/\partial t = -(1/\mu_0)\, \nabla \times (\nabla \times \boldsymbol{\mathcal{E}}). \qquad (7.4.40)$$

Since this equation involves the first time derivative of the electric field, we would normally not expect propagating solutions to exist. This is correct when there is no external magnetic field; however, when such a field is present, the conductivity tensor is not diagonal and, if the field is strong enough in a pure metal so that $\omega_c\tau \gg 1$ (ω_c is the cyclotron frequency and τ is the relaxation time), the off-diagonal elements will be large in magnitude compared to the diagonal elements. In this case a solution appears representing a wave which is only weakly damped.

To have a specific example, let the external magnetic field define the **z** axis, and suppose that there is an electric field in the xy plane which is propagating in the z direction so that the field quantities depend on position and time as $\exp[i(kz - \omega t)]$. In these circumstances (7.4.40) yields a pair of coupled equations

$$i\omega(\sigma_{xx}\mathcal{E}_x + \sigma_{xy}\mathcal{E}_y) = (k^2/\mu_0)\mathcal{E}_x, \qquad i\omega(\sigma_{yx}\mathcal{E}_x + \sigma_{yy}\mathcal{E}_y) = (k^2/\mu_0)\mathcal{E}_y. \tag{7.4.41}$$

The determinant of the coefficients of $\mathcal{E}_x$ and $\mathcal{E}_y$ must vanish if we are to obtain a nontrivial solution. We note that $\sigma_{xx} = \sigma_{yy} = \sigma_d$ and $\sigma_{yx} = -\sigma_{xy} = \sigma_n$ [refer to Eq. (7.3.9)]. The determinantal equation is

$$\omega^2\sigma_n^2 - [i\omega\sigma_d - (k^2/\mu_0)]^2 = 0. \tag{7.4.42}$$

We are interested in the solution in which k^2 is almost real and positive, as this will correspond to a weakly damped field. This case is

$$\omega = (k^2/\mu_0)[1/(\sigma_n + i\sigma_d)] = (k^2/\mu_0)[(\sigma_n - i\sigma_d)/(\sigma_n^2 + \sigma_d^2)]. \tag{7.4.43}$$

This solution is interesting in the high field limit in which

$$\sigma_n \approx ne/B \gg \sigma_d \approx (ne/B)(1/\omega_c\tau). \tag{7.4.44}$$

Then

$$\begin{aligned}\omega &\approx (k^2/\mu_0\sigma_n)[1 - (i\sigma_d/\sigma_n)] \approx k^2B/\mu_0 ne[1 - (i/\omega_c\tau)] \\ &= k^2c^2(\omega_c/\omega_p^2)[1 - (i/\omega_c\tau)]\end{aligned} \tag{7.4.45}$$

in which ω_p is the plasma frequency

$$\omega_p^2 = ne^2/m^*\epsilon_0. \tag{7.4.46}$$

The wave is circularly polarized with $\mathcal{E}_x \approx -i\mathcal{E}_y$.

Electromagnetic waves obeying the dispersion relation (45) are known as *helicons* (Aigrain, 1960). The group velocity $d\omega/dk$ is proportional to k, so that waves of higher frequency travel faster. Disturbances of this type occur in the ionosphere, where they are known as *whistlers* since the frequency observed at a point diminishes with time. They were first observed in metals by Bowers *et al.* (1961) in sodium.

If the metal contains open orbits, we see from (7.3.33) and (7.3.34) that the elements of the conductivity tensor in strong fields acquire a different frequency dependence such that we do not have $|\sigma_{xy}| \gg \sigma_{xx}$. Instead, the reverse situation obtains. Hence, the existence of open orbits tends to destroy the helicon mode.

7.4.3 *Magnetoplasma Waves*

The discussion of helicons applies primarily to simple metals. Electromagnetic phenomena are different in a compensated metal in which there are equal numbers of electrons and holes. In such a case, the plasma is neutral. We shall refer to the waves which can exist in such a case as *magnetoplasma waves*. Here also we may treat the conductivity as independent of the spatial coordinates.

It is convenient to go back to the equations of motion for a charged particle of effective mass m_α^* ($\alpha = 1, 2$, corresponding to electrons and holes) in a steady magnetic field and time-dependent electric field. We will assume that the effective masses are scalars. This is not appropriate for real semimetals (such as bismuth) in which these effects are observed, but it is sufficient to discuss the essential features:

$$m_\alpha^* [d\mathbf{v}_\alpha/dt + \mathbf{v}_\alpha/\tau_\alpha] = e[\mathbf{v}_\alpha \times \mathbf{B} + \mathcal{E}e^{-\iota\omega t}]. \qquad (7.4.47)$$

It is specifically assumed that the concentration of electrons and holes are equal. The Hall effect is small in the high field limit. Equation (7.4.47) leads to an expression for the dielectric tensor. This tensor is diagonal in the absence of a Hall effect. The xx and yy elements have the form

$$\kappa_{xx} = \kappa_{yy} = \kappa_{\mathrm{B}} + \sum_\alpha [\omega_{\mathrm{p}\alpha}{}^2(1 + i/\omega\tau_\alpha)]/[\omega_{\mathrm{c}\alpha}{}^2 - \omega^2(1 + i/\omega\tau_\alpha)^2] \qquad (7.4.48)$$

in which κ_{B} is the background due to the polarizabilities of the tightly bound electrons, $\omega_{\mathrm{p}\alpha}$ is the plasma frequency for carriers of type α, given by Eq. (7.4.46), and $\omega_{\mathrm{c}\alpha}$ is the corresponding cyclotron frequency. Note that $\omega_{\mathrm{c}2}$ is negative here since the sign of the charge of the hole is opposite to that of the electron. Equation (7.4.48) would appear to lead to results similar to those of the previous section; however, when $n_1 = n_2$ and $\omega_{\mathrm{c}\alpha} \gg \omega$, (7.4.48) becomes

$$\kappa_{xx} = \kappa_{yy} = \kappa_{\mathrm{B}} + \sum_\alpha (\omega^2{}_{\mathrm{p}\alpha}/\omega^2{}_{\mathrm{c}\alpha})[1 + (i/\omega\tau_\alpha)]. \qquad (7.4.49)$$

The real parts of these components are positive: thus the plasma tends to behave as a dielectric. On the other hand, the zz component of the dielectric tensor is given by an expression similar to (7.4.48) except that the term

$\omega_{c\alpha}$ is missing. Thus

$$\kappa_{zz} = \kappa_B - \sum (\omega^2_{p\alpha}/\omega^2)[1 - (i/\omega\tau)]. \tag{7.4.50}$$

We are interested in a regime in which $\omega_{p\alpha} \gg \omega$, so that (7.4.50) will be large and negative.

Let us consider a plane electromagnetic wave of the form $\exp[i(\mathbf{k}\cdot\mathbf{r} - \omega t)]$. Maxwell's equations give a general relation connecting $\mathbf{k}$, ω, and the dielectric tensor $\boldsymbol{\kappa}$:

$$\mathbf{k} \times (\mathbf{k} \times \boldsymbol{\mathcal{E}}) = \boldsymbol{\kappa}(\omega^2/c^2)\,\boldsymbol{\mathcal{E}}. \tag{7.4.51}$$

Equation (7.4.51) is equivalent to a set of three simultaneous, linear, homogeneous equations for the components of $\boldsymbol{\mathcal{E}}$. The condition that a nontrivial solution should exist is that a determinantal equation must be satisfied:

$$\det[(\mathbf{k}^2\,\delta_{\alpha\beta} - k_\alpha k_\beta) - (\omega^2/c^2)\,\kappa_{\alpha\beta}] = 0. \tag{7.4.52}$$

Equation (7.4.52) determines the dispersion relation $\omega(\mathbf{k})$ for the electromagnetic wave.

Let us consider a wave which propagates in the plane perpendicular to the magnetic field, with $\mathbf{k} \perp \mathbf{B}$. Specifically suppose $\boldsymbol{\mathcal{E}}$ is in the x direction and $\mathbf{k}$ is in the y direction (propagation across the magnetic field). The magnetic field produced by the wave is then parallel to the external magnetic field.

$$k^2 = \kappa_{xx}\omega^2/c^2. \tag{7.4.53}$$

The wave propagates with a speed

$$v = v_A = c/\kappa_{xx}^{1/2} \approx B[\mu_0 n(m_1^* + |m_2|^*)]^{-1/2} \tag{7.4.54}$$

in which $n = n_1 = n_2$, and we have neglected the background portion of the dielectric function. This wave is known as the *magnetosonic wave*.

If the wave vector $\mathbf{k}$ of the electromagnetic field has a component parallel to $\mathbf{B}$, the situation is slightly different. The magnetic field produced by the wave has a component perpendicular to the external magnetic field. Let $\mathbf{k}$ lie in the xz plane. Put $k_x = k \sin\theta$, $k_z = k\cos\theta$. We obtain from Eq. (7.4.52)

$$\kappa_{xx}\kappa_{zz}(\omega^2/c^2) - \mathbf{k}^2(\kappa_{zz}\cos^2\theta + \kappa_{xx}\sin^2\theta) = 0. \tag{7.4.55}$$

From (7.4.49) and (7.4.50), it follows that the ratio $|\kappa_{zz}|/\kappa_{xx}$ is large (of order ω_c/ω where ω_c is a carrier cyclotron frequency). Hence, except when θ is very close to $\pi/2$, we may obtain an approximate solution of (7.4.55) in the form

$$k^2 = (\kappa_{xx}/\cos^2\theta)(\omega^2/c^2). \tag{7.4.56}$$

Thus the propagation speed is

$$v = c \cos\theta/\kappa_{xx}^{1/2} = v_A \cos\theta \tag{7.4.57}$$

in which v_A is given by (7.4.54). The speed of propagation depends on the angle between **k** and the field. This is the Alfven wave. It disappears when **k** becomes perpendicular to **B**, and is replaced by the magnetosonic wave. The damping of these waves is seen from (7.4.49) to be small when $\omega\tau$ is large. Magnetoplasma waves of this type were proposed to exist in interstellar space by Alfven (1942). In solids, conditions are most favorable for their observation in semimetals. Experimental observations have been made using bismuth (see Isaacson and Williams, 1969).

7.4.4 *Cyclotron Resonance*

In our discussion of helicons and other magnetoplasma waves, we have assumed that the **k** dependence of the dielectric tensor or the conductivity tensor was negligible. We will now investigate phenomena where, as in the theory of the anomalous skin effect, this is not the case. Such an investigation must be based on the Boltzmann equation, as the simple kinetic approach is not sufficient. An adequate starting point is furnished by Chambers' expression for the distribution function, Eq. (7.3.25),

$$f - f_0 = df_0/dE \int_{-\infty}^{t_0} e\mathcal{E}(t)\cdot\mathbf{v}(t) \exp\left[-\int_t^{t_0} dt'/\tau(\mathbf{k}(t'))\right] dt. \tag{7.3.25$'$}$$

In the present case, we put $\mathcal{E}(t) \propto \exp[i(\mathbf{q}\cdot\mathbf{r}' - \omega t)]$ and assume that the relaxation time τ is constant. It is apparent from the argument leading to Chambers' formula that if the electric field is a function of position, this leads to an implicit time dependence as the electron moves along a trajectory

$$\mathbf{r}'(t) = \mathbf{r} + \int_{t_0}^{t} \mathbf{v}(t')\, dt'. \tag{7.4.58}$$

Here $\mathbf{r}$ is the position reached at time t_0, while $\mathbf{r}'(t)$ is the instantaneous position at t, which is the position which must be used in the expression for the field. Thus

$$f - f_0 = (df_0/dE)\, e\mathcal{E}(\mathbf{r}, t_0) \int_{-\infty}^{t_0} dt\, \mathbf{v}(t) \exp\Big[(\tau^{-1} - i\omega)(t - t_0) + i\mathbf{q}\cdot\int_{t_0}^{t} \mathbf{v}(t')\, dt'\Big]. \tag{7.4.59}$$

It is convenient to introduce a new variable $\phi' = \omega_c t$ where ω_c is the cyclotron frequency. We proceed to construct the elements $\sigma_{\alpha\beta}$ of the conductivity tensor in the usual way:

$$\sigma_{\alpha\beta} = (-e^2/4\pi^3) \int d^3k \, df_0/dE \, v_\alpha(\mathbf{k})/\omega_c \int_{-\infty}^{\phi} d\phi' \, v_\beta(\phi')$$

$$\times \exp[(1/\omega_c)(\tau^{-1} - i\omega)(\phi' - \phi) + (i/\omega_c) \int_{\phi}^{\phi'} \mathbf{q}\cdot\mathbf{v}(\phi'') \, d\phi''].$$

(7.4.60)

Under the influence of a strong magnetic field, electrons move in orbits around a cross section of the Fermi surface in a plane perpendicular to the field. We can then interpret ϕ, which measures angular position along such a trajectory, as the azimuthal angle of polar coordinate system. As the temperature dependence of σ is not of particular concern, we may approximate df_0/dE by $-\delta(E - \mu)$. It will also be assumed that the Fermi surface is spherical. It is convenient to use the velocity v as the independent variable in the integration; thus $\mu = m^* v_F{}^2/2$, $E = m^* v^2/2$, and $d^3k = (m^*/\hbar)^3 \, d^3v$. We must also consider the geometry of the system. The first case which will be discussed is that in which the magnetic field is parallel to the surface of the metal. Azbel and Kaner (1956) suggested that cyclotron resonance should be observable in metals in this geometry. It will also be assumed that $\boldsymbol{\mathcal{E}}$ is parallel to $\mathbf{B}$ (which defines the z direction). Then we need only consider σ_{zz}. The normal into the surface of the metal is now taken to define the positive y direction, that is, the direction of the propagation vector $\mathbf{q}$ of the electromagnetic field. Thus

$$\int_{\phi}^{\phi'} \mathbf{q}\cdot\mathbf{v}(\phi'') \, d\phi'' = qv_F \sin\theta(\cos\phi - \cos\phi')$$

and

$$\sigma = \sigma_{zz} = [2m^{*2} v_F{}^3 e^2/(2\pi\hbar)^3 \omega_c] \int d\theta \, d\phi \sin\theta \cos^2\theta \int_{-\infty}^{\phi}$$

$$\times \exp\{-i\bar{\omega}/\omega_c[\phi' - \phi - (qv_F/\bar{\omega}) \sin\theta \, (\cos\phi' - \cos\phi)] \, d\phi'$$

(7.4.61)

in which $\bar{\omega} = \omega + i/\tau$.

The integration of (7.4.61) is fairly difficult. We will sketch the procedure (see Cohen *et al.* (1960) for details). The integral over ϕ' may be expressed as

$$\int_{-\infty}^{\phi} d\phi' \exp\{-i(\bar{\omega}/\omega_c)[\phi' - (qv_F/\bar{\omega}) \sin\theta \cos\phi']\}$$

$$= \sum_n (i^n) J_n[(qv_F/\bar{\omega}) \sin\theta] \int_{-\infty}^{\phi} \exp\{i[n - (\bar{\omega}/\omega_c)]\phi'\} \, d\phi'$$

$$= \sum_{n=-\infty}^{\infty} \{(i)^n/i[n - (\bar{\omega}/\omega_c)]\} J_n[(qv_F/\bar{\omega}) \sin\phi] \exp\{i[n - (\bar{\omega}/\omega_c)]\phi\}$$

in which J_n is a Bessel function, and we have used the identity

$$e^{i\rho\cos\phi} = \sum_{n=-\infty}^{\infty} i^n J_n(\rho) e^{in\phi}.$$

The integral over ϕ may be done in a similar manner. The expression for σ reduces to

$$\sigma = [4\pi m^{*2} v_F^3 e^2/(2\pi\hbar)^3 \omega_c] \sum_n 1/i[n - (\bar{\omega}/\omega_c)]$$

$$\times \int_0^{\pi} d\theta \sin\theta \cos^2\theta \, J_n^2[(qv_F/\omega_c) \sin\theta]. \tag{7.4.62}$$

A reasonably simple result can be obtained in the limit $qv_F/\omega_c \gg 1$. In this case we may employ an asymptotic expansion for the Bessel function

$$J_n(\rho) \approx (2/\pi\rho)^{1/2} \cos(\rho - n\pi/2 - \pi/4).$$

Since $J_n(\rho)$ oscillates rapidly when ρ is large, we may replace $J_n^2(\rho)$ by its average, which will be $1/\pi\rho$. The integration over θ is then trivial. After some further simplifications (7.4.62) reduces to

$$\sigma = \tfrac{3}{4} ne^2/m^* v_F q \sum_{n=-\infty}^{\infty} \omega_c\tau/[1 + i(n\omega_c - \omega)\tau]. \tag{7.4.63}$$

The sum can be performed:

$$\sum_{n=-\infty}^{\infty} 1/[1 + i(n\omega_c - \omega)\tau] = z(1 - i\omega\tau)^{-1}[(1/z) + 2z \sum_{n=1}^{\infty} 1/(z^2 + n^2\pi^2)]$$

$$= z \operatorname{ctnh}[z/(1 - i\omega\tau)]$$

in which $z = \pi(1 - i\omega\tau)/\omega_c\tau$. Thus the final formula for σ is

$$\sigma = (3\pi/4)(ne^2/m^* v_F q) \operatorname{ctnh}[\pi(1 - i\omega\tau)/\omega_c\tau]. \tag{7.4.64}$$

If $\omega\tau \gg 1$, this function oscillates with ω/ω_c. These oscillations will be manifest as resonances in the surface impedance. Note also that if the magnetic field vanishes, the ctnh function goes to unity and the expression for σ agrees with (7.4.33), the result obtained in our discussion of the anomalous skin effect. Since the dependence on the wavevector of the field q is the same as was obtained in the theory of the anomalous skin effect, we may write the surface impedance in the present case as

$$Z(B) = Z(0)\{\tanh[\pi(1 - i\omega\tau)/\omega_c\tau]\}^{1/3}. \tag{7.4.65}$$

Here $Z(0)$ is the surface impedance in the absence of the field, which is given by (7.4.38) under the assumption of specular reflection of electrons at the surface. Structure in the surface impedance will repeat whenever the ratio ω/ω_c changes by an integer. Thus the resonant condition is essentially $\omega_c = \omega/(n + \frac{1}{2})$.

A simple physical picture can be attached to these results: An electron moves in an orbit perpendicular to **B**. Since the electric field is confined to the skin depth, only those electrons whose orbits take them into the skin depth can absorb energy. For a general orbit, the time of arrival of electron in the region of the field will correspond to varying values of the field, and nothing interesting happens. However, if ω/ω_c is an integer, the field makes an integral number of oscillations in a revolution of the electron, and the electron always arrives in the skin layer in phase with the field.

Figure 7.4.2 shows cyclotron resonance oscillations in the surface impedance of antimony, according to Herrod and Goodrich (1970). The structure at low fields is real, and is a manifestation of magnetically induced surface quantum states (Koch, 1968). In such a state, an electron moving approximately parallel to the surface (a "skipping" trajectory) is effectively trapped in a potential well. This well is bounded on one side by the metal–vacuum surface barrier. On the other side, the magnetic field confines the electron to a depth of the order of an orbit radius.

We will now investigate what can happen when the magnetic field is perpendicular to the surface of the metal. The incoming normal to the surface defines the z axis. Let the electric field be circularly polarized in the xy plane. Define

$$\mathcal{E}_\pm = \mathcal{E}_x \pm i\mathcal{E}_y.$$

The current may be treated similarly:

$$j_\pm = j_x \pm ij_y = \sigma_\pm \mathcal{E}_\pm$$

where

$$\sigma_\pm = \sigma_{xx} \pm i\sigma_{yx} = \sigma_{xx} \mp i\sigma_{xy} \tag{7.4.66}$$

and we have used the Onsager relation $\sigma_{xy}(B) = -\sigma_{yx}(B)$, Eq. (7.3.13).

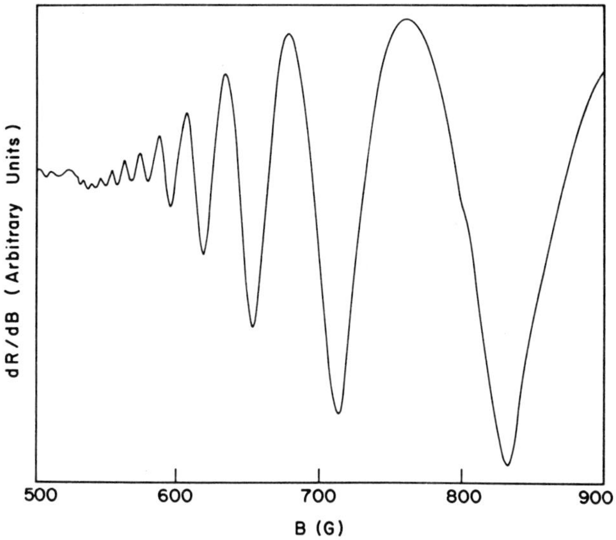

Fig. 7.4.2. Cyclotron resonance in antimony. The measurement is in the Azbel–Kaner geometry. The irregular structure at low fields is a manifestation of magnetic surface states. [Courtesy R. A. Herrod and R. G. Goodrich.]

We consider the calculation of $\sigma_\pm$ from (7.4.60). Put $v_x \pm iv_y = v \cos\theta\, e^{\pm i\phi}$. The situation is similar to the helicon problem. We will calculate σ_-. The wave vector $\mathbf{q}$ is in the z direction. We obtain, in place of (7.4.61)

$$\sigma_- = [2m^{*2}v_F{}^3e^2/(2\pi\hbar)^3\omega_c] \int_0^\pi d\phi \sin^3\phi \int_0^{2\pi} d\phi \cos\phi$$

$$\times \int_{-\infty}^{\phi} \exp\{i\phi' - (i\bar{\omega}/\omega_c)[\phi' - \phi - (qv_F/\bar{\omega}) \cos\theta\,(\phi' - \phi)]\}\, d\phi' \tag{7.4.67}$$

(since $\mathbf{q}\cdot\mathbf{v}$ is independent of ϕ). In this case, the integrals turn out to be elementary, and we obtain

$$i\sigma_- = (3ne^2/4m^*qv_F)\{2\xi + (1 - \xi^2) \ln[(\xi + 1)/(\xi - 1)]\} \tag{7.4.68}$$

in which

$$\xi = (\omega_c - \omega - i/\tau)qv_F. \tag{7.4.69}$$

To obtain σ_+, we change the sign of ω in (7.4.69) and insert an overall negative sign. However, only ξ_- and σ_- are important as only waves with this circular polarization can propagate in the metal.

In the limit $q \to 0$ and $\omega_c \gg \omega \gg 1/\tau$, Eq. (7.4.68) yields $i\sigma_- \approx \sigma_N$, as it should [see (7.4.44)]. This brings us back to the helicon problem. In the present case, we are interested in a different limit, namely ξ near 1 or

$$\omega_c - \omega = qv_F. \tag{7.4.70}$$

For $\xi < 1$, the logarithm acquires a large imaginary part, and the conductivity becomes dissipative. At this point, helicon propagation effectively ceases. Miller and Haering (1962) show that a spike in the surface impedance is obtained. The surface resistance will be proportional to [see (7.4.37)]

$$R \sim \int \sigma_R/[(q^2 + \mu_0\omega\sigma_i)^2 + \mu^2\omega^2\sigma_R{}^2]\, dq \tag{7.4.71}$$

in which $\sigma_i(q)$ and $\sigma_R(q)$ are the imaginary and real parts of σ_-. Below the value of q for which (7.4.70) is satisfied σ_R is small [it comes from the $i/\tau qv_F$ term in (7.4.69)]. The integral is too complicated to do analytically, however, insight into the structure of the results can be obtained by examining the term in the denominator $q^2 + \mu\omega\sigma_i$ near $\xi = 1$. Put $\xi = 1 + x$. We find, for x small and positive,

$$\begin{aligned} q^2 + \mu\omega\sigma_i &= [(\omega_c - \omega)^2/v_F{}^2(1 + x)^2] \\ &\quad \times \{1 - \tfrac{3}{2}[ne^2\mu_0\omega v_F{}^2/m^*(\omega_c - \omega)^3](1 + x)^3\}. \end{aligned} \tag{7.4.72}$$

The spike in the surface resistance occurs close to the value of the frequency such that the term in (7.4.72) independent of x vanishes. Since $\omega_c \gg \omega$, this condition is

$$\omega = \tfrac{2}{3}(m^*\omega_c{}^3/ne^2\mu_0 v_F{}^2) = \tfrac{2}{3}\omega_c(\omega_c/\omega_p)^2(c^2/v_F{}^2). \tag{7.4.73}$$

We note from (7.4.73) that the frequency at which the spike occurs is proportional to B^3. Observations in sodium and potassium give the position of the spike in good agreement with the position predicted by (7.4.73) (Taylor, 1965).

The following physical description is possible: Since $\omega \ll \omega_c$, the condition $qv_F = \omega_c$ corresponds to

$$qR_o = 1 \tag{7.4.74}$$

where R_o is the radius of the orbit of an electron in the magnetic field. When (7.4.74) is satisfied, the wavelength of the field corresponds exactly to a cyclotron orbit, and electrons with velocities v_F parallel to the field move so as always to experience an electric field having the same phase. This leads to resonant absorption of energy, and to the sudden appearance of a large real part of the conductivity.

7.4.5 *Radio-Frequency Size Effects*

The preceding discussion does not by any means exhaust the interesting electromagnetic effects in metals. Let us consider the penetration of a circularly polarized electric field into a metal in circumstances in which (7.4.68) is a good approximation to the conductivity. The general expression for the field at a depth z in the metal can be obtained by a slight generalization of Eq. (7.4.36). We assume specular reflection

$$\mathcal{E}_{\pm}(z) = \mathcal{E}_x(z) \pm i\mathcal{E}_y(z) = -(1/\pi)(d\mathcal{E}_{\pm}/dz)_{0^+}\int_{-\infty}^{\infty} dp\, e^{ipz}/[p^2 - i\mu\omega\sigma_{\pm}(p)]. \tag{7.4.75}$$

We wish to evaluate this expression for large values of z. The procedure is to convert (7.4.75) into a contour integral in which the contour is closed in the upper half of the complex p plane. The singularities of the integrand are, for a fixed magnetic field, poles corresponding to helicon modes, and branch cuts (see Fig. 7.4.3). When the parameters of the problem are such that the poles are further from the real axis than the terminus of the branch cut (which are characterized by $\xi = \pm 1$) the dominant singularities are those due to the branch cuts. Let us consider the specific case of $\mathcal{E}_-$. The contour is to be closed in the upper half of the complex p plane. The branch point occurs in this case (suppose $\omega_c \gg \omega$) when

$$p = p_B = (1/v_F)(-\omega_c + i/\tau) = -(1/R_o) + (i/l) \tag{7.4.76}$$

in which l is the mean free path and R_o is the radius of the orbit. The discontinuity of the integrand across the cut must be integrated along the

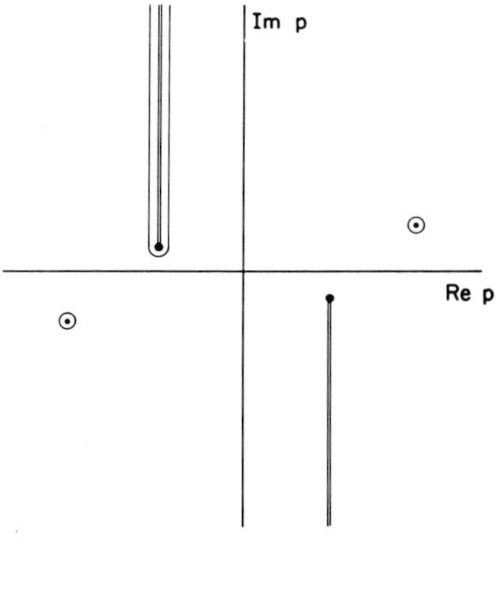

Fig. 7.4.3. Position of singularities in the integrand of Eq. (7.4.75) in the case of σ_-. Solid lines indicate branch cuts; circled points are poles.

cut, parallel to the imaginary axis, from p_B to $i\infty$. The asymptotic form of the result is (Gantmakher and Kaner, 1965)

$$\mathcal{E}_{-} \sim (Ae^{-z/l}/z^2)e^{-i(z/R_0+i\pi/2)} \tag{7.4.77}$$

in which A is a constant we will not evaluate. The field alternates with the mean free path as the characteristic distance, and oscillates with a period determined by the radius of an electron orbit.

Observation of these oscillations is possible if the sample is in the form of a slab of finite thickness L. The necessary electromagnetic theory is somewhat more complicated (Platzman and Buchsbaum, 1963), but solutions can be obtained under the assumption of specular reflection. Baraff (1968) has treated the more physical case of diffuse reflection. The field which emerges from the slab shows the oscillations in L/R_0 inferred from (7.4.77). Observation of such oscillations is evidently possible only when the ratio of slab thickness to mean free path, L/l, is not large.

Note that since the oscillatory term involves $1/R_0 = eB/m^*v_F$, the oscillations observed in transmission should be periodic in B; such is in fact observed [for one example (potassium), see Libchaber *et al.* 1970]. Pictorially, the effect is due to the radio-frequency currents carried through the specimen by electrons moving in helical paths in the magnetic field. Somewhat similar effects are observed when the magnetic field is parallel

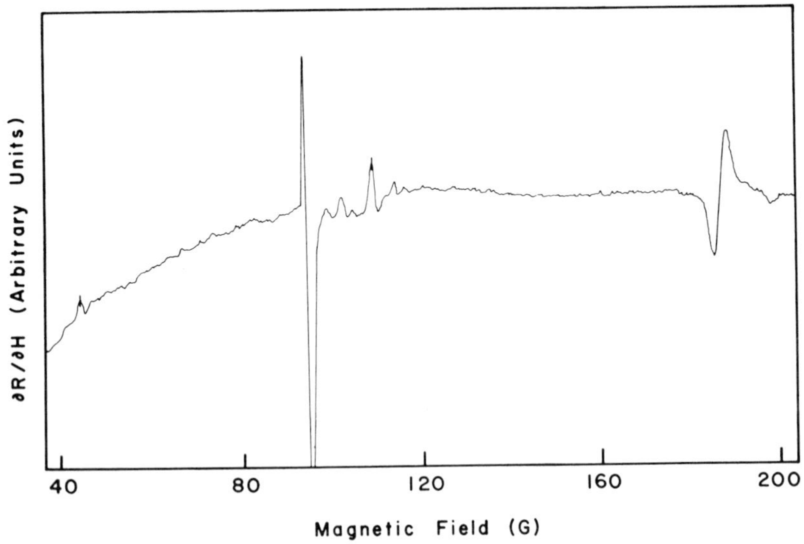

Fig. 7.4.4. Radio-frequency size effect in cadmium. The large structure near 90 G arises from the so-called "lens" portion of the Fermi surface. The reversed structure near 180 G occurs when two lens orbits fit within the sample. [From Jones *et al.* (1968).]

to the surface of the specimen and the electric field is perpendicular to B. (Gantmakher, 1967). In this case, large and sharp oscillations in transmitted power are found with maxima occurring when an integral number of extremal orbits are just contained within the sample. This condition is

$$2nR_o = L \tag{7.4.78}$$

in which n is an integer.

Equation (7.4.78) enables a simple determination of the diameter of a Fermi surface. The radio-frequency size effect has proved to be a valuable tool for the study of Fermi surfaces. Physically, it results from thin sheets of current which are set up inside the material as a result of the orbital motion of the carriers. Such a sheet is, in effect, an image of the surface current. This problem can be analyzed simply in terms of the ineffectiveness concept of Pippard, as was done for the anomalous skin effect in subsection 7.4.1. Only those electrons whose velocities lie in an angle $\delta/2R_o$ can contribute; the nth current sheet being damped by a factor $(\delta/2R_o)^{n/2}$. Figure 7.4.4 shows an example of the size effect observed in cadmium, according to Jones *et al.* (1968).

7.5 Ultrasonic Attenuation by Electrons

There are many processes which cause attenuation of sound waves in crystals, including scattering by dislocations, interactions with the phonons normally present in the crystal, and so on. Attention will be focused here on attenuation by the electron distribution. We recall that point impurities, for example, do not scatter long wavelength sound waves efficiently, but are effective scatterers for electrons at the top of a Fermi distribution. Hence, while the electron distribution tends to be driven out of equilibrium by a sound wave, it moves back toward equilibrium by interaction with impurities and with phonons. In this way, energy is removed from the sound wave and dissipated. It turns out that the attenuation of a sound wave in a metal is substantially affected by the presence of a strong magnetic field. Oscillations in the attenuation are produced which furnish one of the principal means of determining properties of the Fermi surface. The subject of ultrasonic attenuation is a difficult one and agreement with experiment has proved somewhat elusive. Our discussion will follow, in the main, the work of Cohen *et al.* (1960).

7.5.1 *Ultrasonic Attenuation in Zero Magnetic Field*

We will begin by considering the problem in the absence of a magnetic field. Certain limitations will be imposed at the start. We will consider only sound waves of long wavelength ($\lambda \gg a$) where a is a lattice constant. For such waves the discrete atomic structure of the material can be neglected. The displacement of the ions from their equilibrium positions will be described by a continuous velocity field of wave vector $\mathbf{q}$ and frequency ω: $\mathbf{u}(\mathbf{r}, t) \propto \exp[i(\mathbf{q}\cdot\mathbf{r} - \omega t)]$.

We must determine the distribution function for electrons in the presence of the wave. The Boltzmann equation is in the present case

$$\partial f/\partial t + \mathbf{v}(\mathbf{k})\cdot\nabla_{\mathbf{r}} f + (\mathbf{F}/\hbar)\cdot\nabla_{\mathbf{k}} f = (\partial f/\partial t)_{\mathrm{scat}} \tag{7.5.1}$$

in which $\mathbf{F} = -e[\boldsymbol{\mathcal{E}} + \mathbf{v}(\mathbf{k}) \times \mathbf{B}]$. We will proceed by making the usual relaxation time approximation for the right side of (7.5.1):

$$(\partial f/\partial t)_{\mathrm{scat}} = -(f - f_s)/\tau \tag{7.5.2}$$

in which f_s is the distribution to which the electrons relax. In previous sections, we have put $f_s = f_0$, the equilibrium Fermi distribution. In the present case, we assume that the relaxation time of the electrons is small compared to the period of the sound wave, so that the electrons scatter into a Fermi distribution centered about the velocity of the medium $\mathbf{u}$. Thus

$$f_s(\mathbf{v}) = f_0(\mathbf{v} - \mathbf{u}(\mathbf{r}, t), E_{\mathrm{F}}(\mathbf{r}, t)). \tag{7.5.3}$$

The Fermi energy E_{F} adjusts itself so as to maintain the correct electron density in the system. This is important, since local charge neutrality must be maintained throughout.

We will further restrict our considerations to a system with a single parabolic band of effective mass m^*. The function f_s may be expanded, with the ion velocity $\mathbf{u}$ and the difference between the local electron density ρ and the equilibrium density ρ_0 being regarded as small parameters. Let $\Delta\rho = \rho - \rho_0$. We have

$$f_s = f_0(\mathbf{v}, \rho_0) - \mathbf{u}\cdot\nabla_{\mathbf{v}} f_0 + \Delta\rho(\partial f_0/\partial\rho)_{\rho_0}.$$

The equilibrium distribution function depends on the electron density through the local Fermi energy E_{F}:

$$\partial f_0/\partial\rho = \partial E_{\mathrm{F}}/\partial\rho\; \partial f_0/\partial E_{\mathrm{F}} = \tfrac{2}{3}(E_{\mathrm{F}}^{(0)}/\rho)\; \partial f_0/\partial E_{\mathrm{F}} = -\tfrac{2}{3}(E_{\mathrm{F}}^{(0)}/\rho)\; \partial f_0/\partial E$$

since f_0 is a function of the difference $E - E_{\mathrm{F}}$. Thus

$$f_s \approx f_0(\mathbf{v}, \rho_0) - \partial f_0/\partial E\; (m^*\mathbf{u}\cdot\mathbf{v} + \tfrac{2}{3}E_{\mathrm{F}}^{(0)}\, \Delta\rho/\rho_0). \tag{7.5.4}$$

Now we follow the usual procedure and set $f = f_0 + f_1$ with $f_1 \propto \exp[i(\mathbf{q}\cdot\mathbf{r} - \omega t)]$, while f_0 is independent of position and time. Equation (7.5.1) becomes

$$i(\mathbf{q}\cdot\mathbf{v} - \omega)f_1 - e\boldsymbol{\mathcal{E}}\cdot\mathbf{v}\,\partial f_0/\partial E$$
$$= -(f_1/\tau) - (1/\tau)(m^*\mathbf{u}\cdot\mathbf{v} + \tfrac{2}{3}E_F{}^{(0)}\,\Delta\rho/\rho_0)\,\partial f_0/\partial E.$$

The solution is

$$f_1 = -(\partial f_0/\partial E)[-e\tau\boldsymbol{\mathcal{E}}\cdot\mathbf{v} + m^*\mathbf{u}\cdot\mathbf{v} + \tfrac{2}{3}E_F{}^{(0)}\,\Delta\rho/\rho_0][1 + i\tau(\mathbf{q}\cdot\mathbf{v} - \omega)]^{-1}. \tag{7.5.5}$$

We can now calculate the electronic portion j_e of the total current from (7.2.7). The result looks somewhat different from formulas we have previously obtained since there is a new type of term, involving $\Delta\rho/\rho$, which describes electron diffusion:

$$j_{e,\alpha} = \sigma_{\alpha\beta}[\mathcal{E}_\beta - (m^*\mu_\beta/e\tau)] - \Delta\rho\; ev_sR_\alpha, \tag{7.5.6}$$

in which v_s is the speed of sound, $\sigma_{\alpha\beta}$ represents the conductivity tensor, and R_α is the diffusion vector. A summation over repeated Cartesian indices is implied. The quantities are defined as

$$\sigma_{\alpha\beta} = (e^2/4\pi^3)\int v_\alpha v_\beta\tau\,(-\partial f_0/\partial E)/[1 + i\tau(\mathbf{q}\cdot\mathbf{v} - \omega)]\,d^3k \tag{7.5.7}$$

and

$$R_\alpha = (1/6\pi^3)(E_F{}^{(0)}/v_s\rho_0)\int v_\alpha\,(-\partial f_0/\partial E)/[1 + i\tau(\mathbf{q}\cdot\mathbf{v} - \omega)]\,d^3k. \tag{7.5.8}$$

There are two new features of the result (7.5.6). The first is a change in the effective electric field by $m^*\mathbf{u}/e\tau$ which results from the fact that the electrons scatter against moving impurities. This is the collision drag effect. The second is the presence of the diffusion vector $\mathbf{R}$. Since the electron distribution in the presence of a sound wave is not uniform, diffusion occurs as the electrons attempt to redistribute themselves.

There is a relation between the conductivity and the diffusion current. We introduce the identity

$$[1 + i\tau(\mathbf{q}\cdot\mathbf{v} - \omega)]^{-1} = (1 - i\omega\tau)^{-1}\{1 - [i\mathbf{q}\cdot\mathbf{v}\tau/(1 + i\tau(\mathbf{q}\cdot\mathbf{v} - \omega))]\}$$

and obtain

$$i\sigma_{\alpha\beta}q_\beta = -[3v_s\rho_0e^2(1 - i\omega\tau)/(2E_F{}^{(0)})]R_\alpha. \tag{7.5.9}$$

The diffusion current can be interpreted as the response of the system to a density gradient. Suppose that, in analogy to the relation between ordinary current and an external field, we define a diffusion current $\mathbf{j}_D$:

$$\mathbf{j}_D(\mathbf{r}, t) = e \int d^3r' \int dt' \, \mathfrak{D}(\mathbf{r} - \mathbf{r}', t - t') \cdot \nabla_\rho(\mathbf{r}', t'). \qquad (7.5.10)$$

Let $D_{\alpha\beta}(\mathbf{q}, \omega)$ be the Fourier transform of $\mathfrak{D}$. We have

$$j_{D\alpha}(\mathbf{q}, \omega) = ieD_{\alpha\beta}(\mathbf{q}, \omega) q_\beta \, \Delta\rho(\mathbf{q}, \omega). \qquad (7.5.11)$$

However, (7.5.6) implies $j_{D\alpha}(\mathbf{q}, \omega) = -ev_s R_\alpha \, \Delta\rho(\mathbf{q}, \omega)$. Thus

$$v_s R_\alpha = -iD_{\alpha\beta}(\mathbf{q}, \omega) q_\beta. \qquad (7.5.12)$$

It is convenient to relate the conductivity tensor $\sigma_{\alpha\beta}$ to a mobility tensor $\mu_{\alpha\beta}$ by

$$\sigma_{\alpha\beta}(\mathbf{q}, \omega) = e\rho_0(\mathbf{q}, \omega) \mu_{\alpha\beta}(\mathbf{q}, \omega). \qquad (7.5.13)$$

The mobility may now be related to the diffusion tensor through (7.5.9) and (7.5.12):

$$\tfrac{2}{3} E_F^{(0)} \mu_{\alpha\beta}(\mathbf{q}, \omega) = eD_{\alpha\beta}(\mathbf{q}, \omega)(1 - i\omega\tau). \qquad (7.5.14)$$

This is a generalization of the so-called "Einstein relation" (which applies when $\omega = 0$) to time-dependent fields. This relation is unchanged by the inclusion of an external magnetic field (however, μ and D must be recalculated).

The expression (7.5.7) for the conductivity is quite similar to one obtained previously in the case of the anomalous skin effect, Eq. (7.4.30). In the absence of an external magnetic field, the off-diagonal components of σ vanish. The diagonal components can be evaluated as was done in Section 7.4.1. Let us take the direction of q as defining the x axis. We obtain σ_{zz} (and σ_{yy}) from (7.4.32) noting the change in coordinate system:

$$\sigma_{zz} = \sigma_{yy} = \tfrac{3}{2}[\sigma_0/(1 - i\omega\tau)\zeta^3][(1 + \zeta^2) \tan^{-1}\zeta - \zeta] \qquad (7.4.32')$$

in which $\zeta = qv\tau[1 - i\omega\tau]^{-1}$, and σ_{xx} is found by a similar procedure to be

$$\sigma_{xx} = [3\sigma_0/(1 - i\omega\tau)\zeta^3](\zeta - \tan^{-1}\zeta). \qquad (7.5.15)$$

In these equations, $\sigma_0 = ne^2\tau/m^*$.

The only nonvanishing component of $\mathbf{R}$ is R_x. This may be found from Eqs. (7.5.12–7.5.15) or directly by integrating (7.5.8):

$$R_x = (-i)[2E_F^{(0)} \sigma_0 q / e^2 v_s \rho_0 \zeta^3](\zeta - \tan^{-1}\zeta)/(1 - i\omega\tau)^2. \qquad (7.5.16)$$

These results may be combined to yield an expression for the x component of the electronic current:

$$j_{ex} = \sigma_{xx}[\mathcal{E}_x - (mu_\alpha/e\tau) + i(\Delta\rho/3\rho_0)(m^* v\zeta/e\tau)]. \qquad (7.5.17)$$

The total current in the system is the sum of the electronic current $\mathbf{j}_e$ and that due to the background motion of the ions $\rho e\mathbf{u}$:

$$\mathbf{j} = \mathbf{j}_e + \rho e\mathbf{u}. \qquad (7.5.18)$$

The electric field which is present in these formulas arises from the sound wave itself. It must ultimately be eliminated from the expressions. Additional relations between current and field are required, which are furnished by the Maxwell equations. The components of the electric field perpendicular to the wave vector $\mathbf{q}$ of the sound wave satisfy the wave equation

$$\nabla^2 \mathbf{\mathcal{E}}_\perp = \mu\, \partial \mathbf{j}_\perp/\partial t + \mu\epsilon\, \partial^2 \mathbf{\mathcal{E}}_\perp/\partial t^2.$$

We will take free space values for μ and ϵ, and write $\omega = v_s q$. In this way, we obtain

$$\mathbf{\mathcal{E}}_\perp = (i\mu_0/\omega)v_s^2/(1 - v_s^2/c^2)\mathbf{j}_\perp \qquad (7.5.19)$$

in which c is the speed of light in vacuum. The longitudinal components of the field satisfy $\nabla\cdot\mathbf{\mathcal{E}} = e\rho_c/\epsilon_0$ in which ρ_c contains contributions (of opposite sign) from electrons and ions. We also have the equation of continuity

$$\nabla\cdot\mathbf{j} + e\, \partial\rho_c/\partial t = 0. \qquad (7.5.20)$$

We obtain from these equations

$$\mathcal{E}_{||} = (-i/\epsilon_0\omega)j_{||} \qquad (7.5.21)$$

in which $\mathcal{E}_{||}$ and $j_{||}$ refer to components parallel to $\mathbf{q}$. Equation (7.5.20) must also be satisfied by the electronic component of the current. Only the displaced electron density $\Delta\rho$ is time-dependent. Thus

$$\Delta\rho = qj_{||}/e\omega = -j_{e||}/ev_s. \qquad (7.5.22)$$

We are now ready to consider the attenuation. This requires calculation of the net power dissipation per unit volume. The sound wave supplies energy to the electrons as it moves through the material which the electrons dissipate by collisions. The average rate of transfer of electron energy per unit volume is just the work done by the electric field (resistive loss) $\mathbf{j}_e\cdot\mathbf{\mathcal{E}}$. Some of this energy is dissipated as heat, but a part is coherently fed back into the sound wave. This occurs because the impurities with which the electrons collide are moving. The average electron velocity before collision $\langle\mathbf{v}\rangle$ differs from that after collision, which is just $\mathbf{u}$, the velocity of the lattice. The average change in momentum of the electrons, which gives the average force exerted on the lattice (per unit volume), is $\rho m^*(\langle\mathbf{v}\rangle - \mathbf{u})/\tau$. The power that is fed back into the sound wave is

$\mathbf{u}^* \cdot \rho_0 m^*(\langle \mathbf{v} \rangle - \mathbf{u})/\tau$. The net power dissipated per unit volume is

$$Q = \tfrac{1}{2} \operatorname{Re}[\mathbf{j}_e^* \cdot \boldsymbol{\mathcal{E}} - \mathbf{u}^* \cdot \rho_0 m^*(\langle \mathbf{v} \rangle - \mathbf{u})/\tau]. \qquad (7.5.23)$$

It is convenient to work with complex fields and currents; this convenience means that we must write instead of $\mathbf{A} \cdot \mathbf{B}$ involving real fields, $\frac{1}{2} \operatorname{Re} \mathbf{A}^* \cdot \mathbf{B}$ for complex fields. The second term in (7.5.23) describes "collision drag."

The preceding equations (7.5.17)–(7.5.22) allow us to determine the electron current $\mathbf{j}_e$ and the electric field $\boldsymbol{\mathcal{E}}$ in terms of the sound wave velocity $\mathbf{u}$. These relations are linear and can be symbolically represented in terms of tensors $\mathbf{V}$ and $\mathbf{W}$ by

$$\mathbf{j}_e = -\rho_0 e \mathbf{V} \cdot \mathbf{u}; \qquad \boldsymbol{\mathcal{E}} = -(\rho_0 e/\sigma_0) \mathbf{W} \cdot \mathbf{u}. \qquad (7.5.24)$$

Certain components of these tensors will be given explicitly below. It turns out to be permissible to neglect collision drag for frequencies which are not too high. Then the power dissipated can be written as

$$Q = \rho_0 (m^* \mid \mathbf{u} \mid^2 / 2\tau) \hat{\mathbf{u}} \cdot \mathbf{S} \cdot \hat{\mathbf{u}} \qquad (7.5.25)$$

in which the tensor $\mathbf{S}$ is given by

$$\mathbf{S} = \operatorname{Re} \mathbf{V}^* \cdot \mathbf{W} \qquad (7.5.26)$$

and $\hat{\mathbf{u}}$ is a unit vector in the direction of the polarization of the sound wave. We choose the direction of $\mathbf{q}$ as the $\mathbf{z}$ axis and let this coincide with $\mathbf{u}_3$, which thus represents longitudinal polarization. Other independent directions of polarization are taken to define the remaining coordinate directions. For a specific direction of polarization i, the power dissipated is

$$Q_i = \rho_0 m^* \mid \mathbf{u} \mid^2 / 2\tau S_{ii}. \qquad (7.5.27)$$

Hence, our concern will be to calculate the diagonal element of the tensor $\mathbf{S}$. The factor $\rho_0 m^* \mid \mathbf{u} \mid^2 / 2\tau$ defines a relevant energy scale for the attenuation problem; also we expect the diagonal elements of $\mathbf{S}$ not to differ enormously from unity.

Experimentally, one measures the attenuation coefficient α. The sound intensity decays with distance l in the direction of propagation as $e^{-\alpha l}$. As in the case of optical absorption, α is the ratio of the power dissipated per unit volume to the energy flux. The energy flux is $\frac{1}{2} M \mid \mathbf{u} \mid^2 v_s$ in which M is the atomic mass. It is assumed that there is one conduction electron per atom

$$\alpha = Q_i / (\tfrac{1}{2} M \mid \mathbf{u} \mid^2 v_s) = (m^* v / M v_s)(S_{ii}/l) \qquad (7.5.28)$$

in which $l = v\tau$ is the electron mean free path. A reasonable order of magnitude estimate gives $M v_s / m^* v \approx 100$.

We must now determine the tensors $\mathbf{V}$ and $\mathbf{W}$. Consider the third term in (7.5.6) and use (7.5.22):

$$-\Delta\rho\, e v_s R_x = R_x j_{ex} = \mathbf{R}\cdot\mathbf{j}_e. \tag{7.5.29}$$

Equation (7.5.29) is used to define the tensor $\mathbf{R}$, which has only a single nonzero element:

$$R_{ij} = R_x\, \delta_{ix}\, \delta_{jx} \tag{7.5.30}$$

in which R_x is given by (7.5.16). Equation (7.5.6) becomes

$$\mathbf{j}_e = \boldsymbol{\sigma}\cdot[\boldsymbol{\mathcal{E}} - (m^*\mathbf{u}/e\tau)] + \mathbf{R}\cdot\mathbf{j}_e.$$

This equation can be solved in the form

$$\mathbf{j}_e = \sigma_0\sigma'\cdot[\boldsymbol{\mathcal{E}} - (m^*\mathbf{u}/e\tau)] \tag{7.5.31}$$

in which σ' is a dimensionless modified conductivity tensor which includes diffusion

$$\boldsymbol{\sigma}' = [\mathbf{I} - \mathbf{R}]^{-1}\cdot\boldsymbol{\sigma}/\sigma_0. \tag{7.5.32}$$

Next, we must consider the total current. Equation (7.5.31) is substituted into (7.5.18):

$$\mathbf{j} = \sigma_0\boldsymbol{\sigma}'\cdot[\boldsymbol{\mathcal{E}} - (m^*\mathbf{u}/e\tau)] + \rho e\mathbf{u}. \tag{7.5.33}$$

Additional relations between total current and electric field are furnished by (7.5.19) and (7.5.21). These are combined in the form

$$\mathbf{j} = \sigma_0\mathbf{T}\cdot\boldsymbol{\mathcal{E}} \tag{7.5.34}$$

in which the dimensionless tensor $\mathbf{T}$ has elements

$$T_{zz} = T_{yy} = -(i\omega/\mu_0\sigma_0)(1 - v_s^2/c^2)/v_s^2 = -(i\epsilon_0\omega/\sigma_0)[(c^2/v_s^2) - 1] \tag{7.5.35a}$$

$$T_{xx} = i\epsilon_0\omega/\sigma_0. \tag{7.5.35b}$$

We substitute (7.5.34) into (7.5.33) and solve for $\boldsymbol{\mathcal{E}}$. The result has the form (7.5.24) with

$$\mathbf{W} = [\mathbf{T} - \boldsymbol{\sigma}']^{-1}[\boldsymbol{\sigma}' - I]. \tag{7.5.36}$$

Next, insert (7.5.24) into (7.5.31) and determine $\mathbf{V}$:

$$\mathbf{V} = \boldsymbol{\sigma}'\cdot(I + \mathbf{W}) = \boldsymbol{\sigma}'\cdot[\boldsymbol{\sigma}' - \mathbf{T}]^{-1}[I - \mathbf{T}]. \tag{7.5.37}$$

Finally, we obtain $\mathbf{S}$ from (7.5.26):

$$\mathbf{S} = \mathrm{Re}\,\boldsymbol{\sigma}'\cdot\{[\boldsymbol{\sigma}' - \mathbf{T}]^{-1}[\mathbf{I} - \mathbf{T}]\}\cdot[\boldsymbol{\sigma}' - \mathbf{T}]^{-1}[\mathbf{I} - \boldsymbol{\sigma}']. \tag{7.5.38}$$

The diagonal elements of $\mathbf{S}$ determine the attenuation coefficient.

Equation (7.5.38) is relatively difficult to interpret as it stands. It should be noted that all the tensors which appear in it are diagonal (in the case of zero magnetic field) so that there is no combination between longitudinal and transverse components.

Let us consider the longitudinal component S_{zz}. For moderately low frequencies $\sigma_0 \gg \epsilon_0\omega$, and we neglect $\mathbf{T}$ compared to $\boldsymbol{\sigma}'$. In this case $\mathbf{V} \approx \mathbf{I}$, the electron current is small ($\mathbf{j}_e \approx -\rho_0 e\mathbf{u}$), and we find

$$S_{ii} = \mathrm{Re}[(1/\sigma'_{ii}) - 1]. \tag{7.5.39}$$

We use (7.5.32), (7.5.15), and (7.6.16) to obtain, in the low frequency limit in which ζ is real ($\zeta \approx ql$, where l is the electron mean free path),

$$S_{xx} = [\zeta^2/3(\zeta - \tan^{-1}\zeta)]\tan^{-1}\zeta - 1. \tag{7.5.40}$$

Note that, in spite of first appearances, a contribution is obtained from R_x. The attenuation of a transverse wave is determined by S_{zz}, where in this limit

$$S_{zz} = \{2\zeta^3/3[(1 + \zeta^2)\tan^{-1}\zeta - \zeta]\} - 1. \tag{7.5.41}$$

Let us suppose that the q is large. The attenuation coefficient for longitudinal phonons is found from (7.5.40) and (7.5.28) to be

$$\alpha_L = (\pi/6)(m^* v_F/M v_s)q. \tag{7.5.42}$$

This result can also be obtained from ordinary perturbation theory applied to the electron–phonon interaction as described by the deformation potential [Eq. (7.1.33)]. The result for transverse phonons in the same limit of large q is quite similar:

$$\alpha_T = (4/3\pi)(m^* v_F/M v_s)q. \tag{7.5.43}$$

It should be noted that these results are independent of l. In the opposite limit in which $ql \ll 1$, we find

$$\alpha_L = (4/15)(m^* v_F/M v_s)q^2 l \tag{7.5.44a}$$

and

$$\alpha_T = \tfrac{1}{5}(m^* v_F/M v_s)q^2 l. \tag{7.5.44b}$$

These expressions contain an additional power of ql as compared with (7.5.42) and (7.5.43).

7.5.2 *Magnetic Field Dependence*

Oscillations may be observed in the attenuation of sound waves in the presence of an external magnetic field provided that $ql \gg 1$ and $\omega_c t \gg 1$ (ω_c is, as usual, the cyclotron frequency). The theory presented in subsection 7.5.1 can be generalized in a straightforward way to include magnetic effects. The general formula for the attenuation, Eq. (7.5.38) remains valid in the presence of an external field: the essential problem is to obtain an adequate expression for the effective conductivity tensor $\boldsymbol{\sigma}'$. Qualitatively, we expect that the acoustic attenuation will depend on the inverse of the effective conductivity. The conductivity should be a periodic function of $\omega/\omega_c = 2\pi(R/\lambda)$ (where R is the radius of a cyclotron orbit and λ is the wavelength of a sound wave). This quantity measures the number of wavelengths of sound contained in a cyclotron orbit. We expect that when ω/ω_c is an integer, the absorption of sound would be most effective, and peaks in the attenuation coefficient should occur. Observation of such oscillations furnishes another useful method for the determination of Fermi surface parameters.

Oscillations may also be observed when the phonon wavelength is of the order of the radius of an electron orbit. More precisely, maxima are found in the attenuation for integral values of the product qR. These are called *geometric resonances* and occur in circumstances in which $\omega/\omega_c = (qR)v_s/v_F \ll 1$.

The theory of ultrasonic attenuation in a magnetic field can be based on expressions for the elements of the conductivity tensor in a magnetic field similar to Eq. (7.4.62). To be specific, let us consider waves propagating perpendicular to the magnetic field. We will take B to define the z direction. The direction of q will continue to be x. The nonzero components of the conductivity tensor σ (and of the effective conductivity σ') are σ_{11}, σ_{22}, σ_{12} $(= -\sigma_{21})$, and σ_{33}. We find from (7.5.38) in the approximation $\mathbf{T} = \mathbf{I}$ that

$$
\begin{aligned}
S_{11} &= \mathrm{Re}[(\sigma'_{11} + \sigma_{12}'^2/\sigma'_{22})^{-1} - 1], \\
S_{22} &= \mathrm{Re}[(\sigma'_{22} + \sigma_{12}'^2/\sigma'_{11})^{-1} - 1], \qquad (7.5.45) \\
S_{33} &= \mathrm{Re}[(\sigma'_{33})^{-1} - 1].
\end{aligned}
$$

The calculation of the elements of the conductivity follows the procedures described in Section 7.4.4. We will omit the detailed expressions which can be found in Cohen, Harrison, and Harrison (1960). In the limit $\bar{\omega}/\omega_c \ll 1$, we see from (7.4.62) that we may consider only the term with $n = 0$ in the summation over the Bessel functions in this and similar expressions. In this case (7.4.62) reduces to

$$
\sigma_{33} = [3\sigma_0/(1 - i\omega\tau)] \int_0^{\pi/2} J_0^2[(qv_F/\omega_c)\sin\theta] \cos^2\theta \sin\theta \, d\theta. \qquad (7.5.46)
$$

Integrals of this type can exhibit oscillatory structure with respect to

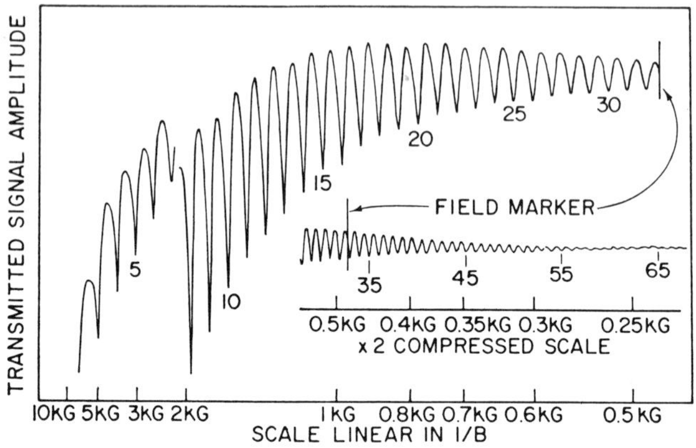

Fig. 7.5.1. Magnetoacoustic oscillations in copper. The wave vector $\mathbf{q}$ is parallel to the (0, 1, 1) axis and B is perpendicular to this axis. The frequency is 476 MHz. [From Kamm (1970).]

(qv_F/ω_c) resulting from the oscillations of J_0^2, superimposed on a smooth background.

For higher fields, absorption of energy from the sound wave can occur through cyclotron resonance. The conductivity can be determined from equations like (7.4.64). The theory requires inclusion of the collision drag term in (7.5.23) and also the elements of $\mathbf{T}$ cannot be neglected compared to those of $\boldsymbol{\sigma}'$. Results are given by Cohen *et al.* (1960). Oscillations in the attenuation are observed as functions of ω/ω_c. Figure 7.5.1 shows observed magnetoacoustic oscillations in copper according to Kamm (1970).

The preceding discussion has considered only magnetic fields weak enough so that quantization of the electron energy levels can be neglected. In the regime in which quantization is important, the very interesting phenomenon of "giant quantum oscillations" (GQO) can be observed in the ultrasonic attenuation (Gurevich *et al.* 1961; Shapira and Lax, 1965). Electrons (or holes) whose velocity is equal to the sound velocity are able to absorb phonons for specific values of the magnetic field for which these electrons have energies within KT of the Fermi energy.

We consider the interaction of a sound wave which is a coherent beam of phonons (wave vector $\mathbf{q}$, frequency ω), with electrons in a pure metal $(\tau \longrightarrow \infty)$. The direction of the magnetic field $\mathbf{B}$ defines the z axis. Let the initial state of the electron be described by the Landau quantum number n, z component of wave vector k_z, and energy $E_n(k_z)$. The final state is specified by l, k_z', and $E_l(k_z')$. Conservation of energy and momentum in the collision requires that

$$k_z' = k_z + q_z, \qquad E_l(k_z') - E_n(k_z) = \hbar\omega(\mathbf{q}). \tag{7.5.47}$$

When the magnetic field is large, the phonon does not have sufficient energy to cause a change in the Landau level. Let v_s be (as before) the velocity of sound. We then have

$$E_n(k_z + q_z) - E_n(k_z) = \hbar v_s q. \tag{7.5.48}$$

The left side of (7.5.48) will depend on k_z, and thus (7.5.48) will determine a specific value of k_z (denoted k_0) at which absorption can take place. If q_z is small, we can expand $E_n(k_z + q)$, and we have through first order

$$(\cos\theta/\hbar)\ (\partial E_n/\partial k_z)_{k_z=k_0} = v_s \tag{7.5.49}$$

in which θ is the angle between q and B. Since $\hbar^{-1}\,\partial E_n/\partial k_z$ is the z component of the group velocity of an electron wave packet centered at $\mathbf{k}_0$, (7.5.49) asserts that those electrons whose velocity along $\mathbf{q}$ is equal to v_s can absorb energy. For this reason, the GQO can be considered as a surf-riding resonance.

There is one vital qualification to this argument. Suppose that (7.5.48) or (7.5.49) is satisfied. Absorption will still only be possible if the states $|\,n, k_z + q_z\rangle$ and $|\,n, k_z\rangle$ are within KT of the Fermi energy. Only if this is true will there be an appreciable probability of having both an electron in the initial state and an unoccupied final state so that a transition can occur. The position of the Landau level is controlled by the external field; hence as $\mathbf{B}$ varies states of a given Landau quantum number n can move so that they are close to the Fermi energy where absorption is large or away from it giving small absorption. The interval between oscillations is determined by the increment in field required to bring successive level pairs through the Fermi surface. The situation is quite similar to the de Haas–van Alphen effect. The argument of Section 6.4.2 can be applied: the periodicity of the absorption is characterized by a frequency (in B) of $\mathcal{a}_0(\hbar k_z)/\hbar eB$, in which $\mathcal{a}_0$ is the area of a cross section of Fermi surface perpendicular to the magnetic field (evaluated at $k_z = k_0$).

7.6 Electrical Resistance Due to Lattice Vibrations

Our calculations of transport properties in previous sections have been based on the relaxation time approximation. It was determined in Section 7.2 that this is correct when electrons are scattered elastically. Scattering by impurities (without internal degrees of freedom) is elastic; however, scattering by lattice vibrations necessarily involves emission and absorption of phonons, and thus is not elastic. In this section, we will study the contribution of lattice scattering to the electrical resistivity of metals.

7.6.1 *The Collision Term in the Boltzmann Equation*

It is necessary to obtain the collision term $(\partial f/\partial t)_c$. This will be done by ordinary first-order time-dependent perturbation theory involving the electron–phonon interaction. To simplify the calculation, we will assume that the electron states involved belong to a single band (so that the band index can be dropped). This assumption will be valid except where band degeneracies occur since the energy changes involved are small. More restrictively, we will assume that only longitudinal phonons are involved and replace the factor $\mathbf{q}\cdot\mathbf{e}_j(\mathbf{q})$, which occurs in the Hamiltonian Eq. (7.1.25), by q for longitudinal phonons and by zero for others. (The phonon wave vector is $\mathbf{q}$ and $\mathbf{e}$ is its polarization vector.) The approximate form of the electron phonon interaction is taken to be

$$H_{\text{e-p}} = \mathfrak{N}^{-1/2}\sum_{\mathbf{q}} D(\mathbf{q})[c^+(\mathbf{k}+\mathbf{q})c(\mathbf{k})a(\mathbf{q}) + c^+(\mathbf{k}-\mathbf{q})c(\mathbf{k})a^+(\mathbf{q})] \tag{7.6.1}$$

with

$$D(\mathbf{q}) = i(\hbar/2M\omega(\mathbf{q}))^{1/2}qC. \tag{7.6.2}$$

In these equations, c^+, c (a^+, a) are electron (phonon) creation or destruction operators, while C is a constant which can be estimated as $-\frac{2}{3}E_F$ from deformation potential theory. In addition, M is the atomic mass and ω is the phonon frequency.

The matrix elements for specific processes may be determined from (7.6.1) and (7.6.2) by inspection, and the transition rate obtained by standard procedures. For example, the rate of transitions involving absorption of a phonon of wave vector $\mathbf{q}$ in which an electron is transferred from a state of wave vector $\mathbf{k}$ to one of wave vector $\mathbf{k}+\mathbf{q}$ is

$$(2\pi/\hbar)\,|\,D(\mathbf{q})\,|^2 n(\mathbf{q})f(\mathbf{k})[1-f(\mathbf{k}+\mathbf{q})]\,\delta[\epsilon(\mathbf{k})+\hbar\omega(\mathbf{q})-\epsilon(\mathbf{k}+\mathbf{q})]. \tag{7.6.3}$$

The function $n(\mathbf{q})$ specifies the probability of finding a phonon of wave-vector $\mathbf{q}$ present at temperature T, while $f(\mathbf{k})$ gives the probability that the initial electron state is occupied.

The quantity $(\partial f/\partial t)_c$ may now be computed. It is the difference between the rate at which electrons in other states ($\mathbf{k}\pm\mathbf{q}$, say) are scattered into $\mathbf{k}$ by phonon emission or absorption, and the rate at which electrons in $\mathbf{k}$

are scattered into other states (see Section 7.2.1) $\mathbf{k} \pm \mathbf{q}$. Expressions for these partial rates are formed by combining terms of the form (7.6.3), and the result is summed over all values of q:

$$(\partial f/\partial t)_c = (\pi C^2/\mathfrak{N}M) \sum_{\mathbf{q}} [q^2/\omega(\mathbf{q})]\{[f(\mathbf{k}+\mathbf{q})(1-f(\mathbf{k}))(n(\mathbf{q})+1)$$
$$- f(\mathbf{k})(1-f(\mathbf{k}+\mathbf{q}))n(\mathbf{q})]\,\delta[\epsilon(\mathbf{k}+\mathbf{q}) - \hbar\omega(\mathbf{q}) - \epsilon(\mathbf{k})]$$
$$+ [f(\mathbf{k}+\mathbf{q})(1-f(\mathbf{k}))n(-\mathbf{q}) - f(\mathbf{k})(1-f(\mathbf{k}+\mathbf{q}))$$
$$\times (n(-\mathbf{q})+1)]\,\delta[\epsilon(\mathbf{k}) - \hbar\omega(\mathbf{q}) - \epsilon(\mathbf{k}+\mathbf{q})]\}. \qquad (7.6.4)$$

In the second group of terms, we have replaced $\mathbf{q}$ by $-\mathbf{q}$. In general, the phonon distribution is not in equilibrium, so that $n(\mathbf{q}) \neq n(-\mathbf{q})$. However, this effect will be neglected, although it must be taken into account in a more refined calculation. Thus we will, in the following, replace $n(-\mathbf{q})$ by $n(\mathbf{q})$.

If both the electrons and the phonons are in equilibrium, $(\partial f/\partial t)_c$ must vanish, since scattering cannot alter an equilibrium situation. Let us see how this occurs. Consider the coefficient of the first delta function in (7.6.4). If $f(\mathbf{k})$ is replaced by the equilibrium form

$$f(\mathbf{k}) = f_0(\mathbf{k}) = \{\exp[\epsilon(\mathbf{k}) - \mu]/KT + 1\}^{-1} \qquad (7.6.5)$$

where μ is the Fermi energy, a similar replacement is made for $f(\mathbf{k}+\mathbf{q})$, and for $n(\mathbf{q})$,

$$n(\mathbf{q}) = \{\exp[\hbar\omega(\mathbf{q})/KT] - 1\}^{-1}, \qquad (7.6.6)$$

it will be seen that the coefficient vanishes, provided the energies of the states are related by the vanishing of the argument of the delta function:

$$f_0(\mathbf{k}+\mathbf{q})[1 - f_0(\mathbf{k})][n(\mathbf{q}) + 1] = f_0(\mathbf{k})[1 - f_0(\mathbf{k}+\mathbf{q})]n(\mathbf{q}) \qquad (7.6.7)$$

if $\epsilon(\mathbf{k}+\mathbf{q}) = \epsilon(\mathbf{k}) + \hbar\omega(\mathbf{q})$. As in Section 7.2.1, we take advantage of this by introducing a function $\Phi(\mathbf{k})$ by

$$f(\mathbf{k}) = f_0(\mathbf{k}) + \Phi(\mathbf{k})\, \partial f_0(\mathbf{k})/\partial E$$
$$= f_0(\mathbf{k}) - [\Phi(\mathbf{k})/KT] f_0(\mathbf{k})[1 - f_0(\mathbf{k})]. \qquad (7.6.8)$$

Equation (7.6.8) and its equivalent involving $\Phi(\mathbf{k}+\mathbf{q})$ are substituted into (7.6.4). The sum over $\mathbf{q}$ is replaced by an integral according to the usual formula

$$\mathfrak{N}^{-1} \sum_{\mathbf{q}} \to \Omega/(2\pi)^3 \int d^3q.$$

The following result is obtained after some algebra:

$$
\begin{aligned}
(\partial f/\partial t)_c = &-(\pi C^2/MKT)[\Omega/(2\pi)^3] \int [q^2 n(\mathbf{q})/\omega(\mathbf{q})] \\
&\times \{ f_0(\mathbf{k})[1 - f_0(\mathbf{k}+\mathbf{q})]\,\delta[\epsilon(\mathbf{k}) - \epsilon(\mathbf{k}+\mathbf{q}) + \hbar\omega(\mathbf{q})] \\
&+ f_0(\mathbf{k}+\mathbf{q})[1 - f_0(\mathbf{k})]\,\delta[\epsilon(\mathbf{k}) - \epsilon(\mathbf{k}+\mathbf{q}) - \hbar\omega(\mathbf{q})]\} \\
&\times [\Phi(\mathbf{k}+\mathbf{q}) - \Phi(\mathbf{k})]\, d^3q. \qquad (7.6.9)
\end{aligned}
$$

This expression is still extremely complicated. One general result can be obtained. If KT is large compared to the maximum phonon energy $K\Theta$ (Θ is the Debye temperature), it will be a legitimate first approximation to neglect $\hbar\omega(\mathbf{q})$. Under these circumstances the scattering will be effectively elastic. This means that the relaxation time approximation will be valid, and a relaxation time can be extracted from (7.6.9). This will be done in Section 7.6.2.

The Boltzmann equation in the absence of a magnetic field is obtained from (7.2.16)

$$(\mathbf{v}_\mathbf{k}\cdot\mathbf{A})\ \partial f^{(0)}/\partial E\ (\mathbf{k}) = (\partial f/\partial t)_c \qquad (7.6.10)$$

in which

$$A = -[e\boldsymbol{\mathcal{E}} + T\ \nabla_\mathbf{r}(\mu/T) + (E/T)\ \nabla T] \qquad (7.6.11)$$

and $(\partial f/\partial t)_c$ is given by (7.6.9). To reduce the complications of this integral equation, we will introduce the assumption that the band of interest is parabolic: $E(\mathbf{k}) = \gamma\mathbf{k}^2$, $\gamma = \hbar^2/2m^*$. Further, let us look for a solution of (7.6.10) of the form

$$\Phi(\mathbf{k}) = -g(E)\,\mathbf{k}\cdot\mathbf{A} \qquad (7.6.12)$$

in which g is a function of energy only.

In the case of a parabolic band, the delta functions in (7.6.9) have the form

$$
\begin{aligned}
&\delta[\gamma(2kq\cos\theta + q^2) \pm \hbar\omega(\mathbf{q})] \\
&\qquad = (1/2\gamma kq)\ \delta\{\cos\theta \pm [\hbar\omega(\mathbf{q})/2\gamma kq] + (q/2k)\}.
\end{aligned}
$$

The + sign refers to the second delta function, the − sign to the first. The conditions under which the argument can vanish so that a contribution will be obtained are

$$-1 \le (q/2k) \mp [\hbar\omega(\mathbf{q})/2\gamma kq] \le 1. \qquad (7.6.13)$$

For the monovalent metals to which the theory is applicable, $\hbar\omega(\mathbf{q})/2\gamma kq \approx K\theta/\mu$. Thus this term is virtually negligible. Also the largest value of q

(a Brillouin zone "radius") is smaller than $2k_F$ and k is constrained to the neighborhood of k_F. Hence (7.6.13) is satisfied in simple metals for all q (although this will not generally be the case in degenerate semiconductors and semimetals). Since

$$\mathbf{q}\cdot\mathbf{A} = qA[\cos\theta_{\mathbf{k}\mathbf{A}}\cos\theta + \sin\theta_{\mathbf{k}\mathbf{A}}\sin\theta\cos\phi]$$

in which θ is the angle between $\mathbf{k}$ and $\mathbf{q}$, we find on integrating over angles

$$\begin{aligned}(\partial f/\partial t)_c = {} & (\pi C^2/MKT)[\Omega A\cos\theta_{\mathbf{k}A}/(2\pi)^2 2\gamma] \\ & \times \int [qn(\mathbf{q})/\omega(\mathbf{q})]\{f_0(E)[1 - f_0(E+\hbar\omega)] \\ & \times [(1 + (\hbar\omega/2E) - (q^2/2k^2))g(E+\hbar\omega) - g(E)] \\ & + f_0(E-\hbar\omega)[1 - f_0(E)] \\ & \times [(1 - (\hbar\omega/2E) - (q^2/2k^2))g(E-\hbar\omega) - g(E)]\}q^2\,dq.\end{aligned} \tag{7.6.14}$$

7.6.2 *The High Temperature Limit*

At high temperatures, $\hbar\omega$ can be neglected as mentioned previously. Equation (7.6.14) simplifies. Since

$$(1/KT)f_0(E)[1 - f_0(E)] = -\partial f_0/\partial E,$$

the Boltzmann equation (7.6.10) reduces to

$$2\gamma k = (\pi C^2/M)[\Omega/2\gamma(2\pi)^2]g(E)\int_0^{q_m}[qn(\mathbf{q})/\omega(\mathbf{q})](q^2/k^2)q^2\,dq. \tag{7.6.15}$$

The upper limit q_m is the radius of a sphere of volume equal to the Brillouin zone. It is evident from the comparison between (7.6.12) and (7.2.26) that $g(E)$ is proportional to the relaxation time $\tau[g = 2\gamma\tau]$. Since temperature is assumed to be high, we approximate $n(q)$ by

$$n(\mathbf{q}) \approx KT/\hbar\omega(\mathbf{q}) = KT/\hbar v_s q$$

in which v_s is the speed of sound and we have assumed a simple Debye model for the phonon spectrum. It is convenient to introduce the Debye temperature Θ through

$$K\Theta = \hbar v_s q_m = \hbar v_s(6\pi^2/\Omega)^{1/3}. \tag{7.6.16}$$

The integral is easily performed, and the resulting equation can be solved for the relaxation time:

$$\tau = (8/3\pi)(M/m^*)(\hbar k^3/C^2 q_m{}^3)[(K\Theta)^2/KT] \qquad (T \gg \Theta). \tag{7.6.17a}$$

This model is applicable to a monovalent metal in which $k = k_F = q_m/2^{1/3}$. Thus

$$\tau = (4/3\pi)(M/m^*)[\hbar(K\Theta)^2/C^2KT]. \tag{7.6.17b}$$

We note that the relaxation time is inversely proportional to the temperature. Consequently, the resistivity will be proportional to temperature, as is observed. The order of magnitude of τ at room temperature is 10^{-14} sec. These results are not to be taken too literally, since we have used a form of the electron–phonon interaction valid only for very long wavelengths, have neglected coupling to transverse waves, the possibility of Umklapp processes, dispersion in the phonon spectrum, and so on. Approximations of this type are more meaningful at low temperatures, which will now be considered.

7.6.3 *The T^5 Law*

In order to discuss the low temperature behavior of the electrical resistivity, it is convenient to combine the emission and absorption terms in (7.6.14). This can be done if we regard formally the emission of a phonon of energy $\hbar\omega$ as the absorption of a phonon of negative energy $-\hbar\omega$. Since, in the simple model we are using, ω is linearly related to the magnitude of the wave vector $\mathbf{q}$, the range of this variable will be extended to negative values $-q_m \leq q \leq q_m$. With the aid of the identity (7.6.7), we obtain

$$\left(\frac{\partial f}{\partial t}\right)_c = -C^2\frac{\Omega A\cos\theta_{kA}}{8\pi\gamma M}\frac{\partial f_0}{\partial E}\int_0^{q_m} dq\,\frac{q^3 n(\mathbf{q})}{\omega(\mathbf{q})}$$
$$\times\left\{\left[\left(1+\frac{\hbar\omega}{2E}-\frac{q^2}{2k^2}\right)g(E+\hbar\omega)-g(E)\right]\frac{f_0(E+\hbar\omega)}{f_0(E)}\,e^{\hbar\omega/KT}\right.$$
$$\left.+\left[\left(1-\frac{\hbar\omega}{2E}-\frac{q^2}{2k^2}\right)g(E-\hbar\omega)-g(E)\right]\frac{f_0(E-\hbar\omega)}{f_0(E)}\right\}.$$

The Boltzmann equation (7.6.10) becomes

$$2\gamma k\left(\frac{\partial f_0}{\partial E}\right) = -\frac{c^2\Omega}{8\pi\gamma M}\left(\frac{\partial f_0}{\partial E}\right)\int_{-q_m}^{q_m} dq\,\frac{q^3}{\omega(\mathbf{q})}$$
$$\times\left[\left(1+\frac{\hbar\omega}{2E}-\frac{q^2}{2k^2}\right)g(\zeta+x)-g(x)\right]\frac{e^{\zeta}+1}{e^{\zeta+x}+1}\,\frac{1}{|1-e^{-x}|}, \tag{7.6.18}$$

in which $\zeta = (E - \mu)/KT$, $x = \hbar\omega/KT$. Equation (7.6.18) replaces (7.6.15). We have not canceled the factors of $\partial f_0/\partial E$ for reasons which will be apparent below.

We require $g(E)$ only for energies close to the Fermi energy. There is no particular reason to expect $g(E)$ to vary rapidly in this region. Let us try as ansatz, $g(E)$ = constant. [Actually, we can expand $g(E)$ in terms of $E - \mu$, and determine subsequently that only the first term is important for our purposes.] To evaluate this constant, we integrate both sides of (7.6.18) with respect to E. On the left side, we simply replace $\partial f_0/\partial E$ by $-\delta(E - \mu)$, while on the right we change the variable of integration to ζ and observe that everything varies slowly with respect to ζ except the Fermi function. The limits of the ζ integration can be made $-\infty$ to $+\infty$ without appreciable error. Thus

$$2\gamma k_F = -(C^2\Omega g/8\pi\gamma M)\int_{-q_m}^{q_m} dq\,[q^3/\omega(\mathbf{q})][(\hbar\omega/2\mu) - (q^2/2k_F^2)][1/|1 - e^{-x}|]\int_{-\infty}^{\infty}[(e^{-\zeta}+1)(e^{\zeta+x}+1)]^{-1}\,d\zeta. \tag{7.6.19}$$

The ζ integral is elementary (substitute $z = e^{\zeta}$), and yields

$$\int d\zeta\,[(e^{-\zeta}+1)(e^{\zeta+x}+1)]^{-1} = x/(e^x - 1). \tag{7.6.20}$$

We substitute (7.6.20) into (7.6.19) and observe that the term with $\hbar\omega/2\mu$ cancels. The remainder can be expressed as

$$g = 2\gamma(2/3\pi)(M/m^*)(\hbar k_F^3/C^2 q_m^3)K\Theta(\Theta/T)^5[1/J_5(\Theta/T)] \tag{7.6.21}$$

in which

$$J_n(\Theta/T) = \int_0^{\Theta/T} x^n/[(e^x - 1)(1 - e^{-x})]\,dx. \tag{7.6.22}$$

For very low temperatures, $T \ll \Theta$, the upper limit may be made infinite. We find

$$J_5(\infty) = \int_0^{\infty} x^5 e^x/(e^x - 1)^2\,dx = 5!\zeta(5) = 124.4 \tag{7.6.23}$$

(ζ is here the Riemann zeta function). We can now obtain the electrical conductivity with the aid of (7.6.8) and the procedures of Section 7.2

[see (7.2.29)–(7.2.31)]:

$$\sigma = (2/3\pi)(Mne^2/m^{*2})(\hbar k_F{}^3/C^2 q_m{}^3) K\Theta(\Theta/T)^5[1/J_5(\Theta/T)]. \quad (7.6.24)$$

The essential conclusion is that the contribution to the electrical resistivity from lattice scattering should go to zero at low temperatures as T^5 (the T^5 law). This is at least approximately what is observed but exact degree of validity of (7.6.24) with respect to experiment remains somewhat controversial. In particular, phonon drag (the distortion of the phonon distribution due to the electron motion) has been neglected, and significant corrections may be required for real metals (Kaveh and Wiser, 1972).

7.7 The Polaron Problem

In this section, we consider phenomena associated with the motion of a slow electron in an ionic crystal. This problem has attracted substantial attention both because of its intrinsic importance and because it furnishes a model for some problems in quantum field theory. Our considerations will be based on the "Fröhlich Hamiltonian," Eq. (7.1.117),

$$H = (p^2/2m^*) + \hbar\omega \sum_{\mathbf{q}} a_{\mathbf{q}}{}^+ a_{\mathbf{q}} + i\,\mathfrak{N}^{-1/2} \sum_{\mathbf{q}} V(\mathbf{q}) \times [a_{\mathbf{q}}{}^+ \exp(-i\mathbf{q}\cdot\mathbf{r}) - a_{\mathbf{q}} \exp(i\mathbf{q}\cdot\mathbf{r})] \quad (7.1.117)$$

in which

$$V(\mathbf{q}) = (4\pi\alpha)^{1/2}[(\hbar\omega/q)(\hbar/2m^*\omega)^{1/4}] \quad (7.1.120)$$

and α is a dimensionless coupling constant

$$\alpha = (e^2/8\pi\epsilon_0)(1/\hbar\omega)(2m^*\omega/\hbar)^{1/2}[(1/\kappa_\infty) - (1/\kappa_0)]. \quad (7.1.119)$$

The electron whose "band" effective mass is m^*, interacts with a field describing longitudinal optical phonons of frequency ω (assumed independent of q). The quantities κ_∞ and κ_0 are the high and low frequency dielectric constants.

If α is small, the coupling between the electrons and the lattice is weak and it is permissible to use ordinary perturbation theory. This is the situation in many common semiconductors. As α increases the problem becomes substantially more complex and special methods (such as the Feynman path integral technique) must be employed. In the limit of very large α—the strong coupling case—a simple variational approach is possible. A survey of different approaches to the polaron problem may be found in Kuper and Whitfield (1963).

7.7.1 *Weak Coupling*

We will calculate the self-energy and effective mass of the polaron in the weak coupling limit using ordinary perturbation theory. We start from a noninteracting system in which the electrons have energies $\epsilon_0(\mathbf{k}) = \hbar^2\mathbf{k}^2/2m^*$. The states of the noninteracting system can be characterized by giving the electron wave vector and the numbers of phonons present in different modes: thus $|\,\mathbf{k}, 1\mathbf{q}\rangle$ indicates that the electron has wave vector $\mathbf{k}$ and there is a single phonon in $\mathbf{q}$. At $T = 0°\text{K}$, there are no phonons present in the ground state of the noninteracting system. This situation changes when the interaction is included.

The computation of the energy of an electron to second order in the interaction requires us to consider the emission of a phonon of wave vector $\mathbf{q}$ followed by its reabsorption. Thus if the initial state is $|\,\mathbf{k}, 0\rangle$, the intermediate state is $|\,\mathbf{k} - \mathbf{q}, 1\mathbf{q}\rangle$. The first-order perturbation vanishes and we have

$$\epsilon(\mathbf{k}) = \epsilon_0(\mathbf{k}) + \sum_{\mathbf{q}} |\langle \mathbf{k} - \mathbf{q}, 1\mathbf{q}\,|\,H_{\text{int}}\,|\,\mathbf{k}, 0\mathbf{q}\rangle|^2/[\epsilon_0(\mathbf{k}) - \epsilon_0(\mathbf{k} - \mathbf{q}) - \hbar\omega] \tag{7.7.1}$$

where H_{int} is the third term in Eq. (7.1.117). The value of the matrix element is just $V(\mathbf{q})$. Thus

$$\epsilon(\mathbf{k}) = \epsilon_0(\mathbf{k}) - (2m^*/\hbar^2\mathfrak{N}) \sum_{\mathbf{q}} |\,V(\mathbf{q})\,|^2/(-2\mathbf{k}\cdot\mathbf{q} + \mathbf{q}^2 + 2m^*\omega/\hbar).$$

The sum is converted to an integral the usual way:

$$\epsilon(\mathbf{k}) - \epsilon_0(\mathbf{k}) = -(2m^*/\hbar^2)(2\pi)^{-3}(4\pi\alpha)(\hbar\omega)^2(\hbar/2m^*\omega)^{1/2} \times \int (d^3q/\mathbf{q}^2)(-2\mathbf{k}\cdot\mathbf{q} + \mathbf{q}^2 + 2m^*\omega/\hbar)^{-1}. \tag{7.7.2}$$

We choose $\mathbf{k}$ as the polar axis, and perform the integration over angles. We are left with an integral over q of the form

$$\epsilon(\mathbf{k}) - \epsilon_0(\mathbf{k}) = -\hbar\omega\alpha(2m^*\omega/\hbar)^{1/2}/\pi \times \int_0 (dq/2kq)\ln[(q^2 + 2m^*\omega/\hbar + 2kq)/(q^2 + 2m^*\omega/\hbar - 2kq)].$$

If Umklapp processes are ignored, the upper limit of the integral should be q_m, the radius of a sphere whose volume is equal to that of the Brillouin zone.

TABLE I

ESTIMATED VALUES OF THE POLARON COUPLING CONSTANT

Substance	α	Substance	α
NaF	6.3	AgBr	1.6
NaCl	5.5	ZnO	0.85
NaBr	5.0	PbS	0.16
NaI	4.8	GaAs	0.06
AgCl	1.7	InSb	0.014

We will, however, follow the conventional practice in which the upper limit of the integral is made infinite and the difference is neglected. Typically, this leads to an error in the value of the integral of the order of 10%. Further, we require $\epsilon(\mathbf{k}) - \epsilon_0(\mathbf{k})$ only for small values of k, since the effective mass approximation employed in the Fröhlich Hamiltonian is reasonably valid only for small k. We expand the logarithm, retaining only the first two nonvanishing terms:

$$\ln[(1 + x)/(1 - x)] = 2(x + \tfrac{1}{3}x^3).$$

In this case, $x = 2kq/(q^2 + 2m^*\omega/\hbar)$. The resulting integrals are simple, and we find

$$\epsilon(\mathbf{k}) - \epsilon_0(\mathbf{k}) = -\alpha[\hbar\omega + \tfrac{1}{6}(\hbar^2k^2/2m^*)]$$

or

$$\epsilon(\mathbf{k}) = -\alpha\hbar\omega + (\hbar^2/2m^*)(1 - \tfrac{1}{6}\alpha)k^2. \qquad (7.7.3)$$

The energy of the lowest band state ($k = 0$) is lowered by $-\alpha\hbar\omega$, and the effective mass is increased. The effective mass in the presence of interaction is, through first order in α,

$$m^*_{\text{pol}} = m^*(1 + \tfrac{1}{6}\alpha). \qquad (7.7.4)$$

It is apparent from the form of the Fröhlich Hamiltonian that perturbation theory will yield some sort of series in powers of α. The numerical factor of $\frac{1}{6}$ which appears in (7.7.4) is somewhat unexpected, and indicates that the results of perturbation theory may have a greater range of validity than might have been anticipated. Values of α for a representative sample of crystals, as estimated by Brown (1963) are given in Table I. Perturbation theory should be adequate in common semiconductors.

Further insight into the significance of the parameter α can be obtained if we calculate, in perturbation theory, the average number of phonons accompanying an electron of wave vector **k** (**k** is assumed to be small). The state vector for an electron including the electron–phonon interaction, is denoted by $| \mathbf{k}, \text{int}\rangle$ and can be expressed as

$$| \mathbf{k}, \text{int}\rangle = | \mathbf{k}, 0\rangle + \sum_{\mathbf{q}} | \mathbf{k} - \mathbf{q}, 1\mathbf{q}\rangle \times \langle \mathbf{k} - \mathbf{q}, 1\mathbf{q} | H_{\text{int}} | \mathbf{k}, 0\mathbf{q}\rangle / [\epsilon_0(\mathbf{k}) - \epsilon_0(\mathbf{k} + \mathbf{q}) - \hbar\omega]. \tag{7.7.5}$$

Equation (7.7.5) is correct through first order in α. The average number of phonons $\langle N\rangle$ included in (7.7.5) is

$$\langle N\rangle = \langle \mathbf{k}, \text{int} | \sum_{\mathbf{p}} a_{\mathbf{p}}^+ a_{\mathbf{p}} | \mathbf{k}, \text{int}\rangle. \tag{7.7.6}$$

We have $\langle \mathbf{k} - \mathbf{q}', 1\mathbf{q}' | a_{\mathbf{p}}^+ a_{\mathbf{p}} | \mathbf{k} - \mathbf{q}, 1\mathbf{q}\rangle = \delta_{\mathbf{pq}}\, \delta_{\mathbf{p},\mathbf{q}'}$. The procedure used before yields

$$\begin{aligned} \langle N\rangle &= \sum_{\mathbf{q}} | \langle \mathbf{k} - \mathbf{q}, 1\mathbf{q} | H_{\text{int}} | \mathbf{k} - \mathbf{q}, 1\mathbf{q}\rangle |^2 / [\epsilon_0(\mathbf{k}) - \epsilon_0(\mathbf{k} + \mathbf{q}) - \hbar\omega]^2 \\ &= (2m^*/\hbar^2)^2 (2\pi)^{-3} (4\pi\alpha)(\hbar\omega)^2 (\hbar/2m^*\omega)^{1/2} \\ &\quad \times \int (d^3q/q^2)[q^2 + (2m^*\omega/\hbar) - 2\mathbf{k}\cdot\mathbf{q}]^{-2} \\ &= (2\alpha/\pi)(2m^*\omega/\hbar)^{3/2} \int_0^\infty dq\, \{[q^2 + (2m^*\omega/\hbar)]^2 - k^2q^2\}^{-1}. \end{aligned} \tag{7.7.7}$$

We have integrated over directions in order to obtain the last line. Again we will confine our attention to small k. The integral will be evaluated in the limit $k = 0$, and we obtain a result correct up to second order. The result is, simply,

$$\langle N\rangle = \alpha/2. \tag{7.7.8}$$

Thus the average number of phonons accompanying a slow electron is just $\alpha/2$, provided α is small.

There have been numerous attempts to extend the range of validity of the theory. A glance at the values of α in Table I indicates that in order to deal with the values of α that are encountered in the alkali halides, something better than perturbation theory is needed. We will not discuss the intermediate coupling region in detail, but will simply note the possibility of applying variational methods (Lee *et al.*, 1953). Moreover, there exists an extremely elegant and apparently quite accurate approach due to

Feynman (1955) (see also Feynman *et al.*, 1962), based on the path integral formulation of quantum mechanics. We will turn to the strong coupling region, in which new physical phenomena should appear.

7.7.2 *Strong Coupling*

The weak coupling and strong coupling regimes may be contrasted by noting that α is inversely proportional to $\omega^{-1/2}$ [see (7.1.119)]. Weak coupling therefore corresponds qualitatively to high frequencies, such that the lattice polarization easily follows the motion of the electron. Conversely, strong coupling implies low frequencies, and we can imagine that the electron follows the motion of the lattice. In effect, the electron becomes bound in a potential well: it digs a hole for itself and falls in.

This problem can be formulated mathematically as follows (for a more extended discussion, see Allcock, 1963): A variational method is employed in which the wave function is assumed to be

$$\psi_0(\mathbf{r} - \mathbf{X}, \mathbf{X}) = \Omega(\mathbf{r} - \mathbf{X})\Phi(\mathbf{X}) \tag{7.7.9}$$

with $\mathbf{r}$ being the electron coordinate. The function $\Omega(\mathbf{r} - \mathbf{X})$ is a bound state wave function for the electron centered at some point $\mathbf{X}$. This point can be regarded as the center of the potential well in which the electron traps itself. The function Φ, which is independent of the electron coordinate, describes the state of the lattice polarization field. The picture here is that of an electron following adiabatically the motion of the lattice polarization. Evidently, all positions $\mathbf{X}$ should give equivalent results and it is a defect of the assumed function (7.7.9) that it is not translationally invariant. This difficulty can be remedied by adopting a slightly more complicated trial function

$$\psi(\mathbf{k}, \mathbf{r}) = (1/V) \int d^3X \exp(i\mathbf{k}\cdot\mathbf{X})\psi_0(\mathbf{r} - \mathbf{X}, \mathbf{X}) \tag{7.7.10}$$

in which V is the volume of the system and ψ_0 is given by (7.7.9). We will, however, consider only the simpler function. The same results are obtained from the polaron self-energy; however, (7.7.10) should be used to calculate the effective mass.

We now proceed to determine the polaron binding energy. The first step is to write a formal expression for the expectation value of the energy. This can be expressed as

$$\langle E \rangle = \langle \Phi \mid H_{\mathrm{R}} \mid \Phi \rangle \tag{7.7.11}$$

in which H_R is a reduced Hamiltonian such that the electronic coordinates have been integrated out:

$$H_R = \int \Omega^*(\mathbf{r} - \mathbf{X})(\mathbf{p}^2/2m^*)\Omega(\mathbf{r} - \mathbf{X})\, d^3r + \hbar\omega \sum_{\mathbf{q}} a_{\mathbf{q}}^+ a_{\mathbf{q}}$$
$$+ i\mathfrak{N}^{-1/2} \sum_{\mathbf{q}} V(\mathbf{q})[f^*(\mathbf{q})a_{\mathbf{q}}^+ - f(\mathbf{q})a_{\mathbf{q}}] \tag{7.7.12}$$

and

$$f(\mathbf{q}) = \exp(i\mathbf{q}\cdot\mathbf{X}) \int |\Omega(\mathbf{y})|^2 \exp(i\mathbf{q}\cdot\mathbf{y})\, d^3y. \tag{7.7.13}$$

The reduced Hamiltonian is a function of the oscillator coordinates $\mathbf{q}$. As far as these coordinates are concerned, the problem is that of a displaced harmonic oscillator, and can be solved simply by the introduction of new creation and annihilation operators $b_{\mathbf{q}}$ and $b_{\mathbf{q}}^+$. Put

$$a_{\mathbf{q}} = b_{\mathbf{q}} - iV(\mathbf{q})f^*(\mathbf{q})/\hbar\omega\mathfrak{N}^{1/2}, \qquad a_{\mathbf{q}}^+ = b_{\mathbf{q}}^+ + iV(\mathbf{q})f(\mathbf{q})/\hbar\omega\mathfrak{N}^{1/2}. \tag{7.7.14}$$

It is easily verified that the new operators $b_{\mathbf{q}}^+$ and $b_{\mathbf{q}}$ obey the same commutation relations as the set $a_{\mathbf{q}}^+$, $a_{\mathbf{q}}$. In fact, it is not difficult to show that the substitution (7.7.14) corresponds to a unitary transformation of the operators $a_{\mathbf{q}}$. Substitution of (7.7.14) into (7.7.12) leads us to a new reduced Hamiltonian $\bar{H}_R$ (with $y = \mathbf{r} - \mathbf{x}$):

$$\bar{H}_R = \int \Omega^*(\mathbf{y})(\mathbf{p}^2/2m^*)\Omega(\mathbf{y})\, d^3y + \hbar\omega \sum_{\mathbf{q}} b_{\mathbf{q}}^+ b_{\mathbf{q}}$$
$$- \mathfrak{N}^{-1} \sum_{\mathbf{q}} V^2(\mathbf{q}) |f(\mathbf{q})|^2/\hbar\omega. \tag{7.7.15}$$

This reduced Hamiltonian is diagonal in the new phonon operators. Since the eigenvalues of $b_{\mathbf{q}}^+ b_{\mathbf{q}}$ are positive integers (or zero), the state of lowest energy for the system must correspond to a zero eigenvalue for this quantity (for each $\mathbf{q}$). Further, (7.7.15) is independent of X. Since Φ_0 is normalized for all X, Eq. (7.7.11) reduces to

$$\langle E \rangle = \int \Omega^*(\mathbf{y})(\mathbf{p}^2/2m^*)\Omega(\mathbf{y})\, d^3y - \sum_{\mathbf{q}} V(\mathbf{q})^2 |f(\mathbf{q})|^2/\hbar\omega. \tag{7.7.16}$$

We now use (7.7.13) to express (7.7.16) in a manner which makes the dependence on Ω in the last term apparent. The sum is converted to an integral in the usual way:

$$\langle E\rangle = \int \Omega^*(\mathbf{y})(\mathbf{p}^2/2m^*)\Omega(\mathbf{y})\, d^3y - (\alpha\hbar\omega/2\pi^2)(\hbar/2m^*\omega)^{1/2}$$

$$\times \int d^3y \mid \Omega(\mathbf{y}) \mid^2 \int d^3z \mid \Omega(\mathbf{z}) \mid^2 \int d^3q\{\exp[i\mathbf{q}\cdot(\mathbf{y}-\mathbf{z})]/q^2\}.$$

The q integral can be done immediately. The result is

$$\langle E\rangle = \int \Omega^*(\mathbf{y})(\mathbf{p}^2/2m)\Omega(\mathbf{y})\, d^3y$$

$$- \alpha\hbar\omega(\hbar/2m^*\omega)^{1/2} \int d^3y \int d^3z \mid \Omega(\mathbf{y}) \mid^2 \mid \Omega(\mathbf{z}) \mid^2/\mid \mathbf{y}-\mathbf{z} \mid. \tag{7.7.17}$$

We may vary the function Ω^* subject to the constraint of normalization and obtain an effective Schrödinger equation:

$$[(\mathbf{p}^2/2m^*) - 2\alpha\hbar\omega(\hbar/2m^*\omega)^{1/2} \int d^3z \mid \Omega(\mathbf{z}) \mid^2/\mid \mathbf{y}-\mathbf{z} \mid - \epsilon]\Omega(\mathbf{y}) = 0. \tag{7.7.18}$$

This equation is quite analogous to the Hartree form of a one-electron self-consistent field equation; note, however, that this "self"-interaction term is attractive. Rather than solve (7.7.18) numerically, we proceed by assuming a trial form for Ω:

$$\Omega(\mathbf{y}) = [2\beta m^*\omega/\pi\hbar]^{3/4} \exp[-(\beta m^*\omega/\hbar)\mathbf{y}^2]. \tag{7.7.19}$$

This is to be substituted into (7.7.17). The integrals can be calculated and the result differentiated with respect to β. We will not give details of this calculation. The result is

$$\beta^{1/2} = (\alpha/3)(2/\pi)^{1/2}, \qquad \langle E\rangle = -(\alpha^2/3\pi)\hbar\omega. \tag{7.7.20}$$

We note that the binding energy in this limit is proportional to α^2.

7.8 Superconductivity

One really would not expect anything sudden and radical to happen to the properties of a metal at liquid helium temperatures. Nature frequently confounds this expectation with the phenomenom of superconductivity. Electrical resistance drops totally to zero, magnetic fields are expelled, the specific heat exhibits a λ anomaly. A second-order phase transition occurs in which a new state is formed which must be described in a very different fashion from that used in normal metals. The existence of a Fermi surface is one of the most definitive features of the metallic state; yet the Fermi surface disappears and an energy gap of order KT_c (where T_c is the transition temperature) appears in the excitation spectrum.

Superconductivity is found in very many metallic systems though apparently not in all. The existence of order is not necessary: numerous disordered alloys become superconductors. The occurrence of superconductivity in quite diverse metals suggests that details of the band structure must be only of secondary importance. The isotope effect—the transition temperature of metal samples formed from different isotopes of the same element satisfies, in many instances,

$$T_c M^{1/2} = \text{const} \tag{7.8.1}$$

in which M is the atomic mass—suggests quite strongly that lattice vibrations must be intimately involved since characteristic vibrational frequencies will tend to vary as $M^{-1/2}$. On the other hand, the small magnitude of the energy difference between the normal and the superconducting states, which is of the order of 10^{-8} eV/atom, implies that the effect must be quite subtle, as the first-order electron–phonon interaction involves energies considerably larger than this. It will be shown below that in second order, the electron–phonon interaction gives rise to an effective electron–electron interaction which is attractive for electrons in states close to the Fermi energy. A normal Fermi system is unstable with respect to such an interaction, and a condensed state with the required unusual properties is formed.

We will develop the theory along the lines first presented by Bardeen *et al.* (1957). General references include Parks (1969), de Gennes (1966), Rickayzen (1965), Schrieffer (1964).

7.8.1 *The Origin of Superconductivity*

Superconductivity results from the operation of an attractive electron–electron interaction produced by phonon exchange. This attraction can be derived by an extension of the argument in Section 7.1.6 in which the effects of the electron–phonon interaction on the self-energy of an electron were considered.

The starting point is the following Hamiltonian for a system of interacting electrons and phonons:

$$\begin{aligned} H = {} & \sum_{k\sigma} E(\mathbf{k}) c^+_{k\sigma} c_{k\sigma} + \sum_{q} \hbar\omega(\mathbf{q}) a_q^+ a_q \\ & + \sum_{kq\sigma} [D(\mathbf{q}) c^+_{k+q\sigma} c_{k\sigma} a_q + D(-\mathbf{q}) c^+_{k-q,\sigma} c_{k\sigma} a_q^+] \\ & + \sum_{kk'q\sigma\sigma'} \mathcal{V}(\mathbf{q}) c^+_{k-q\sigma} c^+_{k'+q\sigma'} c_{k'\sigma'} c_{k\sigma}. \end{aligned} \tag{7.8.2}$$

The first term represents the energy of a noninteracting system of electrons; the second, that of the phonons; the third term describes the electron–phonon coupling; the fourth, the electron–electron interaction. A single-band approach has been adopted in which electrons couple only to one branch of the phonon spectrum; other branches are omitted. To determine the effect of the electron–phonon interaction on the electron system, it is desirable to apply a unitary transformation which will eliminate the third term from H. This cannot be done completely, but it is possible to do this in first order.

The procedure is similar to that of Section 5.3. Let H_0 denote the first two terms of (7.8.2), H_1 the third, and U, the fourth: $H = H_0 + H_1 + U$. Consider the transformation $H_T = e^{-iS} H e^{iS}$ in which S is a Hermitian operator. Equation (5.3.14) shows that, through second order in S,

$$H_T = H + i[H, S] - \tfrac{1}{2}[[H, S], S]$$

where S is chosen to satisfy $i[H_0, S] = -H_1$. The transformed Hamiltonian is

$$H_T = H_0 + U + i[U, S] - \tfrac{1}{2}[[U, S], S] + \tfrac{1}{2}i[H_1, S]. \tag{7.8.3}$$

We will make the approximation of neglecting the commutator $[U, S]$, that is, we ignore the effect of the electron–phonon interaction on the electron–electron terms. Suppose now that eigenstates of H_0 are available, which are denoted temporarily by $|m\rangle$, $|n\rangle$. Then

$$\langle m | S | n \rangle = i\langle m | H_1 | n \rangle / (E_m - E_n).$$

Let n_q represent the number of phonons present with wave vector $\mathbf{q}$. Then H_1 has matrix elements in which n_q changes by ± 1. Note that

$$\langle n_q - 1 | a_q | n_q \rangle = n_q^{1/2}, \qquad \langle n_q + 1 | a_q^+ | n_q \rangle = (n_q + 1)^{1/2}.$$

The transformation operator S will have matrix elements connecting the same states:

$$\begin{aligned} & \langle n_q - 1 | S | n_q \rangle \\ & \quad = i \sum_{k\sigma} D(\mathbf{q}) c^+_{k+q,\sigma} c_{k\sigma} n_q^{1/2} / [E(\mathbf{k} + \mathbf{q}) - E(\mathbf{k}) - \hbar\omega_q] \end{aligned} \tag{7.8.4}$$

and

$$\langle n_{\mathbf{q}} + 1 \mid S \mid n_{\mathbf{q}} \rangle = i \sum_{\mathbf{k}\sigma} D(-\mathbf{q}) c^{+}_{\mathbf{k}-\mathbf{q},\sigma} c_{\mathbf{k}\sigma} (n_q + 1)^{1/2} / [E(\mathbf{k} - \mathbf{q}) - E(\mathbf{k}) + \hbar\omega_{\mathbf{q}}].$$

Interest centers on matrix elements of the transformed Hamiltonian H_{T} which are diagonal in the phonon occupation numbers, since it is these elements which will contain the effective electron–electron interaction. Thus

$$\langle n_{\mathbf{q}} \mid \tfrac{1}{2} i [H_1, S] \mid n_{\mathbf{q}} \rangle = \tfrac{1}{2} i \sum_{\pm} \{ \langle n_{\mathbf{q}} \mid H_1 \mid n_{\mathbf{q}} \pm 1 \rangle \langle n_{\mathbf{q}} \pm 1 \mid S \mid n_{\mathbf{q}} \rangle - \langle n_{\mathbf{q}} \mid S \mid n_{\mathbf{q}} \pm 1 \rangle \langle n_{\mathbf{q}} \pm 1 \mid H_1 \mid n_{\mathbf{q}} \rangle \}.$$

The matrix element products are easily constructed. For example,

$$\tfrac{1}{2} i \langle n_{\mathbf{q}} \mid H_1 \mid n_q - 1 \rangle \langle n_{\mathbf{q}} - 1 \mid S \mid n_{\mathbf{q}} \rangle = -\tfrac{1}{2} \sum_{\mathbf{k}\mathbf{k}',\sigma\sigma'} c^{+}_{\mathbf{k}'-\mathbf{q},\sigma'} c_{\mathbf{k}'\sigma'} c^{+}_{\mathbf{k}+\mathbf{q},\sigma} c_{\mathbf{k},\sigma} \times | D(\mathbf{q}) |^2 n_{\mathbf{q}} / [E(\mathbf{k} + \mathbf{q}) - E(\mathbf{k}) - \hbar\omega_{\mathbf{q}}].$$

We have used the relation $D(-\mathbf{q}) = D^*(\mathbf{q})$. After the matrix element products have been determined, the result may be summed over phonon wave vectors $\mathbf{q}$. The additional term in H_{T} which is produced will be called H_{I}. Some algebra, involving the change of dummy variables and use of the commutation rules yields

$$H_{\mathrm{I}} = \sum_{\mathbf{k}\mathbf{k}'\mathbf{q},\sigma\sigma'} | D(\mathbf{q}) |^2 \{ \hbar\omega_{\mathbf{q}} / [E(\mathbf{k}) - E(\mathbf{k} - \mathbf{q})]^2 - (\hbar\omega_{\mathbf{q}})^2 \} \times c^{+}_{\mathbf{k}-\mathbf{q},\sigma} c^{+}_{\mathbf{k}'+\mathbf{q},\sigma'} c_{\mathbf{k}'\sigma'} c_{\mathbf{k}\sigma}. \tag{7.8.5}$$

The interaction matrix elements given by (7.8.5) can be either attractive or repulsive. If the states $\mathbf{k}$, $\mathbf{k} - \mathbf{q}$ are separated by an energy larger than $\hbar\omega_q$, the effect is repulsive, but if the energy difference is smaller than this, an attraction is present. The system will have to adjust itself to the presence of this interaction, which will occur—as we shall see—through the formation of pairs.

7.8.2 *The Superconducting State*

The electron–electron coupling derived from phonon exchange as expressed by (7.8.5) should be combined with the coulomb interaction [fourth term of (7.8.2)] to form an effective total electron–electron interaction. The Coulomb potential $\mathcal{V}(\mathbf{q})$ should include screening effects described by a dielectric function. A specific expression for $\mathcal{V}(\mathbf{q})$ will not be required.

At this point, we introduce a further approximation. Since the superconducting state, as we will see below, is formed through the association of electrons in pairs with opposite wave vectors and spins (thus $\mathbf{k}\uparrow, -\mathbf{k}\downarrow$) we drop all those terms in the interaction which do not connect such pairs. This we set $\mathbf{k} = -\mathbf{k}'$ in (7.8.5). The interaction term is rewritten as

$$\tfrac{1}{2} \sum_{\mathbf{k}\mathbf{k}',\sigma\sigma'} V_{\mathbf{k}\mathbf{k}'} c^+_{\mathbf{k}'\sigma'} c^+_{-\mathbf{k}'\sigma} c_{-\mathbf{k}\sigma} c_{\mathbf{k}\sigma'} \tag{7.8.6}$$

with

$$V_{\mathbf{k}\mathbf{k}'} = 2\,|D(\mathbf{k}-\mathbf{k}')|^2 \hbar\omega(\mathbf{k}-\mathbf{k}')/\{[E(\mathbf{k}) - E(\mathbf{k}')]^2 - [\hbar\omega(\mathbf{k}-\mathbf{k}')]^2\} + \mathcal{V}(\mathbf{k}-\mathbf{k}') \tag{7.8.7}$$

in which $\mathcal{V}(\mathbf{k}-\mathbf{k}')$ is a matrix element of the Coulomb interaction. The summation over spin states has four terms, two of which ($\sigma = \sigma'$) correspond to interaction of triplet pairs and two of which have singlet components. The theory of superconductivity is constructed in part in (imperfect) analogy with the condensation of a Bose gas of spin zero particles. We retain only the singlet pairs. Since $V_{\mathbf{k}\mathbf{k}'} = V_{-\mathbf{k}-\mathbf{k}'}$ the two terms are the same and the factor of $\frac{1}{2}$ in (7.8.6) disappears. Since we have now in effect fixed the spin ordering, we can delete the index σ; letting $\mathbf{k}$ stand for $\mathbf{k}\uparrow$, say, and $-\mathbf{k}$ for $-\mathbf{k}\downarrow$. We now have the Hamiltonian

$$H = \sum_{\mathbf{k}} E(\mathbf{k}) c_{\mathbf{k}}^+ c_{\mathbf{k}} + \sum_{\mathbf{k}\mathbf{k}'} V_{\mathbf{k}\mathbf{k}'} c^+_{\mathbf{k}'} c^+_{-\mathbf{k}'} c_{-\mathbf{k}} c_{\mathbf{k}}. \tag{7.8.8}$$

We will choose the zero of energy to be at the Fermi energy in the normal state. Equation (7.8.8) is the pair Hamiltonian of the BCS theory. The term "pair" is appropriate since the operator combinations $c_{-\mathbf{k}}c_{\mathbf{k}}$ and $c^+_{\mathbf{k}'}c^+_{-\mathbf{k}}$ can be regarded as destroying or creating an electron pair. Thus (7.8.8) contains an interaction which transfers a pair from $|\mathbf{k}\uparrow, -\mathbf{k}\downarrow\rangle$ to $|\mathbf{k}'\uparrow, -\mathbf{k}'\downarrow\rangle$. Superconductivity results if the potential $V_{\mathbf{k}\mathbf{k}'}$ is attractive for states near the Fermi energy.

Many approximations have been involved in the derivation of the pair Hamiltonian. Equation (7.8.8) should be regarded as a plausible model. It is possible to proceed more rigorously from this point and deduce the consequences of (7.8.8) with some rigor. The success of the results in describing the properties of real superconductors offers some justification of the model employed. It is, in fact, quite surprising that a theory founded on a simple model of pairing as is incorporated in (7.8.8) is in fact able to account for much of the observational information about superconductivity.

We will now consider the diagonalization of the pair Hamiltonian (7.8.8) following the method of Valatin (1958). This is quite similar to the procedure employed in the theory of antiferromagnetism (Section 7.2).

Let us define new operators ξ_{k} by

$$\xi_{\mathrm{k}} = \alpha_{\mathrm{k}} c_{\mathrm{k}} - \beta_{\mathrm{k}} c^{+}{}_{-\mathrm{k}}, \qquad \xi_{-\mathrm{k}} = \alpha_{\mathrm{k}} c_{-\mathrm{k}} + \beta_{\mathrm{k}} c_{\mathrm{k}}{}^{+} \tag{7.8.9a}$$

in which α_{k} and β_{k} are real and positive. The conjugate relations are

$$\xi_{\mathrm{k}}{}^{+} = \alpha_{\mathrm{k}} c_{\mathrm{k}}{}^{+} - \beta_{\mathrm{k}} c_{-\mathrm{k}}, \qquad \xi^{+}{}_{-\mathrm{k}} = \alpha_{\mathrm{k}} c^{+}{}_{-\mathrm{k}} + \beta_{\mathrm{k}} c_{\mathrm{k}}. \tag{7.8.9b}$$

We will require that the ξ_k (like the c_k) obey Fermion rules:

$$\{\xi^{+}{}_{\mathrm{k}'}, \xi_{\mathrm{k}}\} = \delta_{\mathrm{kk}'}, \qquad \{\xi_{\mathrm{k}}, \xi_{\mathrm{k}'}\} = 0. \tag{7.8.10}$$

The brace indicates an anticommutator. Equations (7.8.10) will be satisfied provided that

$$\alpha_{\mathrm{k}}{}^{2} + \beta_{\mathrm{k}}{}^{2} = 1. \tag{7.8.11}$$

The transformation (7.8.9) can be easily inverted with the aid of (7.8.11):

$$\begin{aligned} c_{\mathrm{k}} &= \alpha_{\mathrm{k}}\xi_{\mathrm{k}} + \beta_{\mathrm{k}}\xi^{+}{}_{-\mathrm{k}}, \qquad & c_{-\mathrm{k}} &= \alpha_{\mathrm{k}}\xi_{-\mathrm{k}} - \beta_{\mathrm{k}}\xi_{\mathrm{k}}{}^{+}, \\ c_{\mathrm{k}}{}^{+} &= \alpha_{\mathrm{k}}\xi_{\mathrm{k}}{}^{+} + \beta_{\mathrm{k}}\xi_{-\mathrm{k}}, \qquad & c^{+}{}_{-\mathrm{k}} &= \alpha_{\mathrm{k}}\xi^{+}{}_{-\mathrm{k}} - \beta_{\mathrm{k}}\xi_{\mathrm{k}}. \end{aligned} \tag{7.8.12}$$

We substitute (7.8.12) into (7.8.8). It is convenient to define the number operator η_{k} for the operators ξ_{k}:

$$\eta_{\mathrm{k}} = \xi_{\mathrm{k}}{}^{+}\xi_{\mathrm{k}}. \tag{7.8.13}$$

The first term in (7.8.8) becomes

$$\sum_{\mathrm{k}} E(\mathbf{k})[2\beta_{\mathrm{k}}{}^{2} + (\alpha_{\mathrm{k}}{}^{2} - \beta_{\mathrm{k}}{}^{2})(\eta_{\mathrm{k}} + \eta_{-\mathrm{k}}) + 2\alpha_{\mathrm{k}}\beta_{\mathrm{k}}(\xi_{\mathrm{k}}{}^{+}\xi^{+}{}_{-\mathrm{k}} + \xi_{-\mathrm{k}}\xi_{\mathrm{k}})]. \tag{7.8.14}$$

It is evident that this term is no longer diagonal. The interaction term is more complicated. We find

$$\begin{aligned} \sum_{\mathrm{k}'} V_{\mathrm{kk}'}[&\alpha_{\mathrm{k}'}\alpha_{\mathrm{k}}\beta_{\mathrm{k}}\beta_{\mathrm{k}'}(1 - \eta_{\mathrm{k}'} - \eta_{-\mathrm{k}'})(1 - \eta_{\mathrm{k}} - \eta_{-\mathrm{k}}) \\ &+ \alpha_{\mathrm{k}'}\beta_{\mathrm{k}'}(\alpha_{\mathrm{k}}{}^{2} - \beta_{\mathrm{k}}{}^{2})(1 - \eta_{\mathrm{k}'} - \eta_{-\mathrm{k}'})(\xi_{-\mathrm{k}}\xi_{\mathrm{k}} + \xi_{\mathrm{k}}{}^{+}\xi^{+}{}_{-\mathrm{k}}) \\ &+ \text{fourth-order terms}. \end{aligned} \tag{7.8.15}$$

We have used the symmetry of $V_{\mathrm{kk}'}$ $(=V_{\mathrm{k}'\mathrm{k}})$ in order to obtain the second line of (7.8.15). This equation involves a diagonal part, second-order nondiagonal terms, and fourth-order nondiagonal terms. The latter will be discarded. It can be shown that the omitted terms have a negligible expectation value in the BCS ground state. We may then combine (7.8.14) and (7.8.15). The second-order nondiagonal terms in (7.8.14) and (7.8.15) have the same form and may be eliminated by proper choice of α and β. The operators ξ_k^+ and ξ_k are considered to create (or destroy) an excited state of the system (which consists of a correlated hole electron pair). The ground state $\mathcal{G}$ of the system should therefore satisfy

$$\xi_k \mid \mathcal{G}\rangle = \xi_{-k} \mid \mathcal{G}\rangle = 0 \tag{7.8.16a}$$

and thus

$$\eta_k \mid \mathcal{G}\rangle = \eta_{-k} \mid \mathcal{G}\rangle = 0. \tag{7.8.16b}$$

The ground state $\mid \mathcal{G}\rangle$ is not to be confused with the vacuum state $\mid 0\rangle$. Unfortunately, $\mid \mathcal{G}\rangle$ is not an exact eigenstate of the total particle number. However, the expectation value of this quantity

$$N = \sum_k (c_k{}^+c_k + c^+{}_{-k}c_{-k}) \tag{7.8.17}$$

in the ground state is

$$\langle N\rangle = \langle \mathcal{G} \mid N \mid \mathcal{G}\rangle = 2 \sum_k \beta_k{}^2. \tag{7.8.18}$$

We will have to make sure that (7.8.18) gives the correct number of particles.

In view of (7.8.16a,b), the condition that the second-order off-diagonal terms should vanish is

$$2E(\mathbf{k})\alpha_k\beta_k + (\alpha_k{}^2 - \beta_k{}^2) \sum_{k'} V_{kk'}\alpha_{k'}\beta_{k'} = 0. \tag{7.8.19}$$

Equations (7.8.11) and (7.8.19) must be solved simultaneously to determine α_k and β_k. The solution is facilitated if we introduce

$$\alpha_k = \sin\theta_k, \qquad \beta_k = \cos\theta_k. \tag{7.8.20}$$

This substitution enables us to satisfy (7.8.11) automatically. Equation (7.8.19) then becomes

$$E(\mathbf{k}) \sin 2\theta_k - \tfrac{1}{2} \cos 2\theta_k \sum_{k'} V_{kk'} \sin 2\theta_{k'} = 0. \tag{7.8.21}$$

At this point, a further substitution becomes convenient. We define a quantity Δ_k by

$$\Delta_k = -\tfrac{1}{2} \sum_{k'} V_{kk'} \sin 2\theta_{k'}. \tag{7.8.22}$$

This is inserted into (7.8.21), which immediately yields

$$\tan 2\theta_k = -\Delta_k/E(\mathbf{k}). \tag{7.8.23a}$$

Note that $E(\mathbf{k})$ is negative below the Fermi energy. The angle θ_k should be in the first quadrant for such energies. Hence

$$\sin 2\theta_k = \Delta_k/[\Delta_k^2 + E(\mathbf{k})^2]^{1/2}; \qquad \cos 2\theta_k = -E(\mathbf{k})/[\Delta_k^2 + E(\mathbf{k})^2]^{1/2}. \tag{7.8.23b}$$

Equation (7.8.23b) may now be substituted back into (7.8.22). The result is an integral equation for Δ_k of the form

$$\Delta_k = -\tfrac{1}{2} \sum_{k'} V_{kk'} \Delta_{k'}/[\Delta^2_{k'} + E^2(\mathbf{k}')]^{1/2}. \tag{7.8.24}$$

A simple assumption concerning the function $V_{kk'}$ will be introduced to facilitate solution of (7.8.24). It will be supposed that $V_{kk'}$ is a (negative) constant within a small region of energy centered about the Fermi energy (which is our zero of energy). Specifically, let

$$\begin{aligned} V_{kk'} &= -V_0 \qquad \text{if } |E(\mathbf{k})| \text{ and } |E(\mathbf{k}')| < \hbar\omega_0, \\ &= 0 \qquad \text{otherwise.} \end{aligned} \tag{7.8.25}$$

We would expect [on the basis of Eq. (7.8.4)] that ω_0 would be of the order of magnitude of the Debye frequency for the phonon spectrum. The solution of (7.8.24) for Δ_k has the corresponding property that

$$\begin{aligned} \Delta_k &= \Delta_0 \qquad \text{if } |E(\mathbf{k})| < \hbar\omega_0, \\ &= 0 \qquad \text{otherwise.} \end{aligned} \tag{7.8.26}$$

The equation for Δ_0 is then

$$1 = \tfrac{1}{2} V_0 \sum_{k} [\Delta_0^2 + E(\mathbf{k})^2]^{-1/2}. \tag{7.8.27}$$

The summation is converted into an integral by introducing the density of states $G(E)$:

$$1 = \tfrac{1}{2} V_0 \int_{-\hbar\omega_0}^{\hbar\omega_0} dE\, G(E)/(\Delta_0^2 + E^2)^{1/2}. \tag{7.8.28}$$

The integral extends over only a small range of energies ($2\hbar\omega_0$) near the Fermi energy. It is adequate in first approximation to treat G as a constant G_0 equal to its value at the Fermi energy. The integral is then a simple one. We solve for Δ_0 and obtain finally

$$\Delta_0 = \hbar\omega_0/\sinh(1/V_0G_0) \approx 2\hbar\omega_0 \exp(-1/V_0G_0). \tag{7.8.29}$$

The last step in (7.8.29) follows if V_0G_0 is small, which is frequently the case.

In order to estimate the order of magnitude of Δ_0 we obtain V_0 from (7.8.4) with $\hbar\omega = \hbar\omega_D$ (ω_D being the Debye frequency) and $E(\mathbf{k}) = E(\mathbf{k} - \mathbf{q})$. The result should be an overestimate of V_0, since the contribution from the Coulomb repulsion $\mathcal{V}(\mathbf{k} - \mathbf{k}')$ has been ignored. Thus

$$V_0 \approx |D|^2/\hbar\omega_D. \tag{7.8.30}$$

The coupling constant D is found from (7.1.34) and (7.1.28) in the deformation potential approximation (and $K = k_F$). The result can be inserted in (7.8.30), which becomes

$$V_0 \approx \tfrac{4}{9}[E_F{}^3/(\hbar\omega_D)^2](m/M).$$

The density of states is more difficult to estimate accurately since band structure effects will contribute. We will make a rough guess: $G_0 \approx 1/E_F$. Thus

$$V_0 G_0 \approx \tfrac{4}{9}(E_F/\hbar\omega_D)^2(m/M). \tag{7.8.31}$$

The ratio $E_F/\hbar\omega_D$ is typically of the order 100–200 while $m/M \approx 10^{-5}$. Hence a reasonable range for V_0G_0 is between 1 and $\frac{1}{10}$. For order of magnitude purposes in the following, we will choose $V_0G_0 \approx 0.2$. Then (with $\hbar\omega_0 \approx 0.03$ eV) we see that $\Delta_0 \approx 4 \times 10^{-4}$ eV (or 5°K). The actual value is extremely sensitive to V_0G_0, but our argument indicates that we should not expect $K\Delta_0$ to exceed a few degrees at most.

We will proceed to consider the ground state energy of the system. From (7.8.14) and (7.8.15) we see that this is given by

$$E_g = 2\sum_{\mathbf{k}} E(\mathbf{k})\beta_{\mathbf{k}}{}^2 + \sum_{\mathbf{k}\mathbf{k}'} V_{\mathbf{k}\mathbf{k}'}\alpha_{\mathbf{k}'}\alpha_{\mathbf{k}}\beta_{\mathbf{k}}\beta_{\mathbf{k}'}. \tag{7.8.32}$$

Equations (7.8.20) and (7.8.22) are substituted into (7.8.32) which then becomes

$$E_g = 2\sum_{\mathbf{k}} E(\mathbf{k})\cos^2\theta_{\mathbf{k}} - \tfrac{1}{2}\sum_{\mathbf{k}} \Delta_{\mathbf{k}}\sin 2\theta_{\mathbf{k}}. \tag{7.8.33}$$

The quantity of physical interest is the change in the ground state energy produced by the interaction. If V were zero, the one-particle states of the system would be occupied up to $k = k_F$. The change in the energy is therefore

$$\delta E_g = 2\sum_{k<k_F} E(\mathbf{k})[\cos^2\theta_{\mathbf{k}} - 1] + 2\sum_{k>k_F} E(\mathbf{k})\cos^2\theta_{\mathbf{k}} - \tfrac{1}{2}\sum_{\mathbf{k}} \Delta_{\mathbf{k}}\sin 2\theta_{\mathbf{k}}. \tag{7.8.34}$$

The solutions (7.8.23b) for the trigonometric functions are inserted into

(7.8.34). We obtain

$$\delta E_{\rm g} = -\sum_{k<k_{\rm F}} E(\mathbf{k})\left(1 + \frac{E(\mathbf{k})}{[\Delta^2 + E^2(\mathbf{k})]^{1/2}}\right)$$

$$+ \sum_{k>k_{\rm F}} E(\mathbf{k})\left(1 - \frac{E(\mathbf{k})}{[\Delta^2 + E^2(\mathbf{k})]^{1/2}}\right) - \tfrac{1}{2}\sum_{\mathbf{k}} \frac{\Delta^2}{[\Delta^2 + E^2(\mathbf{k})]^{1/2}}$$

$$= 2\sum_{k>k_{\rm F}}\left(E(\mathbf{k}) - \frac{2E^2(\mathbf{k}) + \Delta^2}{2[\Delta^2 + E^2(\mathbf{k})]^{1/2}}\right).$$

The second term is obtained by observing that $E(\mathbf{k})\ (k < k_{\rm F}) = -E(\mathbf{k})\ (k > k_{\rm F})$ to sufficient accuracy, since $\Delta \neq 0$ over only a small range of energies. The sum is converted into an integral. We find, using the approximation of a constant density of states,

$$\delta E_{\rm g} = 2G_0 \int_0^{\hbar\omega_0} E - [(2E^2 + \Delta^2)/2(E^2 + \Delta^2)^{1/2}]\, dE$$

$$= (\hbar\omega_0)^2 G_0[1 - (1 + \Delta^2/\hbar^2\omega_0^2)^{1/2}]$$

$$= (\hbar\omega_0)^2 G_0[1 - \coth(1/V_0G_0)]. \qquad (7.8.35)$$

In the weak coupling approximation, the square root may be expanded. The result is

$$\delta E_{\rm g} = -\tfrac{1}{2}G_0\Delta^2. \qquad (7.8.36)$$

This is a very small energy change. With our previous estimate for Δ (and $G_0 \approx 0.2\ \text{eV}^{-1}$), we have $\delta E_{\rm g} = 1.6 \times 10^{-8}$ eV. The change is small because the coupling is weak and, moreover, affects only states near the very top of the Fermi distribution. It is remarkable that an interaction which produces so small a change in the energy of the system can produce such dramatic effects. It should also be noted, in view of (7.8.29), that the change in the ground state energy is not an analytic function of the interaction strength V_0. When V_0 is small, $\delta E_{\rm g}$ vanishes more rapidly than any power of V_0. Although the effect of the interaction on the ground state energy is exceedingly small, it could not have been predicted by any finite-order perturbation theory.

The state vector for the lowest state $|\,\mathcal{G}\rangle$ is defined by (7.8.16). It follows from the definition of the operators $\xi_{\rm k}$ and the properties of the single Fermion operators $c_{\rm k}$ that $\xi_{\rm k}\xi_{\rm k} = \xi_{-\rm k}\xi_{-\rm k} = 0$. Consider $\mathbf{k}$ fixed for the moment. A state satisfying (7.8.16) is

$$\xi_{\rm k}\xi_{-\rm k}\,|\,0\rangle = (\alpha_{\rm k}\beta_{\rm k} + \beta_{\rm k}^2 c_{\rm k}^{+} c^{+}_{-\rm k})\,|\,0\rangle. \qquad (7.8.37)$$

We have used the fact that $c_{\mathbf{k}}$ annihilates the vacuum state $|0\rangle$, $c_{\mathbf{k}}|0\rangle = 0$. The state (7.8.37) is not normalized. However, it is easily found with the aid of (7.8.11) that the normalization constant is simply $1/\beta_{\mathbf{k}}$. The ground state may now be constructed as

$$|\mathcal{G}\rangle = [\prod_{\mathbf{k}} (\alpha_{\mathbf{k}} + \beta_{\mathbf{k}} c_{\mathbf{k}}^{+} c^{+}_{-\mathbf{k}})]\,|0\rangle. \tag{7.8.38}$$

This is the state vector employed by Bardeen, Cooper, and Schrieffer (1957) in a variational calculation.

It was observed previously [Eq. (7.8.18)] that this ground state vector is not eigenstate of the total number of particles. However, from (7.8.18) and (7.8.23b), the average number is

$$\langle N\rangle = 2\sum_{\mathbf{k}} \beta_{\mathbf{k}}^2 = \sum_{\mathbf{k}} 1 - \{E(\mathbf{k})/[E^2(\mathbf{k}) + \Delta_{\mathbf{k}}^2]^{1/2}\} = 2\sum_{k<k_F} (1) = N. \tag{7.8.39}$$

We have again used the fact that $E(\mathbf{k})$ changes sign on passing through k_F. The average number of particles is given correctly, and this turns out to be sufficient.

The state $\xi_{\mathbf{k}}^{+}|\mathcal{G}\rangle$ is an excited state of the system. It is apparent from (7.8.9) that $\xi_{\mathbf{k}}^{+}|\mathcal{G}\rangle$ is a state of wave vector $\mathbf{k}$, involving a superposition of an electron of wave vector $\mathbf{k}$ and a hole of wave vector $-\mathbf{k}$. We call this a *quasi-particle state.* The energy of such states will be determined. After the coefficients α, β have been determined in accord with (7.8.19), the Hamiltonian becomes

$$H = E_g + \sum_{\mathbf{k}} \{E(\mathbf{k})(\alpha_{\mathbf{k}}^2 - \beta_{\mathbf{k}}^2) - 2\alpha_{\mathbf{k}}\beta_{\mathbf{k}} \sum_{\mathbf{k}'} V_{\mathbf{k}\mathbf{k}'}\alpha_{\mathbf{k}'}\beta_{\mathbf{k}'}$$
$$\times (1 - \tfrac{1}{2}\eta_{\mathbf{k}'} - \tfrac{1}{2}\eta_{-\mathbf{k}'})\}(\eta_{\mathbf{k}} + \eta_{-\mathbf{k}}) + \cdots \text{ nondiagonal terms} \tag{7.8.40}$$

in which E_g is the ground state energy just discussed. We wish to find the energy required to create an excitation. At $T = 0$, we may ignore $\eta_{\mathbf{k}'}$ and $\eta_{-\mathbf{k}'}$ compared to unity. We will denote the excitation energy by $\mathcal{E}(\mathbf{k})$. This is

$$\mathcal{E}(\mathbf{k}) = (\alpha_{\mathbf{k}}^2 - \beta_{\mathbf{k}}^2)E(\mathbf{k}) - 2\alpha_{\mathbf{k}}\beta_{\mathbf{k}} \sum_{\mathbf{k}'} V_{\mathbf{k}\mathbf{k}'}\alpha_{\mathbf{k}'}\beta_{\mathbf{k}'}$$
$$= -\cos 2\theta_{\mathbf{k}} E(\mathbf{k}) + \sin 2\theta_{\mathbf{k}}\,\Delta_{\mathbf{k}} = [E(\mathbf{k})^2 + \Delta_{\mathbf{k}}^2]^{1/2}. \tag{7.8.41}$$

To interpret (7.8.41), recall that we measure energies from the Fermi level $\mathcal{E}_F$, where $\mathcal{E}(\mathbf{k})$ vanishes. Equation (7.8.41) indicates that the excitation energy $\mathcal{E}(\mathbf{k})$ never vanishes. In other words, the excitation spectrum contains a gap. This gap is characteristic of the superconducting state, and

Δ is known as the *gap parameter*. It should be observed, however, that external fields do not create these excitations singly. Since $\xi_{\mathbf{k}}^{+}$ is a linear combination of single Fermion operators, the ξ's always occur in pairs in an interaction Hamiltonian which can scatter particles, but must conserve the number. As a result, the excitations are created in pairs, and the gap observed experimentally (for instance, in absorption of electromagnetic waves) is 2Δ.

It is interesting to consider the density of states of these excitations. We will assume for this purpose that the band structure is isotropic: $E(\mathbf{k}) = E(|\mathbf{k}|)$, and that the gap parameter Δ is independent of $\mathbf{k}$. Let the density of quasi-particle states be denoted by $\mathcal{G}(\mathcal{E})$. We have

$$\mathcal{G}(\mathcal{E}) \approx dk/d\mathcal{E} = dk/dE\, dE/d\mathcal{E}. \tag{7.8.42}$$

However, $d\mathcal{E}/dE = E/|\mathcal{E}| = (\mathcal{E}^2 - \Delta^2)^{1/2}/|\mathcal{E}|$, and the density of band states is proportional to dk/dE. Excitations produced by $\xi_{\mathbf{k}}^{+}$ are degenerate with those produced by $\xi^{+}_{-\mathbf{k}}$. The spin degeneracy of the normal metal is replaced by this degeneracy. As a result

$$\begin{aligned} \mathcal{G}(\mathcal{E}) &\approx G_0(|\mathcal{E}|/[\mathcal{E}^2 - \Delta^2]^{1/2}), \qquad & |\mathcal{E}| > \Delta, \\ &= 0, & |\mathcal{E}| < \Delta. \end{aligned} \tag{7.8.43}$$

It is to be noted that the density of states is singular at $\mathcal{E} = \Delta$, in a fashion similar to that encountered in a one-dimensional system. This singularity

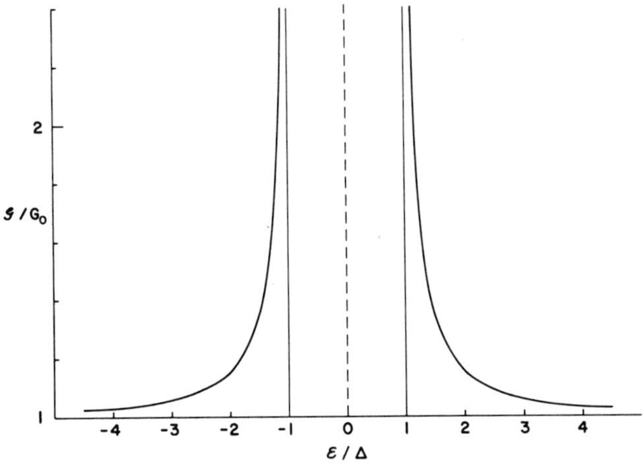

Fig. 7.8.1. Ratio of the density of states in the superconducting state to that in the normal state.

would presumably disappear if the assumption of isotropic $E(\mathbf{k})$ and constant gap were removed. The relative change in the density of states, $\mathcal{G}(\mathcal{E})/G_0$ is illustrated in Fig. 7.8.1.

7.8.3 *The Transition Temperature*

The excitations we have discussed are created and annihilated by operators obeying the anticommutation rules of Fermi statistics. Thus at finite temperatures the population of the excited states should be described by the usual Fermi function. Since the number of excitations is not conserved, no chemical potential appears in this function. There is an important complication, however. At finite temperatures, we should not neglect the terms in (7.8.40) involving products of number operators η. It is apparent from (7.8.40) that the energy required to create an excitation depends on the numbers of other excitations present; and thus on the temperature. We will therefore consider the excitation energy ϵ, and the gap parameter Δ to be functions of temperature.

It is necessary to define the excitation energy for $T \neq 0$. To this end, let $\langle H \rangle$ and $\langle \eta_\mathbf{k} \rangle$ denote thermal averages of the Hamiltonian (7.8.40) and the number operator $\eta_\mathbf{k}$ for temperature T. The basis states used in constructing this average are eigenstates of the number operator $\eta_\mathbf{k}$. Our definition is

$$\mathcal{E}(T, \mathbf{k}) = \partial \langle H \rangle / \partial \langle \eta_\mathbf{k} \rangle. \tag{7.8.44}$$

We will also approximate the terms involving averages of products of η's as

$$\langle \eta_\mathbf{k} \eta_{\mathbf{k}'} \rangle \approx \langle \eta_\mathbf{k} \rangle \langle \eta_{\mathbf{k}'} \rangle. \tag{7.8.45}$$

We then find

$$\mathcal{E}(T, \mathbf{k}) = E(\mathbf{k})(\alpha_\mathbf{k}^2 - \beta_\mathbf{k}^2) - 2\alpha_\mathbf{k}\beta_\mathbf{k} \sum_{\mathbf{k}'} V_{\mathbf{k}\mathbf{k}'}\alpha_{\mathbf{k}'}\beta_{\mathbf{k}'}(1 - \langle \eta_{\mathbf{k}'} \rangle - \langle \eta_{-\mathbf{k}'} \rangle). \tag{7.8.46}$$

The excitation energy for a given $\mathbf{k}$ is reduced by the presence of other excitations. According to our argument above, we set

$$\langle \eta_{\mathbf{k}'} \rangle = f[\mathcal{E}(T, \mathbf{k}')] = [\exp(\mathcal{E}(T, \mathbf{k}')/KT) + 1]^{-1}. \tag{7.8.47}$$

The factor $1 - \langle \eta_{\mathbf{k}'} \rangle - \langle \eta_{-\mathbf{k}'} \rangle = \tanh \mathcal{E}(T, \mathbf{k}')/2KT$, which occurs in (7.8.46), is conveniently combined with $V_{\mathbf{k}\mathbf{k}'}$ in defining a temperature-dependent gap parameter

$$\Delta_\mathbf{k}(T) = -\tfrac{1}{2} \sum_{\mathbf{k}} V_{\mathbf{k}\mathbf{k}'} \sin 2\theta_{\mathbf{k}'} \tanh \mathcal{E}(T, \mathbf{k}')/2KT. \tag{7.8.48}$$

The relations (7.8.20) defining $\sin \theta_\mathbf{k}$, etc. are unaltered. The definition of a temperature-dependent gap parameter is useful since from (7.8.14) and (7.8.15), we should have for $T \neq 0$ a modified equation replacing (7.8.19),

$$2E(\mathbf{k})\alpha_k\beta_k + (\alpha_k{}^2 - \beta_k{}^2) \sum V_{kk'}\alpha_{k'}\beta_{k'}(1 - \langle\eta_{k'}\rangle - \langle\eta_{-k'}\rangle) = 0. \tag{7.8.49}$$

As a result, Eqs. (7.8.23) are modified only by the substitution of $\Delta_k(T)$ for Δ_k. Then we obtain from (7.8.48), in place of (7.8.24)

$$\Delta_k(T) = -\tfrac{1}{2} \sum_{k'} \{V_{kk'}\, \Delta_{k'}(T)/[\Delta_{k'}(T)^2 + E^2(\mathbf{k}')]^{1/2}\} \tanh \mathcal{E}(T, \mathbf{k}')/2KT. \tag{7.8.50}$$

Moreover, the extension of (7.8.23) to finite temperatures enables us to rewrite (7.8.46) as

$$\mathcal{E}(T, \mathbf{k}) = [E^2(\mathbf{k}) + \Delta_k(T)^2]^{1/2}. \tag{7.8.51}$$

The solution of (7.8.50) is evidently a difficult task in general. We use the simplifying approximations of (7.8.25) and (7.8.26), introduce the constant density of states G_0, and find in place of (7.8.28)

$$1 = V_0G_0 \int_0^{\hbar\omega_0} \tanh\{[E^2 + \Delta^2(T)]^{1/2}/2KT\}\, dE/(E^2 + \Delta^2)^{1/2}. \tag{7.8.52}$$

The gap parameter $\Delta(T)$, determined from (7.8.52), is a decreasing function of temperature which vanishes at the transition temperature T_c. Thus we can substitute $\Delta = 0$ into (7.8.52) and determine T_c directly:

$$1/V_0G_0 = \int_0^{\hbar\omega_0} (\tanh E/2KT_c)/E\, dE = \int_0^{\hbar\omega_0/2KT_c} (\tanh x)/x\, dx \tag{7.8.53}$$

in which $x = E/2KT_c$.

It is desirable to integrate by parts once:

$$1/V_0G_0 = \ln(\hbar\omega_0/2KT_c) \tanh(\hbar\omega_0/2KT_c) - \int_0^{\hbar\omega_0/2KT_c} \ln x \operatorname{sech}^2 x\, dx. \tag{7.8.54}$$

In fact, $\hbar\omega_0/2KT_c$ is quite large. It is a good approximation to make the upper limit of the remaining integral infinite. Thus we have

$$KT_c = \tfrac{1}{2}A\hbar\omega_0 \exp(-1/G_0V_0)$$

where the numerical constant A is given by (Rickayzen, 1965)

$$\ln A = -\int_0^\infty \ln x \operatorname{sech}^2 x\, dx = \gamma + \ln(4/\pi)$$

in which γ is Euler's constant (0.5772). Thus

$$KT_c = 1.13\hbar\omega_0 \exp(-1/V_0G_0). \tag{7.8.55}$$

The transition temperature is typically much smaller than the Debye temperature ($KT_c \approx 0.08\hbar\omega_0$ for $V_0G_0 = 5$). We can also relate T_c to the gap at $T = 0°K$ by the use of (7.8.29)

$$2\Delta(0)/KT_c = 3.53. \tag{7.8.56}$$

This relation is predicted to hold for all superconductors. Although it is not exactly obeyed in reality, agreement with experiment is quite reasonable (Meservey and Schwartz, 1969). See McMillan (1968) for a more quantitative calculation of the transition temperature.

The existence of the isotope effect is an immediate consequence of (7.8.55). The essential fact is that the Debye frequency is proportional to $M^{-1/2}$ where M is the atomic mass (see Section 1.1, Part A). In view of this, the simple expression for V_0G_0, Eq. (7.8.31), is predicted to be independent of M. Then (7.8.56) indicates that $T_c \approx M^{-1/2}$. This is the isotope effect. Also in this case, experimental results tend to be in reasonable, although not exact agreement with theory. Deviations can be attributed to effects of the Coulomb interaction $\mathcal{V}(\mathbf{k} - \mathbf{k}')$ [see Eq. (7.8.7)] (Garland, 1963).

The temperature dependence of the gap parameter is determined by (7.8.52). This requires numerical computation if results are to be obtained for arbitrary temperatures. The calculation can be set up as follows. Integrate (7.8.50) by parts once. The result is, with $\delta = \Delta/KT$,

$$1/G_0V_0 = \sinh^{-1}(\hbar\omega_0/\Delta) \tanh[(\hbar\omega_0)^2 + \Delta^2)^{1/2}/2KT] + F(\delta) \tag{7.8.57}$$

where $F(\delta)$ is specified subsequently. The weak coupling approximation, $\hbar\omega_0 \gg \Delta$, now gives

$$\Delta(T) \simeq 2\hbar\omega_0 \exp(-1/V_0G_0) \exp F(\delta). \tag{7.8.58}$$

This expression can be combined with (7.8.29) to yield

$$\Delta(T)/\Delta_0 = \exp[F(\Delta/KT)]. \tag{7.8.59}$$

The expression for F is

$$F(\delta) = -\delta \int_0^\infty [x \sinh^{-1}x \operatorname{sech}^2\delta(1 + x^2)^{1/2}]/(1 + x^2)^{1/2}\, dx. \tag{7.8.60}$$

This function must be computed numerically in general. A detailed investigation shows that $\Delta \approx (T_c - T)^{1/2}$ for T near T_c. This result is characteristic of a molecular field theory, as we saw in Chapter 2, Part A. The gap parameter Δ is the order parameter of this theory.

7.8.4 *Tunneling*

We have seen that the gap parameter Δ is fundamental to the theory of superconductivity. It is therefore of considerable importance to be able to determine this quantity experimentally. This is possible by means of a tunneling experiment, as was first demonstrated by Giaever (1960a).

Consider a thin sandwich composed of two metals separated by a thin insulating film (perhaps 20 Å thick). The insulating film acts as a barrier to the flow of electrons from one metal to the other. However, if a potential difference is established between the metals, a current may flow in which electrons tunnel through the barrier. It turns out that the voltage dependence of the tunnel current is radically different if one (or both) of the metals is superconducting compared to a situation in which both are in the normal state.

Let us consider an idealized tunneling experiment involving two semi-infinite metal slabs with plane parallel faces. Metal A is on the left, B on the right. The normal from A into B defines the z axis. The tunneling experiment can be described by a transfer Hamiltonian H_T. Let the operators c^+, c refer to metal A, while d^+, d refer to B

$$H_T = \sum_{qk\sigma} [T_{kq}c^+_{k\sigma}d_{q\sigma} + T^*_{qk}d^+_{q\sigma}c_{k\sigma}] \tag{7.8.61}$$

in which T_{kq} is a matrix element which can be treated as approximately independent of energy. This expression for the transfer Hamiltonian will simply be assumed to be valid here: for derivation see Bardeen (1961), Cohen *et al.* (1962), and Prange (1963). An approach not involving the use of a tunneling Hamiltonian has been given by Caroli *et al.* (1971). The first term in (7.8.61) describes the transfer of an electron from B to A; the second, a transfer from A to B.

Suppose that metal A is maintained at a voltage V above B, and for the moment let both metals be normal (i.e., nonsuperconducting). The net current flow from A to B is proportional to

$$R_{AB} = (2\pi/\hbar) \sum_{kq\sigma} | T_{kq} |^2 \{ f(\mathbf{k})[1 - f(\mathbf{q})] - f(\mathbf{q})[1 - f(\mathbf{k})]$$
$$\times \delta[E_A(\mathbf{k}) + eV - E_B(\mathbf{q})]\}. \tag{7.8.62}$$

The f's represent Fermi functions. Equation (7.8.62) is written using a convention that the single particle energies do not include the external potential V. This potential is, instead, incorporated in the Fermi function in order to express the fact that states in A are raised by an energy eV with respect to those in B. The Fermi functions depend on energy only. Thus on introducing the densities of states,

$$R_{AB} \sim \int dE_A \,|\, T \,|^2 [f(E_A) - f(E_A + eV)] G_A(E_A) G_B(E_A + eV) \tag{7.8.63}$$

in which $|\, T \,|$ is a suitable average of the matrix element, and G_A and G_B are densities of states for metals A and B, respectively. The theory of tunneling in normal systems (Harrison, 1961) indicates that $|\, T \,|^2$ is in fact inversely proportional to the product of the densities of states

$$|\, T \,|^2 \approx [G_A G_B]^{-1} \tag{7.8.64}$$

so that if V is small,

$$R_{AB} \sim \int dE_A \,[f(E_A) - f(E_A + eV)] \approx \int dE_A \,[-eV \,(df/dE)_{E_A}] = eV. \tag{7.8.65}$$

Thus the current is proportional to voltage if V is small, and Ohm's law is satisfied.

The situation is quite different if one of the metals (B, say) is superconducting. In this case, we must use (7.8.12) to introduce quasi-particle operators in metal B. The transfer Hamiltonian (7.8.61) is rewritten as

$$\begin{aligned} H_T = \sum_{kq} \{ & T_{kq} c_k^{+}(\alpha_q \xi_q + \beta_q \xi^{+}_{-q}) + T_{-k-q} c^{+}_{-k}(\alpha_q \xi_{-q} - \beta_q \xi_q^{+}) \\ & + T_{qk}(\alpha_q \xi_q^{+} + \beta_q \xi_{-q}) c_k + T^{*}_{-q-k}(\alpha_q \xi^{+}_{-q} - \beta_q \xi_q) c_{-k} \}. \end{aligned} \tag{7.8.66}$$

The calculation of R is to be repeated with this Hamiltonian. The terms can be interpreted as follows: $c_k^{+}\xi_q$ creates an electron in the normal metal A and destroys a quasi particle in the superconductor B, while $c_k^{+}\xi^{+}_{-q}$ creates both an electron and a quasi particle, etc. All terms describe independent processes, so that the corresponding transition probabilities, which are computed in first-order perturbation theory, are additive. The specific **k**, **q** dependence of the matrix element will be ignored. An average $|\, T \,|^2$ is introduced (see Scalapino *et al.* 1966). States in the normal metal A will be raised by an amount eV with respect to those in the superconductor. Contributions to R from half of the terms in (7.8.66) are listed in Table II, in accord with the following conventions: The Fermi functions are denoted by f_A and f_B, with all energies being measured from the chemical potential which is the same in both metals. The energies of electron states in the normal metal are denoted by E_A, while the energies of quasi-particle states in B are labeled $\mathcal{E}_B$. The algebraic sign is chosen to be positive for current flow from left to right. A common factor of $2\pi \,|\, T \,|^2/\hbar$ is omitted. The

TABLE II

Term	Contributions
$c_{\mathbf{k}}^{+}\mathcal{E}_{\mathbf{q}}$	$-\alpha_{\mathbf{q}}^2\,\delta[E_A(\mathbf{k}) + eV - \mathcal{E}_B(\mathbf{q})]\,[1 - f_A(E_A)]\,f_B(\mathcal{E}_B)$
$c_{\mathbf{k}}^{+}\,\mathcal{E}^{+}_{-\mathbf{q}}$	$-\beta_{\mathbf{q}}^2\,\delta[E_A(\mathbf{k}) + eV + \mathcal{E}_B(\mathbf{q})]\,[1 - f_A(E_A)]\,[1 - f_B(\mathcal{E}_B)]$
$\mathcal{E}_{\mathbf{q}}^{+}\,c_{\mathbf{k}}$	$\alpha_{\mathbf{q}}^2\,\delta[E_A(\mathbf{k}) + eV - \mathcal{E}_B(\mathbf{q})]\,f_A(E_A)[1 - f_B(\mathcal{E}_B)]$
$\mathcal{E}_{-\mathbf{q}}\,c_{\mathbf{k}}$	$\beta_{\mathbf{q}}^2\,\delta[E_A(\mathbf{k}) + eV + \mathcal{E}_B(\mathbf{q})]\,f_A(E_A)f_B(\mathcal{E}_B)$

remaining terms differ only by a change of sign for the wave vectors and give identical contributions. All contributions are added.

$$R_{AB} = (4\pi/\hbar)\sum_{\mathbf{kq}} |\,T\,|^2\{\alpha_{\mathbf{q}}^2[f_A(E_A) - f_B(\mathcal{E}_B)]\,\delta[E_A(\mathbf{k}) + eV - \mathcal{E}_B(\mathbf{q})] + \beta_{\mathbf{q}}^2[f_A(E_A) + f_B(\mathcal{E}_B) - 1]\,\delta[E_A(\mathbf{k}) + eV + \mathcal{E}_B(\mathbf{q})]\}. \tag{7.8.67}$$

The sums over wave vectors are converted to integrals over energies. This process introduces a product of density of states factors $G_A(E_A)\mathcal{G}_B(\mathcal{E}_B)$. Note that $\mathcal{G}_B$ is the density of states in the superconductor as given by (7.8.43). The integration over $\mathcal{E}_B$ can then be performed. The first delta function implies that

$$\mathcal{E}_B = (E_B^2 + \Delta^2)^{1/2} = E_A + eV \qquad \text{or} \qquad E_B = \pm[(E_A + eV)^2 - \Delta^2]^{1/2}. \tag{7.8.68}$$

Both signs contribute to the sum. However, it follows from (7.8.20) and (7.8.23) that

$$\alpha_{\mathbf{q}}^2(E_B) + \alpha_{\mathbf{q}}^2(-E_B) = \beta_{\mathbf{q}}^2(E_B) + \beta_{\mathbf{q}}^2(-E_B) = 1. \tag{7.8.69}$$

Thus the factor $\alpha_{\mathbf{q}}^2$ disappears from the sum. Application of the same argument to the second delta function shows that $\beta_{\mathbf{q}}^2$ also disappears:

$$\begin{aligned} R_{AB} &= (4\pi/\hbar)\int |\,T\,|^2 G_A(E_A)\{\mathcal{G}_B(E_A + eV)[f_A(E_A) - f_B(E_A + eV)] \\ &\qquad + \mathcal{G}_B(-E_A - eV)[f_A(E_A) + f_B(-E_A - eV) - 1]\}\,dE_A \\ &= (8\pi/\hbar)\int_{-\infty}^{\infty} |\,T\,|^2 G_A(E - eV)\mathcal{G}_B(E)[f_A(E - eV) - f_B(E)]\,dE. \end{aligned} \tag{7.8.70}$$

We have introduced $E = E_A + eV$. Equation (7.8.64) implies that the factors of the ordinary density of states G_A and G_B also disappear from

(7.8.70). We consider the remainder. At very low temperatures, the Fermi factors may be replaced by unit step functions. Then

$$R_{AB} \approx \int_{\Delta}^{eV} E/(E^2 - \Delta^2)^{1/2}\, dE = [(eV)^2 - \Delta^2]^{1/2}. \qquad (7.8.71)$$

The tunnel current is proportional to R_{AB}. This is zero for $eV < \Delta$, and rises sharply when the voltage difference equals the gap. Note that the ratio of the differential conductance in the superconducting and normal states is

$$(dI/dV)_{sup}/(dI/dV)_{norm} = eV/[(eV)^2 - \Delta^2]^{1/2}. \qquad (7.8.72)$$

Measurement of this ratio gives the density of quasi-particle states in the superconductor. Equation (7.8.72) turns out to be a reasonably good approximation.

Observations of tunneling in some superconductors (notably the so-called "strong coupling" superconductors such as lead) reveal more structure in the voltage dependence of the tunneling current than can be accounted for by (7.8.72). This structure occurs at voltages which are of the order of typical phonon energies. The theory has been extended to describe this situation through the introduction of a complex, energy-dependent, gap parameter $\Delta(\epsilon)$. Equation (7.8.72) is to be replaced by (Schrieffer, 1964),

$$R_{AB}(\text{sup})/R_{AB}(\text{norm}) = \text{Re}\{eV/[(eV)^2 - \Delta^2(V)]^{1/2}\}. \qquad (7.8.73)$$

Structure in the density of phonon states is transferred into $\Delta(V)$; conversely tunneling measurements can be used to determine properties such as the position of van Hove singularities in the phonon density of states (McMillan and Rowell, 1965).

The theory can also be extended in a straightforward way to describe tunneling between two superconductors separated by a barrier. Quasi-particle operators are introduced in both A and B. Equation (7.8.71) becomes

$$R_{AB} \approx \int \mathcal{G}_A(E)\mathcal{G}_B(E_A + eV)[f(E) - f(E + eV)]\, dE$$

$$= \int [|E|/(E^2 - \Delta_A{}^2)^{1/2}][|E + eV|/(|E + eV|^2 - \Delta_B{}^2)^{1/2}]$$

$$\times [f(E) - f(E + eV)]\, dE. \qquad (7.8.74)$$

This integral can be evaluated at $T = 0$ in terms of complete elliptic integrals (Douglass and Falicov, 1964). The tunnel current is zero for voltages less than the sum of the gap parameters ($eV < \Delta_A + \Delta_B$). There is a discontinuity when the inequality becomes an equality. For temperatures greater than zero, the integral must be evaluated numerically. In this case, there is a logarithmic singularity at $eV = |\Delta_A - \Delta_B|$ and a finite discontinuity at $eV = \Delta_A + \Delta_B$. A region of negative resistance occurs between these points. Neither a singularity nor a discontinuity can be observed experimentally; however, a pronounced maximum and a sharp increase are found (Giaever, 1960b).

Tunneling between two superconductors is, however, more complicated than the previous analysis indicates. Equation (7.8.74) describes the tunneling of single quasi particles, but it is also possible to transfer a pair of particles between superconductors (Josephson, 1962). A "supercurrent" can flow thin insulating (or normal metal) barriers between superconductors. In the presence of a steady voltage difference, the current oscillates. These phenomena comprise the Josephson effect (see Anderson, 1967, for a review). We will present here a simple description due to Ferrell and Prange (1963).

The argument is based on an analogy with the tight binding method in band theory. In band theory, an electron may be shifted through any direct lattice vector without a change of energy. Bloch's theorem asserts that such a displacement merely changes the wave function by a phase factor. It is then possible to form linear combinations of displaced states which satisfy Bloch's theorem and are approximate energy eigenfunctions.

In the present case, we observe that no energy is required to transfer a pair of electrons from one side of the barrier to the other. Let ϕ_n denote a state in which n pairs have been transferred from left to right. States with different values of n are degenerate. This degeneracy is resolved by constructing the linear combination

$$\psi_\alpha = \sum_n e^{i\alpha n}\phi_n. \tag{7.8.75}$$

A new quantum number α, which is a phase difference between the two superconductors, is introduced here and is conjugate to the number n in the same way that $\mathbf{k}$ is conjugate to $\mathbf{R}$. The tunneling Hamiltonian (7.8.61) transfers single electrons between the metals; we must, therefore, consider the effect of H_T in second order:

$$H_T^{(2)} = H_T(E - H_0)^{-1} H_T \tag{7.8.76}$$

where $H_T^{(2)}$ connects states ϕ_n which differ by a change of ± 1 in n. The energy of ψ_α is therefore

$$E(\alpha) = \langle\psi_\alpha | H_T^{(2)} | \psi_\alpha\rangle / \langle\psi_\alpha | \psi_\alpha\rangle. \tag{7.8.77}$$

This can be expressed as

$$E(\alpha) = -2\hbar\mathcal{J}\cos\alpha \tag{7.8.78}$$

in which

$$\hbar \mathcal{G} = | \langle \phi_{n+1} | H_T^{(2)} | \phi_n \rangle |. \tag{7.8.79}$$

The negative sign in (7.8.78) is a convention. The dominant matrix elements are expected, from (7.8.76), to be negative. Equation (7.8.78) indicates that the energy is a minimum if $\alpha = 0$. Such a state of definite phase does not correspond to a definite number of pairs in either superconductor.

Hamilton's canonical equations apply to the time rate of change of the average values of canonically conjugate variables, in this case n and $\hbar\alpha$. Thus

$$d\langle n\rangle/dt = \langle \partial E(\alpha)/\partial(\hbar\alpha)\rangle = 2\mathcal{G}\langle \sin \alpha\rangle \approx 2\mathcal{G} \sin \langle\alpha\rangle. \tag{7.8.80}$$

In the last step, we assume that the states of interest correspond to a narrow range of values of α. In addition $\hbar\, \partial\langle\alpha\rangle/dt = 2eV(t)$ in which $2eV(t)$ is the difference in potential energy between states $n + 1$ and n in which an additional pair has been transferred. This potential energy is produced by an assumed applied potential difference $V(t)$ between the superconductors. The rate at which electrons cross the barrier $J(t)$ is twice the rate of transfer of electron pairs

$$J(t) = 2\, d\langle n\rangle/dt = 4\mathcal{G} \sin \langle\alpha\rangle = 4\mathcal{G} \sin \left[(2e/\hbar) \int^t V(t')\, dt' \right]. \tag{7.8.81a}$$

Consider a potential difference containing a constant term V_0 and an alternating part $V(t) = V_0 + V_1 \cos(\omega_a t + \phi_0)$. Equation (7.8.81a) yields

$$J(t) = 4\mathcal{G} \sin[(2eV_0 t/\hbar) + (2eV_1/\hbar\omega a) \sin(\omega_a t + \phi_0) + \theta_0). \tag{7.8.81b}$$

This is a frequency modulated alternating current. If only the constant potential difference exists, the frequency of the alternating current is 484 MHz-μV.

Equations (7.8.81) describe the Josephson effect. Both the dc and the ac effects have been observed (see, for instance, Langenberg *et al.*, 1965). The current as given by (7.8.81b) can be expanded in Bessel functions with the aid of the identity

$$e^{ir\sin\phi} = \sum_{n=-\infty}^{\infty} J_n(r) e^{in\phi}$$

in which J_n is an nth order Bessel function. The result is

$$J(t) = 4\mathcal{G} \sum_n (-1)^n J_n(2eV_1/\hbar\omega a) \sin[(2eV_0 t/\hbar) - n\omega_a t - n\phi_0 + \theta_0]. \tag{7.8.81c}$$

If $2eV_0/\hbar = l\omega_a$, for some integral l, the current has a dc component which

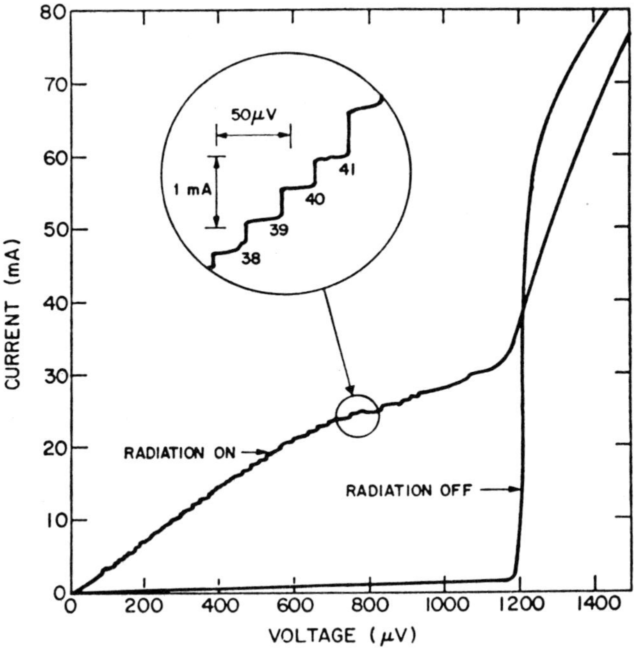

Fig. 7.8.2. Voltage–current curve of a Sn–Sn oxide–Sn tunnel junction irradiated with 10-GHz microwave radiation at 1.2°K. The constant voltage–current steps occur when the radiation is present, and disappear when it is absent. [From Parker *et al.* (1969).]

is independent of V_0:

$$J_{dc} = 4\mathcal{J}(-1)^l J_l(2eV_1/\hbar\omega)\sin(\theta_0 - l\phi_0).$$

This is, in a sense, indeterminante, as it depends on an arbitrarily assigned phase. The dc current can vary between limits $\pm 4\mathcal{J}J_l(2eV_1/\hbar\omega_a)$ while the dc voltage is constant. At a series of dc voltages $V_l = l\hbar\omega_a/2e$, a change in the dc current can occur which is manifested as a sharp step in the current–voltage characteristic of the junction. This effect is shown in Fig. 7.8.2, according to the measurements of Parker *et al.* (1969).

The discovery of Josephson tunneling has made possible experimental observation of macroscopic quantum interference effects. See, for example, Jaklevic *et al.* (1965).

7.8.5 *The Meissner Effect*

Superconductors tend to exclude applied magnetic fields. For small applied fields, $B = 0$ inside a superconductor of macroscopic size. The magnetic field penetrates only a small distance (the penetration depth) being attenuated in a characteristic distance λ which is of the order of 10^{-5} cm. This is the Meissner effect. If the external field is increased, superconductivity will ultimately be destroyed. There are two ways in which this may occur. In a type I superconductor, the entire material becomes normal sharply at an applied field H_c, the critical field; and the flux density inside approaches closely $\mu_0 H_{app}$. In contrast, the applied field begins to penetrate a type II superconductor at fields greater than H_{c1}, but superconductivity is not destroyed until a higher field H_{c2} is reached, when $B \approx u_0 H_{app}$. A thin surface layer may remain superconducting up to a still higher field H_{c3}.

This subsection will be concerned with the theoretical treatment of the Meissner effect on the basis of the BCS theory. The essential feature is that in a superconductor, in contrast with the situation in a normal metal, the current produced by the system in response to an external field is proportional to the vector potential, rather than to the field. Let us begin by supposing that this is so, and write

$$\mathbf{j} = -(\lambda^2 \mu_0)^{-1} \mathbf{A}. \tag{7.8.82}$$

A relation of this type was proposed by London (see London, 1950) long before the development of the BCS theory. Then

$$\nabla \times \mathbf{j} = -(\lambda^2 \mu_0)^{-1} \nabla \times \mathbf{A} = -\lambda^{-2} \mathbf{H}.$$

Since $\nabla \times \mathbf{H} = \mathbf{j}$, we have, from Maxwell's equation,

$$\nabla \times \nabla \times \mathbf{H} = -\lambda^{-2} \mathbf{H} \quad \text{or} \quad \nabla^2 \mathbf{H} = \lambda^{-2} \mathbf{H}. \tag{7.8.83}$$

This equation has solutions which indicate that the magnetic field decays exponentially in a characteristic distance λ, which is referred to as the *penetration depth.*

This argument indicates the manner in which a microscopic theory of the Meissner effect can be developed. The current, which is the response of the system to an external electromagnetic field, is to be calculated. If one finds that in the limit of a static field

$$\mathbf{j} = -(\text{positive constant}) \mathbf{A},$$

the system will expel the field and exhibit a Meissner effect. The calculation employs the procedure of Section 6.5.4. It should be noted that in the case of a normal metal, Eq. (6.5.60) indicates that a Meissner effect will not be obtained.

The starting point is Eq. (6.5.54) which is a formal expression for the time-dependent current produced by a wave vector and time-dependent vector potential. We take the Fourier transform of this equation with respect to time and obtain

$$\mathcal{J}_\alpha(\mathbf{q}, \omega) = -(Ne^2/m)\mathcal{A}_\alpha(\mathbf{q}, \omega) + (i/\hbar)\int_{-\infty}^{0} dt$$

$$\times \langle[j_\alpha(\mathbf{q}, 0), j_\beta(-\mathbf{q}, t)]\rangle e^{-i\omega t}\mathcal{A}_\beta(\mathbf{q}, \omega). \quad (7.8.84)$$

The current operator $j_\beta(\mathbf{q}, t)$ is to be expressed in the interaction picture (Section 8.1.4):

$$j_\beta(t) = \exp(iH_0t/\hbar)j_\beta\exp(-iH_0t/\hbar)$$

in which H_0 is the Hamiltonian for the system neglecting the external electromagnetic field, but including the electron–phonon interaction which produces superconductivity. The angular brackets $\langle\cdots\rangle$ indicate a thermal average. Let us introduce a complete set of states for the system. These states are denoted for the time being by $|m\rangle$, $|n\rangle$, and are eigenstates of H_0:

$$\langle j_\alpha(\mathbf{q}, 0), j_\beta(-\mathbf{q}, t)\rangle$$

$$= Z^{-1}\sum_{m,n}\exp(-\beta E_m)\langle m\,|\,j_\alpha(\mathbf{q}, 0)\,|\,n\rangle\langle n\,|\,j_\beta(-\mathbf{q}, t)\,|\,m\rangle$$

$$= Z^{-1}\sum_{m,n}\exp(-\beta E_m)\exp[i(E_n - E_m)t/\hbar]$$

$$\times \langle m\,|\,j_\alpha(\mathbf{q}, 0)\,|\,n\rangle\langle n\,|\,j_\alpha(-\mathbf{q}, 0)\,|\,m\rangle$$

in which $Z = \sum_m \exp(-\beta E_m)$. The integration over time can be performed. The frequency ω is considered to possess an infinitesimal positive imaginary part which is allowed to go to zero where appropriate. We will be concerned here with low frequencies at which real absorption cannot take place; hence a delta function, which is present when absorption or emission occurs, will be ignored:

$$\mathcal{J}_\alpha(\mathbf{q}, \omega) = (-Ne^2/m)\mathcal{A}_\alpha(\mathbf{q}, \omega) + Z^{-1}\mathcal{A}_\beta(\mathbf{q}, \omega)\sum_{m,n}\exp(-\beta E_m)$$

$$\times\left[\frac{\langle m\,|\,j_\alpha(\mathbf{q})\,|\,n\rangle\langle n\,|\,j_\beta(-\mathbf{q})\,|\,m\rangle}{E_m - E_n + \hbar\omega}\right.$$

$$\left.+\frac{\langle m\,|\,j_\beta(-\mathbf{q})\,|\,n\rangle\langle n\,|\,j_\alpha(\mathbf{q})\,|\,m\rangle}{E_m - E_n - \hbar\omega}\right]. \quad (7.8.85)$$

The current operator $\mathbf{j}$ is given in position space by Eq. (6.5.41). Only the terms independent of $\mathbf{A}$ are required in (7.8.85). The Fourier transformed operator $\mathbf{j}(\mathbf{q})$ is

$$\mathbf{j}(\mathbf{q}) = (e\hbar/2m) \sum_{\mathbf{k}\sigma} (2\mathbf{k} + \mathbf{q}) c^+_{\mathbf{k}\sigma} c_{\mathbf{k}+\mathbf{q}\sigma}. \tag{7.8.86}$$

In order to deal with the superconducting state, we must introduce the operators ξ according to (7.8.12):

$$\begin{aligned}\mathbf{j}(\mathbf{q}) &= (e\hbar/2m) \sum_{\mathbf{k}} (2\mathbf{k} + \mathbf{q}) [c^+_{\mathbf{k}\uparrow} c_{\mathbf{k}+\mathbf{q}\uparrow} - c^+_{-(\mathbf{k}+\mathbf{q})\downarrow} c_{-\mathbf{k}\downarrow}] \\ &= (e\hbar/2m) \sum_{\mathbf{k}} (2\mathbf{k} + \mathbf{q}) [(\alpha_{\mathbf{k}}\alpha_{\mathbf{k}+\mathbf{q}} + \beta_{\mathbf{k}}\beta_{\mathbf{k}+\mathbf{q}}) (\xi_{\mathbf{k}}^+ \xi_{\mathbf{k}+\mathbf{q}} - \xi^+_{-(\mathbf{k}+\mathbf{q})} \xi_{-\mathbf{k}}) \\ &\quad + (\alpha_{\mathbf{k}}\beta_{\mathbf{k}+\mathbf{q}} - \beta_{\mathbf{k}}\alpha_{\mathbf{k}+\mathbf{q}}) (\xi_{\mathbf{k}}^+ \xi^+_{-(\mathbf{k}+\mathbf{q})} - \xi_{-\mathbf{k}} \xi_{\mathbf{k}+\mathbf{q}})].\end{aligned} \tag{7.8.87}$$

The matrix elements of the current operator can now be obtained. Consider the second and third lines of (7.8.87); let m be the "initial state." The intermediate state n can differ from m by the scattering of a quasi particle from $\mathbf{k} + \mathbf{q}$ to $\mathbf{k}$ [or from $-\mathbf{k}$ to $-(\mathbf{k} + \mathbf{q})$], by the appearance of an additional quasi-particle pair in $\mathbf{k}$, $-(\mathbf{k} + \mathbf{q})$, or by destruction of one already present in $-\mathbf{k}$, $\mathbf{k} + \mathbf{q}$. The energy differences $E_m - E_n$ are $\mathcal{E}(\mathbf{k} + \mathbf{q}) - \mathcal{E}(\mathbf{k})$, $\mathcal{E}(\mathbf{k}) - \mathcal{E}(\mathbf{k} + \mathbf{q})$, $-\mathcal{E}(\mathbf{k}) - \mathcal{E}(\mathbf{k} + \mathbf{q})$, and $\mathcal{E}(\mathbf{k}) + \mathcal{E}(\mathbf{k} + \mathbf{q})$, respectively. The thermal average in (7.8.85) implies that the matrix elements are to be replaced by occupation factors of $f[\mathcal{E}(\mathbf{k})]$, etc. It is also convenient, when writing $\mathbf{j}(-\mathbf{q})$ to change the sign of the summation variable which appears in (7.8.87). Finally, we obtain

$$\begin{aligned}\mathcal{J}_\alpha(\mathbf{q}, \omega) = {} & (-Ne^2/m)\mathcal{A}_\alpha(\mathbf{q}, \omega) + (e^2\hbar^2/4m^2) \\ & \times \sum_{\mathbf{k}} (2\mathbf{k} + \mathbf{q})_\alpha (2\mathbf{k} + \mathbf{q})_\beta \mathcal{A}_\beta(\mathbf{q}, \omega) \\ & \times \{(\alpha_{\mathbf{k}}\alpha_{\mathbf{k}+\mathbf{q}} + \beta_{\mathbf{k}}\beta_{\mathbf{k}+\mathbf{q}})^2 \{ f[\mathcal{E}(\mathbf{k})] - f[\mathcal{E}(\mathbf{k} + \mathbf{q})]\} \\ & \times [(\mathcal{E}(\mathbf{k} + \mathbf{q}) - \mathcal{E}(\mathbf{k}) + \hbar\omega)^{-1} \\ & + (\mathcal{E}(\mathbf{k} + \mathbf{q}) - \mathcal{E}(\mathbf{k}) - \hbar\omega)^{-1}] + (\alpha_{\mathbf{k}}\beta_{\mathbf{k}+\mathbf{q}} - \beta_{\mathbf{k}}\alpha_{\mathbf{k}+\mathbf{q}})^2 \\ & \times \{1 - f[\mathcal{E}(\mathbf{k})] - f[\mathcal{E}(\mathbf{k} + \mathbf{q})]\} [(\mathcal{E}(\mathbf{k} + \mathbf{q}) + \mathcal{E}(\mathbf{k}) + \hbar\omega)^{-1} \\ & + (\mathcal{E}(\mathbf{k} + \mathbf{q}) + \mathcal{E}(\mathbf{k}) - \hbar\omega)^{-1}]\}.\end{aligned} \tag{7.8.88}$$

Equation (7.8.88) applies to an arbitrary position- and time-dependent electromagnetic field. For the purposes of the Meissner effect, we may take the static ($\omega = 0$) limit of (7.8.88). We may also neglect $\mathbf{q}$ compared to

$2\mathbf{k}$ in the factors multiplying the braces. We will also restrict attention to a cubic crystal in which the relation between $\mathcal{J}$ and $\mathcal{A}$ is scalar. In accord with (7.8.82) we have

$$\mathcal{J}_\alpha(\mathbf{q}) = -\lambda^{-2}(\mathbf{q})\mathcal{A}_\alpha(\mathbf{q})/\mu_0 \tag{7.8.89}$$

with

$$\lambda^{-2} = (Ne^2\mu_0/m) - \tfrac{2}{3}(e^2\hbar^2\mu_0/m^2)\sum_{\mathbf{k}}(\mathbf{k})^2\Bigg\{(\alpha_{\mathbf{k}}\alpha_{\mathbf{k+q}} + \beta_{\mathbf{k}}\beta_{\mathbf{k+q}})^2$$
$$\times\frac{f[\mathcal{E}(\mathbf{k})] - f[\mathcal{E}(\mathbf{k}+\mathbf{q})]}{\mathcal{E}(\mathbf{k}+\mathbf{q}) - \mathcal{E}(\mathbf{k})} + (\alpha_{\mathbf{k}}\beta_{\mathbf{k+q}} - \beta_{\mathbf{k}}\alpha_{\mathbf{k+q}})^2$$
$$\times\frac{1 - f[\mathcal{E}(\mathbf{k})] - f[\mathcal{E}(\mathbf{k}+\mathbf{q})]}{\mathcal{E}(\mathbf{k}) + \mathcal{E}(\mathbf{k}+\mathbf{q})}\Bigg\}. \tag{7.8.90}$$

The Meissner effect requires $\lambda(0)$. To obtain this limit we may set $q = 0$ in α and β:

$$\lambda^{-2}(0) = (Ne^2\mu_0/m) + (2e^2\hbar^2\mu_0/3m^2)$$
$$\times\sum_{\mathbf{k}}(\mathbf{k})^2\lim_{q\to 0}\{f[\mathcal{E}(\mathbf{k})] - f[\mathcal{E}(\mathbf{k}-\mathbf{q})]/[\mathcal{E}(\mathbf{k}) - \mathcal{E}(\mathbf{k}+\mathbf{q})]\}. \tag{7.8.91}$$

The limit yields $\partial f/\partial\mathcal{E}$. The sum may be converted to an integral in the usual way:

$$\lambda^{-2}(0) = (Ne^2\mu_0/m) + \tfrac{2}{3}(e^2\hbar^2/m^2)\mu_0\int \mathcal{G}(\mathcal{E})k^2\,\partial f/\partial\mathcal{E}\,d\mathcal{E}. \tag{7.8.92}$$

Note that the integral involves the single particle energy $\mathcal{E}$. The slowly varying factors $\mathcal{G}_0(\mathcal{E})k^2$ may be taken outside the integrand. We use the relation, valid for a free electron system, $\mathcal{G}_0(\mathcal{E}_F) = 3Nm/2\hbar^2k_F^2$ and have, with the aid of (7.8.41),

$$\lambda^{-2}(0) = (Ne^2\mu_0/m)\left\{1 + \int_{-\infty}^{-\Delta} + \int_{\Delta}^{\infty} |\mathcal{E}|/(\mathcal{E}^2 - \Delta^2)^{1/2}\,\partial f/\partial\mathcal{E}\,d\mathcal{E}\right\}. \tag{7.8.93}$$

At temperatures greater than T_c, $\Delta = 0$, and the integral gives -1; hence λ^{-2} vanishes in this limit and no Meissner effect is obtained. Below T_c, the cancellation is not complete. At $T = 0$, $\partial f/\partial\mathcal{E}$ is different from zero only at $\mathcal{E} = 0$, so that the integral over $\mathcal{E}$ vanishes. Thus we find a penetration depth

$$\lambda = (m/Ne^2\mu_0)^{1/2} = c/\omega_p \tag{7.8.94}$$

in which c is the speed of light and ω_p is the plasma frequency. This is of the correct order of magnitude.

The destruction of superconductivity by a magnetic field can be understood in terms of the Meissner effect. The expulsion of flux from the interior of a sample can be considered as raising the energy of the system by an amount equal to the energy of the field displaced from the material. For a volume $\mathcal{V}$ of superconductor whose dimensions are large compared to the penetration depth, this energy is $\mu_0\mathcal{V}\mathbf{H}^2/2$. If this exceeds the energy gained by condensation into the superconducting state, the material will make a transition to the normal state (or remain in the normal state). The condensation energy is given by (7.8.36). The critical field H_c therefore satisfies

$$\mu_0\mathcal{V}\mathbf{H}_c^2/2 = \tfrac{1}{2}\mathcal{V}G_0\,\Delta^2.$$

(A factor of volume has been included in the condensation energy; G_0 must be interpreted as the density of states per unit volume.) Thus

$$\mathbf{H}_c^2 = G_0\,\Delta^2/\mu_0. \tag{7.8.95}$$

This equation applies at $T = 0$°K. At finite temperatures, it is necessary to consider the free energies of the normal and superconducting states.

7.8.6 *Zero Resistance*

The results of the preceding subsection also show that currents can flow in superconductors without electrical resistance. Experimentally, this was the first property of a superconductor to be observed and is responsible for the name "superconductor."

Equation (7.8.88) can be readily generalized to time-dependent fields:

$$\mathcal{J}_\alpha(\mathbf{q}, \omega) = -\mathcal{A}_\alpha(\mathbf{q}, \omega)/\mu_0\lambda^2(\mathbf{q}, \omega). \tag{7.8.96}$$

Since the electric field $\mathbf{E}(t)$ is related to the vector potential $\mathbf{A}(t)$ by $\mathbf{E}(t) = -\partial\mathbf{A}/\partial t$, their Fourier transforms satisfy $\boldsymbol{\mathcal{E}}(\mathbf{q}, \omega) = i\omega\boldsymbol{\mathcal{A}}(\mathbf{q}, \omega)$. Thus the relation between current and field is

$$\boldsymbol{\mathcal{J}}(\mathbf{q}, \omega) = i\boldsymbol{\mathcal{E}}(\mathbf{q}, \omega)/\mu_0\lambda^2(\mathbf{q}, \omega)\omega. \tag{7.8.97}$$

To discuss static conductivity, we should let q vanish and then allow ω to tend to zero. However, it can be shown that under rather general conditions the results are independent of the order in which the limits are taken. For very small frequencies, we may therefore use $\lambda^2(0)$ as given by (7.8.91)–(7.8.94). Since λ^2 may be taken constant in (7.8.97), we recover a simple relation between time-dependent quantities:

$$d\mathbf{J}(t)/dt = (1/\mu_0\lambda^2)\mathbf{E}(t). \tag{7.8.98}$$

This relation describes freely accelerating electrons. Note that at $T = 0$, (7.8.94) implies that

$$(\mu_0\lambda^2)^{-1} = Ne^2/m. \tag{7.8.99}$$

This is exactly the value to be expected from a simple kinetic derivation of (7.8.98).

From a microscopic point of view, we may describe a persistent current as modifying the basic pairing of electrons in a superconductor. Thus instead of combining $\mathbf{k}\uparrow$ and $-\mathbf{k}\downarrow$, we pair $\mathbf{k} + \mathbf{q}/2\uparrow$ with $-\mathbf{k} + \mathbf{q}/2\downarrow$. The same value of $\mathbf{q}$ will be used for all pairs. A state of the entire system with some $\mathbf{q} \neq 0$ is metastable. Individual particle scattering could tend to change the value of $\mathbf{q}$ for a single particle, but is ineffective in changing $\mathbf{q}$ for the entire system. Hence, $\mathbf{q}$ is in fact unchanged, and currents flow without resistance.

7.8.7 *Flux Quantization*

One of the remarkable phenomena associated with the superconducting state is the quantization of magnetic flux through a superconducting ring. This effect was originally suggested by London (1950) and Onsager (1954). The first observations were made by Deaver and Fairbank (1961) and by Doll and Nabäuer (1961). The basic quantum unit of flux is $\phi_0 = h/2e$. The effect is closely related to the Meissner effect and gives clear evidence for the pairing of electrons in a superconductor. It should be emphasized that flux quantization is determined by the properties of superconductors and is not the result of some new physical principle concerning electromagnetic fields. We present first the simple argument of London (1950) and Onsager (1961).

Consider a ring of superconducting material such that some flux passes through the center. At a distance greater than the penetration depth λ in the material of the ring, the Meissner effect implies that the magnetic field $\mathbf{B}$ vanishes. However, the vector potential $\mathbf{A}$ will not vanish. This follows since the flux Φ through the ring is

$$\Phi = \int \mathbf{B}\cdot d\mathbf{S} = \int \nabla \times \mathbf{A}\cdot d\mathbf{S} = \oint \mathbf{A}\cdot d\mathbf{l} \tag{7.8.100}$$

in which the line integral runs along any path inside the superconductor which goes around the hole. On the other hand, the current density in the presence of a vector potential $\mathbf{A}$ is proportional to the average of $\mathbf{p} - q\mathbf{A}$, in which q is the charge of the particles which carry the current. The current will be confined to the surface of the ring and will vanish in the interior. In order that the wave function of the superconductor be single-valued whenever the coordinate of any electron is taken around the ring, it is necessary that

$$\oint \mathbf{p}\cdot d\mathbf{l} = -2\pi n\hbar. \tag{7.8.101}$$

This condition, combined with the vanishing of the current, implies that

$$\Phi = \oint \mathbf{A}\cdot d\mathbf{l} = -2\pi n\hbar/q.$$

In the superconductor, we should take $q = -2e$. Thus

$$\Phi = \pi n\hbar/e, \tag{7.8.102}$$

which is the observed result.

A more detailed discussion can be based on the work of Byers and Yang (1961). The wave function for a system of particles of charge $-q$ in the superconductor satisfies the equation

$$\sum_n (1/2m)[(\hbar/i)\,\nabla_n + qA(\mathbf{r}_n)]^2\psi + V\psi = E\psi \tag{7.8.103}$$

in which all interactions have been placed in the potential energy term V. Since $B = 0$ inside the ring, $\nabla \times \mathbf{A} = 0$ and $\mathbf{A}$ can be represented as the gradient of a scalar, $\mathbf{A} = \nabla\chi$. However, χ is not single-valued, since if we go around the ring along some closed path,

$$\Delta\chi = \oint \nabla\chi\cdot d\mathbf{l} = \oint \mathbf{A}\cdot d\mathbf{l} = \Phi. \tag{7.8.104}$$

The vector potential can be eliminated from (7.8.103) by making the transformation

$$\psi' = \psi \exp[\sum_n (iq/\hbar)\chi(\mathbf{r}_n)]. \tag{7.8.105}$$

The transformed function ψ' satisfies

$$\sum_n [-(\hbar^2/2m)\,\nabla_n{}^2]\psi' + V\psi' = E\psi'. \tag{7.8.106}$$

The elimination of the vector potential has not been achieved without price: the boundary conditions for ψ' are more complicated than for ψ. Let us take the coordinate of one electron all the way around the ring, while the others are held fixed. The original function ψ was single-valued, so that it returns to its original value, but, on account of (7.8.104), ψ' does not. Instead

$$\psi' \rightarrow \psi' \exp[(iq/\hbar)\Phi]. \tag{7.8.107}$$

The eigenvalues E are determined by the differential equation (7.8.106) and the boundary condition (7.8.107). It is seen from this that the energy levels of the system are periodic in the magnetic flux Φ with period $2\pi\hbar/q$. Furthermore, for real potentials V, we see by taking the complex conjugate of (7.8.105) and (7.8.106) (or more generally, by using the time reversal operator) that the energy levels of the system are even functions of Φ.

These considerations can be illustrated by a collection of noninteracting electrons in a cylindrical box. We may choose the vector potential $\mathbf{A}$ to be in the $\hat{\phi}$ direction (around the cylinder) and to be constant, such that if r is the radius of the cylinder, $\Phi = 2\pi rA$. The cylinder is assumed to be thin enough so that r is essentially constant. The energy of a particle is

$$\epsilon_l = (1/2m)\{p_z^2 + (\hbar^2/r^2)[l + (q\Phi/2\pi\hbar)]^2\} \qquad (7.8.108)$$

in which l is the angular momentum quantum number. We will ignore the z coordinate. The total energy is then (for $T = 0°K$)

$$E = (\hbar^2/2mr^2) \sum_l [l + (q\Phi/2\pi\hbar)]^2. \qquad (7.8.109)$$

The sum is to be carried out over all occupied l.

In a normal metal, it may be supposed that the three-dimensional problem reduces to many one-dimensional problems with varying values of the upper limit of summation on l. Cancellations can then occur, so that the energy is essentially flat as a function of Φ. In a superconductor, states connected by time reversal symmetry should be paired. In the present

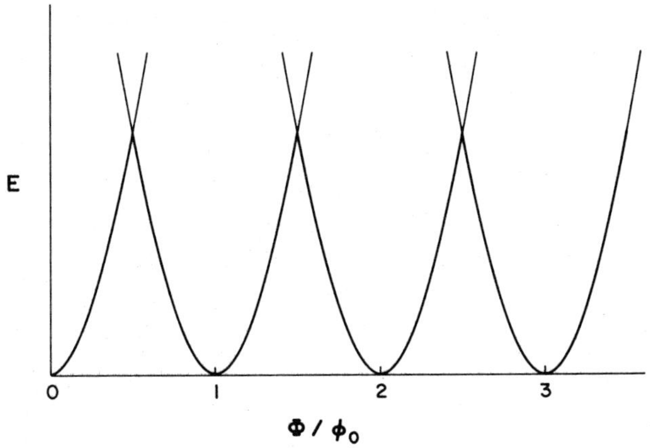

Fig. 7.8.3. The energy of a superconducting ring at zero temperature is shown as a function of the magnetic flux Φ through it (ϕ_0 is the flux quantum $\pi\hbar e$).

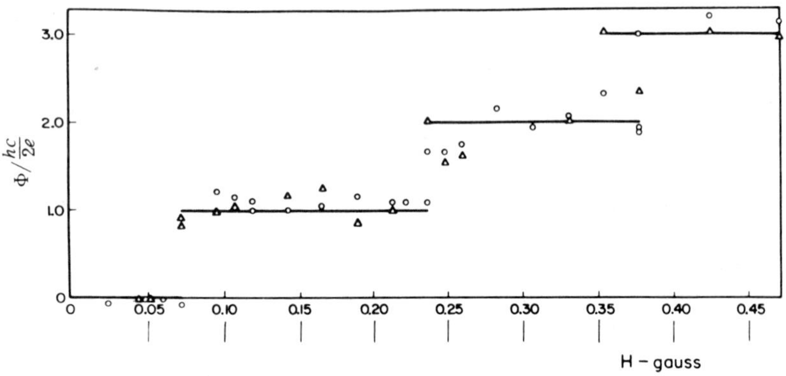

Fig. 7.8.4. The trapped flux inside a superconducting cylinder is shown as a function of the magnetic field in which the cylinder is cooled. [From Deaver and Fairbank (1961).]

instance, this means that states l and $-l$ must either be both occupied or both empty. The total energy of the system in this case varies with Φ as

$$E = \text{const} + (N\hbar^2/2mr^2)(q\Phi/2\pi\hbar)^2. \tag{7.8.110}$$

We interpret q as equal to $2e$ and we see that in this region the energy of the system will be a minimum if Φ is held at $\pi\hbar/e$. The energy is shown schematically as a function of flux in Fig. 7.8.3. In view of the periodicity of the energy, the system will maintain itself at the lowest possible energy, if it keeps the flux through it equal to an integral number of flux quanta. Experimental evidence for flux quantization is shown in Fig. 7.8.4, according to Deaver and Fairbank (1961).

7.8.8 *The Ginzburg–Landau Equations*

The description of superconductivity which has been presented here is incomplete in one respect of considerable practical importance. We have considered the system to be either entirely superconducting or entirely normal. In fact, a given sample can, in the presence of a magnetic field, have both normal and superconducting regions coexisting simultaneously. This enables the system to reduce the energy associated with the expulsion of the magnetic field, while retaining some of the condensation energy associated with the superconducting transition. The situation is somewhat similar to the existence of domains in macroscopic ferromagnets.

In a superconductor of type I, a state of this sort (the intermediate state) is formed only when geometric factors tend to increase the energy associated with the excluded magnetic field, as in the case of a thin disk held perpendicular to an applied field. A type II superconductor of whatever shape allows the penetration of some field (forming a mixed state) when the field is sufficiently large. The two types of superconductors can be distinguished according to the value of a parameter κ as will be discussed subsequently.

The Ginzburg–Landau equations describe spatial variation of superconductivity. These equations were presented phenomenologically before the microscopic theory was developed (Ginzburg and Landau, 1950). They have subsequently been derived on the basis of the BCS theory by Gor'kov (1959a, b). We shall use here the phenomenological approach, which fits in the framework of the Landau theory of phase transitions presented in Section 2.7, Part A. For a review, see Werthamer (1969).

The theory is based on an expansion for the free energy in terms of an order parameter, conventionally denoted $\Psi(\mathbf{r})$. Although Ψ will be found to obey an equation resembling the Schrödinger equation, it should not be confused with a wave function: in fact, it is proportional to the ratio of the gap parameter Δ, which should be regarded as a complex function of spatial coordinates in the circumstances of interest, to its value at zero temperature and field.

The assumed expression for the free energy is

$$F_s = F_n + \int d^3r \, \{ f(|\Psi(\mathbf{r})|^2) + (\mu_0/2) \, |\mathbf{H}(\mathbf{r}) - \mathbf{H}_a|^2$$
$$+ (1/2m^*) \, |[-i\hbar \nabla + q\mathbf{A}(\mathbf{r})]\Psi(\mathbf{r})|^2\}. \tag{7.8.111}$$

The term F_n represents the free energy of the normal state of the material at the same temperature. The normal state could be obtained by increasing the magnetic field above the critical field. The function f describes the change in the free energy produced by the transition to superconductivity. This must decrease if the superconducting state is to be stable. The phase transition is of second order so that f will be continuous. According to the general principles of the Landau theory of second-order phase transitions, we suppose that near the transition

$$f = -\alpha |\Psi|^2 - \tfrac{1}{2}\beta |\Psi|^4. \tag{7.8.112}$$

This approximation should be valid when $|\Psi|$ is small, that is, either near the critical temperature or the critical field.

The next term in F represents the change in the energy of the magnetic field due to the expulsion of flux: $\mathbf{H}_a$ is the externally applied field. The last term describes the increase in the free energy of the superconductor which results from spatial variations of the order parameter, and from current flow. The quantity q represents an effective charge which will be determined later.

One principal difference between the Ginzburg–Landau equations and the theory of second-order phase transitions presented in Section 2.7, Part A, is that the order parameter Ψ is complex. This is useful in that it allows a discussion of current flow. To see how such effects arise, we proceed to determine equations for Ψ and $\mathbf{A}$ by minimizing F_s with respect to variations of Ψ^*, Ψ, and $\mathbf{A}(\mathbf{r})$. The boundary condition which is imposed at the surface of the specimen is

$$\hat{\mathbf{n}}\cdot[(\hbar/i)\ \nabla + q\mathbf{A}]\Psi = 0. \tag{7.8.113}$$

Variation of Ψ^* yields an equation for Ψ, which is

$$(1/2m)[(\hbar/i)\ \nabla + q\mathbf{A}]^2\Psi - (\alpha + \beta \mid \Psi \mid^2)\Psi = 0. \tag{7.8.114}$$

An additional equation is obtained by varying $\mathbf{A}$. After some manipulation, which we will not give in detail, one finds

$$\nabla \times \mathbf{H} = (-q^2/m) \mid \Psi \mid^2\mathbf{A} - (\hbar q/2im)(\Psi^* \nabla\Psi - \Psi \nabla\Psi^*). \tag{7.8.115}$$

This equation has a relatively simple interpretation. According to Maxwell's equations,

$$\nabla \times \mathbf{H} = \mathbf{J}.$$

The right side of (7.8.115) must therefore represent the current. Suppose first that there is no position dependence to the superconductivity, so that the second term on the right vanishes. Thus,

$$\mathbf{J} = (-q^2/m) \mid \Psi \mid^2\mathbf{A}$$

and comparison with (7.8.89) indicates that the penetration depth λ^2 is related to $\mid \Psi \mid^2$ by

$$\lambda^{-2} = (q^2\mu_0/m) \mid \Psi \mid^2. \tag{7.8.116}$$

The constants appearing in (7.8.114) and (7.8.115) may be determined as follows: Far from all boundaries in a homogeneous specimen, we have in view of the Meissner effect, simply

$$\alpha + \beta\Psi_h^2 = 0. \tag{7.8.117}$$

The order parameter should be real and independent of position under these circumstances: its value is there denoted by Ψ_h. Also, in view of the discussion in subsection 7.8.5 concerning the critical field,

$$\alpha\Psi_h^2 + \tfrac{1}{2}\beta\Psi_h^4 = \mu_0 H_c^2/2. \tag{7.8.118}$$

These equations determine α and β, while (7.8.116) can be used to express

Ψ_h^2 in terms of the penetration depth λ. Then it becomes convenient to replace Ψ in (7.8.114) by

$$\Psi' = \Psi/\Psi_h. \tag{7.8.119}$$

The resulting equation can be rewritten as

$$[(\hbar/iq)\nabla + \mathbf{A}]^2\Psi' - 2\mu_0^2 H_c^2\lambda^2(1 - |\Psi'|^2)\Psi' = 0.$$

We may now measure $\mathbf{A}$ and $\mathbf{H}$ in units of

$$A_0 = \sqrt{2}\mu_0 H_c\lambda, \qquad H_0 = \sqrt{2}H_c \tag{7.8.120}$$

by defining $\mathbf{A}'$, $\mathbf{H}'$ such that

$$\mathbf{A}' = \mathbf{A}/A_0, \qquad \mathbf{H}' = \mathbf{H}/H_0. \tag{7.8.121}$$

Finally, it is useful to measure coordinates in terms of the penetration depth: Introduce $\mathbf{r}'$:

$$\mathbf{r}' = \mathbf{r}/\lambda. \tag{7.8.122}$$

These measures enable us to write (7.8.114) and (7.8.115) with all constants contained in a single parameter κ

$$\kappa = \sqrt{2}\mu_0 H_c\lambda^2 q/\hbar. \tag{7.8.123}$$

Equation (7.8.114) now becomes

$$[(1/i\kappa)\nabla' + \mathbf{A}']^2\Psi' - (1 - |\Psi'|^2)\Psi' = 0 \tag{7.8.124}$$

while (7.8.115) simplifies

$$\nabla' \times \mathbf{H}' = -\mathbf{A}' - (1/2i\kappa)(\Psi'^*\nabla'\Psi' - \Psi'\nabla'\Psi'^*). \tag{7.8.125}$$

Moreover, the boundary condition (7.8.113) is

$$\hat{\mathbf{n}}\cdot[(1/i\kappa)\nabla' - \mathbf{A}']\Psi' = 0. \tag{7.8.126}$$

Equations (7.8.124)–(7.8.126) show that the Ginzburg–Landau theory depends on the single dimensionless parameter κ. This parameter can be written as the ratio of two lengths,

$$\kappa = \lambda/\xi_0 \tag{7.8.127}$$

with

$$\xi_0 = \hbar/\sqrt{2}\mu_0 H_c\lambda q. \tag{7.8.128}$$

The quantity ξ_0 is called the *coherence length* (however, other definitions exist which differ from this by a numerical factor). It was shown by Gor'kov (1959a, b) that the effective charge q should be taken as that of an electron pair $2e$. Then ξ_0 can be related to microscopic parameters using (7.8.94) and (7.8.95). At $T = 0°\mathrm{K}$

$$\xi_0 = [1/2\sqrt{3}]\hbar v_\mathrm{F}/\Delta \tag{7.8.129}$$

in which v_F is the speed of an electron at the Fermi surface. A free electron band structure has been assumed.

The coherence length can be regarded as a crude measure of the range of strong correlations between electrons. This is developed by an argument based on the uncertainty principle. Electrons with energies in the range

$$E_\mathrm{F} - \Delta \leq \hbar^2k^2/2m \leq E_\mathrm{F} + \Delta$$

will be strongly affected by the transition to the superconducting state. Now put $k = k_\mathrm{F} + \delta k_\mathrm{F}$. It is easy to see that $\delta k \approx k_\mathrm{F}\Delta/E_\mathrm{F}$. Then the spatial range is $\delta r \approx 1/\delta k \approx \hbar v_\mathrm{F}/2\Delta$ which agrees qualitatively with (7.8.129).

Superconductors of type I are characterized by $\kappa < 1/\sqrt{2}$; those of type II by $\kappa > 1/\sqrt{2}$. This turns also out to characterize the energy associated with formation of a surface separating normal and superconducting regions. Type I superconductors have positive surface energies so that formations of distinct regions is inhibited; type II superconductors have negative surface energies, favoring a mixed state.

Let us consider finally the onset of superconductivity in the presence of a magnetic field. Let H_m be the highest field for which superconductivity can exist, and H_m' similarly refer to H'. For fields close to this, the scaled-order parameter Ψ' is small, so that we can drop the $|\Psi'|^2$ term in (7.8.124). Further, it will be legitimate to take the actual field equal to the applied field, since the screening effect of supercurrents is of order Ψ^2, and thus is small. Thus set $A_x' = H'y'$. Equation (7.8.124) becomes

$$((1/i\kappa)\,\partial/\partial x' + H'y')^2\Psi' - [(1/\kappa^2)\,\partial^2/\partial y'^2 + (1/\kappa^2)\,\partial^2/\partial z'^2 + 1]\Psi' = 0. \tag{7.8.130}$$

This equation resembles the Schrödinger equation for the motion of a charged particle in a magnetic field. The largest field for which superconductivity can occur is determined by the lowest eigensolution of (7.8.130): this means we may take $\Psi' = f(y)$. The equation becomes

$$d^2f/dy^2 + \kappa^2(1 - H_\mathrm{m}'^2y^2)f = 0. \tag{7.8.131}$$

The solution of interest is of the form $f \approx \exp(-a^2y^2)$. Substitution shows that this requires

$$H_\mathrm{m}' = \kappa \tag{7.8.132a}$$

or, from (7.8.120) and (7.8.121)

$$H_m = \kappa\sqrt{2}H_c. \qquad (7.8.132b)$$

It is seen from this result that for type II superconductors ($\kappa > 1/\sqrt{2}$), superconductivity appears at a field $H_m = H_{c2} > H_c$. In this situation, the material is in a mixed state. Type I superconductors, with $\kappa < 1/\sqrt{2}$, may actually remain normal in decreasing fields down to field $H_m < H_c$. Under such conditions, the sample is in a metastable state. These arguments ignore surface effects.

Additional applications of the Ginzburg–Landau equations will not be presented here. For further information, see Parks (1969) and Saint-James *et al.* (1969).

Problems

1 Apply the pseudopotential method to the calculation of the conductivity of a liquid metal. The total pseudopotential can be assumed to be the sum of contributions from each atom, and the scattering cross section can be calculated using the Born approximation. Show that the resistivity ρ is given by (Ziman, 1961)

$$\rho = (3m/16\hbar e^2)(N\Omega_0/E_F) \int_0^2 |S(q)|^2 |V_A(q)|^2 (q/k_F)^3 \, dq/k_F$$

where $S(q)$ is the structure factor defined by Eq. (4.4.26), and $V_A(q)$ is the Fourier coefficient of an atomic pseudopotential.

2 Determine the conductivity of a thin film of metal of thickness a. The mean free path l_0 (in a bulk sample under the same conditions) is greater than a. Show that if the electrons hitting the surface are scattered randomly, the conductivity of the thin film σ is related to that of the bulk metal σ_0 by

$$\sigma/\sigma_0 = (3a/4l_0) + (a/2l_0)\ln(l_0/a).$$

3 Estimate the speed of sound in potassium in a free electron model, and compare the results with more complete calculations (Shyu and Gespari, 1969), and with experiment.

4 Verify Eqs. (7.4.67) and (7.4.68).

5 Find the $\omega(\mathbf{k})$ relation for the propagation of electromagnetic waves in a metal in a magnetic field if the Fermi surface permits an open orbit in the $k_x - k_y$ plane. The magnetic field is applied in the z direction which is

also the direction of wave propagation. The conductivity tensor is to be obtained from Section (7.3.4).

6 Consider a semiconductor in which carriers are predominately scattered by acoustic phonons. The temperature is larger than the Debye temperature. Show that the electron mobility depends on temperature as $T^{-3/2}$.

7 Calculate the energy of the lowest state of the bound polaron in the strong coupling limit variationally using Eq. (7.7.17) with the trial function given by Eq. (7.7.20).

8 BCS define creation and annihilation operators for pairs by $b_{\mathbf{k}}^{+} = c^{+}_{\mathbf{k}\uparrow}c_{-\mathbf{k}\downarrow}$; $b_{\mathbf{k}} = c_{-\mathbf{k}\downarrow}c_{\mathbf{k}\uparrow}$. Evaluate the commutators $[b_{\mathbf{k}}, b^{+}_{\mathbf{k}'}]$, $[b_{\mathbf{k}}, b_{\mathbf{k}'}]$ and the anticommutator $\{b_{\mathbf{k}}, b_{\mathbf{k}'}\}$.

9 Obtain the fourth-order terms discarded in going from Eq. (7.8.8) to Eq. (7.8.15). Why is it legitimate to neglect these terms?

10 Assume that the entropy of electrons in a superconductor is the same as that of a set of noninteracting quasiparticles obeying Fermi statistics with energies $\mathcal{E}(\mathbf{k})$ given by Eq. (7.8.51). Show that the electronic specific heat is given by

$$C = 2\kappa\beta^2 \sum_{\mathbf{k}} f(\mathcal{E})[1 - f(\mathcal{E})] \left[\mathcal{E}(\mathbf{k}, T)^2 + \frac{\beta}{2}\frac{d}{d\beta}\Delta_{\mathbf{k}}^2(T) \right]$$

in which $\beta = 1/KT$. Show that the specific heat is discontinuous at T_c and is proportional to $e^{-\Delta(0)/\kappa T}$ as $T \to 0$.

References

Abrikosov, A. A. (1972). "Introduction to the Theory of Normal Metals." Academic Press, New York.

Aigrain, P. (1960). *Proc. Int. Conf. Semiconductor Phys., Prague* p. 225.

Alfven, H. (1942). *Nature (London)* **150,** 405.

Allcock, G. R. (1963). Strong Coupling Polaron Theory. *In* "Polarons and Excitons" (C. G. Kuper and G. D. Whitefield, eds.), p. 45. Plenum Press, New York.

Anderson, P. W. (1967). *Progr. Low Temp. Phys.* **5,** 1.

Azbel, M. Ia., and Kaner, E. (1956). *J. Exp. Theoret. Phys. (USSR)* **30,** 811 (*English Transl.: Sov. Phys. JETP* **3,** 772).

Baraff, G. A. (1968). *Phys. Rev.* **167,** 625.

Bardeen, J. (1937). *Phys. Rev.* **52,** 688.

Bardeen, J. (1961). *Phys. Rev. Lett.* **6,** 57.

Bardeen, J., and Shockley, W. (1950). *Phys. Rev.* **80,** 72.

Bardeen, J., Cooper, L. N., and Schrieffer, J. R. (1957). *Phys. Rev.* **108,** 1175.

Baynham, A. C., and Boardman, A. D. (1970). *Advan. Phys.* **19,** 575.

Blatt, F. J. (1968). "Physics of Electronic Conduction in Solids." McGraw-Hill, New York.
Bohm, D., and Staver, T. (1951). *Phys. Rev.* **84,** 836.
Born, M., and Oppenheimer, J. R. (1927). *Ann. Phys.* **84,** 457.
Bowers, R., Legendy, C., and Rose, F. (1961). *Phys. Rev. Lett.* **7,** 339.
Brockhouse, B. N., Arase, T., Caglioti, G., Rao, K. R., and Woods, A. D. B. (1962). *Phys. Rev.* **128,** 1099.
Brown, F. C. (1963). Experiments on the Polaron. *In* "Polarons and Excitons" (C. G. Kuper and G. D. Whitfield, eds.), p. 323. Plenum Press, New York.
Budd, H. (1962). *Phys. Rev.* **127,** 4.
Byers, N., and Yang, C. N. (1961). *Phys. Rev. Lett.* **7,** 46.
Caroli, C., Combescot, R., Nozieres, P., and Saint-James, D. (1971). *J. Phys. C* **4,** 916.
Chambers, R. G. (1952). *Proc. Phys. Soc.* **A65,** 458.
Chambers, R. G. (1960). Magnetoresistance. *In* "The Fermi Surface" (W. A. Harrison and M. B. Webb, eds.), p. 100. Wiley, New York.
Chester, G. V. (1961). *Advan. Phys.* **10,** 357.
Cohen, M. H., Harrison, M. J., and Harrison, W. A. (1960). *Phys. Rev.* **117,** 937.
Cohen, M. H., Falicov, L. M., and Phillips, J. C. (1962). *Phys. Rev. Lett.* **8,** 316.
Coleman, R. V., Funes, A. J., Plaskett, J. S., and Tapp, C. M. (1964). *Phys. Rev.* **133,** A521.
Deaver, B. S., and Fairbank, W. M. (1961). *Phys. Rev. Lett.* **7,** 43.
Doll, R., and Nabauer, M. (1961). *Phys. Rev. Lett.* **7,** 51.
Douglass, D. H., and Falicov, L. M. (1964). The Superconducting Energy Gap. *In* "Progress in Low Temperature Physics" (C. J. Gorter, ed.), Vol. 4, p. 102. North-Holland Publ., Amsterdam.
Fawcett, E. (1964). *Advan. Phys.* **13,** 139.
Ferrell, R. A., and Prange, R. E. (1963). *Phys. Rev. Lett.* **10,** 479.
Feynman, R. P. (1955). *Phys. Rev.* **97,** 660.
Feynman, R. P., Hellwarth, R. W., Iddings, C. K., and Platzmann, P. M. (1962). *Phys. Rev.* **127,** 1004.
Fröhlich, H. (1954). *Advan. Phys.* **3,** 325.
Fry, J. L. (1969). *Phys. Rev.* **179,** 892.
Gantmakher, V. F. (1967). Radio Frequency Size Effects in Metals. *In* "Progress in Low Temperature Physics" (C. J. Gorter, ed.), Vol. V., p. 18. North-Holland Publ., Amsterdam.
Gantmakher, V. F., and Kaner, E. A. (1965). *J. Exp. Theoret. Phys. USSR* **48,** 1572 (*English Transl.: Sov. Phys. JETP* **21,** 1053).
Garland, J. W. (1963). *Phys. Rev. Lett.* **11,** 114.
Gennes, P. G. de (1966). "Superconductivity of Metals and Alloys" (P. A. Pincus, trans.). Benjamin, New York.
Giaever, I. (1960a). *Phys. Rev. Lett.* **5,** 147.
Giaever, I. (1960b). *Phys. Rev. Lett.* **5,** 464.
Ginzburg, V. L., and Landau, L. D. (1950). *J. Exp. Theoret. Phys.* (*USSR*) **20,** 1064.
Gor'kov, L. P. (1959a). *J. Exp. Theoret. Phys.* (*USSR*) **36,** 1918 (*English Transl.: Sov. Phys. JETP* **9,** 1364.
Gor'kov, L. P. (1959b). *J. Exp. Theoret. Phys.* (*USSR*) **37,** 1407 (*English Transl.: Sov. Phys. JETP* **10,** 998.
Gurevich, V. L., Skobov, V. G., and Firsov, Yu. A. (1961). *J. Exp. Theoret. Phys.* (*USSR*) **40,** 786 (*English Transl.: Sov. Phys. JETP* **13,** 552).

Harrison, W. A. (1961). *Phys. Rev.* **123,** 85.
Herrod, R. A., and Goodrich, R. G. (1970). *Phys. Lett.* **33A,** 331.
Isaacson, R. T., and Williams, G. A. (1969). *Phys. Rev.* **177,** 738.
Jaclevic, R. C., Lambe, J., Mercereau, J. E., and Silver, A. H. (1965). *Phys. Rev.* **140,** A1628.
Jones, H. (1956). Theory of Electrical and Thermal Conductivity in Metals. *In* "Encyclopedia of Physics" (S. Flügge, ed.), Vol. 19, p. 227. Springer-Verlag, Berlin.
Jones, H., and Zener, C. (1934a). *Proc. Roy. Soc.* (*London*) **144,** 101.
Jones, H., and Zener, C. (1934b). *Proc. Roy. Soc.* (*London*) **145,** 268.
Jones, R. C., Goodrich, R. G., and Falicov, L. M. (1968). *Phys. Rev.* **174,** 672.
Josephson, B. D. (1962). *Phys. Lett.* **1,** 251.
Joshi, S. K., and Rajagopal, A. K. (1968). *Solid State Phys.* **22,** 159.
Kamm, G. N. (1970). *Phys. Rev. B* **1,** 554.
Kaner, E. A., and Skobov, V. G. (1968). *Advan. Phys.* **17,** 605.
Kaveh, M., and Wiser, N. (1972). *Phys. Rev. Lett.* **29,** 1374.
Kittel, C. (1963). "Quantum Theory of Solids." Wiley, New York.
Klemens, P. G. (1956). Thermal Conductivity of Solids at Low Temperatures. *In* "Encyclopedia of Physics" (S. Flügge, ed.), Vol. 14, p. 198. Springer-Verlag, Berlin.
Koch, J. F. (1968). "Magnetic Field Induced Surface Quantum States in Metals. *In* "Solid State Physics" (J. F. Cochran and R. R. Haering, eds.), Vol. 1, p. 253. Gordon and Breach, New York.
Kohler, M. (1948). *Z. Phys.* **124,** 772.
Kohn, W. (1959). *Phys. Rev. Lett.* **2,** 393.
Kuper, C. G., and Whittfield, G. O. (1963). "Polarons and Excitons." Plenum Press, New York.
Langenberg, D. N., Scalapino, D. J., Taylor, B. N., and Eck, R. E. (1965). *Phys. Rev. Lett.* **15,** 294.
Langer, J. S., and Vosko, S. H. (1959). *J. Phys. Chem. Solids* **12,** 196.
Lee, M. J. G. (1970). *Phys. Rev. B* **2,** 250.
Lee, T. D., Low, F. E., and Pines, D. (1953). *Phys. Rev.* **90,** 297.
Libchaber, A., Adams, G., and Grimes, C. C. (1970). *Phys. Rev. B* **1,** 361.
Lighthill, M. J. (1958). "Introduction to Fourier Analysis and Generalized Functions." Cambridge Univ. Press, London and New York.
Lindhard, J. (1954). *Kgl. Danske Videnskab. Selskab, Mat.-Fys. Medd.* **28,** No. 8.
London, F. (1950). "Superfluids," Vol. 1. Wiley, New York.
MacDonald, D. K. C. (1962). "Thermoelectricity. An Introduction to Its Principles." Wiley, New York.
McMillan, W. L. (1968). *Phys. Rev.* **167,** 331.
McMillan, W. L., and Rowell, J. M. (1965). *Phys. Rev. Lett.* **14,** 108.
Mattis, D. C., and Bardeen, J. (1958). *Phys. Rev.* **111,** 412.
Mendelssohn, K., and Rosenberg, H. M. (1961). *Solid State Phys.* **12,** 223.
Meservey, R., and Schwartz, B. B. (1969). "Superconductivity" (R. D. Parks, ed.), Vol. 1, p. 117. Dekker, New York.
Migdal, A. B. (1958). *J. Exp. Theoret. Phys.* (*USSR*) **34,** 1438 (*English Transl.: Sov. Phys. JETP* **7,** 996).
Miller, P. B., and Haering, R. (1962). *Phys. Rev.* **128,** 126.
Onsager, L. (1954). *Proc. Int. Conf. Theoret. Phys., Kyoto and Tokyo* p. 935. Science Council of Japan, Tokoyo.
Onsager, L. (1961). *Phys. Rev. Lett.* **7,** 50.

Parker, W. H., Langenberg, D. N., Denenstein, A., and Taylor, B. (1969). *Phys. Rev.* **177,** 639.

Parks, R. D. (ed.) (1969). "Superconductivity." Dekker, New York.

Penn, D. R. (1962). *Phys. Rev.* **128,** 2093.

Pippard, A. B. (1947). *Proc. Roy. Soc.* **A191,** 385.

Platzman, P. M., and Buchsbaum, S. J. (1963). *Phys. Rev.* **132,** 2.

Platzman, P. M., and Wolff, P. A. (1973). "Waves and Interactions in Solid State Plasmas." Academic Press, New York.

Prange, R. E. (1963). *Phys. Rev.* **131,** 1083.

Reuter, G. E. H., and Sondheimer, E. H. (1948). *Proc. Roy. Soc.* (*London*) **A195,** 336.

Rickayzen, G. (1965). "Theory of Superconductivity." Wiley (Interscience), New York.

Saint James, D., Thomas, E. D., and Sarma, G. (1964). "Type II Superconductivity." Pergamon, Oxford.

Scalapino, P. J., Schrieffer, J. R., and Wilkins, J. W. (1966). *Phys. Rev.* **148,** 263.

Schrieffer, J. R. (1964). "Theory of Superconductivity." Benjamin, New York.

Sham, L. J., and Ziman, J. M. (1963). *Solid State Phys.* **15,** 221.

Shapira, Y., and Lax, B. (1965). *Phys. Rev.* **138,** A119.

Shyu, W. M., and Gaspari, G. (1969). *Phys. Rev.* **177,** 1041.

Smith, A. C., Janak, J. F., and Adler, R. B. (1967). "Electronic Conduction in Solids." McGraw-Hill, New York.

Stedman, R., and Nilsson, G. (1965). *Phys. Rev. Lett.* **15,** 634.

Taylor, M. T. (1965). *Phys. Rev.* **137,** A1145.

Toigo, F., and Woodruff, T. O. (1970). *Phys. Rev. B* **2,** 3958.

Toigo, F., and Woodruff, T. O. (1971). *Phys. Rev. B* **4,** 4312.

Valatin, D. G. (1958). *Nuovo Cimento* **7,** 843.

Vashishta, P., and Singwi, K. S. (1972). *Phys. Rev. B* **6,** 875.

Vosko, S. H., Taylor, R., and Keech, G. H. (1965). *Can. J. Phys.* **43,** 1187.

Walsh, W. M. (1968). *Solid State Phys.* **1,** 39.

Walter, J. P., and Cohen, M. L. (1970). *Phys. Rev. B* **2,** 1821.

Wannier, G. H. (1972). *Phys. Rev. B* **5,** 3836.

Werthamer, N. R. (1969). The Ginzburg-Landau Equations and Their Extensions. *In* "Superconductivity" (R. D. Parks, ed.), Vol. 1, p. 321. Dekker, New York.

Wilson, A. H. (1953). "The Theory of Metals." Cambridge Univ. Press, London and New York.

Woll, E. J., and Kohn, W. (1962). *Phys. Rev.* **126,** 1693.

Ziman, J. M. (1960). "Electrons and Phonons." Oxford Univ. Press, London and New York.

Ziman, J. M. (1961). *Phil. Mag.* [8] **6,** 1013.

CHAPTER 8

Aspects of the Electron–Electron Interaction

In this chapter we will discuss some of the important concepts and methods used in the study of the effects of the electron–electron interaction in solids. Some applications will be considered, in particular, the problems of the correlation energy of a free electron gas and of magnetic order in metals.

Every topic in solid state physics involves, at some level of investigation, the question of electron interactions. It is fortunate, in view of the complexity of this problem, that simple, independent particle models work satisfactorily for many problems. Examples include optical properties, particularly in semiconductors, effects of static fields, transport theory, and the characteristics of Fermi surfaces. The question as to why single-particle theories enjoy a large degree of success is most significant; we will attempt to provide some tentative answers.

Our study will begin with the method of Green's functions, which is a vital tool of many-body theory. The properties of an electron gas will be considered. We will then discuss the Landau theory of Fermi liquids. This chapter will conclude with an investigation of itinerant electron magnetism. Useful general references include: Zubarev (1960), Pines (1962, 1963), Kadanoff and Baym (1962), Abrikosov *et al.* (1963), Nozieres (1964), Schultz (1964) Pines and Nozieres (1966), Kirzhnits (1967), Mattuck (1967), March *et al.* (1969), Mills (1969), Hedin and Lundqvist (1969), Ziman (1969), Fetter and Walecka (1971), Thouless (1972). This list is not exhaustive.

8.1 Properties of Green's Functions

Green's functions furnish a powerful tool for the investigation of the properties of many-body systems. These objects are generalizations of the ordinary Green's functions which appear in the theory of partial differential equations.

8.1.1 *The Single-Particle Green's Function at* $T = 0°K$

Let $\psi_\sigma(\mathbf{r}, t)$ be the electron field operator in the Heisenberg picture: $\psi_\sigma(\mathbf{r}, t)$ destroys an electron of spin σ at point $\mathbf{r}$ and time t; while $\psi^+_{\sigma'}(\mathbf{r}', t')$ creates one with spin σ' at $\mathbf{r}', t'$. The single-particle Green's function is defined by

$$\mathcal{G}(\mathbf{r}, t, \sigma; \mathbf{r}', t', \sigma') = -i\langle N \mid T\{\psi_\sigma(\mathbf{r}, t)\psi^+_{\sigma'}(\mathbf{r}', t')\} \mid N\rangle \qquad (8.1.1)$$

in which $\mid N\rangle$ represents the actual ground state of an N electron system, all interactions being included. The operator T arranges the operators within the braces in chronological order in such a way that time increases from right to left. A minus sign is supplied if the order of operators established by T differs from that already present in the braces. [This remark applies to Fermion systems. The factor of (-1) is omitted when the corresponding definition is made for Boson systems.] Thus

$$\begin{aligned} T\{\psi_\sigma(\mathbf{r}, t)\psi^+_{\sigma'}(\mathbf{r}', t')\} &= \psi_\sigma(\mathbf{r}, t)\psi^+_{\sigma'}(\mathbf{r}', t') \qquad &\text{if} \quad t' < t, \\ &= -\psi^+_{\sigma'}(\mathbf{r}', t')\psi_\sigma(\mathbf{r}, t) \qquad &\text{if} \quad t' > t. \end{aligned} \qquad (8.1.2)$$

Suppose that $t > t'$, so that the first line of (8.1.2) applies. An electron has been inserted at $\mathbf{r}', t'$ in the system and one has been removed subsequently at $\mathbf{r}, t$. Specifically, $\mathcal{G}$ can be interpreted as a probability amplitude for finding an electron at $\mathbf{r}, t$ when one has been added at $\mathbf{r}', t'$. In this case, $\mathcal{G}$ describes the "propagation" of an extra particle and is referred to as a "propagator."

If $t < t'$, a particle is first removed, then added. In this situation, $\mathcal{G}$ describes the propagation of a hole.

If the system is such that the total spin is a good quantum number, $\mathcal{G}$ will differ from zero only if $\sigma = \sigma'$. Further, if there is no spin order, $\mathcal{G}$ will be independent of spin and in this case, the spin indices may be dropped, which we will frequently do.

Note that $\mathcal{G}$ is discontinuous at $t = t'$: Let δ be a small positive quantity,

$$\begin{aligned} &\mathcal{G}(\mathbf{r}, t; \mathbf{r}'t - \delta) - \mathcal{G}(\mathbf{r}, t - \delta; \mathbf{r}', t) \\ &\quad = -i\langle N \mid \psi(\mathbf{r}, t)\psi^+(\mathbf{r}', t - \delta) + \psi^+(\mathbf{r}', t)\psi(\mathbf{r}, t - \delta) \mid N\rangle \\ &\quad = -i\,\delta(\mathbf{r} - \mathbf{r}') \qquad \text{as} \quad \delta \to 0. \end{aligned} \qquad (8.1.3)$$

We have used the anticommutation relations obeyed by the field operators.

Let us now introduce a complete set of energy eigenstates for the $N + 1$ particle system. We will denote these by an index m: thus $| N + 1, m\rangle$. Then, for $t > t'$,

$$\mathcal{G}(\mathbf{r}, t; \mathbf{r}', t') = -i \sum_m \langle N \mid \psi(\mathbf{r}, t) \mid N + 1; m\rangle\langle N + 1, m \mid \psi^+(\mathbf{r}', t') \mid N\rangle. \tag{8.1.4}$$

Since

$$\psi(\mathbf{r}, t) = e^{iHt/\hbar}\psi(\mathbf{r})e^{-iHt/\hbar}, \tag{8.1.5}$$

we have

$$\langle N \mid \psi(\mathbf{r}, t) \mid N + 1, m\rangle = \exp\{i[E_N - E_{N+1,m}]t/\hbar\}\langle N \mid \psi(\mathbf{r}) \mid N + 1, m\rangle \tag{8.1.6}$$

in which $E_{N+1,m}$ is the energy of the $N + 1$ particle system in state m. Let

$$E_{N+1,m} - E_N = \epsilon_m. \tag{8.1.7}$$

We see that $\epsilon_m \geq \mu$, where μ is the chemical potential. Also define

$$f_m(\mathbf{r}) = \langle N \mid \psi(\mathbf{r}) \mid N + 1, m\rangle. \tag{8.1.8}$$

In a Hartree–Fock approximation in which a fixed basis set is used (that is, the N and $N + 1$ particle Slater determinants employ the same one-particle functions except for the added particle), $f_m(\mathbf{r})$ is the one-electron wave function for state m.

$$\mathcal{G}(\mathbf{r}, t; \mathbf{r}', t') = -i \sum_m f_m(\mathbf{r})f_m^*(\mathbf{r}') \exp[-i\epsilon_m(t - t')/\hbar] \qquad (t > t'). \tag{8.1.9}$$

We proceed in similar fashion for $t < t'$. Let

$$\epsilon_m = E_N - E_{N-1,m} \qquad (\epsilon_m < \mu) \tag{8.1.10}$$

and

$$f_m(\mathbf{r}) = \langle N - 1, m \mid \psi(\mathbf{r}) \mid N\rangle. \tag{8.1.11}$$

We can express the result in a form valid for all times:

$$\mathcal{G}(\mathbf{r}, t; \mathbf{r}', t') = -i \sum_m f_m(\mathbf{r})f_m^*(\mathbf{r}') \exp[-i\epsilon_m(t - t')/\hbar] \times [\theta(\epsilon_m - \mu)\theta(t - t') - \theta(\mu - \epsilon_m)\theta(t' - t)] \tag{8.1.12}$$

in which θ is a unit step function

$$\theta(x) = 1 \quad \text{for} \quad x \geq 0, \qquad = 0 \quad \text{otherwise}.$$

Equation (8.1.12) shows that $\mathcal{G}$ is a function of time through the difference $t - t'$ only. In a system which is spatially uniform, we would similarly expect that $\mathcal{G}$ depends on coordinates through $\mathbf{r} - \mathbf{r}'$ only. This will, however, not be true in the presence of a periodic potential.

It is convenient to introduce a Fourier transform with respect to time:

$$\mathcal{G}(\mathbf{r}, \mathbf{r}', \epsilon) = \int_{-\infty}^{\infty} d(t - t')\mathcal{G}(\mathbf{r}, t; \mathbf{r}', t') \exp[i\epsilon(t - t')/\hbar]. \quad (8.1.13)$$

The inverse relation is

$$\mathcal{G}(\mathbf{r}, t, \mathbf{r}', t') = (1/2\pi\hbar) \int_{-\infty}^{\infty} \exp[-i\epsilon(t - t')/\hbar]\mathcal{G}(\mathbf{r}, \mathbf{r}', \epsilon)\, d\epsilon. \quad (8.1.14)$$

In order that the integral should exist, it is convenient to consider ϵ_m as having a positive infinitesimal imaginary part when $\epsilon_m < \mu(\epsilon_m \rightarrow \epsilon_m + i\delta)$ and a negative imaginary part $(\epsilon_m \rightarrow \epsilon_m - i\delta)$ for $\epsilon_m > \mu$. Then we have

$$\mathcal{G}(\mathbf{r}, \mathbf{r}', \epsilon) = \hbar \sum_m f_m(\mathbf{r}) f_m^*(\mathbf{r}')[(\epsilon + i\delta - \epsilon_m)^{-1}\theta(\epsilon_m - \mu)$$
$$+ (\epsilon - i\delta - \epsilon_m)^{-1}\theta(\mu - \epsilon_m)]. \quad (8.1.15a)$$

Equation (8.1.15a) simplifies if we suppress δ keeping the convention concerning ϵ_m in mind,

$$\mathcal{G}(\mathbf{r}, \mathbf{r}', \epsilon) = \hbar \sum_m f_m(\mathbf{r}) f_m^*(\mathbf{r}')/(\epsilon - \epsilon_m). \quad (8.1.15b)$$

This form is reminiscent of the usual expression for a single-electron Green's function as the matrix element $\langle \mathbf{r}' \mid [\epsilon - H]^{-1} \mid \mathbf{r} \rangle$. There is an important difference in that $f_m(\mathbf{r})$ is not the solution of a one-particle Schrödinger equation. However, note that these functions are complete:

$$\sum_m f_m(\mathbf{r}) f_m^*(\mathbf{r}') = \langle N \mid \psi(\mathbf{r})\psi^+(\mathbf{r}') + \psi^+(\mathbf{r}')\psi(\mathbf{r}) \mid N \rangle = \delta(\mathbf{r} - \mathbf{r}'). \quad (8.1.16)$$

The summation in Eq. (8.1.16) includes all states—those with energies below as well as those above μ. Equation (8.1.15a) expresses an important result. The single-particle energy-dependent Green's function $\mathcal{G}(\mathbf{r}, \mathbf{r}', \epsilon)$ has poles near the energies which characterize the propagation of an extra particle (if $\epsilon_m > \mu$) or a hole (if $\epsilon_m < \mu$).

It is convenient to define a new function $A(\mathbf{r}, \mathbf{r}', \epsilon)$ called the *spectral weight function* by

$$A(\mathbf{r}, \mathbf{r}', \epsilon) = \sum_m f_m(\mathbf{r}) f_m^*(\mathbf{r}')\, \delta(\epsilon - \epsilon_m). \quad (8.1.17)$$

When the f_m are normalized (as in a noninteracting or Hartree–Fock system) A can be related to the density of states

$$G(\epsilon) = \int A(\mathbf{r}, \mathbf{r}, \epsilon)\, d^3r. \tag{8.1.18}$$

We see from (8.1.16) that

$$\int_{-\infty}^{\infty} A(\mathbf{r}, \mathbf{r}', \epsilon)\, d\epsilon = \delta(\mathbf{r} - \mathbf{r}').$$

The Green's function is related to the spectral weight function by

$$\mathcal{G}(\mathbf{r}, \mathbf{r}', \epsilon) = \hbar \int_c A(\mathbf{r}, \mathbf{r}', \epsilon')/(\epsilon - \epsilon')\, d\epsilon'. \tag{8.1.19}$$

We can reproduce (8.1.15a) for the Green's function through use of (8.1.17) and (8.1.19), if the path of integration in (8.1.19) is just above the real ϵ' axis for $\epsilon < \mu$, and just below it for $\epsilon > \mu$.

Suppose we next introduce a complete set of single-particle wave functions [for example, Bloch functions $\psi_n(\mathbf{k}, \mathbf{r})$] in terms of which the field operators may be expanded:

$$\psi(\mathbf{r}) = \sum_{n\mathbf{k}} \psi_n(\mathbf{k}, \mathbf{r}) c_{n\mathbf{k}}. \tag{8.1.20}$$

For convenience, we treat $\mathbf{k}$ as discrete. We may then represent the Green's function in the crystal momentum representation as

$$\mathcal{G}_{nn'}(\mathbf{k}, \mathbf{k}', \epsilon) = \int \psi_n^*(\mathbf{k}, \mathbf{r}) \mathcal{G}(\mathbf{r}, \mathbf{r}', \epsilon) \psi_n'(\mathbf{k}', \mathbf{r}')\, d^3r\, d^3r'. \tag{8.1.21}$$

This object can be directly expressed in terms of the c's. We find

$$\mathcal{G}_{nn'}(\mathbf{k}, \mathbf{k}', \epsilon) = -i \int_{-\infty}^{\infty} e^{i\epsilon t/\hbar} \langle N \mid T[c_{n\mathbf{k}}(t) c^+_{n'\mathbf{k}'}(0)] \mid N \rangle\, dt \tag{8.1.22}$$

in which $c_{n\mathbf{k}}(t)$ is defined by an equation similar to (8.1.5):

$$c_{n\mathbf{k}}(t) = e^{iHt/\hbar} c_{n\mathbf{k}} e^{-iHt/\hbar}. \tag{8.1.23}$$

Analogous equations apply to the spectral weight functions:

$$A_{nn'}(\mathbf{k}, \mathbf{k}', \epsilon) = \int \psi_n^*(\mathbf{k}, \mathbf{r}) A(\mathbf{r}, \mathbf{r}', \epsilon) \psi_n'(\mathbf{k}', \mathbf{r}')\, d^3r\, d^3r' \tag{8.1.24}$$

and

$$\mathcal{G}_{nn'}(\mathbf{k}, \mathbf{k}', \epsilon) = \hbar \int_c d\epsilon'\, A_{nn'}(\mathbf{k}, \mathbf{k}', \epsilon')/(\epsilon - \epsilon'). \tag{8.1.25}$$

The normalization is

$$\int_{-\infty}^{\infty} A_{nn'}(\mathbf{k}, \mathbf{k}', \epsilon)\, d\epsilon = \delta_{nn'}\, \delta_{\mathbf{k}\mathbf{k}'}. \tag{8.1.26}$$

The integration in (8.1.19) and (8.1.25) may be simplified with use of the identity $(x \pm i\epsilon)^{-1} = P(1/x) \mp i\pi\, \delta(x)$. The result is

$$\mathcal{G}_{nn'}(\mathbf{k}, \mathbf{k}', \epsilon) = \hbar\, P \int d\epsilon\, A_{nn'}(\mathbf{k}, \mathbf{k}', \epsilon)/(\epsilon - \epsilon') + i\hbar\pi A_{nn'}(\mathbf{k}, \mathbf{k}', \epsilon)[\theta(\mu - \epsilon) - \theta(\epsilon - \mu)]. \tag{8.1.27a}$$

The spectral weight function can be regarded as a Hermitian matrix. In many cases, it will be found to be real. When this holds, (8.1.27a) yields

$$\operatorname{Im} \mathcal{G}_{nn'}(\mathbf{k}, \mathbf{k}', \epsilon) = \mp\pi\hbar A_{nn'}(\mathbf{k}, \mathbf{k}', \epsilon) \tag{8.1.27b}$$

in which the $+(-)$ sign applies if ϵ is below (above) the Fermi energy.

In the limit in which the interaction of electrons is neglected, the N particle wave functions are simply Slater determinants of Bloch functions. In such a case, $\mathcal{G}_{ll'}(\mathbf{k}, \mathbf{k}', \epsilon)$ is diagonal in the indices of the states. A superscript (0) will be added to designate this important function. We find from (8.1.12) and (8.1.21) that

$$\begin{aligned} \mathcal{G}_{ll'}{}^{(0)}(\mathbf{k}, \mathbf{k}', t) &= \delta_{ll'}\, \delta_{\mathbf{k}\mathbf{k}'}\, \{-i[1 - n_l(\mathbf{k})]\} \exp[-i\epsilon_l(\mathbf{k})t/\hbar] \qquad (t > 0) \\ &= \delta_{ll'}\, \delta_{\mathbf{k}\mathbf{k}'}\, i n_l(\mathbf{k}) \exp[-i\epsilon_l(\mathbf{k})t/\hbar] \qquad (t < 0) \end{aligned} \tag{8.1.28}$$

in which $n_l(\mathbf{k})$ is the occupation number of the single-particle state. Thus, for $t > 0$, $\mathcal{G}$ describes the propagation of electrons in excited states (energy greater than the Fermi energy), and describes holes for $t < 0$. The energy-dependent Green's function is simply [from (8.1.15a) and (8.1.21)]

$$\mathcal{G}_{ll'}{}^{(0)}(\mathbf{k}, \mathbf{k}', \epsilon) = \hbar\delta_{ll'}\, \delta_{\mathbf{k}\mathbf{k}'}\, (\epsilon - \epsilon_l(\mathbf{k}) + i \operatorname{sgn}[\epsilon_l(\mathbf{k}) - \mu]\delta)^{-1}. \tag{8.1.29}$$

The Green's function has poles close to the energies of the single-particle states. The spectral weight function is simply

$$A_{ll'}(\mathbf{k}, \mathbf{k}', \epsilon) = \delta_{ll'}\, \delta_{\mathbf{k}\mathbf{k}'}\, \delta(\epsilon_l(\mathbf{k}) - \epsilon).$$

These results will not hold in a real system in which electrons interact. Since the Coulomb interaction is translationally invariant, the wave vector is conserved, and we may expect $\mathcal{G}$ to remain diagonal in $\mathbf{k}$ (but not in the bands). We will frequently consider from this point a one-band model, however, in which the band index may be discarded.

As an example of what can be expected to happen in realistic circumstances, suppose that $\mathcal{G}(\mathbf{k}, \epsilon)$ has a pole at a complex energy $\epsilon_c(\mathbf{k})$,

$$\mathcal{G}(\mathbf{k}, \epsilon) = \hbar z(\mathbf{k})/[\epsilon - \epsilon_c(\mathbf{k})] + \phi(\mathbf{k}, \epsilon) \qquad (8.1.30)$$

where ϕ is a smooth function. It follows from (8.1.14) that, if the imaginary part of ϵ_c is negative,

$$\epsilon_c = \epsilon_1 - i\Gamma \qquad (\Gamma > 0), \qquad (8.1.31a)$$

the time-dependent Green's function contains a term

$$-iz \exp(-i\epsilon_1 t/\hbar - \Gamma t/\hbar) \qquad (t > 0) \qquad (8.1.31b)$$

which describes the decay of an excited electron of energy ϵ_1 and lifetime Γ. On the other hand, if the imaginary part is positive,

$$\epsilon_c = \epsilon_1 + i\Gamma, \qquad (8.1.32a)$$

then the time-dependent Green's function has a term

$$iz \exp(-i\epsilon_1 t/\hbar + \Gamma t/\hbar) \qquad (t < 0) \qquad (8.1.32b)$$

describing a hole state. The spectral function is obtained from (8.1.27b). Let $z = z_1 + iz_2$. Then

$$A(\mathbf{k}, \epsilon) = (1/\pi)\{[\Gamma z_1 \mp z_2(\epsilon - \epsilon_1)]/((\epsilon - \epsilon_1)^2 + \Gamma^2) + \operatorname{Im} \phi\} \qquad (8.1.33)$$

in which the upper sign corresponds to (8.1.31) and the lower to (8.1.32). This is a resonance formula of the asymmetric Breit–Wigner type. Note that

$$1/\pi \int_{-\infty}^{\infty} [\Gamma/((\epsilon - \epsilon_1)^2 + \Gamma^2)]\, d\epsilon = 1.$$

The diagonal elements of the spectral weight function must be positive definite, and we expect that $\operatorname{Im} \phi(\epsilon)$, which represents a background contribution, is also positive. Then the sum rule (8.1.26) indicates that $z_1 \leq 1$. The Green's function (8.1.30) describes a single-particle excitation (electron or hole) in the medium. Such an excitation is called a *quasi particle.* The state decays because it can scatter against other particles in the medium and make a transition to another state. However, the number of particles available for scattering is restricted by the Pauli principle, and vanishes at the Fermi energy. A detailed calculation, to be described subsequently (Section 8.2) shows that $\Gamma \approx (\epsilon - \mu)^2$ when ϵ is close to μ. If we have a situation in which (8.1.30) applies with $\Gamma(k) < \epsilon_1(k)$, and in which values of the energy parameter ϵ being considered are close to ϵ_c, we refer to the excitation as a *quasi-particle state.*

8.1.2 The Two-Particle Green's Function

The single-particle Green's function just described is useful in discussions of the spectrum of quasi particles. In order to calculate the total energy of the system and response functions as well, we require the two-particle Green's function. Let 1 denote the coordinate, spin, and time labels $\mathbf{r}_1$, $\mathbf{t}_1$, σ_1, etc. The two-particle Green's function is defined by

$$\mathcal{G}_2(1, 2, 1', 2') = (-i)^2 \langle N \mid T\{\psi(1)\psi(2)\psi^+(2')\psi^+(1')\} \mid N\rangle. \tag{8.1.34}$$

We will apply a subscript 2 to the two-particle Green's function. This function is considerably more complex than the single-particle Green's function. We note a few of its properties. It follows from the properties of the T operator that

$$\mathcal{G}_2(1, 2, 1', 2') = -\mathcal{G}_2(2, 1, 1', 2') = -\mathcal{G}_2(1, 2, 2', 1') = \mathcal{G}_2(2, 1, 2', 1'). \tag{8.1.35}$$

Further, $\mathcal{G}_2$ will vanish unless

$$\sigma_1 + \sigma_2 = \sigma_1' + \sigma_2'. \tag{8.1.36}$$

There will be two cases to consider, corresponding to singlet and triplet states for the pair of particles being propagated. We note that if t_2' and t_1' are both earlier than t_2 and t_1, $\mathcal{G}$ will describe the motion of a pair of added electrons, and thus involves intermediate states for a $N = 2$ particle system. Similarly, if t_2' and t_1' are later than t_2 and t_1, a pair of holes is being propagated. If $t_1' < t_1$, but $t_2' > t_2$ (or interchanged), a particle–hole pair in the N particle system is involved.

Ground state expectation values of one- and two-particle operators can be calculated if the one- and two-particle Green's functions are known. Consider first a one-particle operator $\mathcal{O}$ which can be expressed in second quantized form as

$$\mathcal{O} = \int d^3r\, \psi^+(\mathbf{r})\, \mathcal{O}(\mathbf{r})\psi(\mathbf{r}). \tag{8.1.37}$$

The integration over $\mathbf{r}$ includes implicitly summation over spin components. The average value of $\mathcal{O}$ in the ground state $\mid N\rangle$ is

$$\begin{aligned}\langle N \mid \mathcal{O} \mid N\rangle &= \int d^3r\, \{\mathcal{O}(\mathbf{r})\, \langle N \mid \psi^+(\mathbf{r}')\psi(\mathbf{r}) \mid N\rangle\}_{\mathbf{r}'=\mathbf{r}} \\ &= -i \int d^3r\, \{\mathcal{O}(\mathbf{r})\mathcal{G}(\mathbf{r}t, \mathbf{r}'t_+)\}_{\mathbf{r}'=\mathbf{r}} \end{aligned}\tag{8.1.38}$$

in which $t^+ = t + \delta$ where δ is a positive infinitesimal quantity. Similarly, we have for a two-electron operator

$$\mathfrak{F} = \int d^3r\, d^3r'\, \psi^+(\mathbf{r})\psi^+(\mathbf{r}')\mathfrak{F}(\mathbf{r}, \mathbf{r}')\psi(\mathbf{r}')\psi(\mathbf{r}) \qquad (8.1.39)$$

$$\langle N \mid \mathfrak{F} \mid N\rangle = -\int d^3r\, d^3r'\, \mathfrak{F}(\mathbf{r}, \mathbf{r}')\mathcal{G}_2(\mathbf{r}t, \mathbf{r}'t, \mathbf{r}t_+, \mathbf{r}'t_+). \qquad (8.1.40)$$

These formulas can be applied to the Hamiltonian

$$H = \int \psi^+(\mathbf{r})h(\mathbf{r})\psi(\mathbf{r})\, d^3r + \tfrac{1}{2}\int \psi^+(\mathbf{r})\psi^+(\mathbf{r}')V(\mathbf{r}, \mathbf{r}')\psi(\mathbf{r}')\psi(\mathbf{r}')\, d^3r\, d^3r' \qquad (8.1.41)$$

in which $h(\mathbf{r})$ contains the kinetic energy and one-particle potential energy, and $V(\mathbf{r}, \mathbf{r}')$ represents the two-body interaction, to give the energy

$$E = -i\int d^3r\, \{h(\mathbf{r})\mathcal{G}_1(\mathbf{r}t, \mathbf{r}'t_+)\}_{\mathbf{r}'=\mathbf{r}}$$

$$-\tfrac{1}{2}\int d^3r\, d^3r'\, V(\mathbf{r}, \mathbf{r}')\mathcal{G}_2(\mathbf{r}t, \mathbf{r}'t, \mathbf{r}t_+, \mathbf{r}'t_+). \qquad (8.1.42)$$

8.1.3 *Equations of Motion*

The field operators ψ and ψ^+ obey the equations of motion characteristic of operators in the Heisenberg picture,

$$i\hbar\, \partial/\partial t\, \psi_\sigma(\mathbf{r}, t) = [\psi_\sigma(\mathbf{r}, t), H] \qquad (8.1.43)$$

in which H is the Hamiltonian (8.1.41) for the system. In order to compute the commutator, recall the fundamental anticommutation rule for the field operators

$$\{\psi_\sigma(\mathbf{r}, t), \psi_\sigma{}^+(\mathbf{r}', t)\} = \psi_\sigma(\mathbf{r}, t)\psi^+{}_{\sigma'}(\mathbf{r}', t) + \psi^+{}_{\sigma'}(\mathbf{r}', t)\psi_\sigma(\mathbf{r}, t)$$
$$= \delta_{\sigma\sigma'}\, \delta(\mathbf{r} - \mathbf{r}'). \qquad (8.1.44)$$

The times are the same in the operators in (8.1.44). It would be necessary to solve the equations of motion in order to obtain the anticommutation rules at different times. Equation (8.1.43) yields

$$i\hbar\, \partial/\partial t\, \psi_\sigma(\mathbf{r}, t) = [h(\mathbf{r}) + \sum_{\sigma'}\int d^3r'\, \psi^+{}_{\sigma'}(\mathbf{r}', t)V(\mathbf{r}, \mathbf{r}')\psi_{\sigma'}(\mathbf{r}', t)]\psi_\sigma(\mathbf{r}, t). \qquad (8.1.45)$$

This is the fundamental dynamical equation. It resembles the equation obeyed by a time-dependent wave function in the Hartree approximation, but it is actually far more complicated, in that operators, rather than functions, are involved. In order to obtain an equation for the single-particle Green's function, we apply the operator $i\hbar\, \partial/\partial t$ to (8.1.1), use (8.1.45) and the equation for the derivative of a step function $d\theta(t)/dt = \delta(t)$. A straightforward calculation yields

$$[i\hbar\, \partial/\partial t - h(\mathbf{r})]\mathcal{G}(\mathbf{r}t\sigma, \mathbf{r}'t'\sigma')$$
$$= \hbar\delta_{\sigma\sigma'}\, \delta(\mathbf{r} - \mathbf{r}')\, \delta(t - t') - i \sum_{\sigma''} \int d^3r''\, V(\mathbf{r}, \mathbf{r}'')$$
$$\times \langle N \mid T\{\psi^+_{\sigma''}(\mathbf{r}'', t)\psi_{\sigma''}(\mathbf{r}'', t)\psi_\sigma(\mathbf{r}, t)\psi^+_{\sigma'}(\mathbf{r}', t')\} \mid N\rangle. \quad (8.1.46)$$

The last term in (8.1.46) contains four field operators and therefore involves the two-body Green's function. This equation may be rewritten with $\mathcal{G}_2$ explicitly introduced:

$$[i\hbar\, \partial/\partial t - h(\mathbf{r})]\mathcal{G}(\mathbf{r}t\sigma, \mathbf{r}'t'\sigma')$$
$$= \hbar\delta_{\sigma\sigma'}\, \delta(\mathbf{r} - \mathbf{r}')\, \delta(t - t') + i \sum_{\sigma''} \int d^3r''\, V(\mathbf{r}, \mathbf{r}'')$$
$$\times \mathcal{G}_2(\mathbf{r}''t\sigma'', \mathbf{r}t\sigma, \mathbf{r}'t'\sigma', \mathbf{r}''t_+\sigma''). \quad (8.1.47)$$

This is the equation of motion for the one-particle Green's function. It is possible to derive similar equations for the two-particle Green's function, which will connect this object with a three-body Green's function. In general, an n particle Green's function obeys an equation relating it to an $n + 1$ particle function. An hierarchy of equations is obtained. Unless special circumstances occur, this hierarchy must be terminated in a possibly arbitrary manner by an ansatz which expresses the $n + 1$ particle function as a combination of n particle functions. Such an ansatz is frequently referred to as a *decoupling procedure*.

Equation (8.1.47) can be used to rewrite the formula for the energy of the system in terms of the single-particle Green's function

$$E = (-i/2) \int d^3r\, \{[i\hbar\partial/\partial t + h(\mathbf{r})]\mathcal{G}(\mathbf{r}t, \mathbf{r}'t_+)\}_{\mathbf{r}=\mathbf{r}', t=t'}. \quad (8.1.48)$$

This formula can be expressed in terms of the spectral weight function. To do this, we first introduce the Fourier transform of the Green's function through (8.1.14). Next, use (8.1.19), recalling carefully the location of the contour of integration in (8.1.19). The contour must be closed in the lower half plane. The result is

$$E = \tfrac{1}{2} \int_{-\infty}^{\mu} d\epsilon \, d^3r \, [\epsilon + h(\mathbf{r})][A(\mathbf{r}, \mathbf{r}', \epsilon) \,|_{\mathbf{r}=\mathbf{r}'}. \tag{8.1.49}$$

Note that the upper limit of the ϵ integration is specified by the chemical potential. This results from the behavior of the prescribed contour in (8.1.19).

We continue to examine the equation of motion (8.1.47). It is convenient to introduce a function $K(1, 2, 1', 2')$ by writing (8.1.34) in the form

$$\mathcal{G}_2(1, 2, 1', 2') = \mathcal{G}(1, 2')\mathcal{G}(2, 1') - \mathcal{G}(1, 1')\mathcal{G}(2, 2') + K(1, 2, 1', 2'). \tag{8.1.50}$$

The first two terms in (8.1.50) describe the propagation of two independent particles, taking due account of antisymmetry; the third term contains specific interaction effects. Equation (8.1.50) may be regarded as defining K. It is desirable to express K as (the four-dimensional integration includes time and a summation over spin indices)

$$K(1, 2, 1'2') = \int \cdots \int d^4x_3 \, d^4x_{3'} \, d^4x_4 \, d^4x_{4'} \, \mathcal{G}(1, 3)\mathcal{G}(2, 4)$$
$$\times \gamma(3, 4, 3', 4')\mathcal{G}(3', 1')\mathcal{G}(4', 2'). \tag{8.1.51}$$

The function γ describes the interaction of two electrons.

Equation (8.1.50) is substituted into (8.1.47). The result is written as

$$[i\hbar \, \partial/\partial t - h(1) - V(1)]\mathcal{G}(1, 1') - \int M(1, 2)\mathcal{G}(2, 1') \, d^4x_2 = \hbar\delta(1 - 1') \tag{8.1.52}$$

in which

$$\delta(1 - 1') = \delta_{\sigma\sigma'} \, \delta(t - t') \, \delta(\mathbf{r} - \mathbf{r}'). \tag{8.1.53}$$

Two new objects have been introduced in (8.1.52). The first is the term

$$V(1) = V(\mathbf{r}) = -i \sum_{\sigma} \int d^3r' \, V(\mathbf{r}, \mathbf{r}')\mathcal{G}_1(\mathbf{r}'t\sigma, \mathbf{r}'t_+\sigma). \tag{8.1.54a}$$

This term is analogous to an ordinary potential. This observation will be clear if we substitute (8.1.12) into (8.1.54a):

$$V(\mathbf{r}) = \sum_{m\sigma} \int d^3r' \, | f_{m\sigma}(\mathbf{r}') \,|^2 V(\mathbf{r}, \mathbf{r}')\theta(\mu - \epsilon_{m\sigma}). \tag{8.1.54b}$$

In a Hartree–Fock approximation, f_m is a one-particle wave function for state m. States with energies below the Fermi energy are included in the sum. It is evident in this case that $V(\mathbf{r})$ is just the average electrostatic potential of the electron distribution.

The other new object in (8.1.52a) is the mass operator (or self-energy) M. An expression for this operator can be found (however, it involves the unknown interaction function γ):

$$M(1, 2) = iV(1, 2)\mathcal{G}(1, 2_+) - i \iiiint V(1, 3)\mathcal{G}(3, 4)\mathcal{G}(1, 5)\gamma(4, 5, 2, 6)\mathcal{G}(6, 2_+)\, d^4x_3\, d^4x_4\, d^4x_5\, d^4x_6 \tag{8.1.55}$$

in which $V(1, 2) = V(\mathbf{r}_1, \mathbf{r}_2)\, \delta(t_1 - t_2)$ and the time associated with "2_+" is $t_2 + i\delta$. The Hartree–Fock approximation is obtained by dropping the complicated second term of (8.1.55):

$$M_{\mathrm{HF}}(\mathbf{r}t\sigma, \mathbf{r}''t_+\sigma'') = iV(\mathbf{r}, \mathbf{r}'')\mathcal{G}(\mathbf{r}t\sigma, \mathbf{r}''t_+\sigma''). \tag{8.1.56a}$$

If (8.1.12) is inserted in (8.1.56), the mass operator has a typical exchange form,

$$M_{\mathrm{HF}}(\mathbf{r}t\sigma, \mathbf{r}''t_+\sigma) = -V(\mathbf{r}, \mathbf{r}'') \sum_m f_{m\sigma}(\mathbf{r})f^*_{m\sigma}(\mathbf{r}'')\theta(\mu - \epsilon_{m\sigma}). \tag{8.1.56b}$$

A slight simplification of (8.1.52) can be achieved if we introduce the Fourier transform of the Green's function with respect to time according to (8.1.13). Since the physical system must be invariant with respect to a choice of the origin of the time coordinate, it follows that M can depend only on the difference of the time coordinates. Thus we can introduce the Fourier transform of M,

$$M(\mathbf{r}\sigma, \mathbf{r}'\sigma', \epsilon) = \int_{-\infty}^{\infty} d(t - t')\, M(\mathbf{r}t\sigma, \mathbf{r}'t'\sigma')e^{i\epsilon(t-t')/\hbar}$$

and we obtain

$$[\epsilon - h(\mathbf{r}) - V(\mathbf{r})]\mathcal{G}(\mathbf{r}, \mathbf{r}', \epsilon) - \sum_{\sigma''} \int d\epsilon\, M(\mathbf{r}\sigma, \mathbf{r}''\sigma'', \epsilon)\mathcal{G}(\mathbf{r}''\sigma'', \mathbf{r}'\sigma', \epsilon) = \delta(\mathbf{r} - \mathbf{r}')\, \delta_{\sigma\sigma'}. \tag{8.1.57}$$

For further discussion of the equations of motion, see Martin and Schwinger (1959) and Kadanoff and Baym (1962).

8.1.4 *The Interaction Picture: Perturbation Theory*

The preceding discussion has presented some of the important general properties of Green's functions leading to the development of equations of motion. These equations of motion form a maze of increasing complexity as the number of particles studied becomes larger. We will not pursue this path further. Instead, we will develop the perturbation procedure which leads to the diagrammatic expansion of the Green's function. Although the convergence of perturbation theory is sometimes questionable, experience indicates that the procedure is useful.

Let H be the Hamiltonian for the system of interacting electrons. We write

$$H = H_0 + V \tag{8.1.58}$$

in which H_0 contains the single-particle terms (kinetic energy plus periodic potential due to the ions) and V contains the electron interaction. The discussion in Sections 8.1.1–8.1.3 employed the Heisenberg picture in which the state vectors are constant in time, and the operators obey equations of motion governed by the full Hamiltonian. The perturbation calculation is most conveniently performed in the interaction (or Dirac) picture (see Section 6.5.4): If $\mathcal{O}_S$ is an operator in the Schrödinger picture, the corresponding operator in the interaction picture is

$$\mathcal{O}_I = \exp(iH_0t/\hbar)\,\mathcal{O}_S\exp(-iH_0t/\hbar) \tag{8.1.59}$$

while a state vector $|\psi_s\rangle$ in the Schrödinger picture becomes in the interaction picture

$$|\psi_I\rangle = \exp(iH_0t/\hbar)\,|\psi_s\rangle. \tag{8.1.60}$$

It follows from (8.1.59) and (8.1.60) that the time dependence of $|\psi_I\rangle$ is determined by

$$i\hbar\, d/dt\,|\psi_I\rangle = V_I(t)\,|\psi_I\rangle \tag{8.1.61}$$

in which V_I is related to V which appears in (8.1.60) by

$$V_I(t) = \exp(iH_0t/\hbar)\,V\exp(-iH_0t/\hbar). \tag{8.1.62}$$

The operator V_I will usually depend on time even if V is independent of time. Note that in the case in which the Hamiltonian is independent of time, an operator $\mathcal{O}_H$ in the Heisenberg picture is related to the Schrödinger picture by

$$\mathcal{O}_H = \exp(iHt/\hbar)\,\mathcal{O}_S\exp(-iHt/\hbar) \tag{8.1.63}$$

so that the relation between $\mathcal{O}_H$ and $\mathcal{O}_I$ is

$$\mathcal{O}_I = \exp(iH_0t/\hbar)\exp(-iHt/\hbar)\,\mathcal{O}_H\exp(iHt/\hbar)\exp(-iH_0t/\hbar). \tag{8.1.64}$$

If H_0 and V do not commute, as is usually the case, the exponents in (8.1.64) cannot be combined directly. The development of $| \psi_I(t) \rangle$ in time is described by a time development operator $U(t, t')$ such that

$$| \psi_I(t) \rangle = U(t, t') \, | \psi_I(t') \rangle. \tag{8.1.65}$$

The operator U is unitary and satisfies two important relations as a consequence of its definition by (8.1.65):

$$U(t, t') U(t't'') = U(t, t'') \tag{8.1.66}$$

and

$$U(t, t) = 1. \tag{8.1.67}$$

An equation for $U(t, t')$ can be deduced by combining (8.1.61) and (8.1.65):

$$i\hbar \, d/dt \, U(t, t') = V_I(t) U(t, t'). \tag{8.1.68}$$

The solution of this equation is subject to the initial condition (8.1.67). If t'' is set equal to t in (8.1.66) and the result compared with (8.1.67), we see that

$$U(t, t') = U(t', t)^{-1}. \tag{8.1.69}$$

The differential equation (8.1.68) and the initial condition (8.1.67) can be combined in an integral equation

$$U(t, t') = 1 - (i/\hbar) \int_{t'}^{t} V(t'') U(t''t') \, dt''. \tag{8.1.70}$$

A formal solution of (8.1.70) can be obtained by iteration

$$U(t, t') = 1 + \sum_{n=1}^{\infty} U_n(t, t') \tag{8.1.71}$$

in which the nth-order term is given by

$$U_n(t, t') = (-i/\hbar)^n \int_{t'}^{t} dt_1 \int_{t'}^{t_1} dt_2 \cdots \int_{t'}^{t_{n-1}} dt_n \, V_I(t_1) \cdots V_I(t_n). \tag{8.1.72}$$

This expression is actually quite complicated, since the operators V_I at different times do not commute. A more symmetric way of writing (8.1.72) exists in which the limits are the same on each integral:

$$U_n(t, t') = (1/n!)(-i/\hbar)^n \int_{t'}^{t} dt_1 \int_{t'}^{t} dt_2 \cdots \int_{t'}^{t} dt_n \times P\{V_I(t_1) \cdots V_I(t_n)\}. \tag{8.1.73}$$

In this equation, $P\{\cdots\}$ represents Dyson's chronological product which arranges the operators $V_{\rm I}$ in order so that time increases from right to left, without [in contrast to the operator T introduced previously] causing any change of sign. The equivalence between (8.1.72) and (8.1.73) is discussed in detail by Schweber (1961, Section 11f). Essentially, the factor of $1/n!$ in (8.1.73) arises because there are $n!$ possible time orderings of the $V_{\rm I}$, each of which gives the same contribution to U_n.

The electron–electron interaction can be expressed in second quantized form in the Schrödinger picture. We will suppress band indices in writing this equation in order to keep from making the notation too cumbersome:

$$V_{\rm S} = \tfrac{1}{2} \sum_{\mathbf{kqk}',\sigma\sigma'} v(\mathbf{q}) c^+{}_{\mathbf{k}+\mathbf{q}\sigma} c^+{}_{\mathbf{k}'+\mathbf{q}\sigma'} c_{\mathbf{k}'\sigma'} c_{\mathbf{k}\sigma}. \tag{8.1.74}$$

The effect of transforming to the interaction picture is to introduce time-dependent creation and annihilation operators:

$$\begin{aligned} c_{\mathbf{k}\sigma}(t) &= \exp(iH_0t/\hbar) c_{\mathbf{k}\sigma} \exp(-iH_0t/\hbar) = \exp[-iE_\sigma(\mathbf{k})t/\hbar] c_{\mathbf{k}\sigma} \\ c^+{}_{\mathbf{k}\sigma}(t) &= \exp[iE_\sigma(\mathbf{k})t/\hbar] c^+{}_{\mathbf{k}\sigma}. \end{aligned} \tag{8.1.75}$$

Thus

$$V_{\rm I}(t) = \tfrac{1}{2} \sum_{\mathbf{kqk}',\sigma\sigma'} v(\mathbf{q}) c^+{}_{\mathbf{k}+\mathbf{q}\sigma}(t) c^+{}_{\mathbf{k}'-\mathbf{q}\sigma'}(t) c_{\mathbf{k}'\sigma'}(t) c_{\mathbf{k}\sigma}(t). \tag{8.1.76}$$

Each term in U_n will involve $2n$ creation operators and $2n$ annihilation operators. Since the number of creation and annihilation operators in (8.1.73) is even, the operator P can be replaced by the time-ordering operator T used earlier.

8.1.5 *Relation between the Time Development Operator and Green's Functions*

Before we consider the specific procedure for the evaluation of contributions to U, it is useful to see how the ground state wave function, the ground state energy, and the Green's functions are related to U. Let the state vector for the lowest state of the noninteracting system in the interaction picture be denoted $|0\rangle$. Its energy will be denoted W_0.

$$H_0 |0\rangle = W_0 |0\rangle.$$

This vector is independent of time. It is useful to define the so-called "vacuum" amplitude

$$R(t, t_0) = \langle 0 | U(t, t_0) | 0\rangle. \tag{8.1.77}$$

We will show that the ground state energy can be determined in principal from $R(t, t_0)$. To this end, we will let t_0 vanish, and allow the time t, which will be taken to be positive, to have a small negative imaginary part. Let

E_g be the ground state energy of the interacting system. It will be proved that

$$E_g = W_0 + i \lim_{t\to\infty(1-i\eta)} [\hbar\, d/dt \ln R(t, 0)]. \tag{8.1.78}$$

Use of a complex time coordinate ensures that there will be no problem with the existence of integrals with infinite limits. To establish (8.1.78), it is convenient to return briefly to the Schrödinger picture. The time development operator in the interaction picture $U(t, t_0)$ is related to the corresponding quantity in the Schrödinger picture by

$$U_S(t, t_0) = \exp(-iH_0t/\hbar)\, U(t, t_0) \exp(iH_0t_0/\hbar). \tag{8.1.79}$$

This relation may be established by using (8.1.60), or by differentiating with respect to t, from which we see that

$$i\hbar\, d/dt\, U_S(t, t_0) = HU_S(t, t_0)$$

where H is the full Hamiltonian. If H_S is independent of time, we may write

$$U_S(t, t_0) = \exp[-iH(t - t_0)/\hbar]. \tag{8.1.80}$$

It follows from (8.1.79) and (8.1.80) that

$$U(t, t_0) = \exp(iH_0t/\hbar) \exp[-iH(t - t_0)/\hbar] \exp(-iH_0t_0/\hbar). \tag{8.1.81}$$

Thus

$$R(t, t_0) = \exp[iW_0(t - t_0)/\hbar]\langle 0 \mid \exp[-iH(t - t_0)/\hbar] \mid 0\rangle. \tag{8.1.82}$$

It is now convenient to introduce a complete set of time-independent eigenstates of the full Hamiltonian H. These states will be denoted simply by $\mid n\rangle$. In order to avoid confusion, the full ground state will be denoted $\mid g\rangle$. These states are the complete state vectors in the Heisenberg picture. Note that the ground state of the noninteracting system is also independent of time. Thus $H \mid n\rangle = E_n \mid n\rangle$, and

$$\begin{aligned} R(t, t_0) &= \exp[iW_0(t - t_0)/\hbar] \sum_n \langle 0 \mid n\rangle\langle n \mid \exp[-iH(t - t_0)/\hbar] \mid 0\rangle \\ &= \exp[iW_0(t - t_0)/\hbar] \sum_n \exp[-iE_n(t - t_0)/\hbar] \mid \langle 0 \mid n\rangle \mid^2. \end{aligned} \tag{8.1.83}$$

Now consider

$$\hbar \frac{d}{dt} \ln R(t, 0) = iW_0 + \frac{\sum_n (-iE_n) \exp(-iE_nt/\hbar) \mid \langle 0 \mid n\rangle \mid^2}{\sum_n \exp(-iE_nt/\hbar) \mid \langle 0 \mid n\rangle \mid^2}. \tag{8.1.84}$$

If we now allow t to become large in the combination $t(1 - i\eta)$, the contribution from the ground state dominates the sum, and we have

$$\lim_{t\to\infty(1-i\eta)} \hbar\, d/dt \ln R(t, 0) = -i(E_g - W_0). \tag{8.1.85}$$

Equation (8.1.78) follows immediately.

It is obviously required for this argument that the true ground state and the ground state of the noninteracting system must not be orthogonal. Difficulty can arise here if the ground state of the interacting system has a different symmetry than that of the noninteracting system. This can occur if the interaction leads to a phase transition.

It will now be shown that the single-particle Green's function is related to U as

$$\mathcal{G}(\mathbf{r}t\sigma, \mathbf{r}'t'\sigma') = -i \lim_{\substack{T_1\to-\infty(1-i\eta)\\ T_2\to\infty(1-i\eta)}} \frac{\langle 0 \mid T\{U(T_2, T_1)\; {}^{(\mathrm{I})}\psi_\sigma(\mathbf{r}, t)\; {}^{(\mathrm{I})}\psi^+{}_{\sigma'}(\mathbf{r}', t')\} \mid 0\rangle}{\langle 0 \mid U(T_2, T_1) \mid 0\rangle}. \tag{8.1.86}$$

The operators ${}^{(\mathrm{I})}\psi$, ${}^{(\mathrm{I})}\psi^+$ carry a superscript I to indicate that they refer to the interaction picture, the time development operators refer to the same picture, and T is the time ordering operator introduced earlier.

The procedure used in establishing (8.1.86) is quite similar to that employed in discussing the ground state energy. We will consider specifically the case in which $t' < t$: the reverse situation may be handled in a similar manner. The times in (8.1.86) are thus ordered as follows: $T_2 > t > t' > T_1$. It is convenient to use (8.1.66) to express U in the form

$$U(T_2, T_1) = U(T_2, t)\, U(t, t')\, U(t', T_1).$$

The numerator in (8.1.86) is then written as

$$\langle 0 \mid U(T_2, t)\; {}^{(\mathrm{I})}\psi_\sigma(\mathbf{r}, t)\, U(t, t')\; {}^{(\mathrm{I})}\psi^+{}_{\sigma'}(\mathbf{r}', t')\, U(t', T_1) \mid 0\rangle. \tag{8.1.87}$$

Since the Green's function was defined with operators in the Heisenberg picture, we must introduce the relevant transformations which are described in Eqs. (8.1.59)–(8.1.64). Note that the noninteracting ground state vector is the same in both pictures. Substitution of (8.1.81) in (8.1.87) yields

$$\exp[iW_0(T_2 - T_1)/\hbar]\langle 0 \mid \exp[-iH(T_2 - t)/\hbar]\; {}^{(\mathrm{S})}\psi_\sigma(\mathbf{r}, t)$$
$$\times \exp[-iH(t - t')/\hbar]\; {}^{(\mathrm{S})}\psi_\sigma{}^+(\mathbf{r}', t') \exp[-iH(t' - T_1)/\hbar] \mid 0\rangle.$$

The operators ${}^{(\mathrm{S})}\psi$, ${}^{(\mathrm{S})}\psi^+$ are in the Schrödinger picture. Use of (8.1.63) gives finally the Heisenberg operators and we drop the left superscript at this point:

$$\exp[iW_0(T_2 - T_1)/\hbar] \times \langle 0 \mid \exp(-iHT_2/\hbar)\psi_\sigma(\mathbf{r}, t)\psi^+_{\sigma'}(\mathbf{r}', t') \exp(iHT_1/\hbar) \mid 0\rangle. \quad (8.1.88)$$

Now we set up the quotient demanded by (8.1.86). The denominator has a similar form to (8.1.88) except that there are no creation or destruction operators in it. The exponential prefactor cancels, so that we consider the quantity

$$\frac{\langle 0 \mid \exp(-iHT_2/\hbar)\psi_\sigma(\mathbf{r}, t)\psi^+_{\sigma'}(\mathbf{r}', t') \exp(iHT_1/\hbar) \mid 0\rangle}{\langle 0 \mid \exp(-iHT_2/\hbar) \exp(iHT_1/\hbar) \mid 0\rangle}. \quad (8.1.89)$$

As before, two complete sets of eigenstates of H are introduced. In the Heisenberg picture, these are independent of time. The expression (8.1.89) becomes

$$\frac{\sum_{n,l} \exp[-i(E_nT_2 - E_lT_1)/\hbar]\langle 0 \mid n\rangle\langle n \mid \psi_\sigma(\mathbf{r}, t)\psi^+_{\sigma'}(\mathbf{r}', t') \mid l\rangle\langle l \mid 0\rangle}{\sum_n \exp[-iE_n(T_2 - T_1)/\hbar] \mid \langle 0 \mid n\rangle \mid^2}. \quad (8.1.90)$$

We now proceed to extract the limits as required by (8.1.86). As before, only the ground state survives when the limit is taken and the double sum contains only a single term. The scalar products cancel as well, and we have

$$-i \lim_{\substack{T_2\to\infty(1-i\eta)\\ T_1\to-\infty(1-i\eta)}} \frac{\langle 0 \mid T\{U(T_2, T_1)\, {}^{(\mathrm{I})}\psi_\sigma(\mathbf{r}, t)\, {}^{(\mathrm{I})}\psi_\sigma{}^+(\mathbf{r}', t')\} \mid 0\rangle}{\langle 0 \mid U(T_2, T_1) \mid 0\rangle} = -i\langle N \mid T\{\psi_\sigma(\mathbf{r}, t)\psi^+_{\sigma'}(\mathbf{r}', t')\} \mid N\rangle. \quad (8.1.91)$$

The time ordering operator has been reinserted. The right side of (8.1.91) is just the definition of the Green's function according to (8.1.1).

Similar results for the two-particle Green's function are obtained by a straightforward extension of the argument just presented:

$$\mathcal{G}(1, 2, 1'2') = (-i)^2 \lim_{\substack{T_1\to-\infty(1-i\eta)\\ T_2=\infty(1-i\eta)}} \langle 0 \mid T\{U(T_2, T_1)\, {}^{(\mathrm{I})}\psi(1)\, {}^{(\mathrm{I})}\psi(2) \times {}^{(\mathrm{I})}\psi^+(2')\, {}^{(\mathrm{I})}\psi^+(1')\} \mid 0\rangle / \langle 0 \mid U(T_2, T_1) \mid 0\rangle. \quad (8.1.92)$$

The utility of the results (8.1.78), (8.1.86), and (8.1.92) is that they enable us to use (8.1.71) to generate a perturbation expansion for the ground state energy and the Green's functions. Each term in the perturbation expansion of U generates a class of terms in the expansion for the Green's function.

8.1.6 *Wick's Theorem*

Each term in the perturbation expansion for the time development operator (8.1.73) can be described in terms of specific physical processes, which may be real or virtual. What is required is a systematic procedure for extracting matrix elements from the general formalism. This is accomplished with the aid of Wick's theorem (Wick, 1950). This result enables us to express the time ordered product $T\{V_{\rm I}(t_1)\cdots V_{\rm I}(t_n)\}$ in a form suitable for practical evaluation in which virtual processes are represented explicitly. To do this, the time ordered product is decomposed into so-called "normal" products.

The normal product of a set of free particle creation and annihilation operators is defined to be an arrangement in which all of the creation operators stand to the left of all destruction operators. Then if states of the system are specified by a definite number of noninteracting particles with given momenta and spins, there exists one and only one normal product with a nonzero matrix element connecting a pair of states.

In order to make this concept useful in a solid, it is convenient from this point to work with the Green's function in the crystal momentum representation as is specified by (8.1.22). The Fermi distribution function for the system at $T = 0°{\rm K}$ in the absence of interactions has all states occupied for energies less than μ, and all vacant for energies greater than μ. Then, let

$$\begin{aligned} b_{n\mathbf{k}} &= c_{n\mathbf{k}} \qquad \text{if} \quad E_n(\mathbf{k}) > \mu, \\ &= c^+{}_{n\mathbf{k}} \qquad \text{if} \quad E_n(\mathbf{k}) < \mu \end{aligned} \tag{8.1.93}$$

(and correspondingly for $b^+{}_{n\mathbf{k}}$). Thus $b_{n\mathbf{k}}$ ($b^+{}_{n\mathbf{k}}$) destroys (creates) an electron if $E_n(\mathbf{k})$ is above the Fermi energy and destroys (creates) a hole by filling (emptying) the state if $E_n(\mathbf{k})$ is below the Fermi energy. This convention means that the noninteracting ground state $|\,0\rangle$ satisfies

$$b_{n\mathbf{k}}\,|\,0\rangle = 0, \tag{8.1.94}$$

so that $|\,0\rangle$ is analogous to the vacuum in quantum field theory. An excited state of the noninteracting system, $|\,1\rangle$ say, is formed by

$$|\,l\rangle = b^+{}_{n_1\mathbf{k}}b^+{}_{n_2\mathbf{k}}\cdots b^+{}_{n_l\mathbf{k}}\,|\,0\rangle. \tag{8.1.95}$$

This state contains some specified arrangement of excited electrons and holes. Note that

$$\begin{aligned} \{b_{n\mathbf{k}}(t), b_{n'\mathbf{k}'}(t')\} &= 0 \\ \{b^+{}_{n\mathbf{k}}(t), b^+{}_{\hbar'\mathbf{k}'}(t')\} &= 0 \\ \{b_{nk}(t), b^+{}_{n'k'}(t')\} &= \delta_{nn'}\,\delta_{\mathbf{k}\mathbf{k}'}\,\exp[iE_n(\mathbf{k})(t'-t)/\hbar]. \end{aligned} \tag{8.1.96}$$

The braces $\{\cdots\}$ denote the anticommutator.

We may now ready to be more specific about the normal product. Consider a set of factors $AB\cdots Z$ which represent some arrangement of operators of the type $b_{n\mathbf{k}}$ and $b^+{}_{n\mathbf{k}}$. The normal product $N(AB\cdots Z)$ is defined to be an arrangement in which all the $b^+{}_{n\mathbf{k}}$ are to the left of the $b_{n\mathbf{k}}$ and the product has the sign of the permutation of the original order required to give the final arrangement. The order of the creation or annihilation operators among themselves is of no significance except in regard to sign.

Consider a specific normal product of the form

$$\Theta = (\prod_i b^+{}_{n_i\mathbf{k}_i})(\prod_j b_{n_j\mathbf{k}_j}). \tag{8.1.97}$$

The expectation value of this normal product in the noninteracting ground state must vanish,

$$\langle 0 \mid \Theta \mid 0 \rangle = 0. \tag{8.1.98}$$

Further, for specific excited states $\mid l \rangle$, $\mid l' \rangle$ of the type (8.1.95), the matrix element $\langle l' \mid \Theta \mid l \rangle$

$$\langle l' \mid \Theta \mid l \rangle = \langle 0 \mid (\prod_m b_{n_m\mathbf{k}_m})\Theta \prod_p b^+{}_{n_p\mathbf{k}_p} \mid 0 \rangle \tag{8.1.99}$$

will vanish unless all the $n_i\mathbf{k}_i$ are contained in the $n_m\mathbf{k}_m$ and the $n_j\mathbf{k}_j$ are contained in the $n_p\mathbf{k}_p$.

It is necessary to decompose a given time ordered product into a sum of normal products. We shall study this process first for a product of two operators, A and B. We define the difference between the time ordered product of two operators and the normal product to be the contraction, which is denoted by $\overline{AB}$. Thus

$$\overline{AB} = T(AB) - N(AB). \tag{8.1.100}$$

From the definitions of T and N it is seen that if A and B anticommute, $\{A, B\} = 0$, the contraction $\overline{AB} = 0$. The contraction of two creation or annihiliation operators in the interaction picture is an ordinary function, so that for such operators,

$$\overline{AB} = \langle 0 \mid \overline{AB} \mid 0 \rangle = \langle 0 \mid T(AB) \mid 0 \rangle. \tag{8.1.101}$$

The contraction of an annihilation operator and a creation operator is of particular importance. This can be evaluated using (8.1.101) and (8.1.75). The result can be summarized succinctly by introducing the single-particle Green's function $\mathcal{G}^{(0)}$ for a noninteracting system which is given by (8.1.28)

$$\overline{c^+{}_{n\mathbf{k}}(t)c_{n'\mathbf{k}'}(t')} = -i\,\mathcal{G}_{nn'}{}^{(0)}(\mathbf{k}, \mathbf{k}', t - t'). \tag{8.1.102}$$

The decomposition of an arbitrary time ordered product of creation and annihiliation operators into a sum of normal products and contractions is described by *Wick's theorem.* This theorem states that the time ordered product of N operators $F_1\cdots F_n$ is related to the normal product by

$$T(F_1F_2\cdots F_N) = N(F_1F_2\cdots F_N) + \sum_{i<j} \lambda_{ij}\overline{F_iF_j}N_{ij}(F_1\cdots F_N) + \sum_{i<j,k<l} \lambda_{ijkl}\overline{F_iF_j}\,\overline{F_kF_l}N_{ijkl}(F_1\cdots F_N) + \cdots + \sum_{i_1<i_2,i_{N-1}<i_N} \lambda_{i_1\cdots i_N}\overline{F_{i_1}F_{i_2}}\cdots\overline{F_{i_{N-1}}F_{i_N}}. \qquad (8.1.103)$$

In this expression, the coefficients $\lambda_{i\cdots}$ are ±1 according as the permutation required to bring the contracted operators together is even or odd, and $N_{ijkl}(F_1\cdots F_N)$ is the normal product with the operators F_i, F_j, F_k, and F_l deleted. We will not give the proof here (see Wick, 1950, Schweber, 1961, or Nozières, 1964). In words, the theorem states that a time ordered product is a sum of normal products and contractions: one contracts all possible different pairs of operators, first single pairs, then two pairs at a time and so on. Each contraction is multiplied by the normal product of the remaining uncontracted operators, and carries the sign of the permutation of the original order which associates the contracted operators. It follows that

$$\langle 0 \mid T(F_1F_2\cdots F_N) \mid 0\rangle = \sum \lambda_{i_1\cdots i_N}\overline{F_{i_1}F_{i_2}}\cdots\overline{F_{i_{N-1}}F_{i_N}}, \qquad (8.1.104)$$

the expectation value in the noninteracting ground state is simply the sum of all possible contractions of distinct pairs of operators, taken with the appropriate sign.

8.1.7 *Diagrammatic Representation*

Wick's theorem forms the basis for the diagrammatic representation of matrix elements. Let us consider the nth order term $\langle 0 \mid U_n(t, t') \mid 0\rangle$. According to Eq. (8.1.72) this term contains n factors V_{I}; and each factor of V_{I} involves two creation operators, two annihilation operators, and one matrix element of the interaction. Ultimately there are summations over momenta and integrations over time to be performed. As a consequence of Wick's theorem, pairs involving a creation and an annihilation operator are to be contracted: each such contraction will introduce a factor of the free particle Green's function according to (8.1.102).

In order to bring order out of chaos, we proceed to construct diagrams as follows. Select a vertical axis to represent increasing time. An interaction V_{I} occurs at a specific time. This will be drawn as a dashed horizontal line. The states connected by this interaction are represented by directed solid

Fig. 8.1.1. An interaction vertex.

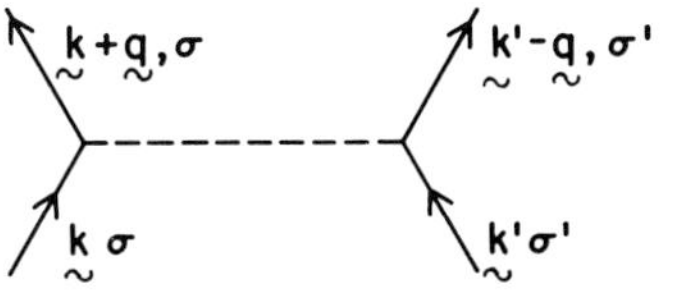

lines with a vertical component entering (in the case of an annihilation operator) or leaving (creation operator) the points connected by the dashed line. The result is a "vertex."

An interaction vertex is shown in Fig. 8.1.1. Lines are labeled with wave vector and spin indices according to Eq. (8.1.76). A band index could be added to complete the specification of states. The spin index is unchanged at a vertex, the wave vector changes in accord with the Fourier component of the potential involved, however, the total wave vector of states entering and leaving the vertex is constant.

There will be n vertices of this general type in an nth-order diagram. The lines entering and leaving the vertices must be coupled together so as to leave (in the case of diagrams contributing to $\langle 0 \mid U \mid 0 \rangle$) no unconnected lines. Diagrams contributing to $\mathcal{G}$ are constructed similarly, except that there will be two open ends or unconnected lines corresponding to the state

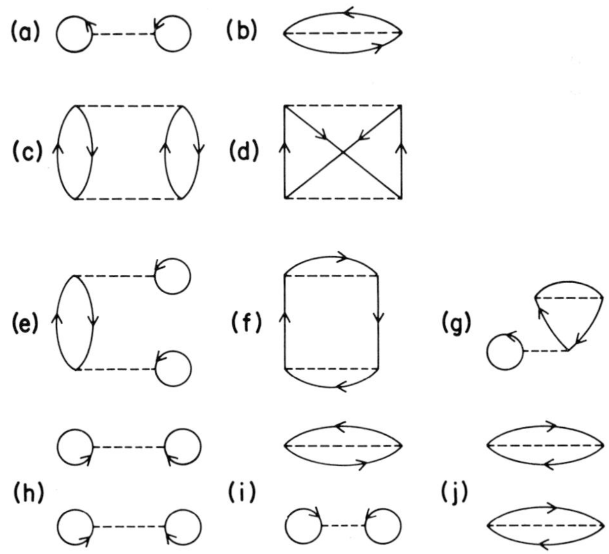

Fig. 8.1.2. Some diagrams contributing to $\langle 0 \mid U \mid 0 \rangle$. All first- and second-order diagrams are included.

being propagated by the Green's function which is being calculated. Each connection represents a contraction of a creation and an annihilation operator according to Wick's theorem. Since all indices on the contracted pair must be the same in order to obtain a nonzero result [Eq. (8.1.102)], each line can be characterized by a single set of indices. Each line has a direction, which is indicated by an arrow, from the creation operator to the annihilation operator. The sense of the arrows need not be up (as in the case of Fig. 8.1.1). If the creation operator is at the earlier time, the arrow points up, and an electron is propagated. Conversely, if the annihilation operator is earlier, the arrow points down and a hole is propagated. It may happen that a pair of creation and annihilation operators at the same time have the same indices. In such a case, we join the electron and hole lines to make a bubble.

Some graphs which contribute $\langle 0 \mid U \mid 0 \rangle$ are shown in Fig. 8.1.2. Diagrams need not be completely connected: instead they may contain disconnected pieces, as is the case for (g) and (h) in the figure. Some diagrams contributing to the Green's function are shown in Fig. 8.1.3. These also involve connected and disconnected graphs.

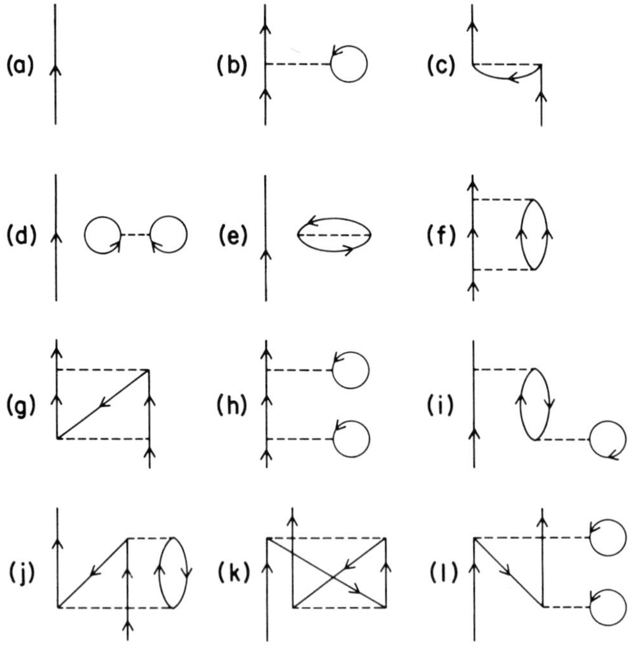

Fig. 8.1.3. Some diagrams contributing to the single-particle Green's function: (a) zeroth order, (b)–(e) first order; (f)–(l) some (not all) second-order terms.

Some reflection on the construction of diagrams should lead to the conclusion that for each basic connected diagram (such as Fig. 8.1.3f) there will exist, in higher order, diagrams in which the basic one is supplemented by disconnected pieces; these disconnected pieces include *all* diagrams contributing to $\langle 0 \mid U \mid 0 \rangle$. Each diagram corresponds to some term in the relevant perturbation expansion. Therefore it has a value, and we may legitimately speak of the addition of diagrams. In the case of a disconnected diagram, the values of the pieces are multiplied to obtain the total value. As a result, it is easy to see that in the series for the Green's function, each connected diagram is multiplied by the full $\langle 0 \mid U \mid 0 \rangle$. However, the denominator of (8.1.92) is just $\langle 0 \mid U \mid 0 \rangle$; and thus these factors cancel. We arrive at the very important result that the Green's function can be expressed as a sum of connected (or linked) diagrams only.

A similar result is valid for the ground state energy. To see this, we must investigate the structure of $\langle 0 \mid U \mid 0 \rangle$ in more detail. Let us focus attention on a particular linked diagram, for example, Fig. 8.1.2c. In higher orders, we will find that this diagram is repeated; and it will occur in conjunction with other linked diagrams to which it is not connected. It can be shown (see Nozières, 1964; Mattuck, 1967) that, including all numerical factors, the quantity $R(t)$, Eq. (8.1.77) is given as the exponential of the sum of all linked diagrams with no free ends

$$R(t) = \exp\Big[\sum_{\mathrm{L}} \langle 0 \mid U_{\mathrm{L}}(t, 0) \mid 0 \rangle\Big] \tag{8.1.105}$$

in which U_{L} is the contribution to U from a specific linked diagram. The ground state energy requires the logarithm of $R(t)$, hence the exponential disappears, and we are left with just the sum of contributions from linked diagrams.

We are now ready to investigate the evaluation of the contribution of a diagram, and to specify more precisely what diagrams must be drawn and how they are to be drawn. It would appear that in the nth order there would be an overall factor of $(1/n!)$ [from Eq. (8.1.73)]; however, there are normally $n!$ ways of assigning the factors t_i to the vertices of a diagram. The factor of $(n!)^{-1}$ thus cancels (unless permutation of some of the vertices should fail to yield a distinct graph—as this occurs only for diagrams with unlinked parts, this possibility can now be ignored. Hence, we need draw only graphs which are not topologically equivalent with respect to the time order of the vertices.

Equation (8.1.76) suggests that each vertex should contribute a factor of $\frac{1}{2}V(\mathbf{q})$, where $\mathbf{q}$ is the wave vector transfer. However, the factor of $\frac{1}{2}$ can be shown to be canceled by a diagram in which the particular interaction line has been twisted (or rotated) through 180°. Thus, we may discard twisted diagrams if we simply replace $V/2$ by V. It is seen from (8.1.73) that a factor of $-i/\hbar$ should accompany each V.

Each electron or hole line represents a contraction. If the line runs from time t_1 to time t_2, (8.1.102) indicates that we should insert a factor $i\mathcal{G}^{(0)}(\mathbf{k}, t_2 - t_1)$. [We have dropped the $\mathbf{k}'$ symbol in (8.1.102) since $\mathbf{k} = \mathbf{k}'$. Band and spin indices are understood.] The momenta, spins, and band indices of internal lines (lines which both begin and end at vertices in the diagram) are to be summed, and the times are to be integrated from $-\infty(1 - i\eta)$ to $\infty(1 - i\eta)$. Finally, we must specify how to treat a line which begins and ends at the same time (same vertex). Such a line may be closed on itself: a "bubble." Since the line "begins" with a creation and "ends" in an annihilation, it follows from (8.1.2), for example, that we require an infinitesimal negative time. Thus a bubble gives rise to a factor $i\mathcal{G}^{(0)}(\mathbf{k}, -\delta)$, in which δ is small. Finally, we must consider the question of possible minus signs generated in rearranging the operators to form contractions in accord with Wick's theorem. This can be shown (but we will not do it) to require a factor of (-1) for each closed Fermion loop (and has no other effect). Diagrams 8.1.3b, d–f, h–l have closed loops.

As an example of the preceding considerations, we write down the contribution from the diagram 8.1.3b. Bands and spins are neglected. Let the wave vector of the incoming and outgoing lines in this diagram be $\mathbf{k}$; that associated with the bubble $\mathbf{k}'$. The result gives a contribution to $\mathcal{G}(\mathbf{k}, t_2 - t_1)$ of the form

$$(-1) \sum_{k'<k_F} \int_{-\infty}^{\infty} [i\mathcal{G}^{(0)}(\mathbf{k}, t - t_1)][(-i/\hbar)V(q = 0)] \times [i\mathcal{G}^{(0)}(\mathbf{k}', -\delta)][i\mathcal{G}^{(0)}(\mathbf{k}, t_2 - t)]\, dt. \tag{8.1.106}$$

The initial factor of (-1) occurs because the bubble counts as a closed loop; the summation over $\mathbf{k}'$ is restricted to $k' < k_F$ (k_F is the radius of the Fermi sphere) by (8.1.28), and $\mathbf{q}$ is set equal to zero in $V(\mathbf{q})$ because there is no wave vector transfer.

It is normally more useful to work with the Fourier transform of the Green's function with respect to time than with the time-dependent function itself. This is desirable since the spectral weight function can then be obtained simply via (8.1.27b). Furthermore there is considerable convenience in calculations, since the time integrations are typically of the folding type, such as

$$\int_{-\infty}^{\infty} f_1(t_1 - t) f_2(t)\, dt.$$

The Fourier transform of such an integral with respect to t_1 is just the product of the Fourier transforms of the factors in the integrand.

The transformation procedure is as follows. The Fourier transformed free particle Green's function, according to (8.1.14), is

$$\mathcal{G}^{(0)}(\mathbf{k}, t) = (2\pi\hbar)^{-1} \int_{-\infty}^{\infty} d\epsilon\, \mathcal{G}^{(0)}(\mathbf{k}, \epsilon) \exp(-i\epsilon t/\hbar).$$

Four lines are involved at each vertex, two entering and two leaving. Each such line will contribute a time dependence of the form $\exp(-i\epsilon_i t/\hbar)$ [each line will have a specific parameter ϵ_i]. In the case of a bubble, where a line enters and leaves a vertex, a factor $e^{i\epsilon\delta}$ is included. The time integration at a vertex is now of the form

$$\int_{-\infty}^{\infty} dt \exp[i(\epsilon_1 + \epsilon_2 - \epsilon_3 - \epsilon_4)t/\hbar] = 2\pi\hbar\, \delta(\epsilon_1 + \epsilon_2 - \epsilon_3 - \epsilon_4). \tag{8.1.107}$$

This relation recalls a delta function of energy conservation which occurs in the calculation of transition rates and we will say that the "energy parameter" is conserved at each vertex. However, it is important not to confuse the energy parameter with the single-particle energy of the state propagated by the line. Instead of integrating over the times of the individual vertices, we must now integrate over the energy parameter associated with each internal line. In accord with (8.1.14), each such integration is to be accompanied by a factor of $(2\pi)^{-1}$. It is convenient to simplify the notation by introducing units in which $\hbar = 1$.

From this point, we will work exclusively with the Fourier transformed diagrams for the Green's function. The rules for the evaluation of diagrams are summarized in Table I ("Goldstone rules"). The + and − signs on the free particle Green's function apply to electron and hole lines, respectively: the values of these contributions are taken from Eq. (8.1.29). Use of the contributions given in this table for the nonpropagating line removes the necessity to introduce and integrate over the energy parameter associated with such a line. This factor is obtained if one proceeds as mentioned earlier by considering a contribution

$$\begin{aligned} i\int_{-\infty}^{\infty} \mathcal{G}(\mathbf{k}, \epsilon) e^{i\epsilon\delta}(d\epsilon/2\pi) &= (i/2\pi)\int_{\infty}^{\infty} 1/[\epsilon - E(\mathbf{k}) \pm i\eta] e^{i\epsilon\delta}\, d\epsilon \\ &= -1 \qquad \text{if} \quad E(\mathbf{k}) > \mu, \\ &= 0 \qquad \text{if} \quad E(\mathbf{k}) < \mu. \end{aligned} \tag{8.1.108}$$

TABLE I

Name	Part	Contribution		
		Goldstone Rules	Feynman Rules	Finite Temperature
Electron line		$i\mathcal{G}^{(+)}_{l\sigma}(\underset{\sim}{k},\epsilon) = \frac{i}{\epsilon - E_{l\sigma}(\underset{\sim}{k}) + i\delta}$	$\frac{i}{\epsilon - E_{l\sigma}(\underset{\sim}{k}) + i\delta_{l\sigma}(\underset{\sim}{k})}$	$-\frac{1}{i\omega_n + \mu - E_{l\sigma}(\underset{\sim}{k})}$
Hole line		$i\mathcal{G}^{(-)}_{l\sigma}(\underset{\sim}{k},\epsilon) = \frac{i}{\epsilon - E_{l\sigma}(\underset{\sim}{k}) - i\delta}$	$\delta_{l\sigma}(\underset{\sim}{k}) = +\delta$ for $E_{l\sigma}(\underset{\sim}{k}) > \mu$ $= -\delta$ for $E_{l\sigma}(\underset{\sim}{k}) < \mu$	$\omega_n = (2n+1)\frac{\pi}{\beta}$
Non propagating lines (bubble)	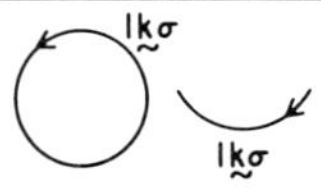	-1 if $E_{l\sigma}(\underset{\sim}{k}) < \mu$ 0 if $E_{l\sigma}(\underset{\sim}{k}) > \mu$	same	$-f[E_{l\sigma}(\underset{\sim}{k})] = -[e^{\beta(E_{l\sigma}(\underset{\sim}{k})-\mu)}+1]^{-1}$
Vertex	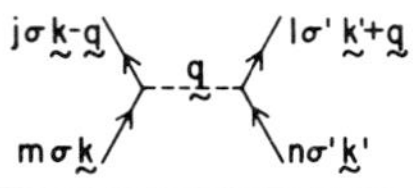	$-iV_{jlmn}\ (q)$	same	$-V_{jlmn}\ (q)$
Closed loop	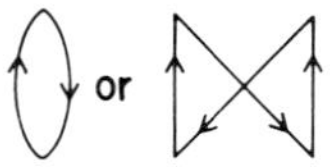	-1	same	same
Each intermediate	$l\underset{\sim}{k}\sigma\omega$	$\sum_{l\underset{\sim}{k}\sigma} \int \frac{d\epsilon}{2\pi}$	same	$\sum_{l\underset{\sim}{k}\sigma} \frac{1}{\beta} \sum_{n=-\infty}^{\infty}$

The exponential factor has forced us to close the contour of integration in the upper half plane.

Certain apparent violations of the exclusion principle are to be overlooked in the diagrammatic procedure. For example, in Fig. 8.1.3j, the initial (or final) state $\mathbf{k}, \epsilon$ apparently contains two particles; however, the diagram is legitimate and must be evaluated. The situation is as follows. Before the unlinked diagrams were eliminated, this diagram was exactly canceled by an unlinked diagram Fig. 8.1.4 whose value differs only by an algebraic sign, due to the presence of an additional closed loop in the unlinked diagram (Goldstone, 1957). The elimination of unlinked diagrams leaves a diagram which violates the exclusion principle: it cannot be discarded if the linked cluster expansion is to be maintained.

It is desirable to present a few examples of the applications of these rules. Accordingly, we give below the contributions to the Green's function from the diagrams of Figs. 8.1.3b, c, f, g. We will continue to suppress band and spin indices. In each case, the incoming line will be characterized by wave vector $\mathbf{k}$ and energy parameter ϵ.

$$i\mathcal{G}^{(+)}(\mathbf{k}, \epsilon)\Big[\sum_{k<k_F} V(0)\Big]\mathcal{G}^{(+)}(\mathbf{k}, \epsilon) \qquad \text{(Fig. 8.1.3b)} \qquad (8.1.109)$$

$$-i\mathcal{G}^{(+)}(\mathbf{k}, \epsilon)\Big[\sum_{k'<k_F} V(\mathbf{k}-\mathbf{k}')\Big]\mathcal{G}^{(+)}(\mathbf{k}, \epsilon) \qquad \text{(Fig. 8.1.3c)} \qquad (8.1.110)$$

$$i\mathcal{G}^{(+)}(\mathbf{k}, \epsilon)\left[\sum_{\mathbf{p},\mathbf{q}} \iint d\omega/(2\pi)\, d\omega'/(2\pi)\, V(\mathbf{q})^2 \mathcal{G}^{(+)}(\mathbf{k}-\mathbf{q}, \epsilon-\omega)\right.$$
$$\left.\times\, \mathcal{G}^{(-)}(\mathbf{p}, \omega')\mathcal{G}^{(+)}(\mathbf{p}+\mathbf{q}, \omega'+\omega)\right]\mathcal{G}^{(+)}(\mathbf{k}, \epsilon)$$
$$\text{(Fig. 8.1.3f)} \qquad (8.1.111)$$

$$-i\mathcal{G}^{(+)}(\mathbf{k}, \epsilon)\left[\sum_{\mathbf{p},\mathbf{q}} \iint d\omega/(2\pi)\, d\omega'/(2\pi)\, V(\mathbf{q}) V(\mathbf{q}+\mathbf{p}-\mathbf{k})\right.$$
$$\left.\times\, \mathcal{G}^{(+)}(\mathbf{k}-\mathbf{q}, \epsilon-\omega)\mathcal{G}^{(-)}(\mathbf{p}, \omega')\mathcal{G}^{(+)}(\mathbf{p}+\mathbf{q}, \omega+\omega')\right]\mathcal{G}^{+}(\mathbf{k}, \epsilon)$$
$$\text{(Fig. 8.1.3g).} \qquad (8.1.112)$$

These expressions actually give the contribution to $(i\mathcal{G})$, since the values assigned to the lines include a factor of i. A single factor of i may be discarded if we desire the contribution to $\mathcal{G}$ alone. The integrals over ω and ω' in (8.1.111) and (8.1.112) are not difficult. They may be made into

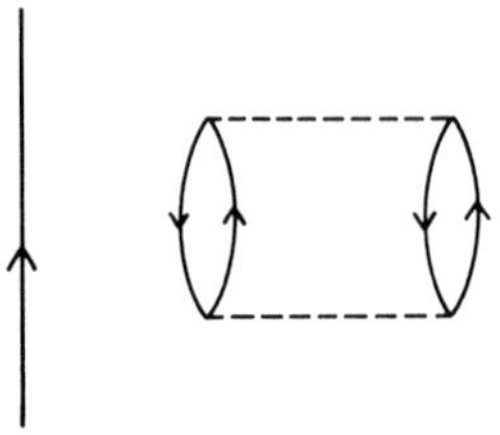

Fig. 8.1.4. Unlinked graph which cancels Fig. 8.1.3 before unlinked graphs are removed.

contour integrals and evaluated by the residue theorem. It is necessary to pay careful attention to the location of the poles: For example, (8.1.111) simplifies to

$$i\mathcal{G}^{(+)}(\mathbf{k}, \epsilon)\Big[\sum_{\mathbf{p}<k_F,\mathbf{q}} [V(\mathbf{q})]^2/\{\epsilon + E(\mathbf{p}) - E(\mathbf{p}+\mathbf{q}) - E(\mathbf{k}-\mathbf{q})\}\Big] \times \mathcal{G}^{(+)}(\mathbf{k}, \epsilon). \tag{8.1.113}$$

The structure of these results (8.1.109)–(8.1.113) should be noted. We obtain a function involving the potential sandwiched between two $\mathcal{G}$'s. It is useful to consider now the set of all diagrams contributing to $\mathcal{G}$ which cannot be cut into two separate pieces by removing a single particle line. Relevant examples are shown in Fig. 8.1.5. Such diagrams are called *proper self-energy diagrams.* Let us imagine summing all the diagrams of this type, and deleting the factors of $\mathcal{G}(\mathbf{k}, \epsilon)$ in front and rear. We are thus performing the sum of factors of the sort found in the square brackets in (8.1.109)–(8.1.113). The result of this summation is designated $M(\mathbf{k}, \epsilon)$, and is denoted graphically in Fig. 8.1.6 by a large circle. We can now obtain the series for $\mathcal{G}$ by repeating M shown in that drawing. The series can be

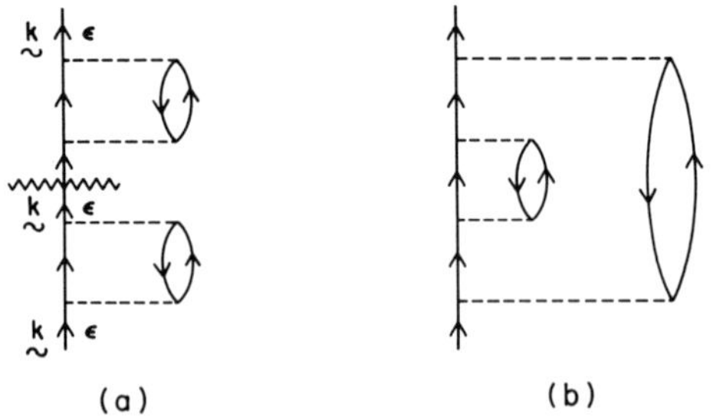

Fig. 8.1.5. (a) A diagram which can be separated into two pieces by cutting a single-particle line. (b) A diagram which cannot be separated.

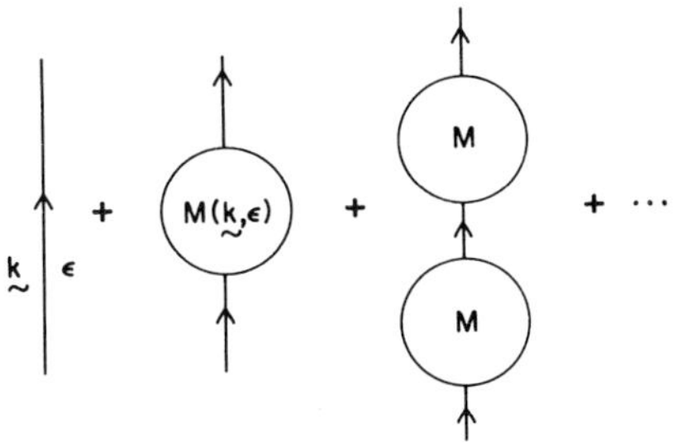

Fig. 8.1.6. The series for $\mathcal{G}$ in terms of M.

summed

$$\begin{aligned}\mathcal{G}(\mathbf{k}, \epsilon) &= \mathcal{G}^{(0)}(\mathbf{k}, \epsilon) + \mathcal{G}^{(0)}(\mathbf{k}, \epsilon)M(\mathbf{k}, \epsilon)\mathcal{G}^{(0)}(\mathbf{k}, \epsilon) \\ &\quad + \mathcal{G}^{(0)}(\mathbf{k}, \epsilon)M(\mathbf{k}, \epsilon)\mathcal{G}^{(0)}(\mathbf{k}, \epsilon)M(\mathbf{k}, \epsilon)\mathcal{G}^{(0)}(\mathbf{k}, \epsilon) + \cdots \\ &= \mathcal{G}^{(0)}(\mathbf{k}, \epsilon)\{1/[1 - M(\mathbf{k}, \epsilon)\mathcal{G}^{(0)}(\mathbf{k}, \epsilon)]\} \\ &= \{1/[\epsilon - E(\mathbf{k}) - M(\mathbf{k}, \epsilon) + i\delta]\}. \end{aligned} \quad (8.1.114)$$

This result is known as *Dyson's equation*. The poles of the exact propagator, which gives the physical single-particle energies, are solutions of the equation

$$\epsilon = E(\mathbf{k}) + M(\mathbf{k}, \epsilon). \quad (8.1.115)$$

Unfortunately, this is a transcendental equation since M depends on ϵ.

The simplest approximation to M employs just the first-order terms from (8.1.109) and (8.1.110). In this case, $M(\mathbf{k}, \epsilon)$ is independent of ϵ

$$M(\mathbf{k}, \epsilon) = M(\mathbf{k}) = \sum_{k'<k_F} [V(0) - V(\mathbf{k} - \mathbf{k}')]. \quad (8.1.116)$$

The first term of (8.1.116) corresponds to the average potential of (8.1.55) in a plane wave approximation, the second corresponds to (8.1.58). The energies which are obtained from (8.1.115) are those of the Hartree–Fock approximation

$$E_{HF} = E(\mathbf{k}) + \sum_{k'<k_F} [V(0) - V(\mathbf{k} - \mathbf{k}')]. \quad (8.1.117)$$

Diagrams which have been drawn in accord with the prescriptions presented above are known as *Goldstone diagrams*. Certain simplifications are possible. Specifically, it is possible to avoid drawing separate diagrams for electron and hole processes. To do this we simply employ the free particle Green's function $i/[\epsilon - E(\mathbf{k}) + i\,\delta(\mathbf{k})]$ in which $\delta(\mathbf{k})$ is a small positive quantity for states above the Fermi energy and a small negative quantity for hole states. Summation over all values of **k** is required.

The following rules govern the construction of diagrams by these rules (the Feynman rules):

All nth-order Green's function diagrams have n vertices and two external lines.

Vertices are joined to each other by directed lines in all possible ways subject to the condition that one line must enter and one must leave on each side of a vertex. Only diagrams which cannot be deformed into each other are by stretching or twisting are considered distinct and are included. Wave vector and energy parameter are conserved at each vertex. Diagrams with a particle and a hole in the same state are not allowed.

The evaluation of such diagrams is made in accord with Feynman rules presented in Table I.

8.1.8 *The Ground State Energy*

The graphs which contribute to the ground state energy have a time dependence such that $\lim_{t\to\infty(1-i\eta)} d \ln R(t)/dt$ exists. What happens can be illustrated by an example. Let us consider the specific case of Fig. 8.1.2c. The time-dependent formalism will be considered according to the rules described in the previous subsection. Denote the value of the diagram by $R_{2c}(t)$. (The labeling is shown in Fig. 8.1.7.)

$$R_{2c}(t) = \tfrac{1}{2}\int_0^t dt_1 \int_0^t dt_2 \sum_{\mathbf{qkp}} [-iV(\mathbf{q})]^2 \mathcal{G}^{(+)}(\mathbf{k}-\mathbf{q}, t_2-t_1)$$

$$\times\, \mathcal{G}^{(+)}(\mathbf{p}+\mathbf{q}, t_2-t_1)\mathcal{G}^{(-)}(\mathbf{p}, t_1-t_2)\mathcal{G}^{(-)}(\mathbf{k}, t_1-t_2). \quad (8.1.118)$$

The initial factor of $\frac{1}{2}$ occurs because the counting procedures for diagrams contributing to the ground state energy yield slightly different results than in the case of the Green's function. We saw there that factors of $\frac{1}{2}$ associated with the potential are eliminated by counting the effect of twisted diagrams.

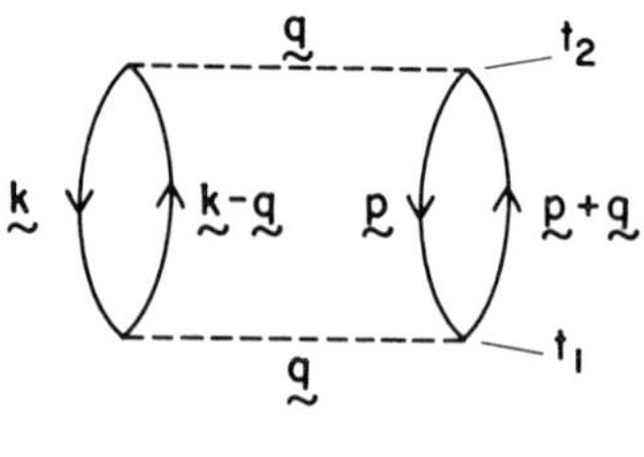

Fig. 8.1.7. A second-order contribution to the ground state energy.

If, as is possible in the case of the ground state energy, the diagram is completely symmetric left and right, rotating all interactions does not produce a new diagram, and an overall factor of $\frac{1}{2}$ remains. The values of the Green's functions are obtained from (8.1.28). Let us single out the time-dependent factors and the time integration. This contributes

$$\int_0^t dt_1 \int_0^t dt_2 \exp[-iD(t_2 - t_1)] = \int_0^t dt_1 \int_0^{t-t_1} d(t_2 - t_1) \exp[-iD(t_2 - t_1)] \tag{8.1.119}$$

in which $D = E(\mathbf{k} - \mathbf{q}) + E(\mathbf{p} - \mathbf{q}) - E(\mathbf{k}) - E(\mathbf{p})$. The integral gives

$$-(i/D)[t - (iD)^{-1}(1 - e^{-iDt})]. \tag{8.1.120}$$

We differentiate with respect to time, then take the limit as $t \to \infty(1 - i\eta)$. The quantity D is greater than zero since the energies of excited electrons are above the Fermi energy, while those of holes are below it. Thus the exponential term vanishes and the contribution of R_{2c} to E_g is ΔE, where, according to (8.1.78),

$$\Delta E = \frac{1}{2} \sum_{\mathbf{kpq}} \frac{[V(\mathbf{q})]^2[1 - n(\mathbf{k} - \mathbf{q})][1 - n(\mathbf{p} + \mathbf{q})]n(\mathbf{k})n(\mathbf{p})}{E(\mathbf{k}) + E(\mathbf{p}) - E(\mathbf{k} - \mathbf{q}) - E(\mathbf{p} + \mathbf{q})}. \tag{8.1.121}$$

This is just the result which is obtained by computing the second-order direct term according to ordinary Raleigh–Schrödinger perturbation theory.

There are many general features about this result. Suppose the vertices of each graph are ordered so that $t > t_n > t_{n-1} > \cdots > t_1$. The propagators will combine so that in the time interval between t_{r-1} and t_r, the time dependence in the integration is of the form $\exp[-iD_{r-1}(t_r - t_{r-1})]$, where D_{r-1} is the difference between the sum of the energies of the electron states and that of the hole states. The integration gives simply a factor of $(-i/D_{r-1})$. The leftover exponential factors ultimately disappear with all derivatives since the upper time limit has an infinite negative imaginary part. The final time integration $\int_0^t dt_1$ does not have an associated exponential factor, and leads to $\ln R \propto t$. The coefficient of t determines the interaction contribution to the ground state energy.

We have already eliminated all unlinked graphs from the series for $\ln R$ and for the Green's functions. In addition, we may discard all those graphs like Figs. 8.1.2e, f, g, 8.1.3i, l which have in effect a particle and a hole in

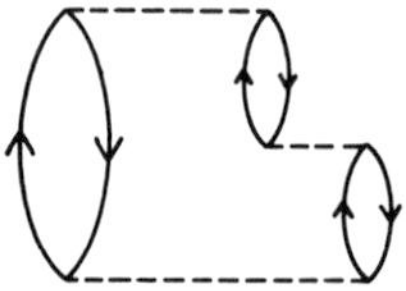

Fig. 8.1.8. A ground state energy diagram which is not completely symmetric.

the same state. Thus, through second order, the calculation of the ground state energy requires only Figs. 8.1.2a–d.

The following rules are obtained for ground state energy diagrams after the time integrations have been performed:

(1) For each time interval between two vertices, insert an energy denominator which is [sum of hole energies − sum of particle energies]$^{-1}$.
(2) Each vertex contributes a matrix element of the interaction between the states which enter and leave.
(3) A factor of (-1) is required for each closed loop.
(4) An additional factor of (-1) is contributed by each hole line.
(5) If the diagram is completely symmetric (as is true for Figs. 8.1.2a–d, but not in the case of Fig. 8.1.8), insert a factor of $\frac{1}{2}$.
(6) Finally, sum over all wave vectors, restricting particle lines to energies greater than the Fermi energy and hole lines to energies below it. For convenience, we will frequently omit factors of $\mathfrak{N}^{-1/2}$ and $\mathfrak{N}^{-1}$ ($\mathfrak{N}$ is the number of unit cells) and simply convert the sums to integrals according to

$$\sum_{\mathbf{k}} \rightarrow (2\pi)^{-3} \int d^3k.$$

8.1.9 *Finite Temperatures*

We now consider the extension of the previous discussion to apply at finite temperatures (Luttinger and Ward, 1960). It turns out that the formalism can be carried over to nonzero temperatures with remarkably little change. The basic idea is to alter the definition of the Green's function (8.1.1) from an average of a certain operator in the ground state to a statistical average involving the density matrix.

It is desirable to work with the grand canonical ensemble. This means that instead of considering just the usual Hamiltonian H for the system in defining the density matrix ρ, we introduce the grand Hamiltonian $\mathcal{H}$,

$$\mathcal{H} = H - \mu N \tag{8.1.122}$$

in which μ is the chemical potential and N is the operator which specifies the total number of particles in the system. Let $\beta = 1/KT$. The density matrix is

$$\rho = \exp(-\beta\mathcal{H}). \tag{8.1.123}$$

The grand partition function $\mathcal{Q}$ is related to ρ by

$$\mathcal{Q} = \mathrm{Tr}(\rho). \tag{8.1.124}$$

The average value of an operator $\mathcal{O}$ is

$$\langle\mathcal{O}\rangle = \mathcal{Q}^{-1}\,\mathrm{Tr}(\mathcal{O}\rho). \tag{8.1.125}$$

It is important to note that the trace is taken over states with an arbitrary number of particles present. However, we are interested in a system with a specified average number of particles N such that

$$\bar{N} = \langle N\rangle. \tag{8.1.126}$$

The chemical potential μ is to be adjusted so that (8.1.126) holds.

The density matrix ρ satisfies Bloch's equation,

$$\partial\rho/\partial\beta = -\mathcal{H}\rho = -(H - \mu N)\rho, \tag{8.1.127}$$

as is seen by differentiating (8.1.123). This equation is analogous to the time-dependent Schrödinger equation if one makes the correspondence $\beta \to it/\hbar$, $\mathcal{H} \to H$, $\rho \to \Psi$. The resemblance between the Bloch equation and the time-dependent Schrödinger equation (with an imaginary time) is basic to the development of a finite temperature theory.

We wish to define the single-particle Green's function. In doing this, we will focus attention on the Green's function in **k** space, and suppress band and spin indices. It will be further supposed that $\mathcal{G}$ is diagonal in **k** as a consequence of the conservation of wave vector in the interactions. The temperature-dependent single-particle Green's function is then

$$\mathcal{G}^{(T)}(\mathbf{k}, t_2 - t_1) = -i\langle T\{c_{\mathbf{k}}(t_2)c_{\mathbf{k}}^{+}(t_1)\}\rangle \tag{8.1.128}$$

in which the average is defined by (8.1.125). It will also be important to introduce the imaginary time Green's function

$$\mathcal{G}(\mathbf{k}, \tau_2 - \tau_1) = -i\langle T\{c_{\mathbf{k}}(\tau_2)c_{\mathbf{k}}^{+}(\tau_1)\}\rangle. \tag{8.1.129}$$

The real quantity τ corresponds to $it/\hbar$. This replacement is made in all expressions involving time: thus for any operator $\mathcal{O}$, the analog of the Heisenberg representation is

$$\mathcal{O}(\tau) = \exp(\mathcal{H}\tau)\,\mathcal{O}\,\exp(-\mathcal{H}\tau). \tag{8.1.130}$$

The operators $c_{\mathbf{k}}(\tau)$ have been transformed in this way. The operator T now orders the operators so that the values of τ_2, τ_1 increase from right to left.

As we will see subsequently, the range of variation of τ_1 and τ_2 is confined to $0 < \tau_1, \tau_2 < \beta$. This causes substantial complications.

In the absence of interactions, we are concerned with the operator

$$\mathcal{H}_0 = H_0 - \mu N = \sum_{\mathbf{k}} [E(\mathbf{k}) - \mu] c_{\mathbf{k}}^+ c_{\mathbf{k}}. \tag{8.1.131}$$

The propagators for the noninteracting system are defined with respect to $\mathcal{H}_0$. We find, with the aid of (8.1.75),

$$\begin{aligned}\mathcal{G}^{(0)}(\mathbf{k}, t_2 - t_1) &= -i \exp\{-i[E(\mathbf{k}) - \mu](t_2 - t_1)/\hbar\} \langle c_{\mathbf{k}} c_{\mathbf{k}}^+ \rangle_0 \qquad (t_2 > t_1),\\ &= i \exp\{-i[E(\mathbf{k}) - \mu](t_2 - t_1)/\hbar\} \langle c_{\mathbf{k}}^+ c_{\mathbf{k}} \rangle_0 \qquad (t_2 < t_1).\end{aligned}$$

The subscript 0 indicates that the thermal averages are taken with $\mathcal{H}_0$. Since the states of different $\mathbf{k}$ are independent of each other in the absence of interactions, the average can actually be constructed with respect to states of a single $\mathbf{k}$ only. Thus

$$\begin{aligned}\langle c_{\mathbf{k}}^+ c_{\mathbf{k}} \rangle_0 &= \sum_{n=0}^{1} n \exp\{-\beta n[E(\mathbf{k}) - \mu]\} \Big/ \sum_{n=0}^{1} \exp\{-\beta n[E(\mathbf{k}) - \mu]\}\\ &= \{1 + \exp\{\beta[E(\mathbf{k}) - \mu]\}\}^{-1} = f(\mathbf{k}).\end{aligned} \tag{8.1.132}$$

The function $f(\mathbf{k})$ is the usual Fermi distribution. Thus we have

$$\begin{aligned}\mathcal{G}^{(0)}(\mathbf{k}, t) &= -i \exp\{-i[E(\mathbf{k}) - \mu]t/\hbar\}[1 - f(\mathbf{k})] \qquad (t > 0),\\ &= i \exp\{-i[E(\mathbf{k}) - \mu]t/\hbar\} f(\mathbf{k}) \qquad (t < 0).\end{aligned} \tag{8.1.133}$$

This result is a straightforward generalization of (8.1.28).

We wish to obtain a diagrammatic expansion for the Green's function. This will be done using the imaginary time Green's function, defined by (8.1.129). The development of perturbation theory in subsection 8.1.4 can be taken over almost directly with $\mathcal{H}_0$ replacing H_0. We are concerned with the imaginary time development operator $U(\beta, 0) \equiv U(\beta)$. This is expanded in a perturbation series

$$U(\beta) = 1 + \sum_{n=1}^{\infty} U_n(\beta) \tag{8.1.134}$$

in which

$$U_n(\beta) = (1/n!)(-1)^n \int_0^\beta d\tau_1 \cdots \int_0^\beta d\tau_n \, T[V_I(\tau_1) \cdots V_I(\tau_n)]. \tag{8.1.135}$$

We may now proceed to relate the imaginary time Green's function to the operator $U(\beta)$ as was done in subsection 8.1.5. The procedure is simpler here than before in one sense: for finite β, there are no convergence problems associated with integrals over an infinite time interval. Equation (8.1.86) is replaced by

$$\mathcal{G}(\mathbf{k}, \tau_2 - \tau_1) = -i\langle T\{U(\beta)c_{\mathbf{k}}(\tau_2)c_{\mathbf{k}}^+(\tau_1)\}\rangle_0/\langle U(\beta)\rangle_0. \qquad (8.1.136)$$

It is also useful to define the temperature-dependent vacuum amplitude

$$R(\beta) = \langle U(\beta)\rangle_0. \qquad (8.1.137)$$

Equations (8.1.136) and (8.1.137) are combined with (8.1.135) to give the perturbation expansion for $\mathcal{G}$ and R.

The principal difference between the zero temperature and finite temperature procedures arises from the fact that the quantities τ_i are restricted to the finite interval 0 to β_i Therefore, use of a Fourier transform to introduce the energy parameter ϵ is not permitted. Instead, we must employ a Fourier series representation. Since nothing yet concerns values of $\tau_2 - \tau_1$ outside the range from $-\beta$ to β, we are free to extend $\mathcal{G}$ by defining it as a periodic function in a suitable manner.

Consider Eq. (8.1.129) in a case in which $\tau_1 > \tau_2$:

$$\begin{aligned}\langle T\{c_k(\tau_2)c_k^+(\tau_1)\}\rangle &= -\mathrm{Tr}[\exp(-\beta\mathcal{H})c_{\mathbf{k}}^+(\tau_1)c_{\mathbf{k}}(\tau_2)] \\ &= -\mathrm{Tr}[c_{\mathbf{k}}(\tau_2)\exp(-\beta\mathcal{H})c_{\mathbf{k}}^+(\tau_1)] \\ &= -\mathrm{Tr}[\exp(-\beta\mathcal{H})\exp(\beta\mathcal{H})c_{\mathbf{k}}(\tau_2)\exp(-\beta\mathcal{H})c_{\mathbf{k}}^+(\tau_1)] \\ &= -\mathrm{Tr}[\exp(-\beta\mathcal{H})c_{\mathbf{k}}(\tau_2+\beta)c_{\mathbf{k}}^+(\tau_1)].\end{aligned}$$

We have used the invariance of the trace with respect to a cyclic permutation of the operators: $\mathrm{Tr}(ABC) = \mathrm{Tr}(CAB)$. Note that since $\tau_1, \tau_2 < \beta$, $\tau_2 + \beta > \tau_1$. Thus we should have

$$\mathcal{G}(\mathbf{k}, \tau_2 - \tau_1) = -\mathcal{G}(\mathbf{k}, \tau_2 + \beta - \tau_1). \qquad (8.1.138)$$

We now define a transformed Green's function by

$$\mathcal{G}(\mathbf{k}, \tau) = (1/\beta)\sum_{l=-\infty}^{\infty}\exp(-i\omega_l\tau)\mathcal{G}(k, \omega_l). \qquad (8.1.139)$$

In order to be consistent with (8.1.138), we require

$$\omega_l = (2l+1)\pi/\beta. \qquad (8.1.140)$$

Then the inverse to (8.1.139) is

$$\mathcal{G}(\mathbf{k}, \omega_l) = \int_0^\beta \exp(i\omega_l\tau)\mathcal{G}(\mathbf{k}, \tau)\, d\tau. \qquad (8.1.141)$$

Wick's theorem remains valid for imaginary times. We can thus proceed to set up the diagrammatic procedure just as was done at zero temperature. The free particle propagators are obtained immediately from (8.1.133):

$$\mathcal{G}^{(0)}(\mathbf{k}, \tau_2 - \tau_1) = -i \exp\{-[E(\mathbf{k}) - \mu](\tau_2 - \tau_1)\}\{[1 - f(\mathbf{k})]\theta(\tau_2 - \tau_1) + f(\mathbf{k})\theta(\tau_1 - \tau_2)\}. \quad (8.1.142)$$

The transform of this object according to (8.1.141) is obtained by a straightforward integration

$$\mathcal{G}^{(0)}(\mathbf{k}, \omega_l) = i/[i\omega_l + \mu - E(\mathbf{k})]. \quad (8.1.143)$$

We note that the same free propagator is used for both electron and hole lines.

For finite temperatures, the Fermi function is not a step function, but is rounded, or smeared over an energy interval of order β^{-1} on each side of μ. States in this energy region are neither completely occupied nor completely empty. As a result, it is possible to have both electrons and holes in the same states. Diagrams containing such situations which were excluded previously must now be included in the summation (Examples: Figs. 8.1.3i, l). We still include only linked diagrams. The series (8.1.134), (8.1.135) for $U(\beta)$ now contains no factor of i^n: hence, the factor of $(-i)$ previously included with each vertex is to be replaced by (-1). The integration over the energy parameter is to be replaced by a summation over ω_l. Nonpropagating lines or bubbles require a factor of $-f(\mathbf{k})$ which is the direct generalization of the zero-temperature prescription.

The argument leading to the Dyson equation (8.1.114), which expresses the Green's function in terms of the proper self-energy M, is still valid as long as one adheres to the new rules

$$\mathcal{G}(\mathbf{k}, \omega_l) = i/[i\omega_l - E(\mathbf{k}) + \mu - M(\mathbf{k}, \omega_l)]. \quad (8.1.144)$$

The self-energy now depends on temperature. The lowest approximation, that of Hartree–Fock theory (Figs. 8.1.3b, c), becomes, in place of (8.1.116),

$$M(\mathbf{k}, \omega_l) = M(\mathbf{k}) = \sum_{\mathbf{k}'} [V(0) - V(\mathbf{k} - \mathbf{k}')]f(\mathbf{k}'). \quad (8.1.145)$$

We obtain from this temperature-dependent single-particle energies

$$E_{\mathrm{HF}} = E(\mathbf{k}) + \sum_{\mathbf{k}'} [V(0) - V(\mathbf{k} - \mathbf{k}')]f(\mathbf{k}'). \quad (8.1.146)$$

Diagram rules for finite temperatures are also summarized in Table I.

The function $R(t)$, defined by Eq. (8.1.77), is expressed by (here we refer to the "zero temperature problem")

$$R(t) = \langle 0 \mid \exp(iH_0t/\hbar) \exp(-iHt/\hbar) \mid 0 \rangle \qquad (8.1.147)$$

[see Eqs. (8.1.80) and (8.1.82)]. Our finite temperature procedures generalize this to

$$R(\beta) = \langle \exp(\beta\mathcal{H}_0) \exp(-\beta\mathcal{H}) \rangle_0. \qquad (8.1.148)$$

This function can be related to the grand partition function

$$\mathcal{Q} = \mathrm{Tr} \exp(-\beta\mathcal{H}) = \mathrm{Tr} \exp(-\beta\mathcal{H}_0) U(\beta) \qquad (8.1.149)$$

in which $U(\beta) = \exp(\beta\mathcal{H}_0) \exp(-\beta\mathcal{H})$. The states used in evaluating the trace in (8.1.148) are eigenstates of $\mathcal{H}_0$. The same states can be used in (8.1.149). The grand partition function for the system neglecting interactions is $\mathcal{Q}_0 = \mathrm{Tr} \exp(-\beta\mathcal{H}_0)$. Thus

$$\mathcal{Q}/\mathcal{Q}_0 = \mathrm{Tr}[\exp(-\beta\mathcal{H})]/\mathrm{Tr}[\exp(-\beta\mathcal{H}_0)] = R(\beta). \qquad (8.1.150)$$

The thermodynamic free energy is given by $F = -KT \ln \mathcal{Q}$. This gives the relation

$$F = F_0 - KT \ln R(\beta) \qquad (8.1.151)$$

in which F_0 is the free energy in the absence of interactions. The problem of calculating the ground state energy generalizes to the calculation of the free energy. The graphical form of the perturbation series for $R(\beta)$ is the same as that obtained at zero temperatures. The linked cluster theorem is still valid, but, as in the case of diagrams for the Green's function, we must include diagrams in which both an electron and a hole are present in the same state. There are some complications in the evaluation of these diagrams, which will not be discussed here (see Bloch, 1962, for further details).

8.2 Some Properties of an Electron Gas

Much work on the electron interaction problem has focused on the so-called "free electron gas," in which one considers a system of free electrons moving in a uniform background of positive charge which keeps the system electrically neutral. This model (jellium) represents a very considerable idealization with respect to real metals, in that the periodic potential which gives rise to the band structure is smeared out. However, it does allow one to focus attention on the characteristic many-body aspects of the electron–electron interaction without the complications posed by anisotropic energies and many-band effects. Since many simple metals can be adequately described by weak effective pseudopotentials there is some hope that results for the free electron gas could be used in some actual metallic systems. Unfortunately, it turns out that the problem is tractable only in the limit of high densities in which the kinetic energy is large compared to the interaction energy. This limit does not apply to real metals, and as a result, approximations of uncertain validity have to be applied to obtain results of physical interest.

Previously we have discussed the Hartree–Fock approximation in considerable detail (see Sections 4.9, 4.10, Part A, and 8.1.7). It has been observed that the Hartree–Fock approximation corresponds to a first-order perturbation treatment of the electron interaction. In this section we will investigate how this approximation can be improved. Unfortunately, we will quickly see that it is not possible to proceed on an order-by-order basis in perturbation theory because divergencies arise in the evaluation of certain graphs in second order and in each higher order. Those divergencies result from the long range of the Coulomb potential. This problem is surmounted by the technique of selective summation: it turns out to be possible to sum all graphs of certain restricted classes, including those which give rise to the characteristic divergencies. Finite results are then obtained. Useful general references include Pines (1962), Brout and Carruthers (1963), Pines and Nozières (1966), and Hedin and Lundquist (1969).

The characteristic parameter of the electron gas problem is r_s, the radius of a sphere whose volume is the average volume available for each electron,

$$r_s = [(3/4\pi)(\mathcal{V}/N)]^{1/3} \tag{8.2.1}$$

in which $\mathcal{V}$ is the volume of the entire system and N is the number of electrons. Small values of r_s correspond to high densities. The Fermi wave vector k_F is related to r_s by $k_F^3 = 3\pi^2\rho$, where ρ is the density, or

$$k_F = (\tfrac{9}{4}\pi)^{1/3}(1/r_s) = 1.919/r_s. \tag{8.2.2}$$

The average kinetic energy per particle is

$$E_K = \tfrac{3}{5}E_F = 3\hbar^2 k_F^2/10m = 2.210/r_s^2 \quad \text{Ry} \tag{8.2.3}$$

in which E_F is the Fermi energy. The assumption of electrical neutrality implies that the average potential [$V(q = 0)$ in Eq. (8.1.76)] is zero. The average exchange energy per particle is given by Eq. (4.10.16), Part A,

$$E_x = -(3e^2/2)(3\rho/8\pi)^{1/3} = -(3/2)(9/32\pi^2)^{1/3}(e^2/r_s) = 0.916/r_s. \tag{8.2.4}$$

The first-order term in the total energy of the system is smaller than the zeroth-order term (the kinetic energy) by one power of r_s.

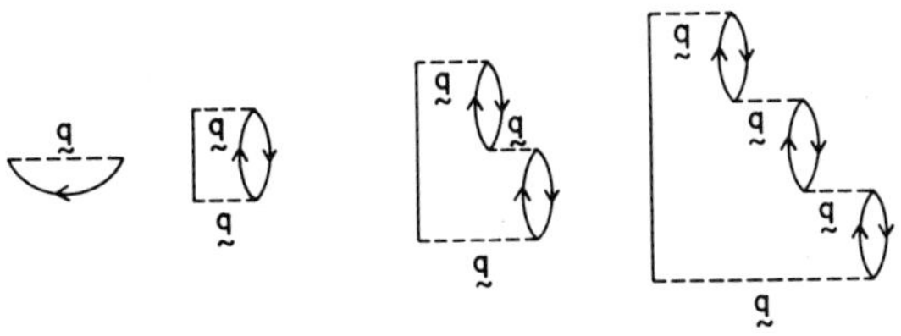

Fig. 8.2.1. Dominant self-energy diagrams for the free electron gas.

8.2.1 *The Random Phase Approximation*

Let us examine the calculation of the single-particle Green's function for the system. To do this, we wish to compute the self-energy function $M(\mathbf{k}, \epsilon)$ which appears in (8.1.116). The first-order term in the self-energy, which is the Hartree–Fock approximation (8.1.119), was already computed in Section 4.10, Part A, and is given by Eq. (4.10.9). We will refer to this as $E_1(k)$:

$$E_1(k) = -(2k_F/\pi)[1 + ((k^2 - k_F^2)/2kk_F) \ln |(k - k_F)/(k + k_F)|]. \tag{8.2.5}$$

We now turn to the second-order terms. The second-order direct contribution to $M(\mathbf{k}, \epsilon)$ is contained in the square brackets in Eq. (8.1.113). This term will be denoted M_{2R}:

$$M_{2R}(\mathbf{k}, \epsilon) = \sum_{\mathbf{p}<k_F, \mathbf{q}} [V(\mathbf{q})]^2/\{\epsilon + E(\mathbf{p}) - E(\mathbf{p} + \mathbf{q}) - E(\mathbf{k} - \mathbf{q})\}. \tag{8.2.6}$$

This looks harmless. The problem is that

$$V(\mathbf{q}) = 4\pi e^2/q^2 \tag{8.2.7}$$

for a Coulomb potential. As a result, the sum over $\mathbf{q}$ involves $1/q^4$ and diverges as $q \to 0$. Similar divergences occur in higher order. The dominant diagrams, which are the most highly divergent ones, are the so-called *ring diagrams* (see Fig. 8.2.1). These diagrams have the essential feature that the wave vector transfer from loop to loop is the same in each case. As a result, a ring diagram with n loops contains a contribution $[V(\mathbf{q})]^{n+1} \approx q^{-2(n+1)}$ and the divergencies pile up.

A little thought about the evaluation of the diagrams of this series will convince one that the terms of the series evolve by the multiplication of factors shown diagrammatically in Fig. 8.2.2. This factor will be called

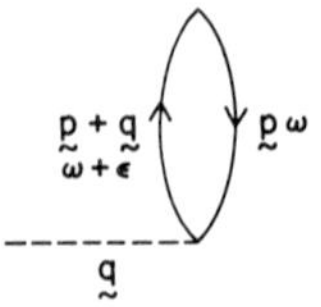

Fig. 8.2.2. Polarization part.

$P(\mathbf{q}, \epsilon)$. It will be evaluated using the Feynman rules:

$$P(\mathbf{q}, \epsilon) = iV(\mathbf{q}) \sum_{\mathbf{p}} \int (d\omega/2\pi)[iG(\mathbf{p}, \omega)][iG(\mathbf{p} + \mathbf{q}, \omega + \epsilon)]$$

$$= -iV(\mathbf{q}) \sum_{\mathbf{p}} \int (d\omega/2\pi)[\omega - E(\mathbf{p}) + i\,\delta_{\mathbf{p}}]^{-1}$$

$$\times [\omega + \epsilon - E(\mathbf{p} + \mathbf{q}) + i\,\delta_{\mathbf{p}+\mathbf{q}}]^{-1}. \qquad (8.2.8)$$

We must include a factor of 2 in the sum over $\mathbf{p}$ to account for the two directions of spin. The integral is to be done as a contour integral. If both p and $|\,\mathbf{p} + \mathbf{q}\,|$ are greater or less than k_F, the poles in the integral are on the same side of the real axis. The contour can be closed on the opposite side, so that the value of the integral is zero. Two cases give a nonvanishing contribution.

Case (1): $p < k_F$; $|\,\mathbf{p} + \mathbf{q}\,| > k_F$; $\delta_{\mathbf{p}} = -\delta_1$; $\delta_{\mathbf{p}+\mathbf{q}} = \delta_2$.

Close the contour above the real axis. The region of allowed values of p and $\mathbf{p} + \mathbf{q}$ is conveniently indicated by Fermi functions $f(\mathbf{p})$, $f(\mathbf{p} + \mathbf{q})$ which are in this case, simply step functions. We obtain

$$V(\mathbf{q}) \sum_{\mathbf{p}} f(\mathbf{p})[1 - f(\mathbf{p} + \mathbf{q})]/[\epsilon + E(\mathbf{p}) - E(\mathbf{p} + \mathbf{q}) + i\delta]$$

$$(\delta = \delta_1 + \delta_2). \qquad (8.2.9)$$

Case (2): $p > k_F$; $|\,\mathbf{p} + \mathbf{q}\,| < k_F$; $\delta_{\mathbf{p}} = \delta_1$; $\delta_{\mathbf{p}+\mathbf{q}} = -\delta_2$.

The contour will be closed below the real axis. This introduces an additional negative sign. The result is

$$-V(\mathbf{q}) \sum_{\mathbf{p}} f(\mathbf{p} + \mathbf{q})[1 - f(\mathbf{p})]/[\epsilon + E(\mathbf{p}) - E(\mathbf{p} + \mathbf{q}) - i\delta].$$

$$(8.2.10)$$

The significance of $P(\mathbf{q}, \epsilon)$ may be appreciated if we pass immediately

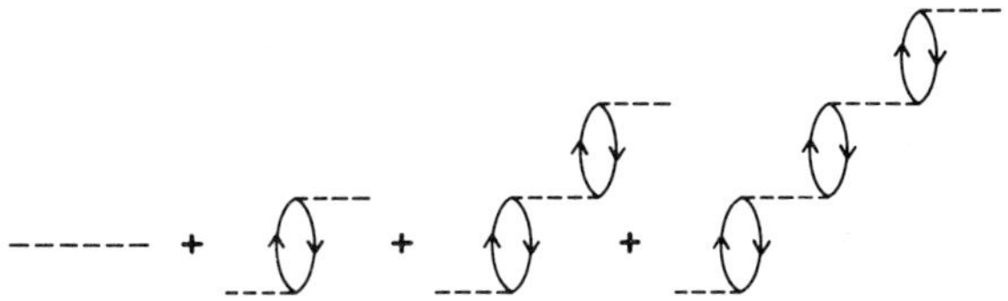

Fig. 8.2.3. Series of polarization parts and attached interaction lines. The first term is the bare interaction.

to the limit $\epsilon = 0$, in which case δ may be set equal to zero. This gives

$$P(\mathbf{q}, 0) = V(\mathbf{q}) \sum_{\mathbf{p}} [f(\mathbf{p}) - f(\mathbf{p}+\mathbf{q})]/[E(\mathbf{p}) - E(\mathbf{p}+\mathbf{q})]. \tag{8.2.11}$$

This result can be compared with Eq. (7.1.58). It is seen that $S(\mathbf{q}) = -P(\mathbf{q}, 0)/V(\mathbf{q})$ is the object which specifies the static dielectric function of the system. It is convenient to generalize this function to nonzero values of ϵ by defining $S(\mathbf{q}, \epsilon)$ through

$$P(\mathbf{q}, \epsilon) = -V(\mathbf{q})S(\mathbf{q}, \epsilon). \tag{8.2.12}$$

The series of polarization parts, with attached interaction lines, and including the bare interaction (see Fig. 8.2.3) gives

$$-iV(\mathbf{q})[1 - V(\mathbf{q})S(\mathbf{q}, \epsilon) + V^2(\mathbf{q})S^2(\mathbf{q}, \epsilon) - V^3(\mathbf{q})S^3(\mathbf{q}, \epsilon) + \cdots]$$
$$= -iV(\mathbf{q})/[1 + V(\mathbf{q})S(\mathbf{q}, \epsilon)]. \tag{8.2.13}$$

The effect of summing the polarization parts is to introduce a screened effective interaction in which $V(\mathbf{q})$ is replaced by

$$V(\mathbf{q}) \to V(\mathbf{q})/\kappa(\mathbf{q}, \epsilon) \tag{8.2.14a}$$

in which

$$\kappa(\mathbf{q}, \epsilon) = 1 - P(\mathbf{q}, \epsilon) = 1 + V(\mathbf{q})S(\mathbf{q}, \epsilon). \tag{8.2.14b}$$

The simplest approximation to S, which is to take $\epsilon = 0$, leads to the explicit result given in Eq. (7.1.59).

In order to obtain an expression valid beyond the static ($\epsilon = 0$) limit, we let $\delta \to 0$ in (8.2.9) and (8.2.10), and use the identity

$$\lim_{\eta \to 0} (x + i\eta)^{-1} = P(1/x) - i\pi\, \delta(x).$$

The results can be simplified somewhat by interchanging the variables $\mathbf{p}$ and $\mathbf{p} + \mathbf{q}$ in (8.2.10). Define

$$E_{\mathbf{pq}} = E(\mathbf{p} + \mathbf{q}) - E(\mathbf{p}).$$

The result of these considerations is the following expression for the dielectric function:

$$\kappa(\mathbf{q}, \epsilon) = 1 - V(\mathbf{q}) \Big\{ 4P \int [d^3p/(2\pi)^3][f(\mathbf{p})E_{\mathrm{pq}}/(\epsilon^2 - E_{\mathrm{pq}}{}^2)]$$

$$- 2\pi i \int [d^3p/(2\pi)^3] f(\mathbf{p})[1 - f(\mathbf{p}+\mathbf{q})]$$

$$\times [\delta(\epsilon - E_{\mathrm{pq}}) + \delta(\epsilon + E_{\mathrm{pq}})] \Big\}. \tag{8.2.15}$$

A factor of 2 for spins has been inserted. It will be observed the κ is an even function of ϵ.

Let us consider evaluation of (8.2.15) for small q and arbitrary ϵ in the case of a parabolic band $E(p) = \hbar^2p^2/2m^*$. Then

$$E_{\mathrm{pq}} = \hbar^2(2\mathbf{p}\cdot\mathbf{q} + \mathbf{q}^2)/2m,$$

so that this quantity vanishes as $q \to 0$. Hence, for finite ϵ, there can be no contribution from the delta functions. In the second term, we may neglect $E^2{}_{\mathrm{pq}}$ in the denominator, and the term in the numerator involving $\mathbf{p}\cdot\mathbf{q}$ vanishes by symmetry. Thus we find

$$\kappa(\mathbf{q}, \epsilon) = 1 - [2\hbar^2q^2V(\mathbf{q})/m^*\epsilon^2] \int [d^3p/(2\pi)^3] f(\mathbf{p}) = 1 - (4\pi Ne^2/m^*\nu^2) \tag{8.2.16}$$

in which $\nu = \epsilon/\hbar$, and N is the electron density. The dielectric function vanishes if

$$\nu^2 = \omega^2{}_{\mathrm{pl}} = 4\pi Ne^2/m^* = Ne^2/m^*\epsilon_0 \tag{8.2.17}$$

(the last form being in mks units). The quantity ω_{pl} is the plasma frequency. The vanishing of κ indicates that the system possesses an undamped oscillation at this frequency (see Section 1.3, Part A). This excitation is called the *plasmon*. As q increases, the delta functions remain unable to contribute, while the contribution from $E^2{}_{\mathrm{pq}}$ in the second term indicates that the plasma frequency increases with q. For sufficiently large q, the delta functions come in and the plasmon becomes damped through the transfer of energy to electron hole pairs.

The role of plasma oscillations in the theory of the electron gas was emphasized by Bohm and Pines (1953). We will return to this topic in Section 8.3.6. We continue with the calculation of the electron self-energy. The considerations leading to Eq. (8.2.14) can be generalized. We may

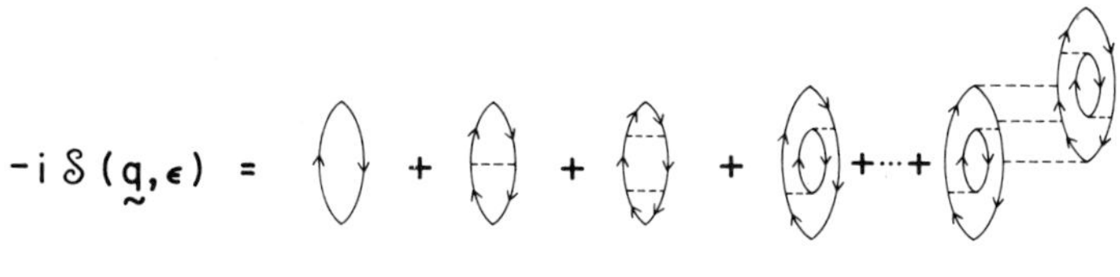

Fig. 8.2.4. Series of proper polarization parts.

consider all sorts of processes happening within a single bubble. Let us define a proper polarization part as any diagram without external interaction lines which cannot be broken into two disconnected pieces by breaking one interaction line. We may consider the sum of all possible proper polarization parts in a formal sense. The diagrammatic series is illustrated in Fig. 8.2.4. This sum will be denoted as $-i\mathcal{S}(\mathbf{q}, \epsilon)$.

Next, we may replace the contribution from each simple polarization part, which was $-iS(\mathbf{q}, \epsilon)$, by the complete sum of proper parts $-i\mathcal{S}$. The series which gives the full screened potential is found by generalizing (8.2.12). Let the screened potential be $V_s(\mathbf{q}, \epsilon)$:

$$V_s(\mathbf{q}, \epsilon) = V(\mathbf{q})/\kappa(\mathbf{q}, \epsilon) \tag{8.2.18a}$$

in which

$$\kappa(\mathbf{q}, \epsilon) = 1 + V(\mathbf{q})\mathcal{S}(\mathbf{q}, \epsilon). \tag{8.2.18b}$$

These equations extend (8.2.14a) and (8.2.14b). We will refer to the dielectric function (8.2.16), computed in the limit in which $\mathcal{S}$ is replaced by S, as the *random phase approximation* (RPA).

The screened interaction (8.2.18a) can now be used to simplify the perturbation series for the Green's function. The diagrammatic series for the self-energy, which is represented by Fig. 8.2.1, with the addition of more complicated loops as in Fig. 8.2.4, is summed simply by replacing $V(\mathbf{q})$ by the screened interaction $V_s(\mathbf{q}, \epsilon)$. The resulting approximation for the self-energy part is (Feynman rules)

$$\begin{aligned} M(\mathbf{k}, \epsilon) &= \sum_{\mathbf{q}} \int (d\omega/2\pi) V_s(\mathbf{q}, \omega) iG^{(0)}(\mathbf{k} - \mathbf{q}, \epsilon - \omega) \\ &= i \sum_{\mathbf{q}} \int (d\omega/2\pi) [V(\mathbf{q})/\kappa(\mathbf{q}, \omega)][\epsilon - \omega - E(\mathbf{k} - \mathbf{q}) + i\,\delta_{\mathbf{k}-\mathbf{q}}]^{-1}. \end{aligned} \tag{8.2.18c}$$

Introduction of the dielectric function leads to a finite result for the self-energy even though the individual ring diagrams were infinite. Such are the advantages of selective summation to infinite order.

The integration of (8.2.18) is complicated. The calculation is given in detail by Quinn and Ferrell (1958). The result is to be substituted into the Dyson equation (8.1.115), and the result solved for the renormalized energy, which is given by the location of the pole of the actual Green's function. The energy is complex: the real part has the usual significance as specifying the energy of a single, quasi-particle state, while the imaginary part gives the lifetime of such a state. Their results are, for an electron near the Fermi surface,

$$\text{real part:}\quad E(k) = (\hbar^2k^2/2m) - 0.166r_s(\hbar^2k_F{}^2/2m) \times [(k/k_F)(\ln r_s + 0.203) + \ln r_s - 1.80]; \tag{8.2.19a}$$

$$\text{imaginary part:}\quad E_i(k) = \pm\hbar/\tau = 0.252r_s{}^{1/2}[\hbar^2(k - k_F)^2/2m]. \tag{8.2.19b}$$

In these equations, r_s is measured in atomic units. The plus sign in (8.2.19b) refers to holes, the negative to electrons.

In the Hartree–Fock approximation, the single-particle energy (8.2.5) contains a logarithmic term gives rise to a singularity in dE/dk at k_F. This pecularity, which is in violent disagreement with experiment, is removed by the screening of the interaction potential. It is seen that the real part of the energy behaves smoothly at $k = k_F$. The effective mass of an electron near the Fermi surface is found from the derivative of the energy. We put

$$1/m^* = 1/\hbar^2k_F\,(dE/dk)_{k_F} = (1/m)[(1 - 0.083r_s(\ln r_s + 0.203)]. \tag{8.2.20}$$

This is the density of states mass. The specific heat of an interacting electron system is still proportional to the density of states at the Fermi energy (Luttinger and Ward, 1960), and the density of states is proportional to m^*. Thus the ratio of the specific heat of an interacting electron system to that of a noninteracting system is

$$\frac{C(\text{interacting})}{C(\text{noninteracting})} = 1 + 0.083r_s(\ln r_s + 0.203). \tag{8.2.21}$$

This result was first obtained by Gell-Mann (1957).

The imaginary part of the self-energy results physically from electron–electron scattering. A single-particle state of wave vector **k** is not a stationary state of the system. Instead, the particle interacts with others in the Fermi sea, and is scattered out of the original state. If the original state is close to the Fermi surface, the number of particles available for real scattering is small, since a real transition requires the excitation of some electron from a state below the Fermi surface to a state above it. An electron exactly at the Fermi surface cannot produce a real excitation: hence, the lifetime must become infinite.

Let us apply this argument to determine that the dependence of $1/\tau$ is in fact proportional to $(k - k_F)^2$ as asserted by (8.2.19b). It follows from the Dyson equation that it will be correct to leading order to set $\epsilon = E(\mathbf{k})$ in the computation of $\text{Im}(M)$. In order to have a specific situation, let us suppose that $k > k_F$: the propagation of an excited electron is being considered.

The major complications of the ω integral in (8.2.18) result from $\kappa(k, \omega)$, which has a branch cut along the real axis. However, it can be shown (Quinn and Ferrell, 1958) that the imaginary part of the self-energy is obtained from the residue at the pole of the free Green's function:

$$\text{Im}\, M(\mathbf{k}, E(\mathbf{k})) = \sum_{\mathbf{q}} V(\mathbf{q})\, \text{Im}\{\kappa[\mathbf{q}, E(\mathbf{k}) - E(\mathbf{k} - \mathbf{q})]\}^{-1}. \qquad (8.2.22)$$

The imaginary portion of the dielectric function can be obtained from (8.2.15). Note that ϵ is positive, as is $E_{\mathbf{pq}}$; thus only the first delta function can contribute:

$$\text{Im}\, \kappa(\mathbf{q}, \epsilon) = [V(\mathbf{q})/(2\pi)^2] \int d^3p\, f(\mathbf{p})[1 - f(\mathbf{p} + \mathbf{q})]\, \delta[\epsilon - E_{\mathbf{pq}}]. \qquad (8.2.23)$$

Integrals of this sort are difficult because of the complicated nature of the region of integration. Small values of q are dominant in (8.2.22). Equation (8.2.23) simplifies when terms of order q^2 are neglected. A parabolic band $E(p) = \gamma p^2$ is used. We find

$$\text{Im}\, \kappa(\mathbf{q}, \epsilon) = V(\mathbf{q})\epsilon/8\pi\gamma^2 q. \qquad (8.2.24)$$

For fixed $\mathbf{q}$ and sufficiently small ϵ, $\text{Im}\, \kappa$ is small compared to $\text{Re}\, \kappa$. Thus we may approximate

$$\text{Im}[1/\kappa] = [-\text{Im}\, \kappa(\mathbf{q}, \epsilon)]/[\text{Re}\, \kappa(\mathbf{q}, 0)]^2. \qquad (8.2.25)$$

Equations (8.2.24) and (8.2.25) are inserted into (8.2.22). The following expression is obtained:

$$\text{Im}\, M(\mathbf{k}, E(\mathbf{k})) = (-e^4/4\pi^2\gamma^2) \int d^3q\, [E(\mathbf{k}) - E(\mathbf{k} - \mathbf{q})]/q^5[\text{Re}\, \kappa(\mathbf{q}, 0)]^2. \qquad (8.2.26)$$

It is necessary once again to be careful about the limits of integration. These are determined by the fact that we are interested in a real scattering process, involving states above the Fermi surface. Thus

$$k > |\mathbf{k} - \mathbf{q}| > k_F.$$

These inequalities determine the range of the angular integration: If x is the angle between $\mathbf{k}$ and $\mathbf{q}$, $\mathbf{k}\cdot\mathbf{q} = kqx$, then x lies in the range

$$0 \leq x \leq x_m = (k^2 - k_F{}^2)/2kq \leq 1.$$

Terms of order q^2 are neglected. Evidently $(k - k_F)$ must be small. Equation (8.2.26) now yields

$$\operatorname{Im} M(\mathbf{k}, E(\mathbf{k})) = (-e^4/2\pi\gamma)k_F(k - k_F)^2 \int dq/q^4[\operatorname{Re} \kappa(\mathbf{q}, 0)]^2. \quad (8.2.27)$$

The essential role played by Re κ should be noticed. If this factor were not present, the integral would be highly divergent at the lower limit. This would imply an infinite damping rate, and a single-electron picture would never be appropriate. However [from Eq. (7.1.60)],

$$\operatorname{Re} \kappa(\mathbf{q}, 0) = 1 + (4\pi e^2/q^2)(3/2E_F\Omega)$$

and the integral is convergent. The behavior for small q dominates the integral. To leading order in r_s, we may set the upper limit in (8.2.27) to be ∞. The integral can now be readily performed. The result can be expressed as (in atomic units)

$$\operatorname{Im} M(\mathbf{k}, E(\mathbf{k})) = (-\pi^2/32)(4\alpha r_s/\pi)^{1/2}(k - k_F)^2 \quad (8.2.28)$$

in which $\alpha = (\frac{4}{9}\pi)^{1/3}$. This result yields (8.2.19b). Note that, in the case of hole states the sign of (8.2.28) should be reversed. This results from the fact that a hole bubbles upward toward the Fermi surface as it loses energy through real scattering processes. (In contrast, an electron falls down onto the surface.)

The essential feature of this result is the dependence on $(k - k_F)^2$. This is an exact result. Its importance results from the fact that it validates the single-particle description for states close to the Fermi surface. States with a single excited particle are nearly eigenstates in the sense that the ratio of the imaginary to the real part of the excitation energy vanishes as $k \to k_F$. More will be said on this point in Section 8.2.3.

8.2.2 *The Ground State Energy*

This subsection is devoted to a calculation of the ground state energy of the free electron gas according to the method of Gell-Mann and Brueckner

Fig. 8.2.5. Second-order ground state energy diagrams: (a) direct, (b) exchange.

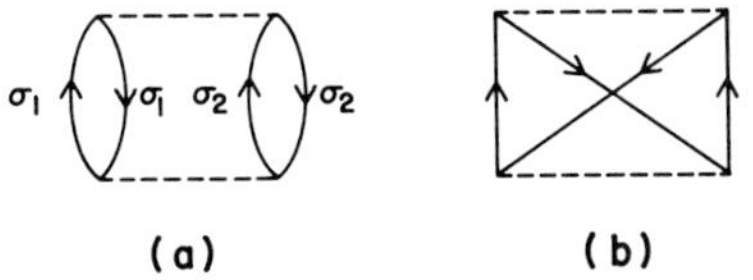

(1957). Although the ground state energy can be obtained by further calculations involving the single-particle Green's function [see Eqs. (8.1.50) and (8.1.51)] that was considered in the previous subsection, the procedure of these authors is informative and direct. The diagrams making the dominant—of the highest apparent divergence—contribution to the ground state energy are summed.

The zero- and first-order terms in the ground state energy are of order r_s^{-2} and r_s^{-1}, respectively; the values are given explicitly in (8.2.3) and (8.2.4). If the perturbation series were a simple power series, we would expect the next term to be a constant, independent of r_s. Let us consider the second-order ground state energy diagrams shown in Fig. 8.2.5. The direct term (a) can be obtained from Eq. (8.1.121). Note that each line must be assigned a spin index, and a sum over spins must be performed. Since the spin associated with each loop in (a) is independent, the spin sum gives a factor of 4 in this case; but only a factor of 2 in (b). Note also that the signs of (a) and (b) are different. It is further convenient to introduce scaled momenta by dividing each **k**, **q**, **p** by k_F. Each summation over **k** is replaced by an integration over **k**. This brings in a factor of $[\Omega/2\pi]^3$, in which Ω is the volume of the system. Finally, Ω is replaced by k_f using the standard relation $k_F{}^3 = 3\pi^2 N/V$ (with N the number of particles). The direct contribution to the total energy per particle is, in rydbergs,

$$(E_2/N) \quad \text{(direct)}$$

$$= (-3/8\pi^5) \int d^3q/q^4 \int_{\substack{k<1 \\ |\mathbf{q}-\mathbf{k}|>1}} d^3k \int_{\substack{p<1 \\ |\mathbf{q}-\mathbf{p}|>1}} d^3p \, [\mathbf{q}^2 + \mathbf{q}\cdot(\mathbf{p}-\mathbf{k})]^{-1}. \tag{8.2.29}$$

It is convenient to label the exchange diagram as similarly as possible to the direct diagram. Its contribution is

$$(E_2/N) \quad \text{(exchange)}$$

$$= (3/16\pi^5) \int d^3q/q^2 \int_{\substack{k<1 \\ |\mathbf{q}-\mathbf{k}|>1}} d^3k \int_{\substack{p<1 \\ |\mathbf{q}-\mathbf{p}|>1}} d^3p \, [\mathbf{q}^2 + \mathbf{q}\cdot(\mathbf{p}-\mathbf{k})]^{-1} \times |\mathbf{q}+\mathbf{p}-\mathbf{k}|^{-2}. \tag{8.2.30}$$

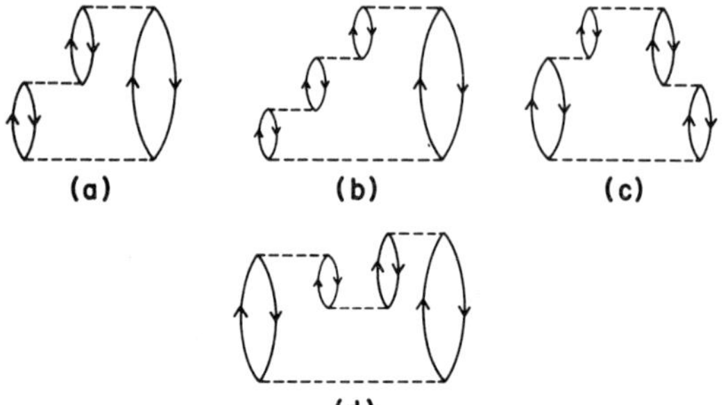

Fig. 8.2.6. Third- and fourth-order ring diagrams contributing to the self-energy.

These expressions are independent of r_s as expected. The difficulty is that (8.2.29) is divergent for small q. To see this, we note that the restrictions on the integration over intermediate wave vectors **k** and **p** are such that the allowed volume, in each case, is of order q for small q. The energy denominator is of order q^{-1}, so we face a q integral $\int d^3q/q^3$ which is logarithmically divergent. On the other hand, the exchange integral (8.2.30) behaves like $\int d^3qq$ for small q and $\int d^3q/q^6$ for large q, and thus is convergent. The result of doing the integration in (8.2.30) will be needed later. The calculation is complicated, and we simply quote the result

$$E_2/N \quad \text{(exchange)} = 0.046 \quad \text{(Ry/electron)}. \tag{8.2.31}$$

Let us see what happens in higher order. The dominant diagrams are the ring diagrams in which the wave vector **q** transferred is the same at each vertex. This is the same situation as was encountered in the case of the self-energy. The relevant diagrams are shown in Fig. 8.2.6. Diagrams of this type have a factor of $(1/q^2)$ for each vertex and are clearly the most highly divergent ones. If the wave vector exchange is different, the diagram will be ignored.

The contribution of the third-order diagram 8.2.6a differs from 8.2.29 by an additional intermediate wave vector integration with an associated energy denominator and an additional matrix element of potential. The wave vectors are scaled as before, and a numerical factor $-2(\alpha r_s/\pi^2)$ appears, in which $\alpha = (\frac{4}{9}\pi)^{1/3}$. Note that the diagram is not completely symmetric

$$E_3/N = 2(\alpha r_s/\pi^2)(3/8\pi^5) \int d^3q/q^6 \int d^3p_1 \int d^3p_2 \int d^3p_3$$
$$\times [\mathbf{q}^2 + \mathbf{q}\cdot(\mathbf{p}_1 - \mathbf{p}_2)]^{-1}[\mathbf{q}^2 + \mathbf{q}\cdot(\mathbf{p}_1 - \mathbf{p}_3)]^{-1}. \tag{8.2.32}$$

The same restrictions on intermediate wave vectors obtain as in second order. The fourth-order terms are more complicated because there are three different energy denominators. The argument by which the small q dependence was determined for second order shows that the third-order term diverges as $\int d^3q/q^5$, the fourth-order as $\int d^3q/q^7$, and so on.

In spite of the divergence of the individual terms, a finite result can be obtained if we add the terms *prior* to integration over q. This addition is accomplished by the following trick: Let us write the series beginning with second order as

$$\Delta E = (1/N)(E_2 + E_2 + E_4 + \cdots)$$

$$= (-3/8\pi^5) \int d^3q/q^4 \sum_{n=2}^{\infty} (-1)^n (\alpha r_s/\pi^2 q^2)^{n-2} A_n. \qquad (8.2.33)$$

The A_n contain all the integration over intermediate momenta. Thus, for example,

$$A_3 = 2 \int d^3p_1 \int d^3p_2 \int d^3p_3 \, [\mathbf{q}^2 + \mathbf{q}\cdot(\mathbf{p}_1 + \mathbf{p}_2)]^{-1}[\mathbf{q}^2 + \mathbf{q}\cdot(\mathbf{p}_1 + \mathbf{p}_3)]^{-1}. \qquad (8.2.34)$$

Note that it is permissible to change the signs of the intermediate momenta and this has been done in (8.2.34). The energy denominators can be combined by introducing an auxiliary variable t and an extra integration. Each internal momentum will be associated with the function

$$F_q(t_i) = \int d^3p_i \exp[- | t_i | (\tfrac{1}{2}q^2 + p_i^2)]. \qquad (8.2.35)$$

This expression is used as a propagator. We simply multiply factors of this type, one for each intermediate momentum, and multiply this contribution by an overall delta function containing the sum of all t_i, and a numerical factor of $1/n$; then integrate over all t_i

$$A_n = (1/n) \int_{-\infty}^{\infty} F_q(t_1)F_q(t_2)\cdots F_q(t_n) \, \delta(t_1 + t_2 + \cdots t_n) \, dt_1\cdots dt_n. \qquad (8.2.36)$$

This can be confirmed by comparison with the specific formulas. For example, in the case of A_2, we have simply

$$A_2 = \tfrac{1}{2} \int d^3p_1 \int d^3p_2 \int_{-\infty}^{\infty} dt_2 \int_{-\infty}^{\infty} dt_1\, \delta(t_1 + t_2)$$

$$\times \exp[-\mid t_1 \mid (\tfrac{1}{2}\mathbf{q}^2 + \mathbf{q}\cdot\mathbf{p}_1) - \mid t_2 \mid (\tfrac{1}{2}\mathbf{q}^2 + \mathbf{q}\cdot\mathbf{p}_2)]$$

$$= \int d^3p_1 \int d^3p_2 \int_0^{\infty} dt \exp[-t(\mathbf{q}^2 + \mathbf{q}\cdot(\mathbf{p}_1 + \mathbf{p}_2))]$$

$$= \int d^3p_1 \int d^3p_2\, [\mathbf{q}^2 + \mathbf{q}\cdot(\mathbf{p}_1 + \mathbf{p}_2)]^{-1}. \tag{8.2.37}$$

This is the correct expression. Higher-order terms follow, and it turns out that all the complications in fourth and higher orders arising from different energy denominators are contained properly.

Next, introduce the Fourier transform of the delta function in the expression for A_n:

$$\delta(t_1 + t_2 + \cdots + t_n) = (q/2\pi) \int_{-\infty}^{\infty} \exp[iuq(t_1 + t_2 + \cdots + t_n)]\, du.$$

This enables us to write the integral (8.2.36) for A_n as the product of n identical parts $Q_q(u)$, followed by an overall integral

$$A_n = (q/2\pi n) \int_{-\infty}^{\infty} [Q_q(u)]^n\, du \tag{8.2.38}$$

in which

$$Q_q(u) = \int d^3p \int_{-\infty}^{\infty} dt \exp(ituq) \exp\{-\mid t \mid [\tfrac{1}{2}\mathbf{q}^2 + \mathbf{q}\cdot\mathbf{p}]\}\, dt. \tag{8.2.39}$$

The series (8.2.33) for ΔE is now

$$\Delta E = (-3/16\pi^6) \int d^3q/q^3 \int_{-\infty}^{\infty} du \sum_{n=2}^{\infty} [(-1)^n/n][Q_q(u)]^n (\alpha r_s/\pi^2 q^2)^{n-2}. \tag{8.2.40}$$

The sum can be performed

$$\sum_{n=2}^{\infty} [(-1)^n/n] a^n b^{n-2} = (1/b^2) \sum_{n=1}^{\infty} [(-1)^n/n] a^n b^n + (a/b)$$

$$= (a/b) - (1/b^2) \ln(1 + ab). \tag{8.2.41}$$

Thus we have

$$\Delta E = (-3/16\pi^6) \int d^3q/q^3 \int du \, \{(\pi^2 q^2/\alpha r_s) Q_q(u) - (\pi^2 q^2/\alpha r_s)^2 \ln[1 + Q_q(u)(\alpha r_s/\pi^2 q^2)]\}. \tag{8.2.42}$$

We now face the problem of evaluating the integral (8.2.39). In third and higher orders of perturbation theory, we require the contribution only for small values of q; the most highly divergent parts. For small q, we may drop the $\frac{1}{2}q^2$ in the exponential. The restrictions on the intermediate wave vectors which are spelled out in Eq. (8.2.30) are transferred to (8.2.39). Specifically these are $p < 1$ and $|\mathbf{p} + \mathbf{q}| > 1$. Let θ be the angle between $\mathbf{p}$ and $\mathbf{q}$. For small q, such that terms of order q^2 are neglected, these conditions require that $0 \leq \cos\theta \leq 1$ and $(1 - q\cos\theta) \leq p \leq 1$. These approximations enable us to simplify (8.2.39) (with $x = \cos\theta$).

$$Q_q(u) = 2\pi p \int_0^1 x \, dx \int_{-\infty}^{\infty} dt \exp(ituq) \exp(-|t| qx)$$

$$= 4\pi \int_0^1 x^2 \, dx/(x^2 + u^2) = 4\pi R(u) \tag{8.2.43a}$$

in which

$$R(u) = 1 - u \tan^{-1}(1/u). \tag{8.2.43b}$$

Note that $Q_q(u)$ is actually independent of q for small q.

We have already seen that the contributions from the ring diagrams converge readily when q is large. It will be sufficient to approximate $Q_q(u)$ for all q as

$$\begin{aligned} Q_q(u) &= 4\pi R(u) \qquad (0 \leq q \leq 1), \\ &= 0 \qquad (q > 1). \end{aligned} \tag{8.2.44}$$

However, a correction must be supplied which restores the exact value of the second-order term if we are to obtain correctly the value of the term in the correlation energy which is independent of r_s. The large q contributions from higher-order graphs do not affect the correlation energy to the order of interest.

The angular integration in (8.2.42) can be done immediately and the result now simplifies to

$$\Delta E = (-12/\pi^3) \int_{-\infty}^{\infty} du \int_0^1 dq/q \times \{q^2[R(u)/\lambda] - (q^2/\lambda)^2 \ln[1 + (\lambda R(u)/q^2)]\} + \delta \tag{8.2.45}$$

in which $\lambda = 4\alpha r_s/\pi$. The correction δ is found as the difference between the exact second-order perturbation (8.2.29) and the second-order term appearing in (8.2.45). Thus

$$\delta = (1/N)E_2 - \left[(-12/\pi^3)\int_{-\infty}^{\infty} du \int_0^1 (dq/q)\ \tfrac{1}{2}R^2\right]. \qquad (8.2.46)$$

The second term in (8.2.46) has been constructed in such a way that the logarithmic divergences must cancel. Evaluation yields $\delta = -0.0508$.

The q integral in (8.2.45) is convergent at $q = 0$. In the result, we may discard terms which vanish as $r_s \to 0$ ($\lambda \to 0$). We find

$$\Delta E = (3/\pi^3)\int_{-\infty}^{\infty} du\, R^2(u)[\ln \lambda + \ln R(u) - \tfrac{1}{2}] + \delta$$

$$= (2/\pi^2)(1 - \ln 2)\ln(4\alpha r_s/\pi) + B \qquad (8.2.47)$$

in which B is a numerical constant, independent of r_s. The apparent divergence of the second-order perturbation term has been replaced by a logarithmic dependence on r_s. The correlation energy can now be obtained by adding to ΔE, above, the value of the second-order exchange correction given by (8.2.31). After all constants have been evaluated, the correlation energy E_c, which is the difference between the ground state energy per electron as calculated here and that obtained in the Hartree–Fock equation, is, in rydbergs,

$$E_c = 0.0622 \ln r_s - 0.094. \qquad (8.2.48)$$

This expression is valid for small r_s. A more complete theory than that presented here will correct this result by adding terms of order $r_s \ln r_s$ and higher (Du Bois, 1959, 1960).

8.2.3 *Quasi Particles and the Fermi Surface*

In a system of noninteracting particles at zero temperature, the number of particles in a state of wave vector $\mathbf{k}$ and spin σ, $n_\sigma(\mathbf{k})$, is a simple step function: $n_\sigma(\mathbf{k}) = 1$ if $E(\mathbf{k}) < \mu$, $n_\sigma(\mathbf{k}) = 0$ if $E(\mathbf{k}) > \mu$. This discontinuity in the occupation number leads to the existence of a sharp Fermi surface, and is vitally involved in many characteristic properties of metals. Real metals are not systems of noninteracting particles, but they exhibit sharp Fermi surfaces. It is therefore of considerable interest to see what happens to the occupation number in the presence of interaction. In fact, a discontinuity persists, although its magnitude is reduced. The general theory is due to Luttinger (1960).

The occupation number $n_\sigma(\mathbf{k})$ is not diagonal in the presence of interactions. Instead, we must ask for the expectation value in the ground state

$$\bar{n}_\sigma(\mathbf{k}) = \langle n_\sigma(\mathbf{k}) \rangle = \langle N \mid c_\sigma^+(\mathbf{k}) c_\sigma(\mathbf{k}) \mid N \rangle. \tag{8.2.49}$$

This will be recognized as a limiting form of the single-particle Green's function:

$$\bar{n}_\sigma(\mathbf{k}) = -i\mathcal{G}(\mathbf{k}t\sigma, \mathbf{k}t_+\sigma) = -i \lim_{t\to 0^-} \mathcal{G}_\sigma(\mathbf{k}, t). \tag{8.2.50}$$

We will investigate the properties of the Green's function near the Fermi energy. The Fourier transform $\mathcal{G}(\mathbf{k}, \omega)$ has, according to (8.1.114), the form

$$\mathcal{G}(\mathbf{k}, \omega) = [\omega - \epsilon(\mathbf{k}) - M(\mathbf{k}, \omega)]^{-1}.$$

We will assume that the system has no magnetic order and so suppress the spin index. The function $\epsilon(\mathbf{k})$ denotes the unperturbed energy. The mass operator $M(\mathbf{k}, \omega)$ is found from (8.2.18). We separate real and imaginary parts

$$M(\mathbf{k}, \omega) = M_r(\mathbf{k}, \omega) + iM_i(\mathbf{k}, \omega).$$

It is consistent with Eq. (8.2.28) to write the imaginary part as

$$M_i(\mathbf{k}, \omega) = B(\mathbf{k}) \operatorname{sgn}(\mu - \omega)(\omega - \mu)^2$$

in which B is a function to be determined by a more elaborate calculation. The imaginary part vanishes on the Fermi surface. The energies of excited states of the system are found by solving (8.1.115), which becomes

$$\omega - \epsilon(\mathbf{k}) - M_r(\mathbf{k}, \omega) - iB(\mathbf{k}) \operatorname{sgn}(\mu - \omega)(\omega - \mu)^2 = 0. \tag{8.2.51}$$

This equation can be solved formally when $\omega = \mu$. We have

$$\mu - \epsilon(\mathbf{k}) - M_r(\mathbf{k}, \mu) = 0. \tag{8.2.52}$$

This equation defines the Fermi surface in an interacting system: $\mathbf{k} = \mathbf{k}_F = \mathbf{k}_F(\mu)$. For ω close to μ and k close to k_F, we may expand M_r:

$$M_r(\mathbf{k}, \omega) = M_r(\mathbf{k}_F, \mu) + (\omega - \mu)\, \partial M_r/\partial\omega + (\mathbf{k} - \mathbf{k}_F) \cdot \nabla M_r.$$

This is inserted into (8.2.51). Real and imaginary parts are separated. Put $\omega = \omega_r + i\omega_i$. Since ω_i vanishes on the Fermi surface, it may be treated as small. The solution for the real part will be denoted $E(\mathbf{k})$:

$$E(\mathbf{k}) = \omega_r = \mu + \frac{(\mathbf{k} - \mathbf{k}_F) \cdot \{\nabla_k[\epsilon(\mathbf{k}) + M_r(\mathbf{k}, \mu)]\}_{\mathbf{k}=\mathbf{k}_F}}{[1 - \partial M_r/\partial\omega\, (\mathbf{k}_F, \omega)]_{\omega=\mu}}. \tag{8.2.53}$$

Then we have, on solving for the imaginary part,

$$\omega_i = E(\mathbf{k}) + i\,\{\operatorname{sgn}[\mu - E(\mathbf{k})]B(\mathbf{k})[E(\mathbf{k}) - \mu]^2/[1 - \partial M_r/\partial\omega]_{\omega=E(\mathbf{k})}\}. \tag{8.2.54}$$

It is convenient to define a quantity $Z(\mathbf{k})$,

$$Z(\mathbf{k}) = [1 - (\partial M_r/\partial\omega)_{\omega=E(\mathbf{k})}]^{-1}. \tag{8.2.55}$$

These results can now be substituted into (8.1.114) in order to determine the Green's function for values of ω which are not solutions of (8.2.51). We consider values of ϵ which are close to ω as defined by (8.2.51) so that it is possible to expand quantities to first order in $\epsilon - \omega$. The result is

$$\mathcal{G}(\mathbf{k}, \epsilon) = Z(\mathbf{k})/[\epsilon - E(\mathbf{k}) - i\Gamma(\mathbf{k})] + \phi(\mathbf{k}, \epsilon) \tag{8.2.56}$$

in which

$$\Gamma(\mathbf{k}) = \operatorname{sgn}[\mu - E(\mathbf{k})]Z(\mathbf{k})B(\mathbf{k})[E(\mathbf{k}) - \mu]^2 \tag{8.2.57}$$

and ϕ is a substantially unknown correction for higher-order effects not included here. It will be seen that (8.2.56) is in exact accord with (8.1.30), and thus describes the propagation of a quasi particle. Note that $Z(\mathbf{k})$ is real in this instance.

The time-dependent Green's function can be obtained from the Fourier transform of (8.2.56). The essential results were already given in Eqs. (8.1.31) and (8.1.32):

$$\begin{aligned}\mathcal{G}(\mathbf{k}, t) = {} & -iZ(\mathbf{k})\{\theta(t)\theta(E(\mathbf{k}) - \mu)\exp[-iE(\mathbf{k})t - \Gamma(\mathbf{k})t] \\ & - \theta(-t)\theta(\mu - E(\mathbf{k}))\exp[-iE(\mathbf{k})t + \Gamma(\mathbf{k})t]\} + \phi(\mathbf{k}, t)\end{aligned} \tag{8.2.58}$$

in which the last term is the Fourier transform of the function $\phi(\mathbf{k}, \epsilon)$ in (8.2.56). We can now determine $\bar{n}(\mathbf{k})$ from (8.2.50):

$$\bar{n}(\mathbf{k}) = Z(\mathbf{k})\theta(\mu - E(\mathbf{k})) + \phi(\mathbf{k}, 0_-). \tag{8.2.59}$$

It is assumed that $\phi(\mathbf{k}, 0_-)$ is continuous across the Fermi surface. Then there is a discontinuity of magnitude $Z(\mathbf{k})$ in the occupation number on crossing the Fermi surface. It was shown following Eq. (8.1.33) that $Z(\mathbf{k}) \leq 1$. Thus the discontinuity in the occupation number characteristic of a system of noninteracting particles persists when the interaction is present, provided perturbation theory converges, but its magnitude is decreased. The actual occupation number function is shown schematically

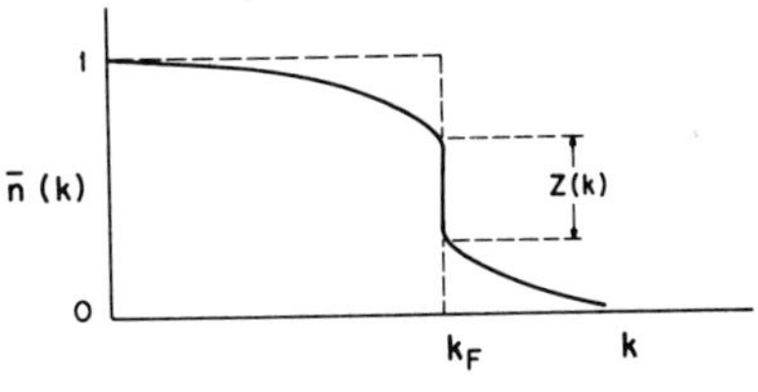

Fig. 8.2.7. Average occupation number in an interacting system (—) and in a non-interacting system (– – –).

in Fig. 8.2.7. If perturbation theory fails to converge, the Fermi surface may disappear, as we have seen to be the case in superconductivity.

The magnitude of the discontinuity has been estimated for a free electron gas by Daniel and Vosko (1960). Their work is a straightforward extension of the calculation of the correlation energy by Gell-Mann and Brueckner as presented in the previous subsection. A term is added to the Hamiltonian whose expectation value gives (8.2.49):

$$H = H_0 + H' = H_0 + \xi \sum_{p\sigma} \delta_{\mathbf{p},\mathbf{k}}\, c^+_{\mathbf{p}\sigma} c_{\mathbf{k}\sigma}. \tag{8.2.60}$$

In the absence of spin order, the sum over spins in (8.2.53) is permissible. The quantity ξ is an arbitrary small parameter. The Hamiltonian H_0 contains the complete electron–electron interaction. The energy of the system is evaluated to all orders in the electron–electron interaction, but only to first order in ξ. This first-order term is just ξ times the required expectation value. Their calculation has been improved and extended to metallic densities by Geldart *et al.* (1964), who find that a substantial discontinuity persists for densities in the range of interest.

One method of measuring the average occupation number, interpreted as a momentum distribution function, is through observation of x-ray Compton scattering. In this process, a high energy photon interacts with an electron in the Fermi sea and transfers it to an excited state. The differential cross section for this process is proportional to

$$d^2\sigma/d\omega\, d\Omega \approx \int \delta[\hbar\omega - (\hbar^2\mathbf{p}^2/2m) - (\hbar^2\mathbf{k}\cdot\mathbf{p}/m)]\bar{n}(\mathbf{k})\, d^3k \tag{8.2.61}$$

in which $\hbar\omega$ is the change in photon energy, $\mathbf{k}$ is the initial wave vector of the electron, and $\mathbf{p}$ is the momentum transferred to the electron (Platzman and Tzoar, 1965). The delta function simply expresses conservation of energy in the process. If the initial and final photon momenta are specified experimentally, $\mathbf{p}$ is known, and energy conservation fixes a unique value of $\mathbf{k}\cdot\mathbf{p}$. As a result, the experiment measures a plane cross section of the momentum distribution, perpendicular to $\mathbf{p}$. Such observations have been made by Phillips and Weiss (1968), who find the expected discontinuity.

8.3 The Landau Theory of Fermi Liquids

It is a remarkable fact of nature that a single-particle description is applicable, at least in a qualitative sense, to real solids. The success of band theory requires explanation since the interaction between electrons is not weak. A general description of interacting electrons can be based on the Landau theory of Fermi liquids (Landau, 1956, 1957, 1959) which was extended to charged systems by Silin (1958).

8.3.1 *Quasi Particles*

The essential idea of the Landau theory is that the energy levels of the interacting system can be classified according to the same quantum numbers used for noninteracting particles. In other words, we start from some definite state in the absence of interaction, then turn on the interaction slowly. The interaction representation matrix elements of the potential can be regarded as containing a factor $e^{\alpha t}$ for $t < 0$, in which α is a small positive quantity. Under the influence of the interaction, our state evolves in some way, but remains characterized by the same wave vector. Wave vector is conserved at each vertex when the two-body interaction is translationally invariant and there is a series for calculating the full single-particle Green's function for a state of wave vector $\mathbf{k}$, starting from the free particle Green's function.

The states of the interacting system are called "quasi-particle states" (or "elementary excitations"), each of which has a definite $\mathbf{k}$. Physically, a quasi particle may be pictured as a single particle surrounded by a self-consistent distribution of other particles. Quasi-particle states include quasi electrons and quasi holes. The states are occupied according to Fermi statistics. The energy of the entire system is not simply the sum of the energies of individual quasi-particle states: it depends on the distribution function as well. This is already evident in the Hartree–Fock approximation.

A single quasi-particle state is not, in general, an eigenstate of the Hamiltonian for the interacting system. Such a state has a lifetime, which is determined from the imaginary part of the self-energy function. However, close to the Fermi energy, the lifetime becomes very long (it is infinite exactly at the Fermi energy). Thus the notion of quasi particle is most useful close to the Fermi energy.

The distribution of quasi particles is described by a function $n(\mathbf{k})$. If this function were specified, it would be possible, in principle, to compute the total energy. Thus the total energy of the system is a "functional" of $n(\mathbf{k})$, which we denote as $\mathcal{E}[n(\mathbf{k})]$. [For an interesting extension of these ideas to position space, see Hohenberg and Kohn (1964)]. In general, we do not know $n(\mathbf{k})$ explicitly. Suppose that $n(\mathbf{k})$ is changed by a small amount $\delta n(\mathbf{k})$. The change in the energy of the system is $\delta\mathcal{E}$. To first order in δn, this is

$$\delta\mathcal{E} = [V/(2\pi)^3] \int E(\mathbf{k})\, \delta n(\mathbf{k})\, d^3k \tag{8.3.1a}$$

where V is the volume of the system. In a situation in which it is necessary to take account of spin explicitly, we consider a separate distribution for each spin state: $n \to n_\sigma(\mathbf{k})$, and replace (8.3.1a) by

$$\delta\mathcal{E} = [V/(2\pi)^3] \sum_\sigma \int E_\sigma(\mathbf{k})\, \delta n_\sigma(\mathbf{k})\, d^3k. \tag{8.3.1b}$$

The quantity $E_\sigma(\mathbf{k})$ is the effective energy of the quasi particle. It is to be regarded, according to (8.3.1b), as the first functional derivative of the energy with respect to the distribution function

$$E_\sigma(\mathbf{k}) = \delta\mathcal{E}/\delta n_\sigma(\mathbf{k}). \tag{8.3.2}$$

If $\mathbf{k}$ is on the Fermi surface (we denote this by k_F without implying that the Fermi surface is spherical), the addition of one particle in k_F, with the original system in its ground state of energy $\mathcal{E}_\mathrm{g}$, gives us an $(N + 1)$ particle system in its ground state. Thus

$$E(k_\mathrm{F}) = \mathcal{E}_\mathrm{g}(N + 1) - \mathcal{E}_\mathrm{g}(N) = \mu \tag{8.3.3}$$

in which μ is the chemical potential

$$\mu = \partial\mathcal{E}_\mathrm{g}/\partial N. \tag{8.3.4}$$

It is frequently necessary to obtain the change in energy of the system to second order in δn. This is written as

$$\delta\mathcal{E} = [V/(2\pi)^3] \sum_\sigma \int \epsilon_\sigma(\mathbf{k})\, \delta n_\sigma(\mathbf{k})\, d^3k$$

$$+ \tfrac{1}{2}[V/(2\pi)^6] \sum_{\sigma'\sigma} \int f_{\sigma\sigma'}(\mathbf{k}, \mathbf{q})\, \delta n_\sigma(\mathbf{k})\, \delta n_{\sigma'}(\mathbf{q})\, d^3k\, d^3q. \tag{8.3.5}$$

The function $f_{\sigma\sigma'}(\mathbf{k}, \mathbf{q})$ is the second functional derivative of the energy with respect to the distribution function. This function is a fundamental quantity which characterizes the Fermi liquid. It is symmetric,

$$f_{\sigma\sigma'}(\mathbf{k}, \mathbf{q}) = f_{\sigma'\sigma}(\mathbf{q}, \mathbf{k}). \tag{8.3.6}$$

It can be shown that f is the forward amplitude for quasi electron, quasi-hole scattering when both $\mathbf{k}$ and $\mathbf{q}$ are on the Fermi surface.

We can interpret $\epsilon_\sigma(\mathbf{k})$ as the energy of an isolated quasi particle, that is, if there are no other quasi particles present. In the presence of other particles, we have the energy $E_\sigma(\mathbf{k})$, which is, according to (8.3.1),

$$E_\sigma(\mathbf{k}) = \epsilon_\sigma(\mathbf{k}) + [1/(2\pi)^3] \sum_{\sigma'} \int f_{\sigma\sigma'}(\mathbf{k}, \mathbf{q})\, \delta n_{\sigma'}(\mathbf{q})\, d^3q. \quad (8.3.7)$$

If the system has no magnetic order and no external magnetic field is present, the energies must be independent of σ and $f_{\sigma\sigma'}$ can depend only on $\sigma \cdot \sigma'$. It is then convenient to express $f_{\sigma\sigma'}$ as a sum of direct and exchange terms $f_{\rm d}$ and $f_{\rm x}$, respectively,

$$f_{\sigma\sigma'}(\mathbf{k}, \mathbf{q}) = f_{\rm d}(\mathbf{k}, \mathbf{q}) + f_{\rm x}(\mathbf{k}, \mathbf{q})\, \delta_{\sigma\sigma'}. \quad (8.3.8)$$

Quasi particles are fermions. It will be assumed that the expression for the entropy of noninteracting fermions is applicable here:

$$S = [V/(2\pi)^3] \sum_{\sigma} \int \{n_\sigma(\mathbf{k}) \ln n_\sigma(\mathbf{k}) + [1 - n_\sigma(\mathbf{k})] \ln[1 - n_\sigma(\mathbf{k})]\}\, d^3k. \quad (8.3.9)$$

The entropy is to be maximized subject to the constraints that the total number of particles N and the total energy $\mathcal{E}$ are fixed. The number is

$$N = [V/(2\pi)^3] \sum_{\sigma} \int n_\sigma(\mathbf{k})\, d^3k. \quad (8.3.10)$$

It is required that

$$\delta N = 0 \quad \text{and} \quad \delta \mathcal{E} = 0. \quad (8.3.11)$$

The constraints are imposed using the method of Lagrange multipliers. The resulting distribution function is the usual Fermi function

$$n(\mathbf{k}) = n[E(\mathbf{k})] = \{\exp[(E(\mathbf{k}) - \mu)/KT] + 1\}^{-1}. \quad (8.3.12)$$

It follows from (8.3.12) and (8.3.7) that the energies of quasi-particle states depend on temperature.

8.3.2 *Specific Heat, Compressibility, and Speed of Sound*

We will examine in this subsection some of the thermodynamic properties of an electrically neutral Fermi liquid. The theory of the specific heat proceeds exactly as in Section 4.1.5, Part A, except that the particle energy is to be interpreted as referring to a quasi particle. In particular, Eq. (4.1.59) is valid for the low temperature specific heat

$$C_V = (\pi^2/3)K^2G(\mu_0)T \tag{8.3.13}$$

in which K is Boltzmann's constant and $G(\mu_0)$ is the density of quasi-particle states at the zero-temperature value of the chemical potential. The density of states at energy E is obtained from the usual formula

$$G(E) = \sum_\sigma G_\sigma(E) = [\Omega/(2\pi)^3] \sum_\sigma \int d^3k\, \delta(E - E_\sigma(\mathbf{k})). \tag{8.3.14}$$

It is frequently convenient to introduce the thermal, or density of states, effective mass by the relation

$$G(\mu) = \Omega m^* k_F/\pi^2\hbar^2 \tag{8.3.15}$$

in which k_F is given by the usual expression for a free electron system

$$k_F{}^3 = 3\pi^2\rho \tag{8.3.16}$$

and ρ $(=N/V)$ is the number density.

We now turn to the compressibility. Suppose the system is confined in a volume V. The total energy of the system in the ground state is proportional to the volume and can be written as

$$\mathcal{E} = Vf(\rho) \tag{8.3.17}$$

and the pressure is

$$P = -\partial\mathcal{E}/\partial V. \tag{8.3.18}$$

The compressibility κ is defined by

$$\kappa = -1/V\, \partial V/\partial P. \tag{8.3.19}$$

The compressibility is related to the energy by

$$\kappa^{-1} = -V\, \partial^2\mathcal{E}/\partial V^2 = \rho^2\, d^2f/d\rho^2. \tag{8.3.20}$$

The chemical potential μ is the derivative of the energy with respect to the number of particles

$$\mu = \partial\mathcal{E}/\partial N = df/d\rho. \tag{8.3.21}$$

Equations (8.3.20) and (8.3.21) can be combined to yield

$$\kappa^{-1} = N\rho\, \partial\mu/\partial N. \tag{8.3.22}$$

The speed of sound can be found, once the compressibility is known. It follows from the basic macroscopic equations of fluid dynamics (see, for instance Serrin, 1959), that v_s, the speed of sound in a fluid, is given by

$$v_s^2 = \partial P/\partial \rho_m \tag{8.3.23}$$

in which $\rho_m = m\rho$ is the mass density. This leads to

$$v_s^2 = 1/m\rho\kappa = N/m\ \partial\mu/\partial N. \tag{8.3.24}$$

We will now calculate $d\mu/dN$. If the number of particles in the system is changed by an amount δN, μ changes as a result of two effects: (1) a change in density produces a change in k_F through (8.3.16), and (2) there is a contribution from the interaction according to the second term of (8.3.7):

$$\delta\mu = (\partial\epsilon/\partial k)_{k_F}\,\delta k_F + [2/(2\pi)^3]\int d^3q\, f(\mathbf{k}, \mathbf{q})\,\delta n(\mathbf{q}). \tag{8.3.25}$$

The factor of 2 in the second term accounts for summation over spins. Spin indices are not explicitly written. In terms of the direct and exchange contributions defined according to (8.3.8), the function f appearing in (8.3.25) is

$$f = f_d + \tfrac{1}{2}f_x. \tag{8.3.26}$$

It is possible to add or subtract particles only near the top of the Fermi distribution. Thus a change in the number of particles is related to a change in k_F by

$$\delta N = 8\pi V k_F^2\,\delta k_F/(2\pi)^3.$$

Further, since $\delta n(\mathbf{q})$ is different from zero only for values of $\mathbf{q}$ close to k_F, it is possible to integrate over the magnitude of $\mathbf{q}$ first. For simplicity, it will be assumed that $\epsilon(\mathbf{k})$ is parabolic, characterized by an effective mass m^*. We can obtain an expression for $\partial\mu/\partial N$ in the form

$$\partial\mu/\partial N = (\pi^2\hbar^2/Vm^*k_F) + (1/4\pi)\int d\Omega_q\, f(\mathbf{k}, \mathbf{q}). \tag{8.3.27}$$

The integral includes solid angles on the surface of the Fermi sphere. The speed of sound can now be obtained from (8.3.24)

$$v_s^2 = (\hbar^2 k_F^2/3mm^*) + (1/4\pi m\Omega)\int d\Omega_q\, f(\mathbf{k}, \mathbf{q}). \tag{8.3.28}$$

This result, although rigorous, is not directly applicable to solids, since no account has been taken of the motion of the ions. Note that a contribution is obtained from the particle interaction.

8.3.3 *Current*

We will now consider the effect of particle interactions on the calculation of the current in a system. The wave functions for the system are not known, and it is not practical to work directly with the Kubo formalism of Section 6.5.6. Instead, we adopt a kinetic approach. Denote the current in the system by **J**. It is supposed that

$$\mathbf{J} = \mathrm{Tr}(n\mathbf{v}) = V \sum_{\sigma} \int [d^3k/(2\pi)^3] n_\sigma(\mathbf{k}) \mathbf{v}_\sigma(\mathbf{k}) \tag{8.3.29}$$

in which n_σ specifies the quasi-particle occupation probability and $\mathbf{v}(\mathbf{k})$ is the velocity. In the ground state of the system, $n_\sigma = n_\sigma^{(0)}$, which is a function of energy only, and **J** vanishes by symmetry. The system will be assumed to be disturbed from its equilibrium state so that $n_\sigma(\mathbf{k}) - n_\sigma^{(0)}(\mathbf{k}) = \delta n_\sigma(\mathbf{k})$. The velocity of a quasi particle in state **k** will differ from the equilibrium value $\mathbf{v}_\sigma^{(0)}(\mathbf{k})$ by a correction $\delta\mathbf{v}_\sigma(\mathbf{k})$ which results from the change in the distribution function. The total current $\delta\mathbf{J}$ is, to first order in the departures from equilibrium values,

$$\delta\mathbf{J} = V \sum_{\sigma} \int [d^3k/(2\pi)^3][\mathbf{v}_\sigma^{(0)}(\mathbf{k})\ \delta n_\sigma(\mathbf{k}) + n_\sigma^{(0)}(\mathbf{k})\ \delta\mathbf{v}_\sigma(\mathbf{k})]. \tag{8.3.30}$$

In this equation,

$$\mathbf{v}^{(0)}(\mathbf{k}) = (1/\hbar)\ \nabla_{\mathbf{k}}\epsilon_\sigma(\mathbf{k}) \tag{8.3.31}$$

and $\delta\mathbf{v}_\sigma(\mathbf{k})$ can be obtained from (8.3.7) as

$$\delta\mathbf{v}_\sigma = (1/\hbar)\ \nabla_{\mathbf{k}} \left\{ \sum_{\sigma'} \int [d^3q/(2\pi)^3] f_{\sigma\sigma'}(\mathbf{k}, \mathbf{q})\ \delta n_{\sigma'}(\mathbf{q}) \right\}. \tag{8.3.32}$$

The second term in (8.3.30) can be transformed through integration by parts and a relabeling of the variables which makes use of the symmetry of f:

$$\sum_{\sigma} \int [d^3q/(2\pi)^3] n_\sigma^{(0)}(\mathbf{k})\ \delta\mathbf{v}_\sigma(\mathbf{k}) = -(1/\hbar) \sum_{\sigma',\sigma} \iint [d^3k/(2\pi)^3][d^3q/(2\pi)^3] \times f_{\sigma\sigma'}(\mathbf{k}, \mathbf{q})\ \nabla_{\mathbf{q}} n_{\sigma'}^{(0)}(\mathbf{q})\ \delta n_\sigma(\mathbf{k}). \tag{8.3.33}$$

For low temperatures, we have

$$\nabla_{\mathbf{q}} n_\sigma(\mathbf{q}) = -\nabla_{\mathbf{q}}\epsilon_\sigma(\mathbf{q})\ \delta[\epsilon_\sigma(\mathbf{q}) - \mu].$$

An integration over energies can be done in (8.3.33). The usual parabolic band is assumed. The result is

$$(1/\hbar) \sum_{\sigma',\sigma} \iint [d^3k/(2\pi)^3] k_F{}^2 [d\Omega_q/(2\pi)^3] \hat{\mathbf{q}} f_{\sigma\sigma'}(\mathbf{k}, \mathbf{q})$$

in which $\hat{\mathbf{q}}$ is a unit vector. In an isotropic system, symmetry requires that $f_{\sigma\sigma'}(\mathbf{k}, \mathbf{q})$ can depend on angles only through the angle θ between the vectors $\mathbf{k}$ and $\mathbf{q}$. We will suppose that this applies with sufficient accuracy in the system under study. Then $\mathbf{q}$ may be replaced by $\cos\theta$ in the integral. The expression (8.3.30) for the current can be written as

$$\delta J = V \sum_{\sigma} \int [d^3k/(2\pi)^3]\, \delta n(\mathbf{k})\, j_\sigma(\mathbf{k}) \tag{8.3.34}$$

with

$$\hbar j_\sigma(\mathbf{k}) = \nabla_{\mathbf{k}}\epsilon(\mathbf{k}) + [k_F{}^2/(2\pi)^3] \sum_{\sigma'} \int d\Omega_q \cos\theta\, f_{\sigma\sigma'}(\cos\theta). \tag{8.3.35}$$

The magnitudes of $\mathbf{k}$ and $\mathbf{q}$ are equal to k_F. The first term in (8.3.35) gives the usual expression for the current; the second represents the contribution from the interaction. This term is usually pictured as describing the back flow of the Fermi liquid around a moving quasi particle. In a translationally invariant system in which $\mathbf{k}$ is proportional to the actual mechanical momentum, the current must be related to this momentum by $j_{\mathbf{k}} = \hbar\mathbf{k}/m$ in which m is the real mass in contrast to the effective mass. Since the interactions between particles are translationally invariant, they cannot affect the value of the current. It is then possible to obtain a relation between the real and the effective mass (spin effects being ignored)

$$1/m = (1/m^*) + [2k_F/(2\pi)^3\hbar^2] \int d\Omega_q \cos\theta\, f(\cos\theta). \tag{8.3.36}$$

This relation is not directly applicable to solids in which $\mathbf{k}$ does not specify momentum.

8.3.4 *Spin Susceptibility*

We will consider the contribution to the magnetic susceptibility of a Fermi liquid from spin alignment. Diamagnetic contributions from Landau level formation in a charged system will be ignored. Suppose the system is placed in a magnetic field $\mathbf{B}$ which defines the "z" direction. The energy of a free spin in this field is $\beta s_z B$ in which $\beta = \hbar e/m$, and $s_z = \pm\frac{1}{2}$. However, it is necessary to consider a change in the distribution function in the presence of the field. The change in energy of a state of spin σ can be written as

$$\delta E_\sigma(\mathbf{k}) = -\beta s_z B + \sum_{\sigma'} \int [d^3q/(2\pi)^3] f_{\sigma\sigma'}(\mathbf{k}, \mathbf{q})\, \delta n_{\sigma'}(\mathbf{q}). \quad (8.3.37)$$

The change in the distribution function δn has to be considered. At first glance, this may appear to present a complicated problem, since the presence of a field will produce a change in the chemical potential μ. However, this change is of second order in B (see Section 6.4.1) and thus does not contribute to the calculation of the magnetization to first order in the field, as is required to determine the susceptibility. Therefore, it is sufficient to write

$$\delta n_\sigma(\mathbf{q}) = (\partial n_\sigma/\partial E)\, \delta E_\sigma(\mathbf{q}) = -\delta(E_\sigma - \mu)\, \delta E_\sigma(\mathbf{q}).$$

This expression is valid at low temperatures. It is to be substituted into (8.3.37). We look for a solution of the resulting equation of the form $\delta E_\sigma = -\gamma s_z B$ with γ a constant to be determined. The following equation is obtained:

$$\begin{aligned}\gamma &= \beta - \gamma \sum_{\sigma'} \int [d^3q/(2\pi)^3](s_{z'}/s_z) f_{\sigma\sigma'}(\mathbf{k}, \mathbf{q})\, \delta(E_{\sigma'} - \mu) \\ &= \beta - \gamma(m^* k_F/\hbar^2) \int [d\Omega_q/(2\pi)^3][f_{\sigma\sigma}(\mathbf{k}, \mathbf{q}) - f_{\sigma,-\sigma}(\mathbf{k}, \mathbf{q})] \\ &= \beta - \gamma[m^* k_F/(2\pi)^3\hbar^2] \int d\Omega_q\, f_x(\mathbf{k}, \mathbf{q}). \end{aligned} \quad (8.3.38)$$

In the last step of (8.3.38) we have introduced the "exchange" component of the function f [see (8.3.8)]. This equation is easily solved for γ. The magnetization per unit volume M is

$$M = -\beta \sum_\sigma \int [d^3k/(2\pi)^3]\, s_z \delta n_\sigma(\mathbf{k}) = (m^* k_F/4\pi^2\hbar^2)\beta\gamma B. \quad (8.3.39)$$

The susceptibility χ is defined by $M = \mu_0 \chi H$, but as discussed in Section 6.4, as long as the susceptibility is small, it is legitimate to set $B = \mu_0 H$ in (8.3.39). Then

$$\chi = (m^* k_F/4\pi^2\hbar^2)\beta\gamma. \quad (8.3.40)$$

We have from (8.3.40) and (8.3.38)

$$1/\chi = (1/\beta^2)\left[(4\pi^2\hbar^2/m^* k_F) + (1/2\pi) \int d\Omega_q\, f_x(\mathbf{k}, \mathbf{q})\right]. \quad (8.3.41)$$

We have now obtained, through Eqs. (8.3.28), (8.3.36), and (8.3.41), expressions for the speed of sound, the effective mass, and the spin susceptibility in terms of averages of the function f over the Fermi surface. As was pointed out previously, for an isotropic system, and approximately in a simple metal, $f(\mathbf{k}, \mathbf{q})$ is a function of the angle θ between the vectors $\mathbf{k}$ and $\mathbf{q}$. Since $\mathbf{k}$ and $\mathbf{q}$ are constrained to have magnitude k_F, θ is in fact the only relevant variable. It now becomes convenient to introduce an expansion of f in terms of Legendre polynomials $P_l(\cos\theta)$:

$$f(\mathbf{k}, \mathbf{q})\,|_{k=k_F,\, q=q_F} = \sum_l A_l P_l(\cos\theta). \tag{8.3.42}$$

Our results can be simply expressed in terms of the coefficient $A_l(k_F)$. The speed of sound is, according to (8.3.28),

$$v_s^2 = (\hbar^2 k_F^2/3mm^*) + (1/m\Omega)A_0. \tag{8.3.43}$$

The effective mass, from (8.3.19), is

$$1/m^* = (1/m) - (k_F/3\pi^2\hbar^2)A_1. \tag{8.3.44}$$

The paramagnetic susceptibility involves only the exchange component of f. Obviously, this can be expanded in a manner identical to (8.3.42), with coefficients $A_l^{(x)}$. Thus we have from (8.3.41)

$$1/\chi = (1/\beta^2)[(4\pi^2\hbar^2/m^* k_F) + 2A_0^{(x)}]. \tag{8.3.45}$$

8.3.5 *Boltzmann Equation and Collective Oscillations*

The procedures of Section 7.1 can be generalized to apply to the transport properties of quasi particles. The quasi-particle distribution function n is allowed to depend on position and time as well as on $\mathbf{k}$: $n = n_\sigma(\mathbf{k}, \mathbf{r}, t)$. The general arguments of Section 7.1 show that n should obey an equation similar to the Boltzmann equation:

$$\partial n_\sigma(\mathbf{k}, \mathbf{r}, t)/\partial t + \mathbf{v}_\sigma(\mathbf{k})\cdot\nabla_{\mathbf{r}} n_\sigma(\mathbf{k}, \mathbf{r}, t) + d\mathbf{k}/dt\cdot\nabla_{\mathbf{k}} n_\sigma(\mathbf{k}, \mathbf{r}, t) = P_\sigma(n) \tag{8.3.46}$$

in which $P_\sigma(n)$ represents the rate of change of n_σ due to collisions. It will always be assumed here that the departures from equilibrium are small:

$$n_\sigma(\mathbf{k}, \mathbf{r}, t) = n_\sigma^{(0)}(\mathbf{k}) + \delta n_\sigma(\mathbf{k}, \mathbf{r}, t). \tag{8.3.47}$$

It is then legitimate to construct a theory which is of first order in δn. To this end, we set

$$\hbar\, d\mathbf{k}/dt = -\nabla_{\mathbf{r}} E_\sigma(\mathbf{k}, n); \qquad \hbar\mathbf{v}_\sigma(\mathbf{k}) = \nabla_{\mathbf{k}}\epsilon_\sigma(\mathbf{k}). \tag{8.3.48}$$

The linearized Boltzmann equation is then

$$\partial \delta n_\sigma/\partial t + (1/\hbar)[\nabla_{\mathbf{k}}\epsilon_\sigma \cdot \nabla_{\mathbf{r}}\, \delta n_\sigma - \nabla_{\mathbf{r}} E_\sigma \cdot \nabla_{\mathbf{k}} n_\sigma^{(0)}(\mathbf{k})] = P(n). \quad (8.3.49)$$

The position dependence of E_σ is a consequence of the second term in (8.3.7):

$$\nabla_{\mathbf{r}} E_\sigma(\mathbf{k}, n) = [1/(2\pi)^3] \sum_\sigma \int f_{\sigma\sigma'}(\mathbf{k}, \mathbf{q})\, \nabla \delta n_{\sigma'}(\mathbf{q}, \mathbf{r}, t)\, d^3q. \quad (8.3.50)$$

Since $n^{(0)}$ depends on $\mathbf{k}$ only through the energy, we have at low temperatures

$$\nabla_{\mathbf{k}} n_\sigma^{(0)}(\mathbf{k}) = -\hbar v_\sigma(\mathbf{k})\, \delta(\epsilon_\sigma(\mathbf{k}) - \mu). \quad (8.3.51)$$

Equations (8.3.50) and (8.3.51) are substituted into (8.3.49), which becomes

$$\partial \delta n_\sigma/\partial t + \mathbf{v}_\sigma(\mathbf{k}) \cdot \nabla_{\mathbf{r}}\, \delta n_\sigma + \mathbf{v}_\sigma(\mathbf{k}) \cdot \delta(\epsilon_\sigma(\mathbf{k}) - \mu)$$

$$\times [1/(2\pi)^3] \sum_{\sigma'} \int f_{\sigma\sigma'}(\mathbf{k}, \mathbf{q})\, \nabla_{\mathbf{r}}\, \delta n_{\sigma'}(\mathbf{q}, \mathbf{r}, t)\, d^3q = P(n). \quad (8.3.52)$$

The preceding discussion applies in the absence of external forces. Now suppose that the system is disturbed by an external electric field $\boldsymbol{\mathcal{E}}$. A term must be added to the expression for $d\mathbf{k}/dt$, Eq. (8.3.48), to represent the force:

$$\hbar\, d\mathbf{k}/dt = -\nabla_{\mathbf{r}} E_\sigma(\mathbf{k}, n) + \mathbf{F} \quad (8.3.53)$$

where $\mathbf{F}$ represents the total force applied to a quasi particle. This is not simply the external force $\mathbf{F}_e = e\boldsymbol{\mathcal{E}}_e$. The field inside the medium is modified by the response of the particles, which contributes an induced field $\boldsymbol{\mathcal{E}}_I$:

$$\mathbf{F} = e(\boldsymbol{\mathcal{E}}_e + \boldsymbol{\mathcal{E}}_I). \quad (8.3.54)$$

The induced field is related to the density variation δn by

$$\nabla_{\mathbf{r}} \cdot \boldsymbol{\mathcal{E}}_I(\mathbf{r}, t) = (e/\epsilon_0)\, \delta n(\mathbf{r}, t) = [e/\epsilon_0(2\pi)^3] \sum_\sigma \int \delta n_\sigma(\mathbf{k}, \mathbf{r}, t)\, d^3k. \quad (8.3.55)$$

Evidently we have a coupled set of equations for δn_σ which must be solved self-consistently.

We will not investigate the details of quasi-particle scattering. It will simply be supposed that a time of relaxation exists:

$$P(n) = -\delta n_\sigma(\mathbf{k}, \mathbf{r}, t)/\tau(\mathbf{k}). \quad (8.3.56)$$

The relaxation time will be assumed to be independent of position and time. It will be recalled from Section 7.2.2 that a time of relaxation is expected to exist whenever the processes which scatter quasi particles are elastic.

Let us consider the specific case of a periodic applied field

$$\mathbf{\mathcal{E}}_e = \mathbf{\mathcal{E}}_0 \exp[i(\mathbf{p}\cdot\mathbf{r} - \omega t)]. \tag{8.3.57}$$

The coupled set of equations for δn_σ is linear, and the solutions will have the position and time dependence of the applied field:

$$\delta n_\sigma(\mathbf{k}, \mathbf{r}, t) = \delta n_\sigma(\mathbf{k}, \mathbf{p}, \omega) \exp[i(\mathbf{p}\cdot\mathbf{r} - \omega t)]. \tag{8.3.58}$$

Our procedure is equivalent to the introduction of a Fourier transform for an applied field which is an arbitrary function of position and time. Equations (8.3.57) and (8.3.58) are substituted into (8.3.52) and (8.3.55), and the force term from (8.3.53) is added. The result is a set of equations for $\delta n_\sigma(\mathbf{k}, \mathbf{p}, \omega)$:

$$[\mathbf{v}_\sigma(\mathbf{k})\cdot\mathbf{p} - \omega]\,\delta n_\sigma(\mathbf{k}, \mathbf{p}, \omega) + \mathbf{v}_\sigma(\mathbf{k})\,\delta[\varepsilon_\sigma(\mathbf{k}) - \mu]\cdot\Big[ie(\mathbf{\mathcal{E}}_e + \mathbf{\mathcal{E}}_I)$$

$$+ \mathbf{p}[1/(2\pi)^3] \sum_\sigma \int f_{\sigma\sigma'}(\mathbf{k}, \mathbf{q})\,\delta n_{\sigma'}(\mathbf{q}, \mathbf{p}, \omega)\,d^3q\Big] = i\delta n_\sigma(\mathbf{k}, \mathbf{p}, \omega)/\tau \tag{8.3.59}$$

and

$$i\mathbf{p}\cdot\mathbf{\mathcal{E}}_I(p, \omega) = [e/(2\pi)^3\epsilon_0] \sum_\sigma \int \delta n_\sigma(\mathbf{k}, \mathbf{p}, \omega)\,d^3k. \tag{8.3.60}$$

These equations are obviously quite complex, and we will confine our attention to certain limiting situations. If the relaxation time is short $(\omega\tau \ll 1)$, the quasiparticle system remains in equilibrium, and the deformation δn proceeds adiabatically. In this limit, the ordinary sound waves previously described are obtained. In the opposite limit in which $\omega\tau \ll 1$, collisions are of no importance. We will investigate some phenomena which occur under these circumstances.

For simplicity, we will first suppose that there are no driving forces and that the system is electrically neutral, as in the case of liquid ^{3}He. Then only (8.3.59) is to be considered. It simplifies to

$$[\mathbf{v}_\sigma(\mathbf{k})\cdot\mathbf{p} - \omega]\,\delta n_\sigma(\mathbf{k}, \mathbf{p}, \omega) + \mathbf{p}\cdot\mathbf{v}_\sigma(\mathbf{k})\,\delta[\epsilon_\sigma(\mathbf{k}) - \mu]$$

$$\times [1/(2\pi)^3] \sum_{\sigma'} \int f_{\sigma\sigma'}(\mathbf{k}, \mathbf{q})\,\delta n_{\sigma'}(\mathbf{q}, \mathbf{p}, \omega)\,d^3q = 0. \tag{8.3.61}$$

We will first consider disturbances which are independent of σ. The σ index is dropped, and $\sum_{\sigma'}$ is replaced by a factor of 2. Equation (8.3.61) is a homogenous integral equation which has nontrivial solutions only for certain values—eigenvalues—of the frequency ω. These frequencies are those of collective oscillations of the system.

Equation (8.3.61) can be simplified as follows. The density variation δn will be different from zero only at the Fermi energy (providing that the temperature is low enough). Thus δn should contain a delta function, and it becomes convenient to introduce a function $\xi(\mathbf{k}, \mathbf{p}, \omega)$ by

$$\delta n(\mathbf{k}, \mathbf{p}, \omega) = \hbar \mid \mathbf{v}(\mathbf{k}) \mid \delta[\epsilon(\mathbf{k}) - \mu]\xi(\mathbf{k}, \mathbf{p}, \omega). \tag{8.3.62}$$

It will be convenient to drop the arguments $\mathbf{p}$, ω in the specification of ξ. It is also useful to let $\mathbf{p}$ define the polar axis of a spherical coordinate system, with θ the angle between $\mathbf{k}$ and $\mathbf{p}$, and α the angle between $\mathbf{p}$ and $\mathbf{q}$. A parameter u and a function $F(\alpha)$ are introduced by the definitions

$$u = \omega/pv_{\mathrm{F}} = m^*\omega/\hbar k_{\mathrm{F}}p, \tag{8.3.63a}$$

$$F(\alpha) = [1/(2\pi)^3](m^*k_{\mathrm{F}}/\hbar p)f(\alpha). \tag{8.3.63b}$$

In (8.3.63b), $f(\alpha) = f_{\mathrm{d}} + \frac{1}{2}f_{\mathbf{x}}$, where f_{d} and $f_{\mathbf{x}}$ are defined by (8.3.8). Equation (8.3.61) reduces to

$$[\cos\theta - u]\xi(\mathbf{k}) + 2\cos\theta \int F(\alpha)\xi(\mathbf{q})\, d\Omega_q = 0. \tag{8.3.64}$$

This equation evidently describes oscillations of the Fermi surface. The nature of the solutions depend on the angular dependence of F. To obtain some insight into what is going on, we will consider the case of constant F: $F(\alpha) = F_0$. A solution of (8.3.64) can then be obtained in the form

$$\xi(\mathbf{k}) = [(\cos\theta)/u - \cos\theta]C \tag{8.3.65}$$

in which C is a constant. Substitution of (8.3.65) into (8.3.64) shows that u and F_0 must be related by

$$1 = 2F_0 \int (\cos\alpha)/(u - \cos\alpha)\, d\Omega_q. \tag{8.3.66}$$

It should be noticed that if $u < 1$, (8.3.66) is singular. In this case, a resonant transfer of energy between the wave and the individual quasi particles is possible, and leads to a damping of the oscillation. This situation may be treated in the usual way by allowing the frequency to have a negative imaginary part. We will, however, confine out attention to $u > 1$, for which the waves are not damped. The integral is easily evaluated

$$1 = 8\pi F_0[(u/2)\ln\{(u+1)/(u-1)\} - 1]. \qquad (8.3.67)$$

This equation must be solved for u in terms of F_0.

Solutions of (8.3.67) exist for positive values of F_0. They must be obtained numerically for general values of F_0; however, analytic results are easily obtained in two limits

$$\text{large } F_0: \quad u = (8\pi F_0/3)^{1/2}, \qquad (8.3.68a)$$

$$\text{small } F_0: \quad u = 1 + (2/e^2)\exp(-\tfrac{1}{4}\pi F_0). \qquad (8.3.68b)$$

The mode of oscillation described by (8.3.68) can be pictured as an egg-shaped deformation of the Fermi surface with the elongation in the direction of **p**. This mode is known as "zero sound." Equation (8.3.68b) indicates that when the coupling is weak, $\omega \approx pv_F$; zero sound propagates with the Fermi velocity.

The procedure we have employed can be generalized to permit the study of oscillations when more general expressions [such as Eq. (8.3.42)] are used for F. It is particularly interesting that oscillations of a spin wave character exist, even though the system is not magnetically ordered. To obtain these waves, we return to Eq. (8.3.61) and consider a deformation of the Fermi surface which depends on the spin:

$$\delta n_\sigma(\mathbf{k}, \mathbf{p}, \omega) = \hbar\,|\,\mathbf{v}(\mathbf{k})\,|\,\delta[\epsilon(\mathbf{k}) - \mu]\,\sigma\xi(\mathbf{k}, \mathbf{p}, \omega). \qquad (8.3.69)$$

It is supposed here that $\epsilon_\sigma(\mathbf{k})$ and $\mathbf{v}_\sigma(\mathbf{k})$ are independent of σ to a satisfactory degree of accuracy. In (8.3.69) the quantity σ takes the value $+1$ or -1. We can now repeat the argument which leads to an equation of the form of (8.3.63). The results can be taken over without change except for a redefinition of F. Instead of (8.3.63b) we require

$$F(\alpha) = [1/(2\pi)^3](m^* k_F/\hbar p)\cdot\tfrac{1}{2}[f_{++}(\alpha) - f_{+-}(\alpha)]$$

$$= [1/(2\pi)^3](m^*/2\hbar p) f_x(\alpha) \qquad (8.3.70)$$

in which f_x is the exchange contribution to the Fermi liquid interaction function.

Spin waves in nonferromagnetic metals (specifically sodium and potassium) have been observed (Schultz and Dunifer, 1967). The analysis is somewhat more complex than we have indicated here: for the details, see Platzman and Wolff (1967).

We will now discuss briefly the effects of the Coulomb interactions in a charged Fermi liquid on the transport equation. We may proceed by solving Eq. (8.3.60) for the induced field $\mathcal{E}_I$. For convenience, we again ignore spin, and introduce

$$\rho(\mathbf{p}, \omega) = [2/(2\pi)^3] \int \delta n(\mathbf{k}, \mathbf{p}, \omega)\, d^3k. \tag{8.3.71}$$

Then we find $\mathcal{E}_I$ to be given by

$$\mathcal{E}_I(p, \omega) = -ie(\mathbf{p}/\epsilon_0\mathbf{p}^2)\rho(\mathbf{p}, \omega). \tag{8.3.72}$$

Equation (8.3.72) may now be inserted in (8.3.59). The result is, when we discard spin indices,

$$\begin{aligned} &[\mathbf{v}(\mathbf{k})\cdot\mathbf{p} - \omega]\,\delta n(\mathbf{k}, \mathbf{p}, \omega) + ie\mathbf{v}(\mathbf{k})\,\delta[\epsilon(\mathbf{k}) - \mu]\cdot\mathcal{E}_e \\ &\quad + \delta[\epsilon(\mathbf{k}) - \mu]\mathbf{p}\cdot\mathbf{v}(\mathbf{k})[2/(2\pi)^3] \\ &\quad \times \int [f(\mathbf{k}, \mathbf{q}) + (e^2/\epsilon_0\mathbf{p}^2)]\,\delta n(\mathbf{q}, \mathbf{p}, \omega)\, d^3q = 0. \end{aligned} \tag{8.3.73}$$

This result displays an essential difference with respect to (8.3.61): the effective interaction between charged quasi particles is not simply $f(\mathbf{k}, \mathbf{q})$, but $f(\mathbf{k}, \mathbf{q}) + e^2/\epsilon_0\mathbf{p}^2$. This additional term appears even if the external field $\mathcal{E}_e$ is absent, since the fluctuations in a charged Fermi liquid displace charge and so must induce an electric field. The consequences of this additional term will now be investigated.

8.3.6 *Plasma Oscillations*

The collective oscillations of a charged Fermi liquid, such as an electron gas, are different in character from those of a neutral system. This occurs because of the presence of the Coulomb interaction term $e^2/\epsilon_0\mathbf{p}^2$ in (8.3.73) which becomes dominant in the long wavelength limit in which p is small. These modes are the plasma oscillations which have been encountered previously (Section 8.2.1). In this subsection, these oscillations will be obtained from the solution of the transport equation (8.3.73).

It is assumed that no external electric field is present. Attention is concentrated on the small p limit. In this case, the interaction function $f(\mathbf{k}, \mathbf{q})$ in (8.3.73) is negligible compared to the coulomb term. Equation (8.3.73) may be rewritten as

$$\mathbf{p}\cdot\mathbf{v}(\mathbf{k})\{\delta n(\mathbf{k}, \mathbf{p}, \omega) + [e^2\rho(\mathbf{p}, \omega)/\epsilon_0\mathbf{p}^2]\delta[\epsilon(\mathbf{k}) - \mu]\} - \omega\,\delta n(\mathbf{k}, \mathbf{p}, \omega) = 0 \tag{8.3.74}$$

in which ρ is defined by (8.3.71). We follow the procedures of the previous subsection, and introduce ξ through Eq. (8.3.62). The quantity ρ may be expressed in terms of ξ as

$$\rho(\mathbf{p}, \omega) = k_F^2/4\pi^3 \int d\Omega_q \, \xi(\mathbf{q}) \tag{8.3.75}$$

in which an assumption of a parabolic band has been introduced. Equation (8.3.74) becomes

$$\mathbf{p}\cdot\mathbf{v}(\mathbf{k}) \left[\xi(\mathbf{k}) + (e^2 k_F/4\pi^3 \epsilon_0 \mathbf{p}^2 \hbar v_F) \int d\Omega_q \, \xi(\mathbf{q})\right] - \omega\xi = 0. \tag{8.3.76}$$

The constant multiplying the integral can be expressed as

$$e^2 k_F^2/4\pi^3 \epsilon_0 \mathbf{p}^2 \hbar v_F = (3/4\pi)[ne^2/m^*\epsilon_0 (pv_F)^2] = (3/4\pi)(\omega_p/pv_F)^2$$

in which n is the electron density and ω_p is the plasma frequency. Let θ be the angle between $\mathbf{p}$ and $\mathbf{k}$ and define $u = \omega/pv_F$. Equation (8.3.76) becomes

$$\cos\theta \left[\xi(\mathbf{k}) + (3/4\pi)(\omega_p/pv_F)^2 \int d\Omega_q \, \xi(\mathbf{q})\right] - u\xi = 0. \tag{8.3.77}$$

This equation is equivalent to (8.3.64) in the case in which $F(\alpha)$ is a positive constant. Consequently, the solution (8.3.65), (8.3.67) may be used provided that we identify F_0 in (8.3.66) and (8.3.67) as

$$F_0 = (3/8\pi)(\omega_p/pv_F)^2. \tag{8.3.78}$$

We are interested in the limit of long wavelengths: $pv_F \ll \omega_p$. In this case, (8.3.68a) is valid and we find simply

$$\omega = \omega_p. \tag{8.3.79}$$

The frequency of the oscillation is just the plasma frequency, as is expected. The Coulomb interaction between the electrons has caused a major modification of the collective modes. Zero sound has been replaced by a plasma oscillation. Although the treatment we have given is idealized in many respects, it is important to note that plasma oscillations can exist in real metals and are observed in studies of the energy lost by fast electrons passing through a metal.

8.3.7 *Derivation of the Interaction Function*

We will now consider the derivation of the interaction function $f_{\sigma\sigma'}(\mathbf{k}, \mathbf{q})$ for a free electron gas according to many-body theory. If $E_\sigma(\mathbf{k})$ is the single-particle energy for a state of wave vector **k** and spin σ, we require the variational derivative [see Eq. (8.3.7)]

$$f_{\sigma\sigma'}(\mathbf{k}, \mathbf{q}) = \delta E_\sigma(\mathbf{k})/\delta n_{\sigma'}(\mathbf{q}) = \delta^2 \mathcal{E}/\delta n_\sigma(\mathbf{k})\, \delta n_{\sigma'}(\mathbf{q}). \qquad (8.3.80)$$

Our calculation is based on the treatments of Rice (1965) and Herring (1966a). The single-particle energy $E_\sigma(\mathbf{k})$ will be specified by

$$E_\sigma(\mathbf{k}) = \epsilon(\mathbf{k}) + M_\sigma(\mathbf{k}, \epsilon(\mathbf{k})) \qquad (8.3.81)$$

in which M is the self-energy. The self-energy function is discussed for the free electron gas in Section 8.2. We will use the expression (8.2.18) which was derived in the random phase approximation

$$M_\sigma(\mathbf{k}, \epsilon) = \sum_{\mathbf{p}} \int (d\omega/2\pi)\, [V(\mathbf{p})/\kappa(\mathbf{p}, \omega)][i\mathcal{G}_\sigma^{(0)}(\mathbf{k} - \mathbf{p}, \epsilon - \omega)]. \qquad (8.3.82)$$

It is convenient to express the Green's function explicitly in terms of the occupation numbers

$$\mathcal{G}_\sigma^{(0)}(\mathbf{k}, \omega) = [n_\sigma(\mathbf{k})/(\omega - \epsilon(\mathbf{k}) - i\delta)] + [(1 - n_\sigma(\mathbf{k}))/(\omega - \epsilon(\mathbf{k}) + i\delta)]. \qquad (8.3.83)$$

Consider the variational derivative of an integral involving the Green's function. Suppose $f_{\sigma'}(\mathbf{k}, \omega)$ is an arbitrary function which is independent of occupation numbers. We wish to evaluate

$$\begin{aligned}
&[\delta/\delta n_\sigma(\mathbf{p})] \sum_{\mathbf{k}\sigma'} \int (d\omega/2\pi) i\mathcal{G}_{\sigma'}^{(0)}(\mathbf{k}, \omega) f_{\sigma'}(\mathbf{k}, \omega) \\
&\quad = \sum_{\mathbf{k}\sigma'} \int (d\omega/2\pi) i[(\omega - \epsilon(\mathbf{k}) - i\delta)^{-1} - (\omega - \epsilon(\mathbf{k}) + i\delta)^{-1}] \\
&\qquad \times \delta_{\mathbf{kp}}\, \delta_{\sigma\sigma'}\, f_{\sigma'}(\mathbf{k}, \omega) \\
&\quad = \sum_{\mathbf{k}\sigma'} \int (d\omega/2\pi) i[2\pi i\, \delta(\omega - \epsilon(\mathbf{k}))]\, \delta_{\mathbf{kp}}\, \delta_{\sigma\sigma'}\, f_{\sigma'}(\mathbf{k}, \omega) \\
&\quad = -f_\sigma(\mathbf{p}, \epsilon(\mathbf{p})).
\end{aligned} \qquad (8.3.84)$$

This result is to be applied to (8.3.82); however, we must take account of the dependence of the dielectric function on the occupation numbers. We have from Eq. (8.2.18)

$$\begin{aligned} f_{\sigma\sigma'}(\mathbf{k}, \mathbf{q}) &= \delta M_\sigma(\mathbf{k}, \epsilon(\mathbf{k}))/\delta n_{\sigma'}(\mathbf{q}) \\ &= -[V(\mathbf{k}-\mathbf{q})/\kappa(\mathbf{k}-\mathbf{q}, \epsilon(\mathbf{k}) - \epsilon(\mathbf{q}))]\,\delta_{\sigma\sigma'} \\ &\quad - \sum_{\mathbf{p}} \int (d\omega/2\pi)[V(\mathbf{k}-\mathbf{p})/\kappa^2(\mathbf{k}-\mathbf{p}, \epsilon(\mathbf{k}) - \omega)] \\ &\quad \times [i\mathcal{G}^{(0)}(\mathbf{p}, \omega)][\delta\kappa(\mathbf{k}-\mathbf{p}, \epsilon(\mathbf{k}) - \omega)/\delta n_{\sigma'}(\mathbf{q})]. \end{aligned} \tag{8.3.85}$$

The interaction function is of interest for states very close to the Fermi surface. Hence, we may set $\epsilon(\mathbf{k}) = \epsilon(\mathbf{q}) = \mu$. It is also useful to recall the decomposition of $f_{\sigma\sigma'}$ into direct and exchange parts:

$$f_{\sigma\sigma'}(\mathbf{k}, \mathbf{q}) = f_{\mathrm{d}}(\mathbf{k}, \mathbf{q}) + f_{\mathrm{x}}(\mathbf{k}, \mathbf{q})\,\delta_{\sigma\sigma'}. \tag{8.3.86}$$

We have for the exchange part

$$f_{\mathrm{x}}(\mathbf{k}, \mathbf{q}) = -V(\mathbf{k}-\mathbf{q})/\kappa(\mathbf{k}-\mathbf{q}, 0). \tag{8.3.87}$$

This is simply a screened Coulomb potential. The direct part is more complicated. Equations (8.2.8) and (8.2.14b) can be combined to give

$$\begin{aligned} \kappa(\mathbf{k}-\mathbf{p}, \epsilon(\mathbf{k}) - \omega) &= 1 - iV(\mathbf{k}-\mathbf{p}) \sum_{\mathbf{k}'\sigma'} \int (d\omega'/2\pi)[i\mathcal{G}_{\sigma'}(\mathbf{k}', \omega')] \\ &\quad \times [i\mathcal{G}_{\sigma'}(\mathbf{k}' + \mathbf{k} - \mathbf{p}, \omega' + \epsilon(\mathbf{k}) - \omega)]. \end{aligned} \tag{8.3.88}$$

The variation of this quantity can be obtained according to (8.3.84):

$$\begin{aligned} &\delta\kappa(\mathbf{k}-\mathbf{p}, \epsilon(\mathbf{k}) - \omega)/\delta n_\sigma(\mathbf{q}) \\ &\quad = -V(\mathbf{k}-\mathbf{p})[\mathcal{G}_\sigma(\mathbf{q} + \mathbf{k} - \mathbf{p}, \epsilon(\mathbf{q}) + \epsilon(\mathbf{k}) - \omega) \\ &\qquad + \mathcal{G}_\sigma(\mathbf{q} + \mathbf{p} - \mathbf{k}, \omega + \epsilon(\mathbf{q}) - \epsilon(\mathbf{k})]. \end{aligned} \tag{8.3.89}$$

This formula is substituted into (8.3.85). We obtain, after some relabeling of variables,

$$\begin{aligned} f_{\mathrm{d}}(\mathbf{k}, \mathbf{q}) &= \sum_{\mathbf{p}} \int (d\omega/2\pi)[V^2(\mathbf{p})/\kappa^2(\mathbf{p}, \omega)][i\mathcal{G}_{\sigma'}(\mathbf{k}-\mathbf{p}, \epsilon(\mathbf{k}) - \omega] \\ &\quad \times [\mathcal{G}_\sigma(\mathbf{q}+\mathbf{p}, \epsilon(\mathbf{q}) + \omega) + \mathcal{G}_\sigma(\mathbf{q}-\mathbf{p}, \epsilon(\mathbf{q}) - \omega)]. \end{aligned} \tag{8.3.90}$$

We now have formal expressions for the interaction function in the random phase approximation. We will not attempt to evaluate the integral (8.3.90). Herring (1966a) points out that if one first averages over directions of $\mathbf{p}$, the integral with respect to ω diverges when $\mathbf{k} = -\mathbf{q}$. The function f_d turns out to possess a logarithmic singularity in this case.

The expressions obtained for the interaction function can now be used to calculate the properties described in preceding subsections: effective mass, compressibility, magnetic susceptibility, etc. These calculations are rather involved, and will not be described here. For further information, see Rice (1965).

8.4 Electron Interactions and Magnetic Order

The problem of understanding the origin and properties of magnetic order in metals has been one of the most challenging in the theory of solids. In many respects, the difficulties have turned out to be greater than are encountered in the theory of superconductivity, and much controversy remains. Several features of this question will be discussed in this section. A detailed and critical review has been given by Herring (1966a).

An essential part of the problem is that quite differing forms of magnetic order occur in metals whose electronic structure apparently varies widely. It does not seem to be possible to analyze the magnetic order problem in a single theoretical framework as is furnished for superconductivity by the BCS theory and its extensions. It is true that magnetism can be attributed to exchange interaction and thus to electron interactions: the problem is to find an adequate description of exchange effects in the systems of interest. A complete theory must account for the ferromagnetism of transition metals (Fe, Co, Ni, and related alloys) and of some rare earth metals (Gd, Tb, Dy), the antiferromagnetism of some transition metals (Cr, Mn) and the oscillatory structures (linear or helical) of certain rare earths (Tb, Dy, Ho, En, Tm). It is in fact possible for a single metal to exhibit differing types of magnetic order in different temperature ranges, and several rare earth metals do this (Cooper, 1968). In addition, there are a few metallic ferromagnets, such as $ZrZn_2$ which do not involve a 3d, 4f, or 5f magnetic constituent.

The theory of magnetic order in insulators is usually based on the Heisenberg Hamiltonian, as discussed in Chapter 2, Part A. To a limited extent the same approach can be applied as a phenomenological description of magnetism in metals. Serious objections can, however, be raised to its use on a fundamental basis: The derivation of the Heisenberg Hamiltonian was considered in Section 2.1 following the procedure of Herring (1966b). The effective Hamiltonian is an asymptotic formula, valid only for a collection of well-separated atoms. Specifically, it cannot be applied to electrons in a shell which participate in establishing metallic properties. There is direct evidence that the d electrons in the transition metals contribute to the Fermi surface (see, for example, Gold *et al.*, 1971) and so cannot be described as localized on atomic sites as required to justify the use of atomic spin operators. Moreover, the magnetic moment per atom in iron, cobalt, and nickel, is not integral, as would be expected in a localized model. A theory of the magnetic properties of these materials should be based on band theory.

The situation in rare earth metals is different. The 4f electrons which produce the magnetic moment are localized. The magnetic properties of these materials can be described using an indirect exchange model. The essential idea here is that localized 4f electrons are coupled by an exchange interaction with itinerant electrons in s-d bands. This coupling tends to align the spins of the itinerant electrons parallel to those of localized electrons. In second order, it produces an alignment of the localized spins as well.

We will begin our detailed discussion with indirect exchange as this is somewhat simpler than the full band theory problem, which follows. This section concludes with a study of antiferromagnetism in metals.

8.4.1 *Indirect Exchange*

The indirect exchange mechanism was introduced by Vonsovski (1946) and Zener (1951). The interaction between localized and itinerant electron spins was considered in Section 5.6 in connection with the Kondo effect. The coupling is described by a Hamiltonian

$$U = -J \sum_{\mu,i} f(\mathbf{r}_i - \mathbf{R}_\mu)\mathbf{S}_\mu \cdot \mathbf{s}_i \tag{8.4.1}$$

in which $\mathbf{S}_\mu$ is a localized spin located at sites $\mathbf{R}_\mu$, $\mathbf{s}_i$ is the spin operator for an itinerant electron, and $f(\mathbf{r})$ is a form factor. The sign of J is determined by the competition between direct exchange (Kasuya, 1956) and a localized–itinerant mixing briefly described in Section 5.6. It will be assumed here that the ferromagnetic direct exchange is dominant. We can express (8.4.1) in second quantized form according to (5.6.4) and (5.6.5):

$$\begin{aligned} U = (-1/2N) \sum_{\mu,\mathbf{k},\mathbf{q}} &\exp[i(\mathbf{q} - \mathbf{k})\cdot\mathbf{R}_\mu]J(\mathbf{k}, \mathbf{q}) \\ &\times \{S_\mu^{(z)}[c_\uparrow^+(\mathbf{k} + \mathbf{q})c_\uparrow(\mathbf{k}) - c_\downarrow^+(\mathbf{k} + \mathbf{q})c_\downarrow(\mathbf{k})] \\ &+ S_\mu^{(+)}c_\downarrow^+(\mathbf{k} + \mathbf{q})c_\uparrow(\mathbf{k}) + S_\mu^{(-)}c_\uparrow^+(\mathbf{k} + \mathbf{q})c_\downarrow(\mathbf{k})\} \end{aligned} \tag{8.4.2}$$

in which

$$J(\mathbf{k}, \mathbf{q}) = J \int \psi^*(\mathbf{k}, \mathbf{r})f(\mathbf{r})\psi(\mathbf{q}, \mathbf{r})\, d^3r. \tag{8.4.3}$$

A single band is considered. We will approximate $J(\mathbf{k}, \mathbf{q})$ by a positive constant independent of $\mathbf{k}$ and $\mathbf{q}$. This approximation must be removed if quantitative application to real materials is considered.

The change in the energy of the system will be calculated through second order in perturbation theory. The first-order change in energy of a state $\mathbf{k}$, $\Delta\epsilon_1(\mathbf{k})$, is

$$\Delta\epsilon_1(\mathbf{k}) = (-J/2N) \sum_\mu S_\mu^{(z)}[n_\uparrow(\mathbf{k}) - n_\downarrow(\mathbf{k})]. \tag{8.4.4}$$

This vanishes if there is no difference between $n_\uparrow(\mathbf{k})$ and $n_\downarrow(\mathbf{k})$. A subtle point arises: perhaps the localized spins could cause some polarization. Suppose a number σ of electrons of down spin are changed to spin up. The change in total energy, obtained by summing (8.4.4) over $\mathbf{k}$, is

$$\Delta E_1 = \sum_\mathbf{k} \Delta\epsilon_1(\mathbf{k}) = (-J\sigma/N) \sum_\mu S_\mu^{(z)}. \tag{8.4.5}$$

There will, however, be an increase in the energy of the system caused by the fact that some $\mathbf{k}$ states near the Fermi energy are singly rather than doubly occupied. Let $G(\mu)$ denote the density of states per atom for a single spin direction at the Fermi energy. Straightforward calculation shows that this change is

$$\Delta E_\mathrm{B} = \sigma^2/NG(\mu). \tag{8.4.6}$$

We may add (8.4.5) and (8.4.6) and then minimize to determine σ. This yields

$$\sigma = \sigma_0 = (J/2)G(\mu) \sum_\mu S_\mu^{(z)} \tag{8.4.7a}$$

and

$$\Delta E_1' = (\Delta E_1 + \Delta E_\mathrm{B})_{\sigma_0} = [-J^2G(\mu)/4N](\sum S_\mu^{(z)})^2. \tag{8.4.7b}$$

This formula indicates that a uniform magnetization of the electron system should be established even by a single localized spin. Such a result is inconsistent with the expectation that in the case of a single localized spin, the magnetization should be local, that is, confined to the vicinity of the spin. It is necessary to proceed to second order in perturbation theory to obtain a satisfactory result. The second-order calculation is due to Ruderman and Kittel (1954), Kasuya (1956), and Yosida (1957).

The calculation can be made using ordinary perturbation theory as was done in Section 5.6. The second-order change in energy of a state of wave vector $\mathbf{k}$ and spin s is $\Delta\epsilon_2$ and this must be summed over $\mathbf{k}$ to obtain ΔE_2, the second-order change in energy of the system. There is one complication: According to the rules of perturbation theory, intermediate states which are the same as the initial state are to be excluded. This means that if we consider an initial state $|\mathbf{k}s\rangle$, and intermediate states $|\mathbf{q}s'\rangle$; in summing over q for $s' = s$, we must exclude $\mathbf{q} = \mathbf{k}$; however, if $s' = -s$, there is no such exclusion. We are interested in the interaction between local spins on different sites; hence, we ignore the term giving rise to the Kondo effect. Our result is

$$\Delta E_2 = 2(J/2N)^2 \sum_{\mu\nu} [S_\mu^{(z)} S_\nu^{(z)} \sum_{\mathbf{k}} \sum_{\mathbf{q}\neq\mathbf{k}} + \tfrac{1}{2}(S_\mu^{(+)} S_\nu^{(-)} + S_\mu^{(-)} S_\nu^{(+)}) \sum_{\mathbf{k}} \sum_{\mathbf{q}}]$$
$$\times f[\epsilon(\mathbf{k})]\{\exp[i(\mathbf{q}-\mathbf{k})\cdot(\mathbf{R}_\mu - \mathbf{R}_\nu)]/[\epsilon(\mathbf{k}) - \epsilon(\mathbf{q})]\}. \quad (8.4.8)$$

The initial factor of 2 allows for the two directions of electron spin with reference to the initial state $|\mathbf{k}s\rangle$, and the factor $f[\epsilon(\mathbf{k})]$ ensures that we count only occupied initial states. All other Fermi functions disappear.

We will evaluate (8.4.8) for a parabolic band $\epsilon(\mathbf{k}) = \gamma\mathbf{k}^2$ and zero temperature: Set $\mathbf{q} - \mathbf{k} = \mathbf{p}$. We have

$$(1/N) \sum_{\mathbf{kq}} f[\epsilon(\mathbf{k})]\{\exp[i(\mathbf{q}-\mathbf{k})\cdot\mathbf{R}]/[\epsilon(\mathbf{k}) - \epsilon(\mathbf{q})]\}$$
$$= (1/N) \sum_{\mathbf{kq}} f(\mathbf{k})\{\exp(i\mathbf{p}\cdot\mathbf{R})/[\epsilon(\mathbf{k}) - \epsilon(\mathbf{k}+\mathbf{p})]\}.$$

Spin directions are not to be included in the sum. The sum on $\mathbf{k}$ is done first:

$$(1/N) \sum_{\mathbf{k}} f(\mathbf{k})/[\epsilon(\mathbf{k}) - \epsilon(\mathbf{k}+\mathbf{p})]$$
$$= [\Omega/(2\pi)^3] \int d^3k\, f(\mathbf{k})/[\epsilon(\mathbf{k}) - \epsilon(\mathbf{k}+\mathbf{p})] = -S(\mathbf{p})/4 \quad (8.4.9)$$

in which $S(\mathbf{p})$ is given by (8.1.58) and (for a parabolic band) by Eq. (7.1.59). This can be written as

$$S(\mathbf{p}) = G(\mu)[1 + (4k_F^2 - p^2)/4k_F p \ln|(p + 2k_F)/(p - 2k_F)|]. \quad (8.4.10)$$

Here, $G(\mu)$ is the density of states for a single spin direction as considered above. This result is to be inserted in (8.4.8) and the sum on $\mathbf{q}$ performed. The contribution from $p = 0$ is to be excluded from the first term. This contribution is

$$(-)(J^2/2N)(\sum_\mu S_\mu^{(z)})^2 S(0)/4.$$

Since $S(0) = 2G(\mu)$, it is seen that this term equals the first-order result of (8.4.7b). Hence, we obtain the correct sum of first- and second-order terms simply by dropping the restriction excluding $p = 0$ in the first term of (8.4.8), so that the sum includes all p. We may now rewrite (8.4.8) in the form

$$\Delta E = \Delta E_1' + \Delta E_2 = \sum_{\mu\nu} (\mathbf{S}_\mu \cdot \mathbf{S}_\nu) J(|\mathbf{R}_\mu - \mathbf{R}_\nu|). \tag{8.4.11}$$

This is an effective exchange interaction of the Heisenberg-type coupling the local spins. The exchange parameter J is given by

$$J(R) = -J^2/N \sum_{\mathbf{p}} \exp(i\mathbf{p}\cdot\mathbf{R}) S(\mathbf{p}).$$

The sum is expressed as an integral

$$(1/N) \sum_{\mathbf{p}} \exp(i\mathbf{p}\cdot\mathbf{R}) S(\mathbf{p}) = [\Omega/(2\pi)^3] \int \exp(i\mathbf{p}\cdot\mathbf{R}) S(\mathbf{p})\, d^3p$$

$$= (\Omega/4\pi^2 iR) \int_{-\infty}^{\infty} pS(\mathbf{p}) e^{ipR}\, dp. \tag{8.4.12}$$

The integral can be evaluated by an elegant procedure due to Van Vleck (1962). Its value will be changed only infinitesimally by choosing a contour which is infinitesimally above the real p axis in the region $-2k_F \leq p \leq 2k_F$, and lies along the real axis elsewhere (see Fig. 8.4.1). Then, if the logarithm in (8.4.10) did not contain the absolute value sign, the integral would be zero. This would occur because we could close the contour with an infinite semicircle in the upper half of the p plane. Such a contour contains no singularities, since the deformation of the contour along the real axis now avoids the branch points of the logarithm. Further the function $pS(p)$ vanishes like $1/p$ for large p, so that the integral converges. As a result of these considerations, the value of the integral in (8.4.12) is equal to the

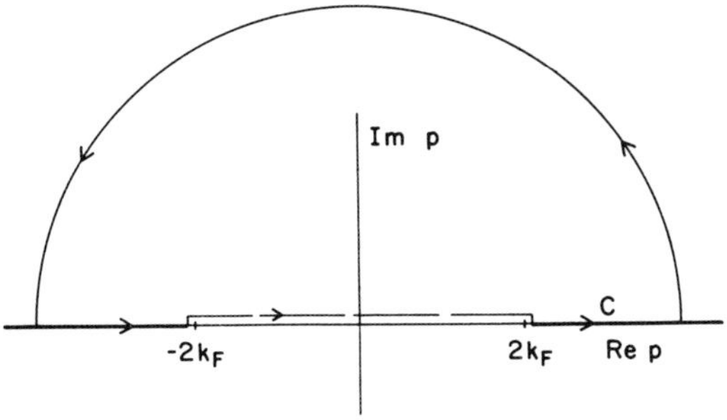

Fig. 8.4.1. Contour of integration for Eq. (8.4.12).

difference between the integral as it stands, and the integral with the absolute value signs in the logarithm removed. The difference arises from the imaginary part of the logarithm in the second case, which is just πi. Equation (8.4.12) becomes

$$[-\Omega G(\mu)/4\pi R] \int_{-2k_F}^{2k_F} [(4k_F^2 - p^2)/4k_F p] e^{ipR}\, dp.$$

This integral is elementary, and we finally obtain for $J(R)$

$$J(R) = [3\pi Z G(\mu) J^2/(2k_F R)^3][\cos 2k_F R - ((\sin 2k_F R)/2k_F R)]. \tag{8.4.13}$$

We have used the relation $k_F^3\Omega = 3\pi^2 Z$, in which Z is the valence.

Equation (8.4.13) shows that the effective exchange interaction is of relatively long range, decreasing like R^{-3} for large R, and oscillating in R with a period of $2k_F$. This interaction can also be applied to nuclear spins in a metal, and is significant in determining the width of nuclear magnetic resonance lines in some metals (Ruderman and Kittel, 1954).

8.4.2 *The Band Theory of Ferromagnetism*

We now turn to the problem of the transition metal ferromagnets: specifically, iron, cobalt, nickel, and their alloys. In these materials, it is not adequate to treat the d electron spins as localized, since, as was pointed out earlier, there is direct evidence that these electrons contribute to the Fermi surface.

The fundamental concept of the energy band theory of ferromagnetism is that the d electrons in transition metals must be regarded as itinerant, that is, occupying Bloch states. The question of whether ferromagnetism occurs in a particular metal can be put in the following terms: Is the decrease in the energy of electrostatic repulsion produced by aligning the electron spins large enough to compensate for the increase in energy which results from the promotion of electrons into states of higher energy in the band structure? The decrease in energy due to magnetization is due to the fact that electrons avoid each other more efficiently when their spins are aligned because the Pauli principle forces the total wave function of the system to vanish when two electrons are in the same place.

The formulation is generally sound. Several difficulties arise in attempts to estimate the relevant quantities: In the ferromagnetism problem, we are concerned with the behavior of electrons in narrow energy bands. A simple single-determinant wave function may be a good approximation in the ferromagnetic state, but it is likely to be a poor one in the nonmagnetic state. In such a case, a single-determinant wave function contains contributions from situations in which some atoms contain the wrong number of electrons. These states are of high energy, and the energy of the nonmagnetic state is predicted to be too large. This problem can be illustrated in the simple case of the hydrogen molecule (see, for example, Slater, 1953). Let a(1), b(1) denote one-electron wave functions in which electron 1 is bound to atom a or b, respectively. We form molecular orbitals [a ± b] (neglecting normalization) which are the analogs of Bloch functions. Consider the singlet determinantal function

$$\psi_s(1,2) \approx \begin{vmatrix} [\mathrm{a}(1)+\mathrm{b}(1)]\alpha(1) & [\mathrm{a}(1)+\mathrm{b}(1)]\beta(1) \\ [\mathrm{a}(2)+\mathrm{b}(2)]\alpha(2) & [\mathrm{a}(2)+\mathrm{b}(2)]\beta(2) \end{vmatrix}$$

$$= \{[\mathrm{a}(1)\mathrm{b}(2)+\mathrm{a}(2)\mathrm{b}(1)]+[\mathrm{a}(1)\mathrm{a}(2)+\mathrm{b}(1)\mathrm{b}(2)]\} \times [\alpha(1)\beta(2)-\beta(1)\alpha(2)]. \tag{8.4.14}$$

Here α and β are spin functions. This wave function contains equal components of the Heitler–London wave function a(1)b(2) + a(2)b(1) which is nearly correct at moderately (but not extremely) large distances, and an ionized function a(1)a(2) + b(1)b(2) which is of high energy. The triplet function is

$$\psi_T(1,2) \approx \begin{vmatrix} [\mathrm{a}(1)+\mathrm{b}(1)]\alpha(1) & [\mathrm{a}(1)-\mathrm{b}(1)]\alpha(1) \\ [\mathrm{a}(2)+\mathrm{b}(2)]\alpha(2) & [\mathrm{a}(2)-\mathrm{b}(2)]\alpha(2) \end{vmatrix}$$

$$= [\mathrm{b}(1)\mathrm{a}(2)-\mathrm{a}(1)\mathrm{b}(2)]\alpha(1)\alpha(2). \tag{8.4.15}$$

This function does not contain any components corresponding to ionized states. Use of (8.4.14) and (8.4.15) will lead to an incorrect prediction that the triplet state has a lower energy than the singlet state at large distances.

This tendency will also be present in the energy band problem. A Hartree–Fock calculation of energies based on simple determinantal wave functions will produce a spurious ferromagnetism at large interatomic spacings, and is suspect at actual separations. We have to find an approximation which is better than Hartree–Fock, particularly for the nonmagnetic state. In this way, the problem of ferromagnetism is intimately connected with that of electron correlation.

8.4.3 *Ferromagnetism in the Hubbard Model*

In Sections 8.1 and 8.2, the properties of a system of interacting electrons was studied using a formalism based on plane waves (or the crystal momentum representation). Ferromagnetism in the transition metals involves electron states with wave functions which overlap rather weakly. It seems plausible to treat the problem of interactions in such a system in terms of localized functions. Accordingly, we consider a Hamiltonian which is expressed on a basis of Wannier functions $a_n(\mathbf{r} - \mathbf{R}_\mu)$ (n is the band index, and $\mathbf{R}_\mu$ is the lattice site):

$$H = \sum_{n\mu\nu\sigma} \mathcal{E}_n(\mathbf{R}_\mu - \mathbf{R}_\nu) c^+_{n\mu\sigma} c_{n\nu\sigma} + \tfrac{1}{2} \sum_{\mu\nu\rho\tau} \sum_{\sigma\sigma'} \sum_{ijln} V_{ijln}(\mathbf{R}_\mu, \mathbf{R}_\nu, \mathbf{R}_\rho, \mathbf{R}_\tau) c^+_{i\mu\sigma} c^+_{j\nu\sigma'} c_{n\tau\sigma'} c_{l\rho\sigma}. \tag{8.4.16}$$

The operator $c^+_{i\mu\sigma}$ creates an electron in the state of spin σ whose space wave function is $a_i(\mathbf{r} - \mathbf{R}_\mu)$. The quantity $\mathcal{E}_n(\mathbf{R}_\mu - \mathbf{R}_\nu)$ is a Fourier coefficient of the energy band function $E_n(\mathbf{k})$ defined by Eq. (5.1.47)

$$\mathcal{E}_n(\mathbf{R}_\mu) = [\Omega/(2\pi)^3] \int E_n(\mathbf{k}) \exp(i\mathbf{k}\cdot\mathbf{R}_\mu)\, d^3k \tag{8.4.17}$$

and V_{ijln} is a matrix element of the Coulomb interaction:

$$V_{ijln}(\mathbf{R}_\mu, \mathbf{R}_\nu, \mathbf{R}_\rho, \mathbf{R}_\tau) = \int a_i^*(\mathbf{r}_1 - \mathbf{R}_\mu) a_j^*(\mathbf{r}_2 - \mathbf{R}_\nu)(e^2/r_{12}) a_l(\mathbf{r}_1 - \mathbf{R}_\rho) \times a_n(\mathbf{r}_2 - \mathbf{R}_\tau)\, d^3r_1\, d^3r_2. \tag{8.4.18}$$

Equations (8.4.16)–(8.4.18) furnish the general basis for a fundamental theory. The complications of keeping track of the matrix elements are enormous. In order to simplify matters, the following approximations are usually made.

(1) Only those matrix elements involving electrons on the same site ($\mathbf{R}_\mu = \mathbf{R}_\nu = \mathbf{R}_\rho = \mathbf{R}_\tau$) are retained. Elements of this type are certainly the largest; however, terms with $\mathbf{R}_\mu = \mathbf{R}_\rho$, $\mathbf{R}_\nu = \mathbf{R}_\tau \neq \mathbf{R}_\mu$ will not be negligible. However, we may argue qualitatively that in a ferromagnet the interactions of the d electrons on different atoms should be screened by mobile s electrons so that perhaps one should use in the matrix element the interaction $e^2 \exp(-\kappa r_{12})/r_{12}$. The screening constant κ is assumed to be large enough so that interactions between different atoms can be discarded.

(2) A further approximation is to consider only a single band. At first sight, this appears to be quite unrealistic for magnetic problems because ferromagnetism occurs in materials with complex, overlapping band structures. However, there is some evidence that the complications of

overlapping bands are not essential to a qualitative treatment, and it is certainly interesting to consider the implications of the simplest possible model. We will label the single matrix element

$$V_0 = V_{iiii}(\mathbf{R}_\mu, \mathbf{R}_\mu, \mathbf{R}_\mu, \mathbf{R}_\mu)$$

and we have the simple Hamiltonian

$$H = \sum_{\mu\nu\sigma} \mathcal{E}(\mathbf{R}_\mu - \mathbf{R}_\nu) c^+{}_{\mu\sigma} c_{\nu\sigma} + \tfrac{1}{2} V_0 \sum_{\mu\sigma} n_{\mu\sigma} n_{\mu-\sigma} \tag{8.4.19}$$

in which $n_{\mu\sigma} = c^+{}_{\mu\sigma} c_{\mu\sigma}$, and the band index has been dropped. This is the Hubbard Hamiltonian which was previously discussed in Section 3.4.8, Part A.

Herring (1966a) has estimated some of the most important matrix elements of the type (8.4.18) using an unscreened Coulomb interaction and replacing the Wannier functions by atomic wave functions. This yields $V_0 \approx 20$ eV, which is quite large compared to typical d band widths. It is quite probable that V_0 is substantially smaller than this, as a result of dielectric screening and other effects; however, it remains interesting to consider the model defined by (8.4.19) for large V_0, even if quantitative applications to real metals require inclusion of interband elements.

The Hubbard Hamiltonian describes a system of electrons interacting through a strong, short range, repulsion. It is then appropriate to apply the techniques of many-body theory developed by Brueckner and collaborators (Brueckner and Levinson, 1955; Bethe, 1956; Brueckner and Gammel, 1958; Brueckner, 1958) in the study of nuclear matter. The essential idea of this approach is that if a system of particles interacts through short range forces, it is possible to set up an expansion in which the density, or a quantity related to it, rather than the strength of the interaction, is the small parameter. The low density limit may actually be relevant to the ferromagnetism problem in nickel and nickel–copper alloys: in nickel, it is believed that there are 0.56 holes per atom in the d band, and the number appears to decrease linearly with increasing concentration of copper.

The low density concept can be formulated in the language of diagrammatic perturbation theory if we observe that each hole line in a Goldstone diagram will lead to an integration over the occupied portion of **k** space, and thus will tend to bring in a factor of the density. Consequently, we select the set of relevant diagrams with the smallest number of hole lines. In the case of the ground state energy, two hole lines must always be present. The present problem is actually slightly simpler than nuclear matter as considered by Brueckner, since the Hamiltonian includes only interaction between particles of opposite spins. Consequently exchange diagrams are not present, and we may restrict consideration to the series shown in Fig.

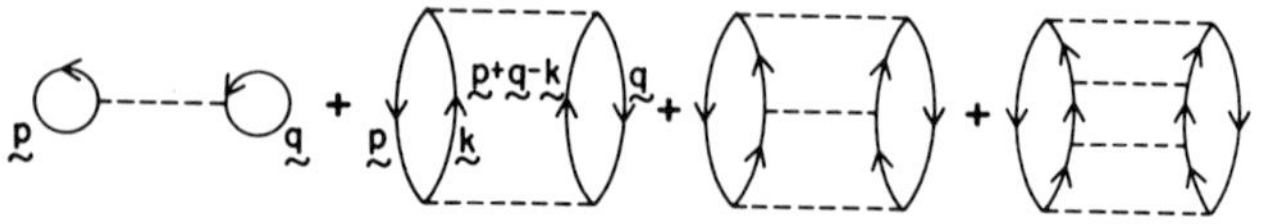

Fig. 8.4.2. Self-energy ladder diagrams.

8.4.2: the so-called *ladder diagrams.* These diagrams have only two hole lines and an arbitrary number of electron–electron interactions. In order to evaluate these, it is convenient to transform (8.4.19) to wave vector space, using

$$c_{\mu\sigma} = \mathfrak{N}^{-1/2} \sum_{\mathbf{k}} \exp(i\mathbf{k}\cdot\mathbf{R}_\mu) c_\sigma(\mathbf{k}) \tag{8.4.20}$$

in which $\mathfrak{N}$ is the number of sites in the system. The result is

$$H = \sum_{\mathbf{k}\sigma} E_n(\mathbf{k}) c_\sigma^+(\mathbf{k}) c_\sigma(\mathbf{k}) + (V_0/\mathfrak{N}) \sum_{\mathbf{kpq}} c_\uparrow^+(\mathbf{k}) c_\downarrow^+(\mathbf{p}) c_\downarrow(\mathbf{q}) c_\uparrow(\mathbf{k}+\mathbf{p}-\mathbf{q}). \tag{8.4.21}$$

The spin states have been indicated explicitly in the second term of (8.4.21). If the spin of one of the lines is fixed, the spin of the remaining lines is uniquely determined. Therefore we do not need to indicate spins in evaluating the diagrams for a nonmagnetic state of the system.

These diagrams may be evaluated according to the rules given in Table I. We recall that each completely symmetric diagram brings in a factor of $\frac{1}{2}$. The series of Fig. 8.4.2 is of the form

$$\begin{aligned}(1/2\mathfrak{N}) \sum_{\mathbf{q},\mathbf{p}<k_F} \{V_0 &+ (V_0^2/\mathfrak{N}) \sum_{\mathbf{k}}{}' [\epsilon(\mathbf{q}) + \epsilon(\mathbf{p}) - \epsilon(\mathbf{k}) - \epsilon(\mathbf{p}+\mathbf{q}-\mathbf{k})]^{-1} \\ &+ (V_0^3/\mathfrak{N}^2) \sum_{\mathbf{k}s}{}' [\epsilon(\mathbf{q}) + \epsilon(\mathbf{p}) - \epsilon(\mathbf{k}) - \epsilon(\mathbf{p}+\mathbf{q}-\mathbf{k})]^{-1} \\ &\times [\epsilon(\mathbf{q}) + \epsilon(\mathbf{p}) - \epsilon(\mathbf{s}) - \epsilon(\mathbf{p}+\mathbf{q}-\mathbf{s})]^{-1} + \cdots\}.\end{aligned} \tag{8.4.22}$$

The prime on the summations indicates that the electron states must all be above the Fermi energy.

The series in the braces can be represented symbolically in the form

$$t = V_0 + V_0\mathcal{G}V_0 + V_0\mathcal{G}V_0\mathcal{G}V_0 + \cdots = V_0(1 - \mathcal{G}V_0)^{-1} \tag{8.4.23}$$

in which

$$\mathcal{G}(\mathbf{p},\mathbf{q}) = (1/N) \sum_{\mathbf{k}}{}' [\epsilon(\mathbf{q}) + \epsilon(\mathbf{p}) - \epsilon(\mathbf{k}) - \epsilon(\mathbf{p}+\mathbf{q}-\mathbf{k})]^{-1}. \tag{8.4.24}$$

The sum of the series will be denoted t on account of its resemblance to the two-body t matrix discussed in Section 5.2. A significant difference exists, however, with respect to the discussion in the theory of scattering. As a result of the restriction that electron states which are included in (8.4.24) must be unoccupied, the energy denominator does not vanish, and $\mathcal{G}$ is real.

The ground state energy per atomic site will be

$$E_g = E_0 + (1/2\mathfrak{N}) \sum_{\mathbf{q},\mathbf{p}<k_F} t(\mathbf{p}, \mathbf{q}). \tag{8.4.25}$$

The sum over $\mathbf{k}$ in (8.4.24) is difficult for a realistic band structure. A parabolic model is not appropriate. We will make only a rough estimate, which should be reasonably valid for a low density system: neglect the wave vectors $\mathbf{p}$, $\mathbf{q}$ and energies of the hole states compared to the wave vector $\mathbf{k}$ and energy of the electron states. Then $\mathcal{G}(\mathbf{p}, \mathbf{q})$ is independent of $\mathbf{p}$ and $\mathbf{q}$:

$$G = -\tfrac{1}{2}I(\mu) \tag{8.4.26}$$

where

$$I(\mu) = 1/\mathfrak{N} \sum_{\mathbf{k}}{}' 1/\epsilon(\mathbf{k}) = \int_{\mu} G(E)\, dE/E \tag{8.4.27}$$

[in which $G(E)$ is the density of states for a single spin]. In this case, t is also independent of $\mathbf{p}$ and $\mathbf{q}$ so that we have

$$t = V_0/(1 + \tfrac{1}{2}IV_0). \tag{8.4.28}$$

Note that t saturates at a value $2/I$ no matter how large V_0 becomes.

We wish to use (8.4.25) to determine a criterion for ferromagnetism by considering whether an infinitesimal spin excess will lead to a state of lower energy. In order to do this, we must note that $\mathbf{q}$ and $\mathbf{p}$ in (8.4.25) refer to states of different spin. Let $\rho_\uparrow = n_\uparrow/n$ be the density of electrons of $\uparrow$ spin, and put

$$\rho_{\uparrow(\downarrow)} = (\rho_0/2)(1 \pm \zeta). \tag{8.4.29}$$

The ground state energy from (8.4.25) is, for constant t,

$$E_g = E_0(\zeta) + t\rho_\uparrow\rho_\downarrow = E_0(\zeta) + (t\rho_0^2/4)(1 - \zeta^2). \tag{8.4.30}$$

Equation (8.4.30) indicates that the potential energy of the system is decreased by having some spin alignment.

We must now consider the dependence of E_0 on ζ. In the nonmagnetic state of the system, the single-particle states are occupied up to the Fermi energy μ_0:

$$2\mathfrak{N} \int_0^{\mu_0} G(E)\, dE = n.$$

For convenience, the bottom of the band is chosen as the zero of energy. In the ferromagnetic state, it is useful to regard the electrons of ↑ and ↓ spins as having different Fermi energies, the difference being maintained by an exchange splitting of the bands. Let these Fermi energies be denoted by $\mu_\pm$ such that

$$\int_0^{\mu_\pm} G(E)\,dE = (\rho_0/2)(1 \pm \zeta). \tag{8.4.31}$$

It will be supposed that ζ is small, so that the integrals may be expanded about μ_0. Through second order in ζ, we find

$$\mu_\pm - \mu_0 = \pm[\rho_0\zeta/2G(\mu_0)] - [\rho_0{}^2\zeta^2 G'(\mu_0)/8G^3(\mu_0)]. \tag{8.4.32}$$

In (8.4.32), $G'(\mu_0)$ is the derivative of the density of states evaluated at μ_0.

We can now evaluate E_0:

$$E_0 = \left[\int_0^{\mu_+} + \int_0^{\mu_-}\right] EG(E)\,dE. \tag{8.4.33}$$

Let E_a be the value of E_0 in the absence of ferromagnetism, i.e., if $\mu_+ = \mu_- = \mu_0$. A short calculation yields, correct through second order in ζ,

$$E_0 = E_a + [\rho_0{}^2\zeta^2/4G(\mu_0)]. \tag{8.4.34}$$

The ground state energy is

$$E_g = E_a + (t\rho_0{}^2/4) + (\rho_0{}^2\zeta^2/4)\{[1/G(\mu_0)] - t\}. \tag{8.4.35}$$

The ferromagnetic state will have lower energy if

$$tG(\mu_0) > 1. \tag{8.4.36}$$

This is the criterion for the occurrence of ferromagnetism. This result was first obtained by Kanamori (1963). The effective interaction t is restricted to a quantity of the order of the band width by (8.4.28). Ferromagnetism will be favored when the Fermi energy falls in a region where the density of states is greater than average. We also note that, for very low densities $G(\mu_0) \simeq \mu_0{}^{1/2}$. Since t is finite as $\mu \to 0$, whatever V_0 is, this means that ferromagnetism cannot occur for sufficiently low densities, no matter how strong the interaction. The most favorable situation for the occurrence of ferromagnetism has the Fermi energy near the maximum of a sharp peak in the density of states. In this way $G(\mu_0)$ can be large without making $I(\mu_0)$ too large.

The criterion (8.4.36) applies to unsaturated ferromagnetism: it does not imply that all electrons will align. Furthermore, it must be regarded with some caution, since the approximation (8.4.26) may produce an underestimate of I and thus an overestimate of the tendency to ferromagnetism.

8.4.4 *Spin Waves*

The prediction of spin waves was a major success of the Heisenberg model (see Section 2.4, Part A). For many years it was generally believed that the question of the validity of a localized or a band model in transition metals might be settled in favor of the former if convincing evidence for the existence of spin waves in these materials was forthcoming. It is now understood that spin waves are present as elementary excitations in the band theory of ferromagnetism, and it is possible to make moderately quantitative calculations of the spin wave spectrum.

Spin waves of nonzero wave vector are only approximate eigenstates of the Hamiltonian for interacting electrons. For small wave vectors, the lifetime becomes long fairly rapidly, so that spin waves are well defined. Spin waves of wave vector 0 are exact eigenstates, if the Hamiltonian is independent of spin, since they are formed by applying the operator $S_x - iS_y$ (where S_x is the x component of the total spin) to the ground state vector and have the same energy as the ferromagnetic state. Pictorially, spin waves can be considered to be quasi-bound states of a minority spin electron and a majority spin hole which differ by the wave vector of the spin wave.

An operator which creates a spin disturbance of wave vector $\mathbf{q}$ is

$$S(\mathbf{q}) = \sum_{\mathbf{k}} c_{\downarrow}^{+}(\mathbf{k} + \mathbf{q})c_{\uparrow}(\mathbf{q}). \qquad (8.4.37)$$

We will consider the calculation of spin wave energies for the one-band Hubbard Hamiltonian (8.4.19). The case of a saturated ferromagnet (all spins aligned) will be considered. Then, since only electrons of opposite spin can interact, the ferromagnetic state is known exactly. If the band is less than half full, the n electron ferromagnetic state is

$$| F \rangle = c_{\uparrow}^{+}(\mathbf{k}_n) \cdots c_{\uparrow}^{+}(\mathbf{k}_1) \ | 0 \rangle \qquad (8.4.38)$$

where $| 0 \rangle$ is the vacuum state. If the band is more than half full, we introduce hole operators by $b_{\uparrow}^{+}(\mathbf{k}) = c_{\uparrow}(\mathbf{k})$. The substitution has the effect of changing the sign of the band energy $[E(\mathbf{k})]$ term in the Hamiltonian (8.4.21), and adds a constant. If we take the change in sign of the single-particle energy into proper account and discard the constant term, the problem is symmetric between electrons and holes, so that we may restrict our attention to electrons in a band less than half full ($n \leq N$).

Spin wave energies may be calculated as follows: If the operator $S(\mathbf{q})$ given by (8.4.37) acts on the ferromagnetic state $| F \rangle$, a spin disturbance is created. Suppose this is done at $t = 0$. The probability amplitude that it is

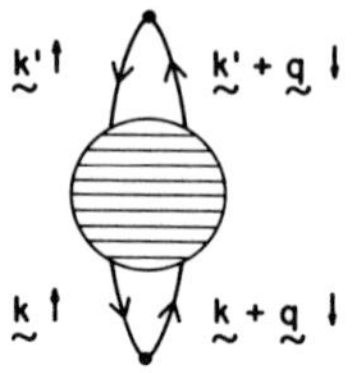

Fig. 8.4.3. Diagram representing $\chi(\mathbf{q},\omega)$.

found at a later time t is given by

$$\chi(\mathbf{q}, t) = -\langle F \mid T\{S^{+}(\mathbf{q}, t) S(\mathbf{q}, 0)\} \mid F\rangle. \tag{8.4.39}$$

The negative sign is a convention. The Fourier transform of $\chi(\mathbf{q}, t)$ is defined by

$$\chi(\mathbf{q}, \omega) = \int_{-\infty}^{\infty} dt\, e^{i\omega t} \chi(\mathbf{q}, t). \tag{8.4.40}$$

The energy and lifetime of the spin wave can be found from the poles of $\chi(\mathbf{q}, \omega)$. The treatment follows that of Young and Callaway (1970).

The function $\chi(\mathbf{q}, \omega)$ will be calculated using the rules of diagrammatic perturbation theory (Section 8.1) in application to the Hamiltonian (8.4.21). Each line must be labeled with a spin index. There are no down spin hole lines. The graphs contributing to $\chi(\mathbf{q}, \omega)$ have the general structure shown in Fig. 8.4.3. The central blob represents an interaction of arbitrary complexity. Evaluations will be made according to the Feynman rules presented in Table I. However, since there are no $\downarrow$ spin holes,

$$\mathcal{G}_{\downarrow}^{(0)}(\mathbf{k}) = [\epsilon - E(\mathbf{k}) + i\delta]^{-1} \qquad (\delta > 0). \tag{8.4.41}$$

Also since only particles of opposite spin interact, the exact $\uparrow$ spin Green's function is identical to the free particle Green's function:

$$\mathcal{G}_{\uparrow}(\mathbf{k}) = \mathcal{G}_{\uparrow}^{(0)}(\mathbf{k}).$$

The self-energy correction to the $\downarrow$ spin Green's function will be found in the same approximation as used previously in the calculation of the ground state energy. The ladder series is shown in Fig. 8.4.4. The self-energy

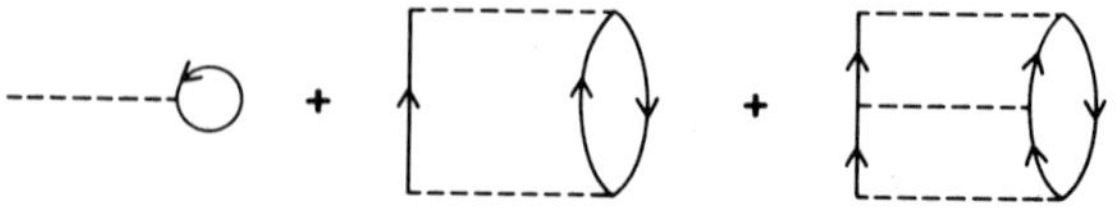

Fig. 8.4.4. Ladder series for the self-energy.

$M_\downarrow(\mathbf{k}, \epsilon)$ (see Section 8.1) is given by

$$M_\downarrow(\mathbf{k}, \epsilon) = (1/\mathfrak{N}) \sum_{\mathbf{p}} \int (d\omega/2\pi)[-i\mathcal{G}_\uparrow^{(0)}(\mathbf{p}, \omega)t(\mathbf{k}\epsilon, \mathbf{p}\omega)] \quad (8.4.42)$$

in which the two-body t matrix is here expressed by the series

$$\begin{aligned} t(\mathbf{k}\epsilon, \mathbf{p}\omega) &= V_0 + i(V_0^2/\mathfrak{N}) \sum_{\mathbf{q}} \int (d\omega'/2\pi)\mathcal{G}_\downarrow^{(0)}(\mathbf{q}, \omega') \\ &\quad \times \mathcal{G}_\uparrow^{(0)}(\mathbf{k}+\mathbf{p}-\mathbf{q}, \epsilon+\omega-\omega') \\ &\quad - (V_0^3/\mathfrak{N}^2) \sum_{\mathbf{qs}} \int [d\omega'\, d\omega''/(2\pi)^2]\mathcal{G}_\downarrow^{(0)}(\mathbf{q}, \omega') \\ &\quad \times \mathcal{G}_\uparrow^{(0)}(\mathbf{k}+\mathbf{p}-\mathbf{q}, \epsilon+\omega-\omega')\mathcal{G}_\downarrow^{(0)}(\mathbf{s}, \omega'') \\ &\quad \times \mathcal{G}_\uparrow^{(0)}(\mathbf{k}+\mathbf{p}-\mathbf{s}, \epsilon+\omega-\omega'') + \cdots \\ &= V_0\Bigg[1 - (iV_0/\mathfrak{N}) \sum_{\mathbf{q}} \int (d\omega'/2\pi)\mathcal{G}_\downarrow^{(0)}(\mathbf{q}, \omega') \\ &\quad \times \mathcal{G}_\uparrow^{(0)}(\mathbf{k}+\mathbf{p}-\mathbf{q}, \epsilon+\omega-\omega')\Bigg]^{-1}. \end{aligned} \quad (8.4.43a)$$

This t matrix reduces to the object previously considered if ϵ and ω are replaced by appropriate energies. We now turn to the calculation of χ. This is somewhat delicate because of a question of consistency. The diagrams summed in calculating χ must be consistent with those used in obtaining M, otherwise a gap may appear erroneously separating the spin wave spectrum from the ground state.

In order to alleviate the cumbersome nature of the notation, we will henceforth treat wave vectors as four vectors, the fourth component being the energy parameter: $k = (\mathbf{k}, \epsilon)$, etc. It is also useful to introduce a convention that subscripted variables are to be summed and integrated as appropriate. The presence of k_1 in an expression indicates that we are to perform the operations: $(1/\mathfrak{N}) \sum_{\mathbf{k}_1} \int d\epsilon_1/(2\pi)$. As an example we may

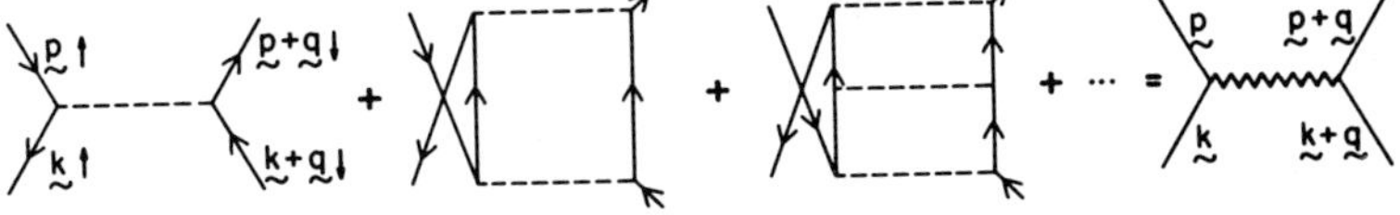

Fig. 8.4.5. Series replacing V by t in the interaction of two particles.

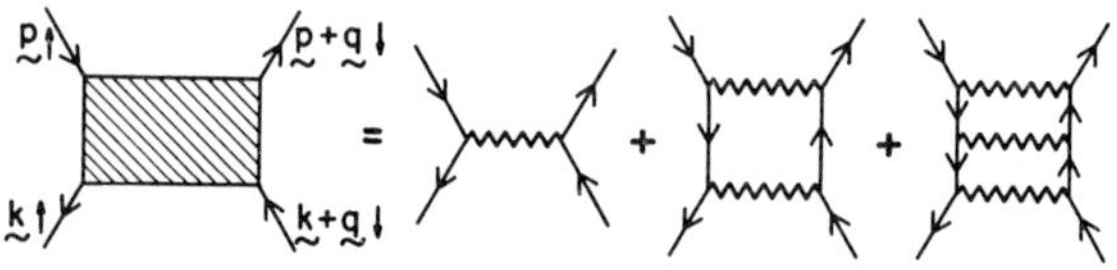

Fig. 8.4.6. Series for the iterated interaction.

rewrite (8.4.34a) as

$$t(k, p) = V_0[1 - iV_0\mathcal{G}_\downarrow(q_1)\mathcal{G}_\uparrow(k + p - q_1)]^{-1}. \qquad (8.4.43b)$$

In the calculation of $\chi(q)$ we may replace the free particle ↓ spin Green's function $\mathcal{G}_\downarrow^{(0)}(q)$ everywhere by the corrected Green's function to which it is related by Dyson's equation (8.1.114):

$$\mathcal{G}_\downarrow^{-1}(k) = \mathcal{G}_\downarrow^{(0)-1}(k) - M_\downarrow(k) \qquad (8.4.44)$$

[with $M_\downarrow(k)$ being given by (8.4.42)].

It is necessary to introduce a set of diagrams which bring the t matrix into the susceptibility calculation. Such diagrams are shown in Fig. 8.4.5. Evaluation of the series presented there indicates that we may introduce an interaction characterized by a wavy line and described by $t(p, k + q)$. Specifically, the sum of the series is

$$i\mathcal{G}_\uparrow^{(0)}(p)\mathcal{G}_\downarrow(p + q)t(p, k + q)\mathcal{G}_\downarrow^{(0)}(k)\mathcal{G}_\downarrow(k + q) \qquad (8.4.45)$$

in which t is given by (8.4.43b). Note that in considering this series it is not possible to have diagrams of the sort shown in Fig. 8.4.5 with similar structure on the right (↓ spin) side since down spin holes are not present.

Next, it is necessary to iterate the t interaction. This is done by constructing a series of diagrams, similar to the ladder series, in which V is replaced by t. The relevant series is shown in Fig. 8.4.6. We will call the shaded box $\Gamma(p, k + q)$. The series for Γ has the form

$$\begin{aligned}\Gamma(p, k + q) &= t(p, k + q) + it(p, k_1 + q)\mathcal{G}_\uparrow(k_1)\mathcal{G}_\downarrow(k_1 + q)t(k_1, k + q)\\ &\quad - t(p, k_1 + q)\mathcal{G}_\uparrow(k_1)\mathcal{G}_\downarrow(k_1 + q)t(k_1, k_2 + q)\mathcal{G}_\uparrow(k_2)\\ &\quad \times \mathcal{G}_\downarrow(k_2 + q)t(k_2, k_2 + q) + \cdots. \end{aligned} \qquad (8.4.46)$$

This series is equivalent to an integral equation

$$\Gamma(p, k + q) = t(p, k + q) + it(p, k_1 + q)\mathcal{G}_\uparrow(k_1)\mathcal{G}_\downarrow(k_1 + q)\Gamma(k_1, k + q). \qquad (8.4.47)$$

The quantity $\Gamma(p, k + q)$ is referred to as a *vertex function*. In order to complete the formal calculation of $\chi(q, \omega)$ it is only necessary to close off

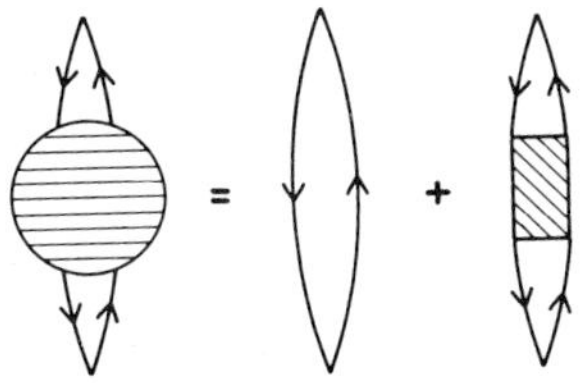

Fig. 8.4.7. Determination of $\chi(q)$. The shaded box represents the series shown in Fig. 8.4.6.

the lines at the top and bottom, and add the zero-order "no interaction" diagram. This is shown in Fig. 8.4.7. The result is

$$\chi(q) = \mathcal{G}_\uparrow(p_1)\mathcal{G}_\downarrow(p_1+q) + i\mathcal{G}_\uparrow(p_1)\mathcal{G}_\downarrow(p_1+q)\Gamma(p_1,k_1+q)\mathcal{G}_\uparrow(k_1)\mathcal{G}_\downarrow(k_1+q). \tag{8.4.48}$$

We now have a formal procedure for the determination of the susceptibility. Unfortunately, it involves the complicated integral equation (8.4.47). The approximation of a constant t matrix which was already introduced in (8.4.28) will be introduced at this point. It should be noted that the function $I(\mu)$ defined in (8.4.27) and appearing in (8.4.28) now refers to the chemical potential in the ferromagnetic state, rather than the paramagnetic state. A constant t matrix leads to a constant self-energy: Eq. (8.4.42) becomes

$$M_\downarrow = t\rho \tag{8.4.49}$$

in which $\rho = n/N$ is the number density.

It is convenient to introduce $\chi_0(q)$ the susceptibility if Γ is ignored:

$$\chi_0(q) = \mathcal{G}_\uparrow(p_1)\mathcal{G}_\downarrow(p_1+q). \tag{8.4.50}$$

The series (8.4.46) for Γ can be summed directly to yield, when t is constant,

$$\Gamma(p, k+q) = t/[1 - it\chi_0(q)]. \tag{8.4.51}$$

This result is substituted into (8.4.48) which yields

$$\chi(q) = \chi_0(q)/[1 - it\chi_0(q)]. \tag{8.4.52}$$

The poles of $\chi(q)$ determine the spin wave energies

$$1 = it\chi_0(q). \tag{8.4.53}$$

Equations (8.4.44), (8.4.49), and (8.4.50) are introduced so that (8.4.53) becomes

$$1 = (1/\mathfrak{N}) \sum_{\mathbf{p}(\mathrm{occ})} t/[t\rho + E(\mathbf{p}+\mathbf{q}) - E(\mathbf{p}) - \omega] \tag{8.4.54}$$

in which ω is the spin wave energy to be determined as a function of $\mathbf{q}$.

Solution to this transcendental equation generally requires numerical calculation. However, for small q we may expand the denominator to second order:

$$E(\mathbf{p}+\mathbf{q}) - E(\mathbf{p}) = (\mathbf{q}\cdot\nabla)E(\mathbf{p}) + \tfrac{1}{2}(\mathbf{q}\cdot\nabla)^2E(\mathbf{p}) + \cdots. \quad (8.4.55)$$

After a straightforward calculation, we find

$$\omega = (1/2\rho\mathfrak{N}) \sum_{\mathbf{p}(\mathrm{occ})} \{(\mathbf{q}\cdot\nabla_{\mathbf{p}})^2E(\mathbf{p}) - 2[\mathbf{q}\cdot\nabla E(\mathbf{p})]^2/t\rho\}. \quad (8.4.56)$$

We have used the fact that $\sum_{\mathbf{p}} \nabla_{\mathbf{p}}E(\mathbf{p})$ vanishes by time reversal symmetry. For cubic crystals (8.4.56) simplifies to

$$\omega = q^2/6\rho\mathfrak{N} \sum_{\mathbf{p}(\mathrm{occ})} [\nabla^2E(\mathbf{p}) - 2\,|\,\nabla E(\mathbf{p})\,|^2/t\rho]. \quad (8.4.57)$$

This result is an explicit expression for the spin wave energy to second order in the wave vector. This result is not exact: a formally correct expression for the spin wave energy to second order in $\mathbf{q}$ was given by Edwards (1967). The differences between his result and Eq. (8.4.57) appear not to be large when ρ is small (Callaway, 1968).

The first term of (8.4.57), involving ∇^2E may be either positive or negative: the contribution of the second is always negative. It is necessary that the spin wave energy be positive for the ferromagnetic state to be stable. Otherwise, spin waves would occur spontaneously at $T = 0$, and the assumed saturated ferromagnetism will be destroyed.

If there are just enough electrons (one per atom) to fill the band half full in the paramagnetic state, the saturated ferromagnetic state has all ↑ spin states in the band occupied. Under these circumstances,

$$(1/\mathfrak{N}) \sum_{\mathbf{p}(\mathrm{occ})} \nabla^2E(\mathbf{p}) = [\Omega/(2\pi)^3] \int d^3p\, \nabla^2E(\mathbf{p})$$

$$= [\Omega/(2\pi)^3] \int d\mathbf{S}\cdot\nabla E(\mathbf{p}) = 0. \quad (8.4.58)$$

It follows that the half-filled band cannot be ferromagnetic. The most probable kind of magnetic order under these circumstances is antiferromagnetic. However, if the band is slightly less than half full, the ground state of the Hubbard Hamiltonian (8.4.21) was shown rigorously by Nagaoka (1966) to be ferromagnetic.

At lower densities, (8.4.57) indicates that the ferromagnetic state may be stable if $t\rho$ is large enough. This quantity, which appears in the denominator of the second term, is just the spin splitting of the energy bands in the ferromagnetic state. As a result of (8.4.49), the energies of states of ↓

spin are raised with respect to those of ↑ spin by $t\rho$. In fact, it follows from the work of Edwards (1967), that for higher densities than those for which (8.4.49) is valid, one can obtain a good approximation by simply replacing $t\rho$ in the second term of (8.4.57) by $M_{\downarrow}(\mathbf{p})$.

It is easily determined from (8.4.57) that the ferromagnetic state cannot be stable for sufficiently small densities. Assume for simplicity that the band is parabolic: $E(\mathbf{p}) = \gamma \mathbf{p}^2$. This assumption is not realistic for application to actual ferromagnets. We find that

$$\omega/\gamma q^2 = 1 - (4\mu/5t\rho) \tag{8.4.59}$$

in which $\mu = \gamma k_F^2$ is the Fermi energy ($\mu \approx \rho^{2/3}$). Thus there is a minimum density for which ferromagnetism can occur, regardless of the strength of the interaction. This is consistent with our conclusion following (8.4.36).

8.4.5 *Spin Density Waves*

Ferromagnetism is obviously not the only kind of magnetic order that may exist in a metal. Antiferromagnetism is a possible alternative. We will examine briefly here the interesting possibility of spin density wave antiferromagnetism, as introduced by Overhauser (1962). The possibility is realized in chromium.

The term "simple spin density wave" describes an electron distribution which is uniform in space, but has an oscillatory spin density. Overhauser (1962) showed that the ground state of an interacting system of free electrons in the Hartree–Fock approximation contains a spin density wave. We will therefore consider the Hartree–Fock equations (4.9.13), Part A, in the symbolic form

$$[H_0 + A]\psi(\mathbf{k}, \mathbf{r}) = E(\mathbf{k})\psi(\mathbf{k}, \mathbf{r}) \tag{8.4.60}$$

in which H_0 contains the kinetic energy and ordinary potential terms and A is the exchange operator. The wave function ψ is a two-component spinor which is not necessarily an eigenfunction of s_z. The exchange operator A may, and in this case will, have off-diagonal terms connecting different spin components. Let us assume with Overhauser that $A = A_0 + A_1$,

$$A_1 = -g\boldsymbol{\sigma}\cdot(\hat{\mathbf{i}} \cos \mathbf{q}\cdot\mathbf{r} + \hat{\mathbf{j}} \sin \mathbf{q}\cdot\mathbf{r}) \tag{8.4.61}$$

where $\hat{\mathbf{i}}$, $\hat{\mathbf{j}}$ are the usual unit vectors in the x and y directions, g is a constant, the direction of $\mathbf{q}$ defines the z axis, σ is the Pauli spin operator, and A_0 is the portion of the exchange operator which is diagonal in the spin indices. An A of this form will occur for a system possessing a fractional spin polarization at every point but with the direction of the polarization being a continuous function of position [see Eq. (4.9.19), Part A]. If the explicit expressions for the spin operators are used, A_1 has the form

$$A_1 = -g\begin{pmatrix} 0 & e^{-i\mathbf{q}\cdot\mathbf{r}} \\ e^{i\mathbf{q}\cdot\mathbf{r}} & 0 \end{pmatrix}. \tag{8.4.62}$$

We suppose for simplicity that the eigenfunctions of the operator $H_0 + A_0$ are plane waves $\exp(i\mathbf{k}\cdot\mathbf{r})$, as is correct for a free electron gas, and denote the eigenvalues of this operator by $\epsilon(\mathbf{k})$. The operator A_1 connects plane waves of different wave vectors: $\mathbf{k} - \mathbf{q}/2$ and $\mathbf{k} + \mathbf{q}/2$. The eigenvalues of H may be found by diagonalizing the simple matrix

$$\begin{pmatrix} \epsilon(\mathbf{k} - \mathbf{q}/2) & -g \\ -g & \epsilon(\mathbf{k} + \mathbf{q}/2) \end{pmatrix}$$

to be

$$E(\mathbf{k}) = \tfrac{1}{2}[\epsilon(\mathbf{k} - \mathbf{q}/2) + \epsilon(\mathbf{k} + \mathbf{q}/2)] \pm \{\tfrac{1}{4}[\epsilon(\mathbf{k} - \mathbf{q}/2) - \epsilon(\mathbf{k} + \mathbf{q}/2)]^2 + g^2\}^{1/2}. \tag{8.4.63}$$

The eigenfunctions are

$$\psi_1 = \Omega^{-1/2}\begin{pmatrix} \cos\theta_{\mathbf{k}/2}\exp[i(\mathbf{k} - \mathbf{q}/2)\cdot\mathbf{r}] \\ \sin\theta_{\mathbf{k}/2}\exp[i(\mathbf{k} + \mathbf{q}/2)\cdot\mathbf{r}] \end{pmatrix} \tag{8.4.64a}$$

and

$$\psi_2 = \Omega^{-1/2}\begin{pmatrix} -\sin\theta_{\mathbf{k}/2}\exp[i(\mathbf{k} - \mathbf{q}/2)\cdot\mathbf{r}] \\ \cos\theta_{\mathbf{k}/2}\exp[i(\mathbf{k} + \mathbf{q}/2)\cdot\mathbf{r}] \end{pmatrix} \tag{8.4.64b}$$

in which $\theta_{\mathbf{k}}$ is given for the lower branch by

$$\cos\theta_{\mathbf{k}/2} = g/\{g^2 + [\epsilon(\mathbf{k} - \mathbf{q}/2) - E(\mathbf{k})]^2\}^{1/2}. \tag{8.4.65}$$

The square moduli of ψ_1 and ψ_2 are constant, as is the case for a single plane wave. Hence, no fluctuation in the charge density is associated with the wave and no alternating term need be included in H_0. The state ψ_1 is the product of $\exp(i\mathbf{k}\cdot\mathbf{r})$ with a spin function which at point $\mathbf{r}$ has eigenvalue $+\frac{1}{2}$ for the spin operator referring to a direction making an angle $\theta_{\mathbf{k}}$ (which is independent of $\mathbf{r}$) with the z axis, and having an azimuthal angle $\phi = \mathbf{q}\cdot\mathbf{r}$ (independent of $\mathbf{k}$) with respect to the xz plane. The function ψ_2 has spin eigenvalue $-\frac{1}{2}$ with respect to the same direction, and can be obtained from ψ_1 through the substitution $\theta_{\mathbf{k}} \to \pi - \theta_{\mathbf{k}}$, $\phi = \phi + \pi$ (followed by the suppression of a common phase factor).

The energy spectrum described by (8.4.63) has two branches. This is shown schematically in Fig. 8.4.8. If the off-diagonal exchange were to be

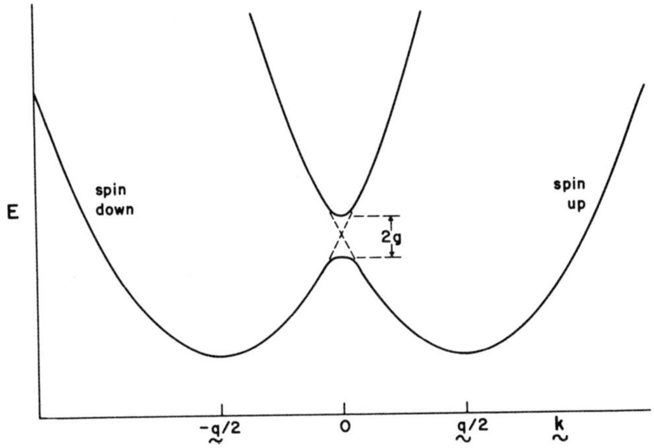

Fig. 8.4.8. Energy of a spin density wave.

neglected ($g = 0$), the angle θ_k as given by (8.4.65) would be zero. In this case, the two functions (8.4.64a, b) would represent pure spin up and spin down states, respectively. The energy levels in this limit are just simple band functions, shown as parabolas centered around $\mathbf{k} = \pm\mathbf{q}/2$. The interaction which couples the spin states causes a gap of magnitude $2g$ to appear in the spectrum. In the region of small values of $\mathbf{k}\cdot\mathbf{q}$, the spin will describe a wide angle cone around the z axis, collapsing to a spiral in the xy plane when $\mathbf{k}\cdot\mathbf{q} = 0$ ($\theta_k = \pi/2$). The spins of the two bands are everywhere opposite. In the ground state of the system, states belonging to the lower branch of the spectrum will be preferentially occupied.

Formation of a spin density wave state will occur only if the total energy of the system is lowered. We would expect this to occur under conditions which permit the Fermi energy to fall in the gap between upper and lower bands. A reduction in the energy of the system can then be obtained by populating just the states in the lower band. However, note now that the area of the Fermi surface is reduced. To see this, we note that the Fermi surface will be truncated, in a manner similar to that which occurs when Brillouin zones form, for directions of $\mathbf{k}$ such that the Fermi energy falls in the gap between bands.

We now wish to investigate how a spin density wave state arises in the Hartree–Fock approximation. This discussion is based on the review of Herring (1966a). The general expression [Eq. (4.9.12), Part A] for the total energy can be employed. Specifically, we delete the single-particle potential $V(\mathbf{r}_1)$ and consider instead of the actual electron–electron interaction e^2/r_{12}, a general function $V(\mathbf{r}_1 - \mathbf{r}_2)$:

$$E = \sum_{\mathbf{k}} \left\{ \int \psi^*(\mathbf{k}, \mathbf{r})(-\nabla^2)\psi(\mathbf{k}, \mathbf{r})\, d^3r \right.$$

$$+ \tfrac{1}{2} \sum_{\mathbf{k}'} \iint |\psi(\mathbf{k}, \mathbf{r}_1)|^2 |\psi(\mathbf{k}', \mathbf{r}_2)|^2 V(\mathbf{r}_1 - \mathbf{r}_2)\, d^3r_1\, d^3r_2$$

$$- \tfrac{1}{2} \sum_{\mathbf{k}'} \iint \psi^*(\mathbf{k}, \mathbf{r}_1)\psi^*(\mathbf{k}', \mathbf{r}_2) V(\mathbf{r}_1 - \mathbf{r}_2)\psi(\mathbf{k}, \mathbf{r}_2)\psi(\mathbf{k}', \mathbf{r}_1)\, d^3r_1\, d^3r_2. \tag{8.4.66}$$

Integrations include sums over spin components. This expression will be considered using the wave function (8.4.64). In order to simplify the calculation, we shall assume that the number of particles is such that only states belonging to the lower band in Fig. 8.4.8 are occupied. The total energy as given by (8.4.66) can be regarded as the sum of kinetic, average potential, and exchange parts:

$$\text{kinetic energy} = \sum_{\mathbf{k}} [\mathbf{k}^2 + \tfrac{1}{4}\mathbf{q}^2 - \mathbf{k}\cdot\mathbf{q} \cos\theta_{\mathbf{k}}] \tag{8.4.67}$$

$$\text{average potential energy} = \tfrac{1}{2} N V_{\text{Av}} \tag{8.4.68}$$

in which V_{Av} is

$$V_{\text{Av}} = \sum_{\mathbf{k}'} \int |\psi(\mathbf{k}', \mathbf{r}_2)|^2 V(\mathbf{r}_1 - \mathbf{r}_2)\, d^3r_2 = (N/\Omega) \int V(\mathbf{r})\, d^3r. \tag{8.4.69}$$

The potential matrix element involved in the calculation of the exchange energy is

$$\int \psi^*(\mathbf{k}_1, \mathbf{r}_1)\psi^*(\mathbf{k}_2, \mathbf{r}_2) V(\mathbf{r}_1 - \mathbf{r}_2)\psi(\mathbf{k}_3, \mathbf{r}_1)\psi(\mathbf{k}_4, \mathbf{r}_2)\, d^3r_1\, d^3r_2$$

$$= \Omega^{-1} \delta_{\mathbf{k}_1+\mathbf{k}_2, \mathbf{k}_3+\mathbf{k}_4} V(\mathbf{k}_1 - \mathbf{k}_3) S(\mathbf{k}_1, \mathbf{k}_3) S(\mathbf{k}_2, \mathbf{k}_4) \tag{8.4.70}$$

in which $V(\mathbf{k}_1 - \mathbf{k}_3)$ is a Fourier coefficient of $V(\mathbf{r})$:

$$V(\mathbf{k}) = \int \exp(i\mathbf{k}\cdot\mathbf{r}) V(\mathbf{r})\, d^3r \tag{8.4.71}$$

and

$$S(\mathbf{k}, \mathbf{k}') = \cos[\tfrac{1}{2}(\theta_{\mathbf{k}} - \theta_{\mathbf{k}'})]. \tag{8.4.72}$$

Thus

$$\text{exchange energy} = (-1/2\Omega) \sum_{\mathbf{k}\mathbf{k}'} V(\mathbf{k} - \mathbf{k}') \cos^2[\tfrac{1}{2}(\theta_{\mathbf{k}} - \theta_{\mathbf{k}'})]. \tag{8.4.73}$$

Equations (8.4.67), (8.4.68), and (8.4.73) are to be added. The result gives the total energy as a function of the wave vector $\mathbf{q}$ of the spin density wave:

$$E(\mathbf{q}) = \sum_{\mathbf{k}} \{\mathbf{k}^2 + \tfrac{1}{4}\mathbf{q}^2 - \mathbf{k}\cdot\mathbf{q}\cos\theta_{\mathbf{k}} - (1/2\Omega)\sum_{\mathbf{k}'} V(\mathbf{k}-\mathbf{k}')\cos^2[\tfrac{1}{2}(\theta_{\mathbf{k}}-\theta_{\mathbf{k}'})]\} + \tfrac{1}{2}NV_{\text{Av}}. \quad (8.4.74)$$

The evaluation of the total energy for a three-dimensional system with Coulomb interactions is quite complicated. Overhauser (1962) showed in this case that, as was mentioned above, the Hartree–Fock ground state contains a spin density wave. We shall confine our attention here to a much simpler model: a one-dimensional system with a repulsive delta function interaction between particles. Thus we set $V(\mathbf{k}) = V_0$. For a one-dimensional system, the Fermi wave vector k_F is related to the number of particles by

$$(a/\pi)\int_{-k_F}^{k_F} dk = 2ak_F/\pi = N. \quad (8.4.75)$$

The energy per particle is

$$E(q)/N = \tfrac{1}{3}k_F^2 + \tfrac{1}{4}q^2 - q\langle k\cos\theta_k\rangle + \tfrac{1}{2}(NV_0/a) - (V_0/4aN)\sum_{\mathbf{k}}\sum_{\mathbf{k}'}[1 + \cos\theta_{\mathbf{k}}\cos\theta_{\mathbf{k}'} + \sin\theta_{\mathbf{k}}\sin\theta_{\mathbf{k}'}] \quad (8.4.76)$$

in which $\langle\cdots\rangle$ denotes an average over the occupied states. For this system, $\Omega = a$. We shall assume that the system is not ferromagnetic. This means, in view of Eq. (8.4.64), $\sum_{\mathbf{k}}\cos\theta_k = 0$. Equation (8.4.76) becomes

$$E(q)/N = \tfrac{1}{3}k_F^2 + \tfrac{1}{4}q^2 - q\langle k\cos\theta_k\rangle - (NV_0/4\Omega)[\langle\sin\theta_k\rangle]^2 + \tfrac{1}{4}(NV_0/\Omega). \quad (8.4.77)$$

This expression is to be minimized with respect to (each) θ_k: This yields

$$\operatorname{ctn}\theta_k = k/\alpha \quad (8.4.78)$$

in which

$$\alpha = NV_0\langle\sin\theta_k\rangle/2q\Omega. \quad (8.4.79)$$

We have from (8.4.78)

$$\langle\sin\theta_{\mathbf{k}}\rangle = 1/k_F\int_0^{k_F}\alpha/(k^2+\alpha^2)^{1/2}\,dk = (\alpha/k_F)\sinh^{-1}(k_F/\alpha). \quad (8.4.80)$$

Equations (8.4.80) and (8.4.79) are combined to yield

$$\alpha = k_F/\sinh(2qak_F/NV_0). \tag{8.4.81}$$

Equations (8.4.78)–(8.4.81) are substituted into (8.4.77) to determine the total energy. Note that

$$\langle k \cos\theta_k \rangle = (1/k_F) \int_0^{k_F} k^2/(k^2 + \alpha^2)^{1/2}\, dk$$

$$= \tfrac{1}{2}[(k_F{}^2 + \alpha^2)^{1/2} - (\alpha^2/k_F)\sinh^{-1}(k_F/\alpha)]. \tag{8.4.82}$$

Equation (8.4.77) reduces to

$$E(q) = \tfrac{1}{3}k_F{}^2 + \tfrac{1}{4}q^2 - \tfrac{1}{2}q(k_F{}^2 + \alpha^2)^{1/2}. \tag{8.4.83}$$

We can now determine q by minimizing $E(q)$. If the interaction V_0 and (thus) α vanish, we find $q = k_F$. Then (8.4.78) indicates $\theta_k = 0$ or π, and we choose $\cos\theta_k = \operatorname{sign}(k)$. This describes the normal paramagnetic state. If the interaction is weak, a solution is obtained with q slightly smaller than k_F. We then have

$$\alpha \approx 2k_F \exp[-2q\Omega k_F/NV_0]. \tag{8.4.84}$$

Put $q = k_F(1 - y)$. Minimization of the energy yields, to leading order in α,

$$y \approx 2\Omega\alpha^2/NV_0. \tag{8.4.85}$$

In obtaining (8.4.85), we have assumed $\Omega/NV_0 \gg 1/k_F{}^2$. Thus, q differs from k_F by terms which are exponentially small when the interaction is weak. The energy is

$$E = (k_F{}^2/12) - (\alpha^2/4) = (k_F{}^2/12) - k_F{}^2 \exp[-2\Omega k_F{}^2/NV_0]. \tag{8.4.86}$$

The energy depends on V_0 in a manner that is not analytic for small V_0. This is the same kind of behavior found in the theory of superconductivity. Equation (8.4.86) tells us that the uniform "normal" state of a one-dimensional system is never the ground state in the Hartree–Fock approximation, no matter how weak is the repulsion between particles. This conclusion is due to Overhauser (1960). For strong interactions, q decreases, and ultimately vanishes, leading to a ferromagnetic state of the system.

The situation in three dimensions is not so clear cut. It can then be shown that for short range interactions, such as the repulsive delta function considered here, the spin density wave state is not the ground state. The Coulomb interaction yields a spin density wave ground state for free electrons in the Hartree–Fock approximation as pointed out earlier, however, this disappears when account is taken of the dielectric screening of the electron–electron interaction (Rajagopal, 1966; Fedders and Martin, 1966; Hamann and Overhauser, 1966). It is possible for special band structure effects to bring about a situation favorable to the formation of spin density waves or peculiar ordering of local magnetic moments (as occurs in rare earth metals). We will turn to a brief investigation of these.

8.4.6 *The Wave Vector Dependence of the Susceptibility*

The essential quantity involved in the determination of magnetic order is the wave vector-dependent spin susceptibility. This function was studied in Section 8.4.1 in the limit of free electrons. We must include band structure effects in the present discussion, as these can lead to significant qualitative modifications of the results. Consider the Hamiltonian

$$H = H_0 + H_1 = H_0 + (e\hbar/m_0) \sum_j \mathbf{s}_j \cdot \mathbf{B}(\mathbf{q}) \cos \mathbf{q}\cdot\mathbf{r}_j. \qquad (8.4.87)$$

A magnetic field of wave vector $\mathbf{q}$ is included. Here $\mathbf{s}_j$ is spin operator for the jth electron and H_0 contains all periodic potential terms. It is convenient to take the magnetic field $\mathbf{B}$ in the x direction $\mathbf{B}(\mathbf{q}) = B_q\hat{\mathbf{x}}$. This avoids a redistribution of spin population between up and down spin states (which refer to the z axis) even in the limit $q = 0$. Diamagnetic effects are ignored. The unperturbed states are the usual Bloch states $\psi_{n\sigma}(\mathbf{k}, \mathbf{r})$, in which n is the band index and σ $(= \uparrow, \downarrow)$ denotes the spin. The perturbation can be written as

$$H_1 = (e\hbar/4m_0)B_{\mathrm{q}} \sum_j (s_j^+ + s_j^-)[\exp(i\mathbf{q}\cdot\mathbf{r}_j) + \exp(-i\mathbf{q}\cdot\mathbf{r}_j)]. \qquad (8.4.88)$$

The first-order change in the energy is zero.

The second-order change in the energy of the state $| n\mathbf{k}\sigma\rangle$ will be denoted $\Delta E_{n\sigma}^{(2)}(\mathbf{k})$. To determine this quantity, we require the matrix element

$$\langle n\mathbf{k} \mid \exp(-i\mathbf{q}\cdot\mathbf{r}) \mid l\mathbf{p}\rangle = \sum \delta_{\mathbf{k}+\mathbf{q},\mathbf{p}+\mathbf{K}_t} M_{nl}(\mathbf{k}, \mathbf{K}_t, \mathbf{p}) \qquad (8.4.89)$$

in which the sum runs over reciprocal lattice vectors and the matrix element M_{nl} is

$$M_{nl}(\mathbf{k}, \mathbf{K}_t, \mathbf{p}) = [(2\pi)^3/\Omega] \int_{\mathrm{cell}} \exp(i\mathbf{K}_t\cdot\mathbf{r}) u_n^*(\mathbf{k}, \mathbf{r}) u_l(\mathbf{p}, \mathbf{r})\, d^3r. \qquad (8.4.90)$$

Then $\Delta E^{(2)}$ is given by

$$\Delta E_{n\sigma}{}^{(2)}(\mathbf{k}) = (e\hbar B_q/4m_0)^2 \sum_{tl\mathbf{p}} |M_{nl}(\mathbf{k}, \mathbf{K}_t, \mathbf{p})|^2 \times [1 - f(E_l)]/[E_n(\mathbf{k}) - E_l(\mathbf{p})][\delta_{\mathbf{k}+\mathbf{q},\mathbf{p}+\mathbf{K}_t} + \delta_{\mathbf{k}-\mathbf{q},\mathbf{p}+\mathbf{K}_t}] \tag{8.4.91}$$

in which $f(E)$ is the usual Fermi function. Only unoccupied intermediate states occur. The sum over the spins of these states has been performed. The change in the total energy of the system $\Delta\mathcal{E}$ is found by summing (8.4.91) over occupied states $|n\mathbf{k}\sigma\rangle$:

$$\Delta\mathcal{E} = \left(\frac{e\hbar B_q}{4m_0}\right)^2 \sum_{ln,\mathbf{k}t,\sigma} \left\{\frac{f(E_n)[1 - f(E_l)]\,|M_{nl}(\mathbf{k}, \mathbf{K}_t, \mathbf{k}+\mathbf{q})|^2}{E_n(\mathbf{k}) - E_l(\mathbf{k}+\mathbf{q})} + \frac{f(E_n)[1 - f(E_l)]\,|M_{nl}(\mathbf{k}, \mathbf{K}_t, \mathbf{k}-\mathbf{q}|^2}{E_n(\mathbf{k}) - E_l(\mathbf{k}-\mathbf{q})}\right\}$$

$$= \left(\frac{e\hbar B_q}{4m_0}\right)^2 \sum_{ln,\mathbf{k}t,\sigma} \frac{\{f[E_n(\mathbf{k})] - f[E_l(\mathbf{k}+\mathbf{q})]\}}{E_n(\mathbf{k}) - E_l(\mathbf{k}+\mathbf{q})}\,|M_{nl}(\mathbf{k}, \mathbf{K}_t, \mathbf{k}+\mathbf{q})|^2. \tag{8.4.92}$$

Dummy indices have been altered to achieve the simplified form in the second line. It is convenient to introduce the generalized susceptibility function $S(\mathbf{q})$

$$S(\mathbf{q}) = -\sum_{ln,\mathbf{k}t,\sigma} \frac{\{f[E_n(\mathbf{k})] - f[E_l(\mathbf{k}+\mathbf{q})]\}}{E_n(\mathbf{k}) - E_l(\mathbf{k}+\mathbf{q})}\,|M_{nl}(\mathbf{k}, \mathbf{K}_t, \mathbf{k}+\mathbf{q})|\}^2 \tag{8.4.93}$$

so that

$$\Delta\mathcal{E} = -(e\hbar B_{\mathbf{q}}/4m_0)^2 S(\mathbf{q}). \tag{8.4.94}$$

The spin susceptibility can now be determined according to (6.3.25c):

$$\chi(\mathbf{q})/[1 + \chi(\mathbf{q})]^2 = -2\mu_0\, \partial^2(\Delta\mathcal{E})/\partial B_{\mathbf{q}}{}^2. \tag{8.4.95}$$

The additional factor of 2 in (8.4.95) in comparison with Eq. (6.3.25c) results from the fact that we have considered an oscillatory field $B_q \cos^2 \mathbf{q}\cdot\mathbf{r}$. The energy in this field is proportional to $B_{\mathbf{q}}{}^2 \cos^2 \mathbf{q}\cdot\mathbf{r}$, and the spatial average of $\cos^2 \mathbf{q}\cdot\mathbf{r}$ is $\frac{1}{2}$. We compensate for the factor of $\frac{1}{2}$ by inserting a factor of 2 in (8.4.95). Thus

$$\chi(\mathbf{q})/[1 + \chi(\mathbf{q})]^2 = \mu_0(e\hbar/2m_0)^2 S(\mathbf{q}) = \mu_0\mu_\beta^2 S(\mathbf{q}) \qquad (8.4.96)$$

(μ_0 is the permittivity of free space and μ_β is the Bohr magneton). Under ordinary circumstances, $\chi(\mathbf{q})$ is small, and it is an adequate approximation to write

$$\chi(\mathbf{q}) = \mu_0\mu_\beta^2 S(\mathbf{q}). \qquad (8.4.97)$$

The generalized susceptibility function $S(\mathbf{q})$ is essential to the study of magnetic order. If this function is evaluated in the free electron approximation, the result given in (8.4.10) is obtained. We are interested in band structure and interaction effects on this function.

The essential result is Eq. (8.4.94). In the spirit of molecular field theory (Section 2.3, Part A), we consider an exchange interaction between members of a spin system as equivalent to a magnetic field acting on those spins. A magnetic structure which is characterized by a wave vector $\mathbf{q}$ will give rise to an effective magnetic field described by the same wave vector $\mathbf{q}$. From (8.4.94), the energy of such a structure is proportional to $-S(\mathbf{q})$. Under such circumstances the stable magnetic configuration will be determined by the maximum of the response function $S(\mathbf{q})$.

The response function for a free electron gas with interactions neglected, as given by (8.4.10), has its maximum at $q = 0$ and decreases monotonically with increasing q. Band structure effects can modify this situation. Equation (8.4.93) indicates that transitions between occupied and unoccupied states contribute to S: if $|n\mathbf{k}\rangle$ and $|l\mathbf{q}\rangle$ are both full or both empty, the contribution vanishes. A maximum contribution will result if the energy denominator is small. For a simple parabolic band, small values of q dominate. However, if the Fermi surface is complicated in a suitable way, other values of q may be important. Specifically, suppose the Fermi surface has components (which may be in different bands) that are nearly plane parallel and are separated by $\mathbf{q}_0$ (nesting portions). Then $S(\mathbf{q})$ will tend to have a maximum for $\mathbf{q} = \mathbf{q}_0$. Actually, if portions of the surface were exactly parallel, a logarithmic singularity would occur for $\mathbf{q} = \mathbf{q}_0$. Deviations from parallelism will convert the singularity into a local maximum.

If the local maximum of $S(\mathbf{q})$ is the absolute maximum, the system will respond by establishing spin order of the required periodicity. This may be a spin density wave. If local spin moments exist in the system, they will tend to order through the indirect exchange mechanism in a structure with the appropriate period. Arguments of this type can be employed to account for the spin density wave antiferromagnetism of chromium (Lomer, 1962), and to explain the peculiar magnetic structure of some rare earth elements (Keeton and Loucks, 1968; Evenson and Liu, 1968, 1969).

In the case of chromium, a large part of the electron-filled Fermi surface centered at Γ can be made nearly to coincide over a substantial area with a large hole-filled portion of Fermi surface, centered at H by a translation $\mathbf{q}$ in the (100) direction. The length of this translation is slightly less than the distance from Γ to H, and appears to correspond with the wave vector $q = 0.96(2\pi/a)$ of the sinusoidal spin density waves which are established in chromium.

The preceding discussion is incomplete in that it does not take into account the effect of electron interactions on the susceptibility. It turns out that the susceptibility is modified in a significant way by Coulomb interactions. The existence of a spin density wave ground state in the Hartree–Fock approximation implies that the susceptibility has a maximum not at $q = 0$: in fact, there is a pole for $q \simeq 2k_F$ (Fedders and Martin, 1966). When a screened electron–electron interaction is employed, the pole disappears, but there is an increase in the susceptibility. In the limit that the screened interaction is of very short range, the enhancement takes a simple form. The theory was worked out by Wolff (1959) in the random phase approximation.

A detailed theory of interaction effects on the susceptibility could be built along the lines of the procedure followed in Section 8.4.4 to treat spin waves. We will not develop the details here. The essential features of the result can be understood through a simple argument due to Watson and Freeman (1966). In the absence of interactions, the system responds to a magnetic field $H(\mathbf{q})$ by producing a magnetization $M(\mathbf{q})$ determined by

$$M(\mathbf{q}) = \chi_0(\mathbf{q})H(\mathbf{q}) \tag{8.4.98}$$

where χ is assumed to be a scalar so that M and H are in the same direction. The vector nature of M and H can then be neglected. Let us further assume that this response induces, by exchange coupling, an effective field $H_{\text{ind}}(\mathbf{q})$ proportional to $M(\mathbf{q})$. This is in line with the molecular field arguments presented earlier. A quantity $V(\mathbf{q})$ is defined through the relation

$$H_{\text{ind}} = V(\mathbf{q})M(\mathbf{q})/\mu_0\mu_B{}^2.$$

The constant $\mu_0\mu_B{}^2$ is included for convenience. Then we write

$$M(\mathbf{q}) = \chi_0(\mathbf{q})[H(\mathbf{q}) + V(\mathbf{q})M(\mathbf{q})/\mu_0\mu_B{}^2]. \tag{8.4.99}$$

This equation is solved for $M(\mathbf{q})$, which is related to $H(\mathbf{q})$ through Eq. (8.4.98) with χ_0 replaced by the enhanced susceptibility $\chi_e(\mathbf{q})$. We suppose that (8.4.97) applies, and obtain

$$\chi_e(\mathbf{q}) = \chi_0(\mathbf{q})/[1 - V(\mathbf{q})S(\mathbf{q})]. \qquad (8.4.100)$$

The susceptibility is increased if V is positive. In the calculation of Wolff, $V(\mathbf{q})$ is the Fourier transform of the screened interaction potential between electrons. If the interaction is of short range, then $V(\mathbf{q}) = V_0$, a constant. We note from (8.4.93), using the argument leading to Eq. (7.1.53), that $S(0) = G(\mu)$ (including both spin directions) since if $q = 0$, $K_t = 0$, and the matrix element M_{nl} (8.4.90) simply becomes δ_{nl}. Thus, put

$$S(\mathbf{q}) = G(\mu)U(\mathbf{q}) \qquad (8.4.101)$$

in which $U(0) = 1$. Note that an instability develops at $q = 0$ if $1 = V_0G(\mu)$. This instability corresponds to ferromagnetism, and the criterion is similar to (8.4.36). The comparison indicates that at low densities V_0 should be replaced by t_0, the two-body t matrix. The exchange enhancement of the susceptibility given by (8.4.100) can be fairly large for metals, such as palladium (Giovannini *et al.*, 1964).

8.5 Many-Body Effects in Semiconductors

In Chapters 5 and 6, the effects of impurities and external fields on the energy levels of a system of electrons in a periodic potential were studied. The entire discussion was based on a one-particle theory, no account being taken of the electron interactions. In this section, we will reconsider one aspect of this problem from the point of view of many-body theory. It will be shown, following the work of Kohn (1957, 1958), that an effective mass equation is valid for the description of the change in energy levels of an insulator in response to a slowly varying perturbation. This work has been extended by several authors: see, for example, Klein (1959), Ambegaokar and Kohn (1960), Sham (1966).

Consider a system consisting of an insulator plus one extra electron. It can be proved that the response of such a system to a long wavelength, low frequency electric field is the same as that of a free electron of effective mass m^* moving in a medium having the dielectric constant κ, of the perfect insulator. The point impurity is included as a special case in that the bound states which occur when a small positive charge q is embedded in the system have a hydrogenic spectrum specified by

$$E_n = -(m^*/2\kappa^2)(e^2q^2/n^2\hbar^2). \qquad (8.5.1)$$

In addition, it has been shown that the low lying levels in a constant external magnetic field B are characterized by a cyclotron frequency

$$\omega_c = eB/m^* \tag{8.5.2}$$

in which m^* is the same effective mass as in (8.5.1).

8.5.1 *Quasi Particles in Semiconductors*

We begin by considering a perfect semiconductor (or insulator) which contains N electrons. The Hamiltonian for this system contains the kinetic energy of the electrons, the potential of each electron in the field of the fixed ions, and the Coulomb interaction energy between the electrons. The eigenstates of this system will be denoted by $| N, \mathbf{K}, v\rangle$ in which $\mathbf{K}$ designates the total wave vector of the system and v corresponds to the set of quantum numbers other than $\mathbf{K}$ required to specify a state. Specifically, if $T(\mathbf{R}_\mu)$ is a translation operator which increases all electron coordinates by the direct lattice vector $\mathbf{R}_\mu$, then

$$T(\mathbf{R}_\mu) \mid N, \mathbf{K}, v\rangle = \exp(i\mathbf{K}\cdot\mathbf{R}_\mu) \mid N, \mathbf{K}, v\rangle \tag{8.5.3}$$

(see also Section 4.9, Part A). The energy of a state will be denoted $E_{N,v}(\mathbf{K})$, and it will be convenient to refer to v as the band index. The ground state of the system will be assumed to have $\mathbf{K} = 0$ and we shall set $v = 0$ for this state.

A perfect semiconductor is distinguished from a metal by the following property of the energy spectrum. All excited states of the system with $K = 0$ have energies which differ from the ground state by a finite amount ΔE:

$$E_{N,v}(0) - E_{N,0}(0) \geq \Delta E. \tag{8.5.4}$$

This property is characteristic of the independent particle model of a semiconductor, but not of a metal: it is found to be true experimentally for real semiconductors since electromagnetic radiation, which causes transitions between states of the same $\mathbf{K}$, is not absorbed if the photons have less than a definite energy. The energy $E_{N,0}(0)$ is assumed to be an absolute minimum with respect to $\mathbf{K}$.

For the purposes of investigating the shallow impurity problem, we require statements similar to (8.5.4) concerning the $N + 1$ and the $N - 1$ particle system. In such cases, we have, for fixed $\mathbf{K}$ and $v \neq 0$,

$$E_{N+1,v}(\mathbf{K}) - E_{N+1,0}(\mathbf{K}) > \Delta \tag{8.5.5}$$

$$E_{N-1,v}(\mathbf{K}) - E_{N-1,0}(\mathbf{K}) > \Delta \tag{8.5.6}$$

in which Δ is some finite energy. Also, suppose an electron is added to (or removed) from the N particle system:

$$E_{N+1,v}(\mathbf{K}) - E_{N,0}(0) \geq \mu + E_g/2$$
$$E_{N-1,v}(\mathbf{K}) - E_{N,0}(0) \geq \mu - E_g/2. \qquad (8.5.7)$$

Here μ is the chemical potential and E_g is an energy which corresponds in the single particle picture to the band gap.

Let us now consider some of the formal properties of the single-particle Green's function, $\mathcal{G}(\mathbf{r}, \mathbf{r}', \epsilon)$. We have, by a slight rewriting of Eq. (8.1.15a),

$$\mathcal{G}(\mathbf{r}, \mathbf{r}', \epsilon) = \sum_{v,\mathbf{K}} \{ f_v^{\mathrm{e}}(\mathbf{K}, \mathbf{r}) f_v^{\mathrm{e}*}(\mathbf{K}, \mathbf{r}')/[\epsilon - E_v^{\mathrm{e}}(\mathbf{K})]$$
$$+ f_v^{\mathrm{h}}(\mathbf{K}, \mathbf{r}) f_v^{\mathrm{h}*}(\mathbf{K}, \mathbf{r}')/[\epsilon - E_v^{\mathrm{h}}(\mathbf{K})]\} \qquad (8.5.8)$$

in which the functions f^{e} and f^{h} are

$$f_v^{\mathrm{e}}(K, \mathbf{r}) = \langle N, 0, 0 \mid \psi(\mathbf{r}) \mid N+1, v, \mathbf{K} \rangle$$
$$f_v^{\mathrm{h}}(K, \mathbf{r}) = \langle N, 0, 0 \mid \psi^{+}(\mathbf{r}) \mid N-1, v, -\mathbf{K} \rangle, \qquad (8.5.9)$$

and ψ is the usual electron field operator. We will use units in which $\hbar = 1$. The energies in the denominator are

$$E_v^{\mathrm{e}}(\mathbf{K}) = E_{N+1,v}(\mathbf{K}) - E_{N,0}(0)$$
$$E_v^{\mathrm{h}}(\mathbf{K}) = E_{N,0}(0) - E_{N-1,v}(-\mathbf{K}). \qquad (8.5.10)$$

Let us introduce a Green's function referring to states of fixed total $\mathbf{K}$:

$$\mathcal{G}(\mathbf{r}, \mathbf{r}', \epsilon) = \sum_{\mathbf{K}} \mathcal{G}_{\mathbf{K}}(\mathbf{r}, \mathbf{r}', \epsilon). \qquad (8.5.11)$$

The function $\mathcal{G}_{\mathbf{K}}$ obeys a form of Bloch's theorem. The electron field operators transform under a translation $T(\mathbf{R}_\mu)$ as

$$\psi(\mathbf{r} + \mathbf{R}_\mu) = T^{+}(\mathbf{R}_\mu)\psi(\mathbf{r})T(\mathbf{R}_\mu). \qquad (8.5.12)$$

From this and (8.5.3), we have

$$\mathcal{G}_{\mathbf{K}}(\mathbf{r} + \mathbf{R}_\mu, \mathbf{r}', \epsilon) = \exp(i\mathbf{k}\cdot\mathbf{R}_\mu)\mathcal{G}_{\mathbf{K}}(\mathbf{r}, \mathbf{r}', \epsilon) = \mathcal{G}_{\mathbf{K}}(\mathbf{r}, \mathbf{r}' - \mathbf{R}_\mu, \epsilon). \qquad (8.5.13)$$

Equation (8.5.7) implies that the properties of the Green's function in a semiconductor are somewhat different than in a metal. Suppose that a single electron is added to the lowest conduction band in a state whose energy is greater than that of the lowest minimum of this band by an amount less than the magnitude of the band gap. It is then not possible energetically for such an electron to produce a real electron–hole pair. In other words, such a state cannot decay through the electron–electron interaction (of course, in reality, it can decay by emission of phonons) and thus has an indefinitely long lifetime. In a metal, an electron which has been excited across the Fermi surface can always lose energy through the creation of electron–hole pairs. The excited state in the semiconductor corresponds to a quasi particle of infinite lifetime. The single-particle Green's function $\mathcal{G}_{\mathbf{K}}(\mathbf{r}, \mathbf{r}', \epsilon)$ will thus have a simple pole on the real axis at the energy $E_0^{\mathrm{e}}(\mathbf{K})$, provided $E_0^{\mathrm{e}}(\mathbf{K}) < E_{\mathrm{g}}$. Correspondingly, the spectral weight function $A_{\mathbf{K}}(\mathbf{r}, \mathbf{r}', \epsilon)$ will have an isolated delta function spike at $E_0^{\mathrm{e}}(\mathbf{K})$. Similar considerations apply to holes: A single hole state is also an eigenstate of the system provided $E_0^{\mathrm{h}}(\mathbf{K}) < E_{\mathrm{g}}$.

These observations form the basis for an explanation of the remarkable success of band theory in explaining the properties of semiconductors. When the number of excited particles, holes, and electrons is sufficiently small, and their energies are smaller than the band gap, band theory furnishes an adequate description. The electron–electron interaction can alter the values of the parameters involved. Use of the single-particle language is, however, appropriate.

The preceding discussion implies that the Green's function has the form

$$\begin{aligned}\mathcal{G}_{\mathrm{k}}(\mathbf{r}, \mathbf{r}', \epsilon) &= f_0^{\mathrm{e}}(\mathbf{K}, \mathbf{r}) f_0^{\mathrm{e}*}(\mathbf{K}, \mathbf{r}')/[\epsilon - E_0^{\mathrm{e}}(\mathbf{K})] \\ &\quad + f_0^{\mathrm{h}}(\mathbf{K}, \mathbf{r}) f_0^{\mathrm{h}*}(\mathbf{K}, \mathbf{r}')/[\epsilon - E_0^{\mathrm{h}}(\mathbf{K})] \\ &\quad + \left(\int_{-\infty}^{E_0^{\mathrm{h}}(\mathrm{K})-E_{\mathrm{g}}} d\omega + \int_{E_0^{\mathrm{e}}(\mathrm{K})+E_{\mathrm{g}}}^{\infty} d\omega\right) A_{\mathbf{K}}(\mathbf{r}, \mathbf{r}', \omega)/(\epsilon - \omega)\end{aligned} \tag{8.5.14}$$

in which $A_{\mathbf{K}}$ is the spectral weight function.

8.5.2 *Impurity States*

Suppose that a single donor impurity is inserted into the crystal so that it replaces an atom of the perfect crystal. The site of the donor will be used as the origin of coordinates. Most of the preceding analysis remains valid except that additional simple poles will be present in the Green's function. These poles, which lie within the band gap, correspond to the bound states associated with the donor.

We may obtain the generalized effective mass equation as follows (Kohn, 1957, 1958): Let $\Psi(1, \ldots, N+1)$ be the exact many-electron wave function for the system in the presence of the donor, while $\psi_v(\mathbf{K}, \mathbf{r}_1 \cdots \mathbf{r}_{N+1})$ refers to states of the perfect crystal. We expand Ψ as follows:

$$\Psi(1, \ldots, N+1) = \sum_v \int d^3K\, \phi_v(\mathbf{K}) \psi_v(\mathbf{K}, \mathbf{r}_1 \cdots \mathbf{r}_{N+1}). \tag{8.5.15}$$

This procedure generalizes the crystal momentum representation described in Section 5.1. The formal development leading to Eq. (5.1.18) is still valid, and we have

$$[E_v(\mathbf{K}) - E]\phi_v(\mathbf{K}) + \sum_l \int d^3Q \, \langle v\mathbf{K} \mid U \mid l\mathbf{Q}\rangle \phi_l(\mathbf{Q}) = 0. \quad (8.5.16)$$

The matrix element is now given by

$$\langle v\mathbf{K} \mid U \mid l\mathbf{Q}\rangle = \int \psi_v^*(\mathbf{K}, \mathbf{r}_1 \cdots \mathbf{r}_{N+1}) U(1 \cdots N+1) \times \psi_l(\mathbf{Q}, \mathbf{r}_1 \cdots \mathbf{r}_{N+1}) \, d^3r_1 \cdots d^3r_{N+1}. \quad (8.5.17)$$

The perturbation U represents the change in the potential energy of the system of $N + 1$ particles due to the presence of the donor. It is assumed that this can be expressed as a sum of identical perturbing potentials acting on each electron

$$U(1 \cdots N+1) = \sum_{i=1}^{N+1} U(\mathbf{r}_i). \quad (8.5.18)$$

The matrix element must be considered in detail. It is convenient to introduce the Fourier transform $V(\mathbf{q})$ of the single-electron perturbation $U(\mathbf{r}_i)$ in (8.5.17):

$$U(\mathbf{r}_i) = \int V(\mathbf{q}) \exp(-i\mathbf{q}\cdot\mathbf{r}_i) \, d^3q. \quad (8.5.19)$$

The total perturbation can be expressed in terms of $V(\mathbf{q})$:

$$U(1 \cdots N+1) = \int V(\mathbf{q})\rho(\mathbf{q}) \, d^3q \quad (8.5.20)$$

where

$$\rho(\mathbf{q}) = \sum_i \exp(-i\mathbf{q}\cdot\mathbf{r}_i). \quad (8.5.21)$$

The quantity $\rho(\mathbf{q})$ may be interpreted as a Fourier coefficient of the electron density since if we have $(N + 1)$ electrons located at points $\mathbf{r}_i$, the electron density is $\sum_i \delta(\mathbf{r} - \mathbf{r}_i)$ and the Fourier transform of this function is $\rho(\mathbf{q})$. Then we have for the matrix element of the potential

$$\langle 0\mathbf{K} \mid V \mid 0\mathbf{Q}\rangle = (2\pi)^{-3} \int d^3q \, V(\mathbf{q}) \int \psi_0^*(\mathbf{K}, \mathbf{r}_1 \cdots \mathbf{r}_{N+1}) \times \sum_i \exp(-i\mathbf{q}\cdot\mathbf{r}_i)\psi_0(\mathbf{Q}, \mathbf{r}_1 \cdots \mathbf{r}_{N+1}) \, d^3r_1 \cdots d^3r_{N+1}$$

$$= \int d^3q \, \delta(\mathbf{Q} - \mathbf{q} - \mathbf{K}) \, V(\mathbf{q}) \langle 0\mathbf{K} \mid \rho(\mathbf{q}) \mid 0\mathbf{K} + \mathbf{q}\rangle. \quad (8.5.22)$$

The last step follows from the Bloch form of the total wave function, since we have in the previous line the Fourier transform of a periodic function. As in Section 5.3, the perturbing potential is assumed to be slowly varying so that only small values of q are important. Kohn has shown that

$$\lim_{q\to 0} \langle 0\mathbf{K} \mid \rho(\mathbf{q}) \mid 0\mathbf{K} + \mathbf{q}\rangle = 1/\kappa \tag{8.5.23}$$

in which κ is the static dielectric function. This interpretation follows directly from consideration of the significance of (8.5.22). Suppose we wish to study the scattering of a low energy "band 0" electron by a small external charge q embedded in the medium. The first Born approximation requires us to consider the matrix element

$$M_{\mathbf{KQ}} = \left\langle 0\mathbf{K} \left| \sum_{i=1}^{N+1} (-eq/r_i) \right| 0\mathbf{Q} \right\rangle .$$

The procedure of (8.5.22), combined with (8.5.23), leads to the expression

$$M_{\mathbf{KQ}} = 4\pi(-e)q/\kappa \mid \mathbf{K} - \mathbf{Q} \mid^2 \tag{8.5.24}$$

which clearly describes a screened Coulomb potential. Hence, provided the limit specified by (8.5.23) exists, κ must be the dielectric constant. A detailed diagrammatic analysis shows that this interpretation is correct.

A less formal approach is also possible. We may follow the procedure of Section 7.1.4 to calculate the screening charge density produced in the system by the external perturbation. The system then responds to the total potential, which is the sum of the original potential plus that induced in the system. In this way, we arrive at the general expressions (7.1.46) and (7.1.47) which describe the screening. We also recall the discussion following Eq. (7.1.52) in which it is shown that in a semiconductor, the dielectric function must remain finite as the wave vector goes to zero.

As a result of the preceding arguments, we may rewrite (8.5.16) in the form

$$[E_0(\mathbf{K}) - E]\phi_0(\mathbf{K}) + (1/\kappa) \int d^3Q\, V(\mathbf{K} - \mathbf{Q})\phi_0(\mathbf{Q}) = 0. \tag{8.5.25}$$

The contributions from terms of higher order in the expansion of $\rho(\mathbf{q})$ will produce departures from the simple static screening which appears in (8.5.25).

We now have a very straightforward problem. For small values of $\mathbf{K}$, it is possible to replace $E_0(\mathbf{K})$ by the leading terms of its power series expansion. In accord with our fundamental assumption, these are

$$E_0(\mathbf{K}) = E_0(0) + (\hbar^2\mathbf{K}^2/2m^*) \tag{8.5.26}$$

in which m^* is an effective mass. This is substituted into (8.5.25) and the result is transformed back into position space exactly as was done in Section 5.3. The result is just the effective mass equation

$$[-(\hbar^2/2m^*)\nabla^2 + (1/\kappa)U(\mathbf{r})]F(\mathbf{r}) = EF(\mathbf{r}) \tag{8.5.27}$$

in which $F(\mathbf{r})$ is the envelope function

$$F(\mathbf{r}) = \int \exp(i\mathbf{K}\cdot\mathbf{r})\phi_0(\mathbf{K})\, d^3K \tag{8.5.28}$$

which was introduced in Section 5.3.

Problems

1 Let the electron density be expressed as $n(\mathbf{x}) = \psi^+(\mathbf{x})\psi(\mathbf{x})$ and define $\rho(\mathbf{k}) = \int d^3x \exp[i\mathbf{k}\cdot\mathbf{x}]n(\mathbf{x})$. Show that

$$\rho(\mathbf{q}) = \sum_{n\mathbf{q}\sigma} c^+_{n\sigma}(\mathbf{k})c_{n\sigma}(\mathbf{k}+\mathbf{q}).$$

Then show that the Coulomb energy operator for an electron gas can be expressed as

$$V = \sum_{\mathbf{q}} (2\pi e^2/q^2)(\rho^+(\mathbf{q})\rho(\mathbf{q}) - N).$$

2 Label and write down the contributions of the following diagrams to the one-particle Green's function using Goldstone rules. Perform the frequency integrations.

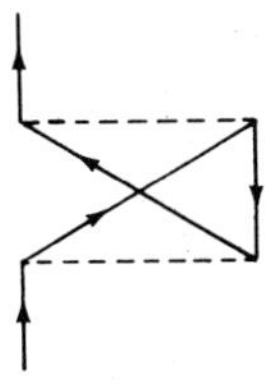

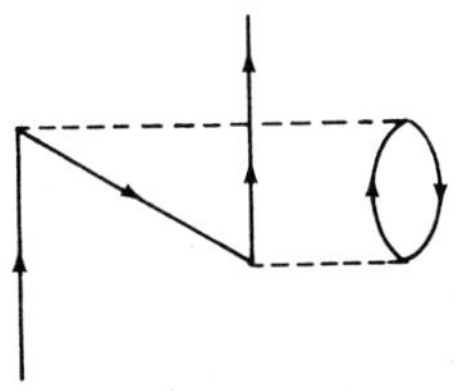

3 Suppose that a Fermi system of spin $\frac{1}{2}$ particles has the interaction potential between pairs $V(r) = V_0 e^{-ar}/r$. Evaluate the self-energy in the Hartree–Fock approximation.

4 Evaluate the dielectric function for the free electron gas in the RPA for all frequencies.

5 For a free electron gas show that

$$[[H_0, \rho(\mathbf{q})], \rho(-\mathbf{q})] = (N/m)q^2$$

(see Problem 1). Then let $|n\rangle$ be an excited state of H_0. Show that

$$\sum_n [\epsilon_n - \epsilon_0]\{|\langle 0 | \rho(\mathbf{q}) | n\rangle|^2 + |\langle 0 | \rho(-\mathbf{q}) | n\rangle|^2\} = (N/m)q^2.$$

6 Suppose that a test charge $e\rho_t \cos(\mathbf{k}_0\mathbf{r} - \omega t)$ is present in an electron gas. Compute the rate of transfer of energy to the gas and show that this can be expressed in terms of the imaginary part of the reciprocal dielectric function by

$$d\epsilon/dt = -(e^2/2\epsilon_0 k^2)\omega\rho_t^2 \operatorname{Im}[(1/\chi(\mathbf{k}, \omega))]$$

(see Nozières and Pines, 1958).

7 The Stoner–Wohlfarth theory of itinerant electron ferromagnetism assumes that the exchange splitting of the electron states is proportional to the magnetization:

$$E_\downarrow(\mathbf{k}) - E_\uparrow(\mathbf{k}) = 2K\theta\sigma \tag{1}$$

in which $\sigma = (\rho\uparrow - \rho\downarrow)/\rho$ and θ is an effective interaction measured in temperature units. (a) Show that (1) can be obtained from the Hubbard model in the low density limit. (b) Suppose that the system is subject to an external magnetic field H. Treat σ as small (a weak ferromagnet) and show that the relation between B and σ can be written as

$$B/\sigma = a_1 + a_2T^2 + a_3\sigma^2$$

correct through terms of order σ^3 and T^3. Obtain expressions for the constants a_1, a_2, and a_3 in terms of the density of states at the Fermi energy and its derivatives (see Edwards and Wohlfarth, 1968).

8 Calculate the Curie temperature of the weak itinerant electron ferromagnet described in Problem 7. Show that below this temperature, but close to it, there is a magnetic contribution to the specific heat $C_m = \gamma_m T$ and that the ratio of the magnetic specific heat to the ordinary electronic specific heat is

$$C_m/C_0 = 2[1 - (\rho/2GK\theta)][(G'/G)^2 - (G''/G)]^2[(G'/G)^2 - (G''/3G)]^{-1}$$

in which ρ is the average electron density, and G, G', and G'' are the density of states and its first and second derivatives at the (paramagnetic) Fermi energy.

9 Let E_0 be the ground state energy of a noninteracting electron gas ($e^2 = 0$) and let E be the ground state energy in the presence of interactions. Suppose ψ_0 is the ground state wave function, which can be considered to be a function of e^2. Show that

$$E = E_0 + \int_0^{e^3} \frac{de'^2}{e'^2} \langle \psi_0(e'^2) \mid V(e'^2) \mid \psi_0(e'^2) \rangle$$

in which V is the interaction including all pairs of particles.

10 Show, by a diagrammatic analysis, that $\langle \psi_0 \mid V \mid \psi_0 \rangle$, the matrix element which appears in Problem 9, can be expressed as

$$\langle \psi_0 \mid V \mid \psi_0 \rangle = [V/(2\pi)^3] \int d^3q \frac{d\omega}{2\pi} \operatorname{Im}[1/\kappa(\mathbf{q}, \omega)]$$

in which $\kappa(\mathbf{q}, \omega)$ is the exact dielectric function for the system.

References

Abrikosov, A. A., Gor'kov, L. P., and Dzyaloshinski, E. I. (1963). "Methods of Quantum Field Theory in Statistical Physics." Prentice-Hall, Englewood Cliffs, New Jersey.

Ambegaokar, V., and Kohn, W. (1960). *Phys. Rev.* **117,** 423.

Bethe, H. A. (1956). *Phys. Rev.* **103,** 1353.

Bloch, C. (1962). General Perturbation Formalism for the Many Body Problem at Non Zero Temperatures. *In* "Lectures on the Many Body Problem" (E. R. Caianello, ed.), Vol. 1, p. 31. Academic Press, New York.

Bohm, D., and Pines, D. (1953). *Phys. Rev.* **92,** 609.

Brout, R., and Carruthers, P. (1963). "Lectures on the Many Electron Problem." Wiley (Interscience), New York.

Brueckner, K. A. (1958). *Phys. Rev.* **110,** 597.

Brueckner, K. A. (1959). "Theory of Nuclear Structure in the Many Body Problem" (C. DeWitt, ed.), p. 47. Wiley, New York.

Brueckner, K. A., and Gammel, J. L. (1958). *Phys. Rev.* **109,** 1023.

Brueckner, K. A., and Levinson, C. A. (1955). *Phys. Rev.* **97,** 1344.

Callaway, J. (1968). *Phys. Rev.* **170,** 576.

Cooper, B. R. (1968). *Solid State Phys.* **21,** 393.

Daniel, E., and Vosko, S. H. (1960). *Phys. Rev.* **120,** 2041.

DuBois, D. F. (1959). *Ann. Phys. N.Y.* **7,** 174.

DuBois, D. F. (1960). *Ann. Phys. N.Y.* **8,** 24.

Edwards, D. M. (1967). *Proc. Roy. Soc.* (*London*) **A300,** 373.

Edwards, D. M., and Wohlfarth, E. P. (1968). *Proc. Roy. Soc.* (*London*) **A303,** 127.

Evenson, W. E., and Liu, S. H. (1968). *Phys. Rev. Lett.* **21,** 432.

Evenson, W. E., and Liu, S. H. (1969). *Phys. Rev.* **178,** 783.

Fedders, P. A., and Martin, P. C. (1966). *Phys. Rev.* **143,** 245.

Fetter, A. L., and Walecka, J. D. (1971). "Quantum Theory of Many Body Systems." McGraw-Hill, New York.

Geldart, D. J. W., Houghton, A., and Vosko, S. H. (1964). *Can. J. Phys.* **42,** 1938.
Gell-Mann, M. (1957). *Phys. Rev.* **106,** 369.
Gell-Mann, M., and Brueckner, K. A. (1957). *Phys. Rev.* **106,** 364.
Giovannini, B., Peter, M., and Schrieffer, J. R. (1964). *Phys. Rev. Lett.* **12,** 736.
Gold, A. V., Hodges, L., Panousis, P. T., and Stone, D. R. (1971). *Int. J. Magn.* **2,** 357.
Goldstone, J. (1957). *Proc. Roy. Soc. (London)* **A239,** 267.
Hamann, D. R., and Overhauser, A. W. (1966). *Phys. Rev.* **143,** 183.
Hedin, L., and Lundqvist, S. (1969). *Solid State Phys.* **23,** 1.
Herring, C. (1966a). Exchange interactions among itinerant electrons. *In* "Magnetism" (G. Rado and H. Suhl, eds.), Vol. 4. Academic Press, New York.
Herring, C. (1966b). Direct Exchange between Well Separated Atoms. *In* "Magnetism" (G. Rado and H. Suhl, eds.), Vol 2B. Academic Press, New York.
Hohenberg, P., and Kohn, W. (1964). *Phys. Rev.* **136,** B864.
Kadanoff, L. P., and Baym, G. (1962). "Quantum Statistical Mechanics." Benjamin, New York.
Kanamori, J. (1963). *Progr. Theor. Phys.* **30,** 275.
Kasuya, T. (1956). *Prog. Theor. Phys.* **16,** 45.
Keeton, S. C., and Loucks, T. L. (1968). *Phys. Rev.* **168,** 672.
Kirzhnits, D. A. (1967). "Field Theoretical Methods in Many Body Systems." Pergamon, Oxford.
Klein, A. (1959). *Phys. Rev.* **115,** 1136.
Kohn, W. (1957). *Phys. Rev.* **105,** 509.
Kohn, W. (1958). *Phys. Rev.* **110,** 857.
Landau, L. D. (1956). *J. Exp. Theoret. Phys. (USSR)* **30,** 158 (English Transl. *Sov. Phys. JETP* **3,** 920).
Landau, L. D. (1957). *J. Exp. Theoret. Phys. (USSR)* **32,** 59 (English Transl. *Sov. Phys. JETP* **5,** 101).
Landau, L. D. (1959). *J. Exp. Theoret. Phys. (USSR)* **35,** 97 (English Transl. *Sov. Phys. JETP* **8,** 70).
Lomer, W. M. (1962). *Proc. Phys. Soc. (London)* **80,** 489.
Luttinger, J. M. (1960). *Phys. Rev.* **119,** 1153.
Luttinger, J. M., and Ward, J. C. (1960). *Phys. Rev.* **118,** 1417.
March, N. H., Young, W. H., and Sampanthar, S. (1969). "The Many Body Problem in Quantum Mechanics." Cambridge Univ. Press. London and New York.
Martin, P. C., and Schwinger, J. (1959). *Phys. Rev.* **115,** 1342.
Mattuck, R. D. (1967). "A Guide to Feynman Diagrams in the Many Body Problem." McGraw-Hill, New York.
Mills, R. (1969). "Propagators for Many Particle Systems." Gordon and Breach, New York.
Nagaoka, Y. (1966). *Phys. Rev.* **147,** 392.
Nozières, P. (1964). "Theory of Interacting Fermi Systems." Benjamin, New York.
Nozières, P., and Pines, D. (1958). *Nuovo Cimento* [X] **9,** 470.
Overhauser, A. W. (1960). *Phys. Rev. Lett.* **4,** 462.
Overhauser, A. W. (1962). *Phys. Rev.* **128,** 1437.
Phillips, W. C., and Weiss, R. J. (1968). *Phys. Rev.* **171,** 790.
Pines, D. (1962). "The Many Body Problem." Benjamin, New York.
Pines, D. (1963). "Elementary Excitations in Solids." Benjamin, New York.
Pines, D., and Nozieres, P. (1966). "The Theory of Quantum Liquids." Benjamin, New York.

Platzman, P. M., and Tzoar, W. (1965). *Phys. Rev.* **139,** 410.
Platzman, P. M., and Wolff, P. A. (1967). *Phys. Rev. Lett.* **18,** 280.
Quinn, J. J., and Ferrell, R. (1958). *Phys. Rev.* **112,** 812.
Rajagopal, A. K. (1966). *Phys. Rev.* **142,** 152.
Rice, T. M. (1965). *Ann. Phys. N.Y.* **31,** 100.
Ruderman, M. A., and Kittel, C. (1954). *Phys. Rev.* **96,** 99.
Schultz, T. D. (1964). "Quantum Field Theory and the Many Body Problem." Gordon and Breach, New York.
Schultz, S., and Dunifer, G. (1967). *Phys. Rev. Lett.* **18,** 283.
Schweber, S. S. (1961). "An Introduction to Relativistic Quantum Field Theory." Harper, New York.
Serrin, J. (1959). Mathematical Principles of Classical Fluid Mechanics. *In* "Encyclopedia of Physics" (S. Flügge, ed.), Vol. 8. p. 1. Springer-Verlag, Berlin.
Sham, L. J. (1966). *Phys. Rev.* **150,** 720.
Silin, V. P. (1958). *J. Exp. Theoret. Phys.* (*USSR*) **33,** 495 (English Transl. *Sov. Phys. JETP* **6,** 387).
Slater, J. C. (1953). *Rev. Mod. Phys.* **25,** 199.
Thouless, D. J. (1972). "The Quantum Mechanics of Many Body Systems," 2nd ed. Academic Press, New York.
Van Vleck, J. H. (1962). *Rev. Mod. Phys.* **34,** 681.
Vonsovski, S. V. (1946). *J. Exp. Theoret. Phys.* (*USSR*) **16,** 981.
Watson, R. E., and Freeman, A. J. (1966). *Phys. Rev.* **152,** 566.
Wick, G. C. (1950). *Phys. Rev.* **80,** 268.
Wolff, P. A. (1959). *Phys. Rev.* **120,** 814.
Yosida, K. (1957). *Phys. Rev.* **106,** 893.
Young, W., and Callaway, J. (1970). *J. Phys. Chem. Solids* **31,** 865.
Zener, C. (1951). *Phys. Rev.* **81,** 440.
Ziman, J. M. (1969). "Elements of Advanced Quantum Theory." Cambridge Univ. Press, London and New York.
Zubarev, D. N. (1960). *Usp. Fiz. Nauk.* **71,** 71 (English transl., *Sov. Phys. Uspekhi* **3,** 320).

Author Index for Part A

Numbers in italics refer to the pages on which the complete references are listed.

Author Index for Part B

Numbers in italics refer to the pages on which the complete references are listed.

S

T

U

V

W

Y

Z

Subject Index

A 6
B 7
C 8
D 9
E 0
F 1
G 2
H 3
I 4
J 5